PROTECTIVE GROUPS IN ORGANIC SYNTHESIS

PROTECTIVE GROUPS IN ORGANIC SYNTHESIS

THIRD EDITION

Theodora W. Greene
The Rowland Institute for Science

and

Peter G. M. Wuts
Pharmacia and Upjohn Company

A WILEY-INTERSCIENCE PUBLICATION
JOHN WILEY & SONS, INC.
New York / Chichester / Weinheim / Brisbane / Toronto / Singapore

This book is printed on acid-free paper. 

Library of Congress Cataloging in Publication Data:

Protective groups in organic synthesis. — 3rd ed. / editors, Theodora W. Greene and Peter G. M. Wuts.

p. cm.

Rev. ed. of: Protective groups in organic synthesis. 2nd ed. / Theodora W. Greene and Peter G. M. Wuts. c1991.

Includes index.

ISBN 0-471-16019-9 (cloth)

1. Organic compounds—Synthesis. 2. Protective groups (Chemistry) I. Greene, Theodora W., 1931– . II. Wuts, Peter G. M. III. Greene, Theodora W., 1931– Protective groups in organic synthesis.

QD262. G665 1999

547.2—dc21 98-38182

Printed in the United States of America.

10 9 8 7 6 5 4 3 2

PREFACE TO THE THIRD EDITION

Organic synthesis has not yet matured to the point where protective groups are not needed for the synthesis of natural and unnatural products; thus, the development of new methods for functional group protection and deprotection continues. The new methods added to this edition come from both electronic searches and a manual examination of all the primary journals through the end of 1997. We have found that electronic searches of *Chemical Abstracts* fail to find many new methods that are developed during the course of a synthesis, and issues of selectivity are often not addressed. As with the second edition, we have attempted to highlight unusual and potentially useful examples of selectivity for both protection and deprotection. In some areas the methods listed may seem rather redundant, such as the numerous methods for THP protection and deprotection, but we have included them in an effort to be exhaustive in coverage. For comparison, the first edition of this book contains about 1500 references and 500 protective groups, the second edition introduces an additional 1500 references and 206 new protective groups, and the third edition adds 2349 new citations and 348 new protective groups.

Two new sections on the protection of phosphates and the alkyne-CH are included. All other sections of the book have been expanded, some more than others. The section on the protection of alcohols has increased substantially, reflecting the trend of the nineties to synthesize acetate- and propionate-derived natural products. An effort was made to include many more enzymatic methods of protection and deprotection. Most of these are associated with the protection of alcohols as esters and the protection of carboxylic acids. Here we have not attempted to be exhaustive, but hopefully, a sufficient number of cases are provided that illustrate the true power of this technology, so that the reader will examine some of the excellent monographs and review articles cited in the references. The Reactivity Charts in Chapter 10 are identical to those in the first edition. The chart number appears beside the name of each protective group when it is first introduced. No attempt was made to update these Charts, not only because of the sheer magnitude of the task, but because it is nearly impossible in a two-dimensional table to address adequately the effect that electronic and steric controlling elements have on a particular instance of protection or deprotection. The concept of fuzzy sets as outlined by Lofti Zadeh would be ideally suited for such a task.

The completion of this project was aided by the contributions of a number of people. I am grateful to Rein Virkhaus and Gary Callen, who for many years forwarded me references when they found them, to Jed Fisher for the information he contributed on phosphate protection, and to Todd Nelson for providing me a preprint of his excellent review article on the deprotection of silyl ethers. I heartily thank Theo Greene for checking and rechecking the manuscript—all 15 cm of it—for spelling and consistency and for the arduous task of checking all the references for accuracy. I thank Fred Greene for reading the manuscript, for his contribution to Chapter 1 on the use of protective groups in the synthesis of himastatin, and for his contribution to the introduction to Chapter 9, on phosphates. I thank my wife, Lizzie, for encouraging me to undertake the third edition, for the hours she spent in the library looking up and photocopying hundreds of references, and for her understanding while I sat in front of the computer night after night and numerous weekends over a two-year period. She is the greatest!

Kalamazoo, Michigan
June 1998

PETER G. M. WUTS

PREFACE TO THE SECOND EDITION

Since publication of the first edition of this book in 1981, many new protective groups and many new methods of introduction or removal of known protective groups have been developed: 206 new groups and approximately 1500 new references have been added. Most of the information from the first edition has been retained. To conserve space, generic structures used to describe Formation/Cleavage reactions have been replaced by a single line of conditions, sometimes with explanatory comments, especially about selectivity. Some of the new information has been obtained from on-line searches of *Chemical Abstracts*, which have limitations. For example, *Chemical Abstracts* indexes a review article about protective groups only if that word appears in the title of the article. References are complete through 1989. Some references, from more widely circulating journals, are included for 1990.

Two new sections on the protection for indoles, imidazoles, and pyrroles and protection for the amide –NH are included. They are separated from the regular amines because their chemical properties are sufficiently different to affect the chemistry of protection and deprotection. The Reactivity Charts in Chapter 8 are identical to those in the first edition. The chart number appears beside the name of each protective group when it is first discussed.

A number of people must be thanked for their contributions and help in completing this project. I am grateful to Gordon Bundy, who loaned me his card file, which provided many references that the computer failed to find, and to Bob Williams, Spencer Knapp, and Tohru Fukuyama for many references on amine and amide protection. I thank Theo Greene who checked and rechecked the manuscript for spelling and consistency and for the herculean task of checking all the references to make sure that my 3's and 8's and 7's and 9's were not interchanged—all done without a single complaint. I thank Fred Greene who read the manuscript and provided valuable suggestions for its improvement. My wife Lizzie was a major contributor to getting this project finished, by looking up and photocopying references, by turning on the computer in an evening ritual, and by typing many sections of the original book, which made the changes and additions much easier. Without her understanding and encouragement, the volume probably would never have been completed.

Kalamazoo, Michigan
May 1990

PETER G. M. WUTS

PREFACE TO THE FIRST EDITION

The selection of a protective group is an important step in synthetic methodology, and reports of new protective groups appear regularly. This book presents information on the synthetically useful protective groups (~500) for five major functional groups: –OH, –NH,–SH,–COOH, and >C=O. References through 1979, the best method(s) of formation and cleavage, and some information on the scope and limitations of each protective group are given. The protective groups that are used most frequently and that should be considered first are listed in Reactivity Charts, which give an indication of the reactivity of a protected functionality to 108 prototype reagents.

The first chapter discusses some aspects of protective group chemistry: the properties of a protective group, the development of new protective groups, how to select a protective group from those described in this book, and an illustrative example of the use of protective groups in a synthesis of brefeldin. The book is organized by functional group to be protected. At the beginning of each chapter are listed the possible protective groups. Within each chapter protective groups are arranged in order of increasing complexity of structure (e.g., methyl, ethyl, *t*-butyl, . . . , benzyl). The most efficient methods of formation or cleavage are described first. Emphasis has been placed on providing recent references, since the original method may have been improved. Consequently, the original reference may not be cited; my apologies to those whose contributions are not acknowledged. Chapter 8 explains the relationship between reactivities, reagents, and the Reactivity Charts that have been prepared for each class of protective groups.

This work has been carried out in association with Professor Elias J. Corey, who suggested the study of protective groups for use in computer-assisted synthetic analysis. I appreciate his continued help and encouragement. I am grateful to Dr. J. F. W. McOmie (Ed., *Protective Groups in Organic Chemistry*, Plenum Press, New York and London, 1973) for his interest in the project and for several exchanges of correspondence, and to Mrs. Mary Fieser, Professor Frederick D. Greene, and

Professor James A. Moore for reading the manuscript. Special thanks are also due to Halina and Piotr Starewicz for drawing the structures, and to Kim Chen, Ruth Emery, Janice Smith, and Ann Wicker for typing the manuscript.

THEODORA W. GREENE

Harvard University
September 1980

CONTENTS

ABBREVIATIONS

PROTECTIVE GROUPS

In some cases, several abbreviations are used for the same protective group. We have listed the abbreviations as used by an author in his or her original paper, including capital and lowercase letters. Occasionally, the same abbreviation has been used for two different protective groups. This information is also included.

ABO	2,7,8-trioxabicyclo[3.2.1]octyl
Ac	acetyl
ACBZ	4-azidobenzyloxycarbonyl
AcHmb	2-acetoxy-4-methoxybenzyl
Acm	acetamidomethyl
Ad	1-adamantyl
Adoc	1-adamantyloxycarbonyl
Adpoc	1-(1-adamantyl)-1-methylethoxycarbonyl
Alloc or AOC	allyloxycarbonyl
Als	allylsulfonyl
AMB	2-(acetoxymethyl)benzoyl
AN	4-methoxyphenyl or anisyl
Anpe	2-(4-acetyl-2-nitrophenyl)ethyl
AOC or Alloc	allyloxycarbonyl
p-AOM	*p*-anisyloxymethyl or (4-methoxyphenoxy)methyl
Azb	*p*-azidobenzyl
Bam	benzamidomethyl
BBA	butane-2,3-bisacetal
BDMS	biphenyldimethylsilyl
Bdt	1,3-benzodithiolan-2-yl
Betsyl or Bts	benzothiazole-2-sulfonyl
Bic	5-benzisoxazolylmethoxycarbonyl
Bim	5-benzisoazolylmethylene
Bimoc	benz[*f*]inden-3-ylmethoxycarbonyl
BIPSOP	*N*-2,5-bis(triisopropylsiloxy)pyrrolyl

BMB	*o*-(benzoyloxymethyl)benzoyl
Bmpc	2,4-dimethylthiophenoxycarbonyl
Bmpm	bis(4-methoxyphenyl)-1′-pyrenylmethyl
Bn	benzyl
Bnpeoc	2,2-bis(4′-nitrophenyl)ethoxycarbonyl
BOC	*t*-butoxycarbonyl
BOM	benzyloxymethyl
Bpoc	1-methyl-1-(4-biphenyl)ethoxycarbonyl
BSB	benzoSTABASE
Bsmoc	1,1-dioxobenzo[*b*]thiophene-2-ylmethoxycarbonyl
Bts or Betsyl	benzothiazole-2-sulfonyl
B^{t}SE	2-*t*-butylsulfonylethyl
Bum	*t*-butoxymethyl
t-Bumeoc	1-(3,5-di-*t*-butylphenyl)-1-methylethoxycarbonyl
Bus	*t*-butylsulfonyl
Bz	benzoyl
CAEB	2-[(2-chloroacetoxy)ethyl]benzoyl
Cam	carboxamidomethyl
CAMB	2-(chloroacetoxymethyl)benzoyl
Cbz or Z	benzyloxycarbonyl
CDA	cyclohexane-1,2-diacetal
CDM	2-cyano-1,1-dimethylethyl
CE or Cne	2-cyanoethyl
Cee	1-(2-chloroethoxy)ethyl
cHex	cyclohexyl
Climoc	2-chloro-3-indenylmethoxycarbonyl
Cms	carboxymethylsulfenyl
Cne or CE	2-cyanoethyl
Coc	cinnamyloxycarbonyl
Cpeoc	2-(cyano-1-phenyl)ethoxycarbonyl
CPTr	4,4′,4″-tris(4,5-dichlorophthalimido)triphenylmethyl
CTMP	1-[(2-chloro-4-methyl)phenyl]-4-methoxypiperidin-4-yl
Cys	cysteine
DAM	di-*p*-anisylmethyl or bis(4-methoxyphenyl)methyl
DATE	1,1-di-*p*-anisyl-2,2,2-trichloroethyl
DB-*t*-BOC	1,1-dimethyl-2,2-dibromoethoxycarbonyl
DBD-Tmoc	2,7-di-*t*-butyl[9-(10,10-dioxo-10,10,10,10-tetra=hydrothioxanthyl)]methoxycarbonyl
DBS	dibenzosuberyl
Dde	2-(4,4-dimethyl-2,6-dioxocyclohexylidene)ethyl
Ddz	1-methyl-1-(3,5-dimethoxyphenyl)ethoxycarbonyl
DEM	diethoxymethyl
DEIPS	diethylisopropylsilyl
Desyl	2-oxo-1,2-diphenylethyl
Dim	1,3-dithianyl-2-methyl

Dmab	4-{*N*-[1-(4,4-dimethyl-2,6-dioxocyclohexylidene)-3-methylbutyl]amino}benzyl
DMB	"3′,5′-dimethoxybenzoin"
Dmb	2,4-dimethoxybenzyl
DMIPS	dimethylisopropylsilyl
Dmoc	dithianylmethoxycarbonyl
Dmp	2,4-dimethyl-3-pentyl
Dmp	dimethylphosphinyl
DMPM	3,4-dimethoxybenzyl
DMT or DMTr	di(*p*-methoxyphenyl)phenylmethyl or dimethoxytrityl
DMTr or DMT	di(*p*-methoxyphenyl)phenylmethyl or dimethoxytrityl
DNB	*p,p*′-dinitrobenzhydryl
DNMBS	4-(4′,8′-dimethoxynaphthylmethyl)benzenesulfonyl
DNP	2,4-dinitrophenyl
Dnpe	2-(2,4-dinitrophenyl)ethyl
Dnpeoc	2-(2,4-dinitrophenyl)ethoxycarbonyl
DNs	2,4-dinitrobenzenesulfonyl
Dnseoc	2-dansylethoxycarbonyl
Dobz	*p*-(dihydroxyboryl)benzyloxycarbonyl
Doc	2,4-dimethylpent-3-yloxycarbonyl
DOPS	dimethyl[1,1-dimethyl-3-(tetrahydro-2*H*-pyran-2-yloxy)propyl]silyl
DPA	diphenylacetyl
DPIPS	diphenylisopropylsilyl
DPM or Dpm	diphenylmethyl
DPMS	diphenylmethylsilyl
Dpp	diphenylphosphinyl
Dppe	2-(diphenylphosphino)ethyl
Dppm	(diphenyl-4-pyridyl)methyl
DPSE	2-(methyldiphenylsilyl)ethyl
Dpt	diphenylphosphinothioyl
DPTBS	diphenyl-*t*-butoxysilyl or diphenyl-*t*-butylsilyl
DTBMS	di-*t*-butylmethylsilyl
DTBS	di-*t*-butylsilylene
DTE	2-(hydroxyethyl)dithioethyl or "dithiodiethanol"
Dts	dithiasuccinimidyl
EE	1-ethoxyethyl
EOM	ethoxymethyl
Fcm	ferrocenylmethyl
Fm	9-fluorenylmethyl
Fmoc	9-fluorenylmethoxycarbonyl
GUM	guaiacolmethyl
HBn	2-hydroxybenzyl
HIP	1,1,1,3,3,3-hexafluoro-2-phenylisopropyl

Hoc	cyclohexyloxycarbonyl
HSDIS	(hydroxystyryl)diisopropylsilyl
HSDMS	(hydroxystyryl)dimethylsilyl
hZ or homo Z	homobenzyloxycarbonyl
IDTr	3-(imidazol-1-ylmethyl)-4′,4″-dimethoxytriphenylmethyl
IETr	4,4′-dimethoxy-3″-[*N*-(imidazolylethyl)carbamoyl]trityl
iMds	2,6-dimethoxy-4-methylbenzenesulfonyl
Ipaoc	1-isopropylallyloxycarbonyl
Ipc	isopinocamphenyl
IPDMS	isopropyldimethylsilyl
Lev	levulinoyl
LevS	4,4-(ethylenedithio)pentanoyl
LevS	levulinoyldithioacetal ester
MAQ	2-(9,10-anthraquinonyl)methyl or 2-methylene-anthraquinone
MBE	1-methyl-1-benzyloxyethyl
MBF	2,3,3a,4,5,6,7,7a-octahydro-7,8,8-trimethyl-4,7-methanobenzofuran-2-yl
MBS or Mbs	*p*-methoxybenzenesulfonyl
Mds	2,6-dimethyl-4-methoxybenzenesulfonyl
MEC	α-methylcinnamyl
MEM	2-methoxyethoxymethyl
Menpoc	α-methylnitropiperonyloxycarbonyl
MeOZ or Moz	*p*-methoxybenzyloxycarbonyl
Mes	mesityl or 2,4,6-trimethylphenyl
MIP	methoxyisopropyl or 1-methyl-1-methoxyethyl
MM	menthoxymethyl
MMT or MMTr	*p*-methoxyphenyldiphenylmethyl
MMTr or MMT	*p*-methoxyphenyldiphenylmethyl
MOM	methoxymethyl
MOMO	methoxymethoxy
Moz or MeOZ	*p*-methoxybenzyloxycarbonyl
MP	*p*-methoxyphenyl
MPM or PMB	*p*-methoxyphenylmethyl or *p*-methoxybenzyl
Mps	*p*-methoxyphenylsulfonyl
Mpt	dimethylphosphinothioyl
Ms	methanesulfonyl or mesyl
Msib	4-(methylsulfinyl)benzyl
Msz	4-methylsulfinylbenzyloxycarbonyl
Mtb	2,4,6-trimethoxybenzenesulfonyl
Mte	2,3,5,6-tetramethyl-4-methoxybenzenesulfonyl
MTHP	4-methoxytetrahydropyranyl
MTM	methylthiomethyl
MTMB	4-(methylthiomethoxy)butyryl

MTMECO	2-(methylthiomethoxy)ethoxycarbonyl
MTMT	2-(methylthiomethoxymethyl)benzoyl
Mtpc	4-(methylthio)phenoxycarbonyl
Mtr	2,3,6-trimethyl-4-methoxybenzenesulfonyl
Mts	2,4,6-trimethylbenzenesulfonyl or mesitylenesulfonyl
NBOM	nitrobenzyloxymethyl
Ne	2-nitroethyl
Noc	4-nitrocinnamyloxycarbonyl
Nosyl or Ns	2- or 4-nitrobenzenesulfonyl
Npe or npe	2-(nitrophenyl)ethyl
Npeoc	2-(4-nitrophenyl)ethoxycarbonyl
Npes	2-(4-nitrophenyl)ethylsulfonyl
NPS or Nps	2-nitrophenylsulfenyl
NpSSPeoc	2-[(2-nitrophenyl)dithio]-1-phenylethoxycarbonyl
Npys	3-nitro-2-pyridinesulfenyl
Ns or Nosyl	2- or 4-nitrobenzenesulfonyl
NVOC or Nvoc	3,4-dimethoxy-6-nitrobenzyloxycarbonyl or 6-nitroveratryloxycarbonyl
OBO	2,6,7-trioxabicyclo[2.2.2]octyl
ONB	*o*-nitrobenzyl
PAB	*p*-acylaminobenzyl
PAC_H	2-[2-(benzyloxy)ethyl]benzoyl
PAC_M	2-[2-(4-methoxybenzyloxy)ethyl]benzoyl
Paloc	3-(3-pyridyl)allyloxycarbonyl or 3-(3-pyridyl)prop-2-enyloxycarbonyl
Pbf	2,2,4,6,7-pentamethyldihydrobenzofuran-5-sulfonyl
Peoc	2-phosphonioethoxycarbonyl
Peoc	2-(triphenylphosphonio)ethoxycarbonyl
Pet	2-(2′-pyridyl)ethyl
Pf	9-phenylfluorenyl
Phamc	phenylacetamidomethyl
Phenoc	4-methoxyphenacyloxycarbonyl
Pim	phthalimidomethyl
Pixyl or Px	9-(9-phenyl)xanthenyl
PMB or MPM	*p*-methoxybenzyl or *p*-methoxyphenylmethyl
PMBM	*p*-methoxybenzyloxymethyl
Pmc	2,2,5,7,8-pentamethylchroman-6-sulfonyl
Pme	pentamethylbenzenesulfonyl
PMP	*p*-methoxyphenyl
PMS	*p*-methylbenzylsulfonyl
PNB	*p*-nitrobenzyl
PNP	*p*-nitrophenyl
PNPE	2-(4-nitrophenyl)ethyl
POM	4-pentenyloxymethyl
POM	pivaloyloxymethyl

Pp	2-phenyl-2-propyl
Ppoc	2-triphenylphosphonioisopropoxycarbonyl
Ppt	diphenylthiophosphinyl
PSE	2-(phenylsulfonyl)ethyl
Psec	2-(phenylsulfonyl)ethoxycarbonyl
PTE	2-(4-nitrophenyl)thioethyl
PTM	phenylthiomethyl
Pv	pivaloyl
Px or pixyl	9-(9-phenyl)xanthenyl
Pyet	1-(α-pyridyl)ethyl
Pyoc	2-(2′- or 4′-pyridyl)ethoxycarbonyl
Qm	2-quinolinylmethyl
SATE	*S*-acetylthioethyl
Scm	*S*-carboxymethylsulfenyl
SEE	1-[2-(trimethylsilyl)ethoxy]ethyl
SEM	2-(trimethylsilyl)ethoxymethyl
SES	2-(trimethylsilyl)ethanesulfonyl
Sisyl	tris(trimethylsilyl)silyl
SMOM	(phenyldimethylsilyl)methoxymethyl
Snm	*S*-(*N*′-methyl-*N*′-phenylcarbamoyl)sulfenyl
STABASE	1,1,4,4-tetramethyldisilylazacyclopentane
Tacm	trimethylacetamidomethyl
TBDMS or TBS	*t*-butyldimethylsilyl
TBDPS	*t*-butyldiphenylsilyl
Tbf-DMTr	4-(17-tetrabenzo[*a*,*c*,*g*,*i*]fluorenylmethyl-4′,4″-dimethoxytrityl
Tbfmoc	17-tetrabenzo[*a*,*c*,*g*,*i*]fluorenylmethoxycarbonyl
TBDS	tetra-*t*-butoxydisiloxane-1,3-diylidene
TBMPS	*t*-butylmethoxyphenylsilyl
TBS or TBDMS	*t*-butyldimethylsilyl
TBTr	4,4′,4″-tris(benzyloxy)triphenylmethyl
TCB	2,2,2-trichloro-1,1-dimethylethyl
TcBOC	1,1-dimethyl-2,2,2-trichloroethoxycarbonyl
TCP	*N*-tetrachlorophthalimido
Tcroc	2-(trifluoromethyl)-6-chromonylmethyleneoxycarbonyl
Tcrom	2-(trifluoromethyl)-6-chromonylmethylene
TDE	(2,2,2-trifluoro-1,1-diphenyl)ethyl
TDS	thexyldimethylsilyl
Teoc	2-(trimethylsilyl)ethoxycarbonyl
TES	triethylsilyl
Tf	trifluoromethanesulfonyl
TFA	trifluoroacetyl
Tfav	4,4,4-trifluoro-3-oxo-1-butenyl
Thexyl	2,3-dimethyl-2-butyl
THF	tetrahydrofuranyl

THP	tetrahydropyranyl
TIBS	triisobutylsilyl
TIPDS	1,3-(1,1,3,3-tetraisopropyldisiloxanylidene)
TIPS	triisopropylsilyl
TLTr	4,4′,4″-tris(levulinoyloxy)triphenylmethyl
Tmb	2,4,6-trimethylbenzyl
Tmob	trimethoxybenzyl
TMPM	trimethoxyphenylmethyl
TMS	trimethylsilyl
TMSE or TSE	2-(trimethylsilyl)ethyl
TMSEC	2-(trimethylsilyl)ethoxycarbonyl
TMSP	2-trimethylsilylprop-2-enyl
TMTr	tris(*p*-methoxyphenyl)methyl
Tos or Ts	*p*-toluenesulfonyl
TPS	triphenylsilyl
TPTE	2-(4-triphenylmethylthio)ethyl
Tr	triphenylmethyl or trityl
Tritylone	9-(9-phenyl-10-oxo)anthryl
Troc	2,2,2-trichloroethoxycarbonyl
Ts or Tos	*p*-toluenesulfonyl
TSE or TMSE	2-(trimethylsilyl)ethyl
Tse	2-(*p*-toluenesulfonyl)ethyl
Voc	vinyloxycarbonyl
Z or Cbz	benzyloxycarbonyl

REAGENTS

9-BBN	9-borabicyclo[3.3.1]nonane
bipy	2,2′-bipyridine
BOP Reagent	benzotriazol-1-yloxytris(dimethylamino)phosphonium hexafluorophosphate
BOP-Cl	bis(2-oxo-3-oxazolidinyl)phosphinic chloride
BroP	bromotris(dimethylamino)phosphonium hexafluorophosphate
Bt	benzotriazol-1-yl or 1-benzotriazolyl
BTEAC	benzyltriethylammonium chloride
CAL	*Candida antarctica* lipase
CAN	ceric ammonium nitrate
CMPI	2-chloro-1-methylpyridinium iodide
cod	cyclooctadiene
cot	cyclooctatetraene
CSA	camphorsulfonic acid
DABCO	1,4-diazabicyclo[2.2.2]octane
DBAD	di-*t*-butyl azodicarboxylate
DBN	1,5-diazabicyclo[4.3.0]non-5-ene

DBU	1,8-diazabicyclo[5.4.0]undec-7-ene
DCC	dicyclohexylcarbodiimide
DDQ	2,3-dichloro-5,6-dicyano-1,4-benzoquinone
DEAD	diethyl azodicarboxylate
DIAD	diisopropyl azodicarboxylate
DIBAL-H	diisobutylaluminum hydride
DIPEA	diisopropylethylamine
DMAC	*N,N*-dimethylacetamide
DMAP	4-*N,N*-dimethylaminopyridine
DMDO	2,2-dimethyldioxirane
DME	1,2-dimethoxyethane
DMF	*N,N*-dimethylformamide
DMPU	1,3-dimethyl-3,4,5,6-tetrahydro-2(1*H*)-pyrimidinone
DMS	dimethyl sulfide
DMSO	dimethyl sulfoxide
dppb	1,4-bis(diphenylphosphino)butane
dppe	1,2-bis(diphenylphosphino)ethane
DTE	dithioerythritol
DTT	dithiothreitol
EDC or EDCI	1-ethyl-3-(3-dimethylaminopropyl)carbodiimide (or 1-[3-(dimethylamino)propyl]-3-ethylcarbodimide) hydrochloride
EDCI or EDC	1-ethyl-3-(3-(dimethylaminopropyl)carbodiimide
EDTA	ethylenediaminetetraacetic acid
HATU	*N*-[(dimethylamino)(3*H*-1,2,3-triazolo(4,5-*b*)pyridin-3-yloxy)methylene]-*N*-methylmethanaminium hexafluorophosphate, previously known as *O*-(7-azabenzotriazol-1-yl)-1,1,3,3-tetramethyluronium hexafluorophosphate
HMDS	1,1,1,3,3,3-hexamethyldisilazane
HMPA	hexamethylphosphoramide
HMPT	hexamethylphosphorous triamide
HOAt	7-aza-1-hydroxybenzotriazole
HOBT	1-hydroxybenzotriazole
Im	imidazol-1-yl or 1-imidazolyl
IPA	isopropyl alcohol
IPCF (=IPCC)	isopropenyl chloroformate (isopropenyl chlorocarbonate)
KHMDS	potassium hexamethyldisilazide
LAH	lithium aluminum hydride
LDBB	lithium 4,4′-di-*t*-butylbiphenylide
MAD	methylaluminumbis(2,6-di-*t*-butyl-4-methylphenoxide)
MCPBA	*m*-chloroperoxybenzoic acid
MoOPH	oxodiperoxymolybdenum(pyridine)hexamethylphosphoramide
ms	molecular sieves
MSA	methanesulfonic acid

MTB	methylthiobenzene
MTBE	*t*-butyl methyl ether
NBS	*N*-bromosuccinimide
$Ni(acac)_2$	nickel acetylacetonate
NMM	*N*-methylmorpholine
NMO	*N*-methylmorpholine *N*-oxide
NMP	*N*-methylpyrrolidinone
P	polymer support
Pc	phthalocyanine
PCC	pyridinium chlorochromate
$PdCl_2(tpp)_2$	dichlorobis[tris(2-methylphenyl)phosphine]palladium
$Pd_2(dba)_3$	tris(dibenzylideneacetone)dipalladium
PG	protective group
PhI(OH)OTs	[hydroxy(tosyloxy)iodo]benzene
PPL	porcine pancreatic lipase
PPTS	pyridinium *p*-toluenesulfonate
proton sponge	1,8-bis(dimethylamino)naphthalene
Pyr	pyridine
$Rh_2(pfb)_4$	rhodium perfluorobutyrate
ScmCl	methoxycarbonylsulfenyl chloride
SMEAH	sodium bis(2-methoxyethoxy)aluminum hydride
Su	succinimidyl
TAS-F	tris(dimethylamino)sulfonium difluorotrimethylsilicate
TBAF	tetrabutylammonium fluoride
TEA	triethylamine
TEBA or TEBAC	triethylbenzylammonium chloride
TEBAC or TEBA	triethylbenzylammonium chloride
TESH	triethylsilane
Tf	trifluoromethanesulfonyl
TFA	trifluoroacetic acid
TFAA	trifluoroacetic anhydride
TFMSA or TfOH	trifluoromethanesulfonic acid
TfOH or TFMSA	trifluoromethanesulfonic acid
THF	tetrahydrofuran
THP	tetrahydropyran
TMEDA	*N,N,N′,N′*-tetramethylethylenediamine
TMOF	trimethyl orthoformate
TPAP	tetrapropylammonium perruthenate
TPP	tetraphenylporphyrin
TPPTS	sulfonated triphenylphosphine
TPS	triisopropylbenzensulfonyl chloride
$Tr^+BF_4^-$ or $Ph_3C^+BF_4^-$	triphenylcarbenium tetrafluoroborate
$TrS^-Bu_4N^+$	tetrabutylammonium triphenylmethanethiolate
Ts	toluenesulfonyl

PROTECTIVE GROUPS IN ORGANIC SYNTHESIS

1

THE ROLE OF PROTECTIVE GROUPS IN ORGANIC SYNTHESIS

PROPERTIES OF A PROTECTIVE GROUP

When a chemical reaction is to be carried out selectively at one reactive site in a multifunctional compound, other reactive sites must be temporarily blocked. Many protective groups have been, and are being, developed for this purpose. A protective group must fulfill a number of requirements. It must react selectively in good yield to give a protected substrate that is stable to the projected reactions. The protective group must be selectively removed in good yield by readily available, preferably nontoxic reagents that do not attack the regenerated functional group. The protective group should form a derivative (without the generation of new stereogenic centers) that can easily be separated from side products associated with its formation or cleavage. The protective group should have a minimum of additional functionality to avoid further sites of reaction. All things considered, no one protective group is the best. Currently, the science and art of organic synthesis, contrary to the opinions of some, has a long way to go before we can call it a finished and well-defined discipline, as is amply illustrated by the extensive use of protective groups during the synthesis of multifunctional molecules. Greater control over the chemistry used in the building of nature's architecturally beautiful and diverse molecular frameworks, as well as unnatural structures, is needed when one considers the number of protection and deprotection steps often used to synthesize a molecule.

HISTORICAL DEVELOPMENT

Since a few protective groups cannot satisfy all these criteria for elaborate substrates, a large number of mutually complementary protective groups are needed and, indeed, are available. In early syntheses, the chemist chose a standard derivative known to be stable to the subsequent reactions. In a synthesis of callistephin chloride, the phenolic—OH group in **1** was selectively protected as an acetate.[1] In the presence of silver ion, the aliphatic hydroxyl group in **2** displaced the bromide ion in a bromoglucoside. In a final step, the acetate group was removed by basic hydrolysis. Other classical methods of cleavage include acidic hydrolysis (eq. 1), reduction (eq. 2) and oxidation (eq. 3):

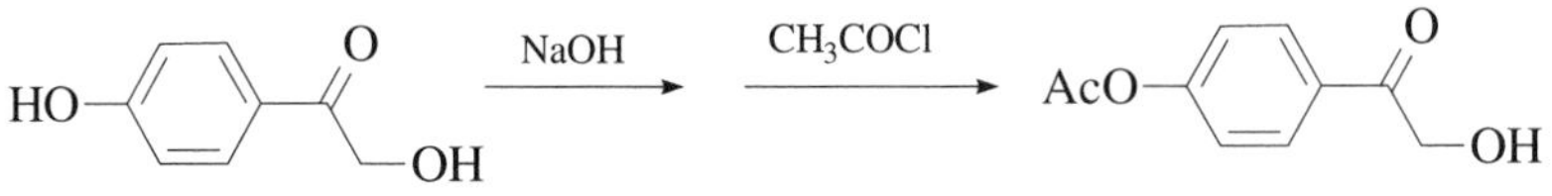

HO–C₆H₄–C(=O)–CH₂OH (**1**) $\xrightarrow{\text{NaOH}}$ $\xrightarrow{\text{CH}_3\text{COCl}}$ AcO–C₆H₄–C(=O)–CH₂OH (**2**)

(1) ArO–R → ArOH

(2) RO–CH_2Ph → ROH

(3) RNH–CHO → [RNHCOOH] → RNH_3^+

Some of the original work in the carbohydrate area in particular reveals extensive protection of carbonyl and hydroxyl groups. For example, a cyclic diacetonide of glucose was selectively cleaved to the monoacetonide.[2] A summary[3] describes the selective protection of primary and secondary hydroxyl groups in a synthesis of gentiobiose, carried out in the 1870s, as triphenylmethyl ethers.

DEVELOPMENT OF NEW PROTECTIVE GROUPS

As chemists proceeded to synthesize more complicated structures, they developed more satisfactory protective groups and more effective methods for the formation and cleavage of protected compounds. At first a tetrahydropyranyl acetal was prepared,[4] by an acid-catalyzed reaction with dihydropyran, to protect a hydroxyl group. The acetal is readily cleaved by mild acid hydrolysis, but formation of this acetal introduces a new stereogenic center. Formation of the 4-methoxytetrahydropyranyl ketal[5] eliminates this problem.

Catalytic hydrogenolysis of an *O*-benzyl protective group is a mild, selective method introduced by Bergmann and Zervas[6] to cleave a benzyl carbamate (>NCO–$OCH_2C_6H_5$ → >NH) prepared to protect an amino group during peptide syntheses. The method has also been used to cleave alkyl benzyl ethers, stable compounds prepared to protect alkyl alcohols; benzyl esters are cleaved by catalytic hydrogenolysis under neutral conditions.

Three selective methods to remove protective groups have received attention: "assisted," electrolytic, and photolytic removal. Four examples illustrate "assisted removal" of a protective group. A stable allyl group can be converted to a labile vinyl ether group (eq. 4)[7]; a β-haloethoxy (eq. 5)[8] or a β-silylethoxy (eq. 6)[9] derivative is cleaved by attack at the β-substituent; and a stable *o*-nitrophenyl derivative can be reduced to the *o*-amino compound, which undergoes cleavage by nucleophilic displacement (eq. 7):[10]

$$\text{(4)}\quad ROCH_2CH{=}CH_2 \xrightarrow{t\text{-}BuO^-} [ROCH{=}CHCH_3] \xrightarrow{H_3O^+} ROH$$

$$\text{(5)}\quad RO{-}CH_2{-}CCl_3 + Zn \longrightarrow RO^- + CH_2{=}CCl_2$$

$$\text{(6)}\quad RO{-}CH_2{-}CH_2{-}SiMe_3 \xrightarrow{F^-} RO^- + CH_2{=}CH_2 + FSiMe_3$$

R = alkyl, aryl, R′CO–, or R′NHCO–

(7)

NO2 → NH2 → H N O + NH

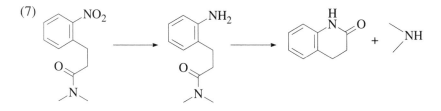

The design of new protective groups that are cleaved by "assisted removal" is a challenging and rewarding undertaking.

Removal of a protective group by electrolytic oxidation or reduction is useful in some cases. An advantage is that the use and subsequent removal of chemical oxidants or reductants (e.g., Cr or Pb salts; Pt– or Pd–C) are eliminated. Reductive cleavages have been carried out in high yield at −1 to −3 V (vs. SCE), depending on the group; oxidative cleavages in good yield have been realized at 1.5–2 V (vs. SCE). For systems possessing two or more electrochemically labile protective groups, selective cleavage is possible when the half-wave potentials, $E_{1/2}$, are sufficiently different; excellent selectivity can be obtained with potential differences on the order of 0.25 V. Protective groups that have been removed by electrolytic oxidation or reduction are described at the appropriate places in this book; a review article by Mairanovsky[11] discusses electrochemical removal of protective groups.[12]

Photolytic cleavage reactions (e.g., of *o*-nitrobenzyl, phenacyl, and nitrophenylsulfenyl derivatives) take place in high yield on irradiation of the protected compound for a few hours at 254–350 nm. For example, the *o*-nitrobenzyl group, used to protect alcohols,[13] amines,[14] and carboxylic acids,[15] has been removed by irradiation. Protective groups that have been removed by photolysis are described at the appropriate places in this book; in addition, the reader may wish to consult five review articles.[16–20]

One widely used method involving protected compounds is solid-phase synthesis[21–24] (polymer-supported reagents). This method has the advantage of requiring only a simple workup by filtration such as in automated syntheses, especially of polypeptides, oligonucleotides, and oligosaccharides.

Internal protection, used by van Tamelen in a synthesis of colchicine, may be appropriate:[25]

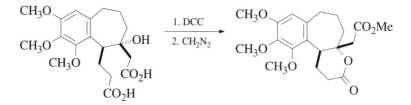

CH_3O, CH_3O, CH_3O, OH, CO_2H, CO_2H —(1. DCC; 2. CH_2N_2)→ CH_3O, CH_3O, CH_3O, CO_2Me, O, O

SELECTION OF A PROTECTIVE GROUP FROM THIS BOOK

To select a specific protective group, the chemist must consider in detail all the reactants, reaction conditions, and functionalities involved in the proposed synthetic scheme. First, he or she must evaluate all functional groups in the reactant to determine those that will be unstable to the desired reaction conditions and that, accordingly, require protection. Then the chemist should examine the reactivities of possible protective groups, listed in the Reactivity Charts, to determine whether the protective group and the reaction conditions are compatible. A guide to these considerations is found in Chapter 10. (The protective groups listed in the Reactivity Charts in that chapter were the most widely used groups at the time the charts were prepared in 1979 in a collaborative effort with other members of Professor Corey's research group.) The chemist should consult the complete list of protective groups in the relevant chapter and consider their properties. It will frequently be advisable to examine the use of one protective group for several functional groups (e.g., a 2,2,2-trichloroethyl group to protect a hydroxyl group as an ether, a carboxylic acid as an ester, and an amino group as a carbamate). When several protective groups are to be removed simultaneously, it may be advantageous to use the same protective group to protect different functional groups (e.g., a benzyl group, removed by hydrogenolysis, to protect an alcohol and a carboxylic acid). When selective removal is required, different classes of protection must be used (e.g., a benzyl ether cleaved by hydrogenolysis, but stable to basic hydrolysis, to protect an alcohol, and an alkyl ester cleaved by basic hydrolysis, but stable to hydrogenolysis, to protect a carboxylic acid). One often overlooked issue in choosing a protective group is that the electronic and steric environments of a given functional group will greatly influence the rates of formation and cleavage. As an obvious example, a tertiary acetate is much more difficult to form or cleave than a primary acetate.

If a satisfactory protective group has not been located, the chemist has a number of alternatives available: rearrange the order of some of the steps in the synthetic scheme, so that a functional group no longer requires protection or a protective group that was reactive in the original scheme is now stable; redesign the synthesis, possibly making use of latent functionality[26] (i.e., a functional group in a precursor form, e.g., anisole as a precursor of cyclohexanone); include the synthesis of a new protective group in the overall plan; or, better yet, design new chemistry that avoids the use of a protective group.

Several books and chapters are associated with protective group chemistry. Some of these cover the area;[27,28] others deal with more limited aspects. Protective groups continue to be of great importance in the synthesis of three major classes of naturally occurring substances—peptides,[22] carbohydrates,[23] and oligonucleotides[24]—and significant advances have been made in solid-phase synthesis,[22–24] including automated procedures. The use of enzymes in the protection and deprotection of functional groups has been reviewed.[29] Special attention is also called to a review on selective deprotection of silyl ethers.[30]

SYNTHESIS OF COMPLEX SUBSTANCES: TWO EXAMPLES (AS USED IN THE SYNTHESIS OF HIMASTATIN AND PALYTOXIN) OF THE SELECTION, INTRODUCTION, AND REMOVAL OF PROTECTIVE GROUPS

Synthesis of Himastatin

Himastatin, isolated from an actinomycete strain (ATCC) from the Himachal Pradesh State in India and active against gram-positive microorganisms and a variety of tumor probe systems, is a $C_{72}H_{104}N_{14}O_{20}$ compound, **1**.[31] It has a novel bisindolyl structure in which the two halves of the molecule are identical. Each half contains a cyclic peptidal ester that contains an L-tryptophanyl unit, D-threonine, L-leucine, D-[(*R*)-5-hydroxy]piperazic acid, (*S*)-2-hydroxyisovaleric acid, and D-valine. The synthesis of himastatin,[32] which illustrates several important aspects of protective group usage, involved the preparation of the pyrroloindoline moiety **A**, its conversion to the bisindolyl unit **A′**$_2$, synthesis of the peptidal ester moiety **B**, the subsequent joining of these units (**A′**$_2$ and two **B** units), and cyclization leading to himastatin. The following brief account focuses on the protective-group aspects of the synthesis.

Unit A (Scheme 1)

The first objective was the conversion of L-tryptophan into a derivative that could be converted to pyrroloindoline **3**, possessing a *cis* ring fusion and a *syn* relationship of the carboxyl and hydroxyl groups. This was achieved by the conversions shown in Scheme 1. A critical step was ***e***. Of many variants tried, the use of the trityl group on the NH_2 of tryptophan and the *t*-butyl group on the carboxyl resulted in stereospecific oxidative cyclization to afford **3** of the desired *cis–syn* stereochemistry in good yield.

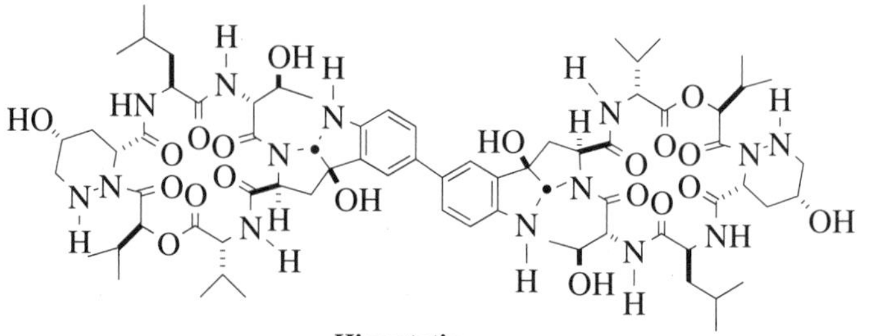

Himastatin
1

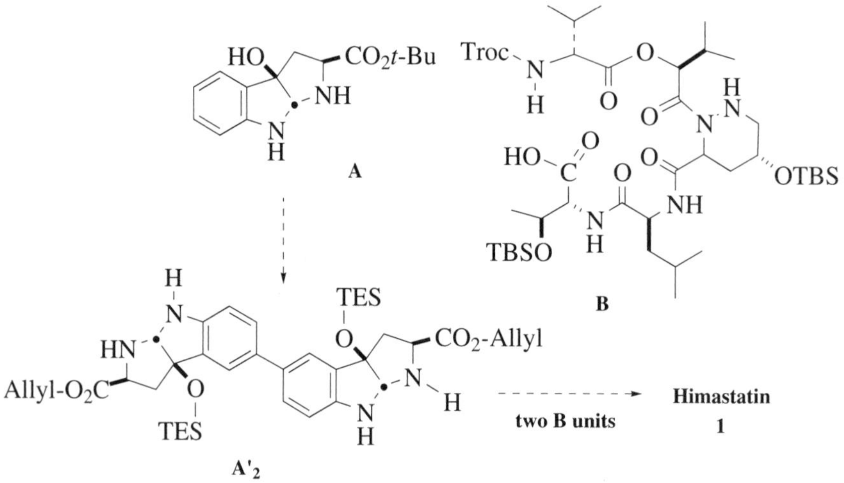

A

B

A'₂

two B units → **Himastatin** **1**

Bisindolyl Unit A′$_2$ (Schemes 2 and 3)

The conversion of **3** to **8** is summarized in Scheme 2. The trityl group (too large and too acid sensitive for the ensuing steps) was removed from N, and both N's were protected by Cbz (benzyloxycarbonyl) groups. Protection of the tertiary OH specifically as the robust TBS (*t*-butyldimethylsilyl) group was found to be necessary for the sequence involving the electrophilic aromatic substitution step, **5** to **6**, and the Stille coupling steps (**6** + **7** → **8**).

The TBS group then had to be replaced (two steps, Scheme 3: **a** and **b**) by the more easily removable TES (triethylsilyl) group to permit deblocking at the last step in the synthesis of himastatin. Before combination of the bisindolyl unit with the peptidal ester unit, several additional changes in the state of protection at the two nitrogens and the carboxyl of **8** were needed (Schemes 2 and 3). The Cbz protective groups were removed from both N's, and the more reactive pyrrolidine N was protected as the FMOC (fluorenylmethoxycarbonyl) group. At the carboxyl, the *t*-butyl group was replaced by the allyl group. [The smaller allyl group was needed for the later condensation of the adjacent pyrrolidine nitrogen of **15** with the threonine carboxyl of **24** (Scheme 5); also, the allyl group can be

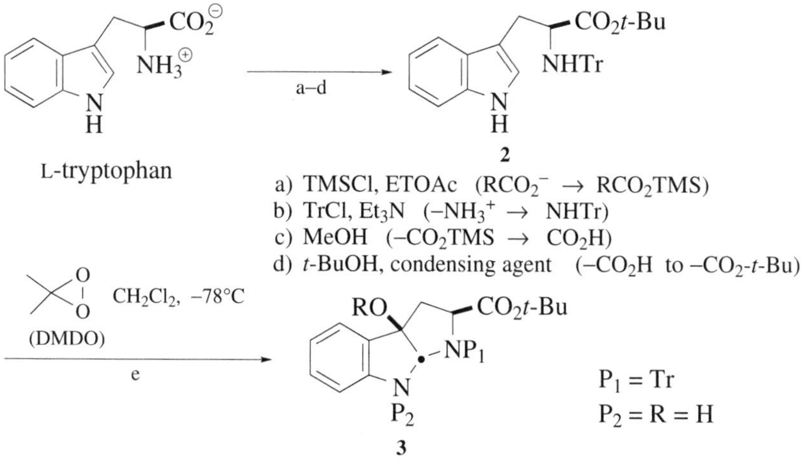

L-tryptophan → (a–d) → 2

a) TMSCl, ETOAc ($RCO_2^- \rightarrow RCO_2TMS$)
b) TrCl, Et_3N ($-NH_3^+ \rightarrow$ NHTr)
c) MeOH ($-CO_2TMS \rightarrow CO_2H$)
d) t-BuOH, condensing agent ($-CO_2H$ to $-CO_2$-t-Bu)

(DMDO) CH_2Cl_2, −78°C (e) → 3 P_1 = Tr; P_2 = R = H

Scheme 1

3 P_1 = Tr; P_2 = R = H
(a) ↓
4 P_1 = P_2 = R = H
(b) ↓
5 P_1 = P_2 = Cbz; R = TBS → (c)

(a) HOAc, MeOH, CH_2Cl_2 (N-Trityl → NH)
(b) (i) CbzCl, pyridine, CH_2Cl_2 (both NH's → N-Cbz)
(ii) TBSCl, DBU, CH_3CN (29% from 2) (−OH → OTBS)
(c) ICl, 2,6-di-t-butylpyridine, CH_2Cl_2 (75%) (X = H → X = I)
(d) Me_6Sn_2, $Pd(Ph_3P)_4$, THF (86%) (X = I → X = $SnMe_3$)
(e) **6**, Pd_2dba_3, Ph_3As, DMF, 45°C, (79%) (**6** + **7** → **8**)

6 R = TBS; P_1 = P_2 = Cbz; X = I
↓ d
7 R = TBS; P_1 = P_2 = Cbz; X = $SnMe_3$
↓ **6**, e
8 R = TBS; P_1 = P_2 = Cbz; X = dimer

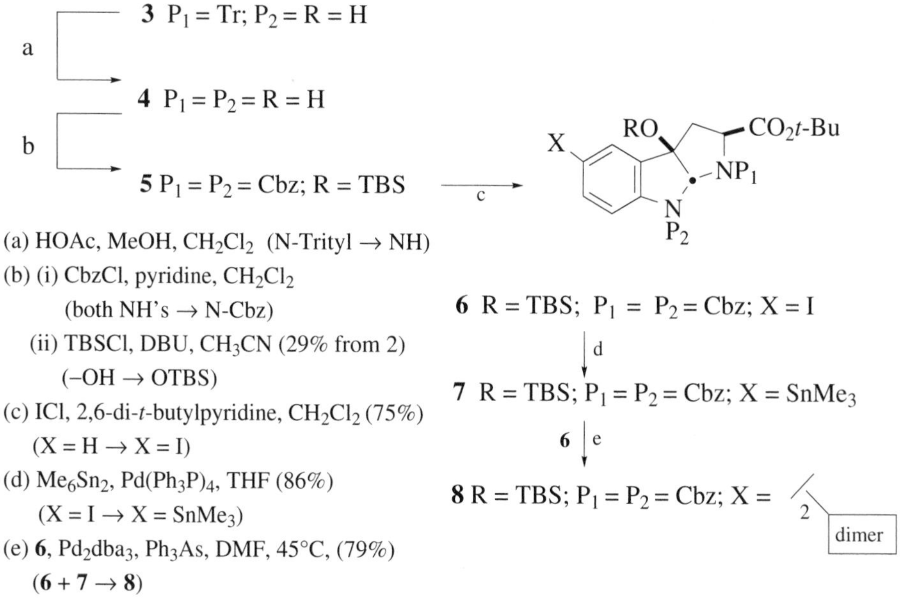

Scheme 2

cleaved by the $Pd(Ph_3P)_4$-$PhSiH_3$ method, conditions under which many protective groups (including, of course, the other protective groups in **25**; see Scheme 6) are stable.] Returning to Scheme 3, the FMOC groups on the two equivalent pyrrolidine N's were then removed, affording **15**.

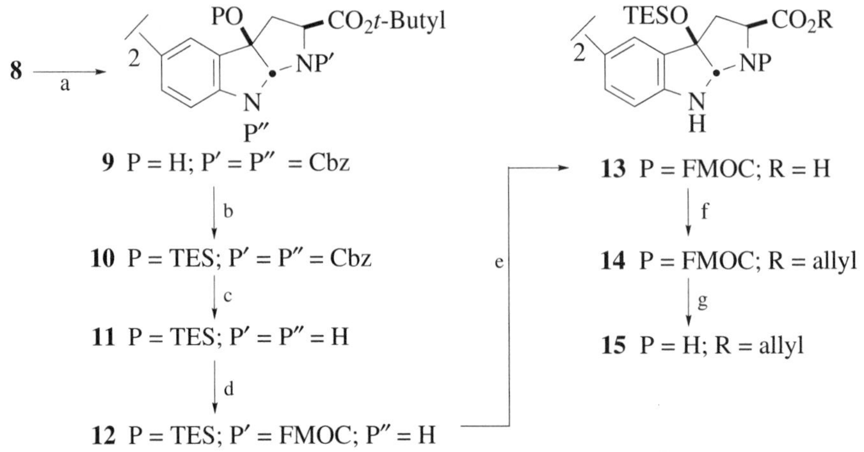

8 →(a)

PO, CO_2t-Butyl, NP′, N, P″, 2

9 P = H; P′ = P″ = Cbz

↓ b

10 P = TES; P′ = P″ = Cbz

↓ c

11 P = TES; P′ = P″ = H

↓ d

12 P = TES; P′ = FMOC; P″ = H

→ e

TESO, CO_2R, NP, N H, 2

13 P = FMOC; R = H

↓ f

14 P = FMOC; R = allyl

↓ g

15 P = H; R = allyl

(a) TBAF, THF (91%) (TBSO– → HO–)
(b) TESCl, DBU, DMF (92%) (HO– → TESO–)
(c) H_2, Pd/C, EtOAc (100%) (both NCbz's → NH)
(d) FMOC-HOSU, pyridine, CH_2Cl_2 (95%) (NH → NFMOC)
(e) TESOTf, lutidine, CH_2Cl_2 ($-CO_2$–t-Bu → $-CO_2H$)
(f) allyl alcohol, DBAD, Ph_3P, CH_2Cl_2 (90% from **12**) ($-CO_2H$ → $-CO_2$–allyl
(g) piperidine, CH_3CN (74%) (NFMOC → NH)

Scheme 3

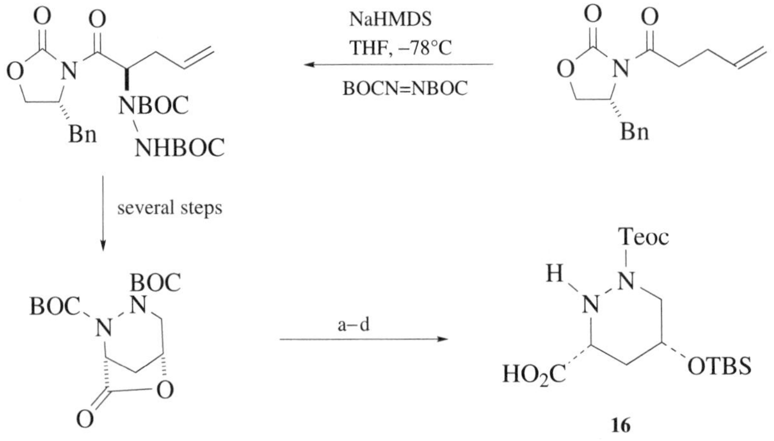

O O O N Bn NBOC NHBOC ← NaHMDS, THF, –78°C, BOCN=NBOC — O O O N Bn

↓ several steps

BOC BOC N N O O →(a–d) Teoc H N N HO_2C OTBS

16

(a) TFA (both –NBOC's → NH)
(b) TeocCl, pyridine (–NH → NTeoc)
(c) LiOH (lactone → $-CO_2^-$ + HO–)
(d) TBSOTf, lutidine (–OH → –OTBS)

Scheme 4

Peptidal Ester Unit B (Schemes 4 and 5)

Several of these steps are common in peptide synthesis and involve standard protective groups. Attention is called to the 5-hydroxypiperazic acid. Its synthesis (Scheme 4) has the interesting feature of the introduction of the two nitrogens in protected form as BOC (*t*-butoxycarbonyl) groups in the same step. Removal of the BOC groups and selective conversion of the nitrogen furthest from the carboxyl group into the N-Teoc (2-trimethylsilylethoxycarbonyl) group, followed by hydrolysis of the lactone and TBS protection of the hydroxyl,

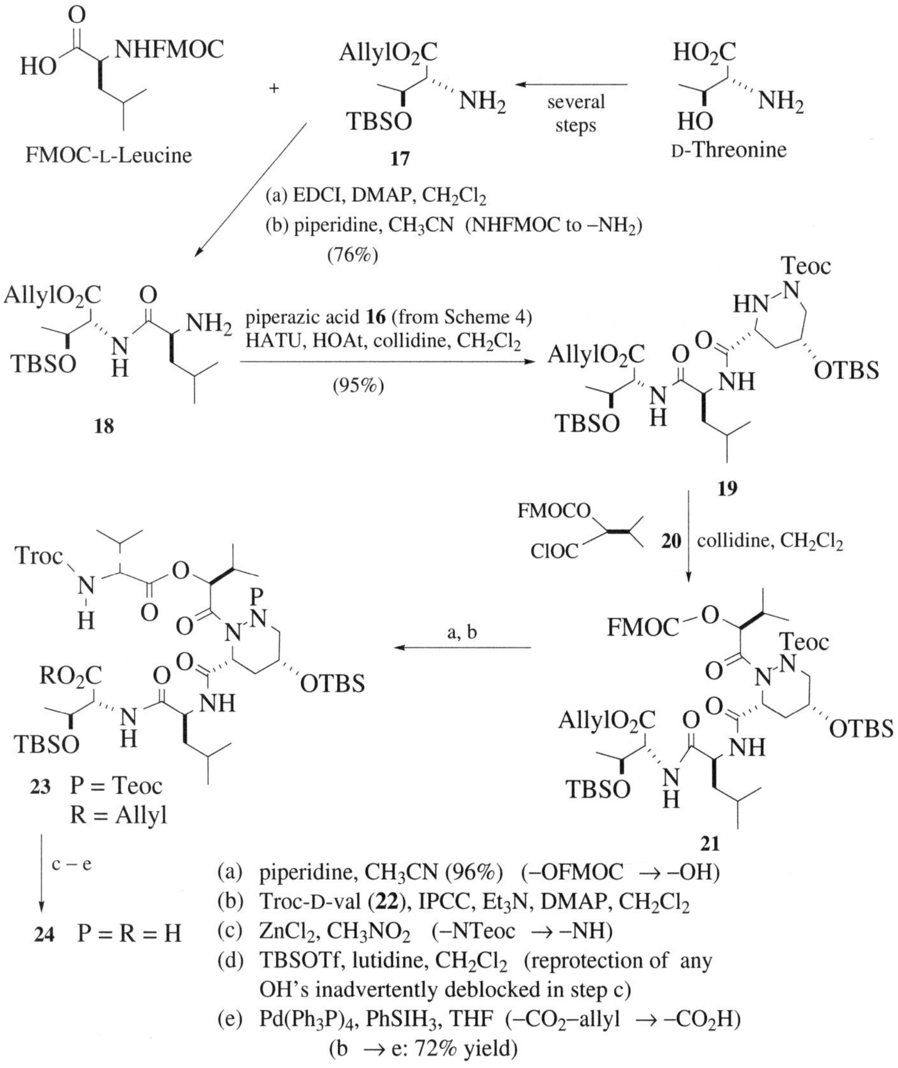

FMOC-L-Leucine + 17 ← several steps ← D-Threonine

(a) EDCI, DMAP, CH_2Cl_2
(b) piperidine, CH_3CN (NHFMOC to $-NH_2$)
(76%)

18 → 19: piperazic acid **16** (from Scheme 4), HATU, HOAt, collidine, CH_2Cl_2 (95%)

19 → 21: **20**, collidine, CH_2Cl_2

21 → 23: a, b

23 P = Teoc, R = Allyl

c – e

24 P = R = H

(a) piperidine, CH_3CN (96%) (–OFMOC → –OH)
(b) Troc-D-val (**22**), IPCC, Et_3N, DMAP, CH_2Cl_2
(c) $ZnCl_2$, CH_3NO_2 (–NTeoc → –NH)
(d) TBSOTf, lutidine, CH_2Cl_2 (reprotection of any OH's inadvertently deblocked in step c)
(e) $Pd(Ph_3P)_4$, $PhSIH_3$, THF ($-CO_2$–allyl → $-CO_2H$)
(b → e: 72% yield)

Scheme 5

afforded the piperazic acid entity **16** in a suitable form for combination with dipeptide **18** (Scheme 5). Because of the greater reactivity of the leucyl $-NH_2$ group of **18** in comparison to the piperazyl $-N_{\alpha}H$ group in **16**, it was not necessary to protect this piperazyl –NH in the condensation of **18** and **16** to form **19**. In the following step (**19** + **20** → **21**), this somewhat hindered piperazyl –NH **is** condensed with the acid chloride **20**. Note that the hydroxyl in **20** is protected by the FMOC group—not commonly used in hydroxyl protection. A requirement for the protective group on this hydroxyl was that it be removable (for the next condensation: **21** + Troc-D-valine **22** → **23**) under conditions that would leave unaltered the –COO–allyl, the N-Teoc, and the OTBS groups. The FMOC group (cleavage by piperidine) met this requirement. The choice of the Troc (2,2,2-trichloroethoxycarbonyl) group for N-protection of valine was based on the requirements of removability, without affecting the OTBS and OTES groups, and stability to the conditions of removal of allyl from –COO–allyl [easily met by the use of $Pd(Ph_3P)_4$ for this deblocking].

Himastatin 1 (Scheme 6)

Of special importance to the synthesis was the choice of condensing agents and conditions.[33] HATU-HOAt[34] was of particular value in these final stages. Condensation of the threonine carboxyl of **24** (from Scheme 5) with the pyrrolidine N's of the bisindolyl compound **15** (from Scheme 3) afforded **25**. Removal

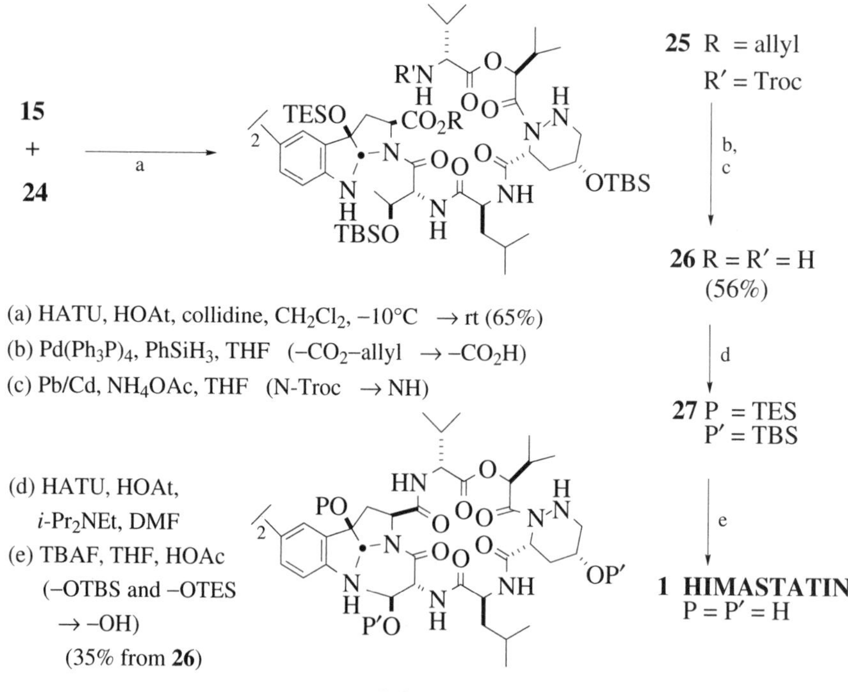

Scheme 6

of the allyl groups from the tryptophanyl carboxyls and the Troc groups from the valine amino nitrogens, followed by condensation (macrolactamization), gave **27**. Removal of the six silyl groups (the two quite hindered TES groups and the four, more accessible, TBS groups) by fluoride ion afforded himastatin.

Synthesis of Palytoxin Carboxylic Acid

Palytoxin carboxylic acid, $C_{123}H_{213}NO_{53}$, Figure 1 ($R^1 - R^8 = H$), derived from palytoxin, $C_{129}H_{223}N_3O_{54}$, contains 41 hydroxyl groups, one amino group, one ketal, one hemiketal, and one carboxylic acid, in addition to some double bonds and ether linkages.

The total synthesis[35] was achieved through the synthesis of eight different segments, each requiring extensive use of protective group methodology, followed by the appropriate coupling of the various segments in their protected forms.

The choice of what protective groups to use in the synthesis of each segment was based on three aspects: (a) the specific steps chosen to achieve the synthesis of each segment; (b) the methods to be used in coupling the various segments, and (c) the conditions needed to deprotect the 42 blocked groups in order to

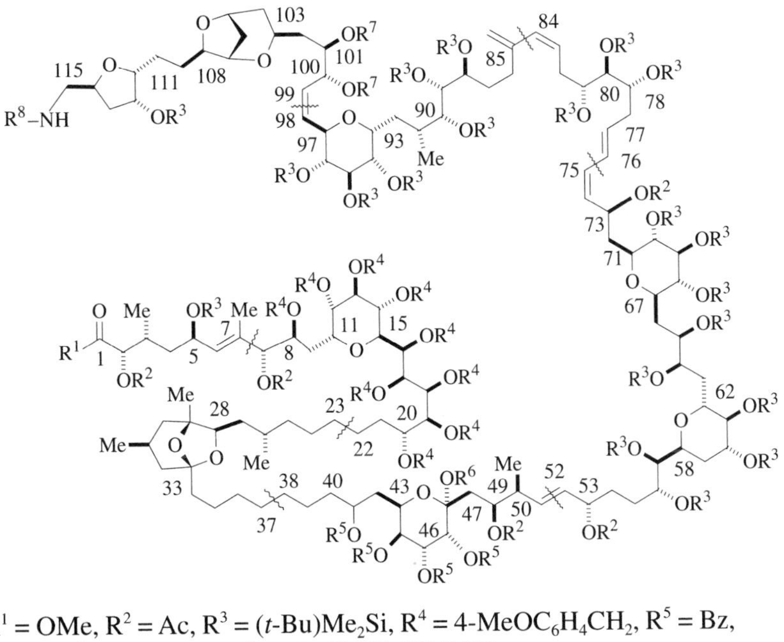

1: $R^1 = OMe$, $R^2 = Ac$, $R^3 = (t\text{-Bu})Me_2Si$, $R^4 = 4\text{-}MeOC_6H_4CH_2$, $R^5 = Bz$, $R^6 = Me$, R^7 = acetonide, $R^8 = Me_3SiCH_2CH_2OCO$

2: Palytoxin carboxylic acid: $R^1 = OH$, $R^2\text{–}R^8 = H$

Figure 1. Palytoxin carboxylic acid.

liberate palytoxin carboxylic acid in its unprotected form. (These conditions must be such that the functional groups already deprotected are stable to the successive deblocking conditions.) Kishi's synthesis employed only eight different protective groups for the 42 functional groups present in the fully protected form of palytoxin carboxylic acid (Figure l, **1**). A few additional protective groups were used for "end group" protection in the synthesis and sequential coupling of the eight different segments. The synthesis was completed by removal of all of the groups by a series of five different methods. The selection, formation, and cleavage of these groups are described next.

For the synthesis of the C.l–C.7 segment, the C.1 carboxylic acid was protected as a methyl ester. The C.5 hydroxyl group was protected as the *t*-butyldimethylsilyl (TBS) ether. This particular silyl group was chosen because it improved the chemical yield and stereochemistry of the Ni(II)/Cr(II)-mediated coupling reaction of segment C.1–C.7 with segment C.8–C.51. Nine hydroxyl groups were protected as *p*-methoxyphenylmethyl (MPM) ethers, a group that was stable to the conditions used in the synthesis of the C.8–C.22 segment. These MPM groups were eventually cleaved oxidatively by treatment with 2,3-dichloro-5,6-dicyano-1,4-benzoquinone (DDQ).

The C.2 hydroxyl group was protected as an acetate, since cleavage of a *p*-methoxyphenylmethyl (MPM) ether at C.2 proved to be very slow. An acetyl group was also used to protect the C.73 hydroxyl group during synthesis of the right-hand half of the molecule (C.52–C.115). Neither a *p*-methoxyphenylmethyl (MPM) nor a *t*-butyldimethylsilyl (TBS) ether was satisfactory at C.73: dichlorodicyanobenzoquinone (DDQ) cleavage of a *p*-methoxyphenylmethyl (MPM) ether at C.73 resulted in oxidation of the *cis-trans* dienol at C.78–C.73 to a *cis-trans* dienone. When C.73 was protected as a *t*-butyldimethylsilyl (TBS) ether, Suzuki coupling of segment C.53–C.75 (in which C.75 was a vinyl iodide) to segment C.76–C.115 was too slow. In the synthesis of segment C.38–C.51, the C.49 hydroxyl group was also protected at one stage as an acetate, to prevent benzoate migration from C.46. The C.8 and C.53 hydroxyl groups were protected as acetates for experimental convenience. A benzoate ester, more electron withdrawing than an acetate ester, was used to protect the C.46 hydroxyl group to prevent spiroketalization of the C.43 and C.51 hydroxyl groups during synthesis of the C.38–C.51 segment. Benzoate protection of the C.46 hydroxyl group also increased the stability of the C.47 methoxy group (part of a ketal) under acidic cleavage conditions. Benzoates, rather than acetates, were used during the synthesis of the C.38–C.51 segment, since they were more stable and were better chromophores in purification and characterization.

Several additional protective groups were employed in the coupling of the eight different segments. A tetrahydropyranyl (THP) group was used to protect the hydroxyl group at C.8 in segment C.8–C.22, and a *t*-butyldiphenylsilyl (TBDPS) group for the hydroxyl group at C.37 in segment C.23–C.37. The TBDPS group at C.37 was later removed by $Bu_4N^+F^-$/THF in the presence of nine *p*-methoxyphenylmethyl (MPM) groups. After the coupling of segment C.8–C.37 with segment C.38–C.51, the C.8 THP ether was hydrolyzed with pyridinium *p*-toluenesulfonate (PPTS) in methanol-ether, 42°, in the presence of the bicyclic ketal at C.28–C.33 and the cyclic ketal at C.43–C.47. (As noted previously, the resistance of this ketal to these acidic conditions was due to the electron-withdrawing effect of the benzoate at C.46.) A cyclic acetonide (a 1,3-dioxane) at C.49–C.51 was also removed by this step and had to be reformed (acetone/PPTS) prior to the coupling of segment C.8–C.51 with segment C.1–C.7. After the coupling of these segments to form segment C.1–C.51, the new hydroxyl group at C.8 was protected as an acetate, and the acetonide at C.49–C.51 was, again, removed without alteration of the bicyclic ketal at C.28–C.33 or the cyclic ketal at C.43–C.47, still stabilized by the benzoate at C.46.

The synthesis of segment C.77–C.115 from segments C.77–C.84 and C.85–C.115 involved the liberation of an aldehyde at C.85 from its protected form as a dithioacetal, $RCH(SEt)_2$, by mild oxidative deblocking ($I_2/NaHCO_3$, acetone, water) and the use of the *p*-methoxyphenyldiphenylmethyl (MMTr) group to protect the hydroxyl group at C.77. The C.77 MMTr ether was subsequently converted to a primary alcohol (PPTS/MeOH-CH_2Cl_2, rt) without affecting the 19 *t*-butyldimethylsilyl (TBS) ethers or the cyclic acetonide at C.100–C.101.

The C.100–C.101 diol group, protected as an acetonide, was stable to the Wittig reaction used to form the *cis* double bond at C.98–C.99 and to all of the conditions used in the buildup of segment C.99–C.115 to fully protected palytoxin carboxylic acid (Figure 1, **1**).

The C.115 amino group was protected as a trimethylsilylethyl carbamate ($Me_3SiCH_2CH_2OCONHR$), a group that was stable to the synthesis conditions, and cleaved by the conditions used to remove the *t*-butyldimethylsilyl (TBS) ethers.

Thus, the 42 functional groups in palytoxin carboxylic acid (39 hydroxyl groups, one diol, one amino group, and one carboxylic acid) were protected by eight different groups:

1 methyl ester	–COOH
5 acetate esters	–OH
20 *t*-butyldimethylsilyl (TBS) ethers	–OH
9 *p*-methoxyphenylmethyl (MPM) ethers	–OH
4 benzoate esters	–OH
1 methyl "ether"	–OH of a hemiketal
1 acetonide	1,2-diol
1 $Me_3SiCH_2CH_2OCO$	$-NH_2$

The protective groups were then removed in the following order by the five methods listed :

(1) To cleave *p*-methoxyphenylmethyl (MPM) ethers: DDQ (dichlorodicyanobenzoquinone)/*t*-BuOH–CH_2Cl_2–phosphate buffer (pH 7.0), 4.5 h.

(2) To cleave the acetonide: 1.18 *N* $HClO_4$–THF, 25°, 8 days.
(3) To hydrolyze the acetates and benzoates: 0.08 *N* LiOH/H_2O–MeOH–THF, 25°, 20 h.
(4) To remove *t*-butyldimethylsilyl (TBS) ethers and the carbamoyl ester ($Me_3SiCH_2CH_2OCONHR$): $Bu_4N^+F^-$, THF, 22°, 18 h → THF–DMF, 22°, 72 h.
(5) To hydrolyze the methyl ketal at C.47, no longer stabilized by the C.46 benzoate: HOAc–H_2O, 22°, 36 h.

This order was chosen so that DDQ (dichlorodicyanobenzoquinone) treatment would not oxidize a deprotected allylic alcohol at C.73 and so that the C.47 hemiketal would still be protected (as the ketal) during basic hydrolysis (Step 3).

And so the skillful selection, introduction, and removal of a total of 12 different protective groups has played a major role in the successful total synthesis of palytoxin carboxylic acid (Figure 1, **2**).

1. A. Robertson and R. Robinson, *J. Chem. Soc.*, 1460 (1928).
2. E. Fischer, *Ber.*, **28**, 1145–1167 (1895); see p. 1165.
3. B. Helferich, *Angew. Chem.*, **41**, 871 (1928).
4. W. E. Parham and E. L. Anderson, *J. Am. Chem. Soc.*, **70**, 4187 (1948).
5. C. B. Reese, R. Saffhill, and J. E. Sulston, *J. Am. Chem. Soc.*, **89**, 3366 (1967).
6. M. Bergmann and L. Zervas, *Chem. Ber.*, **65**, 1192 (1932).
7. J. Cunningham, R. Gigg, and C. D. Warren, *Tetrahedron Lett.*, 1191 (1964).
8. R. B. Woodward, K. Heusler, J. Gosteli, P. Naegeli, W. Oppolzer, R. Ramage, S. Ranganathan, and H. Vorbruggen, *J. Am. Chem. Soc.*, **88**, 852 (1966).
9. P. Sieber, *Helv. Chim. Acta*, **60**, 2711 (1977).
10. I. D. Entwistle, *Tetrahedron Lett.*, 555 (1979).
11. V. G. Mairanovsky, *Angew. Chem., Int. Ed. Engl.*, **15**, 281 (1976).
12. See also M. F. Semmelhack and G. E. Heinsohn, *J. Am. Chem. Soc.*, **94**, 5139 (1972).
13. S. Uesugi, S. Tanaka, E. Ohtsuka, and M. Ikehara, *Chem. Pharm. Bull.*, **26**, 2396 (1978).
14. S. M. Kalbag and R. W. Roeske, *J. Am. Chem. Soc.*, **97**, 440 (1975).
15. L. D. Cama and B. G. Christensen, *J. Am. Chem. Soc.*, **100**, 8006 (1978).
16. V. N. R. Pillai, *Synthesis*, 1–26 (1980).
17. P. G. Sammes, *Quart. Rev., Chem. Soc.*, **24**, 37–68 (1970); see pp. 66-68.
18. B. Amit, U. Zehavi, and A. Patchornik, *Isr. J. Chem.*, **12**, 103–113 (1974).
19. V. N. R. Pillai, "Photolytic Deprotection and Activation of Functional Groups," *Org. Photochem.*, **9**, 225–323 (1987).
20. V. Zehavi, "Applications of Photosensitive Protecting Groups in Carbohydrate Chemistry," *Adv. Carbohydr. Chem. Biochem.*, **46**, 179–204 (1988).

21. (a) R. B. Merrifield, *J. Am. Chem. Soc.*, **85**, 2149 (1963); (b) P. Hodge, *Chem. Ind. (London)*, 624 (1979); (c) C. C. Leznoff, *Acc. Chem. Res.*, **11**, 327 (1978); (d) E. C. Blossey and D. C. Neckers, Eds., *Solid Phase Synthesis*, Halsted, New York, 1975; P. Hodge and D. C. Sherrington, Eds., *Polymer-Supported Reactions in Organic Synthesis*, Wiley-Interscience, New York, 1980 (A comprehensive review of polymeric protective groups by J. M. J. Fréchet is included in this book); (e) D. C. Sherrington and P. Hodge, *Synthesis and Separations Using Functional Polymers*, Wiley-Interscience, New York, 1988.

22. **Peptides**: (a) M. W. Pennington and B. M. Dunn, Eds., *Methods in Molecular Biology, Vol 35: Peptide Synthesis Protocols*, Humana Press, Totowa, NJ, 1994, pp. 91–169; (b) G. Grant, Ed., *Synthetic Peptides*, W. H. Freeman & Co.; New York, 1992; (c) *Novabiochem 97/98, Catalog*, Technical Section S1-S85 (this section contains many valuable details on the use and manipulation of protective groups in the amino acid–peptide area); (d) J. M. Stewart and J. D. Young, *Solid Phase Peptide Synthesis*, 2d ed., Pierce Chemical Company, Rockford, IL, 1984; (e) E. Atherton and R. C. Sheppard, *Solid Phase Peptide Synthesis. A Practical Approach*, Oxford-IRL Press, New York, 1989; (f) R. Epton, Ed., *Innovation and Perspectives in Solid Phase Synthesis: Peptides, Polypeptides and Oligonucleotides; Collected Papers, First International Symposium: Macro-Organic Reagents and Catalysts*, SPCC, U.K., 1990; R. Epton, Ed., *Collected Papers, Second International Symposium*, Intercept Ltd, Andover, U.K., 1992; (g) V. J. Hruby and J.-P. Meyer, "The Chemical Synthesis of Peptides," in *Bioorganic Chemistry: Peptides and Proteins*, S. M. Hecht, Ed., Oxford University Press, New York, 1998, Chapter 2, pp. 27–64.

23. **Oligonucleotides**: (a) S. L. Beaucage and R. P. Iyer, *Tetrahedron*, **48**, 2223 (1992); **49**, 1925 (1993); **49**, 6123 (1993); **49**, 10441 (1993); (b) J. W. Engels and E. Uhlmann, *Angew. Chem., Int. Ed. Engl.*, **28**, 716 (1989); (c) S. L. Beaucage and M. H. Caruthers, "The Chemical Synthesis of DNA/RNA," in *Bioorganic Chemistry: Nucleic Acids*, S. M. Hecht, Ed., Oxford University Press, New York, 1996, Chapter 2, pp. 36–74.

24. **Oligosaccharides**: (a) S. J. Danishefsky and M. T. Bilodeau, "Glycals in Organic Synthesis: The Evolution of Comprehensive Strategies for the Assembly of Oligosaccharides and Glycoconjugates of Biological Consequence," *Angew. Chem., Int. Ed. Engl.*, **35**, 1380 (1996); (b) P. H. Seeberger and S. J. Danishefsky, *Acc. Chem. Res.*, **31**, 685 (1998); (c) P. H. Seeberger, M. T. Bilodeau, and S. J. Danishefsky, *Aldrichchimica Acta*, **30**, 75 (1997); (d) J. Y. Roberge, X. Beebe, and S. J. Danishefsky, "Solid Phase Convergent Synthesis of *N*-Linked Glycopeptides on a Solid Support," *J. Am. Chem. Soc.*, **120**, 3915 (1998); (e) B. Frasier-Reid, R. Madsen, A. S. Campbell, C. S. Roberts, and J. R. Merritt, "Chemical Synthesis of Oligosaccharides," in *Bioorganic Chemistry: Oligosaccharides*, S. M. Hecht, Ed., Oxford University Press, 1999, Chapter 3, pp. 89–133; (f) K. C. Nicolaou and N. J. Bockovich, "Chemical Synthesis of Complex Carbohydrates," in *Bioorganic Chemistry: Oligosaccharides*, S. M. Hecht, Ed., Oxford University Press, 1999, Chapter 4, pp. 134–173.

25. E. E. van Tamelen, T. A. Spencer, Jr., D. S. Allen, Jr., and R. L. Orvis, *Tetrahedron*, **14**, 8 (1961).

26. D. Lednicer, *Adv. Org. Chem.*, **8**, 179–293 (1972).

27. (a) H. Kunz and H. Waldmann, "Protecting Groups," in *Comprehensive Organic Synthesis*, B. M. Trost, Ed., Pergamon Press, Oxford, U.K., 1991, Vol. 6, 631–701;

(b) P. J. Kocienski, *Protecting Groups*, Georg Theime Verlag, Stuttgart and New York, 1994; (c) J. F. W. McOmie, Ed., *Protective Groups in Organic Chemistry*, Plenum, New York and London, 1973.

28. *Organic Syntheses*, Wiley-Interscience, New York, *Collect. Vols. I–IX*, 1941–1998, **75**, 1997; W. Theilheimer, Ed., *Synthetic Methods of Organic Chemistry*, S. Karger, Basel, Vols 1–52, 1946–1997; E. Müller, Ed., *Methoden der Organischen Chemie* (Houben-Weyl), Georg Thieme Verlag, Stuttgart, Vols. 1–21f, 1958–1995; *Spec. Period. Rep.*: *General and Synthetic Methods,* Royal Society of Chemistry, **1–16** (1978–1994); S. Patai, Ed., *The Chemistry of Functional Groups*, Wiley-Interscience, Vols. 1–51, 1964–1997.
29. (a) H. Waldmann and D. Sebastian, "Enzymatic Protecting Group Techniques," *Chem. Rev.*, **94**, 911 (1994); (b) K. Drauz and H. Waldmann, Eds., *Enzyme Catalysis in Organic Synthesis: A Comprehensive Handbook*, VCH, 1995, Vol. 2, 851–889.
30. T. D. Nelson and R. D. Crouch, "Selective Deprotection of Silyl Ethers," *Synthesis*, 1031 (1996).
31. (a) K.-S. Lam, G. A. Hesler, J. M. Mattel, S.W. Mamber, and S. Forenza, *J. Antibiot.*, **43**, 956 (1990); (b) J. E. Leet, D. R. Schroeder, B. S. Krishnan, and J. A. Matson, *ibid.*, **43**, 961 (1990); (c) J. E. Leet, D. R. Schroeder, J. Golik, J. A. Matson, T. W. Doyle, K. S. Lam, S. E. Hill, M. S. Lee, J. L. Whitney, and B. S. Krishnan, *ibid.* **49**, 299 (1996); (d) T. M. Kamenecka and S. J. Danishefsky, "Studies in the Total Synthesis of Himastatin: A Revision of the Stereochemical Assignment," *Angew. Chem., Int. Ed. Engl.*, **37**, 2993 (1998).
32. T. M. Kamenecka and S. J. Danishefsky, "The Total Synthesis of Himastatin: Confirmation of the Revised Stereostructure," *Angew. Chem., Int. Ed. Engl.*, **37**, 2995 (1998). We thank Professor Danishefsky for providing us with preprints of the himastatin communications here and in ref. 31(d).
33. (a) J. M. Humphrey and A. R. Chamberlin, *Chem. Rev.*, **97**, 2241 (1997); (b) A. Ehrlich, H.-U. Heyne, R. Winter, M. Beyermann, H. Haber, L. A. Carpino, and M. Bienert, *J. Org. Chem.*, **61**, 8831 (1996).
34. HOAt, 7-aza-1-hydroxybenzotriazole; HATU (CAS Registry No. 148893-10-1), *N*-[(dimethylamino) (3*H*-1,2,3-triazolo(4,5-*b*)pyridin-3-yloxy)methylene]-*N*-methylmethanaminium hexafluorophosphate, previously known as *O*-(7-azabenzotriazol-1-yl)-1,1,3,3-tetramethyluronium hexafluorophosphate. [Note: Assignment of structure to HATU as a guanidinium species rather than as a uronium species, i.e., attachment of the $(Me_2NC{=}NMe_2)^+$ unit to N_3 of 7-azabenzotriazole 1-*N*-oxide instead of to the *O*, is based on X-ray analysis (ref. 33b)].
35. R. W. Armstrong, J.-M. Beau, S. H. Cheon, W. J. Christ, H. Fujioka, W.-H. Ham, L. D. Hawkins, H. Jin, S. H. Kang, YOSHITO KISHI, M. J. Martinelli, W. W. McWhorter, Jr., M. Mizuno, M. Nakata, A. E. Stutz, F. X. Talamas, M. Taniguchi, J. A. Tino, K. Ueda, J.-i. Uenishi, J. B. White, and M. Yonaga, *J. Am. Chem. Soc.*, **111**, 7530–7533 (1989). [See also *idem., ibid.*, **111**, 7525–7530 (1989).]

2

PROTECTION FOR THE HYDROXYL GROUP, INCLUDING 1,2- AND 1,3-DIOLS

ETHERS

Hydroxyl groups are present in a number of compounds of biological and synthetic interest, including nucleosides, carbohydrates, steroids, macrolides, polyethers, and the side chain of some amino acids.[1a] During oxidation, acylation, halogenation with phosphorus or hydrogen halides, or dehydration reactions of these compounds, a hydroxyl group must be protected. In polyfunctional molecules, selective protection becomes an issue that has been addressed by the development of a number of new methods. Ethers are among the most used protective groups in organic synthesis and vary from the simplest, most stable methyl ether to the more elaborate, substituted trityl ethers developed for use in nucleotide synthesis. Ethers are formed and removed under a wide variety of conditions. Some of the ethers that have been used extensively to protect alcohols are included in Reactivity Chart 1.[1a,b]

1. (a) See references 23 (**Oligonucleotides**) and 24 (**Oligosaccharides**) in Chapter 10; (b) see also C. B. Reese, "Protection of Alcoholic Hydroxyl Groups and Glycol Systems," in *Protective Groups in Organic Chemistry*, J. F. W. McOmie, Ed., Plenum, New York and London, 1973, pp. 95–143; H. M. Flowers, "Protection of the Hydroxyl Group," in *The Chemistry of the Hydroxyl Group*, S. Patai, Ed., Wiley-Interscience, New York, 1971, Vol. 10/2, pp. 1001–1044; C. B. Reese, *Tetrahedron*, **34**, 3143–3179 (1978), see pp. 3145–3150; V. Amarnath and A. D. Broom, *Chem. Rev.*, **77**, 183–217 (1977), see pp. 184–194; M. Lalonde and T. H. Chan, "Use of Organosilicon Reagents as Protective Groups in Organic Synthesis," *Synthesis*, 817 (1985); and P. Kocienski, *Protecting Groups*, Thieme Medical Publishers, New York, 1994, see p. 21.

Methyl Ether: ROMe (Chart 1)

Formation

1. Me_2SO_4, NaOH, $Bu_4N^+I^-$, org. solvent, 60–90% yield.[1]
2. MeI or Me_2SO_4,[2] NaH or KH, THF. This is the standard method for introducing the methyl ether function onto hindered and unhindered alcohols.

3. CH_2N_2, silica gel, 0–10°, 100% yield.[3]

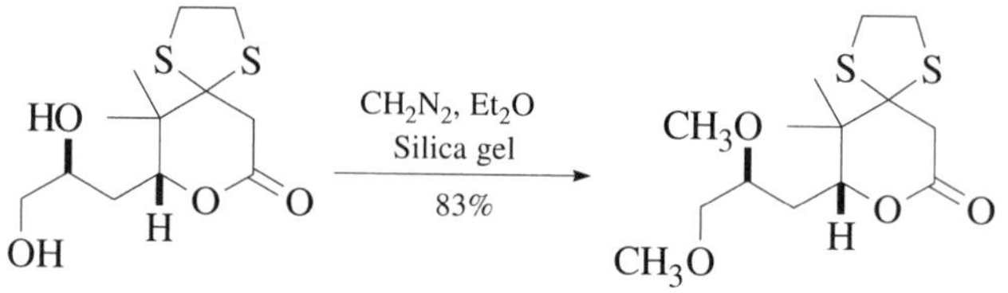

CH_2N_2, Et_2O
Silica gel
83%

Ref. 4

4. CH_2N_2, HBF_4, CH_2Cl_2, Et_3N, 25°, 1 h, 95% yield.[5,6] Hydroxyl amines will *O*-alkylate without the acid catalyst.[7] $TMSCHN_2$ serves as a safe and stable alternative to diazomethane (74–93% yield).[8]
5. MeI, solid KOH, DMSO, 20°, 5–30 min, 85–90% yield.[9]
6. $(MeO)_2POH$, cat. TsOH, 90–100°, 12 h, 60% yield.[10]
7. $Me_3O^+BF_4^-$, 3 days, 55% yield.[11] A simple, large-scale preparation of this reagent has been described.[12]
8. CF_3SO_3Me, CH_2Cl_2, Pyr, 80°, 2.5 h, 85–90% yield.[13,14] The use of 2,6-di-*t*-butyl-4-methylpyridine as a base is also very effective.[15]
9. Because of the increased acidity and reduced steric requirement of the carbohydrate hydroxyl, *t*-BuOK can be used as a base to achieve ether formation.[16]

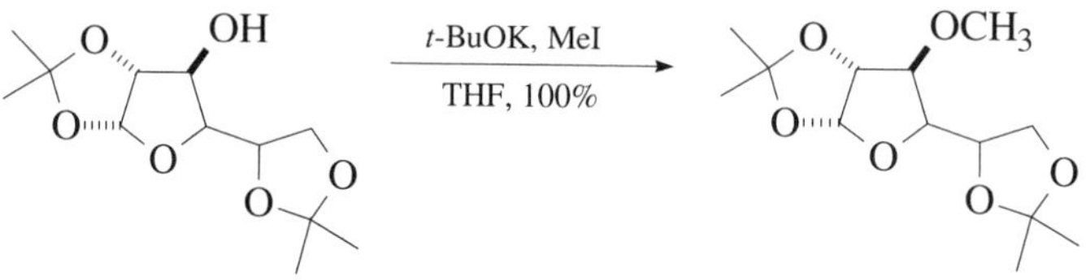

t-BuOK, MeI
THF, 100%

10. MeI, Ag_2O, 93% yield.[17]

MeI, Ag_2O
93%

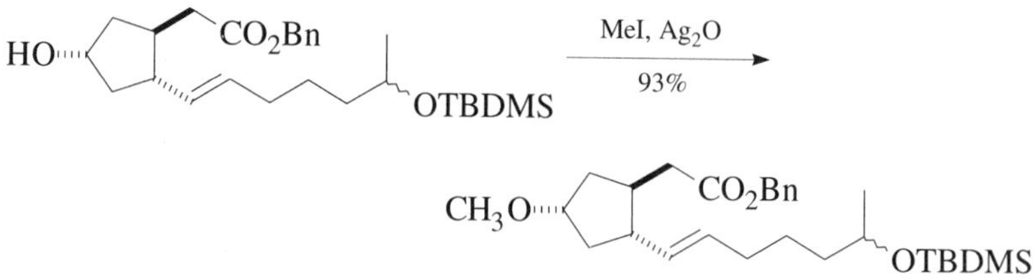

11. Me_2SO_4, DMSO, DMF, $Ba(OH)_2$, BaO, rt, 18 h, 88% yield.[18]

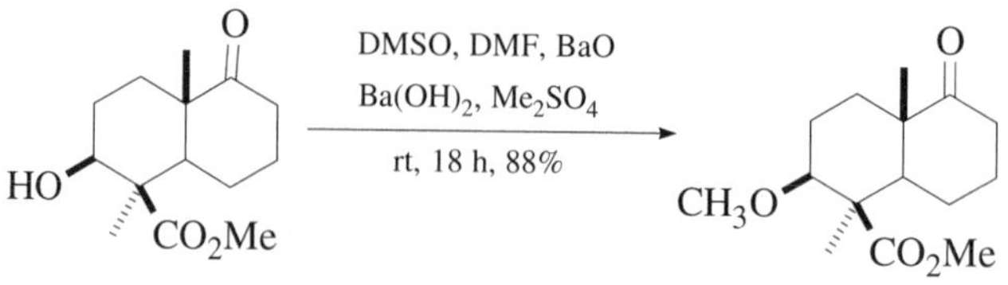

DMSO, DMF, BaO
$Ba(OH)_2$, Me_2SO_4
rt, 18 h, 88%

12. From an aldehyde: MeOH, Pd–C, H_2, 100°, 40 bar, 80–95% yield.[19] Other alcohols can be used to prepare other ethers.
13. AgOTf, MeI, 2,6-di-*t*-butylpyridine, 39–96% yield. This method can be used to prepare alkyl, benzyl, and allyl ethers.[20]
14. From a MOM ether: $Zn(BH_4)_2$, TMSCl, 87% yield.[21]
15. OAc → OCH$_3$ ($BnEt_3N^+$ [$Mo(CO)_5Cl^-$], MeOH, CH_2Cl_2, 30°, AgOTf, 93%)

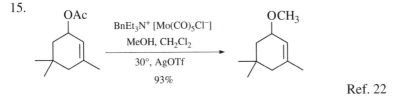

Ref. 22

Cleavage[23]

1. Me_3SiI, $CHCl_3$, 25°, 6 h, 95% yield.[24] A number of methods have been reported in the literature for the *in situ* formation of Me_3SiI,[25] since Me_3SiI is somewhat sensitive to handle. This reagent also cleaves many other ether-type protective groups, but selectivity can be maintained by control of the reaction conditions and the inherent rate differences between functional groups.
2. BBr_3, NaI, 15-crown-5.[26] Methyl esters are not cleaved under these conditions.[27]
3. BBr_3, EtOAc, 1 h, 95% yield.[28]
4. BBr_3, CH_2Cl_2, high yields.[29]

CH_3O, OAc, CH_3O_2C, OAc, CO_2CH_3 → (BBr_3, CH_2Cl_2, −78° → 12°) O, OAc, O, H, OAc, CO_2CH_3

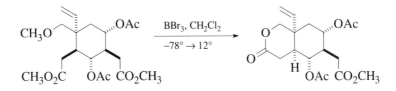

This method is probably the most commonly used one for the cleavage of methyl ethers, because it generally gives excellent yields with a variety of structural types. The solid complex BBr_3–Me_2S that is more easily handled can also be used.[30] BBr_3 will cleave ketals.

5. $BF_3 \cdot Et_2O$, $HSCH_2CH_2SH$, HCl, 15 h, 82% yield.[31,32]
6. $MeSSiMe_3$ or $PhSSiMe_3$, ZnI_2, $Bu_4N^+I^-$.[33] In this case, the 6-*O*-methyl ether was cleaved selectively from permethylated glucose.
7. $SiCl_4$, NaI, CH_2Cl_2, CH_3CN, 80–100% yield.[34]
8. AlX_3 (X = Br, Cl), EtSH, 25°, 0.5–3 h, 95–98% yield.[35]
9. *t*-BuCOCl or AcCl, NaI, CH_3CN, 37 h, rt, 84% yield.[36] In this case, the methyl ether is replaced by a pivaloate or acetate group that can be hydrolyzed with base.
10. Ac_2O, $FeCl_3$, 80°, 24 h.[37] In this case, the methyl ether is converted to an

acetate. The reaction proceeds with complete racemization. Benzyl and allyl ethers are also cleaved.

11. AcCl, NaI, CH_3CN.[38]
12. Me_2BBr, CH_2Cl_2, 0–25°, 3–18 h, 75–93% yield. Tertiary methyl ethers give the tertiary bromide.[39]
13. BI_3–Et_2NPh, benzene, rt, 3–4 h, 94% yield.[40]
14. TMSCl, cat. H_2SO_4, Ac_2O, 71–89% yield.[41]
15. $AlCl_3$, $Bu_4N^+I^-$, CH_3CN, 83% yield.[42,43]

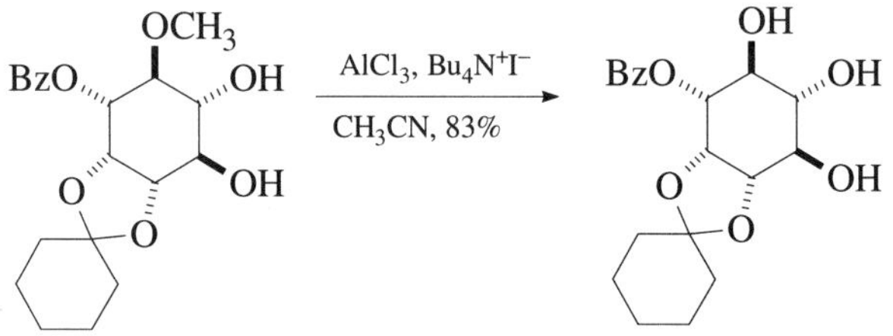

$AlCl_3$, $Bu_4N^+I^-$ / CH_3CN, 83%

16. Treatment of a methyl ether with $RuCl_3$, $NaIO_4$ converts the ether into a ketone.[44]

1. A. Merz, *Angew. Chem., Int. Ed. Engl.*, **12**, 846 (1973).
2. M. E. Jung and S. M. Kaas, *Tetrahedron Lett.*, **30**, 641 (1989).
3. K. Ohno, H. Nishiyama, and H. Nagase, *Tetrahedron Lett.*, 4405 (1979).
4. T. Nakata, S. Nagao, N. Mori, and T. Oishi, *Tetrahedron Lett.*, **26**, 6461 (1985).
5. M. Neeman and W. S. Johnson, *Org. Synth., Collect. Vol. V*, 245 (1973).
6. A. B. Smith, III, K. J. Hale, L. M. Laakso, K. Chen, and A. Riera, *Tetrahedron Lett.*, **30**, 6963 (1989).
7. M. Somei and T. Kawasaki, *Heterocycles*, **29**, 1251 (1989).
8. T. Aoyama and T. Shioiri, *Tetrahedron Lett.*, **31**, 5507 (1990).
9. R. A. W. Johnstone and M. E. Rose, *Tetrahedron*, **35**, 2169 (1979).
10. Y. Kashman, *J. Org. Chem.*, **37**, 912 (1972).
11. H. Meerwein, G. Hinz, P. Hofmann, E. Kroning, and E. Pfeil, *J. Prakt. Chem.*, **147**, 257 (1937).
12. M. J. Earle, R. A. Fairhurst, R. G. Giles, and H. Heaney, *Synlett*, 728 (1991).
13. J. Arnarp and J. Lönngren, *Acta Chem. Scand., Ser. B*, **32**, 465 (1978).
14. R. E. Ireland, J. L. Gleason, L. D. Gegnas, and T. K. Highsmith, *J. Org. Chem.*, **61**, 6856 (1996).
15. J. A. Marshall and S. Xie, *J. Org. Chem.*, **60**, 7230 (1995).
16. P. G. M. Wuts and S. R. Putt, unpublished results.
17. A. E. Greene, C. L. Drian, and P. Crabbe, *J. Am. Chem. Soc.* **102**, 7583 (1980).
18. J. T. A. Reuvers and A. de Groot, *J. Org. Chem.*, **51**, 4594 (1986).
19. V. Bethmont, F. Fache, and M. Lemaive, *Tetrahedron Lett.*, **36**, 4235 (1995).
20. R. M. Burk, T. S. Gac, and M. B. Roof, *Tetrahedron Lett.*, **35**, 8111 (1994).

21. H. Kotsuki, Y. Ushio, N. Yoshimura, and M. Ochi, *J. Org. Chem.*, **52**, 2594 (1987).
22. H. Dvořáková, D. Dvořák, J. Šrogl, and P. Kočovský, *Tetrahedron Lett.*, **36**, 6351 (1995).
23. For a review of alkyl ether cleavage, see B. C. Ranu and S. Bhar, *Org. Prep. Proced. Int.*, **28**, 371 (1996).
24. M. E. Jung and M. A. Lyster, *J. Org. Chem.*, **42**, 3761 (1977).
25. M. E. Jung and T. A. Blumenkopf, *Tetrahedron Lett.*, 3657 (1978); G. A. Olah, A. Husain, B. G. B. Gupta, and S. C. Narang, *Angew. Chem., Int. Ed. Engl.*, **20**, 690 (1981); T.-L. Ho and G. Olah, *Synthesis*, 417 (1977). For a review on the uses of Me_3SiI, see A. H. Schmidt, *Aldrichimica Acta*, **14**, 31 (1981).
26. H. Niwa, T. Hida, and K. Yamada, *Tetrahedron Lett.*, **22**, 4239 (1981).
27. M. E. Kuehne and J. B. Pitner, *J. Org. Chem.*, **54**, 4553 (1989).
28. H. Shimomura, J. Katsuba, and M. Matsui, *Agric. Biol. Chem.*, **42**, 131 (1978).
29. M. Demuynck, P. De Clercq, and M. Vandewalle, *J. Org. Chem.*, **44**, 4863 (1979); P. A. Grieco, M. Nishizawa, T. Oguri, S. D. Burke, and N. Marinovic, *J. Am. Chem. Soc.*, **99**, 5773 (1977).
30. P. G. Williard and C. B. Fryhle, *Tetrahedron Lett.*, **21**, 3731 (1980).
31. G. Vidari, S. Ferrino, and P. A. Grieco, *J. Am. Chem. Soc.*, **106**, 3539 (1984).
32. M. Node, H. Hori, and E. Fujita, *J. Chem. Soc., Perkin Trans. 1*, 2237 (1976).
33. S. Hanessian and Y. Guindon, *Tetrahedron Lett.*, **21**, 2305 (1980); R. S. Glass, *J. Organomet. Chem.*, **61**, 83 (1973); I. Ojima, M. Nihonyangi, and Y. Nagai, *J. Organmet. Chem.*, **50**, C26 (1973).
34. M. V. Bhatt and S. S. El-Morey, *Synthesis*, 1048 (1982).
35. M. Node, K. Nishide, M. Sai, K. Ichikawa, K. Fuji, and E. Fujita, *Chem. Lett.*, 97 (1979).
36. A. Oku, T. Harada, and K. Kita, *Tetrahedron Lett.*, **23**, 681 (1982).
37. B. Ganem and V. R. Small, Jr., *J. Org. Chem.*, **39**, 3728 (1974).
38. T. Tsunoda, M. Amaike, U. S. F. Tambunan, Y. Fujise, S. Ito, and M. Kodama, *Tetrahedron Lett.*, **28**, 2537 (1987).
39. Y. Guindon, C. Yoakim, and H. E. Morton, *Tetrahedron Lett.*, **24**, 2969 (1983).
40. C. Narayana, S. Padmanabhan, and G. W. Kabalka, *Tetrahedron Lett.*, **31**, 6977 (1990).
41. J. C. Sarma, M. Borbaruah, D. N. Sarma, N. C. Barua, and R. P. Sharma, *Tetrahedron*, **42**, 3999 (1986).
42. T. Akiyama, H. Shima, and S. Ozaki, *Tetrahedron Lett.*, **32**, 5593 (1991).
43. E. D. Moher, J. L. Collins, and P. A. Grieco, *J. Am. Chem. Soc.*, **114**, 2764 (1992).
44. L. E. Overman, D. J. Ricca, and V. D. Tran, *J. Am. Chem. Soc.*, **119**, 12031 (1997).

Substituted Methyl Ethers

Methoxymethyl (MOM) Ether: CH_3OCH_2–OR (Chart 1)

Formation

1. CH_3OCH_2Cl,[1] NaH, THF, 80% yield.[2]

2. CH_3OCH_2Cl, *i*-Pr_2NEt, 0°, 1 h → 25°, 8 h, 86% yield.[3] This is the most commonly employed procedure for introducing the MOM group. The reagent chloromethylmethyl ether is reported to be carcinogenic, and dichloromethylmethyl ether, a by-product in its preparation, is considered even more toxic. A preparation that does not produce any of the dichloro ether has been reported.[4]
3. MOMBr, DIPEA, CH_2Cl_2, 0°, 6 h, 72% yield.[3,5]
4. NaI increases the reactivity of MOMCl by the *in situ* preparation of MOMI, which facilitates the protection of tertiary alcohols.[6]

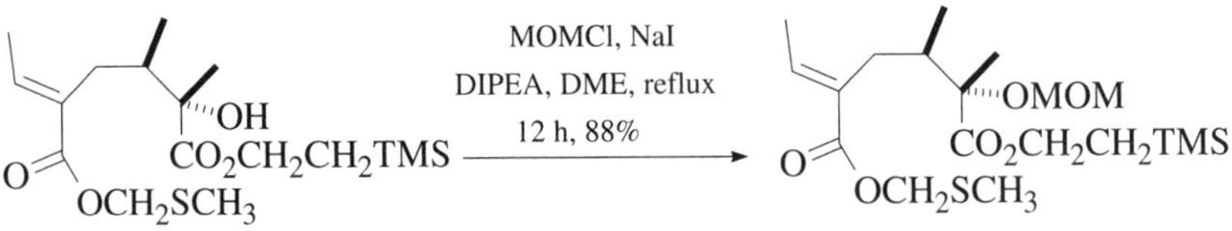

MOMCl, NaI
DIPEA, DME, reflux
12 h, 88%

5. For the selective protection of diols: Bu_2SnO, benzene, reflux; MOMCl, $Bu_4N^+I^-$, rt, 87% yield.[7]
6. $CH_2(OMe)_2$, Nafion H.[8]
7. $CH_2(OMe)_2$, CH_2Cl_2, TfOH, 4 h, 25°, 65% yield.[9] This method is suitable for the formation of primary, secondary, allylic, and propargylic MOM ethers. Tertiary alcohols fail to give a complete reaction. 1,3-Diols give methylene acetals (89% yield).
8. $CH_2(OMe)_2$, CH_2=$CHCH_2SiMe_3$, Me_3SiOTf, P_2O_5, 93–99% yield.[10] This method was used to protect the 2′-OH of ribonucleosides and deoxyribonucleosides, as well as the hydroxyl groups of several other carbohydrates bearing functionality, such as esters, amides, and acetonides.
9. $CH_2(OEt)_2$, Montmorillonite clay (H^+), 72–80% for nonallylic alcohols, 56% for a propargylic alcohol.[11]
10. Selective formation of MOM ethers has been achieved in a diol system.[12]

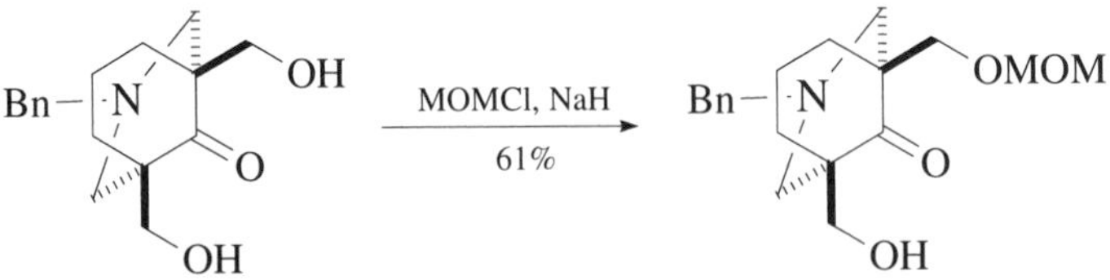

MOMCl, NaH
61%

11. Mono MOM derivatives of diols can be prepared from the ortho esters by diisobutylaluminum hydride reduction (46–98% yield). In general, the most hindered alcohol is protected.[13]

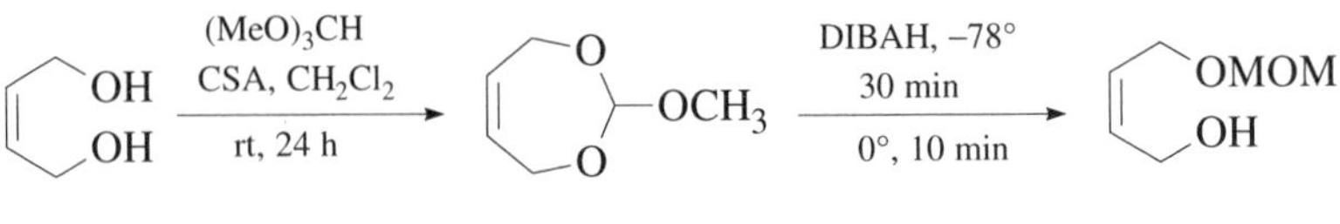

$(MeO)_3CH$
CSA, CH_2Cl_2
rt, 24 h

DIBAH, –78°
30 min
0°, 10 min

In the case of allylic or propargylic diols, the nonallylic (propargylic) alcohol is protected.[14]

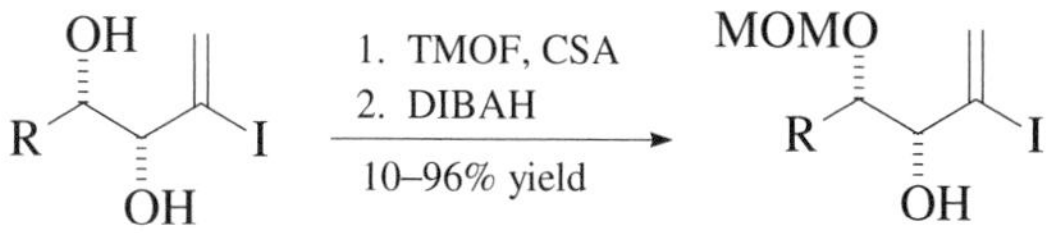

1. TMOF, CSA
2. DIBAH
10–96% yield

12. $CH_3OCH_2OCH_3$, anhydrous $FeCl_3$-ms (3Å), 1–3 h, 70–99% yield.[15]
13. MOMCl, Al_2O_3, ultrasound, 68–92% yield.[16]
14. $CH_2(OMe)_2$, TsOH, LiBr, 9 h, rt, 71–100% yield.[17]
15. $MoO_2(acac)_2$, $CH_2(OMe)_2$, $CHCl_3$, reflux, 63–95% yield.[18]
16. MOMCl, CH_2Cl_2, Na–Y Zeolite, reflux, 70–91% yield.[19]
17. From a PMB ether: $SnCl_2$, CH_2Cl_2, rt, 24 h, 13–81% yield.[20]
18. From a stannylmethyl ether: electrolysis, MeOH, 90% yield.[21]
19. From a trimethylsilyl glycoside: TMSOTf or TFA or BF_3–Et_2O, $CH_3OCH_2OCH_3$, 54–66% yield.[22]
20. $CH_2(OMe)_2$, cat. P_2O_5, $CHCl_3$, 25°, 30 min, 95% yield.[23]
21. $CH_2(OMe)_2$, Me_3SiI or CH_2=$CHCH_2SiMe_3$, I_2, 76–95% yield.[24]
22. $CH_2(OMe)_2$, TsOH, LiBr, 9 h, rt, 71–100% yield.[17]
23. From a PMB ether: $CH_2(OMe)_2$, MOMBr, $SnBr_2$, $ClCH_2CH_2Cl$, rt, 57–81% yield. Phenolic PMB ethers were not converted efficiently. A BOM ether was prepared using this method.[20]

Cleavage

1. Trace concd. HCl, MeOH, 62°, 15 min.[25]
2. 6 *M* HCl, aq. THF, 50°, 6–8 h, 95% yield.[26] An attempt to cleave the MOM group with acid in the presence of a dimethyl acetal resulted in the cleavage of both groups, probably by intramolecular assistance.[27]
3. Concd. HCl, isopropyl alcohol (IPA), 65% yield.[28]

   Other methods attempted for the cleavage of this MOM group were unsuccessful.

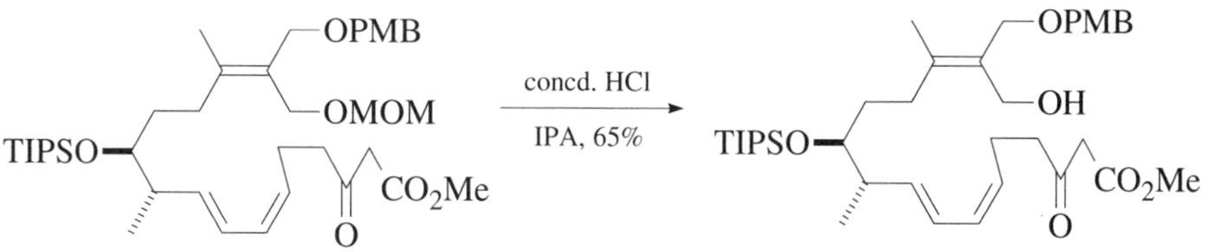

concd. HCl
IPA, 65%

4. 50% AcOH, cat. H_2SO_4, reflux, 10–15 min, 80% yield.[29]
5. PhSH, BF_3·Et_2O, 98% yield.[30]

6. $Ph_3C^+BF_4^-$, CH_2Cl_2, 25°.[31]

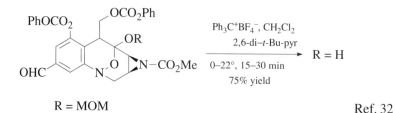

PhOCO2 OCO2Ph OR OHC N O N−CO2Me; $Ph_3C^+BF_4^-$, CH_2Cl_2, 2,6-di-*t*-Bu-pyr, 0–22°, 15–30 min, 75% yield → R = H

R = MOM Ref. 32

7. Pyridinium *p*-toluenesulfonate, *t*-BuOH or 2-butanone, reflux, 80–99% yield.[33] This method is useful for allylic alcohols. MEM ethers are also cleaved under these conditions.

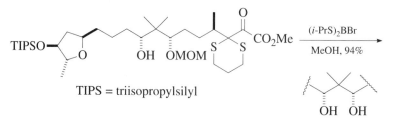

8. (catechol-O2BBr) BBr Catechol boron halides—particularly the bromide—are effective reagents for the cleavage of MOM ethers. The bromide also cleaves the following groups in the order shown: MOMOR ≈ MEMOR > *t*-BOC > Cbz ≈ *t*-BuOR > BnOR > allylOR > *t*-BuO_2CR ≈ 2° alkylOR > BnO_2CR > 1° alkylOR >> $alkylO_2CR$. The *t*-butyldimethylsilyl (TBDMS), *t*-butyldiphenylsilyl (TBDPS), and PMB groups are stable to this reagent.[34] The chloride is less reactive and thus may be more useful for achieving selectivity in multifunctional substrates. Yields are generally > 83%.[35]

9. $(i\text{-PrS})_2BBr$, MeOH, 94% yield.[36] This method has the advantage that 1,2- and 1,3-diols do not give formyl acetals, as is sometimes the case in cleaving MOM groups with neighboring hydroxyl groups.[37] The reagent also cleaves MEM groups and, under basic conditions, affords the *i*-$PrSCH_2OR$ derivatives.

TIPSO O OH OMOM S S O CO2Me → $(i\text{-PrS})_2BBr$, MeOH, 94% → OH OH

TIPS = triisopropylsilyl

10. Me_2BBr, CH_2Cl_2, −78°, then $NaHCO_3/H_2O$, 87–95% yield.[38] This reagent also cleaves the MEM, MTM, and acetal groups. Esters are stable to this reagent.

11. Me_3SiBr, CH_2Cl_2, 0°, 8–9 h, 80–97% yield.[39] This reagent also cleaves the acetonide, THP, trityl, and *t*-$BuMe_2Si$ groups. Esters, methyl and benzyl ethers, *t*-butyldiphenylsilyl ethers, and amides are reported to be stable.

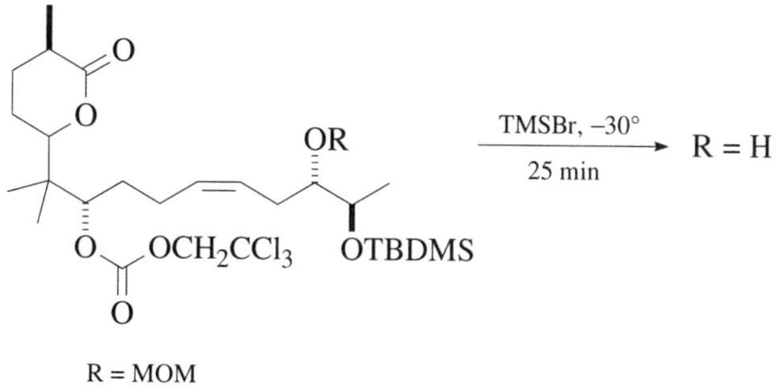

R = MOM

TMSBr, −30°, 25 min → R = H

12. CF_3COOH, CH_2Cl_2, >85% yield.[40]

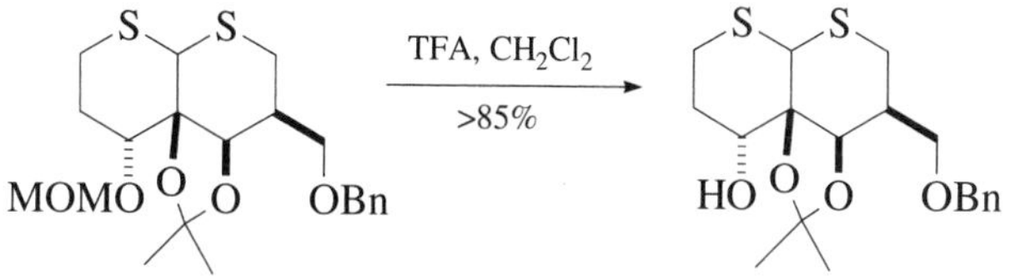

TFA, CH_2Cl_2, >85%

13. $LiBF_4$, CH_3CN, H_2O, 72°, 100% yield.[41]

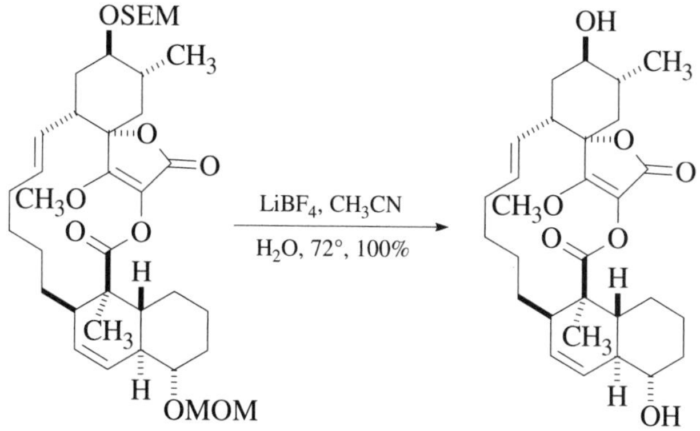

$LiBF_4$, CH_3CN, H_2O, 72°, 100%

14. $MgBr_2$, ether, BuSH, rt, 40–97% yield. Tertiary and allylic MOM derivatives give low yields. MTM and SEM ethers are also cleaved, but MEM ethers are stable.[42]
15. Dowex-50W-X2, aq. MeOH, 42–97% yield.[43]

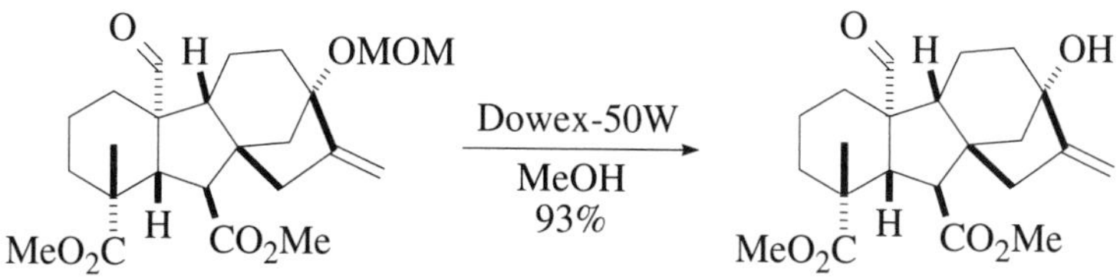

Dowex-50W, MeOH, 93%

Other methods resulted in skeletal rearrangement. This study also showed that the rate of acid-catalyzed MOM cleavage increases in the following order: primary (30 h) < secondary (8 h) < tertiary (0.5–2 h). Tertiary alcohols are cleaved in excellent yields (94–97% yield).

16. $AlCl_3$, NaI, CH_3CN, CH_2Cl_2, 0°, 25 min, >70% yield.[44]

1. For a review of α-monohalo ethers in organic synthesis, see T. Benneche, *Synthesis*, 1 (1995).
2. A. F. Kluge, K. G. Untch, and J. H. Fried, *J. Am. Chem. Soc.*, **94**, 7827 (1972).
3. G. Stork and T. Takahashi, *J. Am. Chem. Soc.*, **99**, 1275 (1977).
4. R. J. Linderman, M. Jaber, and B. D. Griedel, *J. Org. Chem.*, **59**, 6499 (1994).
5. D. Askin, R. P. Volante, R. A. Reamer, K. M. Ryan, and I. Shinkai, *Tetrahedron Lett.*, **29**, 277 (1988).
6. K. Narasaka, T. Sakakura, T. Uchimaru, and D. Guédin-Vuong, *J. Am. Chem. Soc.*, **106**, 2954 (1984).
7. S. David, A. Thieffry, and A. Veyrières, *J. Chem. Soc., Perkin Trans. 1*, 1796 (1981).
8. G. A. Olah, A. Husain, B. G. B. Gupta, and S. C. Narang, *Synthesis*, 471 (1981).
9. M. P. Groziak and A. Koohang, *J. Org. Chem.*, **57**, 940 (1992).
10. S. Nishino and Y. Ishido, *J. Carbohydr. Chem.*, **5**, 313 (1986).
11. U. A. Schaper, *Synthesis*, 794 (1981).
12. M. Ihara, M. Suzuki, K. Fukumoto, T. Kametani, and C. Kabuto, *J. Am. Chem. Soc.*, **110**, 1963 (1988).
13. M. Takasu, Y. Naruse, and H. Yamamoto, *Tetrahedron Lett.*, **29**, 1947 (1988).
14. R. W. Friesen and C. Vanderwal, *J. Org. Chem.*, **61**, 9103 (1996).
15. H. K. Patney, *Synlett*, 567 (1992).
16. B. C. Ranu, A. Majee, and A. R. Das, *Synth. Commun.*, **25**, 363 (1995).
17. J.-L. Gras, Y.-Y. K. W. Chang, and A. Guerin, *Synthesis*, 74 (1985).
18. M. L. Kantam and P. L. Santhi, *Synlett*, 429 (1993).
19. P. Kumar, S. V. N. Raju, R. S. Reddy, and B. Pandey, *Tetrahedron Lett.*, **35**, 1289 (1994).
20. T. Oriyama, M. Kimura, and G. Koga, *Bull. Chem. Soc. Jpn.*, **67**, 885 (1994).
21. J.-i. Yoshida, Y. Ishichi, K. Nishiwaki, S. Shiozawa, and S. Isoe, *Tetrahedron Lett.*, **33**, 2599 (1992).
22. K. Jansson and G. Magnusson, *Tetrahedron*, **46**, 59 (1990).
23. K. Fuji, S. Nakano, and E. Fujita, *Synthesis*, 276 (1975).
24. G. A. Olah, A. Husain, and S. C. Narang, *Synthesis*, 896 (1983).
25. J. Auerbach and S. M. Weinreb, *J. Chem. Soc., Chem. Commun.*, 298 (1974).
26. A. I. Meyers, J. L. Durandetta, and R. Munavu, *J. Org. Chem.*, **40**, 2025 (1975).
27. M. L. Bremmer, N. A. Khatri, and S. M. Weinreb, *J. Org. Chem.*, **48**, 3661 (1983).
28. D. G. Hall and P. Deslogchamps, *J. Org. Chem.*, **60**, 7796 (1995).
29. F. B. Laforge, *J. Am. Chem. Soc.*, **55**, 3040 (1933).
30. G. R. Kieczykowski and R. H. Schlessinger, *J. Am. Chem. Soc.*, **100**, 1938 (1978).
31. T. Nakata, G. Schmid, B. Vranesic, M. Okigawa, T. Smith-Palmer, and Y. Kishi, *J. Am. Chem. Soc.*, **100**, 2933 (1978).

32. J. M. Schkeryantz and S. J. Danishefsky, *J. Am. Chem. Soc.*, **117**, 4722 (1995).
33. H. Monti, G. Léandri, M. Klos-Ringquet, and C. Corriol, *Synth. Commun.*, **13**, 1021 (1983).
34. L. A. Paquette, Z. Gao, Z. Ni, and G. F. Smith, *Tetrahedron Lett.*, **38**, 1271 (1997).
35. R. K. Boeckman, Jr., and J. C. Potenza, *Tetrahedron Lett.*, **26**, 1411 (1985).
36. E. J. Corey, D. H. Hua, and S. P. Seitz, *Tetrahedron Lett.*, **25**, 3 (1984).
37. B. C. Barot and H. W. Pinnick, *J. Org. Chem.*, **46**, 2981 (1981).
38. Y. Guindon, H. E. Morton, and C. Yoakim, *Tetrahedron Lett.*, **24**, 3969 (1983).
39. S. Hanessian, D. Delorme, and Y. Dufresne, *Tetrahedron Lett.*, **25**, 2515 (1984). For *in situ* prepared TMSBr, see R. B. Woodward and 48 co-workers, *J. Am. Chem. Soc.*, **103**, 3213 (note 2) (1981).
40. R. B. Woodward and 48 co-workers, *J. Am. Chem. Soc.*, **103**, 3210 (1981).
41. R. E. Ireland and M. D. Varney, *J. Org. Chem.*, **51**, 635 (1986).
42. S. Kim, I. S. Kee, Y. H. Park, and J. H. Park, *Synlett*, 183 (1991).
43. H. Seto and L. N. Mander, *Synth. Commun.*, **22**, 2823 (1992).
44. E. D. Moher, P. A. Grieco, and J. L. Collins, *J. Org. Chem.*, **58**, 3789 (1993).

Methylthiomethyl (MTM) Ether: CH_3SCH_2OR (Chart 1)

Methylthiomethyl ethers are quite stable to acidic conditions. Most ethers and 1,3-dithianes are stable to the neutral mercuric chloride used to remove the MTM group. One problem with the MTM group is that it is sometimes difficult to introduce.

Formation

1. NaH, DME, CH_3SCH_2Cl, NaI, 0°, 1 h → 25°, 1.5 h, >86% yield.[1]
2. CH_3SCH_2I, DMSO, Ac_2O, 20°, 12 h, 80–90% yield.[2]
3. DMSO, Ac_2O, AcOH, 20°, 1–2 days, 80%.[3]
4. CH_3SCH_2Cl, $AgNO_3$, Et_3N, benzene, 22–80°, 4–24 h, 60–80% yield.[4]
5. DMSO, molybdenum peroxide, benzene, reflux, 7–20 h, ≈60% yield.[5] This method was used to monoprotect 1,2-diols. The method is not general, because oxidation to α-hydroxy ketones and diketones occurs with some substrates. Based on the mechanism and on the results, it would appear that overoxidation has a strong conformational dependence.
6. MTM ethers can be prepared from MEM and MOM ethers by treatment with Me_2BBr to form the bromomethyl ether, which is trapped with MeSH and $(i\text{-}Pr)_2NEt$. This method may have some advantage, since the direct preparation of MTM ethers is not always simple.[6]
7. CH_3SCH_3, CH_3CN, $(PhCO)_2O_2$, 0°, 2 h, 75–95% yield.[7,8] Acetonides, THP ethers, alkenes, ketones, and epoxides all survive these conditions.
8. $(COCl)_2$, DMSO, −78° → −50°; Et_3N, −78° → −15°.[9]

Cleavage

1. $HgCl_2$, CH_3CN, H_2O, 25°, 1–2 h, 88–95% yield.[1] If 2-methoxyethanol is substituted for water, the MTM ether is converted to a MEM ether. Similarly, substitution with methanol affords a MOM ether.[10] If the MTM ether has an adjacent hydroxyl, it is possible to form the formylidene acetal as a by-product of cleavage.[11]
2. $HgCl_2$, $CaCO_3$, MeCN, H_2O.[1] The calcium carbonate is used as an acid scavenger for acid-sensitive substrates.

TBSO MTMO S S —$HgCl_2$, $CaCO_3$ / CH_3CN, H_2O→ TBSO HO S S

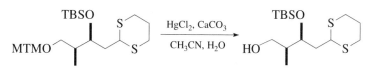

3. MeI, acetone, H_2O, $NaHCO_3$, heat few h, 80–95% yield.[3]
4. Electrolysis: applied voltage = 10 V, AcONa, AcOH; K_2CO_3, MeOH, H_2O, 80–95% yield.[12]
5. Me_3SiCl, Ac_2O, 90%.[13] Treatment of the resulting acetoxymethyl ether with acid or base readily affords the free alcohol.

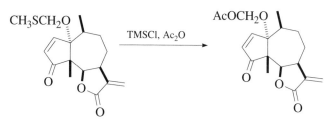

CH_3SCH_2O —TMSCl, Ac_2O→ $AcOCH_2O$

6. $Ph_3C^+BF_4^-$, CH_2Cl_2, 5–30 min, 80–95% yield.[14] The mechanism of this cleavage has been determined to involve complex formation by the trityl cation with the sulfur, followed by hydrolysis, rather than by hydride abstraction.[15]

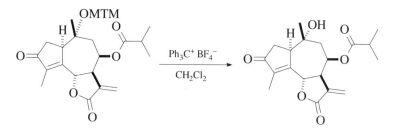

OMTM —$Ph_3C^+ BF_4^-$ / CH_2Cl_2→ OH

In this case, the use of $HgCl_2$, $AgNO_3$, and MeI gave extensive decomposition.

7. $Hg(OTf)_2$, CH_2Cl_2, H_2O, Na_2HPO_4.[16]
8. MgI_2, ether, Ac_2O, rt, 90–100% yield. Cleavage occurs to give a mixture of

acetate and an acetoxymethyl ether that is reported to be very acid and base sensitive.[17]

9. $AgNO_3$, THF, H_2O, 2,6-lutidine, 25°, 45 min, 88–95% yield.[1] These conditions can be used to cleave an MTM ether in the presence of a dithiane.[18]
10. $MgBr_2$, *n*-BuSH, Et_2O, rt, 0.5–3 h, 83–85% yield.[19]

1. E. J. Corey and M. G. Bock, *Tetrahedron Lett.*, 3269 (1975).
2. K. Yamada, K. Kato, H. Nagase, and Y. Hirata, *Tetrahedron Lett.*, 65 (1976).
3. P. M. Pojer and S. J. Angyal, *Aust. J. Chem.*, **31**, 1031 (1978).
4. K. Suzuki, J. Inanaga, and M. Yamaguchi, *Chem. Lett.*, 1277 (1979).
5. Y. Masuyama, M. Usukura, and Y. Kurusu, *Chem. Lett.*, 1951 (1982).
6. H. E. Morton and Y. Guindon, *J. Org. Chem.*, **50**, 5379 (1985).
7. J. C. Medina, M. Salomon, and K. S. Kyler, *Tetrahedron Lett.*, **29**, 3773 (1988).
8. P. Garner and J. U. Yoo, *Tetrahedron Lett.*, **34**, 1275 (1993).
9. D. R. Williams, F. D. Klinger, and V. Dabral, *Tetrahedron Lett.*, **29**, 3415 (1988).
10. P. K. Chowdhury, D. N. Sarma, and R. P. Sharma, *Chem. Ind. (London)*, 803 (1984).
11. M. P. Wachter and R. E. Adams, *Synth. Commun.*, **10**, 111 (1980).
12. T. Mandai, H. Yasunaga, M. Kawada, and J. Otera, *Chem. Lett.*, 715 (1984).
13. D. N. Sarma, N. C. Barua, and R. P. Sharma, *Chem. Ind. (London)*, 223 (1984); N. C. Barur, R. P. Sharma, and J. N. Baruah, *Tetrahedron Lett.*, **24**, 1189 (1983).
14. P. K. Chowdhury, R. P. Sharma, and J. N. Baruah, *Tetrahedron Lett.*, **24**, 4485 (1983).
15. H. Niwa and Y. Miyachi, *Bull. Chem. Soc. Jpn.*, **64**, 716 (1991).
16. G. E. Keck, E. P. Boden, and M. R. Wiley, *J. Org. Chem.*, **54**, 896 (1989).
17. P. K. Chowdhury, *J. Chem. Res., Synop.*, 68 (1992).
18. E. J. Corey, D. H. Hua, B.-C. Pan, and S. P. Seitz, *J. Am. Chem. Soc.*, **104**, 6818 (1982).
19. S. Kim, I. S. Kee, Y. H. Park, and J. H. Park, *Synlett*, 183 (1992).

(Phenyldimethylsilyl)methoxymethyl Ether (SMOM–OR): $C_6H_5(CH_3)_2SiCH_2OCH_2OR$

Formation

1. SMOMCl, *i*-$PrEt_2N$, CH_3CN, 3 h, 40°, 87–91% yield.[1] Diols are selectively protected using the stannylene methodology.

Cleavage

1. AcOOH, KBr, AcOH, NaOAc, 1.5 h, 20°, 82–92% yield.[1] The SMOM group is stable to $Bu_4N^+F^-$; NaOMe/MeOH; 4 *N* NaOH/dioxane/methanol; and *N*-iodosuccinimide, cat. trifluoromethanesulfonic acid.

1. G. J. P. H. Boons, C. J. J. Elie, G. A. van der Marel, and J. H. van Boom, *Tetrahedron Lett.*, **31**, 2197 (1990).

Benzyloxymethyl Ether (BOM–OR): $PhCH_2OCH_2OR$

Formation

1. $PhCH_2OCH_2Cl$, $(i\text{-}Pr)_2NEt$, 10–20°, 12 h, 95% yield.[1] $Bu_4N^+I^-$ can be added to increase the reactivity for protection of more hindered alcohols.
2. $PhCH_2OCH_2Cl$, NaI, proton sponge [1,8-bis(dimethylamino)naphthalene], 84% yield.[2] BOMBr can also be used.[3]

Cleavage

1. Na, NH_3, EtOH.[1]
2. Li, NH_3.[4]
3. PhSH, $BF_3{\cdot}Et_2O$, CH_2Cl_2, −78°, 95% yield.[5,6]

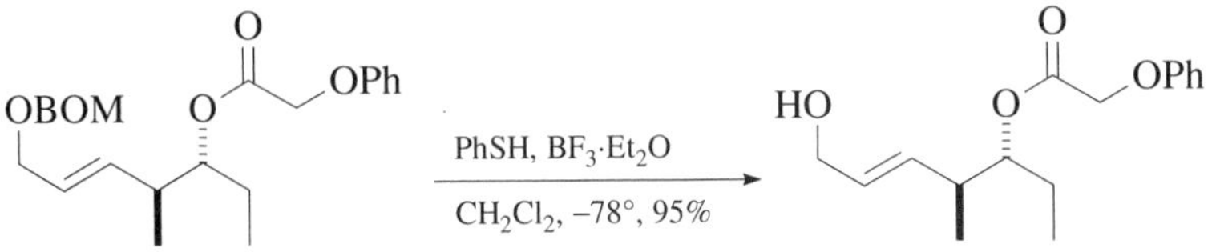

PhSH, $BF_3{\cdot}Et_2O$ / CH_2Cl_2, −78°, 95%

4. H_2, 1 atm, Pd–C, EtOAc–hexane, 68% yield.[7]

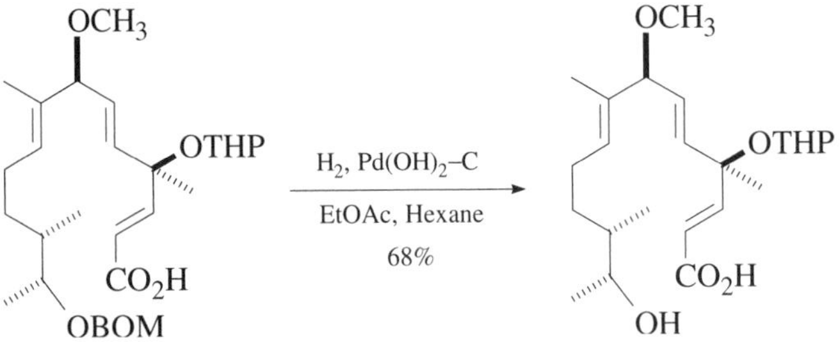

H_2, $Pd(OH)_2$–C / EtOAc, Hexane / 68%

5. H_2, 1 atm, 10% Pd–C, 0.01 *N* $HClO_4$, in 80% THF/H_2O, 25°.[8]
6. HCl, MeOH, 56% yield.[9]
7. MeOH, Dowex 50W–X8, rt, 5–6 days, 90% yield.[10]
8. AlH_2Cl, $AlHCl_2$, or BH_3 in toluene or THF. See the section on SEM ethers for a selectivity study of these reagents with the SEM, MTM, EOM (ethoxymethyl), and *p*-AOM groups.[11]
9. HCl, NaI, 97% yield based on 67% conversion.[12]

1. G. Stork and M. Isobe, *J. Am. Chem. Soc.*, **97**, 6260 (1975).
2. S. F. Martin, W.-C. Lee, G. J. Pacofsky, R. P. Gist, and T. A. Mulhern, *J. Am. Chem. Soc.*, **116**, 4674 (1994).

3. D. A. Evans, S. L. Bender, and J. Morris, *J. Am. Chem. Soc.*, **110,** 2506 (1988).
4. H. Nagaoka, W. Rutsch, G. Schmid, H. Iio, M. R. Johnson, and Y. Kishi, *J. Am. Chem. Soc.*, **102**, 7962 (1980).
5. K. Suzuki, K. Tomooka, E. Katayama, T. Matsumoto, and G.-P. C. Tsuchihashi, *J. Am. Chem. Soc.*, **108**, 5221 (1986).
6. K. C. Nicolaou, C.-K. Hwang, M. E. Duggan, D. A. Nugiel, Y. Abe, K. B. Reddy, S. A. DeFrees, D. R. Reddy, R. A. Awartani, S. R. Conley, F. P. J. T. Rutjes, and E. A. Theodorakis, *J. Am. Chem. Soc.*, **117**, 10227 (1995).
7. D. Tanner and P. Somfai, *Tetrahedron*, **43**, 4395 (1987).
8. D. A. Evans, S. L. Bender, and J. Morris, *J. Am. Chem. Soc.*, **110**, 2506 (1988).
9. R. S. Coleman and E. B. Grant, *J. Am. Chem. Soc.*, **116**, 8795 (1994).
10. W. R. Roush, M. R. Michaelides, D. F. Tai, and W. K. M. Chong, *J. Am. Chem. Soc.*, **109**, 7575 (1987).
11. I. Bajza, Z. Varga, and A. Liptak, *Tetrahedron Lett.*, **34**, 1991 (1993).
12. P. A. Wender, N. F. Badham, S. P. Conway, P. E. Floreancig, T. E. Glass, J. B. Houze, N. E. Krauss, D. Lee, D. G. Marquess, P. L. McGrane, W. Meng, M. G. Natchus, A. J. Shuker, J. C. Sutton, and R. E. Taylor, *J. Am. Chem. Soc.*, **119**, 2757 (1997).

p-Methoxybenzyloxymethyl Ether (PMBM–OR): *p*-$MeOC_6H_4CH_2OCH_2OR$

Formation

1. *p*-$MeOC_6H_4CH_2OCH_2Cl$, (*i*-Pr)$_2$NEt (DIPEA), CH_2Cl_2, 78–100% yield.[1,2]

HO — PMBMCl, DIPEA / CH_2Cl_2, 78% → PMBMO; DDQ, 88% ←

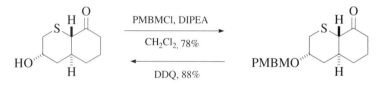

2. Lithium alkoxides react with PMBMCl to form the ethers.[3]

Cleavage

1. DDQ, H_2O, rt, 1–10 h, 63–96% yield.[2]
2. 3:1 THF-6 *M* HCl, 50°, 6 h.[1]

The related [(3,4-dimethoxybenzyl)oxy]methyl group has been used similarly, except that, as expected, it is more easily cleaved (DDQ, CH_2Cl_2, *t*-BuOH, phosphate buffer, pH 6.0, 23°, 110 min, 88% yield).[4]

1. G. Guanti, L. Banfi, E. Narisano, and S. Thea, *Tetrahedron Lett.*, **32**, 6943 (1991).
2. A. P. Kozikowski and J.-P. Wu, *Tetrahedron Lett.*, **28**, 5125 (1987).
3. J. A. Marshall and W. Y. Gung, *Tetrahedron Lett.*, **30**, 7349 (1989).
4. H. Kigoshi, K. Suenaga, T. Mutou, T. Ishigaki, T. Atsumi, H. Ishiwata, A. Sakakura, T. Ogawa, M. Ojika, and K. Yamada, *J. Org. Chem.*, **61**, 5326 (1996).

p-Nitrobenzyloxymethyl Ether: $4\text{-}NO_2C_6H_4CH_2OCH_2OR$

Formation/Cleavage[1]

$$\mathrm{ROH}\ \underset{\mathrm{TBAF}}{\overset{O_2N\text{-}C_6H_4\text{-}CH_2OCH_2\text{-}N^+C_5H_5\ Cl^-,\ \mathrm{DMF},\ Bu_4N^+Br^-,\ 75^\circ}{\rightleftharpoons}}\ O_2N\text{-}C_6H_4\text{-}CH_2OCH_2OR$$

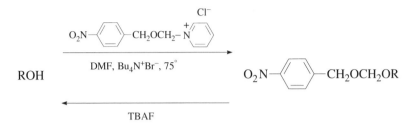

1. G. R. Gough, T. J. Miller, and N. A. Mantick, *Tetrahedron Lett.*, **37**, 981 (1996).

o-Nitrobenzyloxymethyl Ether (NBOM–OR): $2\text{-}NO_2C_6H_4CH_2OCH_2OR$

Formation[1]

1. From a diol: Bu_2SnO, then $2\text{-}NO_2C_6H_4CH_2OCH_2Cl$.

Cleavage[1]

1. *t*-BuOH, H_2O, pH 3.7, long-wave UV for 4.5 h.

1. M. E. Schwartz, R. R. Breaker, G. T. Asteriadis, J. S. deBear, and G. R. Gough, *Biorg. Med. Chem. Lett.*, **2**, 1019 (1992).

(4-Methoxyphenoxy)methyl Ether (*p*-AOM–OR), (*p*-Anisyloxymethyl Ether): $ROCH_2OC_6H_4\text{–}4\text{–}OCH_3$

Formation[1]

1. *p*-AOMCl, $PhCH_2N^+Et_3Cl^-$, CH_3CN, 50% NaOH, rt, 46–91% yield.
2. *p*-AOMCl, $(i\text{-Pr})_2NEt$, CH_2Cl_2, reflux.
3. *p*-AOMCl, DMF, 18-crown-6, K_2CO_3, rt.

Cleavage

1. CAN, CH_3CN, H_2O, 0°, 0.5 h, 60–98% yield.[1] In some cases, the addition of pyridine improves the yields.
2. BH_3, toluene converts the *p*-AOM ether into a methyl ether. For a comparison of the stability of this group with MTM, SEM, BOM, and EOM to various hydride reagents, see the section on SEM ethers.[2]

1. Y. Masaki, I. Iwata, I. Mukai, H. Oda, and H. Nagashima, *Chem. Lett.*, 659 (1989).
2. I. Bajza, Z. Varga, and A. Liptak, *Tetrahedron Lett.*, **34**, 1991 (1993).

Guaiacolmethyl Ether (GUM–OR): 2-$MeOC_6H_4OCH_2OR$

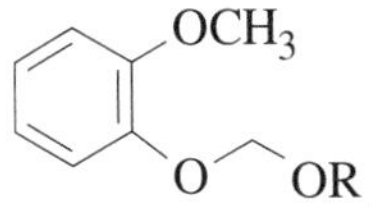

OCH_3 O OR

Formation/Cleavage[1]

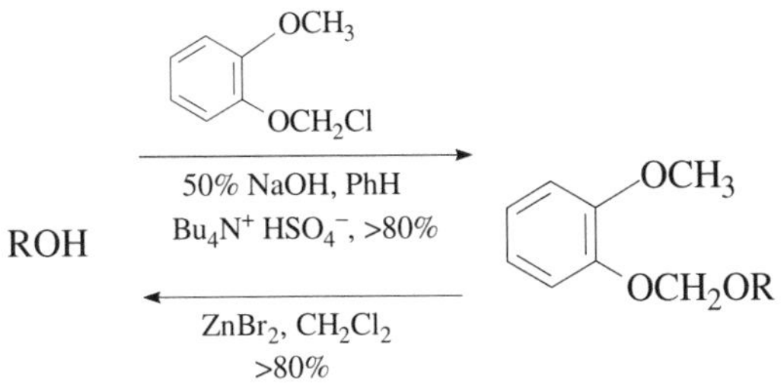

OCH_3 OCH_2Cl

ROH → 50% NaOH, PhH, $Bu_4N^+ HSO_4^-$, >80% → OCH_3 OCH_2OR

← $ZnBr_2$, CH_2Cl_2, >80%

It is possible to introduce this group selectively onto a primary alcohol in the presence of a secondary alcohol. The derivative is stable to $KMnO_4$, *m*-chloroperoxybenzoic acid, $LiAlH_4$, and CrO_3–Pyr. Since this derivative is similar to the *p*-methoxyphenyl ether, it should also be possible to remove it oxidatively. The GUM ethers are less stable than the MEM ethers in acid, but have comparable stability to the SEM ethers. It is possible to remove the GUM ether in the presence of a MEM ether.

1. B. Loubinouz, G. Coudert, and G. Guillaumet, *Tetrahedron Lett.*, **22**, 1973 (1981).

t-Butoxymethyl Ether: *t*-$BuOCH_2OR$

Formation

1. *t*-$BuOCH_2Cl$,[1] Et_3N, −20° → 20°, 3 h, 54–80% yield.[2]
2. *t*-$BuOCH_2SO_2Ph$, LiBr, TEA, toluene, 2–4 days, 70–92% yield.[3]

Cleavage

1. CF_3COOH, H_2O, 20°, 48 h, 85–90% yield.[2] The *t*-butoxymethyl ether is stable to hot glacial acetic acid; aqueous acetic acid, 20°; and anhydrous trifluoroacetic acid.

1. For an improved preparation of this reagent, see J. H. Jones, D. W. Thomas, R. M. Thomas, and M. E. Wood, *Synth. Commun.*, **16**, 1607 (1986).
2. H. W. Pinnick and N. H. Lajis, *J. Org. Chem.*, **43**, 3964 (1978).
3. M. Julia, D. Uguen, and D. Zhang, *Synlett*, 503 (1991).

4-Pentenyloxymethyl Ether (POM–OR):[1] CH_2=$CHCH_2CH_2CH_2OCH_2OR$

Formation

1. POMCl, (*i*-Pr)$_2$NEt, CH_2Cl_2.[2] The related pentenyl glycosides, prepared by the usual methods, were used to protect the anomeric center.[2]

Cleavage

1. NBS, CH_3CN, H_2O, 62–90% yield.[2,3,4] The POM group has been selectively removed in the presence of an ethoxyethyl ether, a TBDMS ether, a benzyl ether, a *p*-methoxybenzyl ether, an acetate, and an allyl group. Because the hydrolysis of a pentenyl 2-acetoxyglycoside was so much slower than a pentenyl 2-benzyloxyglycoside, the 2-benzyl derivative could be cleaved selectively in the presence of the 2-acetoxy derivative.[5] The POM group is stable to 75% AcOH, but is cleaved by 5% HCl.

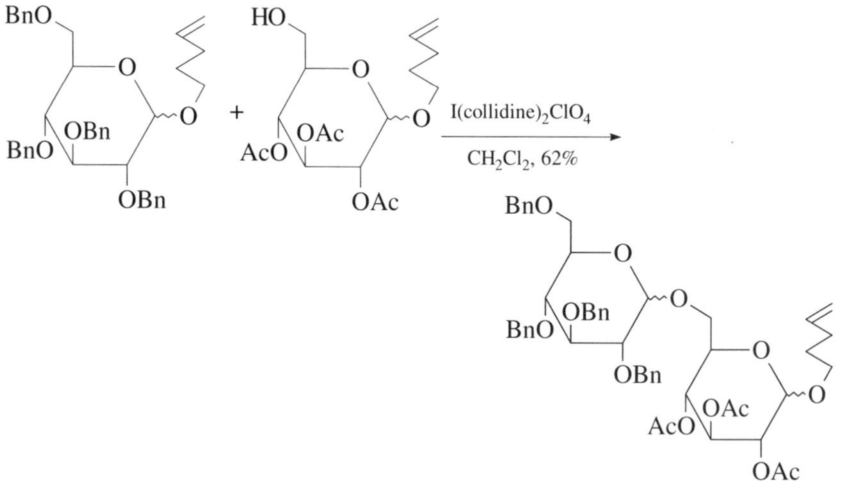

BnO, OBn, BnO, OBn, O + HO, OAc, AcO, OAc, O —I(collidine)$_2$ClO$_4$, CH_2Cl_2, 62%→ BnO, OBn, BnO, OBn, O, OAc, AcO, OAc

Cleavage of the POM group in the presence of neighboring hydroxyls can result in the formation of methylene acetals.[2]

1. The chemistry of the 4-pentenyloxy group has been reviewed by B. Fraser-Reid, U. E. Udodong, Z. Wu, H. Ottosson, J. R. Merritt, C. S. Rao, C. Roberts, and R. Madsen, *Synlett*, 927 (1992).
2. Z. Wu, D. R. Mootoo, and B. Fraser-Reid, *Tetrahedron Lett.*, **29**, 6549 (1988).
3. D. R. Mootoo, V. Date, and B. Fraser-Reid, *J. Am. Chem. Soc.*, **110**, 2662 (1988).
4. For a discussion of the factors that influence the rate of NBS-induced *n*-pentenylglycoside hydrolysis, see C. W. Andrews, R. Rodebaugh, and B. Fraser-Reid, *J. Org. Chem.*, **61**, 5280 (1996).

5. D. R. Mootoo, P. Konradsson, U. Udodong, and B. Fraser-Reid, *J. Am. Chem. Soc.*, **110**, 5583 (1988).

Siloxymethyl Ether: RR'_2SiOCH_2OR'', R′= Me, R = *t*-Bu; R = Thexyl, R′= Me; R = *t*-Bu, R′ = Ph

Formation[1]

1. RR'_2SiOCH_2Cl, $(i\text{-}Pr)_2NEt$, CH_2Cl_2, 73–92% yield.

Cleavage[1]

1. $Bu_4N^+F^-$, THF, 70–80% yield.
2. $Et_4N^+F^-$, CH_3CN, rt, 64–75% yield.
3. AcOH, H_2O.

1. L. L. Gundersen, T. Benneche, and K. Undheim, *Acta Chem. Scand.*, **43**, 706 (1989).

2-Methoxyethoxymethyl Ether (MEM–OR): $CH_3OCH_2CH_2OCH_2OR$ (Chart 1)

Formation

1. NaH or KH, MEMCl, THF or DME, 0°, 10–60 min, >95% yield.[1]
2. $MEMN^+Et_3Cl^-$, CH_3CN, reflux, 30 min, >90% yield.[1]
3. MEMCl, $(i\text{-}Pr)_2NEt$ (DIPEA), CH_2Cl_2, 25°, 3 h, quant.[1]
4. The MEM group has been introduced on one of two sterically similar, but electronically different, alcohols in a 1,2-diol.[2]

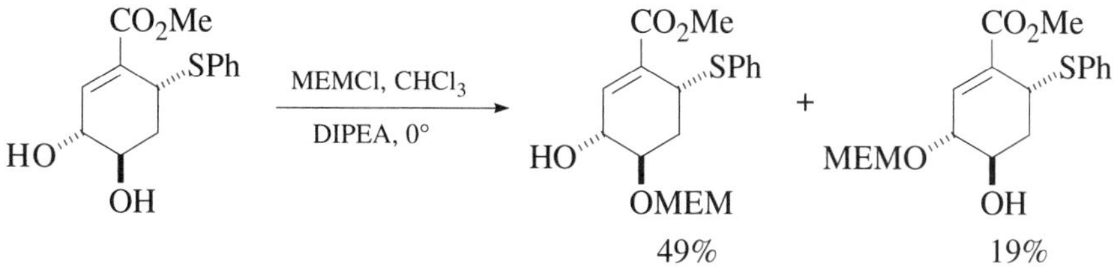

CO_2Me, SPh, HO, OH → (MEMCl, $CHCl_3$; DIPEA, 0°) → CO_2Me, SPh, HO, OMEM (49%) + CO_2Me, SPh, MEMO, OH (19%)

Cleavage

1. $ZnBr_2$, CH_2Cl_2, 25°, 2–10 h, 90% yield.[1] When a MEM-protected diol was cleaved using $ZnBr_2$ in EtOAc, 1,3-dioxolane formation occurred,[3] but this can be prevented by the use of *in situ*–prepared TMSI.[4]
2. $TiCl_4$, CH_2Cl_2, 0°, 20 min, 95% yield.[1,5]

3. Me_2BBr, CH_2Cl_2, −78°; $NaHCO_3$, H_2O, 87–95% yield.[6] This method also cleaves MTM and MOM ethers and ketals.
4. (i-PrS)$_2$BBr, DMAP; K_2CO_3, H_2O.[7] In this case, the MEM ether is converted into the i-PrSCH$_2$ ether that can be cleaved using the same conditions employed to cleave the MTM ether. In one case where the related 2-chloro-1,3,2-dithioborolane was used for MEM ether cleavage, a thiol (–$OCH_2SCH_2CH_2SH$) was isolated as a by-product in 29% yield.[8]
5. Pyridinium p-toluenesulfonate, t-BuOH or 2-butanone, heat, 80–99% yield.[9] This method also cleaves the MOM ether and has the advantage that it cleanly cleaves allylic ethers that could not be cleaved by Corey's original procedure.
6. Me_3SiCl, NaI, CH_3CN, −20°, 79%.[10] Allylic and benzylic ethers tend to form some iodide as a by-product, but less iodide is formed than when Me_3SiI is used directly.
7. (1,3,2-dithioborolane)BBr 2 eq. CH_2Cl_2, −78°.[11] Benzyl, allyl, methyl, THP, TBDMS, and TBDPS ethers are all stable to these conditions. A primary MEM group could be selectively removed in the presence of a hindered secondary MEM group.

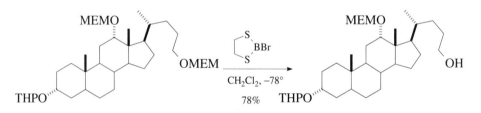

MEMO … OMEM, THPO → (S,S-BBr; CH_2Cl_2, −78°; 78%) → MEMO … OH, THPO

8. HBF_4, CH_2Cl_2, 0°, 3 h, 50–60% yield.[12]
9. In a study of the deprotection of the MEM ethers of hydroxyproline and serine derivatives, it was found that the MEM group was stable to conditions that normally cleave the t-butyl and BOC groups [CF_3COOH, CH_2Cl_2, 1:1 (v:v)]. The MEM group was also stable to 0.2 N HCl, but not stable to 2.0 N HCl or HBr–AcOH.[13]

Removal Time in TFA/CH_2Cl_2 (v/v)

	1:4	1:1	1:0
Z-Hyp(t-Bu)–ONb	45 min	15 min	5 min
Z-Hyp(MEM)OMe	10 h	6 h	2 h

Hyp = hydroxyproline; Nb = 4-nitrobenzoate

10. (a) n-BuLi, THF; (b) $Hg(OAc)_2$, H_2O, THF, 81% yield.[14] In this case, conventional methods to remove the MEM group were unsuccessful.

MEMO … OH, H, H, OTBDMS — 1. BuLi, THF; 2. $Hg(OAc)_2$, H_2O, THF, 81% → HO … OH, H, H, OTBDMS

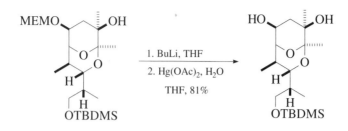

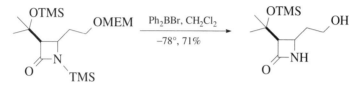

11. (catechol)BBr For a further discussion of this reagent, refer to the section on MOM ethers.[15]
12. Ph_2BBr, CH_2Cl_2, −78°, 71% yield.[16]

OTMS, OMEM, N–TMS, O — Ph_2BBr, CH_2Cl_2, −78°, 71% → OTMS, OH, NH, O

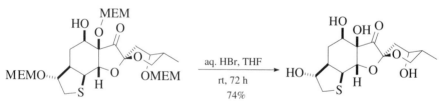

13. $MgBr_2$, Et_2O, 77–95% yield.[17] MOM, SEM, and MTM ethers are also cleaved with this reagent.
14. Aq. HBr, THF, rt, 72 h, 74% yield.[18]

HO, MEM, O, O, MEMO, O, OMEM, H, S — aq. HBr, THF, rt, 72 h, 74% → HO, OH, O, HO, O, OH, H, S

15. $FeCl_3$, Ac_2O, −45°; K_2CO_3, MeOH, 90% yield.[19] A TBDMS group and an acetonide were not affected by these conditions.
16. $H_2ZnCl_2Br_2$, THF, rt, 1 h, 84% or Li_2ZnBr_4, THF, rt, 48 h, 94% yield.[20]

1. E. J. Corey, J.-L. Gras, and P. Ulrich, *Tetrahedron Lett.*, 809 (1976).
2. G. H. Posner, A. Haces, W. Harrison, and C. M. Kinter, *J. Org. Chem.*, **52**, 4836 (1987).
3. J. A. Boynton and J. R. Hanson, *J. Chem. Res., Synop.*, 378 (1992).
4. K. C. Nicolaou, E. W. Yue, S. La Greca, A Nadin, Z. Yang, J. E. Leresche, T. Tsuri, Y. Naniwa, and F. De Riccardis, *Chem.–Eur. J.*, **1**, 467 (1995).
5. O. Miyata, T. Shinada, I. Ninomiya, and T. Naito, *Tetrahedron Lett.*, **32**, 3519 (1991).
6. Y. Quindon, H. E. Morton, and C. Yoakim, *Tetrahedron Lett.*, **24**, 3969 (1983).
7. E. J. Corey, D. H. Hua, and S. P. Seitz, *Tetrahedron Lett.*, **25**, 3 (1984).

8. M. Bénéchie and F. Khuong-Huu, *J. Org. Chem.*, **61**, 7133 (1996).
9. H. Monti, G. Léandri, M. Klos-Ringuet, and C. Corriol, *Synth. Commun.*, **13**, 1021 (1983).
10. J. H. Rigby and J. Z. Wilson, *Tetrahedron Lett.*, **25**, 1429 (1984).
11. D. R. Williams and S. Sakdarat, *Tetrahedron Lett.*, **25**, 3965 (1983).
12. N. Ikota and B. Ganem, *J. Chem. Soc., Chem. Commun.*, 869 (1978).
13. D. Vadolas, H. P. Germann, S. Thakur, W. Keller, and E. Heidemann, *Int. J. Pept. Protein Res.*, **25**, 554 (1985).
14. R. E. Ireland, P. G. M. Wuts, and B. Ernst, *J. Am. Chem. Soc.*, **103**, 3205 (1981).
15. R. K. Boeckman, Jr., and J. C. Potenza, *Tetrahedron Lett.*, **26**, 1411 (1985).
16. M. Shibasaki, Y. Ishida, and N. Okabe, *Tetrahedron Lett.*, **26**, 2217 (1985).
17. S. Kim, Y. H. Park, and I. S. Kee, *Tetrahedron Lett.*, **32**, 3099 (1991).
18. D. R. Williams, P. A. Jass, H.-L. A. Tse, and R. D. Gaston, *J. Am. Chem. Soc.*, **112**, 4552 (1990).
19. R. A. Holton, R. R. Juo, H. B. Kim, A. D. Williams, S. Harusawa, R. E. Lowenthal, and S. Yogai, *J. Am. Chem. Soc.*, **110**, 6558 (1988).
20. J. M. Herbert, J. G. Knight, and B. Sexton, *Tetrahedron*, **52**, 15257 (1996).

2,2,2-Trichloroethoxymethyl Ether: $Cl_3CCH_2OCH_2OR$

Formation

1. $Cl_3CCH_2OCH_2Cl$, NaH or KH, LiI, THF, 5 h, 70–90% yield.[1]
2. $Cl_3CCH_2OCH_2Cl$, $(i\text{-}Pr)_2NEt$, CH_2Cl_2, 30–60% yield.[1]
3. $Cl_3CCH_2OCH_2Br$, 1,8-bis(dimethylamino)naphthalene (proton sponge), CH_3CN, 0–25°, >87% yield.[2]

Cleavage

1. Zn–Cu or Zn–Ag, MeOH, reflux, 97%.[1]
2. Zn, MeOH, Et_3N, AcOH, reflux 4 h, 90–100%.[1]
3. Li, NH_3.[1]
4. SmI_2, THF, 25°, 71% yield.[2]
5. 6% Na(Hg), MeOH, THF, >66% yield.[2]

1. R. M. Jacobson and J. W. Clader, *Synth. Commun.*, **9**, 57 (1979).
2. D. A. Evans, S. W. Kaldor, T. K. Jones, J. Clardy, and T. J. Stout, *J. Am. Chem. Soc.*, **112**, 7001 (1990).

Bis(2-chloroethoxy)methyl Ether: $ROCH(OCH_2CH_2Cl)_2$ (Chart 1)

The mixed ortho ester formed from tri(2-chloroethyl) orthoformate (100°, 10 min to 2 h, 76% yield) is more stable to acid than the unsubstituted derivative, but can be cleaved with 80% AcOH (20°, 1 h).[1]

1. T. Hata and J. Azizian, *Tetrahedron Lett.*, 4443 (1969).

2-(Trimethylsilyl)ethoxymethyl Ether (SEM–OR): $Me_3SiCH_2CH_2OCH_2OR$

Formation

1. $Me_3SiCH_2CH_2OCH_2Cl$, (i-Pr)$_2$NEt (DIPEA), CH_2Cl_2, 35–40°, 1–5 h, 86–100% yield.[1]
2. The preceding conditions failed in this example, unless $Bu_4N^+I^-$ was added to prepare SEMI *in situ*.[2]

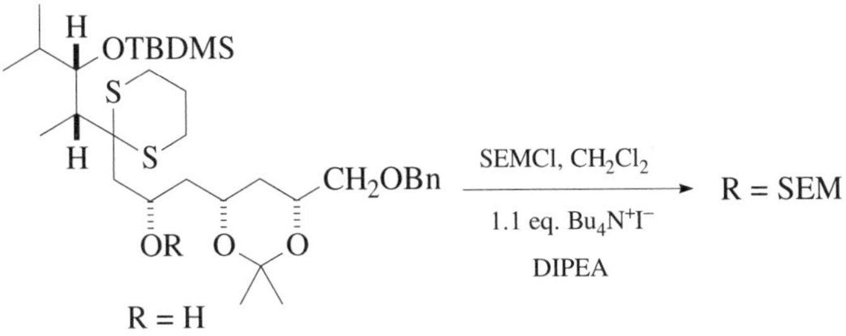

H OTBDMS S S H CH$_2$OBn OR O O R = H —SEMCl, CH_2Cl_2 / 1.1 eq. $Bu_4N^+I^-$ / DIPEA→ R = SEM

3. SEMCl, KH, THF, 0° → rt, 1 h, 87% yield.[3]
4. t-BuMgCl, THF, rt, 5 min, then $Bu_4N^+I^-$, SEMCl, rt, 20–30 h, 78–84% yield. These conditions prevent alkylation of the nitrogen in the nucleoside bases.[4]

Cleavage

SEM ethers are stable to the acidic conditions (AcOH, H_2O, THF, 45°, 7 h) that are used to cleave tetrahydropyranyl and t-butyldimethylsilyl ethers.

1. $Bu_4N^+F^-$, THF or HMPA, 45°, 8–12 h, 85–95% yield.[1,5] The cleavage of 2-(trimethylsilyl)ethyl glycosides is included here because they are functionally equivalent to the SEM group. These glycosides can be prepared by oxymercuration of a glycal with $Hg(OAc)_2$ and $TMSCH_2CH_2OH$, by the reaction of a glycosyl halide using Koenig–Knorr conditions, by a Fischer glycosidation, and by a glycal rearrangement.[4] N,N-Dimethylpropyleneurea can be used to replace the carcinogenic HMPA (45–80% yield).[6] An improved isolation procedure utilizing the insolubility of $Bu_4N^+ClO_4^-$ in water has been developed for isolations in which tetrabutylammonium fluoride is used.[7]
2. TFA, CH_2Cl_2 (2:1, v:v), 0°, 30 min, 93% yield.[8]

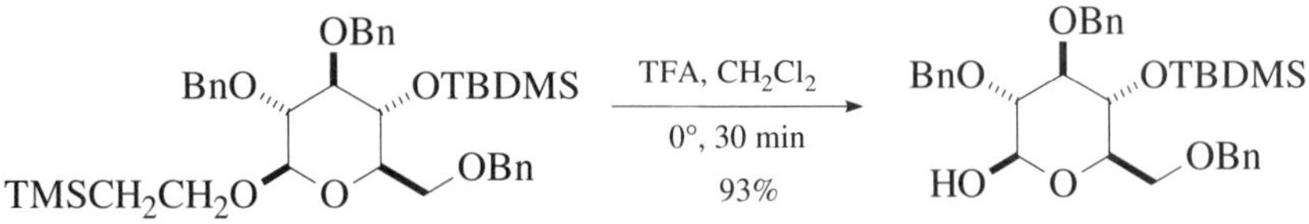

TFA, CH_2Cl_2, 0°, 30 min, 93%

The 4,6-*O*-benzylidene group is also cleaved under these conditions, but the anomeric linkage between sugars is not affected. In addition, anomeric trimethylsilylethyl groups are cleaved with $BF_3 \cdot Et_2O$[9] or $Ac_2O/FeCl_3$. (This reagent also cleaves the BOM group.)[10] The anomeric trimethylsilylethyl group is hydrolyzed much faster than the other alkyl glycosides.[11]

3. $LiBF_4$, CH_3CN, 70°, 3–8 h, 81–90% yield.[12] This system of reagents also cleaves benzylidene acetals. Conventional reagents failed to cleave these glycosides. It is interesting to note that the β-anomers are cleaved more rapidly than the α-anomers and that the furanoside derivatives are not cleaved.
4. CsF, DMF, 130°, >89% yield.[13]
5. $Bu_4N^+F^-$, DMPU, 4-Å molecular sieves, 45–80°, 80–95% yield.[6] These conditions were especially effective in cleaving tertiary SEM derivatives and avoiding the use of the toxic HMPA.
6. $MgBr_2$, *n*-BuSH, Et_2O, rt, 3–24 h, 49–97% yield. MOM and MTM ethers are also cleaved, but MEM and TBDMS ethers are stable. These conditions have resulted in the formation of an ethyl thioether.[14]
7. $ZnCl_2$–Et_2O, 99% yield[15] or $BF_3 \cdot Et_2O$, CH_2Cl_2, 0–25°, 2 h.[16] In these examples, a simple trimethylsilylethyl ether was cleaved, but the method is also applicable to SEM deprotection.[17]
8. 1.5% Methanolic HCl, 16 h, 80–94% yield. These conditions do not cleave the MEM group.[18] 1% Sulfuric acid in methanol has also been used.[19]
9. Concd. HF, CH_3CN, >76% yield.[20] Note that a trimethylsilylethyl ester was not cleaved under these conditions.
10. I_2, sunlamp, 92% yield.[21]

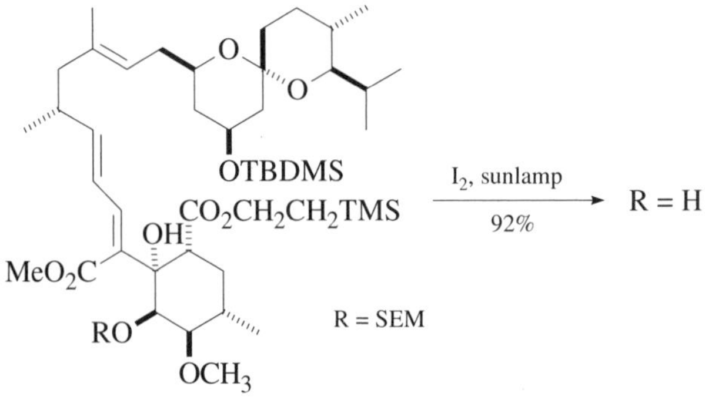

I_2, sunlamp, 92%

R = SEM → R = H

11. Pyridine–HF, THF, 2.5 h, 0–25°, 79% yield.[22]
12. A study of the reductive cleavage of a series of alkoxymethyl ethers using the glucose backbone shows that, depending on the reagent, excellent selectivity can be obtained for deprotection vs. methyl ether formation for most of the common protective groups.[23]

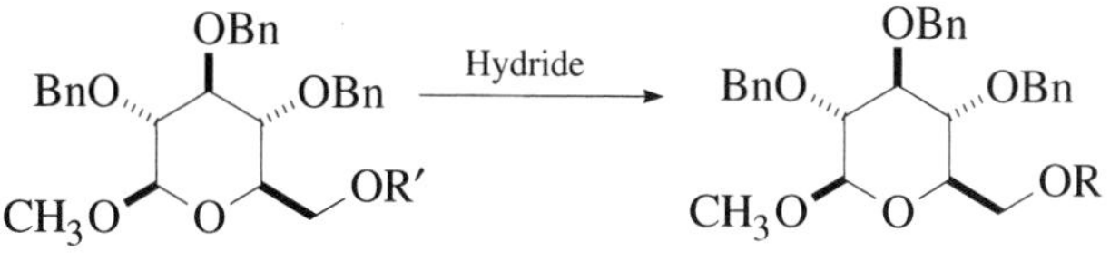

Hydride

1. R = H
2. R = Me

Ether R′=	AlH_2Cl Percent 1	AlH_2Cl Percent 2	$AlHCl_2$ Percent 1	$AlHCl_2$ Percent 2	BH_3/THF Percent 1	BH_3/THF Percent 2	BH_3/Toluene Percent 1	BH_3/Toluene Percent 2
MTM	100	0	100	0	85	15	100	0
SEM	0	0	100	0	100	0	100	0
BOM	0	0	89	11	98	2	100	0
*p*AOM	45	55	32	68	12	86	0	100
EOM	0	0	100	0	0	0	100	0

For secondary derivatives, the selectivity and reactivity vary somewhat. To what extent this depends on the highly functionalized glucose derivative has not been determined. The table that follows gives the cleavage selectivity for the reaction

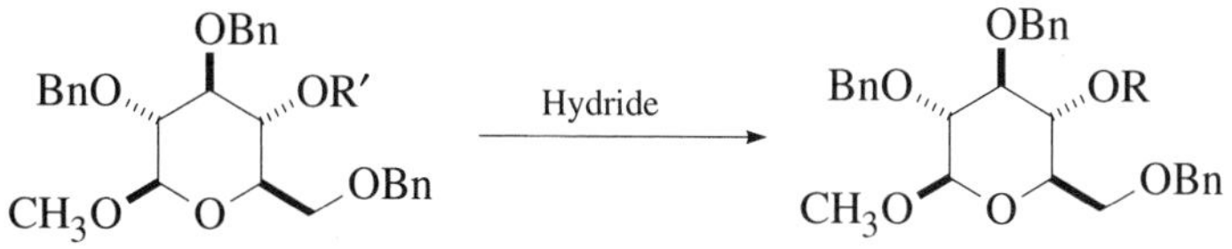

Hydride

1. R = H
2. R = Me

Ether R′=	AlH_2Cl Percent 1	AlH_2Cl Percent 2	$AlHCl_2$ Percent 1	$AlHCl_2$ Percent 2	BH_3/THF Percent 1	BH_3/THF Percent 2	BH_3/Toluene Percent 1	BH_3/Toluene Percent 2
SEM	100	0	82	18	0	0	100	0
BOM	100	0	90	10	0	0	100	0
*p*AOM	0	0	0	0	0	0	0	100
EOM	100	0	100	0	0	0	100	0

1. B. H. Lipshutz and J. J. Pegram, *Tetrahedron Lett.*, **21**, 3343 (1980).
2. B. H. Lipshutz, R. Moretti, and R. Crow, *Tetrahedron Lett.*, **30**, 15 (1989).
3. D. R. Williams, P. A. Jass, H.-L. A. Tse, and R. D. Gaston, *J. Am. Chem. Soc.*, **112**, 4552 (1990).
4. T. Wada, M. Tobe, T. Nagayama, K. Furusawa, and M. Sekine, *Tetrahedron Lett.*, **36**, 1683 (1995).
5. T. Kan, M. Hashimoto, M. Yanagiya, and H. Shirahama, *Tetrahedron Lett.*, **29**, 5417 (1988).
6. B. H. Lipschutz and T. A. Miller, *Tetrahedron Lett.*, **30**, 7149 (1989).
7. J. C. Craig and E. T. Everhart, *Synth. Commun.*, **20**, 2147 (1990).
8. K. Jansson, T. Frejd, J. Kihlberg, and G. Magnusson, *Tetrahedron Lett.*, **29**, 361 (1988). For another case, see R. H. Schlessinger, M. A. Poss, and S. Richardson, *J. Am. Chem. Soc.*, **108**, 3112 (1986).
9. A. Hasegawa, Y. Ito, H. Ishida, and M. Kiso, *J. Carbohydr. Chem.*, **8**, 125 (1989); K. Jansson, T. Frejd, J. Kihlberg, and G. Magnusson, *Tetrahedron Lett.*, **27**, 753 (1986).
10. K. P. R. Kartha, M. Kiso, and A. Hasegawa, *J. Carbohydr. Chem.*, **8**, 675 (1989).
11. K. Jansson, G. Noori, and G. Magnusson, *J. Org. Chem.*, **55**, 3181 (1990).
12. B. H. Lipshutz, J. J. Pegram, and M. C. Morey, *Tetrahedron Lett.*, **22**, 4603 (1981).
13. K. Suzuki, T. Matsumoto, K. Tomooka, K. Matsumoto, and G.-I. Tsuchihashi, *Chem. Lett.,* 113 (1987).
14. S. Kim, I. S. Kee, Y. H. Park, and J. H. Park, *Synlett*, 183 (1991); S. Bailey, A. Teerawutgulrag, and E. J. Thomas, *J. Chem. Soc., Chem. Commun.*, 2521 (1995).
15. H. C. Kolb and H. M. R. Hoffman, *Tetrahedron: Asymmetry*, **1**, 237 (1990).
16. S. D. Burke and G. J. Pacofsky, *Tetrahedron Lett.*, **27**, 445 (1986).
17. F. E. Wincott and N. Usman, *Tetrahedron Lett.*, **35**, 6827 (1994).
18. B. M. Pinto, M. M. W. Buiting, and K. B. Reimer, *J. Org. Chem.*, **55**, 2177 (1990).
19. A. A. Kandil and K. N. Sellsor, *J. Org. Chem.*, **50**, 5649 (1985).
20. J. D. White and M. Kawasaki, *J. Am. Chem. Soc.*, **112**, 4991 (1990).
21. S. Karim, E. R. Parmee, and E. J. Thomas, *Tetrahedron Lett.*, **32**, 2269 (1991).
22. K. Sugita, K. Shigeno, C. F. Neville, H. Sasai, and M. Shibasaki, *Synlett*, 325 (1994).
23. I. Bajza, Z. Varga, and A. Liptak, *Tetrahedron Lett.*, **34**, 1991 (1993).

Menthoxymethyl Ether (MM–OR)

O OR

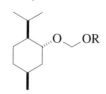

This protective group was developed to determine the enantiomeric excess of chiral alcohols. It is anticipated that many of the methods used to cleave the MOM group would be effective for the MM group as well.

Formation

Menthoxymethyl chloride, DIPEA, CH_2Cl_2, rt, overnight, 77–95% yield.[1]

Cleavage

1. $ZnBr_2$, CH_2Cl_2.[1]
2. TMSOTf, TMSOMe, $ClCH_2CH_2Cl$, 0°–rt, 98% yield. The MM ether is converted to a simple MOM ether. When the TMSOMe was left out of the reaction, neighboring group participation occurred, to give a 1,3-dioxane.[2]

TBDMSO(CH₂)₂CH(CH₃)CH(OTBDMS)CH₂CH(OMM)CH₂SnBu₃ — TMSOTf, $ClCH_2CH_2Cl$, 0°–rt, 81% → 1,3-dioxane (O–CH₂–O) bearing TBDMSO side chain and CH_2SnBu_3

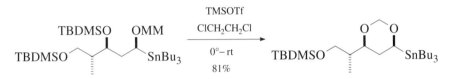

1. D. Dawkins and P. R. Jenkins, *Tetrahedron: Asymmetry*, **3**, 833 (1992).
2. R. J. Linderman, K. P. Cusack, and M. R. Jaber, *Tetrahedron Lett.*, **37**, 6649 (1996).

Tetrahydropyranyl Ether (THP–OR) (Chart 1):

The introduction of a THP ether onto a chiral molecule results in the formation of diastereomers because of the additional stereogenic center present in the tetrahydropyran ring (which can make the interpretation of NMR spectra somewhat troublesome at times). Even so, this is an extensively used protective group in chemical synthesis because of its low cost, ease of installation, general stability to most nonacidic reagents, and ease with which it can be removed. Generally, almost any acidic reagent or any reagent that generates an acid *in situ* can be used to introduce the THP group.

Formation

1. Dihydropyran, TsOH, CH_2Cl_2, 20°, 1.5 h, 100% yield.[1]
2. Pyridinium *p*-toluenesulfonate (PPTS), dihydropyran, CH_2Cl_2, 20°, 4 h, 94–100% yield.[2] The lower acidity of PPTS makes this a very mild method that has excellent compatibility with most functional groups.
3. Reillex 425·HCl, dihydropyran, 86°, 1.5 h, 84–98% yield.[3] The Reillex resin is a macroreticular polyvinylpyridine resin and is thus an insoluble form of the PPTS catalyst.
4. Amberlyst H-15 (SO_3H ion exchange resin), dihydropyran, hexane, 1–2 h, 95% yield.[4]
5. Dihydropyran, K-10 clay, CH_2Cl_2, rt, 63–95% yield.[5] Kaolinitic Clay is

also an effective catalyst, except for phenols, which fail to react.[6] Spanish Speolite clay has also been used.[7]

6. Dihydropyran, $(TMSO)_2SO_2$, CH_2Cl_2, 92–100% yields.[8]
7. Dihydropyran, TMSI, CH_2Cl_2, rt, 80–96% yield.[9]
8. Dihydropyran, $Ph_3P{\cdot}HBr$, 24 h, CH_2Cl_2, 88% yield.[10]
9. Dihydropyran, Dowex 50wx2, toluene, 10–355 min, 78–95% yield. These conditions were developed to monoprotect symmetrical 1,ω-diols.[11]
10. Dihydropyran, $LaCl_3$, CH_2Cl_2, rt, 4 h, 90%.[12]
11. Dihydropyran, sulfonated charcoal, 3Å ms, CH_2Cl_2, 67–98% yield.[13] Sulfated zirconia has also been used as a catalyst with similar effectivness.[14]
12. Polymer-bound dihydropyran, PPTS, 80°.[15]
13. Dihydropyran, $Al(PO_4)_3$, reflux, 15 min, 97% yield.[16]
14. Dihydropyran, DDQ, CH_2Cl_2, 82–100% yield.[17]
15. 2-Tetrahydropyranyl phenyl sulfone, $MgBr_2$–Et_2O, $NaHCO_3$, THF, rt, 47–99% yield.[18]
16. Dihydropyran, H–Y Zeolite, hexane, reflux, 60–95% yield.[19] H–Rho Zeolite can also be used as a catalyst.[20]
17. Dihydropyran, H-MCM-41, ms, 69°, 44–99% yield.[21]
18. Dihydropyran, $H_3[PW_{12}O_{40}]$, CH_2Cl_2, rt, 64–96% yield. The same acid can be used to cleave the THP group if methanol is used as solvent.[22]
19. Tetrahydropyran, $(Bu_4N^+)_2S_2O_8^-$, reflux, 85–95% yield. These oxidative conditions do not affect thioethers.[23]
20. 3,4-$(MeO)_2C_6H_3CH_2OTHF$, DDQ, CH_3CN, 54–94% yield. These conditions can also be used for glycoside synthesis.[24]
21. $Al_2(SO_4)_3$–SiO_2 is a reasonable catalyst for the monotetrahydropyranylation of simple, symmetrical 1,ω-diols.[25]
22. Dihydropyran, Al_2O_3, $ZnCl_2$.[26]
23. Dihydropyran, CAN, CH_3CN, rt, 81–91% yield.[27]
24. Dihydropyran, Zeolite HSZ-330, rt, 1.5 h, 44–100% yield.[28]
25. Dihydropyran, CuCl, CH_2Cl_2, 75–93% yield.[29]

Cleavage

1. AcOH, THF, H_2O, (4:2:1), 45°, 3.5 h.[1] MEM ethers are stable to these conditions.[30]
2. PPTS, EtOH, (pH 3.0), 55°, 3 h, 95–100% yield.[2]
3. Amberlyst H-15, MeOH, 45°, 1 h, 95% yield.[3] Dowex 50W-X8, 25°, 1 h, MeOH, 99% yield.[31]
4. Boric acid, $EtOCH_2CH_2OH$, 90°, 2 h, 80–95% yield.[32]
5. TsOH, MeOH, 25°, 1 h, 94% yield.[33] The use of 2-propanol as solvent was found to enhance the selectivity for THP removal in the presence of a

1,3-TBDPS group.[34] TBDPS ethers are not affected by these conditions.[35]

6. $MgBr_2$, Et_2O, rt, 66–95% yield.[36] *t*-Butyldimethylsilyl and MEM ethers are not affected by these conditions, but the MOM ether is slowly cleaved. The THP derivatives of benzylic and tertiary alcohols give bromides.

THPO OTBDMS CO_2Me —(Me_2AlCl, CH_2Cl_2; −25° to rt, 1 h, 89%)→ HO OTBDMS CO_2Me

7. Me_2AlCl, CH_2Cl_2, −25° → rt, 1 h, 89–100% yield.[37]
8. MeOH, $(TMSO)_2SO_2$, 10–90 min, 93–100% yield.[6]
9. $(NCSBu_2Sn)_2O$ 1%, THF, H_2O.[38] In addition, acetonides and TMS ethers are cleaved under these conditions, but TBDMS, MTM, and MOM groups are stable. This catalyst has also been used to effect transesterifications.[39]
10. The THP ether can be converted directly to an acetate by refluxing in AcOH/AcCl (91% yield).[40] These conditions would probably convert other related acetals to acetates as well.
11. MeOH, reagent prepared by heating Bu_2SnO and Bu_3SnPO_4, heat 2 h, 90% yield.[41] This method is effective for primary, secondary, tertiary, benzylic, and allylic THP derivatives. The MEM group and ketals are inert to this reagent, but TMS and TBDMS ethers are cleaved.
12. $Ph_3P{\cdot}Br_2$, CH_2Cl_2, −50°→ 35°, 85–94% yield.[42] Ethyl acetals and MOM groups are also cleaved with this reagent, but a THP ether can be selectively cleaved in the presence of a MOM ether. The use of this reagent at 0–10° (16 h) will convert a THP ether directly into a bromide,[43] and with a slight modification of the reaction conditions, chlorides, nitriles, methyl ethers, and trifluoroacetates may also be directly produced.[44] Similar processes have been observed using other brominating agents.[45]
13. Bu_3SnSMe, $BF_3{\cdot}Et_2O$, toluene, −20° → 0°, 1.5 h; H_3O^+, 70–97% yield. When treated with various electrophiles, the intermediate stannanes from this reaction form benzyl and MEM ethers, benzoates, and tosylates, and when treated with PCC, they form aldehydes.[46,47]
14. Tonsil, a Mexican Bentonite, acetone, 30 min, rt, 60–95% yield. MOM and MEM groups are stable and phenolic THP ethers were also cleaved.[48]
15. TBDMSOTf, CH_2Cl_2; Me_2S, 95% yield. The THP group is converted directly into a TBDMS ether.[49]
16. $BH_3{\cdot}THF$, 20°, 24 h, 84% yield.[50]
17. DDQ, aq. MeOH, 81–98% yield.[51] DDQ in aqueous CH_3CN has also been used (42–95% yield), but since the medium was reported to be acidic (pH 3), the reaction probably occurs by simple acid catalysis. Benzylic, allylic, and primary THP derivatives are not efficiently cleaved.[52]

18. $NaCNBH_3$, $BF_3{\cdot}Et_2O$, rt, 68–95% yield.[53]
19. LiCl, H_2O, DMSO, 90°, 6 h, 81–92% yield.[54]
20. CAN, MeOH, 0°, 0.5–3 h, 81–95% yield. TBDMS ethers are more easily cleaved; thus, a TBDMS ether is cleaved selectively in the presence of a THP ether (15 min, 95%).[55]
21. $BF_3{\cdot}Et_2O$, $HSCH_2CH_2SH$, CH_2Cl_2, 100% yield. A primary TBDMS ether was not affected.[56]
22. $SnCl_2$, MeOH.[57]
23. THP ethers can be converted directly to TBDMS and TES ethers using the silyl hydride and $Sn(OTf)_2$ or the silyl triflate (70–95% yield). The use of TMSOTf gives the free alcohols upon isolation.[58]
24. Explosions have been reported on distillation of compounds containing a tetrahydropyranyl ether after a reaction with $B_2H_6/H_2O_2–OH^-$, and with 40% CH_3CO_3H:

OTHP → (1. B_2H_6; 2. H_2O_2, NaOH) → OTHP (epoxide O)

OTHP → (40% CH_3CO_3H) → OTHP (epoxide O)

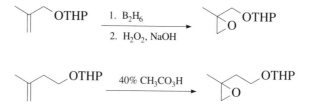

It was thought that the acetal might have reacted with peroxy reagents, forming explosive peroxides. It was suggested that this could also occur with compounds such as tetrahydrofuranyl acetals, 1,3-dioxolanes, and methoxymethyl ethers.[59]

1. K. F. Bernady, M. B. Floyd, J. F. Poletto, and M. J. Weiss, *J. Org. Chem.*, **44**, 1438 (1979).
2. M. Miyashita, A. Yoshikoshi, and P. A. Grieco, *J. Org. Chem.*, **42**, 3772 (1977).
3. R. D. Johnston, C. R. Marston, P. E. Krieger, and G. L. Goe, *Synthesis*, 393 (1988).
4. A. Bongini, G. Cardillo, M. Orena, and S. Sandri, *Synthesis*, 618 (1979).
5. S. Hoyer, P. Laszlo, M. Orlovic, and E. Polla, *Synthesis*, 655 (1986).
6. T. T. Upadhya, T. Daniel, A. Sudalai, T. Ravindranathan, and K. R. Sabu, *Synth. Commun.*, **26**, 4539 (1996).
7. J. M. Campelo, A. Garcia, F. Lafont, D. Luna, and J. M. Marinas, *Synth. Commun.*, **24**, 1345 (1994).
8. Y. Morizawa, I. Mori, T. Hiyama, and H. Nozaki, *Synthesis*, 899 (1981).
9. G. A. Olah, A. Husain, and B. P. Singh, *Synthesis*, 703 (1985).
10. V. Bolitt, C. Mioskowski, D.-S. Shin, and J. R. Falck, *Tetrahedron Lett.*, **29**, 4583 (1988).
11. T. Nishiguchi, M. Kuroda, M. Saitoh, A. Nishida, and S. Fujisaki, *J. Chem. Soc., Chem. Commun.*, 2491 (1995).

12. V. Bhuma and M. L. Kantam, *Synth. Commun.*, **22**, 2941 (1992).
13. H. K. Patney, *Synth. Commun.*, **21**, 2329 (1991).
14. A. Sakar, O. S. Yemul, B. P. Bandgar, N. B. Gaikwad, and P. P. Wadgaonkar, *Org. Prep. Proced. Int.*, **28**, 613 (1996).
15. L. A. Thompson and J. A. Ellman, *Tetrahedron Lett.*, **35**, 9333 (1994).
16. J. M. Campelo, A. Garcia, F. Lafont, D. Luna, and J. M. Marinas, *Synth. Commun.*, **22**, 2335 (1992).
17. K. Tanemura, T. Horaguchi, and T. Suzuki, *Bull. Chem. Soc. Jpn.*, **65**, 304 (1992).
18. D. S. Brown, S. V. Ley, S. Vile, and M. Thompson, *Tetrahedron*, **47**, 1329 (1991).
19. P. Kumar, C. U. Dinesh, R. S. Reddy, and B. Pandey, *Synthesis*, 1069 (1993).
20. D. P. Sabde, B. G. Naik, V. R. Hedge, and S. G. Hedge, *J. Chem. Res., Synop.*, 494 (1996).
21. K. R. Kloetstra and H. van Bekkum, *J. Chem. Res., Synop.*, 26 (1995).
22. A. Molnar and T. Beregszaszi, *Tetrahedron Lett.*, **37**, 8597 (1996).
23. H. C. Choi, K. I. Cho, and Y. H. Kim, *Synlett*, 207 (1995).
24. J. Inanaga, Y. Yokoyama, and T. Hanamato, *Chem. Lett.*, 85 (1993).
25. T. Nishiguchi and K. Kawamine, *J. Chem. Soc., Chem. Commun.*, 1766 (1990).
26. B. C. Ranu and M. Saha, *J. Org. Chem.*, **59**, 8269 (1994).
27. G. Maity and S. C. Roy, *Synth. Commun.*, **23**, 1667 (1993).
28. R. Ballini, F. Bigi, S. Carloni, R. Maggi, and G. Sartori, *Tetrahedron Lett.*, **38**, 4169 (1997).
29. U. T. Bhalerao, K. J. Davis, and B. V. Rao, *Synth. Commun.*, **26**, 3081 (1996).
30. E. J. Corey, R. L. Danheiser, S. Chandrasekaran, P. Siret, G. E. Keck, and J.-L. Gras, *J. Am. Chem. Soc.*, **100**, 8031 (1978).
31. R. Beier and B. P. Mundy, *Synth. Commun.*, **9**, 271 (1979).
32. J. Gigg and R. Gigg, *J. Chem. Soc. C*, 431 (1967).
33. E. J. Corey, H. Niwa, and J. Knolle, *J. Am. Chem. Soc.*, **100**, 1942 (1978).
34. F. Almqvist and T. Frejd, *Tetrahedron: Asymmetry*, **6**, 957 (1995).
35. A. B. Shenvi and H. Gerlach, *Helv. Chim. Acta*, **63**, 2426 (1980).
36. S. Kim and J. H. Park, *Tetrahedron Lett.*, **28**, 439 (1987).
37. Y. Ogawa and M. Shibasaki, *Tetrahedron Lett.*, **25**, 663 (1984).
38. J. Otera and H. Nozaki, *Tetrahedron Lett.*, **27**, 5743 (1986).
39. J. Otera, T. Yano, A. Kawabata, and H. Nozaki, *Tetrahedron Lett.*, **27**, 2383 (1986).
40. M. Jacobson, R. E. Redfern, W. A. Jones, and M. H. Aldridge, *Science*, **170**, 543 (1970); T. Bakos and I. Vincze, *Synth. Commun.*, **19**, 523 (1989).
41. J. Otera, Y. Niibo, S. Chikada, and H. Nozaki, *Synthesis*, 328 (1988).
42. A. Wagner, M.-P. Heitz, and C. Mioskowski, *J. Chem. Soc., Chem. Commun.*, 1619 (1989).
43. M. Schwarz, J. E. Oliver, and P. E. Sonnet, *J. Org. Chem.*, **40**, 2410 (1975).
44. P. E. Sonnet, *Synth. Commun.*, **6**, 21 (1976).
45. A. Tanaka and T. Oritani, *Tetrahedron Lett.*, **38**, 1955 (1997).
46. T. Sato, J. Otera, and H. Nozaki, *J. Org. Chem.*, **55**, 4770 (1990).

47. T. Sato, T. Tada, J. Otera, and H. Nozaki, *Tetrahedron Lett.*, **30**, 1665 (1989).
48. R. Cruz-Almanza, F. J. Peres-Flores, and M. Avila, *Synth. Commun.*, **20**, 1125 (1990).
49. S. Kim and I. S. Kee, *Tetrahedron Lett.*, **31**, 2899 (1990).
50. J. Cossy, V. Bellosta, and M. C. Müller, *Tetrahedron Lett.*, **33**, 5045 (1992).
51. K. Tanemura, T. Suzuki, and T. Horaguchi, *Bull. Chem. Soc. Jpn.*, **67**, 290 (1994).
52. S. Raina and V. K. Singh, *Synth. Commun.*, **25**, 2395 (1995).
53. A. Srikrishna, J. A. Sattigeri, R.Viswajanani, and C. V. Yelamaggad, *J. Org. Chem.*, **60**, 2260 (1995).
54. G. Maiti and S. C. Roy, *J. Org. Chem.*, **61**, 6038 (1996).
55. A. DattaGupta, R. Singh, and V. K. Singh, *Synlett*, 69 (1996).
56. K. P. Nambiar and A. Mitra, *Tetrahedron Lett.*, **35**, 3033 (1994).
57. K. J. Davis, U. T. Bhalerao, and B. V. Rao, *Indian J. Chem.*, *Sect. B*, **36B**, 211 (1997).
58. T. Oriyama, K. Yatabe, S. Sugawara, Y. Machiguchi, and G. Koga, *Synlett,* 523 (1996).
59. A. I. Meyers, S. Schwartzman, G. L. Olson, and H.-C. Cheung, *Tetrahedron Lett.*, 2417 (1976).

3-Bromotetrahydropyranyl Ether: 3-BrTHP–OR

Br
O OR

The 3-bromotetrahydropyranyl ether was prepared from a 17-hydroxy steroid and 2,3-dibromopyran (pyridine, benzene, 20°, 24 h); it was cleaved by zinc/ethanol.[1]

1. A. D. Cross and I. T. Harrison, *Steroids*, **6**, 397 (1965).

Tetrahydrothiopyranyl Ether (Chart 1):

S OR

The tetrahydrothiopyranyl ether was prepared from a 3-hydroxy steroid and dihydrothiopyran (CF_3COOH, $CHCl_3$, 35% yield); it can be cleaved under neutral conditions ($AgNO_3$, aq. acetone, 85% yield).[1]

1. L. A. Cohen and J. A. Steele, *J. Org. Chem.*, **31**, 2333 (1966).

1-Methoxycyclohexyl Ether:[1] A

RO OCH_3

4-Methoxytetrahydropyranyl Ether (MTHP–OR):[1] B (Chart 1)

RO OCH_3
O

4-Methoxytetrahydrothiopyranyl Ether[2] C (Chart 1):

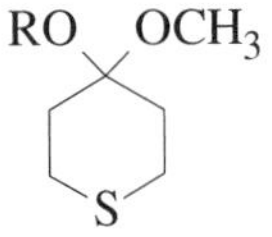

RO OCH$_3$ / S

4-Methoxytetrahydrothiopyranyl Ether *S,S*-Dioxide:[2] D

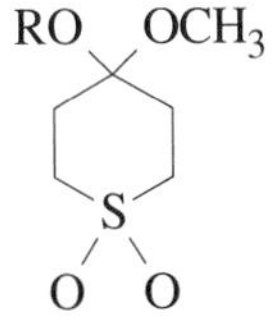

RO OCH$_3$ / S / O O

The preceding ethers have been examined as possible protective groups for the 2′-hydroxyl of ribonucleotides. The following rates of hydrolysis were found: A:B:C:D = 1:0.025:0.005:0.002.[3] These acetals can be prepared by the same methods used for the preparation of the THP derivative. Compounds B and C have been prepared from the vinyl ether and TMSCl as a catalyst.[4] Sulfoxide D was prepared from sulfide C by oxidation with m-ClC$_6$H$_4$CO$_3$H. These ethers have the advantage that they do not introduce an additional stereogenic center into the molecules, as does the THP group. The 4-methoxytetrahydropyranyl group has been employed extensively in nucleoside synthesis, but still suffers from excessive acid lability when the 9-phenylxanthen-9-yl group is used to protect 5′-hydroxy functions in ribonucleotides.[5] The recommended conditions for removal of this group are 0.01 *M* HCl at room temperature. Little, if any, use of these groups has been made by the general synthetic community, but the wide range of selectivities observed in their acidic hydrolysis should make them useful for the selective protection of polyfunctional molecules.

1-[(2-Chloro-4-methyl)phenyl]-4-methoxypiperidin-4-yl Ether (CTMP–OR):[6]

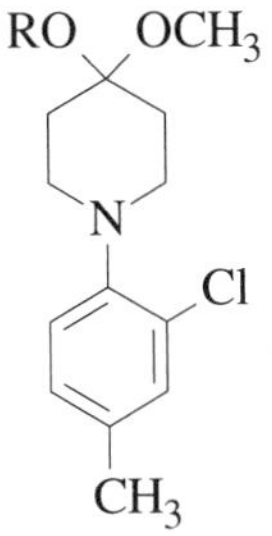

RO OCH$_3$ / N / Cl / CH$_3$

This group was designed to have nearly constant acid stability with decreasing pH ($t_{1/2}$ = 80 min at pH = 3.0, $t_{1/2}$ = 33.5 min at pH = 0.5), which is in contrast to the MTHP group that is hydrolyzed faster as the pH is decreased ($t_{1/2}$ = 125 min at pH = 3, $t_{1/2}$ = 0.9 min at pH = 1.0). The group was reported to have excellent compatibility with the conditions used to remove the 9-phenylxanthen-9-yl group (5.5 eq. CF$_3$COOH, 16.5 eq. pyrrole, CH$_2$Cl$_2$, rt, 30 seconds, 95.5% yield).[3,7,8]

1-(2-Fluorophenyl)-4-methoxypiperidin-4-yl Ether (Fpmp–OR):

F
N
OCH_3
OR

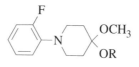

Formation

1. 1-(2-Fluorophenyl)-4-methoxy-1,2,5,6-tetrahydropyridine, mesitylenesulfonic acid or TFA, CH_2Cl_2, 76–91% yield.[9,10,12]

Cleavage

1. Water, pH 2–2.5, 20 h. The $t_{1/2}$ for deblocking the 2′-Fpmp derivative of uridine is 166 min at pH 3 at 25°, whereas it is 75 min for the bis-Fpmp r[UpU] derivative. The increased rate in the latter is assumed to be a result of internal phosphate participation.[12] The Fpmp group is ~1.3 times more stable than the related Ctmp group in the pH range 0.5–1.5. This added stability improves the selectivity for cleavage of the DMTr and pixyl groups in the presence of the Fpmp group during RNA synthesis.[11]

1. C. B. Reese, R. Saffhill, and J. E. Sulston, *J. Am. Chem. Soc.*, **89**, 3366 (1967); *idem*, *Tetrahedron*, **26**, 1023 (1970).
2. J. H. van Boom, P. van Deursen, J. Meeuwse, and C. B. Reese, *J. Chem. Soc., Chem. Commun.*, 766 (1972).
3. C. B. Reese, H. T. Serafinowska, and G. Zappia, *Tetrahedron Lett.*, **27**, 2291 (1986).
4. H. C. P. F. Roelen, G. J. Ligtvoet, G. A. Van der Morel, and J. H. Van Boom, *Recl. Trav. Chim. Pays-Bas*, **106**, 545 (1987).
5. C. B. Reese and P. A. Skone, *Nucleic Acids Res.*, **13**, 5215 (1985).
6. For a large-scale preparation, see M. Faja, C. B. Reese, Q. Song, and P.-Z. Zhang, *J. Chem. Soc., Perkin Trans. 1*, 191 (1997).
7. For an improved preparation of the reagent, see C. B. Reese and E. A. Thompson, *J. Chem. Soc., Perkin Trans. 1*, 2881 (1988).
8. O. Sakatsume, M. Ohtsuki, H. Takaku, and C. B. Reese, *Nucleic Acid Symp. Ser.*, **20**, 77 (1988).
9. B. Beijer, I. Sulston, B. S. Sproat, P. Rider, A. I. Lamond, and P. Neuner, *Nucleic Acids Res.*, **18**, 5143 (1990).
10. A. J. Lawrence, J. B. J. Pavey, I. A. O'Neil, and R. Cosstick, *Tetrahedron Lett.*, **36**, 6341 (1995).
11. D. C. Capaldi and C. B. Reese, *Nucleic Acids Res.*, **22**, 2209 (1994).
12. V. M. Rao, C. B. Reese, V. Schehlmann, and P. S. Yu, *J. Chem. Soc., Perkin Trans. 1*, 43 (1993).

1,4-Dioxan-2-yl Ether:

Formation

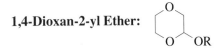

1. 1,4-Dihydrodioxin, $CuBr_2$, THF, rt, 50–88% yield.[1]

Cleavage

1. 6 *N* HCl, EtOH, reflux, 90% yield for cholesterol.[1] Although a direct stability comparison was not made, this group should be more stable than the THP group for the same reasons that the anomeric ethers of carbohydrates are more stable than their 2-deoxy counterparts.

1. M. Fetizon and I. Hanna, *Synthesis*, 806 (1985).

Tetrahydrofuranyl Ether (Chart 1):

Formation

1. 2-Chlorotetrahydrofuran, Et_3N, 30 min, 82–98% yield.[1] 2-Chlorotetrahydrofuran is readily prepared from THF with SO_2Cl_2 (25°, 0.5 h, 85%).
2. Ph_2CHCO_2-2-tetrahydrofuranyl, 1% TsOH, CCl_4, 20°, 30 min, 90–99% yield.[1,2] The authors report that formation of the THF ether by reaction with 2-chlorotetrahydrofuran avoids a laborious procedure[3] that is required when dihydrofuran is used. In addition, the use of dihydrofuran to protect the 2′-OH of a nucleotide gives low yields (24–42%).[4] The tetrahydrofuranyl ester is reported to be a readily available, stable solid. A tetrahydrofuranyl ether can be cleaved in the presence of a THP ether.[1]
3. THF, $[Ce(Et_3NH)_2](NO_3)_6$, 50–100°, 8 h, 30–98% yield.[5] Hindered alcohols give the lower yields. The method was also used to introduce the THP group with tetrahydropyran.
4. THF, $(n\text{-}Bu_4N^+)_2S_2O_8^-$, reflux, 85% yield.[6] These oxidative conditions proved to be compatible with an aromatic thioether.

Cleavage

1. AcOH, H_2O, THF, (3:1:1), 25°, 30 min, 90% yield.[1]
2. 0.01 *N* HCl, THF (1:1), 25°, 10 min, 50% yield.[1]
3. pH 5, 25°, 3 h, 90% yield.[1]

1. C. G. Kruse, F. L. Jonkers, V. Dert, and A. van der Gen, *Recl. Trav. Chim. Pays-Bas*, **98**, 371 (1979).

2. C. G. Kruse, E. K. Poels, F. L. Jonkers, and A. van der Gen, *J. Org. Chem.*, **43**, 3548 (1978).
3. E. L. Eliel, B. E. Nowak, R. A. Daignault, and V. G. Badding, *J. Org. Chem.*, **30**, 2441 (1965).
4. E. Ohtsuka, A. Yamane, and M. Ikehara, *Chem. Pharm. Bull.*, **31**, 1534 (1983).
5. A. M. Maione and A. Romeo, *Synthesis*, 250 (1987).
6. J. C. Jung, H. C. Choi, and Y. H. Kim, *Tetrahedron Lett.*, **34**, 3581 (1993).

Tetrahydrothiofuranyl Ether (Chart 1):

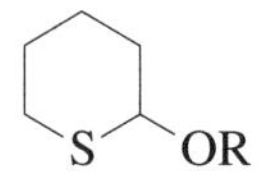

S OR

Formation

1. Dihydrothiofuran, $CHCl_3$, CF_3COOH, reflux, 6 days, 75% yield.[1]
2. (tetrahydrothiofuranyl ester: S, O, C(=O)$CHPh_2$) 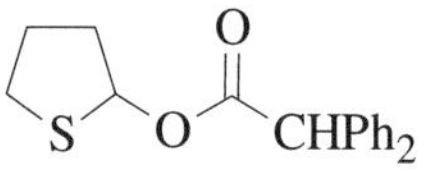 cat. TsOH, $CHCl_3$, 20°, 5 h, 85–95% yield.[2]

Cleavage

1. $AgNO_3$, acetone, H_2O, reflux, 90% yield.[1]
2. $HgCl_2$, CH_3CN, H_2O, 25°, 10 min, quant.[2] Some of the methods used to cleave methylthiomethyl (MTM) ethers should also be applicable to the cleavage of tetrahydrothiofuranyl ethers.

1. L. A. Cohen and J. A. Steele, *J. Org. Chem.*, **31**, 2333 (1966).
2. C. G. Kruse, E. K. Poels, F. L. Jonkers, and A. van der Gen, *J. Org. Chem.*, **43**, 3548 (1978).

2,3,3a,4,5,6,7,7a-Octahydro-7,8,8-trimethyl-4,7-methanobenzofuran-2-yl Ether (RO–MBF)

Formation[1,2]

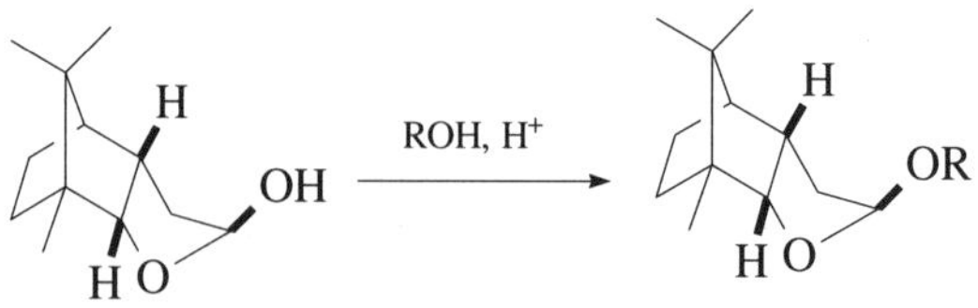

H, H O, OH —ROH, H^+→ H, H O, OR

The advantage of this ketal is that, unlike the THP group, only a single isomer is produced in the derivatization. The disadvantage is that it is not commercially available. Conditions used to hydrolyze the THP group can be used to hydrolyze this acetal.[3] The group may also find applications in the resolution of racemic alcohols.

1. C. R. Noe, *Chem. Ber.*, **115**, 1576, 1591 (1982); C. R. Noe, M. Knollmüller, G. Steinbauer, E. Jangg, and H. Völlenkle, *Chem. Ber.*, **121**, 1231 (1988).
2. U. Girreser and C. R. Noe, *Synthesis*, 1223 (1995).
3. K. Zimmermann, *Synth. Commun.*, **25**, 2959 (1995).

Substituted Ethyl Ethers

1-Ethoxyethyl Ether (EE–OR): $ROCH(OC_2H_5)CH_3$ (Chart 1)

Formation

1. Ethyl vinyl ether, HCl (anhyd).[1]
2. Ethyl vinyl ether, TsOH, 25°, 1 h.[2]
3. Ethyl vinyl ether, pyridinium tosylate (PPTS), CH_2Cl_2, rt, 0.5 h.[3]
4. The ethoxyethyl ether was selectively introduced on a primary alcohol in the presence of a secondary alcohol.[4]

HO ... HO OTBDMS —[CH_2=CHOEt / TsOH, >81%]→ EEO ... HO OTBDMS

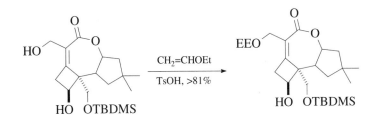

5. $CH_3CH(Cl)OEt$, $PhNMe_2$, CH_2Cl_2, 0°, 10–60 min.[5] These conditions are effective for extremely acid-sensitive substrates or where conditions **1** and **2** fail.
6. CH_2=CHOEt, $CoCl_2$, 65–91% yield.[6]

Cleavage

1. 5% AcOH, 20°, 2 h, 100% yield.[1]
2. 0.5 *N* HCl, THF, 0°, 100% yield.[2] The ethoxyethyl ether is more readily cleaved by acidic hydrolysis than the THP ether, but it is more stable than the 1-methyl-1-methoxyethyl ether. TBDMS ethers are not affected by these conditions.[7]
3. Pyridinium tosylate, *n*-PrOH, 80–85% yield.[8] An acetonide was not affected by these conditions.

1. S. Chládek and J. Smrt, *Chem. Ind. (London)*, 1719 (1964).
2. A. I. Meyers, D. L. Comins, D. M. Roland, R. Henning, and K. Shimizu, *J. Am. Chem. Soc.*, **101**, 7104 (1979).
3. A. Fukuzawa, H. Sato, and T. Masamune, *Tetrahedron Lett.*, **28**, 4303 (1987).
4. M. F. Semmelhack and S. Tomoda, *J. Am. Chem. Soc.*, **103**, 2427 (1981).
5. W. C. Still, *J. Am. Chem. Soc.*, **100**, 1481 (1978).
6. J. Iqbal, R. R. Srivastava, K. B. Gupta, and M. A. Khan, *Synth. Commun.*, **19**, 901 (1989).
7. K. Zimmermann, *Synth. Commun.*, **25**, 2959 (1995).
8. M. A. Tius and A. H. Faug, *J. Am. Chem. Soc.*, **108**, 1035 (1986).

1-(2-Chloroethoxy)ethyl Ether (Cee–OR): $ROCH(CH_3)OCH_2CH_2Cl$

The Cee group was developed for the protection of the 2′-hydroxyl group of ribonucleosides.

Formation

1. CH_2=$CHOCH_2CH_2Cl$, PPTS, CH_2Cl_2, 80–83% yield.[1]

Cleavage

The relative rates of cleavage for a variety of uridine-protected acetals are given in the following table:

	1.5% Cl_2CHCO_2H in CH_2Cl_2		0.01 *N* HCl (pH 2)	
Ether	$T_{1/2}$ (min)	T_∞ (min)	$T_{1/2}$ (min)	T_∞ (min)
$ROCH(CH_3)OCH_2CH_2Cl$	420	960	96	360
$ROCH(CH_3)O$–*i*-Pr	—	30 s	1	4
$ROCH(CH_3)OBu$	2	5	12	34
$ROCH(CH_3)OEt$	20 s	3	5	18
ROTHP	90	273	32	150
ROCTMP[a]	—	—	55	295

[a]CTMP = 1-[(2-chloro-4-methyl)phenyl]-4-methoxypiperidin-4-yl ether

The Cee group is stable under the acidic conditions used to cleave the DMTr group.[3]

1. S.-i. Yamakage, O. Sakatsume, E. Furuyama, and H. Takaku, *Tetrahedron Lett.*, **30**, 6361 (1989).
2. O. Sakatsume, T. Yamaguchi, M. Ishikawa, I. Ichiro, K. Miura, and H. Takaku, *Tetrahedron*, **47**, 8717 (1991).
3. O. Sakatsume, T. Ogawa, H. Hosaka, M. Kawashima, M. Takaki, and H. Takaku, *Nucleosides Nucleotides*, **10**, 141 (1991).

1-[2-(Trimethylsilyl)ethoxy]ethyl Ether (SEE–OR): RO–CH(CH_3)–O–CH_2CH_2–TMS

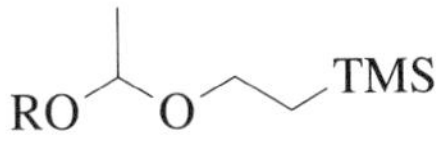

Formation[1]

1. 2-$TMSCH_2CH_2OCH{=}CH_2$, CH_2Cl_2, PPTS, rt, 1–3 h, 76–96% yield. Phenols are readily protected with this reagent.

Cleavage[1]

1. TBAF–H_2O, THF, 45°, 20–24 h, 76–90% yield.
2. TsOH or PPTS, THF, H_2O, 4 h, rt.

1. J. Wu, B. K. Shull, and M. Koreeda, *Tetrahedron Lett.*, **37**, 3647 (1996).

1-Methyl-1-methoxyethyl Ether (MIP–OR): $ROC(OCH_3)(CH_3)_2$ (Chart 1)

Formation

1. $CH_2{=}C(CH_3)OMe$, cat. $POCl_3$, 20°, 30 min, 100% yield.[1]
2. $CH_2{=}C(CH_3)OMe$, neat, 20°, TsOH.[2]

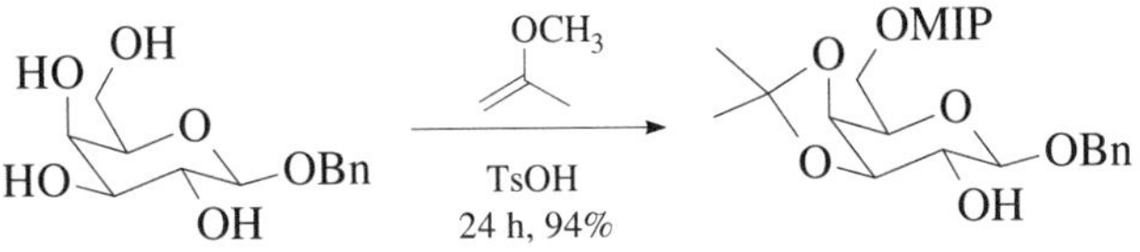

$CH_2{=}C(CH_3)OCH_3$, TsOH, 24 h, 94%

3. $CH_2{=}C(CH_3)OMe$ has been used to protect a hydroperoxide.[3]

Cleavage

1. 20% AcOH, 20°, 10 min.[1]
2. Pyridinium *p*-toluenesulfonate, 5°, 1 h.[4] Similar selectivity can be achieved using a silica–alumina gel prepared by the solgel method.[5]

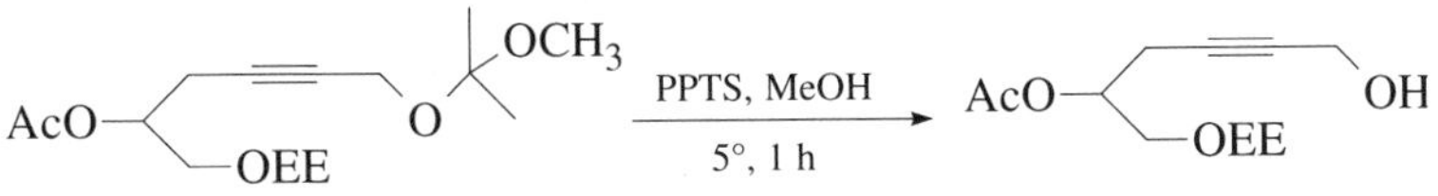

PPTS, MeOH, 5°, 1 h

EE = Ethoxyethyl

Ref. 4

In general, the MIP ether is very labile to acid and silica gel chromatography, unless some TEA is used as part of the eluting solvent. The acid in the NMR solvent, $CDCl_3$, is sufficient to cleave the MIP ether.

1. A. F. Klug, K. G. Untch, and J. H. Fried, *J. Am. Chem. Soc.*, **94**, 7827 (1972).
2. P. L. Barili, G. Berti, G. Catelani, F. Colonna, and A. Marra, *Tetrahedron Lett.*, **27**, 2307 (1986).
3. P. H. Dussault and K. R. Woller, *J. Am. Chem. Soc.*, **119**, 3824 (1997).
4. G. Just, C. Luthe, and M. T. P. Viet, *Can. J. Chem.*, **61**, 712 (1983).
5. Y. Matsumoto, K. Mita, K. Hashimoto, H. Iio, and T. Tokoroyama, *Tetrahedron*, **52**, 9387 (1996).

1-Methyl-1-benzyloxyethyl Ether (MBE–OR): $ROC(OBn)(CH_3)_2$

Formation

1. $CH_2{=}C(OBn)(CH_3)$, $PdCl_2$(1,5-cyclooctadiene) [$PdCl_2$(COD)], 85–95% yield.[1]
2. $CH_2{=}C(OBn)(CH_3)$, $POCl_3$ or TsOH, 61–98% yield.[1] These conditions do not afford a cyclic acetal with a 1,3-diol. This ketal is stable to $LiAlH_4$, diisobutylaluminum hydride, NaOH, alkyllithiums, and Grignard reagents.

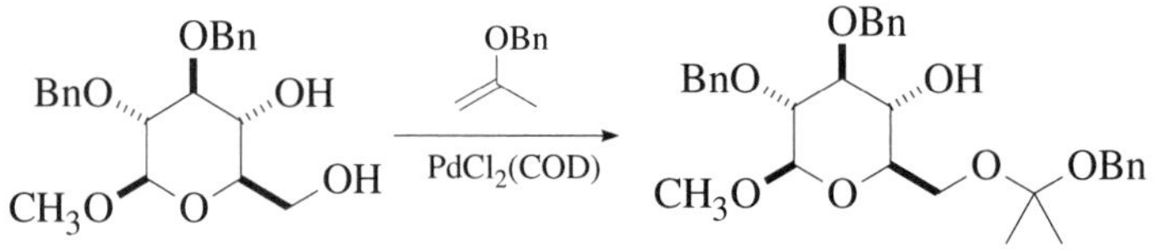

Cleavage

1. H_2, 5% Pd–C, EtOH, rt, 92–99% yield.[1]
2. 3 *M* AcOH, H_2O, THF.[2]

1. T. Mukaiyama, M. Ohshima, and M. Murakami, *Chem. Lett.*, 265 (1984).
2. T. Mukaiyama, M. Ohishima, H. Nagaoka, and M. Murakami, *Chem. Lett.*, 615 (1984).

1-Methyl-1-benzyloxy-2-fluoroethyl Ether: $ROC(OBn)(CH_2F)(CH_3)$

Formation

1. $CH_2{=}C(OBn)CH_2F$, $PdCl_2$(COD), CH_3CN, rt, 24 h, 89–100% yield.[1] Protic acids can also be used to introduce this group, but the yields are sometimes lower. A primary alcohol can be protected in the presence of a secondary alcohol. This reagent does not give cyclic acetals of 1,3-diols with palladium catalysis.

Cleavage

1. H_2, Pd–C, EtOH, 1 atm, 98–100% yield.[1] This group is stable to 3 *M*

aqueous acetic acid at room temperature, conditions that cleave the TBDMS group and the 1-methyl-1-benzyloxyethyl ether.

1. T. Mukaiyama, M. Ohishima, H. Nagaoka, and M. Murakami, *Chem. Lett.*, 615 (1984).

1-Methyl-1-phenoxyethyl Ether: $[ROC(OPh)(CH_3)_2]$

Formation/Cleavage[1]

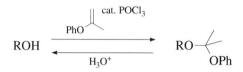

ROH → (PhO–C(=CH$_2$)CH$_3$, cat. $POCl_3$) RO–C(CH$_3$)$_2$–OPh; reverse: H_3O^+

1. P. Zandbergen, H. M. G. Willems, G. A. Van der Marel, J. Brussee, and A. van der Gen, *Synth. Commun.*, **22**, 2781 (1992).

2,2,2-Trichloroethyl Ether: Cl_3CCH_2OR

The anomeric position of a carbohydrate was protected as its trichloroethyl ether. Cleavage is effected with Zn, AcOH, AcONa (3 h, 92%).[1]

1. R. U. Lemieux and H. Driguez, *J. Am. Chem. Soc.*, **97**, 4069 (1975).

1,1-Dianisyl-2,2,2-trichloroethyl Ether (DATE–OR):

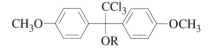

CH_3O–C$_6$H$_4$–C(CCl$_3$)(OR)–C$_6$H$_4$–OCH_3

Formation

1. $An_2(Cl_3C)CCl$, AgOTf, CH_3CN, Pyr, rt, 12–18 h, 92% yield.

Cleavage[1]

1. Li[Co(I)Pc], MeOH, 80–90% yield.
2. Zn, $ZnBr_2$, MeOH, Et_2O or Zn, 80% AcOH–dioxane, 70–80% yield.
3. DATE ethers are stable to concd. HCl–MeOH–dioxane (1:2:2), Cl_2CHCO_2H–CH_2Cl_2 (3:97) and NH_3–dioxane (1:1).

1. R. M. Karl, R. Klösel, S. König, S. Lehnhoff, and I. Ugi, *Tetrahedron*, **51**, 3759 (1995).

1,1,1,3,3,3-Hexafluoro-2-phenylisopropyl Ether (HIP–OR): $Ph(CF_3)_2C–OR$

This group is stable to strong acids and bases, TMSI, $Pd–C/H_2$, DDQ, TBAF, and LAH at low temperatures and thus has the potential to participate in a large number of orthogonal sets.[1]

Formation[1]

1. 1,1,1,3,3,3-Hexafluoro-2-phenylisopropyl alcohol, diethyl azodicarboxylate, PPh_3, benzene, 82–98% yield. Primary alcohols are derivatized effectively, but yields for secondary alcohols are low (46–65%).

Cleavage[1]

1. Lithium naphthalenide, <1 h, −78°.
   The following protective groups can be cleaved in the presence of the HIP group: Tr, THP, MEM, Bn, MPM, TBDPS, Bz; all but the Bz group are stable to the conditions for the cleavage of the HIP group.

1. H.-S. Cho, J. Yu, and J. R. Falck, *J. Am. Chem. Soc.*, **116**, 8354 (1994).

2-Trimethylsilylethyl Ether: $Me_3SiCH_2CH_2OR$

Cleavage

1. $BF_3·Et_2O$, CH_2Cl_2, 0–25°, 79% yield.[1]

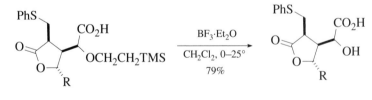

PhS, O, CO_2H, OCH_2CH_2TMS, O, R → ($BF_3·Et_2O$; CH_2Cl_2, 0–25°; 79%) → PhS, O, CO_2H, OH, O, R

2. CsF, DMF, 210°, >65% yield.[2]

TMS, O, OPMB, O, H, CH_3, OCH_3 → (CsF, DMF; 210°, >65%) → OPMB, HO, O, H, CH_3, OCH_3

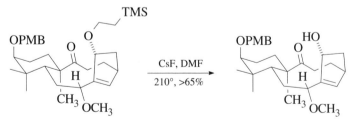

1. S. D. Burke, G. J. Pacofsky, and A. D. Piscopio, *Tetrahedron Lett.*, **27**, 3345 (1986).
2. L. A. Paquette, D. Backhaus, and R. Braun, *J. Am. Chem. Soc.*, **118**, 11990 (1996).

2-(Benzylthio)ethyl Ether: $BnSCH_2CH_2OR$

This ether, developed for the protection of a pyranoside anomeric hydroxyl, is prepared via a Königs–Knorr reaction from the glycosyl bromide and 2-(benzylthio)ethanol in the presence of DIPEA. It is cleaved, after oxidation with dimethyldioxirane, by treatment with LDA or MeONa.[1]

1. T.-H. Chan and C. P. Fei, *J. Chem. Soc., Chem. Commun.*, 825 (1993).

2-(Phenylselenyl)ethyl Ether: $ROCH_2CH_2SePh$ (Chart 1)

This ether was prepared from an alcohol and 2-(phenylselenyl)ethyl bromide ($AgNO_3$, CH_3CN, 20°, 10–15 min, 80–90% yield); it is cleaved by oxidation (H_2O_2, 1 h; ozone; or $NaIO_4$), followed by acidic hydrolysis of the intermediate vinyl ether (dil. HCl, 65–70% yield).[1]

1. T.-L. Ho and T. W. Hall, *Synth. Commun.*, **5**, 367 (1975).

t-Butyl Ether: *t*-BuOR (Chart 1)

Formation

t-Butyl ethers can be prepared from a variety of alcohols, including allylic alcohols. The ethers are stable to most reagents except strong acids. The *t*-butyl ether is probably one of the more underused alcohol protective groups, considering its stability, ease and efficiency of introduction, and ease of cleavage.

1. Isobutylene, $BF_3{\cdot}Et_2O$, H_3PO_4, 100% yield.[1,2]

OH ... O–*t*-Bu

isobutylene / $BF_3{\cdot}Et_2O$, H_3PO_4 / 100%

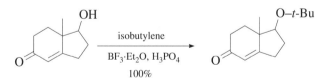

This method has been used for the preparation of the somewhat more hindered 2-methyl-2-butyl ether (*t*-amyl ether); the introduction is selective for primary alcohols.[3]

2. Isobutylene, Amberlyst H-15, hexane.[4] Methylene chloride can also be used as solvent, and in this case a primary alcohol was selectively converted to the *t*-amyl ether in the presence of a secondary alcohol.[5]
3. Isobutylene, H_2SO_4.[6] Acyl migration has been observed using these conditions.[7]
4. Isobutylene, H_3PO_4, $BF_3{\cdot}Et_2O$, $-72°$, 3 h, 0°, 20 h, 79% yield.[8]
5. *t*-BuOC(=NH)CCl_3, $BF_3{\cdot}Et_2O$, CH_2Cl_2, cyclohexane, 59–91% yield.[9]

Cleavage

1. Anhydrous CF_3COOH, 0–20°, 1–16 h, 80–90% yield.[2,4]
2. HBr, AcOH, 20°, 30 min.[10]
3. 4 *N* HCl, dioxane, reflux, 3 h.[11] In this case the *t*-butyl ether was stable to 10 *N* HCl, MeOH, 0–5°, 30 h.
4. HCO_2H, rt, 24 h, >83% yield.[12]
5. Me_3SiI, CCl_4 or $CHCl_3$, 25°, <0.1 h, 100% yield.[13] Under suitable conditions, this reagent also cleaves many other ethers, esters, ketals, and carbamates.[14]
6. Ac_2O, $FeCl_3$, Et_2O, 76–93% yield.[4,15] These conditions give the acetate of the alcohol, which can then be cleaved by simple basic hydrolysis. The method is also effective for the conversion of *t*-butyl glycosides to acetates with retention of configuration (80–100% yield).[16]
7. $TiCl_4$, CH_2Cl_2, 0°, 1 min, 85% yield.[17]
8. TBDMSOTf, CH_2Cl_2, rt, 24 h, 82% yield. The use of a catalytic amount of the triflate will give the alcohol. If the triflate is used stoichiometrically and the reaction worked up with 2,6-lutidine, the TBDMS ether is isolated (98% yield).[18]

1. R. A. Micheli, Z. G. Hajos, N. Cohen, D. R. Parrish, L. A. Portland, W. Sciamanna, M. A. Scott, and P. A. Wehrli, *J. Org. Chem.*, **40**, 675 (1975).
2. H. C. Beyerman and G. L. Heiszwolf, *J. Chem. Soc.*, 755 (1963).
3. B. Figadère, X. Franck, and A. Cavé, *Tetrahedron Lett.*, **34**, 5893 (1993).
4. A. Alexakis and J. M. Duffault, *Tetrahedron Lett.*, **29**, 6243 (1988); A. Alexakis, M. Gardette, and S. Colin, *Tetrahedron Lett.*, **29**, 2951 (1988).
5. X. Franck, B. Figadère, and A. Cavé, *Tetrahedron Lett.*, **38**, 1413 (1997).
6. H. C. Beyerman and J. S. Bontekoe, *Proc. Chem. Soc.*, 249 (1961).
7. N. I. Simirskaya and M. V. Mavrov, *Zh. Org. Khim.*, **31**, 140 (1995); *Chem. Abstr.* **124**: 8220w (1996).
8. N. Cohen, W. F. Eichel, R. J. Lopresti, C. Neukom, and G. Saucy, *J. Org. Chem.*, **41**, 3505 (1976).
9. A. Armstrong, I. Brackenridge, R. F. W. Jackson, and J. M. Kirk, *Tetrahedron Lett.*, **29**, 2483 (1988).
10. F. M. Callahan, G. W. Anderson, R. Paul, and J. E. Zimmerman, *J. Am. Chem. Soc.*, **85**, 201 (1963).
11. U. Eder, G. Haffer, G. Neef, G. Sauer, A. Seeger, and R. Wiechert, *Chem. Ber.*, **110**, 3161 (1977).
12. H. Paulsen and K. Adermann, *Liebigs Ann. Chem.*, 751 (1989).
13. M. E. Jung and M. A. Lyster, *J. Org. Chem.*, **42**, 3761 (1977).
14. A. H. Schmidt, *Aldrichimica Acta*, **14**, 31 (1981).
15. B. Ganem and V. R. Small, Jr., *J. Org. Chem.*, **39**, 3728 (1974).

16. N. Rakotomanomana, J. M. Lacombe, and A. A. Pavia, *J. Carbohydr. Chem.*, **9**, 93 (1990).
17. R. H. Schlessinger and R. A. Nugent, *J. Am. Chem. Soc.*, **104**, 1116 (1982).
18. X. Franck, B. Figadère, and A. Cavé, *Tetrahedron Lett.*, **36**, 711 (1995).

Allyl Ether (Allyl-OR): CH_2=$CHCH_2$-OR (Chart 1)

The use of allyl ethers for the protection of alcohols is common in the literature on carbohydrates because allyl ethers are generally compatible with the various methods for glycoside formation.[1] Obviously, the allyl ether is not compatible with powerful electrophiles such as bromine and catalytic hydrogenation, but it is stable to moderately acidic conditions (1 *N* HCl, reflux, 10 h).[2] The ease of formation, the many mild methods for its cleavage in the presence of numerous other protective groups, and its general stability have made the allyl ether a mainstay of many orthogonal sets. The synthesis of perdeuteroallyl bromide and its use as a protective group in carbohydrates has been reported. The perdeutero derivative has the advantage that the allyl resonances in the NMR no longer obscure other, more diagnostic resonances, such as those of the anomeric carbon in glycosides.[3] The use of the allyl protective group primarily covering carbohydrate chemistry has been reviewed.[4]

Formation

1. CH_2=$CHCH_2Br$, NaOH, benzene, reflux, 1.5 h,[5] or NaH, benzene, 90–100% yield.[6]
2. CH_2=$CHCH_2OC$(=NH)CCl_3, H^+.[7]
3. Allyl carbonates have been converted to allyl ethers.[8] In the following situation, acid- and base-catalyzed procedures failed because of the sensitivity of the $[(i\text{-Pr})_2Si]_2O$ group.

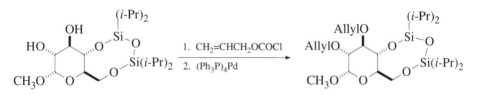

(*i*-Pr)$_2$ OH HO Si O O CH$_3$O O O Si(*i*-Pr)$_2$

1. CH_2=$CHCH_2OCOCl$
2. $(Ph_3P)_4Pd$

(*i*-Pr)$_2$ AllylO AllylO Si O O CH$_3$O O O Si(*i*-Pr)$_2$

4. Bu_2SnO, toluene, THF; CH_2=$CHCH_2Br$, $Bu_4N^+Br^-$, 96% yield.[9]

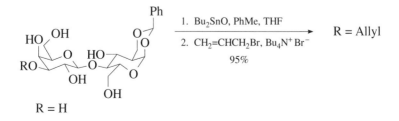

Ph HO OH O O O HO RO O OH O OH

R = H

1. Bu_2SnO, PhMe, THF
2. CH_2=$CHCH_2Br$, Bu_4N^+ Br^-

95%

R = Allyl

The crotyl ether has been introduced using similar methodology.[10]

5. CH_2=$CHCH_2OCO_2Et$, $Pd_2(dba)_3$, THF, 65°, 4 h, 70–97% yield.[11]

$Pd_2(dba)_3$, THF, CH_2=$CHCH_2OCO_2Et$, 65°, 4 h, 70%

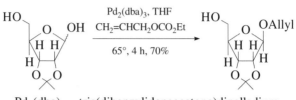

$Pd_2(dba)_3$ = tris(dibenzylideneacetone)dipalladium

Note the preferential reaction at the anomeric hydroxyl. The method is also effective for the protection of primary and secondary alcohols.

6. MeO_2COCH_2CH=CH_2, $Pd_2(dba)_3$, $CHCl_3$, $Ph_2P(CH_2)_4PPh_2$, THF, 65°, 95% yield.[12]
7. Allyl bromide, $(RO)_2Mg$.[13]
8. KF–alumina, allyl bromide, 80% yield. These conditions were developed because the typical strongly basic metal alkoxide-induced alkylation led to Beckmann fragmentation of the isoxazoline.[14]

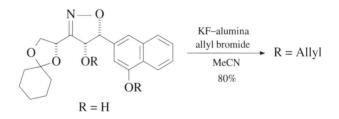

R = H → (KF–alumina, allyl bromide, MeCN, 80%) → R = Allyl

9. Allyl bromide, DMF, BaO, rt.[15]
10. Allyl bromide, Al_2O_3, days. These conditions were developed to alkylate selectively an alcohol in the presence of an amide.[16]

Cleavage

1. $ROCH_2CH$=CH_2 $\xrightarrow[\text{100°, 15 min}]{t\text{-BuOK, DMSO}^{17}}$ $ROCH$=$CHCH_3$ $\xrightarrow{\text{i–x}}$ ROH

i. 0.1 *N* HCl, acetone–water, reflux, 30 min.[8]
ii. $KMnO_4$, NaOH–H_2O, 10°, 100% yield. These basic conditions avoid acid-catalyzed acetonide cleavage.[2]
iii. $HgCl_2/HgO$, acetone–H_2O, 5 min, 100% yield.[18]
iv. Ozonolysis.[17,19]

v. SeO_2, H_2O_2, 92% yield.[20]
vi. 0.1 eq. TsOH, MeOH, 25°, 2.5 h, >86% yield.[21]
vii. Me_3NO, OsO_4, CH_2Cl_2, >76% yield.[22]
viii. MCPBA, MeOH, H_2O.[23]

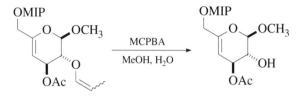

When the OAc group was a hydroxyl, the epoxidation selectivity was not very good, presumably because of the known directing effect of hydroxyl groups in peracid epoxidations.

ix. NIS, CH_2Cl_2, H_2O.[24]
x. $BF_3 \cdot Et_2O$, $Bu_4N^+F^-$, 0°, 52–88% yield.[25]

2. $ROCH_2CH{=}CH_2 \xrightarrow[\text{reflux, 3 h}]{(Ph_3P)_3RhCl,\ DABCO,\ EtOH} \xrightarrow{Hg(II),\ pH\ 2}$ ROH, >90%[6]

Allyl ethers are isomerized by $(Ph_3P)_3RhCl$, and *t*-BuOK/DMSO in the following order:[26]

$(Ph_3P)_3RhCl$: allyl > 2-methylallyl > but-2-enyl

t-BuOK: but-2-enyl > allyl > 2-methylallyl

3. It is possible to remove the allyl group in the presence of an allyloxycarbonyl (AOC, Alloc, or Aloc) group using an $[Ir(COD)(Ph_2MeP)_2]PF_6$–catalyzed isomerization, but the selectivity is not complete. The allyloxycarbonyl group can be removed selectively in the presence of an allyl group using a palladium or rhodium catalyst.[27] Hydrogen-activated $[Ir(COD)(Ph_2MeP)_2]PF_6$ is a better catalyst for allyl isomerization (91–100% yield) because there is no reduction of the alkene, as is sometimes the case with $(Ph_3P)_3RhCl$.[28,29] When Wilkinson's catalyst is prereduced with BuLi, alkene reduction is not observed and high yields of enol ethers are obtained.[30] This method can be used for isomerization of but-2-enyl ethers.[31] The iridium catalyst is also compatible with acetylenes.[32] Because the iridium catalyst can effect isomerization at room temperature, adjacent azides do not cycloadd to the allyl group during the isomerization reaction, as is the case when the isomerization must be performed at reflux.[22]

AllylO OBn; BnOCH$_2$ ... OAOC; OBn —[$Rh(Ph_3P)_3Cl$, 95%]→ AllylO OBn; BnOCH$_2$ ... OH; OBn

$[Ir(COD)(PMePh_2)_2]PF_6$ → O OBn; BnOCH$_2$ ... OAOC; OBn —[$HgCl_2$, H_2O, 67%]→ HO OBn; BnOCH$_2$ ... OAOC; OBn

Ref. 27

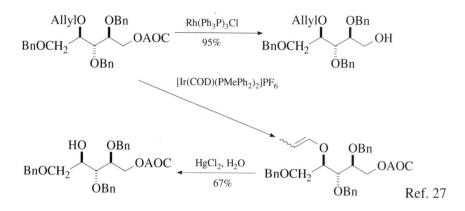

4. Useful selectivity between allyl and 3-methylbut-2-enyl (prenyl) ethers has been achieved.[26]

PrenylO, OBn, OBn, AllylO, O, OBn —[$Rh(PPh_3)_3Cl$, DABCO, EtOH]→ PrenylO, OBn, OBn, O, O, OBn —[t-BuOK, DMSO]→ HO, OBn, OBn, O, O, OBn

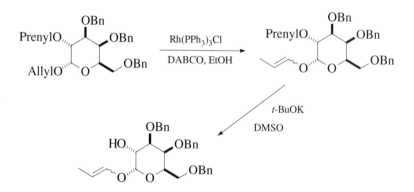

5. *trans*-$Pd(NH_3)_2Cl_2$/*t*-BuOH isomerizes allyl ethers to vinyl ethers that can then be hydrolyzed in 90% yield, but in the presence of an α-hydroxy group the intermediate vinyl ether cyclizes to an acetal.[33] Benzylidene acetals are not affected by this reagent.

6. $$ROCH_2CH{=}CH_2 \xrightarrow[60–80°,\ 24\ h,\ 80–95\%]{Pd/C,\ H_2O,\ MeOH,\ cat.\ TsOH\ or\ HClO_4} ROH$$ [34]

When TsOH is omitted, the reaction gives the vinyl ether.[35]

7. $$ROCH_2CH{=}CH_2 \xrightarrow[reflux,\ 1\ h,\ 50\%]{SeO_2,\ AcOH,\ dioxane} [ROCH(OH)CH{=}CH_2] \longrightarrow ROH$$ [36]

8. $ROCH_2CH{=}CH_2 \xrightarrow[\text{reflux, 2 h, 85–99\%}]{\text{NaTeH, EtOH, AcOH}} ROH$[37]

9. Ac_2O, $BF_3·Et_2O$, then MeONa/MeOH to hydrolyze the acetate.[38]
10. $RhCl_3$, DABCO, EtOH, H_2O; H_3O^+, EtOH.[39]
11. $RhH(Ph_3P)_4$, TFA, EtOH, 50°, 30 min, 98% yield.[40]
12. $PdCl_2$, CuCl, DMF, O_2, 4 h, rt, 88–93% yield.[41]
13. $Pd(Ph_3P)_4$, AcOH, 80°, 10–60 min, 72–98% yield.[42]
14. $TiCl_3$, Mg, THF, 28–96% yield.[43]
15. Electrolysis, DMF, $SmCl_3$, $(n\text{-Bu})_4N^+Br^-$, Mg anode, Ni cathode, 60–90% yield.[44]
16. Electrolysis, $[Ni(bipyr)_3](BF_4)_2$, Mg anode, DMF, rt, 25–99% yield.[45] Aryl halides are reduced.
17. $AlCl_3$–$PhNMe_2$, CH_2Cl_2, 73–100% yield.[46] Benzyl ethers are also cleaved.
18. NBS, $h\nu$, CCl_4; base, 78–99% yield.[47]
19. $CoCl_2$, AcCl, CH_3CN, rt, 8–12 h, 71–84% yield. Benzyl ethers are among those that are also cleaved.[48]
20. Cp_2Zr; H_2O, 50–98% yield. Allyl ethers are cleaved faster than allylamines that are also cleaved (66%).[49]
21. $H_2Ru(PPh_3)_4$, EtOH, 95°; 1.5 h, TsOH, MeOH, 2.5 h, 86% yield.[21]
22. Other catalysts that have been used to isomerize allylic ethers are Polystyrene–CH_2NMe_3-$RhCl_4$ (EtOH, H_2O),[50] $RuCl_2(PPh)_3$ ($NaBH_4$, EtOH),[51] $Rh(diphos)(acetone)_2[ClO_4]_2$ (acetone, 25°),[52] and $Fe(CO)_5$ (xylene, 135°, 8–15 h, 97% yield).[53]
23. $PdCl_2$, AcOH, H_2O, NaOAc, 89% yield.[19]
24. $Ti(O\text{–}i\text{-Pr})_4$, n-BuMgCl, THF, rt, 69–97% yield. Methallyl and other substituted allyl ethers are not cleaved, but ester groups are partially removed, as expected.[54]
25. DDQ, wet CH_2Cl_2, 70–92% yield. Anomeric and secondary allylic ethers could not be cleaved under these conditions.[55]

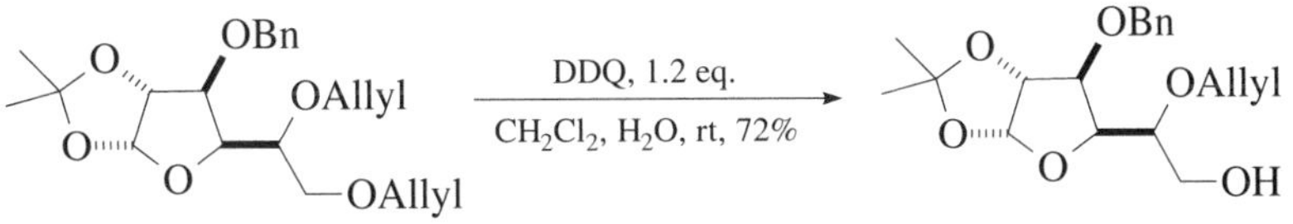

OBn, OAllyl, OAllyl (acetonide furanose) $\xrightarrow[\text{CH}_2\text{Cl}_2\text{, H}_2\text{O, rt, 72\%}]{\text{DDQ, 1.2 eq.}}$ OBn, OAllyl, OH

26. Protection for the double bond in the allyl protecting group may be achieved by epoxidation. Regeneration of the allyl group occurs upon treatment with 3-methylbenzothiazole-2-selone (CH_3, N, S, =Se) and TFA.[56]

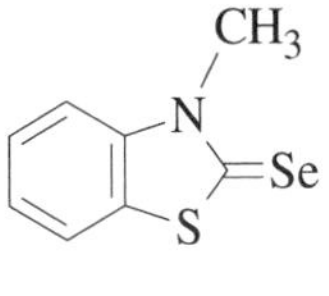

27. Allyl groups are subject to oxidative deprotection with Chromiapillared Montmorillonite Clay, *t*-BuOOH, CH_2Cl_2, isooctane, 85% yield.[57] Allylamines are cleaved in 84–90% yield, and allyl phenyl ethers are cleaved in 80% yield.
28. $NaBH_4$, I_2, THF, 0°, 53–96% yield.[58]
29. $Pd(Ph_3P)_4$, RSO_2Na, CH_2Cl_2 or THF/MeOH, 70–99% yield. These conditions were shown to be superior to the use of sodium 2-ethylhexanoate. Methallyl, crotyl, and cinnamyl ethers, the Alloc group, and allylamines are all efficiently cleaved by this method.[59]
30. DIBAL, $NiCl_2$(dppp), toluene, CH_2Cl_2, THF, or ether, 80–97% yield.[60] These conditions are chemoselective for simple alkyl and phenolic allyl ethers. More highly substituted allyl ethers are unreactive.

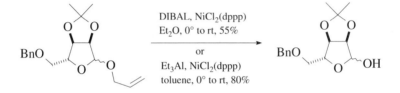

DIBAL, $NiCl_2$(dppp)
Et_2O, 0° to rt, 55%
or
Et_3Al, $NiCl_2$(dppp)
toluene, 0° to rt, 80%

1. R. Gigg, *Am. Chem. Soc. Symp. Ser.*, **39**, 253 (1977); *ibid.*, **77**, 44 (1978); R. Gigg and R. Conant, *Carbohydr. Res.*, **100**, C5 (1982).
2. J. Cunningham, R. Gigg, and C. D. Warren, *Tetrahedron Lett.*, 1191 (1964).
3. J. Thiem, H. Mohn, and A. Heesing, *Synthesis*, 775 (1985).
4. F. Guibé, *Tetrahedron*, **40**, 13509 (1997).
5. R. Gigg and C. D. Warren, *J. Chem. Soc. C*, 2367 (1969).
6. E. J. Corey and W. J. Suggs, *J. Org. Chem.*, **38**, 3224 (1973).
7. T. Iversen and D. R. Bundle, *J. Chem. Soc., Chem. Commun.*, 1240 (1981); H.-P. Wessel, T. Iversen, and D. R. Bundle, *J. Chem. Soc., Perkin Trans. 1*, 2247 (1985).
8. J. J. Oltvoort, M. Kloosterman, and J. H. Van Boom, *Recl: J. R. Neth. Chem. Soc.*, **102**, 501 (1983); F. Guibe and Y. Saint M'Leux, *Tetrahedron Lett.*, **22**, 3591 (1981).
9. S. Sato, S. Nunomura, T. Nakano, Y. Ito, and T. Ogawa, *Tetrahedron Lett.*, **29**, 4097 (1988).
10. A. K. M. Anisuzzaman, L. Anderson, and J. L. Navia, *Carbohydr. Res.*, **174**, 265 (1988).
11. R. Lakhmiri, P. Lhoste, and D. Sinou, *Tetrahedron Lett.*, **30**, 4669 (1989).
12. H. Oguri, S. Hishiyama, T. Oishi, and M. Hirama, *Synlett*, 1252 (1995).
13. J.-M. Lin, H.-H. Li, and A.-M. Zhou, *Tetrahedron Lett.*, **37**, 5159 (1996).
14. H. Yin, R. W. Franck, S.-L. Chen, G. J. Quigley, and L. Todaro, *J. Org. Chem.*, **57**, 644 (1992).
15. J.-C. Jacquinet and P. Sinaÿ, *J. Org. Chem.*, **42**, 720 (1977).
16. I. A. Motorina, F. Parly, and D. S. Grierson, *Synlett*, 389 (1996).
17. J. Gigg and R. Gigg, *J. Chem. Soc. C*, 82 (1966).

18. R. Gigg and C. D. Warren, *J. Chem. Soc. C*, 1903 (1968).
19. A. B. Smith, III, R. A. Rivero, K. J. Hale, and H. A. Vaccaro, *J. Am. Chem. Soc.*, **113**, 2092 (1991).
20. H. Yamada, T. Harada, and T. Takahashi, *J. Am. Chem. Soc.*, **116**, 7919 (1994).
21. K. C. Nicolaou, T. J. Caulfield, H. Kataoka, and N. A. Stylianides, *J. Am. Chem. Soc.*, **112**, 3693 (1990).
22. C. Lamberth and M. D. Bednarski, *Tetrahedron Lett.*, **32**, 7369 (1991).
23. P. L. Barili, G. Berti, D. Bertozzi, G. Catelani, F. Colonna, T. Corsetti, and F. D'Andrea, *Tetrahedron*, **46**, 5365 (1990).
24. K. M. Halkes, T. M. Slaghek, H. J. Vermeer, J. P. Kamerling, and J. F. G. Vliegenthart, *Tetrahedron Lett.*, **36**, 6137 (1995).
25. V. Gevorgyan and Y. Yamamoto, *Tetrahedron Lett.*, **36**, 7765 (1995).
26. P. A. Gent and R. Gigg, *J. Chem. Soc., Chem. Commun.*, 277 (1974); R. Gigg, *J. Chem. Soc., Perkin Trans. 1*, 738 (1980).
27. P. Boullanger, P. Chatelard, G. Descotes, M. Kloosterman, and J. H. Van Boom, *J. Carbohydr. Chem.*, **5**, 541 (1986).
28. J. J. Oltvoort, C. A. A. van Boeckel, J. H. de Koning, and J. H. van Boom, *Synthesis*, 305 (1981).
29. For hydrogenation during isomerization, see C. D. Warren and R. W. Jeanloz, *Carbohydr. Res.*, **53**, 67 (1977); T. Nishiguchi, K. Tachi, and K. Fukuzumi, *J. Org. Chem.*, **40**, 237 (1975); C. A. A. van Boeckel and J. H. van Boom, *Tetrahedron Lett.*, 3561 (1979).
30. G.-J. Boons, A. Burton, and S. Isles, *J. Chem. Soc., Chem. Commun.*, 141 (1996).
31. G.-J. Boons, B. Heskamp, and F. Hout, *Angew. Chem., Int. Ed. Engl.*, **35**, 2845 (1996); G.-J. Boons and S. Isles, *J. Org. Chem.*, **61**, 4262 (1996).
32. J. Alzeer, C. Cai, and A. Vasella, *Helv. Chim. Acta*, **78**, 242 (1995).
33. T. Bieg and W. Szeja, *J. Carbohydr. Chem.*, **4**, 441 (1985).
34. R. Boss and R. Scheffold, *Angew. Chem., Int. Ed. Engl.*, **15**, 558 (1976).
35. A. B. Smith, III, R. A. Rivero, K. J. Hale, and H. A. Vaccaro, *J. Am. Chem. Soc.*, **113**, 2092 (1991).
36. K. Kariyone and H. Yazawa, *Tetrahedron Lett.*, 2885 (1970).
37. N. Shobana and P. Shanmugam, *Indian J. Chem., Sect. B*, **25B**, 658 (1986).
38. C. F. Garbers, J. A. Steenkamp, and H. E. Visagie, *Tetrahedron Lett.*, 3753 (1975).
39. M. Dufour, J.-C. Gramain, H.-P. Husson, M.-E. Sinibaldi, and Y. Troin, *Tetrahedron Lett.*, **30**, 3429 (1989).
40. F. E. Ziegler, E. G. Brown, and S. B. Sobolov, *J. Org. Chem.*, **55**, 3691 (1990).
41. H. B. Mereyala and S. Guntha, *Tetrahedron Lett.*, **34**, 6929 (1993).
42. K. Nakayama, K. Uoto, K. Higashi, T. Soga, and T. Kusama, *Chem. Pharm. Bull.*, **40**, 1718 (1992).
43. S. M. Kadam, S. K. Nayak, and A. Banerji, *Tetrahedron Lett.*, **33**, 5129 (1992).
44. B. Espanet, E. Duñach and J. Périchon, *Tetrahedron Lett.*, **33**, 2485 (1992).
45. S. Olivero and E. Duñach, *J. Chem. Soc., Chem. Commun.*, 2497 (1995).
46. T. Akiyama, H. Hirofuji, and S. Ozaki, *Tetrahedron Lett.*, **32**, 1321 (1991).

47. R. R. Diaz, C. R. Melgarejo, M. T. P. Lopez-Espinosa, and I. I. Cubero, *J. Org. Chem.*, **59**, 7928 (1994).
48. J. Iqbal and R. R. Srivastava, *Tetrahedron*, **47**, 3155 (1991).
49. H. Ito, T. Taguchi, and Y. Hanzawa, *J. Org. Chem.*, **58**, 774 (1993).
50. M. Setty-Fichman, J. Blum, Y. Sasson, and M. Eisen, *Tetrahedron Lett.*, **35**, 781 (1994).
51. H. Frauenrath, T. Arenz, G. Raube, and M. Zorn, *Angew. Chem., Int. Ed. Engl.*, **32**, 83 (1993).
52. S. H. Bergens and B. Bosnich, *J. Am. Chem. Soc.*, **113**, 958 (1991).
53. T. Arnold, B. Orschel, and H.-U. Reissig, *Angew. Chem., Int. Ed. Engl.*, **31**, 1033 (1992).
54. J. Lee and J. K. Cha, *Tetrahedron Lett.*, **37**, 3663 (1996).
55. J. S. Yadav, S. Chandrasekhar, G. Sumithra, and R. Kache, *Tetrahedron Lett.*, **37**, 6603 (1996).
56. G. O. Aspinall, I. H. Ibrahim, and N. K. Khare, *Carbohydr. Res.*, **200**, 247 (1990); H. Paulsen, F. R. Heiker, J. Feldmann, and K. Heyns, *Synthesis*, 636 (1980).
57. B. M. Choudary, A. D. Prasad, V. Swapna, V. L. K. Valli, and V. Bhuma, *Tetrahedron*, **48**, 953 (1992).
58. R. M. Thomas, G. H. Mohan, and D. S. Iyengar, *Tetrahedron Lett.*, **38**, 4721 (1997).
59. M. Honda, H. Morita, and I. Nagakura, *J. Org. Chem.*, **62**, 8932 (1997).
60. T. Taniguchi and K. Ogasawara, *Angew. Chem., Int. Ed. Engl.*, **37**, 1136 (1998).

Propargyl Ethers: $HC{\equiv}CCH_2OR$

Cleavage

1. Propargyl ethers are cleaved with $TiCl_3$–Mg in THF, 54–92% yield. Allyl and benzyl ethers were not cleaved; phenolic propargyl ethers are also cleaved.[1]
2. $(BnNEt_3)_2MoS_4$ (benzyltriethylammonium tetrathiomolybdate).[2]

1. S. K. Nayak, S. M. Kadam, and A. Banerji, *Synlett*, 581 (1993).
2. V. M. Swamy, P. Ilankumaran, and S. Chandrasekaran, *Synlett*, 513 (1997).

p-Chlorophenyl Ether: *p*-ClC_6H_4–OR

Formation / Cleavage[1]

$$ROH \underset{\text{1. Li/NH}_3\text{; 2. H}_3\text{O}^+}{\overset{\text{1. MsCl, Pyr; 2. }p\text{-ClC}_6\text{H}_4\text{ONa}}{\rightleftarrows}} p\text{-ClC}_6\text{H}_4\text{–OR}$$

The *p*-chlorophenyl ether was used in this synthesis to minimize ring sulfonation during cyclization of a diketo ester with concentrated H_2SO_4/AcOH.[1]

1. J. A. Marshall and J. J. Partridge, *J. Am. Chem. Soc.*, **90**, 1090 (1968).

p-Methoxyphenyl Ether: *p*-$MeOC_6H_4OR$

Formation

1. *p*-$MeOC_6H_4OH$, DEAD, Ph_3P, THF, 82–99% yield.[1,2]

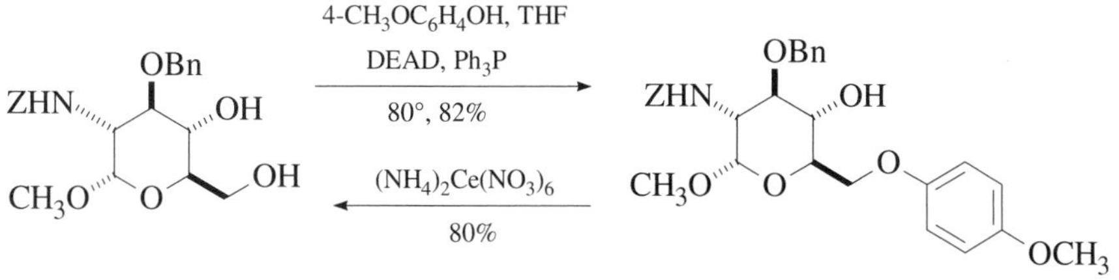

4-$CH_3OC_6H_4OH$, THF, DEAD, Ph_3P, 80°, 82% →

← $(NH_4)_2Ce(NO_3)_6$, 80%

Z = benzyloxycarbonyl, DEAD = diethyl azodicarboxylate

2. From a mesylate: K_2CO_3, 18-crown-6, CH_3CN, reflux, 48 h, 81% yield.[3]
3. From a tosylate: *p*-$MeOC_6H_4OH$, DMF, NaH, 60°, 14 h.[4]

Cleavage

1. Ceric ammonium nitrate, CH_3CN, H_2O (4:1), 0°, 10 min, 80–85% yield;[1,2] or CAN, Pyr, CH_3CN, H_2O, 0°, 0.5 h, 96% yield.[3]
2. Anodic oxidation, CH_3CN, H_2O, $Bu_4N^+PF_6^-$, 20°, 74–100% yield.[5]

This group is stable to 3 *N* HCl, 100°; 3 *N* NaOH, 100°; H_2, 1200 psi; O_3, MeOH, −78°; RaNi, 100°; $LiAlH_4$; Jones reagent and pyridinium chlorochromate (PCC). It has also been used for protection of the anomeric hydroxyl during oligosaccharide synthesis.[6]

1. T. Fukuyama, A. A. Laud, and L. M. Hotchkiss, *Tetrahedron Lett.*, **26**, 6291 (1985).
2. M. Petitou, P. Duchaussoy, and J. Choay, *Tetrahedron Lett.*, **29**, 1389 (1988).
3. Y. Masaki, K. Yoshizawa, and A. Itoh, *Tetrahedron Lett.*, **37**, 9321 (1996).
4. Takano, M. Moriya, M. Suzuki, Y. Iwabuchi, T. Sugihara, and K. Ogasawara, *Heterocycles*, **31**, 1555 (1990).
5. S. Iacobucci, N. Filippova, and M. d'Alarcao, *Carbohydr. Res.*, **277**, 321 (1995).
6. Y. Matsuzaki, Y. Ito, Y. Nakahara, and T. Ogawa, *Tetrahedron Lett.*, **34**, 1061 (1993).

p-Nitrophenyl Ether: (p-$NO_2C_6H_4OR$)

The *p*-nitrophenyl ether was used for the protection of the anomeric position of a pyranoside. It is installed using the Königs–Knorr process and can be cleaved by hydrogenolysis (Pd/C, H_2, Ac_2O), followed by oxidation with ceric ammonium nitrate (81–99% yield).[1]

1. K. Fukase, T. Yasukochi, Y. Nakai, and S. Kusumoto, *Tetrahedron Lett.*, **37**, 3343 (1996).

2,4-Dinitrophenyl Ether (RO–DNP): 2,4-$(NO_2)_2$–C_6H_3OR

Formation

1. 2,4-Dinitrofluorobenzene, DABCO, DMF, 85% yield.[1] When this group was used to protect an anomeric center of a carbohydrate, only the β-isomer was formed, but it could be equilibrated to the α-isomer in 90% yield with K_2CO_3 in DMF.

1. H. J. Koeners, A. J. De Kok, C. Romers, and J. H. Van Boom, *Recl. Trav. Chim. Pays-Bas*, **99**, 355 (1980).

2,3,5,6-Tetrafluoro-4-(trifluoromethyl)phenyl Ether: $CF_3C_6F_4OR$

Treatment of a steroidal alcohol with perfluorotoluene [NaOH, $(n\text{-Bu})_4N^+HSO_4^-$, CH_2Cl_2, 79%] gives the ether, which can be cleaved in 82% yield with NaOMe/DMF.[1]

1. J. J. Deadman, R. McCague, and M. Jarman, *J. Chem. Soc., Perkin Trans. 1*, 2413 (1991).

Benzyl Ether (Bn–OR): $PhCH_2OR$ (Chart 1)

Formation

1. BnCl, powdered KOH, 130–140°, 86% yield.[1]
2. BnCl, $Bu_4N^+HSO_4^-$, 50% KOH, benzene.[2]
3. NaH, THF, BnBr, $Bu_4N^+I^-$, 20°, 3 h, 100%.[3] This method was used to protect a hindered hydroxyl group. Increased reactivity is achieved by the *in situ* generation of benzyl iodide.
4. BnX (X = Cl, Br), Ag_2O, DMF, 25°, good yields.[4] This method is very effective for the monobenzylation of diols.[5]

5. BnCl, $Ni(acac)_2$, reflux, 3 h, 80–90%.[6]
6. BnO–C(=NH)CCl_3, CF_3SO_3H.[7–10]
7. 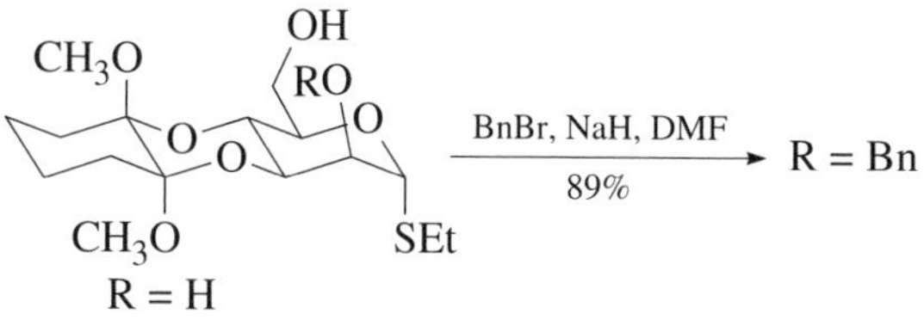 CH_3O, OH, RO, O, O, O, CH_3O, SEt, R = H — BnBr, NaH, DMF, 89% → R = Bn

   Note that in this case the primary alcohol was left unprotected.[11]
8. The following primary alcohol was selectively benzylated using NaH and BnBr at −70°:[12]

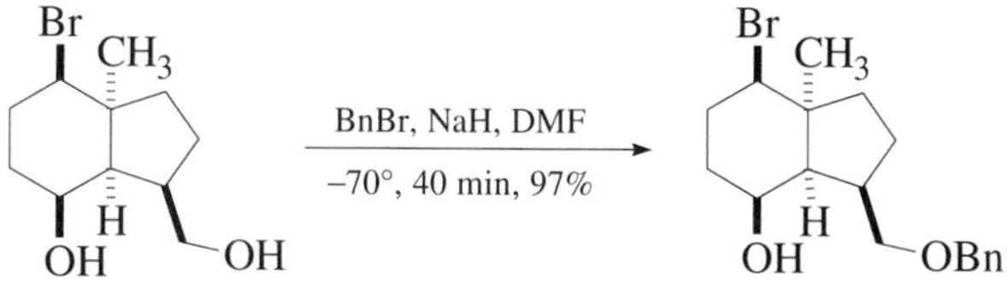

   Br, CH_3, H, OH, OH — BnBr, NaH, DMF, −70°, 40 min, 97% → Br, CH_3, H, OH, OBn

9. Ag_2O, BnBr, DMF, rt, 48 h, 76% yield.[13]

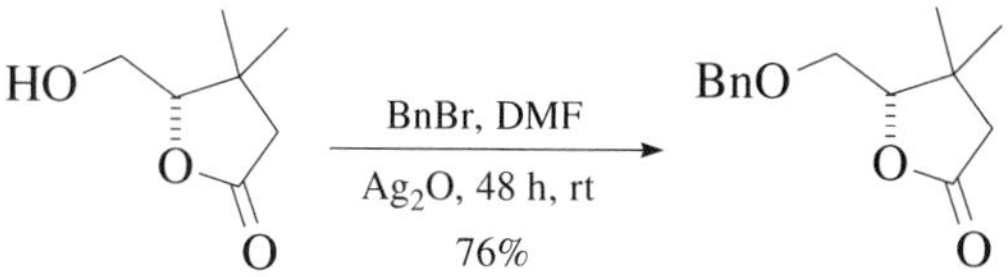

   HO, O, O — BnBr, DMF, Ag_2O, 48 h, rt, 76% → BnO, O, O

10. $(Bu_3Sn)_2O$, toluene, reflux; BnBr, *N*-methylimidazole, 95% yield.[14] Equatorial alcohols are benzylated in preference to axial alcohols in diol-containing substrates.

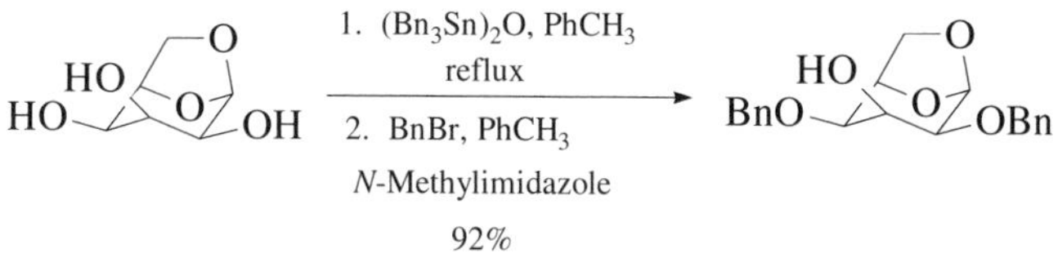

   HO, HO, O, O, OH — 1. $(Bn_3Sn)_2O$, $PhCH_3$ reflux; 2. BnBr, $PhCH_3$ *N*-Methylimidazole 92% → HO, BnO, O, O, OBn

11. Bu_2SnO, benzene; BnBr, DMF, heat, 80% yield.[15] This method has also been used to protect selectively the anomeric hydroxyl in a carbohydrate derivative.[16] The replacement of Bu_2SnO with $Bu_2Sn(OMe)_2$ improves the process procedurally.[17] The use of stannylene acetals for the regioselective manipulation of hydroxyl groups has been reviewed.[18]
12. BnI, NaH, rt, 90% yield.[19] Note that in this case the reaction proceeds without complication of the Payne rearrangement.

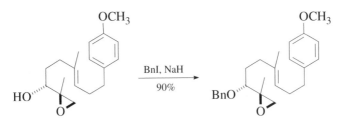

OCH$_3$ OCH$_3$ BnI, NaH 90% HO BnO O O

13. PhCHN$_2$, HBF$_4$, $-40°$, CH_2Cl_2, 66–92% yield.[20] Selective protection of an alcohol in the presence of amines is achieved under these conditions.[21]
14. From a TMS ether: PhCHO, TESH, TMSOTf, 96% yield.[22] This method is effective for the preparation of allyl ethers (85% yield).
15. BnCl, NaH, $CuCl_2$, $Bu_4N^+I^-$, THF, reflux 25 h, 70% yield.[23]

OH HO O Ph O CH$_3$O O BnCl NaH, CuCl$_2$ Bu$_4$N$^+$I$^-$, THF reflux 25 h 70% OBn HO O Ph O CH$_3$O O

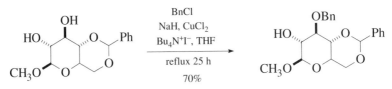

1. H. G. Fletcher, *Methods Carbohydr. Chem.*, **II**, 166 (1963).
2. H. H. Freedman and R. A. Dubois, *Tetrahedron Lett.*, 3251 (1975).
3. S. Czernecki, C. Georgoulis, and C. Provelenghiou, *Tetrahedron Lett.*, 3535 (1976); K. Kanai, I. Sakamoto, S. Ogawa, and T. Suami, *Bull. Chem. Soc. Jpn.*, **60**, 1529 (1987).
4. R. Kuhn, I. Löw, and H. Trishmann, *Chem. Ber.*, **90**, 203 (1957).
5. A. Bouzide and G. Sauvé, *Tetrahedron Lett.*, **38**, 5945 (1997).
6. M. Yamashita and Y. Takegami, *Synthesis*, 803 (1977).
7. T. Iversen and K. R. Bundle, *J. Chem Soc., Chem. Commun.*, 1240 (1981).
8. J. D. White, G. N. Reddy, and G. O. Spessard, *J. Am. Chem. Soc.*, **110**, 1624 (1988).
9. U. Widmer, *Synthesis*, 568 (1987).
10. P. Eckenberg, U. Groth, T. Huhn, N. Richter, and C. Schmeck, *Tetrahedron*, **49**, 1619 (1993).
11. P. Grice, S. V. Ley, J. Pietruszka, H. W. M. Priepke, and S. L. Warriner, *J. Chem. Soc., Perkin Trans. 1*, 351 (1997).
12. A. Fukuzawa, H. Sato, and T. Masamune, *Tetrahedron Lett.*, **28**, 4303 (1987).
13. L. Van Hijfte and R. D. Little, *J. Org. Chem.*, **50**, 3940 (1985).
14. C. Cruzado, M. Bernabe, and M. Martin-Lomas, *J. Org. Chem.*, **54**, 465 (1989).
15. W. R. Roush, M. R. Michaelides, D. F. Tai, B. M. Lesur, W. K. M. Chong, and D. J. Harris, *J. Am. Chem. Soc.*, **111**, 2984 (1989).
16. C. Bliard, P. Herczegh, A. Olesker, and G. Lukacs, *J. Carbohydr. Res.*, **8**, 103 (1989).
17. G. J. Boons, G. H. Castle, J. A. Clase, P. Grice, S. V. Ley, and C. Pinel, *Synlett*, 913 (1993).

18. S. David and S. Hanessian, *Tetrahedron,* **41**, 643 (1985); M. Pereyre, J.-P. Quintard, and A. Rahm, *Tin in Organic Synthesis*, Butterworths, London, 1987, pp. 261–285.
19. E. E. van Tamelen, S. R. Zawacky, R. K. Russell, and J. G. Carlson, *J. Am. Chem. Soc.*, **105**, 142 (1983).
20. L. J. Liotta and B. Ganem, *Tetrahedron Lett.*, **30**, 4759 (1989).
21. L. J. Liotta and B. Ganem, *Isr. J. Chem.*, **31**, 215 (1991).
22. S. Hatakeyama, H. Mori, K. Kitano, H. Yamada, and M. Nishizawa, *Tetrahedron Lett.*, **35**, 4367 (1994).
23. B. Classon, P. J. Garegg, S. Oscarson, and A. K. Tidén, *Carbohydr. Res.*, **216**, 187 (1991).

Cleavage

1. H_2/Pd–C, EtOH, 95% yield.[1,2]
2. Pd is the preferred catalyst, since the use of Pt results in ring hydrogenation.[1] Hydrogenolysis of the benzyl group of threonine in peptides containing tryptophan often results in reduction of tryptophan to the 2,3-dihydro derivative.[3] The presence of nonaromatic amines can retard *O*-debenzylation,[4] and the presence of Na_2CO_3 prevents benzyl group removal, but allows double bond reduction to occur.[5] Although it is possible to effect benzyl ether cleavage in the presence of an isolated olefin (H_2/5% Pd–C, 97% yield),[6] in general, the degree of selectivity is dependent upon the substitution pattern and the degree of steric hindrance. Good selectivity was achieved for hydrogenolysis of a benzyl group in the presence of a trisubstituted olefin conjugated to an ester.[7] Excellent selectivity has been observed in the hydrogenolysis (Pd/C, EtOAc, rt, 18 h) of a benzyl group in the presence of a *p*-methoxybenzyl group.[8] Hydrogenolysis of the benzyl group is solvent dependent, as illustrated in the following table.[9]

Effect of Solvent on the Hydrogenolysis of Benzyl Ether (1.1 bar, 50°, 2% Pd/C)

Solvent	Reaction rate ($mm\ H_2$/min/0.1 g cat)
THF	40
Hexanol	25
Methanol	5
Toluene	2
Hexane	6

3. Pd–C using transfer hydrogenation. A number of methods have been developed in which hydrogen is generated *in situ*. These include the use of cyclohexene (1–8 h, 80–90% yield),[10] cyclohexadiene (25°, 2 h, good yields),[11] HCO_2H,[12] ammonium formate (MeOH, reflux, 91% yield),[13]

and isopropyl alcohol.[14] A benzylidene acetal is not cleaved when ammonium formate is used as the hydrogen source,[13] and a trisubstituted olefin is not affected when formic acid is used as a hydrogen source.[15] In α-methyl 2,3-di-*O*-benzyl-4,6-*O*-benzylideneglucose, the cleavage can be controlled to cleave the 2-benzyl group selectively (83%) when cyclohexene is used as the hydrogen source.[16] Hydrogenation was also shown to cleave only an anomeric benzyl group in perbenzylated galactose.[17]

4. Raney Nickel W2 or W4, EtOH, 85–100% yield.[18] Mono- and dimethoxy-substituted benzyl ethers and benzaldehyde acetals are not cleaved under these conditions, and trisubstituted alkenes are not reduced.
5. Na/ammonia[19,20] or EtOH.[21]

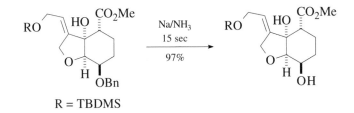

RO HO CO_2Me O H OBn — Na/NH_3, 15 sec, 97% → RO HO CO_2Me O H OH

R = TBDMS

Note that in this example the ester was not reduced. When R = Ac, the benzyl cleavage reaction failed.[22]

6. Li, catalytic naphthalene, −78°, THF, 68–99% yield. In addition, tosyl, benzyl, and mesyl amides are cleaved with excellent efficiency.[23]
7. Electrolytic reduction: −3.1 V, $R_4N^+F^-$, DMF.[24]
8. Me_3SiI, CH_2Cl_2, 25°, 15 min, 100% yield.[25] This reagent also cleaves most other ethers and esters, but selectivity can be achieved with the proper choice of conditions.
9. Lithium aluminum hydride will also cleave benzyl ethers, but this is seldom practical because of the high reactivity of lithium aluminum hydride to other functional groups.[26]
10. Me_2BBr, $ClCH_2CH_2Cl$, 0°–rt, 70–93% yield.[27] The reagent also cleaves phenolic methyl ethers; tertiary ethers and allylic ethers give the bromide rather than the alcohol.
11. $FeCl_3$, Ac_2O, 55–75% yield.[28] The relative rates of cleavage for the 6-, 3- and 2-*O*-benzyl groups of a glucose derivative are 125 : 24 : 1. Sulfuric acid has also been used as a catalyst.[29] $FeCl_3$ (CH_2Cl_2, 0° rt, 64–88% yield) in the absence of acetic anhydride is effective as well and was found to cleave secondary benzyl groups in the presence of a primary benzyl group.[30]
12. CrO_3/AcOH, 25°, 50% yield, [→ ROCOPh (→ ROH + $PhCO_2H$)].[31] This method was used to remove benzyl ethers from carbohydrates that contain functional groups sensitive to catalytic hydrogenation or dissolving metals. Esters are stable, but glycosides or acetals are cleaved.

13. RuO_2, $NaIO_4$, CCl_4, CH_3CN, H_2O, 54–96% yield.[32] The benzyl group is oxidized to a benzoate that can be hydrolyzed under basic conditions.
14. Ozone, 50 min, then NaOMe, 60–88% yield.[33]
15. Electrolytic oxidation: 1.4–1.7 V, Ar_3N, CH_3CN, CH_2Cl_2, $LiClO_4$, lutidine.[34]
16. Ca/NH_3, ether or THF, 2 h; NH_4Cl, H_2O, 90% yield.[35] Acetylenes are **not** reduced under these conditions. One problem with the use of calcium is that the oxide coating makes it difficult to initiate the reaction. This is partially overcome by adding sand to the reaction mixture to abrade the surface of the calcium mechanically.
17. $PhSSiMe_3$, $Bu_4N^+I^-$, ZnI_2, $ClCH_2CH_2Cl$, 60°, 2 h, 75% yield.[36]

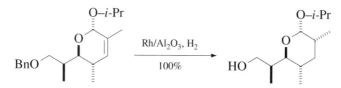

EtO$_2$C, OAc, OBn → PhSTMS, ZnI_2, $Bu_4N^+I^-$, $ClCH_2CH_2Cl$, 60°, 2 h, 75% → EtO$_2$C, OAc, OH

18. Rh/Al_2O_3, H_2, 100%.[37]

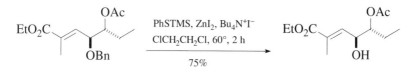

BnO, O, O–*i*-Pr → Rh/Al_2O_3, H_2, 100% → HO, O, O–*i*-Pr

19. $Ph_3C^+BF_4^-$, CH_2Cl_2.[38]

H_3C, O, OBn, SPh, CO_2Me, Br, H → $Ph_3C^+ BF_4^-$ → H_3C, H, O, O, O, SPh, Br, H

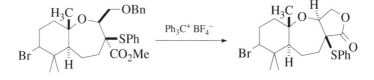

20. *t*-BuMgBr, benzene, 80°, 69%.[39] MeMgI fails in this reaction. In general, benzyl ethers are quite stable to Grignard reagents because these reactions are not usually run at such high temperatures.

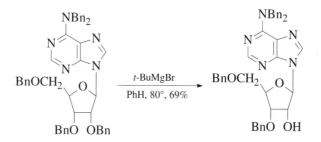

NBn_2, N, N, N, N, $BnOCH_2$, O, BnO, OBn → *t*-BuMgBr, PhH, 80°, 69% → NBn_2, N, N, N, N, $BnOCH_2$, O, BnO, OH

21. EtSH, $BF_3 \cdot Et_2O$, 63% yield.[40] Benzylamines are stable to these conditions, but $BF_3 \cdot Et_2O/Me_2S$ has been used to cleave an allylic benzyl ether.[41]
22. The fungus *Mortierella isabellina* NRRL 1757, 0–100% yield.[42]
23. $BF_3 \cdot Et_2O$, NaI, CH_3CN, 0°, 1 h; rt, 7 h, 80% yield.[43]
24. BCl_3, CH_2Cl_2, −78°→0°; MeOH at −78°, 77% yield.[44]

CO_2Me; N_3; O; O; $(CH_2)_4OBn$ — BCl_3, CH_2Cl_2, − 78° to 0° / then MeOH at − 78° / 77% → $(CH_2)_4OH$

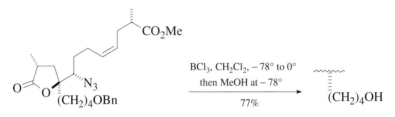

25. BCl_3–DMS, CH_2Cl_2, 5 min → 24 h, rt, 16–100%.[45] A trityl group is cleaved in preference to a benzyl group under these conditions.
26. Me_3SiBr, thioanisole.[46] This reagent combination also cleaves a carbobenzoxy (Z) group and a $4\text{-}MeOC_6H_4CH_2SR$ group and reduces sulf- oxides to sulfides.
27. $SnCl_4$, CH_2Cl_2, rt, 30 min.[47]

BnO, O, O, OBn, OBn — $SnCl_4$, CH_2Cl_2 / rt, 30 min → HO, O, O, OH, OBn (92%) + HO, O, O, OBn, OBn (5%)

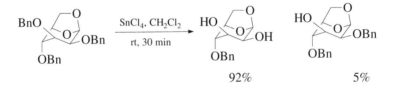

In carbohydrates in which benzyl groups are used extensively for protection, the stability of the benzyl groups toward electrophilic reagents is increased by the presence of electron–withdrawing groups in the ring.[48]

28. Lithium di-*tert*-butylbiphenyl, THF, −78°, 3 h, 95% yield.[49]

OBn, O, O, OH, O, OTIPS — $(t\text{-BuPh})_2^{\bullet -}\,Li^{\bullet +}$ / THF, −78°, 3 h / 95% → OH, O, O, OH, O, OTIPS

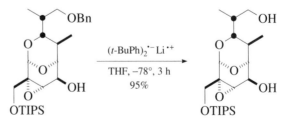

29. Lithium naphthalenide, THF, −25°, 55–80 min, 73–98% yield.[50]
30. DDQ, CH_2Cl_2, 58°, 2 days, 52% yield.[51] In this example, conventional reductive methods failed. Anhydrous DDQ was used to prevent acid-promoted decomposition.

DDQ, CH_2Cl_2, 58°, 2 days, 52%; R = Bn → R = H

31. RO–$CH_2CH_2CH(CH_3)CH_2CH_2$–OBn → RO–$CH_2CH_2CH(CH_3)CH_2CH_2$–OBz (t-BuO_2-benziodoxolone, K_2CO_3, PhH, rt, 47–100 min, 61–90%)

R = MOM, THP, TBDMS, Ac

Ref. 52

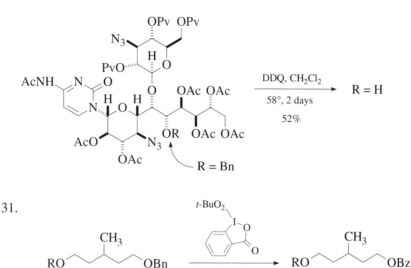

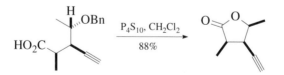

Allyl ethers are oxidized to acrylates with this reagent.

32. NBS, $h\nu$, $CaCO_3$, CCl_4, H_2O, 86% yield.[53]
33. 25% MsOH/$CHCl_3$, 25°, 84% yield.[54]
34. BBr_3, 60% yield.[55]
35. Dimethyldioxirane, acetone, 48 h, rt, 85–93% yield.[56,57] *p*-Bromo-, *p*-cyano- and 2-naphthylmethyl ethers, and benzylidene acetals can also be deprotected.
36. P_4S_{10}, CH_2Cl_2, 88% yield.[58]

P_4S_{10}, CH_2Cl_2, 88%

37. $AlCl_3$–aniline, CH_2Cl_2, rt, 80–96% yield.[59]
38. $PdCl_2$, EtOH, H_2O, H_2, 79–99% yield. These conditions were used for the deprotection of peptides; the $PdCl_2$ was used stochiometrically.[60]
39. TMSOTf, Ac_2O, 10–15°, 85% yield.[61] The acetate is produced that must then be hydrolyzed.
40. AcBr, $SnBr_2$ or $Sn(OTf)_2$, CH_2Cl_2, rt, 1–4 h, 76–97% yield.[62] These conditions convert a benzyl ether into an acetate.
41. $ZnCl_2$, Ac_2O, AcOH, rt, 80–94% yield. These conditions are selective for the cleavage of 6-*O*-benzylpyranosides.[63]

42. PhI(OH)OTs, CH_3CN.[64]

H_3C, O, BnO, BnO — PhI(OH)OTs, CH_3CN, >59% → H_3C, O, BnO, O

1. C. H. Heathcock and R. Ratcliffe, *J. Am. Chem. Soc.*, **93**, 1746 (1971).
2. W. H. Hartung and C. Simonoff, *Org. React.*, **7**, 263 (1953).
3. L. Kisfaludy, F. Korenczki, T. Mohacsi, M. Sajgo, and S. Fermandjian, *Int. J. Pept. Protein Res.*, **27**, 440 (1986).
4. B. P. Czech and R. A. Bartsch, *J. Org. Chem.*, **49**, 4076 (1984).
5. G. R. Cook, L. G. Beholz, and J. R. Stille, *J. Org. Chem.*, **59**, 3575 (1994).
6. J. S. Bindra and A. Grodski, *J. Org. Chem.*, **43**, 3240 (1978).
7. D. Cain and T. L. Smith, Jr., *J. Am. Chem. Soc.*, **102**, 7568 (1980).
8. J. M. Chong and K. K. Sokoll, *Org. Prep. Proced. Int.*, **25**, 639 (1993).
9. S. Hawker, M. A. Bhatti, and K. G. Griffin, *Chim. Oggi*, **10**, 49 (1992).
10. G. M. Anantharamaiah and K. M. Sivanandaiah, *J. Chem. Soc., Perkin Trans. 1*, 490 (1977); S. Hanessian, T. J. Liak, and B. Vanasse, *Synthesis*, 396 (1981).
11. A. M. Felix, E. P. Heimer, T. J. Lambros, C. Tzougraki, and J. Meienhofer, *J. Org. Chem.*, **43**, 4194 (1978).
12. B. ElAmin, G. M. Anantharamaiah, G. P. Royer, and G. E. Means, *J. Org. Chem.*, **44**, 3442 (1979).
13. T. Bieg and W. Szeja, *Synthesis*, 76 (1985).
14. M. Del Carmen Cruzado and M. Martin-Lomias, *Tetrahedron Lett.*, **27**, 2497 (1986).
15. M. E. Jung, Y. Usui, and C. T. Vu, *Tetrahedron Lett.*, **28**, 5977 (1987).
16. D. Beaupere, I. Boutbaiba, G. Demailly, and R. Uzan, *Carbohydr. Res.*, **180**, 152 (1988).
17. T. Bieg and W. Szeja, *Carbohydr. Res.*, **205**, C10 (1990).
18. Y. Oikawa, T. Tanaka, K. Horita, and O. Yonemitsu, *Tetrahedron Lett.*, **25**, 5397 (1984); K. Horita, T. Yoshioka, T. Tanaka, Y. Oikawa, and O. Yonemitsu, *Tetrahedron*, **42**, 3021 (1986).
19. C. M. McCloskey, *Adv. Carbohydr. Chem.*, **12**, 137 (1957); I. Schön, *Chem. Rev.*, **84**, 287 (1984).
20. K. D. Philips, J. Zemlicka, and J. P. Horowitz, *Carbohydr. Res.*, **30**, 281 (1973).
21. E. J. Reist, V. J. Bartuska, and L. Goodman, *J. Org. Chem.*, **29**, 3725 (1964).
22. M. M. Sulikowski, G. E. R. E. Davis, and A. B. Smith, III, *J. Chem. Soc., Perkin Trans. 1*, 979 (1992).
23. E. Alonso, D. J. Ramón, and M. Yus, *Tetrahedron*, **53**, 14355 (1997).
24. V. G. Mairanovsky, *Angew. Chem., Int. Ed. Engl.*, **15**, 281 (1976).
25. M. E. Jung and M. A. Lyster, *J. Org. Chem.*, **42**, 3761 (1977).
26. J. P. Kutney, N. Abdurahman, C. Gletsos, P. LeQuesne, E. Piers, and I. Vlattas, *J. Am. Chem. Soc.*, **92**, 1727 (1970).

27. Y. Guindon, C. Yoakim, and H. E. Morton, *Tetrahedron Lett.*, **24**, 2969 (1983).
28. K. P. R. Kartha, F. Dasgupta, P. P. Singh, and H. C. Srivastava, *J. Carbohydr. Chem.*, **5**, 437 (1986); J. I. Padron and J. T. Vazquez, *Tetrahedron: Asymmetry*, **6**, 857 (1995).
29. J. Sakai, T. Takeda, and Y. Ogihara, *Carbohydr. Res.*, **95**, 125 (1981).
30. R. Rodebaugh, J. S. Debenham, and B. Fraser-Reid, *Tetrahedron Lett.*, **37**, 5477 (1996).
31. S. J. Angyal and K. James, *Carbohydr. Res.*, **12**, 147 (1970).
32. P. F. Schuda, M. B. Cichowicz, and M. R. Heimann, *Tetrahedron Lett.*, **24**, 3829 (1983); P. F. Schuda and M. R. Heimann, *Tetrahedron Lett.*, **24**, 4267 (1983).
33. P. Angibeaud, J. Defaye, A. Gadelle, and J.-P. Utille, *Synthesis*, 1123 (1985).
34. W. Schmidt and E. Steckhan, *Angew. Chem., Int. Ed. Engl.*, **18**, 801 (1979); E. A. Mayeda, L. L. Miller, and J. F. Wolf, *J. Am. Chem. Soc.*, **94**, 6812 (1972).
35. J. R. Hwu, V. Chua, J. E. Schroeder, R. E. Barrans, Jr., K. P. Khoudary, N. Wang, and J. M. Wetzel, *J. Org. Chem.*, **51**, 4731 (1986); J. R. Hwu, Y. S. Wein, and Y.-J. Leu, *J. Org. Chem.*, **61**, 1493 (1996).
36. K. C. Nicolaou, M. R. Pavia, and S. P. Seitz, *J. Am. Chem. Soc.*, **104**, 2027 (1982).
37. Y. Oikawa, T. Tanaka, and O. Yonemitsu, *Tetrahedron Lett.*, **27**, 3647 (1986).
38. T. R. Hoye, A. J. Caruso, J. F. Dellaria, Jr., and M. J. Kurth, *J. Am. Chem. Soc.*, **104**, 6704 (1982).
39. M. Kawana, *Chem. Lett.*, 1541 (1981).
40. S. M. Daly and R. W. Armstrong, *Tetrahedron Lett.*, **30**, 5713 (1989).
41. M. Ishizaki, O. Hoshino, and Y. Iitaka, *Tetrahedron Lett.*, **32**, 7079 (1991).
42. H. L. Holland, M. Conn, P. C. Chenchaiah, and F. M. Brown, *Tetrahedron Lett.*, **29**, 6393 (1988).
43. Y. D. Vankar and C. T. Rao, *J. Chem. Res., Synop.*, 232 (1985).
44. D. R. Williams, D. L. Brown, and J. W. Benbow, *J. Am. Chem. Soc.*, **111**, 1923 (1989).
45. M. S. Congreve, E. C. Davison, M. A. M. Fuhry, A. B. Holmes, A. N. Payne, R. A. Robinson, and S. E. Ward, *Synlett*, 663 (1993).
46. N. Fujii, A. Otaka, N. Sugiyama, M. Hatano, and H. Yajima, *Chem. Pharm. Bull.*, **35**, 3880 (1987).
47. H. Hori, Y. Nishida, H. Ohrui, and H. Meguro, *J. Org. Chem.*, **54**, 1346 (1989).
48. K. Jansson, G. Noori, and G. Magnusson, *J. Org. Chem.*, **55**, 3181 (1990).
49. S. J. Shimshock, R. E. Waltermire, and P. DeShong, *J. Am. Chem. Soc.*, **113**, 8791 (1991).
50. H.-J. Liu, J. Yip, and K.-S. Shia, *Tetrahedron Lett.*, **38**, 2252 (1997).
51. N. Ikemoto and S. L. Schreiber, *J. Am. Chem. Soc.*, **114**, 2524 (1992).
52. M. Ochiai, T. Ito, H. Takahashi, A. Nakanishi, M. Toyonari, T. Sueda, S. Goto, and M. Shiro, *J. Am. Chem. Soc.*, **118**, 7716 (1996).
53. R. W. Binkley and D. G. Hehemann, *J. Org. Chem.*, **55**, 378 (1990).
54. D. S. Matteson, H.-W. Man, and O. C. Ho, *J. Am. Chem. Soc.*, **118**, 4560 (1996).
55. D. E. Ward, Y. Gai, and B. F. Kaller, *J. Org. Chem.*, **60**, 7830 (1995).

56. B. A. Marples, J. P. Muxworthy, and K. H. Baggaley, *Synlett*, 646 (1992).
57. R. Csuk and P. Dörr, *Tetrahedron*, **50**, 9983 (1994).
58. P. A. Jacobi, J. Guo, and W. Zheng, *Tetrahedron Lett.*, **36**, 1197 (1995).
59. T. Akiyama, H. Hirofuji, and S. Ozaki, *Tetrahedron Lett.*, **32**, 1321 (1991).
60. A. J. Pallenberg, *Tetrahedron Lett.*, **33**, 7693 (1992).
61. J. Alzeer and A. Vasella, *Helv. Chim. Acta*, **78**, 177 (1995); P. Angibeaud and J.-P. Utille, *Synthesis*, 737 (1991).
62. T. Oriyama, M. Kimura, M. Oda, and G. Koga, *Synlett*, 437 (1993).
63. G. Yang, X. Ding, and F. Kong, *Tetrahedron Lett.*, **38**, 6725 (1997).
64. A. Kirschning, S. Domann, G. Dräger, and L. Rose, *Synlett*, 767 (1995).

Substituted Benzyl Ethers

Several methoxy-substituted benzyl ethers have been prepared and used as protective groups. Their utility lies in the fact that they are more readily cleaved oxidatively than the unsubstituted benzyl ethers. The following table gives the relative rates of cleavage with dichlorodicyanoquinone (DDQ).[1]

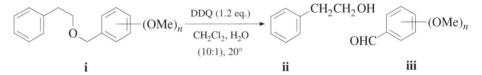

$PhCH_2CH_2OCH_2C_6H_{5-n}(OMe)_n$ (**i**) $\xrightarrow[\text{(10:1), 20°}]{\text{DDQ (1.2 eq.)}, CH_2Cl_2, H_2O}$ $PhCH_2CH_2OH$ (**ii**) + $OHC{-}C_6H_{5-n}(OMe)_n$ (**iii**)

Cleavage of MPM, DMPM, and TMPM Ethers with DDQ in CH_2Cl_2/H_2O at 20°

Protective Group	Time (h)	Yield ii	(%) iii	Protective Group	Time (h)	Yield ii	(%) iii
3,4-DMPM	<0.33	86	84	2-MPM	3.5	93	70
4-MPM	0.33	89	86	3,5-DMPM	8	73	92
2,3,4-TMPM	0.5	60	75	2,3-DMPM	12.5	75	73
3,4,5-TMPM	1	89	89	3-MPM	24	80	94
2,5-DMPM	2.5	95	16	2,6-DMPM	27.5	80	95

From the table, it is clear that there are considerable differences in the cleavage rates of the various ethers, which should prove quite useful.

p-Methoxybenzyl Ether (MPM–OR): *p*-$MeOC_6H_4CH_2OR$

Formation

1. The section on the formation of benzyl ethers should also be consulted.
2. *p*-$MeOC_6H_4CH_2OC(=NH)CCl_3$, H^+, 52–84% yield.[2-4] $BF_3 \cdot Et_2O$.[5] In addition, camphorsulfonic acid[3] and *p*-toluenesulfonic acid[4] have been used as catalysts.

3.

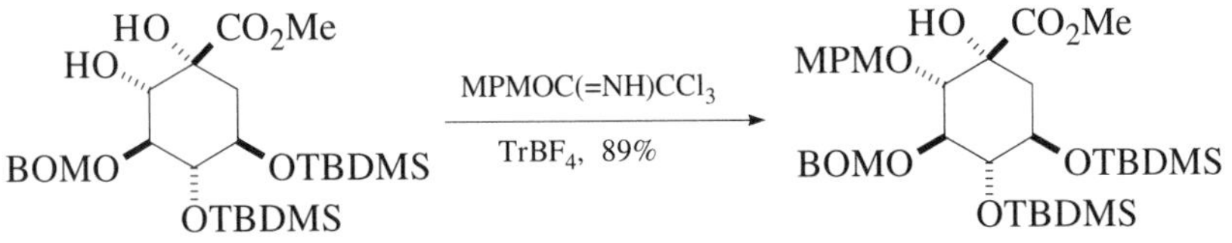

HO CO$_2$Me / HO / BOMO / OTBDMS / OTBDMS $\xrightarrow[\text{TrBF}_4,\ 89\%]{\text{MPMOC(=NH)CCl}_3}$ HO CO$_2$Me / MPMO / BOMO / OTBDMS / OTBDMS

Ref. 6

4. p-MeOC$_6$H$_4$CHN$_2$, SnCl$_2$, ≈50% yield.[7] This method was used to introduce the MPM group at the 2′- and 3′-positions of ribonucleotides without selectivity for either the 2′- or 3′-isomer. The primary 5′-hydroxyl was not affected.
5. NaH, p-MeOC$_6$H$_4$CH$_2$Br, DMF, −5°, 1 h, 65%.[8,9] Other bases, such as BuLi,[10] dimsyl potassium,[11] and NaOH under phase-transfer conditions,[12] have been used to introduce the MPM group. The use of $(n\text{-Bu})_4\text{N}^+\text{I}^-$ for the *in situ* preparation of the very reactive p-methoxybenzyl iodide often improves the protection of hindered alcohols.[13] In the following example, selectivity is probably achieved because of the increased acidity of the 2′-hydroxyl group:

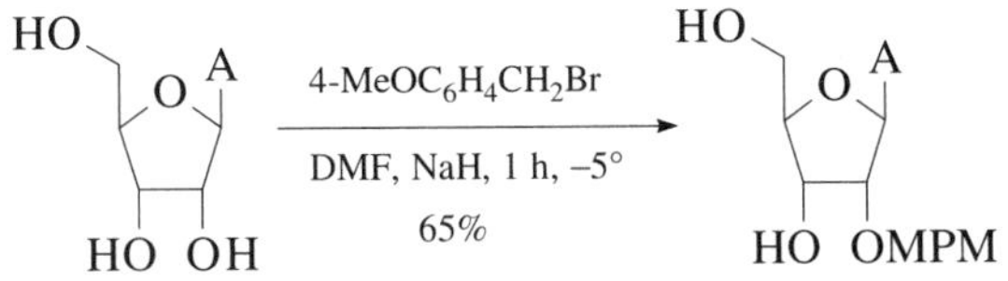

HO / O / A / HO OH $\xrightarrow[\text{DMF, NaH, 1 h, }-5°\text{, }65\%]{4\text{-MeOC}_6\text{H}_4\text{CH}_2\text{Br}}$ HO / O / A / HO OMPM

Ref. 10

6. NaH, p-MeOC$_6$H$_4$CH$_2$Cl, THF, 81% yield.[14]

7.

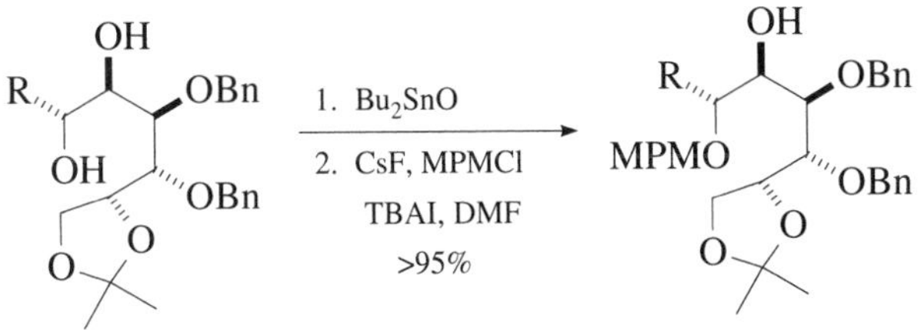

OH / R / OBn / OH / OBn / O / O $\xrightarrow[\text{2. CsF, MPMCl, TBAI, DMF, >95\%}]{\text{1. Bu}_2\text{SnO}}$ OH / R / OBn / MPMO / OBn / O / O

Ref. 15

8. N-(4-Methoxybenzyl)-o-benzenedisulfonimide, NaH, THF, 57–78% yield.[16]

Cleavage

1. The section on the cleavage of benzyl ethers should also be consulted.

2. Electrolytic oxidation: Ar_3N, CH_3CN, $LiClO_4$, 20°, 1.4–1.7 V, 80–90% yield.[17] Benzyl ethers are not affected by these conditions.
3. Dichlorodicyanoquinone (DDQ), CH_2Cl_2, H_2O, 40 min, rt, 84–93% yield.[18–20] This method normally does not cleave simple benzyl ethers, but forcing conditions will result in benzyl ether cleavage.[21] Surprisingly, a glycosidic TMS group was found to survive these conditions.[22] An *O*-MPM group can be cleaved in preference to an *N*-MPM protected amide.[23] The following groups are generally stable to these conditions: ketones, acetals, epoxides, alkenes, acetonides, tosylates, MOM and MEM ethers, THP ethers, acetates, benzyloxymethyl (BOM) ethers, boronates, and TBDMS ethers. Exceptions do occur, however, and will depend on the nature of the reaction conditions.

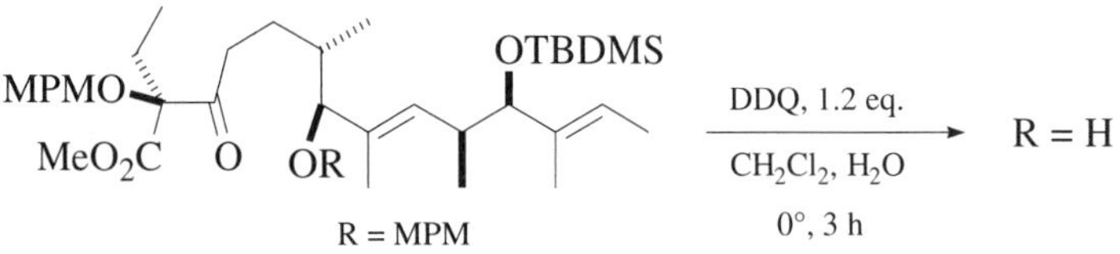

MPMO, MeO_2C, O, OR, OTBDMS; R = MPM —(DDQ, 1.2 eq.; CH_2Cl_2, H_2O; 0°, 3 h)→ R = H

Ref. 24

In this case, the tertiary and electron-deficient MPM group is retained.[24] A very slow cleavage of an MPM protected adenosine was attributed to its reduced electron density as a result of π stacking with the adenine. Typically, these reactions are complete in < 1 h, but in this case complete cleavage required 41 h.[25]

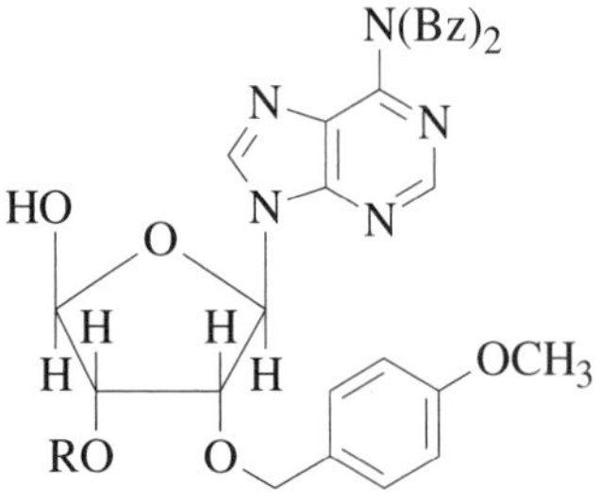

$N(Bz)_2$, N, N, N, N, HO, O, H, H, H, H, RO, O, OCH_3

When MPM ethers bearing a proximal hydroxyl are treated, DDQ acetals are formed.[26,27]

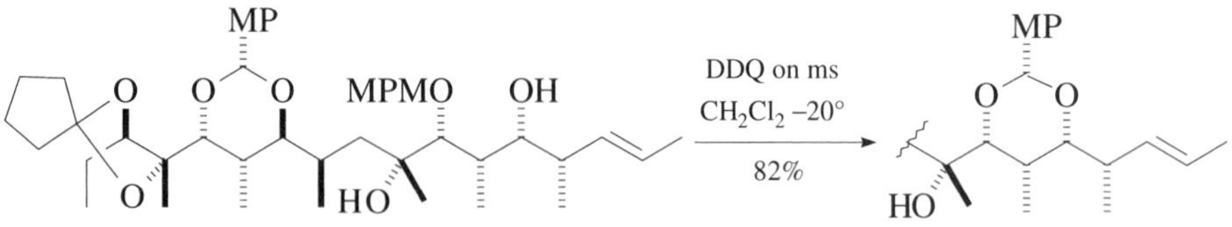

MP, O, O, O, MPMO, OH, O, HO —(DDQ on ms; CH_2Cl_2 −20°; 82%)→ MP, O, O, HO

4. Catalytic DDQ, $FeCl_3$, CH_2Cl_2, H_2O, 62–94% yield.[28]
5. Ozone, acetone, −78°, 42–82% yield.[29,30]

6. Ceric ammonium nitrate (CAN), Br_2 or NBS, CH_2Cl_2, H_2O, 90% yield.[31]

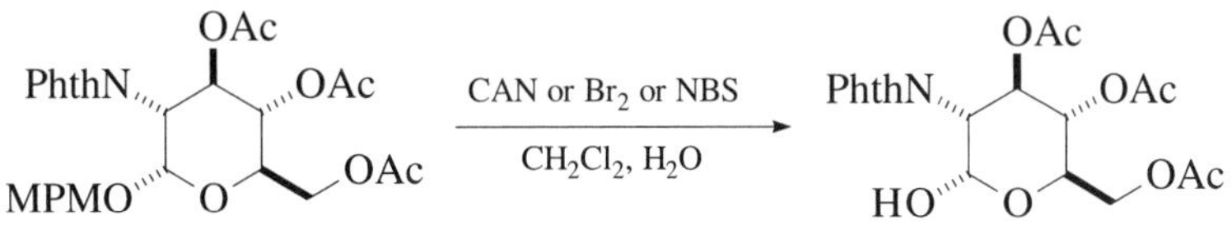

PhthN OAc OAc OAc MPMO O — CAN or Br_2 or NBS, CH_2Cl_2, H_2O → PhthN OAc OAc OAc HO O

Phth = phthalimido

NHBn Ph CO_2Me OPMB — CAN, rt, 1-3 h, CH_3CN, H_2O, 95% → NHBn Ph CO_2Me OH

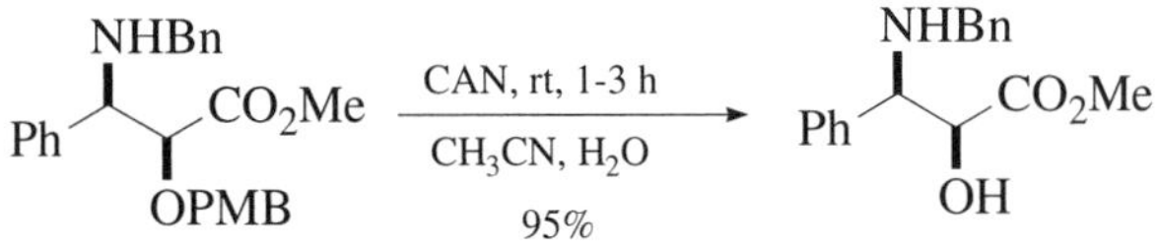

Ref. 32

7. $Ph_3C^+BF_4^-$, CH_2Cl_2 or CH_3CN, H_2O.[1,4] In one case, the reaction with DDQ failed to go to completion. This was attributed to the reduced electron density on the aromatic ring because of its attachment at the more electron-poor anomeric center.
8. $h\nu$ >280 nm, H_2O, 1,4-dicyanonaphthalene, 70–81% yield.[33]
9. $Mg(ClO_4)_2$, $h\nu$, anthraquinone or dicyanoanthracene.[34] These conditions also cleave the DMPM group.
10. This example shows that overoxidation of allylic alcohols[35] may occur with DDQ.

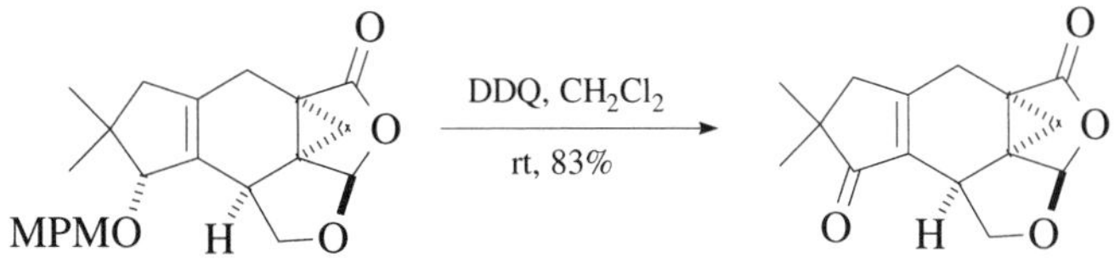

MPMO H O O O — DDQ, CH_2Cl_2, rt, 83% → O H O O O

11. AgO, HNO_3, 74% yield.[36]
12. The following examples illustrate unusual and unexpected cleavage processes because of participation by nearby functionality.

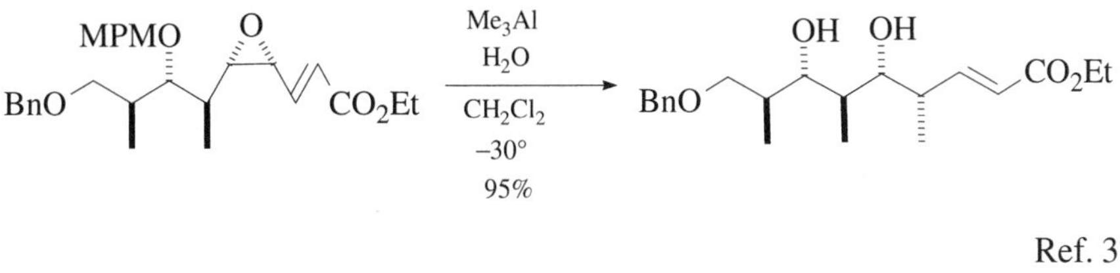

BnO MPMO O CO_2Et — Me_3Al, H_2O, CH_2Cl_2, –30°, 95% → BnO OH OH CO_2Et

Ref. 37

HO MPMO OTBDMS $\xrightarrow[\text{CCl}_4,\ \text{heat}]{\text{PPh}_3}$ Cl + O

Ref. 38

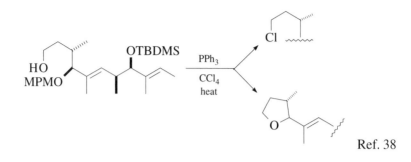

13. An allylic MPM ether has been converted directly to a bromide upon treatment with Me_2BBr (5 min, −78°).[39]
14. I_2, MeOH, reflux, 12–16 h, 75–91% yield. Benzyl ethers are stable to these conditions, but isopropylidenes are cleaved.[40]
15. TMSCl, anisole, $SnCl_2$, CH_2Cl_2, rt, 10–50 min, 78–96% yield.[41]
16. TFA, CH_2Cl_2, rt, 5–30 min, 84–99% yield.[42] An adamantyl glycoside was stable to these conditions. The reaction has also been performed in the presence of anisole to scavenge the liberated benzyl carbenium ion.[43]
17. $BF_3 \cdot Et_2O$, $NaCNBH_3$, THF, reflux 4–24 h, 65–98% yield.[44] Functional groups such aryl ketones and nitro compounds are reduced and electron-rich phenols tend to be alkylated with the released benzyl carbenium ion. The use of $BF_3 \cdot Et_2O$ and triethylsilane as a cation scavenger is also effective.[45]

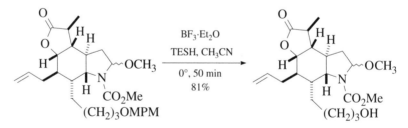

O O H H H H N OCH_3 CO_2Me $(CH_2)_3OMPM$ $\xrightarrow[\text{0°, 50 min, 81\%}]{BF_3 \cdot Et_2O,\ \text{TESH, CH}_3\text{CN}}$ O O H H H H N OCH_3 CO_2Me $(CH_2)_3OH$

18. $AlCl_3$ or $SnCl_2$, EtSH, CH_2Cl_2, 73–97% yield.[46]
19. AcOH, 90°.[47]
20. 1 *M* HCl, EtOH, reflux, 87% yield.[48]
21. BCl_3, dimethyl sulfide.[49]
22. Me_2BBr, CH_2Cl_2, −78°, 5 min, 100% yield.
23. $SnBr_2$, AcBr, CH_2Cl_2, rt, 81–92% yield. These conditions, which also cleave alkyl and aryl benzyl ethers, produce an acetate that must then be hydrolyzed with base to release the alcohol.[50] When $SnCl_2/PhOCH_2COCl$ is used, only MPM ethers are cleaved, leaving benzyl ethers unaffected.
24. TBDMSOTf, TEA, CH_2Cl_2, rt.[51] These conditions result in conversion of the MPM ether into a TBDMS ether.

25. Pd–C, H_2.[52]

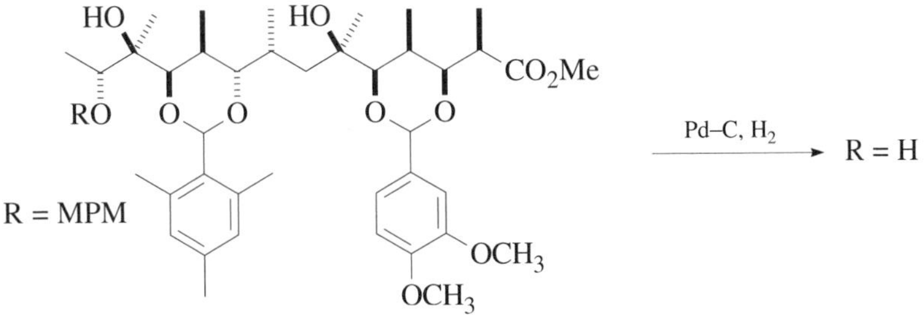

HO HO CO_2Me RO O O O O; R = MPM; OCH_3 OCH_3 —Pd–C, H_2→ R = H

26. TMSI, $CHCl_3$, 0.25 h, 25°.[53]
27. $MgBr_2$–Et_2O, Me_2S, CH_2Cl_2, rt, 75–96% yield.[54] The failure of this substrate to undergo cleavage with the typical conditions was attributed to the presence of the 1,3-diene. Acetonides and TBDMS ethers were found to be stable.

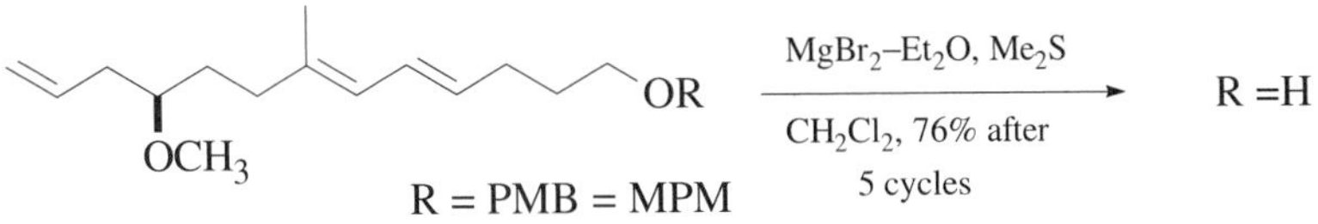

OR OCH_3 R = PMB = MPM —$MgBr_2$–Et_2O, Me_2S; CH_2Cl_2, 76% after 5 cycles→ R =H

28. Na, NH_3, 95% yield.[55]

3,4-Dimethoxybenzyl Ether (DMPM–OR): 3,4-$(MeO)_2C_6H_3CH_2OR$

Formation

1. 3,4-$(MeO)_2C_6H_3CH_2OC(=NH)CCl_3$, TsOH.[15] This ether has properties similar to the *p*-methoxybenzyl (MPM) ether, except that it can be removed from an alcohol with DDQ in the presence of an MPM group with 98% selectivity.[18–20] The selectivity is attributed to the lower oxidation potential of the DMPM group—1.45 V versus 1.78 V for the MPM.

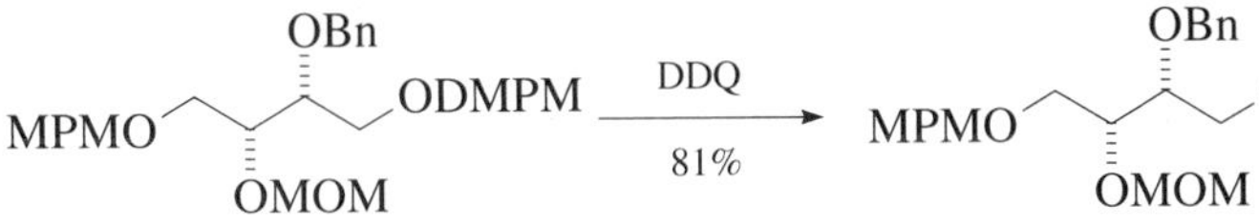

   MPMO OBn ODMPM OMOM —DDQ, 81%→ MPMO OBn OH OMOM Ref. 18

   The dimethoxybenzyl ether has also been used for protection of the anomeric hydroxyl in carbohydrates.[56]
2. NaH, 3,4-$(MeO)_2C_6H_3CH_2Br$, DMF.[57]

1. N. Nakajima, R. Abe, and O. Yonemitsu, *Chem. Pharm. Bull.*, **36**, 4244 (1988).
2. H. Takaku, S. Ueda, and T. Ito, *Tetrahedron Lett.*, **24**, 5363 (1983); N. Nakajima, K. Horita, R. Abe, and O. Yonemitsu, *Tetrahedron Lett.*, **29**, 4139 (1988).

3. R. D. Walkup, R. R. Kane, P. D. Boatman, Jr., and R. T. Cunningham, *Tetrahedron Lett.*, **31**, 7587 (1990).
4. E. Adams, M. Hiegemann, H. Duddeck, and P. Welzel, *Tetrahedron*, **46**, 5975 (1990).
5. N. Hébert, A. Beck, R. B. Lennox, and G. Just, *J. Org. Chem.*, **57**, 1777 (1992).
6. K. K. Reddy, M. Saady, and J. R. Falck, *J. Org. Chem.*, **60**, 3385 (1995).
7. K. Kamaike, H. Tsuchiya, K. Imai, and H. Takaku, *Tetrahedron*, **42**, 4701 (1986).
8. H. Takaku and K. Kamaike, *Chem. Lett.*, 189 (1982).
9. H. Takaku, K. Kamaike, and H. Tsuchiya, *J. Org. Chem.*, **49**, 51 (1984).
10. H. Hoshi, T. Ohnuma, S. Aburaki, M. Konishi, and T. Oki, *Tetrahedron Lett.*, **34**, 1047 (1993).
11. N. Nakajima, T. Hamada, T. Tanaka, Y. Oikawa, and O. Yonemitsu, *J. Am. Chem. Soc.*, **108**, 4645 (1986).
12. P. J. Garegg, S. Oscarson, and H. Ritzen, *Carbohydr. Res.*, **181**, 89 (1988).
13. D. R. Mootoo and B. Fraser-Reid, *Tetrahedron*, **46**, 185 (1990).
14. J. L. Marco and J. A. Hueso-Rodriquez, *Tetrahedron Lett.*, **29**, 2459 (1988).
15. A. Wei and Y. Kishi, *J. Org. Chem.*, **59**, 88 (1994).
16. P. H. J. Carlsen, *Tetrahedron Lett.*, **39**, 1799 (1998).
17. W. Schmidt and E. Steckhan, *Angew. Chem., Int. Ed. Engl.*, **18**, 801 (1979). See also E. A. Mayeda, L. L. Miller, and J. F. Wolf, *J. Am. Chem. Soc.*, **94**, 6812 (1972); S. M. Weinreb, G. A. Epling, R. Comi, and M. Reitano, *J. Org. Chem.*, **40**, 1356 (1975).
18. K. Horita, T. Yoshioka, T. Tanaka, Y. Oikawa, and O. Yonemitsu, *Tetrahedron*, **42**, 3021 (1986); T. Tanaka, Y. Oikawa, T. Hamada, and O. Yonemitsu, *Tetrahedron Lett.*, **27**, 3651 (1986).
19. Y. Oikawa, T. Tanaka, K. Horita, and O. Yonemitsu, *Tetrahedron Lett.*, **25**, 5397 (1984).
20. Y. Oikawa, T. Yoshioka, and O. Yonemitsu, *Tetrahedron Lett.*, **23**, 885 (1982).
21. K. Horita, S. Nagato, Y. Oikawa, and O. Yonemitsu, *Tetrahedron Lett.*, **28**, 3253 (1987); N. Ikemoto and S. L. Schreiber, *J. Am. Chem. Soc.*, **114**, 2524 (1992).
22. K. Hiruma, T. Kajimoto, G. Weitz-Schmidt, I. Ollmann, and C.-H. Wong, *J. Am. Chem. Soc.*, **118**, 9265 (1996).
23. Y. Hamada, Y. Tanada, F. Yokokawa, and T. Shioiri, *Tetrahedron Lett.*, **32**, 5983 (1991).
24. K. J. Hale and J. Cai, *Tetrahedron Lett.*, **37**, 4233 (1996).
25. H. Hotoda, M. Takahashi, K. Tanzawa, S. Takahashi, and M. Kaneko, *Tetrahedron Lett.*, **36**, 5037 (1995).
26. R. Stürmer, K. Ritter, and R. W. Hoffmann, *Angew. Chem., Int. Ed. Engl.*, **32**, 101 (1993).
27. S. Hanessian, N. G. Cooke, B. DeHoff, and Y. Sakito, *J. Am. Chem. Soc.*, **112**, 5276 (1990).
28. S. Chandrasekhar, G. Sumithra, and J. S. Yadav, *Tetrahedron Lett.*, **37**, 1645 (1996).
29. M. Hirama and M. Shimizu, *Synth. Commun.*, **13**, 781 (1983).
30. P. Somfai, *Tetrahedron*, **50**, 11315 (1994).

31. B. Classon, P. J. Garegg, and B. Samuelsson, *Acta Chem. Scand., Ser. B*, **B38**, 419 (1984); R. Johansson and B. Samuelsson, *J. Chem. Soc., Perkin Trans. 1*, 2371 (1984).
32. G. I. Georg, P. M. Mashava, E. Akgün, and M. W. Milstead, *Tetrahedron Lett.*, **32**, 3151 (1991); Y. Wang, S. A. Babirad, and Y. Kishi, *J. Org. Chem.*, **57**, 468 (1992).
33. G. Pandey and A. Krishna, *Synth. Commun.*, **18**, 2309 (1988).
34. A. Nishida, S. Oishi, and O. Yonemitsu, *Chem. Pharm. Bull.*, **37**, 2266 (1989).
35. B. M. Trost and J. Y. L. Chung, *J. Am. Chem. Soc.*, **107**, 4586 (1985).
36. S. Sisko, J. R. Henry, and S. M. Weinreb, *J. Org. Chem.*, **58**, 4945 (1993).
37. M. Miyashita, M. Hoshino, and A. Yoshikoshi, *J. Org. Chem.*, **56**, 6483 (1991).
38. K. J. Hale, J. Cai, S. Manaviazar, and S. A. Peak, *Tetrahedron Lett.*, **36**, 6965 (1995).
39. D. G. Hall and P. Deslongchamps, *J. Org. Chem.*, **60**, 7796 (1995).
40. A. R. Vaino and W. A. Szarek, *Synlett*, 1157 (1995).
41. T. Akiyama, H. Shima and S. Ozaki, *Synlett*, 415 (1992).
42. L. Yan and D. Kahne, *Synlett*, 523 (1995).
43. E. F. De Medeiros, J. M. Herbert, and R. J. K. Taylor, *J. Chem. Soc., Perkin Trans. 1*, 2725 (1991).
44. A. Srikrishna, R. Viswajanani, J. A. Sattigeri, and D. Vijaykumar, *J. Org. Chem.*, **60**, 5961 (1995).
45. Y. Morimoto, M. Iwahashi, K. Nishida, Y. Hayashi, and H. Shirahama, *Angew. Chem., Int. Ed. Engl.*, **35**, 904 (1996).
46. A. Bouzide and G. Sauvé, *Synlett*, 1153 (1997).
47. K. J. Hodgetts and T. W. Wallace, *Synth. Commun.*, **24**, 1151 (1994).
48. D. J. Jenkins, A. M. Riley, and B. V. L. Potter, *J. Org. Chem.*, **61**, 7719 (1996).
49. M. S. Congreve, E. C. Davison, M. A. M. Fuhry, A. B. Holmes, A. N. Payne, R. A. Robinson, and S. E. Ward, *Synlett,* 663 (1993).
50. T. Oriyama, M. Kimura, M. Oda, and G. Koga, *Synlett*, 437 (1993).
51. T. Oriyama , K. Yatabe, Y. Kawada, and G. Koga, *Synlett*, 45 (1995).
52. M. Hikota, H. Tone, K. Horita, and O. Yonemitsu, *J. Org. Chem.*, **55**, 7 (1990).
53. D. M. Gordon and S. J. Danishefsky, *J. Am. Chem. Soc.*, **114**, 659 (1992).
54. T. Onoda, R. Shirai, and S. Iwasaki, *Tetrahedron Lett.*, **38**, 1443 (1997).
55. K. C. Nicolaou, J.-Y. Xu, S. Kim, T. Ohshima, S. Hosokawa, and J. Pfefferkorn, *J. Am. Chem. Soc.*, **119**, 11353 (1997).
56. S. J. Danishefsky, H. G. Selnick, R. E. Zelle, and M. P. DeNinno, *J. Am. Chem. Soc.*, **110**, 4368 (1988); A. De Mesmaeker, P. Hoffmann, and B. Ernst, *Tetrahedron Lett.*, **30**, 3773 (1989).
57. H. Takaku, T. Ito, and K. Imai, *Chem. Lett.*, 1005 (1986).

o-Nitrobenzyl Ether: *o*-$NO_2C_6H_4CH_2OR$ (Chart 1)

p-Nitrobenzyl Ether: *p*-$NO_2C_6H_4CH_2OR$

The *o*-nitrobenzyl and *p*-nitrobenzyl ethers can be prepared and cleaved by many of the methods described for benzyl ethers.[1] The *p*-nitrobenzyl ether is also prepared from an alcohol and *p*-nitrobenzyl alcohol (trifluoroacetic anhydride, 2,6-lutidine, CH_2Cl_2, 67% yield) or with the bromide and Ag_2O.[2,3] In addition, the *o*-nitrobenzyl ether can be cleaved by irradiation (320 nm, 10 min, quant. yield of carbohydrate;[4,5] 280 nm, 95% yield of nucleotide[6]). The *p*-nitrobenzyl ether has been cleaved by electrolytic reduction (−1.1 V, DMF, $R_4N^+X^-$, 60% yield),[7,8] by reduction with $Na_2S_2O_4$ (pH 8–9, 80–95% yield),[9] and by Zn/AcOH followed by acidolysis.[10] These ethers can also be cleaved oxidatively (DDQ or electrolysis) after reduction to the aniline derivative.[2] Clean reduction to the aniline is accomplished with Zn(Cu) (acetylacetone, rt, >93% yield).[3] A polymeric version of the *o*-nitrobenzyl ether has been prepared for oligosaccharide synthesis that is also conveniently cleaved by photolysis.[11] An unusual selective deprotection of a bis-*o*-nitrobenzyl ether has been observed.[12] The photochemical reaction of *o*-nitrobenzyl derivatives has been reviewed.[13]

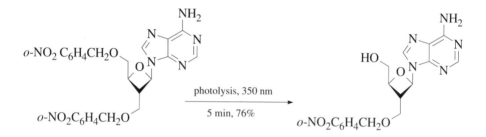

o-$NO_2C_6H_4CH_2O$ / *o*-$NO_2C_6H_4CH_2O$ (adenine nucleoside) → photolysis, 350 nm, 5 min, 76% → HO / *o*-$NO_2C_6H_4CH_2O$ (adenine nucleoside)

1. D. G. Bartholomew and A. D. Broom, *J. Chem. Soc., Chem. Commun.*, 38 (1975).
2. K. Fukase, H. Tanaka, S. Torii, and S. Kusumoto, *Tetrahedron Lett.*, **31**, 389 (1990).
3. K. Fukase, S. Hase, T. Ikenaka, and S. Kusumoto, *Bull. Chem. Soc. Jpn.*, **65**, 436 (1992).
4. U. Zehavi, B. Ami, and A. Patchornik, *J. Org. Chem.*, **37**, 2281 (1972); U. Zehavi and A. Patchornik, *J. Org. Chem.*, **37**, 2285 (1972).
5. For reviews of photoremovable protective groups, see V. N. R. Pillai, *Synthesis*, 1 (1980); V. N. R. Pillai, *Org. Photochem.*, **9**, 225 (1987).
6. E. Ohtsuka, S. Tanaka, and M. Ikehara, *J. Am. Chem. Soc.*, **100**, 8210 (1978).
7. V. G. Mairanovsky, *Angew. Chem., Int. Ed. Engl.*, **15**, 281 (1976).
8. K. Fukase, H. Tanaka, S. Torii, and S. Kusumoto, *Tetrahedron Lett.*, **31**, 381 (1990).
9. E. Guibe-Jampel and M. Wakselman, *Synth. Commun.*, **12**, 219 (1982).
10. T. Abiko and H. Sekino, *Chem. Pharm. Bull.*, **38**, 2304 (1990).
11. K. C. Nicolaou, N. Winssinger, J. Pastor, and F. DeRoose, *J. Am. Chem. Soc.*, **119**, 449 (1997).
12. N. Katagiri, M. Makino, and C. Kaneko, *Chem. Pharm. Bull.*, **43**, 884 (1995).
13. Y. L. Chow, In *The Chemistry of Amino, Nitroso and Nitro Compounds and Their Derivatives*, Supplement F, Part 1, S. Patai, Ed., Wiley, New York, 1982, p. 181.

p-Halobenzyl Ethers: *p*-X-$C_6H_4CH_2OR$, X = Br, Cl

p-Halobenzyl ethers have been prepared to protect side-chain hydroxyl groups in amino acids. More stable to the conditions of acidic hydrolysis (50% CF_3COOH) than the unsubstituted benzyl ether is, they are cleaved by HF (0°, 10 min).[1] Deprotection can also be accomplished with Pearlman's catalyst, Raney nickel W2, or Li/NH_3.[2] These ethers also impart greater crystallinity, which often aids in purification.[3] The electron-withdrawing effect can be used to advantage to stabilize the glycosidic bond toward acid[4] and the benzyl ether bond toward electrophilic reagents, as in the following case, where the BrBn group was used to prevent competition of the ether linkage with the carbonate group for the iodonium intermediate.[5]

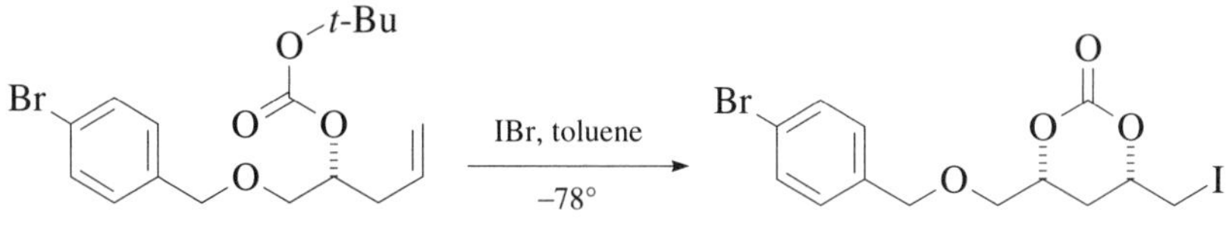

IBr, toluene, −78°

1. D. Yamashiro, *J. Org. Chem.*, **42**, 523 (1977).
2. B. K. Goering, K. Lee, B. An, and J. K. Cha, *J. Org. Chem.*, **58**, 1100 (1993).
3. S. Koto, S. Inada, N. Morishima, and S. Zen, *Carbohydr. Res.*, **87**, 294 (1980).
4. N. L. Pohl and L. L. Kiessling, *Tetrahedron Lett.*, **38**, 6985 (1997).
5. A. B. Smith, III, L. Zhuang, C. S. Brook, Q. Lin, W. H. Moser, R. E. L. Trout, and A. M. Boldi, *Tetrahedron Lett.*, **38**, 8671 (1997).

2,6-Dichlorobenzyl Ether: 2,6-$Cl_2C_6H_3CH_2OR$

Formation[1]

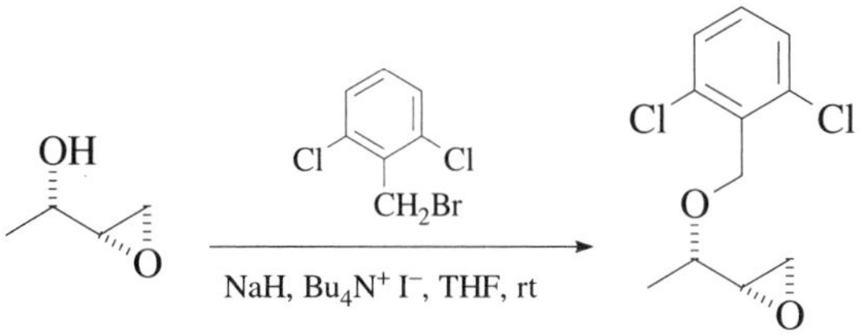

2,6-$Cl_2C_6H_3CH_2Br$, NaH, $Bu_4N^+ I^-$, THF, rt

Cleavage[2]

This group is cleaved during an iodine-promoted tetrahydrofuran synthesis.

1. S. Hatakeyama, K. Sakurai, and S. Takano, *Heterocycles*, **24**, 633 (1986).
2. S. D. Rychnovsky and P. A. Bartlett, *J. Am. Chem. Soc.*, **103**, 3963 (1981).

p-Cyanobenzyl Ether: *p*-CN–$C_6H_4CH_2OR$

The *p*-cyanobenzyl ether, prepared from an alcohol and the benzyl bromide in the presence of sodium hydride (74% yield), can be cleaved by electrolytic reduction (−2.1 V, 71% yield). It is stable to electrolytic removal (−1.4 V) of a tritylone ether [i.e., 9-(9-phenyl-10-oxo)anthryl ether].[1]

1. C. van der Stouwe and H. J. Schäfer, *Tetrahedron Lett.*, 2643 (1979); *idem, Chem. Ber.*, **114**, 946 (1981); J. P. Coleman, Naser-ud-din, H. G. Gilde, J. H. P. Utley, B. C. L. Weedon, and L. Eberson, *J. Chem. Soc., Perkin Trans. 2*, 1903 (1973).

p-Phenylbenzyl Ether: *p*-C_6H_5–$C_6H_4CH_2OR$

Formation

The section on the formation of benzyl ethers should be consulted.

Cleavage

1. $FeCl_3$, CH_2Cl_2, 2–3 min, 68% yield.[1] Benzyl ethers are cleaved in 15–20 min under these conditions. Methyl glycosides, acetates, and benzoates were not affected by this reagent.
2. Pd/C, H_2, EtOAc, >52% yield.[2] The *p*-phenylbenzyl ether is more easily cleaved by hydrogenolysis than are normal benzyl ethers. This property was used to great advantage in the deprotection of the following vineomycinone intermediate:

Pd/C, H_2, EtOAc, >52%

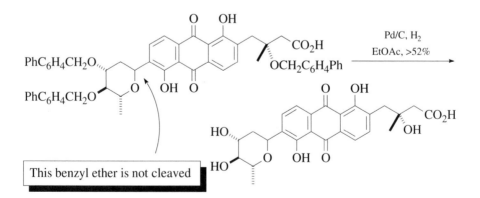

This benzyl ether is not cleaved

1. M. H. Park, R. Takeda, and K. Nakanishi, *Tetrahedron Lett.*, **28**, 3823 (1987).
2. V. Bollitt, C. Mioskowski, R. O. Kollah, S. Manna, D. Rajapaksa, and J. R. Falck, *J. Am. Chem. Soc.*, **113**, 6320 (1991).

2,6-Difluorobenzyl Ether: $C_6H_3F_2CH_2OR$

This group was developed to prevent participation of the BnO bond during cationic reactions. It is formed from the bromide [$C_6H_3F_2CH_2Br$, $Ba(OH)_2 \cdot 8H_2O$, DMF, 25 h, 94% yield][1] and cleaved by dissolving metal reduction (Ca, NH_3, 79% yield).[2]

1. R. Bürli and A. Vasella, *Helv. Chim. Acta*, **79**, 1159 (1996).
2. H. J. Borschberg, *Chimia*, **45**, 329 (1991).

p-Acylaminobenzyl Ethers (PAB–OR): *p*-R′CONH–$C_6H_4CH_2OR$

The pivaloylamidobenzyl group was stable to acetic acid–water–90°, MeOH–NaOMe, iridium-induced allyl isomerization, and to many of the Lewis acids used in glycosylation.[1]

Formation

1. *p*-PvNH–$C_6H_4CH_2Cl$, $Ba(OH)_2$, BaO, DMF, 32 h, 58–99% yield.[1]
2. *p*-PvNH–$C_6H_4CH_2OC(=NH)CCl_3$, TfOH, CH_2Cl_2, 1.5 h, 82% yield.[1]
3. *p*-Acetamidobenzyl ether from a *p*-nitrobenzyl ether: Zn(Cu), acetylacetone; Ac_2O, 93% yield.[2]
4. *p*-Acetamidobenzyl ether from a *p*-nitrobenzyl ether: Pd black, H_2, HCO_2NH_4, or cyclohexadiene, Ac_2O, pyridine.[3]

Cleavage

1. DDQ oxidation.[1,2] Cleavage occurs selectively in the presence of a benzyl and *p*-nitrobenzyl group.
2. Hydrogenolysis.[1]

1. K. Fukase, T. Yoshimura, M. Hashida, and S. Kusumoto, *Tetrahedron Lett.*, **32**, 4019 (1991).
2. K. Fukase, S. Hase, T. Ikenaka, and S. Kusumoto, *Bull. Chem. Soc. Jpn.*, **65**, 436 (1992).
3. K. Fukase, H. Tanaka, S. Torii, and S. Kusumoto, *Tetrahedron Lett.*, **31**, 389 (1990).

p-Azidobenzyl Ether (Azb–OR): 4-$N_3C_6H_4CH_2OR$

Formation[1]

1. *p*-N_3–$C_6H_4CH_2Br$, NaH, DMF, 92–98% yield.

Cleavage[1]

1. (i) H_2, Pd–C, (ii) PPh_3, (iii) DDQ, −5°.

2. DDQ, rt, 90% yield. The reaction is slow.
3. PPh_3, then DDQ, 92% yield.

This benzyl ether is partially stable to $BF_3{\cdot}Et_2O$ as used in glycosylation reactions and NaOMe, but is not stable to TFA at rt for 30 min.

4-Azido-3-chlorobenzyl Ether: 4-N_3–3-Cl–$C_6H_3CH_2OR$

The 3-chloro derivative was developed to impart greater acid stability to the azidobenzyl ether. It is much more stable to $BF_3{\cdot}Et_2O$, but is cleaved in neat TFA. Conditions used to cleave the azidobenzyl ether also cleave the 4-azido-3-chlorobenzyl ether.[2]

1. K. Fukase, M. Hashida, and S. Kusumoto, *Tetrahedron Lett.*, **32**, 3557 (1991).
2. Egusa, K. Fukase, and S. Kusumoto, *Synlett*, 675 (1997).

2-Trifluoromethylbenzyl Ether: 2-$CF_3C_6H_4CH_2OR$

This ether is formed from the alcohol and the benzyl bromide (NaH, DMF, 94% yield). It was used primarily because of its stability toward NBS during the radical-promoted conversion of a 4,6-benzylidenepyranoside to the 6-bromo-4-benzoate. Cleavage occurs quantitatively upon hydrogenolyis (Pd/C, H_2).[1]

1. L. J. Liotta, K. L. Dombi, S. A. Kelley, S. Targontsidis, and A. M. Morin, *Tetrahedron Lett.*, **38**, 7833 (1997).

***p*-(Methylsulfinyl)benzyl Ether (Msib–OR):** *p*-(MeS(O))$C_6H_4CH_2OR$

Formation

1. $CH_3S(O)C_6H_4CH_2Br$, NaH.[1]

Cleavage

The cleavage of this group proceeds by initial reduction of the sulfoxide, which then makes the resulting methylthiobenzyl ether labile to trifluoroacetic acid. Thus, any method used to reduce a sulfoxide could be used to activate this group for deprotection.

1. $SiCl_4$, thioanisole, anisole, TFA, CH_2Cl_2, 25°, 24 h, 82% yield.[2]
2. DMF–SO_3, ethanedithiol, rt, 36 h; 90% aq. TFA, 2-methylindole.[3]

1. S. Futaki, T. Yagami, T. Taike, T. Akita, and K. Kitagawa, *J. Chem. Soc., Perkin Trans.*

1, 653 (1990); Y. Kiso, S. Tanaka, T. Kimura, H. Itoh, and K. Akaji, *Chem. Pharm. Bull.*, **39**, 3097 (1991); S. Futaki, T. Taike, T. Akita, and K. Kitagawa, *J. Chem. Soc., Chem. Commun.*, 523 (1990).

2. Y. Kiso, T. Fukui, S. Tanaka, T. Kimura, and K. Akaji, *Tetrahedron Lett.*, **35**, 3571 (1994).
3. S. Futaki, T. Taike, T. Akita, and K. Kitagawa, *Tetrahedron*, **48**, 8899 (1992).

2- and 4-Picolyl Ether: $C_5H_4NCH_2$–OR

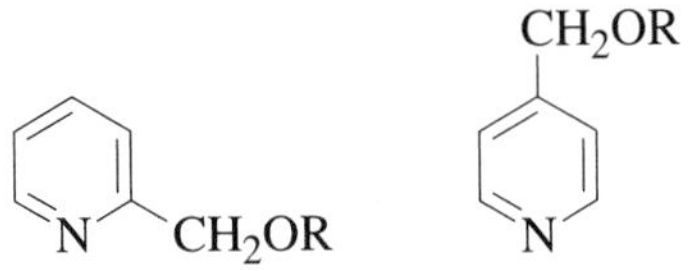

Picolyl ethers are prepared from their chlorides by a Williamson ether synthesis (68–83% yield). Some selectivity for primary vs. secondary alcohols can be achieved (ratios = 4.3–4.6:1). Picolyl ethers are cleaved electrolytically (−1.4 V, 0.5 *M* HBF_4, MeOH, 70% yield). Since picolyl chlorides are unstable as the free base, they must be generated from the hydrochloride prior to use.[1] These derivatives are relatively stable to acid (CF_3CO_2H, HF/anisole). Additionally, cleavage can be effected by hydrogenolysis in acetic acid.[2]

1. S. Wieditz and H. J. Schaefer, *Acta Chem. Scand., Ser. B.*, **B37**, 475 (1983); A. Gosden, R. Macrae, and G. T. Young, *J. Chem. Res., Synop.*, 22 (1977).
2. J. Rizo, F. Albericio, G. Romero, C. G.-Esheverria, J. Claret, C. Muller, E. Giralt, and E. Pedroso, *J. Org. Chem.*, **53**, 5386 (1988).

3-Methyl-2-picolyl *N*-Oxido Ether:

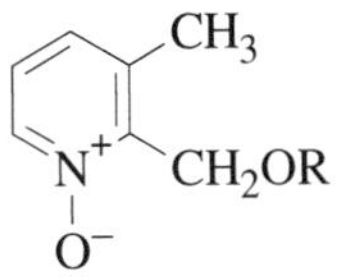

The authors prepared a number of substituted 2-diazomethylene derivatives of picolyl oxide to use for monoprotection of the *cis*-glycol system in nucleosides. The 3-methyl derivative proved most satisfactory.[1]

Formation/Cleavage[1]

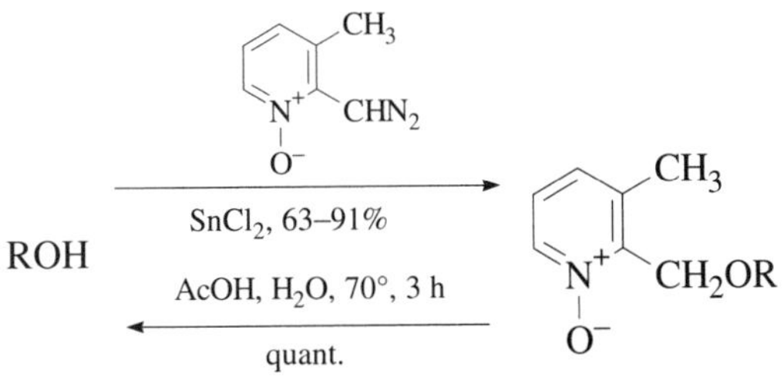

ROH → (3-methyl-2-(diazomethyl)pyridine N-oxide, CHN_2; $SnCl_2$, 63–91%) → 3-methyl-2-picolyl N-oxido ether (CH_2OR)

← AcOH, H_2O, 70°, 3 h, quant.

1. Y. Mizuno, T. Endo, and K. Ikeda, *J. Org. Chem.*, **40**, 1385 (1975); Y. Mizuno, T. Endo, and T. Nakamura, *J. Org. Chem.*, **40**, 1391 (1975).

2-Quinolinylmethyl Ether (Qm–OR)

Formation[1,2]

ROH —(QnCl, KH, THF; 0° –rt, 70–99%)→ 2-quinolinyl-CH_2OR

Cleavage

1. $CuCl_2 \cdot 2H_2O$, DMF, H_2O, air, 65°, 56–80% yield.[1]
2. $h\nu$, 61–85% yield.[2] In this case, cleavage results in simultaneous oxidation of the initially-protected alcohol to give a ketone. The related 6-phenanthridinylmethyl ethers similarly give ketones upon photochemical deprotection.[3]

1. L. Usypchuk and Y. Leblanc, *J. Org. Chem.*, **55**, 5344 (1990).
2. V. Rukachaisirikul, U. Koert, and R. W. Hoffmann, *Tetrahedron*, **48**, 4533 (1992).
3. V. Rukachaisirikul and R. W. Hoffmann, *Tetrahedron*, **48**, 10563 (1992).

1-Pyrenylmethyl Ether: 1-pyrenyl-CH_2OR

This is a fluorescent benzyl ether used for 2′-protection in nucleotide synthesis. It is introduced using 1-pyrenylmethyl chloride (KOH, benzene, dioxane, reflux, 2 h, >65% yield).[1] Most methods used for benzyl ether cleavage should be applicable to this ether.

1. K. Yamana, Y. Ohashi, K. Nunota, M. Kitamura, H. Nakano, O. Sangen, and T. Shimdzu, *Tetrahedron Lett.*, **32**, 6347 (1991).

Diphenylmethyl Ether (DPM–OR): Ph_2CHOR

Formation

1. $(Ph_2CHO)_3PO$, cat. CF_3COOH, CH_2Cl_2, reflux, 4–9 h, 65–92% yield.[1] This methodology has been applied to the protection of amino acid alcohols.[2]

2. Ph_2CHOH, concd. H_2SO_4, 12 h, 70% yield.[3]
3. Ph_2CN_2, CH_3CN or benzene, 79–85% yield.[4]

Cleavage

1. Pd–C, $AlCl_3$, cyclohexene, reflux, 24 h, 91% yield.[5] Simple hydrogenation also cleaves this ether (71–100% yield).[5]
2. Electrolytic reduction: −3.0 V, DMF, $R_4N^+X^-$.[3]
3. 10% CF_3COOH, anisole, CH_2Cl_2.[2] Anisole is present to scavenge the diphenylmethyl cation liberated during the cleavage reaction.

1. L. Lapatsanis, *Tetrahedron Lett.*, 3943 (1978).
2. C. Froussios and M. Kolovos, *Synthesis*, 1106 (1987); M. Kolovos and C. Froussios, *Tetrahedron Lett.*, **25**, 3909 (1984).
3. V. G. Mairanovsky, *Angew. Chem., Int. Ed. Engl.*, **15**, 281 (1976).
4. G. Jackson, H. F. Jones, S. Petursson, and J. M. Webber, *Carbohydr. Res.*, **102**, 147 (1982).
5. G. A. Olah, G. K. S. Prakash, and S. C. Narang, *Synthesis*, 825 (1978).

p,p′-Dinitrobenzhydryl Ether (RO–DNB): $ROCH(C_6H_4\text{-}p\text{-}NO_2)_2$

Formation/Cleavage[1]

$$\mathrm{ROH} \underset{\substack{\text{1. } PtO_2/H_2,\ Fe_3(CO)_{12} \text{ or } NaBH_4\text{-}Ni(OAc)_2 \\ \text{2. pH} < 5, \text{ preferred is 3–4, 81–90\%}}}{\overset{(p\text{-}NO_2\text{–}C_6H_4)_2CN_2,\ BF_3\cdot Et_2O}{\rightleftarrows}} \mathrm{RO\text{–}DNB}$$

The cleavage proceeds by initial reduction of the nitro groups, followed by acid-catalyzed cleavage. The DNB group can be cleaved in the presence of allyl, benzyl, tetrahydropyranyl, methoxyethoxymethyl, methoxymethyl, silyl, trityl, and ketal protective groups.

1. G. Just, Z. Y. Wang, and L. Chan, *J. Org. Chem.*, **53**, 1030 (1988).

5-Dibenzosuberyl Ether:

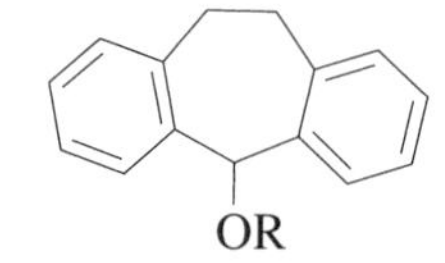

OR

The dibenzosuberyl ether is prepared from an alcohol and the suberyl chloride in the presence of triethylamine (CH_2Cl_2, 20°, 3 h, 75% yield). It is cleaved by acidic hydrolysis (1 *N* HCl/dioxane, 20°, 6 h, 80% yield). This group has also been used to protect amines, thiols, and carboxylic acids. The alcohol derivative can be cleaved in the presence of a dibenzosuberylamine.[1]

1. J. Pless, *Helv. Chim. Acta*, **59**, 499 (1976).

Triphenylmethyl Ether (Tr–OR): Ph_3C–OR (Chart 1)

Formation

1.

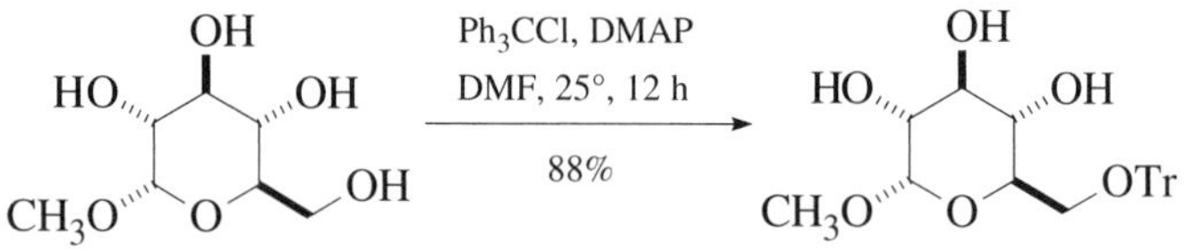

OH, HO, OH, OH, CH_3O, O → (Ph_3CCl, DMAP, DMF, 25°, 12 h, 88%) → OH, HO, OH, OTr, CH_3O, O

DMAP = 4-*N*,*N*-dimethylaminopyridine

A secondary alcohol reacts more slowly (40–45°, 18–24 h, 68–70% yield). In general, excellent selectivity can be achieved for primary alcohols in the presence of secondary alcohols.[1]

2. $C_5H_5N^+CPh_3BF_4^-$, CH_3CN, Pyr, 60–70°, 75–90% yield.[2] Triphenylmethyl ethers can be prepared more readily with triphenylmethylpyridinium fluoroborate than with triphenylmethyl chloride/pyridine.
3. ***P***-*p*-$C_6H_4Ph_2CCl$, Pyr, 25°, 5 days, 90%[3], where ***P*** = styrene-divinylbenzene polymer. Triarylmethyl ethers of primary hydroxyl groups in glucopyranosides have been prepared using a polymeric form of triphenylmethyl chloride. Although the yields are not improved, the workup is simplified.
4. Ph_3CCl, 2,4,6-collidine, CH_2Cl_2, $Bu_4N^+ClO_4^-$, 15 min, 97% yield.[4] This is an improved procedure for installing the trityl group on polymer-supported nucleosides. DBU is also a very effective base, and in this case secondary hydroxyls can be protected in good yield.[5]
5. $Me_2NC_5H_5NCPh_3{}^+Cl^-$, CH_2Cl_2, 25°, 16 h, 95% yield.[6] In this case, a primary alcohol is cleanly protected over a secondary alcohol. The reagent is a stable, isolable salt.[7] If the solvent is changed from CH_2Cl_2 to DMF, the amine of serine can be selectively protected.
6. The trityl group can migrate from one secondary center to another under acid catalysis.[8]

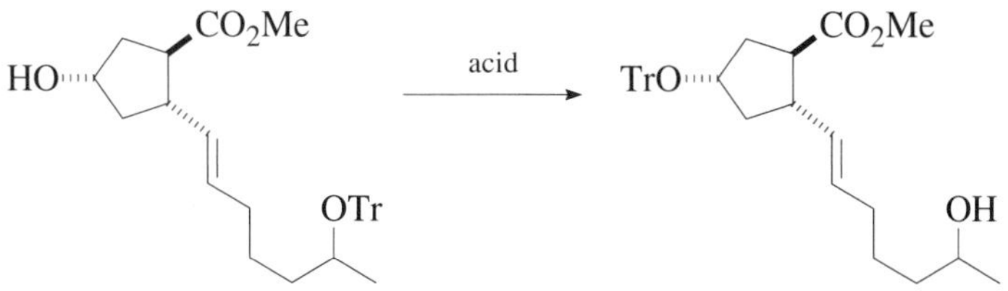

acid

7. $Ph_3COSiMe_3$, Me_3SiOTf, CH_2Cl_2, 0°, 0.5 h, 73–97% yield.[9] These conditions also introduce the trityl group on a carboxyl group. The primary hydroxyl of persilylated ribose was selectively derivatized.
8. TrOTf, 2,6-lutidine, CH_2Cl_2, 0°, >74% yield.[10]

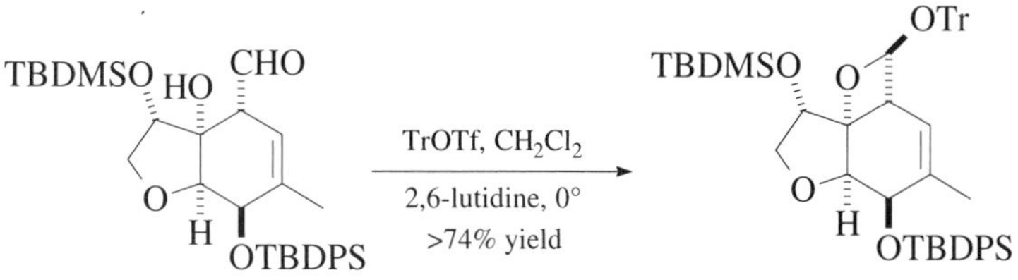

TrOTf, CH_2Cl_2, 2,6-lutidine, 0°, >74% yield

Cleavage

1. Formic acid, ether, 45 min, 88% yield.[11]

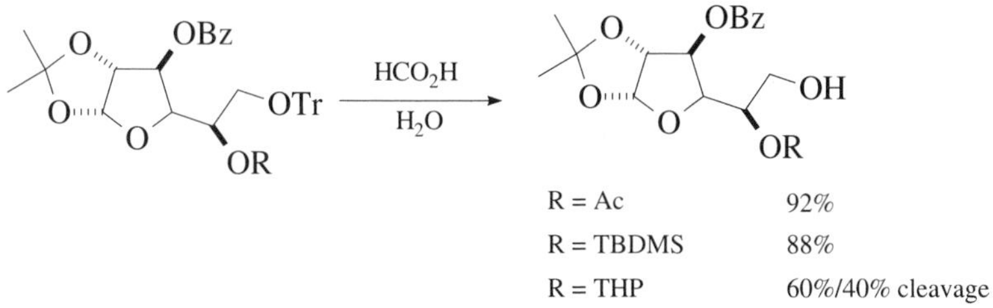

HCO_2H, H_2O

R = Ac	92%
R = TBDMS	88%
R = THP	60%/40% cleavage

2. $CuSO_4$ (anhydrous), benzene, heat, 89–100% yield.[12] In highly acylated carbohydrates, trityl removal proceeds without acyl migration.
3. Amberlyst 15-H, MeOH, rt, 5–10 min, 69–90%.[13]
4. AcOH, 56°, 7.5 h, 96%.[14]
5. 90% CF_3COOH, *t*-BuOH, 20°, 2–30 min, then Bio-Rad 1 × 2(OH^-) resin.[15] These conditions were used to cleave the trityl group from the 5′-hydroxyl of a nucleoside. Bio-Rad resin neutralizes the acid and minimizes cleavage of glycosyl bonds.
6. CF_3COOH, TFAA, CH_2Cl_2. These conditions afford the trifluoroacetate, thus preventing retritylation, which is sometimes a problem when a trityl group is cleaved with acid. A further advantage of these conditions was that a SEM group was completely stable. When TFAA was not used,

traces of moisture resulted in partial SEM cleavage. The TFA group is easily cleaved with methanol and TEA.[16]

7. H_2/Pd, EtOH, 20°, 14 h, 80% yield.[17]
8. HCl(g), $CHCl_3$, 0°, 1 h, 91% yield.[18]
9. Electrolytic reduction: −2.9 V, $R_4N^+X^-$, DMF.[19]

10. $CH_3CH(OCPh_3)(CH_2)_4CH_2OCPh_3 \xrightarrow[20°,\ 15\ min,\ 91\%]{Ph_3C^+BF_4^-,\ CH_2Cl_2\ [20]} CH_3CO(CH_2)_4CH_2OH$

    Since a secondary alcohol is oxidized in preference to a primary alcohol by $Ph_3C^+BF_4^-$, this reaction could result in selective protection of a primary alcohol.

11. $SnCl_2$, Ac_2O, CH_3CN.[21] In this case, a sulfoxide is also reduced.
12. Et_2AlCl, CH_2Cl_2, 3 min, 70–85% yield.[22] This method was used to remove the trityl group from various protected deoxyribonucleotides. The TBDPS group is stable to these conditions.
13. TsOH, MeOH, 25°, 5 h.[23]
14. $BF_3·Et_2O$, $HSCH_2CH_2SH$, 80% yield.[24]

    H₃C, CH₃, CH₃O, O, OTr —[$HSCH_2CH_2SH$ / $BF_3·Et_2O$ / 80%]→ S, S, CH₃, H₃C, HO, OH

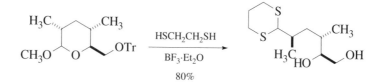

15. $BF_3·Et_2O$, CH_2Cl_2, MeOH, 2 h, rt, 80% yield.[25]
16. Na, NH_3.[26] Additionally, benzyl groups are removed under these conditions.
17. $ZnBr_2$, MeOH, 100% yield.[27,28] TIPS and TBDPS ethers are stable to these conditions.
18. SiO_2, benzene, 25°, 16 h, 81% yield.[29] This cleavage reaction is carried out on a column.
19. K-10 Clay, MeOH, H_2O, 75°, 95% yield.[30]
20. BCl_3, CH_2Cl_2, −10°, 20 min, then cold $NaHCO_3$.[31] TBDMS ethers were stable to these conditions.

α-Naphthyldiphenylmethyl Ether: RO–C(Ph)$_2$–α-$C_{10}H_7$ (Chart 1)

The α-naphthyldiphenylmethyl ether was prepared to protect, selectively, the 5′-OH group in nucleosides. It is prepared from α-naphthyldiphenylmethyl chloride in pyridine (65% yield) and cleaved selectively in the presence of a *p*-methoxyphenyldiphenylmethyl ether with sodium anthracenide, **a** (THF, 97%, yield). The *p*-methoxyphenyldiphenylmethyl ether can be cleaved with acid in the presence of this group.[32]

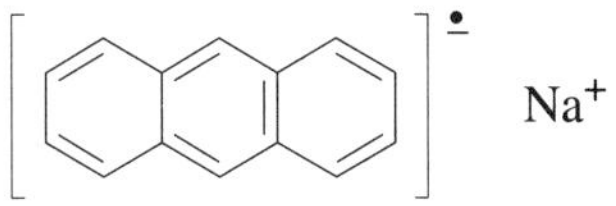

Na^+

a

p-Methoxyphenyldiphenylmethyl Ether (MMTrOR): *p*-$MeOC_6H_4(Ph)_2C$–OR (Chart 1)

Di(p-methoxyphenyl)phenylmethyl Ether (DMTrOR): (*p*-$MeOC_6H_4)_2PhC$–OR

Tri(p-methoxyphenyl)methyl Ether (TMTrOR): (*p*-$MeOC_6H_4)_3C$–OR

These were originally prepared by Khorana from the appropriate chlorotriarylmethane in pyridine,[33] but can also be prepared from the corresponding triaryl tetrafluoroborate salts (80–98% yield for primary alcohols).[34] They were developed to provide a selective protective group for the 5′-OH of nucleosides and nucleotides, which is more acid labile than the trityl group because depurination is often a problem in the acid-catalyzed removal of the trityl group.[35] The introduction of *p*-methoxy groups increases the rate of hydrolysis by about one order of magnitude for each *p*-methoxy substituent. The monomethoxy derivative has been used for the selective protection of a primary allylic alcohol over a secondary allylic alcohol (MMTr, Pyr, −10°).[36] The trimethoxy derivative is too labile for most applications, but the mono- and di-derivatives have been used extensively in the preparation of oligonucleotides and oligonucleosides. A series of triarylcarbinols has been prepared with similar acid stability, which, upon acid treatment, results in different colors. The use of these triarylcarbinols in oligonucleotide synthesis was demonstrated.[37]

Cleavage

1. For 5′-protected uridine derivatives in 80% AcOH, 20°, the times for hydrolysis were as follows:[33]

   $(p\text{-}MeOC_6H_4)_n(Ph)_mCOR$
   $n = 0, m = 3$, 48 h
   $n = 1, m = 2$, 2 h
   $n = 2, m = 1$, 15 min
   $n = 3, m = 0$, 1 min

2. MMTr–OR: 1,1,1,2,2,2-Hexafluoro-2-propanol ($pK_a = 9.3$), 75–90% yield.[38]
3. The following is an example of the use of the MMTr group in a nonnucleoside setting where the usual trityl group was too stable:[39]

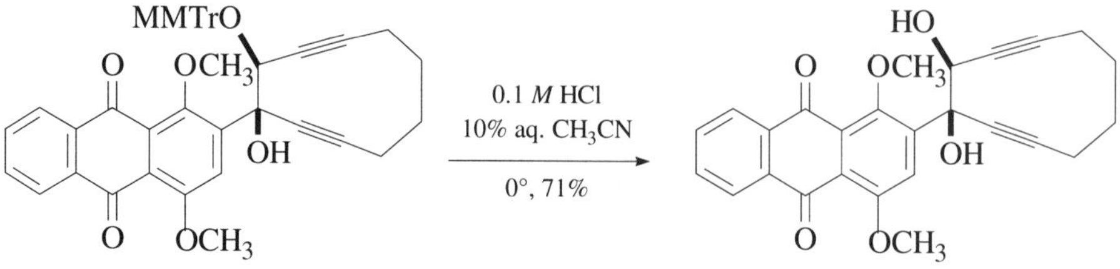

4. MMTr: Cl_2CCO_2H, Et_3SiH.[40]
5. MMTr: Sodium naphthalenide in HMPA (90% yield).[41] The MMTr group is not cleaved by sodium anthracenide, used to cleave α-naphthyldiphenylmethyl ethers.[32]
6. 3% CCl_3CO_2H in 95:5 CH_3NO_2/MeOH is recommended for removal of the DMTr group from the 5′-OH of deoxyribonucleotides because of reduced levels of depurination compared with those of CCl_3CO_2H/CH_2Cl_2, $PhSO_3H/MeOH/CH_2Cl_2$, and $ZnBr_2/CH_3NO_2$.[42]
7. MMTr: MeOH, CCl_4, ultrasound, 25–40°, 1.5–12 h, 69–100% yield.[43]

4-(4′-Bromophenacyloxy)phenyldiphenylmethyl Ether:
p-(*p*-$BrC_6H_4C(O)CH_2O)C_6H_4(Ph)_2C$–OR

This group was developed for protection of the 5′-OH group in nucleosides. The derivative is prepared from the corresponding triarylmethyl chloride and is cleaved by reductive cleavage (Zn/AcOH) of the phenacyl ether to the *p*-hydroxyphenyldiphenylmethyl ether, followed by acidic hydrolysis with formic acid.[44]

4,4′,4″-Tris(4,5-dichlorophthalimidophenyl)methyl Ether (CPTr–OR):

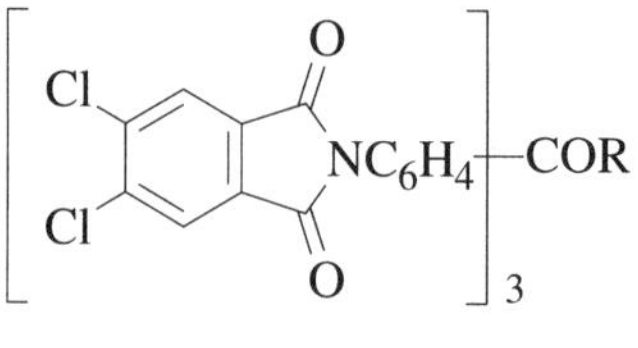

The CPTr group was developed for the protection of the 5′-OH of ribonucleosides. It is introduced with $CPTrBr/AgNO_3/DMF$ (15 min) in 80–96% yield and can be removed by ammonia followed by 0.01 *M* HCl or 80% AcOH.[45] It can also be removed with hydrazine and acetic acid.[46,47]

4,4′,4″-Tris(levulinoyloxyphenyl)methyl Ether (TLTr–OR):

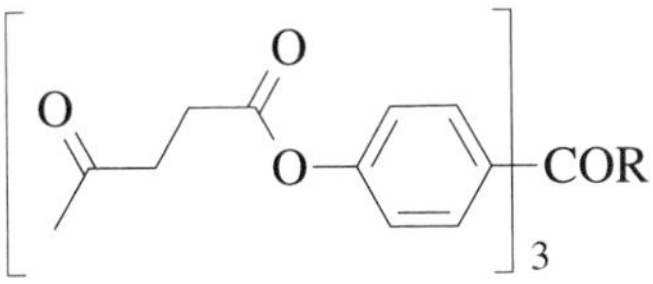

The TLTr group was developed for the protection of the 5′-OH of thymidine. It is introduced in 81% yield with TLTrBr/Pyr and is cleaved with hydrazine (3 min); Pyr–AcOH, 50°, 3 min, 81%. The $t_{1/2}$ in 80% AcOH is 24 h.[48]

4,4′,4″-Tris(benzoyloxyphenyl)methyl Ether (TBTr–OR):

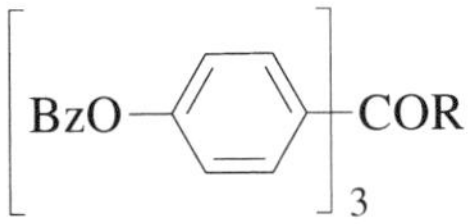

The TBTr group was prepared for 5′-OH protection in oligonucleotide synthesis. The group is introduced in >80% yield with TBTrBr/pyridine at 65°. It is five times more stable to 80% AcOH than the trityl group [$t_{1/2}$ (Tr) = 5 h; $t_{1/2}$ (TBTr) = 25 h]. The TBTr group is removed with 2 *M* NaOH. The di(4-methoxyphenyl)phenylmethyl (DMTr) group can be cleaved without affecting the TBTr derivative (80% AcOH, 95% yield).[49]

4,4′-Dimethoxy-3″-[*N*-(imidazolylmethyl)]trityl Ether (IDTr–OR):

4,4′-Dimethoxy-3″-[*N*-(imidazolylethyl)carbamoyl]trityl Ether (IETr–OR):

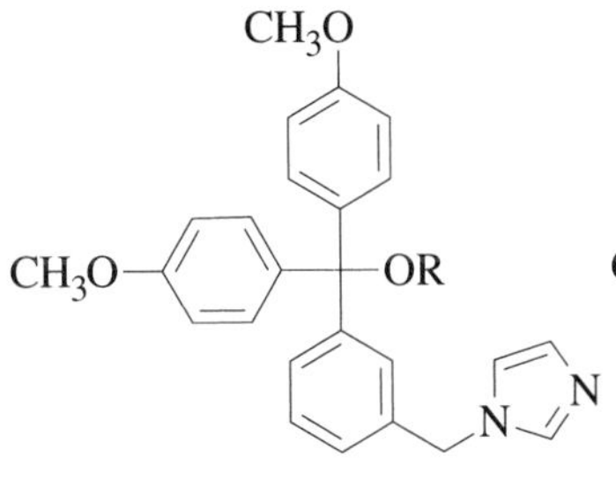

IDTr–OR

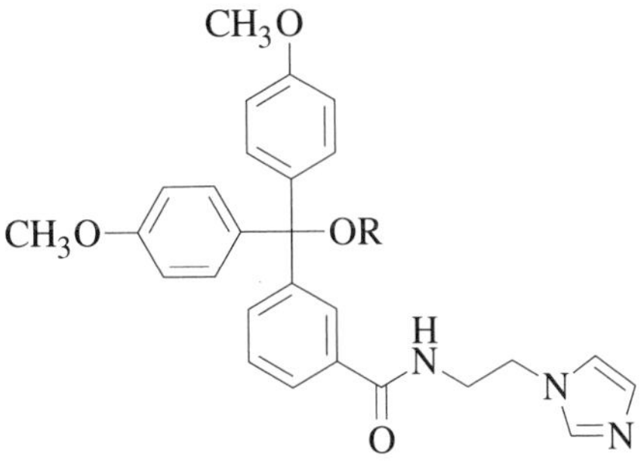

IETr–OR

The IDTr group was developed to protect the 5′-OH of deoxyribonucleotides and to increase the rate of internucleotide bond formation through participation of the pendant imidazole group. Rate enhancements of ≈350 were observed, except when (i-Pr)$_2$EtN was added to the reaction mixture, in which case reactions were complete in 30 seconds, as opposed to the usual 5–6 h without the pendant imidazole group. The group is efficiently introduced with the bistetrafluoroborate salt, IDTr–BBF in DMF (70% yield). It is removed with 0.2 M Cl_2CHCO_2H or 1% CF_3COOH in CH_2Cl_2.[50] The IETr group was developed for the same purpose, but found to be superior in its catalytic activity.[51]

1,1-Bis(4-methoxyphenyl)-1′-pyrenylmethyl Ether (Bmpm–OR):

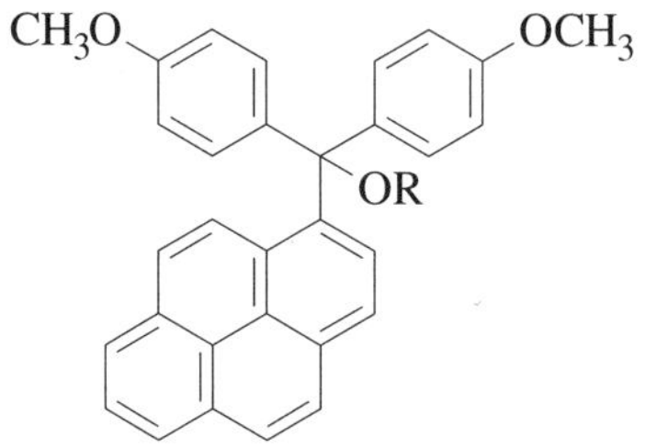

This bulky group was developed as a fluorescent, acid-labile protective group for oligonucleotide synthesis. It has properties very similar to the DMTr group except that it can be detected down to 10^{-10} M on TLC plates with 360-nm ultraviolet light.[52]

4-(17-Tetrabenzo[*a*,*c*,*g*,*i*]fluorenylmethyl)-4′,4″-dimethoxytrityl Ether (Tbf–DMTr–OR):

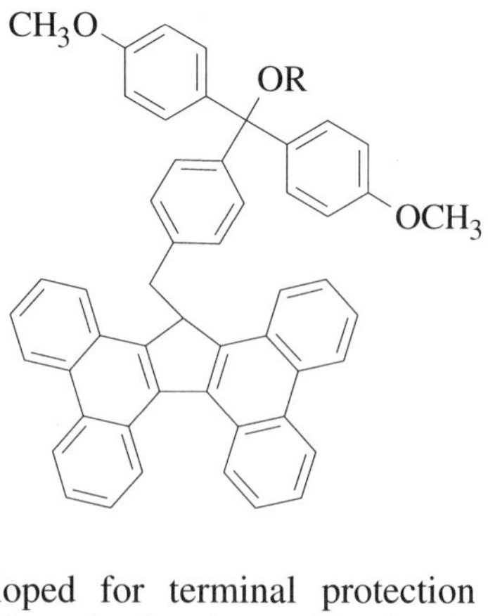

This group was developed for terminal protection of an oligonucleotide sequence for purposes of monitoring the purification by HPLC after a synthesis. It shows characteristic UV maxima at 365 and 380 nm. The group is prepared from the chloride in pyridine and can be bound directly to the support-bound oligonucleotide.[53]

9-Anthryl Ether: 9-anthryl–OR

This group is prepared by the reaction of the anion of 9-hydroxyanthracene and the tosylate of an alcohol. Since the formation of this group requires an S_N2 displacement on the alcohol to be protected, it is best suited for primary alcohols. The group is cleaved by a novel singlet oxygen reaction followed by reduction of the endoperoxide with hydrogen and Raney nickel.[54]

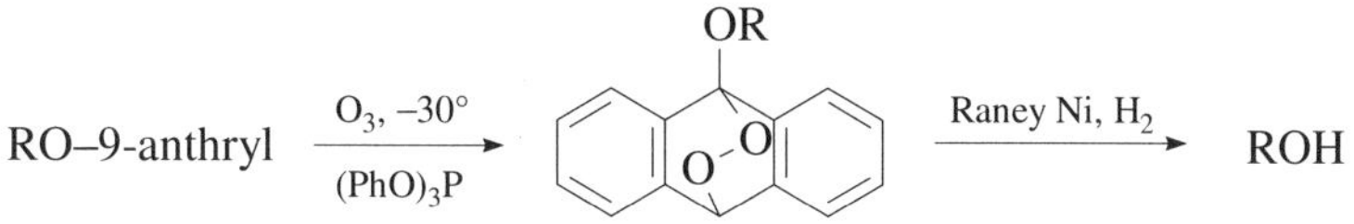

RO–9-anthryl $\xrightarrow[(PhO)_3P]{O_3,\ -30°}$ [endoperoxide, OR] $\xrightarrow{\text{Raney Ni, } H_2}$ ROH

9-(9-Phenyl)xanthenyl Ether (pixyl–OR):

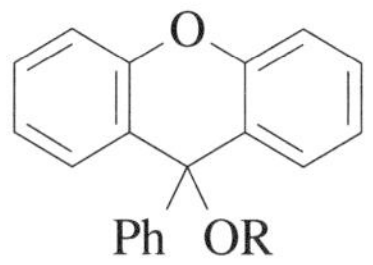

The pixyl ether is prepared from the xanthenyl chloride in 68–87% yield. This group has been used extensively in the protection of the 5′-OH of nucleosides; it is readily cleaved by acidic hydrolysis (80% AcOH, 20°, 8–15 min, 100% yield, or 3% trichloroacetic acid).[55] It can be cleaved under neutral conditions with $ZnBr_2$, thus reducing the extent of the often troublesome depurination of *N*-6-benzyloxyadenine residues during deprotection.[56] In addition, photolysis in CH_3CN/H_2O cleaves the pixyl group.[57] Acidic conditions that remove the pixyl group also partially cleave the THP group ($t_{1/2}$ for THP at 2′-OH of ribonucleoside = 560 sec in 3% Cl_2CHCO_2H/CH_2Cl_2).[58,59] The pixyl group has advantages over the trityl group in that it produces derivatives with a greater tendency to be crystalline and that the UV extinction coefficients are ~100 times greater than for the trityl group.

A series of pixyl derivatives has been prepared and the half-lives of TFA-induced cleavage determined.[60] Reaction conditions for measurement of the half-lives were TFA, CH_2Cl_2, EtOH, 22°.

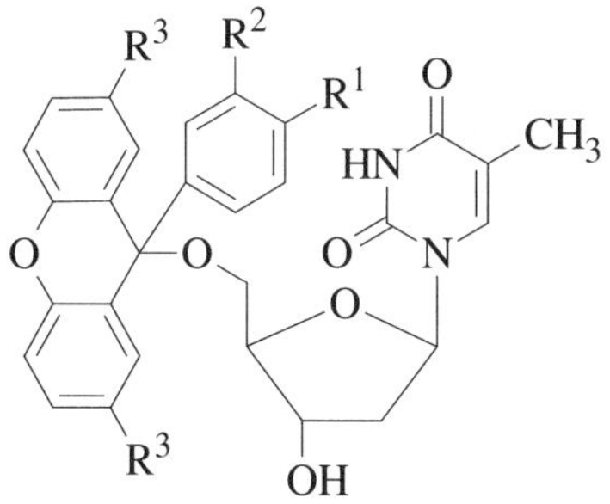

R^1	R^2	R^3	$t_{1/2}$ (min)
OMe	H	H	0.3
Me	H	H	0.55
H	H	H	1.37
H	CF_3	H	8.7
H	H	Br	244
H	CF_3	Br	1560

Under these conditions, the trityl group has an estimated $t_{1/2}$ of ~320 min.

9-(9-Phenyl-10-oxo)anthryl Ether (Tritylone Ether): (Chart 1)

O

Ph OR

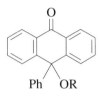

The tritylone ether is used to protect primary hydroxyl groups in the presence of secondary hydroxyl groups. It is prepared by the reaction of an alcohol with 9-phenyl-9-hydroxyanthrone under acid catalysis (cat. TsOH, benzene, reflux, 55–95% yield).[61,62] It can be cleaved under the harsh conditions of the Wolff–Kishner reduction (H_2NNH_2, NaOH, 200°, 88% yield)[32] and by electrolytic reduction (−1.4 V, LiBr, MeOH, 80–85% yield)[42]. It is stable to 10% HCl, 55 h.[32]

1. S. K. Chaudhary and O. Hernandez, *Tetrahedron Lett.*, 95 (1979).
2. S. Hanessian and A. P. A. Staub, *Tetrahedron Lett.*, 3555 (1973).
3. J. M. J. Fréchet and K. E. Haque, *Tetrahedron Lett.*, 3055 (1975); K. Barlos, D. Gatos, J. Kallitsis, G. Papaphotiu, P. Sotiriuc, Y. Wenquig, and W. Schäfer, *Tetrahedron Lett.*, **30**, 3943 (1989).
4. M. P. Reddy, J. B. Rampal, and S. L. Beaucage, *Tetrahedron Lett.*, **28**, 23 (1987).
5. S. Colin-Messager, J.-P. Girard, and J.-C. Rossi, *Tetrahedron Lett.*, **33**, 2689 (1992).
6. O. Hernandez, S. K. Chaudhary, R. H. Cox, and J. Porter, *Tetrahedron Lett.*, **22**, 1491 (1981); R. P. Srivastava and J. Hajdu, *Tetrahedron Lett.*, **32**, 6525 (1991).
7. A. V. Bhatia, S. K. Chaudhary, and O. Hernandez, *Org. Synth.*, **75**, 184 (1997).
8. P. A. Bartlett and F. R. Green, III, *J. Am. Chem. Soc.*, **100**, 4858 (1978).
9. S. Murata and R. Noyori, *Tetrahedron Lett.*, **22**, 2107 (1981).
10. M. Hirama, T. Node, S. Yasuda, and S. Ito, *J. Am. Chem. Soc.*, **113**, 1830 (1991).
11. M. Bessodes, D. Komiotis, and K. Antonakis, *Tetrahedron Lett.*, **27**, 579 (1986).
12. G. Randazzo, R. Capasso, M. R. Cicala, and A. Evidente, *Carbohydr. Res.*, **85**, 298 (1980).

13. C. Malanga, *Chem. Ind. (London)*, 856 (1987).
14. R. T. Blickenstaff, *J. Am. Chem. Soc.*, **82**, 3673 (1960).
15. M. MacCoss and D. J. Cameron, *Carbohydr. Res.*, **60**, 206 (1978).
16. E. Krainer, F. Naider, and J. Becker, *Tetrahedron Lett.*, **34**, 1713 (1993).
17. R. N. Mirrington and K. J. Schmalzl, *J. Org. Chem.*, **37**, 2877 (1972); S. Hanessian and G. Rancourt, *Pure Appl. Chem.*, **49**, 1201 (1977).
18. Y. M. Choy and A. M. Unrau, *Carbohydr. Res.*, **17**, 439 (1971).
19. V. G. Mairanovsky, *Angew. Chem., Int. Ed. Engl.*, **15**, 281 (1976).
20. M. E. Jung and L. M. Speltz, *J. Am. Chem. Soc.*, **98**, 7882 (1976).
21. B. M. Trost and L. H. Latimer, *J. Org. Chem.*, **43**, 1031 (1978).
22. H. Köster and N. D. Sinha, *Tetrahedron Lett.*, **23**, 2641 (1982).
23. A. Ichihara, M. Ubukata, and S. Sakamura, *Tetrahedron Lett.*, 3473 (1977).
24. P.-E. Sum and L. Weiler, *Can. J. Chem.*, **56**, 2700 (1978).
25. D. Cabaret and M. Wakselman, *Can. J. Chem.*, **68**, 2253 (1990).
26. P. Kovác and S. Bauer, *Tetrahedron Lett.*, 2349 (1972); S. Hanessian, N. G. Cooke, B. Dehoff, and Y. Sakito, *J. Am. Chem. Soc.*, **112**, 5276 (1990).
27. V. Kohli, H. Bloecker, and H. Koester, *Tetrahedron Lett.*, **21**, 2683 (1983).
28. T. F. S. Lampe and H. M. R. Hoffmann, *Tetrahedron Lett.*, **37**, 7695 (1996).
29. J. Lehrfeld, *J. Org. Chem.*, **32**, 2544 (1967).
30. J.-i. Asakura, M. J. Robins, Y. Asaka, and T. H. Kim, *J. Org. Chem.*, **61**, 9026 (1996).
31. G. B. Jones, B. J. Chapman, R. S. Huber, and R. Beaty, *Tetrahedron: Asymmetry*, **5**, 1199 (1994).
32. R. L. Letsinger and J. L. Finnan, *J. Am. Chem. Soc.*, **97**, 7197 (1975).
33. H. G. Khorana, *Pure Appl. Chem.*, **17**, 349 (1968); M. Smith, D. H. Rammler, I. H. Goldberg, and H. G. Khorana, *J. Am. Chem. Soc.*, **84**, 430 (1962).
34. C. Bleasdale, S. B. Ellwood, and B. T. Golding, *J. Chem. Soc., Perkin Trans. 1*, 803 (1990).
35. For a review in which the use of various trityl groups in nucleotide synthesis is discussed in the context of the phosphoramididite approach, see S. L. Beaucage and R. P. Iyer, *Tetrahedron*, **48**, 2223 (1992).
36. J. Adams and J. Rokach, *Tetrahedron Lett.*, **25**, 35 (1984).
37. E. F. Fisher and M. H. Caruthers, *Nucleic Acids Res.*, **11**, 1589 (1983).
38. N. J. Leonard and Neelima, *Tetrahedron Lett.*, **36**, 7833 (1995).
39. A. G. Myers and P. S. Dragovich, *J. Am. Chem. Soc.*, **114**, 5859 (1992).
40. V. T. Ravikumar, A. H. Krotz, and D. L. Cole, *Tetrahedron Lett.*, **36**, 6587 (1995).
41. G. L. Greene and R. L. Letsinger, *Tetrahedron Lett.*, 2081 (1975).
42. H. Takaku, K. Morita, and T. Sumiuchi, *Chem. Lett.*, 1661 (1983).
43. Y. Wang and C. McGuigan, *Synth. Commun.*, **27**, 3829 (1997).
44. A. T.-Rigby, Y.-H. Kim, C. J. Crosscup, and N. A. Starkovsky, *J. Org. Chem.*, **37**, 956 (1972).
45. M. Sekine and T. Hata, *J. Am. Chem. Soc.*, **108**, 4581 (1986).
46. M. D. Hagen, C. S.-Happ, E. Happ, and S. Chládek, *J. Org. Chem.*, **53**, 5040 (1988).
47. M. Sekine, J. Heikkilä, and T. Hata, *Bull. Chem. Soc. Jpn.*, **64**, 588 (1991).

48. M. Sekine and T. Hata, *Bull. Chem. Soc. Jpn.*, **58**, 336 (1985).
49. M. Sekine and T. Hata, *J. Org. Chem.*, **48**, 3011 (1983).
50. M. Sekine and T. Hata, *J. Org. Chem.*, **52**, 946 (1987).
51. M. Sekine, T. Mori, and T. Wada, *Tetrahedron Lett.*, **34**, 8289 (1993).
52. J. L. Fourrey, J. Varenne, C. Blonski, P. Dousset, and D. Shire, *Tetrahedron Lett.*, **28**, 5157 (1987).
53. R. Ramage and F. O. Wahl, *Tetrahedron Lett.*, **34**, 7133 (1993).
54. W. E. Barnett and L. L. Needham, *J. Chem. Soc., Chem. Commun.*, 1383 (1970); *idem, J. Org. Chem.*, **36**, 4134 (1971).
55. J. B. Chattopadhyaya and C. B. Reese, *J. Chem. Soc., Chem. Commun.*, 639 (1978).
56. M. D. Matteucci and M. H. Caruthers, *Tetrahedron Lett.*, **21**, 3243 (1980).
57. A. Misetic and M. K. Boyd, *Tetrahedron Lett.*, **39**, 1653 (1998).
58. C. Christodoulou, S. Agrawal, and M. J. Gait, *Tetrahedron Lett.*, **27**, 1521 (1986).
59. H. Tanimura and T. Imada, *Chem. Lett.*, 2081 (1990).
60. P. R. J. Gaffney, L. Changsheng, M. V. Rao, C. B. Reese, and J. C. Ward, *J. Chem. Soc., Perkin Trans. 1*, 1355 (1991); see Errata: *ibid., idem*, 1275 (1992).
61. W. E. Barnett, L. L. Needham, and R. W. Powell, *Tetrahedron*, **28**, 419 (1972).
62. C. van der Stouwe and H. J. Schäfer, *Tetrahedron Lett.*, 2643 (1979).

1,3-Benzodithiolan-2-yl Ether (Bdt–OR):

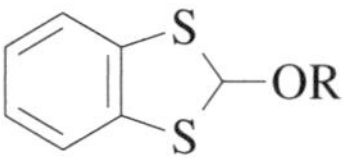

S S OR

Formation

1. BdtO–*i*-Am, H^+, dioxane, rt, 81%.[1]

2. 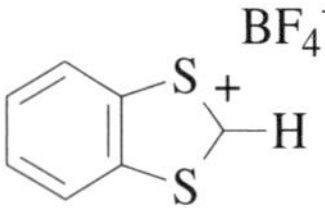 BF_4^- S + S H Pyr, CH_2Cl_2, 95%.[1] The introduction of the Bdt group proceeds under these rather neutral conditions; this proved advantageous for acid-sensitive substrates such as polyenes.[2] The Bdt group can also be reduced with Raney nickel to a methyl group or with Bu_3SnH followed by CH_3I to a [2-(methylthio)phenylthio]methyl ether (MTPM ether)[3,4] that can be cleaved with $AgNO_3$ (DMF : H_2O).[5]

Cleavage

1. 80% AcOH, 100°, 30 min.[1]
2. 2% CF_3COOH, $CHCl_3$, 0°, 20 min, 97% yield.[1]

Half-lives for Cleavage of 5′-Protected Thymidine in 80% AcOH at 15°

	DMTrT	mTHPT	Bdt-5′T	MMTrT	THPT	Bdt-3′T
$t_{1/2}$	3 min	23 min	38 min	48 min	3.5 h	2.5 h
$t_{complete}$	15 min	2.5 h	3 h	3 h	15 h	8 h

DMTrT = 5′-*O*-di-*p*-methoxytritylthymidine
mTHPT = 5′-*O*-(4-methoxytetrahydropyran-4-yl)thymidine
Bdt-5′T = 5′-*O*-(1,3-benzodithiolan-2-yl)thymidine
MMTrT = 5′-*O*-mono-*p*-methoxytritylthymidine
THPT = 5′-tetrahydropyranylthymidine
Bdt-3′T = 3′-*O*-(1,3-benzodithiolan-2-yl)thymidine

3. Dowex W50-1X, MeOH, 1.5 h, rt.[2]

1. M. Sekine and T. Hata, *J. Am. Chem. Soc.*, **105**, 2044 (1983); *idem*, *J. Org. Chem.*, **48**, 3112 (1983).
2. D. Rychnovsky and R. C. Hoye, *J. Am. Chem. Soc.*, **116**, 1753 (1994).
3. M. Sekine and T. Nakanishi, *J. Org. Chem.*, **54**, 5998 (1989).
4. M. Sekine and T. Nakanishi, *Nucleosides Nucleotides*, **11**, 679 (1992).
5. M. Sekine and T. Nakanishi, *Chem. Lett.*, 121 (1991).

Benzisothiazolyl *S,S*-Dioxido Ether

Formation/Cleavage[1]

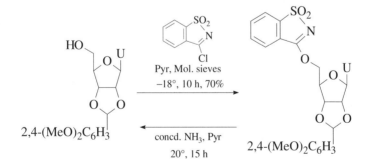

HO ... O U ... O O ... 2,4-(MeO)$_2$C$_6$H$_3$

SO$_2$ N Cl

Pyr, Mol. sieves
−18°, 10 h, 70%

concd. NH$_3$, Pyr
20°, 15 h

SO$_2$ N O ... O U ... O O ... 2,4-(MeO)$_2$C$_6$H$_3$

1. H. Sommer and F. Cramer, *Chem. Ber.*, **107**, 24 (1974).

Silyl Ethers

Silyl ethers are among the most frequently used protective groups for the alcohol function.[1] This stems largely from the fact that their reactivity (both formation and cleavage) can be modulated by a suitable choice of substituents on the silicon atom. Both steric and electronic effects are the basic controlling elements that regulate the ease of cleavage in multiply functionalized substrates. In planning selective deprotection, the steric environment around the silicon atom, as well as the environment of the protected molecular framework, must be considered. For example, it is normally quite easy to cleave a DEIPS group in the presence of a TBDMS group, but examples are known where the reverse is true. In these examples, the backbone structure provides additional steric encumbrance to reverse the selectivity. Differences in electronic factors are also used to achieve selectivity. For two alcohols of similar steric environments that have differing electron densities, the acid-catalyzed deprotection rates vary substantially and can be used to advantage. This is especially true for phenolic vs. alkyl silyl ethers: the alkyl silyl ethers are more easily cleaved by acid, and the phenolic silyl ethers are more easily cleaved by base. The reduced basicity of the silyl oxygen can be used to change the course of Lewis acid-promoted reactions and help to provide selective deprotection.[2] Electron-withdrawing substituents on the silicon atom increase its susceptibility toward basic hydrolysis, but decrease its sensitivity toward acid. For some of the more common silyl ethers, the stability toward acid increases in the order TMS (1) < TES (64) < TBDMS (20,000) < TIPS (700,000) < TBDPS (5,000,000), and the stability toward base increases in the order TMS (1) < TES (10–100) < TBDMS ~ TBDPS (20,000) < TIPS (100,000). Quantitative relationships have been developed[3] to examine the steric factors associated with nucleophilic attack on silicon and the solvolysis of silyl chlorides. Silyl ethers are also considered to be poor donor ligands for chelation-controlled reactions, and thus, their use in reactions where stereoinduction is anticipated must be carefully considered.[4] One of the properties that have made silyl groups so popular is the fact that they are easily cleaved by fluoride ion, which is attributed to the high affinity that fluoride ion has for silicon. The Si–F bond strength is 30 kcal/mol greater than the Si–O bond strength.

An excellent review is available that discusses the selective cleavage of numerous silyl derivatives.[5]

1. For a review on silylating agents, see G. van Look, G. Simchen, and J. Heberle, *Silylating Agents*, Fluka Chemie AG, 1995.
2. M. Oikawa, T. Ueno, H. Oikawa, and A. Ichihara, *J. Org. Chem.*, **60**, 5048 (1995).
3. N. Shimizu, N. Takesue, S. Yasuhara, and T. Inazu, *Chem. Lett.*, 1807 (1993); N. Shimizu, N. Takesue, A. Yamamoto, T. Tsutsumi, S. Yasuhara, and Y. Tsuno, *ibid.*, 1263 (1992).
4. L. Banfi, G. Guanti, and M. T. Zannetti, *Tetrahedron Lett.*, **37**, 521 (1996).
5. T. D. Nelson and R. D. Crouch, *Synthesis*, 1031 (1996).

Migration of Silyl Groups

Silyl groups have found broad appeal as protective groups because their reactivity and stability can be tailored by varying the nature of the substituents on the silicon. Their ability to migrate from one hydroxyl to another is a property that can be used to advantage,[1] but more often, it is a nuisance.[2] The migratory aptitude in nucleosides was found to be solvent dependent, with migration proceeding fastest in protic solvents.[3] Migration usually occurs under basic conditions and proceeds intramolecularly through a pentacoordinate silicon.[4] The TBDMS group has been observed to migrate frequently, [2b,5,6] while migration of the more stable TBDPS and TIPS[7] groups occurs less frequently. The facile migration of the TBDMS residue is a severe problem in the synthesis of oligoribonucleotides.[3,8] Conditions that favor silyl migration are the presence of a strong base in protic solvents, but migrations in aprotic solvents are also observed.[3,9] Both 1,2- [4] and 1,3-migrations[10] have been observed, but if the topological features of a molecule are properly oriented, migrations that span many atoms are observed. Such was the case during the attempted PMB ether formation in a cytovaricin synthesis, where the C-32 DEIPS group migrated to the C-17 hydroxyl. In consonance with the fact that the larger, more stable silyl groups are not as prone to migration, the corresponding TIPS analog gave only the desired C-17 PMB ether.[11]

Although silyl migrations are usually acid- or base-catalyzed, they have been observed to occur thermally.[12]

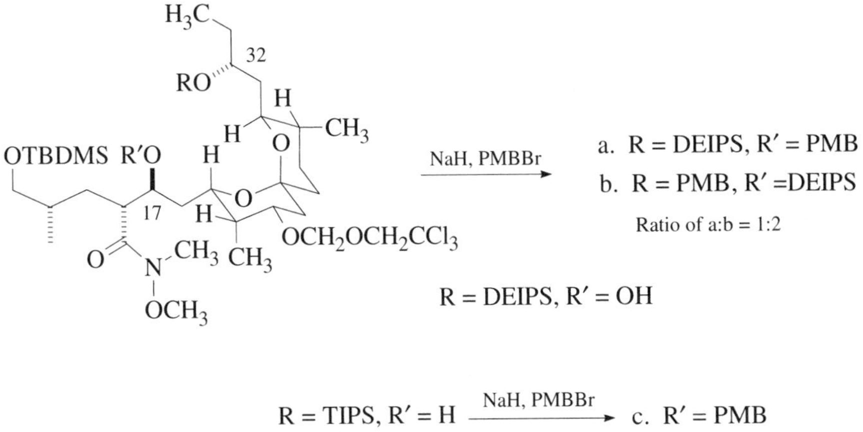

H_3C 32 RO H H CH_3 H O OTBDMS R′O H O 17 H O N CH_3 CH_3 OCH_3 $OCH_2OCH_2CCl_3$

R = DEIPS, R′ = OH $\xrightarrow{\text{NaH, PMBBr}}$ a. R = DEIPS, R′ = PMB; b. R = PMB, R′ = DEIPS

Ratio of a:b = 1:2

R = TIPS, R′ = H $\xrightarrow{\text{NaH, PMBBr}}$ c. R′ = PMB

In the well-known Brook rearrangment,[13] silyl groups migrate from oxygen to carbon, but the following example is less obvious and not necessarily predictable:[14]

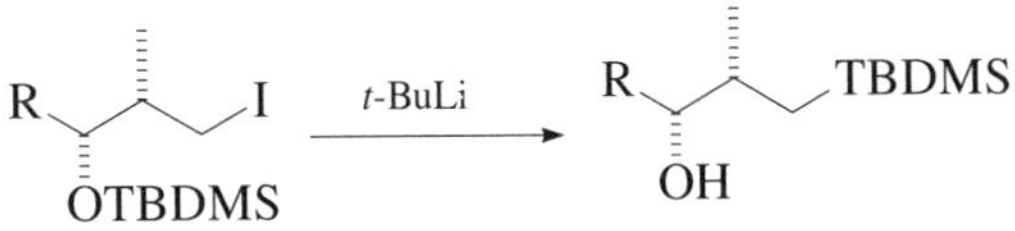

R, OTBDMS, I $\xrightarrow{t\text{-BuLi}}$ R, OH, TBDMS

Other cases of O-to-C migration have been observed.[15] This type of migration has been used to advantage in the preparation of 2-silylated benzyl alcohols.[16]

1. G. A. Molander and S. Swallow, *J. Org. Chem.*, **59**, 7148 (1994); J. M. Lassaletta and R. R. Schmidt, *Synlett*, 925 (1995).
2. (a) C. A. A. Van Boeckel, S. F. Van Aelst, and T. Beetz, *Recl: J. R. Neth. Chem. Soc.*, **102**, 415 (1983); (b) P. G. M. Wuts and S. S. Bigelow, *J. Org. Chem.*, **53**, 5023 (1988); (c) F. Franke and R. D. Guthrie, *Aust. J. Chem.*, **31**, 1285 (1978); (d) Y. Torisawa, M. Shibasaki, and S. Ikegami, *Tetrahedron Lett.*, 1865 (1979); (e) K. K. Ogilvie, S. L. Beaucage, A. L. Schifman, N. Y. Theriaul, and K. L. Sadana, *Can. J. Chem.*, **56**, 2768 (1978); (f) S. S Jones and C. B. Reese, *J. Chem. Soc., Perkin Trans. 1*, 2762 (1979).
3. K. K. Ogilvie and D. W. Entwistle, *Carbohydr. Res.*, **89**, 203 (1981).
4. J. Mulzer and B. Schöllhorn, *Angew. Chem., Int. Ed. Engl.*, **29**, 431 (1990).
5. D. Crich and T. J. Ritchie, *Carbohydr. Res.*, **197**, 324 (1990) and ref. cited therein.
6. R. W. Friesen and A. K. Daljeet, *Tetrahedron Lett.*, **31**, 6133 (1990).
7. F. Seela and T. Fröhlich, *Helv. Chim. Acta*, **77**, 399 (1994).
8. The issue of silyl migration in ribooligonucleotide synthesis has been reviewed: S. L. Beaucage and R. P. Iyer, *Tetrahedron*, **48**, 2223 (1992).
9. S. S. Jones and C. B. Reese, *J. Chem. Soc., Perkin Trans.1*, 2762 (1979); W. Köhler and W. Pfleiderer, *Liebigs Ann. Chem.*, 1855 (1979).
10. U. Peters, W. Bankova, and P. Welzel, *Tetrahedron*, **43**, 3803 (1987).
11. D. A. Evans, S. W. Kaldor, T. K. Jones, J. Clardy, and T. J. Stout, *J. Am. Chem. Soc.*, **112**, 7001 (1990).
12. M. K. Manthey, C. Gonzâlez-Bello, and C. Abell, *J. Chem. Soc., Perkin Trans. 1*, 625 (1997).
13. E. Colvin, *Silicon in Organic Synthesis*, Butterworths, Boston, Chapter 5, 1981.
14. A. B. Smith, III, Y. Qiu, D. R. Jones, and K. Kobayashi, *J. Am. Chem. Soc.*, **117**, 12011 (1995).
15. G. Simchen and J. Pfletschinger, *Angew. Chem., Int. Ed. Engl.*, **15**, 428 (1976); M. H. Hu, P. E. Fanwick, K. Wood, and M. Cushman, *J. Org. Chem.*, **60**, 5905 (1995); J. O. Karlsson, N. V. Nguyen, L. D. Foland, and H. W. Moore, *J. Am. Chem. Soc.*, **107**, 3392 (1985).
16. Y. M. Hijji, P. F. Hudrlik, C. O. Okoro, and A. M. Hudrlik, *Synth. Commun.*, **27**, 4297 (1997).

Trimethylsilyl Ether (TMS–OR): $ROSi(CH_3)_3$ (Chart 1)

A large number of silylating agents exist for the introduction of the trimethylsilyl group onto a variety of alcohols. In general, the sterically least hindered alcohols are the most readily silylated, but are also the most labile to hydrolysis with either acid or base. Trimethylsilylation is used extensively for the derivatization of most functional groups to increase their volatility for gas chromatography and mass spectrometry.

Formation

1. Me_3SiCl, Et_3N, THF, 25°, 8 h, 90% yield.[1]

2. Me_3SiCl, Li_2S, CH_3CN, 25°, 12 h, 75–95% yield.[2] Silylation occurs under neutral conditions with this combination of reagents.
3. $(Me_3Si)_2NH$, Me_3SiCl, Pyr, 20°, 5 min, 100% yield.[3] ROH is a carbohydrate. Hexamethyldisilazane (HMDS) is one of the most common silylating agents and readily silylates alcohols, acids, amines, thiols, phenols, hydroxamic acids, amides, thioamides, sulfonamides, phosphoric amides, phosphites, hydrazines, and enolizable ketones. It works best in the presence of a catalyst such as X–NH–Y, where at least one of the groups X or Y is electron withdrawing.[4] Yttrium-based Lewis acids also serve as catalysts.[5]

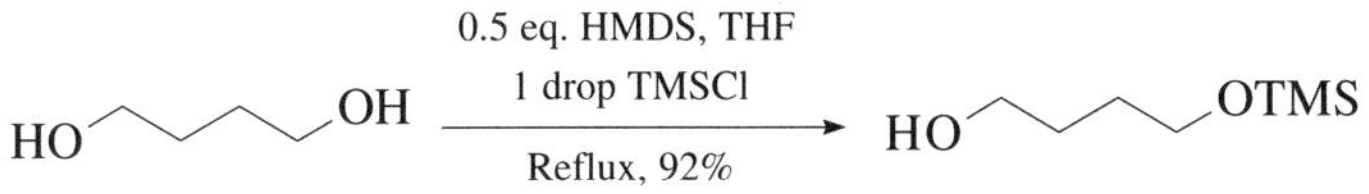

HO~~~~OH → (0.5 eq. HMDS, THF; 1 drop TMSCl; Reflux, 92%) → HO~~~~OTMS   Ref. 6

4. $(Me_3Si)_2O$, PyH^+TsO^-, PhH, mol. sieves, reflux, 4 days, 80–90% yield.[7] These mildly acidic conditions are suitable for acid-sensitive alcohols.
5. Me_3SiNEt_2.[8] Trimethylsilyldiethylamine selectively silylates equatorial hydroxyl groups in quantitative yield (4–10 h, 25°). The report indicated no reaction at axial hydroxyl groups. In the prostaglandin series, the order of reactivity of trimethylsilyldiethylamine is $C_{11} > C_{15} >> C_9$ (no reaction). These trimethylsilyl ethers are readily hydrolyzed in aqueous methanol containing a trace of acetic acid.[9] The reagent is also useful for the silylation of amino acids.[10]

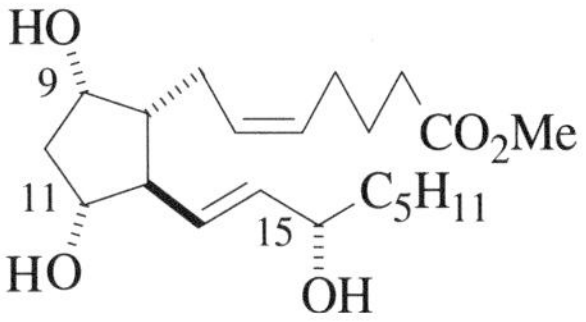

HO, 9, CO_2Me, 11, C_5H_{11}, 15, HO, OH

6. $CH_3C(OSiMe_3){=}NSiMe_3$, DMF, 78°.[11] ROH is a C_{14}-hydroxy steroid. The sterically hindered silyl ether is stable to a Grignard reaction, but is hydrolyzed with 0.1 *N* HCl/10% aq. THF, 25°.[11] The reagent also silylates amides, amino acids, phenols, carboxylic acids, enols, ureas, and imides.[12] Most active hydrogen compounds can be silylated with this reagent.
7. $Me_3SiCH_2CO_2Et$, cat. $Bu_4N^+F^-$, 25°, 1–3 h, 90% yield. This reagent combination allows the isolation of pure products under nonaqueous conditions. The reagent also converts aldehydes and ketones to trimethylsilyl enol ethers.[13] The analogous methyl trimethylsilylacetate has also been used.[14]

8. $Me_3SiNHSO_2OSiMe_3$, CH_2Cl_2, 30°, 0.5 h, 92–98% yield. Higher yields of trimethylsilyl derivatives are realized by reaction of aliphatic, aromatic, and carboxylic hydroxyl groups with *N,O*-bis(trimethylsilyl)sulfamate than by reaction with *N,O*-bis(trimethylsilyl)acetamide.[15]
9. $Me_3SiNHCO_2SiMe_3$, THF, rapid, 80–95% yield. This reagent also silylates phenols and carboxyl groups.[16]
10. $MeCH{=}C(OMe)OSiMe_3$, CH_3CN or CH_2Cl_2, 50°, 30–50 min, 83–95% yield.[17] In addition, this reagent silylates phenols, thiols, amides, and carboxyl groups.
11. $Me_3SiCH_2CH{=}CH_2$, TsOH, CH_3CN, 70–80°, 1–2 h, 90–95% yield.[18] This silylating reagent is stable to moisture. Allylsilanes can be used to protect alcohols, phenols, and carboxylic acids; there is no reaction with thiophenol, except when CF_3SO_3H[19] is used as a catalyst. The method is also applicable to the formation of *t*-butyldimethylsilyl derivatives; the silyl ether of cyclohexanol was prepared in 95% yield from allyl-*t*-butyldimethylsilane. Iodine, bromine, trimethylsilyl bromide, and trimethylsilyl iodide have also been used as catalysts.[20] Nafion-H has been shown to be an effective catalyst.[21]
12. $(Me_3SiO)_2SO_2$.[22] This is a powerful silylating reagent, but has seen little application in organic chemistry.
13. *N,O*-Bis(trimethylsilyl)trifluoroacetamide.[23] This reagent is suitable for the silylation of carboxylic acids, alcohols, phenols, amides, and ureas. It has the advantage over bis(trimethylsilyl)acetamide in that the by-products are more volatile.
14. *N,N′*-Bistrimethylsilylurea, CH_2Cl_2.[24] This reagent readily silylates carboxylic acids and alcohols. The by-product urea is easily removed by filtration.
15. Me_3SiSEt.[25] Alcohols, thiols, amines, and carboxylic acids are silylated.
16. Nafion–TMS, Et_3N, CH_2Cl_2, 100% yield.[26]
17. Isopropenyloxytrimethylsilane.[27] In the presence of an acid catalyst, this reagent silylates alcohols and phenols. It also silylates carboxylic acids without added catalyst.
18. Methyl 3-trimethylsiloxy-2-butenoate.[28] This reagent silylates primary, secondary, and tertiary alcohols at room temperature without added catalyst.
19. *N*-Methyl-*N*-trimethylsilylacetamide.[29] This reagent has been used preparatively to silylate amino acids.[30]
20. Trimethylsilyl cyanide.[31] This reagent readily silylates alcohols, phenols, carboxylic acids, and, more slowly, thiols and amines. Amides and related compounds do not react with it. The reagent has the advantage that a volatile gas (HCN is highly toxic) is the only by-product. In the following case, the use of added base resulted in retro aldol condensation:[32]

21. $Me_3SiOC(O)NMe_2$.[33] This reagent produces only volatile by-products and autocatalytically silylates alcohols, phenols, and carboxylic acids.
22. Trimethylsilylimidazole, CCl_4 or THF, rt.[34] This is a powerful silylating agent for hydroxyl groups. Basic amines are not silylated with it, but as the acidity increases, silylation can occur. TBAF has been used to catalyze trimethylsilylation with this reagent and other silylating agents of the general form $R_3SiNR'_2$.[35]
23. Trimethylsilyl trichloroacetate, K_2CO_3, 18-crown-6, 100–150°, 1–2 h, 80–90% yield.[36] This reagent silylates phenols, thiols, carboxylic acids, acetylenes, urethanes, and β-keto esters, producing CO_2 and chloroform as by-products.
24. 3-Trimethylsilyloxazolidinone.[37] This reagent can be used to silylate most active hydrogen compounds.
25. Trimethylsilyl trifluoromethanesulfonate. This is an extremely powerful silylating agent, but probably is more useful for its many other applications in synthetic chemistry.[38]

Cleavage

Trimethylsilyl ethers are quite susceptible to acid hydrolysis, but acid stability is quite dependent on the local steric environment. For example, the 17α-TMS ether of a steroid is quite difficult to hydrolyze. TMS ethers are readily cleaved with the numerous HF-based reagents. A polymer-bound ammonium fluoride is advantageous for isolation of small polar molecules.[39]

1. $Bu_4N^+F^-$, THF, aprotic conditions.[1]
2. K_2CO_3, anhydrous MeOH, 0°, 45 min, 100% yield.[40]

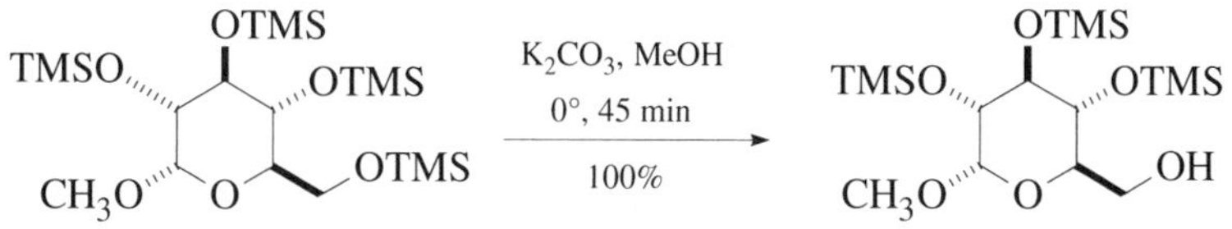

3. Citric acid, MeOH, 20°, 10 min, 100% yield.[41]
4. Rexyn 101 (polystyrenesulfonic acid), 80–91% yield.[42] This method does not cleave the *t*-butyldimethylsilyl ether.

5. $FeCl_3$, CH_3CN, rt, 1 min.[43]
6. $BF_3 \cdot Et_2O$.[44]
7. DDQ, wet EtOAc.[45]
8. RedAl.[46]
9. Direct oxidative cleavage of the TMS ether is possible with $(Ph_3SiO)_2CrO_2$, *t*-BuOOH, CH_2Cl_2, rt, 42–98% yield.[47]

1. E. J. Corey and B. B. Snider, *J. Am. Chem. Soc.*, **94**, 2549 (1972).
2. G. A. Olah, B. G. B. Gupta, S. C. Narang, and R. Malhotra, *J. Org. Chem.*, **44**, 4272 (1979).
3. C. C. Sweeley, R. Bentley, M. Makita, and W. W. Wells, *J. Am. Chem. Soc.*, **85**, 2497 (1963).
4. C. A. Bruynes and T. K. Jurriens, *J. Org. Chem.*, **47**, 3966 (1982).
5. P. Kumar, G. C. G. Pais, and A. Keshavaraja, *J. Chem. Res., Synop.*, 376 (1996).
6. R. K. Kanjolia and V. D. Gupta, *Z. Naturforsch. B*, **35B**, 767 (1980).
7. H. W. Pinnick, B. S. Bal, and N. H. Lajis, *Tetrahedron Lett.*, 4261 (1978); H. Matsumoto, Y. Hoshio, J. Nakabayashi, T. Nakano, and Y. Nagai, *Chem. Lett.*, 1475 (1980).
8. I. Weisz, K. Felföldi, and K. Kovács, *Acta Chim. Acad. Sci. Hung.*, **58**, 189 (1968).
9. E. W. Yankee, U. Axen, and G. L. Bundy, *J. Am. Chem. Soc.*, **96**, 5865 (1974); E. L. Cooper and E. W. Yankee, *J. Am. Chem. Soc.*, **96**, 5876 (1974).
10. K. Rühlmann, *J. Prakt. Chem.*, **9**, 315 (1959); K. Rühlmann, *Chem. Ber.*, **94**, 1876 (1961).
11. M. N. Galbraith, D. H. S. Horn, E. J. Middleton, and R. J. Hackney, *J. Chem. Soc., Chem. Commun.*, 466 (1968).
12. J. F. Klebe, H. Finkbeiner, and D. M. White, *J. Am. Chem. Soc.*, **88**, 3390 (1966).
13. E. Nakamura, T. Murofushi, M. Shimizu, and I. Kuwajima, *J. Am. Chem. Soc.*, **98**, 2346 (1976).
14. L. A. Paquette and T. Sugimura, *J. Am. Chem. Soc.*, **108**, 3841 (1986); T. Sugimura and L. A. Paquette, *J. Am. Chem. Soc.*, **109**, 3017 (1987).
15. B. E. Cooper and S. Westall, *J. Organomet. Chem.*, **118**, 135 (1976).
16. L. Birkofer and P. Sommer, *J. Organomet. Chem.*, **99**, C1 (1975).
17. Y. Kita, J. Haruta, J. Segawa, and Y. Tamura, *Tetrahedron Lett.*, 4311 (1979).
18. T. Morita, Y. Okamoto, and H. Sakurai, *Tetrahedron Lett.*, **21**, 835 (1980).
19. G. A. Olah, A. Husain, B. G. B. Gupta, G. F. Salem, and S. C. Narang, *J. Org. Chem.*, **46**, 5212 (1981).
20. H. Hosomi and H. Sakurai, *Chem Lett.*, 85 (1981).
21. G. A. Olah, A. Husain, and B. P. Singh, *Synthesis*, 892 (1983).
22. L. H. Sommer, G. T. Kerr, and F. C. Whitmore, *J. Am. Chem. Soc.*, **70**, 445 (1948).
23. D. L. Stalling, C. W. Gehrke, and R. W. Zumalt, *Biochem. Biophys. Res. Commun.*,

31, 616 (1968); M. G. Horning, E. A. Boucher, and A. M. Moss, *J. Gas Chromatog.*, **5**, 297 (1967).
24. W. Verboom, G. W. Visser, and D. N. Reinhoudt, *Synthesis*, 807 (1981).
25. E. W. Abel, *J. Chem. Soc.*, 4406 (1960); *idem, ibid.*, 4933 (1961).
26. S. Murata and R. Noyori, *Tetrahedron Lett.*, **21**, 767 (1980).
27. M. Donike and L. Jaenicke, *Angew. Chem., Int. Ed. Engl.*, **8**, 974 (1969).
28. T. Veysoglu and L. A. Mitscher, *Tetrahedron Lett.*, **22**, 1303 (1981).
29. L. Birkofer and M. Donike, *J. Chromatogr.*, **26**, 270 (1967).
30. H. R. Kricheldorf, *Justus Liebigs Ann. Chem.*, **763**, 17 (1972).
31. K. Mai and G. Patil, *J. Org. Chem.*, **51**, 3545 (1986).
32. E. J. Corey and Y.-J. Wu, *J. Am. Chem. Soc.*, **115**, 8871 (1993).
33. D. Knausz, A. Meszticzky, L. Szakacs, B. Csakvari, and K. D. Ujszaszy, *J. Organomet. Chem.*, **256**, 11 (1983); D. Knausz, A. Meszticzky, L. Szakacs, and B. Csakvari, *J. Organomet. Chem.*, **268**, 207 (1984).
34. S. Torkelson and C. Ainsworth, *Synthesis* 722 (1976).
35. Y. Tanabe, M. Murakami, K. Kitaichi, and Y. Yoshida, *Tetrahedron Lett.*, **35**, 8409 (1994); Y. Tanabe, H. Okumura, A. Maeda, and M. Murakami, *ibid.*, **35**, 8413 (1994).
36. J. M. Renga and P.-C. Wang, *Tetrahedron Lett.*, **26**, 1175 (1985).
37. C. Palomo, *Synthesis*, 809 (1981); J. M. Aizpurua, C. Palomo, and A. L. Palomo, *Can. J. Chem.*, **62**, 336 (1984).
38. Review: H. Emde, D. Domsch, H. Feger, U. Frick, A. Götz, H. H. Hergott, K. Hofmann, W. Kober, K. Krägeloh, T. Oesterle, W. Steppan, W. West, and G. Simchen, *Synthesis*, 1 (1982).
39. C. Li, Y. Lu, W. Huang, and B. He, *Synth. Commun.*, **21**, 1315 (1991).
40. D. T. Hurst and A. G. McInnes, *Can. J. Chem.*, **43**, 2004 (1965).
41. G. L. Bundy and D. C. Peterson, *Tetrahedron Lett.*, 41 (1978).
42. R. A. Bunce and D. V. Hertzler, *J. Org. Chem.*, **51**, 3451 (1986).
43. A. D. Cort, *Synth. Commun.*, **20**, 757 (1990).
44. L. Pettersson and T. Frejd, *J. Chem. Soc., Chem. Commun.*, 1823 (1993).
45. A. Oku, M. Kinugasa, and T. Kamada, *Chem. Lett.*, 165 (1993).
46. S.-H. Chen, V. Farina, D. M. Vyas, T. W. Doyle, B. H. Long, and C. Fairchild, *J. Org. Chem.*, **61**, 2065 (1996).
47. J. Muzart and A. N. Ajjou, *Synlett*, 497 (1991).

Triethylsilyl Ether (TES–OR): Et_3SiOR

Formation

1. Et_3SiCl, Pyr. Triethylsilyl chloride is by far the most common reagent for the introduction of the TES group.[1] Silylation also occurs with imidazole and DMF,[2] and with dimethylaminopyridine as a catalyst.[3] Phenols,[4] carboxylic acids,[5] and amines[6] have also been silylated with TESCl.

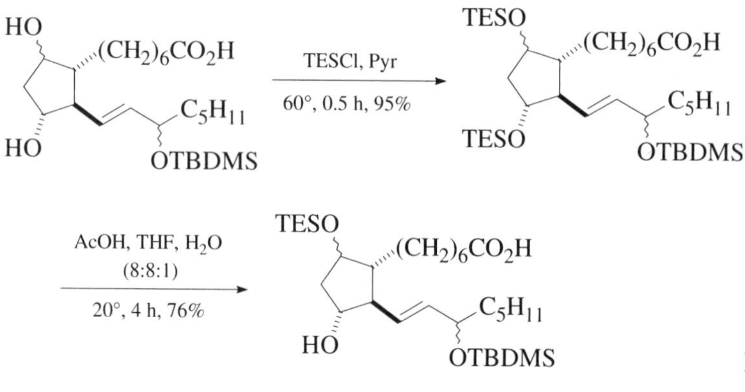

HO, HO, $(CH_2)_6CO_2H$, C_5H_{11}, OTBDMS — TESCl, Pyr, 60°, 0.5 h, 95% → TESO, TESO, $(CH_2)_6CO_2H$, C_5H_{11}, OTBDMS

AcOH, THF, H_2O (8:8:1), 20°, 4 h, 76% → TESO, HO, $(CH_2)_6CO_2H$, C_5H_{11}, OTBDMS

Ref. 3

More acidic conditions [AcOH, THF, H_2O (6:1:3), 45°, 3 h] cleave all the protective groups, 76% yield.

2. *N*-Methyl-*N*-triethylsilyltrifluoroacetamide.[7]
3. Allyltriethylsilane.[8]
4. Triethylsilane, CsF, imidazole.[9]
5. Triethylsilane, CH_2Cl_2, 1% $Rh_2(pfb)_4$ (rhodium perfluorobutyrate), 2 h, 88% yield.[10]
6. *N*-Triethylsilylacetamide.[11]
7. Triethylsilyldiethylamine.[12]
8. 1-Methoxy-1-triethylsiloxypropene.[13]
9. 1-Methoxy-2-methyl-1-triethylsiloxypropene.[14]
10. Triethylsilyl triflate.[15]
11. Triethylsilyl perchlorate.[16] This reagent represents an **explosion** hazard.
12. Triethylsilyl cyanide.[17]

Cleavage

The triethylsilyl ether is approximately 10–100 times more stable[5] than the TMS ether and thus shows a greater stability to many reagents. Although TMS ethers can be cleaved in the presence of TES ethers, steric factors will play an important role in determining selectivity. The TES ether can be cleaved in the presence of a *t*-butyldimethylsilyl ether using 2% HF in acetonitrile.[18] In general, methods used to cleave the TBDMS ether are effective for cleavage of the TES ether.[19]

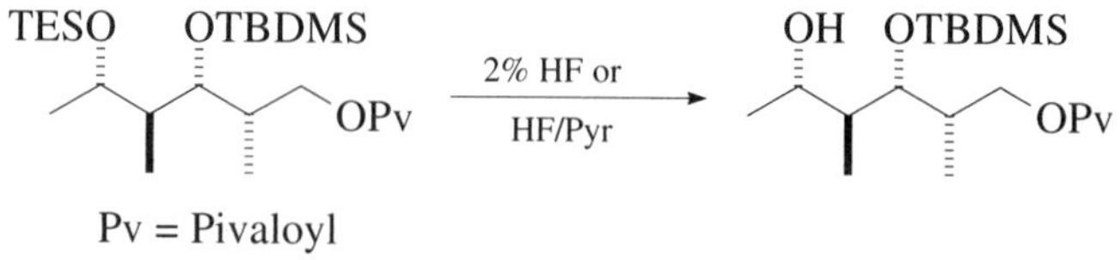

TESO, OTBDMS, OPv — 2% HF or HF/Pyr → OH, OTBDMS, OPv

Pv = Pivaloyl

1. DDQ, CH_3CN or THF, H_2O, 86–100% yield.[20] TBDMS ethers are not cleaved.

1. T. W. Hart, D. A. Metcalfe, and F. Scheinmann, *J. Chem. Soc., Chem. Commun.*, 156 (1979).
2. W. Oppolzer, R. L. Snowden, and D. P. Simmons, *Helv. Chim. Acta*, **64**, 2002 (1981).
3. W. R. Roush and S. Russo-Rodriquez, *J. Org. Chem.*, **52**, 598 (1987).
4. T. L. McDonald, *J. Org. Chem.*, **43**, 3621 (1978).
5. C. E. Peishoff and W. L. Jorgensen, *J. Org. Chem.*, **48**, 1970 (1983).
6. R. West, P. Nowakowski, and P. Boudjouk, *J. Am. Chem. Soc.*, **98**, 5620 (1976).
7. M. Donike and J. Zimmermann, *J. Chromatogr.*, **202**, 483 (1980).
8. A. Hosomi and H. Sakurai, *Chem. Lett.*, 85 (1981).
9. L. Horner and J. Mathias, *J. Organomet. Chem.*, **282**, 175 (1985).
10. M. P. Doyle, K. G. High, V. Bagheri, R. J. Pieters, P. J. Lewis, and M. M. Pearson, *J. Org. Chem.*, **55**, 6082 (1990).
11. J. Dieckman and C. Djerassi, *J. Org. Chem.*, **32**, 1005 (1967); J. Dieckman, J. B. Thompson, and C. Djerassi, *ibid.*, **32**, 3904 (1967).
12. A. R. Bassindale and D. R. M. Walton, *J. Organomet. Chem.*, **25**, 389 (1970).
13. Y. Kita, J. Haruta, J. Segawa, and Y. Tamura, *Tetrahedron Lett.*, 4311 (1979).
14. E. Yoshii and K. Takeda, *Chem. Pharm. Bull.*, **31**, 4586 (1983).
15. C. H. Heathcock, S. D. Young, J. P. Hagen, R. Pilli, and U. Badertscher, *J. Org. Chem.*, **50**, 2095 (1985).
16. T. J. Barton and C. R. Tully, *J. Org. Chem.*, **43**, 3649 (1978); D. B. Collum, J. H. McDonald, III, and W. C. Still, *J. Am. Chem. Soc.*, **102**, 2117 (1980). For *O*-silylation of esters, see C. S. Wilcox and R. E. Babston, *Tetrahedron Lett.*, **25**, 699 (1984).
17. K. Mai and P. Patil, *J. Org. Chem.*, **51**, 3545 (1986).
18. D. Boschelli, T. Takemasa, Y. Nishitani, and S. Masamune, *Tetrahedron Lett.*, **26**, 5239 (1985).
19. For an extensive review on selective silyl ether cleavage, see T. D. Nelson and R. D. Crouch, *Synthesis*, 1031 (1996).
20. K. Tanemura, T. Suzuki, and T. Horaguchi, *J. Chem. Soc., Perkin Trans. 1*, 2997 (1992).

Triisopropylsilyl Ether (TIPS–OR):[1] $(i\text{-Pr})_3SiOR$ (Chart 1)

The greater bulkiness of the TIPS group makes it more stable than the *t*-butyldimethylsilyl (TBDMS) group, but not as stable as the *t*-butyldiphenylsilyl (TBDPS) group, to acidic hydrolysis. The TIPS group is more stable to basic hydrolysis than the TBDMS group or the TBDPS group.[2] Introduction of the TIPS group onto primary hydroxyls proceeds selectively over secondary hydroxyls.[3] The TIPS group has been used to prevent chelation with Grignard reagents during additions to carbonyls.[4]

Formation

1. TIPSCl, imidazole, DMF, 82% yield.[2]
2. TIPSCl, imidazole, DMAP[5] or TEA,[6] CH_2Cl_2.

3. TIPSCl, pyridine, $AgNO_3$ or $Pb(NO_3)_2$, > 90% yield.[7] These conditions cleanly introduce the hindered TIPS group onto the 3′-position of thymidine.
4. TIPSH, CsF, imidazole.[8]
5. $TIPSOSO_2CF_3$, 2,6-lutidine,[9] TEA or DIPEA[10], CH_2Cl_2.
6. The sluggishness of the reaction of TIPSOTf with tertiary alcohols can be exploited to advantage, as was the case in Magnus' synthesis of strychnine. The equilibrium favors the tertiary hemiketal, but silylation favors the primary alcohol.

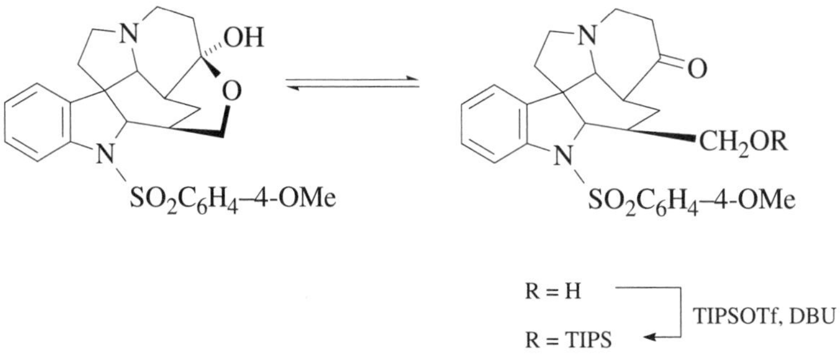

R = H → R = TIPS (TIPSOTf, DBU)

Ref. 11

Cleavage

1. 0.01 *N* HCl, EtOH, 90°, 15 min, 100% yield.[2] HCl in a variety of other concentrations has also been used to cleave the TIPS ether.[11]
2. HCl, EtOAc, −30° → 0°.[12]
3. HF, CH_3CN.[13] In certain sensitive substrates it may be advisable to run this reaction in a polypropylene vessel, as was the case in Schreiber's synthesis of FK-506, where the yield increased from 35% to 73% after switching from the standard glass vessel.[14] This is presumably because of the products formed when HF reacts with glass.
4. 40% aq. HF in THF.[15]
5. Pyr–HF, THF.[16]
6. $Bu_4N^+F^-$, THF.[17]
7. 80% AcOH, H_2O.[18]
8. TFA, THF, H_2O.[19]
9. 40% KOH, MeOH, reflux, 18 h.[20]
10. NO_2BF_4.[21]

1. For an extensive review of the chemistry of the triisopropylsilyl group, see C. Rücker, *Chem. Rev.*, **95**, 1009 (1995).
2. R. F. Cunico and L. Bedell, *J. Org. Chem.*, **45**, 4797 (1980).

3. K. K. Ogilvie, E. A. Thompson, M. A. Quilliam, and J. B. Westmore, *Tetrahedron Lett.*, 2865 (1974)
4. S. V. Frye and E. L. Eliel, *Tetrahedron Lett.*, **27**, 3223 (1986).
5. M. Ohwa and E. L. Eliel, *Chem. Lett.*, 41 (1987).
6. P. R. Maloney and F. G. Fang, *Tetrahedron Lett.*, **35**, 2823 (1994).
7. S. Nishino, Y. Nagato, H. Yamamoto, and Y. Ishido, *J. Carbohydr. Chem.*, **5**, 199 (1986).
8. L. Horner and J. Mathias, *J. Organomet. Chem.*, **282**, 175 (1985).
9. E. J. Corey, H. Cho, C. Rücker, and D. H. Hua, *Tetrahedron Lett.*, **22**, 3455 (1981); K. Tanaka, H. Yoda, Y. Isobe, and A. Kaji, *J. Org. Chem.*, **51**, 1856 (1986).
10. D. W. Knight, D. Shaw, and G. Fenton, *Synlett*, 295 (1994).
11. P. Magnus, M. Giles, R. Bonnert, G. Johnson, L. McQuire, M. Deluca, A. Merritt, C. S. Kim, and N. Vicker, *J. Am. Chem. Soc.*, **115**, 8116 (1993); P. Magnus, M. Giles, R. Bonnert, C. S. Kim, L. McQuire, A. Merritt, and N. Vicker, *ibid.*, **114**, 4403 (1992); H. Yoda, K. Shirakawa, and K. Takabe, *Tetrahedron Lett.*, **32**, 3401 (1991).
12. W.-R. Li, W. R. Ewing, B. D. Harris, and M. M. Joullie, *J. Am. Chem. Soc.*, **112**, 7659 (1990).
13. J. L. Mascareñas, A. Mouriño, and L. Castedo, *J. Org. Chem.*, **51**, 1269 (1986).
14. M. Nakatsuka, J. A. Ragan, T. Sammakia, D. B. Smith, D. B. Uehling, and S. L. Schreiber, *J. Am. Chem. Soc.*, **112**, 5583 (1990).
15. J. Cooper, D. W. Knight, and P. T. Gallagher, *J. Chem. Soc., Chem. Commun.*, 1220 (1987).
16. P. Wipf and H. Kim, *J. Org. Chem.*, **58**, 5592 (1993).
17. J. C.-Y. Cheng, U. Hacksell, and G. P. Daves, Jr., *J. Org. Chem.*, **51**, 4941 (1986).
18. K. K. Ogilvie, S. L. Beaucage, D. W. Entwistle, E. A. Thompson, M. A. Quilliam, and J. B. Westmore, *J. Carbohyd., Nucleosides, Nucleotides*, **3**, 197 (1976).
19. C. Eisenberg and P. Knochel, *J. Org. Chem.*, **59**, 3760 (1994).
20. L. E. Overman and S. R. Angle, *J. Org. Chem.*, **50**, 4021 (1985).
21. N. Hussain, D. O. Morgan, C. R. White, and J. A. Murphy, *Tetrahedron Lett.*, **35**, 5069 (1994).

Dimethylisopropylsilyl Ether (IPDMS–OR): $ROSiMe_2$–*i*-Pr (Chart 1)

Formation

1. (*i*-$PrMe_2Si)_2NH$, *i*-$PrMe_2SiCl$, 25°, 48 h, 98% yield.[1]
2. *i*-$PrMe_2SiCl$, imidazole, DMF, 26°, 2 h, 65% yield.[2]

Cleavage

1. AcOH/H_2O, (3:1), 35°, 10 min, 100% yield.[1] An IPDMS ether is more easily cleaved than a THP ether. It is not, however, stable to Grignard or Wittig reactions or to Jones oxidation.

1. E. J. Corey and R. K. Varma, *J. Am. Chem. Soc.*, **93**, 7319 (1971).
2. K. Toshima, K. Tatsuta, and M. Kinoshita, *Tetrahedron Lett.*, **27**, 4741 (1986).

Diethylisopropylsilyl Ether (DEIPS–OR): ROSiEt$_2$–*i*-Pr

This group is more labile to hydrolysis than the TBDMS group and has been used to protect an alcohol when the TBDMS group was too resistant to cleavage. The DEIPS group is ≈90 times more stable than the TMS group to acid hydrolysis and 600 times more stable than the TMS group to base-catalyzed solvolysis.

Formation

1. Diethylisopropylsilyl chloride, imidazole, CH_2Cl_2, 25°, 1 h.[1]
2. Et$_2$–*i*-PrSiOTf, CH_2Cl_2, 2,6-lutidine, rt.[2]

Cleavage

1. 3:1:3 AcOH, H_2O, THF.[1] Any of the methods used to cleave the TBDMS ether also cleave the DEIPS ether.

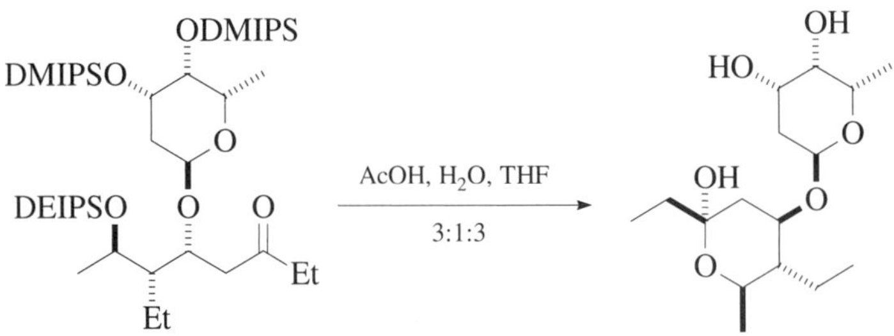

2. AcOH, KF·HF, THF, H_2O, 30°, 46 h, 94% yield.[3] These conditions did not affect a secondary OTBDMS group.
3. H_2, Pd(OH)$_2$.[4] When the cleavage is performed in dioxane, the DEIPS group is stable and benzyl ethers are selectively removed, whereas if MeOH is use as solvent, both the DEIPS and the benzyl ether are cleaved.
4. RMgX.[5]
5. HF–Pyr, Pyr, THF, 74% yield.[6]

1. K. Toshima, K. Tatsuta, and M. Kinoshita, *Tetrahedron Lett.*, **27**, 4741 (1986).
2. K. Toshima, S. Mukaiyama, M. Kinoshita, and K. Tatsuta, *Tetrahedron Lett.*, **30**, 6413 (1989).
3. K. Toshima, M. Misawa, K. Ohta, K. Tatsuta, and M. Kinoshita, *Tetrahedron Lett.*, **30**, 6417 (1989).
4. K. Toshima, K. Yanagawa, S. Mukaiyama, and K. Tatsuta, *Tetrahedron Lett.*, **31**, 6697 (1990).

5. Y. Watanabe, T. Fujimoto, and S. Ozaki, *J. Chem. Soc., Chem. Commun.*, 681 (1992).
6. D. A. Evans, S. W. Kaldor, T. K. Jones, J. Clardy, and T. J. Stout, *J. Am. Chem. Soc.*, **112**, 7001 (1990).

Dimethylthexylsilyl Ether (TDS–OR): $(CH_3)_2CHC(CH_3)_2Si(CH_3)_2OR$

Both TDSCl and $TDSOSO_2CF_3$ are used to introduce the TDS group. In general, conditions similar to those employed to introduce the TBDMS group are effective. This group is slightly more hindered than the TBDMS group, and the chloride has the advantage of being a liquid, which is useful when one handles large quantities of material. Cleavage of the group can be accomplished with the same methods used to cleave the TBDMS group, but is two to three times slower because of the increased steric bulk of the group.[1] A disadvantage is that the NMR spectrum is not as simple as in the case when the similar TBDMS group is used.

1. H. Wetter and K. Oertle, *Tetrahedron Lett.*, **26**, 5515 (1985).

t-Butyldimethylsilyl Ether (TBDMS–OR): *t*-$BuMe_2SiOR$ (Chart 1)

The TBDMS ether has become one of the most popular silyl protective groups used in chemical synthesis. It is easily introduced with a variety of reagents, has the virtue of being quite stable to a variety of organic reactions, and is readily removed under conditions that do not attack other functional groups. It is approximately 10^4 times more stable to basic hydrolysis than the trimethylsilyl (TMS) group. It has excellent stability toward base, but is relatively sensitive to acid. The ease of introduction and removal of the TBDMS ether are influenced by steric factors that often allow for its selective introduction in polyfunctionalized, sterically differentiated molecules. It is relatively easy to introduce a primary TBDMS group in the presence of a secondary alcohol. One problem that has been encountered with the TBDMS group is that it can be metalated on the silyl methyl with *t*-BuLi.[1]

Formation

1. TBDMSCl, imidazole, DMF, 25°, 10 h, high yields.[2] This is the most common method for the introduction of the TBDMS group on alcohols with low steric demand. The method works best when the reactions are run in very concentrated solutions. This combination of reagents also silylates phenols,[3] hydroperoxides,[4] and hydroxylamines.[5] Thiols, amines, and carboxylic acids are not effectively silylated under these conditions.[6] Tertiary alcohols can be silylated with the phosphoramidate

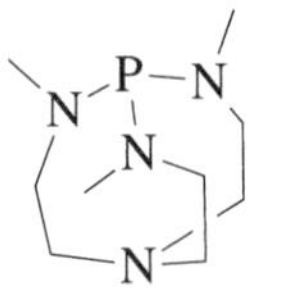

i

catalyst *i*.[7] Although silylation using these conditions normally proceeds uneventfully, the following scheme shows that reactions are not always straight forward.[8]

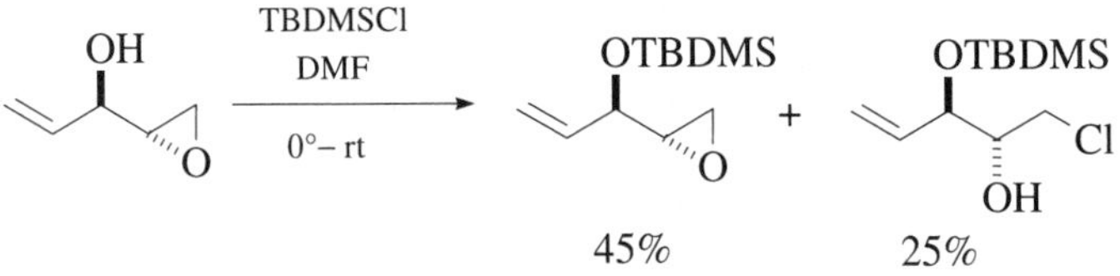

TBDMSCl, DMF, 0°– rt → OTBDMS (45%) + OTBDMS, Cl, OH (25%)

2. TBDMSCl, Li_2S, CH_3CN, 25°, 5–8 h, 75–95% yield.[9] This reaction occurs under nearly neutral conditions.
3. TBDMSCl, DMAP, Et_3N, DMF, 25°, 12 h.[10] These conditions were used to silylate selectively a primary over a secondary alcohol.[11] Besides DMAP, other catalysts, such as 1,1,3,3-tetramethylguanidine,[12] 1,8-diazabicyclo[5.4.0]undec-7-ene (83–99%),[6] 1,5-diazabicyclo[4.3.0]non-5-ene,[13] and ethyldiisopropylamine, have also been used.[14]
4. TBDMSCl, KH, 18-crown-6, THF, 0°–rt, 78% yield.[15] This combination of reagents is very effective in silylating extremely hindered alcohols.
5. 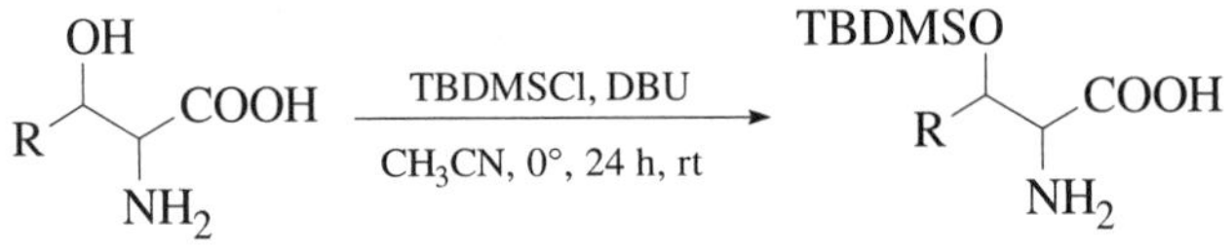 R–CH(OH)–CH(NH_2)–COOH → (TBDMSCl, DBU; CH_3CN, 0°, 24 h, rt) → R–CH(OTBDMS)–CH(NH_2)–COOH

   These conditions were chosen specifically to facilitate the silylation of hydroxylated amino acids.[16]
6. (a) Bu_2SnO, MeOH, (b) TBDMSCl, CH_2Cl_2. These conditions selectively protect the equatorial alcohol of a *cis*-diol on a pyranoside ring.[17] In the case of *β*-lactosides, the primary TBDMS ether is formed in 96% yield.[18] Butane-1,2,4-triol shows unusual selectivity in that the stannylene methods give the 4-TBDMS derivative, whereas benzylation, acetylation, and tosylation give the 1-substituted derivatives.[19]
7. $TBDMSOClO_3$, CH_3CN, Pyr, 20 min, 100% yield.[20] This reagent works well, but has the disadvantage of being **explosive**; it has therefore been supplanted by $TBDMSOSO_2CF_3$.
8. $TBDMSOSO_2CF_3$, CH_2Cl_2, 2,6-lutidine, 0–25°.[21] This is one of the most powerful methods for introducing the TBDMS group. Other bases, such as triethylamine,[22] ethyldiisopropylamine,[23] and pyridine,[24] have also

been used successfully. In the presence of an ester or ketone, it is possible simultaneously to form a silyl enol ether while silylating a hydroxyl group.[20] Not all protections proceed as expected, as illustrated with the following glutarimide:[25]

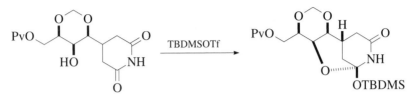

PvO ... HO ... NH ... O —TBDMSOTf→ PvO ... H ... NH ... OTBDMS

9. $TBDMSCH_2CH{=}CH_2$, TsOH, CH_3CN, 70–80°, 2.5 h, 95% yield.[26]
10. 4-*t*-Butyldimethylsiloxy-3-penten-2-one, DMF, TsOH, rt, 83–92% yield.[27]
11. 1-(*t*-Butyldimethylsilyl)imidazole.[28,29]
12. *N*-*t*-Butyldimethylsilyl-*N*-methyltrifluoroacetamide, CH_3CN, 5 min, 97–100% yield.[30] This reagent also silylates thiols, amines, amides, carboxylic acids, and enolizable carbonyl groups.
13. 1-(*t*-Butyldimethylsiloxy)-1-methoxyethene, CH_3CN, 91–100% yield.[31] This reagent also silylates thiols and carboxylic acids.
14. TBDMSCN, 80°, 5 min, 95% yield.[32]
15. A secondary alcohol was selectively protected in the presence of a secondary allylic alcohol with TBDMSOTf, 2,6-lutidine at −78°.[33]

OH, H, OH, OTMS —TBDMSOTf, −78°, 2,6-lutidine, CH_2Cl_2, 91%→ OH, H, OTBDMS, OTMS

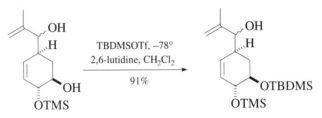

t-Butyl or *t*-amyl ethers are converted to TBDMS ethers with this reagent. If the lutidine is not present, cleavage to the alcohol occurs.[34] Silyl migration has been observed during the protection of an alcohol with a proximal silyl ether when using TBDMSOTf-2,6-lutidine.[35]

TESO OH ... O —TBDMSOTf, 2,6-lutidine→ TESO OTBDMS ... O + TBDMSO TESO ... O

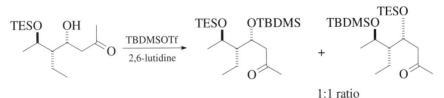

1:1 ratio

16. From a THP ether: TBDMSOTf, Me_2S, CH_2Cl_2, −50°, 24–97% yield. Allylic THP ethers are converted inefficiently.[36]

17. From a THP ether: TBDMSH, CH_2Cl_2, $Sn(OTf)_2$, rt, 1 h, 78% yield. TIPS ethers are prepared analogously.[37]
18. $TBDMSONO_2$.[38]
19. *N,N*-Bis-TBDMSdimethylhydantoin, cat. TBAF.[39] Primary alcohols are selectively protected.
20. $CH_3C(OTBDMS)$=NTBDMS, TBAF, NMP (*N*-methylpyrrolidinone), 76–99% yield.[40]
21. TBDMSH, 10% Pd/C.[41]
22. TBDMSOH, Ph_3P, DEAD, THF, −78°, 68–85% yield.[42]
23. TBDMSH, THF, TBAF, rt, 1 h, 97% yield. Other silanes react similarly.[43]
24. The following schemes represent some interesting examples in which the TBDMS group is introduced selectively on compounds with more than one alcohol:

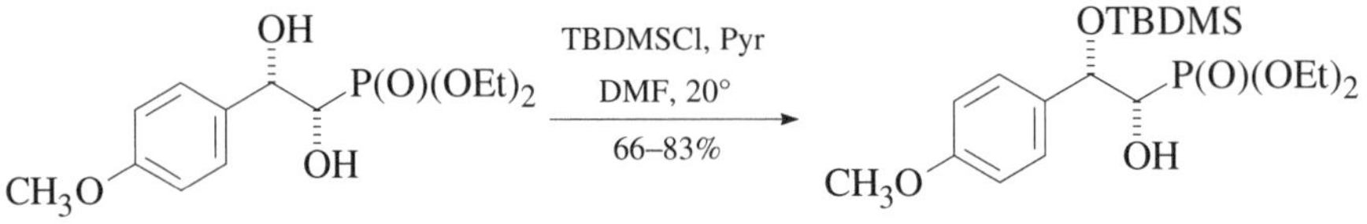

CH_3O–C_6H_4–CH(OH)–CH(OH)–$P(O)(OEt)_2$ —(TBDMSCl, Pyr, DMF, 20°; 66–83%)→ CH_3O–C_6H_4–CH(OTBDMS)–CH(OH)–$P(O)(OEt)_2$

Ref. 44

HO, HO, SO_2Ph —(TBDMSCl, Im, DMF, >79%)→ HO, TBDMSO, SO_2Ph

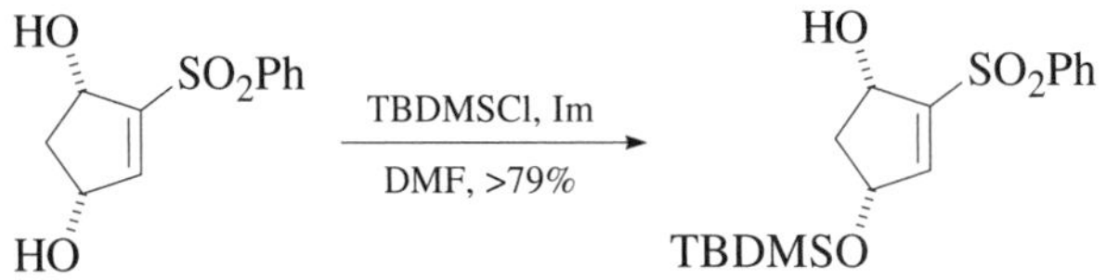

Ref. 45

From these two examples, it appears that, with the reagent TBDMSCl-Im-DMF, the acidity of the alcohol plays an important role in determining the regiochemical preference of hydroxyl protection.

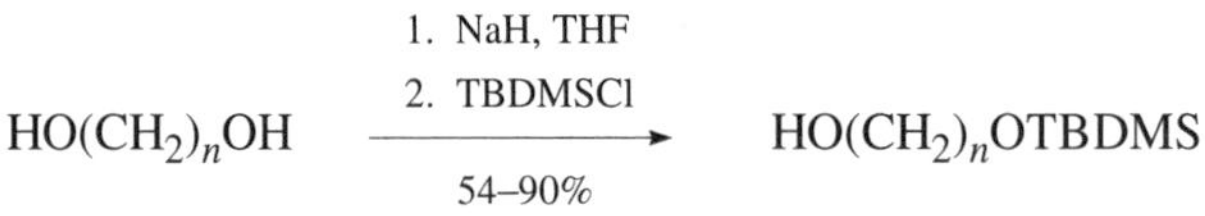

$HO(CH_2)_nOH$ —(1. NaH, THF; 2. TBDMSCl; 54–90%)→ $HO(CH_2)_nOTBDMS$

Ref. 46

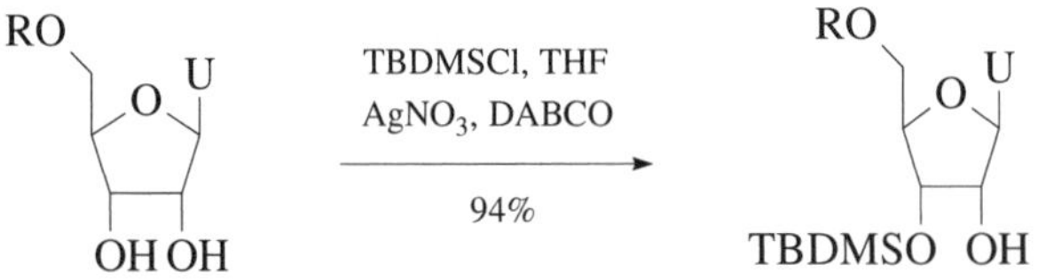

RO, U, O, OH OH —(TBDMSCl, THF, $AgNO_3$, DABCO; 94%)→ RO, U, O, TBDMSO OH

Ref. 47a

R = DMTr or TBDMS

HO–(ribose, U) OH OH —TBDMSCl, Pyr, $AgNO_3$, 90%→ TBDMSO–(ribose, U) OH OTBDMS Ref. 47b

HO–(arabinose, A, HO) OH —TBDMSCl, Pyr / $AgNO_3$→ TBDMSO–(A, RO) OH + 3′,5′-isomer

TBDMSCl, Pyr, $AgNO_3$	94%	3%
3-methylpyridine N-oxide ($N^+–O^-$), TBDMSCl, $AgNO_3$	5%	90%

R = TBDMS

Ref. 48

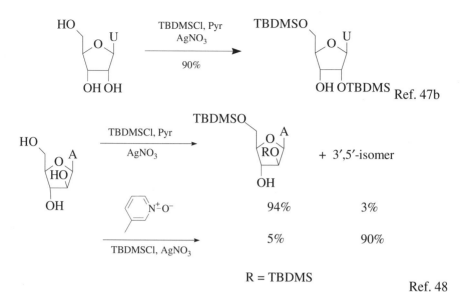

1. R. W. Friesen and L. A. Trimble, *J. Org. Chem.*, **61**, 1165 (1996).
2. E. J. Corey and A. Venkateswarlu, *J. Am. Chem. Soc.*, **94**, 6190 (1972).
3. D. W. Hansen, Jr., and D. Pilipauskas, *J. Org. Chem.*, **50**, 945 (1985).
4. G. R. Clark, M. M. Nikaido, C. K. Fair, and J. Lin, *J. Org. Chem.*, **50**, 1994 (1985).
5. J. F. W. Keana, G. S. Heo, and G. T. Gaughan, *J. Org. Chem.*, **50**, 2346 (1985).
6. J. M. Aizpurua and C. Palomo, *Tetrahedron Lett.*, **26**, 475 (1985).
7. B. A. D'Sa and J. G. Verkade, *J. Am. Chem. Soc.*, **118**, 12832 (1996); B. A. D'Sa, D. McLeod, and J. G. Verkade, *J. Org. Chem.*, **62**, 5057 (1997).
8. J. Jin and S. M. Weinreb, *J. Am. Chem. Soc.*, **119**, 5773 (1997).
9. G. A. Olah, B. G. B. Gupta, S. C. Narang, and R. Malhotra, *J. Org. Chem.*, **44**, 4272 (1979).
10. S. K. Chaudhary and O. Hernandez, *Tetrahedron Lett.*, 99 (1979).
11. K. K. Ogilvie, A. L. Shifman, and C. L. Penney, *Can. J. Chem.*, **57**, 2230 (1979); W. Kinzy and R. R. Schmidt, *Liebigs Ann. Chem.*, 407 (1987).
12. S. Kim and H. Chang, *Synth. Commun.*, **14**, 899 (1984).
13. S. Kim and H. Chang, *Bull. Chem. Soc. Jpn.*, **58**, 3669 (1985).
14. L. Lombardo, *Tetrahedron Lett.*, **25**, 227 (1984).
15. T. F. Braish and P. L. Fuchs, *Synth. Commun.*, **16**, 111 (1986).
16. F. Orsini, F. Pelizzoni, M. Sisti, and L. Verotta, *Org. Prep. Proced. Int.*, **21**, 505 (1989).
17. P. J. Garegg, L. Olsson, and S. Oscarson, *J. Carbohydr. Chem.*, **12**, 955 (1993).
18. A. Glen, D. A. Leigh, R. P. Martin, J. P. Smart, and A. M. Truscello, *Carbohydr. Res.*, **248**, 365 (1993).

19. D. A. Leigh, R. P. Martin, J. P. Smart, and A. M. Truscello, *J. Chem. Soc., Chem. Commun.*, 1373 (1994).
20. T. J. Barton and C. R. Tully, *J. Org. Chem.*, **43**, 3649 (1978).
21. E. J. Corey, H. Cho, C. Rücker, and D. H. Hua, *Tetrahedron Lett.*, **22**, 3455 (1981).
22. L. N. Mander and S. P. Sethi, *Tetrahedron Lett.*, **25**, 5953 (1984).
23. D. Boschelli, T. Takemasa, Y. Nishitani, and S. Masamune, *Tetrahedron Lett.*, **26**, 5239 (1985).
24. P. G. Gassman and L. M. Haberman, *J. Org. Chem.*, **51**, 5010 (1986).
25. W. J. Vloon, J. C. van den Bos, N. P. Willard, G.-J. Koomen, and U. K. Pandit, *Recl. Trav. Chim. Pays-Bas*, **108**, 393 (1989).
26. T. Morita, Y. Okamoto, and H. Sakurai, *Tetrahedron Lett.*, **21**, 835 (1980).
27. T. Veysoglu and L. A. Mitscher, *Tetrahedron Lett.*, **22**, 1299 (1981).
28. M. T. Reetz and G. Neumeier, *Liebigs Ann. Chem.*, 1234 (1981).
29. G. R. Martinez, P. A. Grieco, E. Williams, K.-i. Kanai, and C. V. Srinivasan, *J. Am. Chem. Soc.*, **104**, 1436 (1982).
30. T. P. Mawhinney and M. A. Madson, *J. Org. Chem.*, **47**, 3336 (1982).
31. Y. Kita, J.-i. Haruta, T. Fujii, J. Segawa, and Y. Tamura, *Synthesis*, 451 (1981).
32. K. Kai and G. Patil, *J. Org. Chem.*, **51**, 3545 (1986).
33. D. Askin, D. Angst, and S. Danishefsky, *J. Org. Chem.*, **52**, 622 (1987).
34. X. Franck, B. Figadere, and A. Cavé, *Tetrahedron Lett.*, **36**, 711 (1995).
35. D. Seebach, H. F. Chow, R. F. W. Jackson, M. A. Sutter, S. Thaisrivongs, and J. Zimmermann, *Liebigs Ann. Chem.*, 1281 (1986).
36. S. Kim and I. S. Kee, *Tetrahedron Lett.*, **31**, 2899 (1990).
37. T. Oriyama, K. Yatabe, S. Sugawara, Y. Machiguchi, and G. Koga, *Synlett*, 523 (1996).
38. B. K. Goering, K. Lee, B. An, and J. K. Cha, *J. Org. Chem.*, **58**, 1100 (1993).
39. Y. Tanabe, M. Murakami, K. Kitaichi, and Y. Yoshida, *Tetrahedron Lett.*, **35**, 8409 (1994).
40. D. A. Johnson and L. M. Taubner, *Tetrahedron Lett.*, **37**, 605 (1996).
41. K. Yamamoto and M. Takemae, *Bull. Chem. Soc. Jpn.*, **62**, 2111 (1989).
42. D. L. J. Clive and D. Kellner, *Tetrahedron Lett.*, **32**, 7159 (1991).
43. Y. Tanabe, H. Okumura, A. Maeda, and M. Murakami, *Tetrahedron Lett.*, **35**, 8413 (1994).
44. T. Yokomatsu, K. Suemune, T. Yamagishi, and S. Shibuya, *Synlett*, 847 (1995).
45. R. E. Donaldson and P. L. Fuchs, *J. Am. Chem. Soc.*, **103**, 2108 (1981).
46. P. G. McDougal, J. G. Rico, Y.-I. Oh, and B. D. Condon, *J. Org. Chem.*, **51**, 3388 (1986).
47. (a) G. H. Hakimelahi, Z. A. Proba, and K. K. Ogilvie, *Tetrahedron Lett.*, **22**, 5243 (1981); (b) *idem, ibid.*, **22**, 4775 (1981).
48. K. K. Ogilvie, G. H. Hakimelahi, Z. A. Proba, and D. P. C. McGee, *Tetrahedron Lett.*, **23**, 1997 (1982); K. K. Ogilvie, D. P. C. McGee, S. M. Boisvert, G. H. Hakimelahi, and Z. A. Proba, *Can. J. Chem.*, **61**, 1204 (1983).

Cleavage

1. $Bu_4N^+F^-$, THF, 25°, 1 h, >90% yield.[1] Fluoride ion is very basic, especially under anhydrous conditions, and thus may cause side reactions with base-sensitive substrates.[2] The strong basicity can be moderated by the addition of acetic acid to the reaction, as was the case in the following reaction, after all others methods failed to remove the TBDMS group.[3]

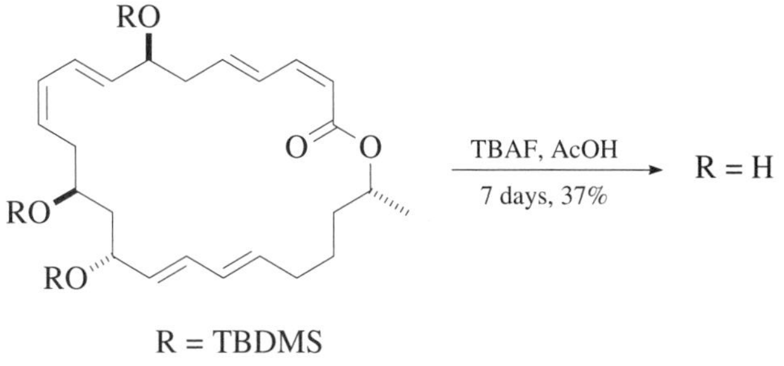

TBAF, AcOH, 7 days, 37% → R = H

R = TBDMS

Commercial TBAF is known to contain water, but the water content seems to vary from lot to lot. This variation in water concentration was determined to be the cause of the often ineffective cleavage of TBDMS groups of ribosyl pyrimidine nucleosides. Interestingly, the cleavage of ribosyl purine nucleoside is not affected by the water content. In order to ensure consistency in deprotection in this case, the reaction should be run with molecular sieve-treated TBAF, which results in a water content of 2.3%.[4] It is also known that the addition of 4Å ms increases the rate of TBAF-induced deprotection.[5] ArOTBDMS ethers can be cleaved in the presence of alkylOTBDMS ethers.[6] Similarly, allyl TBDMS ethers have been cleaved in the presence of alkyl TBDMS ethers.[7] The insolubility of $Bu_4N^+ClO_4^-$ in water has been used to advantage in the workup of reactions that use large quantities of TBAF.[8] Long-range stereoelectronic effects are seen in the rate of silyl ether cleavage, as shown by the TBAF-induced cleavage rates for the following three ethers:[9]

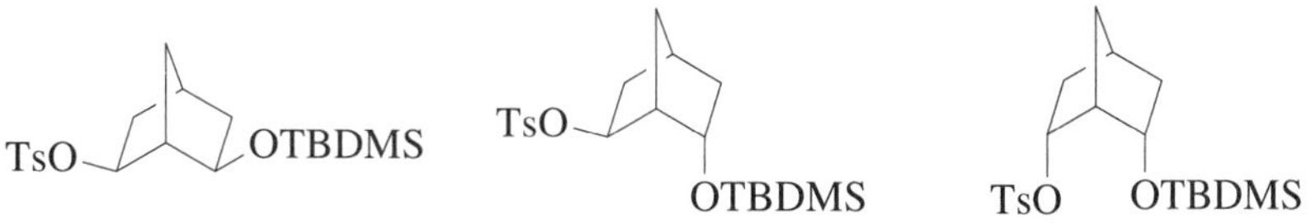

TsO / OTBDMS: 15.4×10^{-3} min^{-1}

TsO / OTBDMS: 4.3×10^{-3} min^{-1}

TsO / OTBDMS: 1.3×10^{-3} min^{-1}

2. KF, 18-crown-6.[10]
3. LiBr, 18-crown-6.[11] Selectivity for primary derivatives was achieved.
4. LiCl, H_2O, DMF, 90°, 81–98% yield.[12]

5. $Bu_4N^+Cl^-$, $KF \cdot H_2O$, CH_3CN, 25°, 4 h, 95% yield.[13] This method generates fluoride ion *in situ* and is reported to be suitable for reactions that normally require anhydrous conditions.
6. Aq. HF, CH_3CN (5:95), 20°, 1–3 h, 90–100% yield.[14] This reagent will cleave ROTBDMS ethers in the presence of ArOTBDMS ethers.[6] The reagent can be used to remove TBDMS groups from prostaglandins.
7. AcOH, H_2O, THF (3:1:1), 25–80°, 15 min to 5 h.[1] Selective cleavage of a primary TBDMS group was achieved with acetic acid in the presence of a secondary TBDMS group.[15]

BOCNH OTBDMS OTBDMS (lactone) —[AcOH, H_2O / THF, 15 h, 30°]→ BOCNH OTBDMS OH (lactone)

8. Dowex 50W-X8, MeOH, 20°.[16] Dowex 50W-X8 is a carboxylic acid resin, H^+ form.
9. $BF_3 \cdot Et_2O$, $CHCl_3$, 0–25°, 15 min to 3 h, 70–90% yield.[17] CH_3CN is also an effective solvent.[18]
10. Pyridine-HF, THF, 0–25°, 70% yield.[19] Cyclic acetals and THP derivatives were found to be stable to these conditions.[20] In the following reaction, if excess pyridine was not included as a buffer, some acyl transfer was observed.[21]

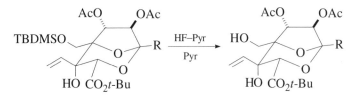

TBDMSO, AcO, OAc, R, HO, CO_2t-Bu —[HF–Pyr / Pyr]→ HO, AcO, OAc, R, HO, CO_2t-Bu

11. 57% HF in urea.[22]
12. $Et_3N \cdot HF$, cyclohexane, rt, 30 min.[23] The use of $Et_3N \cdot 3HF$ was recommended for the desilylation of nucleosides and nucleotides.[24]
13. $NH_4^+F^-$–HF, DMF, NMP, 20°, 90–98% yield. These conditions were developed to remove the TBDMS group from the sensitive carbapenems.[25]
14. $NH_4^+F^-$, MeOH, H_2O, 60–65°, 65% yield.[26,27] Selectivity for primary TBDMS ethers has been observed with this reagent.[28]
15. TsOH (0.1 eq.), THF, H_2O (20:1), 65% yield.[29]
16. 1% concd. HCl in EtOH.[22,30]
17. H_2SO_4.[31]
18. Trifluoroacetic acid, H_2O (9:1), CH_2Cl_2, rt, 96 h.[32] In the accompanying diagram of a riboside, the selectivity is more likely the result of the

reduced basicity of the OTBDMS group adjacent to the carbonyl oxygen, rather than steric differences associated with the two ethers.[33] Similarly, a glycosidic TBDMS group was retained, whereas a primary TBDMS group was cleaved with TFA. In that case also, the glycosidic oxygen is less basic and would be less susceptible to acid-catalyzed cleavage.[34]

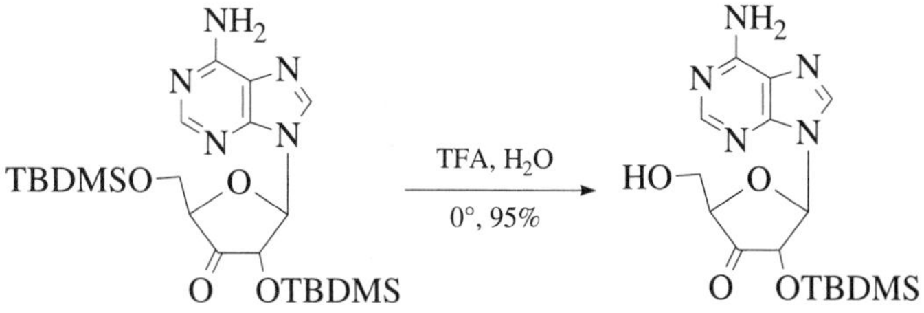

TFA, H_2O, 0°, 95%

19. $Bu_4Sn_2O(NCS)_2$, MeOH, reflux, 16 h, 70% yield.[35] This reagent also cleaves ketals and acetals, 77–97% yield.
20. Me_2BBr.[36]
21. $LiBF_4$, CH_3CN, CH_2Cl_2, 40–86% yield.[37] In this case, $Bu_4N^+F^-$ or acid failed to remove a primary TBDMS group from a steroid.
22. Selectivity in the cleavage of a primary allylic TBDMS group was achieved with HF/CH_3CN in the presence of a more hindered secondary TBDMS group.[38]

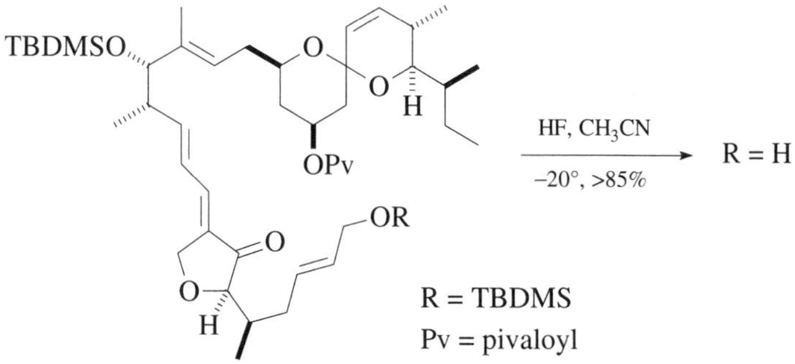

HF, CH_3CN, −20°, >85% → R = H

R = TBDMS
Pv = pivaloyl

23. Selective cleavage of one secondary TBDMS ether in the presence of a somewhat more hindered one was achieved with $Bu_4N^+F^-$ in THF.[39]

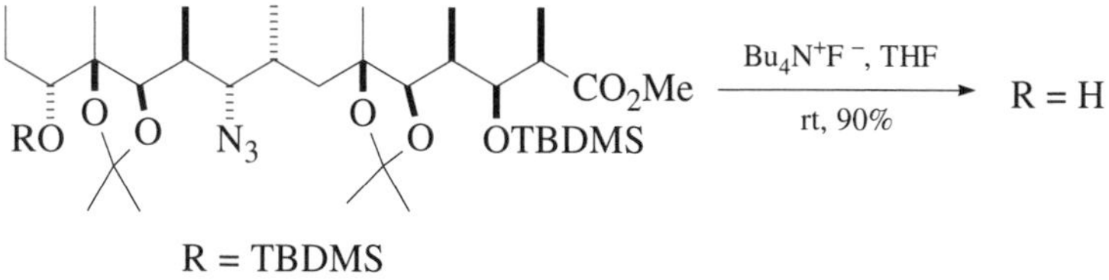

$Bu_4N^+F^-$, THF, rt, 90% → R = H

R = TBDMS

24. NBS, DMSO, H_2O, rt, 17 h.[40] A trisubstituted steroidal alkene was not

affected by these conditions. These conditions have been used to cleave a primary TBDMS ether in the presence of a secondary TBDMS ether.[41]

25. 3 eq. *t*-BuOOH, 1.2 eq. $MoO_2(acac)_2$, CH_2Cl_2, 50–87% yield.[42]
26. 0.01 eq. $PdCl_2(CH_3CN)_2$, acetone, rt, 99% yield.[43,44] Additionally, acetals are cleaved with this reagent, but the TBDPS, MEM, and THP groups are completely stable.
27. Pyridinium *p*-toluenesulfonate, EtOH, 22–55°, 1.2–2 h, 80–92% yield.[45] These conditions were used to remove cleanly a TBDMS group in the presence of a TBDPS group.
28. KO_2, DMSO, DME, 18-crown-6, 50–85% yield.[46]
29. 1 *N* aq. periodic acid in THF was found effective when numerous other methods failed.[47]

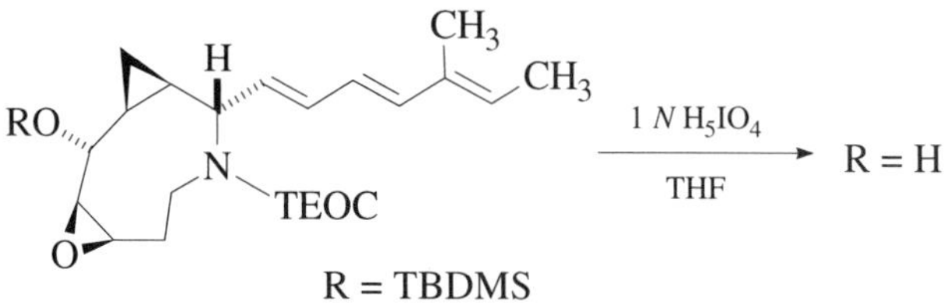

CH3 H CH3 RO N TEOC O $\xrightarrow[\text{THF}]{1\,N\,H_5IO_4}$ R = H

R = TBDMS

30. TMSOTf, CH_2Cl_2, 0°, 5 min, then neutral alumina, 92% yield.[48] TBDPS groups are stable to these conditions.
31. In this case, cleavage of the primary TBDMS group is attributed to the presence of the 2′-hydroxyl, since, in its absence, the cleavage reaction does not proceed.[49]

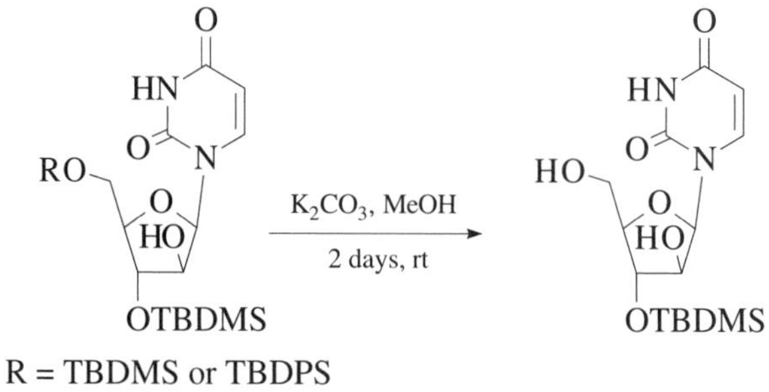

O HN O N RO O HO OTBDMS $\xrightarrow[\text{2 days, rt}]{K_2CO_3,\ MeOH}$ O HN O N HO O HO OTBDMS

R = TBDMS or TBDPS

32. *i*-Bu_2AlH, CH_2Cl_2, 25°, 1–2 h, 84–95% yield.[50]
33. Methanol, CCl_4, ultrasonication, 40–50°, 90–96% yield.[51] Phenolic TBDMS and TBDPS ethers are stable.
34. SiF_4, CH_3CN, 23°, 20 min, 94% yield. This reaction is faster in CH_3CN; tertiary and phenolic TBDMS groups are not cleaved.[52,53]
35. H_2SiF_6, TEA, CH_3CN, >70% yield. TIPS groups are fairly stable to these conditions.[54]

36. BH_3–DMS, TMSOTf, CH_2Cl_2, −78°, 70% yield.[55] Esters and acetals also react with this combination of reagents.
37. Al_2O_3, H_2O, hexanes, 81–98% yield. These conditions are selective for the primary derivative. TBDPS and TMS ethers are also cleaved.[56] The use of alumina in a microwave oven is also effective (68–93% yield).[57]
38. PdO, cyclohexene, methanol, 30 min for a primary ROH, 90–95% yield. Secondary alcohols require longer times. The primary TBDPS and TIPS groups are cleaved much more slowly (18–21 h). Benzylic TBDMS ethers are cleaved without hydrogenolysis.[58]
39. The loss of the TBDMS group during $LiAlH_4$ reductions has been observed in cases where there is an adjacent amine or hydroxyl.[59,60]
40. Ceric ammonium nitrate, MeOH, 0°, 15 min, 82–95% yield.[61] Dioxolanes and some THP ethers are not affected, but in general, with extended reaction times, THP ethers are cleaved.
41. I_2, MeOH, 65°, 12 h, 90% yield.[62] PMB ethers are also cleaved, but benzyl ethers are stable.
42. HCO_2H, THF, H_2O, 82% yield. In this case, all fluoride-based methods failed.[63]

O HN O O RO OR R = TBDMS

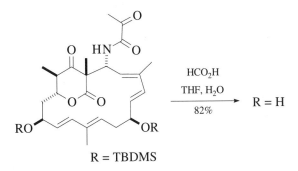

HCO_2H, THF, H_2O, 82% → R = H

In the case of oligonucleotides, the phosphate has been shown to increase the rate of formic-acid-induced TBDMS hydrolysis by internal phosphate participation.[64]

43. AcBr, CH_2Cl_2, rt, 20 min, 90% yield. These conditions convert the TBDMS ether into the acetate. Benzyl and TBDPS ethers are stable, except when $SnBr_2$ is included in the reaction mixture, in which case these groups are also converted to acetates in excellent yield.[65]
44. $(BF_3{\cdot}Et_2O)$–$Bu_4N^+F^-$. This reagent is selective for TBDMS ethers in the presence of TIPS and TBDPS ethers.[66]
45. CsF, CH_3CN, H_2O, reflux.[67]
46. $Ph_3C^+BF_4^-$, CH_3CN, CH_2Cl_2, rt, 60 h.[68]
47. $SnCl_2$, $FeCl_3$, $Cu(NO_3)_2$ or $Ce(NO_3)_3$, CH_3CN, rt, 5 min, 95% yield.[69] TBDPS ethers are also cleaved (3 h, 85–93% yield).

48. During an attempt to metalate a glycal with *t*-BuLi, it was discovered by deuterium labeling that a TBDMS ether can be deprotonated.[70,71]
49. Lewatit 500, MeOH, 96% yield.[72]
50. DMSO, H_2O, 90°, 79–87% yield. These conditions are only effective for primary allylic and homoallylic, primary benzylic, and aryl TBDMS ethers.[73]
51. The oxidative deprotection of silyl ethers, such as the TBDMS ether, has been reviewed.[74]
52. DDQ, CH_3CN, H_2O.[75] These conditions normally cleave the PMB group selectively in the presence of a TBDMS group.[76]
53. Treatment of a primary TBDMS group with Ph_3P and Br_2 converts the group to a primary bromide.[77]
54. The following tables give a comparison of the stability of various silyl ethers to acid, base, and TBAF. The reported half-lives vary as a function of environment and acid or base concentration, but they help define the relative stabilities of these silyl groups.

Half-lives of Hydrolysis of Primary Silyl Ethers[78]

Silyl Ether	Half-lives 5% NaOH–95% MeOH	Half-lives 1% HCl–MeOH, 25°
n-C_6H_{13}OTMS	≤ 1 min	≤ 1 min
n-C_6H_{13}OSi–*i*-BuMe_2	2.5 min	≤ 1 min
n-C_6H_{13}OTBDMS	stable for 24 h	≤ 1 min
n-C_6H_{13}OMDPS	≤ 1 min	14 min
n-C_6H_{13}OTIPS	stable for 24 h	55 min
n-C_6H_{13}OTBDPS	stable for 24 h	225 min

Half-lives of Hydrolysis of Primary Silyl Ethers;[79] Comparison of Trialkylsily vs. Alkoxysilyl Ethers

Ether	Half-lives with $Bu_4N^+F^-$	Half-lives with 0.1 *M* $HClO_4$
n-$C_{12}H_{25}$OTBDMS	140 h	1.4 h
n-$C_{12}H_{25}$OTBDPS	375 h	>200 h
n-$C_{12}H_{25}$OSiPh_2(O–*i*-Pr)	<0.03 h	0.7 h
n-$C_{12}H_{25}$OSiPh_2(O–*t*-Bu)	5.8 h	17.5 h
n-$C_{12}H_{25}$OPh(*t*-Bu)(OMe)	22 h	200 h

55. 4-Methoxysalicylaldehyde–BF_3, CH_2Cl_2, 25°. This method generates HF *in situ*.[80] The following table gives the relative rates of silyl cleavage for three different reagents (TIBS = triisobutylsilyl):

Protective Group	$BF_3 \cdot Et_2O$ CH_2Cl_2, rt	TBAF THF, rt	$BF_3 \cdot Et_2O$ Aldehyde, CH_2Cl_2
TBDMS	45 min	20 min	10 min
TIPS	45 min	15 min	10 min
TIBS	1 h	15 min	15 min
ThxDMS	1.5 h	25 min	15 min
TPS	15 h	2.5 h	20 min
TBDPS	NR	50 min	20 min

1. E. J. Corey and A. Venkateswarlu, *J. Am. Chem. Soc.*, **94**, 6190 (1972).
2. J. H. Clark, *Chem. Rev.*, **80**, 429 (1980).
3. A. B. Smith, III, and G. R. Ott, *J. Am. Chem. Soc.*, **118**, 13095 (1996).
4. R. I. Hogrefe, A. P. McCaffrey, L. U. Borozdina, E. S. McCampbell, and M. M. Vaghefi, *Nucleic Acids Res.*, **21**, 4739 (1993).
5. H. C. Kolb, S. V. Ley, A. M. Z. Slawin, and D. J. Williams, *J. Chem. Soc., Perkin Trans. 1*, 2735 (1992).
6. E. W. Collington, H. Finch, and I. J. Smith, *Tetrahedron Lett.*, **26**, 681 (1985).
7. K. C. Nicolaou, S. Ninkovic, F. Sarabia, D. Vourloumis, Y. He, H. Vallberg, M. R. V. Finlay, and Z. Yang, *J. Am. Chem. Soc.*, **119**, 7974 (1997).
8. J. C. Craig and E. T. Everhart, *Synth. Commun.*, **20**, 2147 (1990).
9. P. M. F. M. Bastiaansen, R. V. A. Orrû, J. B. P. A. Wijnberg, and A. de Groot, *J. Org. Chem.*, **60**, 6154 (1995).
10. G. Stork and P. F. Hudrlik, *J. Am. Chem. Soc.,* **60**, 4462, 4464 (1968); C. L. Liotta and H. P. Harris, *J. Am. Chem. Soc.*, **96**, 2250 (1974).
11. M. Tandon and T. P. Begley, *Synth. Commun.*, **27**, 2953 (1997).
12. J. Farras, C. Serra, and J. Vilarrasa, *Tetrahedron Lett.*, **39**, 327 (1998).
13. L. A. Carpino and A. C. Sau, *J. Chem. Soc., Chem. Commun.*, 514 (1979).
14. R. F. Newton, D. P. Reynolds, M. A. W. Finch, D. R. Kelly, and S. M. Roberts, *Tetrahedron Lett.*, 3981 (1979).
15. A. Kawai, O. Hara, Y. Hamada, and T. Shiari, *Tetrahedron Lett.*, **29**, 6331 (1988).
16. E. J. Corey, J. W. Ponder, and P. Ulrich, *Tetrahedron Lett.*, **21**, 137 (1980).
17. D. R. Kelly, S. M. Roberts, and R. F. Newton, *Synth. Commun.*, **9**, 295 (1979).
18. S. A. King, B. Pipik, A. S. Thompson, A. DeCamp, and T. R. Verhoeven, *Tetrahedron Lett.*, **36**, 4563 (1995).
19. K. C. Nicolaou and S. E. Webber, *Synthesis*, 453 (1986).
20. S. Masamune, L. D.-L. Lu, W. P. Jackson, T. Kaiho, and T. Toyoda, *J. Am. Chem. Soc.*, **104**, 5523 (1982).
21. E. M. Carreira and J. Du Bois, *J. Am. Chem. Soc.*, **117**, 8106 (1995).
22. H. Wetter and K. Oertle, *Tetrahedron Lett.*, **26**, 5515 (1985).
23. J.-E. Nyström, T. D. McCanna, P. Helquist, and R. S. Iyer, *Tetrahedron Lett.*, **26**, 5393 (1985).
24. M. C. Pirrung, S. W. Shuey, D. C. Lever, and L. Fallon, *Biorg. Med. Chem. Lett.*, **4**, 1345 (1994).

25. M. Seki, K. Kondo, T. Kuroda, T. Yamanaka, and T. Iwasaki, *Synlett,* 609 (1995).
26. J. D. White, J. C. Amedio, Jr., S. Gut, and L. Jayasinghe, *J. Org. Chem.*, **54**, 4268 (1989).
27. W. Zhang and M. J. Robins, *Tetrahedron Lett.*, **33**, 1177 (1992).
28. D. Crich and F. Hermann, *Tetrahedron Lett.*, **34**, 3385 (1993).
29. E. J. Thomas and A. C. Williams, *J. Chem. Soc., Chem. Commun.*, 992 (1987).
30. R. F. Cunico and L. Bedell, *J. Org. Chem.*, **45**, 4797 (1980).
31. F. Franke and R. D. Guthrie, *Aust. J. Chem.*, **31**, 1285 (1978).
32. R. Baker, W. J. Cummings, J. F. Hayes, and A. Kumar, *J. Chem. Soc., Chem. Commun.*, 1237 (1986).
33. M. J. Robins, V. Samano, and M. D. Johnson, *J. Org. Chem.*, **55**, 410 (1990).
34. S. F. Martin, J. A. Dodge, L. E. Burgess, and M. Hartmann, *J. Org. Chem.*, **57**, 1070 (1992).
35. J. Otera and H. Nozaki, *Tetrahedron Lett.*, **27**, 5743 (1986).
36. Y. Guindon, C. Yoakim, and H. E. Morton, *J. Org. Chem.*, **49**, 3912 (1984).
37. B. W. Metcalf, J. P. Burkhart, and K. Jund, *Tetrahedron Lett.*, **21**, 35 (1980).
38. S. J. Danishefsky, D. M. Armistead, F. E. Wincott, H. G. Selnick, and R. Hungate, *J. Am. Chem. Soc.*, **109**, 8117 (1987).
39. T. Nakaba, M. Fukui, and T. Oishi, *Tetrahedron Lett.*, **29**, 2219, 2223 (1988).
40. R. J. Batten, A. J. Dixon, R. J. K. Taylor, and R. F. Newton, *Synthesis*, 234 (1980).
41. N. Tsukada, T. Shimada, Y. S. Gyoung, N. Asao, and Y. Yamamoto, *J. Org. Chem.*, **60**, 143 (1995).
42. T. Hanamoto, T. Hayama, T. Katsuki, and M. Yamaguchi, *Tetrahedron Lett.*, **28**, 6329 (1987).
43. B. H. Lipshutz, D. Pollart, J. Monforte, and H. Kotsuki, *Tetrahedron Lett.*, **26**, 705 (1985).
44. N. S. Wilson and B. A. Keay, *J. Org. Chem.*, **61**, 2918 (1996).
45. C. Prakash, S. Saleh, and I. A. Blair, *Tetrahedron Lett.*, **30**, 19 (1989).
46. Y. Torisawa, M. Shibasaki, and S. Ikegami, *Chem. Pharm. Bull.*, **31**, 2607 (1983).
47. G. Kim, M. Y. Chu-Moyer, S. J. Danishefsky, and G. K. Schulte, *J. Am. Chem. Soc.*, **115**, 30 (1993).
48. V. Bou and J. Vilarrasa, *Tetrahedron Lett.*, **31**, 567 (1990).
49. L. L. H. de Fallois, J.-L. Décout, and M. Fontecave, *Tetrahedron Lett.*, **36**, 9479 (1995).
50. E. J. Corey and G. B. Jones, *J. Org. Chem.*, **57**, 1028 (1992).
51. A. S.-Y. Lee, H.-C. Yeh, and M.-H. Tsai, *Tetrahedron Lett.*, **36**, 6891 (1995).
52. E. J. Corey and K. Y. Ki, *Tetrahedron Lett.*, **33**, 2289 (1992).
53. K. C. Nicolaou, K. R. Reddy, G. Skokotas, F. Sato, and X.-Y. Xiao, *J. Am. Chem. Soc.*, **114**, 7935 (1992).
54. A. S. Pilcher, D. K. Hill, S. J. Shimshock, R. E. Waltermire, and P. DeShong, *J. Org. Chem.*, **57**, 2492 (1992); S. J. Shimshock, R. E. Waltermire, and P. Deshong, *J. Am. Chem. Soc.*, **113**, 8791 (1991).
55. R. Hunter, B. Bartels, and J. F. Michael, *Tetrahedron Lett.*, **32**, 1095 (1991).

56. J. Feixas, A. Capdevila, and A. Guerrero, *Tetrahedron*, **50**, 8539 (1994).
57. R. S. Varma, J. B. Lamture, and M. Varma, *Tetrahedron Lett.*, **34**, 3029 (1993).
58. J. F. Cormier, M. B. Isaac, and L.-F. Chen, *Tetrahedron Lett.*, **34**, 243 (1993).
59. J. N. Glushka and A. S. Perlin, *Carbohydr. Res.*, **205**, 305 (1990).
60. E. F. J. De Vries, J. Brussee, and A. van der Gen, *J. Org. Chem.*, **59**, 7133 (1994).
61. A. DattaGupta, R. Singh, and V. K. Singh, *Synlett*, 69 (1996).
62. A. R. Vaino and W. A. Szarek, *J. Chem. Soc., Chem. Commun.*, 2351 (1996).
63. A. S. Kende, K. Liu, I. Kaldor, G. Dorey, and K. Koch, *J. Am. Chem. Soc.*, **117**, 8258 (1995).
64. S.-i. Kawahara, T. Wada, and M. Sekine, *J. Am. Chem. Soc.*, **118**, 9461 (1996).
65. T. Oriyama, M. Oda, J. Gono, and G. Koga, *Tetrahedron Lett.*, **35**, 2027 (1994).
66. S.-i. Kawahara, T. Wada, and M. Sekine, *Tetrahedron Lett.*, **37**, 509 (1996).
67. P. F. Cirillo and J. S. Panek, *J. Org. Chem.*, **55**, 6071 (1990).
68. T. J. Barton and C. R. Tully, *J. Org. Chem.*, **43**, 3649 (1978).
69. A. D. Cort, *Synth. Commun.*, **20**, 757 (1990).
70. R. W. Frieser and L. A. Trimble, *J. Org. Chem.*, **61**, 1165 (1996).
71. J. D. White and M. Kawasaki, *J. Am. Chem. Soc.*, **112**, 4991 (1990).
72. L. F. Tietze, C. Schneider, and A. Grote, *Chem.—Eur. J.*, **2**, 139 (1996).
73. G. Maiti and S. C. Roy, *Tetrahedron Lett.*, **38**, 495 (1997).
74. J. Muzart, *Synthesis*, 11 (1993); S. Chandrasekhar, P. K. Mohanty, and M Takhi, *J. Org. Chem.*, **62**, 2628 (1997).
75. K. Tanemura, T. Suzuki, and T. Horaguchi, *J. Chem. Soc., Perkin Trans. 1*, 2997 (1992); K. Tanemura, T. Suzuki and T. Horaguchi, *Bull. Chem. Soc. Jpn.*, **67** 290 (1994).
76. A. B. Smith, III, Y. Qiu, D. R. Jones, and K. Kobayashi, *J. Am. Chem. Soc.*, **117**, 12011 (1995).
77. P. R. Ashton, R. Königer, J. F. Stoddart, D. Alker, and V. D. Harding, *J. Org. Chem.*, **61**, 903 (1996).
78. J. S. Davies, L. C. L. Higginbotham, E. J. Tremeer, C. Brown, and R. S. Treadgold, *J. Chem. Soc., Perkin Trans. 1*, 3043 (1992).
79. J. W. Gillard, R. Fortin, H. E. Morton, C. Yoakim, C. A. Quesnelle, S. Daignault, and Y. Guindon, *J. Org. Chem.*, **53**, 2602 (1988).
80. S. Mabic and J.-P. Lepoittevin, *Synlett*, 851 (1994).

t-Butyldiphenylsilyl Ether (TBDPS–OR): *t*-$BuPh_2SiOR$ (Chart 1)

The TBDPS group is considerably more stable (≈100 times) than the TBDMS group toward acidic hydrolysis. The TBDPS group is less stable to base than the TBDMS group. The TBDPS group shows greater stability than the TBDMS group to many reagents with which the TBDMS group is incompatible. The TBDMS group is less prone to undergo migration under basic conditions.[1] TBDPS ethers are stable to K_2CO_3/CH_3OH, to 9 *M* NH_4OH, 60°, 2 h, and to $NaOCH_3$ (cat.)/CH_3OH, 25°, 24 h. The ether is stable to 80% AcOH, used to cleave TBDMS, triphenylmethyl, and tetrahydropyranyl ethers. It is also stable to HBr/AcOH, 12°, 2 min, to 25–75% HCO_2H, 25°, 2–6 h, and to 50% aq. CF_3CO_2H, 25°, 15 min (conditions used to cleave acetals).[2]

Formation

1. TBDPSCl, imidazole, DMF, rt.[2] These are the original conditions used to introduce this group and also constitute the most widely employed method.
2. TBDPSCl, DMAP, Pyr.[3] Selective silylation of a primary hydroxyl was achieved under these conditions.
3. TBDPSCl, poly(vinylpyridine), HMPT, CH_2Cl_2.[4]
4. TBDPSCl, DMAP, triethylamine, CH_2Cl_2.[5] This combination of reagents was shown to be very selective for the silylation of a primary hydroxyl in the presence of a secondary hydroxyl.
5. 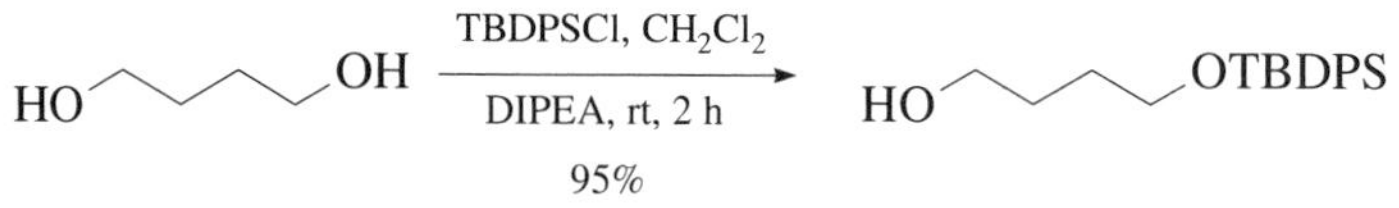 HO~~~~OH → (TBDPSCl, CH_2Cl_2; DIPEA, rt, 2 h; 95%) HO~~~~OTBDPS

Ref. 6

6. TBDPSCl, NH_4NO_3, DMF, 72–96% yield.[7]
7. TBDPSOTf, 2,6-lutidine, CH_2Cl_2.[8]

TBDPSOTf, 2,6-lutidine, CH_2Cl_2

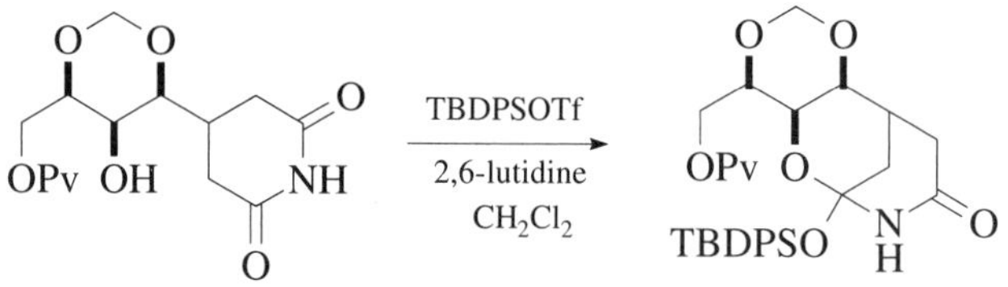

8. TBDPSCl, $AgNO_3$, Pyr, THF, rt, 3 h, 70% yield.[9] The addition of $AgNO_3$ increases the rate of silylation.

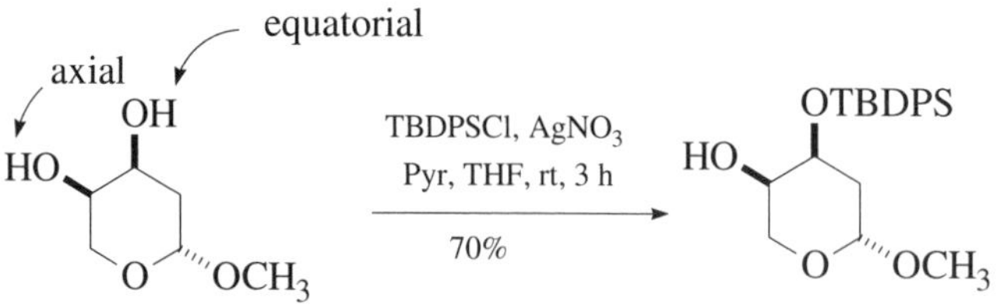

axial, equatorial; TBDPSCl, $AgNO_3$, Pyr, THF, rt, 3 h; 70%

Cleavage

1. $Bu_4N^+F^-$, THF, 25°, 1–5 h, >90% yield.[2]
2. 3% methanolic HCl, 25°, 3 h, 71% yield.[1] In benzoyl-protected carbohydrates, this method gives clean deprotection without acyl migration.[10]
3. 5 *N* NaOH, EtOH, 25°, 7 h, 93% yield.[1] TBDMS ethers are stable.[11]
4. 10% KOH, CH_3OH.[12]

5. Pyr·HF, THF.[13]
6. Amberlite 26 F^- form.[4]
7. HF, CH_3CN.[14]
8. KO_2, DMSO, 18-crown-6.[1]
9. $LiAlH_4$ has resulted in the cleavage of a TBDPS group, but generally,[15,16] TBDPS ethers are not affected by $LiAlH_4$.

$$\text{MeO}_2\text{C}\text{CH}_2\text{CH(OTBDPS)CONH}_2 \xrightarrow{\text{LiAlH}_4} \text{HOCH}_2\text{CH}_2\text{CH(OH)CH}_2\text{NH}_2$$

10. NaH, HMPA, 0°, 5 min; H_2O, 83–84% yield.[17] These conditions selectively cleave a TBDPS ether in the presence of a *t*-butyldimethylsilyl ether.
11. $NH_4^+F^-$.[18]
12. Alumina.[19]
13. $BF_3{\cdot}Et_2O$, 4-methoxysalicylaldehyde.[20] The relative rates of cleavage of the TBDPS ethers of the alcohols $PhCH_2CH_2O$-, propargylO-, BnO-, menthol, and PhO- are 20 min, 45 min, 1.5 h, 5 h, and 8 h, respectively.

1. Y. Torisawa, M. Shibasaki, and S. Ikegami, *Chem. Pharm. Bull.*, **31**, 2607 (1983); W. W. Wood and A. Rashid, *Tetrahedron Lett.*, **28**, 1933 (1987).
2. S. Hanessian and P. Lavallee, *Can. J. Chem.*, **53**, 2975 (1975); *idem, ibid.*, **55**, 562 (1977).
3. R. E. Ireland and D. M. Obrecht, *Helv. Chim. Acta.*, **69**, 1273 (1986); D. M. Clode, W. A. Laurie, D. McHale, and J. B. Sheridan, *Carbohydr. Res.*, **139**, 161 (1985).
4. G. Cardillo, M. Orena, S. Sandri, and C. Tomasihi, *Chem. Ind. (London)*, 643 (1983).
5. S. K. Chaudhary and O. Hernandez, *Tetrahedron Lett.*, 99 (1979); Y. Guindon, C. Yoakim, M. A. Bernstein, and H. E. Morton, *ibid.*, **26**, 1185 (1985).
6. F. Freeman and D. S. H. L. Kim, *J. Org. Chem.*, **57**, 1722 (1992).
7. S. A. Hardinger and N. Wijaya, *Tetrahedron Lett.*, **34**, 3821 (1993).
8. P. A. Grieco, K. J. Henry, J. J. Nunes, and J. E. Matt, Jr., *J. Chem. Soc., Chem. Commun.*, 368 (1992).
9. R. K. Bhatt, K. Chauhan, P. Wheelan, R. C. Murphy, and J. R. Falck, *J. Am. Chem. Soc.*, **116**, 5050 (1994).
10. E. M. Nashed and C. P. J. Glaudemans, *J. Org. Chem.*, **52**, 5255 (1987).
11. S. Hatakeyama, H. Irie, T. Shintani, Y. Noguchi, H. Yamada, and M. Nishizawa, *Tetrahedron*, **50**, 13369 (1994).
12. A. A. Malik, R. J. Cormier, and C. M. Sharts, *Org. Prep. Proced. Int.*, **18**, 345 (1986).
13. K. C. Nicolaou, S. P. Seitz, M. R. Pavia, and N. A. Petasis, *J. Org. Chem.*, **44**, 4011

(1979); K. C. Nicolaou, S. P. Seitz, and M. R. Pavia, *J. Am. Chem. Soc.*, **103**, 1222 (1981).
14. Y. Ogawa, M. Nunomoto, and M. Shibasaki, *J. Org. Chem.*, **51**, 1625 (1986).
15. B. Rajashekhar and E. T. Kaiser, *J. Org. Chem.*, **50**, 5480 (1985).
16. J. C. McWilliams and J. Clardy, *J. Am. Chem. Soc.*, **116**, 8378 (1994).
17. M. S. Shekhani, K. M. Khan, K. Mahmood, P. M. Shah, and S. Malik, *Tetrahedron Lett.*, **31**, 1669 (1990).
18. W. Zhang and M. J. Robins, *Tetrahedron Lett.*, **33**, 1177 (1992).
19. J. Feixas, A. Capdevila, and A. Guerrero, *Tetrahedron*, **50**, 8539 (1994).
20. S. Mabic and J. P. Lepoittevin, *Synlett*, 851 (1994).

Tribenzylsilyl Ether: $ROSi(CH_2C_6H_5)_3$ (Chart 1)

Tri-*p*-xylylsilyl Ether: $ROSi(CH_2C_6H_4\text{-}p\text{-}CH_3)_3$

To control the stereochemistry of epoxidation at the 10,11-double bond in intermediates in prostaglandin synthesis, a bulky protective group was used for the C_{15}–OH group. Epoxidation of the tribenzylsilyl ether yielded 88% α-oxide; epoxidation of the tri-*p*-xylylsilyl ether was less selective.[1]

Formation

1. $ClSi(CH_2C_6H_4\text{-}p\text{-}Y)_3$ (Y = H or CH_3), DMF, 2,6-lutidine, −20°, 24–36 h, 90–100% yield.[1]

Cleavage

1. AcOH, THF, H_2O, (3:1:1), 26°, 6 h → 45°, 3 h, 85% yield.[1]

1. E. J. Corey and H. E. Ensley, *J. Org. Chem.*, **38**, 3187 (1973).

Triphenylsilyl Ether (TPS–OR): $ROSiPh_3$

The stability of the TPS group to basic hydrolysis is similar to that of the TMS group, but its stability to acid hydrolysis is about 400 times greater than that of the TMS group.[1]

Formation

1. Ph_3SiCl, Pyr.[2]
2. Ph_3SiBr, Pyr, −40°, 15 min.[3]
3. Ph_3SiH, cat.[4]

Cleavage

1. AcOH–H_2O-THF (3:1:1), 70°, 3 h, 70% yield.[3]

2. $Bu_4N^+F^-$.[5]
3. NaOH, EtOH.[2]
4. HCl.[6]
5. HF–Pyr, THF, rt, 99% yield.[7]

1. L. H. Sommer, *Stereochemistry, Mechanism and Silicon: An Introduction to the Dynamic Stereochemistry and Reaction Mechanisms of Silicon Centers*, McGraw-Hill, New York, 1965, p. 126.
2. S. A. Barker, J. S. Brimacombe, M. R. Harnden, and J. A. Jarvis, *J. Chem Soc.*, 3403 (1963).
3. H. Nakai, N. Hamanaka, H. Miyake, and M. Hayashi, *Chem Lett.*, 1499 (1979).
4. E. Lukevics and M. Dzintara, *J. Organomet. Chem.*, **271**, 307 (1984); L. Horner and J. Mathias, *J. Organomet. Chem.*, **282**, 175 (1985).
5. K. Maruoka, M. Hasegawa, H. Yamamoto, K. Suzuki, M. Shimazaki, and G.-i. Tsuchihashi, *J. Am. Chem. Soc.*, **108**, 3827 (1986).
6. R. G. Neville, *J. Org. Chem.*, **26**, 3031 (1961).
7. A. Balog, D. Meng, T. Kamenecka, P. Bertinato, D.-S. Su, E. J. Sorensen, and S. J. Danishefsky, *Angew. Chem., Int. Ed. Engl.*, **35**, 2801 (1996).

Diphenylmethylsilyl Ether (DPMS–OR): $Ph_2MeSiOR$

The stability of the DPMS group is intermediate between the stabilities of the TMS and TES (triethylsilyl) groups. The group is incompatible with base, acid, BuLi, $LiAlH_4$, pyridinium chlorochromate, pyridinium dichromate, and CrO_3/pyridine. It is stable to Grignard reagents, Wittig reagents, *m*-chloroperoxybenzoic acid, and silica gel chromatography.[1]

Formation

1. $Ph_2MeSiCl$, DMF, imidazole, 83–92% yield.[1]

Cleavage

1. The group can be cleaved with mild acid, fluoride ion, or base.[1]
2. NaN_3, DMF, 40°, 80–93% yield.[2]
3. Photolysis at 254 nm, CH_3OH, CH_2Cl_2, phenanthrene, 51–84% yield. These conditions are selective for allylic and benzylic alcohols. In the absence of the phenanthrene, TBDMS ethers are also cleaved.[3]

1. S. E. Denmark, R. P. Hammer, E. J. Weber, and K. L. Habermas, *J. Org. Chem.*, **52**, 165 (1987).

2. S. J. Monger, D. M. Parry, and S. M. Roberts, *J. Chem. Soc., Chem. Commun.*, 381 (1989).
3. O. Piva, A. Amougay, and J.-P. Pete, *Synth. Commun.*, **25**, 219 (1995).

Di-*t*-butylmethylsilyl Ether (DTBMS–OR): $(t\text{-Bu})_2MeSiOR$

Formation

1. $DTBMSClO_4$, MeCN, Pyr, 100% yield.[1]
2. DTBMSOTf, 2,6-lutidine, DMAP, 70°, 87% yield.[2,3]

Cleavage

1. $BF_3{\cdot}Et_2O$, CH_2Cl_2; $NaHCO_3$, H_2O, 0°, 30 min, 94% yield. CsF in DMSO fails to cleave this group.[1]
2. 49% Aqueous HF, $MeNO_2$, 0°, 24 h, 30% yield.[2]

1. T. J. Barton and C. R. Tully, *J. Org. Chem.*, **43**, 3649 (1978).
2. K. C. Nicolaou, E. W. Yue, S. La Greca, A. Nadin, Z. Yang, J. E. Leresche, T. Tsari, Y. Naniwa, and F. De Riccardis, *Chem.—Eur. J.*, **1**, 467 (1995).
3. R. S. Bhide, B. S. Levison, R. B. Sharma, S. Ghosh, and R. G. Salomon, *Tetrahedron Lett.*, **27**, 671 (1986).

Sisyl Ether [Tris(trimethylsilyl)silyl Ether]: $[(CH_3)_3Si]_3SiOR$

Formation

$[(CH_3)_3Si]_3SiCl$, CH_2Cl_2, DMAP, 70–97% yield.[1]

Cleavage

1. TBAF, THF.[2]
2. Photolysis, MeOH, CH_2Cl_2, 62–95% yield.[1]

The sisyl ether is stable to Grignard and Wittig reagents, oxidation with Jones reagent, KF/18-crown-6, CsF, and strongly acidic conditions (TsOH, HCl) that cleave most other silyl groups. It is not stable to alkyllithiums or $LiAlH_4$.

1. M. A. Brook, C. Gottardo, S. Balduzzi, and M. Mohamed, *Tetrahedron Lett.*, **38**, 6997 (1997).
2. K. J. Kulicke and B. Giese, *Synlett*, 91 (1990).

(2-Hydroxystyryl)dimethylsilyl Ether (HSDMS–OR) and (2-Hydroxystyryl)diisopropylsilyl Ether (HSDIS–OR)

Formation[1]

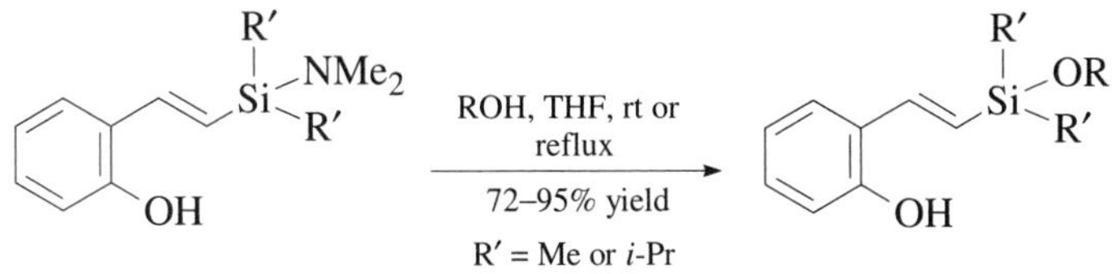

R′ Si NMe$_2$ R′ OH

ROH, THF, rt or reflux
72–95% yield
R′ = Me or *i*-Pr

R′ Si OR R′ OH

Cleavage[1]

1. Photolysis at 254 nm, rt, 30 min, CH_3CN, 75–92% yield.

1. M. C. Pirrung and Y. R. Lee, *J. Org. Chem.*, **58**, 6961 (1993).

t-Butylmethoxyphenylsilyl Ether (TBMPS–OR): *t*-Bu(CH_3O)PhSiOR

The TBMPS group has a greater sensitivity than either the TBDMS or TBDPS group to fluoride ion, which allows for the selective cleavage of the TBMPS group in the presence of the latter two. The TBMPS group is also 140 times more stable to 0.01 *N* $HClO_4$ than is the TBDMS group thus allowing selective hydrolysis of the TBDMS group. The TBMPS group can be introduced onto primary, secondary, and tertiary hydroxyls in excellent yield when DMF is used as the solvent and can be selectively introduced onto primary hydroxyls when CH_2Cl_2 is used as solvent. The main problem with this group is that, when it is introduced onto chiral molecules, diastereomers result that may complicate NMR interpretation.[1]

Formation/Cleavage[1]

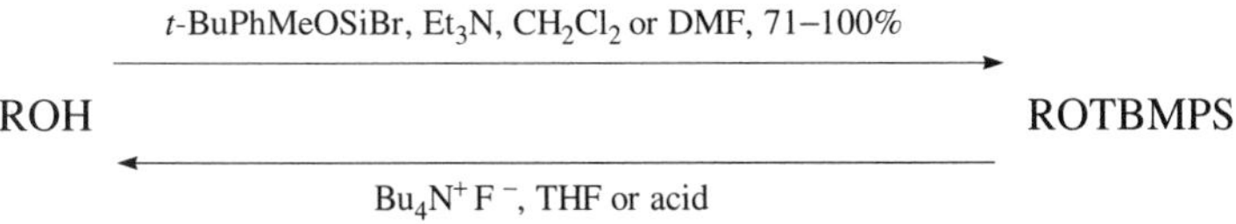

1. Y. Guindon, R. Fortin, C. Yoakim, and J. W. Gillard, *Tetrahedron Lett.*, **25**, 4717 (1984); J. W. Gillard, R. Fortin, H. E. Morton, C. Yoakim, C. A. Quesnelle, S. Daignault, and Y. Guindon, *J. Org. Chem.*, **53**, 2602 (1988).

t-Butoxydiphenylsilyl Ether (DPTBOS–OR): $Ph_2(t\text{-}BuO)SiOR$

The DPTBOS group is considered a low-cost alternative to the TBDMS group with comparable stability.

Formation

1. DPTBOSCl, TEA, CH_2Cl_2, rt, 98% yield.[1]

Cleavage

1. 0.01 *M* $HClO_4$.[2]
2. TBAF.[2]
3. $Na_2S\cdot 9H_2O$, EtOH, rt, 12–72 h, 70–100% yield.[3] TBDMS ethers are stable to these conditions.

1. L. F. Tietze, C. Schneider, and A. Grote, *Chem.—Eur. J.*, **2**, 139 (1996).
2. J. W. Gillard, R. Fortin, H. E. Morton, C. Yoakim, C. A. Quesnelle, S. Daignault, and Y. Guindon, *J. Org. Chem.*, **53**, 2602 (1988).
3. T. Schmittberger and D. Uguen, *Tetrahedron Lett.*, **36**, 7445 (1995).

Conversion of Silyl Ethers to Other Functional Groups

The ability to convert a protective group to another functional group directly, without first performing a deprotection, is a potentially valuable transformation. Silyl-protected alcohols have been converted directly to aldehydes,[1,2] ketones,[3] bromides,[4] acetates,[5] and ethers[6] without first liberating the alcohol in a prior deprotection step.

1. G. A. Tolstikov, M. S. Miftakhov, N. S. Vostrikov, N. G. Komissarova, M. E. Adler, and O. Kuznetsov, *Zh. Org. Khim.*, **24**, 224 (1988); *Chem. Abstr.*, **110**: 7162c (1989).
2. I. Mohammadpoor-Baltork and S. Pouranshirvani, *Synthesis*, 756 (1997).
3. F. P. Cossio, J. M. Aizpurua, and C. Palomo, *Can. J. Chem.*, **64**, 225 (1986).
4. H. Mattes and C. Benezra, *Tetrahedron Lett.*, **28**, 1697 (1987); S. Kim and J. H. Park, *J. Org. Chem.*, **53**, 3111 (1988); J. M. Aizpurua, F. P. Cossio, and C. Palomo, *J. Org. Chem.*, **51**, 4941 (1986).
5. S. J. Danishefsky and N. Mantlo, *J. Am. Chem. Soc.*, **110**, 8129 (1988); B. Ganem and V. R. Small, Jr., *J. Org. Chem.*, **39**, 3728 (1974); S. Kim and W. J. Lee, *Synth. Commun.*, **16**, 659 (1986); E.- F. Fuchs and J. Lehmann, *Chem. Ber.*, **107**, 721 (1974).
6. D. G. Saunders, *Synthesis*, 377 (1988).

ESTERS

See also Chapter 5, on the preparation of esters as protective groups for carboxylic acids.

Formate Ester: ROCHO (Chart 2)

Formation

1. 85% HCOOH, 60°, 1 h, 93% yield.[1] This method can be used to protect selectively only the primary alcohol of a pyranoside.[2]
2. 70% HCOOH, cat. $HClO_4$, 50–55°, good yields.[3]
3. $CH_3COOCHO$, Pyr, −20°, 80–100% yield.[4,5,6]
4. Me_2N^+=CHOBz Cl^-, Et_2O, overnight; dil. H_2SO_4, 60–96% yield.[7]
5. HCO_2H, BF_3·2MeOH, 90% yield.[8]
6. Ethyl formate, $Ce(SO_4)_2$–silica gel, reflux 0.5–24 h, 90–100% yield.[9]

Cleavage

1. $KHCO_3$, H_2O, MeOH, 20°, 3 days.[3]
2. Dil. NH_3, pH 11.2, 22°, 62% yield.[10] A formate ester can be cleaved selectively in the presence of an acetate (MeOH, reflux),[5] dil. NH_3 (formate is 100 times faster than an acetate),[10] or benzoate ester (dil. NH_3)[10].

1. H. J. Ringold, B. Löken, G. Rosenkranz, and F. Sondheimer, *J. Am. Chem. Soc.*, **78**, 816 (1956).
2. L. X. Gan and R. L. Whistler, *Carbohydr. Res.*, **206**, 65 (1990).
3. I. W. Hughes, F. Smith, and M. Webb, *J. Chem. Soc.*, 3437 (1949).
4. F. Reber, A. Lardon, and T. Reichstein, *Helv. Chim. Acta*, **37**, 45 (1954).
5. J. Žemlička, J. Beránek, and J. Smrt, *Collect. Czech. Chem. Commun.*, **27**, 2784 (1962).
6. For a review on acetic formic anhydride, see P. Strazzolini, A. G. Giumanini, and S. Cauci, *Tetrahedron*, **46**, 1081 (1990).
7. J. Barluenga, P. J. Campos, E. Gonzalez-Nuñez, and G. Asensio, *Synthesis*, 426 (1985).
8. M. Dymicky, *Org. Prep. Proced. Int.*, **14**, 177 (1982).
9. T. Nishiguchi and H. Taya, *J. Chem. Soc., Perkin Trans. 1*, 172 (1990).
10. C. B. Reese and J. C. M. Stewart, *Tetrahedron Lett.*, 4273 (1968).

Benzoylformate Ester: ROCOCOPh

The benzoylformate ester can be prepared from the 3′-hydroxy group in a deoxyribonucleotide by reaction with benzoyl chloroformate (anhyd. Pyr, 20°,

12 h, 86% yield); it is cleaved by aqueous pyridine (20°, 12 h, 31% yield), conditions that do not cleave an acetate ester.[1]

1. R. L. Letsinger and P. S. Miller, *J. Am. Chem. Soc.*, **91**, 3356 (1969).

Acetate Ester (ROAc): CH_3CO_2R (Chart 2)

Formation

1. Ac_2O, Pyr, 20°, 12 h, 100% yield.[1] This is one of the most common methods for the introduction of acetate groups. By running the reaction at lower temperatures, good selectivity can be achieved for primary alcohols over secondary alcohols.[2] Tertiary alcohols are generally not acylated under these conditions.
2. CH_3COCl, 25°, 16 h, 67–79% yield.[3]
3. CH_3COCl, CH_2Cl_2, collidine, 91% yield. A primary acetate was formed selectively in the presence of a secondary. These conditions are suitable for a variety of other esters.[4]
4. Ac_2O or AcCl, Pyr, DMAP, 24–80°, 1–40 h, 72–95% yield.[5] The use of DMAP increases the rate of acylation by a factor of 10^4. These conditions acylate most alcohols, including tertiary alcohols. The use of DMAP (4-*N*,*N*-dimethylaminopyridine) as a catalyst to improve the rate of esterification is quite general and works for other esters as well, but it is not effective with hindered anhydrides such as pivaloic anhydride. The phosphine i[6] (48–99% yield) and Bu_3P[7] have been developed as active acylation catalysts for acetates and benzoates.

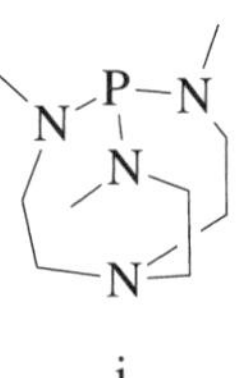

i

5. $AcOC_6F_5$, Et_3N, DMF, 80°, 12–60 h, 72–95% yield.[8] This reagent reacts with amines (25°, no Et_3N) selectively in the presence of alcohols to form *N*-acetyl derivatives in 80–90% yield.
6. The direct conversion of a THP-protected alcohol to an acetate is possible, thus avoiding a deprotection step.[9]

$(CH_2)_8OTHP$ $\xrightarrow[\text{reflux, 91\%}]{\text{AcCl, AcOH}}$ $(CH_2)_8OAc$

7. Ac-Imidazole, $PtCl_2(C_2H_4)$, 23°, 0.5–144 h, 51–87% yield.[10] Platinum(II) acts as a template to catalyze the acetylation of the pyridinyl alcohol, $C_5H_4N(CH_2)_nCH_2OH$. Normally, acylimidazoles are not very reactive acylating agents with alcohols.
8. Ac_2O, CH_2Cl_2, 15 kbar (1.5 GPa), 79–98% yield.[11] This high-pressure technique also works to introduce benzoates and TBDMS ethers onto highly hindered tertiary alcohols.
9. AcOEt, Al_2O_3, 75–80°, 24 h, 45–69% yield.[12] This method is selective for primary alcohols. Phenols do not react under these conditions. The use of $SiO_2{\cdot}NaHSO_4$ as a solid support was also found to be effective.[13]
10. The monoacetylation of alpha–omega diols can be accomplished in excellent yields.[14]

$$HOCH_2(CH_2)_nCH_2OH \xrightarrow[\text{30 h–1 wk; 60–90\%}]{AcOH,\ H_2SO_4,\ H_2O} AcOCH_2(CH_2)_nCH_2OH$$

A monoacetate can be isolated by continuous extraction with organic solvents such as cyclohexane/CCl_4. Monoacylation can also be achieved by ion exchange resin[15] or acid-catalyzed[16] transesterification.

11. Ac_2O, $BF_3{\cdot}Et_2O$, THF, 0°.[17] These conditions give good chemoselectivity for the most nucleophilic hydroxyl group. Alcohols are acetylated in the presence of phenols.
12. $AcOCH_2CF_3$, porcine pancreatic lipase, THF, 60 h, 77% yield.[18] This enzymatic method was used to acetylate selectively the primary hydroxyl group of a variety of carbohydrates. The selective enzymatic acylation of carbohydrates has been partially reviewed.[19]
13. $AcOCH_2CCl_3$, pyridine, porcine pancreatic lipase, 85% yield.[20] These studies examined the selective acylation of carbohydrates. Mannose is acylated at the 6-position in 85% yield in one example.
14. $CH_2{=}C{=}O$, *t*-BuOK, THF.[21] The 17α-hydroxy group of a steroid was acetylated by this method.
15. Bu_2SnO, $PhCH_3$, 110°, 2 h; AcCl, CH_2Cl_2, 0°, 30 min, 84% yield.[22]

HO–CH₂–CH(OH)–CH₂–OBn → (1. Bu_2SnO, $PhCH_3$, 2 h, 110°; 2. AcCl, CH_2Cl_2, 0°, 30 min, 84%) → AcO–CH₂–CH(OH)–CH₂–OBn

16. 3-Acetylthiazolidine-2-thione (O, S, N, S structure), NaH, 93% yield.[23] Primary alcohols are selectively acylated.

17. AcOH, TMSCl, 81% yield.[24]
18. $Sc(OTf)_3$, AcOH, *p*-nitrobenzoic anhydride[25] or $Sc(OTf)_3$, Ac_2O, 66– >95% yield. The lower yields are obtained with allylic alcohols; propargylic alcohols give higher yields. Phenols are effectively acylated with this catalyst, but at a much slower rate than simple aliphatic alcohols.[26] The method was shown to be superior to most other methods for macrolactonization with minimum diolide formation.
19. Ac_2O, $Sc(NTf_2)_3$, CH_3CN, 0°, 1 h, 99% yield. The method is also good for tertiary alcohols.[27,28]
20. Vinyl acetate or 2-propenyl acetate, toluene, $Cp^*{}_2Sm(THF)_2$, rt, 3 h, 88–99% yield. Other esters can also be prepared by this method.[29]
21. Ac_2O, cat. TMSOTf, CH_2Cl_2, 0°, 0.5–60 min, 71–100% yield. This is a more reactive combination of reagents then either DMAP/Ac_2O or $Sc(OTf)_3/Ac_2O$. Phenols are also efficiently acylated by this method.[30]
22. Ethyl acetate, $Ce(SO_4)_2$–silica gel, reflux, 91–99% yield.[31]
23. $Me(OMe)_3$, TsOH, 1.5 h, then H_2O for 30 min.[32]

$C_{11}H_{23}$ OH OH CO_2t-Bu → ($MeC(OMe)_3$, TsOH, 1.5 h, then H_2O) → $C_{11}H_{23}$ OH OAc CO_2t-Bu

HO OH OH EtS O CH_3 → ($MeC(OMe)_3$, TsOH, DMF, 96%) → AcO OH OH EtS O CH_3

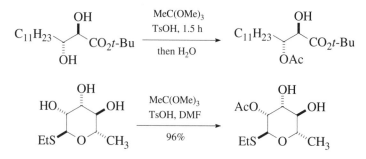

When the reaction was run in CH_3CN, migration of the EtS group to the 2-position was observed. This is attributed to episulfonium salt formation, with the resultant addition of acetate at the anomeric position.[33]

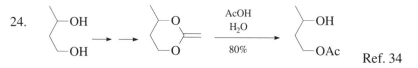

24. OH OH → → O O → (AcOH, H_2O, 80%) → OH OAc Ref. 34

25. Ac_2O, Amberlyst 15, 77% yield. These conditions introduce an acetyl group on oxygen in preference to the normally more reactive primary amine.[35]
26. Ac_2O, cat. I_2, 70–98% yield.[36]
27. The use of biocatalysts for the selective introduction and cleavage of esters is vast and has been extensively reviewed.[37] Therefore only a few examples of the types of transformations that are encountered in this area of protective group chemistry will be illustrated to show some of the basic transformations that have appeared in the literature. The selective

protection or deprotection of symmetrical intermediates to give enantio-enriched products has also been used extensively.

28. *Lipase Fp* from Amano, vinyl acetate, 4 h, 90% yield.[38] This method can also be used for the selective introduction of other esters, such as the methoxyacetyl, phenoxyacetyl, and phenylacetyl groups, in excellent yield.

(a)

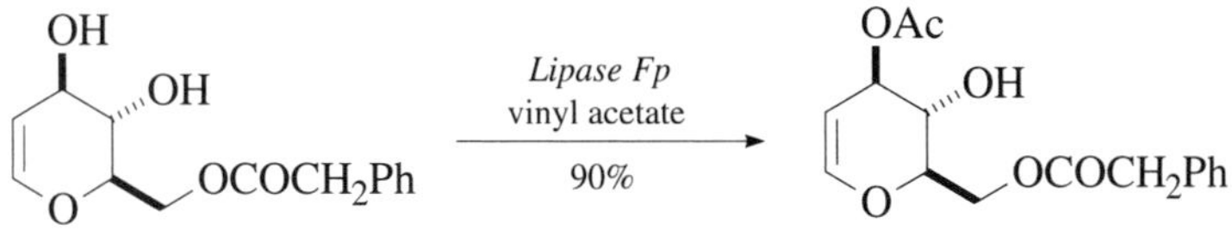

OH, OH, $OCOCH_2Ph$ → *Lipase Fp*, vinyl acetate, 90% → OAc, OH, $OCOCH_2Ph$

(b)

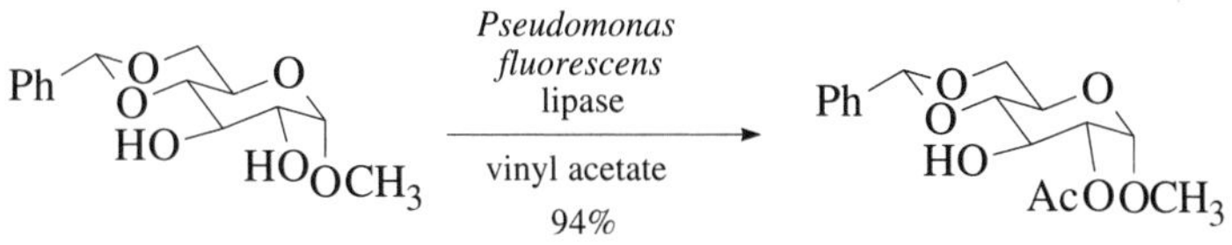

Ph, O, O, O, HO, HO, OCH_3 → *Pseudomonas fluorescens* lipase, vinyl acetate, 94% → Ph, O, O, O, HO, AcO, OCH_3

Ref. 39

(c)

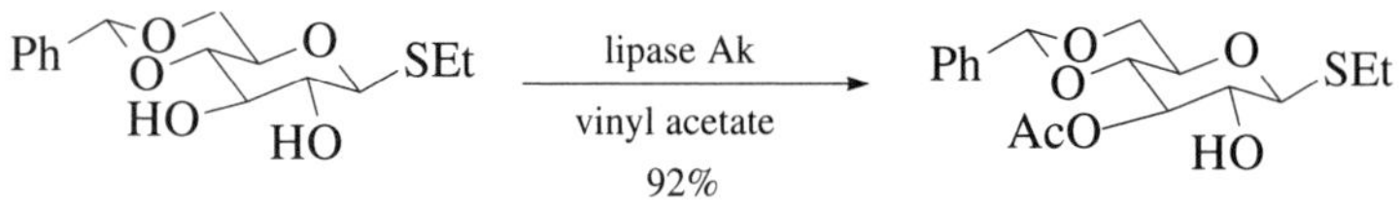

Ph, O, O, O, SEt, HO, HO → lipase Ak, vinyl acetate, 92% → Ph, O, O, O, SEt, AcO, HO

Refs. 40, 41

(d)

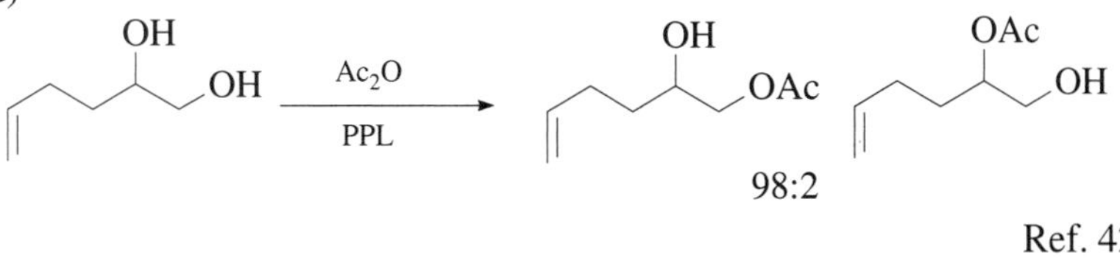

OH, OH → Ac_2O, PPL → OH, OAc ; OAc, OH

98:2

Ref. 42

(e)

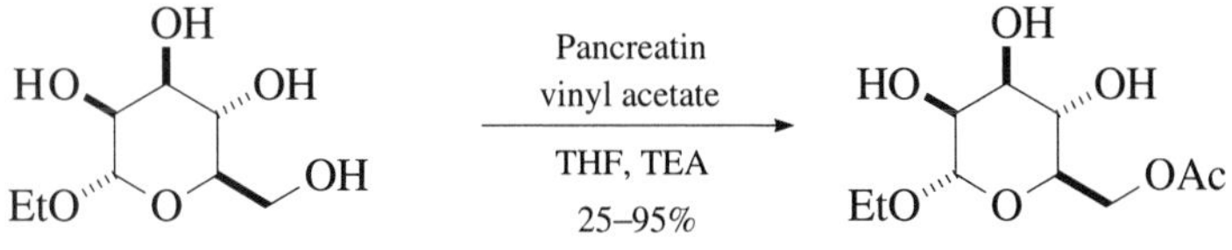

OH, HO, OH, EtO, O, OH → Pancreatin, vinyl acetate, THF, TEA, 25–95% → OH, HO, OH, EtO, O, OAc

Ref. 43

(f)

R′, CH_2OH, R″, CH_2OH → vinyl acetate, lipase, 82–95%, 55–98% ee → R′, CH_2OH, R″, CH_2OAc

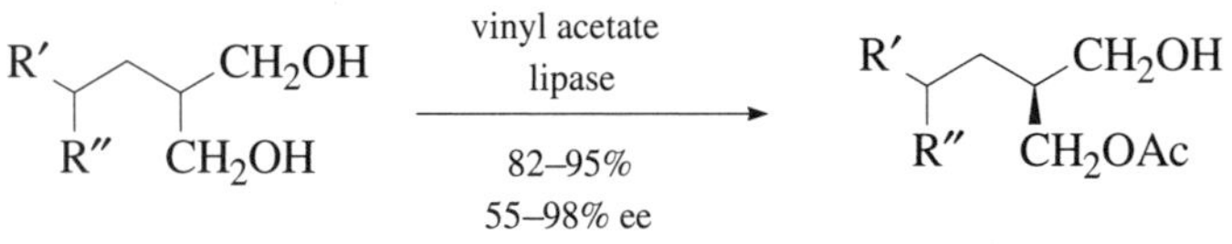

Ref. 44

(g)

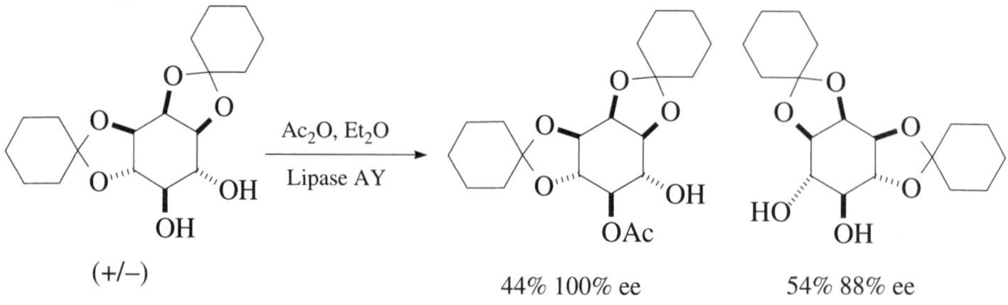

(+/–)

Ac_2O, Et_2O

Lipase AY

44% 100% ee

54% 88% ee

Ref. 45

(h)

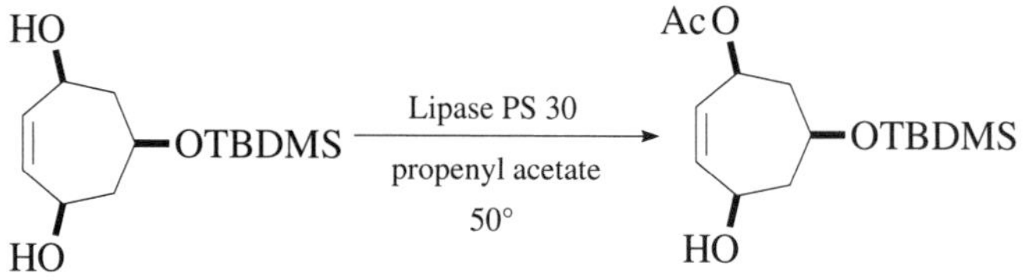

Lipase PS 30

propenyl acetate

50°

Ref. 46

29.

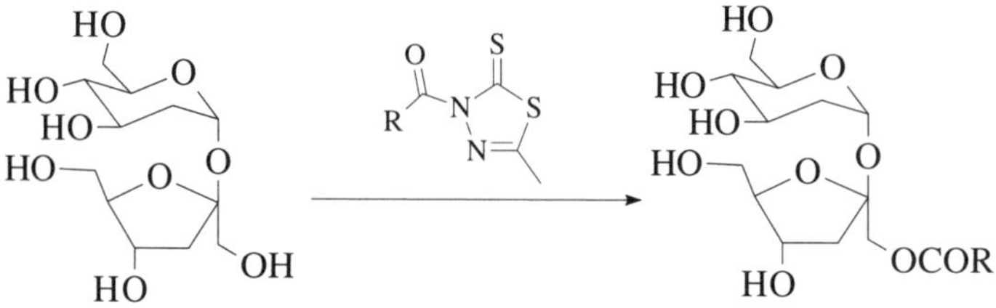

Ref. 47

30. An acyl thiazolidone is also effective for the selective acylation (Ac, Pv, Bz) of primary alcohols.[48]
31. Ac_2O, $CoCl_2$, 69–100% yield. This method does not work for tertiary alcohols.[49]

Cleavage

1. K_2CO_3, MeOH, H_2O, 20°, 1 h, 100% yield.[50] When catalytic NaOMe is used as the base in methanol, the method is referred to as the Zemplén de-*O*-acetylation. Acetyl groups are known to migrate under these conditions, but a recent study indicated that acyl migration is reduced with decreasing solvent polarity (6:1 chloroform/MeOH vs. MeOH).[51]
2. KCN, 95% EtOH, 20° to reflux, 12 h, 93% yield.[52,53] Potassium cyanide is a mild transesterification catalyst, suitable for acid- or base-sensitive compounds. When it is used with 1,2-diol acetates, hydrolysis proceeds slowly until the first acetate is removed.[54]
3. Guanidine, EtOH, CH_2Cl_2, rt, 85–100% yield.[55] Acetamides, benzoates, and pivaloates are stable under these conditions. Phenolic acetates can be removed in the presence of primary and secondary acetates with excellent selectivity.

4. 50% NH_3, MeOH, 20°, 2.5 h, 85% yield.[56] The 3′-acetate is removed from cytosine in the presence of a 5′-benzoate. If the reaction time is extended to two days, both the benzoate and the benzoyl protection on nitrogen are removed.
5. Bu_3SnOMe, $ClCH_2CH_2Cl$, 1 h, 77% yield.[57] These conditions selectively cleave the anomeric acetate of a glucose derivative in the presence of other acetates.
6. $BF_3{\cdot}Et_2O$, wet CH_3CN, 96% yield.[58]

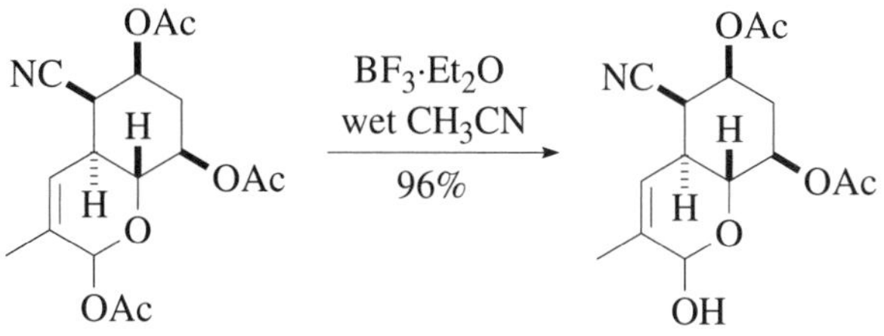

$BF_3{\cdot}Et_2O$, wet CH_3CN, 96%

7. 1,8-Diazabicyclo[5.4.0]undec-7-ene (DBU), benzene, 60°, 45 h, 47–97% yield.[59] Benzoates are not cleaved under these conditions.
8. Sm, I_2, MeOH, rt, 3–60 min, 95–100% yield. Tertiary alcohols were not affected. As the reaction time and temperature are increased, benzoates and carbonates can also be cleaved.[60]
9. HBF_4, MeOH, 23°, 48 h, 83% yield. This system cleaves acetate groups in the presence of benzoate groups.[61] HCl in methanol can also be used.[62]
10. $LiEt_3BH$, THF, −78°, 2 h, 98% yield.[63] An anomeric acetate can be selectively cleaved in the presence of a secondary acetate.
11. Distannoxanes, MeOH or EtOH in $CHCl_3$, CH_2Cl_2, PhH or THF. 1-ω Diacetates are selectively cleaved, but the selectivity goes down as the chain length increases.[64]
12. Bu_2SnO, toluene, 80–110°, 1.5–27 h, 15–92% yield.[65]
13. Mg, MeOH or $Mg(OMe)_2$ in MeOH. The acetate is cleaved in the presence of the benzoate and pivaloate (76–96% yield).[66] The relative rates of cleavage are as follows: *p*-nitrobenzoate > acetate > benzoate > pivaloate >> acetamide. Tertiary acetates are not cleaved.[67]
14. H_2O_2, $NaHCO_3$, THF. The 10-acetate, which is an α-keto acetate, is cleaved in the presence of the taxol side chain that is prone to hydrolysis with other reagents.[68]
15. H_2NNH_2, MeOH, 92% yield. An anomeric acetate was cleaved selectively in the presence of an axial secondary acetate.[69]
16. Deprotection using enzymes can be quite useful. An added benefit is that a racemic or meso substrate can often be resolved with excellent enantioselectivity.[70] Numerous examples of this process are described in the literature. Although acetates are the most common substrates in enzymatic

reactions, other aliphatic esters have been examined with good success.[37]

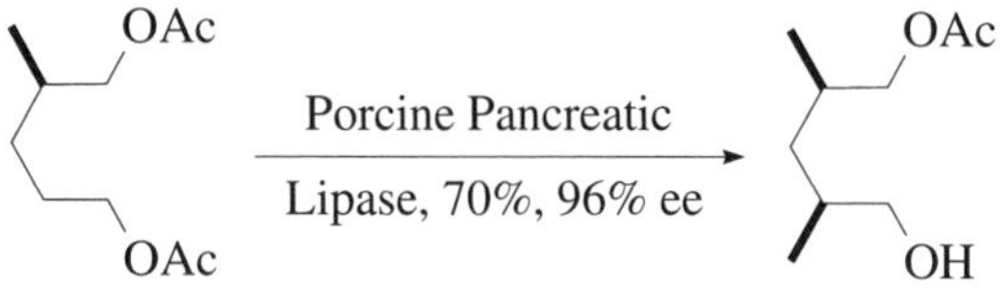

Porcine Pancreatic Lipase, 70%, 96% ee

17. *Candida cylindracea,* phosphate buffer pH 7, Bu_2O.[71] The 6-*O*-acetyl of α-methylperacetylglucose was selectively removed. Porcine pancreatic lipase also hydrolyzes acetyl groups from other carbohydrates. These lipases are not specific for acetate, since they hydrolyze other esters as well. In general, selectivity is dependent upon the ester and the substrate.[18]

18. $AcO(CH_2)_nOAc \xrightarrow[\text{pH 6.9 buffer}]{\text{PPL}} AcO(CH_2)_nOH$ 48–95%

Larger *n* gives lower yield

Ref. 72

19. 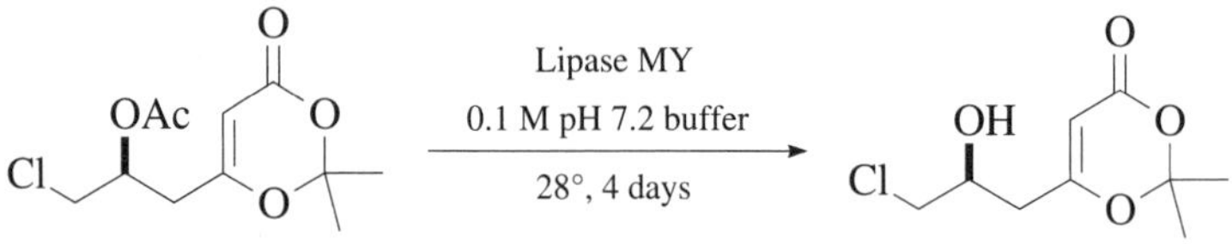 Lipase MY, 0.1 M pH 7.2 buffer, 28°, 4 days

In this case, chemical methods were unsuccessful.[73]

20. 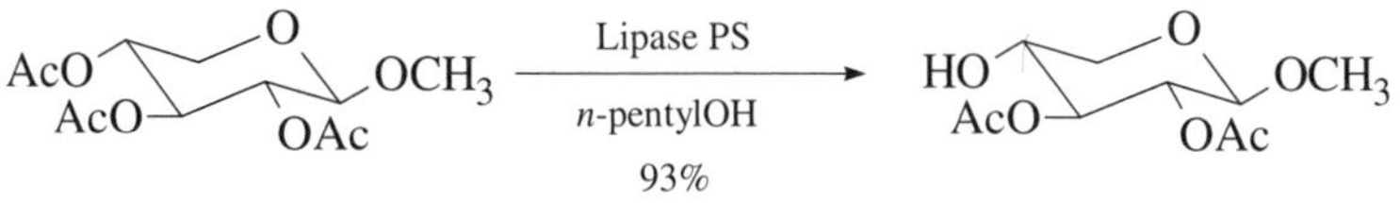 Lipase PS, *n*-pentylOH, 93%

Ref. 74

21. 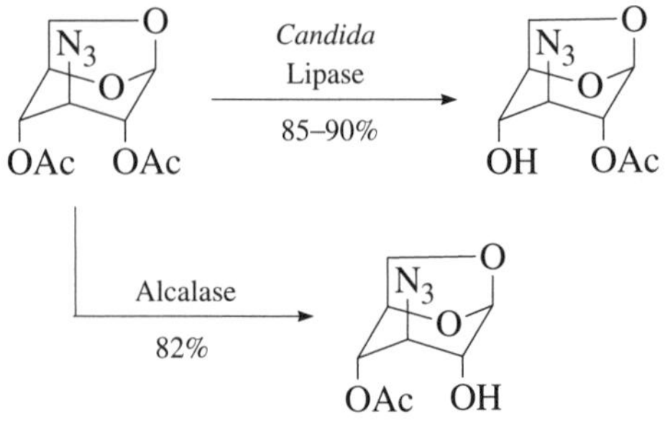 *Candida* Lipase, 85–90%

Alcalase, 82%

Ref. 75

22.

AcO–CH₂–C(R)=CH–CH₂–OAc → HO–CH₂–C(R)=CH–CH₂–OAc

Lipase Al, *Achromobacter sp.*, pH 7.2

Selectivity depends upon R

Ref. 76

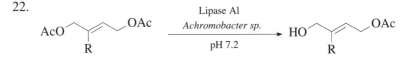

23. Guanidine, guanidinium nitrate, MeOH, CH_2Cl_2, 91–99% yield. These conditions were designed to be compatible with the *N*-Troc group. The tetrachlorophthalimido, *N*-Fmoc, and *O*-Troc groups were unstable in the presence of this reagent. Benzoates are cleaved, but 20 times more slowly.[77]

1. H. Weber and H. G. Khorana, *J. Mol. Biol.*, **72**, 219 (1972); R. I. Zhdanov and S. M. Zhenodarova, *Synthesis*, 222 (1975).
2. G. Stork, T. Takahashi, I. Kawamoto, and T. Suzuki, *J. Am. Chem. Soc.*, **100**, 8272 (1978).
3. D. Horton, *Org. Synth., Collect. Vol. V*, 1 (1973).
4. K. Ishihara, H. Kurihara, and H. Yamamoto, *J. Org. Chem.*, **58**, 3791 (1993).
5. G. Höfle, W. Steglich, and H. Vorbrüggen, *Angew. Chem., Int. Ed. Engl.*, **17**, 569 (1978).
6. B. A. D'Sa and J. G. Verkade, *J. Org. Chem.*, **61**, 2963 (1996).
7. E. Vedejs and S. T. Diver, *J. Am. Chem. Soc.*, **115**, 3358 (1993).
8. L. Kisfaludy, T. Mohacsi, M. Low, and F. Drexler, *J. Org. Chem.*, **44**, 654 (1979).
9. M. Jacobson, R. E. Redfern, W. A. Jones, and M. H. Aldridge, *Science*, **170**, 542 (1970).
10. J. C. Chottard, E. Mulliez, and D. Mansuy, *J. Am. Chem. Soc.*, **99**, 3531 (1977).
11. W. G. Dauben, R. A. Bunce, J. M. Gerdes, K. E. Henegar, A. F. Cunningham, Jr., and T. B. Ottoboni, *Tetrahedron Lett.*, **24**, 5709 (1983).
12. G. H. Posner and M. Oda, *Tetrahedron Lett.*, **22**, 5003 (1981); S. S. Rana, J. J. Barlow, and K. L. Matta, *Tetrahedron Lett.*, **22**, 5007 (1981).
13. T. Nishiguchi and H. Taya, *J. Am. Chem. Soc.*, **111**, 9102 (1989).
14. J. H. Babler and M. J. Coghlan, *Tetrahedron Lett.*, 1971 (1979).
15. T. Nishiguchi, S. Fujisaki, Y. Ishii, Y. Yano, and A. Nishida, *J. Org. Chem.*, **59**, 1191 (1994).
16. T. Nishiguchi and H. Taya, *J. Am. Chem. Soc.*, **111**, 9102 (1989).
17. Y. Nagao, E. Fujita, T. Kohno, and M. Yagi, *Chem. Pharm. Bull.*, **29**, 3202 (1981).
18. W. J. Hennen, H. M. Sweers, Y.-F. Wang, and C.-H. Wong, *J. Org. Chem.*, **53**, 4939 (1988). See also E. W. Holla, *Angew. Chem., Int. Ed. Engl.*, **28**, 220 (1989).
19. N. B. Bashir, S. J. Phythian, A. J. Reason, and S. M. Roberts, *J. Chem. Soc., Perkin Trans. 1*, 2203 (1995).
20. M. Therisod and A. M. Klibanov, *J. Am. Chem. Soc.*, **108**, 5638 (1986); H. M. Sweers and C.-H. Wong, *J. Am. Chem. Soc.*, **108**, 6421 (1986).

21. J. N. Cardner, T. L. Popper, F. E. Carlon, O. Gnoj, and H. L Herzog, *J. Org. Chem.*, **33**, 3695 (1968).
22. F. Aragozzini, E. Maconi, D. Potenza, and C. Scolastico, *Synthesis*, 225 (1989). For a review of the use of Sn–O derivatives to direct regioselective acylation and alkylation, see S. David and S. Hanessian, *Tetrahedron*, **41**, 643 (1985).
23. S. Yamada, *J. Org. Chem.*, **57**, 1591 (1992).
24. R. Nakao, K. Oka, and T. Fukomoto, *Bull. Chem. Soc. Jpn.*, **54**, 1267 (1981).
25. I. Shiina and T. Mukaiyama, *Chem. Lett.*, 677 (1994); J. Izumi, I. Shiina, and T. Mukaiyama, *Chem. Lett.*, 141 (1995).
26. K. Ishihara, M. Kubota, H. Kurihara, and H. Yamamoto, *J. Org. Chem.*, **61**, 4560 (1996).
27. K. Ishihara, M. Kubota, and H. Yamamoto, *Synlett*, 265 (1996).
28. W. R. Roush and D. A. Barda, *Tetrahedron Lett.*, **38**, 8785 (1997).
29. Y. Ishii, M. Takeno, Y. Kawasaki, A. Muromachi, Y. Nishiyama, and S. Sakaguchi, *J. Org. Chem.*, **61**, 3088 (1996).
30. P. A. Procopiou, S. P. D. Baugh, S. S. Flack, and G. G. A. Inglis, *J. Chem. Soc., Chem. Commun.*, 2625 (1996); *idem, J. Org. Chem.*, **63**, 2342 (1998).
31. T. Nishiguchi and H. Taya, *J. Chem. Soc., Perkin Trans. 1*, 172 (1990).
32. M. Oikawa, A. Wada, F. Okazaki, and S. Kusumoto, *J. Org. Chem.*, **61**, 4469 (1996).
33. F. I. Auzanneau and D. R. Bundle, *Carbohydr. Res.*, **212**, 13 (1991).
34. P. C. Zhu, J. Lin, and C. U. Pittman, Jr., *J. Org. Chem.*, **60**, 5729 (1995).
35. V. Srivastava, A.Tandon, and S. Ray, *Synth. Commun.*, **22**, 2703 (1992).
36. K. P. R. Kartha and R. A. Field, *Tetrahedron*, **53**, 11753 (1997).
37. (a) C.-S. Chen and C. J. Sih, "General Aspects and Optimization of Enantioselective Biocatalysis in Organic Solvents—The Use of Lipases," *Angew. Chem., Int. Ed. Engl.*, **28**, 695 (1989). (b) D. H. G. Crout and M. Christen, "Biotransformations in Organic Synthesis," *Mod. Synth. Methods*, **5**, 1 (1989). (c) H. G. Davies, R. H. Green, D. R. Kelly, and S. M. Roberts, *Biotransformations in Preparative Organic Chemistry: The Use of Isolated Enzymes and Whole Cell Systems,* Academic Press, New York, 1989. (d) M. Ohno and M. Otsuka, "Chiral Synthons by Ester Hydrolysis Catalysed by Pig Liver Esterase," *Org. React.*, **37**, 1 (1989). (e) C.-H. Wong, "Enzymatic Catalysts in Organic Synthesis," *Science*, **244**, 1145 (1989). (f) C. J. Sih and S. H. Wu, "Resolution of Enantiomers via Biocatalysis," *Top. Stereochem.*, **19**, 63 (1989). (g) N. Turner, "Recent Advances in the Use of Enzyme-Catalysed Reactions in Organic Synthesis," *Nat. Prod. Rep.*, **6**, 625 (1989). (h) L. Zhu and M. C. Tedford, "Applications of Pig Liver Esterases (PLE) in Asymmetric Synthesis," *Tetrahedron*, **46**, 6587 (1990). (i) A. M. Klibanov, "Asymmetric Transformations Catalysed by Enzymes in Organic Solvents," *Acc. Chem. Res.*, **23**, 114 (1990). (j) D. G. Drueckhammer, W. J. Hennen, R. L. Pederson, C. F. Barbas, III, C. M. Gautheron, T. Krach, and C.-H. Wong, "Enzyme Catalysis in Synthetic Carbohydrate Chemistry," *Synthesis*, 499 (1991). (k) W. Boland, C. Frössl, and M. Lorenz, "Esterolytic and Lipolytic Enzymes in Organic Synthesis," *Synthesis*, 1049 (1991). (l) S. David, C. Augé, and C. Gautheron, "Enzymic Methods in Preparative Carbohydrate Chemistry," *Adv. Carbohydr. Chem. Biochem.*, **49**, 175 (1992). (m) H. Waldmann, "Enzymic Protecting Group Techniques," *Kontakte (Darmstadt)*, **2**, 33 (1991). (n) E. Santaniello, P. Ferraboschi, P. Grisenti, and A. Manzocchi, "The

Biocatalytic Approach to the Preparation of Enantiomerically Pure Chiral Building Blocks," *Chem. Rev.*, **92**, 1071 (1992). (o) L. Poppe and L. Novak, *Selective Biocatalysis: A Synthetic Approach,* VCH, Weinheim 1992. (p) K. Farber, *Biotransformations in Organic Chemistry*, Springer-Verlag, Berlin 1992. (q) A. Reidel and H. Waldmann, "Enzymic Protecting Group Techniques in Bioorganic Synthesis," *J. Prakt. Chem./Chem.–Ztg.*, **335**, 109 (1993). (r) H. Waldmann and D. Sebastian, "Enzymatic Protecting Group Techniques," *Chem. Rev.*, **94**, 911 (1994). (s) K. Drauz and H. Waldmann, Eds., *Enzyme Catalysis in Organic Chemistry: A Comprehensive Handbook,* VCH, Weinheim, 1995.

38. E. W. Holla, *J. Carbohydr. Chem.*, **9**, 113 (1990).
39. G. Iacazio and S. M. Roberts, *J. Chem. Soc., Perkin Trans. 1*, 1099 (1993); M. J. Chinn, G. Iacazio, D. G. Spackman, N. J. Turner, and S. H. Roberts, *J. Chem. Soc., Perkin Trans. 1*, 661 (1992).
40. I. Matsuo, M. Isomura, R. Walton, and K. Ajisaka, *Tetrahedron Lett.*, **37**, 8795 (1996).
41. J. J. Gridley, A. J. Hacking, H. M. I. Osborn, and D. Spackman, *Synlett*, 1397 (1997).
42. S. Ramaswamy, B. Morgan, and A. C. Oehlschager, *Tetrahedron Lett.*, **31**, 3405 (1990).
43. F. Theil and H. Schick, *Synthesis*, 533 (1991).
44. Y. Terao, M. Akamatsu, and K. Achiwa, *Chem. Pharm. Bull.*, **39**, 823 (1991).
45. L. Ling, Y. Watanabe, T. Akiyama, and S. Ozaki, *Tetrahedron Lett.*, **33**, 1911 (1992).
46. C. R. Johnson, A. Golebiowski, T. K. McGill, and D. H. Steensma, *Tetrahedron Lett.*, **32**, 2597 (1991).
47. C. Chauvin and D. Plusquellec, *Tetrahedron Lett.*, **32**, 3495 (1991).
48. S. Yamada, *J. Org. Chem.*, **57**, 1591 (1992).
49. J. Iqbal and R. R. Srivastava, *J. Org. Chem.*, **57**, 2001 (1992).
50. J. J. Plattner, R. D. Gless, and H. Rapoport, *J. Am. Chem. Soc.*, **94**, 8613 (1972).
51. B. Reinhard and H. Faillard, *Liebigs Ann. Chem.*, 193 (1994).
52. K. Mori, M. Tominaga, T. Takigawa, and M. Matsui, *Synthesis*, 790 (1973).
53. K. Mori and M. Sasaki, *Tetrahedron Lett.*, 1329 (1979).
54. J. Herzig, A. Nudelman, H. E. Gottlieb, and B. Fischer, *J. Org. Chem.*, **51**, 727 (1986).
55. N. Kunesch, C. Meit, and J. Poisson, *Tetrahedron Lett.*, **28**, 3569 (1987).
56. T. Neilson and E. S. Werstiuk, *Can. J. Chem.*, **49**, 493 (1971).
57. A. Nudelman, J. Herzig, H. E. Gottlieb, E. Keinan, and J. Sterling, *Carbohydr. Res.*, **162**, 145 (1987).
58. D. Askin, C. Angst, and S. Danishefsky, *J. Org. Chem.*, **52**, 622 (1987).
59. L. H. B. Baptistella, J. F. dos Santos, K. C. Ballabio, and A. J. Marsaioli, *Synthesis*, 436 (1989).
60. R. Yanada, N. Negoro, K. Bessho, and K. Yanada, *Synlett,* 1261 (1995).
61. V. Pozsgay, *J. Am. Chem. Soc.*, **117**, 6673 (1995).
62. N. Yamamoto, T. Nishikawa, and M. Isobe, *Synlett*, 505 (1995).
63. S. V. Ley, A. Armstrong, D. Diez-Martin, M. J. Ford, P. Grice, J. G. Knight, H. C. Kolb, A. Madin, C. A. Marby, S. Mukherjee, A. N. Shaw, A. M. Z. Slawin, S. Vile, A. D. White, D. J. Williams, and M. Woods, *J. Chem. Soc., Perkin Trans. 1*, 667 (1991).

64. J. Otera, N. Dan-oh, and H. Nozaki, *Tetrahedron*, **49**, 3065 (1993).
65. M. G. Perez and M. S. Maier, *Tetrahedron Lett.*, **36**, 3311 (1995).
66. Y.-C. Xu, A. Bizuneh, and C. Walker, *Tetrahedron Lett.*, **37**, 455 (1996).
67. Y.-C. Xu, A. Bizuneh, and C. Walker *J. Org. Chem.*, **61**, 9086 (1996).
68. Q. Y. Zheng, L. G. Darbie, X. Cheng, and C. K. Murray, *Tetrahedron Lett.*, **36**, 2001 (1995).
69. W. R. Roush and X.-F. Lin, *J. Am. Chem. Soc.*, **117**, 2236 (1995).
70. Y.-F. Wang, C.-S. Chen, G. Girdaukas and C. J. Sih, in *Enzymes in Organic Synthesis (Ciba Foundation Symposium,* Vol. **111**), 128 (1985); K. Tsuji, Y. Terao, and K. Achiwa, *Tetrahedron Lett.*, **30**, 6189 (1989); R. Csuk and B. I. Glaenzer, *Z. Naturforsch. B, Chem. Sci.*, **43**, 1355 (1988). For examples in a cyclic series, see K. Laumen and M. Schneider, *Tetrahedron Lett.*, **26**, 2073 (1985); K. Naemura, N. Takahashi, and H. Chikamatsu, *Chem. Lett.*, 1717 (1988); C. R. Johnson and C. H. Senanayake, *J. Org. Chem.*, **54**, 735 (1989); D. R. Deardorff, A. J. Matthews, D. S. McMeekin, and C. L. Craney, *Tetrahedron Lett.*, **27**, 1255 (1986); N. W. Boaz, *Tetrahedron Lett.*, **30**, 2061 (1989).
71. M. Kloosterman, E. W. J. Mosuller, H. E. Schoemaker, and E. M. Meijer, *Tetrahedron Lett.*, **28**, 2989 (1987).
72. O. Houille, T. Schmittberger, and D. Uguen, *Tetrahedron Lett.*, **37**, 625 (1996).
73. J. Sakaki, H. Sakoda, Y. Sugita, M. Sato, and C. Kaneto, *Tetrahedron: Asymmetry*, **2**, 343 (1991).
74. R. Lopez, E. Montero, F. Sanchez, J. Cañada, and A. Fernandez-Mayoralas, *J. Org. Chem.*, **59**, 7027 (1994).
75. E. W. Holla, V. Sinnwell, and W. Klaffke, *Synlett*, 413 (1992).
76. T. Itoh, A. Uzu, N. Kanda, and Y. Takagi, *Tetrahedron Lett.*, **37**, 91 (1996).
77. U. Ellervik and G. Magnusson, *Tetrahedron Lett.*, **38**, 1627 (1997).

Chloroacetate Ester: $ClCH_2CO_2R$

Formation

1. $(ClCH_2CO)_2O$, Pyr, 0°, 70–90% yield.[1]
2. $ClCH_2COCl$, Pyr, ether, 87% yield.[2]
3. PPh_3, DEAD, $ClCH_2CO_2H$, 73% yield.[3] In this case, the esterification proceeds with inversion of the configuration at the alcoholic center.
4. Vinyl chloroacetate, $Cp^*{}_2Sm(THF)_2$, toluene, rt, 99% yield. With SmI_2 as catalyst, the yield is 79%.[4]

Cleavage

The chloroacetate group has been observed to migrate during silica gel chromatography.[5] In general, cleavage of chloroacetates can be accomplished in the presence of other esters such as acetates and benzoates because of the large difference in the hydrolysis rates for esters bearing electron-withdrawing groups.

A study comparing the half-lives for hydrolysis of a variety of esters of 5′-*O*-acyl-uridines gave the following results:[6]

Acyl Group	$t_{1/2}$ min	
	Reagent I	Reagent II
CH_3CO–	191	59
$MeOCH_2CO$–	10.4	2.5
$PhOCH_2CO$–	3.9	<1[a]
Formyl–	0.4	0.22[b]
$ClCH_2CO$–	0.28	0.17[b]

Reagent I = 155 mM NH_3/H_2O; reagent II = NH_3/MeOH.
[a] Reaction is too fast to measure. [b] Time for complete solvolysis of the substrate.

The relative rates of alkaline hydrolysis of acetate, chloro-, dichloro-, and trichloroacetates have been compared and are: $1{:}760{:}1.6\times10^4{:}10$.[5,7]

1. $HSCH_2CH_2NH_2$ or $H_2NCH_2CH_2NH_2$ or *o*-phenylenediamine, Pyr, Et_3N, 1 h, rt.[1]
2. Thiourea, $NaHCO_3$, EtOH, 70°, 5 h, 70% yield.[2]
3. H_2O, Pyr, pH 6.7, 20 h, 100% yield.[8]
4. $NH_2NHC(S)SH$, lutidine, AcOH, 2–20 min, rt, 88–99% yield.[9,10] This method is superior to the use of thiourea in that it proceeds at lower temperatures and affords much higher yields. The reagent also serves to remove the related bromoacetyl esters that are 5–10 times more labile under these conditions. Cleavage occurs cleanly in the presence of an acetate.[11]

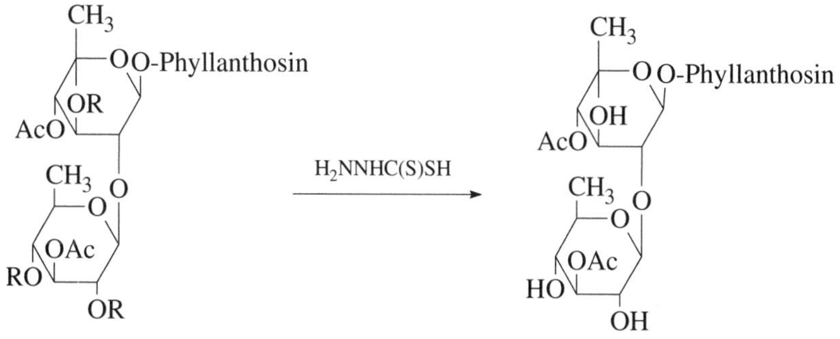

$H_2NNHC(S)SH$

R = $ClCH_2CO$–

5. "Hydrazine acetate."[12]

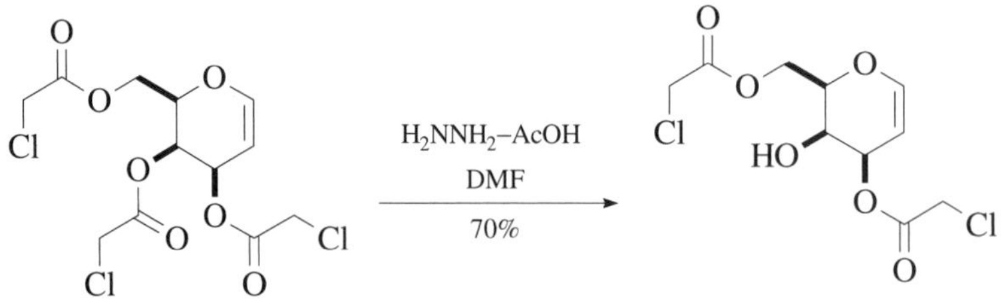

H_2NNH_2–AcOH
DMF
70%

Ref. 13

6. The lipase from *Pseudomonas sp.* K10 has also been used to cleave the chloroacetate, resulting in the resolution of a racemic mixture, since only one enantiomer was cleaved.[14]

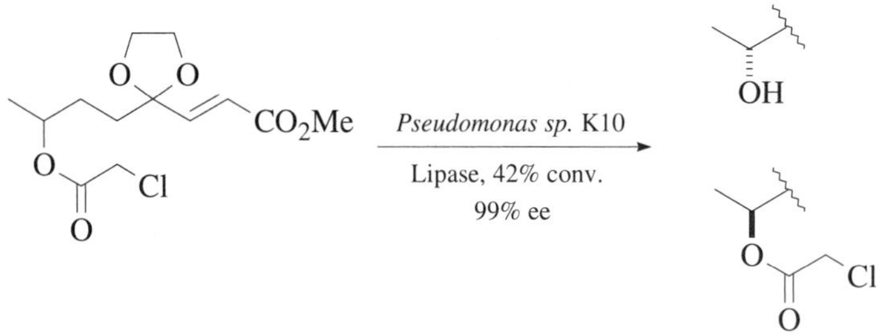

Pseudomonas sp. K10
Lipase, 42% conv.
99% ee

7. *N,N*-Pentamethylenethiourea, TEA, dioxane, 70°, 3 h.[15]
8. NH_3, THF, −50° to −40°, 2.5 h. The use of hydrazine failed in this case.[16]

1. A. F. Cook and D. T. Maichuk, *J. Org. Chem.*, **35**, 1940 (1970).
2. M. Naruto, K. Ohno, N. Naruse, and H. Takeuchi, *Tetrahedron Lett.*, 251 (1979).
3. M. Saiah, M. Bessodes, and K. Antonakis, *Tetrahedron Lett.*, **33**, 4317 (1992); B. Lipshutz and T. A. Miller, *Tetrahedron Lett.*, **31**, 5253 (1990).
4. Y. Ishii, M. Takeno, Y. Kawasaki, A. Muromachi, Y. Nishiyama, and S. Sakaguchi, *J. Org. Chem.*, **61**, 3088 (1996).
5. V. Pozsgay, *J. Am. Chem. Soc.*, **117**, 6673 (1995).
6. C. B. Reese, J. C. M. Stewart, J. H. van Boom, H. P. M. de Leeuw, J. Nagel, and J. F. M. de Rooy, *J. Chem Soc., Perkin Trans. 1*, 934 (1975).
7. N. S. Isaacs, *Physical Organic Chemistry*, 2d. ed., Wiley and Sons, New York, 1995, p 515.
8. F. Johnson, N. A. Starkovsky, A. C. Paton, and A. A. Carlson, *J. Am. Chem. Soc.*, **86**, 118 (1964).
9. C. A. A. van Boeckel and T. Beetz, *Tetrahedron Lett.*, **24**, 3775 (1983).
10. A. S. Cambell and B. Fraser-Reid, *J. Am. Chem. Soc.*, **117**, 10387 (1995).
11. A. B. Smith, III, K. J. Hale, and H. A. Vaccaro, *J. Chem. Soc., Chem. Commun.*, 1026 (1987).
12. U. E. Udodong, C. S. Rao, and B. Fraser-Reid, *Tetrahedron*, **48**, 4713 (1992).

13. S. Bouhroum and P. J. A. Vottero, *Tetrahedron Lett.*, **31**, 7441 (1990).
14. T. K. Ngooi, A. Scilimati, Z.-W. Guo, and C. J. Sih, *J. Org. Chem.*, **54**, 911 (1989).
15. U. Schmidt, M. Kroner, and H. Griesser, *Synthesis*, 294 (1991).
16. J. C. McWilliams and J. Clardy, *J. Am. Chem. Soc.*, **116**, 8378 (1994).

Dichloroacetate Ester: Cl_2CHCO_2R

Formation

1. $Cl_2CHCOCl$.[1]
2. $(Cl_2CHCO)_2O$, Pyr, CH_2Cl_2.[2]
3. $Cl_2CHCOCCl_3$, DMF, 56% yield.[3] This reagent was used to acylate selectively the 6-position of an α-methylglucoside.

Cleavage

1. pH 9–9.5, 20°, 30 min.[1]
2. NH_3, MeOH.[3,4]
3. KOH, *t*-BuOH, H_2O, THF.[2]

1. J. R. E. Hoover, G. L. Dunn, D. R. Jakas, L. L. Lam, J. J. Taggart, J. R. Guarini, and L. Phillips, *J. Med. Chem.*, **17**, 34 (1974).
2. S. Masamune, W. Choy, F. A. J. Kerdesky, and B. Imperiali, *J. Am. Chem. Soc.*, **103**, 1566 (1981).
3. A. H. Haines and E. J. Sutcliffe, *Carbohydr. Res.*, **138**, 143 (1985).
4. C. B. Reese, J. C. M. Stewart, J. H van Boom, H. P. M. de Leeuw, J. Nagel, and J. F. M. de Rooy, *J. Chem Soc., Perkin Trans. 1*, 934 (1975).

Trichloroacetate Ester: RO_2CCCl_3 (Chart 2)

Formation

1. Cl_3CCOCl, Pyr, DMF, 20°, 2 days, 60–90% yield.[1]

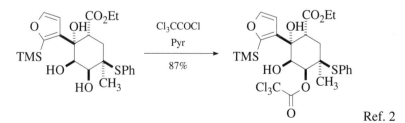

Ref. 2

Cleavage

1. NH_3, EtOH, $CHCl_3$, 20°, 6 h, 81% yield.[1] Cleavage of the trichloroacetate occurs selectively in the presence of an acetate.

2. KOH, MeOH, 72% yield.[1] A formate ester was not hydrolyzed under these conditions.

1. V. Schwarz, *Collect. Czech. Chem. Commun.*, **27**, 2567 (1962).
2. S. Bailey, A. Teerawutgulrag, and E. J. Thomas, *J. Chem. Soc., Chem. Commun.*, 2519 (1995).

Trifluoroacetate Ester (RO–TFA): CF_3CO_2R

Formation

1. $(CF_3CO)_2O$, Pyr.[1]
2. Even with this highly reactive reagent, excellent selectivity was achieved for one of two very similar alcohols.[2]

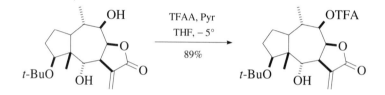

OH, t-BuO, OH → (TFAA, Pyr, THF, – 5°, 89%) → OTFA, t-BuO, OH

3. 2-Pyridyl trifluoroacetate, ether, 20°, 30 min, 99% yield.[3]
4. CF_3CO_3H, 20°, 4 h, 83% yield.[4] In this case, a hindered alcohol was converted to the TFA derivative. The use of TFA failed to give a trifluoroacetate. This method is probably not general.

Cleavage

A series of nucleoside trifluoroacetates was hydrolyzed rapidly in 100% yield at 20°, pH 7.[5]

1. A. Lardon and T. Reichstein, *Helv. Chim. Acta*, **37**, 443 (1954).
2. P. T. Lansbury, T. E. Nickson, J. P. Vacca, R. D. Sindelar, and J. M. Messinger, II, *Tetrahedron*, **43**, 5583 (1987).
3. T. Keumi, M. Shimada, T. Morita, and H. Kitajima, *Bull. Chem. Soc. Jpn.*, **63**, 2252 (1990).
4. G. W. Holbert and B. Ganem, *J. Chem. Soc., Chem. Commun.*, 248 (1978).
5. F. Cramer, H. P. Bär, H. J. Rhaese, W. Sänger, K. H. Scheit, G. Schneider, and J. Tennigkeit, *Tetrahedron Lett.*, 1039 (1963).

Methoxyacetate Ester: $MeOCH_2CO_2R$

Formation

1. $MeOCH_2COCl$, Pyr.[1]

Cleavage

1. $NH_3/MeOH$ or NH_3/H_2O, 78% yield.[1] In nucleoside derivatives, the methoxyacetate is cleaved 20 times faster than an acetate. It can be cleaved in the presence of a benzoate.
2. $Yb(OTf)_3$, MeOH, 0–25°, 92–99% yield. Acetates, benzoates, THP, TBDMS, TBDPS, and MEM ethers are not affected by this reagent.[2]
3. Ethanolamine, IPA, reflux, 21 h, >50% yield. These conditions did not affect the C-10 acetate or the C-13 side chain of a taxol derivative.[3]

1. C. B. Reese and J. C. M. Stewart, *Tetrahedron Lett.*, 4273 (1968).
2. T. Hanamoto, Y. Sugimato, Y. Yokoyama, and J. Inanaga, *J. Org. Chem.*, **61**, 4491 (1996).
3. R. B. Greenwald, A. Pendri, and D. Bolikal, *J. Org. Chem.*, **60**, 331 (1995).

Triphenylmethoxyacetate Ester: $ROCOCH_2OCPh_3$

The triphenylmethoxyacetate was prepared in 53% yield from a nucleoside and the sodium acetate ($Ph_3COCH_2CO_2Na$, *i*-$Pr_3C_6H_2SO_2Cl$, Pyr) as a derivative that could be easily detected on TLC (i.e., it has a distinct orange-yellow color after it is sprayed with ceric sulfate). It is readily cleaved by $NH_3/MeOH$ (100% yield).[1]

1. E. S. Werstiuk and T. Neilson, *Can. J. Chem.*, **50**, 1283 (1972).

Phenoxyacetate Ester: $PhOCH_2CO_2R$ (Chart 2)

Formation

1. $(PhOCH_2CO)_2O$, Pyr.[1,2]
2. $(PhOCH_2CO)_2O$, Pyr, DMAP, CH_2Cl_2, 0°.[3]

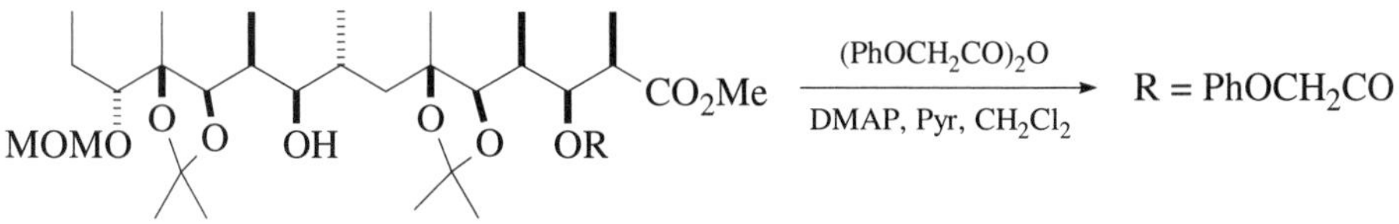

MOMO, O, O, OH, O, O, OR, CO_2Me — $(PhOCH_2CO)_2O$ / DMAP, Pyr, CH_2Cl_2 → R = $PhOCH_2CO$

R = H

Cleavage

1. t-$BuNH_2$, MeOH.[2]
2. NH_3 in H_2O or MeOH.[1] The phenoxyacetate is 50 times more labile to aqueous ammonia than is an acetate.

1. C. B. Reese and J. C. M. Stewart, *Tetrahedron Lett.*, 4273 (1968).
2. T. Kamimura, T. Masegi, and T. Hata, *Chem. Lett.*, 965 (1982).
3. R. B. Woodward and 48 co-workers, *J. Am. Chem. Soc.*, **103**, 3210 (1981).

p-Chlorophenoxyacetate Ester: $ROCOCH_2OC_6H_4$–*p*-Cl

The *p*-chlorophenoxyacetate, prepared to protect a nucleoside by reaction with the acetyl chloride, is cleaved by 0.2 *M* NaOH, dioxane-H_2O, 0°, 30 sec.[1]

1. S. S. Jones and C. B. Reese, *J. Am. Chem. Soc.*, **101**, 7399 (1979).

Phenylacetate Ester: $PhCH_2CO_2R$

Formation

1. *Lipase Fp*, $PhCH_2CO_2CH{=}CH_2$, 84–88% yield.[1]

Cleavage

1. Penicillin G Acylase.[1,2]

1. E. W. Holla, *J. Carbohydr. Chem.*, **9**, 113 (1990).
2. H. Waldmann, A. Heuser, P. Braun, M. Schulz, and H. Kunz, in *Microbial Reagents in Organic Synthesis*, S. Servi, Ed., Kluwer Academic, Dordrecht, 1992, pp 113–122.

p-*P*-Phenylacetate Ester: $ROCOCH_2C_6H_4$–*p*-***P***

Monoprotection of a symmetrical diol can be effected by reaction with a polymer-supported phenylacetyl chloride. The free hydroxyl group is then converted to an ether and the phenylacetate cleaved by aqueous ammonia–dioxane, 48 h.[1]

$$HO(CH_2)_nOH + p\text{-}\boldsymbol{P}\text{-}C_6H_4CH_2COCl \xrightarrow{\text{Pyr}} HO(CH_2)_nOCOCH_2C_6H_4\text{–}p\text{-}\boldsymbol{P}$$

1. J. Y. Wong and C. C. Leznoff, *Can. J. Chem.*, **51**, 2452 (1973).

Diphenylacetate Ester (DPA–OR): Ph_2CHCO_2R

The DPA ester is formed from the acid chloride in pyridine (40–96% yield). It is cleaved oxidatively by treatment with NBS followed by thiourea (40–88% yield).[1]

1. F. Santoyo-González, F. Garcia-Calvo-Flores, J. Isac-Garcia, R. Robles-Diaz, and A. Vargas-Berenguel, *Synthesis*, 97 (1994).

Nicotinate Ester: (3-pyridyl-CO_2R)

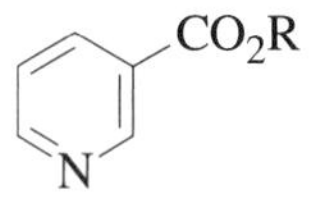

Formation

1. 3-Pyridylcarboxylic acid anhydride, 93–99% yield.[1]

Cleavage

1. MeI followed by hydroxide, 55–98% yield. Quaternization of the pyridine increases the rate of hydrolysis of the ester.

1. S. Ushida, *Chem. Lett.*, 59 (1989).

3-Phenylpropionate Ester: $ROCOCH_2CH_2Ph$

The 3-phenylpropionate ester has been used in nucleoside synthesis.[1] It is cleaved by α-chymotrypsin (37°, 8–16 h, 70–90% yield)[2] and can be cleaved in the presence of an acetate.[3]

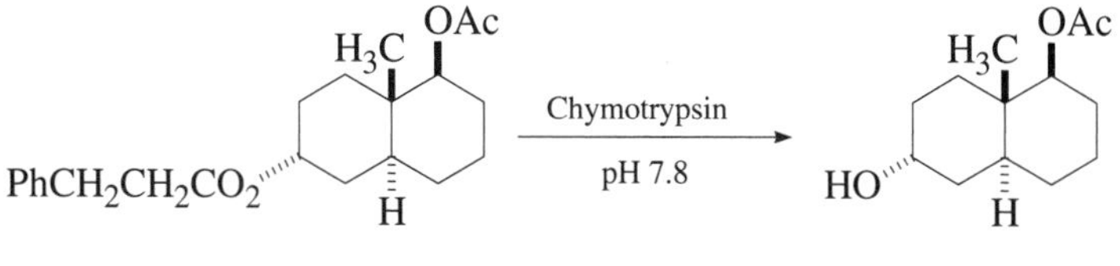

$PhCH_2CH_2CO_2$ / H_3C / OAc / H —Chymotrypsin, pH 7.8→ HO / H_3C / OAc / H

1. H. S. Sachdev and N. A. Starkovsky, *Tetrahedron Lett.*, 733 (1969).
2. A. T.-Rigby, *J. Org. Chem.*, **38**, 977 (1973).
3. Y. Y. Lin and J. B. Jones, *J. Org. Chem.*, **38**, 3575 (1973).

4-Pentenoate Ester: $CH_2{=}CHCH_2CH_2CO_2R$

Formation

1. $CH_2{=}CHCH_2CH_2COCl$.[1] This group was used for the protection of anomeric hydroxyl groups.

Cleavage

1. NBS, 1% H_2O, CH_3CN. 36–85%.[1]

1. J. C. Lopez and B. Fraser-Reid, *J. Chem. Soc., Chem. Commun.*, 159 (1991).

4-Oxopentanoate (Levulinate) Ester (Lev–OR): $ROCOCH_2CH_2COCH_3$

Formation

1. $(CH_3COCH_2CH_2CO)_2O$, Pyr, 25°, 24 h, 70–85% yield.[1]
2. $CH_3COCH_2CH_2CO_2H$, DCC, DMAP, 96% yield.[2]
3. 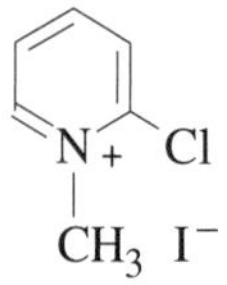 CMPI, $CH_3COCH_2CH_2CO_2H$, DABCO, 86% yield.[3]

(CMPI = 2-chloro-1-methylpyridinium iodide)

Cleavage

The levulinate is less prone to migrate than the benzoate and acetate.[4]

1. $NaBH_4$, H_2O, pH 5–8, 20°, 20 min, 80–95% yield.[1] The by-product, 5-methyl-γ-butyrolactone, is water soluble and thus easily removed.
2. 0.5 *M* H_2NNH_2, H_2O, Pyr, AcOH, 2 min, 100% yield.[2] Normal esters are not cleaved under these conditions.[5]
3. MeMgI, 0°, 2 h, 93% yield.[6] A levulinate is cleaved in preference to a benzoate.
4. $NaHSO_3$, THF, CH_3CN or EtOH, 86–90% yield.[7]

4,4-(Ethylenedithio)pentanoate Ester (Levulinoyl Dithioacetal Ester) RO–LevS:

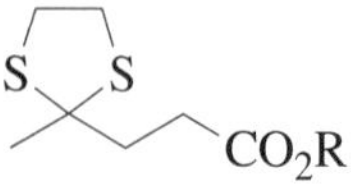

CO_2R

Formation

1. 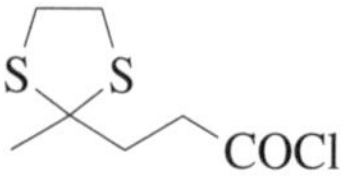 COCl 2,6-lutidine, 0°, 70% yield.[8]

2. 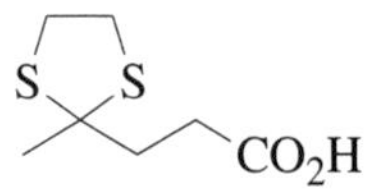 [dithiolane structure] CO_2H CMPI, DABCO, dioxane, 2 h, 20°, 96% yield.[3]

(CMPI = 2-chloro-1-methylpyridinium iodide)

Cleavage

The LevS group is converted to the Lev group with $HgCl_2/HgO$ (acetone/H_2O, 4 h, 20°, 74% yield). It can then be hydrolyzed using the conditions that remove the Lev group.[8] The LevS group is stable to the conditions used for glycoside formation [$HgBr_2$, $Hg(CN)_2$].

1. A. Hassner, G. Strand, M. Rubinstein, and A. Patchornik, *J. Am. Chem. Soc.*, **97**, 1614 (1975).
2. J. H. van Boom and P. M. J. Burgers, *Tetrahedron Lett.*, 4875 (1976).
3. H. J. Koeners, J. Verhoeven, and J. H. van Boom, *Tetrahedron Lett.*, **21**, 381 (1980).
4. J. N. Glushka and A. S. Perlin, *Carbohydr. Res.*, **205**, 305 (1990); R. N. Rej, J. N. Glushka, W. Chew, and A. S. Perlin, *Carbohydr. Res.*, **189**, 135 (1989).
5. N. Jeker and C. Tamm, *Helv. Chim. Acta*, **71**, 1895, 1904 (1988).
6. Y. Watanabe, T. Fujmoto, and S. Ozaki, *J. Chem. Soc., Chem. Commun.*, 681 (1992).
7. M. Ono and I. Itoh, *Chem. Lett.*, 585 (1988).
8. H. J. Koeners, C. H. M. Verdegaal, and J. H. Van Boom, *Recl. Trav. Chim. Pays-Bas*, **100**, 118 (1981).

5-[3-Bis(4-methoxyphenyl)hydroxymethylphenoxy]levulinic Acid Ester

This ester is formed from the anhydride in pyridine and is quantitatively cleaved with H_2NNH_2–H_2O, Pyr–AcOH. The sensitivity of detection of this ester is high with its absorbance maximum of 513 nm and extinction coefficient of 78,600 in 5% Cl_2CHCO_2H/CH_2Cl_2 where it forms the trityl cation.[1]

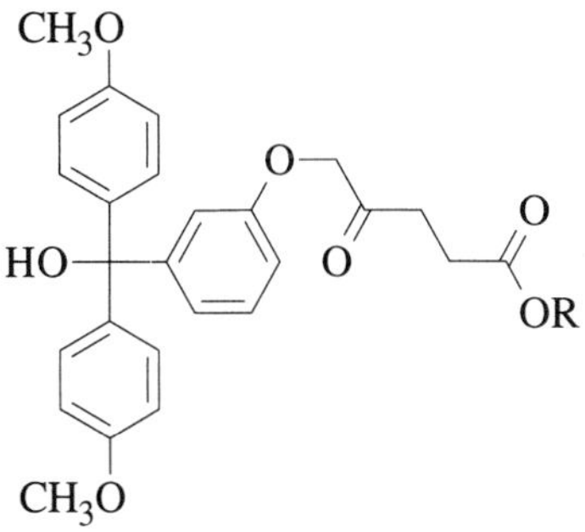

CH_3O

HO — O — O — O — OR

CH_3O

1. E. Leikauf and H. Köster, *Tetrahedron*, **51**, 5557 (1995).

Pivaloate Ester (Pv–OR): $(CH_3)_3CCO_2R$ (Chart 2)

Formation

1. PvCl, Pyr, 0–75°, 2.5 days, 99% yield.[1] In general, such extended reaction times are not required to obtain complete reaction. This is an excellent reagent for selective acylation of a primary alcohol over a secondary alcohol.[2,3,4]

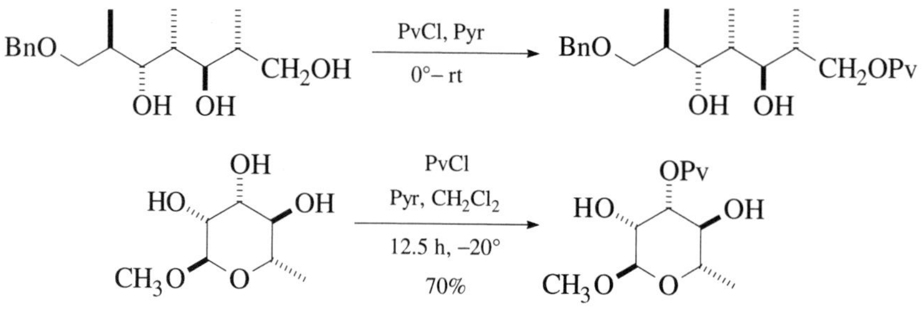

PvCl, Pyr, 0°– rt

PvCl, Pyr, CH_2Cl_2, 12.5 h, –20°, 70%

Ref. 5

2. Selective acylation can be obtained for one of two primary alcohols having slightly different steric environments.[6,7]

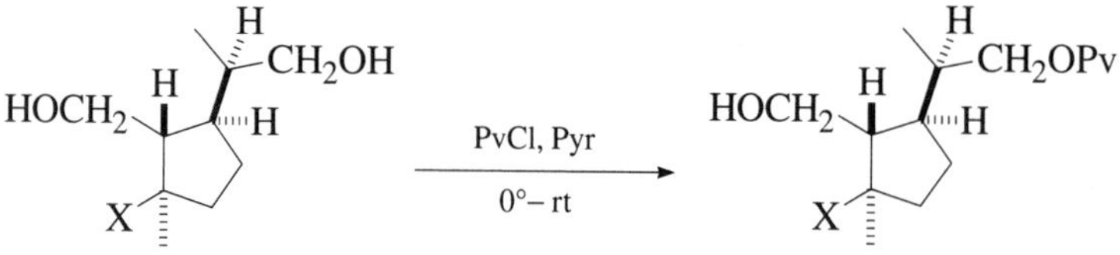

PvCl, Pyr, 0°– rt

α-Methylglucoside can be selectively acylated at the 2,6-positions in 89% yield and α-methyl 4,6-*O*-benzylidineglucoside can be selectively acylated at the 2-position in 77% yield.[8]

3. Vinyl pivaloate, $Cp^*Sm(THF)_2$, toluene, 3 h, 99% yield.[9]
4. Pivaloic anhydride, $Sc(OTf)_3$, CH_3CN, $-20°$, 4 h.[10]
5. Pivaloic anhydride, $MgBr_2$, TEA, CH_2Cl_2, rt, 99% yield.[10]
6.

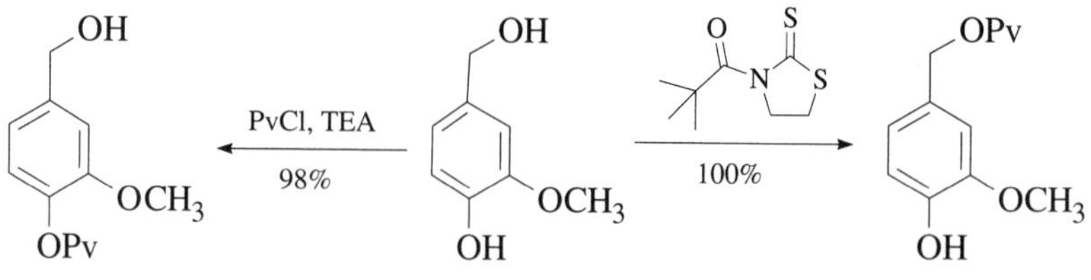

PvCl, TEA, 98% ← → 100%

Thiazolidine-2-thione shows excellent selectivity for primary alcohols over secondary alcohols (>20:1).[11]

Cleavage

1. $Bu_4N^+OH^-$, 20°, 4 h.[12]
2. aq. $MeNH_2$, 20°, $t_{1/2}$ = 3 h.[13] In this case, the 5′-position of uridine was deprotected. Acetates can be cleaved selectively in the presence of a pivaloate with NH_3/MeOH. Pivaloates are not cleaved by hydrazine in refluxing ethanol, a condition that cleaves phthalimides.[14]
3. 0.5 *N* NaOH, EtOH, H_2O, 20°, 12 h, 58% yield.[15]
4. NaOMe, MeOH, 90% yield.[16]
5. MeLi, Et_2O, 20°.[17]
6. *t*-BuOK, H_2O (8:2), 20°, 3 h, 94% yield.[18]
7. *i*-Bu_2AlH, CH_2Cl_2, −78°, 95% yield.[2] *i*-Bu_2AlH, CH_2Cl_2, toluene, 84% yield. Three pivaloates were cleaved from a zaragozic acid intermediate. The use of THF or ether as solvent failed to remove all three.[19]
8. Fungus, *Currulania lunata*, 6 h, 64% yield.[20] In this case, a 21-pivaloate was removed from a steroid.
9. $K^+Et_3BH^-$, THF, −78°, 78% yield.[21]

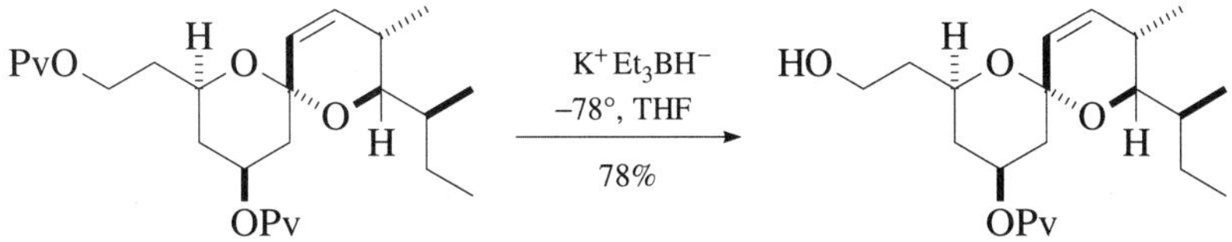

PvO ... OPv —($K^+Et_3BH^-$, −78°, THF, 78%)→ HO ... OPv

10. EtMgBr, Et_2O, 90% yield. Under these conditions silyl migration is not a problem, as it was when the typical hydrolysis conditions were used.[22]
11. PvO–(cyclic phosphate)–OPh —esterase→ [HO–(cyclic phosphate)–OPh] → $(HO)_2P(O)$—OPh

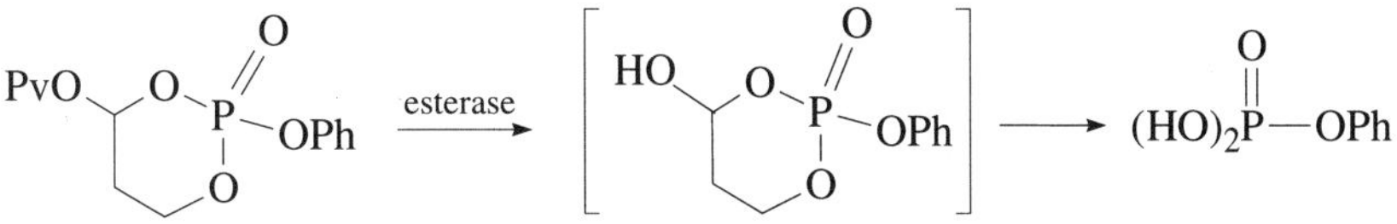

Ref. 23

12. Al_2O_3, microwaves, 12 min, 93% yield.[24] Cleavage of acetates occurs similarly.
13. Esterase from rabbit serum, 53–95% yield.[25]
14. Li, NH_3, Et_2O; NH_4Cl, 70–85% yield.[26]
15. 3 *M* HCl, dioxane, reflux, 18 h, 80% yield.[27]
16. Sm, I_2, MeOH, 24 h reflux, 95% yield.[28] Troc, Ac, and Bz groups are also cleaved.

1. M. J. Robins, S. D. Hawrelak, T. Kanai, J.-M. Siefert, and R. Mengel, *J. Org. Chem.*, **44**, 1317 (1979).

2. K. C. Nicolaou and S. E. Webber, *Synthesis*, 453 (1986).
3. D. Boschelli, T. Takemasa, Y. Nishitani, and S. Masamune, *Tetrahedron Lett.*, **26**, 5239 (1985).
4. H. Nagaoka, W. Rutsch, G. Schmid, H. Ilio, M. R. Johnson, and Y. Kishi, *J. Am. Chem. Soc.*, **102**, 7962 (1980).
5. P. Jütten and H. D. Scharf, *J. Carbohydr. Chem.*, **9**, 675 (1990).
6. N. Kato, H. Kataoka, S. Ohbuchi, S. Tanaka, and H. Takeshita, *J. Chem. Soc., Chem. Commun.*, 354 (1988).
7. P. F. Schuda and M. R. Heimann, *Tetrahedron Lett.*, **24**, 4267 (1983).
8. S. Tomic-Kulenovic and D. Keglevic, *Carbohydr. Res.*, **85**, 302 (1980).
9. Y. Ishii, M. Takeno, Y. Kawasaki, A. Muromachi, Y. Nishiyama, and S. Sakaguchi, *J. Org. Chem.*, **61**, 3088 (1996).
10. E. Vedejs and O. Daugulis, *J. Org. Chem.*, **61**, 5702 (1996).
11. S. Yamada, *Tetrahedron Lett.*, **33**, 2171 (1992); S. Yamada, T. Sugaki, and K. Matsuzaki, *J. Org. Chem.*, **61**, 5932 (1996).
12. C. A. A. van Boeckel and J. H. van Boom, *Tetrahedron Lett.*, 3561 (1979).
13. B. E. Griffin, M. Jarman, and C. B. Reese, *Tetrahedron*, **24**, 639 (1968).
14. T. Nakano, Y. Ito, and T. Ogawa, *Carbohydr. Res.*, **243**, 43 (1993).
15. K. K. Ogilvie and D. J. Iwacha, *Tetrahedron Lett.*, 317 (1973).
16. K. C. Nicolaou, T. J. Caulfield, H. Kataoka, and N. A. Stylianides, *J. Am. Chem. Soc.*, **112**, 3693 (1990).
17. B. M. Trost, S. A. Godleski, and J. L. Belletire, *J. Org. Chem.*, **44**, 2052 (1979).
18. P. G. Gassman and W. N. Schenk, *J. Org. Chem.*, **42**, 918 (1977).
19. E. M. Carreira and J. Du Bois, *J. Am. Chem. Soc.*, **117**, 8106 (1995).
20. H. Kosmol, F. Hill, U. Kerb, and K. Kieslich, *Tetrahedron Lett.*, 641 (1970).
21. S. J. Danishefsky, D. M. Armistead, F. E. Wincott, H. G. Selnick, and R. Hungate, *J. Am. Chem. Soc.*, **111**, 2967 (1989).
22. Y. Watanabe, T. Fujimoto, and S. Ozaki, *J. Chem. Soc., Chem. Commun.*, 681 (1992).
23. D. Farquhar, S. Khan, M. C. Wilkerson, and B. S. Andersson, *Tetrahedron Lett.*, **36**, 655 (1995).
24. S. V. Ley and D. M. Mynett, *Synlett*, 793 (1993).
25. S. Tomic, A. Tresec, D. Ljevakovic, and J. Tomasic, *Carbohydr. Res.*, **210**, 191 (1991); D. Ljevakovic, S. Tomic, and J. Tomasic, *Carbohydr. Res.*, **230**, 107 (1992).
26. H. W. Pinnick and E. Fernandez, *J. Org. Chem.*, **44**, 2810 (1979).
27. A.-M. Fernandez, J.-C. Plaquevent, and L. Duhamel, *J. Org. Chem.*, **62**, 4007 (1997).
28. R. Yanada, N. Negoro, K. Bessho, and K. Yanada, *Synlett*, 1261 (1995).

1-Adamantoate Ester: ROCO-1-adamantyl (Chart 2)

The adamantoate ester is formed selectively from a primary hydroxyl group (e.g., from the 5′-OH in a ribonucleoside) by reaction with adamantoyl chloride, Pyr (20°, 16 h). It is cleaved by alkaline hydrolysis (0.25 *N* NaOH, 20 min), but is stable to milder alkaline hydrolysis (e.g., NH_3, MeOH), conditions that cleave an acetate ester.[1]

1. K. Gerzon and D. Kau, *J. Med. Chem.*, **10**, 189 (1967).

Crotonate Ester: ROCOCH=CHCH$_3$

4-Methoxycrotonate Ester: ROCOCH=CHCH$_2$OCH$_3$

The crotonate esters, prepared to protect a primary hydroxyl group in nucleosides, are cleaved by hydrazine (MeOH, Pyr, 2 h). The methoxycrotonate is 100-fold more reactive to hydrazinolysis and 2-fold less reactive to alkaline hydrolysis than the corresponding acetate.[1]

1. R. Arentzen and C. B. Reese, *J. Chem. Soc., Chem. Commun.*, 270 (1977).

Benzoate Ester (Bz–OR): PHCO$_2$R (Chart 2)

The benzoate ester is one of the more common esters used to protect alcohols. Benzoates are less readily hydrolyzed than acetates, and the tendency for benzoate migration to adjacent hydroxyls, in contrast to that of acetates, is not nearly as strong,[1] but they can be forced to migrate to a thermodynamically more stable position.[2] The *p*-methoxybenzoate is even less prone to migrate than the benzoate.[3] Migration from a secondary to a primary alcohol has also been induced with AgNO$_3$, KF, Pyr, H$_2$O at 100°.[4]

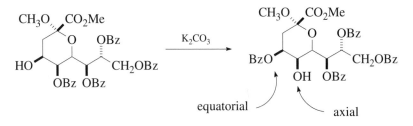

CH$_3$O CO$_2$Me, O, OBz, HO, CH$_2$OBz, OBz OBz —K$_2$CO$_3$→ CH$_3$O CO$_2$Me, O, OBz, BzO, CH$_2$OBz, OH OBz; equatorial, axial

Formation

1. BzCl or Bz$_2$O, Pyr, 0°. Benzoyl chloride is the most common reagent for the introduction of the benzoate group. Reaction conditions vary, depending on the nature of the alcohol to be protected. Cosolvents such as CH$_2$Cl$_2$ are often used with pyridine. Benzoylation in a polyhydroxylated system is much more selective than acetylation.[1] A primary alcohol is selectively protected over a secondary allylic alcohol,[5] and an equatorial alcohol can

be selectively protected in preference to an axial alcohol.[6] A cyclic secondary alcohol was selectively protected in the presence of a secondary acyclic alcohol.[7]

2. Regioselective benzoylation of methyl 4,6-*O*-benzylidene-α-galactopyranoside can be effected by phase transfer catalysis (BzCl, $Bu_4N^+Cl^-$, 40% NaOH, PhH, 69% yield of 2-benzoate; BzCl, $Bu_4N^+Cl^-$, 40% NaOH, HMPA, 62% yield of 3-benzoate).[8]

3. 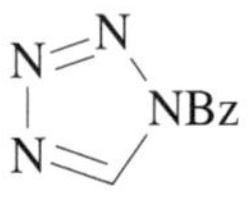 N=N, NBz (1-benzoyl-1,2,4-triazole) Et_3N, DMF, 20°, 15 min, 90% yield.[9] The 2-hydroxyl of methyl 4,6-*O*-benzylidine-α-glucopyranoside was selectively protected.[10]

4. BzCN, Et_3N, CH_3CN, 5 min to 2 h, >80% yield.[11,12] This reagent selectively acylates a primary hydroxyl group in the presence of a secondary hydroxyl group.[13]

5. $BzOCF(CF_3)_2$, TMEDA, 20°, 30 min, 90% yield.[14] This reagent also reacts with amines to form benzamides in high yields.

6. $BzOSO_2CF_3$, −78°, CH_2Cl_2, few min.[15] With acid-sensitive substrates, pyridine is used as a cosolvent. This reagent also reacts with ketals, epoxides,[15] and aldehydes.[16]

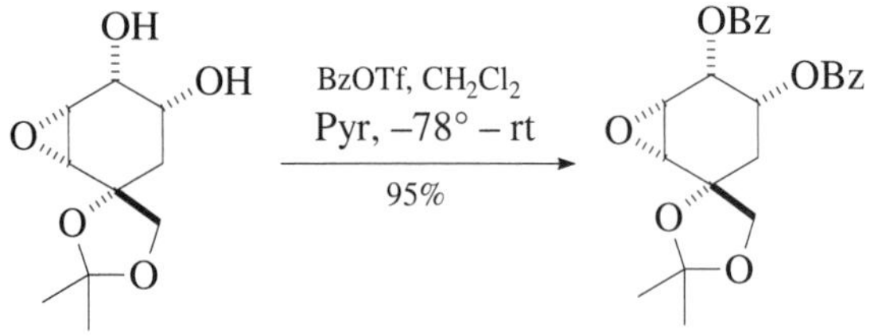

BzOTf, CH_2Cl_2, Pyr, −78° – rt, 95%

7. $PhCO_2H$, DIAD, Ph_3P, THF, 84% yield.[17]

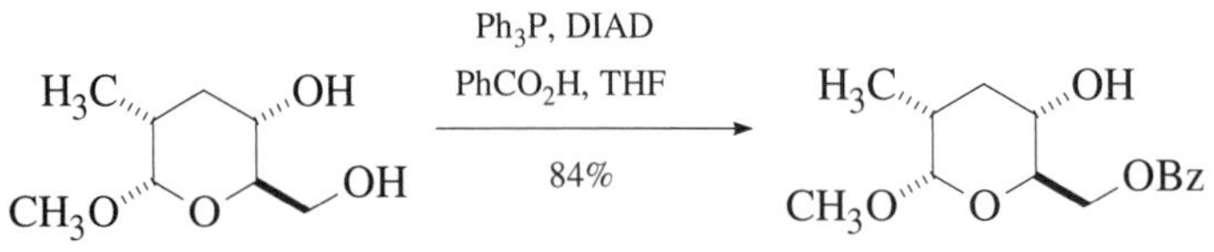

Ph_3P, DIAD, $PhCO_2H$, THF, 84%

DIAD = diisopropyl azodicarboxylate

The Mitsunobu reaction is usually used to introduce an ester with inversion of configuration. The use of this methodology on an anomeric hydroxyl was found to give only the β-benzoate, whereas other methods gave mixtures of anomers.[18] Improved yields are obtained in the Mitsunobu esterification when *p*-nitrobenzoic acid is used as the nucleophile.[19] Bis(dimethylamino) azodicarboxylate as an activating agent was found to be advantageous for hindered esters.[20] Bu_3P=CHCN was introduced as an alternative activating agent for the Mitsunobu reaction.[21]

8.  (N-benzoylimidazole) $CHCl_3$, reflux, 10 h.[22]

9. An alcohol can be selectively benzoylated in the presence of a primary amine.[23]

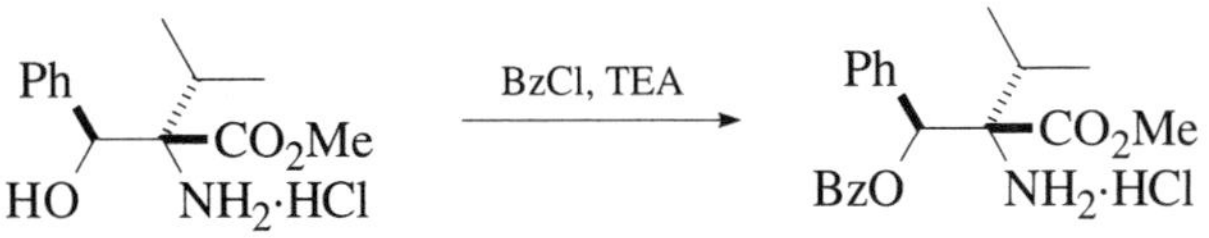

Ph, HO, CO_2Me, $NH_2{\cdot}HCl$ —BzCl, TEA→ Ph, BzO, CO_2Me, $NH_2{\cdot}HCl$

10. BuLi, BzCl; 10% Na_2CO_3, H_2O, 82% yield.[24] These conditions were used to monoprotect 1,4-butanediol.
11. BzOOBz, Ph_3P, CH_2Cl_2, 1 h, rt, ≈80% yield.[25] When these conditions are applied to unsymmetrical 1,2-diols, the benzoate of the kinetically and thermodynamically less stable isomer is formed.
12. $(Bu_3Sn)_2O$; BzCl.[26,27] The use of microwaves accelerates this reaction.[28] $Bu_2Sn(OMe)_2$ is reported to work better than Bu_2SnO in the monoprotection of diols.[29] The monoprotection of diols at the more hindered position can be accomplished through the stannylene if the reaction is quenched with $PhMe_2SiCl$ (45–77% yield).[30] Microwave heating has been found to be effective for this transformation in some cases.[31]

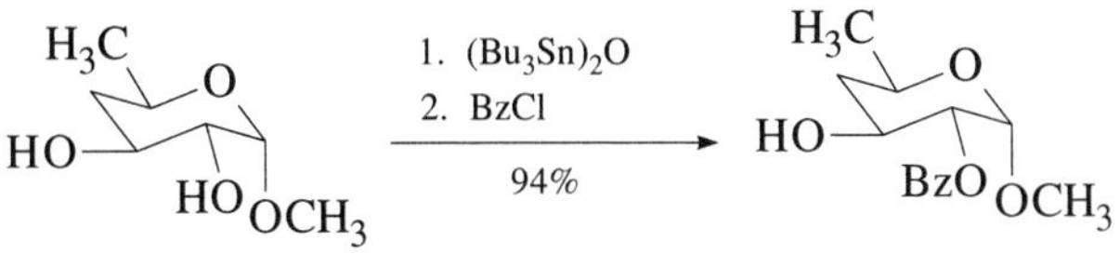

H_3C, O, HO, HO, OCH_3 —1. $(Bu_3Sn)_2O$ 2. BzCl, 94%→ H_3C, O, HO, BzO, OCH_3

13. 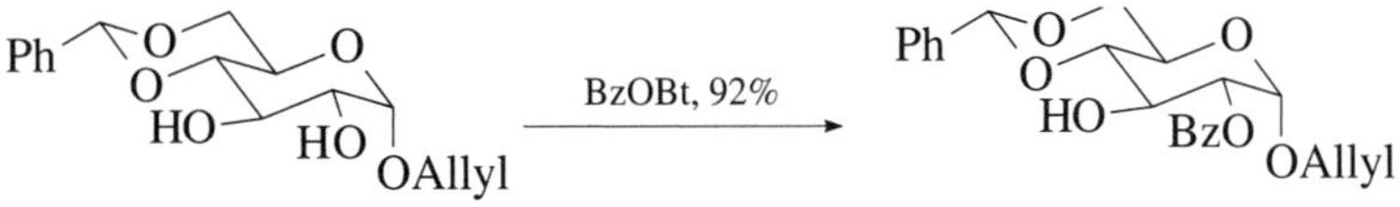 Ph, O, O, O, HO, HO, OAllyl —BzOBt, 92%→ Ph, O, O, O, HO, BzO, OAllyl

Ref. 32

14. 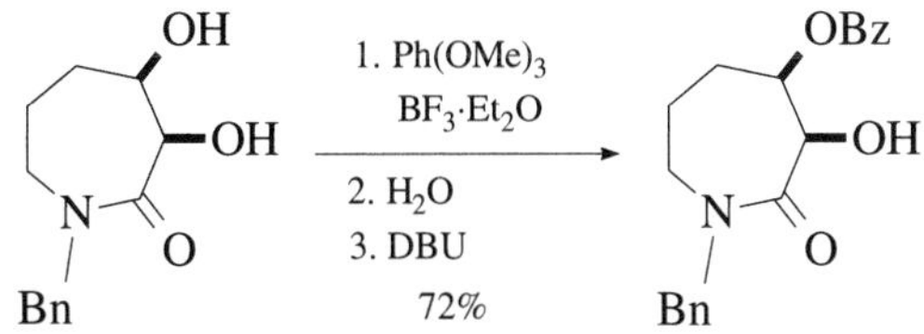 OH, OH, N, O, Bn —1. $Ph(OMe)_3$, $BF_3{\cdot}Et_2O$ 2. H_2O 3. DBU, 72%→ OBz, OH, N, O, Bn

The selectivity here relies on the fact that the β-benzoate is the thermodynamically more stable ester. A mixture of esters is formed upon hydrolysis of the ortho ester and is then equilibrated with DBU.[33] Carbohydrates are selectively protected with this methodology.[34]

15. Bz_2O, $MgBr_2$, TEA, CH_2Cl_2, rt, 95% yield. Tertiary alcohols are readily acylated.[35]
16. Bz_2O, $Sc(NTf_2)_3$, CH_3CN, 25°, 1.5–3 h, 90–98% yield. Phenols are also acylated efficiently.[36]
17. Vinyl benzoate, $Cp^*{}_2Sm(THF)_2$, toluene, rt, 3 h, 99% yield.[37]
18. *N*-Benzoyl-4-(dimethylamino)pyridinium chloride, CH_2Cl_2, TEA.[38]

Cleavage

1. 1% NaOH, MeOH, 20°, 50 min, 90% yield.[39]
2. Et_3N, MeOH, H_2O (1:5:1), reflux, 20 h, 86% yield.[40]
3. MeOH, KCN.[41]
4. A benzoate ester can be cleaved in 60–90% yield by electrolytic reduction at −2.3 V.[42]

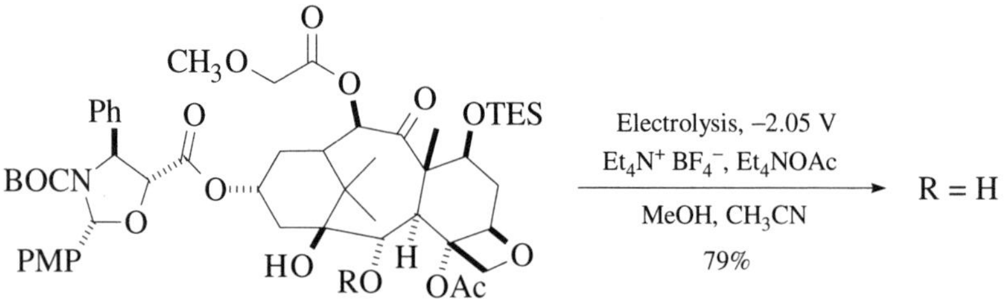

Electrolysis, −2.05 V; $Et_4N^+ BF_4^-$, Et_4NOAc; MeOH, CH_3CN; 79%; R = Bz → R = H

PMP = *p*-methoxyphenyl

Ref. 43

5. The following example illustrates the selective cleavage of a 2′-benzoate in a nucleotide derivative:[44]

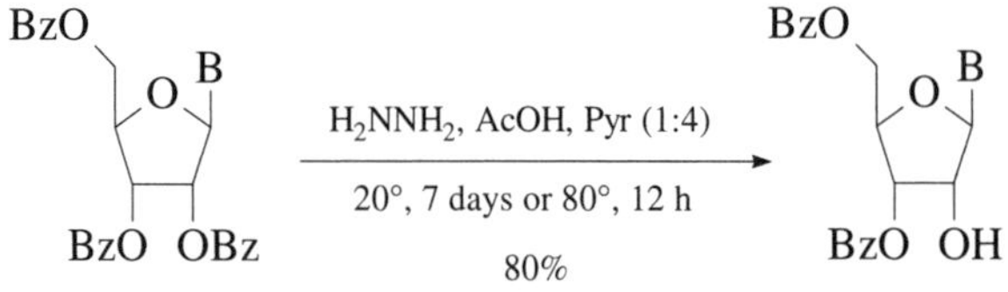

H_2NNH_2, AcOH, Pyr (1:4); 20°, 7 days or 80°, 12 h; 80%

This selectivity is achieved because the hydroxyl at the 2′-position is the most acidic of the three.

6. $BF_3 \cdot Et_2O$, Me_2S.[45]
7. Mg, MeOH, rt, 13 h, 91% yield. Esters are cleaved selectively in the order *p*-nitrobenzoate > acetate > benzoate > pivaloate >> trifluoroacetamide.[46]
8. EtMgBr, Et_2O, rt, 1 h, 90−100% yield.[47,48] These conditions were used to prevent a neighboring silyl ether from migrating. Ethylmagnesium chloride is much more reactive; thus, the reaction can be run at −42°, giving a 90% yield of the alcohol. Acetates and pivaloates are also cleaved.

1. A. H. Haines, *Adv. Carbohydr. Chem. Biochem.*, **33**, 11 (1976).
2. S. J. Danishefsky, M. P. DeNinno, and S.-h. Chen, *J. Am. Chem. Soc.*, **110**, 3929 (1988).
3. E. J. Corey, A. Guzman-Perez, and M. C. Noe, *J. Am. Chem. Soc.*, **117**, 10805 (1995).
4. Z. Zhang and G. Magnusson, *J. Org. Chem.*, **61**, 2383 (1996).
5. R. H. Schlessinger and A. Lopes, *J. Org. Chem.*, **46**, 5252 (1981).
6. A. P. Kozikowski, X. Yan, and J. M. Rusnak, *J. Chem. Soc., Chem. Commun.*, 1301 (1988).
7. K. Furuhata, K. Takeda, and H. Ogura, *Chem. Pharm. Bull.*, **39**, 817 (1991).
8. W. Szeja, *Synthesis*, 821 (1979).
9. J. Stawinski, T. Hozumi, and S. A. Narang, *J. Chem. Soc., Chem. Commun.*, 243 (1976).
10. S. Kim, H. Chang, and W. J. Kim, *J. Org. Chem.*, **50**, 1751 (1985).
11. M. Havel, J. Velek, J. Pospišek, and M. Soucek, *Collect. Czech. Chem. Commun.*, **44**, 2443 (1979).
12. A. Holý and M. Soucek, *Tetrahedron Lett.*, 185 (1971).
13. R. M. Soll and S. P. Seitz, *Tetrahedron Lett.*, **28**, 5457 (1987).
14. N. Ishikawa and S. Shin-ya, *Chem. Lett.*, 673 (1976).
15. L. Brown and M. Koreeda, *J. Org. Chem.*, **49**, 3875 (1984).
16. K. Takeuchi, K. Ikai, M. Yoshida, and A. Tsugeno, *Tetrahedron*, **44**, 5681 (1988).
17. A. B. Smith, III, and K. J. Hale, *Tetrahedron Lett.*, **30**, 1037 (1989).
18. A. B. Smith, III, R. A. Rivero, K. J. Hale, and H. A. Vaccaro, *J. Am. Chem. Soc.*, **113**, 2092 (1991).
19. S. F. Martin and J. A. Dodge, *Tetrahedron Lett.*, **32**, 3017 (1991).
20. T. Tsunodo, Y. Yamamiya, Y. Kawamura, and S. Ito, *Tetrahedron Lett.*, **36**, 2529 (1995).
21. T. Tsunodo, F. Ozaki, and S. Ito, *Tetrahedron Lett.*, **35**, 5081 (1994).
22. C. L. Brewer, S. David, and A. Veyrièrs, *Carbohydr. Res.*, **36**, 188 (1974).
23. Y. Ito, M. Sawamura, E. Shirakawa, K. Hayashizaki, and T. Hayashi, *Tetrahedron*, **44**, 5253 (1988). See also T.-Y. Luh and Y. H. Chong, *Synth. Commun.*, **8**, 327 (1978).
24. A. J. Castellino and H. Rapoport, *J. Org. Chem.*, **51**, 1006 (1986).
25. A. M. Pautard and S. A. Evans, Jr., *J. Org. Chem.*, **53**, 2300 (1988).
26. S. Hanessian and R. Roy, *Can. J. Chem.*, **63**, 163 (1985).
27. For a mechanistic study of the tin-directed acylation, see S. Roelens, *J. Chem. Soc., Perkin Trans. 2*, 2105 (1988).
28. B. Herradón, A. Morcuende, and S. Valverde, *Synlett*, 455 (1995).
29. G. J. Boons, G. H. Castle, J. A. Clase, P. Grice, S. V. Ley, and C. Pinel, *Synlett*, 913 (1993).
30. G. Reginato, A. Ricci, S. Roelens, and S. Scapecchi, *J. Org. Chem.*, **55**, 5132 (1990).
31. A. Morcuende, S. Valverde, and B. Herradón, *Synlett*, 89 (1994).
32. H. Yamda, T. Harada, and T. Takahashi, *J. Am. Chem. Soc.*, **116**, 7919 (1994).

33. J. W. Lampe, P. F. Hughes, C. K. Biggers, S. H. Smith, and H. Hu, *J. Org. Chem.*, **61**, 4572 (1996).
34. F. I. Auzanneau and D. R. Bundle, *Carbohydr. Res.*, **212**, 13 (1991).
35. E. Vedejs and O. Daugulis, *J. Org. Chem.*, **61**, 5702 (1996).
36. K. Ishihara, M. Kubota, and H. Yamamoto, *Synlett*, 265 (1996).
37. Y. Ishii, M. Takeno, Y. Kawasaki, A. Muromachi, Y. Nishiyama, and S. Sakaguchi, *J. Org. Chem.*, **61**, 3088 (1996).
38. M. S. Wolfe, *Synth. Commun.*, **27**, 2975 (1997).
39. K. Mashimo and Y. Sato, *Tetrahedron*, **26**, 803 (1970).
40. K. Tsuzuki, Y. Nakajima, T. Watanabe, M. Yanagiya, and T. Matsumoto, *Tetrahedron Lett.*, 989 (1978).
41. J. Herzig, A. Nudelman, H. E. Gottlieb, and B. Fischer, *J. Org. Chem.*, **51**, 727 (1986).
42. V. G. Mairanovsky, *Angew. Chem., Int. Ed. Engl.*, **15**, 281 (1976).
43. J.-P. Pulicani, D. Bézard, J.-D. Bourzat, H. Bouchard, M. Zucco, D. Deprez, and A. Commercon, *Tetrahedron Lett.*, **35**, 9717 (1994).
44. Y. Ishido, N. Nakazaki, and N. Sakairi, *J. Chem. Soc., Perkin Trans. 1*, 2088 (1979).
45. K. Fuji, T. Kawabata, and E. Fujita, *Chem. Pharm. Bull.*, **28**, 3662 (1980).
46. Y.-C. Xu, E. Lebeau, and C. Walker, *Tetrahedron Lett.*, **35**, 6207 (1994); Y.-C. Xu, A. Bizuneh, and C. Walker, *Tetrahedron Lett.*, **37**, 455 (1996).
47. Y. Watanabe, T. Fujimoto, and S. Ozaki, *J. Chem. Soc., Chem. Commun.*, 681 (1992).
48. Y. Watanabe, T. Fujimoto, T. Shinohara, and S. Ozaki, *J. Chem. Soc., Chem. Commun.*, 428 (1991).

p-Phenylbenzoate Ester: $ROCOC_6H_4$-*p*-C_6H_5

The *p*-phenylbenzoate ester was prepared to protect the hydroxyl group of a prostaglandin intermediate by reaction with the benzoyl chloride (Pyr, 25°, 1 h, 97% yield). It was a more crystalline, more readily separated derivative than 15 other esters that were investigated.[1] It can be cleaved with K_2CO_3 in MeOH in the presence of a lactone.[2]

1. E. J. Corey, S. M. Albonico, U. Koelliker, T. K. Schaaf, and R. K. Varma, *J. Am. Chem. Soc.*, **93**, 1491 (1971).
2. T. V. RaganBabu, *J. Org. Chem.*, **53**, 4522 (1988).

2,4,6-Trimethylbenzoate (Mesitoate) Ester: 2,4,6-$Me_3C_6H_2CO_2R$ (Chart 2)

Formation

1. $Me_3C_6H_2COCl$, Pyr, $CHCl_3$, 0°, 14 h → 23°, 1 h, 95% yield.[1]
2. $Me_3C_6H_2CO_2H$, $(CF_3CO)_2O$, PhH, 20°, 15 min.[2]

Cleavage

1. $LiAlH_4$, Et_2O, 20°, 2 h.[2]
2. *t*-BuOK, H_2O (8:1) "anhydrous hydroxide," 20°, 24–72 h, 50–72% yield.[3] A mesitoate ester is exceptionally stable to base: 2 *N* NaOH, 20°, 20 h; 12 *N* NaOH, EtOH, 50°, 15 min.

1. E. J. Corey, K. Achiwa, and J. A. Katzenellenbogen, *J. Am. Chem. Soc.*, **91**, 4318 (1969).
2. I. J. Bolton, R. G. Harrison, B. Lythgoe, and R. S. Manwaring, *J. Chem. Soc. C*, 2944 (1971).
3. P. G. Gassman and W. N. Schenk, *J. Org. Chem.*, **42**, 918 (1977).

Carbonates

Carbonates, like esters, can be cleaved by basic hydrolysis, but generally are much less susceptible to hydrolysis because of the resonance effect of the second oxygen. In general, carbonates are cleaved by taking advantage of the properties of the second alkyl substituent (e.g., zinc reduction of the 2,2,2-trichloroethyl carbonate). The reagents used to introduce the carbonate onto alcohols react readily with amines as well. As expected, basic hydrolysis of the resulting carbamate is considerably more difficult than basic hydrolysis of a carbonate.

Alkyl Methyl Carbonate: $ROCO_2CH_3$ (Chart 2)

Formation

1.

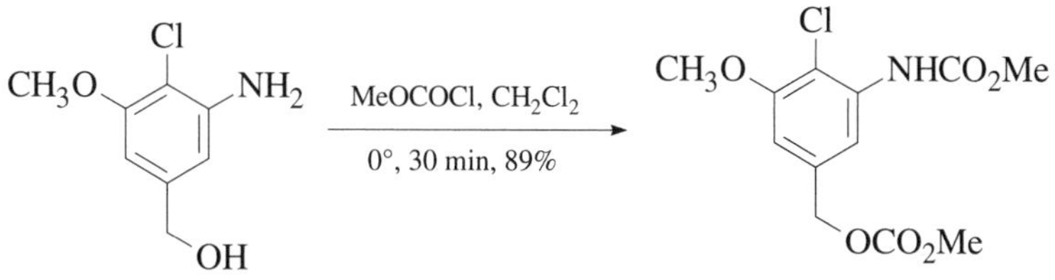

Cl
CH_3O NH_2
OH

$MeOCOCl$, CH_2Cl_2 → 0°, 30 min, 89%

Cl
CH_3O $NHCO_2Me$
OCO_2Me

Ref. 1

2. $(CH_3)_2C{=}NOCO_2CH_3$, CAL, dioxane, 60°, 3 days, 45% yield. Only a primary alcohol is protected.[2]

Cleavage[1]

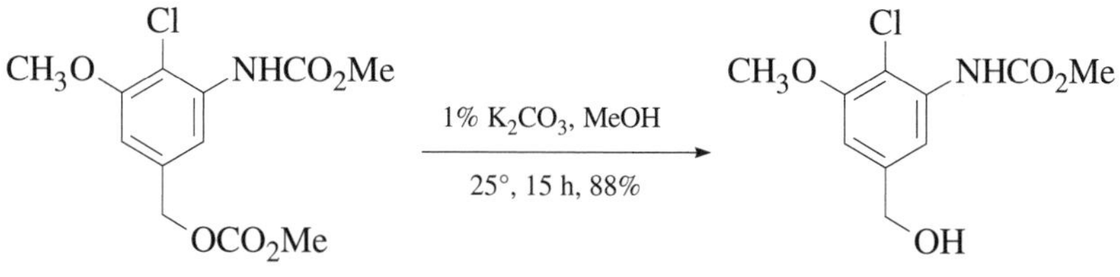

Cl
CH_3O $NHCO_2Me$
OCO_2Me

1% K_2CO_3, MeOH → 25°, 15 h, 88%

Cl
CH_3O $NHCO_2Me$
OH

1. A. I. Meyers, K. Tomioka, D. M. Roland, and D. Comins, *Tetrahedron Lett.*, 1375 (1978).
2. R. Pulido and V. Gator, *J. Chem. Soc., Perkin Trans. 1,* 589 (1993).

Methoxymethyl Carbonate: $CH_3OCH_2OCO_2R$

Formation

1. K_2CO_3, $ClCH_2OMe$, DMF, −20°, 28–95% yield.[1]
2. $AgCO_3$, $ClCH_2OMe$, DMF, −15°, 15–67% yield.[2]

Cleavage

1. K_2CO_3, MeOH, H_2O, 30 min, 20°, 19–93% yield.[2]
2. TFA, MeOH, 30 h, 20°, 79–93% yield.[1,2]

1. K. Teranishi, A. Komoda, M. Hisamatsu, and T. Yamada, *Bull. Chem. Soc. Jpn.*, **68**, 309 (1995).
2. K. Teranishi, H. Nakao, A. Komoda, M. Hisamatsu, and T. Yamada, *Synthesis,* 176 (1995).

Alkyl 9-Fluorenylmethyl Carbonate (Fmoc–OR):

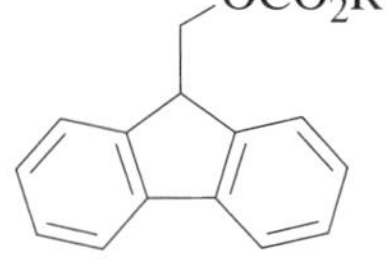

Formation

1. FmocCl, Pyr, 20°, 40 min, 81–96% yield.[1]

2. ROH —(benzotriazol-1-yl OCO_2Fm; DMAP, THF, CH_3CN; 53–95%)→ RO–Fmoc

Ref. 2

Cleavage

1. Et_3N, Pyr, 2 h, 83–96% yield (half-life = 20 min).[1]

1. C. Gioeli and J. B. Chattopadhyaya, *J. Chem. Soc., Chem. Commun.*, 672 (1982).
2. K. Takeda, K. Tsuboyama, M. Hoshino, M. Kishino, and H. Ogura, *Synthesis*, 557 (1987).

Alkyl Ethyl Carbonate: $ROCO_2Et$

An ethyl carbonate, prepared and cleaved by conditions similar to those described for a methyl carbonate, was used to protect a hydroxyl group in glucose.[1]

1. F. Reber and T. Reichstein, *Helv. Chim. Acta*, **28**, 1164 (1945).

Alkyl 2,2,2-Trichloroethyl Carbonate (Troc–OR): $ROCO_2CH_2CCl_3$ (Chart 2)

Formation

1. Cl_3CCH_2OCOCl, Pyr, 20°, 12 h.[1] The trichloroethyl carbonate can be introduced selectively onto a primary alcohol in the presence of a secondary alcohol.[2] DMAP has been used to catalyze this acylation.[3]

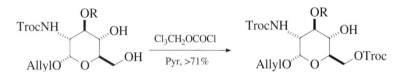

OR TrocNH OH AllylO O OH —(Cl_3CH_2OCOCl, Pyr, >71%)→ OR TrocNH OH AllylO O OTroc

Cleavage

1. Zn, AcOH, 20°, 1–3 h, 80% yield.[1]
2. Zn, MeOH, reflux, short time.[1]
3. Zn–Cu, AcOH, 20°, 3.5 h, 100% yield.[4] A 2,2,2-tribromoethyl carbonate is cleaved by Zn-Cu/AcOH 10 times faster than trichloroethyl carbonate.
4. Electrolysis, −1.65 V, MeOH, $LiClO_4$, 80% yield.[5]
5. Sm, I_2, MeOH, rt, 5 min, 100% yield.[6]

1. T. B. Windholz and D. B. R. Johnston, *Tetrahedron Lett.*, 2555 (1967).
2. M. Imoto, N. Kusunose, S. Kusumoto, and T. Shiba, *Tetrahedron Lett.*, **29**, 2227 (1988).
3. S. Hanessian and R. Roy, *Can. J. Chem.*, **63**, 163 (1985).
4. A. F. Cook, *J. Org. Chem.*, **33**, 3589 (1968).
5. M. F. Semmelhack and G. E. Heinsohn, *J. Am. Chem. Soc.*, **94**, 5139 (1972).
6. R. Yanada, N. Negoro, K. Bessho, and K. Yanada, *Synlett*, 1261 (1995).

1,1-Dimethyl-2,2,2-trichloroethyl Carbonate (TCBOC–OR): $Cl_3CC(CH_3)_2OCO_2R$

Formation

1. $Cl_3CC(CH_3)_2OCOCl$.[1]

Cleavage

1. $(Et_3NH)^+$ $Sn(SPh)_3^-$, tetrabutylammonium cobalt(II)phthalocyanine 5,12,19,26-tetrasulfonate, CH_3CN, MeOH, 20°, 1 h, 90% yield.[1]

1. S. Lehnhoff, R. M. Karl, and I. Ugi, *Synthesis*, 309 (1991).

Alkyl 2-(Trimethylsilyl)ethyl Carbonate (TMSEC–OR): $Me_3SiCH_2CH_2OCO_2R$

Formation

1. $TMSCH_2CH_2OCOCl$, Pyr, 65–97% yield.[1]
2. $TMSCH_2CH_2OCO$-imidazole, DBU, benzene, 54% yield.[2]

Cleavage

1. 0.2 *M* $Bu_4N^+F^-$, THF, 20°, 10 min, 87–94% yield.[1]
2. $ZnCl_2$, CH_2Cl_2 or CH_3NO_2, 20°, 81–90% yield.[1]
3. $ZnBr_2$, CH_2Cl_2 or CH_3NO_2, 20°, 65–92% yield.[1]

1. C. Gioeli, N. Balgobin, S. Josephson, and J. B. Chattopadhyaya, *Tetrahedron Lett.*, **22**, 969 (1981).
2. W. R. Roush and T. A. Blizzard, *J. Org. Chem.*, **49**, 4332 (1984).

Alkyl 2-(Phenylsulfonyl)ethyl Carbonate (Psec–OR): $PhSO_2CH_2CH_2OCO_2R$

Formation

1. $PhSO_2CH_2CH_2OCOCl$, Pyr, 20°, 74–99% yield.[1] 4-Substituted phenylsulfonyl analogs of this protective group have also been prepared and their relative rates of cleavage studied: $T_{1/2}$ (min) (TEA, Pyr, 20°) 4-H, 180; 4-Me, 1140; 4-Cl, 60; 4-NO_2, 10.[2]

Cleavage

1. Et_3N, Pyr, 20 h, rt, 85–99% yield.[1]
2. NH_3, dioxane, H_2O (9:1), 7 min.[1]
3. K_2CO_3 (0.04 *M*) 1 min.[1]

1. N. Balgobin, S. Josephson, and J. B. Chattopadhyaya, *Tetrahedron Lett.*, **22**, 3667 (1981).
2. S. Josephson, N. Balgobin, and J. Chattopadhyaya, *Tetrahedron Lett.*, **22**, 4537 (1981).

Alkyl 2-(Triphenylphosphonio)ethyl Carbonate (Peoc–OR): $Ph_3P^+CH_2CH_2OCO_2R\ Cl^-$

Formation

1. $Ph_3P^+CH_2CH_2OCOCl\ Cl^-$, Pyr, CH_2Cl_2, 4 h, 0°, 65–94% yield.[1]

Cleavage

1. Me_2NH, MeOH, 0°, 75% yield.[1] *t*-Butyl esters could be cleaved with HCl without affecting the Peoc group.

1. H. Kunz and H.-H. Bechtolsheimer, *Synthesis*, 303 (1982).

Alkyl Isobutyl Carbonate: $ROCO_2CH_2CH(CH_3)_2$

An isobutyl carbonate was prepared by reaction with isobutyl chloroformate (Pyr, 20°, 3 days, 73% yield), to protect the 5′-OH group in thymidine. It was cleaved by acidic hydrolysis (80% AcOH, reflux, 15 min, 88% yield).[1]

1. K. K. Ogilvie and R. L. Letsinger, *J. Org. Chem.*, **32**, 2365 (1967).

Alkyl Vinyl Carbonate: $ROCO_2CH{=}CH_2$

Formation

1. $CH_2{=}CHOCOCl$, Pyr, CH_2Cl_2, 93% yield.[1]

Cleavage

1. Na_2CO_3, H_2O, dioxane, warm, 97% yield.[1] Phenols can be protected under similar conditions. Amines are converted by these conditions to carbamates that are stable to alkaline hydrolysis with sodium carbonate. Carbamates are cleaved by acidic hydrolysis (HBr, MeOH, CH_2Cl_2, 8 h), conditions that do not cleave alkyl or aryl vinyl carbonates.

1. R. A. Olofson and R. C. Schnur, *Tetrahedron Lett.*, 1571 (1977).

Alkyl Allyl Carbonate: $ROCO_2CH_2CH{=}CH_2$ (Chart 2)

Formation

1. $CH_2{=}CHCH_2OCOCl$, Pyr, THF, 0° → 20°, 2 h, 90% yield.[1]

2.

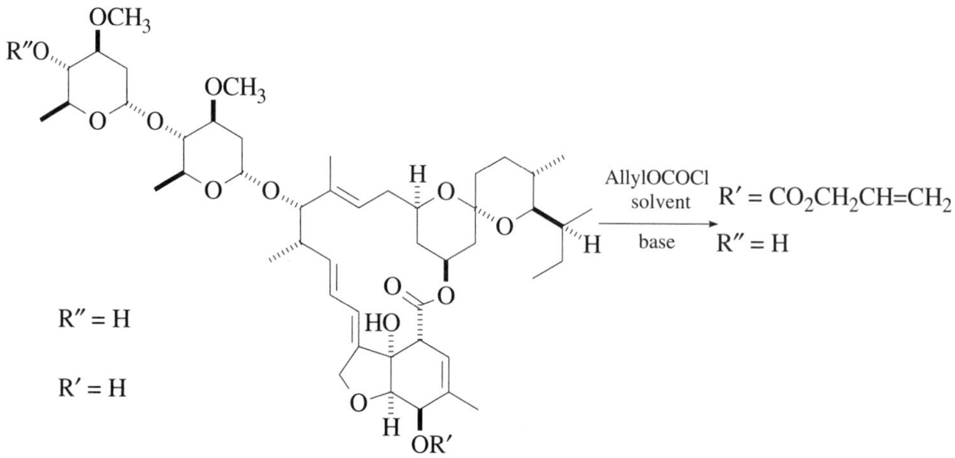

R″ = H

R′ = H

AllylOCOCl, solvent, base → R′ = $CO_2CH_2CH{=}CH_2$, R″ = H

This reaction[2] showed a remarkable selectivity with respect to the solvent and base used. In THF and EtOAc using TEA as the base, a 1:1 mixture of the allylic carbonate and bisacylated products is obtained, but when CH_2Cl_2 is used as solvent, the reaction favors the allylic alcohol by a factor of 97:3 (mono/bis). In THF or MTBE, the use of TMEDA as the base also results in a 97:3 mono/bis ratio.[2]

3. Diallyl carbonate, $Pd(OAc)_2$, Ph_3P. Conventional methods failed to protect this hindered 12-α-hydroxycholestane derivative.[3]
4. $CH_2{=}CHCH_2OCO_2N{=}C(CH_3)_2$, CAL, dioxane, 60°, 3 days.[4]
5. 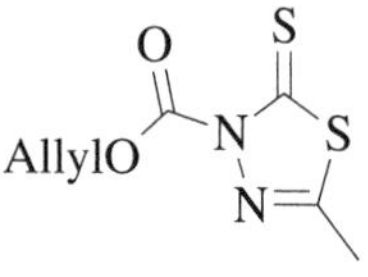 AllylO DMAP, THF, 65% yield. This reaction is selective for primary alcohols.[5] Benzyl, isobutyl, and ethyl carbonates are also prepared using this method (63–85% yield).

Cleavage

1. $Ni(CO)_4$, TMEDA, DMF, 55°, 4 h, 87–95% yield.[1] Because of the toxicity associated with nickel carbonyl, this method is rarely used and has largely been supplanted by palladium-based reagents.
2. $Pd(Ph_3P)_4$, HCO_2NH_4.[6]
3. $Pd(Ph_3P)_4$, Bu_3SnH, 90–100% yield.[7]
4. $PdCl_2(Ph_3P)_2$, dimedone, 91% yield.[8]
5. $Pd(OAc)_2$, TPPTS, Et_2NH, CH_3CN, H_2O, 51–100% yield. If the reaction is run in a biphasic system using butyronitrile as the solvent, a dimethylallyl carbamate can be retained, but in a homogeneous system using CH_3CN, both groups are cleaved quantitatively.[9,10]

6. Pd(dba)$_2$, dppe, Et$_2$NH, THF, 15 min–5 h, 96–100% yield.[11]
7. Pd(Ph$_3$P)$_4$, NaBH$_4$, ethanol, >88% yield.[2]
8. Pd(OAc)$_2$, TPPTS, Et$_2$NH, CH$_3$CN–H$_2$O or Et$_2$O–H$_2$O, 94–98% yield.[12]

1. E. J. Corey and J. W. Suggs, *J. Org. Chem.*, **38**, 3223 (1973).
2. R. J. Cvetovich, D. H. Kelly, L. M. DiMichele, R. F. Shuman, and E. J. J. Grabowski, *J. Org. Chem.*, **59**, 7704 (1994).
3. A. P. Davis, B. J. Dorgan, and E. R. Mageean, *J. Chem. Soc., Chem. Commun.*, 492 (1993).
4. R. Pulido and V. Gotor, *J. Chem. Soc., Perkin Trans. 1*, 589 (1993).
5. M. Allainmat, P. L'Haridon, L. Toupet, and D. Plusquellec, *Synthesis*, 27 (1990).
6. Y. Hayakawa, H. Kato, M. Uchiyama, H. Kajino, and R. Noyori *J. Org. Chem.*, **51**, 2400 (1986).
7. F. Guibe and Y. Saint M'Leux, *Tetrahedron Lett.*, **22**, 3591 (1981).
8. H. X. Zhang, F. Guibe, and G. Balavoine, *Tetrahedron Lett.*, **29**, 623 (1988).
9. S. Lemaire-Audoire, M. Savignac, E. Blart, G. Pourcelot, J. P. Genét, and J. M. Bernard, *Tetrahedron Lett.*, **35**, 8783 (1994).
10. J. P. Genét, E. Blart, M. Savignac, S. Lemeune, S. Lemaire-Audoire, J. M. Paris, and J. M. Bernard, *Tetrahedron*, **50**, 497 (1994).
11. J. P. Genét, E. Blart, M. Savignac, S. Lemeune, S. Lemaire-Audoire, and J.-M. Bernard, *Synlett*, 680 (1993).
12. J. P. Genét, E. Blart, M. Savignac, S. Lemeune, and J.-L. Paris, *Tetrahedron Lett.*, **34**, 4189 (1993).

Alkyl *p*-Nitrophenyl Carbonate: ROCOOC$_6$H$_4$–*p*-NO$_2$ (Chart 2)

Formation/Cleavage[1]

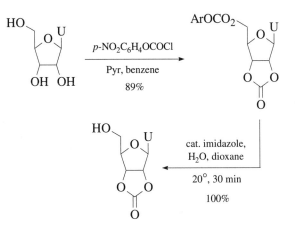

HO, U, O, OH OH

p-NO$_2$C$_6$H$_4$OCOCl
Pyr, benzene
89%

ArOCO$_2$, U, O, O, O, O

cat. imidazole,
H$_2$O, dioxane
20°, 30 min
100%

HO, U, O, O, O, O

Acetates, benzoates, and cyclic carbonates are stable to these hydrolysis conditions. [Cyclic carbonates are cleaved by more alkaline conditions (e.g., dil. NaOH, 20°, 5 min, or aq. Pyr, warm, 15 min, 100% yield).] [1]

1. R. L. Letsinger and K. K. Ogilvie, *J. Org. Chem.*, **32**, 296 (1967).

Alkyl Benzyl Carbonate: $ROCO_2Bn$ (Chart 2)

A benzyl carbonate was prepared in 83% yield from the sodium alkoxide of glycerol and benzyl chloroformate (20°, 24 h).[1] It was also prepared by a lipase-catalyzed ester exchange with allyl benzyl carbonate.[2] It is cleaved by hydrogenolysis (H_2/Pd–C, EtOH, 20°, 2 h, 2 atm, 76% yield)[1] and electrolytic reduction (−2.7 V, $R_4N^+X^-$, DMF, 70% yield).[3] A benzyl carbonate was used to protect the hydroxyl group in lactic acid during a peptide synthesis.[4]

1. B. F. Daubert and C. G. King, *J. Am. Chem. Soc.*, **61**, 3328 (1939).
2. M. Pozo, R. Pulido, and V. Gotor, *Tetrahedron*, **48**, 6477 (1992).
3. V. G. Mairanovsky, *Angew. Chem., Int. Ed. Engl.*, **15**, 281 (1976).
4. G. Losse and G. Bachmann, *Chem. Ber.*, **97**, 2671 (1964).

Alkyl *p*-Methoxybenzyl Carbonate: *p*-$MeOC_6H_4CH_2OCO_2R$

Alkyl 3,4-Dimethoxybenzyl Carbonate: 3,4-$(MeO)_2C_6H_3CH_2OCO_2R$

These groups are readily cleaved with $Ph_3C^+BF_4^-$, 0°, 6 min, 90% yield and 0°, 15 min, 90% yield, respectively. It should also be possible to cleave these carbonates with DDQ like the corresponding methoxy- and dimethoxyphenylmethyl ethers.[1]

1. D. H. R. Barton, P. D. Magnus, G. Smith, G. Streckert, and D. Zurr, *J. Chem. Soc., Perkin Trans. 1*, 542 (1972).

Alkyl *o*-Nitrobenzyl Carbonate: $ROCO_2CH_2C_6H_4$-*o*-NO_2

Alkyl *p*-Nitrobenzyl Carbonate: $ROCO_2CH_2C_6H_4$–*p*-NO_2 (Chart 2)

The nitrobenzyl carbonates were prepared to protect a secondary hydroxyl group in a thienamycin precursor. The *o*-nitrobenzyl carbonate was prepared from the chloroformate (DMAP, CH_2Cl_2, 0°→ 20°, 3 h) and cleaved by irradiation, pH 7.[1] The *p*-nitrobenzyl carbonate was prepared from the chloroformate (−78°, *n*-BuLi, THF, 85% yield) and cleaved by hydrogenolysis (H_2/Pd–C, dioxane, H_2O, EtOH, K_2HPO_4)[2] or by electrolytic reduction.[3]

1. L. D. Cama and B. G. Christensen, *J. Am. Chem. Soc.*, **100**, 8006 (1978).

2. D. B. R. Johnston, S. M. Schmitt, F. A. Bouffard, and B. G. Christensen, *J. Am. Chem. Soc.*, **100**, 313 (1978).
3. V. G. Mairanovsky, *Angew. Chem., Int. Ed. Engl.*, **15**, 281 (1976).

Carbonates Cleaved By β-Elimination

2-Dansylethyl Carbonate (Dnseoc–OR):

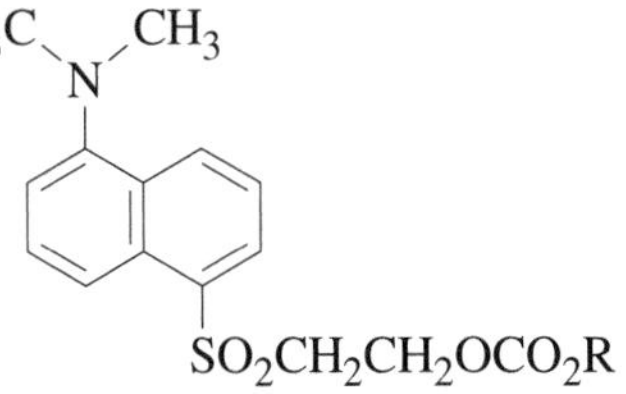

H_3C–N–CH_3 (5-dimethylaminonaphthalene-1-yl) $SO_2CH_2CH_2OCO_2R$

Formation

When the Dnseoc group is used in nucleoside synthesis, the coupling yields are determined by measuring the absorbance at 350 nm of each eluate from the Dnseoc-deprotection steps containing the 5-(dimethylamino)naphthalene-1-yl–vinyl sulfone or by measuring the fluorescence at 530 nm.[1]

Cleavage

DBU, CH_3CN, 140 s.[2] The 2-(4-nitrophenyl)ethyl (Npe) phosphate protective group and the 2-(4-nitrophenyl)ethoxycarbonyl (Npeoc) group are stable to these conditions, but the cyanoethyl group is not.

2-(4-Nitrophenyl)ethyl Carbonate (Npeoc–OR):
$4\text{-}NO_2C_6H_4CH_2CH_2OCO_2R$

Formation

1. $4\text{-}NO_2C_6H_4CH_2CH_2OCOCl$, Pyr, CH_2Cl_2, $-10°$, 3 h, >70% yield.[3]
2. 3-Methyl-1-[2-(4-nitrophenyl)ethoxycarbonyl]-1*H*-imidazol-3-ium chloride, CH_2Cl_2, DMAP, rt, 100% yield.[3]

Cleavage

1. 0.5 *M* DBU in dry pyridine.[3]
2. K_2CO_3, MeOH, 69–75% yield.[4]

2-(2,4-Dinitrophenyl)ethyl Carbonate (Dnpeoc–OR):
$2,4\text{-}(NO_2)_2C_6H_3CH_2CH_2OCO_2R$

Formation

1. $2,4\text{-}(NO_2)_2C_6H_3CH_2CH_2OCOCl$, Pyr, CH_2Cl_2, $-10°$, 3 h, >75% yield.[3]

Cleavage

1. TEA, MeOH, dioxane.[3]

2-Cyano-1-phenylethyl Carbonate (Cpeoc–OR): $NCCH_2CH(C_6H_5)OCO_2R$

In a quest to develop the perfect nucleoside protection scheme, the Cpeoc group was devised for protection of the 5′-OH. It is introduced through the chloroformate in 58–83% yield. Cleavage is achieved with 0.1 *M* DBU with half-lives of 7–14 sec, depending on the nucleoside.[5]

1. F. Bergmann and W. Pfleiderer, *Helv. Chim. Acta*, **77**, 203 (1994).
2. F. Bergmann and W. Pfleiderer, *Helv. Chim. Acta*, **77**, 988 (1994).
3. H. Schirmeister, F. Himmelsbach, and W. Pfleiderer, *Helv. Chim. Acta*, **76**, 385 (1993).
4. M. Wasner, R. J. Suhadolnik, S. E. Horvath, M. E. Adelson, N. Kon, M.-X. Guan, E. E. Henderson, and W. Pfleiderer, *Helv. Chim. Acta*, **79**, 619 (1996).
5. U. Münch and W. Pfeiderer, *Nucleosides Nucleotides,* **16**, 801 (1997)

Alkyl *S*-Benzyl Thiocarbonate: $ROCOSCH_2Ph$ (Chart 2)

Formation

1. $PhCH_2SCOCl$, Pyr, 65–70% yield.[1]

Cleavage

1. H_2O_2, AcOH, AcOK, $CHCl_3$, 20°, 4 days, 50–55% yield.[1]

1. J. J. Willard, *Can. J. Chem.*, **40**, 2035 (1962).

Alkyl 4-Ethoxy-1-naphthyl Carbonate

Formation/Cleavage[1]

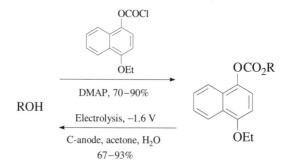

ROH → (4-ethoxy-1-naphthyl chloroformate: OCOCl, OEt; DMAP, 70–90%) → 4-ethoxy-1-naphthyl carbonate (OCO_2R, OEt)

Reverse: Electrolysis, −1.6 V; C-anode, acetone, H_2O; 67–93%

Amines can also be protected by this reagent; cleavage must be carried out in acidic media to avoid amine oxidation. The by-product naphthoquinone can be removed by extraction with basic hydrosulfite. Ceric ammonium nitrate also serves as an oxidant for deprotection, but the yields are much lower.

1. R. W. Johnson, E. R. Grover, and L. J. MacPherson, *Tetrahedron Lett.*, **22**, 3719 (1981).

Alkyl Methyl Dithiocarbonate: CH_3SCSOR

Formation[1]

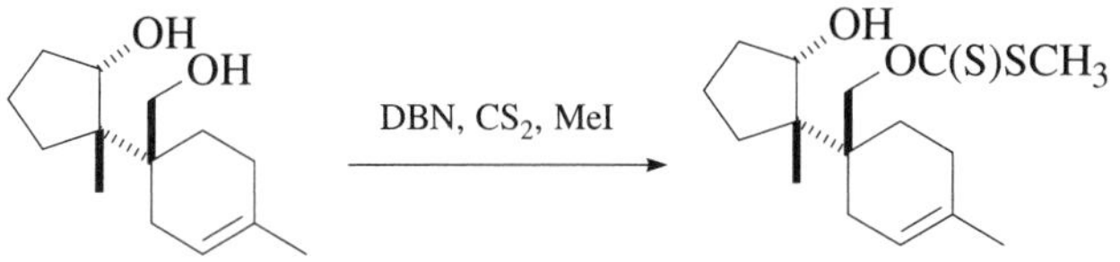

OH OH —— DBN, CS_2, MeI ——→ OH $OC(S)SCH_3$

Most attempts to differentiate these hydroxyl groups with conventional derivatives resulted in the formation of a tetrahydrofuran. The dithiocarbonate can also be prepared by phase transfer catalysis ($Bu_4N^+HSO_4^-$, 50% NaOH/H_2O, CS_2, MeI, rt, 1.5 h).[2]

Cleavage

1. These esters can be deoxygenated with Bu_3SnH[3] or, as in the preceding example, with $LiAlH_4$.[1]

1. R. H. Schlessinger and J. A. Schultz, *J. Org. Chem.*, **48**, 407 (1983).
2. A. W. M. Lee, W. H. Chan, H. C. Wong, and M. S. Wong, *Synth. Commun.*, **19**, 547 (1989).
3. D. H. R. Barton and S. W. McCombie, *J. Chem. Soc., Perkin Trans. 1*, 1574 (1975).

Assisted Cleavage

The following derivatives represent protective groups that contain an auxiliary functionality that, when chemically modified, results in intramolecular, assisted cleavage, thus increasing the rate of cleavage over that of simple basic hydrolysis.

2-Iodobenzoate Ester: 2-I–$C_6H_4CO_2R$

The 2-iodobenzoate is introduced by acylation of the alcohol with the acid (DCC, DMAP, CH_2Cl_2, 25°, 96% yield); it is removed by oxidation with Cl_2 (MeOH, H_2O, Na_2CO_3, pH >7.5).[1]

4-Azidobutyrate Ester: $N_3(CH_2)_3CO_2R$

The 4-azidobutyrate ester is introduced via the acid chloride. Cleavage occurs by pyrrolidone formation after the azide is reduced by hydrogenation, H_2S or Ph_3P.[2,3]

4-Nitro-4-methylpentanoate Ester

Formation/Cleavage[4]

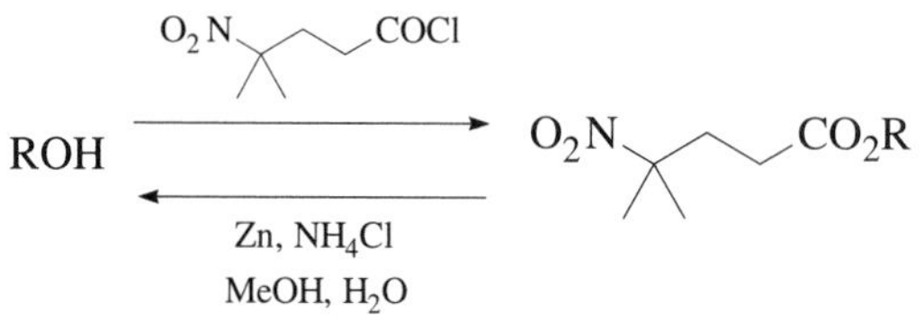

O_2N–C(CH$_3$)$_2$–CH$_2$CH$_2$–COCl; ROH → O_2N–C(CH$_3$)$_2$–CH$_2$CH$_2$–CO_2R

Reverse: Zn, NH_4Cl, MeOH, H_2O

***o*-(Dibromomethyl)benzoate Ester:** *o*-$(Br_2CH)C_6H_4CO_2R$

The *o*-(dibromomethyl)benzoate, prepared to protect nucleosides by reaction with the benzoyl chloride (CH_3CN, 65–90% yield), can be cleaved under nearly neutral conditions. The cleavage involves conversion of the $-CHBr_2$ group to $-CHO$ by silver ion-assisted hydrolysis. The benzoate group, *ortho* to the $-CHO$ group, now is rapidly hydrolyzed by neighboring group participation. (The morpholine and hydroxide ion-catalyzed hydrolyses of methyl 2-formylbenzoate are particularly rapid.)[5,6]

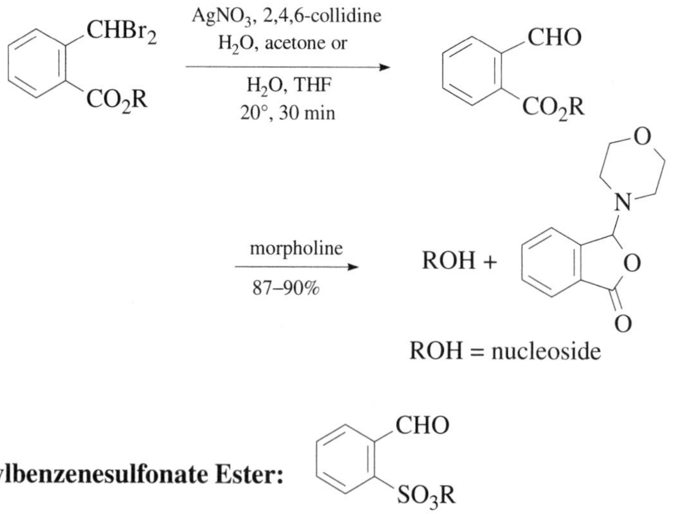

o-$(CHBr_2)C_6H_4CO_2R$ → ($AgNO_3$, 2,4,6-collidine, H_2O, acetone or H_2O, THF, 20°, 30 min) → *o*-$(CHO)C_6H_4CO_2R$

→ (morpholine, 87–90%) → ROH + 3-morpholinophthalide

ROH = nucleoside

2-Formylbenzenesulfonate Ester: *o*-$(CHO)C_6H_4SO_3R$

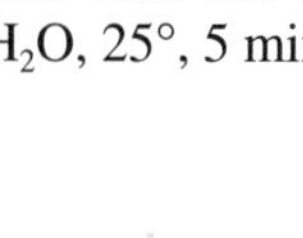

This sulfonate is prepared by reaction with the sulfonyl chloride. Cleavage occurs with 0.05 *M* NaOH (acetone, H_2O, 25°, 5 min, 83–93% yield). Here also, cleavage is facilitated by intramolecular participation through the hydrate of the aldehyde.[7]

Alkyl 2-(Methylthiomethoxy)ethyl Carbonate (MTMEC–OR): $CH_3SCH_2OCH_2CH_2OCO_2R$

Formation

1. $CH_3SCH_2OCH_2CH_2OCOCl$, 1-methylimidazole, CH_3CN, 1 h, >72% yield.[8]

Cleavage

1. $Hg(ClO_4)_2$, 2,4,6-collidine, acetone, H_2O (9:1), 5 h; NH_3, dioxane, H_2O (1:1).[6] In this case Hg(II) is used to cleave the MTM group, liberating a hydroxyl group, which assists in the cleavage of the carbonate upon treatment with ammonia. Cleavage by ammonia is 500 times faster for this hydroxy derivative than for the initial MTM derivative.

4-(Methylthiomethoxy)butyrate Ester (MTMB–OR): $CH_3SCH_2O(CH_2)_3CO_2R$

Formation

1. 4-$(CH_3SCH_2O)(CH_2)_3CO_2H$, 2,6-dichlorobenzoyl chloride, Pyr, CH_3CN, 70% yield.[9] The MTMB group was selectively introduced onto the 5′-OH of thymidine.

Cleavage

1. $Hg(ClO_4)_2$, THF, H_2O, collidine, rt, 5 min; 1 *M* K_2CO_3, 10 min or TEA, 30 min.[7] Hg(II) cleaves the MTM group, liberating a hydroxyl group that assists in the cleavage of the ester.

2-(Methylthiomethoxymethyl)benzoate Ester (MTMT–OR): 2-$(CH_3SCH_2OCH_2)C_6H_4CO_2R$

This group was introduced and removed using the same conditions as those for the MTMB group. The half-lives for ammonolysis of acetate, MTMB, and MTMT, are 5 min, 15 min, and 6 h, respectively.[7]

2-(Chloroacetoxymethyl)benzoate Ester (CAMB–OR):

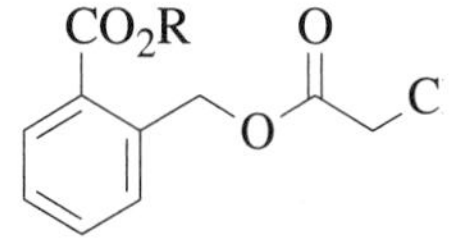

CO_2R O O C

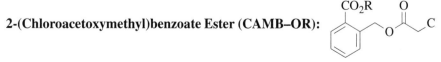

This ester was designed as a protective group for the 2-position in glycosyl donors. It has the stability of the benzoate during glycosylation, but has the ease of removal of the chloroacetate. It is readily introduced through the acid chloride

(CH_2Cl_2, Pyr, 71–88% yield) and is cleaved with thiourea to release the alcohol that closes to the phthalide, releasing the carbohydrate.[10] Its use for nitrogen protection was unsuccessful.

2-[(2-Chloroacetoxy)ethyl]benzoate Ester(CAEB–OR):

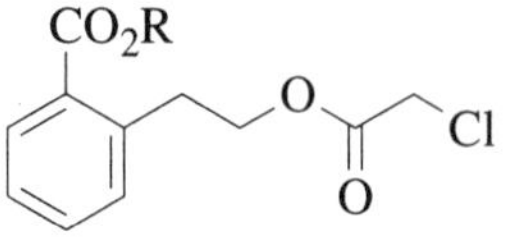

CO_2R O Cl O

The CAEB group is similar to the CAMB group, except that the final deprotection requires acid treatment to initiate ring closure and cleavage.[11] The CAEB group is introduced through the acid chloride (Pyr, CH_2Cl_2, 72 h, 61–91% yield) and is cleaved with thiourea (DMF, 55°, 8–17 h; TsOH, 120 h, 83% yield). This group is reported to be stable to hydrogenolysis.

2-[2-(Benzyloxy)ethyl]benzoate Ester (PAC$_H$–OR) and

2-[2-(4-Methoxybenzyloxy)ethyl]benzoate Ester(PAC$_M$–OR):

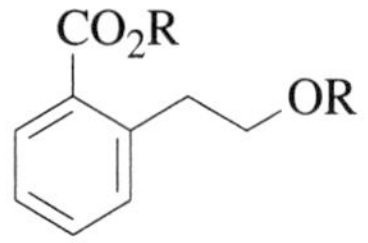

CO_2R OR

R = Bn, MPM

These groups were designed for use in the synthesis of phosphatidylinositol phosphates, where it was desirable to be able to cleave a benzoate without cleaving a glyceryl ester.[12]

Formation

1. PAC–OH, DCC, CH_2Cl_2, DMAP, rt, ~4 h, 87–100% yield.[12]

Cleavage

1. R = H, H_2, Pd/C, AcOEt then *t*-BuOK or *t*-BuMgCl, 85–96% yield.[12]
2. R = OMe, DDQ, CH_2Cl_2, H_2O, 0°C or rt then *t*-BuOK or *t*-BuMgCl, 82–98% yield.[12]
3. R = OMe, $AlCl_3$, $PhNMe_2$, CH_2Cl_2, rt then *t*-BuOK or *t*-BuMgCl, 88–91% yield.[12]

1. R. A. Moss, P. Scrimin, S. Bhattacharya, and S. Chatterjee, *Tetrahedron Lett.*, **28**, 5005 (1987).
2. S. Kusumoto, K. Sakai, and T. Shiba, *Bull. Chem. Soc. Jpn.*, **59**, 1296 (1986).
3. S. Velarde, J. Urbina, and M. R. Peña, *J. Org. Chem.*, **61**, 9541 (1996).
4. T.-L. Ho, *Synth. Commun.*, **10**, 469 (1980).
5. J. B. Chattopadhyaya, C. B. Reese, and A. H. Todd, *J. Chem. Soc., Chem. Commun.*, 987 (1979); J. B. Chattopadhyaya and C. B. Reese, *Nucleic Acids Res.*, **8**, 2039 (1980).

6. K. Zegelaar-Jaarsveld, H. I. Duynstee, G. A. van der Marel, and J. H. van Boom, *Tetrahedron*, **52**, 3575 (1996).
7. M. S. Shashidhar and M. V. Bhatt, *J. Chem. Soc., Chem. Commun.*, 654 (1987).
8. S. S. Jones, C. B. Reese, and S. Sibanda, *Tetrahedron Lett.*, **22**, 1933 (1981).
9. J. M. Brown, C. Christodoulou, C. B. Reese, and G. Sindona, *J. Chem. Soc., Perkin Trans. 1*, 1785 (1984).
10. T. Ziegler and G. Pantkowski, *Liebigs Ann. Chem.*, 659 (1994).
11. T. Ziegler and G. Pantkowski, *Tetrahedron Lett.*, **36**, 5727 (1995).
12. Y. Watanabe, M. Ishimaru, and S. Ozaki, *Chem. Lett.*, 2163 (1994).

Miscellaneous Esters

The following miscellaneous esters have been prepared as protective groups, but have not been widely used. Therefore, they are simply listed for completeness, rather than described in detail.

1. 2,6-Dichloro-4-methylphenoxyacetate ester[1]
2. 2,6-Dichloro-4-(1,1,3,3-tetramethylbutyl)phenoxyacetate ester[1]
3. 2,4-Bis(1,1-dimethylpropyl)phenoxyacetate ester[1]
4. Chlorodiphenylacetate ester[2]
5. Isobutyrate ester[3] (Chart 2)
6. Monosuccinoate ester[4]
7. (*E*)-2-Methyl-2-butenoate (Tigloate) ester[5]
8. *o*-(Methoxycarbonyl)benzoate ester[6]
9. *p*-***P***-Benzoate ester[7]
10. α-Naphthoate ester[8]
11. Nitrate ester[9] (Chart 2)
12. Alkyl *N,N,N′,N′*-tetramethylphosphorodiamidate: $[(CH_3)_2N]_2P(O)OR$[10]
13. 2-Chlorobenzoate ester[11]
14. 4-Bromobenzoate ester[12]
15. 4-Nitrobenzoate ester[13]

1. C. B. Reese, *Tetrahedron*, **34**, 3143 (1978).
2. A. F. Cook and D. T. Maichuk, *J. Org. Chem.*, **35**, 1940 (1970).
3. H. Büchi and H. G. Khorana, *J. Mol. Biol.*, **72**, 251 (1972).
4. P. L. Julian, C. C. Cochrane, A. Magnani, and W. J. Karpel, *J. Am. Chem. Soc.*, **78**, 3153 (1956).
5. S. M. Kupchan, A. D. J. Balon, and E. Fujita, *J. Org. Chem.*, **27**, 3103 (1962).

6. G. Losse and H. Raue, *Chem. Ber.*, **98**, 1522 (1965).
7. R. D. Guthrie, A. D. Jenkins, and J. Stehlicek, *J. Chem. Soc. C*, 2690 (1971).
8. I. Watanabe, T. Tsuchiya, T. Takase, S. Umezawa, and H. Umezawa, *Bull. Chem. Soc. Jpn.*, **50**, 2369 (1977).
9. J. Honeyman and J. W. W. Morgan, *Adv. Carbohydr. Chem.*, **12**, 117 (1957); J. F. W. Keana, in *Steroid Reactions*, C. Djerassi, Ed., Holden-Day, San Francisco, 1963, pp. 75–76; R. Boschan, R. T. Merrow, and R. W. Van Dolah, *Chem. Rev.*, **55**, 485 (1955); R. W. Binkley and D. J. Koholic, *J. Org. Chem.*, **44**, 2047 (1979); R. W. Binkley and D. J. Koholic, *J. Carbohydr. Chem.*, **3**, 85 (1984).
10. R. E. Ireland, D. C. Muchmore, and U. Hengartner, *J. Am. Chem. Soc.*, **94**, 5098 (1972).
11. E. Rozners, R. Renhofa, M. Petrova, J. Popelis, V. Kumpins, and E. Bizdena, *Nucleosides Nucleotides*, **11**, 1579 (1992).
12. K. Ohmori, S. Nishiyama, and S. Yamamura, *Tetrahedron Lett.*, **36**, 6519 (1995).
13. C. Kolar, K. Dehmel, H. Moldenhauer, and M. Gerken, *J. Carbohydr. Chem.*, **9** 873 (1990).

3′,5′-Dimethoxybenzoin Carbonate (DMB–O_2COR)

Formation/Cleavage

ROH → (MeNO$_2$, Pyr 42–95%) → DMB–O$_2$COR

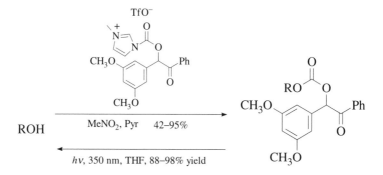

← *hν*, 350 nm, THF, 88–98% yield

The dimethoxybenzoin group has an advantage over the *o*-nitrobenzyl group, because it produces a nonreactive benzofuran upon photolysis, whereas the *o*-nitrobenzyl group gives a reactive nitroso aldehyde upon photolytic cleavage. The DMB group is also cleaved much more rapidly and with greater quantum efficiency than the *o*-nitrobenzyl group.[1] A convenient procedure for the preparation of DMB has been reported.[2]

1. M. C. Pirrung and J.-C. Bradley, *J. Org. Chem.*, **60**, 1116 (1995).
2. M. H. B. Stowell, R. S. Rock, D. C. Rees, and S. I. Chan, *Tetrahedron Lett.*, **37**, 307 (1996).

A Wild and Woolly Photolabile Fluorescent Ester:

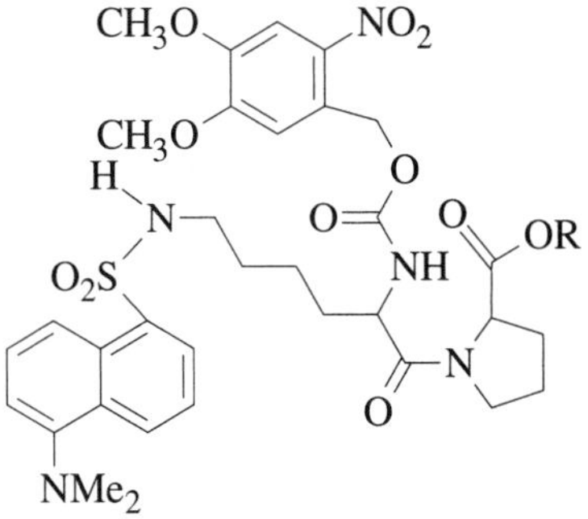

This group was developed as part of a scheme to prepare fluorescent tags to be used in DNA sequencing. Deprotection is accomplished by irradiation at 360 nm to release the NVOC group, which then sets up the system to form a diketopiperazine while releasing the alcohol.[1]

1. K. Burgess, S. E. Jacutin, D. Lim, and A. Shitangkoon, *J. Org. Chem.*, **62**, 5165 (1997).

Alkyl *N*-Phenylcarbamate: ROCONHPh (Chart 2)

Phenyl isocyanates are generally more reactive than alkyl isocyanates in their reactions with alcohols, but with CuCl catalysis, even alkyl isocyanates will react readily with primary, secondary, or tertiary alcohols (45–95% yield).[1]

Formation

1. PhN=C=O, Pyr, 20°, 2–3 h, 100% yield.[2] This method was used to protect selectively the primary hydroxyl group in several pyranosides.[3]

Cleavage

1. MeONa, MeOH, reflux, 1.5 h, good yield.[4]
2. $LiAlH_4$, THF, or dioxane, reflux, 3–4 h, 90% yield.[3]
3. Cl_3SiH, Et_3N, CH_2Cl_2, 4–48 h, 25–80°, 80–95% yield.[5] Primary, secondary, tertiary, allylic, propargylic, or benzylic derivatives are cleaved by this method.

1. M. E. Duggan and J. S. Imagire, *Synthesis*, 131 (1989).
2. K. L. Agarwal and H. G. Khorana, *J. Am. Chem. Soc.*, **94**, 3578 (1972).
3. D. Plusquellec and M. Lefeuvre, *Tetrahedron Lett.*, **28**, 4165 (1987).
4. H. O. Bouveng, *Acta Chem. Scand.*, **15**, 87, 96 (1961).
5. W. H. Pirkle and J. R. Hauske, *J. Org. Chem.*, **42**, 2781 (1977).

Borate Ester: $(RO)_3B$

Formation

1. $BH_3 \cdot Me_2S$, 25°, 1 h, 80–90% yield.[1]
2. $B(OH)_3$, benzene, $-H_2O$, 100% yield.[2,3]

Cleavage

Borate esters are hydrolyzed with aqueous acid or base. More sterically hindered borates, such as pinanediol derivatives, are quite stable to hydrolysis.[4] Some hindered borates are stable to anhydrous acid and base, to HBr/BzOOBz, to NaH, and to Wittig reactions.[3]

1. C. A. Brown and S. Krishnamurthy, *J. Org. Chem.*, **43**, 2731 (1978).
2. W. I. Fanta and W. F. Erman, *J. Org. Chem.*, **37**, 1624 (1972).
3. W. I. Fanta and W. F. Erman, *Tetrahedron Lett.*, 4155 (1969).
4. D. S. Matteson and R. Ray, *J. Am. Chem. Soc.*, **102**, 7590 (1980).

Dimethylphosphinothioyl Ester: $(CH_3)_2P(S)OR$

The dimethylphosphinothioyl group has been used to protect hydroxyl groups in carbohydrates. It is prepared from the alcohol and $Me_2P(S)Cl$ (cat. DMAP, DBU). The group is not prone to undergo "acyl" migration, as are carboxylate esters. It is stable to the acidic conditions used to cleave acetonides and trityl groups, as well as to DBU/MeOH, $Bu_4N^+F^-$, Bu_3SnH, Grignard reagents, and cat. NaOMe/MeOH. The dimethylphosphinothioyl group is cleaved with $BnMe_3N^+OH^-$. It can also be cleaved by $Bu_4N^+F^-$ after conversion to the dimethylphosphonyl group with *m*-chloroperoxybenzoic acid.[1]

1. T. Inazu and T. Yamanoi, *Noguchi Kenkyusho Jiho*, 43–47 (1988); *Chem. Abstr.*, **111**: 7685w (1989).

Alkyl 2,4-Dinitrophenylsulfenate: $ROSC_6H_3$–2,4-$(NO_2)_2$ (Chart 2)

A nitrophenylsulfenate, cleaved by nucleophiles under very mild conditions, was developed as protection for a hydroxyl group during solid-phase nucleotide synthesis.[1] The sulfenate ester is stable to the acidic hydrolysis of acetonides.[2]

Formation

1. 2,4-$(NO_2)_2C_6H_3SCl$, Pyr, DMF or CH_2Cl_2, 20°, 1 h, 70–85% yield.[1]

2. 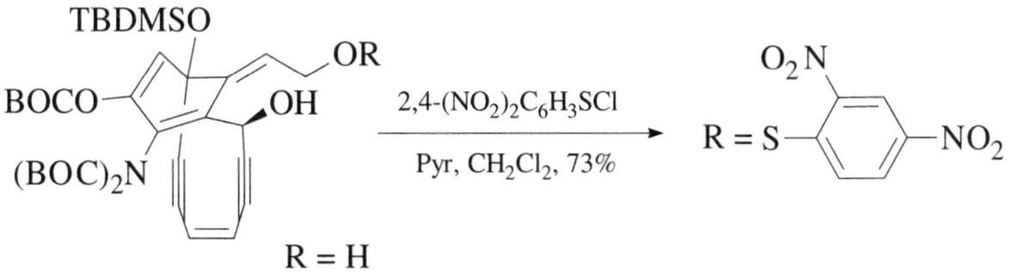 TBDMSO ... OR, BOCO, OH, $(BOC)_2N$, R = H $\xrightarrow[\text{Pyr, } CH_2Cl_2\text{, 73\%}]{2,4\text{-}(NO_2)_2C_6H_3SCl}$ R = S–C₆H₃(O₂N)(NO₂)

Ref. 3

Cleavage

1. Nu^-, MeOH, H_2O, 25°, 4 h, 63–80% yield.[1]
2. Nu^- = $Na_2S_2O_3$, pH 8.9; NaCN, pH 8.9; Na_2S, pH 6.6; PhSH, pH 11.8.[1]
3. H_2, Raney Ni, 54% yield.[1]
4. Al, $Hg(OAc)_2$, MeOH, 5 h, 67% yield.[2]
5. An *o*-nitrophenylsulfenate is cleaved by electrolytic reduction (−1.0 V, DMF, $R_4N^+X^-$).[4]
6. PhSH, Pyr, THF, 83% yield.[3]

1. R. L. Letsinger, J. Fontaine, V. Mahadevan, D. A. Schexnayder, and R. E. Leone, *J. Org. Chem.*, **29**, 2615 (1964).
2. K. Takiura, S. Honda, and T. Endo, *Carbohydr. Res.*, **21**, 301 (1972).
3. P. Magnus, G. F. Miknis, N. J. Press, D. Grandjean, G. M. Taylor, and J. Harling, *J. Am. Chem. Soc.*, **119**, 6739 (1997).
4. V. G. Mairanovsky, *Angew. Chem., Int. Ed. Engl.*, **15**, 281 (1976).

Sulfonates as Protective Groups for Alcohols

Sulfonate protective groups have largely been restricted to carbohydrates, where they serve to protect the 2-OH with a nonparticipating group so that coupling gives predominately 1,2-*cis* glycosides.

Sulfate: $ROSO_3^-$

Formation[1]/Cleavage[2]

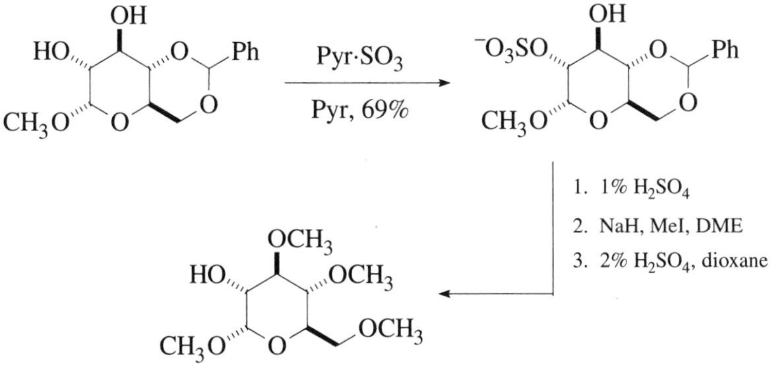

OH, HO, O, Ph, CH_3O, O, O $\xrightarrow[\text{Pyr, 69\%}]{\text{Pyr·}SO_3}$ OH, ^-O_3SO, O, Ph, CH_3O, O, O

1. 1% H_2SO_4
2. NaH, MeI, DME
3. 2% H_2SO_4, dioxane

OCH_3, HO, OCH_3, CH_3O, O, OCH_3

The α-anomer gives better selectivity for the 2-OH than does the β-anomer (3:2). Note that the conditions used to remove the 4,6-*O*-benzylidene group are sufficiently mild to retain the sulfate.[2]

Allylsulfonate (Als–OR): $CH_2{=}CHCH_2SO_2R$

The allylsulfonate was developed for the protection of carbohydrates.

Formation

1. Allylsulfonyl chloride, Pyr, CH_2Cl_2, 55–71% yield.[3]

Cleavage

1. THF, morpholine, 35% aq. formaldehyde, $(Ph_3P)_4Pd$, >85% yield.[3]

Methanesulfonate (Mesylate) (RO–Ms): $MeSO_3R$

Formation

1. MsCl, Et_3N, CH_2Cl_2, 0°, generally >90% yield.[4]

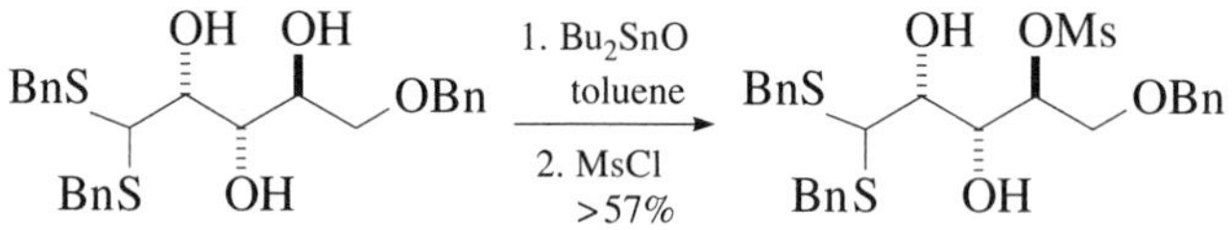

OH OH; BnS; OBn; BnS OH → 1. Bu_2SnO, toluene; 2. MsCl, >57% → OH OMs; BnS; OBn; BnS OH

Ref. 5

Cleavage

1. Na(Hg), 2-propanol, 84–98% yield.[6] The use of methanol or ethanol gives very slow reactions. Benzyl groups are not affected by these conditions.
2. Photolysis, KI, MeOH.[7] The triflates are also cleaved, but the products are partitioned between cleavage and reduction.[8]
3. MeMgBr, THF.[9]

Benzylsulfonate: $ROSO_2Bn$

Formation

1. $BnSO_2Cl$, 2,6-lutidine, CH_2Cl_2, >72% yield.[10]

Cleavage

1. $NaNH_2$, DMF, 67–95% yield.[3]

Tosylate (TsOR): $CH_3C_6H_4SO_3R$

Formation

1. TsCl, Pyr.[11]

2. 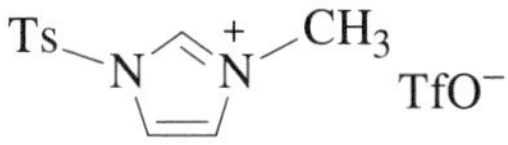 $Ts-N^{\frown}N^{+}-CH_3$ (imidazolium) TfO^- This reagent selectively protects a primary alcohol in the presence of a secondary alcohol.[12]
3. Bu_2SnO, toluene, reflux; TsCl, $CHCl_3$, 36–99% yield. The primary alcohol of a 1,2-diol is selectively tosylated, but when hexamethylenestannylene acetals are used, selectivity is reversed and the secondary diol is preferentially tosylated.[13,14]
4. TsCl, DABCO, CH_2Cl_2, MTBE or AcOEt, 45–97% yield. In many cases, these conditions were found to be superior to the use of pyridine as a base. DABCO is also less toxic than pyridine, which may prove useful in a commercial setting.[15]

Cleavage

1. $h\nu$, 90% CH_3CN/H_2O, 1,5-dimethoxynaphthalene, NH_2NH_2 or $NaBH_4$ or Pyr·BH_3, 59–97% yield.[16]
2. $h\nu$, Et_3N, MeOH, 12 h, 91% yield.[17]
3. The tosyl group has also been removed by reductive cleavage with Na/NH_3 (65–73% yield),[18] Na/naphthalene (50–87% yield),[19] and Na(Hg)/MeOH (96.7% yield).[20]
4. $NaBH_4$, DMSO, 140°, 71% yield.[21]
5. $LiAlH_4$, ether.[22]
6. Mg, MeOH, 4–6 h, 80–95% yield.[23]

2-[(4-Nitrophenyl)ethyl]sulfonate (Npes–OR): $4\text{-}NO_2C_6H_4CH_2CH_2SO_3R$

Formation

1. $4\text{-}NO_2C_6H_4CH_2CH_2SO_2Cl$, Pyr, 70–90% yield.[24]

Cleavage

1. 0.1 *M* DBU, CH_3CN, 2 h.[25] The Npes group is more labile to base than either the Npeoc or the Npe group. It is not very rapidly removed by

fluoride ions. K_2CO_3, MeOH can be used for acetate cleavage in the presence of a Npes ester.[26]

1. A. Liav and M. B. Goren, *Carbohydr. Res.*, **131**, C8 (1984).
2. M. B. Goren and M. E. Kochansky, *J. Org. Chem.*, **38**, 3510 (1973); A. Liav and M. B. Goren, *Carbohydr. Res.*, **127**, 211 (1984).
3. W. K. D. Brill and H. Kunz, *Synlett*, 163 (1991).
4. A. Fürst and F. Koller, *Helv. Chim. Acta*, **30**, 1454 (1947).
5. M. W. Bredenkamp, C. W. Holzapfel, and A. D. Swanepoel, *Tetrahedron Lett.*, **31**, 2759 (1990).
6. K. T. Webster, R. Eby, and C. Schuerch, *Carbohydr. Res.*, **123**, 335 (1983).
7. R. W. Binkley and X. Liu, *J. Carbohydr. Chem.*, **11**, 183 (1992).
8. X. G. Liu, R. W. Binkley, and P. Yeh, *J. Carbohydr. Chem.*, **11**, 1053 (1992).
9. J. Cossy, J.-L. Ranaivosata, V. Bellosta, and R. Wietzke, *Synth. Commun.*, **25**, 3109 (1995).
10. L. F. Awad, El S. H. Ashry, and C. Schuerch, *Bull. Chem. Soc. Jpn.*, **59**, 1587 (1986).
11. L. F. Fieser and M. Fieser, *Reagents for Organic Synthesis*, Vol. 1, Wiley, New York, p. 1179 (1967).
12. M. Gerspacher and H. Rapoport, *J. Org. Chem.*, **56**, 3700 (1991).
13. X. Kong and T. B. Grindley, *Can. J. Chem.*, **72**, 2396 (1994).
14. Y. Tsuda, M. Nishimura, T. Kobayashi, Y. Sato, and K. Kanemitsu, *Chem. Pharm. Bull.*, **39**, 2883 (1991).
15. J. Hartung, S. Hünig, R. Kneuer, M. Schwaz, and H. Wenner, *Synthesis*, 1433 (1997).
16. A. Nishida, T. Hamada, and O. Yonemitsu, *J. Org. Chem.*, **53**, 3386 (1988); *idem.*, *Chem. Pharm. Bull.*, **38**, 2977 (1990).
17. R. W. Binkley and D. J. Koholic, *J. Org. Chem.*, **54**, 3577 (1989).
18. M. A. Miljkovic, M. Pesic, A. Jokic, and E. A. Davidson, *Carbohydr. Res.*, **15**, 162 (1970); J. Kovar, *Can. J. Chem.*, **48**, 2383 (1970).
19. H. C. Jarrell, R. G. S. Ritchie, W. A. Szarek, and J. K. N. Jones, *Can. J. Chem.*, **51**, 1767 (1973); E. Lewandowska, V. Neschadimenko, S. F. Wnuk, and M. J. Robins, *Tetrahedron*, **53**, 6295 (1997).
20. R. S. Tipson, *Methods Carbohydr. Chem.*, **II**, 250 (1963).
21. V. Pozsgay, E. P. Dubois, and L. Pannnell, *J. Org. Chem.*, **62**, 2832 (1997).
22. H. B. Borén, G. Ekborg, and J. Lönngren, *Acta. Chem. Scand., Ser. B*, **B29**, 1085 (1975).
23. M. Sridhar, B. A. Kumar, and R. Narender, *Tetrahedron Lett.*, **39**, 2847 (1998).
24. M. Pfister, H. Schirmeister, M. Mohr, S. Farkas, K.-P. Stengele, T. Reiner, M. Dunkel, S. Gokhale, R. Charubala, and W. Pfleiderer, *Helv. Chim. Acta*, **78**, 1705 (1995).
25. H. Schirmeister and W. Pfleiderer, *Helv. Chim. Acta*, **77**, 10 (1994); R. Charubala, W. Pfleiderer, R. W. Sobol, S. W. Li, and R. J. Suhadolnik, *Helv. Chim. Acta*, **72**, 1354 (1989).
26. C. Hörndler and W. Pfleiderer, *Helv. Chim. Acta*, **79**, 798 (1996).

PROTECTION FOR 1,2- AND 1,3-DIOLS

The prevalence of diols in synthetic planning and in natural sources (e.g., in carbohydrates, macrolides, and nucleosides) has led to the development of a number of protective groups of varying stability to a substantial array of reagents. Dioxolanes and dioxanes are the most common protective groups for diols. The ease of formation follows the order $HOCH_2C(CH_3)_2CH_2OH$ > $HO(CH_2)_2OH$ > $HO(CH_2)_3OH$.

In some cases the formation of a dioxolane or dioxane can result in the generation of a new stereogenic center, either with complete selectivity or as a mixture of the two possible isomers. Although the new stereogenic center is removed on deprotection, this center often causes problems because it complicates NMR interpretation.

Cyclic carbonates and cyclic boronates have also found considerable use as protective groups. In contrast to most acetals and ketals, the carbonates are cleaved with strong base, and sterically unencumbered boronates are readily cleaved by water.

Some of the protective groups for diols are listed in Reactivity Chart 3.

Cyclic Acetals and Ketals

Methylene Acetal (Chart 3)

Methylene acetals are the most stable acetals to acid hydrolysis. Difficulty in their removal is probably the reason that these compounds have not seen much use. Cleavage usually occurs under strongly acidic or Lewis acid conditions.

Formation

1. 40% CH_2O, concd. HCl, 50°, 4 days, 68% yield.[1] The trismethylenedioxy derivative of a carbohydrate was formed.
2. Paraformaldehyde, H_2SO_4, AcOH, 90°, 1 h, good yield.[2]
3. DMSO, NBS, 50°, 12 h, 62% yield.[3]
4. CH_2Br_2, NaH, DMF, 0–30°, 40 h, 46% yield.[4]
5. $(MeO)_2CH_2$, 2,6-lutidine, TMSOTf, 0°, 15 min.[5] Similar conditions have been used to introduce MOM ethers on alcohols.

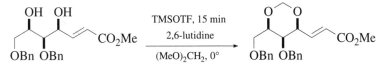

OH OH ... CO_2Me (OBn OBn) → TMSOTF, 15 min; 2,6-lutidine; $(MeO)_2CH_2$, 0° → O O (methylene acetal) ... CO_2Me (OBn OBn)

6. $(MeO)_2CH_2$, LiBr, TsOH, CH_2Cl_2, 23°, 83% yield.[6] In this case, a 1,3-methylene acetal is formed in preference to a 1,2-methylene acetal from a 1,2,3-triol. These conditions also protect simple alcohols as their MOM derivatives.

7. CH_2Br_2, NaOH, CH_2Cl_2, cetyl$N^+Me_3Br^-$, heat, 81% yield.[7] This method is effective for both *cis*- and *trans*-1,2-diols.
8. DMSO, TMSCl, 36–72 h.[8]
9. DMSO, $POCl_3$ or $SOCl_2$, 30–120 min, 10–95% yield.[9]

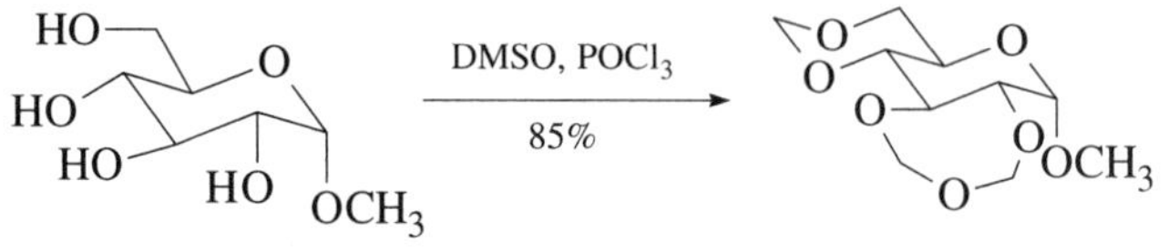

DMSO, $POCl_3$ → 85%

In some examples, the trioxaheptane system could be hydrolyzed with acid to give the diol. The trioxaheptane may also release formaldehyde upon heating.
10. CH_2Br_2, powdered KOH, DMSO, rt, 49% yield.[10]
11. HCHO, cat. SO_2.[11]
12. From a bis MEM ether: $ZnBr_2$, EtOAc, rt.[12]
13. 1,1′-Thiocarbonyldiimidazole, solvent, rt then reduce with Ph_3SnH, AIBN, toluene, reflux, 36–90% yield.[13]

Cleavage

1. BCl_3, CH_2Cl_2, −80°, 30 min, warm to 20°, 61% yield; isolated as the acetate derivative.[1]
2. 2 *N* HCl, 100°, 3 h.[2]
3. AcOH, Ac_2O, H_2SO_4, 2 h, 0°, 91.5% yield.[14]

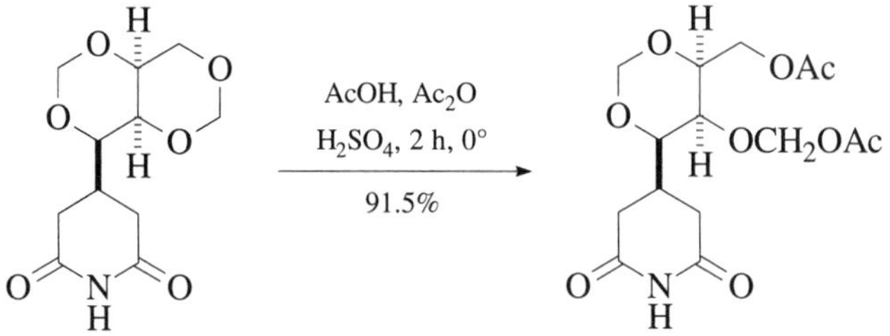

AcOH, Ac_2O, H_2SO_4, 2 h, 0° → 91.5%

4. NaI, $SiCl_4$, rt, 20–60 min, 78% yield. Cleavage results in subsequent formation of a diiodide, but this is not a general process. For the most part, ketals are cleaved to give the ketone, while catechol methylene acetals return the catechol.[15]
5. $Ph_3C^+BF_4^-$, CH_2Cl_2, reflux, 48 h; HCl, rt, 17.5 h, 86% yield.[16] Cleavage occurs by hydride abstraction.
6. $(CF_3CO)_2O$, AcOH, CH_2Cl_2, 21°; MeOH, K_2CO_3, 92% yield.[17]

7. HF, EtOH, THF, 0–5°, 14 h.[18]

HF, EtOH / THF, 0–5° / 14 h

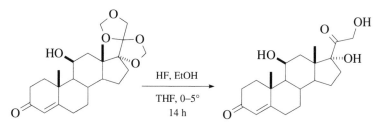

8. AcCl, $ZnCl_2$, Et_2O; ROH, 75–97% yield.[19,20] When methanol is replaced with benzyl alcohol or methoxyethanol the BOM or MEM groups are formed respectively.

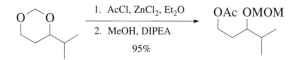

1. AcCl, $ZnCl_2$, Et_2O / 2. MeOH, DIPEA / 95% → OAc OMOM

1. T. G. Bonner, *Methods Carbohydr. Chem.*, **II**, 314 (1963).
2. L. Hough, J. K. N. Jones, and M. S. Magson, *J. Chem. Soc.*, 1525 (1952).
3. S. Hanessian, G. Y.-Chung, P. Lavallee, and A. G. Pernet, *J. Am. Chem. Soc.*, **94**, 8929 (1972).
4. J. S. Brimacombe, A. B. Foster, B. D. Jones, and J. J. Willard, *J. Chem. Soc. C*, 2404 (1967).
5. F. Matsuda, M. Kawasaki, and S. Terashima, *Tetrahedron Lett.*, **26**, 4639 (1985).
6. J. L. Gras, R. Nouguier, and M. Mchich, *Tetrahedron Lett.*, **28**, 6601 (1987).
7. D. G. Norman, C. B. Reese, and H. T. Serafinowska, *Synthesis*, 751 (1985). For a similar method, see K. S. Kim and W. A. Szarek, *Synthesis*, 48 (1978).
8. B. S. Bal and H. W. Pinnick, *J. Org. Chem.*, **44**, 3727 (1979); Z. Gu, L. Zeng, X.-p. Fang, T. Colman-Saizarbitoria, M. Huo, and J. L. Mclaughlin, *J. Org. Chem.*, **59**, 5162 (1994).
9. M. Guiso, C. Procaccio, M. R. Fizzano, and F. Piccioni, *Tetrahedron Lett.*, **38**, 4291 (1997).
10. A. Liptak, V. A. Oláh, and J. Kerékgyártó, *Synthesis*, 421 (1982).
11. B. Burczyk, *J. Prakt. Chem.*, **322**, 173 (1980).
12. J. A. Boynton and J. R. Hanson, *J. Chem. Res., Synop.*, 378 (1992).
13. F. De Angelis, M. Marzi, P. Minetti, D. Misiti, and S. Muck, *J. Org. Chem.*, **62**, 4159 (1997).
14. M. J. Wanner, N. P. Willard, G. J. Kooman, and U. K. Pandet, *Tetrahdron*, **43**, 2549 (1987).
15. S. S. Elmorsy, M. V. Bhatt, and A. Pelter, *Tetrahedron Lett.*, **33**, 1657 (1992).
16. H. Niwa, O. Okamoto, and K. Yamada, *Tetrahedron Lett.*, **29**, 5139 (1988).
17. J.-L. Gras, H. Pellissier, and R. Nouguier, *J. Org. Chem.*, **54**, 5675 (1989).

18. H. Shibasaki, T. Furuta, and Y. Kasuya, *Steroids*, **57**, 13 (1992).
19. W. F. Bailey, L. M. J. Zarcone, and A. D. Rivera, *J. Org. Chem.*, **60**, 2532 (1995).
20. W. F. Bailey, M. W. Carson, and L. M. J. Zarcone, *Org. Synth.*, **75**, 177 (1997).

Ethylidene Acetal (Chart 3)

Formation

1. CH_3CHO, $CH_3CH(OMe)_2$, or paraldehyde, concd. H_2SO_4, 2–3 h, 60% yield.[1]
2. In the following example the ethylidene acetal was used, because attempts to make the acetonide led to the formation of a 1:1 mixture of the 1,3- and 1,4-acetonides.[2]

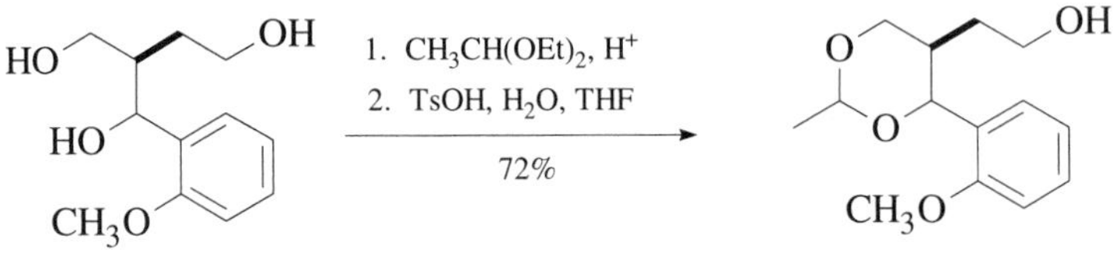

HO, HO, CH_3O, OH — 1. $CH_3CH(OEt)_2$, H^+; 2. TsOH, H_2O, THF; 72% → O, O, CH_3O, OH

3. Diborane reduction of an ortho ester that is prepared from a triol with $CH_3C(OEt)_3$, PPTS.[3]

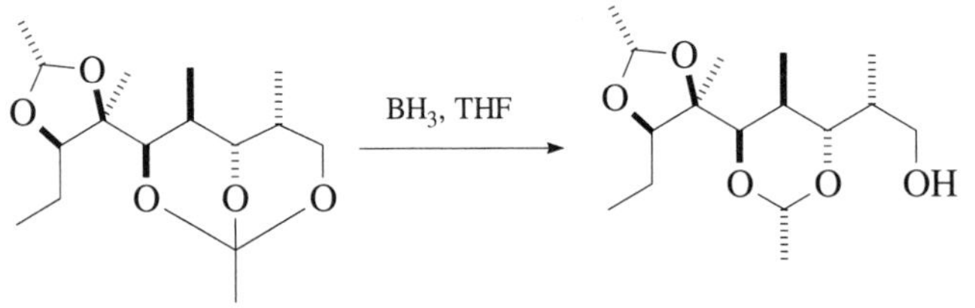

BH_3, THF →

Cleavage

1. 0.67 *N* H_2SO_4, aq. acetone, reflux, 7 h.[1]
2. Ac_2O, cat. H_2SO_4, 20°, 5 min, 60% yield.[1] The ethylidene acetal is cleaved to form an acetate that can be hydrolyzed with base.
3. 80% AcOH, reflux, 1.5 h.[4]
4. O_3, CH_2Cl_2, 75% yield.[3]

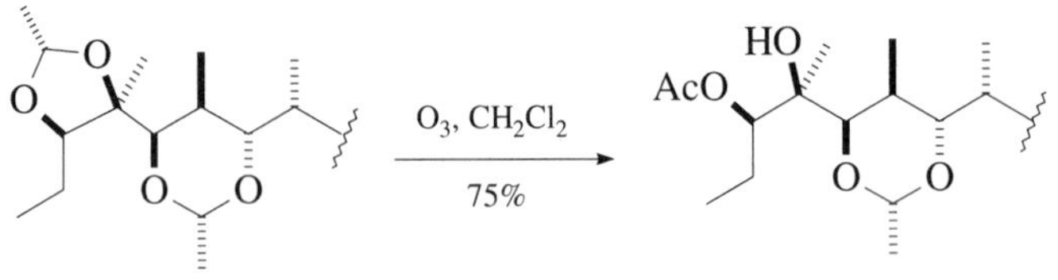

O_3, CH_2Cl_2; 75% → AcO, HO

1. T. G. Bonner, *Methods Carbohydr. Chem.*, **II**, 309 (1963); D. M. Hall, T. E. Lawler, and B. C. Childress, *Carbohydr. Res.*, **38**, 359 (1974).

2. A. G. Brewster and A. Leach, *Tetrahedron Lett.*, **27**, 2539 (1986).
3. G. Stork and S. D. Rychnovsky, *J. Am. Chem. Soc.*, **109**, 1565 (1987).
4. J. W. Van Cleve and C. E. Rist, *Carbohydr. Res.*, **4**, 82 (1967).

t-Butylmethylidene Ketal: *t*-BuCH(OR)$_2$[1]

1-*t*-Butylethylidene Ketal: *t*-BuC(CH$_3$)(OR)$_2$[2]

1-Phenylethylidene Ketal: Ph(CH$_3$)C(OR)$_2$[2]

1-*t*-Butylethylidene and 1-phenylethylidene ketals were prepared selectively from the C_4–C_6, 1,3-diol in glucose by an acid-catalyzed transketalization reaction [e.g., $Me_3CC(OMe)_2CH_3$, TsOH/DMF, 24 h, 79% yield; $PhC(OMe)_2Me$, TsOH, DMF, 24 h, 90% yield, respectively]. They are cleaved by acidic hydrolysis: AcOH, 20°, 90 min, 100% yield, and AcOH, 20°, 3 days, 100% yield, respectively.[2] Ozonolysis of the *t*-butylmethylidene ketal affords hydroxy ester, albeit with poor regiocontrol, but a more sterically differentiated derivative may give better selectivity, as was observed with the ethylidene ketal.[1]

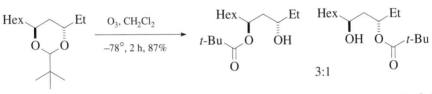

Hex, Et, O, O — O_3, CH_2Cl_2, −78°, 2 h, 87% → Hex, Et, *t*-Bu, O, OH, O + Hex, Et, OH, O, *t*-Bu, O (3:1)

Ref. 1

1. S. D. Rychnovsky and N. A. Powell, *J. Org. Chem.*, **62**, 6460 (1997).
2. M. E. Evans, F. W. Parrish, and L. Long, Jr., *Carbohydr. Res.*, **3**, 453 (1967).

1-(4-Methoxyphenyl)ethylidene Acetal

Formation/Cleavage[1]

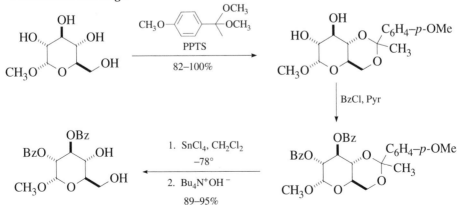

OH, HO, OH, CH_3O, O, OH — CH_3O–C$_6$H$_4$–C(OCH$_3$)$_2$CH$_3$, PPTS, 82–100% → OH, HO, O, C_6H_4-*p*-OMe, CH_3, CH_3O, O, O

BzCl, Pyr ↓

OBz, BzO, O, C_6H_4-*p*-OMe, CH_3, CH_3O, O, O — 1. $SnCl_4$, CH_2Cl_2, −78°; 2. $Bu_4N^+OH^-$, 89–95% → OBz, BzO, OH, CH_3O, O, OH

PPTS = pyridinium *p*-toluensulfonate

1. B. H. Lipshutz and M. C. Morey, *J. Org. Chem.*, **46**, 2419 (1981).

2,2,2-Trichloroethylidene Acetal

Formation

1. Trichloroacetaldehyde (chloral) reacts with glucose in the presence of sulfuric acid to form two mono- and four diacetals.[1]

2. chloral, DCC, $ClCH_2CH_2Cl$, reflux, 4.5 h, 63%

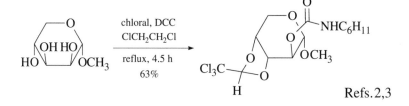

Refs. 2,3

Cleavage

Cleavage occurs by prior conversion to the ethylidene acetal with RaNi or Bu_3SnH and then the normal acid hydrolysis.[2,3] The trichloro acetal is cleaved by reduction (H_2, Raney Ni, 50% NaOH, EtOH, 15 min).[3] The trichloro acetal can probably be cleaved with Zn/AcOH [cf. $ROCH(R')OCH_2CCl_3$, cleaved by Zn/AcOH, AcONa, 20°, 3 h, 90% yield[4]].

1. S. Forsén, B. Lindberg, and B.-G. Silvander, *Acta Chem. Scand.*, **19**, 359 (1965).
2. R. Miethchen and D. Rentsch, *J. Prakt. Chem.*, **337**, 422 (1995).
3. R. Miethchen and D. Rentsch, *Synthesis*, 827 (1994).
4. R. U. Lemieux and H. Driguez, *J. Am. Chem. Soc.*, **97**, 4069 (1975).

Acrolein Acetal: $CH_2{=}CHCH(OR)_2$

Formation

1. Bu_2SnO, toluene, reflux, 4 h; $Pd(Ph_3P)_4$, THF, $CH_2{=}CHCH(OAc)_2$, rt, 1 h, 80–89% yield. In pyranoside protection, selectivity for 1,3-dioxane formation is generally observed, but dioxolanes are often formed.

Cleavage

1. $(Ph_3P)_3RhCl$, EtOH, with or without TFA, 90% yield.
2. 1% H_2SO_4, refluxing dioxane, >80% yield.[1]
3. Reductive cleavage of the acrolein acetal proceeds similarly to that of the benzylidene acetals.[2]

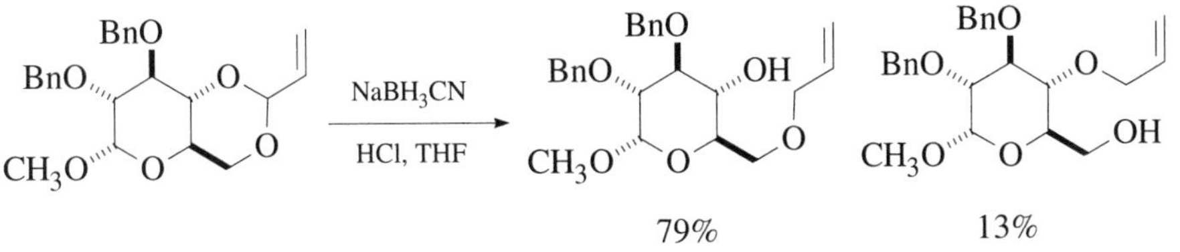

BnO BnO CH$_3$O O O O → NaBH$_3$CN / HCl, THF → BnO BnO CH$_3$O O OH O (79%) + BnO BnO CH$_3$O O O OH (13%)

1. C. W. Holzapfel, J. J. Huyser, T. L. Van der Merwe, and F. R. Van Heerden, *Heterocycles*, **32**, 1445 (1991).
2. P. J. Garegg, *Acc. Chem. Res.*, **25**, 575 (1992).

Acetonide (Isopropylidene Ketal) (Chart 3)

Acetonide formation is the most commonly used protection for 1,2- and 1,3-diols. The acetonide has been used extensively in carbohydrate chemistry to mask selectively the hydroxyls of the many different sugars.[1] In preparing acetonides of triols, the 1,2-derivative is generally favored over the 1,3-derivative which in turn is favored over the 1,4-derivative,[2] but the extent to which the 1,2-acetonide is favored is dependent upon the structure of the triol.[3–6] Note that the 1,2-selectivity for the ketal from 3-pentanone is better than that from acetone.[7] Its greater lipophilicity also improves the isolation of the ketals of small alcohols such as glycerol.[8]

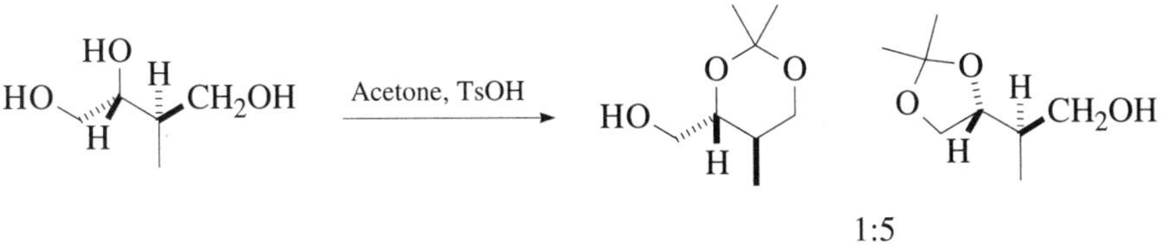

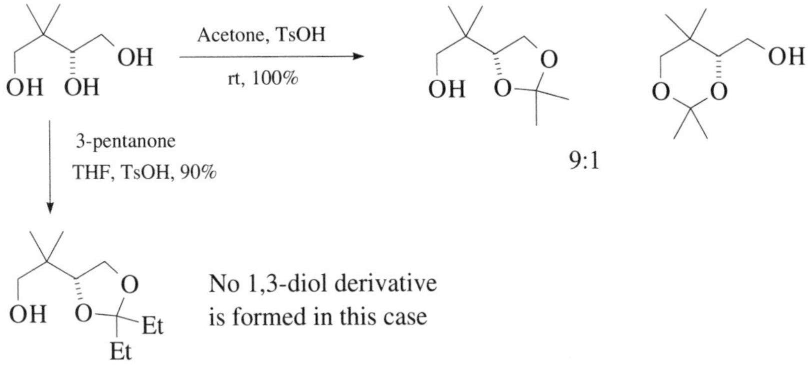

HO HO H H CH$_2$OH → Acetone, TsOH → HO O O H + O O H H CH$_2$OH

1:5

OH OH OH → Acetone, TsOH / rt, 100% → OH O O + O O OH

9:1

↓ 3-pentanone / THF, TsOH, 90%

OH O O Et Et

No 1,3-diol derivative is formed in this case

In cases where two 1,2-acetonides are possible, the thermodynamically more favored one prevails. Secondary alcohols have a greater tendency to form cyclic acetals than do primary alcohols,[7,9] but an acetonide from a primary alcohol is preferred over an acetonide from two *trans*, secondary alcohols.

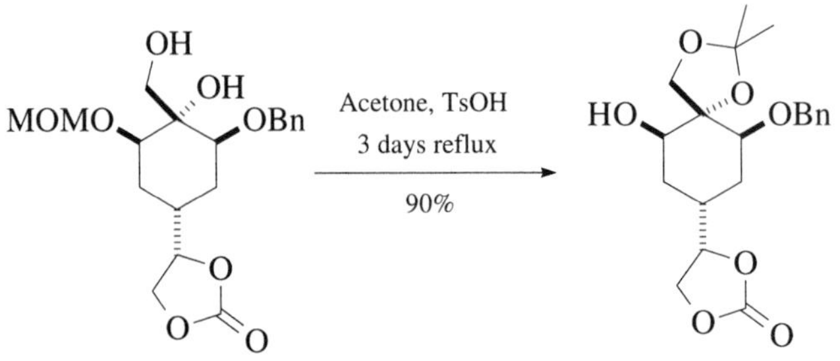

MOMO, OH, OH, OBn → Acetone, TsOH, 3 days reflux, 90% → HO, O, O, OBn

Ref. 10

In the following two situations, **i** is isomerized to **ii**, producing a *trans* derivative, but acetonide **iii** fails to isomerize to the internal derivative because the less favorable *cis* product would be formed.[11,12]

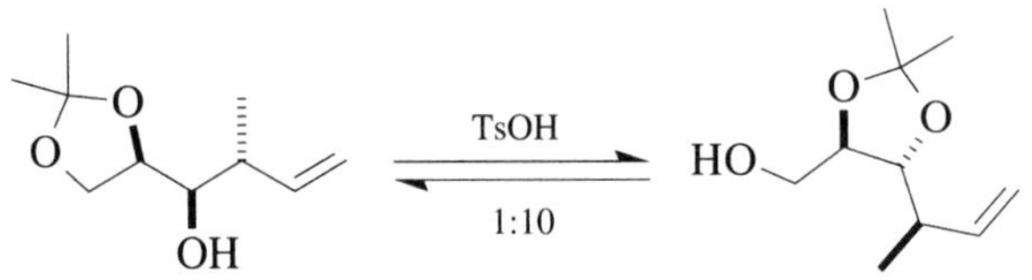

TsOH, 1:10

HO

OH

i

ii

OH

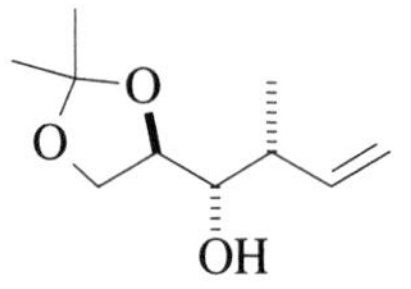

iii

Acetonides may also participate in unexpected reactions, such as the chlorination show below.[13]

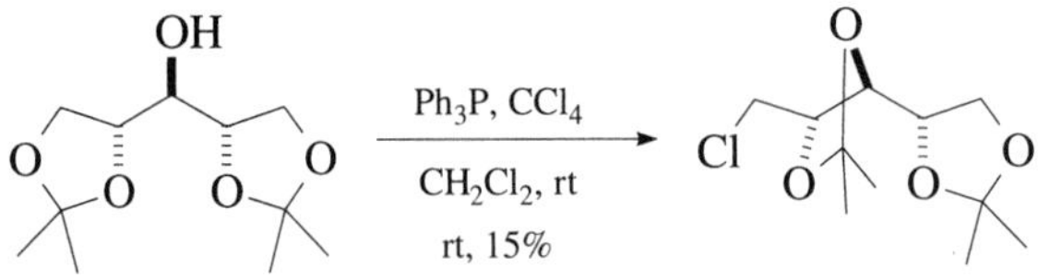

Ph_3P, CCl_4

CH_2Cl_2, rt

rt, 15%

The attempted allylation of the aldehyde shown in the following equation resulted in unanticipated tetrahydrofuran formation:[14]

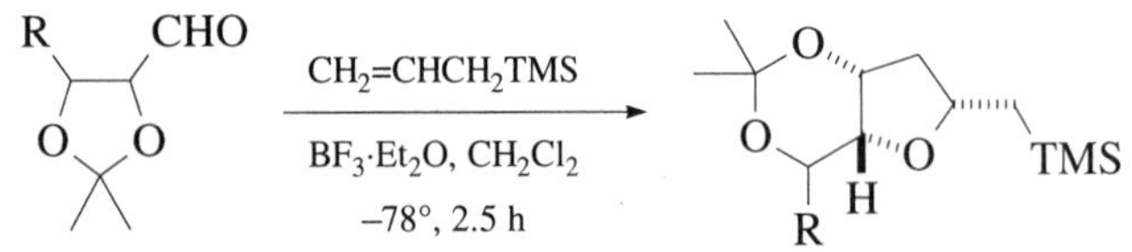

R CHO

CH_2=$CHCH_2TMS$

BF_3·Et_2O, CH_2Cl_2

–78°, 2.5 h

TMS

H

R

These examples serve to illustrate the fact that, in reactions in which carbenium ions are formed in proximity to the acetal lone pairs, unexpected rearrangements may occur.

Formation

1. $CH_3C(OCH_3)=CH_2$, dry HBr, CH_2Cl_2, 0°, 16 h, 75% yield.[15]

2. 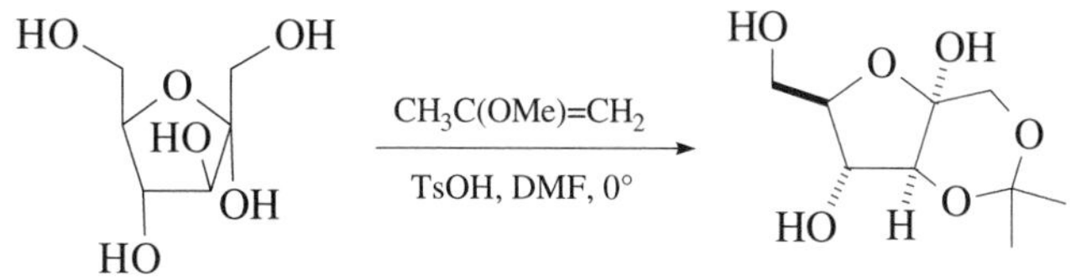 $CH_3C(OMe)=CH_2$, TsOH, DMF, 0°

   Under these conditions, 2-methoxypropene reacts to form the kinetically controlled 1,3-*O*-isopropylidene, instead of the thermodynamically more stable 1,2-*O*-isopropylidene.[16]

3. TsOH, DMF, $Me_2C(OMe)_2$, 24 h.[17,18] This method has become one of the most popular methods for the preparation of acetonides. It generally gives high yields and is compatible with acid-sensitive protective groups such as the TBDMS group.
4. $Me_2C(OMe)_2$, DMF, pyridinium *p*-toluenesulfonate (PPTS).[19] The use of PPTS for acid-catalyzed reactions has been quite successful and is particularly useful when TsOH is too strong an acid for the functionality in a given substrate. TBDMS groups are stable under these conditions.[20]
5. Anhydrous acetone, $FeCl_3$, 36°, 5 h, 60–70% yield.[21]
6. $Me_2C(OMe)_2$, di-*p*-nitrophenyl hydrogen phosphate, 3–5 h, 90–100% yield.[22]
7. $Me_2C(OMe)_2$, $SnCl_2$, DME, 30 min, 54% yield. This reaction has been used to prepare the bisacetonide of mannitol on a 100-kg scale.[23]
8. $MeC(OEt)=CH_2$, cat. HCl, DMF, 25°, 12 h, 90–100% yield.[24] This method is subject to solvent effects. In the formation of a *trans*-acetonide, the use of CH_2Cl_2 did not give the acetonide, but when the solvent was changed to THF, acetonide formation proceeded in 90% yield.[25] These conditions are used to obtain the kinetic acetonide.[26]

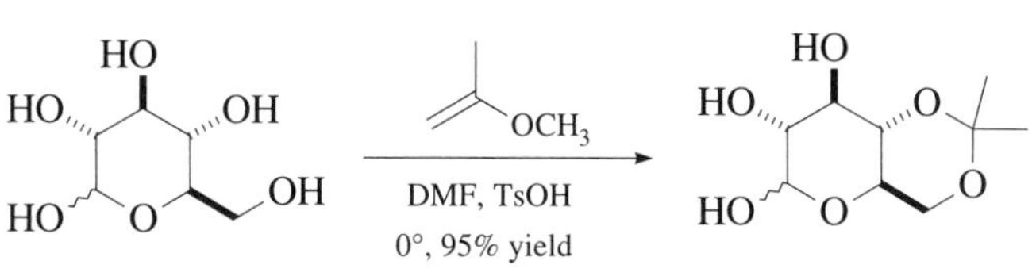

   OCH_3; DMF, TsOH, 0°, 95% yield — Ref. 25

9. $MeC(OTMS)=CH_2$, concd. HCl or TMSCl, 10–30 min, 80–85% yield.[27] This method is effective for the formation of *cis*- or *trans*-acetonides of 1,2-cyclohexanediol.

10. The classical method for acetonide formation is by reaction of a diol with acetone and an acid catalyst.[28,29]
11. Acetone, I_2, 70–85% yield, rt or reflux.[30]
12. Acetone, $CuSO_4$, H_2SO_4, 90% yield.[31] If PPTS replaces H_2SO_4 as the acid, the acetonide can be formed in the presence of a trityl group.[32]

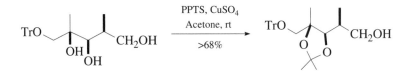

TrO, OH, OH, CH_2OH —(PPTS, $CuSO_4$, Acetone, rt; >68%)→ TrO, O, O, CH_2OH

13. Trimethylsilylated diols are converted to acetonides with acetone and TMSOTf, −78°, 3.5 h, >76% yield.[33]
14. Acetone, $AlCl_3$, Et_2O, rt, 3.5 h, 80% yield.[34] Other methods failed in this sterically demanding case.
15. $CH_3CCl(OMe)CH_3$, DMF, 92% yield.[35]
16. Conversion of silyl ethers to acetonides without prior cleavage of the silyl ether is possible (acetone, AcOH, $CuSO_4$, 81% yield),[36] but is dependent upon the conditions of the reaction.[12] Compare the following examples:

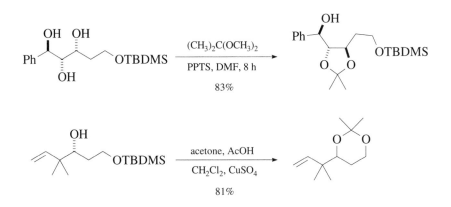

Ph, OH, OH, OH, OTBDMS —($(CH_3)_2C(OCH_3)_2$, PPTS, DMF, 8 h; 83%)→ Ph, OH, O, O, OTBDMS

OH, OTBDMS —(acetone, AcOH, CH_2Cl_2, $CuSO_4$; 81%)→ O, O

17. Lactone methanolysis followed by acetonide formation has also been observed.[37]
18. Conversion of an epoxide directly to an acetonide is accomplished with acetone and $SnCl_4$ (81–86% yield)[38] or with *N*-(4-methoxybenzyl)-2-cyan- opyridinium hexafluoroantimonate [*N*-(4-$MeOC_6H_4CH_2$)-2-CN-$PyrSbF_6$] (59–100% yield).[39]
19. $(CH_3)_2C(OCH_2CH_2CH_2CH{=}CH_2)_2$, NBS, TESOTf, 94% yield.[40]
20. Acetone, K-10 clay.[41]

Cleavage

Cleavage rates for 1,3-dioxanes are greater than for 1,3-dioxolanes,[42] but hydrolysis of a *trans*-fused dioxolane is faster than that of the dioxane. In substrates having more than one acetonide, the least hindered and more electron-rich acetonide can be hydrolyzed selectively.[43] In a classic example, 1,2-5,6-diacetoneglucofuranose is hydrolyzed selectively at the 5,6-acetonide.

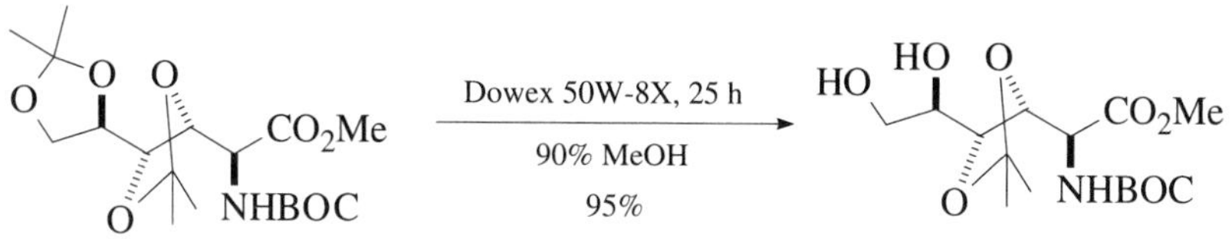

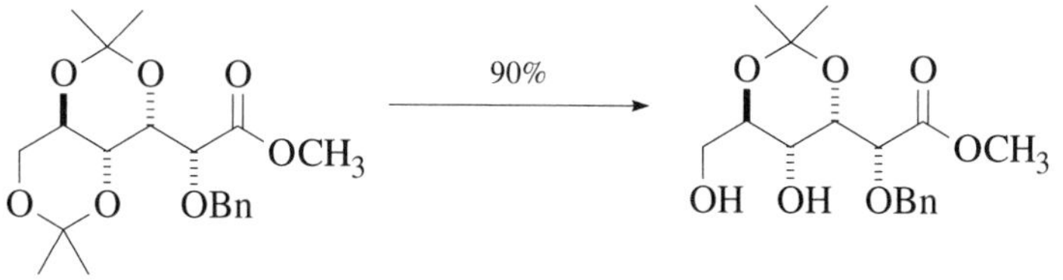

Dowex 50W-8X, 25 h; 90% MeOH; 95%

90%

1. Cat. I_2, MeOH, rt, 24 h, >80% yield.[44] Benzylidene ketals and thioketals are also cleaved under these conditions.
2. 1 *N* HCl, THF (1:1), 20°.[7]
3. 2 *N* HCl, 80°, 6 h.[45]
4. 60–80% AcOH, 25°, 2 h, 92% yield of *cis*-1,2-diol.[46] MOM groups are stable to these conditions.[47]
5. 80% AcOH, reflux, 30 min, 78% yield of *trans*-1,2-diol.[46]
6. TsOH, MeOH, 25°, 5 h.[48] These conditions failed to cleave the acetonide of a 2′,3′-ribonucleoside.[49]
7. Dowex 50-W (H^+), water, 70°, excellent yield.[50]
8. BCl_3, 25°, 2 min, 100% yield.[51]
9. Br_2, Et_2O.[29]
10. $PdCl_2(CH_3CN)_2$, CH_3CN, H_2O, rt.[52] When the solvent is changed to wet acetone, the reagent cleaves an ethylene glycol ketal from ketones in 82–100% yield. TBDPS and MEM groups are stable, but TBDMS and THP groups are cleaved under these conditions.
11. CF_3CO_2H, THF, H_2O, 83% yield.[53]

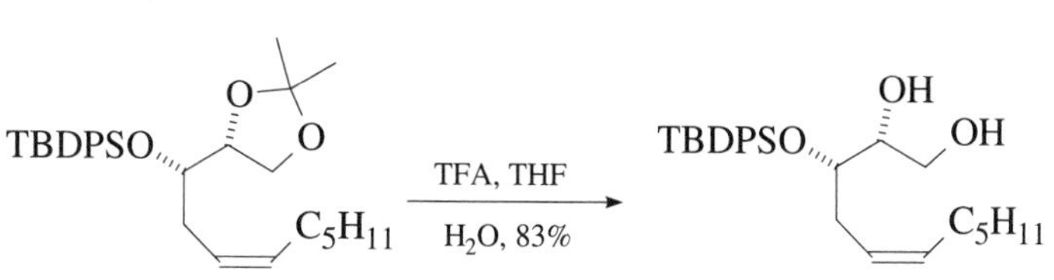

TFA, THF; H_2O, 83%

12. In the following example, the acetonide protective group is selectively converted to one of two *t*-butyl groups. The reaction appears to be general, but the alcohol bearing the *t*-butyl group varies with structure.[54] Benzylidene ketals are also cleaved. The reaction of acetonides with MeMgI proceeds similarly.[55, 56]

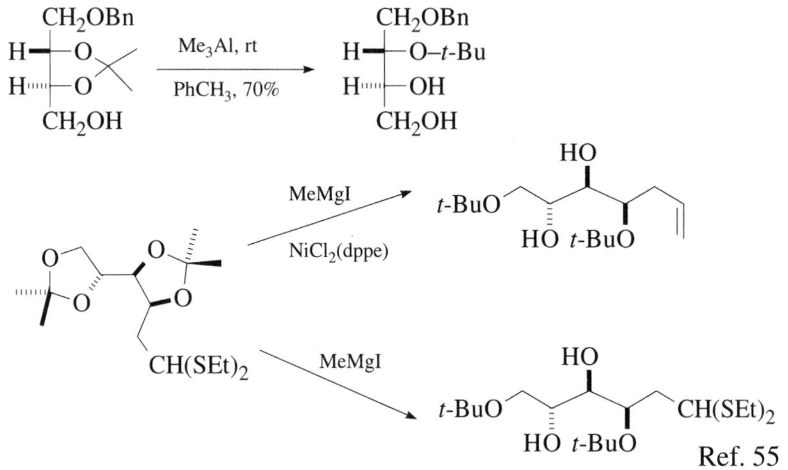

Me_3Al, rt; $PhCH_3$, 70%

MeMgI, $NiCl_2(dppe)$

$CH(SEt)_2$ — MeMgI

Ref. 55

13. Although acetonides are generally considered stable to reagents like BH_3, on occasion they can undergo unexpected side reactions, such as the cleavage observed during a hydroboration.[57,58]

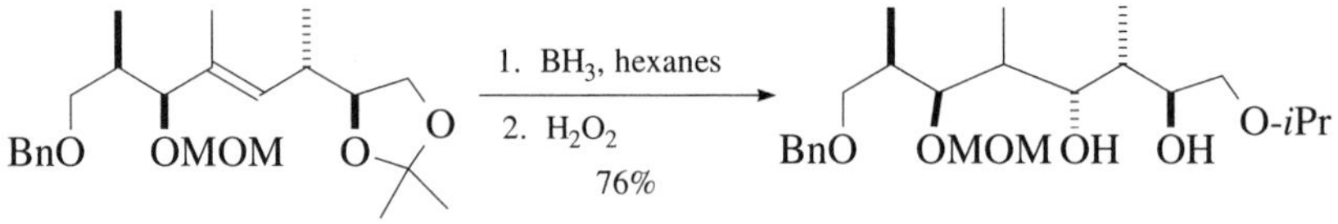

1. BH_3, hexanes 2. H_2O_2 76%

14. The rather unusual conversion of an acetonide to an isopropenyl ether was developed to differentiate a terminal acetonide from several internal ones. The isopropenyl ether was in turn converted to the 1-methylcyclopropyl ether, which was later cleaved with NBS or DDQ.[59,60]

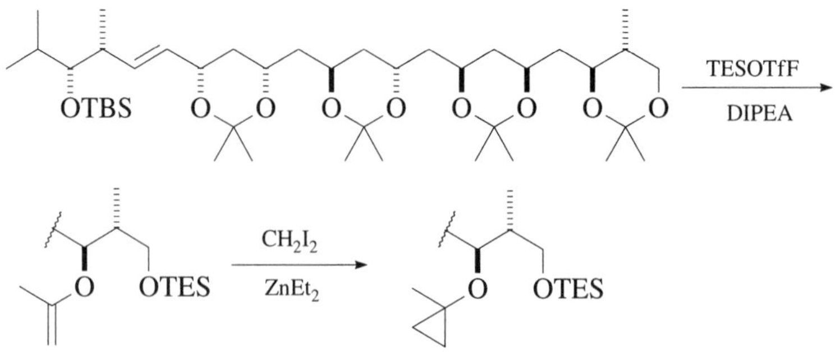

TESOTfF, DIPEA

CH_2I_2, $ZnEt_2$

The intermediate isopropenyl group can be removed with I_2 ($NaHCO_3$, THF, H_2O, rt, 78% yield).

15. $(Bu_2SnNCS)_2O$, diglyme, H_2O, 100°, 82% yield.[61] The THP group is also cleaved by this reagent.
16. $FeCl_3$–SiO_2, $CHCl_3$, 74% yield.[62] When used in acetone, this reagent cleaves the trityl and TBDMS groups. These conditions also cleave THP and TMS groups, but TBDMS, MTM, and MOM groups are not affected when $CHCl_3$ is used as solvent.
17. MeOH, PPTS, heat, high yield.[63]
18. EtSH, TsOH, $CHCl_3$, >76% yield.[3]
19. $HSCH_2CH_2CH_2SH$, $BF_3{\cdot}Et_2O$, CH_2Cl_2, 0°, 89% yield. A primary TBDMS group was not affected.[64]
20. Me_2BBr, CH_2Cl_2, −78°, ~50% yield.[65]
21. SO_2, H_2O, 40°, >67% yield.[66]

1. For a review, see D. M. Clode, *Chem. Rev.*, **79**, 491 (1979).
2. M. R. Kotecha, S. V. Ley, and S. Mantegani, *Synlett*, 395 (1992).
3. D. R. Williams and S.-Y. Sit, *J. Am. Chem. Soc.*, **106**, 2949 (1984).
4. P. Lavallee, R. Ruel, L. Grenier, and M. Bissonnette, *Tetrahedron Lett.*, **27**, 679 (1986).
5. A. I. Meyers and J. P. Lawson, *Tetrahedron Lett.*, **23**, 4883 (1982).
6. S. Hanessian, *Aldrichimica Acta*, **22**, 3 (1989).
7. S. J. Angyal and R. J. Beveridge, *Carbohydr. Res.*, **65**, 229 (1978).
8. C. R. Schmid and D. A. Bradley, *Synthesis*, 587 (1992).
9. P. A. Grieco, Y. Yokoyama, G. P. Withers, F. J. Okuniewicz, and C.-L. J. Wang, *J. Org. Chem.*, **43**, 4178 (1978).
10. S. Nishiyama, Y. Ikeda, S. Yoshida, and S. Yamamura, *Tetrahedron Lett.*, **30**, 105 (1989).
11. J. W. Coe and W. R. Roush, *J. Org. Chem.*, **54**, 915 (1989).
12. C. Mukai, M. Miyakawa, and M. Hanaoka, *J. Chem. Soc., Perkin Trans. 1*, 913 (1997).
13. J. S. Edmonds, Y. Shibata, F. Yang, and M. Morita, *Tetrahedron Lett.*, **38**, 5819 (1997).
14. K. Osumi and H. Sugimura, *Tetrahedron Lett.*, **36**, 5789 (1995).
15. E. J. Corey, S. Kim, S. Yoo, K. C. Nicolaou, L. S. Melvin, Jr., D. J. Brunelle, J. R. Falck, E. J. Trybulski, R. Lett, and P. W. Sheldrake, *J. Am. Chem. Soc.*, **100**, 4620 (1978).
16. E. Fanton, J. Gelas, and D. Horton, *J. Chem. Soc., Chem. Commun.*, 21 (1980).
17. M. E. Evans, F. W. Parrish, and L. Long, Jr., *Carbohydr. Res.*, **3**, 453 (1967).
18. B. H. Lipshutz and J. C. Barton, *J. Org. Chem.* **53**, 4495 (1988).
19. M. Kitamura, M. Isobe, Y. Ichikawa, and T. Goto, *J. Am. Chem. Soc.*, **106**, 3252 (1984).

20. K. Mori and S. Maemoto, *Liebigs Ann. Chem.*, 863 (1987).
21. P. P. Singh, M. M. Gharia, F. Dasgupta, and H. C. Srivastava, *Tetrahedron Lett.*, 439 (1977).
22. A. Hampton, *J. Am. Chem. Soc.*, **83**, 3640 (1961).
23. C. R. Schmid, J. D. Bryant, M. Dowlatzedah, J. L. Phillips, D. E. Prather, R. D. Schantz, N. L. Sear, and C. S. Vianco, *J. Org. Chem.*, **56**, 4056 (1991).
24. S. Chládek and J. Smrt, *Collect. Czech. Chem. Commun.*, **28**, 1301 (1963).
25. J. Cai, B. E. Davison, C. R. Ganellin, and S. Thaisrivongs, *Tetrahedron Lett.*, **36**, 6535 (1995).
26. J. Gelas and D. Horton, *Heterocycles*, **16**, 1587 (1981).
27. G. L. Larson and A. Hernandez, *J. Org. Chem.*, **38**, 3935 (1973).
28. O. Th. Schmidt, *Methods Carbohydr. Chem.*, **II**, 318 (1963).
29. A. N. de Belder, *Adv. Carbohydr. Chem.*, **20**, 219 (1965).
30. K. P. R. Kartha, *Tetrahedron Lett.*, **27**, 3415 (1986).
31. P. Rollin and J.-R. Pougny, *Tetrahedron*, **42**, 3479 (1986).
32. T. Nakata, M. Fukui, and T. Oishi, *Tetrahedron Lett.*, **29**, 2219 (1988).
33. S. D. Rychnovsky, *J. Org. Chem.*, **54**, 4982 (1989).
34. B. Lal, R. M. Gidwani, and R. H. Rupp, *Synthesis*, 711 (1989).
35. A. Kilpala, M. Lindberg, T. Norberg, and S. Oscarson, *J. Carbohydr. Chem.*, **10**, 499 (1991).
36. D. Schinzer, A. Limberg, and O. M. Böhm, *Chem.—Eur. J.*, **2**, 1477 (1996).
37. J. P. Férézou, M. Julia, Y. Li, L. W. Liu, and A. Pancrazi, *Synlett*, 766 (1990).
38. R. Stürmer, *Liebigs Ann. Chem.*, 311 (1991).
39. S. B. Lee, T. Takata, and T. Endo, *Chem. Lett.*, 2019 (1990).
40. R. Madsen and B. Fraser-Reid, *J. Org. Chem.*, **60**, 772 (1995).
41. J.-i. Asakura, Y. Matsubara, and M. Yoshihara, *J. Carbohydr. Chem.*, **15**, 231 (1996).
42. S.-K. Chun and S.-H. Moon, *J. Chem. Soc., Chem. Commun.*, 77 (1992).
43. K.-H. Park, Y. J. Yoon, and S. G. Lee, *Tetrahedron Lett.*, **35**, 9737 (1994); S. D. Burke, K. W. Jung, J. R. Phillips, and R. E. Perri, *Tetrahedron Lett.*, **35**, 703 (1994); M. Gerspacher and H. Rapoport, *J. Org. Chem.*, **56**, 3700 (1991).
44. W. A. Szarek, A. Zamojski, K. N. Tiwari, and E. R. Isoni, *Tetrahedron Lett.*, **27**, 3827 (1986).
45. T. Ohgi, T. Kondo, and T. Goto, *Tetrahedron Lett.*, 4051 (1977).
46. M. L. Lewbart and J. J. Schneider, *J. Org. Chem.*, **34**, 3505 (1969).
47. S. Hanessian, D. Delorme, P. C. Tyler, G. Demailly, and Y. Chapleur, *Can. J. Chem.*, **61**, 634 (1983).
48. A. Ichihara, M. Ubukata, and S. Sakamura, *Tetrahedron Lett.*, 3473 (1977).
49. J. Kimura and O. Mitsunobu, *Bull. Chem. Soc. Jpn.*, **51**, 1903 (1978).
50. P.-T. Ho, *Tetrahedron Lett.*, 1623 (1978).
51. T. J. Tewson and M. J. Welch, *J. Org. Chem.*, **43**, 1090 (1978).
52. B. H. Lipshutz, D. Pollart, J. Monforte, and H. Kotsuki, *Tetrahedron Lett.*, **26**, 705 (1985); C. Schmeck and L. S. Hegedus, *J. Am. Chem. Soc.*, **116**, 9927 (1994).

53. Y. Leblanc, B. J. Fitzsimmons, J. Adams, F. Perez, and J. Rokach, *J. Org. Chem.*, **51**, 789 (1986).
54. S. Takano, T. Ohkawa, and K. Ogasawara, *Tetrahedron Lett.*, **29**, 1823 (1988).
55. T.-Y. Luh, *Synlett,* 201 (1996).
56. W.-L. Cheng, S.-M. Yeh, and T.-Y. Luh, *J. Org. Chem.*, **58**, 5576 (1993).
57. L. D. Coutts, C. L. Cywin, and J. Kallmerten, *Synlett*, 696 (1993).
58. G. Casiraghi, F. Ulgheri, P. Spanu, G. Rassu, L. Pinna, G. Gasparri Fava, M. Belicchi Ferrari, and G. Pelosi, *J. Chem. Soc., Perkin Trans.1*, 2991 (1993).
59. S. D. Rychnovsky and J. Kim, *Tetrahedron Lett.*, **32**, 7219 (1991).
60. S. D. Rychnovsky and R. C. Hoye, *J. Am. Chem. Soc.*, **116**, 1753 (1994).
61. J. Otera and H. Nozaki, *Tetrahedron Lett.*, **27**, 5743 (1986).
62. K. S. Kim, Y. H. Song, B. H. Lee, and C. S. Hahn, *J. Org. Chem.*, **51**, 404 (1986).
63. R. Van Rijsbergen, M. J. O. Anteunis, and A. De Bruyn, *J. Carbohydr. Chem.*, **2**, 395 (1983).
64. T. Konosu and S. Oida, *Chem Pharm. Bull.*, **39**, 2212 (1991).
65. S. E. de Laszlo, M. J. Ford, S. V. Ley, and G. N. Maw, *Tetrahedron Lett.*, **31**, 5525 (1990).
66. A. Dondoni and D. Perrone, *J. Org. Chem.*, **60**, 4749 (1995).

Cyclopentylidene Ketal, i

Cyclohexylidene Ketal, ii

Cycloheptylidene Ketal, iii

Compounds **i**, **ii**, and **iii** can be prepared by an acid-catalyzed reaction of a diol and the cycloalkanone in the presence of ethyl orthoformate and mesitylenesulfonic acid.[1] The relative ease of acid-catalyzed hydrolysis [0.53 *M* H_2SO_4, H_2O, PrOH (65:35), 20°] for compounds **i**, **iii**, acetonide, and **ii** is $C_5 \approx C_7 >$ acetonide>> C_6 (e.g., $t_{1/2}$'s for 1,2-*O*-alkylidene-α-D-glucopyranoses of C_5, C_7, acetonide, and C_6 derivatives are 8, 10, 20, and 124 h, respectively).[1,2] The efficiency of cleavage seems to be dependent upon the electronic environment about the ketal.[3]

The cyclohexylidene ketal has been prepared from dimethoxycyclohexane and TsOH;[4] $HC(OEt)_3$, cyclohexanone, TsOH, EtOAc, heat, 5 h, 78%; 1-(trimethylsiloxy)cyclohexene, concd. HCl, 20°, 10–30 min, 70–75% yield,[5] cyclohexanone, TsOH, $CuSO_4$,[6] and 1-ethoxycyclohexene, TsOH, DMF.[7] The cyclohexylidene derivative of a *trans*-1,2-diol has been prepared.[8] Cyclohexylidene derivatives are cleaved by acidic hydrolysis: 10% HCl, Et_2O, 25°, 5 min;[3] TFA, H_2O, 20°, 6 min to 2 h, 65–85% yield;[9] 0.1 *N* HCl, dioxane;[8] BCl_3, CH_2Cl_2, −80°, 15 h, 90% yield.[10] The cyclohexylidene derivative is also subject to cleavage with Grignard reagents, but under harsh reaction conditions (MeMgI, PhH, 85°, 58% yield).[11] *trans*-Cyclohexylidene ketals are preferentially cleaved in the the presence of *cis*-cyclohexylidene ketals.[12,13] Selective cleavage of the less substituted of two cyclohexylidenes is possible.[14,15]

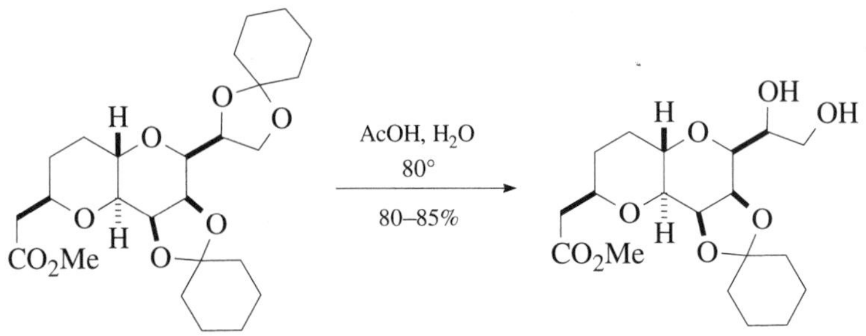

AcOH, H_2O
80°
80–85%

In addition, the cyclopentylidene ketal has been prepared from dimethoxycyclopentane, TsOH, CH_3CN,[16] or cyclopentanone (PTSA, $CuSO_4$ >70% yield)[17] and can be cleaved with 2:1 AcOH/H_2O, rt, 2 h.[18] Certain epoxides can be converted directly to cyclopentylidene derivatives as illustrated in the following reaction: [19]

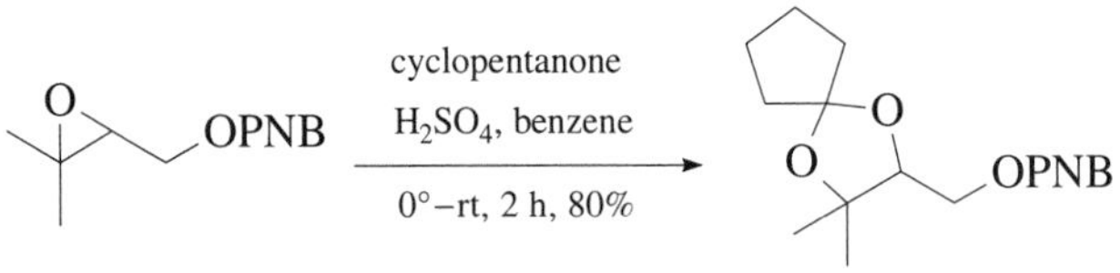

cyclopentanone
H_2SO_4, benzene
0°–rt, 2 h, 80%

PNB = *p*-nitrobenzyl

The 1,2-position of a 6-deoxyglucose derivative has been protected using this reagent, giving primarily the pyranose form, which can be cleaved by alcoholysis with allyl alcohol (benzene, CSA, Δ, 29 h, 82–96%).[20] Methoxycyclopentene (PPTS, CH_2Cl_2, rt, 100%) has been used to introduce this group.[21] The following example shows that a cyclopentylidene can be hydrolyzed in the presence of a *p*-methoxybenzaldehyde ketal. The ketal is first deactivated toward acid hydrolysis by the formation of a charge transfer complex with trinitrotoluene.[22]

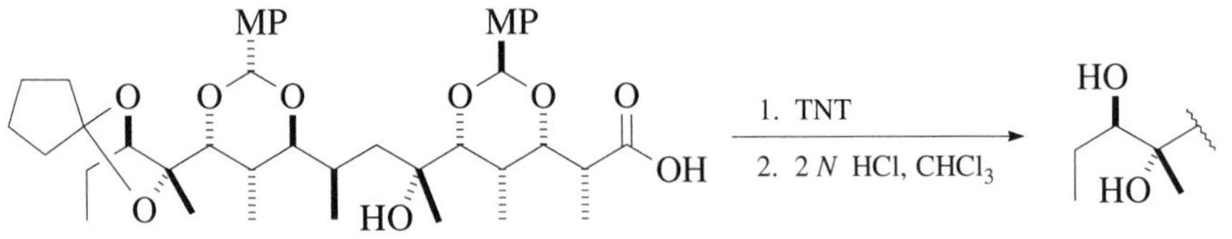

1. TNT
2. 2 *N* HCl, $CHCl_3$

MP = *p*-methoxyphenyl

1. W. A. R. van Heeswijk, J. B. Goedhart, and J. F. G. Vliegenthart, *Carbohydr. Res.*, **58**, 337 (1977).
2. J. M. J. Tronchet, G. Zosimo-Landolfo, F. Villedon-Denaide, M. Balkadjian, D. Cabrini, and F. Barbalat-Rey, *J. Carbohydr. Chem.*, **9**, 823 (1990).
3. J. D. White, J. H. Cammack, K. Sakuma, G. W. Rewcastle, and R. K. Widener, *J. Org. Chem.*, **60**, 3600 (1995).

4. C. Kuroda, P. Theramongkol, J. R. Engebrecht, and J. D. White, *J. Org. Chem.*, **51**, 956 (1986).
5. G. L. Larson and A. Hernandez, *J. Org. Chem.*, **38**, 3935 (1973).
6. W. R. Rousch, M. R. Michaelides, D. F. Tai, B. M. Lesur, W. K. M. Chong, and D. J. Harris, *J. Am. Chem. Soc.*, **111**, 2984 (1989).
7. H. B. Mereyala and M. Pannala, *Tetrahedron Lett.*, **36**, 2121 (1995).
8. D. Askin, C. Angst, and S. Danishefsky, *J. Org. Chem.*, **50**, 5005 (1985).
9. S. L. Cook and J. A. Secrist, *J. Am. Chem. Soc.*, **101**, 1554 (1979).
10. S. D. Géro, *Tetrahedron Lett.*, 591 (1966).
11. M. Kawana and S. Emoto, *Bull. Chem. Soc. Jpn.*, **53**, 230 (1980).
12. Y.-C. Liu and C.-S. Chen, *Tetrahedron Lett.*, **30**, 1617 (1989).
13. J. E. Innes, P. J. Edwards, and S. V. Ley, *J. Chem. Soc., Perkin Trans. 1*, 795 (1997).
14. D. P. Stamos and Y. Kishi, *Tetrahedron Lett.*, **37**, 8643 (1996).
15. M. S. Wolfe, B. L. Anderson, D. R. Borcherding, and R. T. Borchardt, *J. Org. Chem.*, **55**, 4712 (1990).
16. K. Ditrich, *Liebigs Ann. Chem.*, 789 (1990).
17. D. B. Collum, J. H. McDonald, III, and W. C. Still, *J. Am. Chem. Soc.*, **102**, 2118 (1980).
18. C. B. Reese and J. G. Ward, *Tetrahedron Lett.*, **28**, 2309 (1987).
19. A. B. Smith, III, J. Kingery-Wood, T. L. Leenay, E. G. Nolen, and T. Sunazuka, *J. Am. Chem. Soc.*, **114**, 1438 (1992).
20. A. B. Smith, III, R. A. Rivero, K. J. Hale, and H. A. Vaccaro, *J. Am. Chem. Soc.*, **113**, 2092 (1991).
21. R. M. Soll and S. P. Seitz, *Tetrahedron Lett.*, **28**, 5457 (1987).
22. R. Stürmer, K. Ritter, and R. W. Hoffmann, *Angew. Chem., Int. Ed. Engl.*, **32**, 101 (1993).

Benzylidene Acetal (Chart 3)

A benzylidene acetal is a commonly used protective group for 1,2- and 1,3-diols. In the case of a 1,2,3-triol, the 1,3-acetal is the preferred product, in contrast to the acetonide, which gives the 1,2-derivative. The benzylidene acetal has the advantage that it can be removed under neutral conditions by hydrogenolysis or by acid hydrolysis. Benzyl groups[1] and isolated olefins[2] have been hydrogenated in the presence of 1,3-benzylidene acetals. Benzylidene acetals of 1,2-diols are more susceptible to hydrogenolysis than are those of 1,3-diols. In fact, the former can be removed in the presence of the latter.[3] A polymer-bound benzylidene acetal has also been prepared.[4]

Formation

1. PhCHO, $ZnCl_2$, 28°, 4 h.[5]
2. PhCHO, DMSO, concd. H_2SO_4, 25°, 4 h.[6]

3. 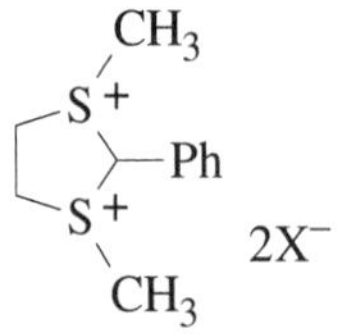 $X = FSO_3^-$, or BF_4^-

K_2CO_3 or Pyr, CH_2Cl_2, 25°, 16 h,45–82% yield.[7] This method is suitable for the protection of 1,2-, 1,3- and 1,4-diols.

4. PhCHO, TsOH, reflux, $-H_2O$, 72% yield.[8]
5. $PhCHBr_2$, Pyr.[9]
6. $PhCH(OMe)_2$, HBF_4, Et_2O, DMF, 97% yield.[10,11] 1,3-Diols are protected in preference to 1,2-diols.[12]
7. $PhC(OMe)_2$, $SnCl_2$, DME, heat, 45 min.[13]

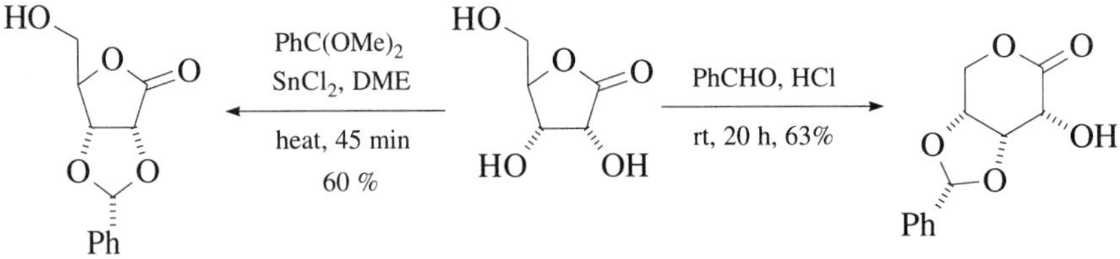

$PhC(OMe)_2$, $SnCl_2$, DME, heat, 45 min, 60 %; PhCHO, HCl, rt, 20 h, 63%

8.

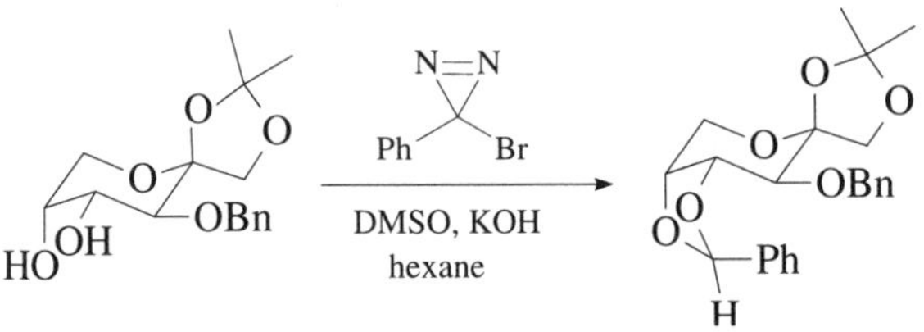

DMSO, KOH, hexane — Ref. 14

9. $PhCH(OCH_2CH_2CH{=}CH_2)_2$, CSA, NBS. Standard methods failed because of cleavage of the dispiroketal (dispoke) protective group.[15,16]

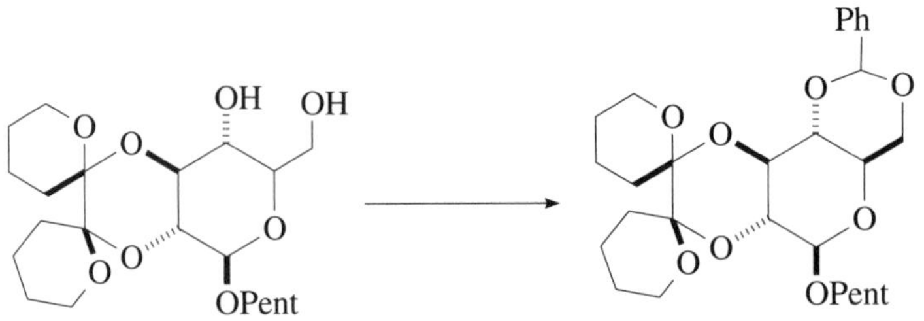

10.

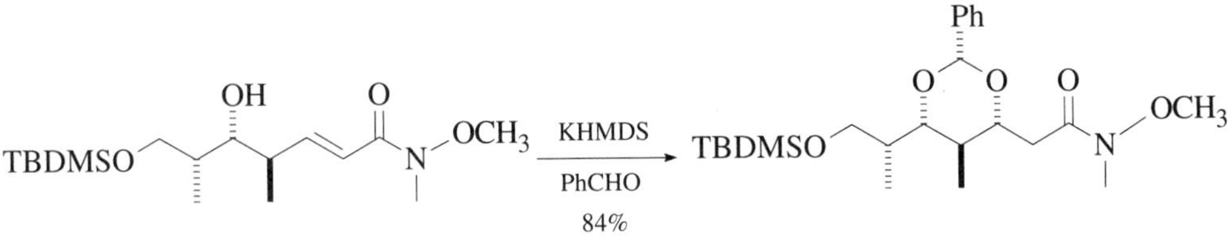

TBDMSO ... OH ... O ... N(CH3)OCH3 —KHMDS, PhCHO, 84%→ TBDMSO ... (Ph benzylidene acetal) ... O ... N(CH3)OCH3

Ref. 17

Cleavage

1. H_2/Pd-C, AcOH, 25°, 30–45 min, 90% yield.[18]
2. Na, NH_3, 85% yield.[19]
3. The benzylidene acetal is cleaved by acidic hydrolysis (e.g., 0.01 *N* H_2SO_4, 100°, 3 h, 92% yield;[20] 80% AcOH, 25°, $t_{1/2}$ for uridine = 60 h[21]), conditions that do not cleave a methylenedioxy group.[20]
4. Electrolysis: −2.9 V, $R_4N^+X^-$, DMF.[22]
5. BCl_3, 100% yield. This reagent also cleaves a number of other ketal-type protective groups.[23]
6. I_2, MeOH, 85% yield.[24]
7. $FeCl_3$, CH_2Cl_2, 3–30 min, 68–85% yield.[25] Benzyl groups are also cleaved by this reagent.
8. $Pd(OH)_2$, cyclohexene, 98% yield.[1]
9. Pd-C, hydrazine, MeOH.[26] In this case a 1,2-benzylidene acetal was cleaved in the presence of a 1,3-benzylidene acetal.
10. Pd–C, HCO_2NH_4, 97% yield.[3]
11. EtSH, $NaHCO_3$, $Zn(OTf)_2$,CH_2Cl_2, rt, 5 h, 90% yield.[27]
12. $SnCl_2$, CH_2Cl_2, rt, 3–12 h, 86–95% yield.[28]

Benzylidene acetals have the useful property that one of the two C–O bonds can be selectively cleaved. The direction of cleavage is dependent upon steric and electronic factors, as well as on the nature of the cleavage reagent. This transformation has been reviewed in the context of carbohydrates.[29]

13. $(i\text{-Bu})_2AlH$, CH_2Cl_2 or $PhCH_3$, 0°–rt, yields generally >80%.[30,31] With this reagent, cleavage occurs to give the least hindered alcohol.

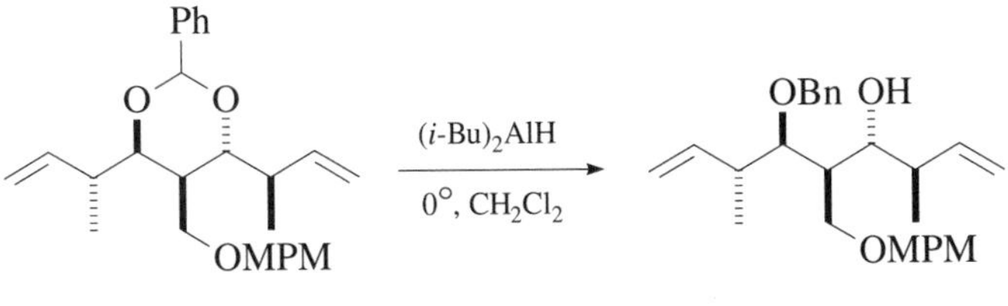

(Ph benzylidene acetal, OMPM) —$(i\text{-Bu})_2AlH$, 0°, CH_2Cl_2→ OBn OH ... OMPM

14.

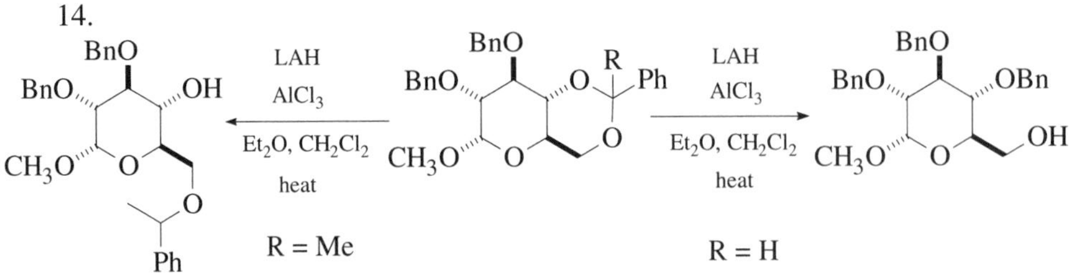

BnO, BnO, OH, CH$_3$O, O, O, Ph ← LAH, $AlCl_3$, Et_2O, CH_2Cl_2, heat, R = Me — BnO, BnO, O, R, Ph, O, CH$_3$O, O — LAH, $AlCl_3$, Et_2O, CH_2Cl_2, heat, R = H → BnO, BnO, OBn, OH, CH$_3$O, O

Ref. 32

15.

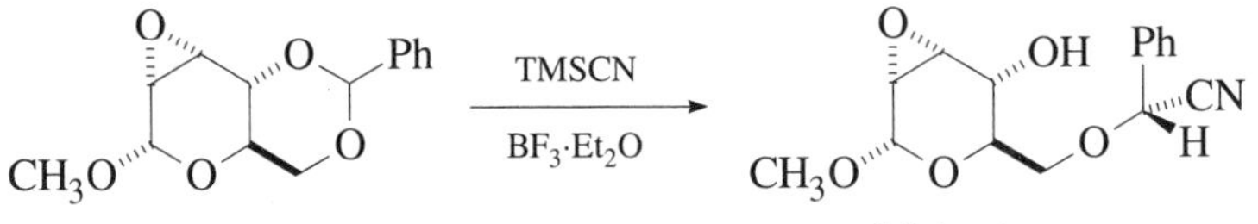

O, O, Ph, O, CH$_3$O, O — TMSCN, $BF_3 \cdot Et_2O$ → O, OH, Ph, CN, H, O, CH$_3$O, O

Major isomer

The regiochemistry of this transformation can be controlled by the choice of Lewis acid. In another substrate the use of $ZnBr_2$/TMSCN gives the cyanohydrin at the more substituted hydroxyl, whereas the use of $TiCl_4$ as a Lewis acid places the cyanohydrin at the least substituted hydroxyl.[34]

16. $Zn(BH_4)_2$, TMSCl, Et_2O, 25°, 45 min, 77–97% yield.[35] Reduction occurs to form a monobenzyl derivative of a diol.
17. NBS, CCl_4, H_2O, 75% yield.[36]

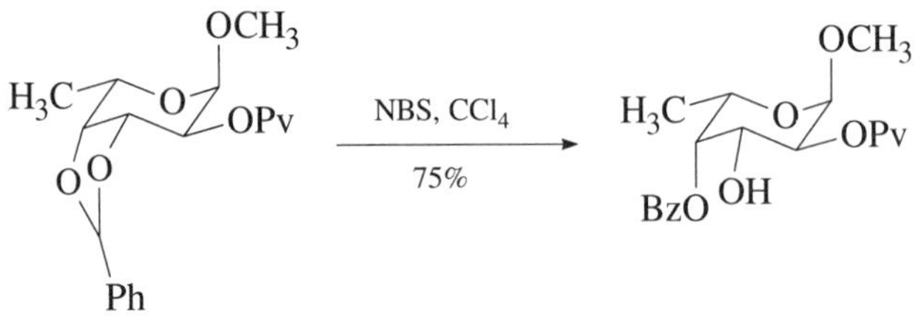

OCH$_3$, H$_3$C, O, OPv, O, O, Ph — NBS, CCl_4, 75% → OCH$_3$, H$_3$C, O, OPv, BzO, OH

In this type of cleavage reaction, it appears that the axial benzoate is the preferred product. If water is excluded from the reaction, a bromo benzoate is obtained.[37] The highly oxidizing medium of 2,2′-bipyridinium chlorochromate and MCPBA in CH_2Cl_2 at rt for 36 h effects a similar conversion of benzylidene acetals to hydroxy benzoates in 25–72% yield.[38]

18. $BrCCl_3$, CCl_4, $h\nu$, 30 min, 100% yield.[39]

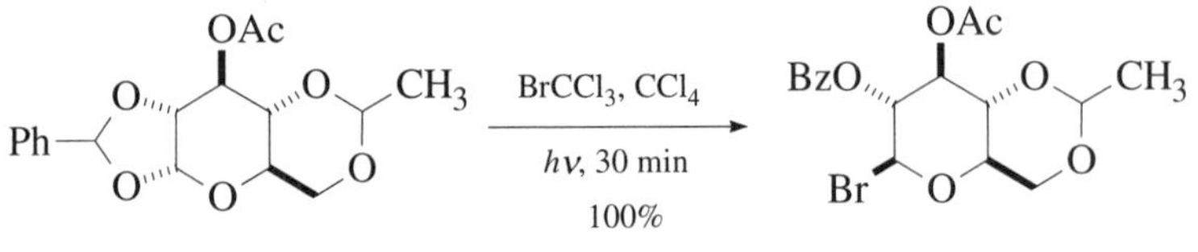

OAc, O, O, CH$_3$, Ph, O, O, O — $BrCCl_3$, CCl_4, $h\nu$, 30 min, 100% → OAc, BzO, O, CH$_3$, Br, O, O

19. $NaBH_3CN$, $TiCl_4$, CH_3CN, rt, 3 h, 83% yield.[40] $NaBH_3CN$, THF, ether/HCl converts a 4,6-benzylidene to a 6-*O*-benzylpyranoside[41]. The use of triflic acid improves this process because the stoichiometry is more conveniently controlled.[42]

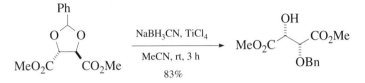

$NaBH_3CN$, $TiCl_4$; MeCN, rt, 3 h; 83%

20. Me_2BBr, TEA, BH_3–THF, −78° warm to −20° over 1 h, 70–97% yield. These conditions cleave the benzylidene acetal, to leave the least hindered alcohol as a free hydroxyl. If diborane is omitted from the reaction mixture and the reaction is quenched with PhSH and TEA, the benzylidene group is cleaved to give an *O,S*-acetal [ROCH(SPh)Ph]. Acetonides are cleaved similarly.[43]
21. $Ph_3C^+BF_4^-$, CH_3CN, 25°, 8 h, 80% yield.[44] A 1:1 mixture of diol monobenzoates is formed.
22. $AlCl_3$, BH_3·TEA, THF, 60°, 96% yield.[45] In a 2-aminoglucose derivative, the 6-*O*-benzyl derivative was formed selectively. The use of $Me_3N–BH_3$ in THF gives the 6-*O*-benzyl derivative, but when the solvent is changed to toluene or CH_2Cl_2, the 4-*O*-benzyl ether is produced.[46] A mechanistic study on the reductive cleavage of acetals has been published.[47]
23. Bu_2BOTf, BH_3–THF, CH_2Cl_2, 0°, 70–91% yield. In a variety of pyranosides, cleavage occurs primarily to give the primary alcohol, with the secondary alcohol protected as the benzyl ether.[48]
24. TFA, Et_3SiH, CH_2Cl_2.[49] 6-*O*-Benzylpyranosides are formed from a 4,6-benzylidenepyranoside in 80–98% yield.
25. *t*-BuOOH, $CuCl_2$, benzene, 50°, 15 h, 87% yield.[50] Additionally, $Pd(OAc)_2$, $FeCl_2$, $PdCl_2$, and $NiCl_2$ were found to be active catalysts in this transformation, but in each case a mixture of benzoates was formed from a 4,6-benzylidene glucose derivative.[51]
26. $BH_3·SMe_2$, CH_2Cl_2, 0°, 1 h, then BF_3, 5 min.[52] Simple benzylidene acetals are cleaved efficiently without hydroboration of alkenes that may be present, and acetonides are converted to the hydroxy isopropyl ethers.

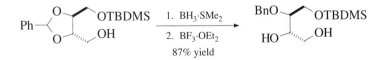

1. $BH_3·SMe_2$; 2. $BF_3·OEt_2$; 87% yield

A related hydroxyl-directed cleavage has been observed using $LiBF_4/BH_3$ or $LiBH_4/BF_3$.[53]

27. Ozonolysis, Ac_2O, NaOAc, −78°, 1 h, 95% yield.[54] In this case, the benzylidene acetal is converted to a diester.
28. $NaBO_3 \cdot 4H_2O$, Ac_2O, 67–95% yield.[55] Cleavage occurs to give a monobenzoate.
29. *t*-BuOOH, Pd(TFA)(*t*-BuOOH), 26–78% yield.[56] Palladium acetate can also be used.[57] Cleavage occurs to give a monobenzoate.

1. S. Hanessian, T. J. Liak, and B. Vanasse, *Synthesis*, 396 (1981).
2. A. B. Smith, III, and K. J. Hale, *Tetrahedron Lett.*, **30**, 1037 (1989).
3. T. Bieg and W. Szeja, *Carbohydr. Res.*, **140**, C7 (1985).
4. J. M. J. M. Fréchet and G. Pellé, *J. Chem. Soc., Chem. Commun.*, 225 (1975).
5. H. G. Fletcher, Jr., *Methods Carbohydr. Chem.*, **II**, 307 (1963).
6. R. M. Carman and J. J. Kibby, *Aust. J. Chem.*, **29**, 1761 (1976).
7. R. M. Munavu and H. H. Szmant, *Tetrahedron Lett.*, 4543 (1975).
8. D. A. McGowan and G. A. Berchtold, *J. Am. Chem. Soc.*, **104**, 7036 (1982).
9. R. N. Russell, T. M. Weigel, O. Han, and H.-w. Liu, *Carbohydr. Res.*, **201**, 95 (1990).
10. R. Albert, K. Dax, R. Pleschko, and A. Stütz, *Carbohydr. Res.*, **137**, 282 (1985); T. Yamanoi, T. Akiyama, E. Ishida, H. Abe, M. Amemiya, and T. Inazu, *Chem. Lett.*, 335 (1989).
11. M. T. Crimmins, W. G. Hollis, Jr., and G. J. Lever, *Tetrahedron Lett.*, **28**, 3647 (1987).
12. Y. Morimoto, A. Mikami, S.-i. Kuwabe, and H. Shirahama, *Tetrahedron Lett.*, **32**, 2909 (1991).
13. S. Y. Han, M. M. Joullie, N. A. Petasis, J. Bigorra, J. Cobera, J. Font, and R. M. Ortuno, *Tetrahedron*, **49**, 349 (1992).
14. C. Li and A. Vasella, *Helv. Chim. Acta*, **76**, 211 (1993).
15. C. W. Andrews, R. Rodebaugh, and B. Fraser-Reid, *J. Org. Chem.*, **61**, 5280 (1996).
16. R. Madsen and B. Fraser-Reid, *J. Org. Chem.*, **60**, 772 (1995).
17. D. A. Evans and J. A. Gauchet-Prunet, *J. Org. Chem.*, **58**, 2446 (1993).
18. W. H. Hartung and R. Simonoff, *Org. React.*, **7**, 263–326 (1953); see pp. 271, 284, 302.
19. M. Zaoral, J. Jezek, R. Straka, and K. Masek, *Collect. Czech Chem. Commun.*, **43**, 1797 (1978).
20. R. M. Hann, N. K. Richtmyer, H. W. Diehl, and C. S. Hudson, *J. Am. Chem. Soc.*, **72**, 561 (1950).
21. M. Smith, D. H. Rammler, I. H. Goldberg, and H. G. Khorana, *J. Am. Chem. Soc.*, **84**, 430 (1962).
22. V. G. Mairanovsky, *Angew. Chem., Int. Ed. Engl.*, **15**, 281 (1976).
23. T. G. Bonner, E. J. Bourne, and S. McNally, *J. Chem. Soc.*, 2929 (1960).
24. W. A. Szarek, A. Zamojski, K. N. Tiwari, and E. R. Ison, *Tetrahedron Lett.*, **27**, 3827 (1986).
25. M. H. Park, R. Takeda, and K. Nakanishi, *Tetrahedron Lett.*, **28**, 3823 (1987).

26. T. Bieg and W. Szeja, *Synthesis*, 317 (1986).
27. K. C. Nicolaou, C. A. Veale, C.-K. Hwang, J. Hutchinson, C. V. C. Prasad, and W. W. Ogilvie, *Angew. Chem., Int. Ed. Engl.*, **30**, 299 (1991).
28. J. Xia and Y. Hui, *Synth. Commun.*, **26**, 881 (1996).
29. S. Hanessian, Ed., *Preparative Carbohydrate Chemistry*, pp. 53–65, Marcel Dekker, Inc., New York, 1997.
30. S. Takano, M. Akiyama, S. Sato, and K. Ogasawara, *Chem. Lett.*, 1593 (1983); S. Hatakeyama, K. Sakurai, K. Saijo, and S. Takano, *Tetrahedron Lett.*, **26**, 1333 (1985).
31. S. L. Schreiber, Z. Wang, and G. Schulte, *Tetrahedron Lett.*, **29**, 4085 (1988).
32. R. Miethchen, J. Holz, and H. Prade, *Tetrahedron*, **48**, 3061 (1992).
33. F. G. De las Heras, A. San Felix, A. Calvo-Mateo, and P. Fernandez-Resa, *Tetrahedron*, **41**, 3867 (1985).
34. R. C. Corcoran, *Tetrahedron Lett.*, **31**, 2101 (1990).
35. H. Kotsuki, Y. Ushio, N. Yoshimura, and M. Ochi, *J. Org. Chem.*, **52**, 2594 (1987).
36. R. W. Binkley, G. S. Goewey, and J. C. Johnston, *J. Org. Chem.*, **49**, 992 (1984).
37. O. Han and H.-w. Liu, *Tetrahedron Lett.*, **28**, 1073 (1987); F. Chretien, M. Khaldi, and Y. Chapleur, *Synth. Commun.*, **20**, 1589 (1990).
38. F. A. Luzzio and R. A. Bobb, *Tetrahedron Lett.*, **38**, 1733 (1997).
39. P. M. Collins, A. Manro, E. C. Opara-Mottah, and M. H. Ali, *J. Chem. Soc., Chem. Commun.*, 272 (1988).
40. G. Adam and D. Seebach, *Synthesis*, 373 (1988).
41. L. Qiao and J. C. Vederas, *J. Org. Chem.*, **58**, 3480 (1993).
42. N. L. Pohl and L. L. Kiessling, *Tetrahedron Lett.*, **38**, 6985 (1997).
43. Y. Guindon, Y. Girard, S. Berthiaume, V. Gorys, R. Lemieux, and C. Yoakim, *Can. J. Chem.*, **68**, 897 (1990).
44. S. Hanessian and A. P. A. Staub, *Tetrahedron Lett.*, 3551 (1973).
45. B. Classon, P. J. Garegg, and A.-C. Helland, *J. Carbohydr. Chem.*, **8**, 543 (1989).
46. P. J. Garegg, *Acc. Chem. Res.*, **25**, 575 (1992).
47. K. Ishihara, A. Mori, and H. Yamamoto, *Tetrahedron*, **46**, 4595 (1990).
48. L. Jiang and T.-H. Chan, *Tetrahedron Lett.*, **39**, 355 (1998).
49. M. P. DeNinno, J. B. Etienne, and K. C. Duplantier, *Tetrahedron Lett.*, **36**, 669 (1995); A. Arasappan and B. Fraser-Reid, *J. Org. Chem.*, **61**, 2401 (1996).
50. K. Sato, T. Igarashi, Y. Yanagisawa, N. Kawauchi, H. Hashimoto, and J. Yoshimura, *Chem. Lett.*, 1699 (1988).
51. F. E. Ziegler and J. S. Tung, *J. Org. Chem.*, **56**, 6530 (1991).
52. S. Saito, A. Kuroda, K. Tanaka, and R. Kimura, *Synlett*, 231 (1996).
53. B. Delpech, D. Calvo, and R. Lett, *Tetrahedron Lett.*, **37**, 1015 (1996).
54. P. Deslongchamps, C. Moreau, D. Fréhel, and R. Chênevert, *Can. J. Chem.*, **53**, 1204 (1975).
55. S. Bhat, A. R. Ramesha, and S. Chandrasekaran, *Synlett*, 329 (1995).
56. T. Hosokawa, Y. Imada, and S. I. Murahashi, *J. Chem. Soc., Chem. Commun.*, 1245 (1983).

57. P. H. G. Wiegerinck, L. Fluks, J. B. Hammink, S. J. E. Mulders, F. M. H. de Groot, H. L. M. van Rozendaal, and H. W. Scheeren, *J. Org. Chem.*, **61**, 7092 (1996).

p-Methoxybenzylidene Acetal (Chart 3)

The *p*-methoxybenzylidene acetal is a versatile protective group for diols that undergoes acid hydrolysis 10 times faster than the benzylidene group.[1] As with the benzylidene derivative, the 1,3-derivative is thermodynamically favored over the 1,2-derivative.[2]

Formation

1. *p*-$MeOC_6H_4CHO$, acid, 70–95% yield.[1]
2. From a trimethylsilylated triol: MeOPhCHO, TMSOTf, CH_2Cl_2, −78°, 5 h, 96% yield.[3]
3. *p*-$MeOC_6H_4CH_2OMe$, DDQ, CH_2Cl_2, rt, 30 min, 49–82% yield.[4]
4. *p*-$MeOC_6H_4CHO$, $ZnCl_2$.[5]
5. *p*-$MeOC_6H_4CH(OMe)_2$, acid.[6] The related *o*-methoxybenzylidene acetal has been prepared by this method.[7] Useful diol selectivity has been achieved, as in the following illustration:[8]

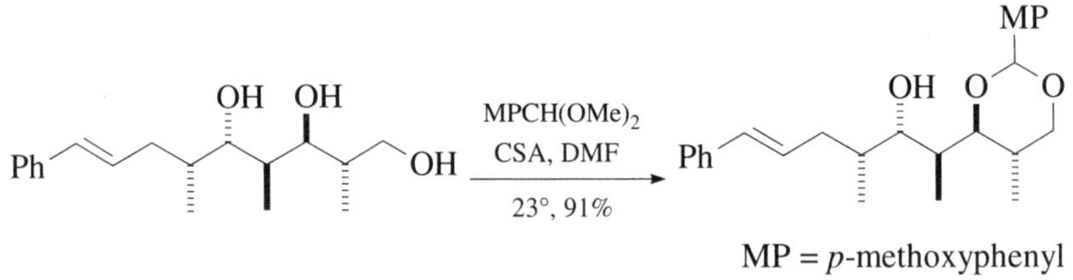

Ph … OH OH … OH —(MPCH(OMe)$_2$, CSA, DMF, 23°, 91%)→ Ph … OH O(MP)O

MP = *p*-methoxyphenyl

6. The *p*-methoxybenzylidene ketal can be prepared by DDQ oxidation of a *p*-methoxybenzyl group that has a neighboring hydroxyl.[9] This methodology has been used to advantage in a number of syntheses.[10,11] In one case, to prevent an unwanted acid-catalyzed acetal isomerization, it was necessary to recrystallize the DDQ and use molecular sieves.[12] The following examples serve to illustrate the reaction:[13,14]

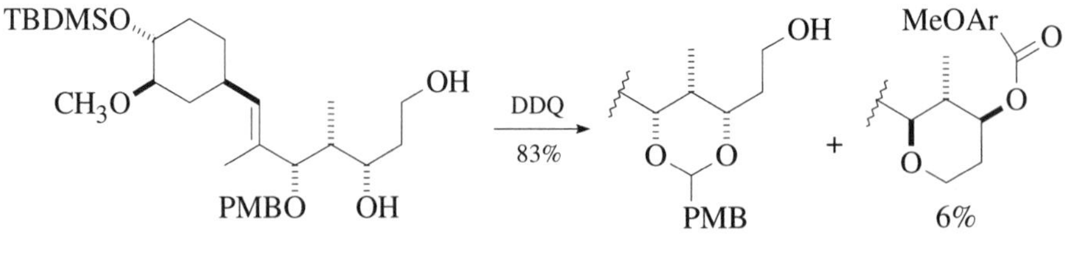

TBDMSO, CH_3O, PMBO, OH, OH —(DDQ, 83%)→ OH, O O, PMB + MeOAr, O, O, O, 6%

Ref. 14

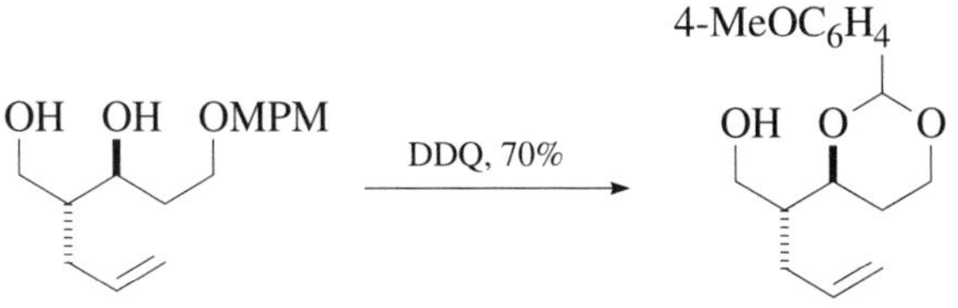

4-$MeOC_6H_4$

OH OH OMPM → (DDQ, 70%) → OH O O

MPM = PMB = *p*-methoxybenzyl

Ref. 13

Cleavage

1. 80% AcOH, 25°, 10 h, 100% yield.[1] Mesitylene acetals have been found to be stable during the acid (pH =1) catalyzed cleavage of *p*-methoxybenzylidene acetals.[15]
2. $Pd(OH)_2$, 25°, 2 h, H_2, >95% yield.[16]
3. $Ce(NH_4)_2(NO_3)_6$, CH_3CN, H_2O.[17] As with the benzylidene group, a variety of methods, shown in the equations that follow, have been developed to effect cleavage of one of the two C–O bonds in this acetal.
4. $(i\text{-}Bu)_2AlH$, $PhCH_3$, 75% yield.[10,18]

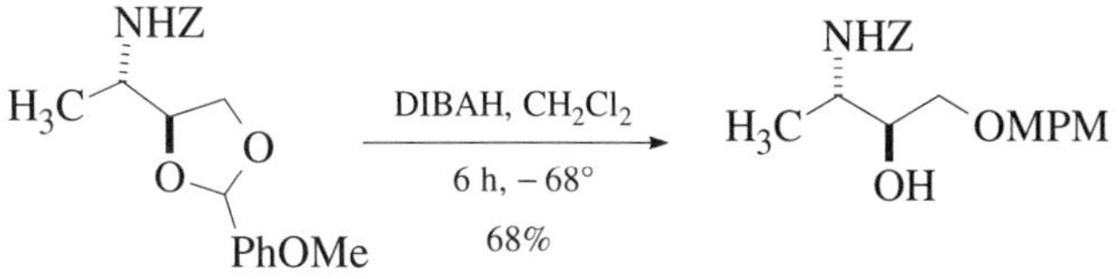

NHZ, H_3C, O, O, PhOMe → (DIBAH, CH_2Cl_2, 6 h, – 68°, 68%) → NHZ, H_3C, OMPM, OH

5. DDQ, water, 87% yield.[7] This method results in the formation of a mixture of the two possible monobenzoates.[19]

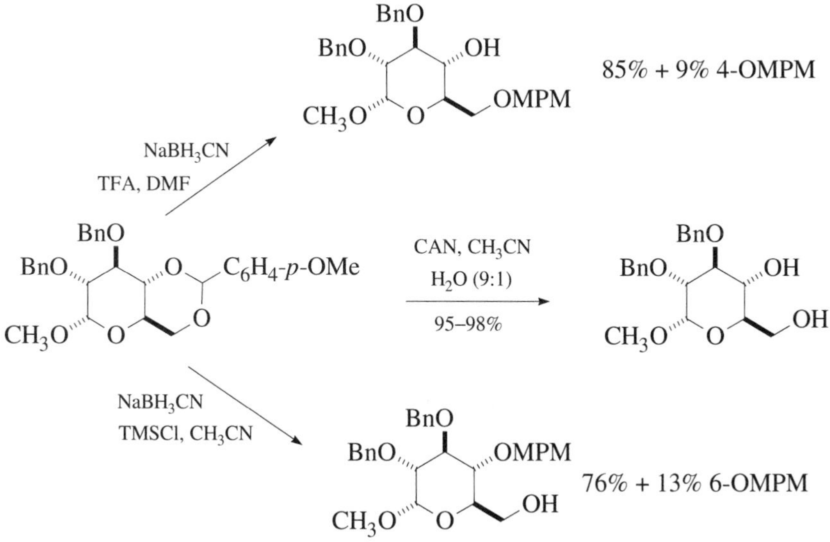

BnO, BnO, OH, CH_3O, O, OMPM — 85% + 9% 4-OMPM

NaBH$_3$CN, TFA, DMF

BnO, BnO, O, C_6H_4-*p*-OMe, CH_3O, O, O → (CAN, CH_3CN, H_2O (9:1), 95–98%) → BnO, BnO, OH, CH_3O, O, OH

NaBH$_3$CN, TMSCl, CH_3CN

BnO, BnO, OMPM, CH_3O, O, OH — 76% + 13% 6-OMPM

Treatment with $LiAlH_4/AlCl_3$,[20,21] $BH_3{\cdot}NMe_3/AlCl_3$,[4] BH_3, THF, heat,[21] or $NaBH_3CN$/TMSCl, CH_3CN,[11] results in cleavage at the least hindered side of the ketal, giving the more hindered ether, whereas treatment with $NaBH_3CN$/HCl[4] or $NaBH_3CN$/TFA/DMF[11] results in the formation of an MPM ether at the least hindered alcohol.

6. DDQ, CH_2Cl_2, $Bu_4N^+Cl^-$, $ClCH_2CH_2Cl$, 96% yield. When $CuBr_2$/ $Bu_4N^+Br^-$ is used, the 6-Br derivative is produced in 93% yield.[19]

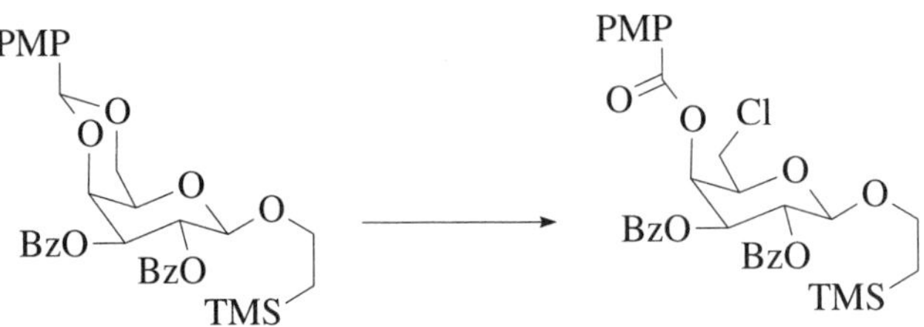

PMP = *p*-methoxyphenyl

7. Ozone.[22] Most acetals are subject to cleavage with ozone, giving a mono ester of the original diol.
8. PDC, *t*-BuOOH, 0°, 4–8 h.[23] Other acetals are similarly cleaved.

1. M. Smith, D. H. Rammler, I. H. Goldberg, and H. G. Khorana, *J. Am. Chem. Soc.*, **84**, 430 (1962).
2. K. Takebuchi, Y. Hamada, and T. Shioiri, *Tetrahedron Lett.*, **35**, 5239 (1994).
3. P. Breuilles, G. Oddon, and D. Uguen, *Tetrahedron Lett.*, **38**, 6607 (1997).
4. Y. Oikawa, T. Nishi, and O. Yonemitsu, *Tetrahedron Lett.*, **24**, 4037 (1983).
5. S. Hanessian, J. Kloss, and T. Sugawara, *ACS Symp. Ser.*, **386**, "Trends in Synth. Carbohydr. Chem.," 64 (1989).
6. M. Kloosterman, T. Slaghek, J. P. G. Hermans, and J. H. Van Boom, *Recl: J. R. Neth. Chem. Soc.*, **103**, 335 (1984).
7. V. Box, R. Hollingsworth, and E. Roberts, *Heterocycles*, **14**, 1713 (1980).
8. D. A. Evans and H. P. Ng, *Tetrahedron Lett.*, **34**, 2229 (1993).
9. Y. Oikawa, T. Yoshioka, and O. Yonemitsu, *Tetrahedron Lett.*, **23**, 889 (1982).
10. A. F. Sviridov, M. S. Ermolenko, D. V. Yaskunsky, V. S. Borodkin, and N. K. Kochetkov, *Tetrahedron Lett.*, **28**, 3835 (1987).
11. J. S. Yadav, M. C. Chander, and B. V. Joshi, *Tetrahedron Lett.*, **29**, 2737 (1988).
12. R. Stürmer, K. Ritter, and R. W. Hoffman, *Angew. Chem., Int. Ed. Engl.*, **32**, 101 (1993).
13. A. B. Jones, M. Yamaguchi, A. Patten, S. J. Danishefsky, J. A Ragan, D. B. Smith, and S. L. Schreiber, *J. Org. Chem.*, **54**, 17 (1989); A. B. Smith, III, K. J. Hale, L. M. Laakso, K. Chen, and A. Riera, *Tetrahedron Lett.*, **30**, 6963 (1989).
14. J. A. Marshall and S. Xie, *J. Org. Chem.*, **60**, 7230 (1995).
15. S. F. Martin, T. Hida, P. R. Kym, M. Loft, and A. Hodgson, *J. Am. Chem. Soc.*, **119**, 3193 (1997).

16. K. Toshima, S. Murkaiyama, T. Yoshida, T. Tamai, and K. Tatsuta, *Tetrahedron Lett.*, **32**, 6155 (1991).
17. R. Johansson and B. Samuelsson, *J. Chem. Soc., Chem. Commun.*, 201 (1984).
18. E. Marotta, I. Pagani, P. Righi, G. Rosini, V. Bortolasi, and A. Medici, *Tetrahedron: Asymmetry*, **6**, 2319 (1995).
19. Z. Zhang and G. Magnusson, *J. Org. Chem.*, **61**, 2394 (1996).
20. I. Sato, Y. Akahori, K.-i. Iida, and M. Hirama, *Tetrahedron Lett.*, **37**, 5135 (1996).
21. T. Tsuri and S. Kamata, *Tetrahedron Lett.*, **26**, 5195 (1985).
22. P. Deslongchamps, P. Atlani, D. Fréhel, A. Malaval, and C. Moreau, *Can. J. Chem.*, **52**, 3651 (1974).
23. N. Chidambaram, S. Bhat, and S. Chandrasekaran, *J. Org. Chem.*, **57**, 5013 (1992).

2,4-Dimethoxybenzylidene Ketal: $2,4\text{-}(CH_3O)_2C_6H_3CH(OR)_2$

Formation

1. $2,4\text{-}(MeO)_2C_6H_3CHO$, benzene, TsOH, heat, >81% yield.[1]

Cleavage

1. This acetal is stable to hydrogenation with W4-Raney Ni, which was used to cleave a benzyl group in 99% yield.[2]

1. M. Smith, D. H. Rammler, I. H. Goldberg, and H. G. Khorana, *J. Am. Chem. Soc.*, **84**, 430 (1962).
2. K. Horita, T. Yoshioka, T. Tanaka, Y. Oikawa, and O. Yonemitsu, *Tetrahedron*, **42** , 3021 (1986).

3,4-Dimethoxybenzylidene Acetal: $3,4\text{-}(CH_3O)_2C_6H_3(OR)_2$

Formation/Cleavage[1]

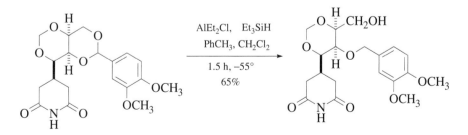

H O O O O O H OCH$_3$ OCH$_3$ O N O H

$AlEt_2Cl$, Et_3SiH
$PhCH_3$, CH_2Cl_2
1.5 h, –55°
65%

H O CH_2OH O O H OCH$_3$ OCH$_3$ O N O H

The acetal can also be cleaved with DDQ (CH_2Cl_2, H_2O, 66% yield) to afford the monobenzoate. Treatment with DIBAL (CH_2Cl_2, 0°, 91% yield) affords the hydroxy ether.[2] Treatment of a 3,4-dimethoxybenzyl ether containing a free hydroxyl with DDQ (benzene, 3Å molecular sieves, rt) affords the 3,4-dimethoxybenzylidene acetal.[3]

1. M. J. Wanner, N. P. Willard, G. J. Koomen, and U. K. Pandet, *Tetrahedron*, **43**, 2549 (1987).
2. A. B. Smith, III, Q. Lin, K. Nakayama, A. M. Boldi, C. S. Brook, M. D. McBriar, W. H. Moser, M. Sobukawa, and L. Zhuang, *Tetrahedron Lett.*, **38**, 8675 (1997).
3. K. Nozaki and H. Shirahama, *Chem. Lett.*, 1847 (1988).

2-Nitrobenzylidene Acetal: 2-$NO_2C_6H_4CH(OR)_2$

The 2-nitrobenzylidene acetal has been used to protect carbohydrates. It can be cleaved by photolysis (45 min, MeOH; CF_3CO_3H, CH_2Cl_2, 0°, 95% yield) to form primarily axial 2-nitrobenzoates from diols containing at least one axial alcohol.[1]

4-Nitrobenzylidene Acetal: 4-$NO_2C_6H_4CH(OR)_2$

This acetal was prepared [4-$NO_2C_6H_4CH(OMe)_2$, TsOH, DMF, benzene, heat] to protect a 4,6-glucopyranoside.[2]

1. P. M. Collins and V. R. N. Munasinghe, *J. Chem. Soc., Perkin Trans. 1*, 921 (1983).
2. W. Guenther and H. Kunz, *Carbohydr. Res.*, **228**, 217 (1992).

Mesitylene Acetal: $MesCH(OR)_2$

Formation

1. $MesCH(OR)_2$, CSA, CH_2Cl_2, 61–91 % yield.[1,2]

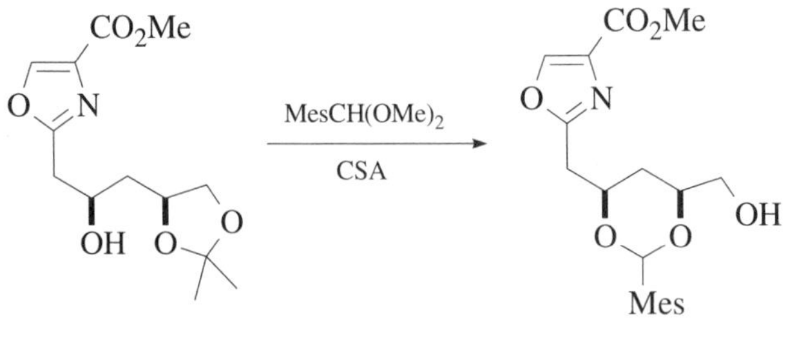

CO_2Me O N OH O O —[$MesCH(OMe)_2$, CSA]→ CO_2Me O N O O OH Mes

Cleavage

1. $Pd(OH)_2$, H_2, EtOH, rt, 12 h.[2] A BOM group can be removed by hydrogenolysis (10% Pd/C, MeOH, THF, 83% yield) in the presence of the mesitylene and 4-methoxyphenyl acetals.[1]
2. 50% aq. AcOH, 35°, >70% yield.[1] In the following illustration, methoxy-

substituted benzylidene acetals could not be hydrolyzed:[3]

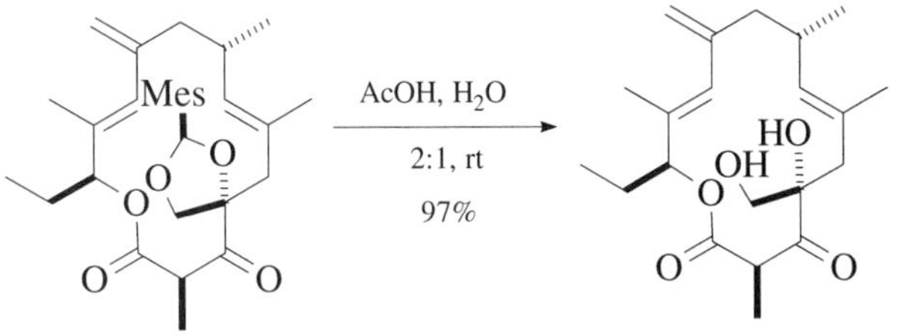

AcOH, H_2O 2:1, rt 97%

1. S. F. Martin, T. Hida, P. R. Kym, M. Loft, and A. Hodgson, *J. Am. Chem. Soc.*, **119**, 3193 (1997).
2. M. Hikota, H. Tone, K. Horita, and O. Yonemitsu, *J. Org. Chem.*, **55**, 7 (1990).
3. B. Tse, *J. Am. Chem. Soc.*, **118**, 7094 (1996).

1-Naphthaldehyde Acetal: $C_{10}H_7CH(OR)_2$

This acetal was prepared to confer crystallinity on the intermediates in the synthesis of the lysocellin antibiotics.[1]

Formation

1. $C_{10}H_7CHO$, trichloroacetic acid, PhH, >74% yield.

Cleavage

1. Pd/C, H_2O, $(COOH)_2$, EtOAc, 0°, 61% yield.
2. 2:1 THF-1 *M* H_2SO_4, 45°, 81% yield.

1. D. A. Evans, R. P. Polniaszek, K. M. DeVries, D. E. Guinn, and D. J. Mathre, *J. Am. Chem. Soc.*, **113**, 7613 (1991).

Benzophenone Ketal: $Ph_2C(OR)_2$

Formation

1. $Ph_2C(OMe)_2$, H_2SO_4.[1]
2. Ph_2CCl_2, Pyr.[2]

Cleavage

1. AcOH, H_2O.[3]
2. Hydrochloric acid, 80% dioxane/water.[4] Cleavage rates for various ring sizes were examined.

1. T. Yoon, M. D. Shair, S. J. Danishefsky, and G. K. Shulte, *J. Org. Chem.*, **59**, 3752 (1994).
2. A. Borbas, J. Hajko, M. Kajtar-Peredy, and A. Liptak, *J. Carbohydr. Chem.*, **12**, 191 (1993).
3. K. S. Feldman and A. Sambandam, *J. Org. Chem.*, **60**, 8171 (1995).
4. T. Oshima, S.-y. Ueno, and T. Nagai, *Heterocyles*, **40**, 607 (1995).

***o*-Xylyl Ether:**

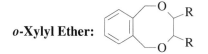

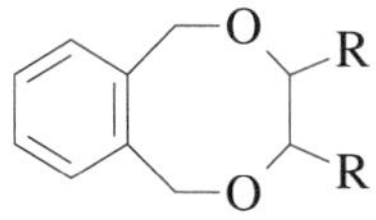

This derivative is formed from the diol and 1,2-di(bromomethyl)benzene (NaH, THF, HMPA, 0°, 66% yield). It is cleaved by hydrogenolysis [$Pd(OH)_2$, EtOH, H_2, 89–99% yield].[1]

1. A. J. Poss and M. S. Smyth, *Synth. Commun.*, **19**, 3363 (1989).

Chiral Ketones

The use of chiral ketones for the protection of diols serves two purposes: first, diol protection is accomplished, and second, symmetrical intermediates are converted to chiral derivatives that can be elaborated further, so that when the diol is deprotected, the molecule retains chirality.[1]

Camphor Ketal

Formation

1. Camphor dimethyl ketal, TMSOTf, DMSO, 90°, 3 h, 25% yield.[2]
2. Camphor, TsOH, 65–70% yield.[3]

Cleavage

1. AcOH, H_2O, > 88% yield.[3]

Menthone

Formation

1. Menthone TMS enol ether, TfOH, THF, −40°, 2 h, 51–91% yield.[4,5]
2. From a TMS protected triol using (−)-menthone.[6]

Cleavage

1. CSA, MeOH, 2 days, rt, 89–90% yield.[6]
2. $CHCl_3$ saturated with 9 *N* HCl, 85% yield.[6]

1. T. Harada and A. Oku, *Synlett*, 95 (1994).
2. K. S. Bruzik and M.-D. Tsai, *J. Am. Chem. Soc.*, **114**, 6361 (1992).
3. G. M. Salamonczyk and K. M. Pietrusiewicz, *Tetrahedron Lett.*, **32**, 4031 (1991).
4. T. Harada, Y. Kagamihara, S. Tanaka, K. Sakamoto, and A. Oku, *J. Org. Chem.*, **57**, 1637 (1992); T. Harada, S. Tanaka, and A. Oku, *Tetrahedron*, **48**, 8621 (1992).
5. T. Harada, T. Shintani, and A. Oku, *J. Am. Chem. Soc.*, **117**, 12346 (1995).
6. N. Adjé, P. Breuilles, and D. Uguen, *Tetrahedron Lett.*, **33**, 2151 (1992).

Cyclic Ortho Esters

A variety of cyclic ortho esters,[1,2] including cyclic orthoformates, have been developed to protect *cis*-1,2-diols. Cyclic ortho esters are more readily cleaved by acidic hydrolysis (e.g., by a phosphate buffer, pH 4.5–7.5, or by 0.005–0.05 *M* HCl)[3] than are acetonides. Careful hydrolysis or reduction can be used to prepare selectively monoprotected diol derivatives.

Methoxymethylene and Ethoxymethylene Acetal (Chart 3)

Formation

1. $HC(OMe)_3$ or $HC(OEt)_3$, acid catalyst, 77% or 45–80% yields, respectively.[4–6]

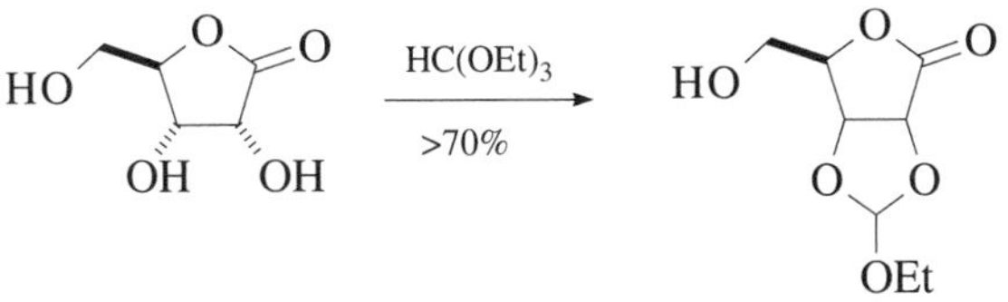

Cleavage

1. 98% formic acid or HCl at pH 2, 20°.[4]
2. Reduction with $(i\text{-Bu})_2AlH$ affords a diol, with one hydroxyl group protected as a MOM group. The more substituted hydroxyl bears the MOM group.[7]

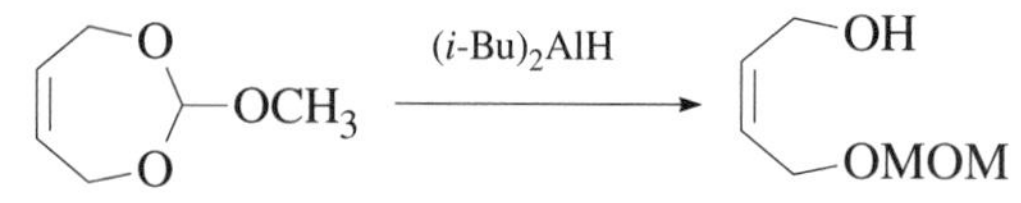

$(i\text{-Bu})_2AlH$

1. C. B. Reese, *Tetrahedron*, **34**, 3143 (1978).
2. V. Amarnath and A. D. Broom, *Chem. Rev.*, **77**, 183 (1977).
3. M. Ahmad, R. G. Bergstrom, M. J. Cashen, A. J. Kresge, R. A. McClelland, and M. F. Powell, *J. Am. Chem. Soc.*, **99**, 4827 (1977).
4. B. E. Griffin, M. Jarman, C. B. Reese, and J. E. Sulston, *Tetrahedron*, **23**, 2301 (1967).
5. J. Zemlicka, *Chem. Ind. (London)*, 581 (1964); F. Eckstein and F. Cramer, *Chem. Ber.*, **98**, 995 (1965).
6. R. M. Ortuño, R. Mercé, and J. Font, *Tetrahedron Lett.*, **27**, 2519 (1986).
7. M. Takasu, Y. Naruse, and H. Yamamoto, *Tetrahedron Lett.*, **29**, 1947 (1988).

The following ortho esters have been prepared to protect the diols of nucleosides. They are readily hydrolyzed with mild acid to afford monoester derivatives, generally as a mixture of positional isomers.

Dimethoxymethylene Ortho Ester[1] (Chart 3)

1-Methoxyethylidene Ortho Ester[2]

1-Ethoxyethylidene Ortho Ester[3]

Formation

1. $CH_2{=}C(OMe)_2$, DMF, TsOH, < 5°.[4] These conditions will completely protect certain triols.[5]

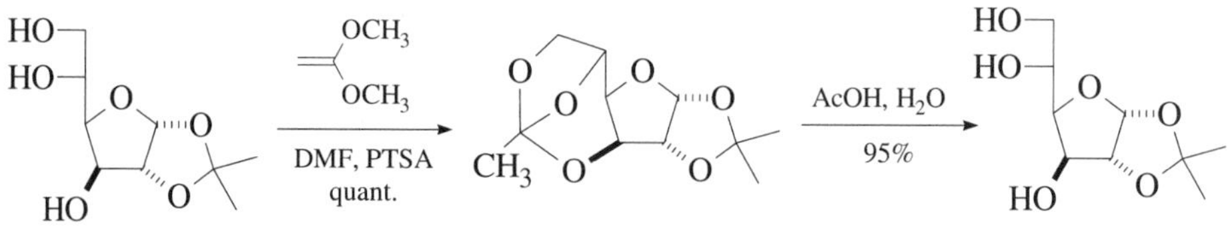

DMF, PTSA quant. → AcOH, H_2O 95%

2. $CH_3C(OEt)_3$.[6] With this ortho ester, good selectivity for the axial alcohol is achieved in the acidic hydrolysis of a pyranoside derivative.[7,4]

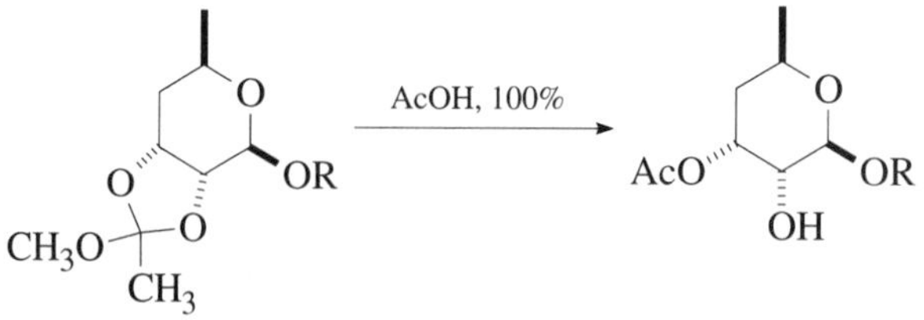

AcOH, 100%

Ref. 7

Methylidene Ortho Ester

Formation[8]

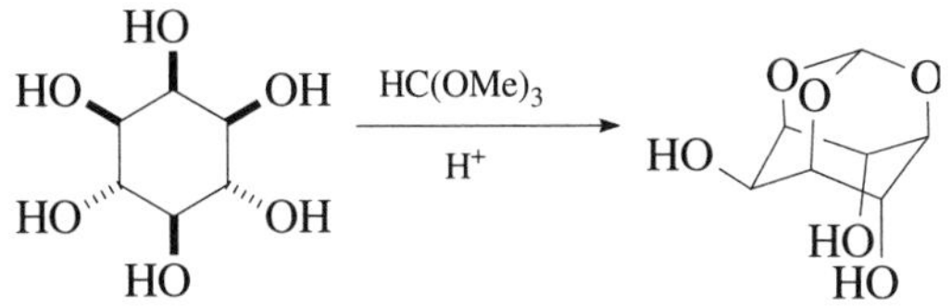

HO, HO, OH, HO, OH, HO — $HC(OMe)_3$, H^+ → HO, O, O, O, HO, HO

Cleavage

1. TFA, H_2O, rt, 40 h, 85% yield.[9]

2.

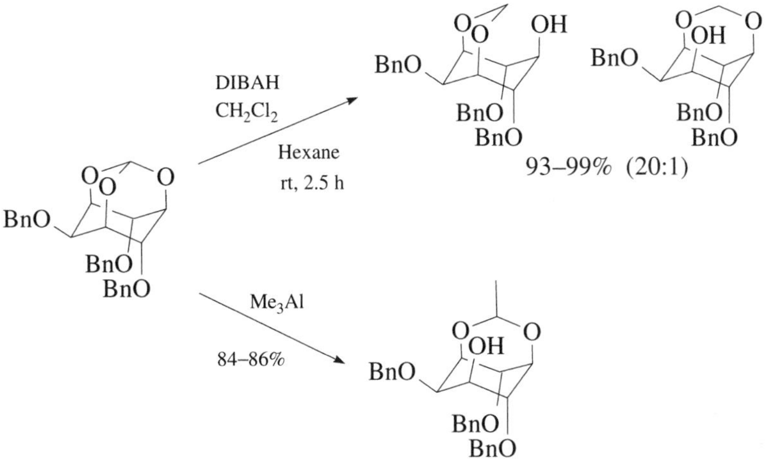

BnO, O, O, O, BnO, BnO — DIBAH, CH_2Cl_2, Hexane, rt, 2.5 h → BnO, O, O, OH, BnO, BnO + BnO, O, OH, O, BnO, BnO — 93–99% (20:1)

Me$_3$Al, 84–86% → BnO, O, OH, O, BnO, BnO

Ref. 10

Phthalide Ortho Ester

Formation/Cleavage[11]

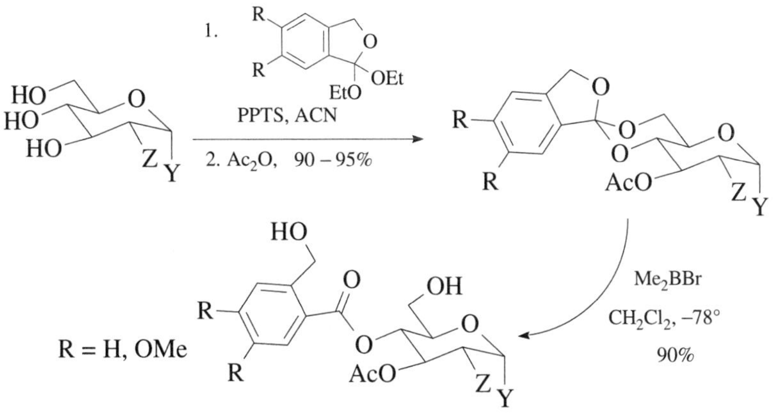

HO, HO, O, HO, Z, Y — 1. (R, R, O, OEt, EtO) PPTS, ACN; 2. Ac_2O, 90 – 95% → R, O, O, O, O, R, AcO, Z, Y

Me_2BBr, CH_2Cl_2, –78°, 90% → HO, O, OH, R, O, O, R, AcO, Z, Y

R = H, OMe

1,2-Dimethoxyethylidene Ortho Ester[12]

α-Methoxybenzylidene Ortho Ester[2]

1-(*N*,*N*-Dimethylamino)ethylidene Derivative[13]

α-(*N*,*N*-Dimethylamino)benzylidene Derivative[13]

1. G. R. Niaz and C. B. Reese, *J. Chem. Soc., Chem. Commun.*, 552 (1969).
2. C. B. Reese and J. E. Sulston, *Proc. Chem. Soc.*, 214 (1964).
3. V. P. Miller, D.-y. Yang, T. M. Weigel, O. Han, and H.-w. Liu, *J. Org. Chem.*, **54**, 4175 (1989).
4. M. Bouchra, P. Calinaud, and J. Gelas, *Carbohydr. Res.*, **267**, 227 (1995).
5. M. Bouchra, P. Calinaud, and J. Gelas, *Synthesis*, 561 (1995).
6. H. P. M. Fromageot, B. E. Griffin, C. B. Reese, and J. E. Sulston, *Tetrahedron*, **23**, 2315 (1967); U. E. Udodong, C. S. Rao, and B. Fraser-Reid, *Tetrahedron*, **48**, 4713 (1992).
7. S. Hanessian and R. Roy, *Can. J. Chem.*, **63**, 163 (1985).
8. J. P. Vacca, S. J. De Solms, S. D. Young, J. R. Huff, D. C. Billington, R. Baker, J. J. Kulagowski, and I. M. Mawer, *ACS Symp. Ser.*, **463**, "Inositol Phosphates Deriv.", 66 (1991).
9. A. M. Riley, M. F. Mahon, and B. V. L. Potter, *Angew. Chem., Int. Ed. Engl.*, **36**, 1472 (1997).
10. S.-M. Yeh, G. H. Lee, Y. Wang, and T.-Y. Luh, *J. Org. Chem.*, **62**, 8315 (1997).
11. A. Arasappan and P. L. Fuchs, *J. Am. Chem. Soc.*, **117**, 177 (1995).
12. J. H. van Boom, G. R. Owen, J. Preston, T. Ravindranathan, and C.B. Reese, *J. Chem. Soc. C*, 3230 (1971).
13. S. Hanessian and E. Moralioglu, *Can. J. Chem.*, **50**, 233 (1972).

2-Oxacyclopentylidene Ortho Ester

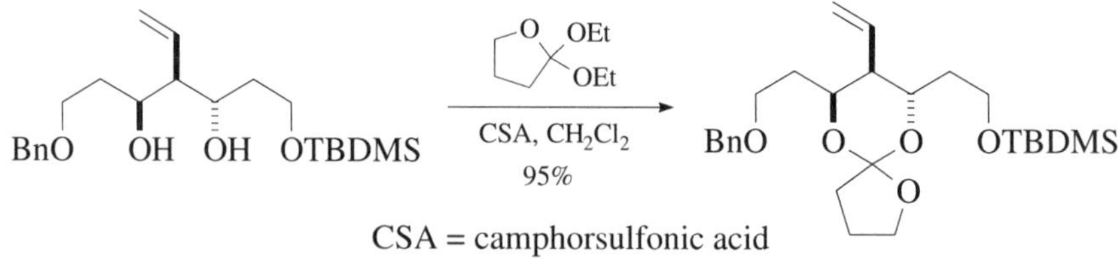

This ortho ester does not form a monoester upon deprotection, as do acyclic ortho esters; thus a hydrolysis step is avoided.[1]

1. R. M. Kennedy, A. Abiko, T. Takemasa, H. Okumoto, and S. Masamune, *Tetrahedron Lett.*, **29**, 451 (1988).

Butane-2,3-bisacetal (BBA):

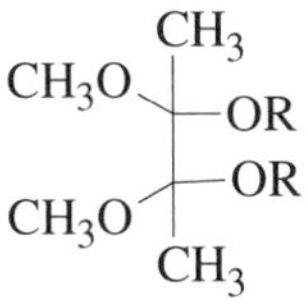

Formation

1. 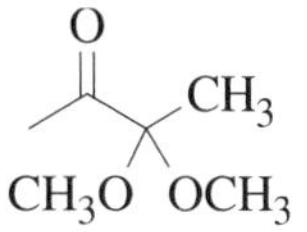 $CH_3C(O)C(CH_3)(OCH_3)_2$ EtOAc, $BF_3{\cdot}Et_2O$, 65% yield.[1]
2. 2,3-Butanedione, TMOF (trimethyl orthoformate), CSA, MeOH, 60–82% yield.[2, 3]
3. 2,2,3,3-Tetramethoxybutane, TMOF, MeOH, CSA, 54–91% yield. *trans*-Diols are protected in preference to *cis*-diols, in contrast to acetonide formation, which prefers protection of *cis*-diols.[4]

1. U. Berens, D. Leckel, and S. C. Oepen, *J. Org. Chem.*, **60**, 8204 (1995).
2. N. L. Douglas, S. V. Ley, H. M. I. Osborn, D. R. Owen, H. W. M. Priepke, and S. L. Warriner, *Synlett*, 793 (1996).
3. A. Hense, S. V. Ley, H. M. I. Osborn, D. R. Owen, J.-R. Poisson, S. L. Warriner, and D. E. Wesson, *J. Chem. Soc., Perkin Trans. 1*, 2023 (1997).
4. J.-L. Montchamp, F. Tian, M. E. Hart, and J. W. Frost, *J. Org. Chem.*, **61**, 3897 (1996).

Cyclohexane-1,2-diacetal (CDA):

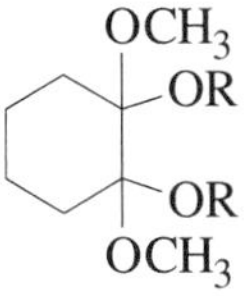

Formation

1. 1,1,2,2-Tetramethoxycyclohexane,[1] CSA, MeOH, trimethyl orthoformate.[2]
2. 1,2-Cyclohexanedione, trimethyl orthoformate, CSA, MeOH, 61 yield.[3] 9,10-Phenanthrenequinone and 2,3-butanedione were similarly converted to diacetals by this method.[4]

Cleavage

1. TFA, H_2O, 5 min, 81% per CDA unit.[2]

1. S. V. Ley, H. M. I. Osborn, H. W. M. Priepke, and S. L. Warriner, *Org. Synth.*, **75**, 170 (1997).

2. P. Grice, S. V. Ley, J. Pietruszka, and H. M. W. Priepke, *Angew. Chem., Int. Ed. Engl.*, **35**, 197 (1996); P. Grice, S. V. Ley, J. Pietruszka, H. M. W. Priepke, and S. L.Warriner, *J. Chem. Soc., Perkin Trans. 1*, 351 (1997).
3. N. L. Douglas, S. V. Ley, H. M. I. Osborn, D. R. Owen, H. W. M. Priepke, and S. L. Warriner, *Synlett*, 793 (1996).
4. A. Hense, S. V. Ley, H. M. I. Osborn, D. R. Owen, J.-F. Poisson, S. L. Warriner, and K. E. Wesson, *J. Chem. Soc., Perkin Trans. 1*, 2023 (1997).

Dispiroketals:

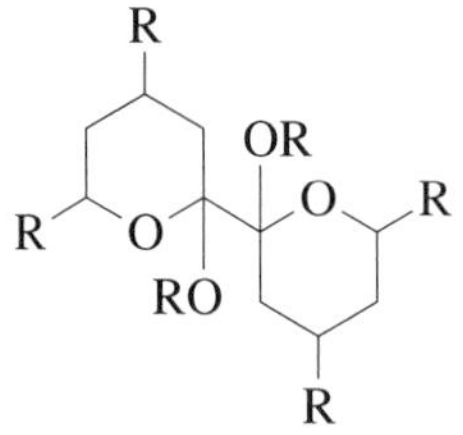

R OR O R R O RO R

Formation

1. Bisdihydropyran, CSA, toluene, reflux, 36–98% yield.[1]
2. 2,2′-Bis(phe nylthiomethyl)dihydropyran, CSA, $CHCl_3$, 54–93% yield. This dihydropyran can be used for the resolution of racemic diols or for regioselective protection, which is directed by the chirality of the dihyropyran.[2] Other 2,2′-substituted bisdihydropyrans that can be cleaved by a variety of methods are available, and their use in synthesis has been reviewed.[3]

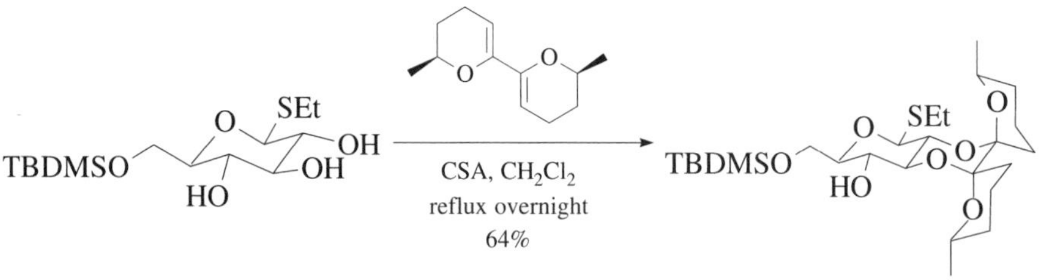

TBDMSO O SEt OH OH HO → (CSA, CH_2Cl_2, reflux overnight, 64%) TBDMSO O SEt HO O O O O

Cleavage

1. The 2,2′-bis(phenylthiomethyl) dispiroketal (dispoke) derivative is cleaved by oxidation to the sulfone, followed by treatment with $LiN(TMS)_2$.[2] The related bromo and iodo derivatives are cleaved reductively with LDBB (lithium 4,4′-di-*t*-butylbiphenylide) or by elimination with the P4-*t*-butylphosphazene base and acid hydrolysis of the enol ether.[3] The 2,2-diphenyl dispiroketal is cleaved with $FeCl_3$ (CH_2Cl_2, rt, overnight).[4] The dimethyl dispiroketal is cleaved with TFA,[5] and the allyl derivative is cleaved by ozonolysis followed by elimination.[1]

1. For a review, see S. V. Ley, R. Downham, P. J. Edwards, J. E. Innes, and M. Woods, *Contemp. Org. Synth.*, **2**, 365 (1995).

2. S. V. Ley, S. Mio, and B. Meseguer, *Synlett*, 791 (1996).
3. S. V. Ley and S. Mio, *Synlett*, 789 (1996).
4. D. A. Entwistle, A. B. Hughes, S. V. Ley, and G. Visentin, *Tetrahedron Lett.*, **35**, 777 (1994).
5. R. Downham, P. J. Edwards, D. A. Entwistle, A. B. Hughes, K. S. Kim, and S. V. Ley, *Tetrahedron: Asymmetry*, **6**, 2403 (1995).

Silyl Derivatives

Di-*t*-butylsilylene Group [DTBS(OR)$_2$]

The DTBS group is probably the most useful of the bifunctional silyl ethers. Dimethylsilyl and diisopropylsilyl derivatives of diols are very susceptible to hydrolysis even in water and therefore are of limited use, unless other structurally imposed steric effects provide additional stabilization.

Formation

1. $(t\text{-Bu})_2SiCl_2$, CH_3CN, TEA, HOBt, 65°.[1,2] Tertiary alcohols do not react under these conditions. The reagent is effective for both 1,2- and 1,3-diols.
2. $(t\text{-Bu})_2Si(OTf)_2$, 2,6-lutidine, 0–25°, $CHCl_3$.[3] This reagent readily silylates 1,2-, 1,3- and 1,4-diols, even when one of the alcohols is tertiary. THP and PMB protected diols are converted to the silylene derivative with the reagent.[4] 1,3-Diols are preferably protected over *cis*- or *trans*-1,2-diols.[5]
3. The di-*t*-butylsilylene group has been used to connect a diene and a dienophile to control the intramolecular Diels–Alder reaction.[6]
4. $(t\text{-Bu})_2SiCl_2$, $AgNO_3$, Pyr, DMF, >84% yield.[7]
5. HO, B, O, OH OH → $(t\text{-Bu})_2Si(OTf)_2$ / DMF, Pyr / 90–95% yield → B, O, O, *t*-Bu—Si, *t*-Bu, O, OH

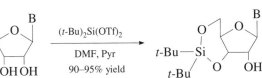

   DMF is the only solvent that works in this transformation.[8]

Cleavage

Derivatives of 1,3- and 1,4-diols are stable to pH 4–10 at 22° for several hours, but derivatives of 1,2-diols undergo rapid hydrolysis under basic conditions (5:1 THF–pH 10 buffer, 22°, 5 min) to form monosilyl ethers of the parent diol.

1. 48% aq. HF, CH_3CN, 25°, 15 min, 95% yield.[3]
2. $Bu_3NH^+F^-$, THF.[9]

3. Pyr·HF, THF, 25°, 88% yield.[1]

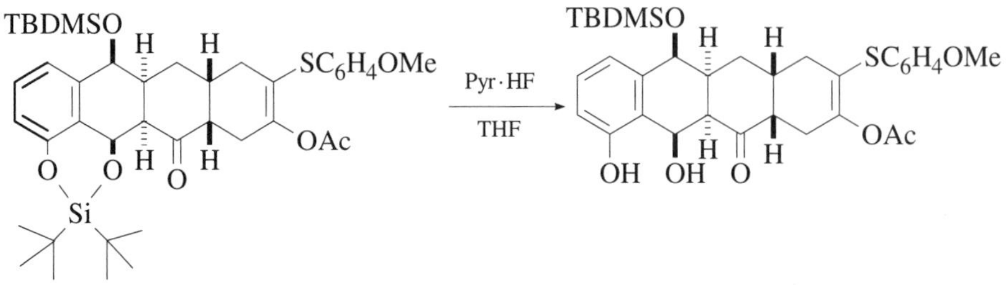

Pyr · HF, THF

4. TBAF, $ZnCl_2$, ms, rt–65°, 3 h.[10]
5. TBAF, THF, rt, 96% yield.[5]
6. TBAF, AcOH, 60°, 12 h, 45% yield.[11]
7. Tris(dimethylamino)sulfonium difluorotrimethylsilicate (TSAF), THF, 0°, 5 h, 64% yield. A TES and two phenolic TIPS groups were also cleaved.[12]

Dialkylsilylenes Groups

1. Three different silylene derivatives were used to achieve selective protection of a more hindered diol during a taxol synthesis. Treatment of the silylene with MeLi opens the ring to afford the more hindered silyl ether.[13]

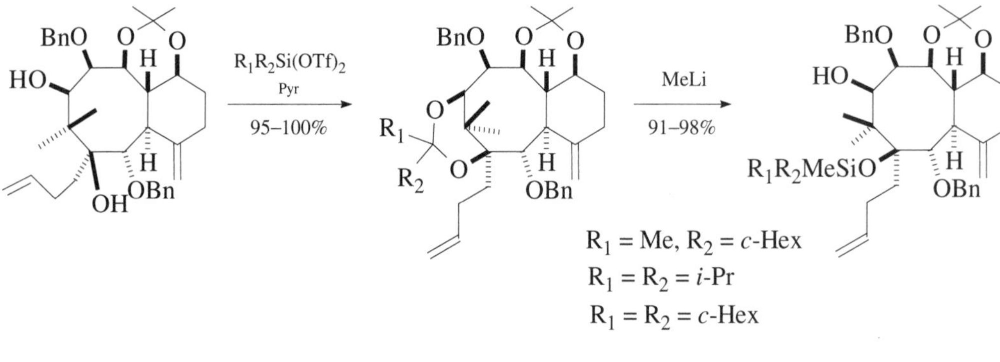

$R_1R_2Si(OTf)_2$, Pyr, 95–100%; MeLi, 91–98%

R_1 = Me, R_2 = *c*-Hex
R_1 = R_2 = *i*-Pr
R_1 = R_2 = *c*-Hex

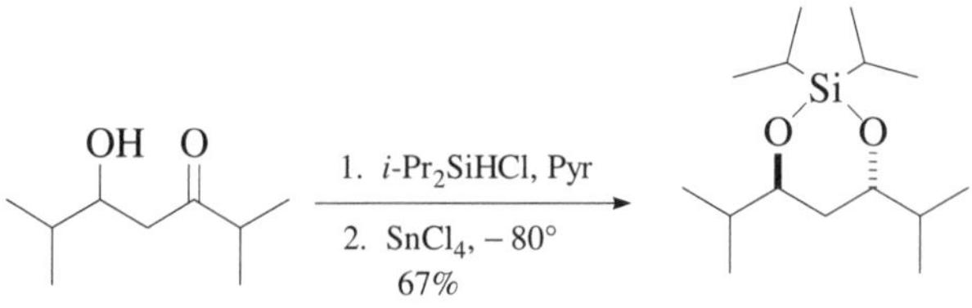

1. *i*-Pr_2SiHCl, Pyr
2. $SnCl_4$, – 80°
67%

Ref. 14

1. B. M. Trost and C. G. Caldwell, *Tetrahedron Lett.*, **22**, 4999 (1981).
2. B. M. Trost, C. G. Caldwell, E. Murayama, and D. Heissler, *J. Org. Chem.*, **48**, 3252 (1983).
3. E. J. Corey and P. B. Hopkins, *Tetrahedron Lett.*, **23**, 4871 (1982).
4. T. Oriyama, K. Yatabe, S. Sugawara, Y. Machiguchi, and G. Koga, *Synlett*, 523 (1996).
5. B. Delpech, D. Calvo, and R. Lett, *Tetrahedron Lett.*, **37**, 1019 (1996).
6. J. W. Gillard, R. Fortin, E. L. Grimm, M. Maillard, M. Tjepkema, M. A. Bernstein, and R. Glaser, *Tetrahedron Lett.*, **32**, 1145 (1991).
7. C. W. Gundlach, IV, T. R. Ryder, and G. D. Glick, *Tetrahedron Lett.*, **38**, 4039 (1997).
8. K. Furusawa, K. Ueno, and T. Katsura, *Chem. Lett.*, 97 (1990).
9. K. Furusawa, *Chem. Lett.*, 509 (1989).
10. R. Van Speybroeck, H. Guo, J. van der Eycken, and M. Vandewalle, *Tetrahedron*, **47**, 4675 (1991).
11. K. Toshima, H. Yamaguchi, T. Jyojima, Y. Noguchi, M. Nakata, and S. Matsumura, *Tetrahedron Lett.*, **37**, 1073 (1996).
12. D. A. Johnson and L. M. Taubner, *Tetrahedron Lett.*, **37**, 605 (1996).
13. I. Shiina, T. Nishimura, N. Ohkawa, H. Sakoh, K. Nishimura, K. Saitoh, and T. Mukaiyama, *Chem. Lett.*, 419 (1997).
14. S. Anwar and A. P. Davis, *J. Chem. Soc., Chem. Commun.*, 831 (1986).

1,3-(1,1,3,3-Tetraisopropyldisiloxanylidene) Derivative [TIPDS$(OR)_2$]

Formation

1. $TIPDSCl_2$, DMF, imidazole.[1–3] This reagent is primarily used in carbohydrate protection, but occasionally it proves valuable in other circumstance s.[4]

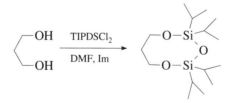

OH, OH —($TIPDSCl_2$ / DMF, Im)→ O–Si, O, O–Si

2. $TIPDSCl_2$, Pyr.[5–7] In polyhydroxylated systems, the regiochemical outcome is determined by the initial reaction at the sterically less hindered alcohol.[8]
3. $TIPDSCl_2$, AgOTf, *sym*-collidine, DMF, 45% yield.[9]

Cleavage

1. $Bu_4N^+F^-$, THF.[1,5,10] When $Bu_4N^+F^-$ is used to remove the TIPDS group, ester groups can migrate because of the basic nature of the fluoride ion. Migration can be prevented by the addition of Pyr·HCl.[11]

2. TEA·HF.[12]
3. 0.2 *M* HCl, dioxane, H_2O or MeOH.[1]
4. 0.2 *M* NaOH, dioxane, H_2O.[1]
5. TMSI, CH_2Cl_2, 0°, 0.5 h, 83% yield.[13]
6. Ac_2O, AcOH, H_2SO_4.[2]
7. The TIPDS derivative can be induced to isomerize from the thermodynamically less stable eight-membered ring to the more stable seven-membered ring derivative.[5,14] The isomerization occurs only in DMF.

DMF, H^+ >82%

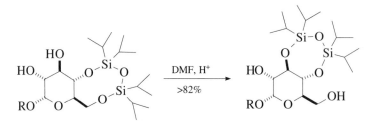

8. $NH_4^+F^-$, MeOH, 60°, 3 h, 99% yield.[15]
9. TBAF, AcOH, THF.[16]
10. CsF, NH_3, MeOH.[17]
11. KF·$2H_2O$, 18-crown-6, DMF or THF, rt, 55–81% yield.[18]
12. Treatment of a TIPDS group with methyl pyruvate (TMSOTf, 0° to rt, 69–99% yield) converts the group to the pyruvate acetal.[9]

1,1,3,3-Tetra-*t*-butoxydisiloxanylidene Derivative [TBDS$(OR)_2$]

Formation

1. 1,3-Dichloro-1,1,3,3-tetra-*t*-butoxydisiloxane, Pyr, rt, 50–87% yield.[19]

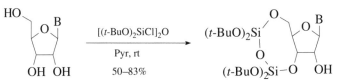

$[(t\text{-BuO})_2SiCl]_2O$, Pyr, rt, 50–83%

B = pyrimidine or purine residue

Cleavage

1. $Bu_4N^+F^-$, THF, 2 min.[19] This group is less reactive toward triethylammonium fluoride than is the TIPDS group. It is stable to 2 *M* HCl, aq. dioxane, overnight. Treatment with 0.2 *M* NaOH, aq. dioxane leads to cleavage of only the Si–O bond at the 5′-position of the uridine derivative. The TBDS derivative is 25 times more stable than the TIPDS derivative to basic hydrolysis.

1. W. T. Markiewicz, *J. Chem. Res. Synop.*, 24 (1979).
2. J. P. Schaumberg, G. C. Hokanson, J. C. French, E. Smal, and D. C. Baker, *J. Org. Chem.*, **50**, 1651 (1985).
3. E. Ohtsuka, M. Ohkubo, A. Yamane, and M. Ikebara, *Chem. Pharm. Bull.*, **31**, 1910 (1983).
4. A. G. Myers, P. M. Harrington, and E. Y. Kuo, *J. Am. Chem. Soc.*, **113**, 694 (1991).
5. C. A. A. van Boeckel and J. H. van Boom, *Tetrahedron*, **41**, 4545, 4557 (1985).
6. J. Thiem, V. Duckstein, A. Prahst, and M. Matzke, *Liebigs Ann. Chem.*, 289 (1987).
7. J. S. Davies, E. J. Tremeer, and R. C. Treadgold, *J. Chem. Soc., Perkin Trans. 1*, 1107 (1987).
8. W. T. Markiewicz, N. Sh. Padyukova, S. Samek, and J. Smrt, *Collect. Czech. Chem. Commun.*, **45**, 1860 (1980).
9. T. Ziegler, E. Eckhardt, K. Neumann, and V. Birault, *Synthesis*, 1013 (1992).
10. M. D. Hagen, C. S.-Happ, E. Happ, and S. Chládek, *J. Org. Chem.*, **53**, 5040 (1988).
11. J. J. Oltvoort, M. Kloosterman, and J. H. Van Boom, *Recl: J. R. Neth. Chem. Soc.*, **102**, 501 (1983).
12. W. T. Markiewicz, E. Biala, and R. Kierzek, *Bull. Pol. Acad. Sci.,Chem.*, **32**, 433 (1984).
13. T. Tatsuoka, K. Imao, and K. Suzuki, *Heterocycles*, **24**, 617 (1986).
14. C. H. M. Verdegaal, P. L. Jansse, J. F. M. de Rooij, and J. H. Van Boom, *Tetrahedron Lett.*, **21**, 1571 (1980).
15. W. Zhang and M. J. Robins, *Tetrahedron Lett.*, **33**, 1177 (1992).
16. W. Pfleiderer, M. Pfister, S. Farkas, H. Schirmeister, R. Charubala, K. P. Stengele, M. Mohr, F. Bergmann, and S. Gokhale, *Nucleosides Nucleotides*, **10**, 377 (1991).
17. J. R. McCarthy, D. P. Matthews, D. M. Stemerick, E. W. Huber, P. Bey, B. J. Lippert, R. D. Snyder, and P. S. Sunkara, *J. Am. Chem. Soc.*, **113**, 7439 (1991).
18. J. N. Kremsky and N. D. Sinha, *Bioorg. Med. Chem. Lett.*, **4**, 2171 (1994).
19. W. T. Markiewicz, B. Nowakowska, and K. Adrych, *Tetrahedron Lett.*, **29**, 1561 (1988); W. T. Markiewicz and A.-R. Katarzyna, *Nucleosides Nucleotides*, **10**, 415 (1991).

Cyclic Carbonates (Chart 3)

Cyclic carbonates[1,2] are very stable to acidic hydrolysis (AcOH, HBr, and H_2SO_4/MeOH) and are more stable to basic hydrolysis than are esters.

Formation

1. Cl_2CO, Pyr, 20°, 1 h.[3]
2. The related thionocarbonate is prepared from thiophosgene (Pyr, DMAP, 78% yield).[4]
3. *p*-$NO_2C_6H_4OCOCl$, Pyr, 20°, 5 days, 72% yield.[5]
4. *N,N'*-Carbonyldiimidazole, PhH, heat, 12 h–4 days, 90% yield.[6,7]

5. Cl_3CCOCl, Pyr, 1 h, rt, >80% yield.[8]

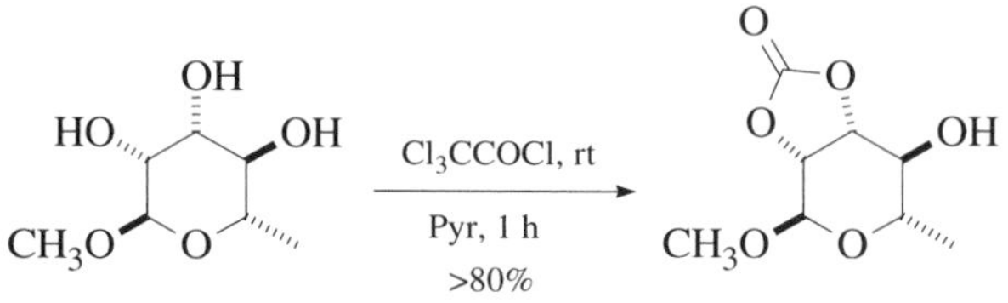

Cl_3CCOCl, rt
Pyr, 1 h
>80%

6. $Cl_3COCO_2CCl_3$ (triphosgene), Pyr, CH_2Cl_2, 84–99% yield.[9] Triphosgene is a much safer source of phosgene and is an easily handled solid. A 1,2,3-triol was selectively protected at the 1,2-position with this reagent.[10] Reactions using triphosgene often need to be run at higher temperatures because triphosgene is not as reactive as phosgene.
7. CO, S, Et_3N, 80°, 4 h; $CuCl_2$, rt, 18 h, 66–100% yield.[11]
8. Ethylene carbonate, $NaHCO_3$, 120°, 80% yield.[12]
9. Cyclic carbonates are prepared directly from epoxides with LiBr, CO_2, NMP (1-methyl-2-pyrrolidinone), 100°.[13]

Cleavage

1. $Ba(OH)_2$, H_2O, 70°.[14]
2. Pyr, H_2O, reflux, 15 min, 100% yield.[4] These conditions were used to remove the carbonate from uridine.
3. 0.5 *M* NaOH, 50% aq. dioxane, 25°, 5 min, 100% yield.[4]
4. As with the benzylidene ketals, the carbonate can be opened to give a monoprotected diol.[15]

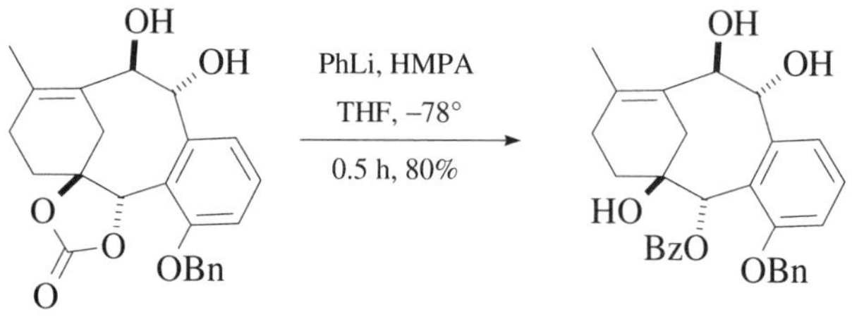

PhLi, HMPA
THF, –78°
0.5 h, 80%

1. L. Hough, J. E. Priddle, and R. S. Theobald, *Adv. Carbohydr. Chem.*, **15**, 91–158 (1960).
2. V. Amarnath and A. D. Broom, *Chem. Rev.*, **77**, 183 (1977).
3. W. N. Haworth and C. R. Porter, *J. Chem. Soc.*, 151 (1930).
4. S. Y. Ko, *J. Org. Chem.*, **60**, 6250 (1995).
5. R. L. Letsinger and K. K. Ogilvie, *J. Org. Chem.*, **32**, 296 (1967).
6. J. P. Kutney and A. H. Ratcliffe, *Synth., Commun.*, **5**, 47 (1975).
7. K. Narasaka, *ACS Symp. Ser.*, **386**, "Trends in Synth. Carbohydr. Chem.," p. 290 (1989).

8. K. Tatsuta, K. Akimoto, M. Annaka, Y. Ohno, and M. Kinoshita, *Bull. Chem. Soc. Jpn.*, **58**, 1699 (1985).
9. R. M. Burk and M. B. Roof, *Tetrahedron Lett.*, **34**, 395 (1993).
10. S. K. Kang, J. H. Jeon, K. S. Nam, C. H. Park, and H. W. Lee, *Synth. Commun.*, **24**, 305 (1994).
11. T. Mizuno, F. Nakamura, Y. Egashira, I. Nishiguchi, T. Hirashima, A. Ogawa, N. Kambe, and N. Sonoda, *Synthesis*, 636 (1989)
12. T. Desai, J. Gigg, and R. Gigg, *Carbohydr. Res.*, **277**, C5 (1995).
13. N. Kihara, Y. Nakawaki, and T. Endo, *J. Org. Chem.*, **60**, 473 (1995).
14. W. G. Overend, M. Stacey, and L. F. Wiggins, *J. Chem. Soc.*, 1358 (1949).
15. K. C. Nicolaou, C. F. Claiborne, K. Paulvannan, M. H. D. Postema, and R. K. Guy, *Chem.—Eur. J.*, **3**, 399 (1997).

Cyclic Boronates

Although boronates are quite susceptible to hydrolysis, they have been useful for the protection of carbohydrates.[1,2] Note that as the steric demands of the diol increase, the rate of hydrolysis decreases. For example, pinacol boronates are rather difficult to hydrolyze; in fact, they can be isolated from aqueous systems with no hydrolysis. The section on the protection of boronic acids should be consulted.

Ethyl Boronate[3] (Chart 3)

Formation

1.

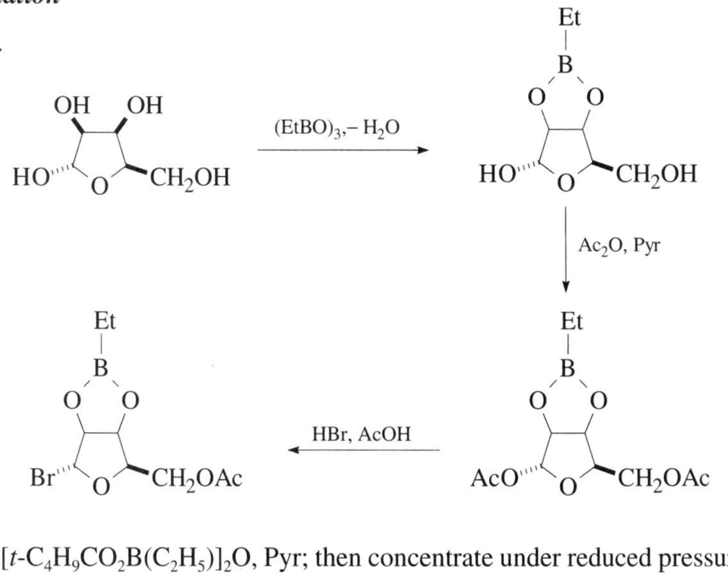

OH OH

HO O CH_2OH

$(EtBO)_3$,– H_2O

Et B O O

HO O CH_2OH

Ac_2O, Pyr

Et B O O

AcO O CH_2OAc

HBr, AcOH

Et B O O

Br O CH_2OAc

Ref. 4

2. [t-$C_4H_9CO_2B(C_2H_5)]_2O$, Pyr; then concentrate under reduced pressure.[5]
3. $EtB(OMe)_2$, ion exchange resin, 85% yield.[6]
4. $LiEt_3BH$, THF, 0°–rt, 98% yield.[7]

Phenyl Boronate

Formation

1. $PhB(OH)_2$, PhH[8] or Pyr.[9] A polymeric version of the phenyl boronate has been developed.[10] Phenyl boronates are stable to the conditions of stannylation and have been used for selective sulfation to produce monosulfated monosaccharides.[11] Phenyl boronates were found to be stable to oxidation with PCC.[12]

Cleavage

1. 1,3-Propanediol, acetone.[1] This method removes the boronate by exchange. 2-Methylpentane-2,5-diol in acetic acid cleaves a phenyl boronate (85% yield).[13]
2. Acetone, H_2O (4:1), 30 min, 83% yield.[10]
3. H_2O_2, EtOAc, >80% yield.[14,15]
4. Ac_2O, Pyr, 99% yield. In this case, the boronate is converted to an acetate.[16]
5. Treatment of the boronate with BuI, AgO affords the monoalkylated diol in a manner similar to stannylene-directed monoalkylation and acylation.[17]

o-Acetamidophenyl Boronate: $[2,6\text{-}(AcNH)_2C_6H_3B(OR)_2]$

This boronate was developed to confer added stability toward hydrolysis. It was shown to be substantially more stable to hydrolysis than the simple phenyl boronate because of coordination of the the ortho acetamide to the boronate.[18]

1. R. J. Ferrier, *Adv. Carbohydr. Chem. Biochem.*, **35**, 31–80 (1978).
2. W. V. Dahlhoff and R. Köster, *Heterocycles*, **18**, 421 (1982).
3. W. V. Dahloff and R. Köster, *J. Org. Chem.*, **41**, 2316 (1976), and references cited therein.
4. W. V. Dahlhoff, A. Geisheimer, and R. Köster, *Synthesis*, 935 (1980); W. V. Dahlhoff and R. Köster, *Synthesis*, 936 (1980).
5. R. Köster, K. Taba, and W. V. Dahlhoff, *Liebigs Ann. Chem.*, 1422 (1983).
6. W. V. Dahlhoff, W. Fenzl, and R. Köster, *Liebigs Ann. Chem.*, 807 (1990).
7. L. Garlaschelli, G. Mellerio, and G. Vidari, *Tetrahedron Lett.*, **30**, 597 (1989).
8. R. J. Ferrier, *Methods Carbohydr. Chem.*, **VI**, 419–426 (1972).
9. J. M. J. Fréchet, L. J. Nuyens, and E. Seymour, *J. Am. Chem. Soc.*, **101**, 432 (1979).
10. N. P. Bullen, P. Hodge, and F. G. Thorpe, *J. Chem. Soc., Perkin Trans. 1*, 1863 (1981).
11. S. Langston, B. Bernet, and A. Vasella, *Helv. Chim. Acta*, **77**, 2341 (1994).
12. C. Lifjebris, B. M. Nilsson, B. Resul, and U. Hacksell, *J. Org. Chem.*, **61**, 4028 (1996).
13. E. Bertounesque, J.-C. Florent, and C. Monneret, *Synthesis*, 270 (1991).

14. D. A. Evans and R. P. Polniaszek, *Tetrahedron Lett.*, **27**, 5683 (1986).
15. D. A. Evans, R. P. Polniaszek, D. M. DeVries, D. E. Guinn, and D. J. Mathre, *J. Am. Chem. Soc.*, **113**, 7613 (1991).
16. A. Flores-Parra, C. Paredes-Tepox, P. Joseph-Nathan, and R. Contreras, *Tetrahedron*, **46**, 4137 (1990).
17. K. Oshima, E.-i. Kitazono, and Y. Aoyama, *Tetrahedron Lett.*, **38**, 5001 (1997).
18. S. X. Cai and J. F. W. Keana, *Bioconjugate Chem.*, **2**, 317 (1991).

3

PROTECTION FOR PHENOLS AND CATECHOLS

The phenolic hydroxyl group occurs widely in plant and animal life, both land based and aquatic, as demonstrated by the vast number of natural products that contain this group. In developing a synthesis of any phenol-containing product, protection is often mandatory to prevent reaction with oxidizing agents and electrophiles or reaction of the nucleophilic phenoxide ion with even mild alkylating and acylating agents. Many of the protective groups developed for alcohol protection are also applicable to phenol protection: thus, the chapter on alcohol protection should also be consulted. Ethers are the most widely used protective groups for phenols, and in general, they are more easily cleaved than the analogous ethers of simple alcohols.[1] Esters are also important protective groups for phenols, but are not as stable to hydrolysis as the related alcohol derivatives. Simple esters are easily hydrolyzed with mild base (e.g., $NaHCO_3$/aq. MeOH, 25°), but more sterically demanding esters (e.g., pivaloate) require harsher conditions to effect hydrolysis. Catechols can be protected in the presence of phenols as cyclic acetals or ketals, or cyclic esters. Some of the more important phenol and catechol protective groups are included in Reactivity Chart 4.[2]

1. For a review on ether cleavage, see M. V. Bhatt and S. U. Kulkarni, *Synthesis,* 249 (1983).
2. See also E. Haslam "Protection of Phenols and Catechols," in *Protective Groups in Organic Chemistry*, J. F. W. McOmie, Ed., Plenum, New York and London, 1973, pp. 145–182.

PROTECTION FOR PHENOLS

Ethers

Historically, simple *n*-alkyl ethers formed from a phenol and a halide or sulfate were cleaved under rather drastic conditions (e.g., refluxing HBr). New ether protective groups have been developed that are removed under much milder conditions (e.g., via nucleophilic displacement, hydrogenolysis of benzyl ethers, or mild acid hydrolysis of acetal-type ethers) that often do not affect other functional groups in a molecule.

Methyl Ether: $ArOCH_3$ (Chart 4)

Deuteromethyl ethers have been used to protect phenols to prevent the methyl hydrogens from participating in free-radical reactions.[1]

Formation

1. MeI, K_2CO_3, acetone, reflux, 6 h.[2,3] This is a very common and often very efficient method for the preparation of phenolic methyl ethers. The method is also applicable to the formation of phenolic benzyl ethers.
2. Me_2SO_4, NaOH, EtOH, reflux, 3 h, 71–74% yield.[2]

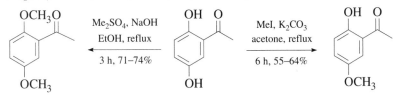

OCH_3 O / OCH_3 ← Me_2SO_4, NaOH, EtOH, reflux, 3 h, 71–74% — OH O / OH — MeI, K_2CO_3, acetone, reflux, 6 h, 55–64% → OH O / OCH_3

3. Li_2CO_3, MeI, DMF, 55°, 18 h, 54–90% yield.[4]

HO / HO (O) — Li_2CO_3, MeI, DMF, 55°, 18 h, 90% → HO / CH_3O (O)

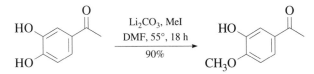

   This method selectively protects phenols with $pK_a \leq 8$ as a result of electron-withdrawing *ortho-* or *para*-substituents.

4. RX, or R'_2SO_4, NaOH, CH_2Cl_2, H_2O, $PhCH_2N^+Bu_3Br^-$, 25°, 2–13 h, 75–95% yield.
   Ar = simple; 2- or 2,6-disubstituted[5,6]

   R = Me, allyl, (epoxide, O)$-CH_2-$, *n*-Bu, *c*-C_5H_{11}, $PhCH_2$, $-CH_2CO_2Et$, R′ = Me, Et
5. Methyl, ethyl, and benzyl ethers have been prepared in the presence of tetraethylammonium fluoride as a Lewis base (alkyl halide, DME, 20°, 3 h, 60–85% yields).[6]

6. $p\text{-}NO_2\text{–}C_6H_4ONa + MeN(NO)CONH_2 \xrightarrow{\text{DME, } 0° \rightarrow 25°\text{, 6 h}}$
$[p\text{-}NO_2\text{–}C_6H_4O^- + CH_2N_2] \longrightarrow p\text{-}NO_2\text{–}C_6H_4OMe$, >90%[7]

7. Phenols protected as *t*-$BuMe_2Si$ ethers can be converted directly to methyl or benzyl ethers (MeI or BnBr, KF, DMF, rt, >90% yield).[8]
8. $TMSCHN_2$, MeOH, MeCN, rt, DIPEA, 31–100% yield.[9]
9. Diazomethane, ether, 80% yield.[10]

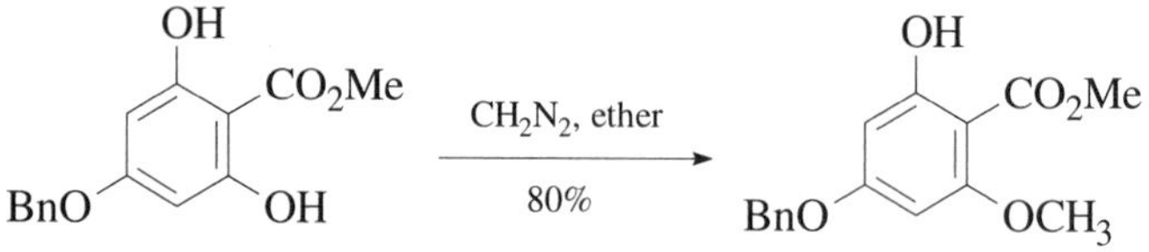

$\xrightarrow[\text{80\%}]{CH_2N_2\text{, ether}}$

10. Dimethyl carbonate, $(Bu_2N)_2C{=}NMe$, 180°, 4.5 h, 54–99% yield.[11] In the presence of this guanidine, aromatic methyl carbonates are converted to methyl ethers with loss of CO_2.

Cleavage

1. Me_3SiI, $CHCl_3$, 25–50°, 12–140 h.[12] Iodotrimethylsilane in quinoline (180°, 70 min) selectively cleaves an aryl methyl group, in 72% yield, in the presence of a methylenedioxy group.[13] Me_3SiI cleaves esters more slowly than ethers and cleaves alkyl aryl ethers (48 h, 25°) more slowly than alkyl alkyl ethers (1.3–48 h, 25°), but benzyl, trityl, and *t*-butyl ethers are cleaved quite rapidly (0.1 h, 25°).[12]
2. Toluene, potassium, 18-crown-6, 100% yield.[14] Tetrahydrofuran can also be used as the solvent in this process.[15]
3. Sodium, liquid ammonia.[16] The utility of this method depends on the nature of the substituents on the aromatic ring. Rings containing electron-withdrawing groups will be reduced, as in the classic Birch reduction.
4. EtSNa, DMF, reflux, 3 h, 94–98% yield.[17,18] Potassium thiophenoxide has been used to cleave an aryl methyl ether without causing migration of a double bond.[19] Sodium benzylselenide ($PhCH_2SeNa$) and sodium thiocresolate ($p\text{-}CH_3C_6H_4SNa$) cleave dimethoxyaryl compounds regioselectively, reportedly because of steric factors in the former case[20] and electronic factors in the latter case.[21]

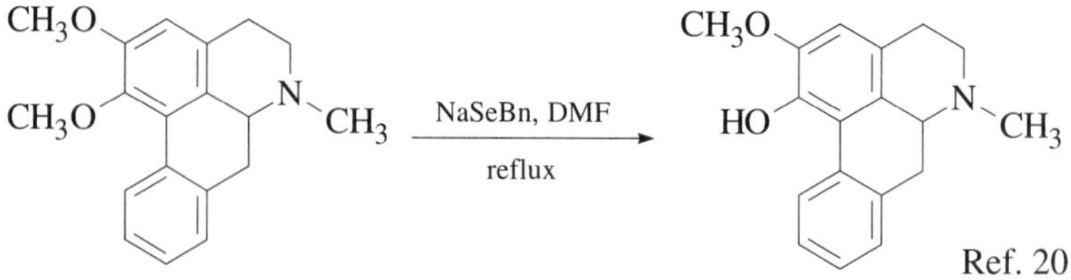

$\xrightarrow[\text{reflux}]{\text{NaSeBn, DMF}}$

Ref. 20

5. Sodium ethanethiolate has been examined for the selective cleavage of aryl methyl ethers. Methyl ethers *para* to an electron-withdrawing group are cleaved preferentially.[22]

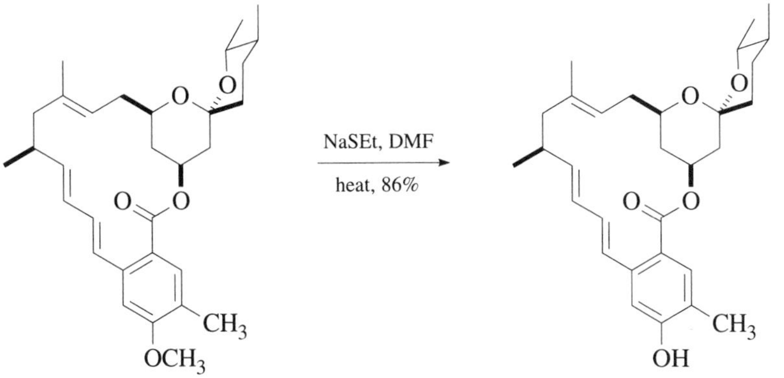

NaSEt, DMF

heat, 86%

Ref. 23

EtMgBr, THF, 0°
EtSNa, DMF, reflux

71%

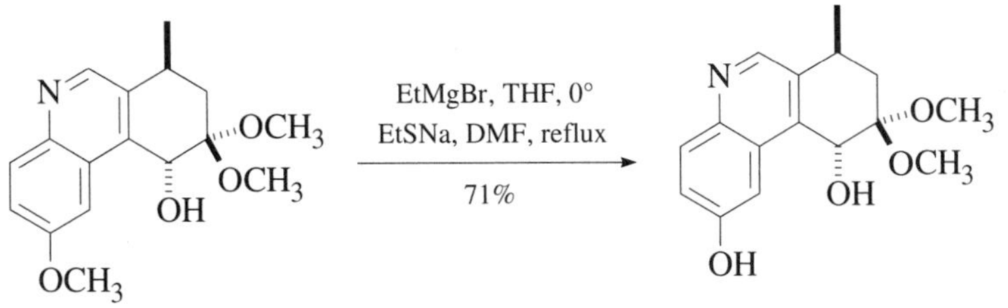

Ref. 24

In this case, the magnesium alkoxide protects the ketal from cleavage.[24]

6. Sodium sulfide in *N*-methylpyrrolidone, NMP, (140°, 2–4 h) cleaves aryl methyl ethers in 78–85% yield.[25]
7. Me_3SiSNa, DMPU, 185°, 78–95% yield.[26]
8. PhSH, catalytic K_2CO_3, NMP, 60–97% yield.[27]
9. Lithium diphenyphosphide (THF, 25°, 2 h; HCl, H_2O, 87% yield) selectively cleaves an aryl methyl ether in the presence of an aryl ethyl ether.[28] It also cleaves a phenyl benzyl ether and a phenyl allyl ether to the phenol in 88% and 78% yield, respectively.[29,30]
10. $(TMS)_2NNa$ or LDA, THF, DMPU, 185°, 80–91% yield.[31]
11. DMSO, NaCN, 125–180°, 5–48 h, 65–90% yield.[32] This cleavage reaction is successful for aromatic systems containing ketones, amides, and carboxylic acids; mixtures are obtained from nitro-substituted aromatic compounds; there is no reaction with 5-methoxyindole (180°, 48 h).
12. LiI, collidine, reflux, 10 h, quant.[33] Aryl ethyl ethers are cleaved more slowly; dialkyl ethers are stable to these conditions.

13. LiI, quinoline, 140–180°, 10–30 min, 65–88% yield.[34]

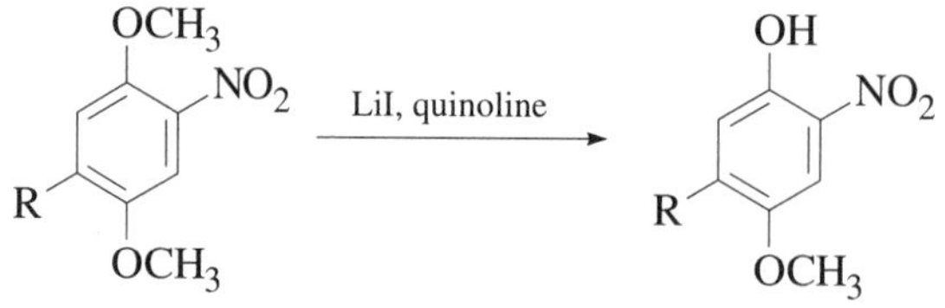

14. $AlBr_3$, EtSH, 25°, <1 h, 94% yield.[35] Both methyl aryl and methyl alkyl ethers are cleaved under these conditions. A methylenedioxy group, used to protect a catechol, is cleaved under similar conditions in satisfactory yields; methyl and ethyl esters are stable (0–20°, 2 h)[35].
15. $AlCl_3$, $HSCH_2CH_2SH$.[36]
16. $AlBr_3$, CH_3CN.[37]

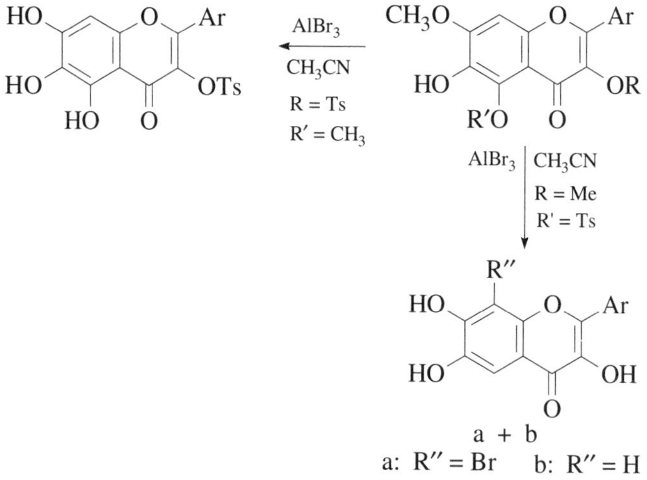

a + b

a: R″ = Br b: R″ = H

17. Regioselective cleavage of dimethoxyaryl derivatives with methanesulfonic acid/methionine has been reported.[38]
18. BBr_3, CH_2Cl_2, −80° → 20°, 12 h, 77–86% yield.[39] Methylenedioxy groups and diphenyl ethers are stable to these cleavage conditions. Benzyloxycarbonyl and *t*-butoxycarbonyl groups, benzyl esters,[40] and 1,3-dioxolanes are cleaved with this reagent. Boron tribromide is reported to be more effective than iodotrimethylsilane for cleaving aryl methyl ethers.[41]
19. Boron triiodide rapidly cleaves methyl ethers of *o*-, *m*-, or *p*-substituted aromatic aldehydes (0°, 25°; 0.5–5 min; 40–86% yield).[42] BI_3 complexed with *N,N*-diethylaniline is similarly effective, but benzyl ethers are converted to the iodide.[43]
20. $BBr_3 \cdot S(CH_3)_2$, $ClCH_2CH_2Cl$, 83°, 50–99% yield.[44] The advantage of this method is that the reagent is a stable, easily handled solid. Methylenedioxy groups are also cleaved by this reagent.

21. 9-Bromo-9-borabicyclo[3.3.0]nonane (9-Br–BBN), CH_2Cl_2, reflux, 87–100% yield.[45] 9-Br–BBN also cleaves dialkyl ethers, allyl aryl ethers, and methylenedioxy groups.
22. BH_2Cl–DMS, toluene, reflux, 95% yield. Acetonides and THP ethers are cleaved, and epoxides are converted to the chlorohydrin.[46]
23. Me_2Br, CH_2Cl_2, 70°, 30–36 h, 72–96% yield.[47] Alkyl methyl ethers are also cleaved, but tertiary methyl ethers are converted to the bromide.
24. 2-Bromo-1,3,2-benzodioxaborole, CH_2Cl_2 (cat. $BF_3 \cdot Et_2O$), 25°, 0.5–36 h, 95–98% yield. Aryl benzyl ethers, methyl esters, and aromatic benzoates are also cleaved.[48]
25. Pyr·HCl, 220°, 6 min, 34% yield of morphine from codeine.[49]
26. Excess MeMgI, 155–165°, 15 min, 80% yield.[50]
27. Sodium *N*-methylanilide, xylene, HMPA, 60–120°, 70–95% yield. Methyl ethers of polyhydric phenols are cleaved to give the monophenol.[51] Benzyl ethers are also cleaved. Halogenated phenols are not effectively cleaved, because of competing aromatic substitution.
28. L-Selectride or Super Hydride, 67°, 88–92% yield.[52] Other methods for converting thebaine to oripavine have not been successful.[53]

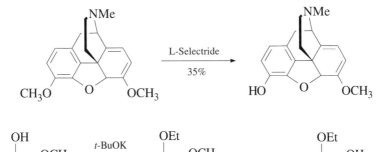

NMe, CH_3O, O, OCH_3 → (L-Selectride, 35%) → NMe, HO, O, OCH_3

29. OH, OCH_3 ← (*t*-BuOK, TDA-1, microwaves, 90%) ← OEt, OCH_3 → (KF, Al_2O_3, 5 h, 210–215°, $HOCH_2CH_2OH$) → OEt, OH

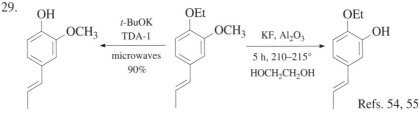

Refs. 54, 55

The loss of the ethyl group probably occurs by an E-2 elimination, whereas methyl cleavage occurs by an S_N-2 process.

30. 48% HBr, AcOH, reflux, 30 min, 85%.[56] The efficiency of this method is significantly improved if a phase transfer catalyst (n-$C_{16}H_{33}P^+Bu_3\ Br^-$) is added to the mixture.[57] Methods that use HBr for ether cleavage can give bromides in the presence of benzylic alcohols.[58]
31. 48% HBr, $Bu_4N^+Br^-$, 100°, 6 h, 80–98% yield.[59]
32. HBr, NaI, 90–94°, sealed tube, 90% yield.[60]
33. TFA, thioanisole, TfOH, 2 h, 0°, 87% yield.[60]

34. BCl_3, CH_2Cl_2, –20°, 94% yield.[61]

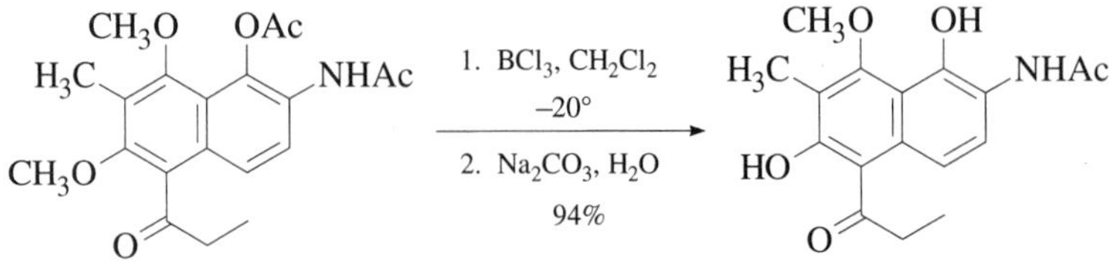

Either an aryl methyl ether or a methylenedioxy group can be cleaved with boron trichloride under various conditions.[62]

35. $AlCl_3$, 3 h, 0°, 75% yield.[63,64] A selectivity study on the demethylation of polymethoxy-substituted acetophenones has been performed using $AlCl_3$ in CH_3CN.[65]

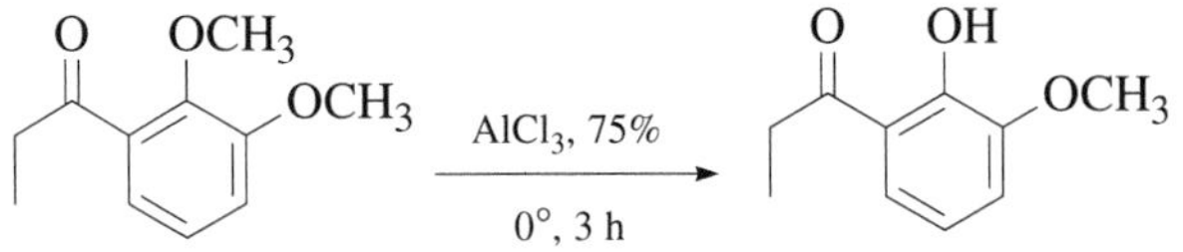

36. LiCl, DMF, heat, 4–72 h.[66]

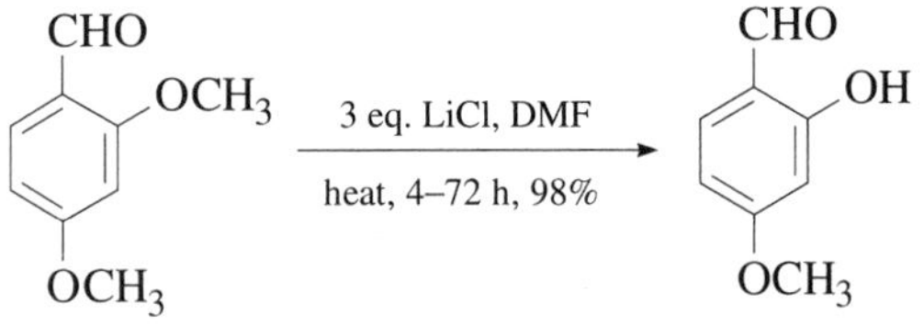

37. CF_3SO_3H, PhSMe, 0–25°.[67,68] In this case *O*-methyltyrosine was deprotected without evidence of O→C migration, which is often a problem when removing protective groups from tyrosine.

38. Ceric ammonium nitrate converts a 1,4-dimethoxy aromatic compound to the quinone, which is reduced with sodium dithionite to give a deprotected hydroquinone.[69]

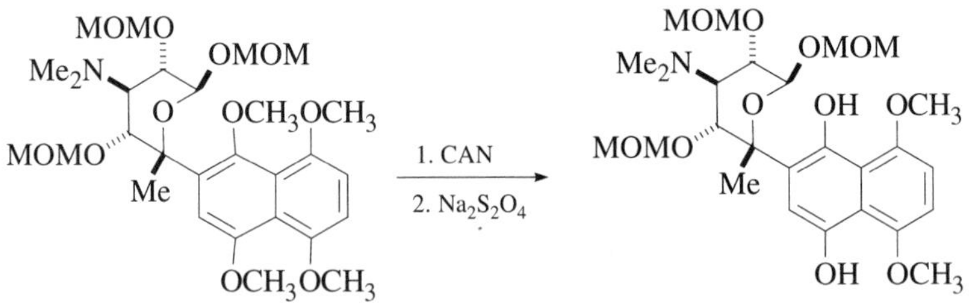

39.

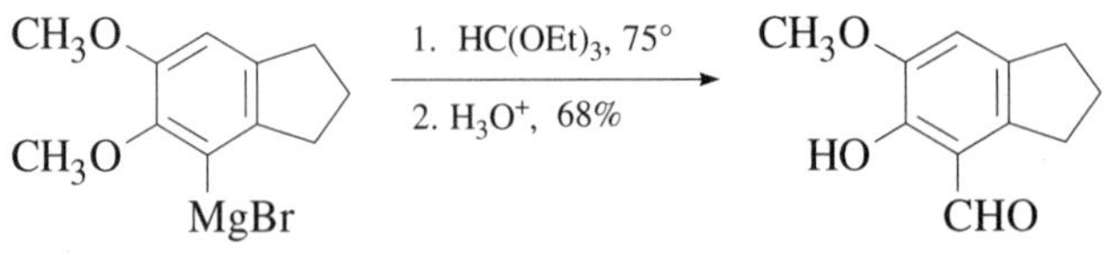

Ref. 70

1. D. L. J. Clive, M. Cantin, A. Khodabocus, X. Kong, and Y. Tao, *Tetrahedron*, **49**, 7917 (1993); D. L. J. Clive, A. Khodabocus, M. Cantin, and Y. Tao, *J. Chem. Soc., Chem. Commun.*, 1755 (1991).
2. G. N. Vyas and N. M. Shah, *Org. Synth., Collect. Vol. IV*, 836 (1963).
3. A. R. MacKenzie, C. J. Moody, and C. W. Rees, *Tetrahedron*, **42**, 3259 (1986).
4. W. E. Wymann, R. Davis, J. W. Patterson, Jr., and J. R. Pfister, *Synth. Commun.*, **18**, 1379 (1988).
5. A. McKillop, J.-C. Fiaud, and R. P. Hug, *Tetrahedron*, **30**, 1379 (1974).
6. J. M. Miller, K. H. So, and J. H. Clark, *Can. J. Chem.*, **57**, 1887 (1979).
7. S. M. Hecht and J. W. Kozarich, *Tetrahedron Lett.*, 1307 (1973).
8. A. K. Sinhababu, M. Kawase, and R. T. Borchardt, *Tetrahedron Lett.*, **28**, 4139 (1987).
9. T. Aoyama, S. Terasawa, K. Sudo, and T. Shioiri, *Chem. Pharm. Bull.*, **32**, 3759 (1984).
10. F. Bracher and B. Schulte, *J. Chem. Soc., Perkin Trans. 1*, 2619 (1996).
11. G. Barcelo, D. Grenouillat, J. P. Senet, and G. Sennyey, *Tetrahedron*, **46**, 1839 (1990).
12. M. E. Jung and M. A. Lyster, *J. Org. Chem.*, **42**, 3761 (1977).
13. J. Minamikawa and A. Brossi, *Tetrahedron Lett.*, 3085 (1978).
14. T. Ohsawa, K. Hatano, K. Kayoh, J. Kotabe, and T. Oishi, *Tetrahedron Lett.*, **33**, 5555 (1992).
15. U. Azzena, T. Denurra, G. Melloni, E. Fenude, and G. Rassa, *J. Org. Chem.*, **57**, 1444 (1992).
16. A. J. Birch, *Quart. Rev.*, **4**, 69 (1950).
17. G. I. Feutrill and R. N. Mirrington, *Tetrahedron Lett.*, 1327 (1970); *idem, Aust. J. Chem.*, **25**, 1719, 1731 (1972).
18. A. S. Kende and J. P. Rizzi, *Tetrahedron Lett.*, **22**, 1779 (1981).
19. J. W. Wildes, N. H. Martin, C. G. Pitt, and M. E. Wall, *J. Org. Chem.*, **36**, 721 (1971).
20. R. Ahmad, J. M. Saá, and M. P. Cava, *J. Org. Chem.*, **42**, 1228 (1977).
21. C. Hansson and B. Wickberg, *Synthesis*, 191 (1976).
22. J. A. Dodge, M. G. Stocksdale, K. J. Fahey, and C. D. Jones, *J. Org. Chem.*, **60**, 739 (1995).
23. A. B. Smith, III, S. R. Schow, J. D. Bloom, A. S. Thompson, and K. N. Winzenberg, *J. Am. Chem. Soc.*, **104**, 4015 (1982).
24. A. G. Myers, N. J. Tom, M. E. Fraley, S. B. Cohen, and D. J. Mader, *J. Am. Chem. Soc.*, **119**, 6072 (1997).
25. M. S. Newman, V. Sankaran, and D. R. Olson, *J. Am. Chem. Soc.*, **98**, 3237 (1976).
26. J. R. Hwu and S.-C. Tsay, *J. Org. Chem.*, **55**, 5987 (1990).
27. M. K. Nayak and A. K. Chakraborti, *Tetrahedron Lett.*, **38**, 8749 (1997).
28. R. E. Ireland and D. M. Walba, *Org. Synth., Collect. Vol. VI*, 567 (1988).
29. F. G. Mann and M. J. Pragnell, *Chem. Ind. (London)*, 1386 (1964).
30. H. Meier and U. Dullweber, *Tetrahedron Lett.*, **37**, 1191 (1996).
31. J. R. Hwu, F. F. Wong, J.-J. Huang, and S.-C. Tsay, *J. Org. Chem.*, **62**, 4097 (1997).
32. J. R. McCarthy, J. L. Moore, and R. J. Crege, *Tetrahedron Lett.*, 5183 (1978).

33. I. T. Harrison, *J. Chem. Soc., Chem. Commun.*, 616 (1969).
34. K. Kirschke and E. Wolff, *J. Prakt. Chem./Chem. Ztg.*, **337**, 405 (1995).
35. M. Node, K. Nishide, K. Fuji, and E. Fujita, *J. Org. Chem.*, **45**, 4275 (1980).
36. T. Inaba, I. Umezawa, M. Yuasa, T. Inoue, S. Mihashi, H. Itokawa, and K. Ogura, *J. Org. Chem.*, **52**, 2957 (1987).
37. T. Horie, T. Kobayashi, Y. Kawamura, I. Yoshida, H. Tominaga, and K. Yamashita, *Bull. Chem. Soc. Jpn.*, **68**, 2033 (1995).
38. N. Fujii, H. Irie, and H. Yajima, *J. Chem . Soc., Perkin Trans. 1*, 2288 (1977).
39. J. F. W. McOmie and D. E. West, *Org. Synth., Collect. Vol. V*, 412 (1973).
40. A. M. Felix, *J. Org. Chem.*, **39**, 1427 (1974).
41. E. H. Vickery, L. F. Pahler, and E. J. Eisenbraun, *J. Org. Chem.*, **44**, 4444 (1979).
42. J. M. Lansinger and R. C. Ronald, *Synth. Commun.*, **9**, 341 (1979).
43. C. Narayana, S. Padmanabhan, and G. W. Kabalka, *Tetrahedron Lett.*, **31**, 6977 (1990).
44. P. G. Williard and C. B. Fryhle, *Tetrahedron Lett.*, **21**, 3731 (1980).
45. M. V. Bhatt, *J. Organomet. Chem.*, **156**, 221 (1978).
46. P. Bovicelli, E. Mincione, and G. Ortaggi, *Tetrahedron Lett.*, **32**, 3719 (1991).
47. Y. Guindon, C. Yoackim, and H. E. Morton, *Tetrahedron Lett.*, **24**, 2969 (1983).
48. P. F. King and S. G. Stroud, *Tetrahedron Lett.*, **26**, 1415 (1985).
49. M. Gates and G. Tschudi, *J. Am. Chem. Soc.*, **78**, 1380 (1956).
50. R. Mechoulam and Y. Gaoni, *J. Am. Chem. Soc.*, **87**, 3273 (1965).
51. B. Loubinoux, G. Coudert, and G. Guillaumet, *Synthesis*, 638 (1980).
52. G. Majetich, Y. Zhang, and K. Wheless, *Tetrahedron Lett.*, **35**, 8727 (1994).
53. A. Coop, J. W. Lewis, and K. C. Rice, *J. Org. Chem.*, **61**, 6774 (1996).
54. A. S. Radhakrishna, K. R. K. P. Rao, S. K. Suri, K. Sivaprakash, and B. B. Singh, *Synth. Commun.*, **21**, 379 (1991).
55. A. Oussaïd, L. N. Thach, and A. Loupy, *Tetrahedron Lett.*, **38**, 2451 (1997).
56. I. Kawasaki, K. Matsuda, and T. Kaneko, *Bull. Chem. Soc. Jpn.*, **44**, 1986 (1971).
57. D. Landini, F. Montanari, and F. Rolla, *Synthesis*, 771 (1978).
58. A. Kamai and N. L. Gayatri, *Tetrahedron Lett.*, **37**, 3359 (1996).
59. K. Hwang and S. Park, *Synth. Commun.*, **23**, 2845 (1993).
60. G. Li, D. Patel, and V. J. Hruby, *Tetrahedron Lett.*, **34**, 5393 (1993).
61. H. Nagaoka, G. Schmid, H. Iio, and Y. Kishi, *Tetrahedron Lett.*, **22**, 899 (1981).
62. M. Gerecke, R. Borer, and A. Brossi, *Helv. Chim. Acta*, **59**, 2551 (1976).
63. K. A. Parker and J. J. Petraitis, *Tetrahedron Lett.*, **22**, 397 (1981).
64. T.-t. Li and Y. L. Wu, *J. Am. Chem. Soc.*, **103**, 7007 (1981).
65. Y. Kawamura, H. Takatsuki, F. Torii, and T. Horie, *Bull. Chem. Soc. Jpn.*, **67**, 511 (1994).
66. A. M. Bernard, M. R. Ghiani, P. P. Piras, and A. Rivoldini, *Synthesis*, 287 (1989).
67. Y. Kiso, S. Nakamura, K. Ito, K. Ukawa, K. Kitagawa, T. Akita, and H. Moritoki, *J. Chem. Soc., Chem. Commun.*, 971 (1979).
68. Y. Kiso, K. Ukawa, S. Nakamura, K. Ito, and T. Akita, *Chem. Pharm. Bull.*, **28**, 673 (1980).
69. M. Kawaski, F. Matsuda, and S. Terashima, *Tetrahedron*, **44**, 5713 (1988).

70. P. Deslongchamps, A. Bélanger, D. J. F. Berney, H. J. Borschberg, R. Brousseau, A. Doutheau, R. Durand, H. Katayama, R. Lapalme, D. M. Leturc, C.-C. Liao, F. N. MacLachan, J.-P. Maffrand, F. Marazza, R. Martino, C. M. L. Ruest, L. Saint-Laurent, and R. S. et P. Soucy, *Can. J. Chem.*, **68**, 115 (1990).

Methoxymethyl (MOM) Ether: $ArOCH_2OCH_3$ (Chart 4)

Formation

1. $ClCH_2OCH_3$, CH_2Cl_2, NaOH–H_2O, Adogen (phase transfer cat.), 20°, 20 min, 80–95% yield.[1,2] This method has been used to protect selectively a phenol in the presence of an alcohol.[3]
2. $ClCH_2OCH_3$, CH_3CN, 18-crown-6, 80% yield.[4]
3. $CH_3OCH_2OCH_3$ TsOH, CH_2Cl_2, mol. sieves, N_2, reflux, 12 h, 60–80% yield.[5] This method of formation avoids the use of the carcinogen chloromethyl methyl ether.
4. $ClCH_2OCH_3$, acetone, K_2CO_3, 86% yield.[6]
5. $ClCH_2OCH_3$, DMF, NaH, 93% yield.[6]

Cleavage

1. HCl, *i*-PrOH, THF, 25°, 12 h, quant.[5]
2. 2 *N* HOAc, 90°, 40 h, high yield.[7] The group has been used in a synthesis of 13-desoxydelphonine from *o*-cresol, a synthesis that required the group to be stable to many reagents.[8]
3. NaI, acetone, cat. HCl, 50°, 85% yield.[9]
4. P_2I_4, CH_2Cl_2, 0° $\rightarrow$ rt, 30 min, 70–90% yield.[10] This method is also effective for removal of the SEM and MEM groups.
5. $(EtO)_3SiCl$, NaI, CH_3CN, CH_2Cl_2, $-5°$, 0.5 h, 74% yield. This method was reported to work better than TMSI.[11] TBDPS groups were not affected by this reagent.
6. TMSBr, CH_2Cl_2, 30° $\rightarrow$ 0°, 87% yield.[12]

1. F. R. van Heerden, J. J. van Zyl, G. J. H. Rall, E. V. Brandt, and D. G. Roux, *Tetrahedron Lett.*, 661 (1978).
2. W. R. Roush, D. S. Coffey, and D. J. Madar, *J. Am. Chem. Soc.*, **119**, 11331 (1997).
3. T. R. Kelly, C. T. Jagoe, and Q. Li, *J. Am. Chem. Soc.*, **111**, 4522 (1989).
4. G. J. H. Rall, M. E. Oberholzer, D. Ferreira, and D. G. Roux, *Tetrahedron Lett.*, 1033 (1976).
5. J. P. Yardley and H. Fletcher, III, *Synthesis*, 244 (1976).
6. M. Süsse, S. Johne, and M. Hesse, *Helv. Chim. Acta*, **75**, 457 (1992).
7. M. A. A.-Rahman, H. W. Elliott, R. Binks, W. Küng, and H. Rapoport, *J. Med. Chem.*, **9**, 1 (1966).
8. K. Wiesner, *Pure Appl. Chem.*, **51**, 689 (1979).

9. D. R. Williams, B. A. Barner, K. Nishitani, and J. G. Phillips, *J. Am. Chem. Soc.*, **104**, 4708 (1982).
10 H. Saimoto, Y. Kusano, and T. Hiyama, *Tetrahedron Lett.*, **27**, 1607 (1986).
11. J. R. Falck, K. K. Reddy, and S. Chandrasekhar, *Tetrahedron Lett.*, **38**, 5245 (1997).
12. J. W. Huffman, X. Zhang, M.-J. Wu, H. H. Joyner, and W. T. Pennington, *J. Org. Chem.*, **56**, 1481 (1991).

Benzyloxymethyl (BOM) Ether: $C_6H_5CH_2OCH_2OAr$

Formation

1. BOMCl, NaH, DMF, >81% yield.[1]

Cleavage

1. MeOH, Dowex 50W-X8 (H^+), 90% yield.[1]

1. W. R. Roush, M. R. Michaelides, D. F. Tai, B. M. Lesur, W. K. M. Chong, and D. J. Harris, *J. Am. Chem. Soc.*, **111**, 2984 (1989).

Methoxyethoxymethyl (MEM)Ether: $ArOCH_2OCH_2CH_2OCH_3$ (Chart 4)

In an attempt to metalate a MEM-protected phenol with BuLi, the methoxy group was eliminated, forming the vinyloxymethyl ether. This was attributed to intramolecular proton abstraction.[1]

A 2-methoxyethoxymethyl ether was used to protect one phenol group during a total synthesis of gibberellic acid.[2]

Formation

1. NaH, THF, 0°; $MeOCH_2CH_2OCH_2Cl$, 0° → 25°, 2 h, 75% yield.[2]
2. $MeOCH_2CH_2OCH_2Cl$, DIPEA.[3]

Cleavage

1. CF_3CO_2H, CH_2Cl_2, 23°, 1 h, 74% yield.[2]
2. $(Ipc)_2BCl$, THF, 0°, 80 h. Cleavage occurred during the reduction of an acetophenone.[3]
3. For other methods of cleavage, the chapter on alcohol protection should be consulted.

1. J. Mayrargue, M. Essamkaoui, and H. Moskowitz, *Tetrahedron Lett.*, **30**, 6867 (1989).
2. E. J. Corey, R. L. Danheiser, S. Chandrasekaran, P. Siret, G. E. Keck, and J.-L. Gras, *J. Am. Chem. Soc.*, **100**, 8031 (1978).
3. E. T. Everhart and J. C. Craig, *J. Chem. Soc., Perkin Trans. 1*, 1701 (1991).

2-(Trimethylsilyl)ethoxymethyl (SEM) Ether:
$(CH_3)_3SiCH_2CH_2OCH_2OAr$

Formation

1. SEMCl, DMAP, Et_3N, benzene, reflux, 3 h, 98% yield.[1]
2. SEMCl, $(i\text{-Pr})_2NEt$, CH_2Cl_2, 97% yield.[3]

Cleavage

1. $Bu_4N^+F^-$, HMPA, 40°, 2 h, >23–51% yield.[2]
2. H_2SO_4, MeOH, THF, 90% yield.[1]
3. P_2I_4, CH_2Cl_2, 0° → rt, 30 min, 62–86% yield.[3,4] These conditions also cleave methoxymethyl and methoxyethoxymethyl ethers.

1. T. L. Shih, M. J. Wyvratt, and H. Mrozik, *J. Org. Chem.*, **52**, 2029 (1987).
2. A. Leboff, A.-C. Carbonnelle, J.-P. Alazard, C. Thal, and A. S. Kende, *Tetrahedron Lett.*, **28**, 4163 (1987).
3. H. Saimoto, Y. Kusano, and T. Hiyama, *Tetrahedron Lett.*, **27**, 1607 (1986).
4. H. Saimoto, S.-i. Ohrai, H. Sashiwa, Y. Shigemasa, and T. Hiyama, *Bull. Chem. Soc. Jpn.*, **68**, 2727 (1995).

Methylthiomethyl (MTM) Ether: $ArOCH_2SCH_3$ (Chart 4)

Formation

1. NaOH, $ClCH_2SMe$, HMPA, 25°, 16 h, 91–94% yield.[1]

Cleavage

1. $HgCl_2$, $CH_3CN–H_2O$, reflux, 10 h, 90–95% yield.[1] Aryl methylthiomethyl ethers are stable to the conditions used to hydrolyze primary alkyl MTM ethers (e.g., $HgCl_2/CH_3CN–H_2O$, 25°, 6 h). They are moderately stable to acidic conditions (95% recovered from HOAc/THF–H_2O, 25°, 4 h).
2. Ac_2O, Me_3SiCl, 25 min, rt, 95% yield.[2]

1. R. A. Holton and R. G. Davis, *Tetrahedron Lett.*, 533 (1977).
2. N. C. Barua, R. P. Sharma, and J. N. Baruah, *Tetrahedron Lett.*, **24**, 1189 (1983).

Phenylthiomethyl (PTM) Ether: $C_6H_5SCH_2OAr$

Formation

1. NaI, $PhSCH_2Cl$, NaH, HMPA, 87–94% yield.[1]

Cleavage

1. $CH_3CN:H_2O$ (4:1), $HgCl_2$, 24 h, 90–94% yield. The methylthiomethyl ether group can be removed in the presence of the phenylthiomethyl ether.[1]

1. R. A. Holton and R. V. Nelson, *Synth. Commun.*, **10**, 911 (1980).

Azidomethyl Ether: N_3CH_2OAr

The azidomethyl ether, used to protect phenols and prepared by the displacement of azide on the chloromethylene group, is cleaved reductively with $LiAH_4$ or by hydrogenolysis (Pd–C, H_2). It is stable to strong acids, permanganate, and free-radical brominations.[1]

1. B. Loubinoux, S. Tabbache, P. Gerardin, and J. Miazimbakana, *Tetrahedron*, **44**, 6055 (1988).

Cyanomethyl Ether: $ArOCH_2CN$

The cyanomethyl ether, formed from bromoacetonitrile (acetone, K_2CO_3, 97–100% yield), is cleaved by hydrogenation of the nitrile with PtO_2 in EtOH, in 98% yield.[1] The method has also been used for the protection of amines and carbamates.

1. A. Benarab, S. Boye, L. Savelon, and G. Guillaumet, *Tetrahedron Lett.*, **34**, 7567 (1993).

2,2-Dichloro-1,1-difluoroethyl Ether: $CHCl_2CF_2OAr$

Formation/Cleavage

$$\text{ArOH} \underset{\text{6\% KOH, H}_2\text{O, DMSO, rt, 85\%}}{\overset{\text{F}_2\text{C=CCl}_2\text{, 40\% KOH, Bu}_4\text{N}^+\text{HSO}_4^-\text{, 92\%}}{\rightleftarrows}} \text{ArOCF}_2\text{CHCl}_2$$

This group decreases the electron density on the aromatic ring and thus inhibits solvolysis of the tertiary alcohol **i** and the derived acetate **ii**.[1]

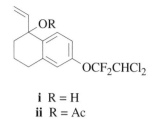

OR

OCF_2CHCl_2

i R = H
ii R = Ac

1. S. G. Will, P. Magriotis, E. R. Marinelli, J. Dolan, and F. Johnson, *J. Org. Chem.*, **50**, 5432 (1985).

2-Chloro- and 2-Bromoethyl Ether: XCH_2CH_2OAr, X=Cl, Br

These ethers can be removed from naphthohydroquinones either by elimination to the vinyl ether followed by hydrolysis or by Finklestein reaction with iodide followed by reduction with zinc.[1]

1. H. Laatsch, *Z. Naturforsch., B: Anorg. Chem., Org. Chem.*, **40B**, 534 (1985).

Tetrahydropyranyl (THP) Ether: ArO–2-tetrahydropyranyl

The tetrahydropyranyl ether, prepared from a phenol and dihydropyran (HCl/EtOAc, 25°, 24 h) is cleaved by aqueous oxalic acid (MeOH, 50–90°, 1–2 h).[1] Tonsil, Mexican Bentonite earth,[2] HSZ Zeolite,[3] and $H_3[PW_{12}O_{40}]$[4] have also been used for the tetrahydropyranylation of phenols. The use of $[Ru(ACN)_3(triphos)](OTf)_2$ in acetone selectively removes the THP group from a phenol in the presence of an alkyl THP group. Ketals of acetophenones are also cleaved.[5]

1-Ethoxyethyl (EE) Ether: $ArOCH(OC_2H_5)CH_3$

The ethoxyethyl ether is prepared by acid catalysis from a phenol and ethyl vinyl ether and is cleaved by acid-catalyzed methanolysis.[6]

1. H. N. Grant, V. Prelog, and R. P. A. Sneeden, *Helv. Chim. Acta*, **46**, 415 (1963).
2. R. Cruz-Almanza, F. J. Pérez-Floress, and M. Avila, *Synth. Commun.*, **20**, 1125 (1990).
3. R. Ballini, F. Bigi, S. Carloni, R. Maggi, and G. Sartori, *Tetrahedron Lett.*, **38**, 4169 (1997).
4. A. Moinar and T. Beregszaszi, *Tetrahedron Lett.*, **37**, 8597 (1996).
5. S. Ma and L. M. Venanzi, *Tetrahedron Lett.*, **34**, 8071 (1993).
6. J. H. Rigby and M. E. Mateo, *J. Am. Chem. Soc.*, **119**, 12, 655 (1997).

Phenacyl Ether: $ArOCH_2COC_6H_5$ (Chart 4)

4-Bromophenacyl Ether: $ArOCH_2COC_6H_4$-4-Br

Formation

1. $BrCH_2COPh$, K_2CO_3, acetone, reflux, 1–2 h, 85–95% yield.[1]

Cleavage

1. Zn, HOAc, 25°, 1 h, 88–96% yield.[1] Phenacyl and *p*-bromophenacyl ethers of phenols are stable to 1% ethanolic alkali (reflux, 2 h), and to 5 *N* sulfuric acid in ethanol–water. The phenacyl ether, prepared from *β*-naphthol, is cleaved in 82% yield by 5% ethanolic alkali (reflux, 2 h).

1. J. B. Hendrickson and C. Kandall, *Tetrahedron Lett.*, 343 (1970).

Cyclopropylmethyl Ether: ArOCH2–*c*-C_3H_5

For a particular phenol, the authors required a protective group that would be stable to reduction (by complex metals, catalytic hydrogenation, and Birch conditions) and that could be easily and selectively removed.

Formation

1. KO–*t*-Bu, DMF, 0°, 30 min; *c*-$C_3H_5CH_2Br$, 20°, 20 min → 40°, 6 h, 80% yield.[1]

Cleavage

1. aq. HCl, MeOH, reflux, 2 h, 94% yield. [1]

1. W. Nagata, K. Okada, H. Itazaki, and S. Uyeo, *Chem. Pharm. Bull.*, **23**, 2878 (1975).

Allyl Ether: $ArOCH_2CH{=}CH_2$ (Chart 4)

Formation

1. Allyl ethers can be prepared by reaction of a phenol and the allyl bromide in the presence of base.[1]
2. AllylOH, $Pd(OAc)_2$, PPh_3, Ti(O–*i*-Pr)$_4$, 73–87% yield.[2]

Cleavage

1. The section on the cleavage of allyl ethers of alcohols should also be consulted.

2. Pd–C, TsOH, H_2O or MeOH; 60–80°, 6 h, > 95% yield.[3]
3. SeO_2/HOAc, dioxane, reflux, 1 h, 40–75% yield.[4]
4. $NaAlH_2(OCH_2CH_2OCH_3)_2$, $PhCH_3$, reflux, 10 h, 62% yield.[5] An aryl allyl ether is selectively cleaved by this reagent (which also cleaves aryl benzyl ethers) in the presence of an *N*-allylamide.
5. Ph_3P/$Pd(OAc)_2$, HCOOH, 90°, 1 h.[6]
6. Pd° cat., Bu_3SnH, AcOH, *p*-NO_2-phenol.[7]
7. $Pd(Ph_3P)_4$, $LiBH_4$, THF, 88% yield.[8] $NaBH_4$ can also be used as an allyl scavenging agent.[9]
8. $Pd(Ph_3P)_4$, $PhSiH_3$, 20–40 min, 74–100% yield.[10]
9. Bis(benzonitrile)palladium(II) chloride, benzene, reflux, 16–20 h, 86% yield.[11]
10. EtOH, $RhCl_3$, reflux, 86% yield. [1]
11. $LiPPh_2$, THF, 4 h, reflux, 78% yield.[12]
12. $SiCl_4$, NaI, CH_2Cl_2, CH_3CN, 8 h, 84% yield.[13]
13. $NaBH_4$, I_2, THF, 0°, 84–95% yield.[14]
14. Electrolysis, $[Ni(bipy)_3](BF_3)$, Mg anode, DMF, rt, 40–99% yield.[15] Aryl bromides and iodides are reduced under these conditions.
15. Electrolysis, DMF, $Bu_4N^+Br^-$, $SmCl_3$, Mg anode, 67–90% yield.[16]
16. $TiCl_3$, Mg, THF, reflux, 3 h, 70% yield.[17]
17. *t*-BuOK, DMSO, 92% yield; MeOH, HCl, >75% yield.[18]
18. Chromium-pillared clay, *t*-BuOOH, CH_2Cl_2, 10 h, 80% yield. Simple allyl ethers are cleaved to give ketones, and allylamines are also deprotected (84–90% yield).[19]
19. Li, naphthalene, THF, 51–91% yield.[20]

1. See, for example; S. F. Martin and P. J. Garrison, *J. Org. Chem.*, **47**, 1513 (1982).
2. T. Satoh, M. Ikeda, M. Miura, and M. Nomura, *J. Org. Chem.*, **62**, 4877 (1997).
3. R. Boss and R. Scheffold, *Angew. Chem., Int. Ed. Engl.*, **15**, 558 (1976).
4. K. Kariyone and H. Yazawa, *Tetrahedron Lett.*, 2885 (1970).
5. T. Kametani, S.-P. Huang, M. Ihara, and K. Fukumoto, *J. Org. Chem.*, **41**, 2545 (1976).
6. H. Hey and H.-J. Arpe, *Angew. Chem., Int. Ed. Engl.*, **12**, 928 (1973).
7. P. Four and F. Guibe, *Tetrahedron Lett.*, **23**, 1825 (1982).
8. M. Bois-Choussy, L. Neuville, R. Beugelmans, and J. Zhu, *J. Org. Chem.*, **61**, 9309 (1996).
9. R. Beugelmans, S. Bourdet, A. Bigot, and J. Zhu, *Tetrahedron Lett.*, **35**, 4349 (1994).
10. M. Dessolin, M.-G. Guillerez, N. Thieriet, F. Guibé, and A. Loffet, *Tetrahedron Lett.*, **36**, 5741 (1995).
11. J. M. Bruce and Y. Roshan-Ali, *J. Chem. Res., Synop.*, 193 (1981).

12. F. G. Mann and M. J. Pragnell, *J. Chem. Soc.*, 4120 (1965).
13. M. V. Bhatt and S. S. El-Morey, *Synthesis*, 1048 (1982).
14. R. M. Thomas, G. H. Mohan, and D. S. Iyengar, *Tetrahedron Lett.*, **38**, 4721 (1997).
15. S. Olivero and E. Duñach, *J. Chem. Soc., Chem. Commun.*, 2497 (1995).
16. B. Espanet, E. Duñach, and J. Perichon, *Tetrahedron Lett.*, **33**, 2485 (1992).
17. S. M. Kadam, S. K. Nayak, and A. Banerji, *Tetrahedron Lett.*, **33**, 5129 (1992).
18. F. Effenberger and J. Jäger, *J. Org. Chem.*, **62**, 3867 (1997).
19. B. M. Choudary, A. D. Prasad, V. Swapna, V. L. K. Valli, and V. Bhuma, *Tetrahedron*, **48**, 953 (1992).
20. E. Alonso, D. J. Ramon, and M. Yus, *Tetrahedron*, **42**, 14355 (1997).

Propargyl Ether: $HC{\equiv}CCH_2OAr$

Cleavage

1. Electrolysis, Ni(II), Mg anode, DMF, rt 77–99% yield. This method is not compatible with halogenated phenols, because of competing halogen cleavage.[1]
2. $TiCl_3$, Mg, THF, 54–92% yield.[2]

1. S. Olivero and E. Duñach, *Tetrahedron Lett.*, **38**, 6193 (1997).
2. S. K. Nayak, S. M. Kadam, and A. Banerji, *Synlett*, 581 (1993).

Isopropyl Ether: $ArOCH(CH_3)_2$

An isopropyl ether was developed as a phenol protective group that would be more stable to Lewis acids than would be an aryl benzyl ether.[1] The isopropyl group has been tested for use in the protection of the phenolic oxygen of tyrosine during peptide synthesis.[2]

Formation

1. Me_2CHBr, K_2CO_3, DMF, acetone, 20°, 19 h.[1]

Cleavage

1. BCl_3, CH_2Cl_2, 0°, rapid; or $TiCl_4$, CH_2Cl_2, 0°, slower. There was no reaction with $SnCl_4$.[1]
2. $SiCl_4$, NaI, 14 h, CH_2Cl_2, CH_3CN, 80% yield.[3]

1. T. Sala and M. V. Sargent, *J. Chem. Soc., Perkin Trans. 1*, 2593 (1979).
2. See cyclohexyl ether in this section: M. Engelhard and R. B. Merrifield, *J. Am. Chem. Soc.*, **100**, 3559 (1978).
3. M. V. Bhatt and S. S. El-Morey, *Synthesis*, 1048 (1982).

Cyclohexyl Ether: ArO–c-C_6H_{11} (Chart 4)

Formation[1]

$$p\text{-}HOC_6H_4CH_2CHCO_2Me\ (NHCOCF_3) \xrightarrow[\text{CH}_2\text{Cl}_2,\ \text{reflux, 24 h}\ \ 60\%]{\text{cyclohexene, BF}_3\cdot\text{Et}_2\text{O}} p\text{-}c\text{-}C_6H_{11}OC_6H_4CH_2CHCO_2Me\ (NHCOCF_3)$$

Cleavage[1]

1. HF, 0°, 30 min, 100% yield.
2. 5.3 *N* HBr/AcOH, 25°, 2 h, 99% yield. An ether that would not undergo rearrangement to a 3-alkyl derivative during acid-catalyzed removal of –NH protective groups was required to protect the phenol group in tyrosine. Four compounds were investigated: *O*-cyclohexyl-, *O*-isobornyl-, *O*-[1-(5-pentamethylcyclopentadienyl)ethyl]- and *O*-isopropyltyrosine.

The *O*-isobornyl- and *O*-[1-(5-pentamethylcyclopentadienyl)ethyl]- derivatives do not undergo rearrangement, but are very labile in trifluoroacetic acid (100% cleaved in 5 min). The cyclohexyl and isopropyl derivatives are more stable to acid, but undergo some rearrangement. The cyclohexyl group combines minimal rearrangement with ready removal.[1] A comparison has been made with several other common protective groups for tyrosine, and the degree of alkylation *ortho* to the phenolic OH decreases in the order Bn > 2-$ClC_6H_4CH_2$ >2,6-$Cl_2C_6H_3CH_2$ > cyclohexyl > *t*-Bu ~ benzyloxycarbonyl ~ 2-Br-benzyloxycarbonyl.[2]

1. M. Engelhard and R. B. Merrifield, *J. Am. Chem. Soc.*, **100**, 3559 (1978).
2. J. P. Tam, W. F. Heath, and R. B. Merrifield, *Int. J. Pept. Protein Res.*, **21**, 57 (1983).

***t*-Butyl Ether:** $ArOC(CH_3)_3$ (Chart 4)

Formation

1. Isobutylene, cat. concd. H_2SO_4, CH_2Cl_2, 25°, 6–10 h, 93% yield.[1] These conditions also convert carboxylic acids to *t*-Bu esters.
2. Isobutylene, CF_3SO_3H, CH_2Cl_2, −78°, 70–90% yield.[2] These conditions will protect a phenol in the presence of a primary alcohol.
3. *t*-Butyl halide, Pyr, 20–30°, few h, 65–95% yield.[3]

Cleavage

The section on *t*-butyl ethers of alcohols should also be consulted.

1. Anhyd. CF_3CO_2H, 25°, 16 h, 81% yield.[1]
2. CF_3CH_2OH, CF_3SO_3H, −5°, 60 sec, 100% yield.[2]

1. H. C. Beyerman and J. S. Bontekoe, *Recl. Trav. Chim. Pays-Bas*, **81**, 691 (1962).
2. J. L. Holcombe and T. Livinghouse, *J. Org. Chem.*, **51**, 111 (1986).
3. H. Masada and Y. Oishi, *Chem. Lett.*, 57 (1978).

Benzyl Ether: $ArOCH_2C_6H_5$ (Chart 4)

Formation

1. In general, benzyl ethers are prepared from a phenol by treating an alkaline solution of the phenol with a benzyl halide.[1]

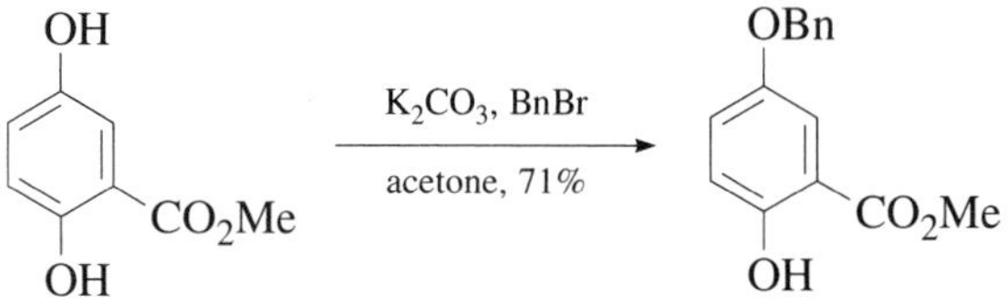

K_2CO_3, BnBr, acetone, 71% — Ref. 2

2. BnCl, NaI, $NaHCO_3$, CH_3CN, 89% — Ref. 3

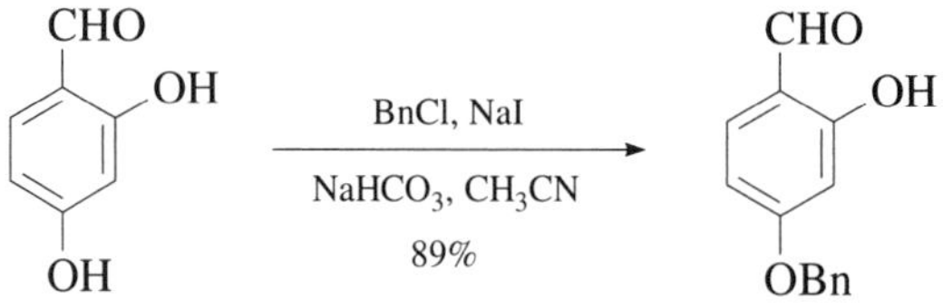

3. $CHCl_3$, MeOH, K_2CO_3, BnBr, 4 h, heat.[4] In this case, some (5:1) selectivity was achieved for a less hindered phenol in the presence of a more hindered one.
4. Benzyl ethers of phenols can also be prepared by reaction with phenyldiazomethane.
5. $(BnO)_2CO$, DMF, 155°, 2 h, 80% yield. Active methylenes are also benzylated.[5]

Cleavage

1. 1. H_2, Pd/C, Ac_2O, AcONa, PhH, 1.5 h; 2. H_2, Pd/C, EtOAc; 3. Ac_2O, Pyr

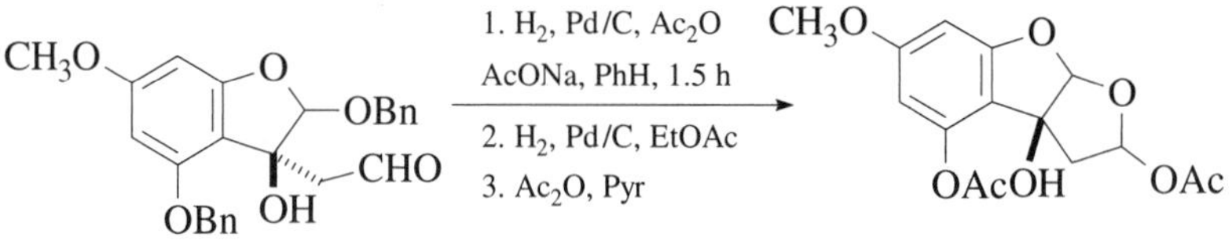

Catalytic hydrogenation in acetic anhydride–benzene removes the aromatic benzyl ether and forms a monoacetate; hydrogenation in ethyl acetate removes the aliphatic benzyl ether to give, after acetylation, the diacetate.[6] Trisubstituted alkenes can be retained during the hydrogenolysis of a phenolic benzyl ether.[7]

2. Pd–C, 1,4-cyclohexadiene, 25°, 1.5 h, 95–100% yield.[8]

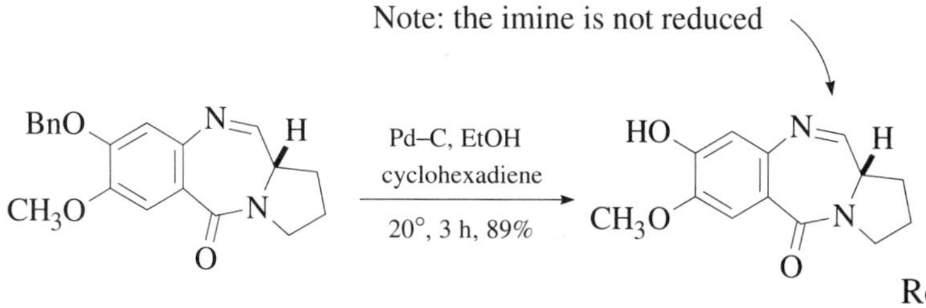

Note: the imine is not reduced

Pd–C, EtOH cyclohexadiene 20°, 3 h, 89%

Ref. 9

3. Palladium black, a more reactive catalyst than Pd–C, must be used to cleave the more stable aliphatic benzyl ethers.[8]
4. Na, *t*-BuOH, 70–80°, 2 h, 78%.[10]

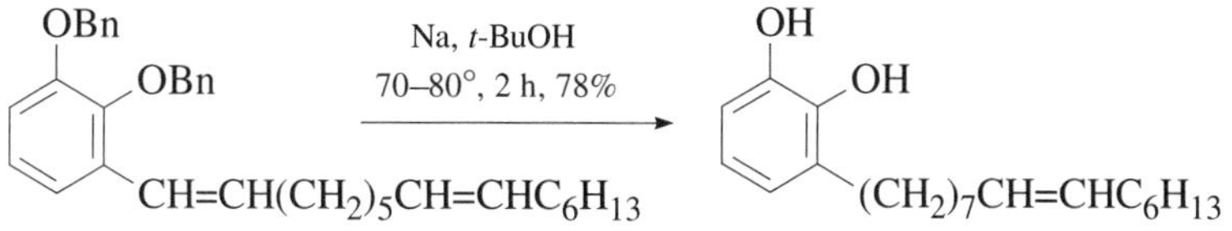

Na, *t*-BuOH 70–80°, 2 h, 78%

$CH{=}CH(CH_2)_5CH{=}CHC_6H_{13}$ → $(CH_2)_7CH{=}CHC_6H_{13}$

In this example, sodium in *t*-butyl alcohol cleaves two aryl benzyl ethers and reduces a double bond that is conjugated with an aromatic ring; non-conjugated double bonds are stable.

5. $BF_3{\cdot}Et_2O$, EtSH, 25°, 40 min, 80–90% yield.[11] Addition of sodium sulfate prevents hydrolysis of a dithioacetal group present in the compound; replacement of ethanethiol with ethanedithiol prevents cleavage of a dithiolane group.
6. CF_3OSO_2F or CH_3OSO_2F, $PhSCH_3$, CF_3CO_2H, 0°, 30 min, 100% yield.[12] Thioanisole suppresses acid-catalyzed rearrangement of the benzyl group to form 3-benzyltyrosine. The more acid-stable 2,6-dichlorobenzyl ether is cleaved in a similar manner.
7. Me_3SiI, CH_3CN, 25–50°, 100% yield.[13] Selective removal of protective groups is possible with this reagent, since a carbamate, $=NCOOCMe_3$, is cleaved in 6 min at 25°; an aryl benzyl ether is cleaved in 100% yield, with no formation of 3-benzyltyrosine, in 1 h at 50°, at which time a methyl ester begins to be cleaved.
8. 2-Bromo-1,3,2-benzodioxaborole, CH_2Cl_2, 95% yield.[14]
9. NaI, $BF_3{\cdot}Et_2O$, 0°, 45 min, rt, 15 min. 75–90% yield.[15]
10. CF_3CO_2H, $PhSCH_3$, 25°, 3 h.[16] The use of dimethyl sulfide or anisole as a cation scavenger was not as effective because of side reactions. Benzyl ethers of serine and threonine were slowly cleaved (30% in 3 h; complete cleavage in 30 h). The use of pentamethylbenzene has been shown to increase the rate of deprotection of *O*-Bn-Tyrosine.[17]
11. $PhNMe_2$, $AlCl_3$, CH_2Cl_2, 78–91% yield.[18]

12. $MgBr_2$, benzene, Et_2O, reflux, 24 h, 63–95% yield.[19]

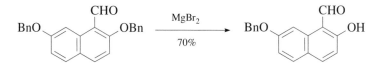

13. Dimethyldioxirane, acetone, 20°, 45 h, 69% yield.[20]
14. Calcium, ammonia, 95% yield.[21]
15. $SnBr_2$, AcBr, CH_2Cl_2, rt, 5–24 h, 76–86% yield. These conditions convert a benzyl ether to the acetate and are effective for alkyl benzyl ethers as well.[22]
16. $TiCl_3$, Mg, THF, reflux, 28–96% yield.[23]
17. TFA, pentamethylbenzene. This method was developed to minimize the formation of 3-benzyltyrosine during the acidolysis of benzyl-protected tyrosine.[24]

1. For example, M. C. Venuti, B. E. Loe, G. H. Jones, and J. M. Young, *J. Med. Chem.*, **31**, 2132 (1988).
2. N. R. Kotecha, S. V. Ley, and S. Montégani, *Synlett*, 395 (1992).
3. W. L. Mendelson, M. Holmes, and J. Dougherty, *Synth. Commun.*, **26**, 593 (1996).
4. H. Schmidhammer and A. Brossi, *J. Org. Chem.*, **48**, 1469 (1983).
5. M. Selva, C. A. Margues, and P. Tundo, *J. Chem. Soc., Perkin Trans. 1*, 1889 (1995).
6. G. Büchi and S. M. Weinreb, *J. Am. Chem. Soc.*, **93**, 746 (1971).
7. A. F. Barrero, E. J. Alvarez-Manzaneda, and R. Chahboun, *Tetrahedron Lett.*, **38**, 8101 (1997).
8. A. M. Felix, E. P. Heimer, T. J. Lambros, C. Tzougraki, and J. Meienhofer, *J. Org. Chem.*, **43**, 4194 (1978).
9. D. E. Thurston, V. S. Murty, D. R. Langley, and G. B. Jones, *Synthesis*, 81 (1990).
10. B. Loev and C. R. Dawson, *J. Am. Chem. Soc.*, **78**, 6095 (1956).
11. K. Fuji, K. Ichikawa, M. Node, and E. Fujita, *J. Org. Chem.*, **44**, 1661 (1979).
12. Y. Kiso, H. Isawa, K. Kitagawa, and T. Akita, *Chem. Pharm. Bull.*, **26**, 2562 (1978).
13. R. S. Lott, V. S. Chauhan, and C. H. Stammer, *J. Chem. Soc., Chem. Commun.*, 495 (1979).
14. P. F. King and S. G. Stroud, *Tetrahedron Lett.*, **26**, 1415 (1985).
15. Y. D. Vankar and C. T. Rao, *J. Chem. Res., Synop.*, 232 (1985).
16. Y. Kiso, K. Ukawa, S. Nakamura, K. Ito, and T. Akita, *Chem. Pharm. Bull.*, **28**, 673 (1980).
17. H. Yoshino, Y. Tsuchiya, I. Saito, and M. Tsujii, *Chem. Pharm. Bull.*, **35**, 3438 (1987).
18. T. Akiyama, H. Hirofuji, and S. Ozaki, *Tetrahedron Lett.*, **32**, 1321 (1991).
19. J. E. Baldwin and G. G. Haraldsson, *Acta. Chem. Scand., Ser. B*, **B40**, 400 (1986).

20. B. A. Marples, J. P. Muxworthy, and K. H. Baggaley, *Synlett*, 646 (1992).
21. J. R. Hwu, Y. S. Wein, and Y.-J. Leu, *J. Org. Chem.*, **61**, 1493 (1996).
22. T. Oriyama, M. Kimura, M. Oda, and G. Koga, *Synlett*, 437 (1993).
23. S. M. Kadam, S. K. Nayak, and A. Banerji, *Tetrahedron Lett.*, **33**, 5129 (1992).
24. H. Yoshino, M. Tsujii, M. Kodama, K. Komeda, N. Niikawa, T. Tanase, N. Asakawa, K. Nose, and K. Yamatsu, *Chem. Pharm. Bull.*, **38**, 1735 (1990).

2,6-Dimethylbenzyl Ether: $2,6\text{-}(CH_3)_2C_6H_3CH_2OAr$

The 2,6-dimethylbenzyl ether is considerably more stable to hydrogenolysis than the benzyl ether. It has a half-life of 15 h at 1 atm of hydrogen in the presence of Pd–C, whereas the benzyl ether has a half-life of ~45 min. This added stability allows hydrogenation of azides, nitro groups, and olefins in the presence of a dimethylbenzyl group.[1]

1. R. Davis and J. M. Muchowski, *Synthesis*, 987 (1982).

4-Methoxybenzyl (MPM–OAr or PMB–OAr) Ether: $4\text{-}CH_3OC_6H_4CH_2OAr$

Formation

1. $MeOC_6H_4CH_2Cl$, $Bu_4N^+I^-$, K_2CO_3, acetone, 55°, 96% yield.[1] Sodium iodide can be used in place of $Bu_4N^+I^-$.[2]
2. $MeOC_6H_4CH_2Br$, $(i\text{-Pr})_2NEt$, CH_2Cl_2, rt, 80% yield.[3]

Cleavage

1. CF_3CO_2H, CH_2Cl_2, 85% yield.[1]
2. Camphorsulfonic acid, $(CH_3)_2C(OCH_3)_2$, rt.[3]
3. $BF_3 \cdot Et_2O$, $NaCNBH_3$, THF, reflux, 6–10 h, 65–77% yield.[4]
4. 18-Crown-6, toluene, K, 2–3 h, 81–96% yield.[5]
5. Acetic acid, 90°, 89–96% yield.[6] Benzyl groups are not affected by these conditions.
6. DDQ, 35% yield.[7] The DDQ-promoted cleavage of phenolic MPM ethers can be complicated by overoxidation, especially with electron-rich phenolic compounds.
7. 5% Pd–C, H_2. In the presence of pyridine, hydrogenolysis of the MPM group is suppressed.[8]

1. J. D. White and J. C. Amedio, Jr., *J. Org. Chem.*, **54**, 736 (1989).
2. I. A. McDonald, P. L. Nyce, M. J. Jung, and J. S. Sabol, *Tetrahedron Lett.*, **32**, 887 (1991).

3. H. Nagaoka, G. Schmid, H. Iio, and Y. Kishi, *Tetrahedron Lett.*, **22**, 899 (1981).
4. A. Srikrishna, R. Viswajanani, J. A. Sattigeri, and D. Vijaykumar, *J. Org. Chem.*, **60**, 5961 (1995).
5. T. Ohsawa, K. Hatano, K. Kayoh, J. Kotabe, and T. Oishi, *Tetrahedron Lett.*, **33**, 5555 (1992).
6. K. J. Hodgetts and T. W. Wallace, *Synth. Commun.*, **24**, 1151 (1994).
7. O. P. Vig, S. S. Bari, A. Sharma, and M. A. Sattar, *Indian J. Chem., Sect. B*, **29B**, 284 (1990).
8. H. Sajiki, H. Kuno, and K. Hirota, *Tetrahedron Lett.*, **38**, 399 (1997).

o-Nitrobenzyl Ether: *o*-NO_2–$C_6H_4CH_2OAr$ (Chart 4)

CO_2Me / NHPG / OH — 1. NaOMe; 2. 2–$NO_2C_6H_4CH_2Cl$, 25°, 15 min, 80% → ; ← 350 nm, MeOH, 25°, 12 h, 98%° — CO_2Me / NHPG / $OCH_2C_6H_4$–2–NO_2

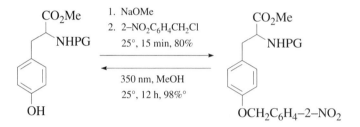

An *o*-nitrobenzyl ether can be cleaved by photolysis. In tyrosine, this avoids the use of acid-catalyzed cleavage and the attendant conversion to 3-benzyltyrosine.[1] (Note that this unwanted conversion can also be suppressed by the addition of thioanisole; see benzyl ether cleavage.)

1. B. Amit, E. Hazum, M. Fridkin, and A. Patchornik, *Int. J. Pept. Protein Res.*, **9**, 91 (1977).

2,6-Dichlorobenzyl Ether: $ArOCH_2C_6H_3$–2,6-Cl_2

This group is readily cleaved by a mixture of CF_3SO_3H, $PhSCH_3$, and CF_3CO_2H.[1,2] Of the common benzyl groups used to protect the hydroxyl of tyrosine, the 2,6-dichlorobenzyl shows a low incidence of alkylation at the 3-position of tyrosine during cleavage with HF/anisole. A comparative study of the deprotection of X-Tyr in HF/anisole gives the following percentages of side reactions for various X groups: Bn, 24.5; 2-ClBn, 9.8; 2,6-Cl_2Bn, 6.5; cyclohexyl, 1.5; *t*-Bu, <0.2; Cbz, 0.5; 2-Br-Cbz, 0.2.[3]

3,4-Dichlorobenzyl Ether: 3,4-$Cl_2C_6H_3CH_2OAr$

The electron-withdrawing chlorine atoms confer greater acid stability to this group than that conferred on the usual benzyl group. It is cleaved by hydrogenolysis (Pd/C–H_2).[4]

1. Y. Kiso, M. Satomi, K. Ukawa, and T. Akita, *J. Chem. Soc., Chem. Commun.*, 1063 (1980).
2. J. Deng, Y. Hamada, and T. Shioiri, *Tetrahedron Lett.*, **37**, 2261 (1996).
3. J. P. Tam, W. F. Heath, and R. B. Merrifield, *Int. J. Pept. Protein Res.*, **21**, 57 (1983).
4. D. A. Evans, C. J. Dinsmore, D. A. Evrard, and K. M. DeVries, *J. Am. Chem. Soc.*, **115**, 6426 (1993).

4-(Dimethylamino)carbonylbenzyl Ether: $(CH_3)_2NCOC_6H_4CH_2OAr$

The 4-(dimethylamino)carbonylbenzyl ether has been used to protect the phenolic hydroxyl of tyrosine. It is stable to CF_3CO_2H (120 h), but not to HBr/AcOH. (Complete cleavage occurs in 16 h.) It can also be cleaved by hydrogenolysis (H_2/Pd–C).[1]

1. V. S. Chauhan, S. J. Ratcliffe, and G. T. Young, *Int. J. Pept. Protein Res.*, **15**, 96 (1980).

4-Methylsulfinylbenzyl (Msib) Ether: $CH_3S(O)C_6H_4CH_2OAr$

The Msib group has been used for the protection of tyrosine. It is cleaved by reduction of the sulfoxide to the sulfide, which is then deprotected with acid. Reduction is achieved with DMF–SO_3/$HSCH_2CH_2SH$ or $Bu_4N^+I^-$[1] or with $SiCl_3$/TFA.[2]

1. S. Futaki, T. Yagami, T. Taike, T. Ogawa, T. Akita, and K. Kitagawa, *Chem. Pharm. Bull.*, **38**, 1165 (1990).
2. Y. Kiso, S. Tanaka, T. Kimura, H. Itoh, and K. Akaji, *Chem. Pharm. Bull.*, **39**, 3097 (1991).

9-Anthrylmethyl Ether: $ArOCH_2$–9-anthryl (Chart 4)

9-Anthrylmethyl ethers, formed from the sodium salt of a phenol and 9-anthrylmethyl chloride in DMF, can be cleaved with CH_3SNa (DMF, 25°, 20 min, 85–99% yield). They are also cleaved by CF_3CO_2H/CH_2Cl_2 (0°, 10 min, 100% yield); they are stable to CF_3CO_2H/dioxane (25°, 1 h).[1]

1. N. Kornblum and A. Scott, *J. Am. Chem. Soc.*, **96**, 590 (1974).

4-Picolyl Ether: $ArOCH_2$–4-pyridyl (Chart 4)

Formation[1]/Cleavage[1,2]

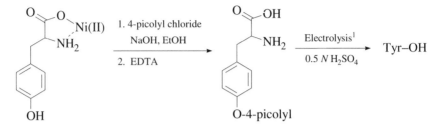

1. 4-picolyl chloride, NaOH, EtOH; 2. EDTA → O-4-picolyl; Electrolysis[1], 0.5 *N* H_2SO_4 → Tyr–OH

An aryl 4-picolyl ether is stable to trifluoroacetic acid, used to cleave an *N-t*-butoxycarbonyl group.[2]

1. A. Gosden, D. Stevenson, and G. T. Young, *J. Chem. Soc., Chem. Commun.*, 1123 (1972).
2. P. M. Scopes, K. B. Walshaw, M. Welford, and G. T. Young, *J. Chem. Soc.*, 782 (1965).

Heptafluoro-*p*-tolyl and Tetrafluoro-4-pyridyl Ethers:
$ArOC_6F_4$–CF_3, $ArOC_5F_4N$

Formation/Cleavage[1–3]

Estradiol —[$C_6F_5CF_3$ (1 eq.), NaOH, CH_2Cl_2, $Bu_4N^+ HSO_4^-$, 95%]→ 4-$CF_3C_6F_4$-estradiol

4-$CF_3C_6F_4$-estradiol —[NaOMe, DMF, 1 h, 10°, 87%[4]]→ Estradiol

If 2 eq. of reagent are used, both hydroxyls can be protected, and the phenolic hydroxyl can be selectively cleaved with NaOMe. The tetrafluoropyridyl derivative is introduced under similar conditions. The use of this methodology has been reviewed.[5]

1. M. Jarman and R. McCague, *J. Chem. Soc., Chem. Commun.*, 125 (1984).
2. M. Jarman and R. McCague, *J. Chem. Res., Synop.*, 114 (1985).
3. J. J. Deadman, R. McCague, and M. Jarman, *J. Chem. Soc., Perkin Trans. 1*, 2413 (1991).
4. S. Singh and R. A. Magarian, *Chem. Lett.*, 1821 (1994).
5. M. Jarman *J. Fluorine Chem.*, **42**, 3 (1989).

Silyl Ethers

Aryl and alkyl trimethylsilyl ethers can often be cleaved by refluxing in aqueous methanol, an advantage for acid- or base-sensitive substrates. The ethers are stable to Grignard and Wittig reactions and to reduction with lithium aluminum hydride at $-15°$. Aryl *t*-butyldimethylsilyl ethers and other sterically more demanding silyl ethers require acid- or fluoride ion-catalyzed hydrolysis for removal. Increased steric bulk also improves their stability to a much harsher set of conditions. An excellent review of the selective deprotection of alkyl silyl ethers and aryl silyl ethers has been published.[1]

1. T. D. Nelson and R. D. Crouch, *Synthesis*, 1031 (1996).

Trimethylsilyl (TMS) Ether: $ArOSi(CH_3)_3$

Formation

1. Me_3SiCl, Pyr, 30–35°, 12 h, satisfactory yield.[1]
2. $(Me_3Si)_2NH$, cat. concd. H_2SO_4, reflux, 2 h, 97% yield.[2]
3. A large number of other silylating agents have been described for the derivatization of phenols, but the first two are among the most common.[3]

Cleavage

Trimethylsilyl ethers are readily cleaved by fluoride ion, mild acids, and mild bases. If the TMS derivative is somewhat hindered, it also becomes less susceptible to cleavage. A phenolic TMS ether can be cleaved in the presence of an alkyl TMS ether [Dowex 1-x8 (HO^-), EtOH, rt, 6 h, 78% yield].[4]

1. Cl. Moreau, F. Roessac, and J. M. Conia, *Tetrahedron Lett.*, 3527 (1970).
2. S. A. Barker and R. L. Settine, *Org. Prep. Proced. Int.*, **11**, 87 (1979).
3. G. van Look, G. Simchen, and J. Heberle, *Silylating Agents*, Fluka Chemie, AG, 1995.
4. Y. Kawazoe, M. Nomura, Y. Kondo, and K. Kohda, *Tetrahedron Lett.*, **28**, 4307 (1987).

t-Butyldimethylsilyl (TBDMS) Ether: $ArOSi(CH_3)_2C(CH_3)_3$ (Chart 4)

The section on alcohol protection should be examined, since many of the methods for the formation and cleavage of TBDMS ethers are similar.

Formation

1. *t*-$BuMe_2SiCl$, DMF, imidazole, 25°, 3 h, 96% yield.[1, 2]

2. 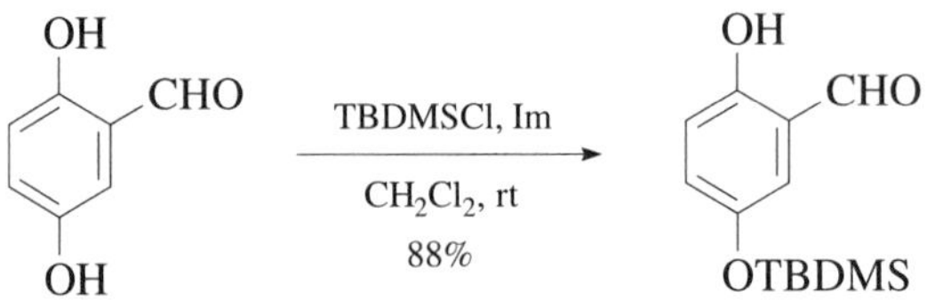 [Scheme: 2,5-dihydroxybenzaldehyde → 5-OTBDMS-2-hydroxybenzaldehyde; TBDMSCl, Im, CH_2Cl_2, rt, 88%]

Ref. 3

3. t-$BuMe_2SiOH$, Ph_3P, DEAD, 86% yield. In this case, the standard methods for silyl ether formation were unsuccessful.[4]

4. 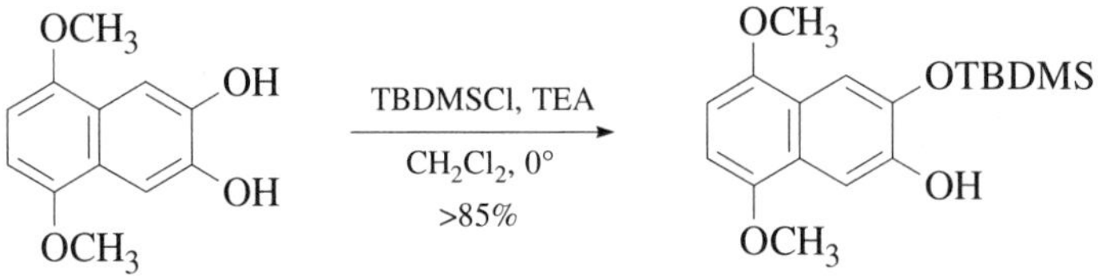 [Scheme: 5,8-dimethoxy-2,3-dihydroxynaphthalene → 3-OTBDMS-2-hydroxy derivative; TBDMSCl, TEA, CH_2Cl_2, 0°, >85%]

Ref. 5

Cleavage

1. 0.1 *M* HF, 0.1 *M* NaF, pH 5, THF, 25°, 2 days, 77% yield.[1] In this substrate, a mixture of products resulted from the attempted cleavage of the *t*-butyldimethylsilyl ether with tetra-*n*-butylammonium fluoride, the reagent generally used.[6]
2. KF, 48% aq. HBr, DMF, rt, 91% yield.[7]

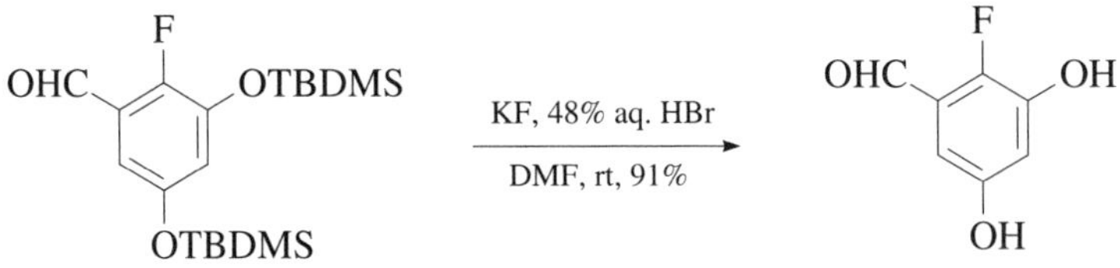

[Scheme: 2-fluoro-3,5-bis(OTBDMS)benzaldehyde → 2-fluoro-3,5-dihydroxybenzaldehyde; KF, 48% aq. HBr, DMF, rt, 91%]

The use of $Bu_4N^+F^-$ results in decomposition of this substrate.

3. 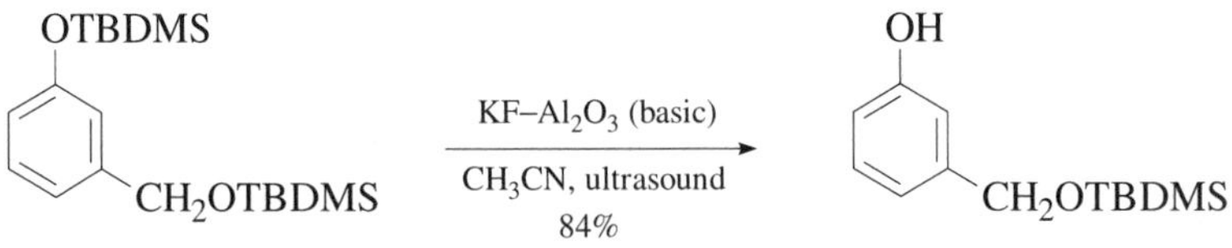 [Scheme: 3-($CH_2OTBDMS$)phenyl OTBDMS → 3-($CH_2OTBDMS$)phenol; KF–Al_2O_3 (basic), CH_3CN, ultrasound, 84%]

Ref. 8

4. $PdCl_2(CH_3CN)_2$, aq. acetone, 75°, 10–96% yield.[9]
5. $BF_3{\cdot}Et_2O$, CH_2Cl_2, rt, 8 h.[10]
6. K_2CO_3, Kriptofix 222, CH_3CN, 55°, 2 h, 70–95% yield.[11,12] Phenolic silyl ethers are cleaved selectively, but when TsOH or $BF_3{\cdot}Et_2O$ is used, alkyl TBDMS groups are cleaved in preference to phenolic derivatives.

7. Amberlite IRA-400 fluoride form, CH_2Cl_2 or DMF; then elute with aq. HCl, 80–90% yield.[13]

Table 1 gives the relative half-life to acid or base hydrolysis of a number of silylated *p*-cresols.[14]

Table 1. Susceptibility of Silylated Cresols to Hydrolysis

	Half-life ($t_{1/2}$, min) at 25°	
Substrate	Acid Hydrolysis 1% HCl in 95% MeOH	Base Hydrolysis 5% NaOH in 95% MeOH
p-$MeC_6H_4OSiEt_3$	≤1[a]	≤1[a]
p-MeC_6H_4OSi–*i*-$BuMe_2$	≤1[a]	≤1[a]
p-MeC_6H_4OSi–*t*-$BuMe_2$	273	3.5
p-MeC_6H_4OSi–*t*-$BuPh_2$	100 (h)	6.5
p-MeC_6H_4OSi–*i*-Pr_3	100 (h)	188

[a] A $t_{1/2}$ of 1 min is a minimum value because of sampling methods.

1. P. M. Kendall, J. V. Johnson, and C. E. Cook, *J. Org. Chem.*, **44**, 1421 (1979).
2. R. C. Ronald, J. M. Lansinger, T. S. Lillie, and C. J. Wheeler, *J. Org. Chem.*, **47**, 2541 (1982).
3. A. Liu, K. Dillon, R. M. Campbell, D. C. Cox, and D. M. Huryn, *Tetrahedron Lett.*, **37**, 3785 (1996).
4. D. L. J. Clive and D. Kellner, *Tetrahedron Lett.*, **32**, 7159 (1991).
5. A. Kojima, T. Takemoto, M. Sodeoka, and M. Shibasaki, *J. Org. Chem.*, **61**, 4876 (1996).
6. E. J. Corey and A. Venkateswarlu, *J. Am. Chem. Soc.*, **94**, 6190 (1972).
7. A. K. Sinhababu, M. Kawase, and R. T. Borchardt, *Synthesis*, 710 (1988).
8. E. A. Schmittling and J. S. Sawyer, *Tetrahedron Lett.*, **32**, 7207 (1991).
9. N. S. Wilson and B. A. Keay, *Tetrahedron Lett.*, **37**, 153 (1996).
10. S. Mabic and J.-P. Lepoittevin, *Synlett*, 851 (1994).
11. C. Prakash, S. Saleh, and I. A. Blair, *Tetrahedron Lett.*, **35**, 7565 (1994).
12. N. S. Wilson and B. A. Keay, *Tetrahedron Lett.*, **38**, 187 (1997).
13. B. P. Bandgar, S. D. Unde, D. S. Unde, V. H. Kulkarni, and S. V. Patil, *Indian J. Chem., Sect. B*, **33B**, 782 (1994).
14. J. S. Davies, C. L. Higginbotham, E. J. Tremeer, C. Brown, and R. C. Treadgold, *J. Chem. Soc., Perkin Trans. 1*, 3043 (1992).

t-Butyldiphenylsilyl (TBDPS) Ether: $(CN_3)_3C(C_6H_5)_2SiOR$

The TBDPS ether has been used for the monoprotection of a catechol (TBDPSCl, Im, DMF, 5 h, 83% yield)[1] or simple phenol protection. It is cleaved with $Bu_4N^+F^-$ (THF, 94% yield).[2]

1. J. C. Kim and W.-W. Park, *Org. Prep. Proced. Int.*, **26**, 479 (1994).
2. A. B. Smith, III, J. Barbosa, W. Wong, and J. L. Wood, *J. Am. Chem. Soc.*, **118**, 8316 (1996).

Triisopropylsilyl (TIPS) Ether: $((CN_3)_2CH)_3SiOR$

The bulk of the TIPS group, introduced with TIPSCl (DMF, Im, 92% yield), directs metallation away from the silyl group as illustrated.[1]

OTIPS
BuLi; DMF
CH_3O OCH_3
OTIPS
CH_3O OCH_3
CHO

1. J. J. Landi, Jr., and K. Ramig, *Synth. Commun.*, **21**, 167 (1991).

Esters

Aryl esters, prepared from the phenol and an acid chloride or anhydride in the presence of base, are readily cleaved by saponification. In general, they are more readily cleaved than the related esters of alcohols, thus allowing selective removal of phenolic esters. 9-Fluorenecarboxylates and 9-xanthenecarboxylates are also cleaved by photolysis. To permit selective removal, a number of carbonate esters have been investigated: aryl benzyl carbonates can be cleaved by hydrogenolysis; aryl 2,2,2-trichloroethyl carbonates by Zn/THF-H_2O. Esters of electron-deficient phenols are good acylating agents for alcohols and amines.

Aryl Formate: HCO_2Ar

The formate ester of phenol is rarely formed, but can be prepared from the phenol, formic acid, and DCC, 94–99% yield. The formate ester is not very stable to basic conditions or to other good nucleophiles.[1]

1. J. Huang and H. K. Hall, Jr., *J. Chem. Res., Synop.*, 292 (1991).

Aryl Acetate: $ArOCOCH_3$ (Chart 4)

Formation

1. AcCl, NaOH, dioxane, $Bu_4N^+HSO_4^-$, 25°, 30 min, 90% yield.[1] Phase-transfer catalysis with tetra-*n*-butylammonium hydrogen sulfate effects acylation of sterically hindered phenols and selective acylation of a phenol in the presence of an aliphatic secondary alcohol.

2. 1-Acetyl-*v*-triazolo[4,5-*b*]pyridine, THF, 1 *N* NaOH, 30 min.[2]

HO–C6H4–CH2OH → (1-acetyl-v-triazolo[4,5-b]pyridine; THF, rt, 1 h, 89%) → AcO–C6H4–CH2OH

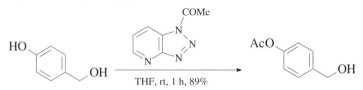

This method is also effective in the selective introduction of a benzoate ester.

3. IPA, NaOH, Ac_2O, pH 7.8. Phenols are selectively esterified in the presence of other alcohols.[3] These authors also showed that an alcohol could be acetylated in the presence of an amine using Ac_2O and Amberlyst 15 resin.
4. *Chromobacterium viscosum* lipase, cyclohexane, vinyl acetate, THF, 40°.[4]

Cleavage

1. $NaHCO_3$/aq. MeOH, 25°, 0.75 h, 94% yield.[5]
2. 3 *N* HCl, acetone, reflux, 2 h.[6]
3. Aq. NH_3, 0°, 48 h.[6]
4. $NaBH_4$, $HO(CH_2)_2OH$, 40°, 18 h, 87% yield.[7] Lithium aluminum hydride can be used to effect efficient ester cleavage if no other functional group is present that can be attacked by this strong reducing agent.[8]
5. $NaBH_4$, LiCl, diglyme. A diacylated guanidine was not deacylated under these conditions, whereas the usual basic conditions for acetate hydrolysis also resulted in guanidine deacylation.[9]
6. Sm, I_2, EtOH, 82–100% yield. Esters of other alcohols are similarly deacylated.[10]

The following conditions selectively remove a phenolic acetate in the presence of a normal alkyl acetate:

7. TsOH, SiO_2, toluene, 80°, 6–40 h, 79–100% yield.[11]
8. $(NH_2)_2C{=}NH$, MeOH, 50°, 95% yield.[12]
9. $Me_2NCH_2C(O)N(OH)Me$, MeOH or THF/H_2O, 84% yield.[13]
10. Zn, MeOH, 91–100% yield.[14]
11. Neutral alumina, microwaves, 82–96% yield.[15]
12. Bi(III)-mandelate, DMSO, 80–125°, 44–96% yield. Phenolic acetates with strong electron-withdrawing groups are hydrolyzed the fastest.[16]
13. Porcine pancreatic lipase, 28–30°, 95% yield.[17]
14. *Candida cylindracea* lipase, BuOH, hexane, 3 h, 25°, 40–100% yield.[18]

1. V. O. Illi, *Tetrahedron Lett.*, 2431 (1979).
2. M. P. Paradisi, G. P. Zecchini, and I. Torrini, *Tetrahedron Lett.*, **27**, 5029 (1986).
3. V. Srivastava, A. Tandon, and S. Ray, *Synth. Commun.*, **22**, 2703 (1992).

4. G. Nicolosi, M. Piattelli, and C. Sanfilippo, *Tetrahedron*, **48**, 2477 (1992).
5. For example, see G. Büchi and S. M. Weinreb, *J. Am. Chem. Soc.*, **93**, 746 (1971).
6. E. Haslam, G. K. Makinson, M. O. Naumann, and J. Cunningham, *J. Chem. Soc.*, 2137 (1964).
7. J. Quick and J. K. Crelling, *J. Org. Chem.*, **43**, 155 (1978).
8. H. Mayer, P. Schudel, R. Rüegg, and O. Isler, *Helv. Chim. Acta*, **46**, 650 (1963).
9. D. Huber, G. Leclerc, and G. Andermann, *Tetrahedron Lett.*, **27**, 5731 (1986).
10. R. Yanada, N. Negoro, K. Bessho, and K. Yanada, *Synlett*, 1261 (1995).
11. G. Blay, M. L. Cardona, M. B. Garcia, and J. P. Pedro, *Synthesis*, 438 (1989).
12. N. Kunesch, C. Miet, and J. Poisson, *Tetrahedron Lett.*, **28**, 3569 (1987).
13. M. Ono and I. Itoh, *Tetrahedron Lett.*, **30**, 207 (1989).
14. A. G. González, Z. D. Jorge, H. L. Dorta, and F. R. Luis, *Tetrahedron Lett.*, **22**, 335 (1981).
15. R. S. Varma, M. Varma, and A. K. Chatterjee, *J. Chem. Soc., Perkin Trans. 1*, 999 (1993).
16. V. Le Boisselier, M. Postel, and E. Duñch, *Tetrahedron Lett.*, **38**, 2981 (1997).
17. V. S. Parmar, A. Kumar, K. S. Bisht, S. Mukherjee, A. K. Prasad, S. K. Sharma, J. Wengel, and C. E. Olsen, *Tetrahedron*, **53**, 2163 (1997).
18. G. Pedrocchi-Fantoni and S. Servi, *J. Chem. Soc., Perkin Trans. 1*, 1029 (1992).

Aryl Levulinate: $CH_3COCH_2CH_2CO_2Ar$

Cleavage[1]

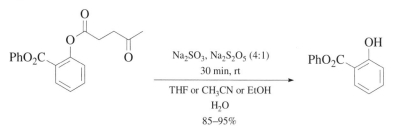

PhO_2C, O, O, O

Na_2SO_3, $Na_2S_2O_5$ (4:1)
30 min, rt
THF or CH_3CN or EtOH
H_2O
85–95%

OH
PhO_2C

1. M. Ono and I. Itoh, *Chem. Lett.*, 585 (1988).

Aryl Pivaloate (ArOPv): $(CH_3)_3CCO_2Ar$ (Chart 4)

Formation/Cleavage[1]

1. Pivaloyl chloride reacts selectively with the less hindered phenol group.

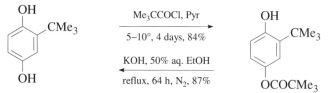

OH, CMe_3, OH

Me_3CCOCl, Pyr
5–10°, 4 days, 84%

KOH, 50% aq. EtOH
reflux, 64 h, N_2, 87%

OH, CMe_3, $OCOCMe_3$

2. Me_3C–CO–N(thiazolidine-2-one, C=O at 2, S ring) NaH, THF, 99% yield.[2] This method works well for the esterification of a phenol in the presence of an aniline. When the thiazolidone is reacted with a hydroxyaniline in the absence of base, only the nitrogen is derivatized to form a pivalamide.[3]

1. L. K. T. Lam and K. Farhat, *Org. Prep. Proced. Int.*, **10**, 79 (1978).
2. K. C. Nicolaou and W.-M. Dai, *J. Am. Chem. Soc.*, **114**, 8908 (1992).
3. W.-M. Dai, Y. K. Cheung, K. W. Tang, P. Y. Choi, and S. L. Chung, *Tetrahedron*, **51**, 12263 (1995).

Aryl Benzoate: $ArOCOC_6H_5$ (Chart 4)

Aryl benzoates, stable to alkylation conditions using K_2CO_3/Me_2SO_4, are cleaved by more basic hydrolysis (KOH).[1] They are stable to anhydrous hydrogen chloride,[2] but are cleaved by hydrochloric acid.[3]

Formation

1. $(ClCO)_2$, Me_2NCHO, PhCOOH; Pyr, 20°, 2 h, 90% yield.[4]
2. 2-(SBz)-1-methylpyridinium (N^+–CH_3) Cl^- aq. $NaHCO_3$ or aq. NaOH, 80% yield.[5] This reagent forms aryl benzoates under aqueous conditions. (It also acylates amines and carboxylic acids.)
3. Monoesterification of a symmetrical dihydroxy aromatic compound can be effected by reaction with polymer-bound benzoyl chloride (Pyr, benzene, reflux, 15 h) to give a polymer-bound benzoate, which can be alkylated with diazomethane to form, after basic hydrolysis (0.5 *M* NaOH, dioxane, H_2O, 25°, 20 h, or 60°, 3 h),[6] a monomethyl ether.
4. $Fe_2(SO_4)_3$–SiO_2, methyl benzoate, 97% yield.[7]

Cleavage

1. Under anhydrous conditions, cesium carbonate or bicarbonate quantitatively cleaves an aryl dibenzoate or diacetate to the monoester; yields are considerably lower with potassium carbonate.[8]

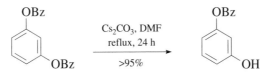

1,3-(OBz)$_2$C$_6$H$_4$ → (Cs_2CO_3, DMF, reflux, 24 h; >95%) → 3-(OBz)C$_6$H$_4$OH

2. $BuNH_2$, benzene, rt, 1–24 h, >85% yield.[9] This method is generally selective for phenolic esters.

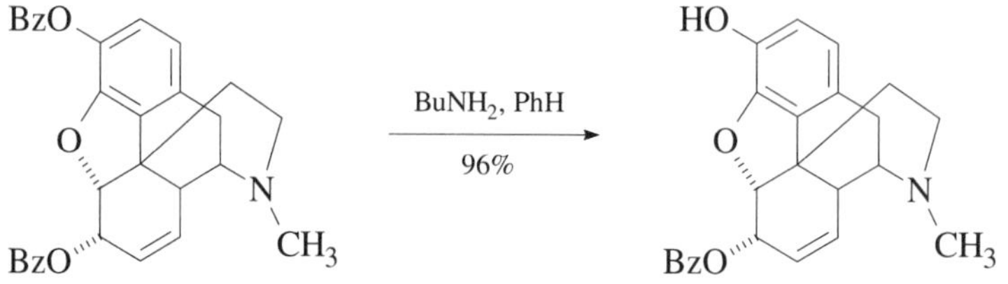

BzO, O, BzO, N, CH_3 → ($BuNH_2$, PhH, 96%) → HO, O, BzO, N, CH_3

3. 2-Bromo-1,3,2-benzodioxaborole, CH_2Cl_2 (cat. $BF_3 \cdot Et_2O$), 25°, 0.25 h, 71% yield.[10]

1. M. Gates, *J. Am. Chem. Soc.*, **72**, 228 (1950).
2. D. D. Pratt and R. Robinson, *J. Chem. Soc.*, 1577 (1922).
3. A. Robertson and R. Robinson, *J. Chem. Soc.*, 1710 (1927).
4. P. A. Stadler, *Helv. Chim. Acta*, **61**, 1675 (1978).
5. M. Yamada, Y. Watabe, T. Sakakibara, and R. Sudoh, *J. Chem. Soc., Chem. Commun.*, 179 (1979).
6. C. C. Leznoff and D. M. Dixit, *Can. J. Chem.*, **55**, 3351 (1977).
7. T. Nishiguchi and H. Taya, *J. Chem. Soc., Perkin Trans. 1*, 172 (1990).
8. H. E. Zaugg, *J. Org. Chem.*, **41**, 3419 (1976).
9. K. H. Bell, *Tetrahedron Lett.*, **27**, 2263 (1986).
10. P. F. King and S. G. Stroud, *Tetrahedron Lett.*, **26**, 1415 (1985).

Aryl 9-Fluorenecarboxylate (Chart 4): CO_2Ar

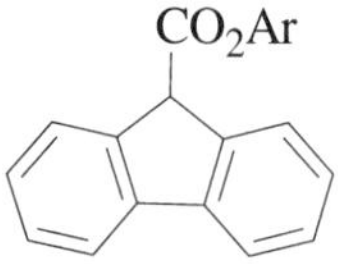

Aryl 9-fluorenecarboxylates (designed to be cleaved photolytically) were prepared from the phenol and the acid chloride (9-fluorenecarbonyl chloride, Pyr, C_6H_6, 25°, 1 h, 65% yield) and cleaved by photolysis (*hν*, Et_2O, reflux, 4 h, 60% yield). The related aryl **xanthenecarboxylates**, **i**, were prepared and cleaved in the same way.[1]

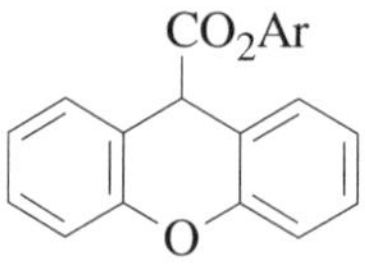

CO_2Ar, O

i

1. D. H. R. Barton, Y. L. Chow, A. Cox, and G. W. Kirby, *J. Chem. Soc.*, 3571 (1965).

Carbonates

Aryl Methyl Carbonate: $ArOCO_2CH_3$ (Chart 4)

In an early synthesis, a methyl carbonate, prepared by reaction of a phenol with methyl chloroformate, was cleaved selectively in the presence of a phenyl ester.[1]

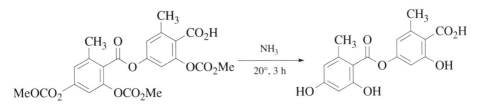

CH3 O CH3 CO2H O MeOCO2 OCO2Me OCO2Me —NH3, 20°, 3 h→ CH3 O CH3 CO2H O OH HO OH

More recently, an ethyl carbonate was cleaved by refluxing in acetic acid for 6 h.[2]

1. E. Fischer and H. O. L. Fischer, *Ber.*, **46**, 1138 (1913).
2. E. Haslam, R. D. Haworth, and G. K. Makinson, *J. Chem. Soc.*, 5153 (1961).

1-Adamantyl Carbonate (Adoc-OAr)

The adamantyl carbonate is prepared from $Adoc_2CO_3$ (DMAP, CH_3CN, >79% yield)[1] or, in the case of electron-deficient phenols, the fluoroformate (THF, Pyr, 54–95% yield).[2] It is somewhat more stable to TFA than the adamantyl carbamate.

1. B. Nyasse and U. Ragnarsson, *Acta Chem. Scand.*, **47**, 374 (1993).
2. I. Niculescu-Duvaz and C. J. Springer, *J. Chem. Res., Synop.*, 242 (1994).

***t*-Butyl Carbonate (BOC-OAr):** $(CH_3)_3COCO_2Ar$

The BOC derivative of phenols can be prepared using a phase transfer protocol (BOC_2O, $Bu_4N^+HSO_4^-$ or 18-crown-6, NaOH, CH_2Cl_2, 80% yield)[1] or by direct acylation with BOC_2O and DMAP as a catalyst (79–100% yield).[2] Cleavage is achieved by refluxing a mixture of the carbonate with 3 *M* HCl in dioxane. The use of TFA for cleavage often results in *t*-butylation of the phenol.[2]

1. F. Houlihan, F. Bouchard, J. M. J. Frechet, and C. G. Willson, *Can. J. Chem.*, **63**, 153 (1985).
2. M. M. Hansen and J. R. Riggs, *Tetrahedron Lett.*, **39**, 2705 (1998).

4-Methylsulfinylbenzyl Carbonate (Msz–OAr): $CH_3S(O)C_6H_4CH_2OCO_2Ar$

Tyrosine-½Cu is protected with 4-methylthiobenzyl 4′-nitrophenyl carbonate ($NaHCO_3$, DMF, H_2O). Release of the copper protection followed by BOC protection of the nitrogen gives a fully protected tyrosine, the sulfide of which is oxidized with $NaBrO_3{\cdot}3H_2O$ to generate the acid stable Msz-protected tyrosine. Cleavage is achieved by reductive acidolysis with $SiCl_4$/TFA.[1]

1. Y. Kiso, S. Tanaka, T. Kimura, H. Itoh, and K. Akaji, *Chem. Pharm. Bull.*, **39**, 3099 (1991).

2,4-Dimethylpent-3-yl Carbonate (Doc–OAr): $(i\text{-Pr})_2COCO_2Ar$

The Doc group, used for the protection of the phenolic hydroxyl group in tyrosine, is introduced with the chloroformate (DIPEA, CH_3CN). It has a half-life in 20% piperidine/DMF of 8 h, compared with 30 sec for the 2-BrZ (2-BrCbz) group. The 2-BrZ group is only slightly more stable to acid than the Doc group. The Doc group is completely cleaved by HF.[1]

1. K. Rosenthal, A. Karlström, and A. Undén, *Tetrahedron Lett.*, **38**, 1075 (1997).

Aryl 2,2,2-Trichloroethyl Carbonate: $ArOCOOCH_2CCl_3$ (Chart 4)

Formation

1. Cl_3CCH_2OCOCl, Pyr or aq. NaOH, 25°, 12 h.[1]

Cleavage

1. Zn, HOAc, 25°, 1–3 h, or Zn, CH_3OH, heat, few min.[1]
2. Zn, THF–H_2O, pH 4.2, 25°, 4 h.[2] The authors suggest that selective cleavage should be possible by this method, since, at pH 4.2, 25°, 2,2,2-trichloroethyl esters are cleaved in 10 min, 2,2,2-trichloroethyl carbamates are cleaved in 30 min, and the 2,2,2-trichloroethyl carbonate of estrone, formed in 87% yield from estrone and the acid chloride, is cleaved in 4 h (97% yield).

1. T. B. Windholz and D. B. R. Johnston, *Tetrahedron Lett.*, 2555 (1967).
2. G. Just and K. Grozinger, *Synthesis*, 457 (1976).

Aryl Vinyl Carbonate: $ArOCO_2CH{=}CH_2$ (Chart 4)

Formation

1. $CH_2{=}CHOCOCl$, Pyr, 95% yield.[1]

Cleavage

1. Na_2CO_3, warm aq. dioxane, 96% yield. Selective protection of an aryl —OH or an amine—NH group is possible by reaction of the compound with vinyl chloroformate. Vinyl carbamates ($RR'NCO_2CH{=}CH_2$) are stable to the basic conditions (Na_2CO_3) used to cleave vinyl carbonates. Conversely, vinyl carbonates are stable to the acidic conditions ($HBr/CH_3OH/CH_2Cl_2$) used to cleave vinyl carbamates. Vinyl carbonates are cleaved by more acidic conditions: 2 *N* anhyd. HCl/dioxane, 25°, 3 h, 10% yield; HBF_4, 25°, 12 h, 30% yield; 2 *N* $HCl/CH_3OH–H_2O$(4:1), 60°, 8 h, 100% yield.[1]

1. R. A. Olofson and R. C. Schnur, *Tetrahedron Lett.*, 1571 (1977).

Aryl Benzyl Carbonate: $ArOCOOCH_2C_6H_5$ (Chart 4)

Formation

1. $PhCH_2OCOCl$, Pyr, CH_2Cl_2, THF.[1]

Cleavage

1. H_2/Pd–C, EtOH, 20°.[1]

o-Bromobenzyl carbonates have been developed for use in solid-phase peptide synthesis. An aryl *o*-bromobenzyl carbonate is stable to acidic cleavage (CF_3CO_2H) of a *t*-butyl carbamate; a benzyl carbonate is cleaved. The *o*-bromo derivative is quantitatively cleaved with hydrogen fluoride (0°, 10 min).[2]

1. M. Kuhn and A. von Wartburg, *Helv. Chim. Acta*, **52**, 948 (1969).
2. D. Yamashiro and C. H. Li, *J. Org. Chem.*, **38**, 591 (1973).

Aryl Carbamate: ArOCONHR

Formation

1. RNCO (R = Ph, *i*-Bu), 60°, 2 h, 65–85% yield.[1]

Cleavage

1. 2 *N* NaOH, 20°, 2 h, 78% yield.[1]
2. $H_2NNH_2 \cdot H_2O$, DMF, 20°, 3 h, 59–87% yield.[1]

1. G. F. Jäger, R. Geiger and W. Siedel, *Chem. Ber.*, **101**, 2762 (1968).

Phosphinates

Dimethylphosphinyl Ester (Dmp–OAr): $(CH_3)_2P(O)OAr$

Formation

1. $Me_2P(O)Cl$, Et_3N, $CHCl_3$, 76% yield.[1] The Dmp group was used to protect tyrosine for use in peptide synthesis. It is stable to 1 *M* HCl/MeOH, 1 *M* HCl/AcOH, CF_3CO_2H, HBr/AcOH, and H_2/Pd–C.

Cleavage

The Dmp group can be cleaved by the following reagents: liq. HF (0°, 1 h); 1 *M* Et_3N/MeOH (rt, 7 h); 0.1 *M* NaOH (rt, < 5 min); 5% aq. $NaHCO_3$ (rt, 5 h); 20% hydrazine/MeOH (rt, < 5 min); 50% pyridine/DMF (rt, 6 h); $Bu_4N^+F^-$ (rt, < 5 min).[1]

Dimethylphosphinothioyl Ester (Mpt–OAr): $(CH_3)_2P(S)OAr$

Formation

1. MptCl, CH_2Cl_2, Et_3N, 66% yield.[2]

Cleavage

The *O*-Mpt group is quite stable to acidic conditions (HBr/AcOH, CF_3CO_2H, 1 *M* HCl/AcOH), but is slowly cleaved under basic conditions (1 *M* NaOH/MeOH, 5 min; 1 *M* Et_3N/MeOH, reflux, 12 h). In contrast, the *N*-Mpt group is readily cleaved with acid (CF_3CO_2H, 60 min; 1 *M* HCl/AcOH, 15 min; HBr/AcOH, 5 min), but not with base. The Mpt group was used to protect tyrosine during peptide synthesis.[2] The Mpt group can be removed with aq. $AgNO_3$ or $Hg(OAc)_2$,[3] or fluoride ion.[4]

Diphenylphosphinothioyl Ester (Dpt–OAr): $(C_6H_5)_2P(S)OAr$

The diphenylphosphinothioyl ester, used to protect a tryptophan, is cleaved with $Bu_4N^+ F^-$-$3H_2O$/DMF.[5]

1. M. Ueki, Y. Sano, I. Sori, K. Shinozaki, H. Oyamada, and S. Ikeda, *Tetrahedron Lett.*, **27**, 4181 (1986).
2. M. Ueki and T. Inazu, *Bull. Chem. Soc. Jpn.*, **55**, 204 (1982).
3. M. Ueki and K. Shinozaki, *Bull. Chem. Soc. Jpn.*, **56**, 1187 (1983).
4. M. Ueki and K. Shinozaki, *Bull. Chem. Soc. Jpn.*, **57**, 2156 (1984).
5. Y. Kiso, T. Kimura, Y. Fujiwara, M. Shimokura, and A. Nishitani, *Chem. Pharm. Bull.*, **36**, 5024 (1988).

Sulfonates

An aryl methane- or toluenesulfonate ester is stable to reduction with lithium aluminum hydride, to the acidic conditions used for nitration of an aromatic ring (HNO_3/HOAc),[1] and to the high temperatures (200–250°) of an Ullmann reaction. Aryl sulfonate esters, formed by reaction of a phenol with a sulfonyl chloride in pyridine or aqueous sodium hydroxide, are cleaved by warming in aqueous sodium hydroxide.[2]

1. E. M. Kampouris, *J. Chem. Soc.*, 2651 (1965).
2. F. G. Bordwell and P. J. Boutan, *J. Am. Chem. Soc.*, **79**, 717 (1957).

Aryl Methanesulfonate: $ArOSO_2CH_3$ (Chart 4)

In a synthesis of decinine, a phenol was protected as a methanesulfonate that was stable during an Ullmann coupling reaction and during condensation, catalyzed by calcium hydroxide, of an amine with an aldehyde. Aryl methanesulfonates are cleaved by warm sodium hydroxide solution.[1,2]

An aryl methanesulfonate was cleaved to a phenol by phenyllithium or phenylmagnesium bromide;[3] it was reduced to an aromatic hydrocarbon by sodium in liquid ammonia.[4]

1. I. Lantos and B. Loev, *Tetrahedron Lett.*, 2011 (1975).
2. J. E. Rice, N. Hussain, and E. J. LaVoie, *J. Labelled Compd. Radiopharm.*, **24**, 1043 (1987).
3. J. E. Baldwin, D. H. R. Barton, I. Dainis, and J. L. C. Pereira, *J. Chem. Soc. C*, 2283 (1968).
4. G. W. Kenner and N. R. Williams, *J. Chem. Soc.*, 522 (1955).

Aryl Toluenesulfonate: $ArOSO_2C_6H_4$-*p*-CH_3

Formation[1]

1\.

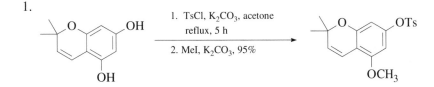

O OH OH → 1. TsCl, K_2CO_3, acetone reflux, 5 h; 2. MeI, K_2CO_3, 95% → O OTs OCH_3

2. *o*-Aminophenol can be selectively protected as a sulfonate or a sulfonamide.[2]

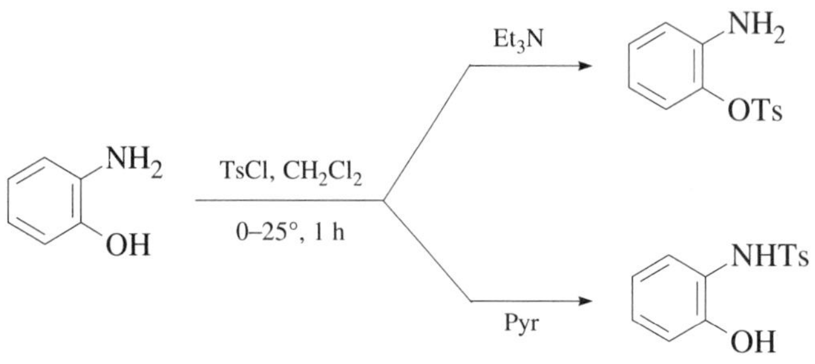

TsCl, CH_2Cl_2, 0–25°, 1 h; Et_3N → 2-OTs aniline; Pyr → NHTs phenol

Cleavage[1]

1. KOH, H_2O, EtOH, reflux, 1 h, 60%

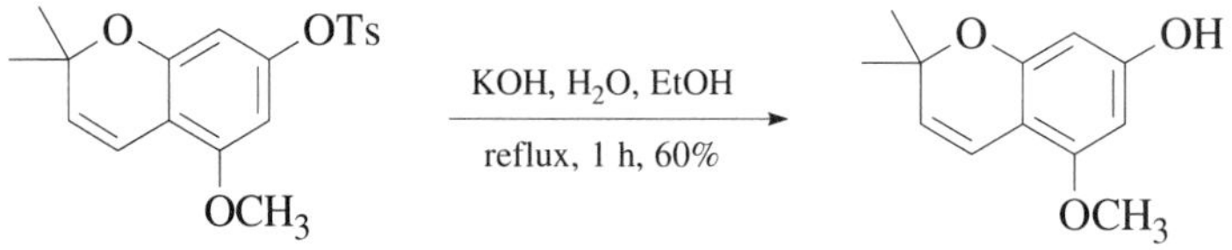

An aryl toluenesulfonate is stable to lithium aluminum hydride (Et_2O, reflux, 4 h) and to *p*-toluenesulfonic acid ($C_6H_5CH_3$, reflux, 15 min).

2. Electrolysis: Hg anode, Pt cathode, DMF, O_2, cyclohexene, $Bu_4N^+Br^-$, 62% yield.[3]

3. Electrolysis, TEAB, CH_3CN, 63%

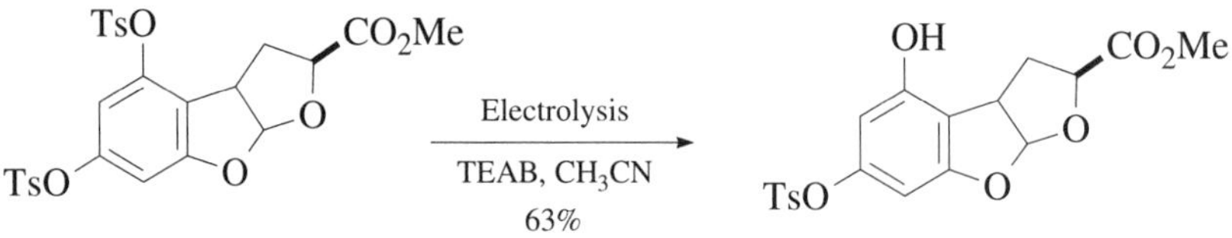

Ref. 4

4. Na(Hg), MeOH, 96.7% yield.[5]
5. Mg, MeOH, 4–6 h, 90–95% yield.[6]

1. M. L. Wolfrom, E. W. Koos, and H. B. Bhat, *J. Org. Chem.*, **32**, 1058 (1967).
2. K. Kurita, *Chem. Ind. (London)*, 345 (1974).
3. S. Dwivedi and R. A. Misra, *Indian J. Chem., Sect. B*, **B31**, 282 (1992).
4. E. R. Civitello and H. Rapoport, *J. Org. Chem.*, **59**, 3775 (1994).
5. R. S. Tipson, *Methods Carbohydr. Chem.*, **2**, 250 (1963).
6. M. Sridhar, B. A. Kumar, and R. Narender, *Tetrahedron Lett.*, **39**, 2847 (1998).

Aryl 2-Formylbenzenesulfonate: [structure: benzene ring with SO_3Ar and CHO substituents]

The formylbenzenesulfonate prepared from a phenol (2-CHO–$C_6H_4SO_2Cl$, Et_3N) can be cleaved with NaOH (aq. acetone, rt, 5 min) in the presence of a hindered acetate.[1]

1. M. S. Shashidhar and M. V. Bhatt, *J. Chem. Soc., Chem. Commun.*, 654 (1987).

PROTECTION FOR CATECHOLS (1,2-Dihydroxybenzenes)

Catechols can be protected as diethers or diesters by methods that have been described to protect phenols. However, formation of cyclic acetals and ketals (e.g., methylenedioxy, acetonide, cyclohexylidenedioxy, and diphenylmethylenedioxy derivatives) or cyclic esters (e.g., borates or carbonates) selectively protects the two adjacent hydroxyl groups in the presence of isolated phenol groups.

Cyclic Acetals and Ketals

Methylene Acetal (Chart 4): [structure: 1,3-benzodioxole]

The methylenedioxy group, often present in natural products, is stable to many reagents. Efficient methods for both formation and removal of the group are available.

Formation

1. CH_2Br_2, NaOH, H_2O, Adogen, reflux, 3 h, 76–86% yield,[1] [Adogen = $R_3N^+CH_3Cl^-$, phase transfer catalyst (R = C_8–C_{10} straight-chain alkyl groups)]. Earlier methods required anhydrous conditions and aprotic solvents.
2. CH_2X_2 (X = Br, Cl), DMF, KF or CsF, 110°, 1.5 h, 70–98% yield.[2]
3. $BrCH_2Cl$, DMF, Cs_2CO_3, 70–110°, 86–97% yield.[3]
4. CH_2Cl_2, CsF, DMF, reflux, 91% yield.[4]

Cleavage

1. $AlBr_3$, EtSH, 0°, 0.5–1 h, 73–78% yield.[5] Aluminum bromide cleaves aryl and alkyl methyl ethers in high yield; methyl esters are stable.

2. PCl_5, CH_2Cl_2, reflux; H_2O; reflux, 3 h, 61% yield.[6]

PCl5, CH2Cl2, reflux → [dichloro intermediate] H2O → [carbonate] H2O, reflux, 3 h → catechol 61%

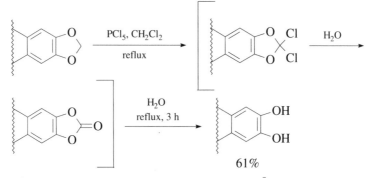

3. BCl_3, CH_3SCH_3, $ClCH_2CH_2Cl$, 83°, 98% yield.[7] Selective cleavage of an aryl methylenedioxy group or an aryl methyl ether by boron trichloride has been investigated.[8–10]
4. 9-Br–BBN, 24 h, 40°, CH_2Cl_2.[11]
5. A 4-nitro-1,2-methylenedioxybenzene has been cleaved to a catechol with 2 *N* NaOH, 90°, 30 min;[12] a similar compound substituted with a 4-nitro or 4-formyl group has been cleaved by $NaOCH_3$/DMSO, 150°, 2.5 min (13–74% catechol, 6–60% recovered starting material).[13]
6. $Pb(OAc)_4$, benzene, 50°, 8 h.[14]
7. $(TMS)_2NNa$ or LDA, THF, DMPU, 93–99% yield.[15]
8. $AlBr_3$, EtSH, 0°, 93% yield.[16]

1. A. P. Bashall and J. F. Collins, *Tetrahedron Lett.*, 3489 (1975).
2. J. H. Clark, H. L. Holland, and J. M. Miller, *Tetrahedron Lett.*, 3361 (1976).
3. R. E. Zelle and W. J. McClellan, *Tetrahedron Lett.*, **32**, 2461 (1991).
4. T. Geller, J. Jakupovic, and H.-G. Schmalz, *Tetrahedron Lett.*, **39**, 1541 (1998).
5. M. Node, K. Nishide, M. Sai, K. Ichikawa, K. Fuji, and E. Fujita, *Chem. Lett.*, 97 (1979).
6. G. L. Trammell, *Tetrahedron Lett.*, 1525 (1978).
7. P. G. Williard and C. B. Fryhle, *Tetrahedron Lett.*, **21**, 3731 (1980).
8. M. Gerecke, R. Borer, and A. Brossi, *Helv. Chim. Acta*, **59**, 2551 (1976).
9. S. Teitel, J. O'Brien, and A. Brossi, *J. Org. Chem.*, **37**, 3368 (1972).
10. F. M. Dean, J. Goodchild, L. E. Houghton, J. A. Martin, R. B. Morton, B. Parton, A. W. Price, and N. Somvichien, *Tetrahedron Lett.*, 4153 (1966).
11. M. V. Bhatt, *J. Organomet. Chem.*, **156**, 221 (1978).
12. E. Haslam and R. D. Haworth, *J. Chem. Soc.*, 827 (1955).
13. S. Kobayashi, M. Kihara, and Y. Yamahara, *Chem. Pharm. Bull.*, **26**, 3113 (1978).
14. Y. Ikeya, H. Taguchi, and I. Yoshioka, *Chem. Pharm. Bull.*, **29**, 2893 (1981).
15. J. R. Hwu, F. F. Wong, J.-J. Huang, and S.-C. Tsay, *J. Org. Chem.*, **62**, 4097 (1997).
16. Y.-Z. Hu and D. L. J. Clive, *J. Chem. Soc., Perkin Trans. 1*, 1421 (1997).

Pivaldehyde Acetal

The acetal is prepared from a catechol and pivaldehyde with TMSCl catalysis.[1]

1. Y. Nishida, M. Abe, H. Ohrui, and H. Meguro, *Tetrahedron: Asymmetry*, **4**, 1431 (1993).

Acetonide Derivative (Chart 4):

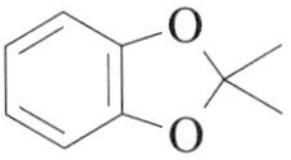

A catechol can be protected as an acetonide (acetone, 70% yield). It is cleaved with 6 *N* HCl (reflux, 2 h, high yield)[1] or by refluxing in acetic acid/H_2O (100°, 18 h, 90% yield).[2]

1. K. Ogura and G.-i. Tsuchihashi, *Tetrahedron Lett.*, 3151 (1971).
2. E. J. Corey and S. D. Hurt, *Tetrahedron Lett.*, 3923 (1977).

Cyclohexylidene Ketal:

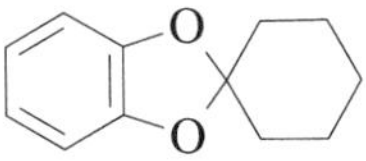

The cyclohexylidene ketal, prepared from a catechol and cyclohexanone (Al_2O_3/TsOH, CH_2Cl_2, reflux, 36 h),[1] is stable to metallation conditions (RX/BuLi) that cleave aryl methyl ethers.[2] The ketal is cleaved by acidic hydrolysis (concd. HCl/EtOH, reflux, 1.5 h, $\rightarrow$ 20°, 12 h); it is stable to milder acidic hydrolysis that cleaves tetrahydropyranyl ethers (1 *N* HCl/EtOH, reflux, 5 h, 91% yield).[3]

1. G. Schill and E. Logemann, *Chem. Ber.*, **106**, 2910 (1973).
2. G. Schill and K. Murjahn, *Chem. Ber.*, **104**, 3587 (1971).
3. J. Boeckmann and G. Schill, *Chem. Ber.*, **110**, 703 (1977).

Diphenylmethylene Ketal (Chart 4):

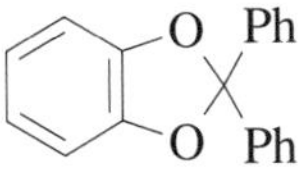

The diphenylmethylene ketal prepared from a catechol (Ph_2CCl_2, Pyr, acetone, 12 h),[1] (Ph_2CCl_2, neat, 170°, 5 min, 59%),[2] or [$Ph_2C(OMe)_2$, H_2SO_4, CH_2Cl_2, 40°, >83% yield][3] can be cleaved by hydrogenolysis (H_2/Pd–C, THF).[4,5] It has also been prepared from a 1,2,3-trihydroxybenzene (Ph_2CCl_2, 160°, 5 min, 80% yield) and cleaved by acidic hydrolysis (HOAc, reflux, 7 h).[6,7]

1. W. Bradley, R. Robinson, and G. Schwarzenbach, *J. Chem. Soc.*, 793 (1930).
2. S. Bengtsson and T. Högberg, *J. Org. Chem.*, **54**, 4549 (1989).
3. M. D. Shair, T. Y. Yoon, K. K. Mosny, T. C. Chou, and S. J. Danishefsky, *J. Am. Chem. Soc.*, **118**, 9509 (1996).
4. E. Haslam, R. D. Haworth, S. D. Mills, H. J. Rogers, R. Armitage, and T. Searle, *J. Chem. Soc.*, 1836 (1961).
5. K. S. Feldman, S. M. Ensel, and R. D. Minard, *J. Am. Chem. Soc.*, **116**, 1742 (1994).
6. L. Jurd, *J. Am. Chem. Soc.*, **81**, 4606 (1959).
7. T. R. Kelly, A. Szabados, and Y.-J. Lee, *J. Org. Chem.*, **62**, 428 (1997).

Ethyl Orthoformate: EtO—(O,O-catechol orthoformate)

The orthoformate is formed by the acid-catalyzed reaction of a catechol with triethyl orthoformate (82% yield) and is cleaved by acid-catalyzed hydrolysis (TsOH, MeOH, H_2O, rt, 16 h, 80–88% yield.).[1]

1. A. Merz and M. Rauschel, *Synthesis*, 797 (1993).

Diisopropylsilylene Derivative: $[(CH_3)_2CH]_2Si(OR)_2$

The diisopropylsilylene, formed from a catechol with $(i\text{-Pr})_2Si(OTf)_2$ and 2,6-lutidine in 96% yield, is cleaved with KF (MeOH, 2 eq. HCl).[1]

1. E. J. Corey and J. O. Link, *Tetrahedron Lett.*, **31**, 601 (1990).

Cyclic Esters

Cyclic Borate (Chart 4): (catechol O,O)BOH

A cyclic borate can be used to protect a catechol group during base-catalyzed alkylation or acylation of an isolated phenol group; the borate ester is then readily hydrolyzed by dilute acid.[1]

Formation[1]

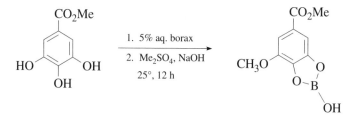

Methyl gallate (CO_2Me; HO, OH, OH) → 1. 5% aq. borax; 2. Me_2SO_4, NaOH, 25°, 12 h → CH_3O-substituted aryl CO_2Me cyclic borate (O–B(OH)–O)

Cleavage[1]

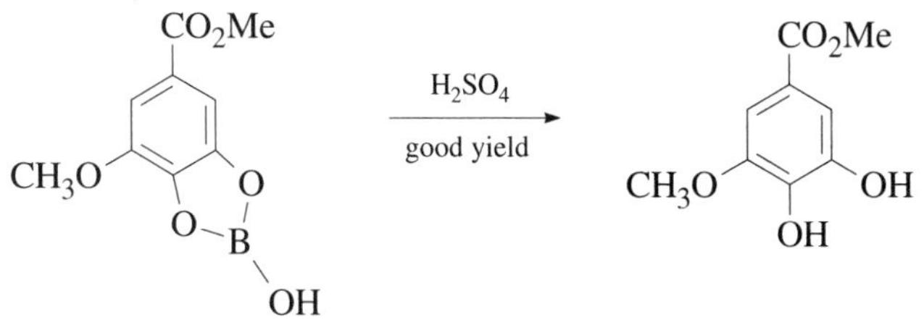

CO_2Me / CH_3O / O–B–OH cyclic borate $\xrightarrow[\text{good yield}]{H_2SO_4}$ CO_2Me / CH_3O / OH / OH

1. R. R. Scheline, *Acta Chem. Scand.*, **20**, 1182 (1966).

Cyclic Carbonate (Chart 4):

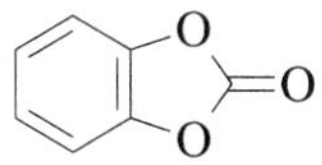

Cyclic carbonates have been used to a limited extent only (since they are readily hydrolyzed) to protect the catechol group in a polyhydroxy benzene.

Formation[1]

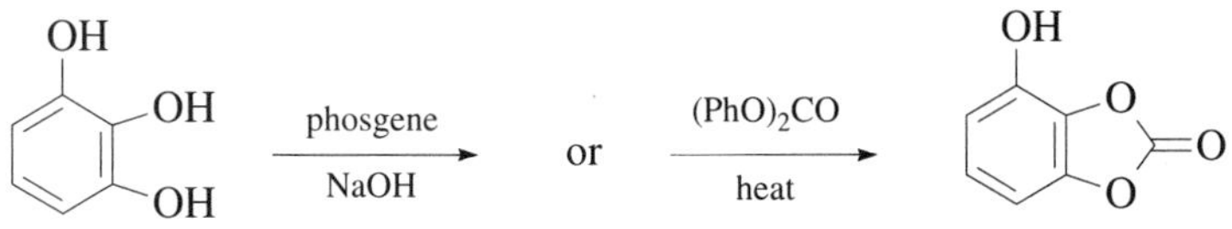

OH / OH / OH $\xrightarrow[\text{NaOH}]{\text{phosgene}}$ or $\xrightarrow[\text{heat}]{(PhO)_2CO}$ OH / O / =O / O

Cleavage

The cyclic carbonate is easily cleaved by refluxing in water for 30 min.[2] It can be converted to the 1,2-dimethoxybenzene derivative (aq. NaOH, Me_2SO_4, reflux, 3 h).[3]

1. A. Einhorn, J. Cobliner, and H. Pfeiffer, *Ber.*, **37**, 100 (1904).
2. H. Hillemann, *Ber.*, **71**, 34 (1938).
3. W. Baker, J. A. Godsell, J. F. W. McOmie, and T. L. V. Ulbricht, *J. Chem. Soc.*, 4058 (1953).

PROTECTION FOR 2-HYDROXYBENZENETHIOLS

Two derivatives have been prepared that may prove useful as protective groups for 2-hydroxybenzenethiols.

SH / OH

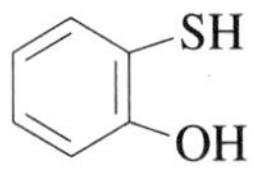

Formation

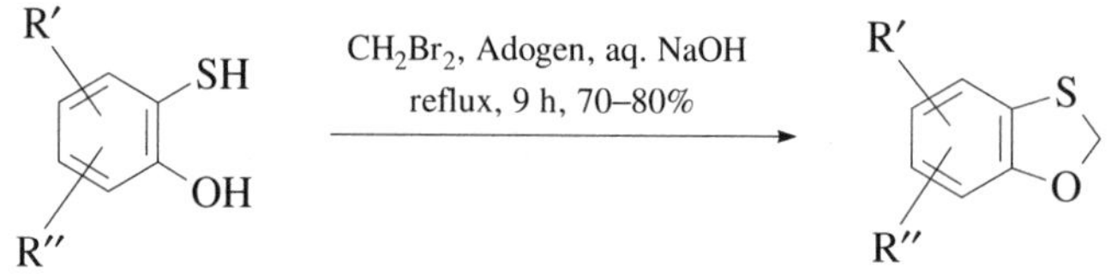

CH_2Br_2, Adogen, aq. NaOH
reflux, 9 h, 70–80%

R′, R″ = H, Me, Cl

Adogen = $MeR_3N^+Cl^-$, phase transfer catalyst

R = C_8–C_{10} straight chain alkyl groups

Ref. 1

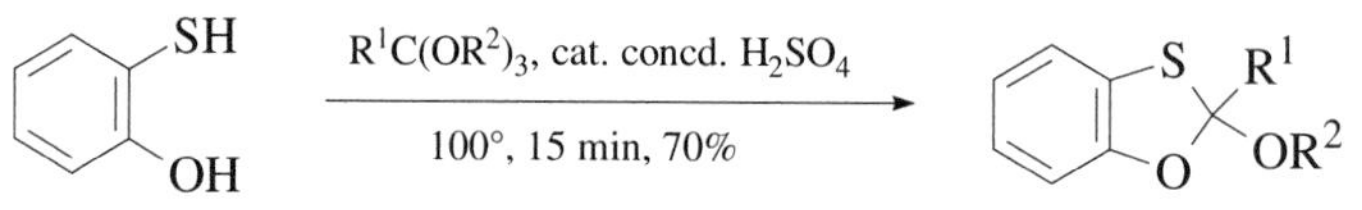

$R^1C(OR^2)_3$, cat. concd. H_2SO_4
100°, 15 min, 70%

R^1 = H, Me, Ph; R^2 = Me, Et

Ref. 2

1. S. Cabiddu, S. Melis, L. Bonsignore, and M. T. Cocco, *Synthesis*, 660 (1975).
2. S. Cabiddu, A. Maccioni, and M. Secci, *Synthesis*, 797 (1976).

4

PROTECTION FOR THE CARBONYL GROUP

During a synthetic sequence, a carbonyl group may have to be protected against attack by various reagents such as strong or moderately strong nucleophiles, including organometallic reagents; acidic, basic, catalytic, or hydride reducing agents; and some oxidants. Because of the order of reactivity of the carbonyl group [e.g., aldehydes (aliphatic > aromatic) > acyclic ketones and cyclohexanones > cyclopentanones > α,β-unsaturated ketones or α,α-disubstituted ketones >> aromatic ketones], it may be possible to protect a reactive carbonyl group selectively in the presence of a less reactive one. In keto steroids, the order of reactivity to ketalization is C_3 or Δ^4-$C_3 > C_{17} > C_{12} > C_{20} > C_{17,21\text{-}(OH)_2}C_{20} > C_{11}$.[1] A review discusses the relative rates of hydrolysis of acetals, ketals, and ortho esters.[2]

The most useful protective groups are the acyclic and cyclic acetals or ketals, and the acyclic or cyclic thio acetals or ketals. The protective group is introduced by treating the carbonyl compound in the presence of acid with an alcohol, diol, thiol, or dithiol. Cyclic and acyclic acetals and ketals are stable to aqueous and nonaqueous bases, to nucleophiles including organometallic reagents, and to hydride reduction. A 1,3-dithiane or 1,3-dithiolane, prepared to protect an aldehyde, is converted by strong base to an anion. The oxygen derivatives are stable to neutral and basic catalytic reduction and to reduction by sodium in ammonia. Although the sulfur analogues poison hydrogenation catalysts, they can be reduced by Raney Ni and by sodium/ammonia. The oxygen derivatives are stable to most oxidants; the sulfur derivatives are cleaved by a wide range of oxidants. The oxygen, but not the sulfur, analogues are readily cleaved by acidic hydrolysis. Sulfur analogues are stable to those conditions. The properties of oxygen and sulfur derivatives are combined in the cyclic 1,3-oxathianes and 1,3-oxathiolanes.

The carbonyl group forms a number of other very stable derivatives that are less used as protective groups because of the greater difficulty involved in their removal. Such derivatives include cyanohydrins, hydrazones, imines, oximes, and semicarbazones. Enol ethers are used to protect one carbonyl group in a 1,2- or 1,3-dicarbonyl compound.

Although IUPAC no longer uses the term "ketal," we have retained it to indicate compounds formed from ketones.

Derivatives of carbonyl compounds that have been used as protective groups in synthetic schemes are described in this chapter; some of the more important protective groups are listed in Reactivity Chart 5.[3–5]

1. H. J. E. Loewenthal, *Tetrahedron*, **6**, 269 (1959).
2. E. H. Cordes and H. G. Bull, *Chem. Rev.*, **74**, 581–603 (1974).
3. See also H. J. E. Loewenthal, "Protection of Aldehydes and Ketones," in *Protective Groups in Organic Chemistry*, J. F. W. McOmie, Ed., Plenum, New York and London, 1973, pp. 323–402.
4. J. F. W. Keana, in *Steroid Reactions*, C. Djerassi, Ed., Holden-Day, San Francisco, 1963, pp. 1–66, 83–87.
5. P. J. Kocienski, *Protecting Groups*, Thieme Medical Publishers, New York, 1994, Chapter 5.

ACETALS AND KETALS

Acyclic Acetals and Ketals

Methods similar to those used to form and cleave dimethyl acetal and ketal derivatives can be used for other dialkyl acetals and ketals.[1]

Dimethyl Acetals and Ketals: $R_2C(OCH_3)_2$ (Chart 5)

Formation

1. MeOH, dry HCl, 2 min.[2]

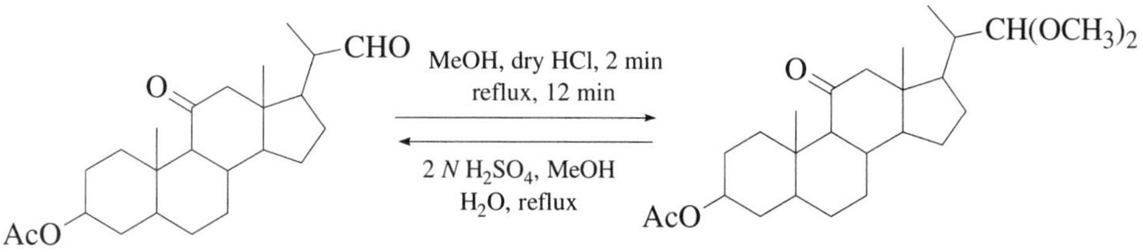

MeOH, dry HCl, 2 min reflux, 12 min
2 *N* H_2SO_4, MeOH H_2O, reflux

2. DCC–$SnCl_4$; ROH, $(CO_2H)_2$, 90% yield.[3]
3. $CH(OMe)_3$, $MeNO_2$, CF_3COOH, reflux, 4 h, 81–93% yield.[4] This procedure was reported to be particularly effective for the preparation of ketals of diaryl ketones.
4. MeOH, $LaCl_3$, $(MeO)_3CH$, 25°, 10 min, 80–100% yield.[5] Dimethyl acetals can be prepared efficiently under neutral conditions by catalysis with lanthanoid halides, but the results of the reaction with ketones are unpredictable.
5. Me_3SiOCH_3, Me_3SiOTf, CH_2Cl_2, −78°, 86% yield.[6] The use of TMSOFs to catalyze this transformation has also been demonstrated.[7] A norbornyl ketone was not ketalized under these conditions.
6. $(MeO)_3CH$, anhydrous MeOH, TsOH, reflux, 2 h.[8] Diethyl ketals have been prepared under similar conditions (EtOH, TsOH, 0–23°, 15 min to 6 h, 80–95% yield) in the presence of molecular sieves to shift the equilibrium by adsorbing water.[9] Amberlyst-15[10] or graphite bisulfate[11] and $(EtO)_3CH$ have been used to prepare diethyl ketals.

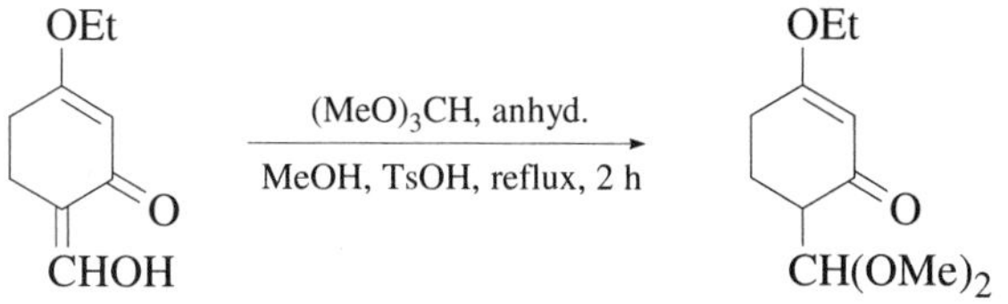

$(MeO)_3CH$, anhyd.
MeOH, TsOH, reflux, 2 h

In the following example, a mixture of the *cis*- and *trans*-decalones is converted completely to the *cis*-isomer—in general, the thermodynamically less favored isomer:[12]

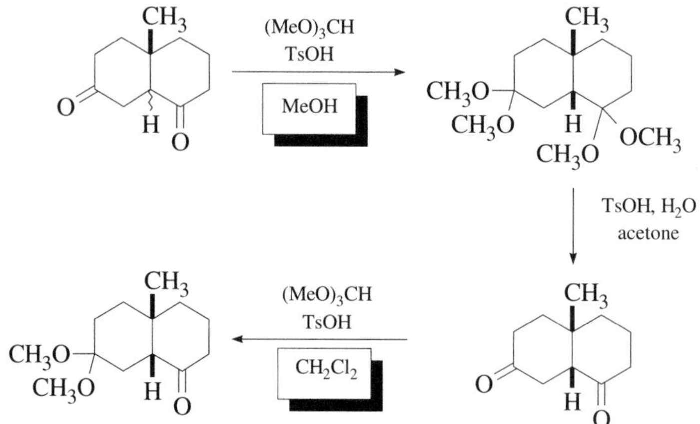

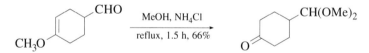

7. MeOH, $(MeO)_4Si$, dry HCl, 25°, 3 days.[13]
8. MeOH, acidic ion-exchange resin, 7–86% yield.[14]
9. $(MeO)_3CH$, Montmorillonite Clay K-10, 5 min to 15 h, > 90% yield.[15] Diethyl ketals have been prepared in satisfactory yield by reaction of the carbonyl compound and ethanol in the presence of Montmorillonite Clay.[16] Kaolinitic clay has also been used.[17]
10. MeOH, NH_4Cl, reflux, 1.5 h, 66% yield.[18]

CH_3O-cyclohexenyl-CHO $\xrightarrow[\text{reflux, 1.5 h, 66\%}]{\text{MeOH, } NH_4Cl}$ O=cyclohexyl-$CH(OMe)_2$

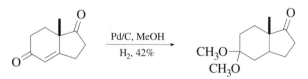

11. Hydrogenation of enones in MeOH with Pd/C resulted in acetal formation. When ethylene glycol/THF is used as solvent, the related dioxolane is formed in 86% yield.[19]

$\xrightarrow[H_2\text{, 42\%}]{\text{Pd/C, MeOH}}$ (CH_3O, CH_3O ketal product)

12. MeOH, $PhSO_2NHOH$, 25°, 15 min, 75–85% yield.[20]
13. Me_2SO_4, 2 *N* NaOH, MeOH, H_2O, reflux, 30 min, 85% yield.[21] In this case, the hemiacetal of phthaldehyde is alkylated with methyl sulfate; such use is probably restricted to cases that are stable to the strongly basic conditions.
14. Allyl bromide, $Sb(OEt)_3$, 80°, 2–6 h, 85–98% yield.[22] This method is chemoselective for aldehydes in the presence of ketones.
15. MeOH, Ce^+-exchanged Montmorillonite Clay, 25°, 0.5–12 h, 18–99% yield. Aldehydes can be selectively protected in the presence of ketones.[23]

16. $Sc(OTf)_3$, $HC(OCH_3)_3$ (TMOF), toluene, 0°, 0.5 h, 92% yield.[24]
17. $CeCl_3 \cdot 7\ H_2O$, MeOH, TMOF.[25]
18. R–CO–CH₂OTs → (KOH, MeOH, 0–5°, 55–83%) → CH_3O OCH_3 R–C–CH₂OH Ref. 26

Cleavage

The acid-catalyzed cleavage of acetals and ketals is greatly influenced by the substitution on the acetal or ketal carbon atom. The following values for k_H^+ illustrate the magnitude of the effect:[27]

Ph—CH(OEt)(OPh)	Ph—CH(OEt)(OEt)	Ph—C(CH₃)(OEt)(OEt)	MeOPh—CH(OEt)(OEt)	CH₂(OEt)(OEt)	CH₃CH(OEt)(OEt)
41	160	6×10^3	5×10^3	1.5×10^{-4}	1.6

1. 50% CF_3COOH, $CHCl_3$, H_2O, 0°, 90 min, 96% yield.[28]

(dimethyl acetal / dithiane / dioxolane substrate) → 50% TFA, $CHCl_3$–H_2O, 0°, 90 min, 96% → OHC (dithiane / dioxolane product)

2. TsOH, acetone.[29]
3. SiO_2 and oxalic or sulfuric acid, 0.5–24 h, 90–95% yield.[30]
4. Me_3SiI, CH_2Cl_2, 25°, 15 min, 85–95% yield.[31] Under these cleavage conditions, 1,3-dithiolanes, alkyl and trimethylsilyl enol ethers, and enol acetates are stable. 1,3-Dioxolanes give complex mixtures. Alcohols, epoxides, trityl, *t*-butyl, and benzyl ethers and esters are reactive. Most other ethers and esters, amines, amides, ketones, olefins, acetylenes, and halides are expected to be stable.
5. $TiCl_4$, LiI, Et_2O, rt, 3 h, 75–90% yield.[32]
6. $LiBF_4$, wet CH_3CN, 96% yield. Unsubstituted 1,3-dioxolanes are hydrolyzed only slowly, but substituted dioxolanes are completely stable.[33] This reagent proved excellent for hydrolysis of the dimethyl ketal in the presence of the acid-sensitive oxazolidine[34] and polyene.[35]

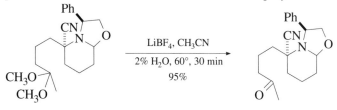

(Ph, CN oxazolidine with CH_3O / CH_3O dimethyl ketal) → $LiBF_4$, CH_3CN, 2% H_2O, 60°, 30 min, 95% → (Ph, CN oxazolidine with ketone O)

TMS, MOMO, $CH(OMe)_2$, $CH_2OTBDMS$ —[$LiBF_4$, H_2O / CH_3CN]→ CHO

7. HCO_2H, pentane, 1 h, 20°.[36] Under these conditions, a β,γ-double bond does not migrate into conjugation.
8. Amberlyst-15, acetone, H_2O, 20 h.[37] Aldehyde acetals conjugated with electron-withdrawing groups tend to be slow to hydrolyze. The use of HCl/THF or PPTS/acetone in the following case was slow and caused considerable isomerization. A TBDMS group is stable under these conditions.[38]

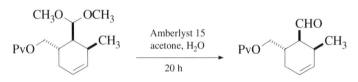

CH_3O, OCH_3, PvO, CH_3 —[Amberlyst 15, acetone, H_2O, 20 h]→ CHO, PvO, CH_3

9. 70% H_2O_2, Cl_3CCO_2H, CH_2Cl_2, *t*-BuOH; dimethyl sulfide, 80% yield.[39] Other methods cleaved the epoxide. This method also cleaves the THP and trityl groups.

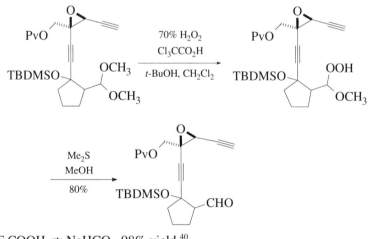

O, PvO, TBDMSO, OCH_3, OCH_3 —[70% H_2O_2, Cl_3CCO_2H, *t*-BuOH, CH_2Cl_2]→ O, PvO, TBDMSO, OOH, OCH_3

—[Me_2S, MeOH, 80%]→ O, PvO, TBDMSO, CHO

10. CF_3COOH, rt; $NaHCO_3$, 98% yield.[40]

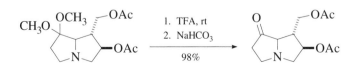

CH_3O, OCH_3, OAc, OAc, N —[1. TFA, rt; 2. $NaHCO_3$; 98%]→ O, OAc, OAc, N

11. AcOH, H_2O, 89% yield.[41] A factor of 400 in the relative rate of hydrolysis is attributed to a conformational effect in which the lone pair on oxygen in the silyl ketals does not overlap with the incipient cation during hydrolysis.

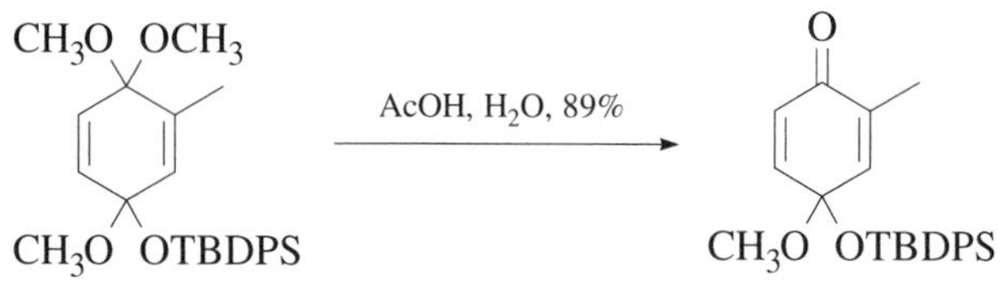

AcOH, H_2O, 89%

12. Oxalic acid, THF, H_2O, rt, 12 min, 72% yield.[42]

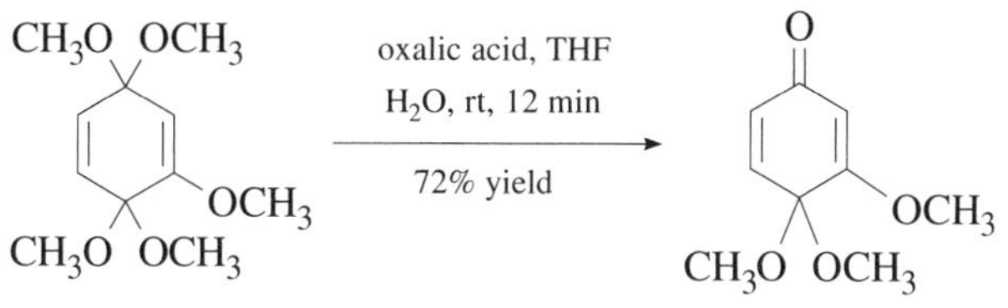

oxalic acid, THF, H_2O, rt, 12 min, 72% yield

13. $BF_3 \cdot Et_2O$, $Et_4N^+I^-$, $CHCl_3$, 69–82% yield.[43]
14. 10% H_2O, silica gel, CH_2Cl_2, 18 h, rt.[44] In this example, attempts to use HCl resulted in THP cleavage followed by cyclization to form a furan.

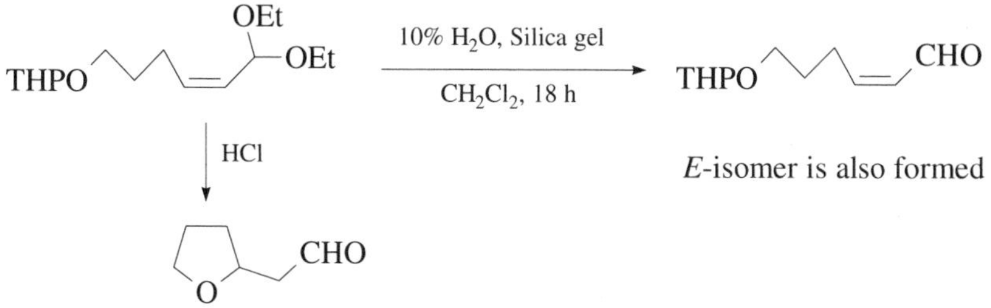

10% H_2O, Silica gel, CH_2Cl_2, 18 h

HCl

E-isomer is also formed

15. DMSO, H_2O, dioxane, reflux, 12 h, 65–99% yield.[45] These conditions cleave a dimethyl ketal in the presence of a *t*-butyldimethylsilyl ether.
16. SiH_2I_2, CH_3CN, $-42°$, 3–40 min, 90–100% yield. Other ketals are also cleaved under these conditions.[46]
17. The direct conversion of dimethyl ketals to other carbonyl protected derivatives is also possible. Treatment of a dimethyl ketal with $HSCH_2CH_2SH$, $TeCl_4$, $ClCH_2CH_2Cl$ gives the dithiolane in 99% yield.[47]
18. $Mo_2(acac)_2$, CH_3CN, rt, 70–91% yield.[48]
19. Acetyl chloride, $SmCl_3$, pentane, rt, 15 min–24 h, 89–96% yield.[49] The efficiency of dioxolane cleavage is very poor under these conditions.
20. $SnCl_2 \cdot 2H_2O$, C_{60}, CH_2Cl_2, 2 h, 84–99% yield.[50] The presence of the buckyball improves the yield.

21. $Ru(CH_3CN)_3(triphos)(OTf)_2$, acetone, rt, 99% yield.[51] Nonphenolic THP groups and dioxolane ketals are stable.

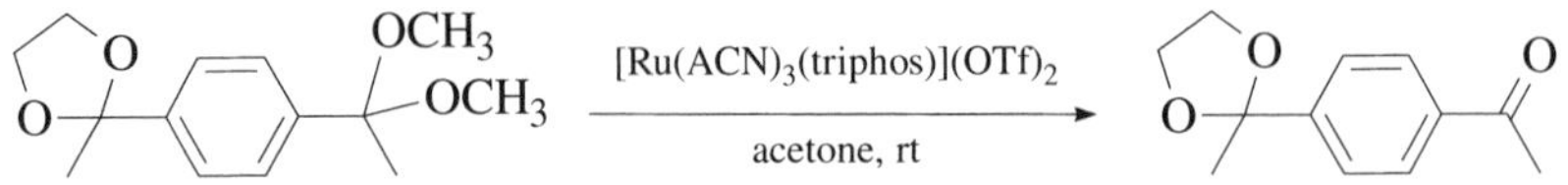

$[Ru(ACN)_3(triphos)](OTf)_2$, acetone, rt

22. DDQ, MeCN, H_2O, rt, 75–92% yield.[52,53] It was shown that this reaction does not proceed through acid catalysis by the hydroquinone.
23. HM-Zeolite, H_2O, PhMe, reflux, 9 h, 89% yield.[54]
24. $ISiCl_3$, rt, 20–30 min, 74–95% yield.[55] Esters and phenolic methyl ethers are reported to survive, whereas with the related TMSI they are cleaved.
25. $ZnCl_2$, Me_2S, AcCl, THF, 89% yield.[56] A dimethyl acetal is chemoselectively cleaved in the presence of a dioxolane acetal.

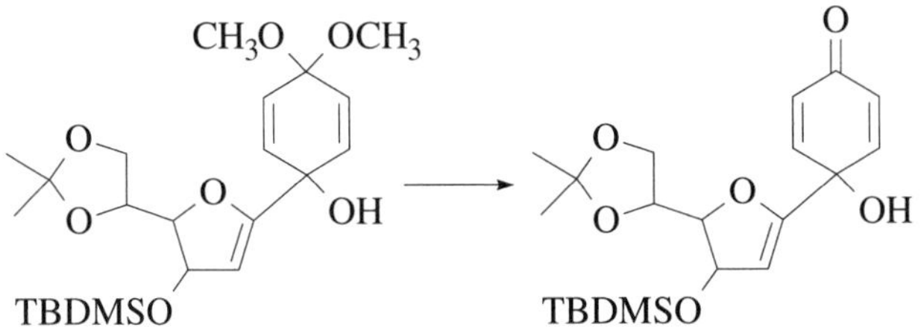

26. $Na_2S_2O_4$, THF, H_2O, 90% yield.[57]
27. Montmorillonite K10, CH_2Cl_2, rt or reflux, 92–100% yield. 1,3-Dioxanes and 1,3-dioxolanes are cleaved similarly.[58]
28. Me_2BBr, CH_2Cl_2, −78°, 45 min, 100% yield. These conditions were chosen when conventional acid-catalyzed hydrolysis resulted in aldehyde epimerization during a kainic acid synthesis.[59]
29. $[Ru(ACN)_3(triphos)](OTf)_2$, acetone, rt, 5 h.[51] Dioxolanes are also cleaved when not conjugated, as in the following case:

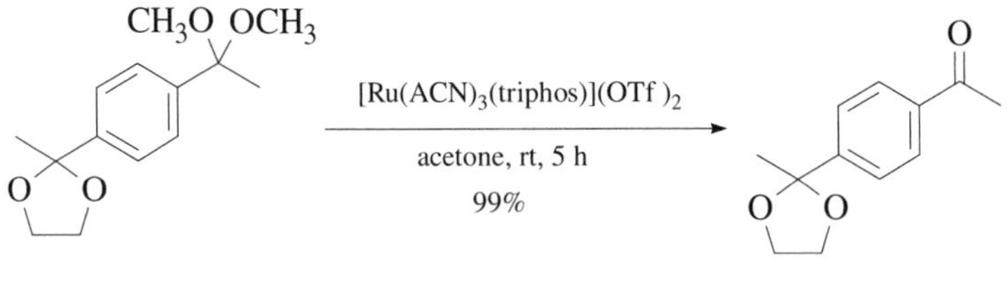

$[Ru(ACN)_3(triphos)](OTf)_2$, acetone, rt, 5 h, 99%

1. F. A. J. Meskens, *Synthesis*, 501 (1981).
2. A. F. B. Cameron, J. S. Hunt, J. F. Oughton, P. A. Wilkinson, and B. M. Wilson, *J. Chem. Soc.*, 3864 (1953).
3. N. H. Andersen and H.-S. Uh, *Synth. Commun.*, **3**, 125 (1973).
4. A. Thurkauf, A. E. Jacobson, and K. C. Rice, *Synthesis*, 233 (1988).
5. A. L. Gemal and J.-L. Luche, *J. Org. Chem.*, **44**, 4187 (1979).

6. M. Vandewalle, J. Van der Eycken, W. Oppolzer, and C. Vullioud, *Tetrahedron*, **42**, 4035 (1986).
7. B. H. Lipshutz, J. Burgess-Henry, and G. P. Roth, *Tetrahedron Lett.*, **34**, 995 (1993).
8. E. Wenkert and T. E. Goodwin, *Synth. Commun.*, **7**, 409 (1977).
9. D. P. Roelofsen, E. R. J. Wils, and H. Van Bekkum, *Recl. Trav. Chim. Pays-Bas*, **90**, 1141 (1971).
10. S. A. Patwardhan and S. Dev, *Synthesis*, 348 (1974).
11. J. P. Alazard, H. B. Kagan, and R. Setton, *Bull. Soc. Chim. Fr.*, 499 (1977).
12. J. B. P. A. Wijnberg, R. P. W. Kesselmans, and A. de Groot, *Tetrahedron Lett.*, **27**, 2415 (1986).
13. W. W. Zajac and K. J. Byrne, *J. Org. Chem.*, **35**, 3375 (1970).
14. N. B. Lorette, W. L. Howard, and J. H. Brown, Jr., *J. Org. Chem.*, **24**, 1731 (1959).
15. E. C. Taylor and C.-S. Chiang, *Synthesis*, 467 (1977). Montmorillonite Clay is activated $Al_2O_3/SiO_2/H_2O$.
16. V. M. Thuy and P. Maitte, *Bull. Soc. Chim. Fr.*, 2558 (1975).
17. D. Ponde, H. B. Borate, A. Sudalai, T. Ravindranathan, and V. H. Deshpande, *Tetrahedron Lett.*, **37**, 4605 (1996).
18. J. I. DeGraw, L. Goodman, and B. R. Baker, *J. Org. Chem.*, **26**, 1156 (1961).
19. P. Hudson and P. J. Parsons, *Synlett*, 867 (1992).
20. A. Hassner, R. Wiederkehr, and A. J. Kascheres, *J. Org. Chem.*, **35**, 1962 (1970).
21. E. Schmitz, *Chem. Ber.*, **91**, 410 (1958).
22. Y. Liao, Y.-Z. Huang, and F.-H. Zhu, *J. Chem. Soc., Chem. Commun.*, 493 (1990).
23. J.-i. Tateiwa, H. Horiuchi, and S. Uemura, *J. Org. Chem.*, **60**, 4039 (1995).
24. K. Ishihara, Y. Karumi, M. Kubota, and H. Yamamoto, *Synlett*, 839 (1996).
25. A. B. Smith, III, M. Fukui, H. A. Vaccaro, and J. R. Empfield, *J. Am. Chem. Soc.*, **113**, 2071 (1991).
26. O. Prakash, N. Saini, and P. K. Sharma, *J. Chem. Res., Synop.*, 430 (1993).
27. D. P. N. Satchell and R. S. Satchell, *Chem. Soc. Rev.*, **19**, 55 (1990).
28. R. A. Ellison, E. R. Lukenbach, and C.-W. Chiu, *Tetrahedron Lett.*, 499 (1975).
29. E. W. Colvin, R. A. Raphael, and J. S. Roberts, *J. Chem. Soc., Chem. Commun.*, 858 (1971).
30. F. Huet, A. Lechevallier, M. Pellet, and J. M. Conia, *Synthesis*, 63 (1978).
31. M. E. Jung, W. A. Andrus, and P. L. Ornstein, *Tetrahedron Lett.*, 4175 (1977).
32. G. Balme and J. Goré, *J. Org. Chem.*, **48**, 3336 (1983).
33. B. H. Lipshutz and D. F. Harvey, *Synth. Commun.*, **12**, 267 (1982).
34. M. Bonin, J. Royer, D. S. Grierson, and H.-P. Husson, *Tetrahedron Lett.*, **27** 1569 (1986).
35. W. R. Roush and R. J. Sciotti, *J. Am. Chem. Soc.*, **116**, 6457 (1994).
36. F. Barbot and P. Miginiac, *Synthesis*, 651 (1983).
37. G. M. Cappola, *Synthesis*, 1021 (1984).
38. A. E. Greene, M. A. Teixeira, E. Barreiro, A. Cruz, and P. Crabbé, *J. Org. Chem.*, **47**, 2553 (1982).

39. A. G. Meyers, M. A. M. Fundy, and P. A. Linstrom, Jr., *Tetrahedron Lett.*, **29**, 5609 (1988).
40. J. J. Tufariello and K. Winzenberg, *Tetrahedron Lett.*, **27**, 1645 (1986).
41. A. J. Stern and J. S. Swenton, *J. Org. Chem.*, **54**, 2953 (1989).
42. D. A. Evans, S. P. Tanis, and D. J. Hart, *J. Am. Chem. Soc.*, **103**, 5813 (1981).
43. A. K. Mandal, P. Y. Shrotri, and A. D. Ghogare, *Synthesis*, 221 (1986).
44. L. Crombie and D. Fisher, *Tetrahedron Lett.*, **26**, 2477 (1985).
45. T. Kametani, H. Kondoh, T. Honda, H. Ishizone, Y. Suzuki, and W. Mori, *Chem. Lett.*, 901 (1989); K. R. Muralidharan, M. K. Mokhallalati, and L. N. Pridgen, *Tetrahedron Lett.*, **35**, 7489 (1994).
46. E. Keinan, D. Perez, M. Sahai, and R. Shvily, *J. Org. Chem.*, **55**, 2927 (1990).
47. H. Tani, K. Masumoto, and T. Inamasu, *Tetrahedron Lett.*, **32**, 2039 (1991).
48. M. L. Kantam, V. Swapna, and P. L. Santhi, *Synth. Commun.*, **25**, 2529 (1995).
49. S.-H. Wu and Z.-B. Ding, *Synth. Commun.*, **24**, 2173 (1994).
50. K. L. Ford and E. J. Roskamp, *J. Org. Chem.*, **58**, 4142 (1993); K. L. Ford and E. J. Roskamp, *Tetrahedron Lett.*, **33**, 1135 (1992).
51. S. Ma and L. M. Venanzi, *Tetrahedron Lett.*, **34**, 8071 (1993).
52. K. Tanemura, T. Suzuki, and T. Horaguchi, *J. Chem. Soc., Chem. Commun.*, 979 (1992).
53. A. Oku, M. Kinugasa, and T. Kamada, *Chem. Lett.*, 165 (1993).
54. M. N. Rao, P. Kumar, A. P. Singh, and R. S. Reddy, *Synth. Commun.*, **22**, 1299 (1992).
55. S. S. Elmorsy, M. V. Bhatt, and A. Pelter, *Tetrahedron Lett.*, **33**, 1657 (1992).
56. C. Chang, K. C. Chu, and S. Yue, *Synth. Commun.*, **22**, 1217 (1992).
57. K. A. Parker and D.-S. Su, *J. Org. Chem.*, **61**, 2191 (1996).
58. E. C. L. Gautier, A. E. Graham, A. McKillop, S. P. Standen, and R. J. K. Taylor, *Tetrahedron Lett.*, **38**, 1881 (1997).
59. S. Hanessian and S. Ninkovic, *J. Org. Chem.*, **61**, 5418 (1996).

Diisopropyl Acetal and Ketals: $(i\text{-PrO})_2CHR$

Formation

$CH(Oi\text{-Pr})_3$, CSA, IPA, removal of MeOH by distillation, 3 h, 68–92% yield.[1,2]

Cleavage

Formic acid, THF, H_2O, 20°, 100% yield. This acetal was chosen to prevent conjugation of a double bond during hydrolysis, which occurred when the corresponding dimethyl acetal was hydrolyzed.[1]

1. J. Sandri and J. Viala, *Synthesis*, 271 (1995).
2. A. Pommier, J.-M. Pons, and P. J. Kocienski, *J. Org. Chem.*, **60**, 7334 (1995).

Bis(2,2,2-trichloroethyl) Acetals and Ketals: $R_2C(OCH_2CCl_3)_2$ (Chart 5)

Formation[1]

$R_2C(OR')_2$ + TsOH, Benzene, reflux; R′ = Me or Et

→ (1.5 eq. Cl_3CCH_2OH) $R_2C(OR')(OCH_2CCl_3)$

→ (4 eq. Cl_3CCH_2OH) $R_2C(OCH_2CCl_3)_2$

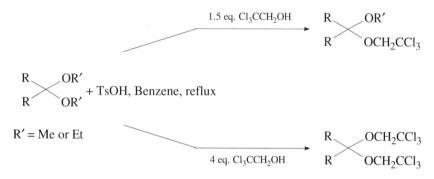

It is more efficient to prepare this ketal by an exchange reaction with the dimethyl or diethyl ketal than directly from the carbonyl compound. Hydrolysis can also be effected by acid catalysis.

Cleavage

1. Zn/EtOAc or THF, reflux, 3–12 h, 40–100% yield.[1]

1. J. L. Isidor and R. M. Carlson, *J. Org. Chem.*, **38**, 554 (1973).

Dibenzyl Acetals and Ketals: $R_2C(OCH_2Ph)_2$

Formation/Cleavage[1]

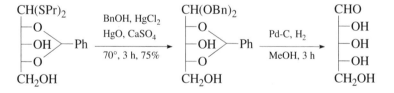

$CH(SPr)_2$ sugar benzylidene (O, OH, O)–Ph, CH_2OH → (BnOH, $HgCl_2$, HgO, $CaSO_4$; 70°, 3 h, 75%) $CH(OBn)_2$ benzylidene (O, OH, O)–Ph, CH_2OH → (Pd-C, H_2; MeOH, 3 h) CHO, OH, OH, OH, CH_2OH

1. J. H. Jordaan and W. J. Serfontein, *J. Org. Chem.*, **28**, 1395 (1963).

Bis(2-nitrobenzyl) Acetals and Ketals: $R_2C(OCH_2C_6H_4\text{–}2\text{-}NO_2)_2$

Formation

$2\text{-}NO_2C_6H_4CH_2OSiMe_3$, Me_3SiOTf, −78°, 78–95% yield.[1]

Cleavage

Photolysis at 350 nm, 85–95% yield.[1]

1. D. Gravel, S. Murray, and G. Ladouceur, *J. Chem. Soc., Chem. Commun.*, 1828 (1985).

Diacetyl Acetals and Ketals: $R_2C(OAc)_2$

Formation

1. Ac_2O, 1 drop concd. H_2SO_4, 20°, 1 h, 95% yield.[1]
2. Ac_2O, $ZnCl_2$.[2]
3. Ac_2O, $FeCl_3$, rt, < 30 min, 60–93% yield.[3] These conditions selectively protect an aldehyde in the presence of a ketone.[4] This combination also converts *t*-butyldimethylsilyl (TBDMS) ethers to acetates.
4. Ac_2O, PCl_3, 20°, 1–24 h, 30–90% yield.[5] Aromatic aldehydes bearing electron-withdrawing groups tend to give low yields under these conditions.
5. Ac_2O, Nafion H, 50–99% yield.[6]
6. Ac_2O, Expansive Graphite, rt, 0.3–6 h, 65–98% yield.[7]
7. Ac_2O, β-Zeolite, 60°, 1.5–5 h, 51–95% yield.[8]
8. Ac_2O, Environcat EPZG, 60–65°, 1.5–8.5 h, 69–99% yield.[9]
9. Ac_2O, HY-Zeolite, rt, CCl_4 or no solvent, 86–95% yield.[10]
10. Ac_2O, I_2, rt, 30 min, 70–99% yield.[11]

Cleavage

1. NaOH or K_2CO_3, THF, H_2O or MeOH.[3] This protective group is stable to MeOH (18 h); 10% HCl (MeOH, 30 min); 10% Na_2CO_3 (H_2O, Et_2O, 70 min); and $NaHCO_3$ (THF, H_2O, 4 h).
2. Alumina, 35°, 30–40 sec, 88–98% yield.[12]
3. Potassium 3-dimethylaminophenoxide, THF, 0°, 10 min, 92–98% yield.[13]
4. Expansive Graphite, CH_2Cl_2 or benzene, reflux, 10–30 min, 95–99% yield.[14]
5. CAN, Silica gel, CH_2Cl_2, 90–95% yield.[15]
6. Montmorillonite Clay K-10 or KSF, CH_2Cl_2, reflux, 86–98% yield.[16]
7. The use of enzymes for the hydrolysis of acylals is effective, and in the case of racemic derivatives some enantioenrichment of the aldehyde is possible.[17]

1. M. Tomita, T. Kikuchi, K. Bessho, T. Hori, and Y. Inubushi, *Chem. Pharm. Bull.*, **11**, 1484 (1963).
2. I. Scriabine, *Bull. Soc. Chim. Fr.*, 1194 (1961).
3. K. S. Kochhar, B. S. Bal, R. P. Deshpande, S. N. Rajadhyaksha, and H. W. Pinnick, *J. Org. Chem.*, **48**, 1765 (1983).

4. J. Kula, *Synth. Commun.*, **16**, 833 (1986).
5. J. K. Michie and J. A. Miller, *Synthesis*, 824 (1981).
6. G. A. Olah and A. K. Mehrotra, *Synthesis*, 962 (1982).
7. T.-S. Jin, G.-Y. Du, Z.-H. Zhang, and T.-S. Li, *Synth. Commun.*, **27**, 2261 (1997).
8. P. Kumar, V. R. Hedge, and J. T. P. Kumar, *Tetrahedron Lett.*, **36**, 601 (1995).
9. B. P. Bandgar, N. P. Mahajan, D. P. Mulay, J. L. Thote, and P. P. Wadgaonkar, *J. Chem. Res., Synop.*, 470 (1995).
10. C. Pereira, B. Gigante, M. J. Marcelo-Curto, H. Carreyre, G. Pérot, and M. Guisnet, *Synthesis*, 1077 (1995).
11. N. Deka, D. J. Kalita, R. Borah, and J. C. Sarma, *J. Org. Chem.*, **62**, 1563 (1997).
12. R. S. Varma, A. K. Chatterjee, and M. Varma, *Tetrahedron Lett.*, **34**, 3207 (1993).
13. Y.-Y. Ku, R. Patel, and D. Sawick, *Tetrahedron Lett.*, **34**, 8037 (1993).
14. T.-S. Jin, Y.-R. Ma, Z.-H. Zhang, and T.-S. Li, *Synth. Commun.*, **27**, 3379 (1997).
15. P. Cotelle and J.-P. Catteau, *Tetrahedron Lett.*, **33**, 3855 (1992).
16. T.-S. Li, Z.-H. Zhang, and C.-G. Fu, *Tetrahedron Lett.*, **38**, 3285 (1997).
17. Y. S. Angelis and I. Smonou, *Tetrahedron Lett.*, **38**, 8109 (1997).

Cyclic Acetals and Ketals

Kinetic studies of acetal/ketal formation from cyclohexanone, and hydrolysis (3×10^{-3} N HCl/dioxane-H_2O, 20°), indicate the following orders of reactivity:[1]

Formation

$HOCH_2C(CH_3)_2CH_2OH > HO(CH_2)_2OH > HO(CH_2)_3OH$

Cleavage

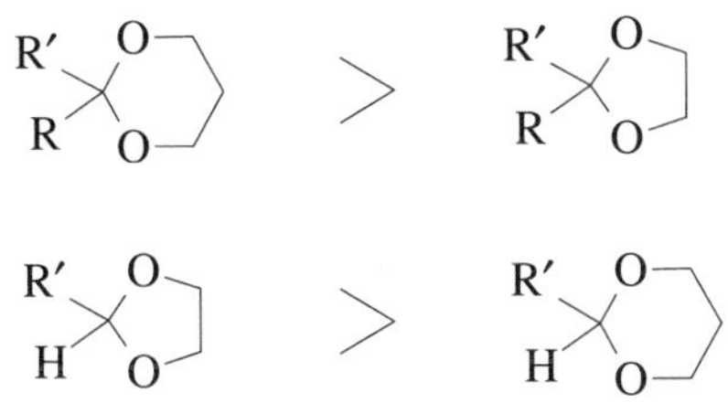

The relative rates of acid-catalyzed hydrolysis of some dioxolanes [dioxolane: aq. HCl (1:1)] are as follows: 2,2-dimethyldioxolane: 2-methyldioxolane: dioxolane, 50,000:5,000:1.[2]

A review[3] discusses the condensation of aldehydes and ketones with glycerol to give 1,3-dioxanes and 1,3-dioxolanes. The chemistry of *O*/*O* and *O*/*S* acetals has been reviewed,[4] and a recent monograph discusses this area of protective groups in a didactic sense.[5]

1. M. S. Newman and R. J. Harper, *J. Am. Chem. Soc.*, **80**, 6350 (1958); S. W. Smith and M. S. Newman, *J. Am. Chem. Soc.*, **90**, 1249, 1253 (1968).
2. P. Salomaa and A. Kankaanperä, *Acta Chem. Scand.*, **15**, 871 (1961).
3. A. J. Showler and P. A. Darley, *Chem. Rev.*, **67**, 427–440 (1967).
4. H. Hagemann and D. Klamann, Eds. *O/O-und O/S-Acetale* [*Methoden der Organischen Chemie*, Houben-Weyl)] 4th ed., G. Thieme, Stuttgart, 1991, Band E 14a/1.
5. P. J. Kocienski, "Carbonyl Protecting Groups," In *Protecting Groups*, Thieme Medical Publishers, New York, 1994, Chapter 5.

1,3-Dioxanes (Chart 5):

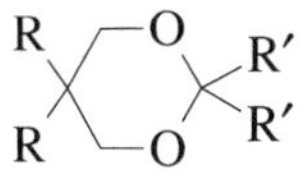

R = H, CH_3

Formation

1. $HO(CH_2)_3OH$, TsOH, benzene, reflux.[1–3]

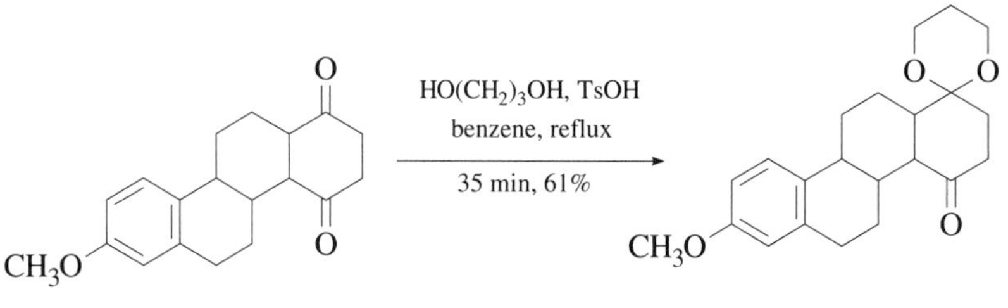

$HO(CH_2)_3OH$, TsOH, benzene, reflux, 35 min, 61%

Ref. 1

TsOH, 91%

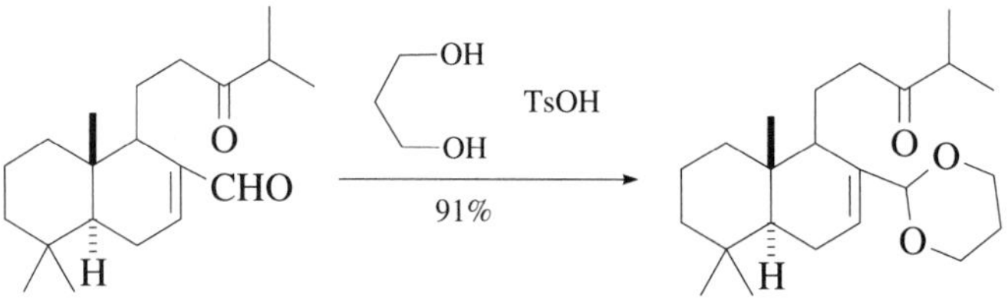

Ref. 2

In the first example, selective protection was more successful with 1,3-propanediol than with ethylene glycol.[1]

2. 1,3-Propanediol, THF, Amberlyst-15, 5 min, 50–70% yield.[4] This method is also effective for the preparation of 1,3-dioxolanes.
3. TMSCl, $SmCl_3$, THF, 71–99% yield. Ketals are cleaved faster than acetals.[5]
4. $HOCH_2C(CH_3)_2CH_2OH$, $Sc(NTf_2)_3$, toluene, 0°, 3 h, 87–92% yield.[6]

5. 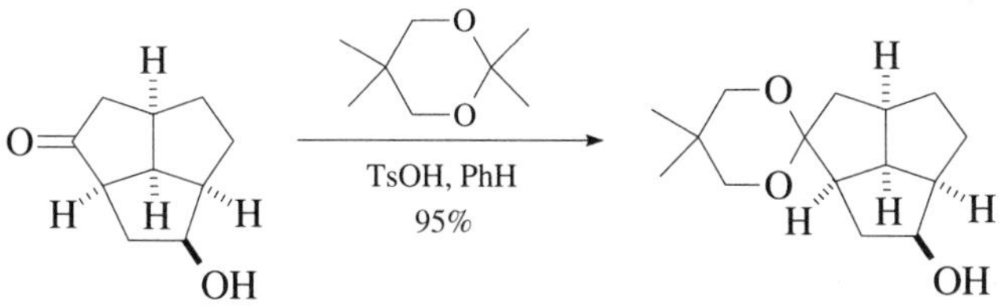 TsOH, PhH 95%

Other methods for ketalization met with failure.[7]

6. $HOCH_2C(CH_3)_2CH_2OH$, *N*-4-methoxybenzyl-2-cyanopyridinium hexafluoroantimonate, toluene, reflux, 1.5–3.7 h, 85–99% yield.[8]
7. $TMSOCH_2C(CH_3)_2CH_2OTMS$, TMSOTf, Pyr, 75% yield.[9]
8. $HOCH_2CH_2CH_2OH$, $Ru(CH_3CN)_3(triphos)(OTf)_2$, 94–99% yield.[10]
9. $HOCH_2C(CH_3)_2CH_2OH$, sulfated Zirconia, benzene, reflux, 88–97% yield.[11]
10. $HOCH_2C(CH_3)_2CH_2OH$, Yittria–Zirconia, rt, $CHCl_3$, 75–96% yield.[12]

Cleavage

The section on the cleavage of 1,3-dioxolanes should be consulted, since a majority of the methods available are applicable to 1,3-dioxanes as well.

1. J. E. Cole, W. S. Johnson, P. A. Robins, and J. Walker, *J. Chem. Soc.*, 244 (1962).
2. H. Okawara, H. Nakai, and M. Ohno, *Tetrahedron Lett.*, **23**, 1087 (1982).
3. For examples of the use of the related 4,4-dimethyl-1,3-dioxane, see E. Piers, J. Banville, C. K. Lau, and I. Nagakura, *Can. J. Chem.*, **60**, 2965 (1982); M. A. Avery, C. Jennings-White, and W. K. M. Chong, *Tetrahedron Lett.*, **28**, 4629 (1987).
4. A. E. Dann, J. B. Davis, and M. J. Nagler, *J. Chem. Soc., Perkin Trans. 1*, 158 (1979).
5. Y. Ukaji, N. Koumoto, and T. Fujisawa, *Chem. Lett.*, 1623 (1989).
6. K. Ishihara, Y. Karumi, M. Kubota, and H. Yamamoto, *Synlett*, 839 (1996).
7. L. A. Paquette and S. Borrelly, *J. Org. Chem.*, **60**, 6912 (1995).
8. S.-B. Lee, S.-D. Lee, T. Takata, and T. Endo, *Synthesis*, 368 (1991).
9. C. K. F. Chiu, L. N. Mander, A. D. Stuart, and A. C. Willis, *Aust. J. Chem.*, **45**, 227 (1992).
10. S. Ma and L. M. Venanzi, *Synlett*, 751 (1993).
11. A. Sakar, O. S. Yemul, B. P. Bandgar, N. B. Gaikwad, and P. P. Wadgaonkar, *Org. Prep. Proced. Int.*, **28**, 613 (1996).
12. G. C. G. Pals, A. Keshavaraja, K. Saravanan, and P. Kumar, *J. Chem. Res., Synop.*, 426 (1996).

5-Methylene-1,3-dioxane (Chart 5):

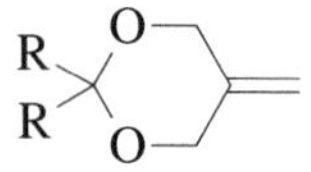

Formation[1]

1. CH_2=C(CH_2OH)$_2$, TsOH, benzene, reflux, 90% yield.

Cleavage[1]

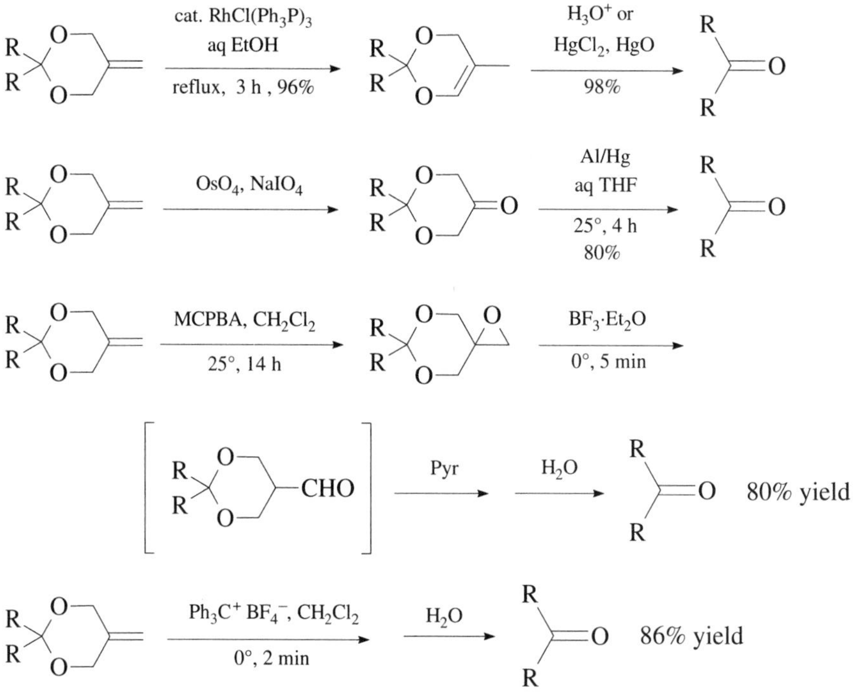

cat. $RhCl(Ph_3P)_3$, aq EtOH, reflux, 3 h, 96%; H_3O^+ or $HgCl_2$, HgO, 98%

OsO_4, $NaIO_4$; Al/Hg, aq THF, 25°, 4 h, 80%

MCPBA, CH_2Cl_2, 25°, 14 h; $BF_3{\cdot}Et_2O$, 0°, 5 min; [CHO intermediate]; Pyr; H_2O; 80% yield

$Ph_3C^+BF_4^-$, CH_2Cl_2, 0°, 2 min; H_2O; 86% yield

The rhodium-catalyzed isomerization can also be carried out with the chiral catalyst, $Ru_2Cl_4(diop)_3$ (H_2, 20–80°, 1–6 h, 47–90% yield). In this case, optically enriched enol ethers are obtained.[2]

1. E. J. Corey and J. W. Suggs, *Tetrahedron Lett.*, 3775 (1975).
2. H. Frauenrath and M. Kaulard, *Synlett*, 517 (1994).

5,5-Dibromo-1,3-dioxane (Chart 5):

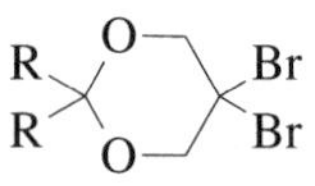

Formation

$Br_2C(CH_2OH)_2$, TsOH, benzene, heat for several hours, 84–94% yield.[1]

Cleavage

Zn–Ag, THF, AcOH, 25°, 1 h, ~90% yield.[1]

1. E. J. Corey, E. J. Trybulski, and J. W. Suggs, *Tetrahedron Lett.*, 4577 (1976).

5-(2′-Pyridyl)-1,3-dioxane:

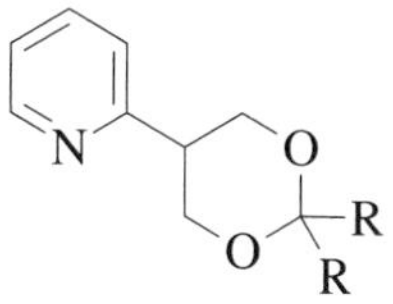

Formation/Cleavage[1]

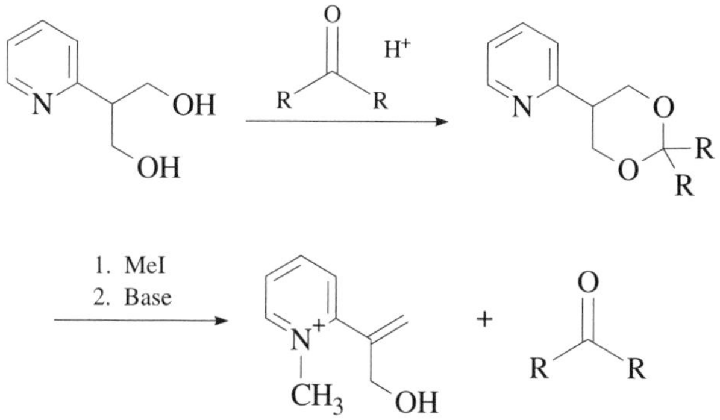

This group is stable to 0.1 *M* HCl.

1. A. R. Katritzky, W.-Q. Fan, and Q.-L. Li, *Tetrahedron Lett.*, **28**, 1195 (1987).

5-Trimethylsilyl-1,3-dioxane:

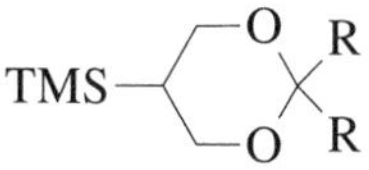

Formation

1. 2-Trimethylsilyl-1,3-propanediol, CH_2Cl_2, rt, 3Å ms, CSA, 45–97% yield.[1]

Cleavage[1]

1. $BF_3 \cdot Et_2O$, THF.
2. $LiBF_4$, THF, 66°, 71–93% yield. 1,3-Dioxolanes of ketones were not affected.

1. B. H. Lipshutz, P. Mollard, C. Lindsley, and V. Chang, *Tetrahedron Lett.*, **38**, 1873 (1997).

Salicylate Acetals

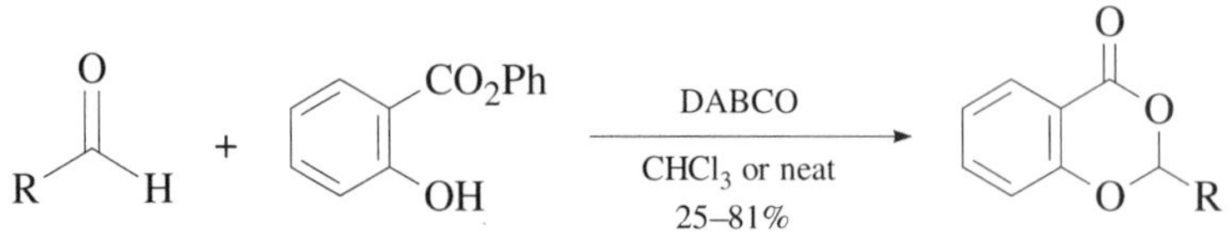

$$\text{RCHO} + \text{2-}(CO_2Ph)C_6H_4OH \xrightarrow[\text{CHCl}_3\text{ or neat, } 25\text{–}81\%]{\text{DABCO}} \text{2-R-4H-1,3-benzodioxin-4-one}$$

Although aromatic aldehydes failed to react, this is one of the few methods available for the preparation of acetals under basic conditions.[1,2]

1. P. Perlmutter and E. Puniani, *Tetrahedron Lett.*, **37**, 3755 (1996).
2. A. A. Khan, N. D. Emslie, S. E. Drewes, J. S. Field, and N. Ramesar, *Chem. Ber.*, **126**, 1477 (1993).

1,3-Dioxolanes (Chart 5):

The 1,3-dioxolane group is probably the most widely used carbonyl protective group. For the protection of carbonyls containing other acid-sensitive functionality, one should use acids of low acidity or pyridinium salts. In general, a molecule containing two similar ketones can be selectively protected at the less hindered carbonyl, assuming that neither or both of the carbonyls are conjugated to an alkene.[1]

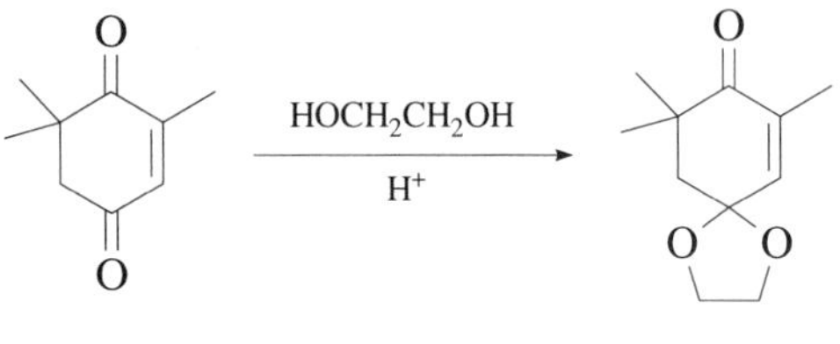

$$\xrightarrow[\text{H}^+]{\text{HOCH}_2\text{CH}_2\text{OH}}$$

Ref. 1b

$$\xrightarrow[\text{PhH, TsOH, reflux, 4 h}]{\text{HOCH}_2\text{CH}_2\text{OH}}$$

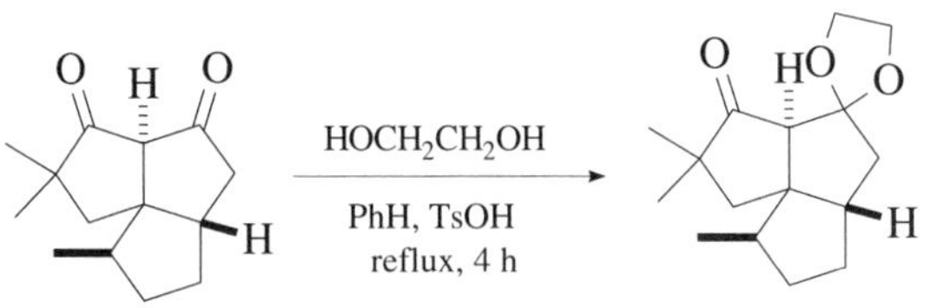

Ref. 1a

If one carbonyl is conjugated with a double bond, the unconjugated carbonyl is selectively protected. This generalization appears to be independent of ring size.[2] Simple aldehydes are generally selectively protected over simple ketones.[3] In the formation of 1,3-dioxolanes of enones, control of the olefin regiochemistry is determined by the acidity of the acid catalyst. Acids of high acidity (pK_a ~1) may cause the double bond to migrate to the β,γ-position, whereas

Table 1[4]

Acid	pK_a	% α,β	% β,γ	% conversion
Fumaric acid	3.03	100	0	90
Phthalic acid	2.89	70	30	90
Oxalic acid[8]	1.23	80	20	93
TsOH acid	<1.0	0	100	100

acids of low acidity (pK_a ~ 3) do not cause double-bond migration (see Table 1).[4] In addition, the use of the bistrimethylsilyl derivative of ethylene glycol and Me_3SiOTf (CH_2Cl_2, −78°, 20 h, pyridine quench, 92%) for the protection of enones proceeds without double-bond migration.[5,6] A similar result was obtained with the Wieland–Miescher ketone using stoichiometric amounts of TsOH.[7]

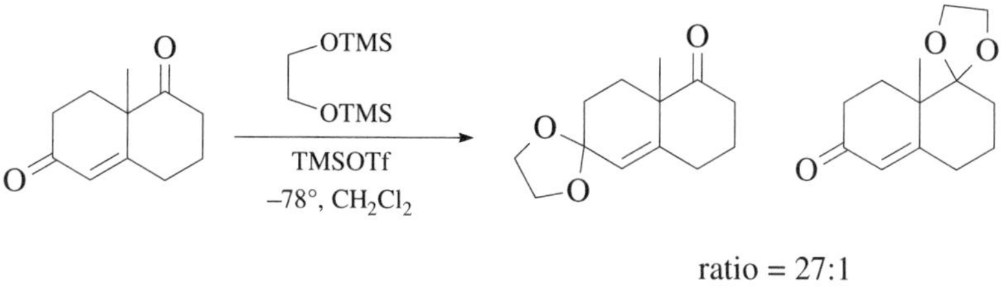

OTMS, OTMS, TMSOTf, −78°, CH_2Cl_2

ratio = 27:1

Ref. 5

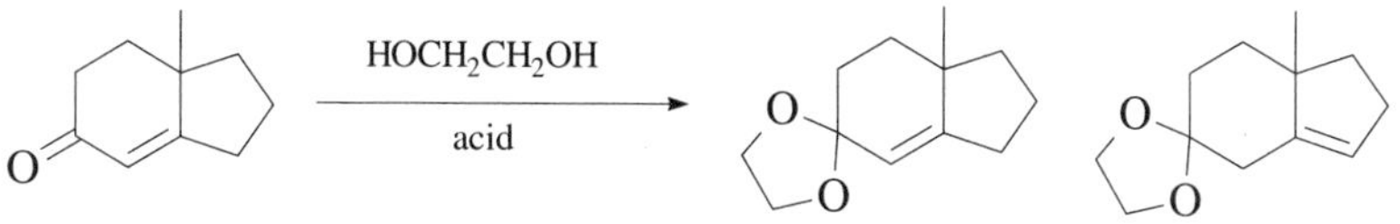

$HOCH_2CH_2OH$, acid

Ref. 4

A polymer-supported 1,2-diol has also been developed for use in carbonyl protection.[9]

Formation

1. $HO(CH_2)_2OH$, TsOH, C_6H_6, reflux, 75–85% yield.[10]
2. $HO(CH_2)_2OH$, TsOH, $(EtO)_3CH$, 25°, 65% yield.[11]
3. $HO(CH_2)_2OH$, $BF_3{\cdot}Et_2O$, HOAc, 35–40°, 15 min, 90% yield.[12]
4. $HO(CH_2)_2OH$, HCl, 25°, 12 h, 55–90% yield.[13]
5. $HO(CH_2)_2OH$, Me_3SiCl, MeOH or CH_2Cl_2.[14]
6. $HO(CH_2)_2OH$, Al_2O_3, $PhCH_3$ or CCl_4, heat, 24 h, 80–100% yield.[3] These conditions are selective for the formation of acetals from aldehydes in the presence of ketones.
7. $Me_3SiOCH_2CH_2OSiMe_3$, Me_3SiOTf, 15 Kbar (1.5 GPa), 40°, 48 h.[15] These conditions were used to prepare the ketal of fenchone, which cannot be done under normal acid-catalyzed conditions.

8. $HO(CH_2)_2OH$, 0.1 eq. $CuCl_2 \cdot H_2O$, 80°, 30 min, 82–100% yield.[16] The use of 5 eq. of $CuCl_2$ results in the formation of the α-chloro ketal.

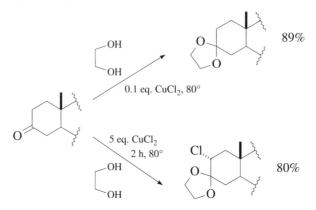

9. $HO(CH_2)_2OH$, oxalic acid, CH_3CN, 25°, 95% yield.[17] Note that ketals prepared with oxalic acid from enones tend to retain the olefin regiochemistry.[8]
10. $HO(CH_2)_2OH$, adipic acid, C_6H_6, reflux, 17–24 h, 10–85% yield.[18]
11. $HO(CH_2)_2OH$, SeO_2, $CHCl_3$, 28°, 4 h, 60% yield.[19]
12. $HO(CH_2)_2OH$, $C_5H_5N^+H\ Cl^-$, C_6H_6, reflux, 6 h, 85% yield.[20]
13. $HO(CH_2)_2OH$, $C_5H_5N^+H\ TsO^-$, C_6H_6, reflux, 1–3 h, 90–95% yield.[21]
14. $HO(CH_2)_nOH$ (n = 2,3)/$MeOCH^+NMe_2\ MeOSO_3^-$, 0–25°, 2 h, 40–95% yield.[22]
15. $HO(CH_2)_nOH$ (n = 2, 3)/column packed with an acid ion-exchange resin, 5 min, 50–90% yield.[23]
16. $HOCH_2CH_2OH$, $(EtO)_3CH$, p-TsOH, 83% yield.[24]
17. 2-Methoxy-1,3-dioxolane/TsOH, C_6H_6, 40–50°, 4 h, 85% yield.[25]
18. 2-Ethoxy-1,3-dioxolane, pyridinium tosylate (PPTS), benzene, heat, 8 h, 89% yield.[26] In this case, protection of an enone proceeds without double-bond migration.
19. 2-Ethyl-2-methyl-1,3-dioxolane/TsOH, reflux, 75% yield.[27,28] These conditions selectively protect a ketone in the presence of an enone.
20. 2,2-Dimethyl-1,3-dioxolane, microwave irradiation, Montmorillonite KSF, 38–95% yield.[29]
21. 2-Dimethylamino-1,3-dioxolane/cat. HOAc, CH_2Cl_2, 83% yield.[30] 2-Dimethylamino-1,3-dioxolane protects a reactive ketone under mild conditions: it reacts selectively with a C_3-keto steroid in the presence of a Δ^4-3-keto steroid. C_{12}- and C_{20}-keto steroids do not react.
22. Diethylene orthocarbonate, $C(-OCH_2CH_2O-)_2$/TsOH or wet $BF_3 \cdot Et_2O$, $CHCl_3$, 20°, 70–95% yield.[31]

23. 1,3-Dioxolanes have been prepared from a carbonyl compound and an epoxide (e.g., ketone/$SnCl_4$, CCl_4, 20°, 4 h, 53% yield[32] or aldehyde/$Et_4N^+Br^-$, 125–220°, 2–4 h, 20–85% yield[33]). Perhalo ketones can be protected by reaction with ethylene chlorohydrin under basic conditions (K_2CO_3, pentane, 25°, 2 h, 85% yield;[34] or NaOH, EtOH–H_2O, 95% yield[35]).
24. When the carbonyl group is very electron-deficient, thus stabilizing the hemiacetal, a dioxolane can be prepared under basic conditions.[34,36]
25. $HOCH_2CH_2OH$, $(i\text{-PrO})_3CH$, $RhCl_3$(triphos), [triphos = $H_3CC(CH_2PPh_2)_3$], rt, reflux, 80–100% yield.[37]
26. $HO(CH_2)_2OH$, PhH, catalyst, quant.[38]

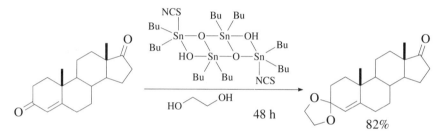

NCS Bu Bu
Bu Sn—O—Sn—OH
Bu Bu
HO—Sn—O—Sn
Bu Bu Bu NCS
HO OH
48 h
82%

4.7% of the 17-ketal and 8.3% of the diketal are also obtained.

27. $HO(CH_2)_2OH$, $ZrOCl_2{\cdot}8\,H_2O$, aq. NaOH, 65–98% yield.[39]
28. $HO(CH_2)_2OH$, PhH, *N*-benzylpyridinium hexafluoroantimonate, 1.5–9 h, reflux, 72–91% yield.[40] It is also possible to form the 4,4-dimethyldioxane (85–99% yield) under these conditions.
29. $HO(CH_2)_2OH$, $[Ru(MeCN)_3(Ph_3P)](OTf)_2$, PhH, azeotropic distillation, 87–99% yield.[41]
30. Ethylene oxide, $BF_3{\cdot}Et_2O$, >120 min, CH_2Cl_2, 25°, 47–95% yield.[42]
31. $HOCH_2CH_2OH$, $BuSnCl_3$, 0°, 10 min, 75–92% yield.[43]
32. From a tosylhydrazone: ethylene glycol, 200°, 89% yield.[44]
33. Selective ketone protection: The –CHO group is converted in Step 1 to a siloxysulfonium salt [$R'CH(OTMS)S^+Me_2\ ^-OTf$] that is reconverted to an aldehyde group in Step 3.[45]

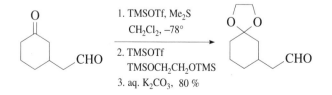

O CHO
1. TMSOTf, Me_2S
CH_2Cl_2, –78°
2. TMSOTf
$TMSOCH_2CH_2OTMS$
3. aq. K_2CO_3, 80 %
O O CHO

34. $HO(CH_2)_nOH$, $n = 2,3$, Fe or Al, rt, 52–99% yield.[46]

35. $TMSOCH_2CH_2OTMS$, TfOH or FsOH (fluorosulfonic acid), BTMSA [bis(trimethylsilyl)acetamide] or BTMSU [bis(trimethylsilyl)urea], 76–97% yield.[47]
36. $HO(CH_2)_nOH$, $n = 2,3$, *i*-PrOTMS, TMSOTf, CH_2Cl_2, −20°, 3 h, 84–99% yield.[48]
37. $HOCH_2CH_2OH$, $MgSO_4$, PhH, L-tartaric acid, reflux, 20 h, 97% yield. These conditions were optimized for the protection of unsaturated aldehydes to prevent double-bond migration.[49]

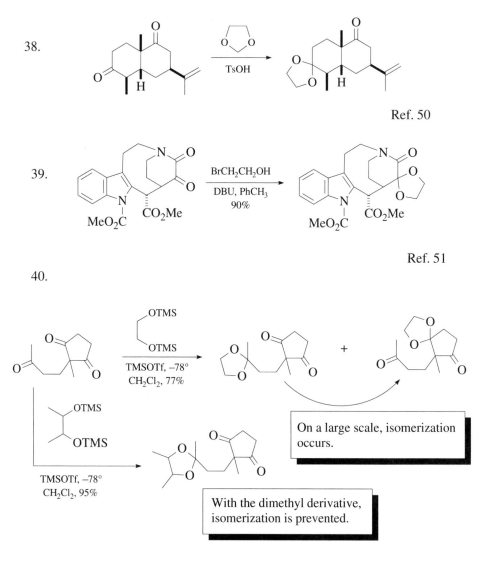

38.

TsOH

Ref. 50

39.

$BrCH_2CH_2OH$

DBU, $PhCH_3$

90%

Ref. 51

40.

TMSOTf, −78°

CH_2Cl_2, 77%

On a large scale, isomerization occurs.

TMSOTf, −78°

CH_2Cl_2, 95%

With the dimethyl derivative, isomerization is prevented.

Ref. 52

Cleavage

1,3-Dioxolanes can be cleaved by acid-catalyzed exchange dioxolanation, acid-catalyzed hydrolysis, or oxidation. Some representative examples are shown in the following list:

1. Pyridinium tosylate (PPTS), acetone, H_2O, heat, 100% yield.[53]

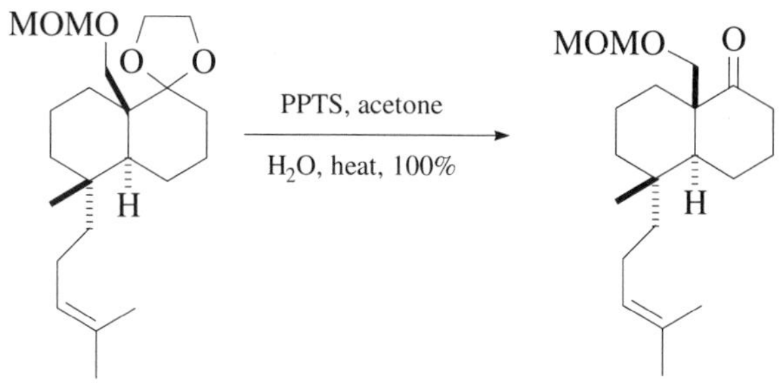

PPTS, acetone
H_2O, heat, 100%

2. Acetone, TsOH, 20°, 12 h.[54] The reactant is a 3,6,17-tris(ethylenedioxy) steroid; the product has carbonyl groups at C-6 and C-17.
3. Acetone, H_2O, PPTS, reflux, 1–3 h, 90–95% yield.[21]
4. 5% HCl, THF, 25°, 20 h.[55]

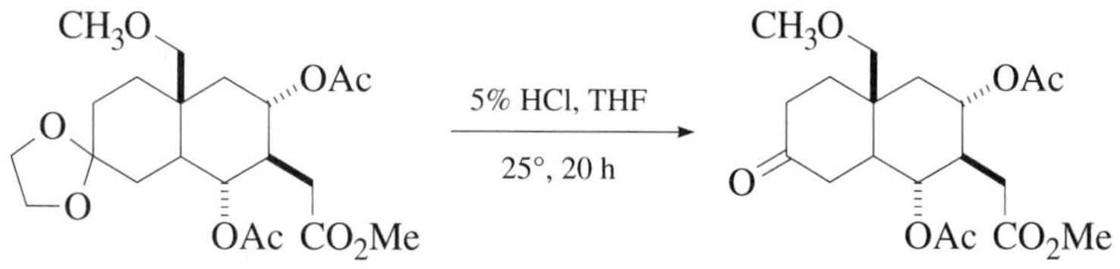

5% HCl, THF
25°, 20 h

5. 1 *M* HCl, THF, 0° → 25°, 13 h, 71% yield. Note that the acetonide survives these conditions.[56] Some variations have been reported in this system (including the use of 30% AcOH, 90°, high yield).[57]

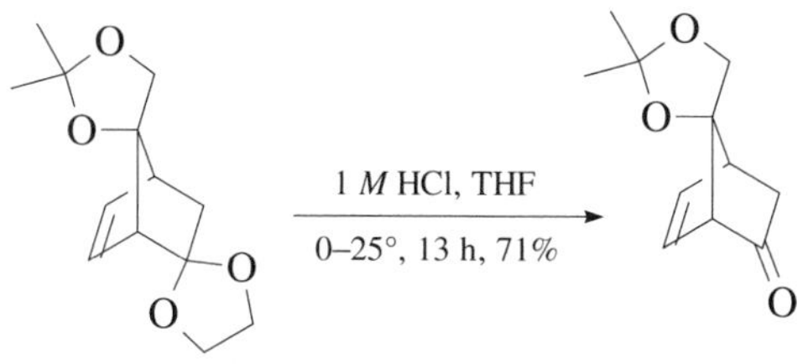

1 *M* HCl, THF
0–25°, 13 h, 71%

6. 80% AcOH, 65°, 5 min, 85% yield.[58]
7. Wet magnesium sulfate (C_6H_6, 20°, 1 h) effects selective, quantitative cleavage of an α,β-unsaturated 1,3-dioxolane in the presence of a 1,3-dioxolane.[18]
8. Perchloric acid (79% $HClO_4/CH_2Cl_2$, 0°, 1 h → 25°, 3 h, 87% yield)[59] and periodic acid (aq. dioxane, 3 h, quant. yield)[60] cleave 1,3-dioxolanes;

the latter drives the reaction to completion by oxidation of the ethylene glycol that forms. Yields are substantially higher from cleavage with perchloric acid (3 *N* $HClO_4$/THF, 25°, 3 h, 80% yield) than with hydrochloric acid (HCl/HOAc, 65% yield).[61]

9. SiO_2, H_2O, CH_2Cl_2, oxalic acid, 90–95% yield.[62] These conditions selectively cleave α,β-unsaturated ketals.

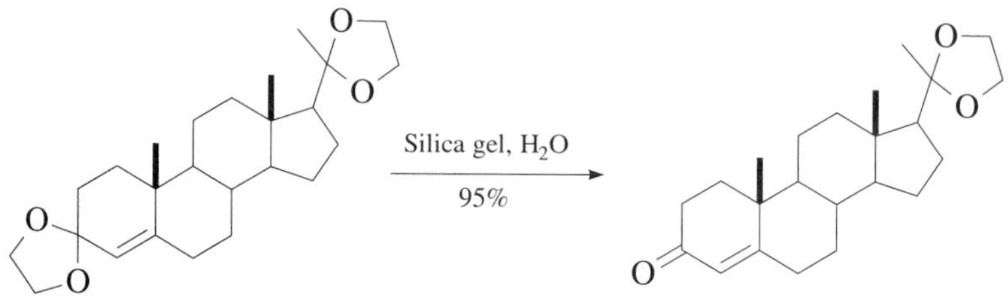

Silica gel, H_2O
95%

10. $Ph_3C^+BF_4^-$, CH_2Cl_2, 25°, 60–100% yield.[63,64]

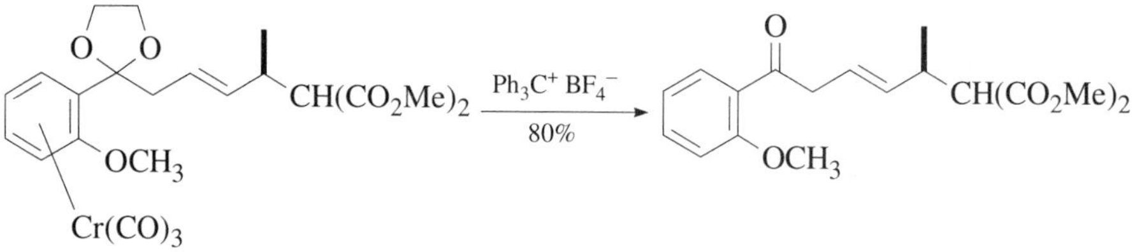

$CH(CO_2Me)_2$ OCH_3 $Cr(CO)_3$ $Ph_3C^+ BF_4^-$ 80% O $CH(CO_2Me)_2$ OCH_3

Ref. 64

1,3-Dithiolanes are not affected by these conditions, but a 1,3-oxathiolane is cleaved (100% yield).[65]

11. Me_2BBr, CH_2Cl_2, −78°, 90–97% yield.[66] This reagent also cleaves MTM, MEM, and MOM ethers (87–95% yield).

12. $PdCl_2(CH_3CN)_2$, acetone, H_2O, 82–100% yield.[67]

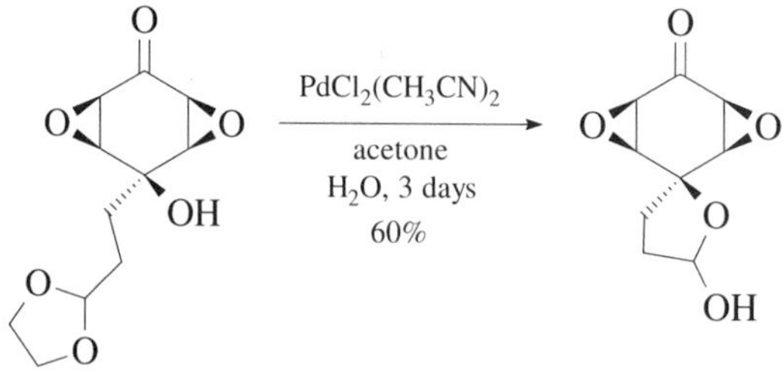

O O O OH O O $PdCl_2(CH_3CN)_2$ acetone H_2O, 3 days 60% O O O O O OH

Ref. 68

13. Me_3SiI.[69]

14. *t*-BuOOH, $Pd(OOCCF_3)(OO\text{-}t\text{-}Bu)$, benzene, 50°, 12 h, 60–80% yield.[70] In this case, an acetal is oxidized to the ester of ethylene glycol ($RCO_2CH_2CH_2OH$).

15. $LiBF_4$, wet CH_3CN.[71] Unsubstituted 1,3-dioxolanes are cleaved slowly

under these conditions (40% in 5 h). The 4,5-dimethyl- and 4,4,5,5-tetramethyldioxolane and the 1,3-dioxane are inert under these conditions. Dimethyl ketals are readily cleaved.

16. $TiCl_4$, Et_2O, LiI, rt, 61–91% yield.[72] A THP ether was stable to these conditions, but methyl ethers can be cleaved.
17. AlI_3, CH_3CN, benzene, 10 min, 70–92% yield.[73] Ethyl ketals are cleaved under these conditions, but thioketals are not affected.
18. Dimethyl sulfoxide, 180°, H_2O, 10 h, 89% yield.[74] A diethyl acetal can be cleaved in the presence of a 1,3-dioxolane under these conditions. TBDMS, THP, and MOM groups are stable.
19. NaTeH, EtOH, 25°, 30 min; air, 80–85% yield.[75]
20. H_2SiI_2, $CDCl_3$, −42°, 1–10 min, 100% yield.[76] Aromatic ketals are cleaved faster than the corresponding aliphatic derivatives, and cyclic ketals are cleaved more slowly than the acyclic analogues, such as dimethyl ketals. Substituted ketals such as those derived from butane-2,3-diol, which react only slowly with Me_3SiI, can also be cleaved with H_2SiI_2. If the reaction is run at 22°, ketals and acetals are reduced to iodides in excellent yield.
21. $CuSO_4$–SiO_2, CH_2Cl_2, 20–80 h, 70–90% yield.[77]
22. Dimethyldioxirane, acetone, CH_2Cl_2, 0°, 24 h, > 95% yield.[78] Ethers are also oxidized under these conditions.
23. DDQ, CH_3CN, H_2O, 68–95% yield.[79]
24. NO_2, silica gel, CCl_4, 30°, 40 min, 88–100% yield.[80]
25. PPh_3, CBr_4, THF, 0°, 96% yield.[81]
26. $SmCl_3$, TMSCl, THF, 92% yield. A ketal is cleaved in preference to an acetal.[82]
27. 2,4,6-Triphenylpyrilium tetrafluoroborate, H_2O, CH_2Cl_2, 3 h, *hν*, 67–88% yield.[83]
28. $RuCl_3 \cdot nH_2O$, *t*-BuOH, PhH, 1 h, rt, 46–86% yield. In this case, the acetal is cleaved with simultaneous oxidation to an ethylene glycol ester.[84]
29. NaI, $CeCl_3 \cdot 7H_2O$, CH_3CN, rt, 0.5–21 h, 84–96% yield.[85] Chemoselective cleavage of ketone derivatives is observed in the presence of aldehyde derivatives, and enone ketals are cleaved in the presence of simple ketone ketals.

1. For two examples, see (a) M. T. Crimmins and J. A. DeLoach, *J. Am. Chem. Soc.*, **108**, 800 (1986); (b) M. G. Constantino, P. M. Donate, and N. Petragnani, *J. Org. Chem.*, **51**, 253 (1986).
2. For a variety of examples with varying ring sizes, see Y. Ohtsuka and T. Oishi, *Tetrahedron Lett.*, **27**, 203 (1986); C. Iwata, Y. Takemoto, M. Doi, and T. Imanishi, *J. Org. Chem.*, **53**, 1623 (1988); S. D. Burke, C. W. Murtiashaw, J. O. Saunders,

and M. S. Dike, *J. Am. Chem. Soc.*, **104**, 872 (1982); P. A. Wender, M. A. Eisenstat, and M. P. Filosa, *J. Am. Chem. Soc.*, **101**, 2196 (1979); A. A. Devreese, P. J. de Clercq, and M. Vandewalle, *Tetrahedron Lett.*, **21**, 4767 (1980); P. G. Baraldi, A. Barco, S. Benetti, G. P. Pollini, E. Polo, and D. Simoni, *J. Org. Chem.*, **50**, 23 (1985); M. P. Bosch, F. Camps, J. Coll, A. Guerrero, T. Tatsuoka, and J. Meinwald, *J. Org. Chem.*, **51**, 773 (1986).

3. Y. Kamitori, M. Hojo, R. Masuda, and T. Yoshida, *Tetrahedron Lett.*, **26**, 4767 (1985).
4. J. W. De Leeuw, E. R. De Waard, T. Beetz, and H. O. Huisman, *Recl. Trav. Chim. Pays-Bas*, **92**, 1047 (1973).
5. J. R. Hwu and J. M. Wetzel, *J. Org. Chem.*, **50**, 3946 (1985); J. R. Hwu, L.-C. Leu, J. A. Robl, D. A. Anderson, and J. M. Wetzel, *J. Org. Chem.*, **52**, 188 (1987).
6. T. Tsunoda, M. Suzuki, and R. Noyori, *Tetrahedron Lett.*, **21**, 1357 (1980).
7. P. Ciceri and F. W. J. Demnitz, *Tetrahedron Lett.*, **38**, 389 (1997).
8. G. H. Posner and G. L. Loomis, *Tetrahedron Lett.*, 4213 (1978).
9. P. Hodge and J. Waterhouse, *J. Chem. Soc., Perkin Trans. 1*, 2319 (1983); Z. H. Xu, C. R. McArthur, and C. C. Leznoff, *Can. J. Chem.*, **61**, 1405 (1983).
10. R. A. Daignault and E. L. Eliel, *Org. Synth., Collect. Vol. V*, 303 (1973).
11. F. F. Caserio, Jr., and J. D. Roberts, *J. Am. Chem. Soc.*, **80**, 5837 (1958).
12. L. F. Fieser and R. Stevenson, *J. Am. Chem. Soc.*, **76**, 1728 (1954).
13. E. G. Howard and R. V. Lindsey, *J. Am. Chem. Soc.*, **82**, 158 (1960).
14. T. H. Chan, M. A. Brook, and T. Chaly, *Synthesis*, 203 (1983).
15. W. G. Dauben, J. M. Gerdes, and G. C. Look, *J. Org. Chem.*, **51**, 4964 (1986); H. Eibisch, *Z. Chem.*, **26**, 375 (1986).
16. J. Y. Satoh, C. T. Yokoyama, A. M. Haruta, K. Nishizawa, M. Hirose, and A. Hagitani, *Chem. Lett.*, 1521 (1974).
17. N. H. Andersen and H.-S. Uh, *Synth. Commun.*, **3**, 125 (1973).
18. J. J. Brown, R. H. Lenhard, and S. Bernstein, *J. Am. Chem. Soc.*, **86**, 2183 (1964).
19. E. P. Oliveto, H. Q. Smith, C. Gerold, L. Weber, R. Rausser, and E. B. Hershberg, *J. Am. Chem. Soc.*, **77**, 2224 (1955).
20. F. T. Bond, J. E. Stemke, and D. W. Powell, *Synth. Commun.*, **5**, 427 (1975).
21. R. Sterzycki, *Synthesis*, 724 (1979).
22. W. Kantlehner and H.-D. Gutbrod, *Liebigs Ann. Chem.*, 1362 (1979).
23. A. E. Dann, J. B. Davis, and M. J. Nagler, *J. Chem. Soc., Perkin Trans. 1*, 158 (1979).
24. M. Koreeda and L. Brown, *J. Org. Chem.*, **48**, 2122 (1983).
25. B. Glatz, G. Helmchen, H. Muxfeldt, H. Porcher, R. Prewo, J. Senn, J. J. Stezowski, R. J. Stojda, and D. R. White, *J. Am. Chem. Soc.*, **101**, 2171 (1979).
26. R. A. Holton, R. M. Kennedy, H.-B. Kim, and M. E. Krafft, *J. Am. Chem. Soc.*, **109**, 1597 (1987).
27. H. J. Dauben, B. Löken, and H. J. Ringold, *J. Am. Chem. Soc.*, **76**, 1359 (1954).
28. H. Hagiwara and H. Uda, *J. Org. Chem.*, **53**, 2308 (1988); Y. Tamai, H. Hagiwara, and H. Uda, *J. Chem. Soc., Perkin Trans. 1*, 1311 (1986).
29. B. Pério, M.-J. Dozias, P. Jacquault, and J. Hamelin, *Tetrahedron Lett.*, **38**, 7867 (1997).

30. H. Vorbrueggen, *Steroids*, **1**, 45 (1963).
31. D. H. R. Barton, C. C. Dawes, and P. D. Magnus, *J. Chem. Soc., Chem. Commun.*, 432 (1975).
32. J. L. E. Erickson and F. E. Collins, *J. Org. Chem.*, **30**, 1050 (1965).
33. F. Nerdel, J. Buddrus, G. Scherowsky, D. Klamann, and M. Fligge, *Liebigs Ann. Chem.*, **710**, 85 (1967).
34. H. E. Simmons and D. W. Wiley, *J. Am. Chem. Soc.*, **82**, 2288 (1960).
35. R. J. Stedman, L. D. Davis, and L. S. Miller, *Tetrahedron Lett.*, 4915 (1967).
36. G. R. Newkome, J. D. Sauer, and C. L. McClure, *Tetrahedron Lett.*, 1599 (1973).
37. J. Ott, G. M. Ramos Tombo, B. Schmid, L. M. Venanzi, G. Wang, and T. R. Ward, *Tetrahedron Lett.*, **30**, 6151 (1989).
38. J. Otera, N. Danoh, and H. Nozaki, *Tetrahedron*, **48**, 1449 (1992).
39. M. Shibagaki, K. Takahashi, H. Kuno, and H. Matsushita, *Bull. Chem. Soc. Jpn.*, **63**, 1258 (1990).
40. S.-B. Lee, S.-D. Lee, T. Takata, and T. Endo, *Synthesis*, 368 (1991).
41. S. Ma and L. M. Venanzi, *Synlett*, 751 (1993).
42. D. S. Torok, J. J. Figueroa, and W. J. Scott, *J. Org. Chem.*, **58**, 7274 (1993).
43. D. Marton, P. Slaviero, and G. Taglianini, *Gazz. Chim. Ital.*, **119**, 359 (1989).
44. Z. Paryzek and J. Martynow, *J. Chem. Soc., Perkin Trans. 1*, 243 (1991).
45. S. Kim, Y. G. Kim, and D.-i. Kim, *Tetrahedron Lett.*, **33**, 2565 (1992).
46. W. Wang, L. Shi, and Y. Huang, *Tetrahedron*, **46**, 3315 (1990).
47. M. El Gihani and H. Heaney, *Synlett*, 433 (1993); *idem, ibid.*, 583 (1993).
48. M. Kurihara and N. Miyata, *Chem. Lett.*, 263 (1995).
49. T.-J. Lu, J.-F. Yang, and L.-J. Sheu, *J. Org. Chem.*, **60**, 2931 (1995).
50. A. A. Haaksma, B. J. M. Jansen, and A. de Groot, *Tetrahedron*, **48**, 3121 (1992).
51. P. Magnus, M. Giles, R. Bonnert, C. S. Kim, L. McQuire, A. Merritt, and N. Vicker, *J. Am. Chem. Soc.*, **114**, 4403 (1992).
52. T. Ohshima, K. Kagechika, M. Adachi, M. Sodeoka, and M. Shibasaki, *J. Am. Chem. Soc.*, **118**, 7108 (1996).
53. H. Hagiwara and H. Uda, *J. Chem. Soc., Chem. Commun.*, 1351 (1987).
54. G. Bauduin, D. Bondon, Y. Pietrasanta, and B. Pucci, *Tetrahedron*, **34**, 3269 (1978).
55. P. A. Grieco, M. Nishizawa, T. Oguri, S. D. Burke, and N. Marinovic, *J. Am. Chem. Soc.*, **99**, 5773 (1977).
56. P. A. Grieco, Y. Yokoyama, G. P. Withers, F. J. Okuniewicz, and C.-L. J. Wang, *J. Org. Chem.*, **43**, 4178 (1978).
57. P. A. Grieco, Y. Ohfune, and G. Majetich, *J. Am. Chem. Soc.*, **99**, 7393 (1977).
58. J. H. Babler, N. C. Malek, and M. J. Coghlan, *J. Org. Chem.*, **43**, 1821 (1978).
59. P. A. Grieco, T. Oguri, S. Gilman, and G. R. DeTitta, *J. Am. Chem. Soc.*, **100**, 1616 (1978).
60. H. M. Walborsky, R. H. Davis, and D. R. Howton, *J. Am. Chem. Soc.*, **73**, 2590 (1951).
61. J. A. Zderic and D. C. Limon, *J. Am. Chem. Soc.*, **81**, 4570 (1959).
62. F. Huet, A. Lechevallier, M. Pellet, and J. M. Conia, *Synthesis*, 63 (1978).

63. D. H. R. Barton, P. D. Magnus, G. Smith, and D. Zurr, *J. Chem. Soc., Chem. Commun.*, 861 (1971).
64. M. Uemura, T. Minami, and Y. Hayashi, *Tetrahedron Lett.*, **29**, 6271 (1988).
65. D. H. R. Barton, P. D. Magnus, G. Smith, G. Streckert, and D. Zurr, *J. Chem. Soc., Perkin Trans. 1*, 542 (1972).
66. Y. Guindon, H. E. Morton, and C. Yoakim, *Tetrahedron Lett.*, **24**, 3969 (1983).
67. B. H. Lipshutz, D. Pollart, J. Monforte, and H. Kotsuki, *Tetrahedron Lett.*, **26**, 705 (1985).
68. A. McKillop, R. J. K. Taylor, R. J. Watson, and N. Lewis, *Synlett*, 1005 (1992).
69. M. E. Jung, W. A. Andrus, and P. L. Ornstein, *Tetrahedron Lett.*, 4175 (1977).
70. T. Hosokawa, Y. Imada, and S.-i. Murahashi, *J. Chem. Soc., Chem. Commun.*, 1245 (1983).
71. B. H. Lipshutz and D. F. Harvey, *Synth. Commun.*, **12**, 267 (1982).
72. G. Balme and J. Goré, *J. Org. Chem.*, **48**, 3336 (1983).
73. P. Sarmah and N. C. Barua, *Tetrahedron Lett.*, **30**, 4703 (1989).
74. T. Kametani, H. Kondoh, T. Honda, H. Ishizone, Y. Suzuki, and W. Mori, *Chem. Lett.*, 901 (1989).
75. P. Lue, W.-Q. Fan, and X.-J. Zhou, *Synthesis*, 692 (1989).
76. E. Keinan, D. Perez, M. Sahai, and R. Shvily, *J. Org. Chem.*, **55**, 2927 (1990).
77. G. M. Caballero and E. G. Gros, *Synth. Commun.*, **25**, 395 (1995).
78. R. Curci, L. D'Accolti, M. Fiorentino, C. Fusco, W. Adam, M. E. González-Nuñez and R. Mello, *Tetrahedron Lett.*, **33**, 4225 (1992).
79. K. Tanemura, T. Suzuki, and T. Horaguchi, *J. Chem. Soc., Chem. Commun.*, 979 (1992).
80. T. Nishiguchi, T. Ohosima, A. Nishida, and S. Fujisaki, *J. Chem. Soc., Chem. Commun.*, 1121 (1995).
81. C. Johnstone, W. J. Kerr, and J. S. Scott, *J. Chem. Soc., Chem. Commun.*, 341 (1996).
82. Y. Ukaji, N. Koumoto, and T. Fujisawa, *Chem. Lett.*, 1623 (1989).
83. H. Garcia, S. Iborra, M. A. Miranda, and J. Primo, *New J. Chem.*, **13**, 805 (1989).
84. S. Murahashi, Y. Oda, and T. Naota, *Chem. Lett.*, 2237 (1992).
85. E. Marcantoni, F. Nobili, G. Bartoli, M. Bosco, and L. Sambri, *J. Org. Chem.*, **62**, 4183 (1997).

4-Bromomethyl-1,3-dioxolane (Chart 5):

CH_2Br
O O
R R

Formation

1. $HOCH_2CH(OH)CH_2Br$, TsOH, benzene, reflux, 5 h, 93–98% yield.[1]

Cleavage[1]

1. Activated Zn, MeOH, reflux, 12 h, 89–96% yield.

This ketal is stable to several reagents that react with carbonyl groups (e.g., m-$ClC_6H_4CO_3H$, NH_3, $NaBH_4$, and MeLi). It is cleaved under neutral conditions.

1. E. J. Corey and R. A. Ruden, *J. Org. Chem.*, **38**, 834 (1973).

4-(3-Butenyl)-1,3-dioxolane: O O / R R

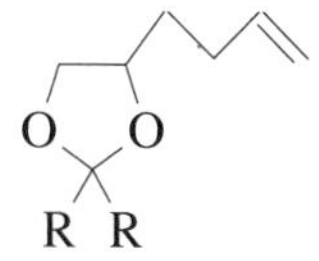

Formation/Cleavage[1]

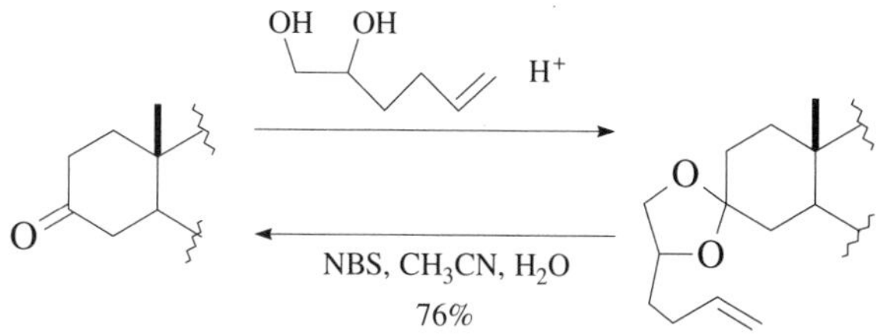

OH OH H$^+$

NBS, CH_3CN, H_2O 76%

1. Z. Wu, D. R. Mootoo, and B. Fraser-Reid, *Tetrahedron Lett.*, **29**, 6549 (1988).

4-Phenyl-1,3-dioxolane: R O Ph / R O

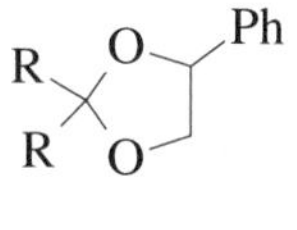

Cleavage

1. Electrolysis: $LiClO_4$, H_2O, Pyr, CH_3CN, *N*-hydroxyphthalimide, 0.85 V SCE, 22–90% yield.[1]
2. Pd/C, H_2.[2]

1. M. Masui, T. Kawaguchi, and S. Ozaki, *J. Chem. Soc., Chem. Commun.*, 1484 (1985).
2. S. Chandrasekhar, B. Muralidhar, and S. Sarkar, *Synth. Commun.*, **27**, 2691 (1997).

4-(4-Methoxyphenyl)-1,3-dioxolane:

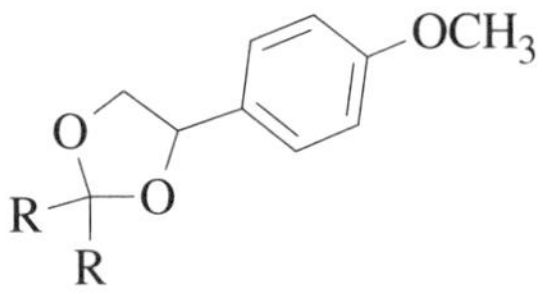

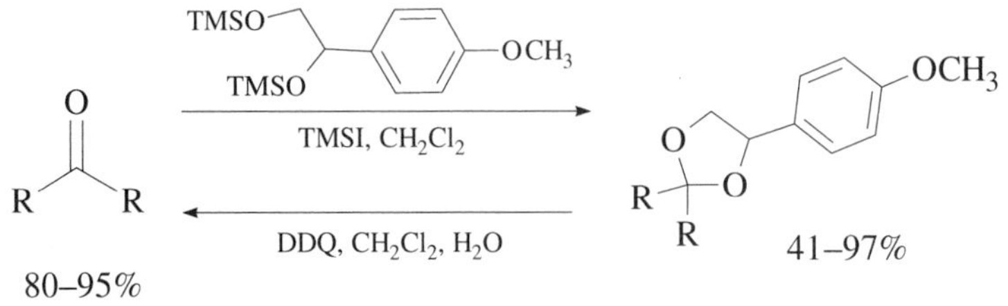

TMSI, CH_2Cl_2 (with 1,2-bis(trimethylsilyloxy)-1-(4-methoxyphenyl)ethane), 41–97%

DDQ, CH_2Cl_2, H_2O, 80–95%

Ref. 1

This protective group can be removed oxidatively in excellent yields.[1]

1. C. E. McDonald, L. E. Nice, and K. E. Kennedy, *Tetrahedron Lett.*, **35**, 57 (1994).

4-(2-Nitrophenyl)-1,3-dioxolane (Chart 5):

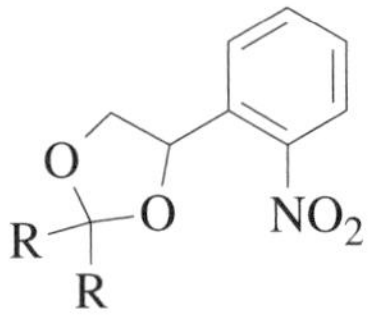

This dioxolane is readily formed from the glycol (TsOH, benzene, reflux, 70–95% yield); it is cleaved by irradiation (350 nm, benzene, 25°, 6 h, 75–90% yield). This group is stable to 5% HCl/THF; 10% AcOH/THF; 2% oxalic acid/THF; 10% aq. H_2SO_4/ THF; and 3% aq. TsOH/THF.[1]

1. J. Hébert and D. Gravel, *Can. J. Chem.*, **52**, 187 (1974); D. Gravel, J. Hébert, and D. Thoraval, *Can. J. Chem.*, **61**, 400 (1983).

4-Trimethylsilylmethyl-1,3-dioxolane:

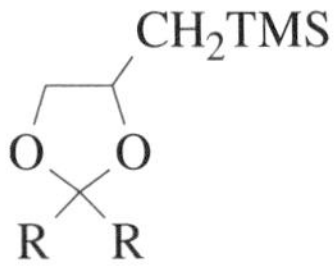

Formation/Cleavage[1]

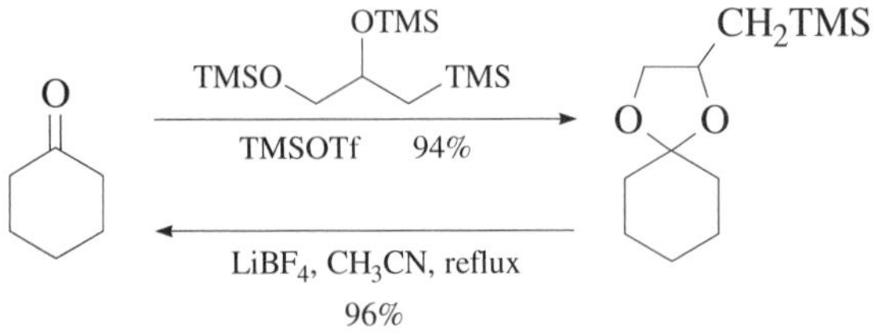

TMSOCH(OTMS)CH₂TMS TMSO / OTMS / TMS, TMSOTf 94%

LiBF$_4$, CH_3CN, reflux 96%

Hindered ketones and enones fail to form the ketal because of competing decomposition of the silyl reagent.

1. B. M. Lillie and M. A. Avery, *Tetrahedron Lett.*, **35**, 969 (1994).

O,O′-Phenylenedioxy Ketal:

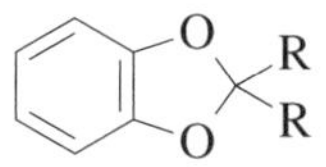

The phenylenedioxy ketal is prepared from catechol (TsOH, 90°, 30 h, 85% yield) and is cleaved with 5 *N* HCl (dioxane, reflux, 6 h). It is more stable to acid than is the ethylene ketal.[1,2]

1. M. Rosenberger, D. Andrews, F. DiMaria, A. J. Duggan, and G. Saucy, *Helv. Chim. Acta*, **55**, 249 (1972).
2. M. Rosenberger, A. J. Duggan, and G. Saucy, *Helv. Chim. Acta*, **55**, 1333 (1972).

1,5-Dihydro-3*H*-2,4-benzodioxepin:

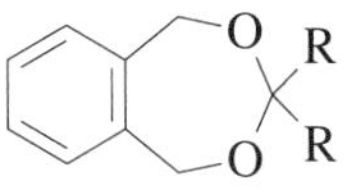

Formation

1.

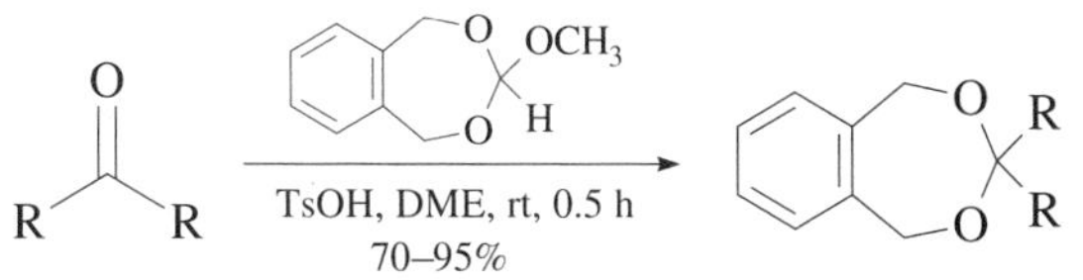

TsOH, DME, rt, 0.5 h
70–95%

Refs. 1, 2

Camphor cannnot be protected with this reagent, indicating that steric factors will prevent its use in very hindered systems.

2. 1,2-Dihydroxymethylbenzene, $CH(OCH_3)_3$, TsOH, 80% yield.[3,4]
3. From a methyl enol ether: 1,2-dihydroxymethylbenzene, Amberlyst H^+, 85% yield.[5]
4. 1,2-Dihydroxymethylbenzene, sulfonated charcoal or TsOH, PhH, reflux, 88–98% yield.[6]
5. 1,2-Ditrimethylsiloxymethylbenzene, TMSOTf, CH_2Cl_2, −78°, 96% yield.[7]
6. 1,2-Dihydroxymethylbenzene, H-Y Zeolite, CH_2Cl_2, reflux, 3–12 h, 46–95% yield.[8]
7. 1,2-Dihydroxymethylbenzene, Environcat EPZG, toluene, reflux, 93–99% yield. Ketones were not reactive under these conditions.[9]

Cleavage

1. H_2, PdO, THF, rt, 0.5 h, 100% yield.[1]

1. N. Machinaga and C. Kibayashi, *Tetrahedron Lett.*, **30**, 4165 (1989).
2. K. Mori, T. Yoshimura, and T. Sugai, *Liebigs Ann. Chem.*, 899 (1988).
3. R. Oi and K. B. Sharpless, *Tetrahedron Lett.*, **33**, 2095 (1992).
4. S. D. Burke and D. N. Deaton, *Tetrahedron Lett.*, **32**, 4651 (1991).
5. L. Schmitt, B. Spiess, and G. Schlewer, *Tetrahedron Lett.*, **33**, 2013 (1992).
6. H. K. Patney, *Tetrahedron Lett.*, **32**, 413 (1992).
7. S. V. D'Andrea, J. P. Freeman, and J. Szmuszkovicz, *Org. Prep. Proced. Int.*, **23**, 432 (1991).
8. T. P. Kumar, K. R. Reddy, and R. S. Reddy, *J. Chem. Res., Synop.*, 394 (1994).
9. B. P. Bandgar, M. M. Kulkarni, and P. P. Wadgaonkar, *Synth. Commun.*, **27**, 627 (1997).

Chiral Acetals and Ketals

Chiral protective groups, although less frequently used in synthesis, provide sought-after protection, diastereochemical control, and enantioselectivity, and can improve the chemical characteristics of a molecule to facilitate a synthesis.[1]

(4*R*,5*R*)-Diphenyl-1,3-dioxolane:

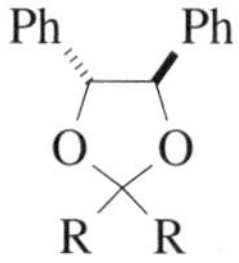

Ph Ph
O O
R R

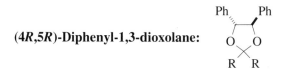

Formation

1. (1*R*,2*R*)-Diphenyl-1,2-ditrimethylsiloxyethane, TMSOTf, 66% yield.[2]
2. (1*R*,2*R*)-Diphenyl-1,2-ethanediol, PPTS, 80°.[3]

Cleavage

1. 2.7 *N* HCl, MeOH, 25°, 90% yield.[3]
2. $Pd(OH)_2$, H_2, EtOAc, quant.[2]

4,5-Dimethyl-1,3-dioxolane

Formation

1. 2,3-Bistrimethylsiloxybutane, TMSOTf, CH_2Cl_2, 66% yield. The double bond of an enone does not migrate out of conjugation.[4]
2. 2,3-Butanediol, benzene, PPTS, reflux, 66% yield.[5]

3.

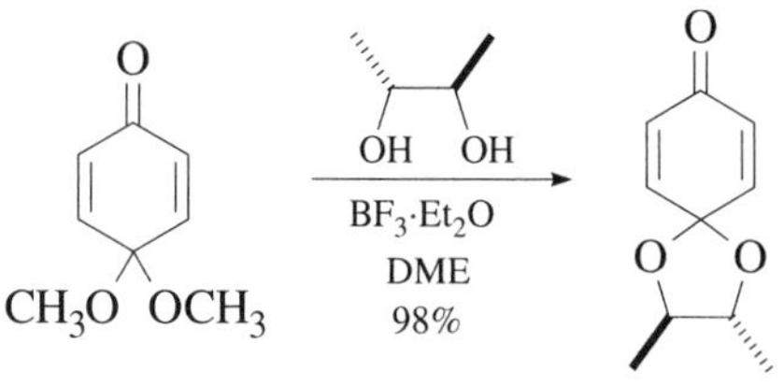

OH OH
$BF_3 \cdot Et_2O$
DME
98%

CH_3O OCH_3

Refs. 6, 7

This reaction also works to form the related dioxane, but the yields are lower.[6]

trans-1,2-Cyclohexanediol Ketal

Formation

1. *trans*-1,2-Cyclohexanediol, *i*-PrOTMS, TMSOTf, CH_2Cl_2, −20°, 3 h, 85% yield.[8]

trans-4,6-Dimethyl-1,3-dioxane

Formation

1. 2,4-Pentanediol, PPTS, >95% yield.[8, 9]

2. R

OH OH
$PdCl_2$, CH_2Cl_2, H_2O
O_2, 50°, 20 h
45–79%

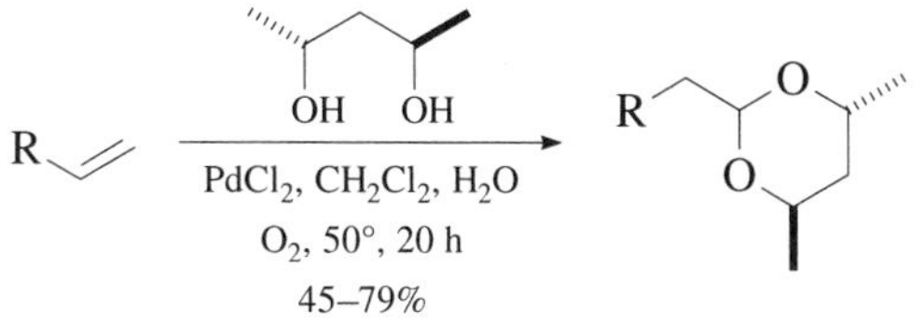

Ref. 10

3. 2,4-Pentanediol, $Sc(OTf)_3$, rt, 13 h–2 days, benzene, THF or CH_2Cl_2, 59–100% yield. This method is also effective for the formation of a 4,5-dimethyldioxolane.[11]

Cleavage

1. NHAloc

OTBDPS

PPTS, TsOH
acetone, H_2O
38°, 48 h
78%

NHAloc

OTBDPS

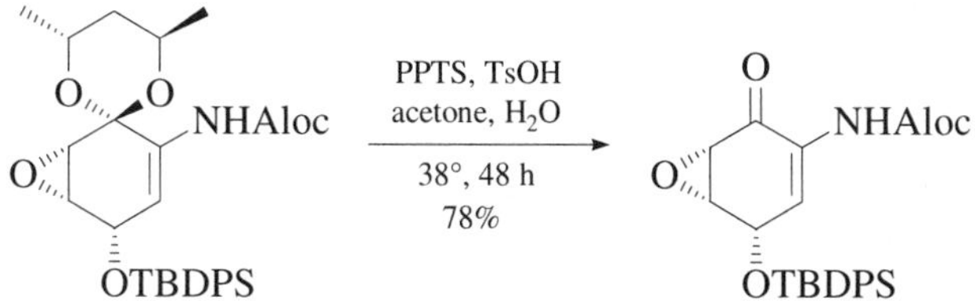

Hydrolysis is facilitated by the increased level of strain imparted by the axial methyl group, thus allowing cleavage under conditions to which the product is stable.[12]

4,5-Bis(dimethylaminocarbonyl)-1,3-dioxolane

Formation[13]

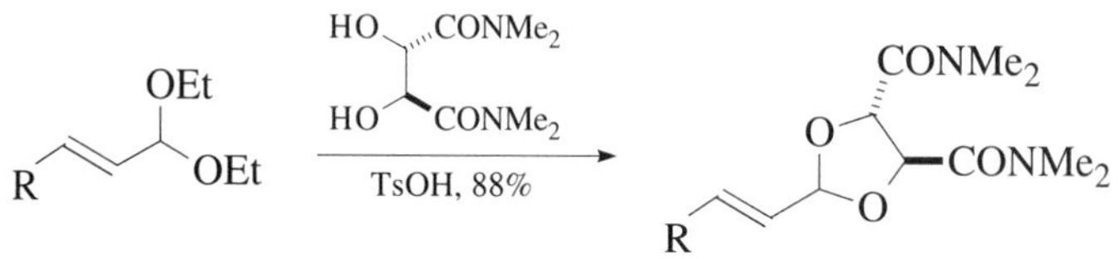

HO, CONMe2 / HO, CONMe2; TsOH, 88%

Cleavage[13]

6 *M* HCl, dioxane, >92% yield.
A chiral protective group was developed for use in the synthesis of optically active alcohols.[13]

4,5-Dicarbomethoxy-1,3-dioxolane

Formation

1. Dimethyl tartrate, $Sc(OTf)_3$, MeCN, rt, 3 h, 95% yield.[14]

2. 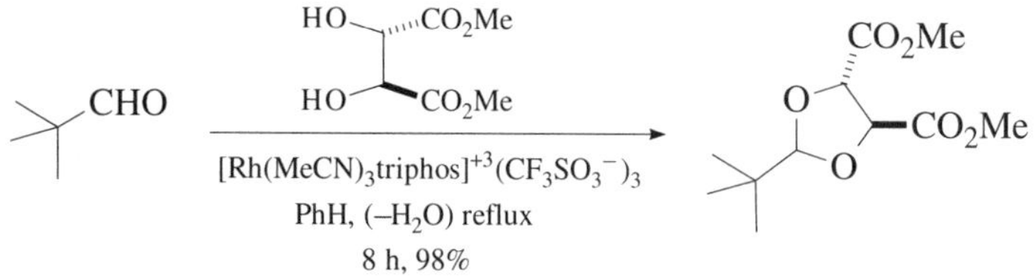 CHO; HO, CO_2Me / HO, CO_2Me; $[Rh(MeCN)_3triphos]^{+3}(CF_3SO_3^-)_3$, PhH, $(-H_2O)$ reflux, 8 h, 98%

Ref. 15

4,5-Dimethoxymethyl-1,3-dioxolane

Formation/Cleavage[16]

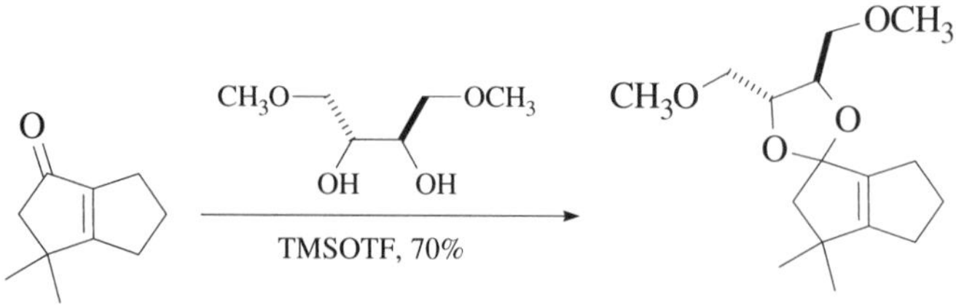

CH_3O, OCH_3, OH OH; TMSOTF, 70%

This protective group was used to direct the selective cyclopropanation of a variety of enones. Hydrolysis (HCl, MeOH, H_2O, rt, 94% yield) affords optically active cyclopropyl ketones.

1. A review: A. Alexakis and P. Mangeney, *Tetrahedron: Asymmetry*, **1**, 477 (1990).
2. C. N. Eid, Jr., and J. P. Konopelski, *Tetrahedron Lett.*, **32**, 461 (1991).
3. J. Cossy and S. BouzBouz, *Tetrahedron Lett.*, **37**, 5091 (1996).
4. E. A. Mash and S. B. Hemperly, *J. Org. Chem.*, **55**, 2055 (1990).
5. M. Toyota, Y. Nishikawa, and K. Fukumoto, *Tetrahedron*, **52**, 10347 (1996).
6. M. C. Pirrung and D. S. Nunn, *Tetrahedron Lett.*, **33**, 6591 (1992).
7. P. de March, M. Escoda, M. Figueredo, J. Font, A. Alvarez-Larena, and J. F. Piniella, *J. Org. Chem.*, **60**, 3895 (1995).
8. M. Kurihara and N. Miyata, *Chem. Lett.*, 263 (1995).
9. A. Mori and H. Yamamoto, *J. Org. Chem.*, **50**, 5444 (1985).
10. T. Hosokawa, T. Ohta, S. Kanayama, and S. I. Murahashi, *J. Org. Chem.*, **52**, 1758 (1987).
11. S.-i. Fukuzawa, T. Tsuchimoto, T. Hotaka, and T. Hiyama, *Synlett*, 1077 (1995).
12. P. Wipf, Y. Kim, and H. Jahn, *Synthesis*, 1549 (1995).
13. J. Fujiwara, Y. Fukutani, M. Hasegawa, K. Maruoka, and H. Yamamoto, *J. Am. Chem. Soc.*, **106**, 5004 (1984).
14. K. Ishihara, Y. Karumi, M. Kubota, and H. Yamamoto, *Synlett*, 839 (1996).
15. J. Ott, G. M. Ramos Tombo, B. Schmid, L. M. Venanzi, G. Wang, and T. R. Ward, *Tetrahedron Lett.*, **30**, 6151 (1989).
16. E. A. Mash, S. K. Math, and C. J. Flann, *Tetrahedron Lett.*, **29**, 2147 (1988).

Dithio Acetals and Ketals

A carbonyl group can be protected as a sulfur derivative—for example, a dithio acetal or ketal, a 1,3-dithiane, or a 1,3-dithiolane—by reaction of the carbonyl compound in the presence of an acid catalyst with a thiol or dithiol. The derivatives are, in general, cleaved by reaction with Hg(II) salts or oxidation; acidic hydrolysis is unsatisfactory. The acyclic derivatives are formed and hydrolyzed much more readily than their cyclic counterparts. Representative examples of formation and cleavage follow.

Acyclic Dithio Acetals and Ketals

***S,S′*-Dimethyl Acetals and Ketals:** $RR'C(SCH_3)_2$ (Chart 5)

***S,S′*-Diethyl Acetals and Ketals:** $RR'C(SC_2H_5)_2$

***S,S′*-Dipropyl Acetals and Ketals:** RR′C(SC_3H_7)$_2$

***S,S′*-Dibutyl Acetals and Ketals:** RR′C(SC_4H_9)$_2$

***S,S′*-Dipentyl Acetals and Ketals:** RR′C(SC_5H_{11})$_2$

***S,S′*-Diphenyl Acetals and Ketals:** RR′C(SC_6H_5)$_2$

***S,S′*-Dibenzyl Acetals and Ketals:** RR′C($SCH_2C_6H_5$)$_2$

General Methods of Formation

1. RSH, concd. HCl, 20°, 30 min.[1] These conditions were used to protect an aldose as the methyl or ethyl thioketal.
2. $RSSiMe_3$, ZnI_2, Et_2O, 0–25°, 70–95% yield.[2] This method is satisfactory for a variety of aldehydes and ketones and is also suitable for the preparation of 1,3-dithianes. Methacrolein gives the product of Michael addition rather than the thioacetal. The less hindered of two ketones is readily protected using this methodology.[3]

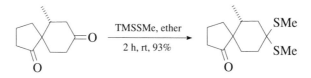

TMSSMe, ether
2 h, rt, 93%
SMe
SMe
O
O
O

3. RSH, Me_3SiCl, $CHCl_3$, 20°, 1 h, >80% yield.[4]
4. $B(SR)_3$, reflux, 2 h or 25°, 18 h, 75–85% yield.[5]
5. $Al(SPh)_3$, 25°, 1 h, 65% yield.[6] This method also converts esters to thioesters.
6. PhSH, $BF_3{\cdot}Et_2O$, $CHCl_3$, 0°, 10 min, 86% yield.[7] $ZnCl_2$[8] and $MgBr_2$[9] have also been used as catalysts. With $MgBr_2$, acetals can be converted to thioacetals in the presence of ketones.
7. RSH, SO_2, benzene, 54–81% yield.[10]
8. EtSH, $TiCl_4$, $CHCl_3$, 6–12 h, rt, 90–98% yield.[11]
9. ***P***-$PPh_2{\cdot}I_2$, RSH, Et_3N, CH_3CN; K_2CO_3, H_2O, 80–98% yield.[12] This method is also effective for the formation of dioxolanes and dithiolanes.
10. RSSR (R = Me, Ph, Bu), Bu_3P, rt, 15–83% yield.[13] This reagent also reacts with epoxides to form 1,2-dithio ethers.
11. H-Y or H-M Zeolite, hexane or CH_2Cl_2, EtSH, reflux, 0.75–144 h, 50–96% yield.[14]

General Methods of Cleavage

1. $AgNO_3$/Ag_2O, CH_3CN–H_2O, 0°, 2 h, 85% yield.[15]

$$\text{BuS,BuS-ketal (OH, C}_5\text{H}_{11}\text{, OTHP)} \xrightarrow[\text{H}_2\text{O, 0°, 2 h, 85\%}]{\text{AgNO}_3\text{, Ag}_2\text{O, CH}_3\text{CN}} \text{lactone (OH, C}_5\text{H}_{11}\text{, OTHP)}$$

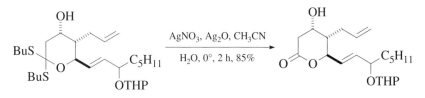

This method has also been used to cleave dithianes and dithiolanes.[16] The *S,S′*-dibutyl group is stable to acids (e.g., HOAc/H_2O–THF, 45°, 3 h; TsOH/CH_2Cl_2, 0°, 0.5 h).[15]

2. $AgClO_4$, H_2O, C_6H_6, 25°, 4 h, 80–100% yield.[17]
3. $HgCl_2$, $CdCO_3$, aq. acetone[18] or $HgCl_2$, $CaCO_3$, CH_3CN, H_2O.[19] In a case where this combination of reagents was not effective, HgO/$BF_3{\cdot}Et_2O$ was found to work.[20]
4. $Me_2CH(CH_2)_2ONO$, CH_2Cl_2, 25°, 15 min; H_2O, 63–93% yield.[21] Isoamyl nitrite cleaves aromatic dithioacetals in preference to aliphatic dithioacetals and dithioacetals in preference to dithioketals. It also cleaves 1,3-oxathiolanes (1 h, 65–90% yield).
5. $Tl(NO_3)_3$, CH_3OH, H_2O, 25°, 5 min, 73–98% yield.[7] These conditions are also effective for the cleavage of dithiolanes and dithianes.
6. SO_2Cl_2, $SiO_2{\cdot}H_2O$, CH_2Cl_2, 25°, 2–3 h, 90–100% yield.[22,23]
7. I_2, $NaHCO_3$, dioxane, H_2O, 25°, 4.5 h, 80–95% yield.[24]
8. I_2, MeOH, reflux, 2 h, 79%; $HClO_4$, H_2O, 25°, 16 h, 87% yield.[25] These conditions also cleave acetonides and benzylidene acetals.[26]
9. H_2O_2, aq. acetone or $NaIO_4/H_2O$, 25°; g HCl/$CHCl_3$, 0°, 50–70% yield.[27]
10. O_2, *hν*, hexane, Ph_2CO, 2–5 h, 60–80% yield.[28] 1,3-Oxathiolanes and dithiolanes are also cleaved by these conditions.
11. CuCl, CuO, H_2O, acetone, 2 h, 20°, 61–73% yield.[29]
12. $HgCl_2$, HgO, 80% CH_3CN, H_2O, 30 min, rt, 96% yield.[30]
13. MCPBA, CF_3COOH, CH_2Cl_2, 0°.[31]
14. Ph_3CClO_4, Ph_3COMe, CH_2Cl_2, −45°, 2.5 h; aq. $NaHCO_3$, 84–96% yield.[32] A diethyl thioketal could be cleaved in the presence of a diphenyl thioketal.
15. DDQ, CH_3CN, H_2O, 80°, 43–95% yield.[33] These conditions also resulted in the cleavage of acetyl groups; a dithiolane was stable to the conditions.
16. $GaCl_3$, CH_2Cl_2, H_2O, rt, 20 min.[34] Thioketals are cleaved in preference to thioacetals and dithianes, which do not react.
17. DMSO, 140–160°, 4–5 h, 79–94% yield.[35]
18. Clay supported NH_4NO_3, CH_2Cl_2, rt, 76–90% yield.[36]
19. The dithioacetal can be converted to an *O,S*-acetal.[37] The mixed acetals were then used to prepare furanosides.

EtS⟍⟋SEt —(AcCl, $BF_3 \cdot Et_2O$)→ EtS⟍⟋Cl —(EtOH, $AgCO_3$)→ EtS⟍⟋OEt

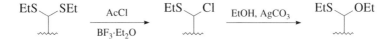

1. H. Zinner, *Chem. Ber.*, **83**, 275 (1950).
2. D. A. Evans, L. K. Truesdale, K. G. Grimm, and S. L. Nesbitt, *J. Am. Chem. Soc.*, **99**, 5009 (1977).
3. D. A. Evans, K. G. Grimm, and L. K. Truesdale, *J. Am. Chem. Soc.*, **97**, 3229 (1975).
4. B. S. Ong and T. H. Chan, *Synth. Commun.*, **7**, 283 (1977).
5. F. Bessette, J. Brault, and J. M. Lalancette, *Can. J. Chem.*, **43**, 307 (1965).
6. T. Cohen and R. E. Gapinski, *Tetrahedron Lett.*, 4319 (1978).
7. E. Fujita, Y. Nagao, and K. Kaneko, *Chem. Pharm. Bull.*, **26**, 3743 (1978).
8. W. E. Truce and F. E. Roberts, *J. Org. Chem.*, **28**, 961 (1963).
9. J. H. Park and S. Kim, *Chem. Lett.*, 629 (1989).
10. B. Burczyk and Z. Kortylewicz, *Synthesis*, 831 (1982).
11. V. Kumar and S. Dev, *Tetrahedron Lett.*, **24**, 1289 (1983).
12. R. Caputo, C. Ferreri, and G. Palumbo, *Synthesis*, 386 (1987).
13. M. Tazaki and M. Takagi, *Chem. Lett.*, 767 (1979).
14. P. Kumar, R. S. Reddy, A. P. Singh, and B. Pandey, *Synthesis*, 67 (1993).
15. E. J. Corey, M. Shibasaki, J. Knolle, and T. Sugahara, *Tetrahedron Lett.*, 785 (1977).
16. C. H. Heathcock, M. J. Taschner, T. Rosen, J. A. Thomas, C. R. Hadley, and G. Popják, *Tetrahedron Lett.*, **23**, 4747 (1982); R. Zamboni and J. Rokach, *Tetrahedron Lett.*, **23**, 4751 (1982).
17. T. Mukaiyama, S. Kobayashi, K. Kamio, and H. Takei, *Chem. Lett.*, 237 (1972).
18. J. English, Jr., and P. H. Griswold, Jr., *J. Am. Chem. Soc.*, **67**, 2039 (1945).
19. A. I. Meyers, D. L. Comins, D. M. Roland, R. Henning, and K. Shimizu, *J. Am. Chem. Soc.*, **101**, 7104 (1979).
20. P. Norris, D. Horton, and B. R. Levine, *Tetrahedron Lett.*, **36**, 7811 (1995).
21. K. Fuji, K. Ichikawa, and E. Fujita, *Tetrahedron Lett.*, 3561 (1978).
22. M. Hojo and R. Masuda, *Synthesis*, 678 (1976).
23. Y. Kamitori, M. Hojo, R. Masuda, T. Kimura, and T. Yoshida, *J. Org. Chem.*, **51**, 1427 (1986).
24. G. A. Russell and L. A. Ochrymowycz, *J. Org. Chem.*, **34**, 3618 (1969).
25. B. M. Trost, T. N. Salzmann, and K. Hiroi, *J. Am. Chem. Soc.*, **98**, 4887 (1976).
26. W. A. Szarek, A. Zamojski, K. N. Tiwari, and E. R. Ison, *Tetrahedron Lett.*, **27**, 3827 (1986).
27. H. Nieuwenhuyse and R. Louw, *Tetrahedron Lett.*, 4141 (1971).
28. T. T. Takahashi, C. Y. Nakamura, and J. Y. Satoh, *J. Chem. Soc., Chem. Commun.*, 680 (1977).
29. B. Cazes and S. Julia, *Tetrahedron Lett.*, 4065 (1978).
30. V. E. Amoo, S. De Bernardo, and M. Weigele, *Tetrahedron Lett.*, **29**, 2401 (1988).
31. J. Cossy, *Synthesis*, 1113 (1987).
32. M. Ohshima, M. Murakami, and T. Mukaiyama, *Chem. Lett.*, 1593 (1986).

33. J. M. Garcia Fernandez, C. Ortiz Mellet, A. M. Marin, and J. Fuentes, *Carbohyd. Res.*, **274**, 263 (1993).
34. K. Saigo, Y. Hashimoto, N. Kihara, H. Umehara, and M. Hasegawa, *Chem. Lett.*, 831 (1990).
35. Ch. S. Rao, M. Chandrasekharam, H. Ila, and H. Junjappa, *Tetrahedron Lett.*, **33**, 8163 (1992).
36. H. M. Meshram, G. S. Reddy, and J. S. Yadav, *Tetrahedron Lett.*, **38**, 8891 (1997).
37. J. C. McAuliffe and O. Hindsgaul, *J. Org. Chem.*, **62**, 1234 (1997).

S,S′-Diacetyl Acetals and Ketals: $R_2C(SCOCH_3)_2$

Formation[1]

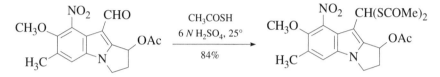

CH_3COSH, 6 *N* H_2SO_4, 25°, 84%

Cleavage[1]

1. NaOMe, MeOH, 25°, 40 h, 96%
2. $CHCl_3$, cat. H_2O or SiO_2

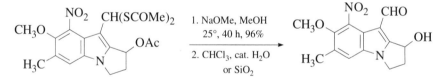

The formyl group was lost during attempted protection with ethylene glycol, TsOH.

1. T. Kametani, Y. Kigawa, K. Takahashi, H. Nemoto, and K. Fukumoto, *Chem. Pharm. Bull.*, **26**, 1918 (1978).

Cyclic Dithio Acetals and Ketals

1,3-Dithiane Derivative ($n = 3$) (Chart 5)

1,3-Dithiolane Derivative ($n = 2$) (Chart 5): $R_2C(S)_2(CH_2)_n$

General Methods of Formation

1. $HS(CH_2)_nSH$, $BF_3{\cdot}Et_2O$, CH_2Cl_2, 25°, 12 h, high yield, $n = 2$,[1] $n = 3$.[2] In α,β-unsaturated ketones, the double bond does not migrate to the β,γ-position, as occurs when an ethylene ketal is prepared.[3] Aldehydes are

selectively protected in the presence of ketones, except when large steric factors disfavor the aldehyde group, as shown in the next example.[4] A TBDMS group is not stable to these conditions.[5] Oxazolidines are converted to the dithiane in 70% yield under these conditions,[6] but the use of methanesulfonic acid as a catalyst is equally effective.[7]

HS(CH2)3SH, $BF_3 \cdot Et_2O$

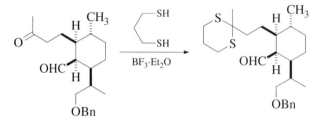

2. 1,3,2-Dithiaborolane B–R R = Cl or Ph

$CHCl_3$, 25°, 2 h, 90–100% yield.[8]

When R = Ph, the reaction is selective for unhindered ketones. Diaryl ketones, generally unreactive compounds, react rapidly when R = Cl.

3. $Me_3SiSCH_2CH_2SSiMe_3$, ZnI_2, Et_2O, 0–25°, 12–24 h, high yields.[9] Less hindered ketones can be selectively protected in the presence of more hindered ketones. *α,β*-Unsaturated ketones are selectively protected (94:1, 94:4) in the presence of saturated ketones by this reagent.[10]
4. $HS(CH_2)_nSH$, $SOCl_2$–SiO_2, 88–100% yield.[11] Aldehydes are selectively protected in the presence of ketones.
5. $HS(CH_2)_2SH$, $TiCl_4$, $-10° \rightarrow 25°$, 96% yield.[12]
6. $HSCH_2CH_2SH$, $Zn(OTf)_2$ or $Mg(OTf)_2$, $ClCH_2CH_2Cl$, heat, 16 h, 85–99% yield.[13,14] Excellent selectivity can be achieved between a hindered and an unhindered ketone.[15] *α,β*-Unsaturated ketones such as carvone are not cleanly converted to ketals, because of Michael addition of the thiol.[13]

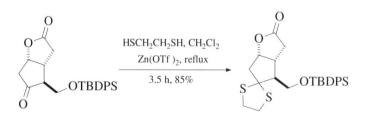

$HSCH_2CH_2SH$, CH_2Cl_2, $Zn(OTf)_2$, reflux, 3.5 h, 85%

In this case, other methods failed because of *β*-elimination.

7. 1,3-Dioxolanes[16,17] and 1,3-dioxanes[18] are readily converted to 1,3-dithiolanes and1,3-dithianes, respectively, in good to excellent yields.

$BF_3·Et_2O$, 84%

Ref. 18

1. $(i\text{-Bu})_2AlS(CH_2)_2SAl(i\text{-Bu})_2$, PhH, rt, 30 min
2. 3 *M* HCl
3. 10% NaOH
80%

Ref. 17

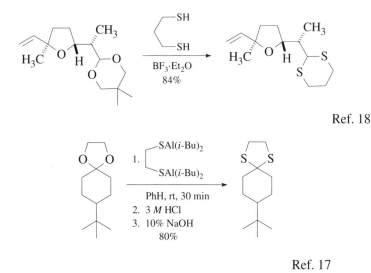

8. 2,2-Dimethyl-2-sila-1,3-dithiane, $BF_3·Et_2O$, CH_2Cl_2, 0°, 82–99% yield.[19] This method was reported to be superior to the conventional synthesis because cleaner products are formed. Aldehydes are selectively protected in the presence of ketones, which do not react competitively with this reagent.
9. 2,2-Dibutyl-2-stanna-1,3-dithiane, $Bu_2Sn(OTf)_2$, $ClCH_2CH_2Cl$, 35°, 1 h, 77–94% yield.[20] TBDMS, TBDPS, THP, and OAc groups are not affected by these conditions.
10. H-Y Zeolite, hexane, or CH_2Cl_2, $HSCH_2CH_2SH$, 0.75–144 h, 50–96% yield.[21]
11. $HS(CH_2)_nSH$, $ClCH_2CH_2Cl$, $TeCl_4$, rt, 80–99% yield.[22] This method is also effective for converting dimethyl acetals to the thioacetal and for selectively protecting an aldehyde in the presence of a ketone.
12. $HSCH_2CH_2SH$, $FeCl_3–SiO_2$, CH_2Cl_2, < 1 min–7 h.[23] Montmorillonite Clay can also be used as a support medium for the ferric ion (75–98% yield). In this case, the reaction is chemoselective for aldehydes.[24]
13. $HSCH_2CH_2SH$, CH_2Cl_2, $(TMSO)_2SO_2$–silica, 75–99% yield.[25]
14. $HSCH_2CH_2SH$, CH_2Cl_2, $CoBr_2$–silica, rt, 3 min–24 h, 87–99% yield.[26]
15. $HSCH_2CH_2SH$, CH_2Cl_2, $LaCl_3$, 1–96 h, 25–93% yield.[27]
16. $HS(CH_2)_nSH$, Montmorillonite KSF Clay, without solvent, 85–90% yield.[28]
17. $HSCH_2CH_2SH$, Amberlyst 15, 83–100% yield.[29]
18. $HSCH_2CH_2SH$, $SnCl_2·H_2O$, THF, reflux, 10–240 min, 51–96% yield.[30] Under these conditions, aldehydes react faster than ketones. Dimethyl ketals, which react faster than dimethyl acetals, are also converted to

dithianes and dithiolanes under these conditions (75–100% yield).[31]

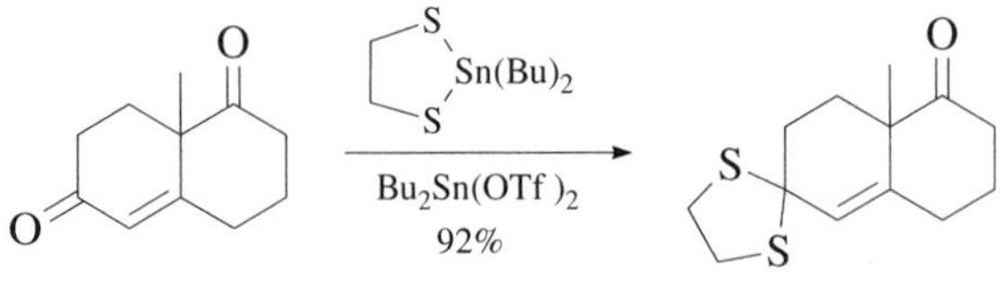

19. $HSCH_2CH_2SH$, MgI_2, Et_2O, rt, 8 h, 95–96% yield.[32] Aryl ketones are not efficiently protected.
20. $HS(CH_2)_nSH$, MeCN, SmI_3, 62–92% yield.[33]
21. $HSCH_2CH_2SH$, Dowex-50W-X8 acidified with HCl, Et_2O, 35–200 min, 60–90% yield.[34]
22. $HSCH_2CH_2SH$, $LiClO_4$, ether, 70–95% yield.[35]
23. $HSCH_2CH_2SH$, THF, $CuSO_4$, 40–96% yield.[36]
24. $HSCH_2CH_2SH$, PhMe, activated Bentonite, 5 h, 99% yield.[37]
25. $HSCH_2CH_2SH$, MeCN, rt, $Bi_2(SO_4)_3$, air, 2.5 h, 93–100% yield.[38]
26. $HSCH_2CH_2SH$, $ZrCl_4$–silica, CH_2Cl_2, rt, 3 h, 98% yield. Unreactive ketones such as benzophenone are efficiently protected.[39]
27. H-Rho–Zeolite, hexane, reflux, 85–94% yield.[40]

General Methods of Cleavage[41]

1. $Hg(ClO_4)_2$, MeOH, $CHCl_3$, 25°, 5 min, 93% yield.[42,43]

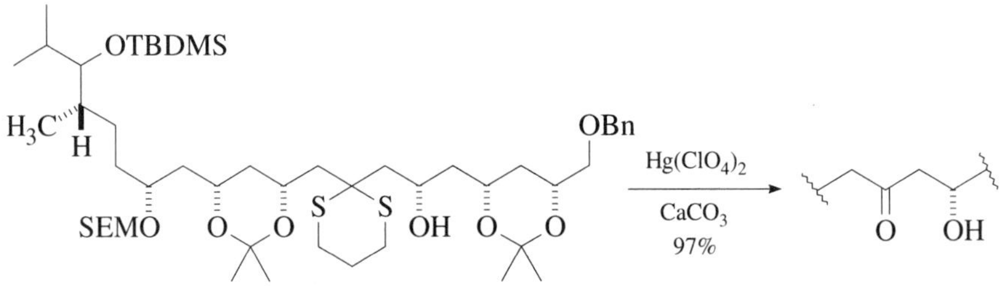

Ref. 43

2. A 1,3-dithiane is stable to the conditions ($HgCl_2$, $CaCO_3$, CH_3CN–H_2O, 25°, 1–2 h) used to cleave a methylthiomethyl (MTM) ether (i.e., a monothio acetal).[44]
3. $CuCl_2$, CuO, acetone, reflux, 90 min, 85% yield.[45]

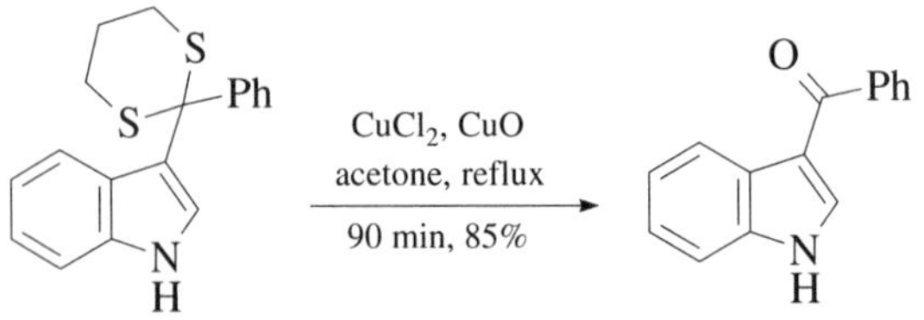

4. $AgNO_3$, EtOH, H_2O, 50°, 20 min, 55% yield.[46]

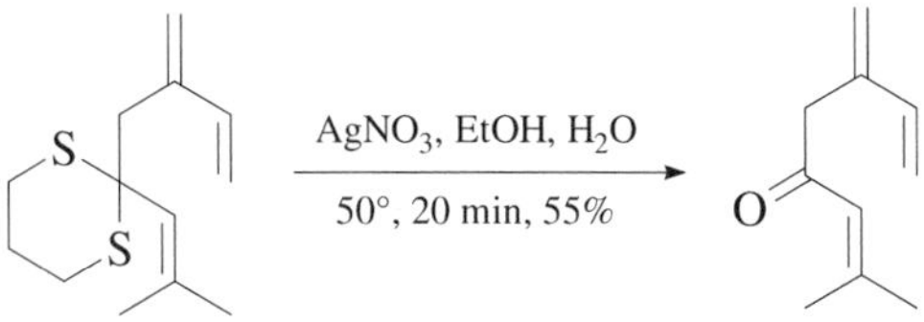

$AgNO_3$, EtOH, H_2O / 50°, 20 min, 55%

Attempted cleavage using Hg(II) salts gave material that could not be distilled. 1,3-Dithiolanes can also be cleaved with Ag_2O (MeOH, H_2O, reflux, 16 h–4 days, 75–85% yield).[47]

5. For ($n = 2$): NBS, aq. acetone, 0°, 20 min, 80% yield.[48]
6. For ($n = 3$): NCS, $AgNO_3$, CH_3CN, H_2O, 25°, 5–10 min, 70–100% yield.[49,50]
7. For ($n = 2,3$): $Tl(NO_3)_3$, CH_3OH, 25°, 5 min, 73–99% yield. These conditions have been used to effect selective cleavage of α,β-unsaturated thioketals.[51] In this case, $Hg(OAc)_2$ was found not to be reliable.

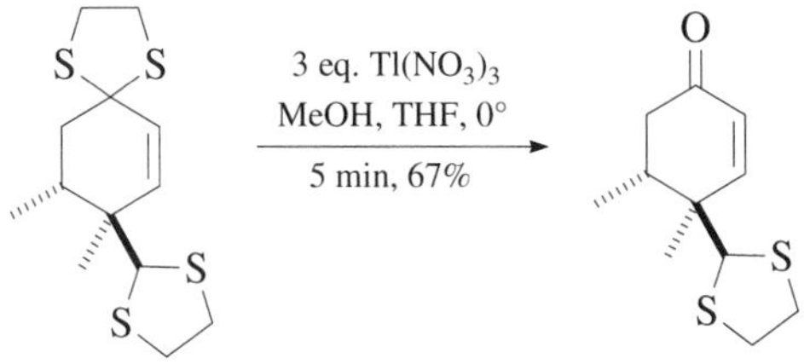

3 eq. $Tl(NO_3)_3$ / MeOH, THF, 0° / 5 min, 67%

8. For ($n = 2,3$): $Tl(OCOCF_3)_3$, THF, 25°, 1 min, 83–95% yield.[52] $Tl(TFA)_3$, Et_2O, H_2O, 94% yield.[53] α,β-Unsaturated 1,3-dithiolanes are selectively cleaved in the presence of saturated 1,3-dithiolanes [$Tl(NO_3)_3$, 5 min, 97% yield].[54]
9. For ($n = 2,3$): SO_2Cl_2, SiO_2, CH_2Cl_2, H_2O, 0–25°, 90–100% yield.[55]
10. For ($n = 2$): I_2, DMSO, 90°, 1 h, 75–85% yield.[56]
11. For ($n = 2$,[57] 3[58]): p-$MeC_6H_4SO_2N(Cl)Na$, aq. MeOH, 75–100% yield.
12. 1,3-Oxathiolanes are also cleaved by Chloramine-T.[58]
13. For ($n = 2,3$): $(PhSeO)_2O$, THF or CH_2Cl_2, 25°, 30 min to 50 h, 63–78% yield.[59]
14. For ($n = 3$): $Me_2CH(CH_2)_2ONO$, CH_2Cl_2, reflux, 2.5 h, 65% yield.[60] 1,3-Oxathiolanes are also cleaved by isoamyl nitrite.
15. For ($n = 2,3$): N-Chlorobenzotriazole, CH_2Cl_2, −80°; NaOH, 50% yield.[61] 1,3-Dithianes and 1,3-dithiolanes, used in this example to protect C_3-keto steroids, were not cleaved by $HgCl_2$–$CdCO_3$.
16. For ($n = 2,3$): $Ce(NH_4)_2(NO_3)_6$, aq. CH_3CN, 3 min, 70–87% yield.[62]
17. For ($n = 2$):O_2, $h\nu$, 4.5 h, 60–80% yield.[63] 1,3-Oxathiolanes are also cleaved by $O_2/h\nu$.

18. Electrolysis: 1.5 V, CH_3CN, H_2O, $LiClO_4$ or $Bu_4N^+ClO_4^-$, 50–75% yield.[64,65] 1,3-Dithiolanes were not cleaved efficiently by electrolytic oxidation. This method has been applied to dithiane deprotection to produce α-diketones.[66]
19. For (n = 2,3): $MeOSO_2F$, C_6H_6, 25°, 1 h, 62–88% yield;[67] or liq. SO_2, 70–85% yield.[68]
20. For (n = 2): MeI, aq. MeOH, reflux, 2–20 h, 60–80% yield.[68]
21. For (n = 3): MeI, aq. CH_3CN, 25°.[69]
22. For (n = 2): $Et_3O^+BF_4^-$, followed by 3% aq. $CuSO_4$, 81% yield.[70]
23. For (n = 2): $Me_2S^+Br\ Br^-$, CH_2Cl_2, 25°, 1 h → reflux, 8 h, followed by H_2O, 55–91% yield.[71]
24. OHCCOOH, HOAc, 25°, 15 min–20 h, 60–90% yield.[72]
25. $NO^+HSO_4^-$, CH_2Cl_2, 25°, 45 min; H_2O, 56–82% yield.[73]
26. Electrolysis: 1 V, $(p\text{-}CH_3C_6H_4)_3N$, CH_3CN, H_2O, $NaHCO_3$, 70–95% yield.[74]
27. Diiodohydantoin, −20°, 5:5:1 acetone:THF:H_2O.[10]
28. $(CF_3CO_2)_2IPh$, H_2O, CH_3CN, 85–99% yield.[75] In the presence of ethylene glycol the dithiane can be converted to a dioxolane (91% yield)[75] or in the presence of methanol to the dimethyl acetal.[76] The reaction conditions are not compatible with primary amides. Thioesters are not affected.[75] A phenylthio ester is stable to these conditions, but amides are not. The hypervalent iodine derivative 1-(*t*-butylperoxy)-1,2-benziodoxol-3(1*H*)-one similarly cleaves thioketals.[77]
29. MCPBA; Ac_2O, Et_3N, H_2O, THF, 28–37% yield.[78]

S S H CH₃ O OH — 1. MCPBA; 2. Ac_2O, TEA, H_2O, THF, 28 – 37% → O H CH_3 H O OH

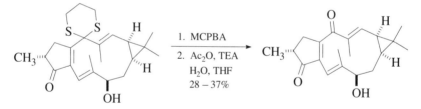

30. $Pyr{\cdot}HBr{\cdot}Br_2$, CH_2Cl_2, pyridine, $Bu_4N^+Br^-$, 0°–rt, 2 h, 80–90% yield.[79] The deprotection proceeds without olefin or aromatic ring bromination.
31. $PhOP(O)Cl_2$, DMF, NaI, 1 h, rt, 71–94% yield.[80]
32. $MeP(Ph)_3^+Br^-$, CH_2Cl_2, H_2O, NaH_2PO_4, Na_2HPO_4, 0–100% yield.[81]
33. For (n = 2): Me_3SiI or Me_3SiBr, DMSO, 65–99% yield.[82]
34. For (n = 3): DMSO, dioxane, 1.8 *M* HCl, 90–96% yield.[83]
35. For (n = 3): $Me_3S^+SbCl_6^-$, −77°; Na_2CO_3, H_2O, 95–97% yield.[83]
36. For (n = 3): MCPBA, TFA, CH_2Cl_2, 0°, 75–96% yield.[84]
37. For (n = 2): $CuCl_2{\cdot}2H_2O$, SiO_2, CH_2Cl_2, H_2O, 50–94% yield.[36]
38. For (n = 2,3): 2,4,6-triphenylpyrylium perchlorate, $h\nu$, O_2, CH_2Cl_2, 13–95% yield.[85,86]

39. TMSOTf, CH_2Cl_2, $NO_2C_6H_4CHO$, rt, 95% yield.[87] Diphenylthio acetals are also cleaved in high yield.
40. DMSO, 140–160°, 4–5 h.[88]
41. For ($n = 2,3$): visible light, methylene green, CH_3CN, H_2O, 86–97% yield.[89]
42. For ($n = 2,3$): nitrogen oxides, CH_2Cl_2, 40–96%, yield.[90]
43. For ($n = 2$): SeO_2, AcOH, rt, 0.5–2 h, 90–98% yield.[91]
44. For ($n = 2, 3$): H_5IO_5, ether, THF, 77–99% yield.[92] This method also cleaves oxathioacetals, but did not affect the acid-sensitive acetonide or 1,3-dioxolane. Note that ethereal periodic acid has been used to cleave terminal acetonides with subsequent glycol cleavage.[93]
45. $AgNO_3$, I_2, THF, H_2O, 53–100% yield[94]
46. An anomolous cleavage of a dithiolane was observed during an attempted hydroboration.[95]

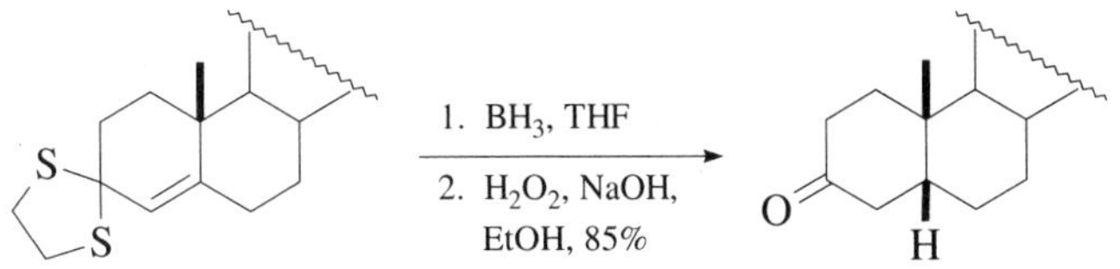

1. BH_3, THF
2. H_2O_2, NaOH, EtOH, 85%

Ref. 95

47. DDQ, BF_3, CH_2Cl_2, air, H_2O, >90% yield.[96]
48. DDQ, CH_3CN, photolysis or reflux, 1.5–2 h, 90–95% yield.[97]
49. DDQ, CH_3CN, H_2O (9:1), 0.5–6 h, 30–88% yield.[98] Dithiane derivatives of aromatic aldehydes give thioesters in low yields; dithiolanes are not effectively cleaved.
50. HgO, BF_3.[99]
51. $HgCl_2$, HgO, MeOH; $LiBF_4$, H_2O, CH_3CN, 89–91% yield.[99]
52. $h\nu$, sen., O_2, CH_3CN or CH_2Cl_2, 62–96% yield.[100,101]
53. NaTeH; H_2O, air, 80–85% yield.[102]
54. $SbCl_5$, N_2, CH_2Cl_2, 0°, 10 min; aq. $NaHCO_3$, 0°, 10 min, 63–100% yield.[103]
55. $GaCl_3$, MeOH, O_2, CH_2Cl_2, rt, 24 h, 71–99% yield.[104]
56. Amberlyst 15, acetone, CH_2O, H_2O, 80°, 10–25 h, 50–80% yield.[105]
57. *N*-Fluoro-2,4,6-trimethylpyridinium trifluormethanesulfonate, −10°, CH_2Cl_2, THF, H_2O, 68–91% yield.[106]
58. Dowex 50W, acetone, paraformaldehyde, reflux, 50–90% yield.[107]
59. $PhI(O_2CCl_3)_2$, CH_3CN, H_2O, rt, 5 min, >95% yield.[108]
60. Oxone, wet alumina, $CHCl_3$, reflux, 15–180 min, 70–96% yield.[109]
61. $Pe(phen)_3(PF_6)$, CH_3CN, H_2O, 43–75% yield. Hydroxyl and THP groups are not compatible with these conditions.[110]
62. Deprotection of a thioketal can occur with HF, which usually does not

affect this group, when neighboring group participation occurs, as in the following case.[111]

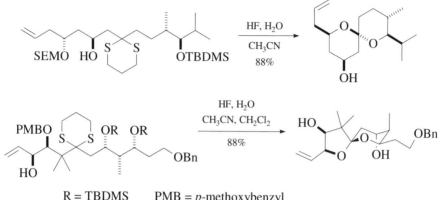

SEMO HO S S OTBDMS → HF, H_2O, CH_3CN, 88% → OH

PMBO S S OR OR HO OBn → HF, H_2O, CH_3CN, CH_2Cl_2, 88% → HO O O OH OBn

R = TBDMS PMB = *p*-methoxybenzyl

Note the unusual cleavage of the PMB ether as well.[112]

63. Clayfen, microwave, 87–97%. The reaction is done in the solid state.[113]
64. $Fe(NO_3)_3$, silica gel, hexane, 40–50°, 3–30 min, 86–100% yield.[114] $Fe(NO_3)_3$ and Montmorillonite K10 Clay in hexane are also effective.[115]

1. R. P. Hatch, J. Shringarpure, and S. M. Weinreb, *J. Org. Chem.*, **43**, 4172 (1978).
2. J. A. Marshall and J. L. Belletire, *Tetrahedron Lett.*, 871 (1971).
3. F. Sondheimer and D. Rosenthal, *J. Am. Chem. Soc.*, **80**, 3995 (1958).
4. W.-S. Zhou, *Pure Appl. Chem.*, **58**, 817 (1986).
5. T. Nakata, S. Nagao, N. Mori, and T. Oishi, *Tetrahedron Lett.*, **26**, 6461 (1985).
6. A. Pasquarello, G. Poli, and C. Scolastico, *Synlett*, 93 (1992).
7. I. Hoppe, D. Hoppe, R. Herbst-Irmer, and E. Egert, *Tetrahedron Lett.*, **31**, 6859 (1990).
8. D. R. Morton and S. J. Hobbs, *J. Org. Chem.*, **44**, 656 (1979).
9. D. A. Evans, L. K. Truesdale, K. G. Grimm, and S. L. Nesbitt, *J. Am. Chem. Soc.*, **99**, 5009 (1977).
10. E. J. Corey, M. A. Tius, and J. Das, *J. Am. Chem. Soc.*, **102**, 7612 (1980).
11. Y. Kamitori, M. Hojo, R. Masuda, T. Kimura, and T. Yoshida, *J. Org. Chem.*, **51**, 1427 (1986).
12. V. Kumar and S. Dev, *Tetrahedron Lett.*, **24**, 1289 (1983).
13. E. J. Corey and K. Shimoji, *Tetrahedron Lett.*, **24**, 169 (1983).
14. M. E. Kuehne, W. G. Bornmann, W. G. Earley, and I. Marko, *J. Org. Chem.*, **51**, 2913 (1986).
15. B. M. Trost and J. R. Parquette, *J. Org. Chem.*, **59**, 7568 (1994).
16. R. A. Moss and C. B. Mallon, *J. Org. Chem.*, **40**, 1368 (1975).
17. T. Satoh, S. Uwaya, and K. Yamakawa, *Chem. Lett.*, 667 (1983).

18. Y. Honda, A. Ori, and G. Tsuchihashi, *Chem. Lett.*, 1259 (1987).
19. J. A. Soderquist and E. I. Miranda, *Tetrahedron Lett.*, **27**, 6305 (1986).
20. T. Sato, E. Yoshida, T. Kobayashi, J. Otera, and H. Nozaki, *Tetrahedron Lett.*, **29**, 3971 (1988).
21. P. Kumar, R. S. Reddy, A. P. Singh, and B. Pandey, *Synthesis*, 67 (1993); *idem, Tetrahedron Lett.*, **33**, 825 (1992).
22. H. Tani, K. Masumoto, T. Inamasu, and H. Suzuki, *Tetrahedron Lett.*, **32**, 2039 (1991).
23. H. K. Patney, *Tetrahedron Lett.*, **32**, 2259 (1991); M. Hirano, K. Ukawa, S. Yakabe, and T. Morimoto, *Org. Prep. Proced. Int.*, **29**, 480 (1997).
24. B. M. Choudary and Y. Sudha, *Synth. Commun.*, **26**, 2993 (1996).
25. H. K. Patney, *Tetrahedron Lett.*, **34**, 7127 (1993).
26. H. K. Patney, *Tetrahedron Lett.*, **35**, 5717 (1994).
27. L. Garlaschelli and G. Vidari, *Tetrahedron Lett.*, **31**, 5815 (1990).
28. D. Villemin, B. Labiad, and M. Hammadi, *J. Chem. Soc., Chem. Commun.*, 1192 (1992).
29. R. B. Perni, *Synth. Commun.*, **19**, 2383 (1989); B. Ku and D. Y. Oh, *ibid.*, **19**, 433 (1989).
30. N. B. Das, A. Nayak, and R. P. Sharma, *J. Chem. Res., Synop.*, 242 (1993).
31. T. Sato, J. Otera, and H. Nozaki, *J. Org. Chem.*, **58**, 4971 (1993).
32. P. K. Chowdhury, *J. Chem. Res., Synop.*, 124 (1993).
33. Y. Zhang, Y. Yu, and R. Lin, *Org. Prep. Proced. Int.*, **25**, 365 (1993).
34. A. K. Maiti, K. Basu, and P. Bhattacharyya, *J. Chem. Res., Synop.*, 108 (1995).
35. V. G. Saraswathy and S. Sankaraman, *J. Org. Chem.*, **59**, 4665 (1994).
36. A. Nayak, B. Nanda, N. B. Das, and R. P. Sharma, *J. Chem. Res., Synop.*, 100 (1994).
37. R. Miranda, H. Cervantes, and P. Joseph-Nathan, *Synth. Commun.*, **20**, 153 (1990).
38. N. Komatsu, M. Uda, and H. Suzuki, *Synlett*, 984 (1995).
39. H. K. Patney and S. Margan, *Tetrahedron Lett.*, **37**, 4621 (1996).
40. D. P. Sabde, B. G. Naik, V. R. Hedge, and S. G. Hedge, *J. Chem. Res., Synop.*, 494 (1996).
41. Mechanisms of hydrolysis of thioacetals: D. P. N. Satchell and R. S. Satchell, *Chem. Soc. Rev.*, **19**, 55 (1990).
42. E. Fujita, Y. Nagao, and K. Kaneko, *Chem. Pharm. Bull.*, **26**, 3743 (1978).
43. B. H. Lipshutz, R. Moretti, and R. Crow, *Tetrahedron Lett.*, **30**, 15 (1989).
44. E. J. Corey and M. G. Bock, *Tetrahedron Lett.*, 2643 (1975).
45. P. Stütz and P. A. Stadler, *Org. Synth., Collect. Vol. VI*, 109 (1988).
46. C. A. Reece, J. O. Rodin, R. G. Brownlee, W. G. Duncan, and R. M. Silverstein, *Tetrahedron*, **24**, 4249 (1968).
47. D. Gravel, C. Vaziri, and S. Rahal, *J. Chem. Soc., Chem. Commun.*, 1323 (1972).
48. E. N. Cain and L. L. Welling, *Tetrahedron Lett.*, 1353 (1975).
49. E. J. Corey and B. W. Erickson, *J. Org. Chem.*, **36**, 3553 (1971).
50. A. V. Rama Rao, G. Venkatswamy, S. M. Javeed, V. H. Deshpande, and B. R. Rao, *J. Org. Chem.*, **48**, 1552 (1983).

51. P. S. Jones, S. V. Ley, N. S. Simpkins, and A. J. Whittle, *Tetrahedron*, **42**, 6519 (1986).
52. T.-L. Ho and C. M. Wong, *Can. J. Chem.*, **50**, 3740 (1972).
53. W. O. Moss, R. H. Bradbury, N. J. Hales, and T. Gallagher, *J. Chem. Soc., Perkin Trans. 1*, 1901 (1992).
54. R. A. J. Smith and D. J. Hannah, *Synth. Commun.*, **9**, 301 (1979).
55. M. Hojo and R. Masuda, *Synthesis*, 678 (1976).
56. J. B. Chattopadhyaya and A. V. Rama Rao, *Tetrahedron Lett.*, 3735 (1973).
57. W. F. J. Huurdeman, H. Wynberg, and D. W. Emerson, *Tetrahedron Lett.*, 3449 (1971).
58. D. W. Emerson and H. Wynberg, *Tetrahedron Lett.*, 3445 (1971).
59. D. H. R. Barton, N. J. Cussans, and S. V. Ley, *J. Chem. Soc., Chem. Commun.*, 751 (1977).
60. K. Fuji, K. Ichikawa, and E. Fujita, *Tetrahedron Lett.*, 3561 (1978).
61. P. R. Heaton, J. M. Midgley, and W. B. Whalley, *J. Chem. Soc., Chem. Commun.*, 750 (1971).
62. T.-L. Ho, H. C. Ho, and C. M. Wong, *J. Chem. Soc., Chem. Commun.*, 791 (1972).
63. T. T. Takahashi, C. Y. Nakamura, and J. Y. Satoh, *J. Chem. Soc., Chem. Commun.*, 680 (1977).
64. Q. N. Porter and J. H. P. Utley, *J. Chem. Soc., Chem. Commun.*, 255 (1978).
65. H. J. Cristau, B. Chabaud, and C. Niangoran, *J. Org. Chem.*, **48**, 1527 (1983).
66. A.-M. Martre, G. Mousset, R. B. Rhlid, and H. Veschambre, *Tetrahedron Lett.*, **31**, 2599 (1990).
67. T.-L. Ho and C. M. Wong, *Synthesis*, 561 (1972).
68. M. Fetizon and M. Jurion, *J. Chem. Soc., Chem. Commun.*, 382 (1972).
69. S. Takano, S. Hatakeyama, and K. Ogasawara, *J. Chem. Soc., Chem. Commun.*, 68 (1977).
70. T. Oishi, K. Kamemoto, and Y. Ban, *Tetrahedron Lett.*, 1085 (1972).
71. G. A. Olah, Y. D. Vankar, M. Arvanaghi, and G. K. S. Prakash, *Synthesis*, 720 (1979).
72. H. Muxfeldt, W.-D. Unterweger, and G. Helmchen, *Synthesis*, 694 (1976).
73. G. A. Olah, S. C. Narang, G. F. Salem, and B. G. B. Gupta, *Synthesis*, 273 (1979).
74. M. Platen and E. Steckhan, *Tetrahedron Lett.*, **21**, 511 (1980); *idem*, *Chem. Ber.*, **117**, 1679 (1984).
75. G. Stork and K. Zhao, *Tetrahedron Lett.*, **30**, 287 (1989).
76. M. Nakatsuka, J. A. Ragan, T. Sammakia, D. B. Smith, D. E. Uehling, and S. L. Schreiber, *J. Am. Chem. Soc.*, **112**, 5583 (1990).
77. M. Ochiai, A. Nakanishi, and T. Ito, *J. Org. Chem.*, **62**, 4253 (1997).
78. A. B. Smith, III, B. D. Dorsey, M. Visnick, T. Maeda, and M. S. Malamas, *J. Am. Chem. Soc.*, **108**, 3110 (1986).
79. G. S. Bates and J. O'Doherty, *J. Org. Chem.*, **46**, 1745 (1981).
80. H.-J. Liu and V. Wiszniewski, *Tetrahedron Lett.*, **29**, 5471 (1988).
81. H.-J. Cristau, A. Bazbouz, P. Morand, and E. Torreilles, *Tetrahedron Lett.*, **27**, 2965 (1986).

82. G. A. Olah, S. C. Narang, and A. K. Mehrotra, *Synthesis*, 965 (1982).
83. M. Prato, U. Quintily, G. Scorrano, and A. Sturaro, *Synthesis*, 679 (1982).
84. J. Cossy, *Synthesis*, 1113 (1987).
85. M. Kamata, Y. Murakami, Y. Tamagawa, M. Kato, and E. Hasegawa, *Tetrahedron*, **50**, 12821 (1994).
86. E. Fasani, M. Freccero, M. Mella, and A. Albini, *Tetrahedron*, **53**, 2219 (1997).
87. T. Ravindranathan, S. P. Chavan, R. B. Tejwani, and J. P. Varghese, *J. Chem. Soc., Chem. Commun.*, 1750 (1991).
88. Ch. S. Rao, M. Chandrasekharam, H. Ila, and H. Junjappa, *Tetrahedron Lett.*, **33**, 8163 (1992).
89. G. A. Epling and Q. Wang, *Synlett*, 335 (1992).
90. G. Mehta and R. Uma, *Tetrahedron Lett.*, **37**, 1897 (1996).
91. S. A. Haroutounian, *Synthesis*, 39 (1995).
92. X.-X. Shi, S. P. Khanapure, and J. Rokach, *Tetrahedron Lett.*, **37**, 4331 (1996).
93. W.-L. Wu and Y.-L. Wu, *J. Org. Chem.*, **58**, 3586 (1993).
94. K. Nishide, K. Yokota, D. Nakamura, T. Sumiya, M. Node, M. Ueda, and K. Fuji, *Tetrahedron Lett.*, **34**, 3425 (1993).
95. C. D'Alessandro, S. Giacopello, A. M. Seldes, and M. E. Deluca, *Synth. Commun.*, **25**, 2703 (1995).
96. J. P. Collman, D. A. Tyvoll, L. L. Chng, and H. T. Fish, *J. Org. Chem.*, **60**, 1926 (1995).
97. L. Mathew and S. Sankararaman, *J. Org. Chem.*, **58**, 7576 (1993).
98. K. Tanemura, H. Dohya, M. Imamura, T. Suzuki, and T. Horaguchi, *Chem. Lett.*, 965 (1994); *idem*, *J. Chem. Soc., Perkin Trans. 1*, 453 (1996).
99. J. A. Soderquist and E. L. Miranda, *J. Am. Chem. Soc.*, **114**, 10078 (1992).
100. M. Kamata, M. Sato, and E. Hasagawa, *Tetrahedron Lett.*, **33**, 5085 (1992); M. Kamata, Y. Murakami, Y. Tamagawa, Y. Kato, and E. Hasegawa, *Tetrahedron*, **50**, 12821 (1994).
101. E. Fasani, M. Freccero, M. Mella, and A. Albini, *Tetrahedron*, **53**, 2219 (1997).
102. P. Lue, W.-Q. Fan, and X.-J. Zhou, *Synthesis*, 692 (1989).
103. M. Kamata, H. Otogawa, and E. Hasegawa, *Tetrahedron Lett.*, **32**, 7421 (1991).
104. K. Saigo, Y. Hashimoto, N. Kihara, H. Umehara, and M. Hasegawa, *Chem. Lett.*, 831 (1990).
105. R. Ballini and M. Petrini, *Synthesis*, 336 (1990).
106. A. S. Kiselyov, L. Strekowski, and V. V. Semenov, *Tetrahedron*, **49**, 2151 (1993).
107. V. S. Giri and P. J. Sankar, *Synth. Commun.*, **23**, 1795 (1993).
108. M. H. B. Stowell, R. S. Rock, D. C. Rees, and S. I. Chan, *Tetrahedron Lett.*, **37**, 307 (1996).
109. P. Ceccherelli, M. Curini, M. C. Marcotullio, F. Epifano, and O. Rosati, *Synlett*, 767 (1996).
110. M. Schmittel and M. Levis, *Synlett*, 315 (1996).
111. P. G. Steet and E. J. Thomas, *J. Chem. Soc., Perkin Trans. 1*, 371 (1997).
112. A. B. Smith, III, J. J.-W. Duan, K. G. Hull, and B. A. Salvatore, *Tetrahedron Lett.*, **32**, 4855 (1991).

113. R. S. Varma and R. K. Saini, *Tetrahedron Lett.*, **38**, 2623 (1997).

114. M. Hirano, K. Ukawa, S. Yakabe, and T. Morimoto, *Synth. Commun.*, **27**, 1527 (1997).

115. M. Hirano, K. Ukawa, S. Yakabe, J. H. Clark, and T. Morimoto, *Synthesis*, 858 (1997).

1,5-Dihydro-3*H*-2,4-benzodithiepin Derivative:

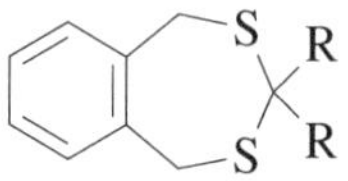

Dithiepin derivatives, prepared in high yield ($FeCl_3$–SiO_3, CH_2Cl_2, rt, 84–99%)[1] from 1,2-bis(mercaptomethyl)benzenes, are cleaved by $HgCl_2$ (80% yield). Neither reagents nor products have unpleasant odors.[2]

1. H. K. Patney, *Synth. Commun.*, **23**, 1829 (1993).

2. I. Shahak and E. D. Bergmann, *J. Chem. Soc. C*, 1005 (1966).

Monothio Acetals and Ketals

Acyclic Monothio Acetals and Ketals

Acyclic monothio acetals and ketals can be prepared directly from a carbonyl compound or by transketalization, a reaction that does not involve a free carbonyl group, from a 1,3-dithiane or 1,3-dithiolane. They are cleaved by acidic hydrolysis or Hg(II) salts.

O-Trimethylsilyl-S-alkyl Acetals and Ketals: $R_2C(SR')OSiMe_3$

Formation

1. $RSSiMe_3$, ZnI_2, 25°, 30 min, 80–90% yield.[1]
2. Me_3SiCl, R′SH, Pyr, 25°, 3 h, 75–90% yield.[2]
3. TMSImidazole, RSH, 90 min, 81–94% yield.[3]

Cleavage

1. dil. HCl.[2]
2. In ether or tetrahydrofuran, organolithium reagents cleave the silicon–oxygen bond; in hexamethylphosphoramide, they react at the carbon atom.[2]

1 D. A. Evans, L. K. Truesdale, K. G. Grimm, and S. L. Nesbitt, *J. Am. Chem. Soc.*, **99**, 5009 (1977).

2. T. H. Chan and B. S. Ong, *Tetrahedron Lett.*, 319 (1976).

3. M. B. Sassaman, G. K. S. Prakash, and G. A. Olah, *Synthesis*, 104 (1990).

O-Alkyl-*S*-alkyl or -*S*-phenyl Acetals and Ketals: $R_2C(OR')SR''$

Formation

1. From a dimethyl acetal: Et_2AlSPh, 0°, 78% yield.[1]
2. From a dimethyl acetal: $BCl_3 \cdot Et_2O$, −45°, CH_3SH, 73% yield.[2]
3. From a dialkyl acetal: Bu_3SnSPh, $BF_3 \cdot Et_2O$, toluene, −78° → 0°, 64–100% yield.[3] These conditions also convert MOM and MEM groups to the corresponding phenylthiomethyl groups in 64–77% yield.
4. From a dialkyl acetal: $MgBr_2$, Et_2O, rt, PhSH, 91% yield.[4] MOM groups are converted to phenylthiomethyl groups in 75% yield, but MEM groups do not react.
5. ROTMS (R = 4-MeBn, 4-MeOBn, 2-butenyl), PhSTMS, $CHCl_3$, TMSOTf, −75°, 37–93%.[5]
6. From a dimethyl ketal: cat. (1,3-dioxolan-2-ylidene)malononitrile [structure: O, O ring =C(C≡N)C≡N], PhSTMS, DMF, 0–60°, 62–90% yield.[6]

Cleavage

1. The mechanisms for hydrolysis of *O,S*-acetals have been reviewed. The following acid-catalyzed cleavage rates show that the *O,S*-acetals have a stability that lies between thioacetals and acetals:[7]

$PhCH(SEt)_2$	$PhCH(OEt)(SEt)$	$PhCH(OMe)(SPh)$	$PhCH(OEt)_2$
3.5×10^{-4}	1.3	41	160

   An extensive review of the chemistry of *O,S*-acetals has been published.[8]
2. Electrolysis: Pt electrode, KOAc, AcOH, 10 V, 18–20°; K_2CO_3, MeOH, 81–91% yield.[9] These cleavage conditions could, in principle, be used to cleave the MTM group.
3. $HgCl_2$, H_2O, $HClO_4$.[10] The section on MTM ethers should be consulted.

1. Y. Masaki, Y. Serizawa, and K. Kaji, *Chem. Lett.*, 1933 (1985).
2. F. Nakatsubo, A. J. Cocuzza, D. E. Keely, and Y. Kishi, *J. Am. Chem. Soc.*, **99**, 4835 (1977).
3. T. Sato, T. Kobayashi, T. Gojo, E. Yishida, J. Otera, and H. Nozaki, *Chem. Lett.*, 1661 (1987).
4. S. Kim, J. H. Park, and S. Lee, *Tetrahedron Lett.*, **30**, 6697 (1989).

5. A. Kusche, R. Hoffmann, I. Münster, P. Keiner, and R. Brückner, *Tetrahedron Lett.*, **32**, 467 (1991).
6. T. Miura and Y. Masaki, *Tetrahedron*, **51**, 10477 (1995); *idem*, *Tetrahedron Lett.*, **35**, 7961 (1994).
7. D. P. N. Satchell and R. S. Satchell, *Chem. Soc. Rev.*, **19**, 55 (1990).
8. P. Wimmer, "O/S Acetale," in *O/O- und O/S-Acetale, [Methoden der Organischen Chemie] (Houben-Weyl)*, Band E14a/1, H. Hagemann and D. Klamann, Eds., 1991, G. Theime Stuttgart, p. 785.
9. T. Mandai, H. Irei, M. Kuwada, and J. Otera, *Tetrahedron Lett.*, **25**, 2371 (1984).
10. J. L. Jensen, D. F. Maynard, G. R. Shaw, and T. W. Smith, Jr., *J. Org. Chem.*, **57**, 1982 (1992).

***O*-Methyl-*S*-2-(methylthio)ethyl Acetals and Ketals:** $R_2C(OMe)SCH_2CH_2SMe$

Formation[1]

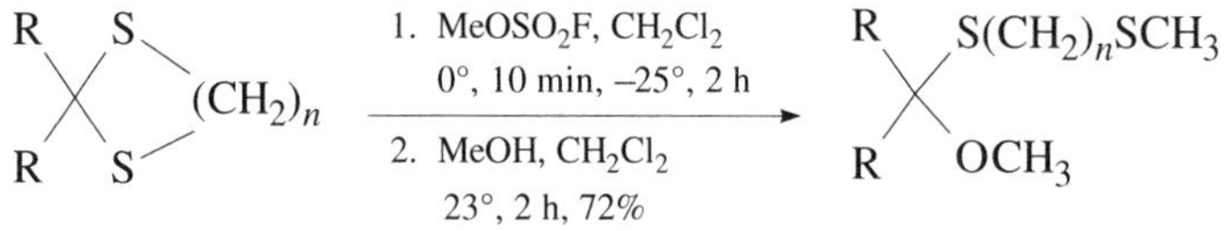

R₂C[S(CH₂)ₙS] → (1. $MeOSO_2F$, CH_2Cl_2, 0°, 10 min, –25°, 2 h; 2. MeOH, CH_2Cl_2, 23°, 2 h, 72%) → $R_2C(S(CH_2)_nSCH_3)(OCH_3)$

$n = 2, 3$

Cleavage[1]

1. $HgCl_2$, $CaCO_3$, THF, H_2O, 0°, rapid.

These derivatives are less susceptible to oxidation and hydrogenolysis than are the 1,3-dithiane and 1,3-dithiolane precursors.

1. E. J. Corey and T. Hase, *Tetrahedron Lett.*, 3267 (1975).

Cyclic Monothio Acetals and Ketals

1,3-Oxathiolanes (Chart 5): R₂C[O–CH₂CH₂–S]

Formation

1. $HSCH_2CH_2OH$, $ZnCl_2$ AcONa, dioxane, 25°, 20 h, 60–90% yield.[1,2]
2. $HSCH_2CH_2OH$, TMSOTf, 10 min, 50–78% yield.[3]

Cleavage

The section on the cleavage of 1,3-dithianes and 1,3-dithiolanes should be consulted, since many of the methods described there are also applicable to the cleavage of oxathiolanes.

1. $HgCl_2$, AcOH, AcOK, 100°, 1 h, 83% yield.[4]
2. $HgCl_2$, NaOH, EtOH, H_2O, 25°, 30 min, 91% yield.[4]
3. Raney Ni, AcOH, AcOK, 100°, 90 min, 92% yield.[4]
4. HCl, AcOH, reflux, 22 h, 60% yield.[5]
5. $AgNO_3$, NCS, 80% CH_3CN, H_2O.[6]
6. Benzyne, $ClCH_2CH_2Cl$, 49–100% yield.[7]
7. 4-Nitrobenzaldehyde, TMSOTf, CH_2Cl_2, rt, 75–97% yield.[8] Dithiolanes are stable to these conditions.

1. J. Romo, G. Rosenkranz, and C. Djerassi, *J. Am. Chem. Soc.*, **73**, 4961 (1951).
2. V. K. Yadav and A. G. Fallis, *Tetrahedron Lett.*, **29**, 897 (1988).
3. T. Ravindranathan, S. P. Chavan, and S. W. Dantale, *Tetrahedron Lett.*, **36**, 2285 (1995).
4. C. Djerassi, M. Shamma, and T. Y. Kan, *J. Am. Chem. Soc.*, **80**, 4723 (1958).
5. R. H. Mazur and E. A. Brown, *J. Am. Chem. Soc.*, **77**, 6670 (1955).
6. S. V. Frye and E. L. Eliel, *Tetrahedron Lett.*, **26**, 3907 (1985).
7. J. Nakayama, H. Sugiura, A. Shiotsuki, and M. Hoshino, *Tetrahedron Lett.*, **26**, 2195 (1985).
8. T. Ravindranathan, S. P. Chaven, J. P. Varghese, S. W. Dantale, and R. B. Tejwani, *J. Chem. Soc., Chem. Commun.*, 1937 (1994); T. Ravindranathan, S. P. Chavan, and M. M. Awachat, *Tetrahedron Lett.*, **35**, 8835 (1994).

Diseleno Acetals and Ketals: $R_2C(SeR')_2$

Selenium compounds are generally HIGHLY TOXIC.

Formation

1. RSeH, $ZnCl_2$, N_2, CCl_4, 20°, 3 h, 70–95% yield.[1]
2. From a ketal: $(PhSe)_3B$, CF_3COOH, $CHCl_3$, 20°, 20 min – 24 h.[2]

Cleavage

1. $HgCl_2$, $CaCO_3$, CH_3CN, H_2O, 20°, 2–4 h, 65–80% yield.[1]
2. $CuCl_2$, CuO, acetone, H_2O, 20°, 5 min–2 h, 73–99% yield.[1]
3. H_2O_2, THF, 0°, 15 min $\rightarrow$ 20°, 3 h, 60–65% yield.[1]
4. $(PhSeO)_2O$, THF, 20° or 60°, 5 min $\rightarrow$ 6 h, 60–90% yield.[1]
5. Clay-supported ferric nitrate (Clayfen) or clay-supported cupric nitrate (Claycop), pentane, rt, 60–97% yield.[3]

Diseleno acetals and ketals are cleaved more rapidly than their dithio counterparts; a methyl derivative is cleaved more rapidly than a phenyl derivative. Methyl iodide or ozone converts diseleno acetals and ketals to vinyl selenides.[1]

1. A. Burton, L. Hevesi, W. Dumont, A. Cravador, and A. Krief, *Synthesis*, 877 (1979).
2. D. L. J. Clive and S. M. Menchen, *J. Org. Chem.*, **44**, 4279 (1979).
3. P. Laszlo, P. Pennetreau, and A. Krief, *Tetrahedron Lett.*, **27**, 3153 (1986).

MISCELLANEOUS DERIVATIVES

O-Substituted Cyanohydrins

_O_-Acetyl Cyanohydrin: $R_2C(CN)OAc$

Formation

1. $Me_2C(CN)OH$, Et_3N, 25°, 2 h, 82% yield; Ac_2O, Pyr, 25°, 40 h, 82% yield.[1]
2. From a cyanohydrin: Ac_2O, $FeCl_3$, 25–92% yield.[2] Other anhydrides are also effective in this conversion.
3. AcCN, K_2CO_3, CH_3CN, 79–96% yield.[3]

Cleavage

1. Li(O-*t*-Bu)$_3$AlH, THF; KOH, CH_3OH, H_2O, 25°, 5 min, 84% yield.[1]

_O_-Trimethylsilyl Cyanohydrin: $R_2C(CN)OSiMe_3$ (Chart 5)

Formation

1. Me_3SiCN, cat. KCN or $Bu_4N^+F^-$, 18-crown-6, 75–95% yield.[4]
2. Me_3SiCN, Ph_3P, CH_3CN, 0°, 1 h, 100% yield.[5]

O, R′, R″, R, O → cat. KCN or $Bu_4N^+F^-$, 18-crown-6, 75–95% / or Ph_3P, CH_3CN, 0°, 1 h, 100% → NC OTMS, R′, R″, R, O

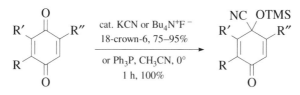

3. $Me_2C(CN)OSiMe_3$, KCN, 130°.[6]
4. Me_3SiCl, KCN, Amberlite XAD-4, CH_3CN, 60°, 8 h, 81–97% yield.[7]
5. Me_3SiCl, KCN, NaI, Pyr, CH_3CN, 50–77% distilled yields, 100% by NMR.[8]
6. R_3SiCl, KCN, ZnI_2, CH_3CN, 86–98% yield.[9] This method was used to prepare the *t*-BuPh$_2$Si, *t*-BuMe$_2$Si, and *i*-Pr$_3$Si cyanohydrins.
7. TMSCN, THF, Yb(CN)$_3$, 0°→ rt, 84–99% yield.[10]
8. TMSCN, CH_2Cl_2, Yb(OTf)$_3$, 55–95% yield. Aromatic ketones fail to react.[11]
9. TMSCN, CH_2Cl_2, −40°, Eu(fod)$_3$, 45–95% yield.[12]

10. TMSCN, TEA, 91–100% yield.[13]
11. TMSCN, CH_3CN, reflux, 2 h, 89–95% yield.[14] These conditions are selective for aldehydes.
12. TMSCN, $MgAlCO_3$, heptane, 90–99% yield.[15]
13. TMSCN, (−)-DIPT [diisopropyl L-tartrate], Ti(*i*-PrO)$_4$, CH_2Cl_2, 0°, 6 h, rt, 12 h, 95% yield. These conditions afford chiral cyanohydrins.[16]
14. (*R*)-BINOL-Ti(O-*i*-Pr)$_2$, TMSCN, CH_2Cl_2. Enantioselectivity of up to 75% is obtained.[17]
15. Chiral (salene)Ti(IV) complexes, TMSCN. This system is selective for aldehydes; the asymmetric induction is dependent upon aldehyde structure.[18,19]
16. Pybox–$AlCl_3$, [(*S,S*)-2,6-bis(4′-isopropyloxazolin-2′-yl)pyridine], TMSCN. Mandelonitrile was formed in 92% yield (>90% ee).[20]
17. Ti(O-*i*-Pr)$_4$, sulfoximines, TMSCN.[21]
18. Bu_2SnCl_2 or Ph_2SnCl_2, TMSCN, 71–97% yield.[22]

Cleavage

1. AgF, THF, H_2O, 25°, 2.5 h, 77% yield.[5]
2. Dilute acid or base.[23]
3. (*S*)-Hydroxynitrile lyase can be used for the decomposition of cyanohydrins with some level of enantioselectivity.[24]

O-1-Ethoxyethyl Cyanohydrin: $R_2C(CN)OCH(OC_2H_5)CH_3$

The ethoxyethyl cyanohydrin was prepared (NaCN, HCl, THF, 0°, 75% yield, followed by $EtOCH{=}CH_2$, HCl, 50% yield) to convert an aldehyde ultimately to a protected ketone. It was cleaved by hydrolysis (0.01 *N* HCl, MeOH, 25°, followed by NaOH, 0°, 85% yield).[25]

O-Tetrahydropyranyl Cyanohydrin: $R_2C(CN)O$–THP

The tetrahydropyranyl cyanohydrin was prepared from a steroid cyanohydrin (dihydropyran, TsOH, reflux, 1.5 h) and cleaved by hydrolysis (cat. concd. HCl, acetone, reflux, 15 min, followed by aq. pyridine, reflux, 1 h).[26]

1. P. D. Klimstra and F. B. Colton, *Steroids*, **10**, 411 (1967).
2. T. Hiyama, H. Oishi, and H. Saimoto, *Tetrahedron Lett.*, **26**, 2459 (1985).
3. M. Okimoto and T. Chiba, *Synthesis*, 1188 (1996).
4. D. A. Evans, J. M. Hoffman, and L. K. Truesdale, *J. Am. Chem. Soc.*, **95**, 5822 (1973).
5. D. A. Evans and R. Y. Wong, *J. Org. Chem.*, **42**, 350 (1977).
6. D. A. Evans and L. K Truesdale, *Tetrahedron Lett.*, 4929 (1973).
7. K. Sukata, *Bull. Chem. Soc. Jpn.*, **60**, 3820 (1987).

8. F. Duboudin, Ph. Cazeau, F. Moulines, and O. Laporte, *Synthesis*, 212 (1982).
9. V. H. Rawal, J. A. Rao, and M. P. Cava, *Tetrahedron Lett.*, **26**, 4275 (1985).
10. S. Matsubara, T. Takai, and K. Utimoto, *Chem. Lett.*, 1447 (1991).
11. Y. Yang and D. Wang, *Synlett*, 1379 (1997).
12. J. H. Gu, M. Okamoto, M. Terada, K. Mikami, and T. Nakai, *Chem. Lett.*, 1169 (1992).
13. S. Kobayashi, Y. Tsuchiya, and T. Mukaiyama, *Chem. Lett.*, 537 (1991).
14. K. Manju and S. Trehan, *J. Chem. Soc., Perkin Trans. 1*, 2383 (1995).
15. B. M. Choudary, N. Narender, and V. Bhuma, *Synth. Commun.*, **25**, 2829 (1995).
16. M. C. Pirrung and S. W. Shuey, *J. Org. Chem.*, **59**, 3890 (1994).
17. M. Mori, H. Imma, and T. Nakai, *Tetrahedron Lett.*, **38**, 6229 (1997).
18. Y. Belokon, M. Flego, N. Ikonnikow, M. Moscalenko, M. North, C. Orizu, V. Tararov, and M. Tasinazzo, *J. Chem. Soc., Perkin Trans. 1*, 1293 (1997).
19. Y. Jiang, X. Zhou, W. Hu, L. Wu, and A. Mi, *Tetrahedron: Asymmetry*, **6**, 405 (1995).
20. I. Iovel, Y. Popelis, M. Fleisher, and E. Lukevics, *Tetrahedron: Asymmetry*, **8**, 1279 (1997).
21. C. Bolm, P. Mueller, and K. Harms, *Acta Chem. Scand.*, **50**, 305 (1996); C. Bolm and P. Mueller, *Tetrahedron Lett.*, **36**, 1625 (1995).
22. J. K. Whitesell and R. Apodaca, *Tetrahedron Lett.*, **37**, 2525 (1996).
23. D. A. Evans, L. K. Truesdale, and G. L. Carroll, *J. Chem. Soc., Chem. Commun.*, 55 (1973).
24. M. Schmidt, S. Herve, N. Klempier, and H. Griengl, *Tetrahedron*, **52**, 7833 (1996).
25. G. Stork and L. Maldonado, *J. Am. Chem. Soc.*, **93**, 5286 (1971).
26. P. deRuggieri and C. Ferrari, *J. Am. Chem. Soc.*, **81**, 5725 (1959).

Substituted Hydrazones

***N,N*-Dimethylhydrazone:** RR′C=NN$(CH_3)_2$ (Chart 5)

Formation

1. H_2NNMe_2, EtOH-HOAc, reflux, 24 h, 90–94% yield.[1]
2. $Me_2AlNHNMe_2$, $PhCH_3$, reflux, 3–5 h, 77–99% yield.[2]
3. H_2NNMe_2, TMSCl, 25°, 36 h, 92% yield.[3]

Cleavage

1. $NaIO_4$, MeOH, pH 7, 2–3 h, 90% yield.[4]
2. $Cu(OAc)_2$, H_2O, THF, pH 5.4, 25°, 15 min, 97% yield.[5]
3. $CuCl_2$, THF, HPO_4^-, → pH 7, 85–100% yield.[5]
4. CH_3I, 95% EtOH, reflux, 80–90% yield.[6]
5. O_3, CH_2Cl_2, −78°, 60–100% yield.[7]
6. O_2, *hν*, Rose Bengal, MeOH, −78° → −20°, followed by Ph_3P or Me_2S, 48–88% yield.[8]

7. CoF_3 ($CHCl_3$, reflux, 67–93% yield);[9] $MoOCl_3$ or MoF_6 (H_2O, THF, 25°, 4 h, 80–90% yield);[10] WF_6 ($CHCl_3$, 0° → 25°, 1 h, 84–95% yield);[11] UF_6 (50–95% yield).[12]
8. N_2O_4, −40° → 0°, CH_3CN, THF, $CHCl_3$, CCl_4, ~10 min, 75–95% yield.[13] This method is also effective for the regeneration of ketones from oximes (45–95% yield).
9. $NaBO_3 \cdot 4H_2O$, *t*-BuOH, pH 7, 60°, 24 h, 70–95% yield.[14]
10. AcOH, THF, H_2O, AcONa, 25°, 24 h, 95% yield.[15]

AcOH, THF, H_2O, NaOAc, 25°, 24 h, 95%

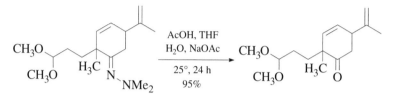

N,N-Dimethylhydrazones are stable to CrO_3/H_2SO_4(0°, 3 min), to $NaBH_4$ (EtOH, 25°), to $LiAlH_4$ (THF, 25°), and to B_2H_6 followed by H_2O_2/OH^-. They are cleaved by CrO_3/Pyr and by *p*-$NO_2C_6H_4CO_3H/CHCl_3$, 25°.[6]
11. Silica gel, THF, H_2O, rt, 3–10 h, 60–74% yield[16] or silica gel, CH_2Cl_2, 77–100% yield.[17]
12. $BF_3 \cdot Et_2O$, acetone, H_2O, 93–100% yield.[18]
13. MCPBA, DMF, −63°, 100% yield.[19] Hydrazones of aldols are cleaved without elimination under these conditions.[20] An axial α-methyl group on a cyclohexanone does not epimerize under these conditions.[19]
14. MMPP·$6H_2O$ (magnesium monoperoxyphthalate), pH 7 buffer, MeOH, 0°, 5–120 min, 76–99% yield.[21] These conditions were used to cleave the related SAMP hydrazone in the presence of two trisubstitued alkenes in 46% yield.[22]
15. Peracetic acid.[23]
16. Dimethyldioxirane, acetone, 89% yield.[24]
17. $NOBF_4$, CH_2Cl_2, Pyr, 59–86% yield. Oximes are cleaved similarly in 55–82% yield.[25]
18. $Pd(OAc)_2$, $SnCl_2$,, DMF, H_2O, 53–100% yield. This is the only catalytic procedure for the cleavage of dimethylhydrazones.[26]
19. $[(n\text{-}Bu)_4N]_2S_2O_8$, $ClCH_2CH_2Cl$, reflux, 0.6 h, 89–97% yield.[27]

1. G. R. Newkome and D. L. Fishel, *Org. Synth., Collect. Vol. VI*, 12 (1988).
2. B. Bildstein and P. Denifl, *Synthesis*, 158 (1994).
3. D. A. Evans, R. P. Polniaszek, K. M. DeVries, D. E. Guinn, and D. J. Mathre, *J. Am. Chem. Soc.*, **113**, 7613 (1991).
4. E. J. Corey and D. Enders, *Tetrahedron Lett.*, 3 (1976).

5. E. J. Corey and S. Knapp, *Tetrahedron Lett.*, 3667 (1976).
6. M. Avaro, J. Levisalles, and H. Rudler, *J. Chem. Soc., Chem. Commun.*, 445 (1969).
7. R. E. Erickson, P. J. Andrulis, J. C. Collins, M. L. Lungle, and G. D. Mercer, *J. Org. Chem.*, **34**, 2961 (1969).
8. E. Friedrich, W. Lutz, H. Eichenauer, and D. Enders, *Synthesis*, 893 (1977).
9. G. A. Olah, J. Welch, and M. Henninger, *Synthesis*, 308 (1977).
10. G. A. Olah, J. Welch, G. K. S. Prakash, and T.-L. Ho, *Synthesis*, 808 (1976).
11. G. A. Olah and J. Welch, *Synthesis*, 809 (1976).
12. G. A. Olah, J. Welch, and T.-L. Ho, *J. Am. Chem. Soc.*, **98**, 6717 (1976).
13. S. B. Shim, K. Kim, and Y. H. Kim, *Tetrahedron Lett.*, **28**, 645 (1987).
14. D. Enders and V. Bhushan, *Z. Naturforsh. B: Chem. Sci.*, **42**, 1595 (1987).
15. E. J. Corey and H. L. Pearce, *J. Am. Chem. Soc.*, **101**, 5841 (1979).
16. R. B. Mitra and G. B. Reddy, *Synthesis*, 694 (1989).
17. H. Kotsuki, A. Miyazaki, I. Kadota, and M. Ochi, *J. Chem. Soc., Perkin Trans. 1*, 429 (1990).
18. D. Enders, H. Dyker, G. Raabe, and J. Runsink, *Synlett*, 901 (1992).
19. M. Duraisamy and H. M. Walborsky, *J. Org. Chem.*, **49**, 3410 (1984).
20. M. M. Claffey and C. H. Heathcock, *J. Org. Chem.*, **61**, 7646 (1996).
21. D. Enders and A. Plant, *Synlett*, 725 (1990).
22. K. C. Nicolaou, F. Sarabia, M. R. V. Finlay, S. Ninkovic, N. P. King, D. Vourloumis, and Y. He, *Chem.—Eur. J.*, **3**, 1971 (1997).
23. L. Horner and H. Fernekess, *Chem. Ber.*, **94**, 712 (1961).
24. A. Altamura, R. Curci, and J. O. Edwards, *J. Org. Chem.*, **58**, 7289 (1993).
25. G. A. Olah and T.-L. Ho, *Synthesis*, 610 (1976).
26. T. Mino, T. Hirota, and M. Yamashita, *Synlett*, 999 (1996).
27. H. C. Choi and Y. H. Kim, *Synth. Commun.*, **24**, 2307 (1994).

Phenylhydrazone: $C_6H_5NHN{=}CR_2$

Formation

1. $PhNHNH_2$, AcOH, EtOH.[1]

Cleavage

1. $PhI(OTFA)_2$, CH_3CN, H_2O, 82–90% yield or PhI(OH)OTs, $CDCl_3$, rt, 2 h, 74–98% yield.[2] Mild oxidative regeneration of ketones occurs in good yields.
2. $(NH_4)_2S_2O_8$, clay, microwaves or ultrasound, 62–90% yield.[3]

1. R. L. Shriner, R. C. Fuson, D. Y. Curtin, and T. C. Morrill, *The Systematic Identification of Organic Compounds: A Laboratory Manual*, 6th ed., Wiley, New York, 1980, p. 165.

2. D. H. R. Barton, J. Cs. Jaszberenyi, and T. Shinade, *Tetrahedron Lett.*, **34**, 7191 (1993).
3. R. S. Varma and H. M. Meshram, *Tetrahedron Lett.*, **38**, 7973 (1997).

2,4-Dinitrophenylhydrazone (2,4-DNP group):
$R_2C{=}NNHC_6H_3$-2,4-$(NO_2)_2$ (Chart 5)

Formation

1. 2,4-$(NO_2)_2C_6H_3NHNH_2{\cdot}H_2SO_4$, EtOH, H_2O, 25°, 10 min, 80% yield.[1] In a synthesis of sativene a carbonyl group was protected as a 2,4-DNP, while a double bond was hydrated with $BH_3/H_2O_2/OH^-$. Attempted protection of the carbonyl group as a ketal caused migration of the double bond; protection as an oxime or oxime acetate was unsatisfactory, since both of these would be reduced with BH_3.

Cleavage

2,4-Dinitrophenylhydrazones are cleaved by various oxidizing and reducing agents, and by exchange reactions.

1. O_3, EtOAc, −78°, 70% yield.[1]
2. $TiCl_3$, DME, H_2O, N_2, reflux, 80–95% yield.[2]
3. Acetone, sealed tube, 75°, 20 h, 80–85% yield.[3]

1. J. E. McMurry, *J. Am. Chem. Soc.*, **90**, 6821 (1968).
2. J. E. McMurry and M. Silvestri, *J. Org. Chem.*, **40**, 1502 (1975).
3. S. R. Maynez, L. Pelavin, and G. Erker, *J. Org. Chem.*, **40**, 3302 (1975).

Tosylhydrazone: $CH_3C_6H_4SO_2NHN{=}CR_2$

Formation

$TsNHNH_2$, AcOH, EtOH.[1]

Cleavage

1. TS-1(titanium silicate molecular sieve), H_2O_2, MeOH, reflux, 4–18 h, 60–64% yield.[2]
2. Dimethyldioxirane, acetone, 95% yield.[3]
3. $Zr(O_3PCH_3)_{1.2}(O_3PC_6H_4SO_3H)_{0.8}$, acetone, H_2O, reflux, 70–95% yield.[4]
4. $KHSO_5$, aq. CH_3CN, 63–99% yield.[5]
5. Dimethyldioxirane, acetone, pH 6 , 10–144 h, 67–99% yield.[6]
6. 70% *t*-Butyl hydroperoxide, CCl_4, reflux, 4–18 h, 50–100% yield.[7] Cleavage is effective only for aromatic tosylhydrazones.

7. Na_2O_2, pentane, H_2O, reflux, 6 h, 69–72% yield.[8]

1. R. H. Shapiro, *Org. React.*, **23**, 405 (1976).
2. P. Kumar, V. R. Hegde, B. Paudey, and T. Ravindranathan, *J. Chem. Soc., Chem. Commun.*, 1553 (1993).
3. A. Altamura, R. Curci, and J. O. Edwards, *J. Org. Chem.*, **58**, 7289 (1993).
4. M. Curini, O. Rosati, and E. Pisani, *Synlett*, 333 (1996).
5. Y. H. Kim, J. C. Jung, and K. S. Kim, *Chem. Ind. (London)*, 31 (1992).
6. J. C. Jung, K. S. Kim and Y. H. Kim, *Synth. Commun.*, **22**, 1583 (1991).
7. N. B. Barhate, A. S. Gajare, R. D. Wakharkar, and A. Sudalai, *Tetrahedron Lett.*, **38**, 653 (1997).
8. T.-L. Ho and G. A. Olah, *Synthesis*, 611 (1976).

Semicarbazone: ($NH_2CONHN{=}CR_2$)

Formation

$NH_2CONHNH_2$, NaOAc, MeOH.[1]

Cleavage

1. $PhI(OAc)_2$, CH_3CN, H_2O, 70–83% yield.[2]
2. $(Bu_4N^+)_2S_2O_8^-$, $ClCH_2CH_2Cl$, reflux, 89–97% yield.[3]
3. Pyruvic acid, acetic acid, $CHCl_3$, 43–61%.[4]
4. $CuCl_2{\cdot}2H_2O$, CH_3CN, reflux, 10–390 min, 7–97% yield.[5]
5. TMSCl, $NaNO_2$ or $NaNO_3$, Aliquat 366, 3–5 h, CH_2Cl_2, 75–95% yield.[6]

Diphenylmethylsemicarbazone: ($Ph_2CHNHCONHN{=}CR_2$)

This derivative was used to improve the solubility characteristic of an argininal semicarbazone for solution-phase peptide synthesis.

Formation

$Ph_2CHNHCONHNH_2$, NaOAc, EtOH, H_2O, reflux, 1 h, 78% yield.[7]

Cleavage

Since hydrogenolysis resulted in only a 20% yield of the free aldehyde, a two-step procedure was developed in which the diphenylmethyl group was first cleaved with HF/anisole and then the unsubstituted semicarbazone was cleaved with formalin in 40–60% overall yield.

1. R. L. Shriner, R. C. Fuson, D. Y. Curtin, and T. C. Morrill, *The Systematic Identification of Organic Compounds*, 6th ed., Wiley, New York, 1980, p. 179.

2. D. W. Chen and Z. C. Chen, *Synthesis*, 773 (1994).
3. H. C. Choi and Y. H. Kim, *Synth. Commun.*, **24**, 2307 (1994).
4. H. Hosoda, K. Osanai, I. Fukasawa, and T. Nambara, *Chem. Pharm. Bull.*, **38**, 1949 (1990).
5. R. N. Ram and K. Varsha, *Tetrahedron Lett.*, **32**, 5829 (1991).
6. R. H. Khan, R. K. Mathur, and A. C. Ghosh, *J. Chem. Res., Synop.*, 506 (1995).
7. R. Dagnino, Jr., and T. R. Webb, *Tetrahedron Lett.*, **35**, 2125 (1994).

Oxime Derivatives: $R_2C{=}NOH$

Formation

1. $H_2NOH{\cdot}HCl$, Pyr, 60°. This is the standard method for the preparation of oximes. Ethanol or methanol can be used as cosolvents.
2. $H_2NOH{\cdot}HCl$, DABCO, MeOH, rt, 87% for a camphor derivative.[1] This method was reported to be better than when pyridine was used as the solvent and base.
3. TMSNHOTMS, KH, 100% yield.[2]
4. $H_2NOH{\cdot}HCl$, Amberlyst A21, EtOH, 1–10 h, 70–97% yield.[3]

Cleavage

Oximes are cleaved by oxidation, reduction, or hydrolysis in the presence of another carbonyl compound. Following are some synthetically useful methods:

1. $CH_3CO(CH_2)_2COOH$, 1 *N* HCl, 25°, 3 h, 94% yield.[4] Pyruvic acid (HOAc, reflux, 1–3 h, 77% yield)[5] and acetone (80–100 h, 72% yield)[6] effect cleavage in a similar manner.
2. $(PhSeO)_2O$/THF, 50°, 1–3 h, 80–95% yield.[7] An *O*-methyl oxime is stable to phenylselenic anhydride.
3. $Na_2S_2O_4$, H_2O, 25°, 12 h or 40°, few hours ~95% yield.[8]
4. $NaHSO_3$, EtOH–H_2O, reflux, 2–16 h; dil. HCl, 30 min, 85% yield.[9,10]
5. Ac_2O, 20°; $Cr(OAc)_2$/THF–H_2O, 25–65°, 75–95% yield.[11] Chromous acetate also cleaves unsubstituted oximes, but the reaction is slow and requires high temperatures.
6. $NaNO_2$, 1 *N* HCl, CH_3OH, H_2O, 0°, 3 h, 76% yield.[12] In the last step of a synthesis of erythronolide A, acid-catalyzed hydrolysis of an acetonide failed because the carbonyl-containing precursor was unstable to acidic hydrolysis (3% MeOH, HCl, 0°, 30 min, conditions developed for the synthesis of erythronolide B). Consequently, the carbonyl group was protected as an oxime, the acetonide was cleaved, and the carbonyl group was regenerated.
7. NOCl, Pyr, −20°; H_2O, reflux, 70–90% yield.[13] Olefins were not affected under these conditions. The related nitrosyl tetrafluoroborate has also been used.[14]

8. $TiCl_3$, H_2O, rt, 1 h, 85% yield.[15] This is an excellent reagent that works when cleavage of a methoxy oxime with chromous ion fails.
9. VCl_2, H_2O, THF, 8 h, rt, 75–92% yield.[16]
10. Et_3NH^+ $ClCrO_3^-$, $ClCH_2CH_2Cl$, 2 h, rt, 60–90% yield.[17] This reagent was reported to work better than PCC (pyridinium chlorochromate). Trimethylsilyl chlorochromate is also effective.[18]
11. Bu_3P, PhSSPh, THF, 85% yield.[19]
12. *t*-BuONO, *t*-BuOK; H_2O, NaOH; acidify, 40°.[20]
13. TMSCl, $NaNO_2$, CCl_4, 5% Aliquat 336, rt, 3–5 h, 64–98% yield.[21]
14. NaOCl, MeCN, rt, 23–99% yield.[22]
15. Zinc bismuthate, $PhCH_3$ or CH_3CN, reflux, 0.5–2 h, 56–85% yield.[23]
16. MnO_2, hexane or CH_2Cl_2, rt, 70–92% yield.[24] The oximes of pyruvates and *O*-alkyl oximes are not cleaved under these conditions.
17. $PhICl_2$, Pyr, $CHCl_3$, 3 h, 10°, 65–80% yield.[25]
18. $TiCl_4$, NaI, CH_3CN, rt, 63–97% yield.[26]
19. Baker's yeast, pH 7.2, H_2O, EtOH, 62–95% yield with sonication.[27]
20. TS-1 Zeolite, H_2O_2, acetone, reflux, 65–86% yield.[28]
21. Dimethyldioxirane, acetone, 0° or rt, 80–100% yield.[29]
22. $Ru_3(CO)_{12}$, CO, 20 atm, 4 h, 100°. These conditions reduce the oxime to an imine that is easily hydrolyzed with water.[30] Aldehyde oximes give low yields of nitriles.
23. $Cu(NO_3)_2$, Bentonite, hexane, acetone, 60–97% yield.[31] When silica gel is used as the support, tosylhydrazones and thioketals are also cleaved in excellent yield.[32]
24. $KMnO_4$, CH_3CN, H_2O, rt, 25-96% yield.[33]
25. $Zr(O_3PCH_3)_{1.2}(O_3PC_6H_4SO_3H)_{0.8}$, acetone, water, reflux 30 min–24 h, 70–95% yield. Semicarbazones, tosylhydrazones, and hydrazones are also cleaved.[34]
26. $(NH_4)_2S_2O_8$–silica gel, microwave irradiation, 59–83% yield.[35]
27. $BiCl_3$, microwave irradiation, 2 min, THF, 70–96% yield. α,β-Unsaturated systems were not effectively cleaved under these conditons.[36]
28. 70% *t*-Butyl hydroperoxide, CCl_4, reflux, 4–18 h, 30–100% yield.[37]
29. NBS, CCl_4, 25°, 80–96% yield.[38]
30. $Mo(CO)_6$, CH_3CN, H_2O, 59–94% yield.[39]
31. $Mn(OAc)_3$, benzene, reflux, 1–2 h, 86–96% yield.[40]
32. Wet $NaIO_4$–silica, microwave, 68–93% yield.[41]
33. $KHSO_5$, AcOH, 70–88% yield.[42]

1. R. V. Stevens, F. C. A. Gaeta, and D. S. Lawrence, *J. Am. Chem. Soc.*, **105**, 7713 (1983).

2. R. V. Hoffman and G. A. Buntain, *Synthesis*, 831 (1987).
3. R. Ballini, L. Barboni, and P. Filippone, *Chem. Lett.*, 475 (1997).
4. C. H. Depuy and B. W. Ponder, *J. Am. Chem. Soc.*, **81**, 4629 (1959).
5. E. B. Hershberg, *J. Org. Chem.*, **13**, 542 (1948).
6. S. R. Maynez, L. Pelavin, and G. Erker, *J. Org. Chem.*, **40**, 3302 (1975).
7. D. H. R. Barton, D. J. Lester, and S. V. Ley, *J. Chem. Soc., Chem. Commun.*, 445 (1977).
8. P. M. Pojer, *Aust. J. Chem.*, **32**, 201 (1979).
9. S. H. Pines, J. M. Chemerda, and M. A. Kozlowski, *J. Org. Chem.*, **31**, 3446 (1966).
10. Y. Watanabe, S. Morimoto, T. Adachi, M. Kashimura, and T. Asaka, *J. Antibiot.*, **46**, 647 (1993).
11. E. J. Corey and J. E. Richman, *J. Am. Chem. Soc.*, **92**, 5276 (1970).
12. E. J. Corey, P. B. Hopkins, S. Kim, S. Yoo, K. P. Nambiar, and J. R. Falck, *J. Am. Chem. Soc.*, **101**, 7131 (1979).
13. C. R. Narayanan, P. S. Ramaswamy, and M. S. Wadia, *Chem. Ind. (London)*, 454 (1977).
14. G. A. Olah and T. L. Ho, *Synthesis*, 609 (1976).
15. G. H. Timms and E. Wildsmith, *Tetrahedron Lett.*, 195 (1971).
16. G. A. Olah, M. Arvanaghi, and G. K. S. Prakash, *Synthesis*, 220 (1980).
17. C. Gundu Rao, A. S. Radhakrishna, B. Bali Singh, and S. P. Bhatnagar, *Synthesis*, 808 (1983).
18. J. M. Aizpurua, M. Juarista, B. L. Lecea, and C. Palomo, *Tetrahedron*, **41**, 2903 (1985).
19. D. H. R. Barton, W. B. Motherwell, E. S. Simon, and S. Z. Zard, *J. Chem. Soc., Chem. Commun.*, 337 (1984).
20. E. J. Corey, M. Narisada, T. Hiraoka, and R. A. Ellison, *J. Am. Chem. Soc.*, **92**, 396 (1970).
21. J. G. Lee, K. H. Kwak, and J. P. Hwang, *Tetrahedron Lett.*, **31**, 6677 (1990).
22. J. M. Khurana, A. Ray, and P. K. Sahoo, *Bull. Chem. Soc. Jpn.*, **67**, 1091 (1994).
23. H. Firouzabadi and I. Mohammadpoor-Baltork, *Synth. Commun.*, **24**, 489 (1994).
24. T. Shinada and K. Yoshihara, *Tetrahedron Lett.*, **36**, 6701 (1995).
25. A. S. Radhakrishna, A. Beena, K. Sivaprakash, and B. B. Singh, *Synth. Commun.*, **21**, 1473 (1991).
26. R. Balicki and L. Kaczmarek, *Synth. Commun.*, **21**, 1777 (1991).
27. A. Kamal, M. V. Rao, and H. M. Meshram, *J. Chem. Soc., Perkin Trans. 1*, 2056 (1991).
28. R. Joseph, A Sudalai, and T. Ravindranathan, *Tetrahedron Lett.*, **35**, 5493 (1994).
29. G. A. Olah, Q. Liao, C. S. Lee, and G. K. S. Prakash, *Synlett*, 427 (1993).
30. M. Akazome, Y. Tsuji, and Y. Watanabe, *Chem. Lett.*, 635 (1990).
31. R. Sanabria, P. Castañeda, R. Miranda, A. Tubón, F. Delgado, and L. Velasco, *Org. Prep. Proced. Int.*, **27**, 480 (1995).
32. J. G. Lee and J. P. Hwang, *Chem. Lett.*, 507 (1995).
33. A. Wali, P. A. Ganeshpure, and S. Satish, *Bull.Chem. Soc. Jpn.*, **66**, 1847 (1993).

34. M. Curini, O. Rosati, and E. Pisani, *Synlett*, 333 (1996).
35. R. S. Varma and H. M. Meshram, *Tetrahedron Lett.*, **38**, 5427 (1997).
36. A. Boruah, B. Baruah, D. Prajapati, and J. S. Sandhu, *Tetrahedron Lett.*, **38**, 4267 (1997).
37. N. B. Barhate, A. S. Gajare, R. D. Wakharkar, and A. Sudalai, *Tetrahedron Lett.*, **38**, 653 (1997).
38. B. P. Bandgar, L. B. Kunde, and J. L. Thote, *Synth. Commun.*, **27**, 1149 (1997).
39. F. Geneste, N. Racelma, and A. Moradpour, *Synth. Commun.*, **27**, 957 (1997).
40. A. S. Demir, C. Tanyeli, and E. Altinel, *Tetrahedron Lett.*, **38**, 7267 (1997).
41. R. S. Varma, R. Dahiya, and R. K. Saini, *Tetrahedron Lett.*, **38**, 8819 (1997).
42. D. S. Bose and P. Srinivas, *Synth. Commun.*, **27**, 3835 (1997).

O-Methyl Oxime: $R_2C{=}NOCH_3$

Formation[1]

1. $MeONH_2{\cdot}HCl$, Pyr, MeOH, 23°, 30 min, 81% yield.

Cleavage[1]

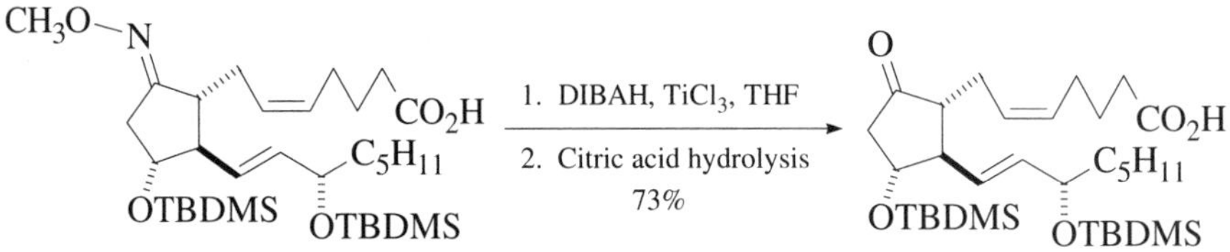

CH_3O—N, CO_2H, C_5H_{11}, OTBDMS, OTBDMS → (1. DIBAH, $TiCl_3$, THF; 2. Citric acid hydrolysis; 73%) O, CO_2H, C_5H_{11}, OTBDMS, OTBDMS

This method was developed because conventional procedures failed to cleave the oxime.

1. E. J. Corey, K. Niimura, Y. Konishi, S. Hashimoto, and Y. Hamada, *Tetrahedron Lett.*, **27**, 2199 (1986).

O-Benzyl Oxime: $R_2C{=}NOCH_2Ph$

The following reactions were used in a synthesis of perhydohistrionicotoxin; the carbonyl groups were protected as an oxime and an *O*-benzyl oxime.[1]

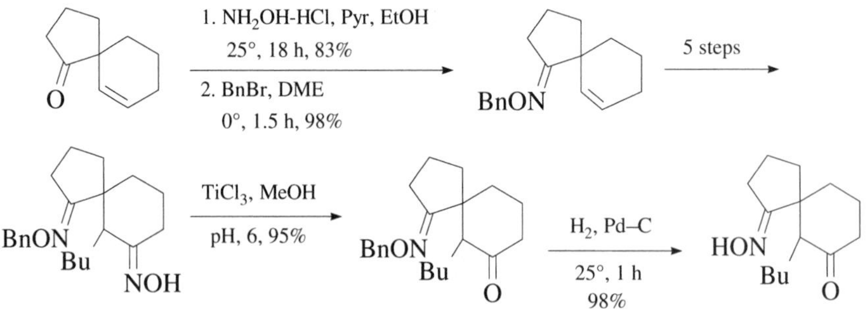

O → (1. NH_2OH-HCl, Pyr, EtOH, 25°, 18 h, 83%; 2. BnBr, DME, 0°, 1.5 h, 98%) BnON → (5 steps)

BnON, Bu, NOH → ($TiCl_3$, MeOH, pH, 6, 95%) BnON, Bu, O → (H_2, Pd–C, 25°, 1 h, 98%) HON, Bu, O

The **2-chlorobenzyl** group has been used in the protection of an oxime during the modification of erythromycin A.[2]

1. E. J. Corey, M. Petrzilka, and Y. Ueda, *Helv. Chim. Acta*, **60**, 2294 (1977).
2. Y. Watanabe, S. Morimoto, T. Adachi, M. Kashimura, and T. Asaka, *J. Antibiot.*, **46**, 647 (1993).

***O*-Phenylthiomethyl Oxime:** $R_2C{=}NOCH_2SC_6H_5$ (Chart 5)

In a prostaglandin synthesis, a carbonyl group was protected as an oxime that had its hydroxyl group protected against Collins oxidation by the phenylthiomethyl group. The phenylthiomethyl group is readily removed to give an oxime that is then cleaved to the carbonyl compound.[1]

Formation[1]

1. $PhSCH_2ONH_2$, Pyr, 25°, 24 h, 100% yield.

Cleavage[1]

1. $HgCl_2$, HgO, AcOH, AcOK, 25–50°, 0.5–48 h, 75% yield; K_2CO_3, MeOH, 25°, 5 min, 100% yield. These conditions remove the $PhSCH_2$– group from the oxime, which is then cleaved with AcOH/$NaNO_2$ (10°, 1 h). This group was also stable to acid, base, and $LiAlH_4$.

1. I. Vlattas, L. Della Vecchia, and J. J. Fitt, *J. Org. Chem.*, **38**, 3749 (1973).

Imines

In general, imines are too reactive to be used to protect carbonyl groups. In a synthesis of juncusol,[1] however, a **bromo**- and an **iodocyclohexylimine** of two identical aromatic aldehydes were coupled by an Ullmann coupling reaction modified by Ziegler.[2] The imines were cleaved by acidic hydrolysis (aq. oxalic acid, THF, 20°, 1 h, 95% yield). Imines of aromatic aldehydes have also been prepared to protect the aldehyde during ring metalation with *s*-BuLi.[3]

1. A. S. Kende and D. P. Curran, *J. Am. Chem. Soc.*, **101**, 1857 (1979).
2. F. E. Ziegler, K. W. Fowler, and S. Kanfer, *J. Am. Chem. Soc.*, **98**, 8282 (1976).
3. B. A. Keay and R. Rodrigo, *J. Am. Chem. Soc.*, **104**, 4725 (1982).

Substituted Methylene Derivatives: RR′C=C(CN)R″ (Chart 5)

RR′ = substituted pyrrole; R″ = –CN,[1] $-CO_2Et$[2]

The substituted methylene derivative, prepared from a 2-formylpyrrole and a malonic acid derivative, was used in a synthesis of chlorophyll.[1] It is cleaved under drastic conditions (concd. alkali).[1,2]

1. R. B. Woodward and 17 co-workers, *J. Am. Chem. Soc.*, **82**, 3800 (1960).
2. J. B. Paine, R. B. Woodward, and D. Dolphin, *J. Org. Chem.*, **41**, 2826 (1976).

Cyclic Derivatives

***N,N′*-Dimethylimidazolidine and *N,N′*-Diarylimidazolidine:**

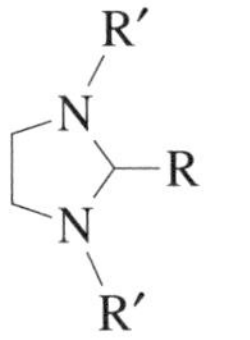

R′ = Me, Ar

The imidazolidine was prepared from an aldehyde with *N,N′*-dimethyl-1,2-ethylenediamine (benzene, heat, 78% yield) and cleaved with MeI (Et_2O; H_2O, 92% yield). Derivatization is chemoselective for aldehydes. The imidazolidine is stable to BuLi and LDA[1,2] and to Li/NH_3.[3] The **diphenylimidazolidine** has been prepared analogously and can be cleaved with aqueous HCl.[4,5] Alternatively, it can be prepared by using thionyl chloride (Pyr, CH_2Cl_2, 0–25°, 7 h, 93% yield).[6] A chiral version using *N,N′*-dimethyl-1*S*,2*S*-diphenyl-1,2-ethylenediamine has been used for protection as well as asymmetric induction.[7,8]

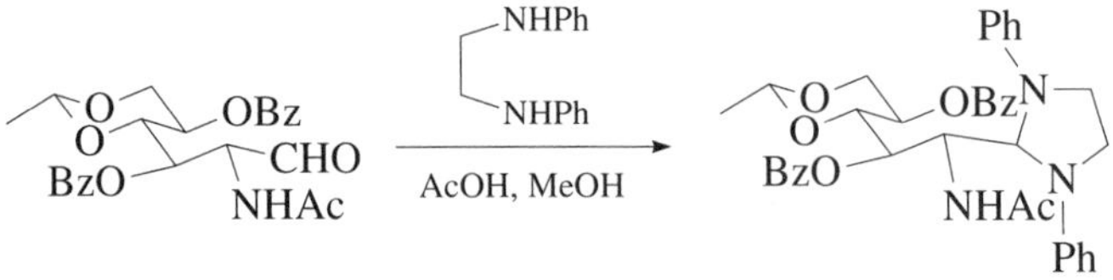

AcOH, MeOH

Ref. 4

The related **bis-*N,N′*-(3,5-dichlorophenyl)imidazolidine** has been used to protect an aldehyde. It is prepared from bis-*N,N′*-(3,5-dichlorophenyl)-1,2 diaminoethane (CSA, DMF, rt, 18 h, 72% yield) and is cleaved with aq. AcOH (rt, overnight, 98% yield).[9]

1. A. J. Carpenter and D. J. Chadwick, *Tetrahedron*, **41**, 3803 (1985).
2. M. Gray and P. J. Parsons, *Synlett*, 729 (1991).
3. L. E. Overman, D. J. Ricca, and V. D. Tran, *J. Am. Chem. Soc.*, **119**, 12031 (1997).

4. H.-W. Wanzlick and W. Löchel, *Chem. Ber.*, **86**, 1463 (1953).
5. A. Giannis, P. Münster, K. Sandhoff, and W. Steglich, *Tetrahedron*, **44**,7177 (1988).
6. J. J. Vanden Eynde, A. Mayence, and A. Maquestiau, *Bull. Soc. Chim. Belg.*, **101**, 233 (1992).
7. A. Alexakis, N. Lensen, and P. Mangeney, *Tetrahedron Lett.*, **32**, 1171 (1991).
8. I. Marek, A. Alexakis, and J.-F. Normant, *Tetrahedron Lett.*, **32**, 5329 (1991).
9. A. Ono, T. Okamoto, M. Inada, H. Nara, and A. Matsuda, *Chem. Pharm. Bull.*, **42**, 2231 (1994).

2,3-Dihydro-1,3-benzothiazole:

S R N R CH_3

The benzothiazole group is introduced by heating 2-methylaminobenzenethiol with a carbonyl compound in ethanol (70–93% yield).[1] An enone is selectively protected over a ketone, and aldehydes react faster than ketones. Cleavage is effected with $AgNO_3$ (CH_3CN, H_2O, pH 7, 83–93% yield)[2] or by heating in Ac_2O followed by aqueous hydrolysis (HCl, $CHCl_3$, 50°, 1 h, 40% yield) of the resulting enamide.[3] **Nonaromatic thiazolidines** have also been used as protective groups. They can be cleaved by basic hydrolysis (NaOH, 25°, 95% yield).[4]

1. H. Chikashita, N. Ishimoto, S. Komazawa, and K. Itoh, *Heterocycles*, **23**, 2509 (1985).
2. H. Chikashita, S. Komazawa, N. Ishimoto, K. Inoue, and K. Itoh, *Bull. Chem. Soc. Jpn.*, **62**, 1215 (1989).
3. G. Trapani, A. Reho, A. Latrofa, and G. Liso, *Synthesis*, 84 (1988).
4. K. Ueno, F. Ishikawa, and T. Naito, *Tetrahedron Lett.*, 1283 (1969).

Diethylamine Adduct: $R_2C[OTi(NEt_2)_3]NEt_2$

Titanium tetrakis(diethylamide) selectively adds to aldehydes in the presence of ketones and to the least hindered ketone in compounds containing more than one ketone. The protection is *in situ*, which thus avoids the usual protection/deprotection sequence. Selective aldol and Grignard additions are readily performed employing this protection methodology.[1]

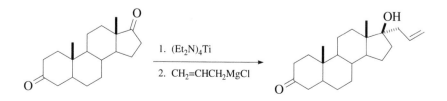

O O

1. $(Et_2N)_4Ti$
2. $CH_2{=}CHCH_2MgCl$

OH O

N-Methoxy-N-methylamine Adduct: $[R_2C(OLi)N(OMe)Me]$

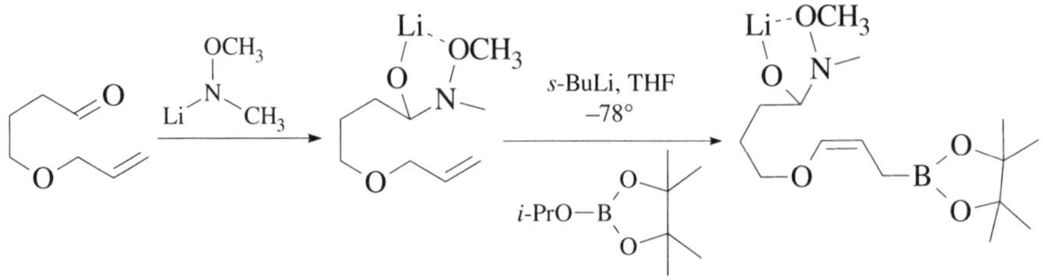

Li(N(OCH₃)CH₃) →; s-BuLi, THF, –78°; i-PrO–B(pinacol)

Refs. 2, 3

The use of various amine adducts of carbonyl compounds as a method of carbonyl protection has been reviewed.[4]

1. M. T. Reetz, B. Wenderoth, and R. Peter, *J. Chem. Soc., Chem. Commun.*, 406 (1983).
2. R. W. Hoffmann and I. Münster, *Tetrahedron Lett.*, **36**, 1431 (1995).
3. D. A. Evans, R. P. Polniaszek, K. M. DeVries, D. E. Guinn, and D. J. Mathre, *J. Am. Chem. Soc.*, **113**, 7613 (1991).
4. D. L. Comins, *Synlett*, 615 (1992).

o-Carborane: R, R′, OH, $B_{10}H_{10}$

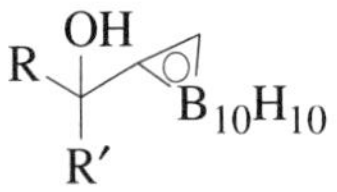

Formation/Cleavage[1]

Li–carborane $B_{10}H_{10}$, –78°, 30 min to rt, 1 h →

← KOH, THF, H_2O, 66–97%

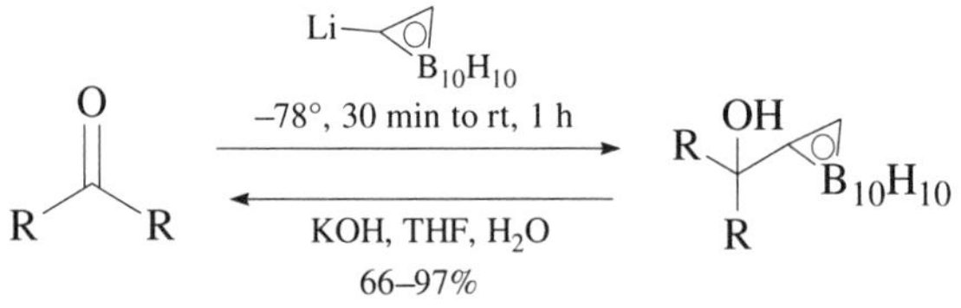

The carboranyl alcohol can also be prepared from the stannyl carborane and an aldehyde using $Pd_2(dba)_3$–$CHCl_3$/dppe. The carborane is stable to Brønsted and Lewis acids and to $LiAlH_4$.

1. H. Nakamura, K. Aoyagi, and Y. Yamamoto, *J. Org. Chem.*, **62**, 780 (1997).

1-Methyl-2-(1′-hydroxyalkyl)imidazoles: R, R, HO, N, N, CH_3

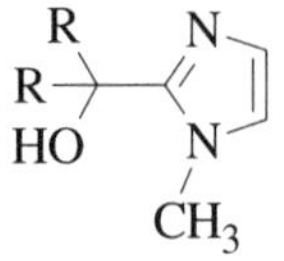

Formation / Cleavage[1]

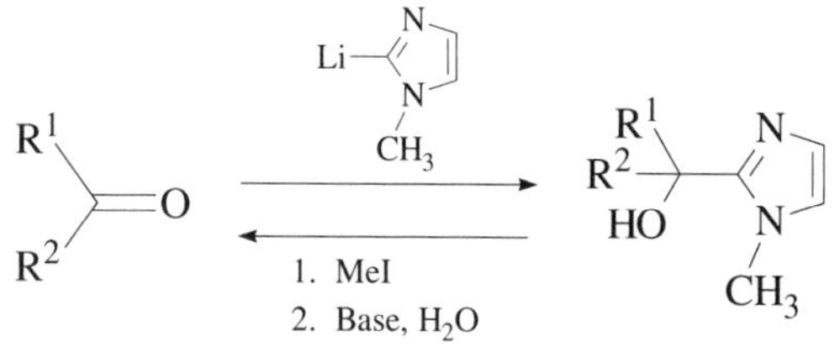

This protective group is stable to 1 *N* KOH/MeOH, 70°, 7 h; 20% H_2SO_4, 70°, 7 h; H_2, Pd–C, EtOH, 1 atm, 18 h; $NaBH_4$, $LiAlH_4$, CF_3COOH, Al_2O_3/MeOH.

1. S. Ohta, S. Hayakawa, K. Nishimura, and M. Okamoto, *Tetrahedron Lett.*, **25**, 3251 (1984).

Protection of the Carbonyl Groups by Conversion to an Enolate Anion

Lithium Diisopropylamide (LDA)

A 17-steroidal ketone was deprotonated by LDA to protect it from reduction during a lithium naphthalenide cleavage of a benzyl ether.[1]

1. H.-J. Liu, J. Yip, and K.-S. Shia, *Tetrahedron Lett.*, **38**, 2253 (1997).

Enamines

The use of enamines as protective groups seems largely to be confined to steroid chemistry, where they serve (in their protonated form) to protect the A–B enone system from bromination[1] and reduction.[2] A large body of literature exists on the preparation and chemistry of enamines,[3] which are easily hydrolyzed with water or aqueous acid.

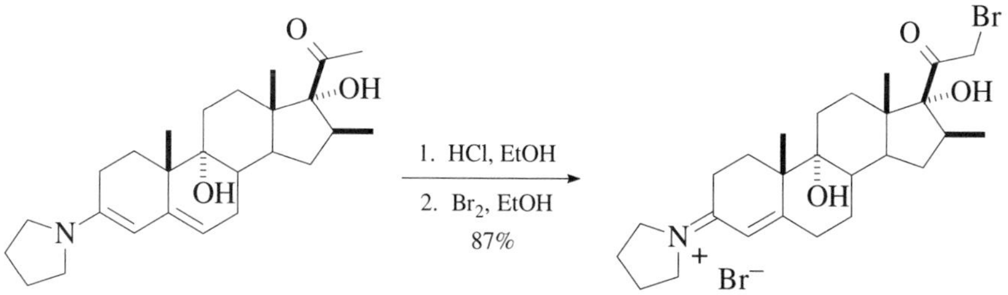

1. N. I. Carruthers, S. Garshasb, and A. T. McPhail, *J. Org. Chem.*, **57**, 961 (1992).
2. J. A. Hogg, *Steroids*, **57**, 593 (1992).

3. Enamine review: A. G. Cook, Ed., *Enamines: Synthesis, Structure and Reactions*, 2d ed., M. Dekker, New York, 1988.

Methylaluminum Bis(2,6-di-*t*-butyl-4-methylphenoxide) (MAD) Complex

This approach to carbonyl protection uses the relative differences in basicity and the differences in steric effects to protect selectively either the more basic carbonyl group or the less hindered carbonyl group from reactions with nucleophiles such as DIBAH[1] and MeLi.[2]

1. K. Maruoka, Y. Araki, and H. Yamamoto, *J. Am. Chem. Soc.*, **110**, 2650 (1988).
2. K. Maruoka, H. Imoto, and H. Yamamoto, *Synlett*, 441 (1994).

MONOPROTECTION OF DICARBONYL COMPOUNDS

Selective Protection of α- and β-Diketones

α- and β-Diketones can be protected as enol ethers, thioenol ethers, enol acetates, and enamines.

Enamines: —C(=O)—C(NR₂)=C— and —C(=O)—C=C(NR₂)—

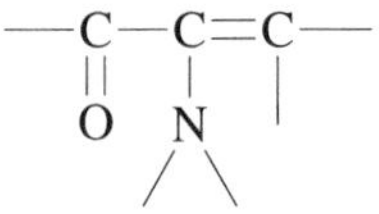

Enol Acetates: —C(=O)—C(OAc)=C— and —C(=O)—C=C(OAc)—

Enol Ethers: —C(=O)—C(OR)=C— and —C(=O)—C=C(OR)—

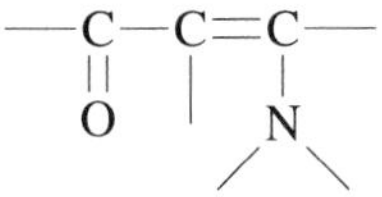

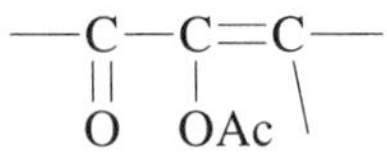

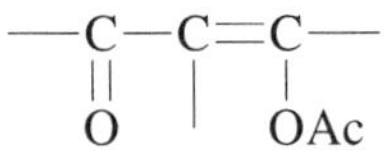

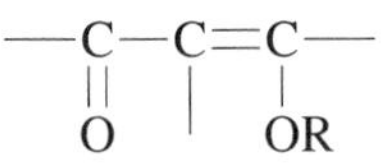

Methyl Enol Ether

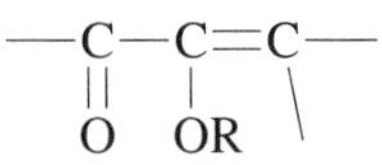

Ethyl Enol Ether

***i*-Butyl Enol Ether**

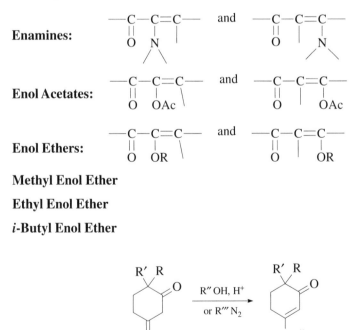

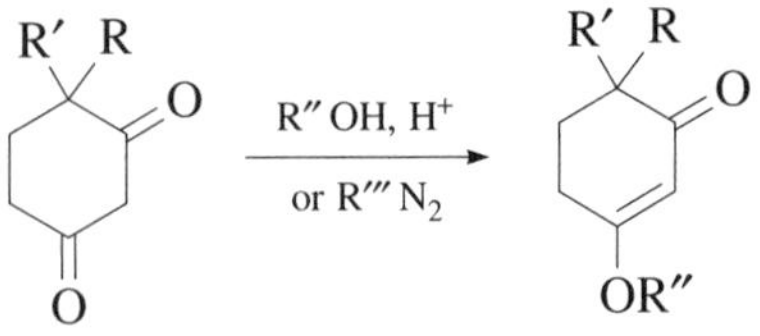

R′ R; O; O → R″OH, H⁺ or R‴N₂ → R′ R; O; OR″

R″ = R‴ + H

R″OH: R″ = Me (HCl, 25°, 8 h, 83% yield)[1]

R″ = Et (TsOH, benzene, reflux, 6–8 h, 70–75% yield)[2]

R‴ = $(CH_3)_2CHCH_2$ (i-BuOH, benzene, reflux, TsOH, 16 h, 100% yield).[3] In this case, 2-methyl-1,3-cyclopentanedione was monoprotected.

Methoxyethoxymethyl (MEM) Enol Ether
Triethylamine, MEMCl, 92% yield[4]

Enamino Derivatives (Vinylogous Amides)

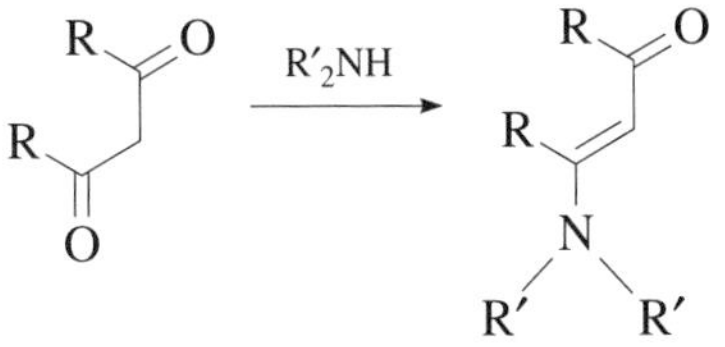

R'_2NH

1. R'_2NH = piperidine, TsOH, benzene, reflux, 92% yield.[5]
2. R'_2NH = morpholine, TsOH, $PhCH_3$, reflux, 4–5 h, 72–80% yield.[6]
3. R'_2NH = various, 300 MPa, with or without $Yb(OTf)_3$, 0–99% yield.[7]
4. R'_2NH = various, K10 clay or SiO_2, 1–10 min, microwave, 35–99% yield.[8]
5. R'_2NH = various, $BF_3{\cdot}Et_2O$, benzene, reflux, 4–6 h, 82–96% yield.[9]
6. R'_2NH = various, Montmorillonite or alumina, 20–100°, 1–5 h, 85–99% yield.[10,11]

4-Methyl-1,3-dioxolanyl Enol Acetate

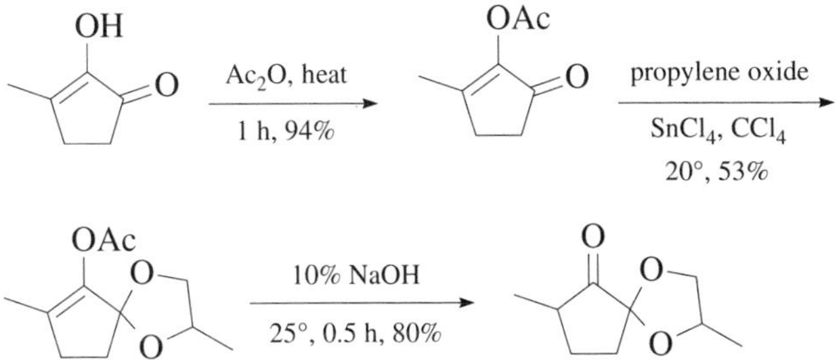

Ac_2O, heat; 1 h, 94%

propylene oxide; $SnCl_4$, CCl_4; 20°, 53%

10% NaOH; 25°, 0.5 h, 80%

Ref. 12

Pyrrolidinyl Enamine

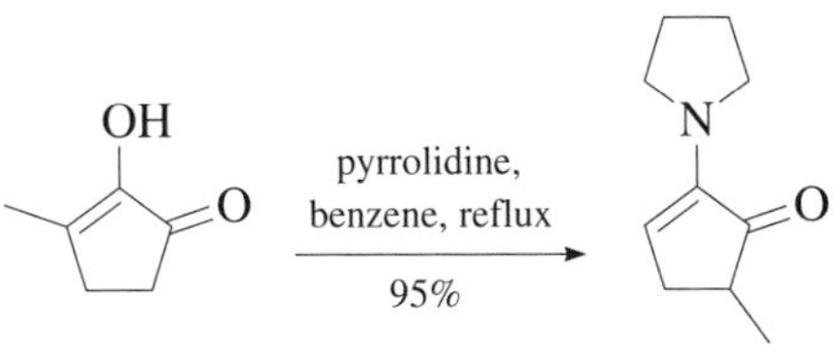

pyrrolidine, benzene, reflux; 95%

Ref. 13

Benzyl Enol Ether

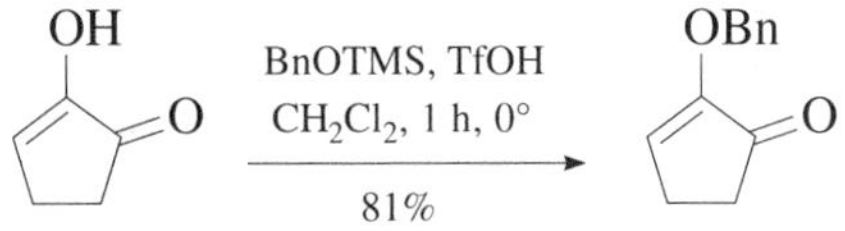

BnOTMS, TfOH
CH_2Cl_2, 1 h, 0°
81%

Ref. 14

Butyl Thioenol Ether

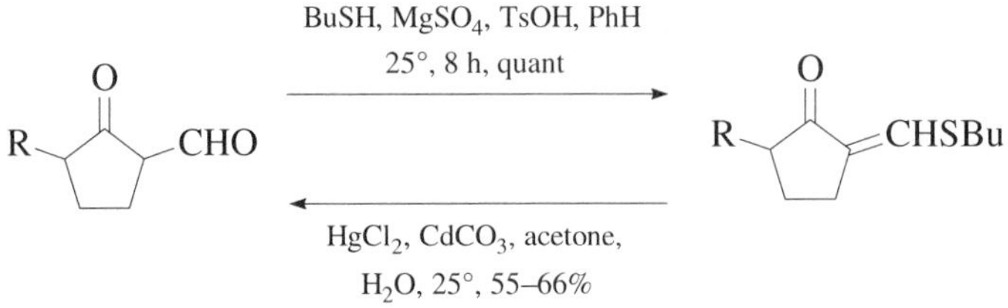

BuSH, $MgSO_4$, TsOH, PhH
25°, 8 h, quant

$HgCl_2$, $CdCO_3$, acetone,
H_2O, 25°, 55–66%

Ref. 15

Protection of Tetronic Acids

1. R′ = Me (MeI, CsF, DMF, 45–81% yield).[16]
2. R′ = Bn, allyl, Me, $TMSCH_2CH_2$, *t*-Bu, etc. (R′OH, Ph_3P, DEAD, 31–100% yield).[17]

1. H. O. House and G. H . Rasmusson, *J. Org. Chem.*, **28**, 27 (1963).
2. W. F. Gannon and H. O. House, *Org. Synth., Collect. Vol. V*, 539 (1973).
3. M. Rosenberger and P. J. McDougal, *J. Org. Chem.*, **47**, 2134 (1982).
4. A. J. H. Klunder, G. J. A. Ariaans, E. A. R. M. v. d. Loop and B. Zwanenburg, *Tetrahedron*, **42**, 1903 (1986).
5. P. Kloss, *Chem. Ber.*, **97**, 1723 (1964).
6. S. Hünig, E. Lücke and W. Brenninger, *Org. Synth., Collect. Vol. V*, 808 (1973).
7. G. Jenner, *Tetrahedron Lett.*, **37**, 3691 (1996).
8. B. Rechsteiner, F. Texier-Boullet, and J. Hamelin, *Tetrahedron Lett.*, **34**, 5071 (1993).
9. M. Azzaro, S. Geribaldi, and B. Videau, *Synthesis*, 880 (1981).
10. F. Texier-Boullet, B. Klein, and J. Hamelin, *Synthesis*, 409 (1986).
11. M. E. F. Braibante, H. S. Braibante, L. Missio, and A. Andricopulo, *Synthesis*, 898 (1994).

12. J. L. E. Erickson and F. E. Collins, Jr., *J. Org. Chem.*, **30**, 1050 (1965).
13. E. Gordon, F. Martens, and H. Gault, *C. R. Hebd. Seances Acad. Sci., Ser. C*, **261**, 4129 (1965).
14. A. A. Ponaras and Md. Y. Meah, *Tetrahedron Lett.*, **27**, 4953 (1986).
15. P. R. Bernstein, *Tetrahedron Lett.*, 1015 (1979).
16. T. Sato, K. Yoshimatsu, and J. Otera, *Synlett*, 845 (1995).
17. J. S. Bajwa and R. C. Anderson, *Tetrahedron Lett.*, **31**, 6973 (1990).

Trimethylsilyl Enol Ethers: R, R, R, OTMS

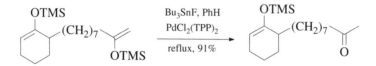

Trimethylsilyl enol ethers can be used to protect ketones, but, in general, are not used for this purpose because they are reactive under both acidic and basic conditions. More highly hindered silyl enol ethers are much less susceptible to acid and base. A less hindered silyl enol ether can be hydrolyzed in the presence of a hindered one. [1]

OTMS, $(CH_2)_7$, OTMS → (Bu_3SnF, PhH; $PdCl_2(TPP)_2$; reflux, 91%) → OTMS, $(CH_2)_7$, O

The preparation of silyl enol ethers has been reviewed.[2–4]

1. H. Urabe, Y. Takano, and I. Kuwajima, *J. Am. Chem. Soc.*, **105**, 5703 (1983).
2. E. Colvin, *Silicon in Organic Synthesis*, Butterworths, Boston, 1981, pp. 198–287.
3. W. P. Weber, *Silicon Reagents for Organic Synthesis*, Springer-Verlag, New York, 1983, pp. 255–272.
4. J. Hydrio, P. Van de Weghe, and J. Collin, *Synthesis*, 68 (1997).

Cyclic Ketals, Monothio and Dithio Ketals

Cyclohexane-1,2-dione reacts with ethylene glycol (TsOH, benzene, 6 h) to form the diprotected compound. Monoprotected 1,3-oxathiolanes and 1,3-dithiolanes are isolated on reaction under similar conditions with 2-mercaptoethanol and ethanedithiol, respectively.[1]

1. R. H. Jaeger and H. Smith, *J. Chem. Soc.*, 160, 646 (1955).

Bismethylenedioxy Derivatives (Chart 5): R, R, O, O, O, O

Formation/Cleavage[1,2]

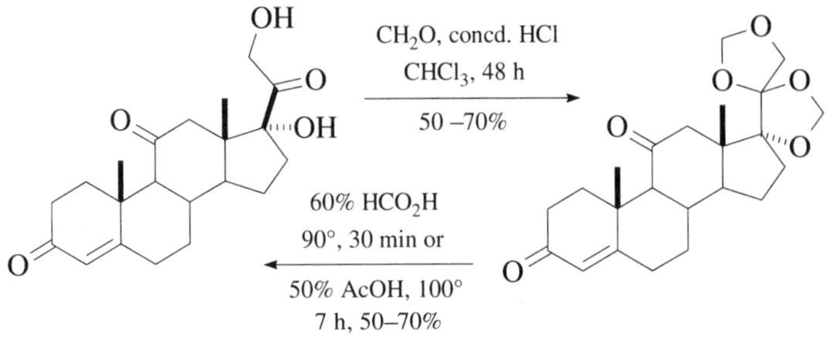

CH_2O, concd. HCl, $CHCl_3$, 48 h, 50–70%

60% HCO_2H, 90°, 30 min or 50% AcOH, 100°, 7 h, 50–70%

This derivative is stable to TsOH/benzene at reflux and to CrO_3/H^+.[3] It is also stable to NBS/*hv*.[4] In the formation of a related derivative, formaldehyde from formalin (containing methanol) converted a C_{11}-hydroxyl group to the C_{11}-methoxymethyl ether. Paraformaldehyde can be used as a source of methanol-free formaldehyde to avoid formation of the ethers.[5]

1. R. E. Beyler, F. Hoffman, R. M. Moriarty, and L. H. Sarett, *J. Org. Chem.*, **26**, 2421 (1961).
2. Y. Nishiguchi, N. Tagawa, F. Watanabe, T. Kiguchi, and I. Ninomiya, *Chem. Pharm. Bull.*, **38**, 2268 (1990).
3. J. F. W. Keana, in *Steroid Reactons*, C. Djerassi, Ed., Holden-Day, San Francisco, 1963, pp. 56–61.
4. D. Duval, R. Condom, and R. Emiliozzi, *C. R. Hebd. Seances Acad. Sci., Ser. C*, **285**, 281 (1977).
5. J. A. Edwards, M. C. Calzada, and A. Bowers, *J. Med. Chem.*, **7**, 528 (1964).

Tetramethylbismethylenedioxy Derivatives

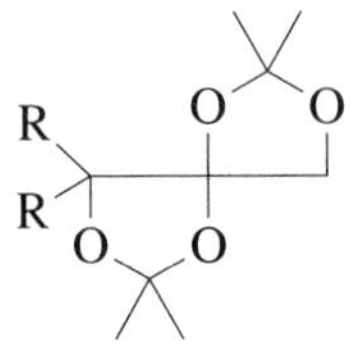

A bismethylenedioxy group in a 4-chloro or 11-keto steroid is stable to cleavage by formic acid or glacial acetic acid (100°, 6 h). The tetramethyl derivative is readily hydrolyzed (50% AcOH, 90°, 3–4 h, 80–90% yield).[1]

1. A. Roy, W. D. Slaunwhite, and S. Roy, *J. Org. Chem.*, **34**, 1455 (1969).

5

PROTECTION FOR THE CARBOXYL GROUP

Carboxylic acids are protected for a number of reasons: (1) to mask the acidic proton so that it does not interfere with base-catalyzed reactions, (2) to mask the carbonyl group to prevent nucleophilic addition reactions, and (3) to improve the handling of the molecule in question (e.g., to make the compound less water soluble, to improve its NMR characteristics, or to make it more volatile so that it can be analyzed by gas chromatography). Besides having stability to a planned set of reaction conditions, the protective group must also be removed without affecting other functionality in the molecule. For this reason, a large number of protective groups for acids have been developed that are removed under a variety of conditions, even though most can readily be cleaved by simple hydrolysis. Hydrolysis is an important means of deprotection, and the rate of hydrolysis is, of course, dependent upon steric and electronic factors that help to achieve differetial deprotection in polyfunctional substrates. These factors are also important in the selective protection of compounds containing two or more carboxylic acids. Hydrolysis using HOO^- is about 400 times faster than simple hydrolysis with hydroxide (phenyl acetate = substrate).[1]

Polymer-supported esters[2] are widely used in solid-phase peptide synthesis, and extensive information on this specialized protection is reported annually.[3] Some activated esters that have been used as macrolide precursors and some that have been used in peptide synthesis are also described in this chapter; the many activated esters that are used in peptide synthesis are discussed elsewhere.[3] A useful list, with references, of many protected amino acids (e.g., $-NH_2$, COOH, and side-chain-protected compounds) has been compiled.[4] Some general methods for the preparation of esters are provided at the beginning of this chapter;[5] conditions that are unique to a protective group are described with that group.[6] Some esters that have been used as protective groups are included in Reactivity Chart 6.

1. W. P. Jencks and M. Gilchrist, *J. Am. Chem. Soc.*, **90**, 2622 (1968).
2. See reference 22 (**Peptides**) in Chapter 1. See also P. Hodge, "Polymer-Supported Protecting Groups," *Chem. Ind.* (*London*), 624 (1979); R. B. Merrifield, G. Barany, W. L. Cosand, M. Engelhard, and S. Mojsov, "Some Recent Developments in Solid Phase Peptide Synthesis," in *Peptides: Proceedings of the Fifth American Peptide Symposium*, M. Goodman and J. Meienhofer, Eds., Wiley, New York, 1977, pp. 488–502; J. M. J. Fréchet, "Synthesis and Applications of Organic Polymers as Supports and Protecting Groups," *Tetrahedron*, **37**, 663 (1981).
3. *Specialist Periodical Reports*: *Amino-Acids, Peptides, and Proteins*, Royal Society of Chemistry, London, Vols. 1–16 (1969–1983); *Amino Acids and Peptides,* Vols. 17–28 (1984–1997).
4. G. A. Fletcher and J. H. Jones, *Int. J. Pept. Protein Res.*, **4**, 347 (1972).
5. For classical methods, see C. A. Buehler and D. E. Pearson, *Survey of Organic Syntheses*, Wiley-Interscience, New York, 1970, Vol. 1, pp. 801–830; 1977, Vol. 2, pp. 711–726.
6. See also E. Haslam, "Recent Developments in Methods for the Esterification and Protection of the Carboxyl Group," *Tetrahedron*, **36**, 2409–2433 (1980); E. Haslam, "Activation and Protection of the Carboxyl Group," *Chem. Ind.* (*London*), 610–617 (1979); E. Haslam, "Protection of Carboxyl Groups," in *Protective Groups in Organic Chemistry*, J. F. W. McOmie, Ed., Plenum, New York and London, 1973, pp.183–215; P. J. Kocienski, *Protecting Groups*, Thieme Medical Publishers, New York 1994, p. 118.

ESTERS

General Preparations of Esters

The preparation of esters can be classified into two main categories: (1) carboxylate activation with a good leaving group and (2) nucleophilic displacement of a carboxylate on an alkyl halide or sulfonate. The latter approach is generally not suitable for the preparation of esters if the halide or tosylate is sterically hindered, but there has been some success with simple secondary halides[1] and tosylates (ROTs, DMF, K_2CO_3, 69–93% yield).[2]

1. RCO_2H + R′OH, MeTHF, Me_3SiCl, (or Me_2SiCl_2, $MeSiCl_3$ or $SiCl_4$), rt, 15 min to 100 h, 90–97% yield.[3] In this case, both R and R′ can be hindered. Since the reaction conditions generate HCl, the substrates should be stable to strong acid.
2. RCO_2H, R′OH, DCC/DMAP, Et_2O, 25°, 1–24 h, 70–95% yield. This method is suitable for a large variety of hindered and unhindered acids and alcohols.[4] Carbodiimide **i** was developed to make the urea by-product water soluble and thus easily washed out.[5]

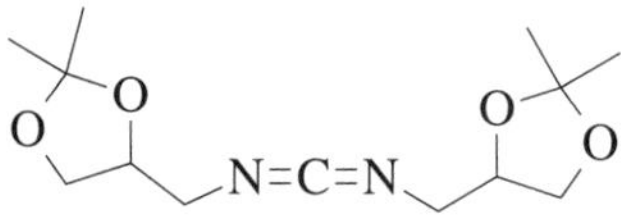

i

3. $(RCO_2)O$, R′OH, Bu_3P, excellent yields.[6] The nearly neutral esterification proceeds without the need for basic additives.
4. RCO_2H, R′OH, BOP–Cl, Et_3N, CH_2Cl_2, 23°, 2 h, 71–99% yield.[7] This is an excellent general method for the preparation of esters.

BOP–Cl =

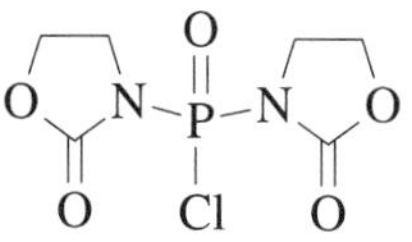

5. $$RCO_2H \xrightarrow[\text{2. R'OH, DMAP, >95\% yield.}]{\text{1. 2,4,6-}Cl_3C_6H_2COCl\text{, }Et_3N\text{, THF}^{8}} RCO_2R'$$

   This method is best suited to the preparation of relatively unhindered esters; otherwise some esterification of the benzoic acid may occur at the expense of the acid to be esterified.

6. $$RCO_2H + R'X \xrightarrow[\text{25–80°, 1–10 h}]{\text{DBU, benzene}^{9}} RCO_2R'\text{, 70–95\% yield}$$

   RCO_2H = alkyl, aryl, hindered acids
   R′ = Et, *n*- and *s*-Bu, CH_3SCH_2, …
   X = Cl, Br, I

   The reaction also proceeds well in acetonitrile, allowing lower temperatures (25°) and shorter times.[10]

7. $$\underset{\text{NHPG}}{RCHCO_2H} \xrightarrow[\text{pH 7}]{Cs_2CO_3} \xrightarrow[\text{6 h}]{\text{R'X, DMF}^{11}} \underset{\text{NHPG}}{RCHCO_2R'}$$

   R′ = Me, 80%; $PhCH_2$, 70–90%; *o*-$NO_2C_6H_4CH_2$, 90%; *p*-$MeOC_6H_4CH_2$, 70%; Ph_3C, 40–60%; *t*-Bu, 14%; PhCOCH(Me), 80%; *N*-phthalimidomethyl, 80% yield.

A study of relative rates of this reaction indicates that $Cs^+ > K^+ > Na^+ > Li^+$; $I^- >> Br^- >> Cl^-$; HMPA > DMSO > DMF.[12]

8. $\mathrm{RCH(NHPG)CO_2H + R'X \xrightarrow[25°,\ 24\ h,\ 90–95\%]{NaHCO_3,\ DMF^{13}} RCH(NHPG)CO_2R'}$

R′ = Et, *n*-Bu, *s*-Bu X = Br, I

9. $\mathrm{RCH(NHPG)CO_2H + R'X \xrightarrow[25°,\ 3–24\ h,\ 70–95\%]{(C_8H_{17})_3N^+MeCl^-,\ aq.\ NaHCO_3,\ CH_2Cl_2^{14}} RCH(NHPG)CO_2R'}$

10. $\mathrm{RCO_2H + R'_3O^+BF_4^- \xrightarrow[25°,\ 1–24\ h]{EtN\text{–}\mathit{i}\text{-}Pr_2,\ CH_2Cl_2^{15}} RCO_2R'}$, 70–95% yield

RCO_2H = hindered acids
R′ = Me, Et

11. $\mathrm{RCO_2H + Me_2NCH(OR')_2 \xrightarrow{25–80°,\ 1–36\ h^{16}} RCO_2R'}$, 80–95%

RCO_2H = Ph, 2,4,6-$Me_3C_6H_2$–, *N*-protected amino acids
R′ = Me, Et, $PhCH_2$, *s*-Bu

12. $\mathrm{RCO_2H + R'OH \xrightarrow[1–20°,\ 24\ h]{\mathit{t}\text{-}BuNC^{17}} RCO_2R'}$, 36–98%

RCO_2H = amino, dicarboxylic acids; ≠ $PhCO_2H$
R′ = Me, Et, *t*-Bu

13. $\mathrm{RCO_2H + R'OH \xrightarrow[25°,12\ h,\ 75–85\%]{Ph_3P(OSO_2CF_3)_2,\ CH_2Cl_2^{18}} RCO_2R'}$

R = aryl
R′ = Et

14. $\mathrm{RCO_2H \xrightarrow[Et_3N,\ DMAP]{ClCO_2R',\ CH_2Cl_2,\ 0°^{19}} RCO_2R'}$, 89–98%

This reaction is not suitable for hindered carboxylic acids, since considerable symmetrical anhydride formation (52% with pivalic acid) results. Symmetrical anhydride formation can sometimes be suppressed by the use of stoichiometric quantities of DMAP.

15. $\mathrm{RCO_2H + R'X \xrightarrow[R''_4N^+X^-,\ rt,\ 80–99\%\ yield]{Electrolysis:\ pyrrolidone,\ DMF^{20}} RCO_2R'}$

This method is based on the generation of the tetraalkylammonium salt of pyrrolidone, which acts as a base. The method is compatible with a large variety of carboxylic acids and alkylating agents. The method is effective for the preparation of macrolides.

16. $\underset{\text{NHPG}}{\text{RCHCO}_2\text{H}} \xrightarrow[\text{DMAP, CH}_2\text{Cl}_2\text{, 0°, R'OH}]{\text{isopropenyl chloroformate}^{21}} \underset{\text{NHPG}}{\text{RCHCO}_2\text{R}'}$, 60–96%

17. $\text{RCO}_2\text{H} + \text{R'OH} \xrightarrow[\text{0.1–48 h, 67–98\%}]{\text{TMSCl, 22–78°}^{22}} \text{RCO}_2\text{R}'$

18. $\text{RCO}_2\text{H} \xrightarrow[\text{2. R'OH, reflux, 5–120 min, 80–90\%}]{\text{1. TsCl, K}_2\text{CO}_3\text{, TEBAC}^{23}\text{, 40°–reflux, 5–60 min}} \text{RCO}_2\text{R}'$

TEBAC = $\text{Et}_3\text{N}^+\text{CH}_2\text{Ph Cl}^-$

19. $\text{RCO}_2\text{H} \xrightarrow[\text{50°, 12–48 h, 50–99\%}^{24}]{\text{R'OH, TiCl(OTf)}_3\text{, (Me}_2\text{SiO)}_4} \text{RCO}_2\text{R}'$

20. $\text{RCO}_2\text{H} \xrightarrow[\text{rt, 0.5–17 h, 90–99\%}^{25}]{\text{R'OH, TiCl}_4\text{,AgClO}_4\text{, (ArCO)}_2\text{O, TMSCl, CH}_2\text{Cl}_2} \text{RCO}_2\text{R}'$

21. $\text{RCO}_2\text{TMS} + \text{R'OTMS} \xrightarrow[\text{rt, 80–99\%}]{\text{TiCl}_4\text{, AgClO}_4\text{, (ArCO)}_2\text{O, CH}_2\text{Cl}_2} \text{RCO}_2\text{R}'$

Sn(OTf)_2 has also been used as an effective catalyst.[26]

22. $\text{RCO}_2\text{H} \xrightarrow[\text{R'OH, 2,6-lutidine}^{27}\text{, 39–84\%}]{\text{4,5-dichloro-1,2,3-dithiazolium chloride}} \text{RCO}_2\text{R}'$

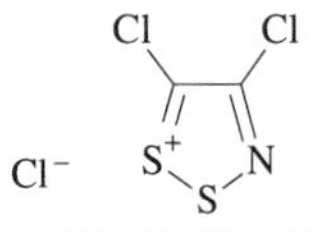

(Reagent structure: Cl, Cl, Cl⁻, S⁺, N, S)

23. $\text{RCO}_2\text{H} \xrightarrow[\text{56–95\%}]{\text{R'OH, EEDQ}^{28}} \text{RCO}_2\text{R}'$

24. 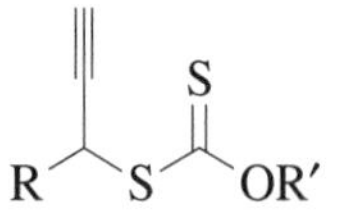 R–CH(C≡CH)–S–C(=S)–OR′ $\xrightarrow[\text{toluene}]{\text{R''CO}_2\text{H}} \text{R''CO}_2\text{R}'$

Esterification proceeds with inversion Ref. 29

25. R–C(=O)–N(imidazolyl) $\xrightarrow[\text{82°, 1–10 h}]{\text{1. MeOTf, ClCH}_2\text{CH}_2\text{Cl; 2. R'OH, NMM}}$ RCO_2R'
70–90%

Ref. 30

The Mitsunobu reaction is used to convert an alcohol and an acid into an ester by the formation of an activated alcohol (Ph_3P, diethyl diazodicarboxylate), which then undergoes displacement with inversion by the carboxylate.[31] Although this reaction works very well, it suffers from the fact that large quantities of by-products are produced, which generally require removal by chromatography.

General Cleavage of Esters[32]

1. $RCO_2R' + Nu^- \xrightarrow{\text{aprotic solvent}^{33}} RCO_2H$

Nu^- = LiS-*n*-Pr: HMPA, 25°, 1 h, ca. quant. yield[34]
= NaSePh: HMPA–THF, reflux, 7 h, 90–100% yield[35]
= LiCl: DMF or Pyr, reflux, 1–18 h, 60–90% yield[36]
= KO-*t*-Bu: DMSO, 50–100°, 1–24 h, 65–95% yield[37]
= NaCN (for decarboxylation of malonic esters): DMSO, 160°, 4 h, 70–80% yield[38]
= NaTeH from Te: DMF, *t*-BuOH, $NaBH_4$, 80–90°, 15 min, 85–98% yield[39]
= KO_2: 18-crown-6, benzene, 25°, 8–72 h, 80–95% yield[40]
= LiI: EtOAc, reflux, 26–98% yield.[41] Bn, PMB, PNB, *t*-Bu, and Me esters are all cleaved.

2. $RCO_2R' \xrightarrow[\text{reflux, 5–35 h, 70–90\%}]{\text{TMSCl, NaI, CH}_3\text{CN}^{42-44}} RCO_2H$

RCO_2H = alkyl, aryl, hindered acids
R′ = Me, Et, *i*-Pr, *t*-Bu, $PhCH_2$

This method generates Me_3SiI *in situ*. The reagent also cleaves a number of other protective groups.

3. $RCO_2R' \xrightarrow[\text{41–96\%}]{\text{MgI}_2\text{, toluene, 1–3 days}^{45}} RCO_2H + R'I$

RCO_2H = alkyl, aryl, hindered acids
R′ = Me, Et, cHex, 1-Ad, 2-Ad, *t*-Bu, $PhCH_2$

4. $RCO_2R' \xrightarrow[\text{15–60 min, 36–98\% yield}]{\text{aq. NaOH, DMF; HCl}^{46}} RCO_2H$

5. $RCO_2R' \xrightarrow[80\text{–}100\%]{\text{KO-}t\text{-Bu/H}_2\text{O (4:1), 25°, 2–48 h}^{47}} RCO_2H$

RCO_2H = Ph, aryl, hindered acids
R′ = Me, *t*-Bu, alkyl
"Anhydrous hydroxide" also cleaves tertiary amides.

6. $\underset{\text{NHPG}}{\text{RCHCO}_2\text{R}'} \xrightarrow[-10°,\ 1\text{ h} \rightarrow 25°,\ 2\text{ h}]{\text{BBr}_3,\ \text{CH}_2\text{Cl}_2\ ^{48}} \underset{\text{NH}_2}{\text{RCHCO}_2\text{H}}$, 60–85%

R′ = Me, Et, *t*-Bu, $PhCH_2$
PG = $-CO_2CH_2Ph$, $-CO_2$–*t*-Bu; OMe, OEt, O–*t*-Bu, OCH_2Ph side-chain ethers

7. $RCO_2R' \xrightarrow[70\text{–}95\%]{\text{AlX}_3,\ \text{R}''\text{SH, 25°, 5–50 h}^{49,50}} RCO_2H$

R = Ph, steroid side chain, ...
R′ = Me, Et, $PhCH_2$–
R″ = Et, $HO(CH_2)_2$–
X = Cl, Br

8. $RCO_2R' \xrightarrow[\text{benzene, 1–30 h}]{\text{xs (Bu}_3\text{Sn)}_2\text{O, 80°}\ ^{51-53}} RCO_2H$, 40–95% yield

R′ = $CH_2O_2CC(CH_3)_3$, Me, Et, Ph

1. T. Shono, O. Ishige, H. Uyama, and S. Kashimura, *J. Org. Chem.*, **51**, 546 (1986).
2. W. L. Garbrecht, G. Marzoni, K. R. Whitten, and M. L. Cohen, *J. Med. Chem.*, **31**, 444 (1988).
3. R. Nakao, K. Oka, and T. Fukomoto, *Bull. Chem. Soc. Jpn.*, **54**, 1267 (1981).
4. A. Hassner and V. Alexanian, *Tetrahedron Lett.*, 4475 (1978).
5. F. S. Gibson, M. S. Park, and H. Rapoport, *J. Org. Chem.*, **59**, 7503 (1994).
6. E. Vedejs, N. S. Bennett, L. M. Conn, S. T. Diver, M. Gingras, S. Lin, P. A. Oliver, and M. J. Peterson, *J. Org. Chem.*, **58**, 7286 (1993).
7. J. Diago-Meseguer, A. L. Palomo-Coll, J. R. Fernández-Lizarbe, and A. Zugaza-Bilbao, *Synthesis*, 547 (1980).
8. J. Inanaga, K. Hirata, H. Saeki, T. Katsuki, and M. Yamaguchi, *Bull. Chem. Soc. Jpn.*, **52**, 1989 (1979).
9. N. Ono, T. Yamada, T. Saito, K. Tanaka, and A. Kaji, *Bull. Chem. Soc. Jpn.*, **51**, 2401 (1978).
10. C. G. Rao, *Org. Prep. Proced. Int.*, **12**, 225 (1980).
11. S.-S. Wang, B. F. Gisin, D. P. Winter, R. Makofske, I. D. Kulesha, C. Tzougraki, and J. Meienhofer, *J. Org. Chem.*, **42**, 1286 (1977).

12. P. E. Pfeffer and L. S. Silbert, *J. Org. Chem.*, **41**, 1373 (1976).
13. V. Bocchi, G. Casnati, A. Dossena, and R. Marchelli, *Synthesis*, 961 (1979).
14. V. Bocchi, G. Casnati, A. Dossena, and R. Marchelli, *Synthesis*, 957 (1979).
15. D. J. Raber, P. Gariano, A. O. Brod, A. Gariano, W. C. Guida, A. R. Guida, and M. D. Herbst, *J. Org. Chem.*, **44**, 1149 (1979).
16. H. Brechbühler, H. Büchi, E. Hatz, J. Schreiber, and A. Eschenmoser, *Helv. Chim. Acta*, **48**, 1746 (1965).
17. D. Rehn and I. Ugi, *J. Chem. Res., Synop.*, 119 (1977).
18. J. B. Hendrickson and S. M. Schwartzman, *Tetrahedron Lett.*, 277 (1975).
19. S. Kim, Y. C. Kim, and J. I. Lee, *Tetrahedron Lett.*, **24**, 3365 (1983).
20. T. Shono, O. Ishige, H. Uyama, and S. Kashimura, *J. Org. Chem.*, **51**, 546 (1986).
21. P. Jouin, B. Castro, C. Zeggaf, A. Pantaloni, J. P. Senet, S. Lecolier, and G. Sennyey, *Tetrahedron Lett.*, **28**, 1661 (1987).
22. M. A. Brook and T. H. Chan, *Synthesis*, 201 (1983).
23. Z. M. Jászay, I. Petneházy, and L. Töke, *Synthesis*, 745 (1989).
24. J. Izumi, I. Shiina, and T. Mukaiyama, *Chem. Lett.*, 141 (1995).
25. I. Shiina, S. Miyoshi, M. Miyashita, and T. Mukaiyama, *Chem. Lett.*, 515 (1994).
26. T. Mukaiyama, I. Shiina, and M. Miyashita, *Chem. Lett.*, 625 (1992).
27. J. J. Folmer and S. M. Weinreb, *Tetrahedron Lett.*, **34**, 2737 (1993).
28. B. Zacharie, T. P. Connolly, and C. L. Penney, *J. Org. Chem.*, **60**, 7072 (1995).
29. J. Boivin, E. Henriet, and S. Z. Zard, *J. Am. Chem. Soc.*, **116**, 9739 (1994).
30. G. Ulibarri, N. Choret, and D. C. H. Bigg, *Synthesis*, 1286 (1996).
31. D. L. Hughes, *Org. React.*, **42**, 335 (1992); O. Mitsunobu, *Synthesis*, 1 (1981).
32. For a review, see C. J. Salomon, E. G. Mata, and O. A. Mascaretti, *Tetrahedron*, **49**, 3691 (1993).
33. J. McMurry, "Ester Cleavages via S_N2-Type Dealkylation," *Org. React.*, **24**, 187–224 (1976).
34. P. A. Bartlett and W. S. Johnson, *Tetrahedron Lett.*, 4459 (1970).
35. D. Liotta, W. Markiewicz, and H. Santiesteban, *Tetrahedron Lett.*, 4365 (1977).
36. F. Elsinger, J. Schreiber, and A. Eschenmoser, *Helv. Chim. Acta*, **43**, 113 (1960).
37. F. C. Chang and N. F. Wood, *Tetrahedron Lett.*, 2969 (1964).
38. A. P. Krapcho, G. A. Glynn, and B. J. Grenon, *Tetrahedron Lett.*, 215 (1967).
39. J. Chen and X. J. Zhou, *Synthesis*, 586 (1987).
40. J. San Filippo, L. J. Romano, C.-I. Chern, and J. S. Valentine, *J. Org. Chem.*, **41**, 586 (1976).
41. J. W. Fisher and K. L. Trinkle, *Tetrahedron Lett.*, **35**, 2505 (1994).
42. M. E. Jung and M. A. Lyster, *J. Am. Chem. Soc.*, **99**, 968 (1977).
43. T. Morita, Y. Okamoto, and H. Sakurai, *J. Chem. Soc., Chem. Commun.*, 874 (1978).
44. G. A. Olah, S. C. Narang, B. G. B. Gupta, and R. Malhotra, *J. Org. Chem.*, **44**, 1247 (1979).
45. A. G. Martinez, J. O. Barcina, G. H. del Veccio, M. Hanack, and L. R. Subramanian, *Tetrahedron Lett.*, **32**, 5931 (1991).
46. J. M. Khurana and A. Sehgal, *Org. Prep. Proced. Int.*, **26**, 580 (1994).

47. P. G. Gassman and W. N. Schenk, *J. Org. Chem.*, **42**, 918 (1977).
48. A. M. Felix, *J. Org. Chem.*, **39**, 1427 (1974).
49. M. Node, K. Nishide, M. Sai, and E. Fujita, *Tetrahedron Lett.*, 5211 (1978).
50. M. Node, K. Nishide, M. Sai, K. Fuji, and E. Fujita, *J. Org. Chem.*, **46**, 1991 (1981).
51. C. J. Salomon, E. G. Mata, and O. A. Mascaretti, *Tetrahedron Lett.*, **32**, 4239 (1991).
52. C. J. Salomon, E. G. Mata, and O. A. Mascaretti, *J. Org. Chem.*, **59**, 7259 (1994).
53. E. G. Mata and O. A. Mascaretti, *Tetrahedron Lett.*, **29**, 6893 (1988).

Transesterification

The process of transesterification is an important way to prepare a large number of esters from more complex or more simple esters without passing through the carboxylic acid. Transesterification can be used to convert one type of ester to another type that can then be removed under a different set of conditions. This section describes many of the methods that have been found effective for ester metathesis.[1]

1. ROH, DBU, LiBr. When a large excess of the alcohol is undesirable, the reaction can be run in THF/CH_2Cl_2 in the presence of 5Å ms. The combination of DBU–LiBr is required, since neither reagent is effective alone.[2]
2. Alkali metal alkoxides, *t*-butyl acetate neat, 45°, 30 min, 98% yield of *t*-butyl ester from methyl benzoate. The rate constant for the reaction increases with increasing ionic radius of the metal and with decreasing polarity of the solvent. Equilibrium for the reaction is achieved in <10 sec. Other examples are presented.[3,4]
3. M(O–*i*-Pr)$_3$; M = La[5], Nd, Gd, Yb.[6]
4. The use of 1,3-disubstituted 1,1,3,3-tetraalkyldistannoxanes for ester metathesis has been reviewed.[7,8]
5. Ti(O–*i*-Pr)$_4$, ROH, 50–90% yield.[9–11]
6. Mg, MeOH.[12]
7. From a *β*-keto ester: ROH, toluene, reflux, 95% yield. The reaction in this case is proposed to proceed through a ketene intermediate.[13]
8. From a *β*-keto ester: ROH, Sulfated-SnO_2, 50–97% yield.[14]
9. $Ce(SO_4)_2$–SiO_2, ROH, reflux, 0.25–2 h.[15]
10. $$RCO_2R' + R''OH \xrightarrow[\text{Toluene >88\% yield}]{Bu_2Sn(OH)OSn(NCS)Bu_2\ \text{cat.}^{16}} RCO_2R'' + R'OH$$

    This method is not effective for tertiary alcohols. It has a strong rate dependence on the polarity of the solvent, with less polar solvents giving faster rates.

1. J. Otera, *Chem. Rev.*, **93**, 1449 (1993).
2. D. Seebach, A. Thaler, D. Blaser, and S. Y. Ko, *Helv. Chim. Acta*, **74**, 1102 (1991).

3. M. G. Stanton and M. R. Gagné, *J. Am. Chem. Soc.*, **119**, 5075 (1997).
4. M. G. Stanton and M. R. Gagné, *J. Org. Chem.*, **62**, 8240 (1997).
5. T. Okano, K. Miyamoto, and J. Kiji, *Chem. Lett.*, 246 (1995).
6. T. Okano, Y. Hayashizaki, and J. Kiji, *Bull. Chem. Soc. Jpn.*, **66**, 1863 (1993).
7. J. Otera, *Adv. Detailed React. Mech.*, **3**, 167 (1994).
8. O. A. Mascaretti and R. L. E. Furlan, *Aldrichimica Acta*, **30**, 55 (1997).
9. D. Seebach, E. Hungerbühler, R. Naef, P. Schnurrenberger, B. Weidmann, and M. Züger, *Synthesis*, 138 (1982).
10. U. D. Lengweiler, M. G. Fritz, and D. Seebach, *Helv. Chim. Acta*, **79**, 670 (1996).
11. For a review of titanium compounds as catalysts for transesterification, see M. I. Siling and T. N. Laricheva, *Russ. Chem. Rev.*, **65**, 279 (1996).
12. Y.-C. Xu, E. Lebeau, and C. Walker, *Tetrahedron Lett.*, **35**, 6207 (1994).
13. A. G. Myers, N. J. Tom, M. E. Fraley, S. B. Cohen, and D. J. Madar, *J. Am. Chem. Soc.*, **119**, 6072 (1997).
14. S. P. Chavan, P. K. Zubaidha, S. W. Dantale, A. Keshavaraja, A. V. Ramaswany, and T. Ravindranathan, *Tetrahedron Lett.*, **37**, 233 (1996).
15. T. Nishiguchi and H. Taya, *J. Chem. Soc., Perkin Trans. 1*, 172 (1990).
16. J. Otera, T. Yano, A. Kawabata, and H. Nozaki, *Tetrahedron Lett.*, **27**, 2383 (1986); J. Otera, S. Ioka, and H. Nozaki, *J. Org. Chem.*, **54**, 4013 (1989).

Enzymatically Cleavable Esters

The enzymatic cleavage of esters is a vast and extensively reviewed area of chemistry.[1] Recently, several new esters have been examined primarily for the preparation of peptides and glycopeptides.

Heptyl Esters: $C_7H_{15}O_2CR$

The heptyl ester was developed as an enzymatically removable protective group for C-terminal amino acid protection.

Formation

Heptyl alcohol, TsOH, benzene, reflux, 66–92% yield.[2]

Cleavage

1. Lipase from *Rhizopus niveus*, pH 7, rt, 50–96% yield.[3]
2. Lipase from *Aspergillus niger*, 0.2 *M* phosphate buffer, acetone, pH 7, 37°, 50–96% yield. This lipase was used in the cleavage of phosphopeptide heptyl esters. These conditions are sufficiently mild to prevent the elimination of phosphorylated serine and threonine residues.[4]
3. Lipase M (*Mucor javanicus*), pH 7, 37°, 70–88% yield. In this case, α- and β-glycosidic peptide derivatives were deprotected. Acetates on the pyranosides were not affected.[5]

2-*N*-(Morpholino)ethyl ($MoC_2H_2O_2CR$) Ester: O(CH₂CH₂)₂N–CH₂CH₂–O_2CR

This ester was developed to impart greater hydrophilicity in C-terminal peptides that contain large hydrophobic amino acids, since the velocity of deprotection with enzymes often was reduced to nearly useless levels. Efficient cleavage is achieved with the lipase from *R. niveus* (pH 7, 37°, 16 h, H_2O, acetone, 78–91% yield).[6]

Choline Ester: $Me_3N^+CH_2CH_2O_2CR\ Br^-$

The choline ester is prepared by treating the 2-bromoethyl ester with trimethylamine. The ester is cleaved with butyrylcholine esterase (pH 6, 0.05 M phosphate buffer, rt, 50–95% yield). As with the morpholinoethyl ester, the choline ester imparts greater solubility to the C-terminal end of very hydrophobic peptides, thus improving the ability to cleave enzymatically the C-terminal ester.[7]

(Methoxyethoxy)ethyl (Mee) Ester: $CH_3OCH_2CH_2OCH_2CH_2O_2CR$

Because *O*-glycoproteins are susceptible to strong base and anomerization with acid, their preparation presents a number of difficulties, among which is the issue of mild and selective deprotection. Although in many cases the heptyl group was found quite useful because of the mild conditions associated with its enzymatic cleavage, in some cases the enzymatic cleavage would not proceed because the high level of hydrophobicity reduced solubility enough that the cleavage velocity approached zero. Increasing the hydrophilicity of the C-terminal protective group by incorporating some oxygen into the chain, as in the Mee ester, allows for reasonably facile cleavage with the lipase M from *M. javanicus* or papain. The pyranosidic acetates were not cleaved with these enzymes, but they could be cleaved with lipase WG.[8]

Methoxyethyl Ester (ME–O_2CR): $CH_3OCH_2CH_2O_2CR$

The advantages of the methoxyethyl ester over some of the other water-solubilizing esters are that many of the amino acid esters are crystalline and thus easily purified, are cleaved with a number of readily available lipases, and are useful for the synthesis of *N*-linked glycopeptides.[9]

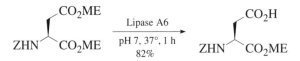

ZHN–CH(CO_2ME)–CH₂–CO_2ME —[Lipase A6, pH 7, 37°, 1 h, 82%]→ ZHN–CH(CO_2ME)–CH₂–CO_2H

1. (a) K. Faber and S. Riva, *Synthesis,* 895 (1992); (b) H. Waldmann and D. Sebastian, *Chem. Rev.*, **94**, 911 (1994); (c) K. Drauz and H. Waldmann, Eds., *Enzyme Catalysis in Organic Synthesis: A Comprehensive Handbook*, VCH, New York, 1995;

(d) T. Pohl, E. Nägele, and H. Waldmann, *Catal. Today*, **22**, 407 (1994); (e) H. Waldmann, P. Braun, and H. Kunz, *Chem. Pept. Proteins*, **5/6** (Pt. A), 227 (1993); (f) A. Reidel and H. Waldmann, *J. Prakt. Chem./Chem.-Ztg.*, **335**, 109 (1993); (g) C.-H. Wong and G. M. Whitesides, *Enzymes in Synthetic Organic Chemistry*, Pergamon, Oxford, U.K. (1994).

2. P. Braun, H. Waldmann, W. Vogt, and H. Kunz, *Synlett*, 105 (1990); *idem*, *Liebigs Ann. Chem.*, 165 (1991).
3. P. Braun, H. Waldmann, W. Vogt, and H. Kunz, *Liebigs Ann. Chem.*, 165 (1991).
4. D. Sebastian and H. Waldmann, *Tetrahedron Lett.*, **38**, 2927 (1997).
5. H. Waldmann, A. Heuser, P. Braun, and H. Kunz, *Indian J. Chem., Sect. B*, **31B**, 799 (1992); H. Waldmann, P. Braun, and H. Kunz, *Biomed. Biochim. Acta*, **50**, S 243 (1991); P. Braun, H. Waldmann, and H. Kunz, *Synlett*, 39 (1992).
6. G. Braum, P. Braun, D. Kowalczyk, and H. Kunz, *Tetrahedron Lett.*, **34**, 3111 (1993).
7. M. Schelhaas, S. Glomsda, M. Hänsler, H.-D. Jakubke, and H. Waldmann, *Angew. Chem., Int. Ed. Engl.*, **35** 106 (1996).
8. J. Eberling, P. Braun, D. Kowalczyk, M. Schultz, and H. Kunz, *J. Org. Chem.*, **61**, 2638 (1996).
9. M. Gewehr and H. Kunz, *Synthesis*, 1499 (1997).

Methyl Ester: RCO_2CH_3 (Chart 6)

Formation

The section on general methods should also be consulted.

1. $H_2NCON(NO)Me$, KOH, DME, H_2O, 0°, 75% yield. This method generates diazomethane *in situ*.[1] *N*-Methyl-*N*-nitrosourea is a proven carcinogen.
2. Me_3SiCHN_2, MeOH, benzene, 20°.[2,3] This reagent does not react with phenols. This is a safe alternative to the use of diazomethane. A detailed, large-scale preparation of this useful reagent has been described.[4] The reagent reacts with various maleic anhydrides in the presence of an alcohol to form diesters (70–96% yield).[5]
3. $Me_2C(OMe)_2$, cat. HCl, 25°, 18 h, 80–95% yield.[6] These reaction conditions were used to prepare methyl esters of amino acids.
4. $(MeO)_2NH$, heat, 98% yield.[7] Amines are also alkylated.
5. MeOH, H_2SO_4, 0°, 1 h; 5°, 18 h, 98% yield.[8]

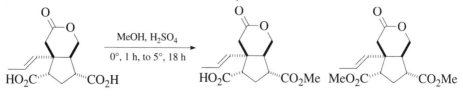

O O O HO₂C CO₂H — MeOH, H_2SO_4 / 0°, 1 h, to 5°, 18 h → HO₂C CO₂Me + MeO₂C CO₂Me

Ratio = 4:1

6. MeOH, HBF_4, Na_2SO_4, 25–60°, 15 h, 45–94% yield.[9] The selectivity observed here is also observed for Et, *i*-Pr, Bn, and cyclohexyl esters ($n = 1, 2$).

$$HO_2C\text{–}(CH_2)_n\text{–}CH(NH_2)\text{–}CO_2H \xrightarrow[45\text{–}94\%]{ROH,\ HBF_4,\ Na_2SO_4,\ 25\text{–}60°,\ 15\ h} RO_2C\text{–}(CH_2)_n\text{–}CH(NH_2)\text{–}CO_2H$$

$n = 1, 2$ R = CH_3, Et, *i*-Pr, Bn, cyclohexyl

7. $NiCl_2 \cdot 6H_2O$, 10 mol%, MeOH, reflux, 9–93% yield.[10] Aromatic and conjugated acids are not effectively esterified under these conditions.

Cleavage

1. LiOH, CH_3OH, H_2O (3:1), 5°, 15 h.[11]
2. $AlBr_3$, tetrahydrothiophene, rt, 62 h, 99% yield.[12]
3. NaCN, HMPA, 75°, 24 h, 75–92% yield.[13] Ethyl esters are not cleaved under these conditions.
4. Cs_2CO_3, PhSH, DMF, 85°, 3 h, 91% yield. A methyl carbonate was cleaved simultaneously.[14]
5. LiI, Pyr, reflux, 91% yield.[15]

EtO$_2$C, NH, CO$_2$Me, N–R $\xrightarrow[91\%]{LiI,\ Pyr,\ reflux}$ EtO$_2$C, NH, CO$_2$H, N–R

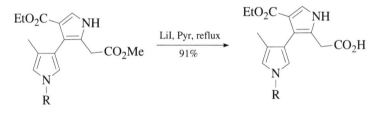

6. $(CH_3)_3SiOK$, ether[16] or THF, 4 h, 61–95% yield as the acid salt.[17]
7. $Ba(OH)_2 \cdot 8H_2O$, MeOH, rt, 7 h, 72% yield.[18]

O, CO$_2$Me, CO$_2$Me $\xrightarrow[MeOH,\ rt,\ 7\ h;\ 72\%]{Ba(OH)_2 \cdot 8H_2O}$ O, CO$_2$Me, CO$_2$H

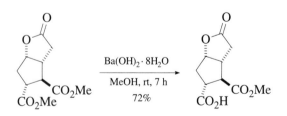

These conditions gave excellent selectivity for an external methyl dienoate in the presence of a more hindered internal dienoate during a

synthesis of the complex macrolide swinholide.[19] These conditions are also mild enough to prevent retroaldol condensation during ester hydrolysis.[20]

8. 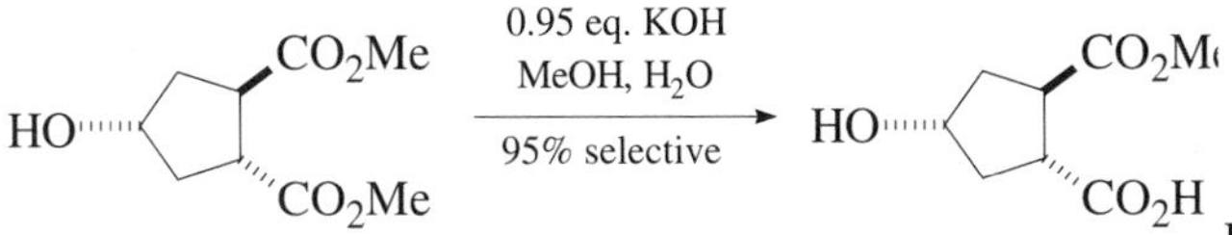 HO⋯cyclopentane diester (CO_2Me, CO_2Me) → 0.95 eq. KOH, MeOH, H_2O, 95% selective → HO⋯cyclopentane (CO_2Me, CO_2H)

Ref. 21

The authors suggest that the selectivity is due to participation of the hydroxyl group.

9. $H_2NC_6H_4SH$, Cs_2CO_3, DMF, 85°, 1–3 h.[22]
10. Pig liver esterase is particularly effective in cleaving one ester of a symmetrical pair.[23–25]
11. 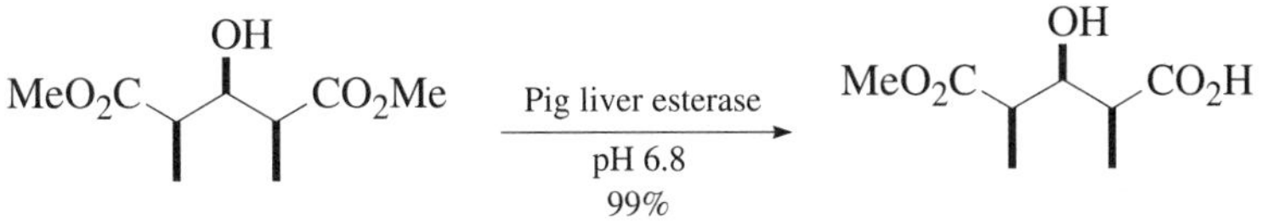 MeO_2C–CH(Me)–CH(OH)–CH(Me)–CO_2Me → Pig liver esterase, pH 6.8, 99% → MeO_2C–CH(Me)–CH(OH)–CH(Me)–CO_2H

Ref. 26

12. 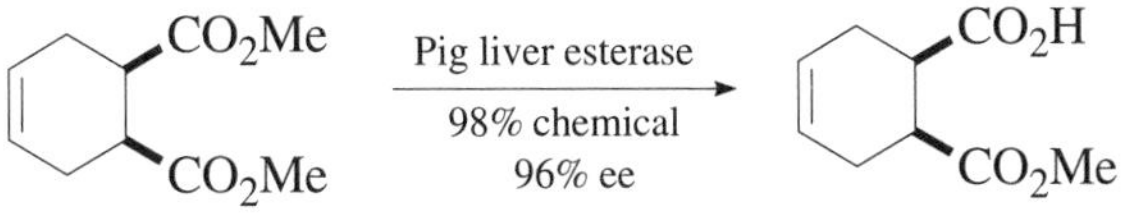 cyclohexene-4,5-dicarboxylate (CO_2Me, CO_2Me) → Pig liver esterase, 98% chemical, 96% ee → (CO_2H, CO_2Me)

Ref. 27

13. MeO_2C–epoxide–CO_2Me → Pig liver esterase → MeO_2C–epoxide–CO_2Me + MeO_2C–epoxide–CO_2H

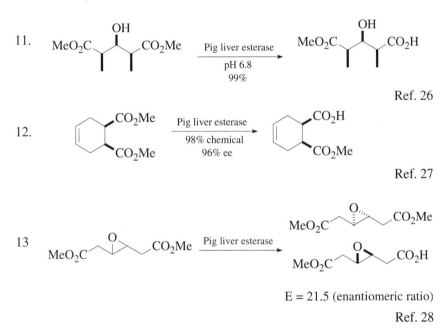

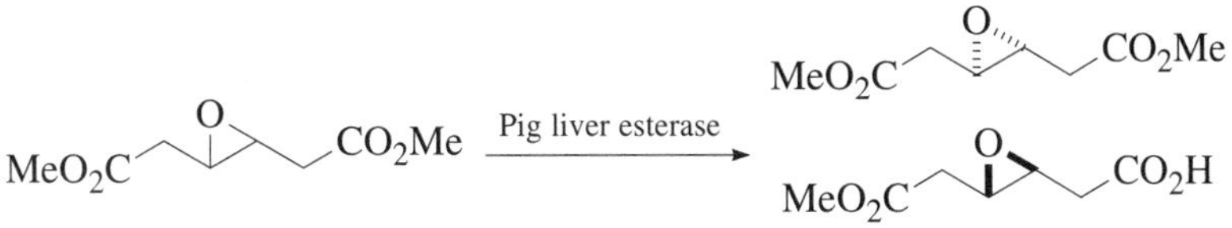

E = 21.5 (enantiomeric ratio)

Ref. 28

14. Carbonic anhydrase, H_2O, 23–83% yield. This enzyme was used for the selective hydrolysis of the D-form of methyl *N*-acetyl α-amino acids.[29]
15. Porcine pancreatic lipase, pH 7.5, 23° 4.5 h, 55% yield. These conditions were used to suppress facile racemization of 2-chlorocyclohexenone.[30]
16. Thermitase, pH 7.5, 55°, 50% DMSO, 3–140 min. This method was used to avoid the degradation of base-sensitive side chains during peptide synthesis. The method is compatible with the Fmoc group.[31]
17. BCl_3, 0°, 5–6 h, 90% yield.[32] In this example, a phenolic methyl group, normally cleaved with boron trichloride, was not affected.
18. $NaBH_4$, I_2, 3 h, rt.[33]

19.

NaH, THF, 0–25°, 68%; H_2O

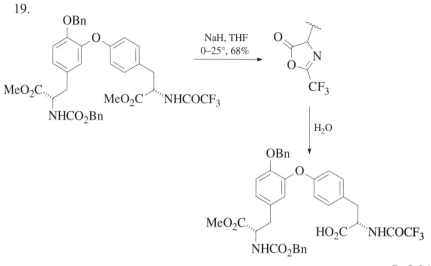

Ref. 34

20. $(Bu_3Sn)_2O$, benzene, 80°, 2–24 h, 73–100% yield.[35] Only relatively unhindered esters are cleaved with this reagent. Acetates of primary and secondary alcohols and phenols are also cleaved efficiently.[36]

21. $CuCO_3$, $Cu(OH)_2$; H_2S workup, 50–60°.[37]

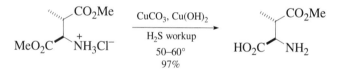

$CuCO_3$, $Cu(OH)_2$; H_2S workup, 50–60°, 97%

1. For example, see S. M. Hecht and J. W. Kozarich, *Tetrahedron Lett.*, 1397 (1973).
2. N. Hashimoto, T. Aoyama, and T. Shioiri, *Chem. Pharm. Bull.*, **29**, 1475 (1981).
3. Y. Hirai, T. Aida, and S. Inoue, *J. Am. Chem. Soc.*, **111**, 3062 (1989).
4. T. Shioiri, T. Aoyama, and S. Mori, *Org. Synth. Collect. Vol. VIII*, 612 (1993).
5. S. C. Fields, W. H. Dent, III, F. R. Green, III, and E. G. Tromiczak, *Tetrahedron Lett.*, **37**, 1967 (1996).
6. J. R. Rachele, *J. Org. Chem.*, **28**, 2898 (1963).
7. V. F. Rudchenko, S. M. Ignator, and R. G. Kostyanovsky, *J. Chem. Soc., Chem. Commun.*, 261 (1990).
8. S. Danishefsky, M. Hirama, K. Gombatz, T. Harayama, E. Berman, and P. Schuda, *J. Am. Chem. Soc.*, **100**, 6536 (1978); *idem, ibid.*, **101**, 7020 (1979).
9. R. Albert, J. Danklmaier, H. Hönig, and H. Kandolf, *Synthesis*, 635 (1987).
10. R. N. Ram and I. Charles, *Tetrahedron*, **53**, 7335 (1997).
11. E. J. Corey, I. Székely, and C. S. Shiner, *Tetrahedron Lett.*, 3529 (1977).
12. A. E. Greene, M.-J. Luche, and J.-P. Deprés, *J. Am. Chem. Soc.*, **105**, 2435 (1983).
13. P. Müller and B. Siegfried, *Helv. Chim. Acta*, **57**, 987 (1974).

14. D. Eren and E. Keinan, *J. Am. Chem. Soc.*, **110**, 4356 (1988); S. Bouzbouz and B. Kirschleger, *Synthesis*, 714 (1994).
15. P. Magnus and T. Gallagher, *J. Chem. Soc., Chem. Commun.*, 389 (1984).
16. E. D. Laganis and B. L. Chenard, *Tetrahedron Lett.*, **25**, 5831 (1984).
17. C. Rasset-Deloge, P. Martinez-Fresneda, and M. Vaultier, *Bull. Soc. Chim. Fr.*, **129**, 285 (1992).
18. K. Inoue and K. Sakai, *Tetrahedron Lett.*, 4063 (1977).
19. I. Paterson, K.-S. Yeung, R. A. Ward, J. D. Smith, J. G. Cumming, and S. Lamboley, *Tetrahedron*, **51**, 9467 (1995).
20. M. Nambu and J. D. White, *J. Chem. Soc., Chem. Commun.*, 1619 (1996).
21. M. Honda, K. Hirata, H. Sueoka, T. Katsuki, and M. Yamaguchi, *Tetrahedron Lett.*, **22**, 2679 (1981).
22. E. Keinan and D. Eren, *J. Org. Chem.*, **51**, 3165 (1986).
23. M. Ohno, Y. Ito, M. Arita, T. Shibata, K. Adachi, and H. Sawai, *Tetrahedron*, **40**, 145 (1984).
24. E. Alvarez, T. Cuvigny, C. Hervé du Penhoat, and M. Julia, *Tetrahedron*, **44**, 119 (1988).
25. K. Adachi, S. Kobayashi, and M. Ohno, *Chimia*, **40**, 311 (1986).
26. D. S. Holmes, U. C. Dyer, S. Russell, J. A. Sherringham, and J. A. Robinson, *Tetrahedron Lett.*, **29**, 6357 (1988).
27. S. Kobayashi, K. Kamiyama, T. Iimori and M. Ohno, *Tetrahedron Lett.*, **25**, 2557 (1984).
28. P. Mohr, L. Rösslein, and C. Tamm, *Tetrahedron Lett.*, **30**, 2513 (1989).
29. R. Chênevert, R. B. Rhlid, M. Lêtourneau, R. Gagnon, and L. D'Astous, *Tetrahedron: Asymmetry*, **4**, 1137 (1993).
30. H. Wild, *J. Org. Chem.*, **59**, 2748 (1994).
31. S. Reissmann and G. Greiner, *Int. J. Pept. Protein Res.*, **40**, 110 (1992).
32. P. S. Manchand, *J. Chem. Soc., Chem. Commun.*, 667 (1971).
33. D. H. R. Barton, L. Bould, D. L. J. Clive, P. D. Magnus, and T. Hase, *J. Chem. Soc. C*, 2204 (1971).
34. D. L. Boger and D. Yohannes, *J. Org. Chem.*, **54**, 2498 (1989).
35. E. G. Mata and O. A. Mascaretti, *Tetrahedron Lett.*, **29**, 6893 (1988).
36. C. J. Salomon, G. E. Mata, and O. A. Mascaretti, *J. Org. Chem.*, **59**, 7259 (1994).
37. J. M. Humphrey, J. B. Aggen and A. R. Chamberlin, *J. Am. Chem. Soc.*, **118**, 11759 (1996).

Substituted Methyl Esters

9-Fluorenylmethyl (Fm) Ester: O_2CR

9-Fluorenylmethyl esters of *N*-protected amino acids were prepared using the DCC/DMAP method (50–89% yield)[1] or by imidazole-catalyzed transesterification of protected amino acid active esters with FmOH.[2] Cleavage is accomplished either with diethylamine or piperidine in CH_2Cl_2 at rt for 2 h. No racemization was observed during formation or cleavage of the Fm esters.[1] The Fm ester is cleaved slowly by hydrogenolysis,[3] but complete selectivity for hydrogenolysis of the benzyloxycarbonyl group could not be obtained. Fm esters also improved the solubility of protected peptides in organic solvents.[2]

1. H. Kessler and R. Siegmeier, *Tetrahedron Lett.*, **24**, 281 (1983).
2. M. A. Bednarek and M. Bodanszky, *Int. J. Pept. Protein Res.*, **21**, 196 (1983).
3. A. Lender, W. Yao, P. A. Sprengeler, R. A. Spanevello, G. T. Furst, R. Hirschmann, and A. B. Smith, III, *Int. J. Pept. Protein Res.*, **42**, 509 (1993).

Methoxymethyl (MOM) Ester: $RCOOCH_2OCH_3$ (Chart 6)

Formation

The section on the formation of MOM ethers should be consulted, since many of the methods described there should also be applicable to the formation of MOM esters.

1. CH_3OCH_2Cl, Et_3N, DMF, 25°, 1 h.[1]
2. $CH_3OCH_2OCH_3$, Zn/$BrCH_2CO_2Et$, 0°; CH_3COCl, 0–20°, 2 h, 75–85%.[2] A number of methoxymethyl esters were prepared by this method, which avoids the use of the carcinogen chloromethyl methyl ether.

Cleavage

1. R'_3SiBr, trace MeOH. Methoxymethyl ethers are stable to these cleavage conditions.[3] Methoxymethyl esters are unstable to silica gel chromatography, but are stable to mild acid (0.01 *N* HCl, EtOAc, MeOH, 25°, 16 h).[4]
2. $MgBr_2$, Et_2O. MEM, MTM, and SEM ethers are cleaved as well.[5]
3. Solvolysis in MeOH/H_2O at 21°. This method was developed for a series of penicillin derivatives where conventional cleavage methods resulted in partial β-lactam cleavage.[6]
4. $AlCl_3$, $PhNMe_2$, 80–99% yield. MEM, MTM, Me, Bn, and SEM esters are cleaved similarly.[7]
5. Pyr, H_2O.[8]

1. A. B. A. Jansen and T. J. Russell, *J. Chem. Soc.*, 2127 (1965).
2. F. Dardoize, M. Gaudemar, and N. Goasdoue, *Synthesis*, 567 (1977).
3. S. Masamune, *Aldrichimica Acta*, **11**, 23–30 (1978); see p. 30.
4. L. M. Weinstock, S. Karady, F. E. Roberts, A. M. Hoinowski, G. S. Brenner, T. B. K. Lee, W. C. Lumma, and M. Sletzinger, *Tetrahedron Lett.*, 3979 (1975).

5. S. Kim, Y. H. Park, and I. S. Kee, *Tetrahedron Lett.*, **32**, 3099 (1991).
6. S. Vanwetswinkel, V. Carlier, J. Marchand-Brynaert, and J. Fastrez, *Tetrahedron Lett.*, **37**, 2761 (1996).
7. T. Akiyama, H. Hirofuji, A. Hirose, and S. Ozaki, *Synth. Commun.*, **24**, 2179 (1994).
8. M. Shimano, H. Nagaoka, and Y. Yamada, *Chem. Pharm. Bull.*, **38**, 276 (1990).

Methylthiomethyl (MTM) Ester: $RCOOCH_2SCH_3$ (Chart 6)

Formation

1. From RCO_2K: CH_3SCH_2Cl, NaI, 18-crown-6, C_6H_6, reflux, 6 h, 85–97% yield.[1]
2. $Me_2S^+ClX^-$, Et_3N, 0.5 h, $-70° \rightarrow 25°$, 80–85% yield.[2]
3. *t*-BuBr, DMSO, $NaHCO_3$, 62–98% yield.[3,4] This method was used to prepare the MTM esters of *N*-protected amino acids.

Cleavage

1. $HgCl_2$, CH_3CN, H_2O, reflux, 6 h; H_2S, 20°, 30 min, 82–98% yield.
2. MeI, acetone, reflux, 24 h; 1 *N* NaOH, 87–97% yield.[5]
3. CF_3COOH, 25°, 15 min, 80–90% yield.[6]
4. HCl, Et_2O, 6 h, 83–88% yield.[4] Acidic deprotection of the BOC group could not be achieved with complete selectivity in the presence of an MTM ester. The trityl and NPS (2-nitrophenylsulfenyl) groups were the preferred nitrogen protective groups.
5. H_2O_2, $(NH_4)_6Mo_7O_{24}$; NaOH, pH 11, 97% yield.[5] The MTM ester is converted to the much-more-base labile methylsulfonylmethyl ester. It is possible to hydrolyze the methylsulfonylmethyl ester in the presence of the MTM ester.
6. MCPBA converts the MTM ester to a methylsulfonylmethyl ester (78–98% yield), which can be hydrolyzed enzymatically with rabbit serum (pH 4.5 phosphate buffer, EtOH, 25–28°, 1 h, 84% yield).[7]

1. L. G. Wade, J. M. Gerdes, and R. P. Wirth, *Tetrahedron Lett.*, 731 (1978).
2. T.-L. Ho, *Synth. Commun.*, **9**, 267 (1979).
3. A. Dossena, R. Marchelli, and G. Casnati, *J. Chem. Soc., Perkin Trans. 1*, 2737 (1981).
4. A. Dossena, G. Palla, R. Marchelli, and T. Lodi, *Int. J. Pept. Protein Res.*, **23**, 198 (1984).
5. J. M. Gerdes and L. G. Wade, *Tetrahedron Lett.*, 689 (1979).
6. T.-L. Ho and C. M. Wong, *J. Chem. Soc., Chem. Commun.*, 224 (1973).
7. A. Kamal, *Synth. Commun.*, **21**, 1293 (1991).

Tetrahydropyranyl (THP) Ester: RCOO-2-tetrahydropyranyl (Chart 6)

Formation

1. Dihydropyran, TsOH, CH_2Cl_2, 20°, 1.5 h, quant.[1]

Cleavage

1. AcOH, THF, H_2O (4:2:1), 45°, 3.5 h.[1]

1. K. F. Bernady, M. B. Floyd, J. F. Poletto, and M. J. Weiss, *J. Org. Chem.*, **44**, 1438 (1979).

Tetrahydrofuranyl Ester: RCO_2-2-tetrahydrofuranyl

Formation/Cleavage[1]

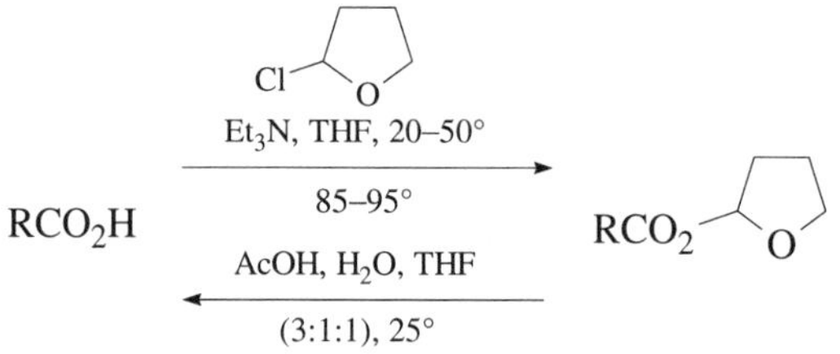

$$RCO_2H \underset{\text{AcOH, } H_2O\text{, THF (3:1:1), 25°}}{\overset{\text{2-chlorotetrahydrofuran, } Et_3N\text{, THF, 20–50°, 85–95°}}{\rightleftarrows}} RCO_2\text{-2-tetrahydrofuranyl}$$

1. C. G. Kruse, N. L. J. M. Broekhof, and A. van der Gen, *Tetrahedron Lett.*, 1725 (1976).

Methoxyethoxymethyl (MEM) Ester: $RCO_2CH_2OCH_2CH_2OCH_3$

Formation/Cleavage[1]

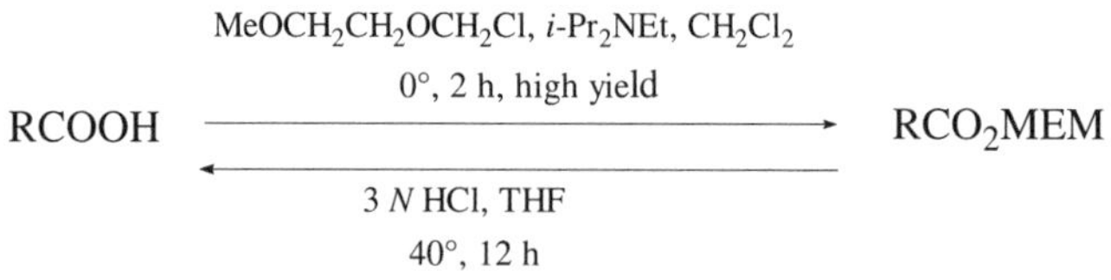

$$RCOOH \underset{3\,N\text{ HCl, THF, 40°, 12 h}}{\overset{MeOCH_2CH_2OCH_2Cl,\ i\text{-}Pr_2NEt,\ CH_2Cl_2,\ 0°\text{, 2 h, high yield}}{\rightleftarrows}} RCO_2MEM$$

In an attempt to synthesize the macrolide antibiotic chlorothricolide, an unhindered –COOH group was selectively protected, in the presence of a hindered –COOH group, as a MEM ester that was then reduced to an alcohol group.[2] $MgBr_2{\cdot}Et_2O$[3] and $AlCl_3$–dimethylaniline[4] efficiently cleave the MEM ester.

1. A. I. Meyers and P. J. Reider, *J. Am. Chem. Soc.*, **101**, 2501 (1979).

2. R. E. Ireland and W. J. Thompson, *Tetrahedron Lett.*, 4705 (1979).
3. A. J. Pearson and H. Shin, *J. Org. Chem.*, **59**, 2314 (1994).
4. T. Akiyama, H. Hirofuji, A. Hirose, and S. Ozaki, *Synth. Commun.*, **24**, 2179 (1994).

2-(Trimethylsilyl)ethoxymethyl (SEM) Ester: $RCO_2CH_2OCH_2CH_2Si(CH_3)_3$

The SEM ester was used to protect a carboxyl group where DCC-mediated esterification caused destruction of the substrate. It was formed from the acid and SEM chloride (THF, 0°, 80% yield) and was removed solvolytically. The ease of removal in this case was attributed to anchimeric assistance by the phosphate group.[1] Normally, SEM groups are cleaved by treatment with fluoride ion. Note that in this case the SEM group is removed considerably faster than the phenyl groups from the phosphate. Additionally, cleavage is effected with $MgBr_2$ in ether (61–100% yield),[2] HF in acetonitrile,[3] or neat HF.[4]

CO_2SEM, $OP(O)(OPh)_2$ —— MeOH, half-life = 12 h, 65% ——→ CO_2H, $OP(O)(OH)_2$

1. E. W. Logusch, *Tetrahedron Lett.*, **25**, 4195 (1984).
2. W.-C. Chen, M. D. Vera, and M. M. Joullié, *Tetrahedron Lett.*, **38**, 4025 (1997).
3. W.-R. Li, W. R. Ewing, B. D. Harris, and M. M. Joullié, *J. Am. Chem. Soc.*, **112**, 7659 (1990).
4. G. Jou, I. Gonzalez, F. Albericio, P. Lloyd-Williams, and E. Giralt, *J. Org. Chem.*, **62**, 354 (1997).

Benzyloxymethyl (BOM) Ester: $RCOOCH_2OCH_2C_6H_5$ (Chart 6)

Formation[1]

$$RCOONa + PhCH_2OCH_2Cl \xrightarrow{\text{HMPA, 25°, 70\%}} RCO_2BOM$$

Cleavage[1]

1. H_2/Pd–C, EtOH, 25°, 70–100% yield.
2. Aqueous HCl, THF, 25°, 2 h, 75–95% yield.

1. P. A. Zoretic, P. Soja, and W. E. Conrad, *J. Org. Chem.*, **40**, 2962 (1975).

Pivaloyloxymethyl Ester (POM–O_2CR): $(CH_3)_3CCO_2CH_2O_2CR$

The ester is prepared from the acid with $PvOCH_2I$ and Ag_2CO_3 in DMF.[1]

Cleavage

$(Bu_3Sn)_2O$, Et_2O, 3 h, 25°, 56% yield.[2–4]

1. D. V. Patel, E. M. Gordon, R. J. Schmidt, H. N. Weller, M. G. Young, R. Zahler, M. Barbacid, J. M. Carboni, J. L. Gullo-Brown, L. Hunihan, C. Ricca, S. Robinson, B. R. Seizinger, A. V. Tuomari, and V. Manne, *J. Med. Chem.*, **38**, 435 (1995).
2. C. J. Salomon, E. G. Mata, and O. A. Mascaretti, *Tetrahedron Lett.*, **32**, 4239 (1991).
3. C. J. Salomon, E. G. Mata, and O. A. Mascaretti, *J. Org. Chem.*, **59**, 7259 (1994).
4. E. G. Mata and O. A. Mascaretti, *Tetrahedron Lett.*, **29**, 6893 (1988).

Phenylacetoxymethyl Ester: $PhCH_2CO_2CH_2O_2CR$

This ester is conveniently formed from a penicillinic acid with $PhCH_2CO_2CH_2Cl$ and TEA. Cleavage is accomplished by enzymatic hydrolysis with penicillin G. acylase in 70–90% yield.[1,2]

1. E. Baldaro, C. Fuganti, S. Servi, A. Tahliani, and M. Terreni, in *Microbial Reagents in Organic Synthesis,* S. Servi, Ed., Kluwer Academic Pubs., Dordrecht (1992), pp. 175ff.
2. E. Baldaro, D. Faiardi, C. Fuganti, P. Grasselli, and A. Lazzzarini, *Tetrahedron Lett.*, **29**, 4623 (1988).

Triisopropylsilylmethyl Ester: $i\text{-}Pr_3SiCH_2O_2CR$

Formation

1. $i\text{-}Pr_3SiCHN_2$, 76–96% yield.[1] In contrast, when $TMSCHN_2$ is used to prepare an ester, the methyl ester is formed.

Cleavage

1. 3 *N* NaOH, EtOH, 6 h, reflux.

1. J. A. Soderquist and E. I. Miranda, *Tetrahedron Lett.*, **34**, 4905 (1993).

Cyanomethyl Ester: RCO_2CH_2CN

Formation

1. $ClCH_2CN$, TEA, 78–96% yield.[1]

2. 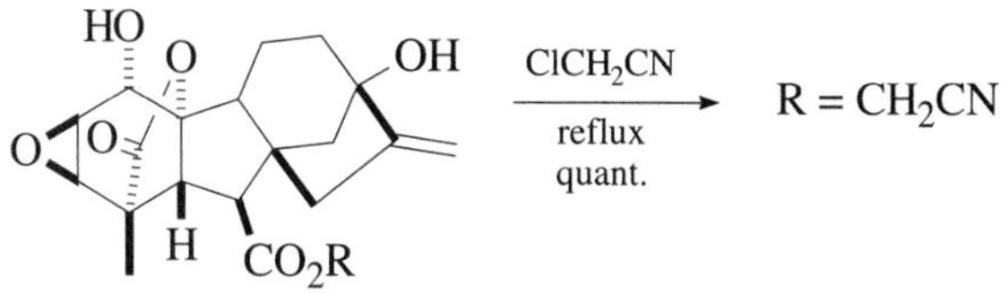 [Structure: R = H] $\xrightarrow[\text{reflux, quant.}]{ClCH_2CN}$ R = CH_2CN

R = H Ref. 2

Cleavage

Na_2S, acetone, water, 74–90% yield.[1]

1. H. M. Hugel, K. V. Bhaskar, and R. W. Longmore, *Synth. Commun.*, **22**, 693 (1992).
2. S. Findlow, P. Gaskin, P. A. Harrison, J. R. Lenton, M. Penny, and C. L. Willis, *J. Chem. Soc., Perkin Trans. 1*, 751 (1997).

Acetol Ester: $CH_3COCH_2O_2CR$

Developed as a carboxyl protective group for peptide synthesis because of its stability to hydrogenolysis and acidic conditions, the acetol (hydroxy acetone) ester is prepared by DCC coupling (68–92% yield) of the acid with acetol. It is cleaved with TBAF in THF.[1]

1. B. Kundu, *Tetrahedron Lett.*, **33**, 3193 (1992).

Phenacyl Ester: $RCOOCH_2COC_6H_5$ (Chart 6)

Formation

1. $PhCOCH_2Br$, Et_3N, EtOAc, 20°,12 h, 83% yield.[1]
2. $PhCOCH_2Br$, KF/DMF, 25°, 10 min, 90–99% yield.[2] Hindered acids are protected at 100°.
3. From the K salt: $PhCOCH_2Br$, $Bu_4N^+\ Br^-$, CH_3CN, rt, dibenzo-18-crown-6, 86–98% yield.[3]

Cleavage

1. Zn/HOAc, 25°, 1 h, 90% yield.[4]
2. Zn, acetylacetone, Pyr, DMF, 35°, 0.6 h, 90–98% yield.[5]
3. H_2/Pd–C, aq. MeOH, 20°, 1 h, 72% yield.[1]
4. PhSNa, DMF, 20°, 30 min, 72% yield.[1]
5. $CuCl_2$, O_2, DMF, H_2O, 23–92% yield.[6]
6. Photolysis, sensitizer, CH_3CN, 2 h, 76–100% yield.[7]

7. Irradiation of buffered solutions of *p*-hydroxyphenacyl esters releases the acid.[8]
8. PhSeH, DMF, rt, 48 h, 79% yield.[9] Under basic coupling conditions, an aspartyl peptide that has a β-phenacyl ester is converted to a succinimide.[10] The use of PhSeH prevents the α,β-rearrangement of the aspartyl residue during deprotection.
9. TBAF, THF or DMSO or DMF, 72–98% yield. 4-Nitrobenzyl and trichloroethyl esters of amino acids are also cleaved.[11]

A phenacyl ester is much more readily cleaved by nucleophiles than are other esters, such as the benzyl ester. Phenacyl esters are stable to acidic hydrolysis (e.g., concd. HCl;[1] HBr/HOAc;[1] 50% CF_3COOH/CH_2Cl_2;[12] HF, 0°, 1 h[12]).

1. G. C. Stelakatos, A. Paganou, and L. Zervas, *J. Chem. Soc. C*, 1191 (1966).
2. J. H. Clark and J. M. Miller, *Tetrahedron Lett.*, 599 (1977).
3. S. J. Jagdale, S. V. Patil, and M. M. Salunkhe, *Synth. Commun.*, **26**, 1747 (1996).
4. J. B. Hendrickson and C. Kandall, *Tetrahedron Lett.*, 343 (1970).
5. D. Hagiwara, M. Neya, and M. Hashimoto, *Tetrahedron Lett.*, **31**, 6539 (1990).
6. R. N. Ram and L. Singh, *Tetrahedron Lett.*, **36**, 5401 (1995).
7. A. Banerjee and D. E. Falvey, *J. Org. Chem.*, **62**, 6245 (1997).
8. R. S. Givens, A. Jung, C.-H. Park, J. Weber, and W. Bartlett, *J. Am. Chem. Soc.*, **119**, 8369 (1997).
9. J. L. Morell, P. Gaudreau, and E. Gross, *Int. J. Pept. Protein Res.*, **19**, 487 (1982).
10. M. Bodanszky and J. Martinez, *J. Org. Chem.*, **43**, 3071 (1978).
11. M. Namikoshi, B. Kundu, and K. L. Rinehart, *J. Org. Chem.*, **56**, 5464 (1991).
12. C. C. Yang and R. B. Merrifield, *J. Org. Chem.*, **41**, 1032 (1976).

p-Bromophenacyl Ester: $RCOOCH_2COC_6H_4$–*p*-Br

In a penicillin synthesis, the carboxyl group was protected as a *p*-bromophenacyl ester that was cleaved by nucleophilic displacement (PhSK, DMF, 20°, 30 min, 64% yield). Hydrogenolysis of a benzyl ester was difficult (perhaps because of catalyst poisoning by sulfur); basic hydrolysis of methyl or ethyl esters led to attack at the β-lactam ring.[1]

1. P. Bamberg, B. Eckström, and B. Sjöberg, *Acta Chem. Scand.*, **21**, 2210 (1967).

α-Methylphenacyl Ester: $RCO_2CH(CH_3)COC_6H_5$

p-Methoxyphenacyl Ester: $RCO_2CH_2COC_6H_4$–*p*-OCH_3

Phenacyl esters can be prepared from the phenacyl bromide, a carboxylic acid, and potassium fluoride as base.[1] These esters can be cleaved by irradiation (313 nm, dioxane or EtOH, 20°, 6 h, 80–95% yield, R = amino acids;[2] >300 nm, 30°, 8 h, R = a gibberellic acid, 36–62% yield[3]). Another phenacyl derivative, $RCO_2CH(COC_6H_5)C_6H_3$–3,5-$(OCH_3)_2$, cleaved by irradiation, has also been reported.[4]

1. F. S. Tjoeng and G. A. Heavner, *Synthesis*, 897 (1981).
2. J. C. Sheehan and K. Umezawa, *J. Org. Chem.*, **38**, 3771 (1973).
3. E. P. Serebryakov, L. M. Suslova, and V. K. Kucherov, *Tetrahedron*, **34**, 345 (1978).
4. J. C. Sheehan, R. M. Wilson, and A. W. Oxford, *J. Am. Chem. Soc.*, **93**, 7222 (1971).

Desyl ester: $RCO_2CH(Ph)C(=O)Ph$

Formation

Desyl bromide, DBU, benzene, reflux, 57–95% yield.[1] A polymer-supported version of this ester has been prepared.[2]

Cleavage

Photolysis, 350 nm, CH_3CN, H_2O. The by-product from the reaction is 2-phenylbenzo[*b*]furan. Cleavage with TBAF and $PhCH_2SH$ has been demonstrated (70–94% yield).[3] The related 3,5-dimethoxybenzoin analogue is cleaved with a rate constant of $>10^{10}$ sec^{-1}. [4] Photolytic cleavage occurs by heterolytic bond dissociation.[5]

1. K. R. Gee, L. W. Kueper, III, J. Barnes, G. Dudley, and R. S. Givens, *J. Org. Chem.*, **61**, 1228 (1996).
2. A. Routledge, C. Abell, and S. Balasubramanian, *Tetrahedron Lett.*, **38**, 1227 (1997).
3. M. Ueki, H. Aoki, and T. Katoh, *Tetrahedron Lett.*, **34**, 2783 (1993).
4. M. H. B. Stowell, R. S. Rock, D. C. Rees, and S. I. Chan, *Tetrahedron Lett.*, **37**, 307 (1996).
5. Y. Shi, J. E. T. Corrie, and P. Wan, *J. Org. Chem.*, **62**, 8278 (1997).

Carboxamidomethyl (Cam) Ester: $RCO_2CH_2CONH_2$

The carboxamidomethyl ester was prepared for use in peptide synthesis. It is formed from the cesium salt of an *N*-protected amino acid and α-chloroacetamide (60–85% yield). It is cleaved with 0.5 *M* NaOH or $NaHCO_3$ in DMF/H_2O. It is stable to the conditions required to remove BOC, Cbz, Fmoc, and *t*-butyl esters. It cannot be selectively cleaved in the presence of a benzyl ester of aspartic acid.[1]

1. J. Martinez, J. Laur, and B. Castro, *Tetrahedron*, **41**, 739 (1985); *idem.*, *Tetrahedron Lett.*, **24**, 5219 (1983).

p-Azobenzenecarboxamidomethyl Ester: $C_6H_5N{=}NC_6H_4NHC(O)CH_2O_2CR$

This ester was developed for C-terminal amino acids during solution-phase peptide synthesis. Purification of intermediates can be monitored colorimetrically or visually. Protection is achieved by reacting the sodium salt of the *N*-protected amino acid with the bromoacetamide derivative to give the ester in 70–95% yield. Cleavage is effected by simple hydrolysis with K_2CO_3 or NH_4OH.[1] A related chromogenic ester, the ***p*-(*p*-(dimethylamino)phenylazo)benzyl ester**, has also been used for the same purpose, except that it can be cleaved by hydrogenolysis.[2]

1. V. G. Zhuravlev, A. A. Mazurov, and S. A. Andronati, *Collect. Czech. Chem. Commun.*, **57**, 1495 (1992).
2. G. D. Reynolds, D. R. K. Harding, and W. S. Hancock, *Int. J. Pept. Protein Res.*, **17**, 231 (1981).

N-Phthalimidomethyl Ester (Chart 6):

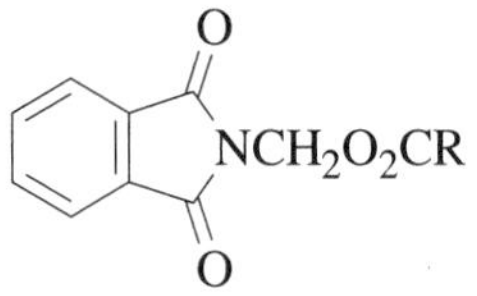

O
NCH_2O_2CR
O

Formation

1. $RCO_2H + XCH_2$-*N*-phthalimido

   X = OH: Et_2NH, EtOAc, 37°, 12 h, 70–80% yield.[1]
   X = Cl: $(c\text{-}C_6H_{11})_2NH$, DMF or DMSO, 60°, few minutes, 70–80% yield.[1]
   X = Cl, Br: KF, DMF, 80°, 2 h, 65–75% yield.[2]

Cleavage

1. H_2NNH_2/MeOH, 20°, 3 h, 90% yield.[1]
2. Et_2NH/MeOH, H_2O, 25°, 24 h or refux, 2 h, 82% yield.[1]
3. NaOH/MeOH, H_2O, 20°, 45 min, 77% yield.[1]
4. Zn/HOAc, 25°, 12 h, 80% yield.[3]
5. g HCl/EtOAc, 20°, 16 h, 83% yield.[1]
6. HBr/HOAc, 20°, 10–15 min, 80% yield.[1]

1. G. H. L. Nefkens, G. I. Tesser, and R. J. F. Nivard, *Recl. Trav. Chim. Pays-Bas*, **82**, 941 (1963).
2. K. Horiki, *Synth. Commun.*, **8**, 515 (1978).
3. D. L. Turner and E. Baczynski, *Chem. Ind. (London)*, 1204 (1970).

2-Substituted Ethyl Esters

2,2,2-Trichloroethyl Ester: $RCO_2CH_2CCl_3$ (Chart 6)

Formation

1. CCl_3CH_2OH, DCC, Pyr.[1]
2. CCl_3CH_2OH, TsOH, toluene, reflux.[1,2]
3. CCl_3CH_2OCOCl, THF, Pyr, >60% yield.[3]

Cleavage

1. Zn, AcOH, 0°, 2.5 h.[1]
2. Zinc, THF buffered at pH 4.2–7.2 (20°, 10 min, 75–95% yield).[4]
3. Zinc dust, 1 *M* NH_4OAc, 66% yield.[5]
4. Electrolysis: −1.65 V, $LiClO_4$, MeOH, 87–91% yield.[6] A **tribromoethyl ester** is cleaved by electrolytic reduction at −0.70 V (85% yield); a **dichloroethyl ester** is cleaved at −1.85 V (78% yield).[6]
5. Cat. Se, $NaBH_4$, DMF, 40–50°, 1 h, 77–93% yield.[7]
6. SmI_2, THF, rt, 2 h, quantitative.[8]
7. Cd, DMF, AcOH, 25°, 15 h, 82% yield.[9]

1. R. B. Woodward, K. Heusler, J. Gosteli, P. Naegeli, W. Oppolzer, R. Ramage, S. Ranganathan, and H. Vorbrüggen, *J. Am. Chem. Soc.*, **88**, 852 (1966).
2. J. F. Carson, *Synthesis*, 24 (1979).
3. R. R. Chauvette, P. A. Pennington, C.W. Ryan, R. D. G. Cooper, F. L. José, I. G. Wright, E. M. Van Heyningen, and G. W. Huffman, *J. Org. Chem.*, **36**, 1259 (1971).
4. G. Just and K. Grozinger, *Synthesis*, 457 (1976).
5. G. Jou, I. Gonzalez, F. Albericio, P. Lloyd-Williams, and E. Giralt, *J. Org. Chem.*, **62**, 354 (1997).
6. M. F. Semmelhack and G. E. Heinsohn, *J. Am. Chem. Soc.*, **94**, 5139 (1972).
7. Z.-Z. Huang and X.-J. Zhou, *Synthesis*, 693 (1989).
8. A. J. Pearson and K. Lee, *J. Org. Chem.*, **59**, 2304 (1994).
9. Y. Génisson, P. C. Tyler, and R. N. Young, *J. Am. Chem. Soc.*, **116**, 759 (1994).

2-Haloethyl Ester: $RCOOCH_2CH_2X$, X = I, Br, Cl (Chart 6)

Cleavage

2-Haloethyl esters have been cleaved under a variety of conditions, many of which proceed by nucleophilic addition.

1. Li^+ or Na^+ Co(I)phthalocyanine/MeOH, 0–20°, 40 min–60 h, 60–98% yield.[1]
2. Electrolysis: Co(I)phthalocyanine, $LiClO_4$, EtOH, H_2O, −1.95 V, 95% yield.[2]
3. $NaS(CH_2)_2SNa/CH_3CN$, reflux, 2 h, 80–85% yield.[3]
4. NaSeH/EtOH, 25°, 1 h → reflux, 6 min, 92–99% yield.[4,5]
5. $(NaS)_2CS/CH_3CN$, reflux, 1.5 h, 75–86% yield.[6]
6. Me_3SnLi/THF, 3 h → $Bu_4N^+F^-$, reflux, 15 min, 78–86% yield.[7]
7. NaHTe, EtOH, 2–60 min, 80–92% yield.[8]
8. Na_2S, 40–68% yield.[9]
9. Li(cobalt phthalocyanine).[10]
10. Cobalt phthalocyanine, $NaBH_4$.[11]
11. SmI_2, THF, rt, 2 h, 88–100% yield.[12] These conditions were found effective when many of the preceding reagents failed to give clean deprotection.

1. H. Eckert and I. Ugi, *Angew. Chem., Int. Ed. Engl.*, **15**, 681 (1976).
2. R. Scheffold and E. Amble, *Angew. Chem., Int. Ed. Engl.*, **19**, 629 (1980).
3. T.-L. Ho, *Synthesis*, 510 (1975).
4. T.-L. Ho, *Synth. Commun.*, **8**, 301 (1978).
5. Z.-Z. Huang and X.-J. Zhou, *Synthesis*, 633 (1990).
6. T.-L. Ho, *Synthesis*, 715 (1974).
7. T.-L. Ho, *Synth. Commun.*, **8**, 359 (1978).
8. J. Chen and X. Zhou, *Synth. Commun.*, **17**, 161 (1987).
9. M. Joaquina, S. A. Amaral Trigo, and M. I. A. Oliveira Sartos, in *Peptides 1986*, D. Theodoropoulos, Ed., Walter de Gruyter & Co., Berlin, 1987, p. 61.
10. P. Lemmen, K. M. Buchweitz, and R. Stumpf, *Chem. Phys. Lipids*, **53**, 65 (1990).
11. H. Eckert, *Z. Naturforch., B: Chem. Sci.*, **45**, 1715 (1990).
12. A. J. Pearson and K. Lee, *J. Org. Chem.*, **59**, 2257, 2304 (1994); *idem, ibid.*, **60**, 7153 (1995).

ω-Chloroalkyl Ester: $RCOO(CH_2)_nCl$

ω-Chloroalkyl esters ($n = 4, 5$) have been cleaved by sodium sulfide (reflux, 4 h, 58–85% yield). The reaction proceeds by sulfide displacement of the chloride ion, followed by intramolecular displacement of the carboxylate group by the (now) sulfhydryl group.[1]

1. T.-L. Ho and C. M. Wong, *Synth. Commun.*, **4**, 307 (1974).

2-(Trimethylsilyl)ethyl (TMSE) Ester: $RCO_2CH_2CH_2Si(CH_3)_3$

Formation

1. $Me_3SiCH_2CH_2OH$, DCC, Pyr, CH_3CN, 0°, 5–15 h, 66–97% yield.[1]
2. From an acid chloride: $Me_3SiCH_2CH_2OH$, Pyr, 25°, 3 h.[2]
3. $Me_3SiCH_2CH_2OH$, Me_3SiCl, THF, reflux, 12–36 h.[3] This method of esterification is also effective for the preparation of other esters.
4. From an anhydride: $Me_2AlOCH_2CH_2SiMe_3$, benzene, heat, >85% yield.[4]

OMOM → $Me_2AlOCH_2CH_2TMS$, benzene, heat, >85% → OMOM, HO_2C $CO_2CH_2CH_2TMS$

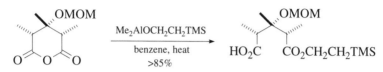

5. $Me_3SiCH_2CH_2OH$, 2-chloro-1-methylpyridinium iodide, Et_3N, 90% yield.[5]
6. From a methyl ester: $Me_3SiCH_2CH_2OH$, Ti(O*i*-Pr)$_4$, 120°, 4 h, 85% yield.[6]

I, AcO, CO_2Me, O, H, OTBDMS → $TMSCH_2CH_2OH$, Ti(O–*i*-Pr)$_4$, 120°, 4 h, 85% → I, HO, $CO_2CH_2CH_2TMS$, O, H, OTBDMS

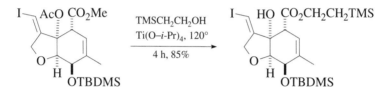

7. $Me_3SiCH_2CH_2OH$, EDC, DMAP, Pyr.[7]
8. $Me_3SiCH_2CH_2OH$, DEAD, Ph_3P, THF, >75% yield.[8]

Cleavage

1. $Et_4N^+F^-$ or $Bu_4N^+F^-$, DMF or DMSO, 20–30°, 5–60 min, quant. yield.[1,9]
2. DMF, $Bu_4N^+Cl^-$, $KF \cdot 2H_2O$, 42–62% yield (substrate = polypeptide).[10]
3. TBAF, SiO_2, 100% yield[7] or TBAF, DMF, 20 min.[11]
4. TBAF, TsOH, THF, 20°. Other conditions in this sensitive ivermectin analogue led to decomposition.[6]
5. Tris(dimethylamino)sulfonium difluorotrimethylsilicate (TAS–F), DMF, >76% yield.[8]

1. P. Sieber, *Helv. Chim. Acta*, **60**, 2711 (1977).
2. H. Gerlach, *Helv. Chim. Acta*, **60**, 3039 (1977).
3. M. A. Brook and T. H. Chan, *Synthesis*, 201 (1983).
4. E. Vedejs and S. D. Larsen, *J. Am. Chem. Soc.*, **106**, 3030 (1984).
5. J. D. White and L. R. Jayasinghe, *Tetrahedron Lett.*, **29**, 2139 (1988).
6. J.-P. Férézou, M. Julia, Y. Li, L. W. Liu, and A. Pancrazi, *Bull. Soc. Chim. Fr.*, **132**, 428 (1995).

7. A. M. Sefler, M. C. Kozlowski, T. Guo, and P. A. Bartlett, *J. Org. Chem.*, **62**, 93 (1997).
8. W. R. Roush, D. S. Coffey, and D. J. Madar, *J. Am. Chem. Soc.*, **119**, 11331 (1997).
9. P. Sieber, R. H. Andreatta, K. Eisler, B. Kamber, B. Riniker, and H. Rink, in *Peptides: Proceedings of the Fifth American Peptide Symposium*, M. Goodman and J. Meienhofer, Eds., Halsted Press, New York, 1977, pp. 543–545.
10. R. A. Forsch and A. Rosowsky, *J. Org. Chem.*, **49**, 1305 (1984).
11. C. K. Marlowe, *Bioorg. Med. Chem. Lett.*, **3**, 437 (1993).

2-Methylthioethyl Ester: $RCO_2CH_2CH_2SCH_3$

The 2-methylthioethyl ester is prepared from a carboxylic acid and methylthioethyl alcohol or methylthioethyl chloride ($MeSCH_2CH_2OH$, TsOH, benzene, reflux, 55 h, 55% yield; $MeSCH_2CH_2Cl$, Et_3N, 65°, 12 h, 50–70% yield).[1] It is cleaved by oxidation [H_2O_2, $(NH_4)_6Mo_7O_{24}$, acetone, 25°, 2 h, 80–95% yield → pH 10–11, 25°, 12–24 h, 85–95% yield][2,3] and by alkylation followed by hydrolysis (MeI, 70–95% yield → pH 10, 5–10 min, 70–95% yield).[1]

1. M. J. S. A. Amaral, G. C. Barrett, H. N. Rydon, and J. E. Willet, *J. Chem. Soc. C*, 807 (1966).
2. P. M. Hardy, H. N. Rydon, and R. C. Thompson, *Tetrahedron Lett.*, 2525 (1968).
3. S. Inoue, K. Okada, H. Tanino, K. Hashizume, and H. Kakoi, *Tetrahedron*, **50**, 2729 (1994).

1,3-Dithianyl-2-methyl (Dim) Ester: (1,3-dithian-2-yl)$-CH_2O_2CR$

The Dim ester was developed for the protection of the carboxyl function during peptide synthesis. It is prepared by transesterification of amino acid methyl esters with 2-(hydroxymethyl)-1,3-dithiane and Al(*i*-PrO)$_3$ (reflux, 4 h, 75°, 12 torr, 75% yield). It is removed by oxidation [H_2O_2, $(NH_4)_2MoO_4$; pH 8, H_2O, 60 min, 83% yield]. Since it must be removed by oxidation it is not compatible with sulfur-containing amino acids such as cysteine and methionine. Its suitability for other, easily oxidized amino acids (e.g., tyrosine and tryptophan) must also be questioned. The Dim ester is stable to CF_3CO_2H and HCl/ether.[1,2]

1. H. Kunz and H. Waldmann, *Angew. Chem., Int. Ed. Engl.*, **22**, 62 (1983).
2. H. Waldmann and H. Kunz, *J. Org. Chem.*, **53**, 4172 (1988).

2-(*p*-Nitrophenylsulfenyl)ethyl Ester: $RCO_2CH_2CH_2SC_6H_4$-*p*-NO_2

This ester is similar to the 2-methylthioethyl ester in that it is prepared from a thioethyl alcohol and cleaved by oxidation [H_2O_2, $(NH_4)_6Mo_7O_{24}$].[1]

1. M. J. S. A. Amaral, *J. Chem. Soc. C*, 2495 (1969).

2-(*p*-Toluenesulfonyl)ethyl (Tse) Ester: $RCO_2CH_2CH_2SO_2C_6H_4$–*p*-CH_3 (Chart 6)

Formation

1. $TsCH_2CH_2OH$, DCC, Pyr, 0°, 1 h → 20°, 16 h, 70–90% yield.[1]

Cleavage

1. Na_2CO_3, dioxane, H_2O, 20°, 2 h, 95% yield.[1]
2. 1 *N* NaOH, dioxane, H_2O, 20°, 3 min, 60–95% yield.[1]
3. KCN, dioxane, H_2O, 20°, 2.5 h, 60–85% yield.[1]
4. DBN, benzene, 25°, quant.[2]
5. DBU, benzene, 11 h, 100% yield.[3]

O O OH $CO_2CH_2CH_2SO_2C_6H_4CH_3$ O O —DBU, benzene / 11 h, rt, 100%→ CO_2H

6. $Bu_4N^+F^-$, THF, 0°, 1 h, 52–95% yield.[4] A primary alcohol protected as the *t*-butyldimethylsilyl ether was cleaved under these conditions, but a similarly protected secondary alcohol was stable.

1. A. W. Miller and C. J. M. Stirling, *J. Chem. Soc. C*, 2612 (1968).
2. E. W. Colvin, T. A. Purcell, and R. A. Raphael, *J. Chem. Soc., Chem. Commun.*, 1031 (1972).
3. H. Tsutsui and O. Mitsunobo, *Tetrahedron Lett.*, **25**, 2163 (1984).
4. H. Tsutsui, M. Muto, K. Motoyoshi, and O. Mitsunobo, *Chem. Lett.*, 1595 (1987).

2-(2′-Pyridyl)ethyl (Pet) Ester: $RCO_2CH_2CH_2$–2-C_5H_4N

Formation

1. DCC, HOBt, $HOCH_2CH_2$–2-C_5H_4N, 0° → rt, CH_2Cl_2 or DMF, overnight, 50–92% yield.[1,2]
2. DCC, DMAP, $HOCH_2CH_2$–2-C_5H_4N, CH_2Cl_2, 61–92% yield.[3]
3. The related **2-(4′-pyridyl)ethyl** ester has also been prepared from the acid chloride and the alcohol.[4]

Cleavage

1. MeI, CH_3CN; morpholine or diethylamine, methanol, 76–95% yield.[1,3] These conditions also cleave the 4′-pyridyl derivative.[4]

The Pet ester is stable to the acidic conditions required to remove the BOC and *t*-butyl ester groups, to the basic conditions required to remove the Fmoc and Fm groups, and to hydrogenolysis. It is not recommended for use in peptides that contain methionine or histidine, since these are susceptible to alkylation with methyl iodide.

1. H. Kessler, G. Becker, H. Kogler, and M. Wolff, *Tetrahedron Lett.*, **25**, 3971 (1984).
2. H. Kessler, G. Becker, H. Kogler, J. Friesse, and R. Kerssebaum, *Int. J. Pept. Protein Res.*, **28**, 342 (1986).
3. H. Kunz and M. Kneip, *Angew. Chem., Int. Ed. Engl.*, **23**, 716 (1984).
4. A. R. Katritsky, G. R. Khan, and O. A. Schwarz, *Tetrahedron Lett.*, **25**, 1223 (1984).

2-(*p*-Methoxyphenyl)ethyl Ester: *p*-$CH_3OC_6H_4CH_2CH_2O_2CR$

Formation of this ester proceeds under standard DCC coupling conditions (DMAP, THF, 28–93%). The ester is cleaved with 1% TFA[1] or dichloroacetic acid in CH_2Cl_2 by DDQ (reflux, CH_2Cl_2, H_2O, 5–15 h, 47–92% yield).[2] Hydrogenolysis (Pd/C, EtOAc, MeOH) cleaves the ester in 23 h, whereas a benzyl ester is cleaved in 10 min under these conditions.[1]

1. M. S. Bernatowicz, H.-G. Chao, and G. R. Matsueda, *Tetrahedron Lett.*, **35**, 1651 (1994).
2. S.-E. Yoo, H. R. Kim, and K. Y. Yi, *Tetrahedron Lett.*, **31**, 5913 (1990).

2-(Diphenylphosphino)ethyl (Dppe) Ester: $(C_6H_5)_2PCH_2CH_2O_2CR$

The Dppe group was developed for carboxyl protection in peptide synthesis. It is formed from an *N*-protected amino acid and the alcohol (DCC, DMAP, 3–12 h, 0°, rt). It is most efficiently cleaved by quaternization with MeI, followed by treatment with fluoride ion or K_2CO_3. The ester is stable to HBr/AcOH, $BF_3{\cdot}Et_2O$, and CF_3CO_2H.[1]

1. D. Chantreux, J.-P. Gamet, R. Jacquier, and J. Verducci, *Tetrahedron*, **40**, 3087 (1984).

1-Methyl-1-phenylethyl (Cumyl) Ester: $RCO_2C(CH_3)_2C_6H_5$

Formation

1. $C_6H_5C(CH_3)_2OC(=NH)CCl_3$, $BF_3{\cdot}Et_2O$, CH_2Cl_2, cHex, 82% yield.[1,2]

Cleavage

1. TFA/CH_2Cl_2, rt, 15 min, 86% yield. BOC and *t*-BuO groups were stable.[1]

2. 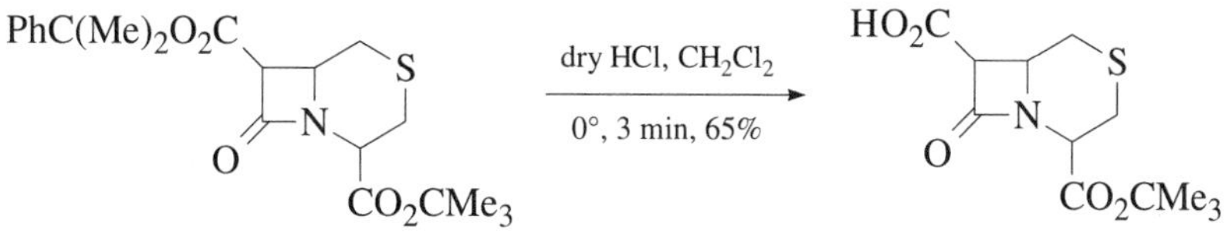 $PhC(Me)_2O_2C$ ... S ... N ... O ... CO_2CMe_3 $\xrightarrow[0°, 3 min, 65\%]{dry HCl, CH_2Cl_2}$ HO_2C ... S ... N ... O ... CO_2CMe_3

Note that a cumyl ester can be selectively cleaved in the presence of a *t*-butyl ester and a *β*-lactam.[3]

1. C. Yue, J. Thierry, and P. Potier, *Tetrahedron Lett.*, **34**, 323 (1993).
2. J. Thierry, C. Yue, and P. Potier, *Tetrahedron Lett.*, **39**, 1557 (1998).
3. D. M. Brunwin and G. Lowe, *J. Chem. Soc., Perkin Trans. 1*, 1321 (1973).

2-(4-Acetyl–2-nitrophenyl)ethyl (Anpe) Ester: RCO_2 ... O ... NO_2

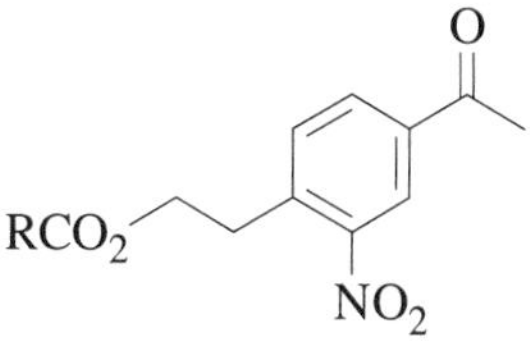

This ester was designed as a base-labile protective group. Monoprotection of aspartic acid was achieved using the DCC/DMAP protocol. Cleavage is

Relative Lability of Aspartic Acid *β*-Carboxyl Protective Groups

Carboxyl Protective Group	Abbreviation	Deprotection time
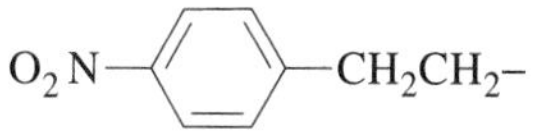 O_2N–C_6H_4–CH_2CH_2–	Npe	1.5–2 h
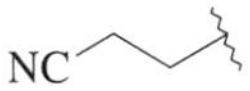 NC (structure)	Cne	45 min
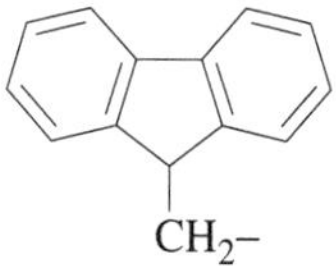 (fluorenyl) CH_2–	Fm	<5 min
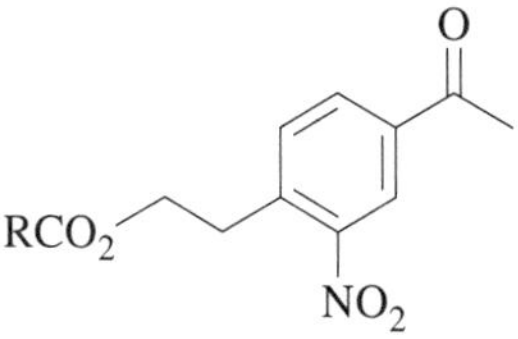 RCO_2 (structure) O, NO_2	Anpe	< 5 min
$O_2NCH_2CH_2$–	Ne	*a*
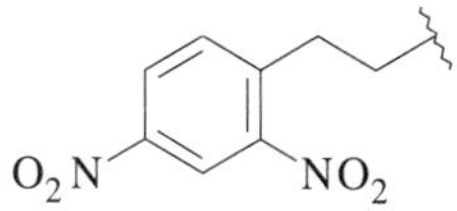 O_2N (structure) NO_2	Dnpe	*a*

a. Not prepared because of a lack of stability.

promoted with 0.1 *M* TBAF. A comparison with other base-labile esters for the β-carboxyl group of aspartic acid to 0.1 *M* TBAF is provided in the preceding table.[1]

1. J. Robles, E. Pedroso, and A. Grandas, *Synthesis*, 1261 (1993).

2-Cyanoethyl Ester: $NCCH_2CH_2O_2CR$

Formation

$HOCH_2CH_2CN$, DCC, DMAP, CH_2Cl_2, 86–97% yield.[1]

Cleavage

1. TBAF, DMF/THF, 64–100% yield. Cleavage occurs in the presence of TMSE and benzyl esters and acetates.[1]
2. K_2CO_3, MeOH, H_2O.[2] Acetates and most other simple esters are cleaved under these conditions.
3. Na_2S, MeOH, 67–91% yield.[3]

1. Y. Kita, H. Maeda, F. Takahashi, S. Fukui, T. Ogawa, and K. Hatayama, *Chem. Pharm. Bull.*, **42**, 147 (1994).
2. P. K. Misra, S. A. N. Hashmi, W. Haq, and S. B. Katti, *Tetrahedron Lett.*, **30**, 3569 (1989).
3. T. Ogawa, K. Hatayama, H. Maeda, and Y. Kita, *Chem. Pharm. Bull.*, **42**, 1579 (1994).

***t*-Butyl Ester:** $RCO_2C(CH_3)_3$ (Chart 6)

Formation

The *t*-butyl ester is a relatively hindered ester, and many of the following methods for its formation should be, and in many cases are, equally effective for the preparation of other hindered esters. The related **1- and 2-adamantyl esters** have been used for the protection of aspartic acid[1] and other amino acids (1-AdOH, toluene, dimethyl sulfate, cat. TsOH, 70–80% yield).[2]

1. Isobutylene, concd. H_2SO_4, Et_2O, 25°, 2–24 h, 50–60% yield.[3] This method works for the preparation of *t*-Bu esters of alkyl acids, amino acids,[4,5] and penicillins.[6]
2. Isobutylene, CH_2Cl_2, H_3PO_4 (P_2O_5), $BF_3 \cdot Et_2O$, $-78°$, 2 h, $\rightarrow$ 0°, 24 h.[7]

3. $(COCl)_2$, benzene, DMF, 7–10°, 45 min; *t*-BuOH, Et_3N, CH_2Cl_2, 0°, 3 h, 75% yield.[8]
4. From an aromatic acid chloride: LiO–*t*-Bu, 25°, 15 h, 79–82% yield.[9]
5. 2,4,6-$Cl_3C_6H_2COCl$, Et_3N, THF; *t*-BuOH, DMAP, benzene, 25°, 20 min, 90% yield.[10]
6. *t*-BuOH, Pyr, $(Me_2N)(Cl)C{=}N^+Me_2Cl^-$, 77% yield.[11] This method is also effective for the preparation of other esters.
7. $(Im)_2CO$ (*N,N′*-carbonyldiimidazole), *t*-BuOH, DBU, 54–91% yield.[12]
8. Bu_3PI_2, Et_2O, HMPA; *t*-BuOH, 73% yield.[13]
9. *t*-BuOH, EDCI (EDCI = 1-ethyl-3-[3-(dimethylamino)propyl]carbodiimide hydrochloride, DMAP, CH_2Cl_2, 88% yield.[14] Cbz-Proline was protected without racemization.
10. *i*-PrN=C(O–*t*-Bu)NH–*i*-Pr, toluene, 60°, 4 h, 90% yield.[15]

O–*t*-Bu
i-Pr–N=C–NH–*i*-Pr
toluene, 60°, 4 h
90% yield

R = H → R = *t*-Bu

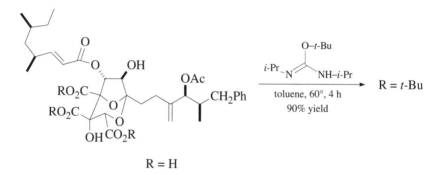

11. $Cl_3C(t\text{-BuO})C{=}NH$, $BF_3{\cdot}Et_2O$, CH_2Cl_2, cyclohexane, 70–92% yield.[16] This reagent also forms *t*-butyl ethers from alcohols.
12. $(t\text{-BuO})_2CHNMe_2$, toluene, 80°, 30 min, 82% yield.[17,18]
13. From an acid chloride: *t*-BuOH, AgCN, benzene, 20–80°, 60–100% yield.[19] Alumina also promotes the conversion of an acid chloride to a *t*-Bu ester in 79–96% yield.[20]
14. 2-Cl-3,5-$(NO_2)_2C_5H_2N$, Pyr, rt → 115°, *t*-BuOH.[21] Other esters are also prepared effectively using this methodology.
15. *t*-BuOCOF, Et_3N, DMAP, CH_2Cl_2, *t*-BuOH, rt, 82–96% yield.[22]
16. $(BOC)_2O$, *t*-BuOH or THF, DMAP, 99% yield. This methodology is effective for the preparation of allyl, methyl, ethyl, and benzyl esters as well.[23]
17. *t*-BuBr, K_2CO_3, BTEAC, DMAC, 55°, 72–100% yield.[24]
18. By transesterification of a methyl ester with *t*-BuOH and sulfated SiO_2.[25]

19.

TrNH–CH(CH₂Ph)–CO_2H → (BOP, TEA, CH_2Cl_2, 80%) → TrNH–CH(CH₂Ph)–CO_2Bt ⇌ TrNH–CH(CH₂Ph)–C(O)–N^+ (benzotriazole N-oxide, O^-)

Thermodynamically favored

→ (t-BuOK, THF, 0°, 90% yield) → TrNH–CH(CH₂Ph)–CO_2t-Bu

Ref. 26

Cleavage

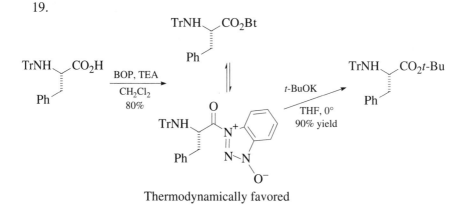

t-Butyl esters are stable to mild basic hydrolysis, to hydrazine, and to ammonia; they are cleaved by moderately acidic hydrolysis, with the release of the t-Bu cation that often must be scavenged to prevent side reactions.

1. HCO_2H, 20°, 3 h.[27]
2. CF_3COOH, CH_2Cl_2, 25°, 1 h.[28] The addition of Et_3SiH to the deprotection step improves the yields over that obtained with the normal cation scavengers.[29]
3. AcOH, HBr, 10°, 10 min, 70% yield.[4] Phthaloyl or trifluoroacetyl groups on amino acids are stable to these conditions; benzyloxycarbonyl (Cbz) or t-butoxycarbonyl (BOC) groups are cleaved.
4. HCl, AcOH, CH_2Cl_2, 5°, 2 h. A t-butyl ether and an Fmoc group were not affected.[30]
5. TsOH, benzene, reflux, 30 min, 76% yield.[4] A t-butyl ester is stable to the conditions needed to convert an α,β-unsaturated ketone to a dioxolane ($HOCH_2CH_2OH$, TsOH, benzene, reflux).[31]
6. KOH, 18-crown-6, toluene, 100°, 5 h, 94% yield.[32] These conditions were used to cleave the t-butyl ester from an aromatic ester; they are probably too harsh to be used on more highly functionalized substrates.
7. 190–200°, 15 min, 100% yield.[33] A thermolysis in quinoline was found advantageous when acid-catalyzed cleavage resulted in partial debenzylation of a phenol.[34] Thermolytic conditions also cleave the BOC group from amines.
8. Bromocatecholborane.[35] Ethyl esters are not affected by this reagent, but it does cleave other groups; see the section on methoxymethyl (MOM) ethers.
9. TMSOTf, TEA, 53–90% yield. t-Butyl esters are cleaved in preference to t-butyl ethers.[36]

10. MgI_2, toluene, 46–111°, 1–3 days, 41–96% yield.[37]
11. Thermitase, pH 7.5, 45°, 20% DMF, 70–89% yield.[38]
12. Pig liver esterase.[39]
13. LiI, EtOAc, reflux,[40]
14. $TiCl_4$, CH_2Cl_2, −10° to 0°, 54–91% yield. These conditions were developed for use with cephalosporin t-butyl esters.[41]

1. Y. Okada and S. Iguchi, *J. Chem. Soc., Perkin Trans. 1*, 2129 (1988).
2. S. M. Iossifidou and C. C. Froussios, *Synthesis*, 1355 (1996).
3. A. L. McCloskey, G. S. Fonken, R. W. Kluiber, and W. S. Johnson, *Org. Synth., Collect. Vol. IV*, 261 (1963).
4. G. W. Anderson and F. M. Callahan, *J. Am. Chem. Soc.*, **82**, 3359 (1960).
5. R. M. Valerio, P. F. Alewood, and R. B. Johns, *Synthesis*, 786 (1988).
6. R. J. Stedman, *J. Med. Chem.*, **9**, 444 (1966).
7. C.-Q. Han, D. DiTullio, Y.-F. Wang, and C. J. Sih, *J. Org. Chem.*, **51**, 1253 (1986).
8. C. F. Murphy and R. E. Koehler, *J. Org. Chem.*, **35**, 2429 (1970).
9. G. P. Crowther, E. M. Kaiser, R. A. Woodruff, and C. R. Hauser, *Org. Synth., Collect. Vol. VI*, 259 (1988).
10. J. Inanaga, K. Hirata, H. Saeki, T. Katsuki, and M. Yamaguchi, *Bull. Chem. Soc. Jpn.*, **52**, 1989 (1979).
11. T. Fujisawa, T. Mori, K. Fukumoto, and T. Sato, *Chem. Lett.*, 1891 (1982).
12. S. Ohta, A. Shimabayashi, M. Aona, and M. Okamoto, *Synthesis*, 833 (1982).
13. R. K. Haynes and M. Holden, *Aust. J. Chem.*, **35**, 517 (1982).
14. M. K. Dhaon, R. K. Olsen, and K. Ramasamy, *J. Org. Chem.*, **47**, 1962 (1982).
15. R. M. Burk, G. D. Berger, R. L. Bugianesi, N. N. Girotra, W. H. Parsons, and M. M. Ponpipom, *Tetrahedron Lett.*, **34**, 975 (1993).
16. A. Armstrong, I. Brackenridge, R. F. W. Jackson, and J. M. Kirk, *Tetrahedron Lett.*, **29**, 2483 (1988).
17. U. Widmer, *Synthesis*, 135 (1983).
18. J. Deng, Y. Hamada, and T. Shioiri, *J. Am. Chem. Soc.*, **117**, 7824 (1995).
19. S. Takimoto, J. Inanaga, T. Katsuki, and M. Yamaguchi, *Bull. Chem. Soc. Jpn.*, **49**, 2335 (1976).
20. K. Nagasawa, S. Yoshitake, T. Amiya, and K. Ito, *Synth. Commun.*, **20**, 2033 (1990); K. Nagasawa, K. Ohhashi, A. Yamashita, and K. Ito, *Chem. Lett.*, 209 (1994).
21. S. Takimoto, N. Abe, Y. Kodera, and H. Ohta, *Bull. Chem. Soc. Jpn.*, **56**, 639 (1983).
22. A. Loffet, N. Galeotti, P. Jouin, and B. Castro, *Tetrahedron Lett.*, **30**, 6859 (1989).
23. K. Takeda, A. Akiyama, H. Nakamura, S.-i. Takizawa, Y. Mizuno, H. Takayanagi, and Y. Harigaya, *Synthesis*, 1063 (1994).
24. P. Chevallet, P. Garrouste, B. Malawska, and J. Martinez, *Tetrahedron Lett.*, **34**, 7409 (1993).
25. S. P. Chavan, P. K. Zubaidha, S. W. Dantale, A. Keshavaraja, A. V. Ramaswamy, and T. Ravindranathan, *Tetrahedron Lett.*, **37**, 233 (1996).

26. K. M. Sliedregt, A. Schouten, J. Kroon, and R. M. J. Liskamp, *Tetrahedron Lett.*, **37**, 4237 (1996).
27. S. Chandrasekaran, A. F. Kluge, and J. A. Edwards, *J. Org. Chem.*, **42**, 3972 (1977).
28. D. B. Bryan, R. F. Hall, K. G. Holden, W. F. Huffman, and J. G. Gleason, *J. Am. Chem. Soc.*, **99**, 2353 (1977).
29. A. Mehta, R. Jaouhari, T. J. Benson, and K. T. Douglas, *Tetrahedron Lett.*, **33**, 5441 (1992).
30. G. M. Makara and G. R. Marshall, *Tetrahedron Lett.*, **38**, 5069 (1997).
31. A. Martel, T. W. Doyle, and B.-Y. Luh, *Can. J. Chem.*, **57**, 614 (1979).
32. C. J. Pedersen, *J. Am. Chem. Soc.*, **89**, 7017 (1967).
33. L. H. Klemm, E. P. Antoniades, and D. C. Lind, *J. Org. Chem.*, **27**, 519 (1962).
34. J. W. Lampe, P. F. Hughes, C. K. Biggers, S. H. Smith, and H. Hu, *J. Org. Chem.*, **61**, 4572 (1996).
35. R. K. Boeckman, Jr., and J. C. Potenza, *Tetrahedron Lett.*, **26**, 1411 (1985).
36. A. Trzeciak and W. Bannwarth, *Synthesis*, 1433 (1996).
37. A. G. Martinez, J. O. Bardina, G. H. del Veccio, M. Hanack, and L. R. Subramanian, *Tetrahedron Lett.*, **32**, 5931 (1991).
38. M. Schultz, P. Hermann, and H. Kunz, *Synlett*, 37 (1992).
39. K. A. Stein and P. L. Toogood, *J. Org. Chem.*, **60**, 8110 (1995).
40. J. W. Fisher and K. L. Trinkle, *Tetrahedron Lett.*, **35**, 2505 (1994).
41. M. Valencic, T. van der Does, and E. de Vroom, *Tetrahedron Lett.*, **39**, 1625 (1998).

3-Methyl-3-pentyl (Mpe) Ester: $(C_2H_5)_2CCH_3O_2CR$

This tertiary ester was developed to reduce aspartimide and piperidide formation during the Fmoc-based peptide synthesis by increasing the steric bulk around the carboxyl carbon. A twofold improvement was achieved over the the standard *t*-butyl ester. The Mpe ester is prepared from the acid chloride and the alcohol and can be cleaved under conditions similar to those used for the *t*-butyl ester.[1]

1. A. Karlström and A. Undén, *Tetrahedron Lett.*, **37**, 4243 (1996).

Dicyclopropylmethyl (Dcpm) Ester: RCO_2–CH(c-C_3H_5)$_2$

The Dcpm group can be removed in the presence of a *t*-butyl or *N*-trityl group with 1% TFA in CH_2Cl_2.[1]

1. L. A. Carpino, H.-G. Chao, S. Ghassemi, E. M. E. Mansour, C. Riemer, R. Warrass, D. Sadat-Aalaee, G. A. Truran, H. Imazumi, A. El-Faham, D. Ionescu, M. Ismail, T. L. Kowaleski , C. H. Han, H. Wenschuh, M. Beyermann, M. Bienert, H. Shroff, F. Albericio, S. A. Triolo, N. A. Sole, and S. A. Kates, *J. Org. Chem.*, **60**, 7718 (1995).

2,4-Dimethyl-3-pentyl (Dmp) Ester: $(i\text{-Pr})_2CHO_2CR$

Formation

1. 2,4-Dimethyl-3-pentanol, DCC, DMAP, CH_2Cl_2, 4 h. This group was developed as an improvement over the use of cylcohexanol for aspartic acid protection during peptide synthesis.[1]

Cleavage

Cleavage is effected with acid. The following table compares the acidolysis rates with Bn and cyclohexyl esters in TFA/phenol at 43°. The Dmp group reduces aspartimide formation during Fmoc-based peptide synthesis.

Protective Group	$t_{1/2}$ (h)
Bn	6
Dmp	40
cHex	500

1. A. H. Karlström and A. E. Unden, *Tetrahedron Lett.*, **36**, 3909 (1995).

Cyclopentyl Ester: $RCO_2\text{–}c\text{-}C_5H_9$

Cyclohexyl Ester: $RCO_2\text{–}c\text{-}C_6H_{11}$

Cycloalkyl esters have been used to protect the β-CO_2H group in aspartyl peptides to minimize aspartimide formation during acidic or basic reactions.[1] Aspartimide formation is limited to 2–3% in TFA (20 h, 25°), 5–7% with HF at 0°, and 1.5–4% TfOH (thioanisole in TFA). Cycloalkyl esters are also stable to Et_3N, whereas use of the benzyl ester leads to 25% aspartimide formation during Et_3N treatment. Cycloalkyl esters are stable to CF_3COOH, but are readily cleaved with HF or TfOH.[2–4]

1. For an improved synthesis of cyclohexyl aspartate, see G. K. Toth and B. Penke, *Synthesis*, 361 (1992).
2. J. Blake, *Int. J. Pept. Protein Res.*, **13**, 418 (1979).
3. J. P. Tam, T.-W. Wong, M. W. Riemen, F.-S. Tjoeng, and R. B. Merrifield, *Tetrahedron Lett.*, 4033 (1979).
4. N. Fujii, M. Nomizu, S. Futaki, A. Otaka, S. Funakoshi, K. Akaji, K. Watanabe, and H. Yajima, *Chem. Pharm. Bull.*, **34**, 864 (1986).

Allyl Ester: $RCO_2CH_2CH{=}CH_2$

The use of various allyl protective groups in complex molecule synthesis has been reviewed.[1]

Formation

1. Allyl bromide, Aliquat 336, $NaHCO_3$, CH_2Cl_2, 83% yield.[2] The carboxylic acid group of Z-serine (Z = Cbz = benzyloxycarbonyl) is selectively esterified without affecting the alcohol.
2. R′R″C=$CHCH_2OH$, NaH, THF, 1–3 days, 80–95% yield.[3] A methyl ester is exchanged for an allyl ester under these conditions.
3. Allyl bromide, Cs_2CO_3, DMF, 84% yield.[4]
4. Allyl alcohol, TsOH, benzene, $-H_2O$.[5] These conditions were used to prepare esters of amino acids.
5. By transesterification of an ethyl ester: AllylOH, DBU, LiBr, 0°, 12 h, >54% yield.[6]
6. $AllylOCO_2CO_2$allyl, THF, DMAP.[7]
7. 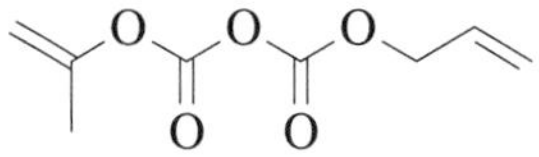 DMAP, 81–100% yield.[8]
8. AllylOC=NH(CCl_3), $BF_3{\cdot}Et_2O$, CH_2Cl_2, cyclohexane, 67–96% yield.[9]
9. Vinyldiazomethane, CH_2Cl_2, 80–92% yield.[10]
10. From the Oppolzer sultam by exchange: AllylOH, $Ti(OR)_4$, 67–95% yield.[11]

Cleavage

1. $Pd(OAc)_2$, sodium 2-methylhexanoate, Ph_3P, acetone.[12] Triethyl phosphite could be used as the ligand for palladium.[13]
2. $(Ph_3P)_3RhCl$ or $Pd(Ph_3P)_4$, 70°, EtOH, H_2O, 91% yield.[14]
3. $Pd(Ph_3P)_4$, pyrrolidine, 0°, 5–15 min, CH_3CN, 70–90% yield.[15] Morpholine has also been used as an allyl scavenger in this process.[2,4] Allylamines are not affected by these conditions.[16]
4. $PdCl_2(Ph_3P)_2$, dimedone, THF, 95% yield.[17] This method is also effective for removing the allyloxycarbonyl group from alcohols and amines.
5. $Pd(Ph_3P)_4$, 2-ethylhexanoic acid[18] or barbituric acid (THF, 3 h, 93% yield).[19] Tributylstannane can serve as an allyl scavenger.[20]
6. Me_2CuLi, Et_2O, 0°, 1 h; H_3O^+, 75–85% yield.[21]
7. $PhSiH_3$, $Pd(Ph_3P)_4$, CH_2Cl_2, 74–100% yield.[22] $CF_3CON(SiMe_3)CH_3$ was also used to scavenge the allyl group from the Alloc and allyl ether protected deriviatives.
8. $Pd(Ph_3P)_4$, $BnONH_2$, CH_2Cl_2, 80% yield.[23]
9. $Pd(OAc)_2$, Ph_3P, TEA, HCO_2H, dioxane, 96% yield.[24,25]
10. Papain, DTT, DMF.[26]

11. $TiCl_4$, Mg–Hg, THF, 40–70% yield.[27] Benzyl esters are also cleaved.
12. $Pd(Ph_3P)_4$, RSO_2Na, CH_2Cl_2 or THF/MeOH, 70–99% yield. These conditions were shown to be superior to the use of sodium 2-ethylhexanoate. Methallyl, allyl, crotyl, and cinnamyl ethers, the Alloc group, and allylamines are all efficiently cleaved by this method.[28]

Methallyl Ester: $CH_2{=}C(CH_3)CH_2O_2CR$

Cleavage of the methallyl ester is achieved in 80–95% yield by solvolysis in refluxing 90% formic acid. Cinnamyl and crotyl alcohols are similarly cleaved.[29]

2-Methylbut-3-en-2-yl Ester: $CH_2{=}CHC(CH_3)_2O_2CR$

This ester is cleaved with $Pd(OAc)_2$, Ph_3P, $Et_3N–H_2CO_2H$, rt, 30 min.[30]

3-Methylbut-2-enyl (Prenyl) Ester: $(CH_3)_2C{=}CHCH_2O_2CR$

Cleavage

1. I_2 in cyclohexane, rt, 75–97% yield.[31]
2. $Pd(OAc)_2$, TPPTS, CH_3CN, H_2O, 75 min, 100% yield. The Alloc group is readily released in the presence of this ester.[32]

3-Buten-1-yl Ester: $CH_2{=}CHCH_2CH_2O_2CR$

This ester, formed from the acid ($COCl_2$, toluene; then $CH_2{=}CHCH_2CH_2OH$, acetone, −78° warm to rt, 70–94% yield), can be cleaved by ozonolysis followed by Et_3N or DBU treatment (79–99% yield). The ester is suitable for the protection of enolizable and base-sensitive carboxylic acids.[33]

4-(Trimethylsilyl)-2-buten-1-yl Ester: $RCO_2CH_2CH{=}CHCH_2Si(CH_3)_3$

This ester is formed by standard procedures and is readily cleaved with $Pd(Ph_3P)_4$ in CH_2Cl_2 to form trimethylsilyl esters that readily hydrolyze on treatment with water or alcohol or on chromatography on silica gel (73–98% yield). Amines can be protected using the related carbamate.[34]

Cinnamyl Ester: $RCO_2CH_2CH{=}CHC_6H_5$ (Chart 6)

The cinnamyl ester can be prepared from an activated carboxylic acid derivative and cinnamyl alcohol or by transesterification with cinnamyl alcohol in the presence of the H-Beta Zeolite (toluene, reflux, 8 h, 59–96% yield).[35] It is cleaved under nearly neutral conditions [$Hg(OAc)_2$, MeOH, 23°, 2–4 h; KSCN, H_2O, 23°, 12–16 h, 90% yield][36] or by treatment with Sulfated-SnO_2, toluene, anisole, reflux.[37] The latter conditions also cleave crotyl and prenyl esters.

α-Methylcinnamyl (MEC) Ester: $RCO_2CH(CH_3)CH{=}CHC_6H_5$

Formation

1. $PhCH{=}CHCH(CH_3)OH$, DCC, DMAP, THF, 98% yield.[38]
2. From an acid chloride: $PhCH{=}CHCH(CH_3)OH$, Pyr, DMAP, 75–88% yield.[38]

Cleavage

$Me_2Sn(SMe)_2$, $BF_3{\cdot}Et_2O$, $PhCH_3$, 0°, 3–24 h; AcOH, 75–100% yield.[33,38] An ethyl ester can be hydrolyzed in the presence of an MEC ester with 1 *N* aqueous NaOH–DMSO (1 : 1), and MEC esters can be cleaved in the presence of ethyl, benzyl, cinnamyl, and *t*-butyl esters, as well as the acetate, TBDMS, and MEM groups.

1. F. Guibé, *Tetrahedron*, **54**, 2967–3041 (1998).
2. S. Friedrich-Bochnitschek, H. Waldmann, and H. Kunz, *J. Org. Chem.*, **54**, 751 (1989).
3. N. Engel, B. Kübel, and W. Steglich, *Angew. Chem., Int. Ed. Engl.*, **16**, 394 (1977).
4. H. Kunz, H. Waldmann, and C. Unverzagt, *Int. J. Pept. Protein Res.*, **26**, 493 (1985).
5. H. Waldmann and H. Kunz, *Liebigs Ann. Chem.*, 1712 (1983).
6. M. J. I. Andrews and A. B. Tabor, *Tetrahedron Lett.*, **38**, 3063 (1997); D. Seebach, A. Thaler, D. Blaser, and S. Y. Ko, *Helv. Chim. Acta*, **74**, 1102 (1991).
7. K. Takeda, A. Akiyama, H. Nakamura, S.-i. Takizawa, Y. Mizuno, H. Takayamagi, and Y. Harigaya, *Synthesis*, 1063 (1994).
8. K. Takeda, A. Akiyama, Y. Konda, H. Takayanagi, and Y. Harigaya, *Tetrahedron Lett.*, **36**, 113 (1995).
9. G. Kokotos and A. Chiou, *Synthesis*, 168 (1997).
10. S. T. Waddell and G. M. Santorelli, *Tetrahedron Lett.*, **37**, 1971 (1996).
11. W. Oppolzer and P. Lienard, *Helv. Chim. Acta*, **75**, 2572 (1992).
12. L. N. Jungheim, *Tetrahedron Lett.*, **30**, 1889 (1989).
13. M. Seki, K. Kondo, T. Kuroda, T. Yamanaka, and T. Iwasaki, *Synlett*, 609 (1995).
14. H. Kunz and H. Waldmann, *Helv. Chim. Acta*, **68**, 618 (1985).
15. R. Deziel, *Tetrahedron Lett.*, **28**, 4371 (1987).
16. J. E. Bardaji, J. L. Torres, N. Xaus, P. Clapés, X. Jorba, B. G. de la Torre, and G. Valencia, *Synthesis*, 531 (1990).
17. H. X. Zhang, F. Guibé, and G. Balavoine, *Tetrahedron Lett.*, **29**, 623 (1988).
18. P. D. Jeffrey and S. W. McCombie, *J. Org. Chem.*, **47**, 587 (1982).
19. H. Kunz and J. März, *Synlett*, 591 (1992).
20. B. G. de la Torre, J. L. Torres, E. Bardají, P. Clapés, N. Xaus, X. Jorba, S. Calvet, F. Albericio, and G. Valencia, *J. Chem. Soc., Chem. Commun.*, 965 (1990).
21. T.-L. Ho, *Synth. Commun.*, **8**, 15 (1978).

22. M. Dessolin, M.-G. Guillerez, N. Thieriet, F. Guibé, and A. Loffet, *Tetrahedron Lett.*, **36**, 5741 (1995).
23. B. T. Lotz and M. J. Miller, *J. Org. Chem.*, **58**, 618 (1993).
24. G. Casy, A. G. Sutherland, R. J. K. Taylor, and R. G. Urben, *Synthesis*, 767 (1989).
25. E. J. Corey and S. Choi, *Tetrahedron Lett.*, **34**, 6969 (1993).
26. N. Xaus, P. Clapés, E. Bardají, J. L. Torres, X. Jorba, J. Mata, and G. Valencia, *Tetrahedron*, **45**, 7421 (1989).
27. K. Satyanarayana, N. Chidambaram, and S. Chandrasekaran, *Synth. Commun.*, **19**, 2159 (1989).
28. M. Honda, H. Morita, and I. Nagakura, *J. Org. Chem.*, **62**, 8932 (1997).
29. C. R. Schmid, *Tetrahedron Lett.*, **33**, 757 (1992).
30. M. Yamaguchi, T. Okuma, A. Horiguchi, C. Ikeura, and T. Minami, *J. Org. Chem.*, **57**, 1647 (1992).
31. J. Cossy, A. Albouy, M. Scheloske, and D. G. Pardo, *Tetrahedron Lett.*, **35**, 1539 (1994).
32. S. Lemaire-Audoire, M. Savignac, E. Blart, G. Pourcelot, J. P. Genét, and J-M. Bernard, *Tetrahedron Lett.*, **35**, 8783 (1994).
33. A. G. M. Barrett, S. A. Lebold, and X.-an Zhang, *Tetrahedron Lett.*, **30**, 7317 (1989).
34. H. Mastalerz, *J. Org. Chem.*, **49**, 4092 (1984).
35. B. S. Balaji, M. Sasidharan, R. Kumar, and B. Chandra, *J. Chem. Soc., Chem. Commun.*, 707 (1996).
36. E. J. Corey and M. A. Tius, *Tetrahedron Lett.*, 2081 (1977).
37. S. P. Chavan, P. K. Zubaidha, S. W. Dantale, A. Keshavaraja, A. V. Ramaswamy, and T. Ravindranathan, *Tetrahedron Lett.*, **37**, 237 (1996).
38. T. Sato, J. Otera, and H. Nozaki, *Tetrahedron Lett.*, **30**, 2959 (1989).

Prop-2-ynyl (Propargyl) Ester: $HC{\equiv}CCH_2O_2CR$

Cleavage

1. Benzyltriethylammonium tetrathiomolybdate in CH_3CN in 61–97% yield. Deprotection is compatible with esters such as benzyl, allyl, acetate, and *t*-butyl esters.[1]
2. $Pd(Ph_3P)_2Cl_2$ (Bu_3SnH, benzene)[2] or cobalt carbonyl.[3] The palladium method cleaves allyl esters, propargyl phosphates, and propargyl carbamates as well.
3. SmI_2.[4,5]
4. Hydrogenolysis.[6]
5. Electrolysis, Ni(II), Mg anode, DMF, rt, 77–99% yield. This method is not compatible with halogenated phenols because of competing halogen cleavage.[7]

1. P. Ilankumaran, N. Manojans, and S. Chandrasekaran, *J. Chem. Soc., Chem. Commun.*, 1957 (1996).

2. H. X. Zhang, F. Guibé, and G. Balavoine, *Tetrahedron Lett.*, **29**, 619, 623 (1988).
3. B. Alcaide, J. Perez-Castels, B. Sanchez-Vigo, and M. A. Sierra, *J. Chem. Soc., Chem. Commun.*, 587 (1994).
4. J. Inanaga, Y. Sugimoto, and T. Hanamoto, *Tetrahedron Lett.*, **33**, 7035 (1992).
5. J. M. Aurrecoechea and R. F.-S. Anton, *J. Org. Chem.*, **59**, 702 (1994).
6. J. Tsuji and T. Mandai, *Synthesis*, 1 (1996); J. Tsuji and T. Mandai, *Angew. Chem., Int. Ed. Engl.*, **34**, 2589 (1995).
7. S. Olivero and E. Duñach, *Tetrahedron Lett.*, **38**, 6193 (1997).

Phenyl Ester: $RCO_2C_6H_5$

Phenyl esters can be prepared from *N*-protected amino acids (PhOH, DCC, CH_2Cl_2, −20°→ 20°, 12 h, 86% yield[1]; PhOH, BOP, Et_3N, CH_2Cl_2, 25°, 2 h, 73–97% yield[2]). Phenyl esters are readily cleaved under basic conditions (H_2O_2, H_2O, DMF, pH 10.5, 20°, 15 min).[3]

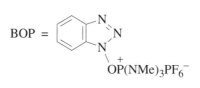

BOP = [benzotriazol-1-yl] $OP^+(NMe)_3PF_6^-$

1. I. J. Galpin, P. M. Hardy, G. W. Kenner, J. R. McDermott, R. Ramage, J. H. Seely, and R.G. Tyson, *Tetrahedron*, **35**, 2577 (1979).
2. B. Castro, G. Evin, C. Selve, and R. Seyer, *Synthesis*, 413 (1977).
3. G. W. Kenner and J. H Seely, *J. Am. Chem. Soc.*, **94**, 3259 (1972).

2,6-Dialkylphenyl Esters

2,6-Dimethylphenyl Ester

2,6-Diisopropylphenyl Ester

2,6-Di-*t*-butyl-4-methylphenyl Ester

2,6-Di-*t*-butyl-4-methoxyphenyl Ester

These esters were prepared from the phenol and the acid chloride, plus DMAP (or from the acid plus trifluoroacetic anhydride). Although the diisopropyl derivative can be cleaved with hot aqueous NaOH, the di-*t*-butyl derivatives could only be cleaved with NaOMe in a mixture of toluene and HMPA.[1] The related 2,6-di-*t*-butyl-4-methoxyphenyl ester can be cleaved oxidatively with ceric ammonium nitrate.[2] These hindered esters have been used to direct the aldol condensation.[3]

1. T. Hattori, T. Suzuki, N. Hayashizaka, N. Koike, and S. Miyano, *Bull. Chem. Soc. Jpn.*, **66**, 3034 (1993).

2. M. P. Cooke, Jr., *J. Org. Chem.*, **51**, 1637 (1986); C. H. Heathcock, M. C. Pirrung, S. H. Montgomery, and J. Lampe, *Tetrahedron,* **37**, 4087 (1981).
3. C. H. Heathcock, in *Asymmetric Synthesis*, J. D. Morrison, Ed., Academic, New York, 1984, Vol. 3, pp. 111–212; D. A. Evans, J. V. Nelson, and T. R. Tabor, in *Top. Stereochem.*, N. L. Allinger, E. L. Eliel, and S. H. Wilen, Eds., Wiley Interscience, New York, **13**, 1 (1982); C. H. Heathcock, in *Comprehensive Organic Synthesis*, B. M. Trost and I. Fleming, Eds., Pergamon, Oxford, U.K. 1991, Vol. 2, pp. 133–238.

p-(Methylthio)phenyl Ester: $RCO_2C_6H_4$-*p*-SCH_3

The *p*-(methylthio)phenyl ester has been prepared from an *N*-protected amino acid and 4-$CH_3SC_6H_4OH$ (DCC, CH_2Cl_2, 0°, 1 h → 20°, 12 h, 60–70% yield). The *p*-(methylthio)phenyl ester serves as an unactivated ester that is activated on oxidation to the sulfone (H_2O_2, AcOH, 20°, 12 h, 60–80% yield), which then serves as an activated ester in peptide synthesis.[1]

1. B. J. Johnson and T. A. Ruettinger, *J. Org. Chem.*, **35**, 255 (1970).

Pentafluorophenyl (Pfp) Ester: $C_6F_5O_2CR$

This active ester was used for carboxyl protection of Fmoc-serine and Fmoc-threonine during glycosylation.[1,2] The esters are then used as active esters in peptide synthesis.

Formation

1. $C_6F_5O_2CCF_3$, Pyr, DMF, rt, 45 min, 92–95% yield.[3] This reagent converts amines to the trifluoroacetamide.[4]
2. C_6F_5OH, DCC, dioxane or EtOAc and DMF, 0°, 1 h then rt, 1 h, 75–99% yield.[5]

1. M. Meldal and K. Bock, *Tetrahedron Lett.*, **31**, 6987 (1990).
2. M. Meldal and K. J. Jensen, *J. Chem. Soc., Chem. Commun.*, 483 (1990).
3. M. Green and J. Berman, *Tetrahedron Lett.*, **31**, 5851 (1990).
4. L. M. Gayo and M. J. Suto, *Tetrahedron Lett.*, **37**, 4915 (1996).
5. L. Kisfaludy and I. Schön, *Synthesis,* 325 (1983); I. Schön and L. Kisfaludy, *ibid.*, 303 (1986).

Benzyl Ester: $RCO_2CH_2C_6H_5$, RCO_2Bn (Chart 6)

Formation

Benzyl esters are readily prepared by many of the classical methods (see the introduction to this chapter), as well as by many newer methods, since benzyl alcohol is unhindered and relatively acid stable.

1. BnOCOCl, Et_3N, 0°, DMAP, CH_2Cl_2, 30 min, 97% yield.[1] In the case of very hindered acids the yields are poor, and formation of the symmetrical anhydride is observed. Useful selectivity can be achieved for a less hindered acid in the presence of a more hindered one.[2]

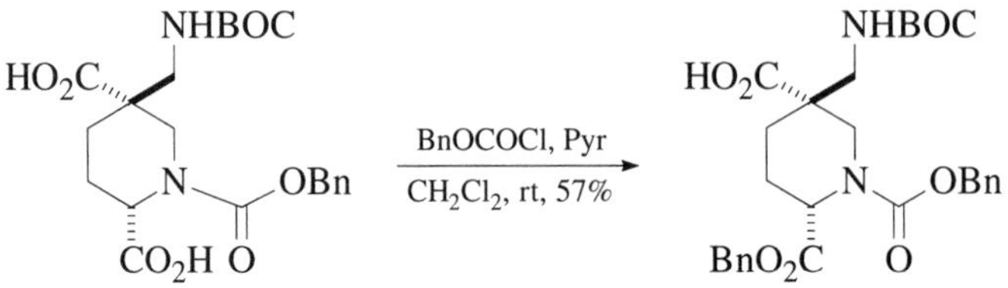

BnOCOCl, Pyr / CH_2Cl_2, rt, 57%

2. A methyl ester can be exchanged thermally (185°, 1.25 h, –MeOH) for a benzyl ester.[3]

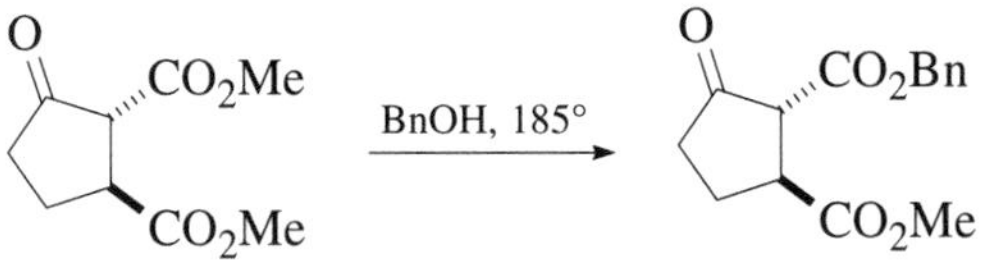

BnOH, 185°

3. For amino acids: DCC, DMAP, BnOH, 92% yield.[4]
4. BnOC=NH(CCl_3), $BF_3 \cdot Et_2O$, CH_2Cl_2, cyclohexane, 60–98% yield.[5,6]
5. 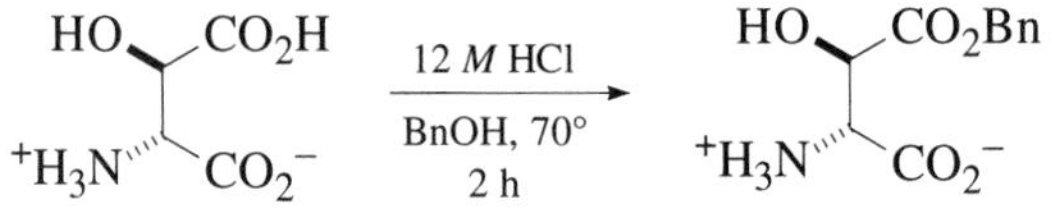 12 *M* HCl / BnOH, 70° / 2 h — Ref. 7
6. $(BnO)_2CHNMe_2$.[8]
7. BnBr, DBU, CH_3CN, 75% yield.[9]
8. BnBr, Cs_2CO_3, CH_3CN, reflux, 93–100% yield.[10] Other esters are prepared similarly.
9. cHexN=C(OBn)NHcHex.[6]
10. From an anhydride, BnOH, Bu_3P, CH_2Cl_2.[11]

Cleavage

The most useful property of benzyl esters is that they are readily cleaved by hydrogenolysis.

1. H_2/Pd–C, 25°, 45 min–24 h, high yields.[12] Catalytic transfer hydrogenation (see entries 2 and 3) can be used to cleave benzyl esters in some compounds that contain sulfur, a poison for hydrogenolysis catalysts.
2. Pd–C, cyclohexene[13] or 1,4-cyclohexadiene,[14] 25°, 1.5–6 h, good yields. Some alkenes,[6] benzyl ethers, BOM groups, and benzylamines[15] are compatible with these conditions.
3. Pd–C, 4.4% HCOOH, MeOH, 25°, 5–10 min in a column, 100% yield.[16]

4. K_2CO_3, H_2O, THF, 0° → 25°, 1 h, 75% yield.[17]

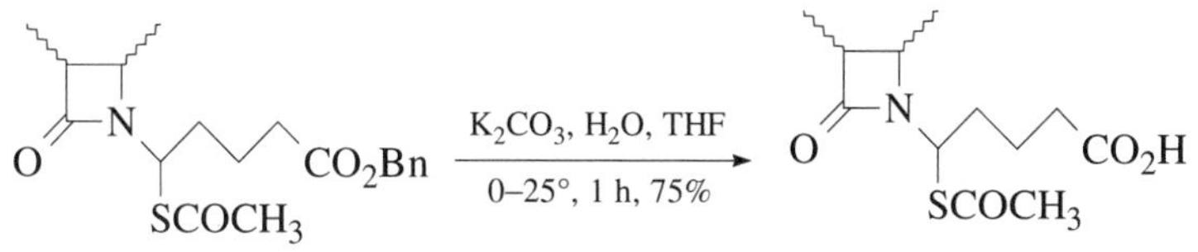

5. $AlCl_3$, anisole, CH_2Cl_2, CH_3NO_2, 0° → 25°, 5 h, 80–95% yield.[18] These conditions were used to cleave the benzyl ester in a variety of penicillin derivatives.
6. BCl_3, CH_2Cl_2, −10°→ rt, 3 h, 90% yield.[19]
7. Na, ammonia, 50% yield.[20] These conditions were used to cleave the benzyl ester of an amino acid; the Cbz and benzylsulfenamide derivatives were also cleaved.
8. Aq. $CuSO_4$, EtOH, pH 8, 32°, 60 min; pH 3; EDTA (ethylenediaminetetraacetic acid), 75% yield.[21]

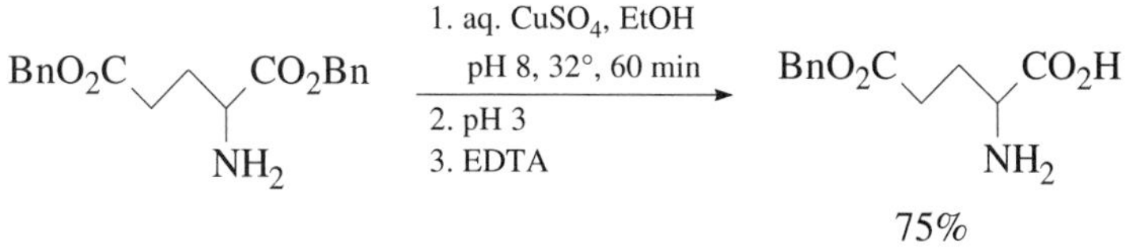

9. Benzyl esters can be cleaved by electrolytic reduction at −2.7 V.[22]
10. t-BuMe$_2$SiH, $Pd(OAc)_2$, CH_2Cl_2, Et_3N, 100% yield.[23] Cbz groups and Alloc groups are also cleaved, but benzyl ethers are stable. $PdCl_2$ and Et_3SiH have also been used to cleave a benzyl ester.[24]

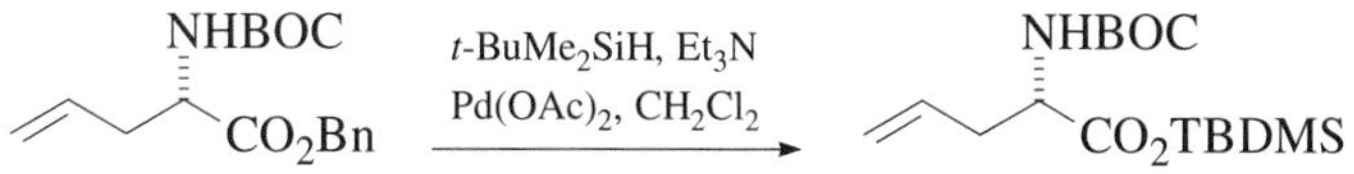

11. NaHTe, DMF, t-BuOH, 80–90°, 5 min, 98% yield.[25] Methyl and propyl esters are also cleaved (13–97% yield).
12. W2 Raney nickel, EtOH, Et_3N, rt, 0.5 h, 75–85% yield.[26] A disubstituted olefin was not reduced.
13. Acidic alumina, microwaves, 7 min, 90% yield.[27]
14. NBS, CCl_4, Bz_2O, reflux, 61–97% yield.[28] Substituted benzyl esters are cleaved similarly.
15. Alcatase, t-BuOH, pH 8.2, 35°, 0.5 h, 91% yield.[29]
16. *P. Fluorescens*, ROH, MTBE convert a benzyl ester by transesterification to Me, Et, and Bu esters.[30]
17. Pronase, 25°, pH 7.2, aq. EtOH, 70–73% yield.[31]

18. Alkaline protease from *Bacillus subtilis* DY, pH 8, 37°, 80–85% yield.[32] Methyl esters are cleaved similarly.
19. Bis(tributyltin) oxide, toluene, 70–90°, 36–96 h, 60–69% yield.[33]

1. S. Kim, Y. C. Kim, and J. I. Lee, *Tetrahedron Lett.*, **24**, 3365 (1983); S. Kim, J. I. Lee, and Y. C. Kim, *J. Org. Chem.*, **50**, 560 (1985).
2. J. E. Baldwin, M. Otsuka, and P. M. Wallace, *Tetrahedron*, **42**, 3097 (1986).
3. W. L. White, P. B. Anzeveno, and F. Johnson, *J. Org. Chem.*, **47**, 2379 (1982).
4. B. Neises, T. Andries, and W. Steglich, *J. Chem. Soc., Chem. Commun.*, 1132 (1982).
5. G. Kokotos and A. Chiou, *Synthesis*, 169 (1997).
6. K. C. Nicolaou, E. W. Yue, Y. Naniwa, F. D. Riccardis, A. Nadin, J. E. Leresche, S. La Greca, and Z. Yang, *Angew. Chem., Int. Ed.. Engl.*, **33**, 2184 (1994).
7. J. F. Okonya, T. Kolasa, and M. J. Miller, *J. Org. Chem.*, **60**, 1932 (1995).
8. G. Emmer, M. A. Grassberger, J. G. Meingassner, G. Schulz, and M. Schaude, *J. Med. Chem.*, **37**, 1908 (1994).
9. M. J. Smith, D. Kim, B. Horenstein, K. Nakanishi, and K. Kustin, *Acc. Chem. Res.*, **24**, 117 (1991).
10. J. C. Lee, Y. S. Oh, S. H. Cho, and J. I. Lee, *Org. Prep. Proced. Int.*, **28**, 480 (1996).
11. E. Vedejs, N. S. Bennett, L. M. Conn, S. T. Diver, M. Gingras, S. Lin, P. A. Oliver, and M. J. Peterson, *J. Org. Chem.*, **58**, 7286 (1993).
12. W. H. Hartung and R. Simonoff, *Org. React.*, **VII**, 263–326 (1953).
13. G. M. Anantharamaiah and K. M. Sivanandaiah, *J. Chem. Soc., Perkin Trans. 1*, 490 (1977).
14. A. M. Felix, E. P. Heimer, T. J. Lambros, C. Tzougraki, and J. Meienhofer, *J. Org. Chem.*, **43**, 4194 (1978).
15. J. S. Bajwa, *Tetrahedron Lett.*, **33**, 2299 (1992).
16. B. ElAmin, G. M. Anantharamaiah, G. P. Royer, and G. E. Means, *J. Org. Chem.*, **44**, 3442 (1979).
17. W. F. Huffman, R. F. Hall, J. A. Grant, and K. G. Holden, *J. Med. Chem.*, **21**, 413 (1978).
18. T. Tsuji, T. Kataoka, M. Yoshioka, Y. Sendo, Y. Nishitani, S. Hirai, T. Maeda, and W. Nagata, *Tetrahedron Lett.*, 2793 (1979).
19. U. Schmidt, M. Kroner, and H. Griesser, *Synthesis*, 294 (1991).
20. C. W. Roberts, *J. Am. Chem. Soc.*, **76**, 6203 (1954).
21. R. L. Prestidge, D. R. K. Harding, J. E. Battersby, and W. S. Hancock, *J. Org. Chem.*, **40**, 3287 (1975).
22. V. G. Mairanovsky, *Angew. Chem., Int. Ed. Engl.*, **15**, 281 (1976).
23. M. Sakaitani, N. Kurokawa, and Y. Ohfune, *Tetrahedron Lett.*, **27**, 3753 (1986).
24. K. M. Rupprecht, R. K. Baker, J. Boger, A. A. Davis, P. J. Hodges, and J. F. Kinneary, *Tetrahedron Lett.*, **39**, 233 (1998).
25. J. Chen and X. J. Zhou, *Synthesis*, 586 (1987).
26. S.-i. Hashimoto, Y. Miyazaki, T. Shinoda, and S. Ikegami, *Tetrahedron Lett.*, **30**, 7195 (1989).

27. R. S. Varma, A. K. Chatterjee, and M. Varma, *Tetrahedron Lett.*, **34**, 4603 (1993).
28. M. S. Anson and J. G. Montana, *Synlett*, 219 (1994).
29. S. T. Chen, S. C. Hsiao, C. H. Chang, and K. T. Wang, *Synth. Commun.*, **22**, 391 (1992).
30. A. L. Gutman, E. Shkolnik, and M. Shapira, *Tetrahedron*, **48**, 8775 (1992).
31. M. Pugniere, B. Castro, N. Domerque, and A. Previero, *Tetrahedron: Asymmetry*, **3**, 1015 (1992).
32. B. Aleksiev, P. Schamlian, G. Videnov, S. Stoev, S. Zachariev, and E. Golovinskii, *Hoppe-Seylers Z. Physiol. Chem.*, **362**, 1323 (1981).
33. C. J. Salomon, E. G. Mata, and O. A. Mascaretti, *J. Chem. Soc., Perkin Trans. 1*, 995 (1996).

Substituted Benzyl Esters

Triphenylmethyl Ester: $RCO_2C(C_6H_5)_3$ (Chart 6)

Triphenylmethyl esters are unstable in aqueous solution, but are stable to oxymercuration.[1]

Formation

$$RCO_2^-M^+ + Ph_3CBr \xrightarrow[\text{3–5 h, 85–95\%}]{\text{Benzene, reflux}^2} RCO_2CPh_3$$

$M^+ = Ag^+, K^+, Na^+$

$$RCO_2SiMe_3 \xrightarrow[\text{0°, 0.5 h, 86\%}]{\text{Ph}_3\text{COTMS, TMSOTf, CH}_2\text{Cl}_2{}^3} RCO_2CPh_3$$

Cleavage

1. $$HCl{\cdot}H_2NCH_2CO_2CPh_3 \xrightarrow[\text{18°, 5 h, 72\%; 18°, 24 h, 98\%; 100°, 1 min, 98\%}^4]{\text{MeOH or H}_2\text{O/dioxane}} HCl{\cdot}H_2NCH_2CO_2H$$

2. Trityl esters have been cleaved by electrolytic reduction at –2.6 V.[5]

1. W. A. Slusarchyk, H. E. Applegate, C. M. Cimarusti, J. E. Dolfini, P. Funke, and M. Puar, *J. Am. Chem. Soc.*, **100**, 1886 (1978).
2. K. D. Berlin, L. H. Gower, J. W. White, D. E. Gibbs, and G. P. Sturm, *J. Org. Chem.*, **27**, 3595 (1962).
3. S. Murata and R. Noyori, *Tetrahedron Lett.*, **22**, 2107 (1981).
4. G. C. Stelakatos, A. Paganou, and L. Zervas, *J. Chem. Soc. C*, 1191 (1966).
5. V. G. Mairanovsky, *Angew. Chem., Int. Ed. Engl.*, **15**, 281 (1976).

Diphenylmethyl (Dpm) Ester: $RCO_2CH(C_6H_5)_2$

Diphenylmethyl esters are similar in acid lability to *t*-butyl esters and can be cleaved by acidic hydrolysis from *S*-containing peptides that poison hydrogenolysis catalysts.

Formation

1. Ph_2CN_2, acetone, 0°, 30 min → 20°, 4 h, 70%.[1,2]
2. $Ph_2C{=}NNH_2$, I_2, AcOH, >90% yield.[3]
3. $Ph_2C{=}NNH_2$, Oxone supported on wet Al_2O_3, cat. I_2, 0°, 66–95% yield.[4]
4. $(Ph_2CHO)_3PO$, CF_3COOH, CH_2Cl_2, reflux, 1–5 h, 70–87% yield.[5] Free alcohols are converted to the corresponding Dpm ethers. This reaction has also been used for the selective protection of amino acids as their tosylate salts (CCl_4, 15 min–3 h, 63–91% yield).[6]
5. $Ph_2C{=}NNH_2$, $PhI(OAc)_2$, CH_2Cl_2, cat. I_2, −10° → 0°, 1 h, 73–93% yield.[7]
6. Ph_2CHOH, cat. TsOH, benzene, azeotropic removal of water, 78–83% yield.[8]
7. $Ph_2CH{=}NNH_2$, AcOOH, 91% yield.[9]

Cleavage

1. H_2/Pd black, MeOH, THF, 3 h, 90% yield.[1,10]
2. CF_3COOH, PhOH, 20°, 30 min, 82% yield.
3. AcOH, reflux, 6 h.[11]
4. $BF_3{\cdot}Et_2O$, AcOH, 40°, 0.5 h → 10°, several hours, 65% yield.[12] The sulfur–sulfur bond in cystine is stable to these conditions.
5. H_2NNH_2, MeOH, reflux, 60 min, 100% yield.[13] In this case, the ester is converted to a hydrazide.
6. Diphenylmethyl esters are cleaved by electrolytic reduction at −2.6 V.[14]
7. HF, CH_3NO_2, AcOH (12:2:1), 91% yield.[15]

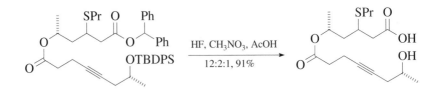

HF, CH_3NO_3, AcOH, 12:2:1, 91%

8. HCl, CH_3NO_2, < 5 min, 25°.[16]
9. 98% HCOOH, 40–50°, 70–97% yield[2].
10. 1 *N* NaOH, MeOH, rt.[6]
11. $AlCl_3$, CH_3NO_2, anisole, 3–6 h, 73–95% yield.[17,18] These conditions also

cleaved the p-$MeOC_6H_4CH_2$ ester and ether in penam- and cephalosporin-type intermediates.

12. 1 eq. TsOH, benzene, reflux, 78–95% yield.[8]

1. G. C. Stelakatos, A. Paganou, and L. Zervas, *J. Chem. Soc. C*, 1191 (1966).
2. T. Kametani, H. Sekine, and T. Hondo, *Chem. Pharm. Bull.*, **30**, 4545 (1982).
3. R. Bywood, G. Gallagher, G. K. Sharma, and D. Walker, *J. Chem. Soc., Perkin Trans. 1*, 2019 (1975).
4. M. Curini, O. Rosati, E. Pisani, W. Cabri, S. Brusco, and M. Riscazzi, *Tetrahedron Lett.*, **38**, 1239 (1997).
5. L. Lapatsanis, *Tetrahedron Lett.*, 4697 (1978).
6. C. Froussios and M. Kolovos, *Synthesis*, 1106 (1987).
7. L. Lapatsanis, G. Milias, and S. Paraskewas, *Synthesis*, 513 (1985).
8. R. Paredes, F. Agudelo, and G. Taborda, *Tetrahedron Lett.*, **37**, 1965 (1996).
9. R. G. Micetich, S. N. Maiti, P. Spevak, M. Tanaka, T. Yamazaki, and K. Ogawa, *Synthesis*, 292 (1986).
10. S. De Bernardo, J. P. Tengi, G. J. Sasso, and M. Weigele, *J. Org. Chem.*, **50**, 3457 (1985).
11. E. Haslam, R. D. Haworth, and G. K. Makinson, *J. Chem. Soc.*, 5153 (1961).
12. R. G. Hiskey and E. L. Smithwick, *J. Am. Chem. Soc.*, **89**, 437 (1967).
13. R. G. Hiskey and J. B. Adams, *J. Am. Chem. Soc.*, **87**, 3969 (1965).
14. V. G. Mairanovsky, *Angew. Chem., Int. Ed. Engl.*, **15**, 281 (1976).
15. L. R. Hillis and R. C. Ronald, *J. Org. Chem.*, **50**, 470 (1985).
16. R. C. Kelly, I. Schletter, S. J. Stein, and W. Wierenga, *J. Am. Chem. Soc.*, **101**, 1054 (1979).
17. T. Tsuji, T. Kataoka, M. Yoshioka, Y. Sendo, Y. Nishitani, S. Hirai, T. Maeda, and W. Nagata, *Tetrahedron Lett.*, 2793 (1979).
18. M. Ohtani, F. Watanabe, and M. Narisada, *J. Org. Chem.*, **49**, 5271 (1984).

Bis(*o*-nitrophenyl)methyl Ester: $RCOOCH(C_6H_4$–o-$NO_2)_2$ (Chart 6)

Bis(o-nitrophenyl)methyl esters are formed and cleaved by the same methods used for diphenylmethyl esters. They can also be cleaved by irradiation ($h\nu = 320$ nm, dioxane, THF, ..., 1–24 h, quant. yield).[1]

1. A. Patchornik, B. Amit, and R. B. Woodward, *J. Am. Chem. Soc.*, **92**, 6333 (1970).

9-Anthrylmethyl Ester: $RCOOCH_2$–9-anthryl (Chart 6)

Formation

1. 9-Anthrylmethyl chloride, Et_3N, MeCN, reflux, 4–6 h, 70–90% yield.[1]

2. N_2CH–9-anthryl, hexane, 25°, 10 min, 80% yield.[2,3]

Cleavage

1. 2 *N* HBr/HOAc, 25°, 10–30 min, 100% yield.[1]
2. 0.1 *N* NaOH/dioxane, 25°, 15 min, 97% yield.[1]
3. MeSNa, THF–HMPA, −20°, 1 h, 90–100% yield.[4]

1. F. H. C. Stewart, *Aust. J. Chem.*, **18**, 1699 (1965).
2. M. G. Krakovyak, T. D. Amanieva, and S. S. Skorokhodov, *Synth. Commun.*, **7**, 397 (1977).
3. K. Hör, O. Gimple, P. Schreier, and H.-U. Humpf, *J. Org. Chem.*, **63**, 322 (1998).
4. N. Kornblum and A. Scott, *J. Am. Chem. Soc.*, **96**, 590 (1974).

2-(9,10-Dioxo)anthrylmethyl Ester (Chart 6):

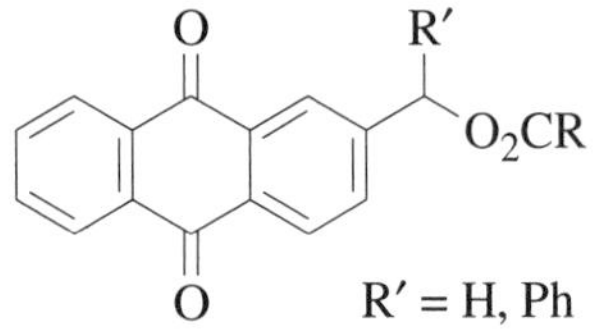

O R′ O_2CR O R′ = H, Ph

This derivative is prepared from an *N*-protected amino acid and the anthrylmethyl alcohol in the presence of DCC/hydroxybenzotriazole.[1] It can also be prepared from 2-(bromomethyl)-9,10-anthraquinone (Cs_2CO_3).[2] It is stable to moderately acidic conditions (e.g., CF_3COOH, 20°, 1 h; HBr/HOAc, $t_{1/2}$ = 65 h; HCl/CH_2Cl_2, 20°, 1 h)[1]. Cleavage is effected by reduction of the quinone to the hydroquinone **i**; in the latter, electron release from the –OH group of the hydroquinone results in facile cleavage of the methylene–carboxylate bond.

The related **2-phenyl-2-(9,10-dioxo)anthrylmethyl** ester has also been prepared, but is cleaved by electrolysis (−0.9 V, DMF, 0.1 *M* $LiClO_4$, 80% yield).[3]

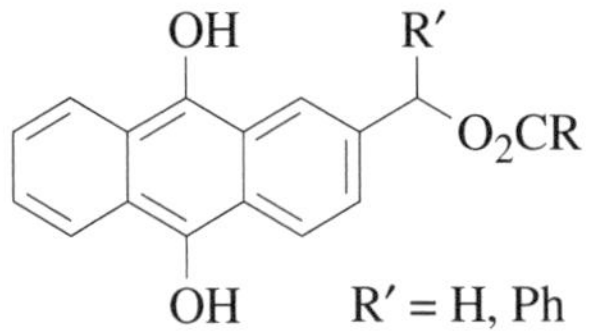

OH R′ O_2CR OH R′ = H, Ph

i

Cleavage[1]

This derivative is cleaved by hydrogenolysis and by the following conditions:

1. $Na_2S_2O_4$, dioxane–H_2O, pH 7–8, 8 h, 100% yield.
2. Irradiation, *i*-PrOH, 4 h, 99% yield.
3. 9-Hydroxyanthrone, Et_3N/DMF, 5 h, 99% yield.

4. 9,10-Dihydroxyanthracene/polystyrene resin, 1.5 h, 100% yield.

1. D. S. Kemp and J. Reczek, *Tetrahedron Lett.*, 1031 (1977).
2. P. Hoogerhout, C. P. Guis, C. Erkelens, W. Bloemhoff, K. E. T. Kerling, and J. H. Boom, *Recl. Trav. Chim. Pays-Bas*, **104**, 54 (1985).
3. R. L. Blankespoor, A. N. K. Lau, and L. L. Miller, *J. Org. Chem.*, **49**, 4441 (1984).

O_2CR

5-Dibenzosuberyl Ester:

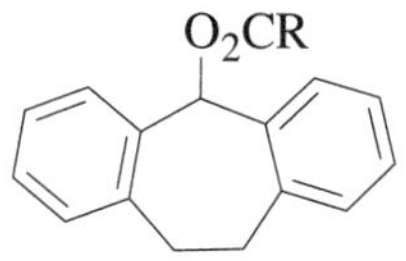

The dibenzosuberyl ester is prepared from dibenzosuberyl chloride (which is also used to protect –OH, –NH, and –SH groups) and a carboxylic acid (Et_3N, reflux, 4 h, 45% yield). It can be cleaved by hydrogenolysis and, like *t*-butyl esters, by acidic hydrolysis (aq. HCl/THF, 20°, 30 min, 98% yield).[1]

1. J. Pless, *Helv. Chim. Acta*, **59** 499 (1976).

R′ O_2CR

1-Pyrenylmethyl Ester (R′= H, Me, Ph):

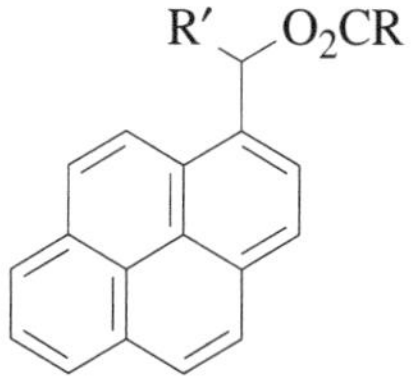

These esters are prepared from the diazomethylpyrenes and carboxylic acids in DMF (R′ = H, 60% yield, R′ = Me, 80% yield, R′ = Ph, 20% yield for 4-methylbenzoic acid). They are cleaved by photolysis at 340 nm (80–100% yield, R′ = H).[1,2] The esters are very fluorescent.

1. M. Iwamura, T. Ishikawa, Y. Koyama, K. Sakuma, and H. Iwamura, *Tetrahedron Lett.*, **28**, 679 (1987).
2. M. Iwamura, C. Hodota, and M. Ishibashi, *Synlett*, 35 (1991).

2-(Trifluoromethyl)-6-chromonylmethyl (Tcrom) Ester:

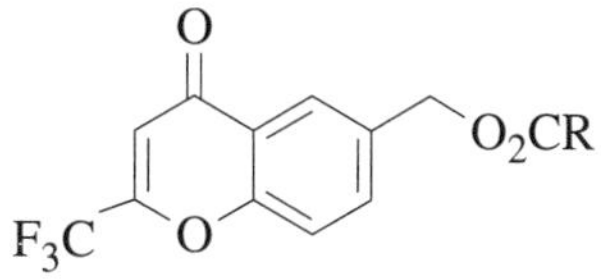

O

O_2CR

F_3C O

The Tcrom ester is prepared from the cesium salt of an *N*-protected amino acid by reaction with 2-(trifluoromonethyl)-6-chromonylmethyl bromide (DMF, 25°,

4 h, 53–89% yield). Cleavage of the Tcrom group is effected by brief treatment with *n*-propylamine (2 min, 25°, 96% yield). The group is stable to HCl/dioxane, used to cleave a BOC group.[1]

1. D. S. Kemp and G. Hanson, *J. Org. Chem.*, **46**, 4971 (1981).

2,4,6-Trimethylbenzyl Ester: $RCOOCH_2C_6H_2$–2,4,6-$(CH_3)_3$

The 2,4,6-trimethylbenzyl ester has been prepared from an amino acid and the benzyl chloride (Et_3N, DMF, 25°, 12 h, 60–80% yield); it is cleaved by acidic hydrolysis (CF_3COOH, 25°, 60 min, 60–90% yield; 2 *N* HBr/HOAc, 25°, 60 min, 80–95% yield) and by hydrogenolysis. It is stable to methanolic hydrogen chloride, used to remove *N*-*o*-nitrophenylsulfenyl groups or triphenylmethyl esters.[1]

1. F. H. C. Stewart, *Aust. J. Chem.*, **21**, 2831 (1968).

***p*-Bromobenzyl Ester:** $RCOOCH_2C_6H_4$–*p*-Br

The *p*-bromobenzyl ester has been used to protect the *β*-COOH group in aspartic acid. It is cleaved by strong acidic hydrolysis (HF, 0°, 10 min, 100% yield), but is stable to 50% CF_3COOH/CH_2Cl_2, used to cleave *t*-butyl carbamates. The *p*-bromobenzyl ester is five to seven times more stable toward acid than is a benzyl ester.[1]

1. D. Yamashiro, *J. Org. Chem.*, **42**, 523 (1977).

***o*-Nitrobenzyl Ester:** $RCOOCH_2C_6H_4$–*o*-NO_2

***p*-Nitrobenzyl Ester:** $RCOOCH_2C_6H_4$–*p*-NO_2

The *o*-nitrobenzyl ester, used in this example to protect penicillin precursors, can be cleaved by irradiation (H_2O/dioxane, pH 7). Reductive cleavage of benzyl or *p*-nitrobenzyl esters occurred in lower yields.[1,2]

p-Nitrobenzyl esters have been prepared from the Hg(I) salt of penicillin precursors and the phenyldiazomethane.[3] They are much more stable to acidic hydrolysis (e.g., HBr) than are *p*-chlorobenzyl esters and are recommended for terminal –COOH protection in solid-phase peptide synthesis.[4]

p-Nitrobenzyl esters of penicillin and cephalosporin precursors have been cleaved by alkaline hydrolysis with Na_2S (0°, aq. acetone, 25–30 min, 75–85% yield).[5] They are also cleaved by electrolytic reduction at –1.2 V,[6] by reduction with $SnCl_2$ (DMF, phenol, AcOH),[7] by reduction with sodium dithionite, or by hydrogenolysis.[8]

1. L. D. Cama and B. G. Christensen, *J. Am. Chem. Soc.*, **100**, 8006 (1978).
2. For a review covering the photolytic removal of protective groups, see V. N. R. Pillai, *Synthesis*, 1 (1980).
3. W. Baker, C. M. Pant, and R. J. Stoodley, *J. Chem. Soc., Perkin Trans. 1*, 668 (1978).
4. R. L. Prestidge, D. R. K. Harding, and W. S. Hancock, *J. Org. Chem.*, **41**, 2579 (1976).
5. S. R. Lammert, A. I. Ellis, R. R. Chauvette, and S. Kukolja, *J. Org. Chem.*, **43**, 1243 (1978).
6. V. G. Mairanovsky, *Angew Chem., Int. Ed. Engl.*, **15**, 281 (1976).
7. M. D. Hocker, C. G. Caldwell, R. W. Macsata, and M. H. Lyttle, *Pept. Res.*, **8**, 310 (1995).
8. J. W. Perich, P. F. Alewood, and R. B. Johns, *Aust. J. Chem.*, **44**, 233 (1991).

p-Methoxybenzyl Ester: $RCOOCH_2C_6H_4$–*p*-OCH_3

Formation

1. *p*-Methoxybenzyl esters have been prepared from the Ag(I) salt of amino acids and the benzyl halide (Et_3N, $CHCl_3$, 25°, 24 h, 60% yield).[1]
2. *p*-Methoxybenzyl alcohol, $Me_2NCH(OCH_2$–*t*-Bu$)_2$, CH_2Cl_2, 90% yield.[2]
3. Isopropenyl chloroformate, $MeOC_6H_4CH_2OH$, DMAP, 0°, CH_2Cl_2, 91%.[3]
4. *p*-Methoxyphenyldiazomethane in CH_2Cl_2, 80–96% yield.[4]
5. *p*-Methoxybenzyl chloride, $NaHCO_3$, DMF, 45°, 89% yield.[5]
6. $MeOC_6H_4CHN_2$.[4]

Cleavage

1. CF_3COOH/PhOMe, 25°, 3 min, 98% yield.[6]
2. HCOOH, 22°, 1 h, 81% yield.[1]
3. TFA, phenol, 1 h, 45°, 73–93% yield.[7,8] These conditions were developed for the mild cleavage of acid-sensitive esters of β-lactam-related antibiotics. Diphenylmethyl and *t*-butyl esters were cleaved with similarly high efficiency.
4. $AlCl_3$, anisole, CH_2Cl_2 or CH_3NO_2, –50°; $NaHCO_3$, –50°, 73–95% yield.[9,10]
5. $CF_3CO_2H/B(OTf)_3$.[11]

1. G. C. Stelakatos and N. Argyropoulos, *J. Chem. Soc. C*, 964 (1970).
2. J. A. Webber, E. M. Van Heyningen, and R. T. Vasileff, *J. Am. Chem. Soc.*, **91**, 5674 (1969).
3. P. Jouin, B. Castro, C. Zeggaf, A. Pantaloni, J. P. Senet, S. Lecolier, and G. Sennyey, *Tetrahedron Lett.*, **28**, 1661 (1987).
4. S. T. Waddell and G. M. Santorelli, *Tetrahedron Lett.*, **37**, 1971 (1996).
5. D. L. Boger, M. Hikota, and B. M. Lewis, *J. Org. Chem.*, **62**, 1748 (1997).

6. F. H. C. Stewart, *Aust. J. Chem.*, **21**, 2543 (1968).
7. H. Tanaka, M. Taniguchi, Y. Kameyama, S. Torii, M. Sasaoka, T. Shiroi, R. Kikuchi, I. Kawahara, A. Shimabayashi, and S. Nagao, *Tetrahedron Lett.*, **31**, 6661 (1990).
8. S. Torii, H. Tanaka, M. Taniguchi, Y. Kameyama, M. Sasaoka, T. Shiroi, R. Kikuchi, I. Kawahara, A. Shimabayashi, and S. Nagao, *J. Org. Chem.*, **56**, 3633 (1991).
9. M. Ohtani, F. Watanabe, and M. Narisada, *J. Org. Chem.*, **49**, 5271 (1984).
10. T. Tsuji, T. Kataoka, M. Yoshioka, Y. Sendo, Y. Nishitani, S. Hirai, T. Maeda, and W. Nagata, *Tetrahedron Lett.*, 2793 (1979).
11. S. D. Young and P. P. Tamburini, *J. Am. Chem. Soc.*, **111**, 1933 (1989).

2,6-Dimethoxybenzyl Ester: 2,6-$(CH_3O)_2C_6H_3CH_2OOCR$

2,6-Dimethoxybenzyl esters prepared from the acid chloride and the benzyl alcohol are readily cleaved oxidatively by DDQ (CH_2Cl_2, H_2O, rt, 18 h, 90–95% yield). A 4-methoxybenzyl ester was found not to be cleaved by DDQ. The authors have also explored the oxidative cleavage (ceric ammonium nitrate, CH_3CN, H_2O, 0°, 4 h, 65–97% yield) of a variety of 4-hydroxy- and 4-amino-substituted phenolic esters.[1]

1. C. U. Kim and P. F. Misco, *Tetrahedron Lett.*, **26**, 2027 (1985).

4-(Methylsulfinyl)benzyl (Msib) Ester: 4-$CH_3S(O)C_6H_4CH_2O_2CR$

The 4-(methylsulfinyl)benzyl ester was recommended as a selectively cleavable carboxyl protective group for peptide synthesis. It is readily prepared from 4-(methylsulfinyl)benzyl alcohol (EDCI, HOBt, $CHCl_3$, 78–100% yield) or from 4-methylthiobenzyl alcohol followed by oxidation of the derived ester with MCPBA or H_2O_2/AcOH. The Msib ester is exceptionally stable to CF_3COOH (cleavage rate = 0.000038% ester cleaved/min) and undergoes only 10% cleavage in HF (anisole, 0°, 1 h). Anhydrous HCl/dioxane rapidly reduces the sulfoxide to the sulfide (Mtb ester), which is completely cleaved in 30 min with CF_3CO_2H. A number of reagents readily reduce the Msib ester to the Mtb ester, with $(CH_3)_3SiCl/Ph_3P$ as the reagent of choice.[1]

1. J. M. Samanen and E. Brandeis, *J. Org. Chem.*, **53**, 561 (1988).

4-Sulfobenzyl Ester: $Na^{+-}O_3SC_6H_4CH_2O_2CR$

4-Sulfobenzyl esters were prepared (cesium salt or dicyclohexylammonium salt, $NaO_3SC_6H_4CH_2Br$, DMF, 37–95% yield) from *N*-protected amino acids. They are cleaved by hydrogenolysis (H_2/Pd), or hydrolysis (NaOH, dioxane/water). Treatment with ammonia or hydrazine results in the formation of the amide or hydrazide. The ester is stable to 2 *M* HBr/AcOH and to CF_3SO_3H in CF_3CO_2H. The relative rates of hydrolysis and hydrazinolysis for different esters are as follows:

Hydrolysis: $NO_2C_6H_4CH_2O– >> C_6H_4CH_2O– > {}^-O_3SC_6H_4CH_2O– > MeO–$
Hydrazinolysis: $NO_2C_6H_4CH_2O– > {}^-O_3SC_6H_4CH_2O– > C_6H_5CH_2O– > MeO–$

A benzyl ester can be cleaved in the presence of the 4-sulfobenzyl ester by CF_3SO_3H.[1,2]

1. R. Bindewald, A. Hubbuch, W. Danho, E. E. Büllesbach, J. Föhles, and H. Zahn, *Int. J. Pept. Protein Res.*, **23**, 368 (1984).
2. A. Hubbuch, R. Bindewald, J. Föhles, V. K. Naithani, and H. Zahn, *Angew. Chem., Int. Ed. Engl.*, **19**, 394 (1980).

4-Azidomethoxybenzyl Ester: $N_3CH_2OC_6H_4CH_2O_2CR$

This ester, developed for peptide synthesis, is prepared by the standard DCC coupling protocol and is cleaved reductively with $SnCl_2$ (MeOH, 25°, 5 h) followed by treatment with mild base to effect quinonemethide formation with release of the acid in 75–95% yield.[1]

1. B. Loubinoux and P. Gerardin, *Tetrahedron*, **47**, 239 (1991).

4-{*N*-[1-(4,4-Dimethyl-2,6-dioxocyclohexylidene)-3-methylbutyl]amino}benzyl (Dmab) Ester:

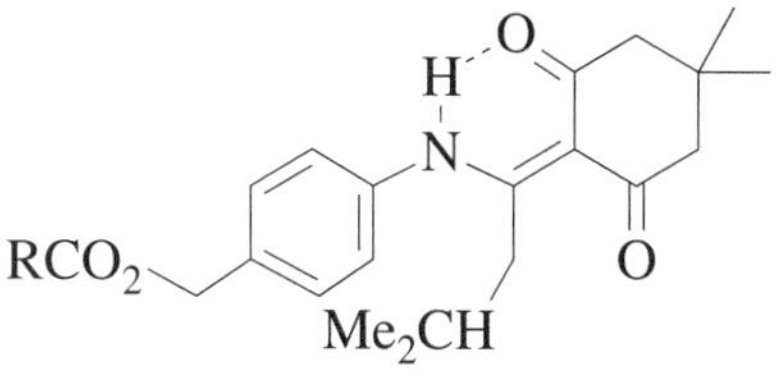

H O N RCO2 O Me2CH

The Dmab group was developed for glutamic acid protection during Fmoc/*t*-Bu based peptide synthesis. The group shows excellent acid stability and stability toward 20% piperidine in DMF. It is formed from the alcohol using the DCC protocol for ester formation and is cleaved with 2% hydrazine in DMF at rt.[1]

1. W. C. Chan, B. W. Bycroft, D. J. Evans, and P. D. White, *J. Chem. Soc., Chem. Commun.*, 2209 (1995).

Piperonyl Ester (Chart 6): RCO2 O O

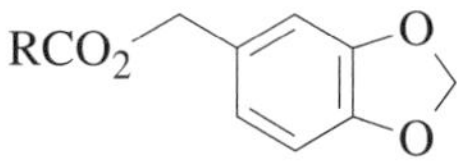

The piperonyl ester can be prepared from an amino acid ester and the benzyl alcohol (imidazole/dioxane, 25°, 12 h, 85% yield) or from an amino acid and the benzyl chloride (Et_3N, DMF, 25°, 57–95% yield). It is cleaved, more readily than a *p*-methoxybenzyl ester, by acidic hydrolysis (CF_3COOH, 25°, 5 min, 91% yield).[1]

1. F. H. C. Stewart, *Aust. J. Chem.*, **24**, 2193 (1971).

4-Picolyl Ester: RCO_2CH_2–4-pyridyl

The picolyl ester has been prepared from amino acids and picolyl alcohol (DCC/CH_2Cl_2, 20°, 16 h, 60% yield) or picolyl chloride (DMF, 90–100°, 2 h, 50% yield). It is cleaved by reduction (H_2/Pd–C, aq. EtOH, 10 h, 98% yield; Na/NH_3, 1.5 h, 93% yield) and by basic hydrolysis (1 *N* NaOH, dioxane, 20°, 1 h, 93% yield). The basic site in a picolyl ester allows its ready separation by extraction into an acidic medium.[1]

1. R. Camble, R. Garner, and G. T. Young, *J. Chem. Soc. C*, 1911 (1969).

p*-*P*-Benzyl Ester:** $RCOOCH_2C_6H_4$–*p*-P***

The first,[1] and still widely used, polymer-supported ester is formed from an amino acid and a chloromethylated copolymer of styrene–divinylbenzene. Originally, it was cleaved by basic hydrolysis (2 *N* NaOH, EtOH, 25°, 1 h). Subsequently, it has been cleaved by hydrogenolysis (H_2/Pd–C, DMF, 40°, 60 psi, 24 h, 71% yield)[2] and by HF, which concurrently removes many amine protective groups.[3]

Monoesterification of a symmetrical dicarboxylic acid chloride can be effected by reaction with a hydroxymethyl copolymer of styrene–divinylbenzene to give an ester; a mono salt of a diacid was converted into a dibenzyl polymer.[4]

1. R. B. Merrifield, *J. Am. Chem. Soc.*, **85**, 2149 (1963).
2. J. M. Schlatter and R. H. Mazur, *Tetrahedron Lett.*, 2851 (1977).
3. J. Lenard and A. B. Robinson, *J. Am. Chem. Soc.*, **89**, 181 (1967).
4. D. D. Leznoff and J. M. Goldwasser, *Tetrahedron Lett.*, 1875 (1977).

Silyl Esters

Silyl esters are stable to nonaqueous reaction conditions. A trimethylsilyl ester is cleaved by refluxing in alcohol; the more substituted, and therefore more stable, silyl esters are cleaved by mildly acidic or basic hydrolysis.

Trimethylsilyl Ester: $RCOOSi(CH_3)_3$ (Chart 6)

Some of the more common reagents for the conversion of carboxylic acids to trimethylsilyl esters follow. For additional methods that can be used to silylate acids, the section on alcohol protection should be consulted, since many of the methods presented there are also applicable to carboxylic acids. Trimethylsilyl esters are cleaved in aqueous solutions.

Formation

1. Me_3SiCl/Pyr, CH_2Cl_2, 30°, 2 h.[1]
2. $MeC(OSiMe_3)=NSiMe_3$, HBr, dioxane, α-picoline, 6 h, 80% yield.[2]
3. $MeCH=C(OMe)OSiMe_3/CH_2Cl_2$, 15–25°, 5–40 min, quant.[3]
4. $Me_3SiNHSO_2OSiMe_3/CH_2Cl_2$, 30°, 0.5 h, 92–98% yield.[4]

1. B. Fechti, H. Peter, H. Bickel, and E. Vischer, *Helv. Chim. Acta*, **51**, 1108 (1968).
2. J. J. de Koning, H. J. Kooreman, H. S. Tan, and J. Verweij, *J. Org. Chem.*, **40**, 1346 (1975).
3. Y. Kita, J. Haruta, J. Segawa, and Y. Tamura, *Tetrahedron Lett.*, 4311 (1979).
4. B. E. Cooper and S. Westall, *J. Organomet. Chem.*, **118**, 135 (1976).

Triethylsilyl Ester: $RCOOSi(C_2H_5)_3$

Formation/cleavage[1]

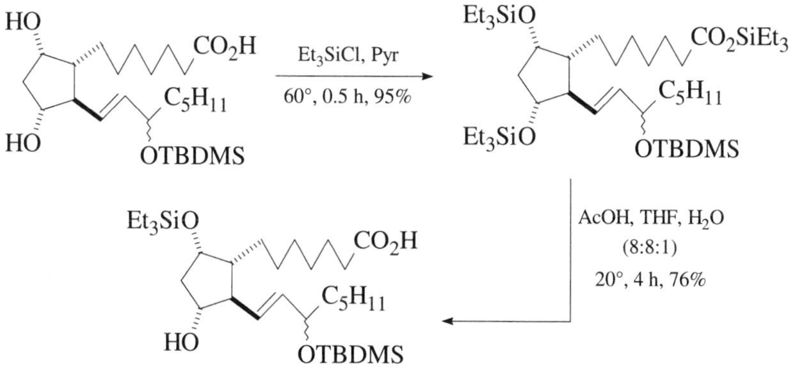

HO, HO, CO_2H, C_5H_{11}, OTBDMS

Et_3SiCl, Pyr; 60°, 0.5 h, 95%

Et_3SiO, Et_3SiO, CO_2SiEt_3, C_5H_{11}, OTBDMS

AcOH, THF, H_2O (8:8:1); 20°, 4 h, 76%

Et_3SiO, HO, CO_2H, C_5H_{11}, OTBDMS

1. T. W. Hart, D. A. Metcalfe, and F. Scheinmann, *J. Chem. Soc., Chem. Commun.*, 156 (1979).

t-Butyldimethylsilyl (TBDMS) Ester: $RCOOSi(CH_3)_2C(CH_3)_3$ (Chart 6)

Formation

1. *t*-$BuMe_2SiCl$, imidazole, DMF, 25°, 48 h, 88%.[1]
2. Morpholine, TBDMSCl, THF, 2 min, 20°, >80% yield.[2] In this case, the ester was formed in the presence of a phenol.
3. *t*-$BuMe_2SiH$, Pd/C, benzene, 70°.[3]

Cleavage

1. AcOH, H_2O, THF, (3 : 1 : 1), 25°, 20 h.[1]

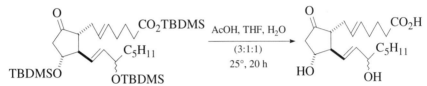

AcOH, THF, H_2O (3:1:1) 25°, 20 h

2. $Bu_4N^+F^-$, DMF, 25°.[1]
3. K_2CO_3, MeOH, H_2O, 25°, 1 h, 88% yield.[4]

K_2CO_3, MeOH, H_2O 25°, 1 h, 88%

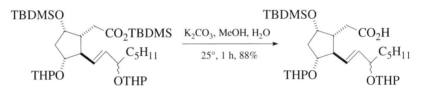

4. The TBDMS ester can be converted directly to an acid chloride [DMF, $(COCl)_2$, rt, CH_2Cl_2] and then converted to another ester, with different properties, by standard means. This procedure avoids the generation of HCl during the acid chloride formation and is thus suitable for acid-sensitive substrates.[5]

1. E. J. Corey and A. Venkateswarlu, *J. Am. Chem. Soc.*, **94**, 6190 (1972).
2. J. W. Perich and R. B. Johns, *Synthesis*, 701 (1989).
3. K. Yamamoto and M. Takemae, *Bull. Chem. Soc. Jpn.*, **62**, 2111 (1989).
4. D. R. Morton and J. L. Thompson, *J. Org. Chem.*, **43**, 2102 (1978).
5. A. Wissner and G. V. Grudzinskas, *J. Org. Chem.*, **43**, 3972 (1978).

i-Propyldimethylsilyl Ester: $RCOOSi(CH_3)_2CH(CH_3)_2$

The *i*-propyldimethylsilyl ester is prepared from a carboxylic acid and the silyl chloride (Et_3N, 0°). It is cleaved at pH 4.5 by conditions that do not cleave a tetrahydropyranyl ether (HOAc–NaOAc, acetone–H_2O, 0°, 45 min → 25°, 30 min, 91% yield).[1]

1. E. J. Corey and C. U. Kim, *J. Org. Chem.*, **38**, 1233 (1973).

Phenyldimethylsilyl Ester: $RCOOSi(CH_3)_2C_6H_5$

The phenyldimethylsilyl ester has been prepared from an amino acid and phenyldimethylsilane (Ni/THF, reflux, 3–5 h, 62–92% yield).[1]

1. M. Abe, K. Adachi, T. Takiguchi, Y. Iwakura, and K. Uno, *Tetrahedron Lett.*, 3207 (1975).

Di-*t*-butylmethylsilyl (DTBMS) Ester: $(t\text{-Bu})_2CH_3SiO_2CR$

The DTBMS ester was prepared (THF, DTBMSOTf, Et_3N, rt) to protect an ester so that a lactone could be reduced to an aldehyde. The ester is cleaved with aq. HF/THF or $Bu_4N^+F^-$ in wet THF. A THP derivative can be deprotected (pyridinium *p*-toluenesulfonate, warm ethanol) in the presence of a DTBMS ester.[1]

1. R. S. Bhide, B. S. Levison, R. B. Sharma, S. Ghosh, and R. G. Salomon, *Tetrahedron Lett.*, **27**, 671 (1986).

Triisopropylsilyl (TIPS) Ester

A TIPS ester, prepared by silylation with TIPSCl, TEA, and THF, is cleaved with HF–Pyr (Pyr, THF, 0°).[1]

1. D. A. Evans, B. W. Trotter, B. Côté, P. J. Coleman, L. C. Dias, and A. N. Tyler, *Angew. Chem., Int. Ed. Engl.*, **36**, 2744 (1997).

Activated Esters

Thiol Esters

Thiol esters, more reactive to nucleophiles than the corrresponding oxygen esters, have been prepared to activate carboxyl groups, for both lactonization and peptide bond formation. For lactonization, *S*-*t*-butyl[1] and *S*-2-pyridyl[2] esters are widely used. Some methods used to prepare thiol esters are as follows (the *S*-*t*-butyl ester is included in Reactivity Chart 6):

1. $RCOOH + R'SH \xrightarrow[0°,\ 5\ min \rightarrow 20°,\ 3\ h]{DCC,\ DMAP,\ CH_2Cl_2{}^{3}} RCOSR'$ 85–92%

   R′ = Et, *t*-Bu

DMAP = 4-dimethylaminopyridine (10^4 times more effective than pyridine)

2. 2-Fluoro-1-methylpyridinium tosylate (N^+–CH_3, F, TsO^-) + RCOOH $\xrightarrow[-15°,\ 1\ h]{Et_3N,\ CH_2Cl_2}$ $\xrightarrow[2\ h,\ 75–95\%]{R'H,\ Et_3N,\ CH_2Cl_2}$ RCOSR′[4]

R′ = *n*-Bu, *s*-Bu, *t*-Bu, Ph, 2-pyridyl

3. RCOOH + R′SH $\xrightarrow[25°,\ 1\ h,\ 70–100\%]{Me_2NPOCl_2,\ Et_3N,\ DME}$ RCOSR′

R′ = Et, *i*-Pr, *t*-Bu, *c*-C_6H_{11}, Ph

These neutral conditions can be used to prepare thiol esters of acid- or base-sensitive compounds, including penicillins.[5]

4.

RCH(NHPG)COOH + Ph_2POCl $\xrightarrow[0°,\ 30\ min]{Et_3N,\ CH_2Cl_2}$ $\xrightarrow{R'SH,\ Et_3N}$ $\xrightarrow[25°,\ 1\ h]{or\ R'STl,\ Et_3N}$ RCH(NHPG)COSR

R′ = *t*-Bu, Ph, $PhCH_2$ 70–100%[6]

5. RCOOH + R′SH $\xrightarrow[Et_3N,\ DMF,\ 25°,\ 3\ h,\ 70–85\%]{(EtO)_2POCN\ or\ (PhO)_2PON_3}$ RCOSR′[7]

R = alkyl, aryl, benzyl, amino acids; penicillins

R′ = Et, *i*-Pr, *n*-Bu, Ph, $PhCH_2$

6. RCOCl + *n*-Bu_3SnSR' $\xrightarrow{CHCl_3}$ RCOSR′[8]

R′ = *t*-Bu: 60°, 0.5 h, 90–95% yield

R′ = Ph: 25°, 12 h, 92–95% yield

R′ = $PhCH_2$: 25°, 0.5–1 h, 87–96% yield

7. RCOOR′ + Me_2AlS–*t*-Bu $\xrightarrow[25°,\ 75–100\%]{CH_2Cl_2}$ RCOS–*t*-Bu[9]

R′ = Me, Et

This reaction avoids the use of toxic thallium compounds.

8. RCOOH + PhSCN $\xrightarrow[25°,\ 30\ min,\ 80–95\%]{Bu_3P,\ CH_2Cl_2}$ RCOSPh[10]

9. RCOOH + ClCOS-2-pyridyl $\xrightarrow[0.5\ h,\ 95–100\%]{Et_3N,\ 0°}$ RCOS-2-pyridyl + $Et_3N{\cdot}HCl$[11]

10. RCO_2H + hydroxybenzotriazole $\xrightarrow{\text{DCC}}$ $\xrightarrow[\text{70–100\%}^{6}]{\text{R'SH, Et}_3\text{N}}$ or $\xrightarrow{\text{R' STl}}$ RCOSR′

R′ = *t*-Bu, Ph, $PhCH_2$

Cleavage

1. $AgNO_3$, H_2O, dioxane, (1:4), 2 h.[12]
2. ROH, $Hg(O_2CCF_3)_2$, 90% yield.[1]
3. Electrolysis, $Bu_4N^+Br^-$, H_2O, CH_3CN, $NaHCO_3$.[13] This method is unsatisfactory for primary and secondary alcohols, aldehydes, olefins, and amines.
4. MeI, ROH (R = *t*-Bu, PhSH, etc.), 68–97% yield.[14]
5. Treatment of the phenylthio ester with Pd/C and TESH results in reduction to the aldehyde.[15]

1. S. Masamune, S. Kamata, and W. Schilling, *J. Am. Chem. Soc.*, **97**, 3515 (1975).
2. T. Mukaiyama, R. Matsueda, and M. Suzuki, *Tetrahedron Lett.*, 1901 (1970); E. J. Corey, P. Ulrich, and J. M. Fitzpatrick, *J. Am. Chem. Soc.*, **98**, 222 (1976).
3. B. Neises and W. Steglich, *Angew. Chem., Int. Ed. Engl.*, **17**, 522 (1978).
4. Y. Watanabe, S.-i. Shoda, and T. Mukaiyama, *Chem. Lett.*, 741 (1976).
5. H. -J. Liu, S. P. Lee, and W. H. Chan, *Synth. Commun.*, **9**, 91 (1979).
6. K. Horiki, *Synth. Commun.*, **7**, 251 (1977).
7. S. Yamada, Y. Yokoyama, and T. Shiori, *J. Org. Chem.*, **39**, 3302 (1974).
8. D. N. Harpp, T. Aida, and T. H. Chan, *Tetrahedron Lett.*, 2853 (1979).
9. R. P. Hatch and S. M. Weinreb, *J. Org. Chem.*, **42**, 3960 (1977).
10. P. A. Grieco, Y. Yokoyama, and E. Williams, *J. Org. Chem.*, **43**, 1283 (1978).
11. E. J. Corey and D. A. Clark, *Tetrahedron Lett.*, 2875 (1979).
12. A. B. Shenvi and H. Gerlach, *Helv. Chim. Acta*, **63**, 2426 (1980).
13. M. Kimura, S. Matsubara, and Y. Sawaki, *J. Chem. Soc., Chem. Commun.*, 1619 (1984).
14. D. Ravi and H. B. Mereyala, *Tetrahedron Lett.*, **30**, 6089 (1989).
15. Fukuyama, S.-C. Lin, and L. Li, *J. Am. Chem. Soc.*, **112**, 7050 (1990).

Miscellaneous Derivatives

Oxazoles: R–(oxazole ring: N, O; substituents R′, R″)

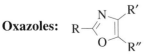

Oxazoles, prepared from carboxylic acids (benzoin, DCC; NH_4OAc, AcOH, 80–85% yield), have been used as carboxylic acid protective groups in a variety of synthetic applications. They are readily cleaved by singlet oxygen followed by hydrolysis (ROH, TsOH, benzene[1] or K_2CO_3, MeOH[2]).

2-Alkyl-1,3-oxazoline (Chart 6):

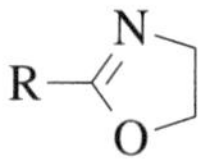

2-Alkyl-1,3-oxazolines are prepared to protect both the carbonyl and hydroxyl groups of an acid. They are stable to Grignard reagents[3] and to lithium aluminum hydride (25°, 2 h).[4]

Formation

1. $HOCH_2C(CH_3)_2NH_2$, $PhCH_3$, reflux, 70–80% yield.[5]
2. From an acid chloride: $HOCH_2C(CH_3)_2NH_2$; $SOCl_2$, CH_2Cl_2, 25°, 30 min, >80% yield.[6]
3. Dimethylaziridine, DCC; 3% H_2SO_4, Et_2O or CH_2Cl_2, rt, 6–16 h, 50–80% yield.[4]
4. $H_2NCH_2CH_2OH$, Ph_3P, Et_3N, CCl_4, CH_3CN, Pyr, rt, 70% yield.[7]
5. From an acid chloride: $BrCH_2CH_2NH_3^+Br^-$; Et_3N, benzene, reflux, 24 h, 46–67% yield.[8]

Cleavage

1. 3 *N* HCl, EtOH, 90% yield.[3]
2. MeI, 25°, 12 h; 1 *N* NaOH, 25°, 15 h, 94% yield.[9]
3. (a) TFA, H_2O; (b) Ac_2O, Pyr; (c) *t*-BuOK, H_2O, THF, quantitative.[10]
4. (a) TFAA; (b) H_2O; (c) diazomethane; (d) KOH, DMSO, 56–88% yield.[11]

1. H. H. Wasserman, K. E. McCarthy, and K. S. Prowse, *Chem. Rev.*, **86**, 845 (1986).
2. M. A. Tius and D. P. Astrab, *Tetrahedron Lett.*, **30**, 2333 (1989).
3. A. I. Meyers and D. L. Temple, *J. Am. Chem. Soc.*, **92**, 6644 (1970).
4. D. Haidukewych and A. I. Meyers, *Tetrahedron Lett.*, 3031 (1972).
5. H. L. Wehrmeister, *J. Org. Chem.*, **26**, 3821 (1961).
6. S. R. Schow, J. D. Bloom, A. S. Thompson, K. N. Winzenberg, and A. B. Smith, III, *J. Am. Chem. Soc.*, **108**, 2662 (1986).
7. H. Vorbrüggen and K. Krolikiewicz, *Tetrahedron Lett.*, **22**, 4471 (1981).
8. C. Kashima and H. Arao, *Synthesis*, 873 (1989).
9. A. I. Meyers, D. L. Temple, R. L. Nolen, and E. D. Mihelich, *J. Org. Chem.*, **39**, 2778 (1974).
10. T. D. Nelson and A. I. Meyers, *J. Org. Chem.*, **59**, 2577 (1994).
11. D. P. Phillion and J. K. Pratt, *Synth. Commun.*, **22**, 13 (1992).

4-Alkyl-5-oxo-1,3-oxazolidine:

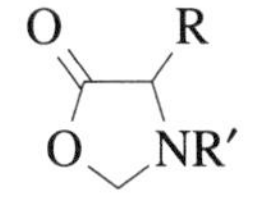

1,3-Oxazolidines are prepared to allow selective protection of the α- or ω-CO_2H groups in aspartic and glutamic acids and α-hydroxy acids.

Formation/Cleavage[1]

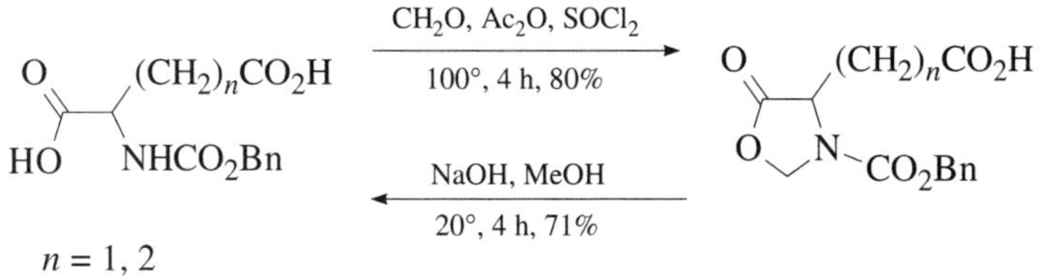

CH_2O, Ac_2O, $SOCl_2$; 100°, 4 h, 80%

NaOH, MeOH; 20°, 4 h, 71%

$n = 1, 2$

The use of paraformaldehyde and acid is equally effective (80–94% yield).[2] The related **2-*t*-butyl** derivative has been prepared and used to advantage as a temporary protective group for the stereogenic center of amino acids during alkylations.[3]

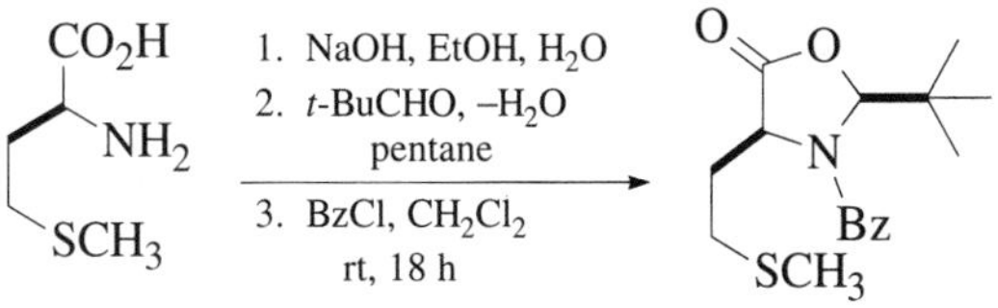

1. NaOH, EtOH, H_2O
2. *t*-BuCHO, $-H_2O$ pentane
3. BzCl, CH_2Cl_2 rt, 18 h

2,2-Bistrifluoromethyl-4-alkyl-5-oxo-1,3-oxazolidine:

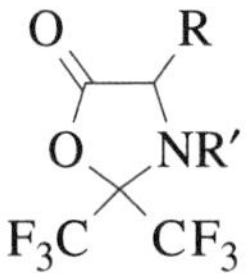

Formation[4,5]

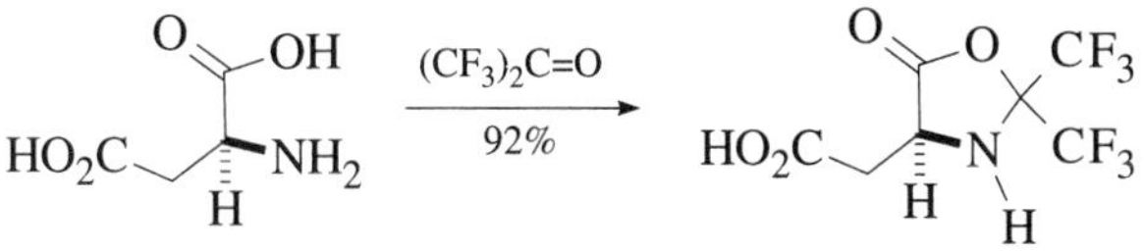

$(CF_3)_2C{=}O$; 92%

Cleavage is achieved with H_2O, IPA, or MeOH.[5] These derivatives also serve as active esters in peptide bond formation.[6]

1. M. Itoh, *Chem. Pharm. Bull.*, **17**, 1679 (1969).
2. M. R. Paleo and F. J. Sardina, *Tetrahedron Lett.*, **37**, 3403 (1996); M. W. Walter, R. M. Adlington, J. E. Baldwin, J. Chuhan, and C. J. Schofield, *Tetrahedron Lett.*, **36**, 7761 (1995).

3. D. Seebach and A. Fadel, *Helv. Chim. Acta*, **68**, 1243 (1985).
4. K. Burger, M. Rudolph, and S. Fehn, *Angew. Chem., Int. Ed. Engl.*, **32,** 285 (1993).
5. K. Burger, E. Windeisen, and R. Pires, *J. Org. Chem.*, **60**, 7641 (1995).
6. K. Burger, M. Rudolph, E. Windeisen, A. Worku, and S. Fehn, *Monatsh. Chem.*, **124**, 453 (1993).

5-Alkyl-4-oxo-1,3-dioxolane:

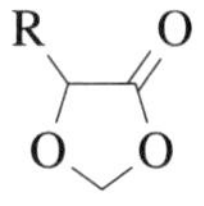

These derivatives are prepared to protect α-hydroxy carboxylic acids; they are cleaved by acidic hydrolysis of the acetal structure (HCl, DMF, 50°, 7 h, 71% yield) or basic hydrolysis of the lactone.[1]

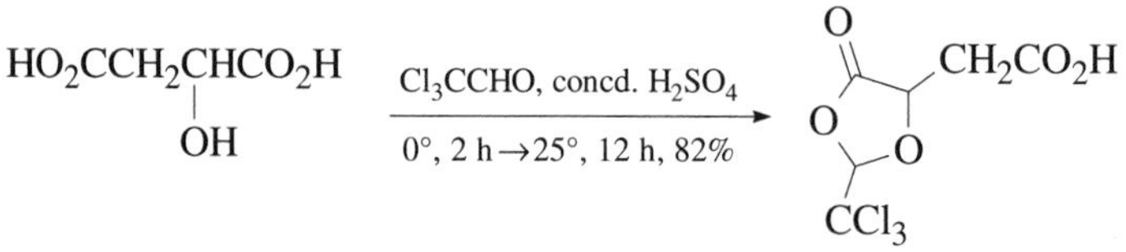

$HO_2CCH_2CH(OH)CO_2H$ → (Cl_3CCHO, concd. H_2SO_4; 0°, 2 h→25°, 12 h, 82%) → 2-CCl_3-5-CH_2CO_2H-4-oxo-1,3-dioxolane

The **2-alkyl** derivatives have been prepared to protect the stereogenic center of the α-hydroxy acid during alkylations.[2]

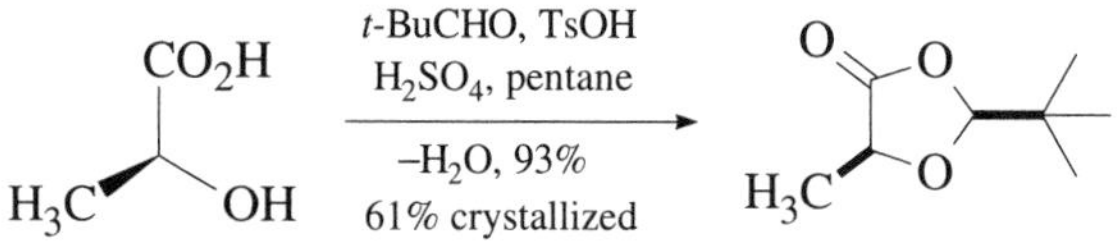

t-BuCHO, TsOH, H_2SO_4, pentane; $-H_2O$, 93%; 61% crystallized

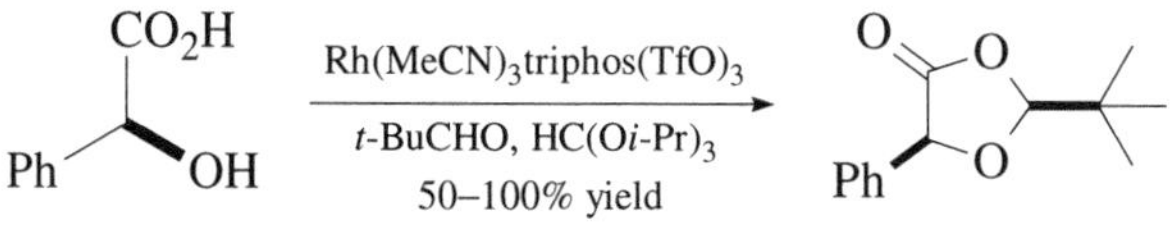

$Rh(MeCN)_3triphos(TfO)_3$; t-BuCHO, $HC(Oi\text{-}Pr)_3$; 50–100% yield

Ref. 3

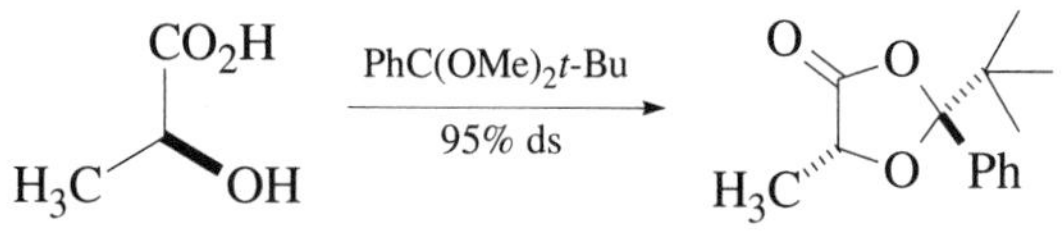

$PhC(OMe)_2t\text{-Bu}$; 95% ds

Ref. 4

This methodology is also effective for the protection of β-hydroxy acids.[5]

1. H. Eggerer and C. Grünewälder, *Justus Liebigs Ann. Chem.*, **677**, 200 (1964).
2. D. Seebach, R. Naef, and G. Calderari, *Tetrahedron*, **40**, 1313 (1984).

3. J. Ott, G. M. R. Tombo, B. Schmid, L. M. Venanzi, G. Wang, and T. R. Ward, *Tetrahedron Lett.*, **30**, 6151 (1989).
4. A. Greiner and J.-Y. Ortholand, *Tetrahedron Lett.*, **31**, 2135 (1990).
5. D. Seebach, R. Imwinkelried, and T. Weber, "EPC Synthesis with C,C Bond Formation via Acetals and Enimines," in *Modern Synthetic Methods 1986*, Vol. 4, R. Scheffold, Ed., Springer-Verlag, New York, 1986, p. 125.

Dioxanones:

O
O $(CH_2)_n$
R O $n = 0, 1$

Dioxanones have been prepared to protect α- or β-hydroxy acids.

Formation

1. RR′C=O, $Sc(NTf_2)_3$ or $Sc(OTf)_3$, CH_2Cl_2, $MgSO_4$ or azeotropic water removal, 54–96% yield. In the case of aldehydes, better stereoselectivity is achieved using $MgSO_4$ as a water scavenger.[1]
2. From a silylated hydroxy acid: RCHO, TMSOTf, 2,6-di-*t*-butylpyridine, 77% yield.[2–4]
3. From a hydroxy acid: pivaldehyde, acid catalyst.[5,6]
4. From a hydroxy acid: $RCH(OR)_2$, PPTS, 20–62% yield.[7,8]

1. K. Ishihara, Y. Karumi, M. Kubota, and H. Yamamoto, *Synlett*, 839 (1996).
2. W. H. Pearson and M.-C. Cheng, *J. Org. Chem.*, **51**, 3746 (1986); *idem, ibid.*, **52**, 1353 (1987).
3. S. L. Schreiber and J. Reagan, *Tetrahedron Lett.*, **27**, 2945 (1986).
4. T. R. Hoye, B. H. Peterson, and J. D. Miller, *J. Org. Chem.*, **52**, 1351 (1987).
5. D. Seebach and J. Zimmerman, *Helv. Chim. Acta*, **69**, 1147 (1986).
6. D. Seebach, R. Imwinkelried, and G. Stucky, *Helv. Chim. Acta*, **70**, 448 (1987).
7. N. Chapel, A. Greiner, and J.-Y. Ortholand, *Tetrahedron Lett.*, **32**, 1441 (1991).
8. J.-Y. Ortholand, N. Vicart, and A. Greiner, *J. Org. Chem.*, **60**, 1880 (1995).

Ortho Esters: $RC(OR')_3$

Ortho esters are one of the few derivatives that can be prepared from acids and esters that protect the carbonyl against nucleophilic attack by hydroxide or other strong nucleophiles such as Grignard reagents. In general, ortho esters are difficult to prepare directly from acids and are therefore more often prepared from the nitrile.[1,2] Simple ortho esters derived from normal alcohols are the least stable in terms of acid stability and stability toward Grignard reagents, but as the ortho ester becomes more constrained, its stability increases.

Formation

1. 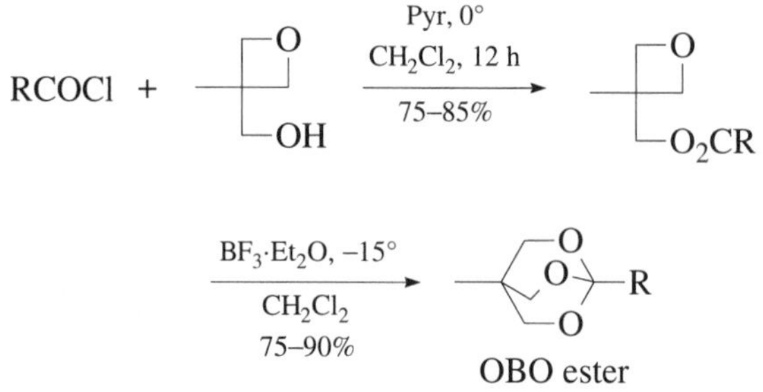 RCOCl + [oxetane alcohol] $\xrightarrow[\text{75–85\%}]{\text{Pyr, 0°, } CH_2Cl_2\text{, 12 h}}$ [oxetane]–O_2CR $\xrightarrow[CH_2Cl_2\text{, 75–90\%}]{BF_3\cdot Et_2O\text{, –15°}}$ OBO ester

Ref. 3

This is one of the few methods available for the direct and efficient conversion of an acid, via the acid chloride, to an ortho ester. The preparation of the oxetane is straightforward, and a large number of oxetanes have been prepared [triol, $(EtO)_2CO$, KOH].[4] In addition, the ***t*-butyl** analogue has been used for the protection of acids.[5] During the course of a borane reduction, the ortho ester was reduced to form a ketal. This was attributed to an intramolecular delivery of the hydride.[6]

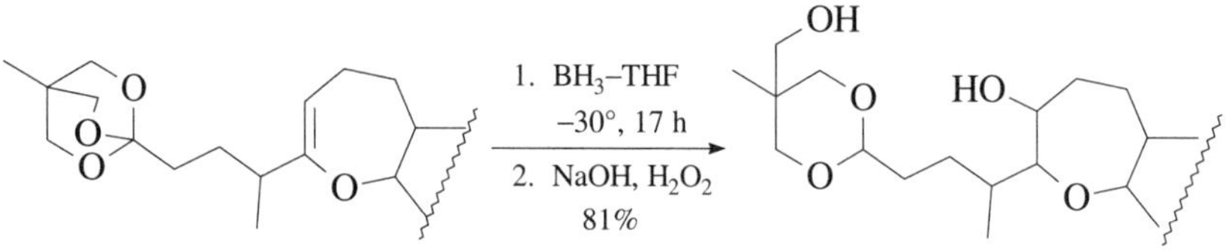

1. BH_3–THF, –30°, 17 h
2. NaOH, H_2O_2
81%

The OBO ester (2,6,7-trioxabicyclo[2.2.2]octyl ester) can also be prepared from a secondary or tertiary amide (Tf_2O, CH_2Cl_2, Pyr then 2,2-bis(hydroxymethyl)-1-propanol, 10–88% yield).[7]

2. The complementary ABO ester (2,7,8-trioxabicyclo[3.2.1]octyl ester) is prepared from the epoxy ester by rearrangement with $Cp_2ZrCl_2/AgClO_4$. The OBO ester is more easily cleaved by Brønsted acids than is the ABO ester, but the ABO ester is cleaved more easily by Lewis acids, thus forming an orthogonal set. The ABO ester can be cleaved with PPTS (MeOH, H_2O, 22°, 2 h; LiOH); the OBO ester is cleaved at 0° in 2 min.[8]

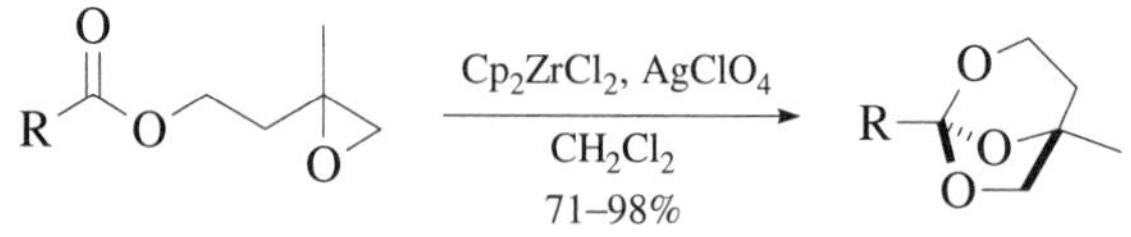

$\xrightarrow[CH_2Cl_2\text{, 71–98\%}]{Cp_2ZrCl_2\text{, } AgClO_4}$

3. 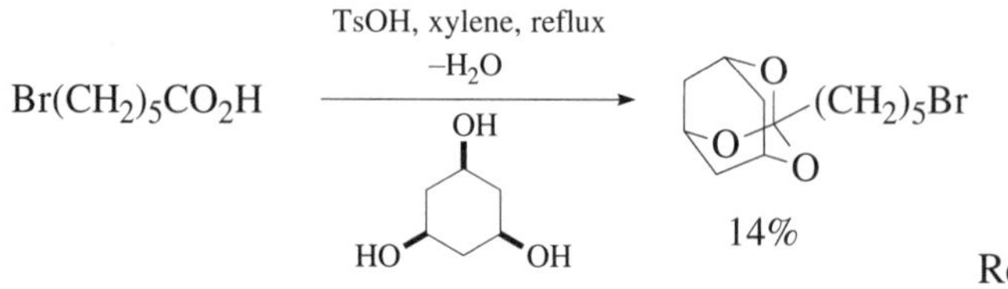 $Br(CH_2)_5CO_2H$ $\xrightarrow{\text{TsOH, xylene, reflux, } -H_2O}$ (with cis-1,3,5-cyclohexanetriol) ortho ester–$(CH_2)_5Br$ 14%

Refs. 9, 10

4. $Br(CH_2)_5CN$ → 1. HCl, MeOH; 2. (cis-1,3,5-cyclohexanetriol, HO, OH, OH) → $(CH_2)_5Br$ ortho ester, 68% Ref. 9

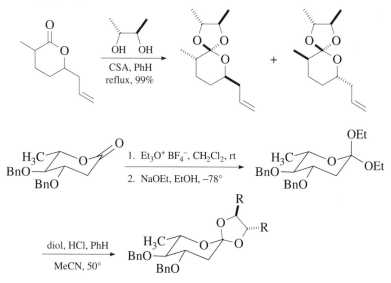

5. HOCH$_2$CH$_2$OH, H$^+$, PhH, reflux, >88% Refs. 11, 12

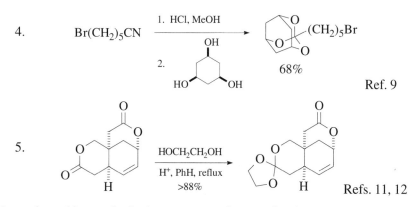

Note that this method does not work on simple esters. In addition, TMSOCH$_2$CH$_2$OTMS/TMSOTF has been used to effect this conversion.[13] The same process was used to introduce the **cyclohexyl** version of this ortho ester in a quassinoid synthesis. Its cleavage was effected with DDQ in aqueous acetone.[14] (*R*,*R*)-2,3-Butanediol can be used to resolve the lactone.[15]

OH OH; CSA, PhH, reflux, 99% → +

1. $Et_3O^+ BF_4^-$, CH_2Cl_2, rt; 2. NaOEt, EtOH, –78° (H$_3$C, BnO, BnO; OEt, OEt)

diol, HCl, PhH, MeCN, 50° (R, R)

2-Substituted gulonolactones failed to react with Meerwein's salt.[16]

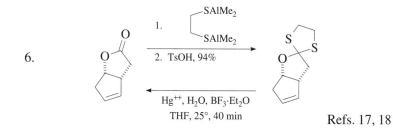

6. 1. $SAlMe_2$ / $SAlMe_2$; 2. TsOH, 94% →
← Hg^{++}, H_2O, $BF_3{\cdot}Et_2O$, THF, 25°, 40 min Refs. 17, 18

7. S, OR′ —R′ONa, R″I, 41–94%→ SR″, OR′, OR′; R, R Ref. 19

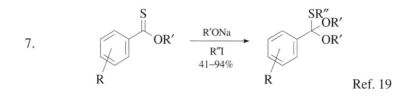

Cleavage

Oxygen ortho esters are readily cleaved by mild aqueous acid (TsOH·Pyr, H_2O;[20] $NaHSO_4$, 5:1 DME, H_2O, 0°, 20 min[21]) to form esters that are then hydrolyzed with aqueous base to give the acid. Note that a trimethyl ortho ester is readily hydrolyzed in the presence of an acid-sensitive ethoxyethyl acetal.[20] The order of acid stability is as follows:

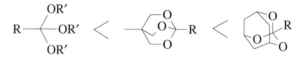

R—C(OR′)3 < (OBO ester)–R < (bicyclic ortho ester)R

The Braun Ortho Ester

Formation/Cleavage[22]

RCO_2H + HO, OH, Ph, Ph, Ph ⟶ R, O, O, OH, Ph, Ph, Ph —LDA, TMSCl, 70–89%→ R, OTMS, O, O, Ph, Ph, Ph

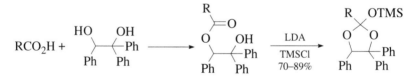

The derivative is stable to *n*-BuLi, *t*-BuLi (−78°), and pH 6–8. It is cleaved with NaOH, MeOH/H_2O at reflux (96% yield).

1. S. M. McElvain and J. W. Nelson, *J. Am. Chem. Soc.*, **64**, 1825 (1942).
2. The synthesis and interconversion of simple ortho esters has been reviewed: R. H. DeWolfe, *Synthesis*, 153 (1974).
3. E. J. Corey and N. Raju, *Tetrahedron Lett.*, **24**, 5571 (1983).
4. C. J. Palmer, L. M. Cole, J. P. Larkin, I. H. Smith, and J. E. Casida, *J. Agric. Food Chem.*, **39**, 1329 (1991).
5. B. R. DeCosta, A. Lewin, J. A. Schoenheimer, P. Skolnick, and K. C. Rice, *Heterocycles*, **32**, 2343 (1991); B. R. DeCosta, A. H. Lewin, K. C. Rice, P. Skolnick, and J. A. Schoenheimer, *J. Med. Chem.*, **34**, 1531 (1991).
6. K. C. Nicolaou, E. A. Theodorakis, F. P. J. T. Rutjes, M. Sato, J. Tiebes, X.-Y. Xiao, C.-K. Hwang, M. E. Duggan, Z. Yang, E. A. Couladouros, F. Sato, J. Shin, H.-M. He, and T. Bleckman, *J. Am. Chem. Soc.*, **117**, 10239 (1995).

7. A. B. Charette and P. Chua, *Tetrahedron Lett.*, **38**, 8499 (1997).
8. P. Wipf, W. Xu, H. Kim, and H. Takahashi, *Tetrahedron*, **53**, 16575 (1997).
9. G. Voss and H. Gerlach, *Helv. Chim. Acta*, **66**, 2294 (1983).
10. C. Müller, G. Voss, and H. Gerlach, *Liebigs Ann. Chem.*, 673 (1995).
11. T. Wakamatsu, H. Hara, and Y. Ban, *J. Org. Chem.*, **50**, 108 (1985).
12. J. D. White, S.-c. Kuo, and T. R. Vedananda, *Tetrahedron Lett.*, **28**, 3061 (1987).
13. D. J. Collins, L. M. Downes, A. G. Jhingran, S. B. Rutschman, and G. J. Sharp, *Aust. J. Chem.*, **42**, 1235 (1989).
14. S. Vasudevan and D. S. Watt, *J. Org. Chem.*, **59**, 361 (1994).
15. A. B. Smith, III, J. W. Leahy, I. Noda, S. W. Remiszewski, N. J. Liverton, and R. Zibuck, *J. Am. Chem. Soc.*, **114**, 2995 (1992).
16. C. Zagar and H. D. Scharf, *Liebigs Ann. Chem.*, 447 (1993).
17. E. J. Corey and D. J. Beames, *J. Am. Chem. Soc.*, **95**, 5829 (1973).
18. M. Nakada, S. Kobayashi, S. Iwasaki, and M. Ohno, *Tetrahedron Lett.*, **34**, 1035 (1993).
19. J. Voss and B. Wollny, *Synthesis*, 684 (1989).
20. G. Just, C. Luthe, and M. T. P. Viet, *Can. J. Chem.*, **61**, 712 (1983).
21. E. J. Corey, K. Niimura, Y. Konishi, S. Hashimoto, and Y. Hamada, *Tetrahedron Lett.*, **27**, 2199 (1986).
22. D. Waldmüller, M. Braun, and A. Steigel, *Synlett*, 160 (1991).

Pentaaminecobalt(III) Complex: $[RCO_2Co(NH_3)_5](BF_4)_3$

The pentaaminecobalt(III) complex has been prepared from amino acids to protect the carboxyl group during peptide synthesis [$(H_2O)Co(NH_3)_5(ClO_4)_3$, 70°, H_2O, 6 h; cool to 0°; filter; HBF_4, 60–80% yield]. It is cleaved by reduction [$NaBH_4$, NaSH, or $(NH_4)_2S$, Fe(II)EDTA]. These complexes do not tend to racemize and are stable to CF_3CO_2H that is used to remove BOC groups.[1–3] The related **bisethylenediamine** complex of amino acids has been prepared. It is stable to strong acids and is cleaved with ammonium sulfide.[4]

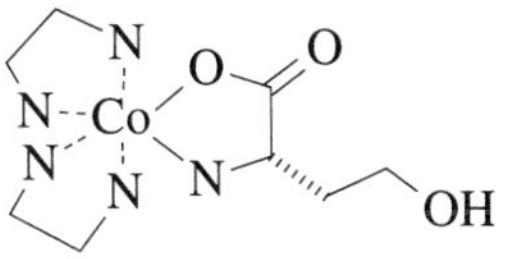

1. S. Bagger, I. Kristjansson, I. Soetofte, and A. Thorlacius, *Acta Chem. Scand. Ser. A*, **A39**, 125 (1985).
2. S. S. Isied, A. Vassilian, and J. M. Lyon, *J. Am. Chem. Soc.*, **104**, 3910 (1982).
3. S. S. Isied, J. Lyon, and A. Vassilian, *Int. J. Pept. Protein Res.*, **19**, 354 (1982).
4. P. M. Angus, B. T. Golding, and A. M. Sargeson, *J. Chem. Soc., Chem. Commun.*, 979 (1993).

Stannyl Esters

Triethylstannyl Ester: $RCOOSn(C_2H_5)_3$

Tri-*n*-butylstannyl Ester: $RCOOSn(n\text{-}C_4H_9)_3$

Stannyl esters have been prepared to protect a –COOH group in the presence of an $-NH_2$ group [$(n\text{-}Bu_3Sn)_2O$ or $n\text{-}Bu_3SnOH$, C_6H_6, reflux, 88%].[1] Stannyl esters of *N*-acylamino acids are stable to reaction with anhydrous amines and to water and alcohols;[2] aqueous amines convert them to ammonium salts.[2] Stannyl esters of amino acids are cleaved in quantitative yield by water or alcohols (PhSK, DMF, 25°, 15 min, 63% yield or HOAc, EtOH, 25°, 30 min, 77% yield).[2]

1. P. Bamberg, B. Ekström, and B. Sjöberg, *Acta Chem. Scand.*, **22**, 367 (1968).
2. M. Frankel, D. Gertner, D. Wagner, and A. Zilkha, *J. Org. Chem.*, **30**, 1596 (1965).

AMIDES AND HYDRAZIDES

To a limited extent, carboxyl groups have been protected as amides and hydrazides, derivatives that complement esters in the methods used for their cleavage. Amides and hydrazides are stable to the mild alkaline hydrolysis that cleaves esters. Esters are stable to nitrous acid, effective in cleaving amides, and to the oxidizing agents [including $Pb(OAc)_4$, MnO_2, SeO_2, CrO_3, and $NaIO_4$;[1] $Ce(NH_4)_2(NO_3)_6$;[2] Ag_2O;[3] and $Hg(OAc)_2$[4]] that have been used to cleave hydrazides.

Formation

Classically, amides and hydrazides have been prepared from an ester or an acid chloride and an amine or hydrazine, respectively; they can also be prepared directly from the acid as shown in the following equations:

1. $RCH(NHPG)COOH + R'NH_2 \xrightarrow[20°,\ 4\ h,\ 70\text{–}90\%]{DCC,\ THF\ or\ CH_2Cl_2\ [5]} RCH(NHPG)CONHR'$

2. $RCH(NHPG)COOH \xrightarrow{H_2NNH_2,\ N\text{-hydroxybenzotriazole}\ [6]} RCH(NHPG)CONHNH_2$

3. $RCOOH + R'R''NH \xrightarrow{Ph_3P\ [7]\ or\ Bu_3P/o\text{-}NO_2\text{–}C_6H_4SCN\ [8]} RCONR'R''$

4. $RCOOH + R'NH_2 \xrightarrow{C_6H_2F_3B(OH)_2,\ \text{toluene, heat}^{9}} RCONHR'$

5. $RCO_2R' + NaEt_2Al(NR''_2)_2 \xrightarrow[83\text{–}96\%^{10}]{} RCONR''_2$

6. $RCOOH \xrightarrow[\text{dioxane, Pyr, 4–16 h}^{11}]{NH_4OCO_2H,\ (BOC)_2O} RCONH_2$

7. $RCOOH + R'OCOCl + TEA \longrightarrow RCO_2CO_2R' \xrightarrow{R''_2NH} RCONR''_2$

This is a very general and mild method for the preparation of amides, applicable to large structural variations in both the acid and the amine. A variety of chloroformates can be employed, but isobutyl chloroformate is used most often. The solvent is not critical, but generally, THF is used.

Cleavage

Equations 1–10 illustrate some mild methods that can be used to cleave amides. Equations 1 and 2 indicate the conditions that were used by Woodward[12] and Eschenmoser,[13] respectively, in their synthesis of vitamin B_{12}. Butyl nitrite,[14] nitrosyl chloride,[15] and nitrosonium tetrafluoroborate ($NO^+BF_4^-$)[16] have also been used to cleave amides. Since only tertiary amides are cleaved by potassium *t*-butoxide (eq. 3), this method can be used to effect selective cleavage of tertiary amides in the presence of primary or secondary amides.[17] (Esters, however, are cleaved by similar conditions.) [18] Photolytic cleavage of nitro amides (eq. 4) is discussed in a review.[19]

1. $RCONH_2 \xrightarrow{N_2O_4/CCl_4{}^{12,20}} RCOOH$

2. $RCONH_2 \xrightarrow{[ClCH_2CH{=}N(\rightarrow O)\text{-}c\text{-}C_6H_{11} + AgBF_4]^{13}} \xrightarrow{H_3O^+} RCOOH$

3. $RCONR'R'' \xrightarrow[24°,\ 2\text{–}48\ \text{h},\ 88\text{–}96\%]{KO\text{-}t\text{-}Bu/H_2O\ (6:2),\ Et_2O^{17}} RCOOH$

$R', R'' \neq H$

4. a, b, or c $\xrightarrow[5\text{–}10\ \text{h},\ 70\text{–}100\%]{350\ \text{nm}^{19}} RCOOH$

a = *o*-nitroanilides[21]
b = *N*-acyl-7-nitroindoles[22]
c = *N*-acyl-8-nitrotetrahydroquinolines[23]

5. $RCONH_2$ → (1,2-dichloro-1,4-dimethoxy-2-cyclopentene / Amberlyst A-21, 45–66°, CH_3CN) → RCO-pyrrol-1-yl → (NaOMe, quant.) → RCO_2Me

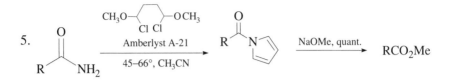

Treatment of acyl pyrroles with primary and secondary amines affords amides.[24]

6. The following cleavage proceeds via intramolecular assistance from the alkoxide formed on base treatment:[25,26]

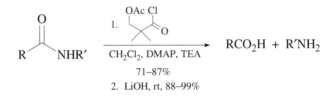

$RCONHR'$ → (1. OAc/Cl acid chloride, CH_2Cl_2, DMAP, TEA, 71–87%; 2. LiOH, rt, 88–99%) → $RCO_2H + R'NH_2$

7. For primary and secondary amides: $CuCl_2$, glyoxal, H_2O, pH 3.5, reflux, 92% yield.[27]

8. For primary amides: DMF dimethyl acetal, MeOH, 92–100% yield. The methyl ester is formed, but if MeOH is replaced with another alcohol, other esters can be prepared with similar efficiency.[28]

9. $NaNO_2$, AcOH, Ac_2O, 30 min, 0°–rt,[29] then hydrolysis with LiOOH. These conditions were developed as a mild method to cleave an amide that was prone to decomposition under the more basic conditions.[30]

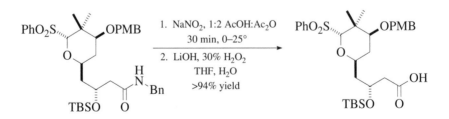

N-Bn amide (PhO₂S, OPMB, TBSO substituted tetrahydropyran) → (1. $NaNO_2$, 1:2 AcOH:Ac_2O, 30 min, 0–25°; 2. LiOH, 30% H_2O_2, THF, H_2O, >94% yield) → carboxylic acid

10. N_2O_4, −20°, CH_3CN, 66–100% yield. Additionally, these conditions cleave hydroxamic acids, anilides, and sulfonamides.[31,32]

Hydrazides have been used in penicillin and peptide syntheses. In the latter syntheses, they are converted by nitrous acid to azides to facilitate coupling.

Some amides and hydrazides that have been prepared to protect carboxyl groups are included in Reactivity Chart 6.

1. M. J. V. O. Baptista, A. G. M. Barrett, D. H. R. Barton, M. Girijavallabhan,

R. C. Jennings, J. Kelly, V. J. Papadimitriou, J. V. Turner, and N. A. Usher, *J. Chem. Soc., Perkin Trans. 1*, 1477 (1977).

2. T.-L. Ho, H. C. Ho, and C. M. Wong, *Synthesis*, 562 (1972).
3. Y. Wolman, P. M. Gallop, A. Patchornik, and A. Berger, *J. Am. Chem. Soc.*, **84**, 1889 (1962).
4. J. B. Aylward and R. O. C. Norman, *J. Chem. Soc. C*, 2399 (1968).
5. J. C. Sheehan and G. P. Hess, *J. Am. Chem. Soc.*, **77**, 1067 (1955).
6. For example, see S. S. Wang, I. D. Kulesha, D. P. Winter, R. Makofske, R. Kutny, and J. Meienhofer, *Int. J. Pept. Protein Res.*, **11**, 297 (1978).
7. L. E. Barstow and V. J. Hruby, *J. Org. Chem.*, **36**, 1305 (1971).
8. P. A. Grieco, D. S. Clark, and G. P. Withers, *J. Org. Chem.*, **44**, 2945 (1979).
9. K. Ishihara, S. Ohara, and H. Yamamoto, *J. Org. Chem.*, **61**, 4196 (1996).
10. T. B. Sim and N. M. Yoon, *Synlett*, 827 (1994).
11. V. F. Posdnev, *Tetrahedron Lett.*, **36**, 7115 (1995).
12. R. B. Woodward, *Pure Appl. Chem.*, **33**, 145 (1973).
13. U. M. Kempe, T. K. Das Gupta, K. Blatt, P. Gygax, D. Felix, and A. Eschenmoser, *Helv. Chim. Acta*, **55**, 2187 (1972).
14. N. Sperber, D. Papa, and E. Schwenk, *J. Am. Chem. Soc.*, **70**, 3091 (1948).
15. M. E. Kuehne, *J. Am. Chem. Soc.*, **83**, 1492 (1961).
16. G. A. Olah and J. A. Olah, *J. Org. Chem.*, **30**, 2386 (1965).
17. P. G. Gassman, P. K. G. Hodgson, and R. J. Balchunis, *J. Am. Chem. Soc.*, **98**, 1275 (1976).
18. P. G. Gassman and W. N. Schenk, *J. Org. Chem.*, **42**, 918 (1977).
19. B. Amit, U. Zehavi, and A. Patchornik, *Isr. J. Chem.*, **12**, 103 (1974).
20. T. Itoh, K. Nagata, Y. Matsuya, M. Miyazaki, and A. Ohsawa, *Tetrahedron Lett.*, **38**, 5017 (1997).
21. B. Amit and A. Patchornik, *Tetrahedron Lett.*, 2205 (1973).
22. B. Amit, D. A. Ben-Efraim, and A. Patchornik, *J. Am. Chem. Soc.*, **98**, 843 (1976).
23. B. Amit, D. A. Ben-Efraim, and A. Patchornik, *J. Chem. Soc., Perkin Trans. 1*, 57 (1976).
24. S. D. Lee, M. A. Brook, and T. H. Chan, *Tetrahedron Lett.*, **24**, 1569 (1983).
25. T. Tsunoda, O. Sasaki, and S. Itô, *Tetrahedron Lett.*, **31**, 731 (1990).
26. T. Tsunoda, O. Sasaki, O. Takenchi, and S. Ito, *Tetrahedron,* **47**, 3925 (1991).
27. L. Singh and R. N. Ram, *J. Org. Chem.*, **59**, 710 (1994).
28. P. L. Anelli, M. Brocchetta, D. Palano, and M. Visigalli, *Tetrahedron Lett.*, **38**, 2367 (1997).
29. I. R. Vlahov, P. I. Vlahova, and R. R. Schmidt, *Tetrahedron Lett.*, **33**, 7503 (1992).
30. D. A. Evans, P. H. Carter, C. J. Dinsmore, J. C. Barrow, J. L. Katz, and D. W. Kung, *Tetrahedron Lett.*, **38**, 4535 (1997).
31. Y. H. Kim, K. Kim, and Y. J. Park, *Tetrahedron Lett.*, **31**, 3893 (1990).
32. D. S. Karanewsky, M. F. Malley, and J. Z. Gougoutas, *J. Org. Chem.*, **56**, 3744 (1991).

Amides

N,N-Dimethylamide: $RCON(CH_3)_2$ (Chart 6)

Formation/Cleavage[1]

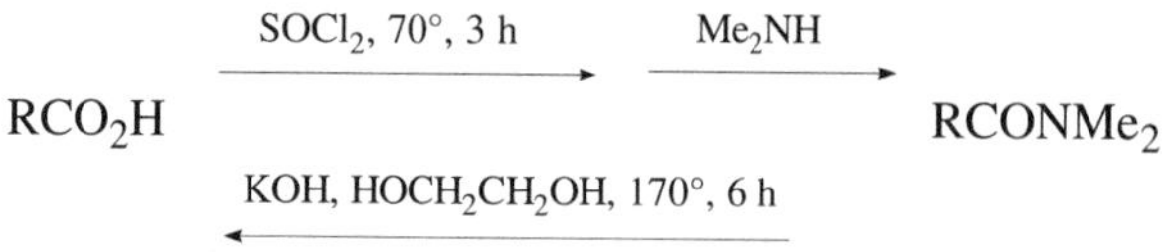

RCO_2H —($SOCl_2$, 70°, 3 h; Me_2NH)→ $RCONMe_2$

$RCONMe_2$ —(KOH, $HOCH_2CH_2OH$, 170°, 6 h)→ RCO_2H

In these papers, the carboxylic acid to be protected was a stable, unsubstituted compound. Harsh conditions were acceptable for both formation and cleavage of the amide. Typically, a simple secondary amide is very difficult to cleave. As the pK_a of the conjugate acid of an amide decreases, the rate of hydrolysis of amides derived from these amines increases. The dimethylamide of a cephalosporin was prepared as follows using 2,2′-dipyridyl disulfide.[2]

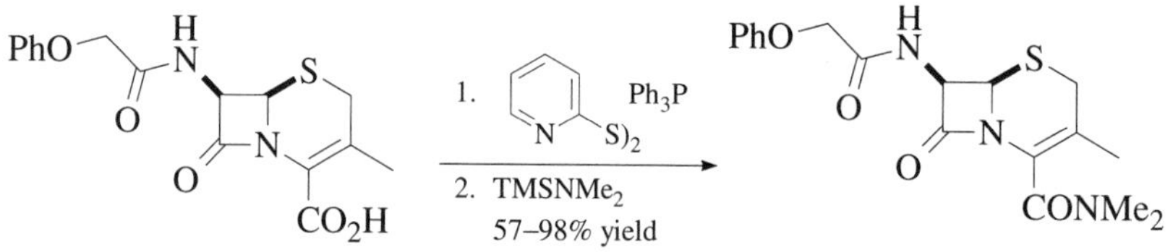

1. (2-pyridyl-S)$_2$, Ph_3P; 2. $TMSNMe_2$; 57–98% yield (cephalosporin CO_2H → $CONMe_2$)

1. D. E. Ames and P. J. Islip, *J. Chem. Soc.*, 351 (1961); *idem, ibid.*, 4363 (1963).
2. R. DiFabio, V. Summa, and T. Rossi, *Tetrahedron*, **49**, 2299 (1993).

Pyrrolidinamide: RCONR′R″, [R′R″= $(-CH_2-)_4$]

Formation/Cleavage[1]

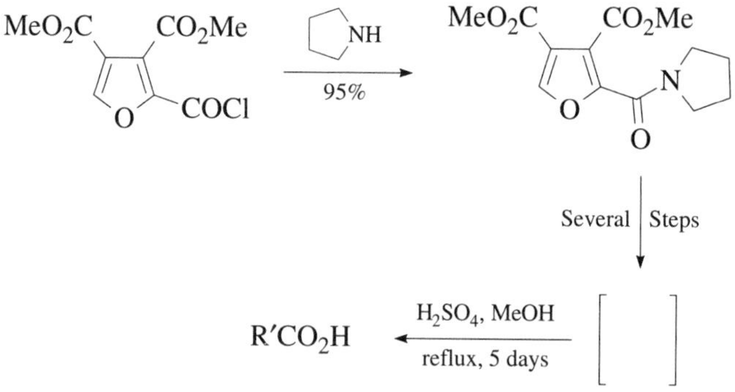

MeO_2C, CO_2Me, COCl (furan) —(pyrrolidine NH, 95%)→ MeO_2C, CO_2Me (furan pyrrolidinamide) —Several Steps→ [ ] —(H_2SO_4, MeOH, reflux, 5 days)→ $R'CO_2H$

$R'CO_2H$ = precursor to DL-camptothecin

1. A. S. Kende, T. J. Bentley, R. W. Draper, J. K. Jenkins, M. Joyeux, and I. Kubo, *Tetrahedron Lett.*, 1307 (1973).

Piperidinamide: RCONR′R″, [R′R″ = $(-CH_2-)_5$]

Formation/Cleavage[1]

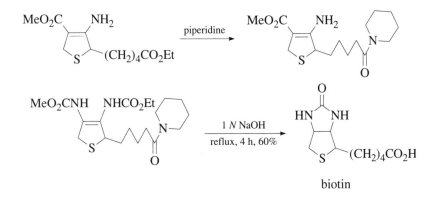

MeO_2C NH_2 S $(CH_2)_4CO_2Et$ —piperidine→ MeO_2C NH_2 S N O

MeO_2CNH $NHCO_2Et$ S N O —1 *N* NaOH, reflux, 4 h, 60%→ O HN NH S $(CH_2)_4CO_2H$

biotin

1. P. N. Confalone, G. Pizzolato, and M. R. Uskokovic, *J. Org. Chem.*, **42**, 1630 (1977).

5,6-Dihydrophenanthridinamide: R N O

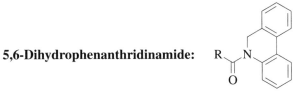

Formation/Cleavage[1]

This amide is stable to HCl or KOH (THF, MeOH, H_2O, 70°, 10 h) and MeMgI, THF, HMPA, −78°. It can also be formed directly from the acid chloride.

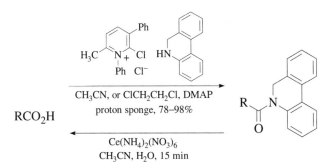

RCO_2H —[Ph, H_3C, N+, Cl, Ph Cl^-; HN], CH_3CN, or $ClCH_2CH_2Cl$, DMAP, proton sponge, 78–98%→ R N O

←$Ce(NH_4)_2(NO_3)_6$, CH_3CN, H_2O, 15 min—

1. T. Uchimaru, K. Narasaka, and T. Mukaiyama, *Chem. Lett.*, 1551 (1981).

o-Nitroanilide: $RCONR'C_6H_4$–o-NO_2, R′≠ H

***N*-7-Nitroindolylamide** (Chart 6):

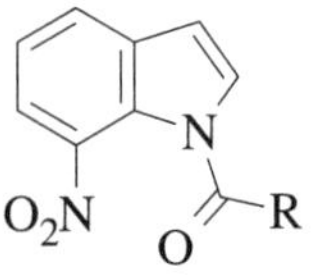

***N*-8-Nitro-1,2,3,4-tetrahydroquinolylamide:**

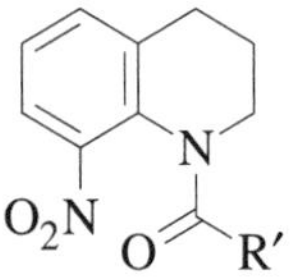

o-Nitroanilides (R′ = Me, n-Bu, c-C_6H_{11}, Ph, $PhCH_2$; ≠ H),[1] nitroindolylamides,[2] and tetrahydroquinolylamides[3] are cleaved in high yields under mild conditions by irradation at 350 nm (5–10 h).

1. B. Amit and A. Patchornik, *Tetrahedron Lett.*, 2205 (1973).
2. B. Amit, D. A. Ben-Efraim, and A. Patchornik, *J. Am. Chem. Soc.*, **98**, 843 (1976).
3. B. Amit, D. A. Ben-Efraim, and A. Patchornik, *J. Chem. Soc., Perkin Trans. 1*, 57 (1976).

2-(2-Aminophenyl)acetaldehyde Dimethyl Acetal Amide:

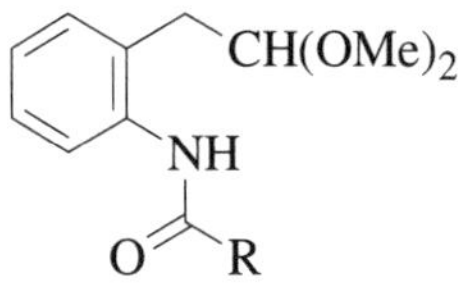

Formation

This amide is readily prepared from the acid chloride (Pyr, rt, 1 h, 77–86% yield) or the acid (DCC, DMAP, CH_2Cl_2, rt, 1 h, 88% yield). Treatment of the amide with camphorsulfonic acid forms an *N*-acylindole. The acid can be regenerated from the *N*-acylindole by LiOH/H_2O_2/THF/H_2O or NaOH/MeOH. Alternatively, it can be transesterified with MeOH/TEA, converted to an amide, by heating with an amine or converted to an aldehyde by DIBAH (62–85% yield).[1]

1. E. Arai, H. Tokuyama, M. S. Linsell, and T. Fukuyama, *Tetrahedron Lett.*, **39**, 71 (1998).

***p-P*-Benzenesulfonamide:** $RCONHSO_2C_6H_4$–*p-P*

A polymer-supported sulfonamide, prepared from an amino acid activated ester and a polystyrene–sulfonamide, is stable to acidic hydrolysis (CF_3COOH; HBr/HOAc). It is cleaved by the "safety-catch" method as shown.[1]

$$RCONHSO_2\text{-}p\text{-}\boldsymbol{P} \xrightarrow{CH_2N_2,\ Et_2O\text{-acetone}} \underset{\text{reactive}}{RCON(CH_3)SO_2C_6H_4\text{-}\boldsymbol{P}} \xrightarrow{0.5\ N\ NaOH} RCO_2H$$

1. G. W. Kenner, J. R. McDermott, and R. C. Sheppard, *J. Chem. Soc., Chem. Commun.*, 636 (1971).

Hydrazides

Hydrazides: $RCONHNH_2$ (Chart 6)

Cleavage

1. NBS/H_2O, 25°, 10 min, 74% yield.[1]
2. 60% $HClO_4$, 48°, 24 h, 100% yield.[2]
3. $POCl_3$, H_2O, 94% yield.[2]
4. HBr/HOAc or HCl/HOAc, 94% yield.[2]
5. $CuCl_2$, H_2O, THF.[3] If an alcohol such as ethanol is substituted for H_2O in this reaction, the ester is produced instead of the acid.

1. H. T. Cheung and E. R. Blout, *J. Org. Chem.*, **30**, 315 (1965).
2. J. Schnyder and M. Rottenberg, *Helv. Chim. Acta*, **58**, 521 (1975).
3. O. Attanasi and F. Serra-Zannetti, *Synthesis*, 314 (1980).

***N*-Phenylhydrazide:** $RCONHNHC_6H_5$ (Chart 6)

Formation

Phenylhydrazides have been prepared from amino acid esters and phenylhydrazine in 70% yield.[1]

Cleavage

1. $Cu(OAc)_2$, 95°, 10 min, 67% yield.[2]
2. $FeCl_3$/1 *N* HCl, 96°, 14 min, 85% yield.[3]

3. Dioxane, DMF, 1 *M* aq. Pyr-AcOH buffer, AcOH, $CuCl_2$, 48 h, air, 86% yield.[4]
4. Horse radish peroxidase, H_2O_2 or Laccase, pH 4, 2% DMSO or DMF. Cleavage occurs by the formation of a phenyldiimide, which decomposes to the acid, nitrogen, and benzene. The laccase method is compatible with the readily oxidized tryptophan and methionine because it does not use peroxide.[5]

1. R. B. Kelly, *J. Org. Chem.*, **28**, 453 (1963).
2. E. W.-Leitz and K. Kühn, *Chem. Ber.*, **84**, 381 (1951).
3. H. B. Milne, J. E. Halver, D. S. Ho, and M. S. Mason, *J. Am. Chem. Soc.*, **79**, 637 (1957).
4. A. N. Semenov and I. V. Lomonosova, *Int. J. Pept. Protein Res.*, **43**, 113 (1994).
5. A. N. Semenov, I. V. Lomonosova, V. I. Berezin, and M. I. Titov, *Biotechnol. Bioengin.*, **42**, 1137 (1993).

N,N'-Diisopropylhydrazide: $RCON(i\text{-}C_3H_7)NH\text{–}i\text{-}C_3H_7$ (Chart 6)

The *N,N'*-diisopropylhydrazide, prepared to protect penicillin derivatives, is cleaved oxidatively by the following methods:[1]

1. $Pb(OAc)_4$/Pyr, 25°, 10 min, 90% yield.
2. $NaIO_4/H_2O$–THF, H_2SO_4, 20°, 5 min, 89% yield.
3. Aq. NBS/THF–Pyr, 20°, 10 min, 90% yield.
4. CrO_3/HOAc, 25°, 10 min, 65% yield.

A number of di- and trisubstituted hydrazides of penicillin and cephalosporin derivatives were prepared to study the effect of *N*-substitution on the ease of oxidative cleavage.[2]

1. D. H. R. Barton, M. Girijavallabhan, and P. G. Sammes, *J. Chem. Soc., Perkin Trans. 1*, 929 (1972).
2. D. H. R. Barton and eight co-workers, *J. Chem. Soc., Perkin Trans. 1*, 1477 (1977).

Phenyl Group: C_6H_5–

The phenyl group became a practical "protective" group for carboxylic acids when Sharpless published a mild, effective one-step method for its conversion to a carboxylic acid.[1] Recently, the group has been used in a synthesis of the amino acid statine, in which it served as a masked or carboxylic acid equivalent.[2]

$$\text{BOCN-oxazolidinone-CH}_2\text{Ph} \xrightarrow[\text{85\% yield}]{\text{RuCl}_3,\ \text{NaIO}_4,\ \text{CH}_3\text{CN},\ \text{H}_2\text{O},\ \text{CCl}_4} \text{BOCN-oxazolidinone-CH}_2\text{CO}_2\text{H}$$

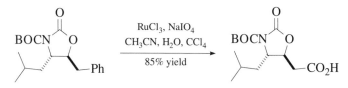

The furan group also serves as a protected carboxylic acid.[3]

1. P. H. J. Carlsen, T. Katsuki, V. S. Martin, and K. B. Sharpless, *J. Org. Chem.*, **46**, 3936 (1981).
2. S. Kano, Y. Yuasa, T. Yokomatsu, and S. Shibuya, *J. Org. Chem.*, **53**, 3865 (1988).
3. S. Sasaki, Y. Hamada, and T. Shioiri, *Tetrahedron Lett.*, **38**, 3013 (1997).

Tetraalkylammonium Salts: $R'_4N^+\ ^-O_2CR$

In a rather nontraditional approach to acid protection, the tetraalkylammonium salts of amino acids allow for coupling of HOBt-activated amino acids in yields of 55–84%.[1]

1. S.-T. Chen and K.-T. Wang, *J. Chem. Soc., Chem. Commun.*, 1045 (1990).

PROTECTION OF SULFONIC ACIDS

Few methods exist for the protection of sulfonic acids. Imidazolides and phenolic esters are too base labile to be useful in most cases. Simple sulfonate esters often cannot be used because these are obviously quite susceptible to nucleophilic reagents.

Neopentyl Ester: $(CH_3)_3CCH_2OSO_2R$

The neopentyl alcohol, prepared from the sulfonyl chloride (Pyr, 95% yield), is cleaved nucleophilically under rather severe conditions ($Me_4N^+\ Cl^-$, DMF, 160°, 16 h, 100% yield).[1]

N-BOC-4-Amino-2,2-dimethylbutyl Sulfonate: $BOCNHCH_2CH_2C(CH_3)_2CH_2OSO_2R$

This sulfonate, prepared from $BOCNHCH_2CH_2C(CH_3)_2CH_2OH$ and the sulfonyl chloride (Pyr, 100% yield) is cleaved by initial BOC cleavage to release the free amine after pH adjustment to 7–8. Intramolecular displacement occurs to release the sulfonate and a pyrrolidine.[1]

Isobutyl Sulfonate: $(CH_3)_2CHCH_2OSO_2R$

The isobutyl sulfonate was examined as a replacement for the isopropyl sulfonate, which had undesirable stability properties. Cleavage occurs with 2 eq. of $Bu_4N^+ I^-$ and proceeds much more readily than cleavage of the isopropyl sulfonate.[2]

Isopropyl Sulfonate: $(CH_3)_2CHOSO_2R$

This sulfonate is cleaved with $Bu_4N^+ I^-$ or ammonia. [3] The group has been reported to suffer from stability problems upon storage and use.[2]

1. J. C. Roberts, H. Gao, A. Gopalsamy, A. Kongsjahju, and R. J. Patch, *Tetrahedron Lett.*, **38**, 355 (1997).
2. M. Xie and T. S. Widlanski, *Tetrahedron Lett.*, **37**, 4443 (1996).
3. B. Musicki and T. S. Widlanski, *Tetrahedron Lett.*, **32**, 1267 (1991); B. Musicki and T. S. Widlanski, *J. Org. Chem.*, **55**, 4231 (1990).

PROTECTION OF BORONIC ACIDS

Boronic esters are easily prepared from a diol and the boronic acid with removal of water, either chemically or azeotropically. (See Chapter 2 on the protection of diols.) Sterically hindered boronic esters, such as those of **pinacol**, can be prepared in the presence of water. Boronic esters of simple unhindered diols are quite sensitive to water and hydrolyze readily. On the other hand, very hindered esters, such as the **pinacol** and **pinanediol** derivatives, are exceedingly difficult to hydrolyze and often require rather harsh conditions to achieve cleavage.

Cleavage

1. Ether, water, phenylboronic acid. Cleavage occurs by transesterification.[1,2]
2. (a) $NaIO_4$, NH_4OAc, acetone, water, 24–48 h; (b) pH 3 with HCl, 55–71% yield.[1]
3. BCl_3, −78°, CH_2Cl_2, 8 h, 83% yield.[3]

CH_3S / AcNH / B / O / O → CH_3S / AcNH / $B(OH)_2$

4. $LiAlH_4$, Et_2O; MeONa, 1,3-propanediol.[4] These conditions reduce the boronate to the hydride.
5. $HN(CH_2CH_2OH)_2$, ether; 1 *N* HCl, ~80% yield.[5,6]

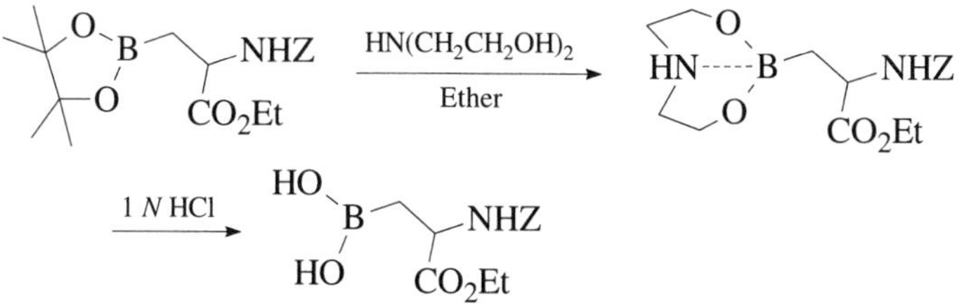

HN(CH$_2$CH$_2$OH)$_2$, Ether; 1 N HCl

6.

NaOH, (HOCH$_2$)$_3$CCH$_2$OH, Ether, H_2O

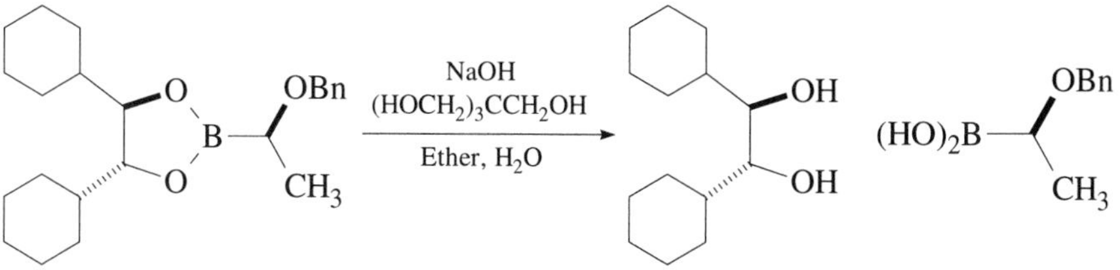

This method was only partially successful with the pinanediol boroanate.[7]

1,2-Benzenedimethanol

This ester is formed quantitatively in THF from the diol in the presence of a dehydrating agent such as sodium sulfate. It can be cleaved by hydrogenolysis, but it is also quite susceptible to hydrolytic cleavage.[8]

1,3-Diphenyl-1,3-propanediol

Esterification is readily achieved in THF in the presence of a dehydrating agent.[9] The boronate is stable to chromatography, has good stability to 2 *M* TFA/CH_2Cl_2, but is not stable to aqueous 1 *M* NaOH. Cleavage is also achieved by hydrogenolysis.[8]

1. S. J. Coutts, J. Adams, D. Krolikowski, and R. J. Snow, *Tetrahedron Lett.*, **35**, 5109 (1994).
2. J. Wityak, R. A. Earl, M. M. Abelman, Y. B. Bethel, B. N. Fisher, G. S. Kauffman, C. A. Kettner, P. Ma, J. L. McMillan, L. J. Mersinger, J. Pesti, M. E. Pierce, F. W. Rankin, R. J. Chorvat, and P. N. Confalone, *J. Org. Chem.*, **60**, 3717 (1995).
3. D. S. Matteson, T. J. Michnick, R. D. Willett, and C. D. Patterson, *Organometallics*, **8**, 726 (1989).
4. M. V. Rangaishenvi, B. Singaram, and H. C. Brown, *J. Org. Chem.*, **56**, 3286 (1991).
5. D. H. Kinder and M. M. Ames, *J. Org. Chem.*, **52**, 2452 (1987).
6. D. S. Matteson and R. Ray, *J. Am. Chem. Soc.*, **102**, 7590 (1980).
7. D. S. Matteson and H.-W. Man, *J. Org. Chem.*, **61**, 6047 (1996).
8. C. Malan, C. Morin, and G. Preckher, *Tetrahedron Lett.*, **37**, 6705 (1996).
9. C. Malan and C . Morin, *Synlett*, 167 (1996).

6

PROTECTION FOR THE THIOL GROUP

Protection for the thiol group is important in many areas of organic research, particularly in peptide and protein syntheses, which often involve the amino acid cysteine, $HSCH_2CH(NH_2)CO_2H$, CySH.[1] Protection of the thiol group in β-lactam chemistry has been reviewed.[2] The synthesis[3] of coenzyme A, which converts a carboxylic acid into a thioester, an acyl transfer agent in the biosynthesis or oxidation of fatty acids, also requires the use of thiol protective groups. A free –SH group can be protected as a thioether or a thioester, or oxidized to a symmetrical disulfide, from which it is regenerated by reduction. Thioethers are, in general, formed by reaction of the thiol, in a basic solution, with a halide; they are cleaved by reduction with sodium/ammonia, by acid-catalyzed hydrolysis, or by reaction with a heavy metal ion such as silver(I) or mercury(II), followed by hydrogen sulfide treatment. Some groups, including *S*-diphenylmethyl and *S*-triphenylmethyl thioethers and *S*-2-tetrahydropyranyl and *S*-isobutoxymethyl hemithioacetals, can be oxidized by thiocyanogen, $(SCN)_2$, iodine, or a sulfenyl chloride to a disulfide that is subsequently reduced to the thiol. Thioesters are formed and cleaved in the same way as oxygen esters; they are more reactive to nucleophilic substitution, as indicated by their use as "activated esters." Several miscellaneous protective groups, including thiazolidines, unsymmetrical disulfides, and *S*-sulfenyl derivatives, have been used to a more limited extent. This chapter discusses some synthetically useful thiol protective groups.[4,5] Some of the more useful groups are included in Reactivity Chart 7.

1. For a review on cysteine protection, see F. Cavelier, J. Daunis, and R. Jacquier, *Bull. Soc. Chim. Fr.*, 210 (1990); see also reference 22 (**Peptides**) in Chapter 1.
2. H. Wild, "Protective Groups in β-Lactam Chemistry," in *The Organic Chemistry of β-Lactams*, G. I. Georg, Ed., VCH Publishers, 1993, pp. 1–48.
3. J. G. Moffatt and H. G. Khorana, *J. Am. Chem. Soc.*, **83**, 663 (1961).
4. See also Y. Wolman, "Protection of the Thiol Group," in *The Chemistry of the Thiol Group*, S. Patai, Ed., Wiley-Interscience, New York, 1974, Vol. 15/2, pp. 669–684; R. G. Hiskey, V. R. Rao, and W. G. Rhodes, "Protection of Thiols," in *Protective Groups in Organic Chemistry*, J. F. W. McOmie, Ed., Plenum Press, New York and London, 1973, pp. 235–308.
5. R. G. Hiskey, "Sulfhydryl Group Protection in Peptide Synthesis," in *The Peptides*, E. Gross and J. Meienhofer, Eds., Academic Press, New York, 1981, Vol. 3, pp. 137–167.

THIOETHERS

S-Benzyl and substituted *S*-benzyl derivatives, readily cleaved with sodium/ammonia, are the most frequently used thioethers. *n*-Alkyl thioethers are difficult to cleave and have not been used extensively as protective groups. Alkoxymethyl or alkylthiomethyl hemithio- or dithioacetals ($RSCH_2OR'$ or $RSCH_2SR'$) can be cleaved by acidic hydrolysis and by reaction with silver or mercury salts, respectively. Mercury(II) salts also cleave dithioacetals ($RS-CH_2SR'$), *S*-triphenylmethyl thioethers ($RS-CPh_3$), *S*-diphenylmethyl thioethers ($RS-CHPh_2$), *S*-acetamidomethyl derivatives ($RS-CH_2NHCOCH_3$), and *S*-(*N*-ethylcarbamates) (RS–CONHEt). *S*-*t*-Butyl thioethers (RS–*t*-Bu) are cleaved if refluxed with mercury(II); *S*-benzyl thioethers ($RS-CH_2Ph$) are cleaved if refluxed with mercury (II)/1 *N* HCl. Some β-substituted *S*-ethyl thioethers are cleaved by reactions associated with the β-substituent.

S-Alkyl Thioethers: $C_nH_{2n+1}SR$

Formation

1. *S*,*S*-Diphenyl-*S*-methoxythiazyne, benzene, 30° was used to prepare the methyl thioether.[1]
2. One of the simplest methods for preparation is by reaction of the thiol with KOH and RX in ethanol as solvent.
3. In many cases, a thiol group is introduced into a substrate through the use of a thiol (e.g., monoprotected H_2S), by a simple displacement or an addition reaction.[2]

Cleavage

1. Na/NH_3, >54% yield. Methyl thioether cleavage of BOC-protected methionine.[3]
2. Na/NH_3[4].

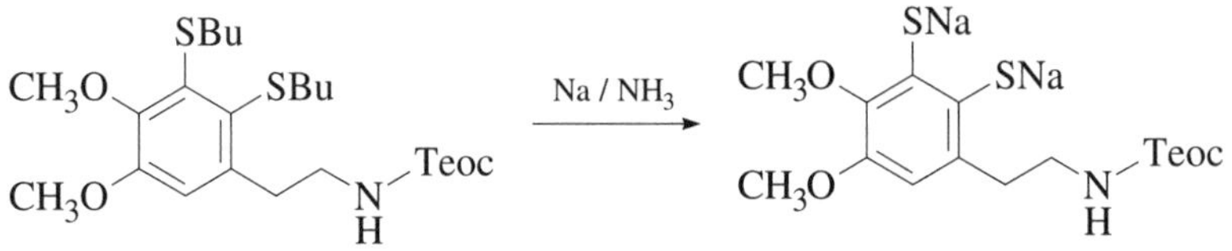

1. T. Yoshimura, E. Tsukurimichi, Y. Sugiyama, H. Kita, C. Shimasaki, and K. Hasegawa, *Bull. Chem. Soc. Jpn.*, **64**, 3176 (1991).
2. For a review, see J. L. Wardell, in *The Chemistry of the Thiol Group, Pt. 1,* S. Patai, Ed., Wiley, New York, 1974, pp. 179–211.
3. R. Lutgring, K. Sujatha, and J. Chmielewski, *Bioorg. Med. Chem. Lett.*, **3**, 739 (1993).
4. P. W. Ford, M. R. Narbut, J. Belli, and B. S. Davidson, *J. Org. Chem.*, **59**, 5955 (1994).

S-Benzyl Thioether: $RSCH_2Ph$ (Chart 7)

For the most part, cysteine and its derivatives have been protected by the reactions that follow.

Formation

1. $PhCH_2Cl$, 2 *N* NaOH or NH_3, EtOH, 30 min, 25°, 90% yield.[1]
2. $PhCH_2Cl$, Cs_2CO_3, DMF, 20°.[2]
3. $PhCH_2Br$, *n*-BuLi, THF, 0°–rt, 30 min, 85% yield.[3]

Cleavage

1. Na, NH_3, 10 min.[4]
2. Sodium in boiling butyl alcohol[5] or in boiling ethyl alcohol[6] can be used if the benzyl thioether is insoluble in ammonia.
3. Li, NH_3, THF, −78°.[7]

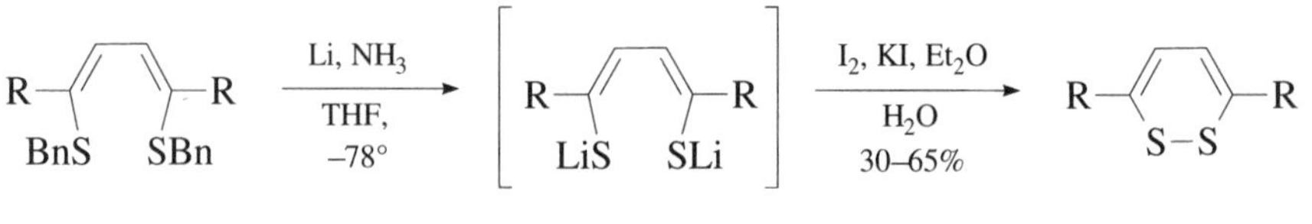

   In this case, the use of Na/NH_3 was slow.
4. HF, anisole, 25°, 1 h.[8] The authors list 15 protective groups that are cleaved by this method, including some branched-chain carbonates and esters,

benzyl esters and ethers, the nitro-protective group in arginine, and *S*-benzyl and *S-t*-butyl thioethers. They report that 12 protective groups, including some straight-chain carbonates and esters, *N*-benzyl derivatives, and *S*-methyl, *S*-ethyl, and *S*-isopropyl thioethers, are stable under these conditions.

5. 5% Cresol, 5% thiocresol, 90% HF.[9] In the HF deprotection of thioethers and many other protective groups, anisole serves as a scavenger for the liberated cation formed during the deprotection process. If cations liberated during this deprotection are not scavenged, they can react with other amino acid residues, especially tyrosine. Dimethyl sulfide, thiocresol, cresol, and thioanisole have also been used as scavengers when strong acids are used for deprotection. A mixture of 5% cresol, 5% *p*-thiocresol, and 90% HF is recommended for benzyl thioether deprotection.[9] These conditions cause cleavage by an S_N1 mechanism. The use of low concentrations of HF in dimethyl sulfide (1:3), which has been recommended for the deprotection of other peptide protective groups, does not cleave the ***S*-4-methylbenzyl** group. Reactions that use low HF concentrations are considered to proceed via an S_N2 mechanism. The use of low HF concentrations with thioanisole results in some methylation of free thiols. The use of HF in anisole can also result in alkylation of methionine.
6. Electrolysis, NH_3, 90 min.[10]
7. Electrolysis, –2.8 V, DMF, $R_4N^+X^-$, 82% yield.[11,12]
8. Ph_2SO, $MeSiCl_3$, TFA, 4°, 4 h, 94% yield. The disulfide is formed.[13]
9. Bu_3SnH, AIBN, PhH, 3 h, Δ, >72% yield. The thiol is released as a stannyl sulfide that was used directly in a glycosylation.[14]

1. M. Frankel, D. Gertner, H. Jacobson, and A. Zilkha, *J. Chem. Soc.*, 1390 (1960); M. Dymicky and D. M. Byler, *Org. Prep. Proced. Int.*, **23**, 93 (1991).
2. F. Vögtle and B. Klieser, *Synthesis*, 294 (1982).
3. J. Yin and C. Pidgeon, *Tetrahedron Lett.*, **38**, 5953 (1997).
4. J. E. T. Corrie, J. R. Hlubucek, and G. Lowe, *J. Chem. Soc., Perkin Trans. 1*, 1421 (1977).
5. W. I. Patterson and V. du Vigneaud, *J. Biol. Chem.*, **111**, 393 (1935).
6. K. Hofmann, A. Bridgwater, and A. E. Axelrod, *J. Am. Chem. Soc.*, **71**, 1253 (1949).
7. M. Koreeda and W. Yang, *Synlett*, 201 (1994).
8. S. Sakakibara, Y. Shimonishi, Y. Kishida, M. Okada, and H. Sugihara, *Bull. Chem. Soc. Jpn.*, **40**, 2164 (1967).
9. W. F. Heath, J. P. Tam, and R. B. Merrifield, *Int. J. Pept. Protein Res.*, **28**, 498 (1986).
10. D. A. J. Ives, *Can. J. Chem.*, **47**, 3697 (1969).
11. V. G. Mairanovsky, *Angew. Chem., Int. Ed. Engl.*, **15**, 281 (1976).
12. C. M. Delerue-Matos, A. M. Freitas, H. L. S. Maia, M. J. Medeiros, M. I. Montenegro, and D. Pletcher, *J. Electroanal. Chem. Interfacial Electrochem.*, **315**, 1 (1991).

13. T. Koide, A. Otaka, H. Suzuki, and N. Fujii, *Synlett*, 345 (1991); K. Akaji, T. Tatsumi, M. Yoshida, T. Kimura, Y. Fujiwara, and Y. Kiso, *J. Chem. Soc., Chem. Comm.*, 167 (1991).
14. W. P. Neumann, *Synthesis*, 665 (1987); H.-S. Byun and R. Bittman, *Tetrahedron Lett.*, **36**, 5143 (1995).

S-p-Methoxybenzyl Thioether: $RSCH_2C_6H_4$–*p*-OCH_3 (Chart 7)

Formation

1. 4-$MeOC_6H_4CH_2Cl$, NH_3, 78% yield.[1]
2. 4-$MeOC_6H_4CH_2Cl$, Na/NH_3, 87% yield.[2]
3. 4-$MeOC_6H_4CH_2OH$, TFA, CH_2Cl_2, 37–81% yield.[3]
4.

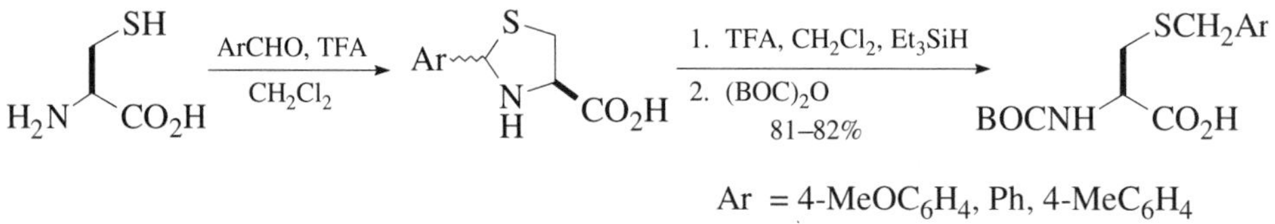

SH / H_2N CO_2H —(ArCHO, TFA / CH_2Cl_2)→ Ar (thiazolidine: S, NH) CO_2H —(1. TFA, CH_2Cl_2, Et_3SiH; 2. $(BOC)_2O$; 81–82%)→ SCH_2Ar / BOCNH CO_2H

Ar = 4-$MeOC_6H_4$, Ph, 4-MeC_6H_4

Ref. 3

5. 4-$MeOC_6H_4CH_2Cl$, NaH, THF, 60°, 1 h.[4]

Cleavage

An *S*-4-methoxybenzyl thioether is stable to HBr/AcOH[1] and to I_2/MeOH.[5] The latter reagent cleaves *S*-trityl and *S*-diphenylmethyl groups.

1. $Hg(OAc)_2$, CF_3COOH, 0°, 10–30 min, or $Hg(OCOCF_3)_2$, aq. AcOH, 20°, 2–3 h, followed by H_2S or $HSCH_2CH_2OH$, 100% yield.[6,7] An *S*-*t*-butyl thioether is cleaved in quantitive yield under these conditions.
2. $Hg(OCOCF_3)_2$, CF_3COOH, anisole.[8] The **dimethoxybenzyl** thioether is also cleaved with this reagent.[9]

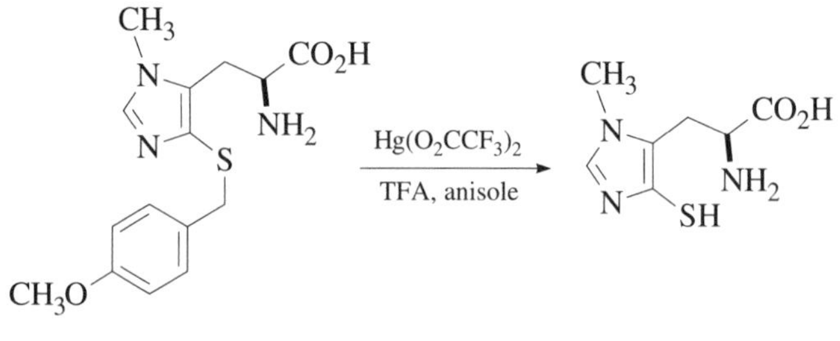

CH_3, N, N, S, CH_3O, CO_2H, NH_2 —($Hg(O_2CCF_3)_2$ / TFA, anisole)→ CH_3, N, N, SH, CO_2H, NH_2

3. CF_3COOH, reflux.[1]
4. CF_3COOH, *o*-cresol, reflux, 24 h, >52% yield.[10]
5. Anhydrous HF, anisole, 25°, 1 h, quant.[11]

6.

BOC–Cys-OMe with 4-$MeOC_6H_4CH_2$ on S —[(4-$BrC_6H_4)_3N^{\cdot+}$ $SbCl_6^-$, 75%[5] or *in situ* electrolysis, CH_3CN, $NaHCO_3$, cat. (4-$BrC_6H_4)_3N^{\cdot+}$ $SbCl_6^-$, 91%]→ BOC–Cys-OMe / BOC–Cys-OMe (disulfide)

7. During the synthesis of peptides that contain 4-methoxybenzyl-protected cysteine residues, sulfoxide formation may occur. These sulfoxides, when treated with HF/anisole, form thiophenyl ethers that cannot be deprotected; therefore, the peptides should be subjected to a reduction step prior to deprotection.[12]

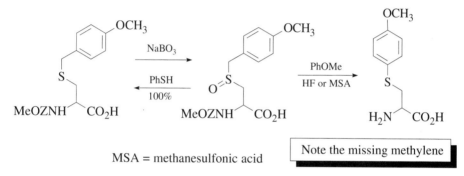

MeOZNH–Cys(4-MeOBn)–CO_2H —[$NaBO_3$]→ sulfoxide (O=S) ; reverse: PhSH, 100% ; sulfoxide —[PhOMe, HF or MSA]→ H_2N–CH(CO_2H)–CH_2–S–C_6H_4–OCH_3

MSA = methanesulfonic acid

Note the missing methylene

8. $AgBF_4$, anisole, TFA, 4°, 1 h, 87% conversion.[13]
9. $MeSiCl_3$, Ph_2SO, TFA, 4°, 10 min, 95% conversion to cystine.[14]

1. S. Akabori, S. Sakakibara, Y. Shimonishi, and Y. Nobuhara, *Bull. Chem. Soc. Jpn.*, **37**, 433 (1964).
2. M. D. Bachi, S. Sasson, and J. Vaya, *J. Chem. Soc., Perkin Trans.1*, 2228 (1980).
3. L. S. Richter, J. C. Marster, Jr., and T. R. Gadek, *Tetrahedron Lett.*, **35**, 1631 (1994).
4. A. W. Tayor and D. K. Dean, *Tetrahedron Lett.*, **29**, 1845 (1988).
5. M. Platen and E. Steckhan, *Liebigs Ann. Chem.*, 1563 (1984).
6. O. Nishimura, C. Kitada, and M. Fujino, *Chem. Pharm. Bull.*, **26**, 1576 (1978).
7. E. M. Gordon, J. D. Godfrey, N. G. Delaney, M. M. Asaad, D. Von Langen, and D. W. Cushman, *J. Med. Chem.*, **31**, 2199 (1988).
8. T. P. Holler, A. Spaltenstein, E. Turner, R. E. Klevit, B. M. Shapiro, and P. B. Hopkins, *J. Org. Chem.*, **52**, 4420 (1987).
9. N. Shibata, J. E. Baldwin, A. Jacobs, and M. E. Wood, *Tetrahedron*, **52**, 12839 (1996).
10. R. Lutgring, K. Sujatha, and J. Chmielewski, *Bioorg. Med. Chem. Lett.*, **3**, 739 (1993).
11. S. Sakakibara and Y. Shimonishi, *Bull. Chem. Soc. Jpn.*, **38**, 1412 (1965).
12. S. Funakoshi, N. Fujii, K. Akaji, H. Irie, and H. Yajima, *Chem. Pharm. Bull.*, **27**, 2151 (1979).

13. M. Yoshida, T. Tatsumi, Y. Fujiwara, S. Iinuma, T. Kimura, K. Akaji, and Y. Kiso, *Chem. Pharm. Bull.*, **38**, 1551 (1990).
14. K. Akaji, T. Tatsumi, M. Yoshida, T. Kimura, Y. Fujiwara and Y. Kiso, *J. Chem. Soc., Chem. Commun.*, 167 (1991).

S-o- or *p*-Hydroxy- or Acetoxybenzyl Thioether:
$RSCH_2C_6H_4$–*o*-(or *p*-)-OR′: R′ = H or Ac

Formation/Cleavage[1]

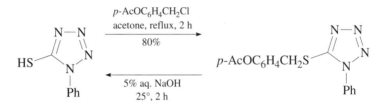

HS–(1-phenyltetrazol-5-yl) → (*p*-$AcOC_6H_4CH_2Cl$, acetone, reflux, 2 h, 80%) → *p*-$AcOC_6H_4CH_2S$–(1-phenyltetrazol-5-yl); reverse: 5% aq. NaOH, 25°, 2 h

The cleavage process occurs by *p*-quinonemethide formation after acetate hydrolysis.

1. L. D. Taylor, J. M. Grasshoff, and M. Pluhar, *J. Org. Chem.*, **43**, 1197 (1978); J. B. Christensen, *Org. Prep. Proced. Int.*, **26**, 471 (1994); C. Gemmell, G. C. Janairo, J. D. Kilburn, H. Ueck, and A. E. Underhill, *J. Chem. Soc., Perkin Trans. 1*, 2715 (1994).

S-p-Nitrobenzyl Thioether: $RSCH_2C_6H_4$–*p*-NO_2 (Chart 7)

Formation

1. 4-$NO_2C_6H_4CH_2Cl$, 1 *N* NaOH, 0°, 1 h → 25°, 0.5 h[1] or NaH, $PhCH_3$, 68% yield.[2]

Cleavage

1. H_2, Pd–C, HCl or AcOH, 7–8 h, 60–68%; $HgSO_4$, H_2SO_4, 20 h, 60%; H_2S, 15 min., 60% yield[2] or RSSR, 76% yield after air oxidation.[1] Hydrogenation initially produces the *p*-amino derivative that is then cleaved with Hg(II).

1. M. D. Bachi and K. J. Ross-Petersen, *J. Org. Chem.*, **37**, 3550 (1972).
2. M. D. Bachi and K. J. Ross-Petersen, *J. Chem. Soc., Chem. Commun.*, 12 (1974).

S-2,4,6-Trimethylbenzyl Thioether: 2,4,6-$Me_3C_6H_2$ CH_2SR

Formation

1. From cysteine: Na/NH_3, 2,4,6-$Me_3C_6H_2CH_2Cl$, 57% yield.[1]

Cleavage

1. HF, anisole, 0°, 30 min or TfOH, TFA, anisole, 30 min. This group is stable to refluxing TFA, whereas the more frequently used 4-methoxybenzyl group is not.[1]
2. Me_2Se, HF, *m*-cresol, 0°, 60 min. These conditions are also excellent for reduction of methionine sulfoxide [Met(O)].[2]
3. $AgBF_4$, anisole, TFA, 4°, 1 h, 73% conversion.[3]

1. F. Brtnik, M. Krojidlo, T. Barth, and K. Jost, *Collect. Czech. Chem. Commun.*, **46**, 286 (1981).
2. M. Yoshida, M. Shimokura, Y. Fujiwara, T. Fujisaki, K. Akaji, and Y. Kiso, *Chem. Pharm. Bull.*, **38**, 382 (1990).
3. M. Yoshida, T. Tatsumi, Y. Fujiwara, S. Iinuma, T. Kimura, K. Akaji, and Y. Kiso, *Chem. Pharm. Bull.*, **38**, 1551 (1990).

S-2,4,6-Trimethoxybenzyl Thioether (Tmob–SR): 2,4,6-$(MeO)_3C_6H_2CH_2SR$

Formation

1. 2,4,6-$(MeO)_3C_6H_2CH_2OH$, TFA, CH_2Cl_2, 84% yield.[1]

Cleavage

1. 5% H_2O, 5% phenol, 5% thioanisol in TFA / CH_2Cl_2 (30% v/v).[1]
2. TFA, CH_2Cl_2, triisopropylsilane or triethylsilane, 30 min, 25°.[1]
3. $Tl(TFA)_3$, DMF, anisole, 0°, 90 min.[1,2]

1. M. C. Munson, C. Garcia-Escheverria, F. Albericio, and G. Barany, *J. Org. Chem.*, **57**, 3013 (1992).
2. M. C. Munson, C. Garcia-Echeverría, F. Albericio, and G. Barany, *Pept.: Chem. Biol., Proc. Am. Pept. Symp., 12th*, 605 (1992).

S-4-Picolyl Thioether: $RSCH_2$–4-pyridyl (Chart 7)

Formation

1. 4-Picolyl chloride, 60% yield.[1]

Cleavage

1. Electrolytic reduction, 0.25 *M* H_2SO_4, 88% yield. *S*-4-Picolylcysteine is stable to CF_3COOH (7 days), to HBr / AcOH, and to 1 *M* NaOH. References for the electrolytic removal of seven other protective groups are included.[1,2]

1. A. Gosden, R. Macrae, and G. T. Young, *J. Chem. Res., Synop.*, 22 (1977).

2. C. M. Delerue-Matos, A. M. Freitas, H. L. S. Maia, M. J. Medeiros, M. I. Montenegro, and D. Pletcher, *J. Electroanal. Chem. Interfacial Electrochem.*, **315**, 1 (1991).

S-2-Quinolinylmethyl Thioether (Qm–SR):

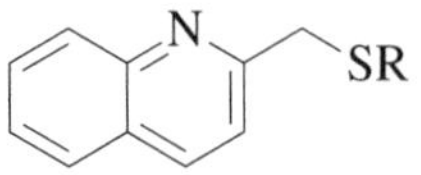

Formation

1. QmCl, NaH or NaOH or TEA, EtOH, 74% from cysteine.[1]

Cleavage

1. $FeCl_3$ or $CuCl_2$, DMF, H_2O, 61–99% yield, isolated as the disulfide. The quinoline group is isolated as the aldehyde.[1]

1. H. Yoshizawa, A. Otaka, H. Habashita, and N. Fujii, *Chem. Lett.*, 803 (1993).

S-2-Picolyl N-Oxide Thioether: $RSCH_2$–2-pyridyl *N*-Oxide (Chart 7)

Formation

1. 2-Picolyl chloride *N*-oxide, aq. NaOH, moderate yields.[1]

Cleavage

1. Ac_2O, reflux, 7 min or 25°, 1.5 h followed by hydrolysis; aq. NaOH, 25°, 3–12 h, 79% yield.[1]
2. Electrolysis on a glassy carbon electrode, DMF, $Bu_4N^+BF_4^-$, 85% yield.[2]

1. Y. Mizuno and K. Ikeda, *Chem. Pharm. Bull.*, **22**, 2889 (1974).
2. M. D. Geraldo and M. J. Medeiros, *Port. Electrochim. Acta*, **9**, 175 (1991).

S-9-Anthrylmethyl Thioether: $RSCH_2$–9-anthryl (Chart 7)

Formation

1. 9-Anthrylmethyl chloride, DMF, –20°, N_2.[1]

Cleavage

1. CH_3SNa, DMF or HMPA, 0–25°, 2–5 h, 68–92% yield.[1]

1. N. Kornblum and A. Scott, *J. Am. Chem. Soc.*, **96**, 590 (1974).

S-9-Fluorenylmethyl Thioether (Fm–SR):

CH_2SR

Formation

1. $Et(i\text{-}Pr)_2N$, DMF, FmCl.[1]
2. FmOTs, DMF, 0°–25°, 71%. This procedure has the advantage that FmOTs is prepared in 83% yield from FmOH, whereas the chloride, FmCl, is produced in only 30% yield from the alcohol and $SOCl_2$.[2]

Cleavage

1. 50% Piperidine, DMF or NH_4OH, 2 h.[3] The *S*-fluorenylmethyl group is stable to 95% HF/5% anisole for 1 h at 0°, to trifluoroacetic acid, to 12 *N* HCl, to 0.1 *M* I_2 in DMF, and to CF_3SO_3H in CF_3COOH.[2]
2. $(Me_2N)_2C{=}N\text{-}t\text{-}Bu$, 23°.[4]

1. M. Bodanszky and M. A. Bednarek, *Int. J. Pept. Protein Res.*, **20**, 434 (1982).
2. F. Albericio, E. Nicolas, J. Rizo, M. Ruiz-Gayo, E. Pedroso, and E. Giralt, *Synthesis*, 119 (1990).
3. M. Ruiz-Gayo, F. Albericio, E. Pedroso, and E. Giralt, *J. Chem. Soc., Chem. Commun.*, 1501 (1986).
4. E. J. Corey, D. Y. Gin, and R. S. Kania, *J. Am. Chem. Soc.*, **118**, 9202 (1996).

S-Xanthenyl Thioether (Xan–SR):

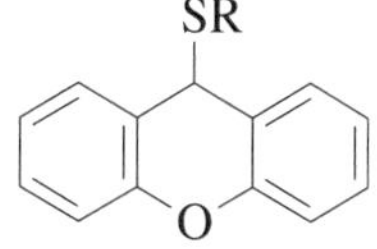

Formation

1. 9*H*-Xanthen-9-ol, TFA, CH_2Cl_2, 25°, 30 min.[1] The 2-methoxy analogue can be prepared similarly, and it is cleaved only slightly faster than the unsubstituted derivative.

Cleavage[1]

1. 0.2% TFA, CH_2Cl_2, Et_3SiH. Other scavengers are not nearly as effective, but when the xanthenyl group is used on a solid phase, more acid is required to get efficient cleavage.
2. I_2, MeOH, DMF or AcOH. AcOH is the most effective solvent, 67–100% yield.
3. $Tl(TFA)_3$, DMF, MeOH, CH_2Cl_2, or acetic acid, 94–100% yield.

1. Y. Han and G. Barany, *J. Org. Chem.*, **62**, 3841 (1997).

S-Ferrocenylmethyl Thioether (Fcm–SR):

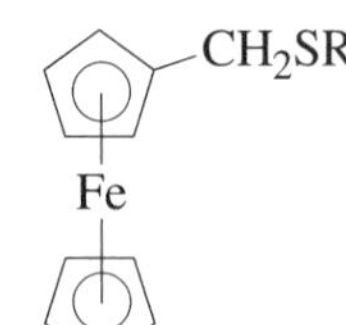

Formation

1. Cp–Fe–$CpCH_2OH$, TFA, acetone, H_2O, rt, overnight, 96% yield.[1]

Cleavage

The Fcm group can be removed with TFA, Ag(I), or Hg(II). The use of scavengers such as thiophenol and anisole is recommended. The Fcm group is stable to mild acid and base, but it is not stable to electrophilic reagents such as $(SCN)_2$, I_2/AcOH, or carboxymethylsulfenyl chloride (CmsCl).[1]

1. A. S. J. Stewart and C. N. C. Drey, *J. Chem. Soc., Perkin Trans. 1*, 1753 (1990).

S-Diphenylmethyl, Substituted *S*-Diphenylmethyl, and *S*-Triphenylmethyl Thioethers

S-Diphenylmethyl, substituted *S*-diphenylmethyl, and *S*-triphenylmethyl thioethers have often been formed or cleaved by the same conditions, although sometimes in rather different yields. As an effort has been made to avoid repetition in the sections that describe these three protective groups, the reader should glance at all the sections.

***S*-Diphenylmethyl Thioether:** $RSCH(C_6H_5)_2$ (Chart 7)

Formation

1. Ph_2CHOH, CF_3COOH, 25°, 15 min or Ph_2CHOH, HBr, AcOH, 50°, 2 h, >90% yield.[1]
2. Boron trifluoride etherate (in HOAc, 60–80°, 15 min, high yields)[2] also catalyzes the formation of *S*-diphenylmethyl and *S*-triphenylmethyl thioethers from aralkyl alcohols.
3. Yields of thioethers, formed under nonacidic conditions (Ph_2CHCl or Ph_3CCl, DMF, 80–90°, 2 h, N_2) are not as high ($RSCHPh_2$, 50% yield; $RSCPh_3$, 75% yield)[3] as the yields obtained under the acidic conditions described in items 1 and 2.

Cleavage

1. CF_3COOH, 2.5% phenol, 30°, 2 h, 65% yield.[1] Zervas and co-workers tried many conditions for the acid-catalyzed formation and removal of the *S*-diphenylmethyl, ***S*-4,4′-dimethoxydiphenylmethyl**, and *S*-triphenylmethyl thioethers. The best conditions for the *S*-diphenylmethyl thioether are those shown here. Phenol or anisole act as cation scavengers.
2. Na, NH_3, 97% yield.[3] Sodium/ammonia is an efficient, but nonselective, reagent. (RS–Ph, RS–CH_2Ph, RS–CPh_3, and RS–SR are also cleaved.)
3. 2-$NO_2C_6H_4SCl$, AcOH (results in disulfide formation), followed by $NaBH_4$ or $HS(CH_2)_2OH$ or dithioerythritol, quant.[4] *S*-Triphenylmethyl, *S*-4,4′-dimethoxydiphenylmethyl, and *S*-acetamidomethyl groups are also removed by this method.

1. I. Photaki, J. T.-Papadimitriou, C. Sakarellos, P. Mazarakis, and L. Zervas, *J. Chem. Soc. C*, 2683 (1970).
2. R. G. Hiskey and J. B. Adams, Jr., *J. Org. Chem.*, **30**, 1340 (1965).
3. L. Zervas and I. Photaki, *J. Am. Chem. Soc.*, **84**, 3887 (1962).
4. A. Fontana, *J. Chem. Soc., Chem. Commun.*, 976 (1975).

***S*-Bis(4-methoxyphenyl)methyl Thioether:** RSCH(C_6H_4–4-OCH_3)$_2$ (Chart 7)

***S*-Bis(4-methoxyphenyl)phenylmethyl Thioether (DMTr)**

Formation

1. DMTrCl (dimethoxytrityl chloride), TEA, 80% aq. AcOH, 91% yield.[1]

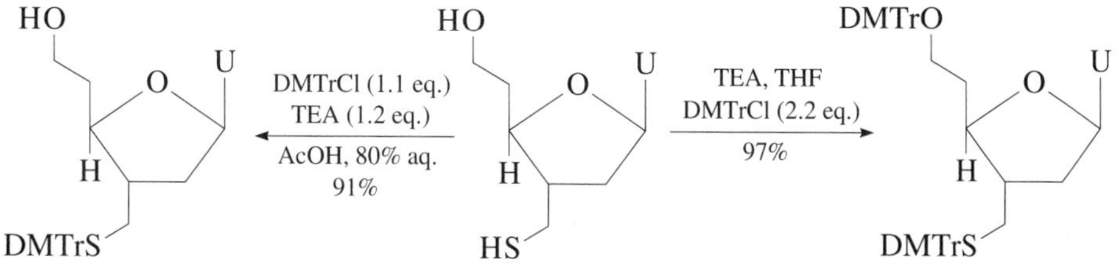

HO O U H DMTrS ← DMTrCl (1.1 eq.) TEA (1.2 eq.) / AcOH, 80% aq. 91% — HO O U H HS → TEA, THF DMTrCl (2.2 eq.) / 97% — DMTrO O U H DMTrS

2. (4-MeOC_6H_4)$_2$CHCl, DMF, 25°, 2 days, 96% yield.[2]

Cleavage

1. Selective cleavage of the DMTr group from oxygen is accomplished with 80% aq. AcOH (rt, 10 min), whereas selective cleavage of the DMTr group from the thiol is effected with $AgNO_3$/NaOAc buffer (rt, 1 min).[1]
2. HBr, AcOH, 50–60°, 30 min, or CF_3COOH, phenol, reflux, 30 min, quant.[2]

1. Z. Huang and S. A. Benner, *Synlett*, 83 (1993).
2. R. W. Hanson and H. D. Law, *J. Chem. Soc.*, 7285 (1965).

S-5-Dibenzosuberyl Thioether:

SR

5-Dibenzosuberyl alcohol reacts in 60% yield with cysteine to give a thioether that is cleaved by mercury(II) acetate or oxidized by iodine to cystine. The dibenzosuberyl group has also been used to protect –OH, $-NH_2$, and $-CO_2H$ groups.[1]

1. J. Pless, *Helv. Chim. Acta*, **59**, 499 (1976).

S-Triphenylmethyl Thioether: $RSC(C_6H_5)_3$ (Chart 7)

S-Triphenylmethyl thioethers have been formed by reaction of the thiol with triphenylmethyl alcohol/anhydrous CF_3COOH (85–90% yield) or with triphenylmethyl chloride (75% yield). Glycosidic triphenylmethyl thioethers are prepared by displacement of the chloride with TrS^- $^+N(Bu)_4$ (tetrabutylammonium triphenylmethanethiolate).[1]

Cleavage

1. HCl, aq. AcOH, 90°, 1.5 h.[2]
2. $Hg(OAc)_2$, EtOH, reflux, 3 h, → 25°, 12 h; H_2S, 61% yield.[2]
3. $AgNO_3$, EtOH, Pyr, 90°, 1.5 h; H_2S, 47% yield.[2] DTE (dithioerythritol) and NaOAc in MeOH/THF can be used in place of H_2S (97% yield).[3] An *S*-triphenylmethyl thioether can be selectively cleaved in the presence of an *S*-diphenylmethyl thioether by acidic hydrolysis or by heavy-metal ions. As a result of the structure of the substrate, the relative yields of cleavage by $AgNO_3$ and $Hg(OAc)_2$ can be reversed.[4]
4. Thiocyanogen [$(SCN)_2$, 5°, 4 h, 40% yield] selectively oxidizes an *S*-triphenylmethyl thioether to the disulfide (RSSR) in the presence of an *S*-diphenylmethyl thioether.[5]
5. *S*-Triphenylmethylcysteine is readily oxidized by iodine (MeOH, 25°) to cystine.[6,7] The *S*-triphenylmethylcysteine group can be selectively cleaved in the presence of a –Cys(Acm)– group (Acm = acetamidomethyl).[8] *S*-Benzyl and *S*-*t*-butyl thioethers are stable to the action of iodine.
6. Electrolysis, –2.6 V, DMF, $R_4N^+X^-$.[9]
7. Et_3SiH, 50% TFA, CH_2Cl_2, 1 h, rt.[10]
8. PhHgOAc 1.2 eq., MeOH–CH_2Cl_2 (4:1), 96% yield. The resulting Hg salt is liberated with H_2S.[1]

1. M. Blanc-Muesser, L. Vigne, and H. Driquez, *Tetrahedron Lett.*, **31**, 3869 (1990).

2. R. G. Hiskey, T. Mizoguchi, and H. Igeta, *J. Org. Chem.*, **31**, 1188 (1966).
3. Z. Huang and S. A. Benner, *Synlett*, 83 (1993).
4. R. G. Hiskey and J. B. Adams, *J. Org. Chem.*, **31**, 2178 (1966).
5. R. G. Hiskey, T. Mizoguchi, and E. L. Smithwick, *J. Org. Chem.*, **32**, 97 (1967).
6. B. Kamber, *Helv. Chim. Acta*, **54**, 398 (1971).
7. K. W. Li, J. Wu, W. Xing, and J. A. Simon, *J. Am. Chem. Soc.*, **118**, 7237 (1996).
8. B. Kamber, A. Hartmann, K. Eisler, B. Riniker, H. Rink, P. Sieber, and W. Rittel, *Helv. Chim. Acta*, **63**, 899 (1980).
9. V. G. Mairanovsky, *Angew. Chem., Int. Ed. Engl.*, **15**, 281 (1976).
10. D. A. Pearson, M. Blanchette, M. L. Baker, and C. A. Guindon, *Tetrahedron Lett.*, **30**, 2739 (1989).

S-Diphenyl-4-pyridylmethyl Thioether: $RSC(C_6H_5)_2$–4-pyridyl

Formation

1. $Ph_2(4\text{-}C_5H_4N)COH$, $BF_3{\cdot}Et_2O$, AcOH, 60°, 48 h.[1]

Cleavage

1. $Hg(OAc)_2$, AcOH, pH 4, 25°, 15 min.[1]
2. Zn, 80% AcOH, H_2O.[2]

The diphenylpyridylmethyl thioether is stable to acids (e.g., CF_3COOH, 21°, 48 h; 45% HBr/AcOH, 21°); it is oxidized by iodine to cystine (91%) or reduced by electrolysis at a mercury cathode.[1]

1. S. Coyle and G. T. Young, *J. Chem. Soc., Chem. Commun.*, 980 (1976).
2. S. Coyle, A. Hallett, M. S. Munns, and G. T. Young, *J. Chem. Soc., Perkin Trans. 1*, 522 (1981).

S-Phenyl Thioether: RSC_6H_5

Although a sulfhydryl group generally is not converted to an *S*-phenyl thioether, the conversion can be accomplished through the use of a Pd-catalyzed arylation with an aryl iodide.[1] Thiophenol can be used to introduce sulfur into molecules by simple displacement or by Michael additions, and thus, the phenyl group serves as a suitable protective group that can be removed by electrolysis (–2.7 V, DMF, $R_4N^+X^-$).[2]

1. P. G. Ciattini, E. Morera, and G. Ortar, *Tetrahedron Lett.*, **36**, 4133 (1995).
2. V. G. Mairanovsky, *Angew. Chem., Int. Ed., Engl.*, **15**, 281 (1976).

S-2,4-Dinitrophenyl Thioether: RSC_6H_3–2,4-$(NO_2)_2$ (Chart 7)

Formation

1. 2,4-$(NO_2)_2$–C_6H_3F, base.[1] The sulfhydryl group in cysteine can be selectively protected in the presence of the amino group by reaction with 2,4-dinitrophenol at pH 5–6.[2]

Cleavage

1. $HSCH_2CH_2OH$, pH 8, 22°, 1 h, quant.[1]

1. S. Shaltiel, *Biochem. Biophys. Res. Commun.*, **29**, 178 (1967).
2. H. Zahn and K. Traumann, *Z. Naturforsch.*, **9B**, 518 (1954).

S-*t*-Butyl Thioether: $RSC(CH_3)_3$ (Chart 7)

Formation

1. Isobutylene, H_2SO_4, CH_2Cl_2, 25°, 12 h, 73% yield.[1] The *S*-*t*-butyl derivative of cysteine is stable to HBr/AcOH and to CF_3COOH.
2. *t*-BuOH, 2 *N* HCl, reflux, 90% yield.[2]
3. *t*-BuOH, H_2SO_4, H_2O, 0°, 0.5 h and rt, 2 h, 98%.[3] A carboxylic acid was left unprotected under these conditions.

Cleavage

1. $Hg(OAc)_2$, CF_3COOH, anisole, 0°, 15 min; H_2S, quant.[4]
2. $Hg(OCOCF_3)_2$, aq. AcOH, 25°, 1 h; H_2S, quant.[4]
3. HF, anisole, 20°, 30 min.[5] No cleavage is observed with HF, *m*-cresol.[6]
4. 2-$NO_2C_6H_4SCl$; $NaBH_4$.[2] Treatment of the thioether with the sulfenyl chloride initially produces a disulfide, which is then reduced to afford the free thiol.
5. Tetramethylene sulfoxide, TMSOTf, 4°, 4 h, 87% yield or Ph_2SO, $MeSiCl_3$ or $SiCl_4$, TFA, 90–96% yield. The latter conditions also cleave the Acm, Bn, MeOBn, and MeBn groups. In all cases, disulfides are isolated.[7]

1. F. M. Callahan, G. W. Anderson, R. Paul, and J. E. Zimmerman, *J. Am. Chem. Soc.*, **85**, 201 (1963).
2. J. J. Pastuszak and A. Chimiak, *J. Org. Chem.*, **46**, 1868 (1981).
3. R. Breitschuh and D. Seebach, *Synthesis*, 83 (1992).
4. O. Nishimura, C. Kitada, and M. Fujino, *Chem. Pharm. Bull.*, **26**, 1576 (1978).
5. S. Sakakibara, Y. Shimonishi, Y. Kishida, M. Okada, and H. Sugihara, *Bull. Chem. Soc. Jpn.*, **40**, 2164 (1967).
6. K. Akaji, K. Fujino, T. Tatsumi, and Y. Kiso, *J. Am. Chem. Soc.*, **115**, 11384 (1993).

7. T. Koide, A. Otaka, H. Suzuki and N. Fujii, *Synlett*, 345 (1991).

S-1-Adamantyl Thioether: RS-1-adamantyl

Formation

1. 1-Adamantyl alcohol, CF_3COOH, 25°, 12 h, 90% yield.[1]
2. From a disulfide: $ArI(OCOAd)_2$, Hg *hν*, CH_2Cl_2.[2]

Cleavage

1. $Hg(OAc)_2$, CF_3COOH, 0°, 15 min, 100% yield.[1]
2. $Hg(OCOCF_3)_2$, aq. AcOH, 20°, 60 min, 100% yield.[1]
3. 1 *M* CF_3SO_3H, $PhSCH_3$ or $Tl(OCOCF_3)_3$.[3]

The *S*-adamantyl group is less prone to sulfoxide formation than the *S*-4-methoxybenzyl group. It is also more stable to CF_3COOH.

1. O. Nishimura, C. Kitada, and M. Fujino, *Chem. Pharm. Bull.*, **26**, 1576 (1978).
2. H. Togo, T. Muraki, and M. Yokoyama, *Synthesis*, 155 (1995).
3. N. Fujii, A. Otaka, S. Funakoshi, H. Yajima, O. Nishimura, and M. Fujino, *Chem. Pharm. Bull.*, **34**, 869 (1986); N. Fujii, H. Yajima, A. Otaka, S. Funakoshi, M. Nomizu, K. Akaji, I. Yamamoto, K. Torizuka, K. Kitagawa, T. Akita, K. Ando, T. Kawamoto, Y. Shimonishi, and T. Takao, *J. Chem. Soc., Chem. Commun.*, 602 (1985).

Substituted *S*-Methyl Derivatives

Monothio, Dithio, and Aminothio Acetals

S-Methoxymethyl Monothioacetal: $RSCH_2OCH_3$

Formation

1. Zn, $(CH_3O)_2CH_2$, $BrCH_2CO_2Et$, 80–82% yield. Formation of the methoxymethyl thioether with dimethoxymethane[1] avoids the use of the carcinogen chloromethyl methyl ether.[2] The reaction forms an intermediate zinc thiolate, which then forms the monothioacetal.
2. $ClCH_2Br$, KOH, $BnN^+Et_3Cl^-$, MeOH, 70–90% yield.[3]

1. F. Dardoize, M. Gaudemar, and N. Goasdoue, *Synthesis*, 567 (1977).
2. T. Fukuyama, S. Nakatsuka, and Y. Kishi, *Tetrahedron Lett.*, 3393 (1976).
3. F. D. Toste and I. W. J. Still, *Synlett*, 159 (1995).

S-Isobutoxymethyl Monothioacetal: $RSCH_2OCH_2CH(CH_3)_2$ (Chart 7)

Formation

1. $ClCH_2OCH_2CH(CH_3)_2$, 82% yield.[1]

Cleavage

1. 2 *N* HBr, AcOH, rapid.[1]

The *S*-isobutoxymethyl monothioacetal is stable to 2 *N* hydrochloric acid and to 50% acetic acid; some decomposition occurs in 2 *N* sodium hydroxide.[1] The monothioacetal is also stable to 12 *N* hydrochloric acid in acetone (used to remove an *N*-triphenylmethyl group) and to hydrazine hydrate in refluxing ethanol (used to cleave an *N*-phthaloyl group). It is cleaved by boron trifluoride etherate in acetic acid, by silver nitrate in ethanol, and by trifluoroacetic acid. The monothioacetal is oxidized to a disulfide by thiocyanogen, $(SCN)_2$.[2]

1. P. J. E. Brownlee, M. E. Cox, B. O. Handford, J. C. Marsden, and G. T. Young, *J. Chem. Soc.*, 3832 (1964).
2. R. G. Hiskey and J. T. Sparrow, *J. Org. Chem.*, **35**, 215 (1970).

S-Benzyloxymethyl Monothioacetal (BOM–SR): $BnOCH_2SR$

Formation

1. $BnOCH_2Cl$, 4 *N* NaOH, 2 h, 0°, 69% yield.[1]

Cleavage

1. AgOTf, TFA.[1]

1. A. Otaka, H. Morimoto, N. Fujii, T. Koide, S. Funakoshi, and H. Yajima, *Chem. Pharm. Bull.*, **37**, 526 (1989).

S-2-Tetrahydropyranyl Monothioacetal: RS-2-tetrahydropyranyl (Chart 7)

Formation

1. Dihydropyran, $BF_3 \cdot Et_2O$, Et_2O, 0°, 0.5 h → 25°, 1 h, satisfactory yields.[1]
2. Dihydropyran, PPTS (pyridinium *p*-toluenesulfonate), 4 hr, 25°, 92% yield.[2]

Cleavage

1. Aqueous $AgNO_3$, 0°, 10 min, quant.[3]
2. HBr, CF_3COOH, 90 min, 100% yield.[4]

An *S*-tetrahydropyranyl monothioacetal is stable to 4 *N* HCl/CH_3OH, 0° and to reduction with Na/NH_3. (An *O*-tetrahydropyranyl acetal is cleaved by 0.1 *N* HCl,

22°, $t_{1/2}$ = 4 min.)[5] An *S*-2-tetrahydropyranyl monothioacetal is oxidized to a disulfide by iodine[3] or thiocyanogen, $(SCN)_2$.[6]

1. R. G. Hiskey and W. P. Tucker, *J. Am. Chem. Soc.*, **84**, 4789 (1962).
2. E. Block, V. Eswarakrishnan, M. Gernon, G. O.-Okai, C. Saha, K. Tang, and J. Zubieta, *J. Am. Chem. Soc.*, **111**, 658 (1989).
3. G. F. Holland and L. A. Cohen, *J. Am. Chem. Soc.*, **80**, 3765 (1958).
4. K. Hammerström, W. Lunkenheimer, and H. Zahn, *Makromol. Chem.*, **133**, 41 (1970).
5. B. E. Griffin, M. Jarman, and C. B. Reese, *Tetrahedron*, **24**, 639 (1968).
6. R. G. Hiskey and W. P. Tucker, *J. Am. Chem. Soc.*, **84**, 4794 (1962).

S-Benzylthiomethyl Dithioacetal: $RSCH_2SCH_2C_6H_5$

S-Phenylthiomethyl Dithioacetal: $RSCH_2SC_6H_5$

Formation

1. $ClCH_2SCH_2Ph$, NH_3, 91% yield.[1]

Cleavage

1. $Hg(OAc)_2$, H_2O, 80% AcOH, $HSCH_2CH_2SH$, 25°, 5–20 min; H_2S, 2 h, high yield.[1]

   The removal of an *S*-benzylthiomethyl protective group from a dithioacetal with mercury(II) acetate avoids certain side reactions that occur when an *S*-benzyl thioether is cleaved with sodium/ammonia. The dithioacetal is stable to hydrogen bromide/acetic acid used to cleave benzyl carbamates.

S-Phenylthiomethyl dithioacetals ($RSCH_2SC_6H_5$) were prepared and cleaved by similar methods.[1]

The dithioacetal is stable to catalytic reduction (H_2/Pd–C, CH_3OH–HOAc, 12 h, the conditions used to cleave a *p*-nitrobenzyl carbamate).[2]

1. P. J. E. Brownlee, M. E. Cox, B. O. Handford, J. C. Marsden, and G. T. Young, *J. Chem. Soc.*, 3832 (1964).
2. R. Camble, R. Purkayastha, and G. T. Young, *J. Chem. Soc. C*, 1219 (1968).

Thiazolidine Derivative: R R S NH R′ R′

Thiazolidines have been prepared from *β*-aminothiols—for example, cysteine—to protect the –SH and –NH groups during syntheses of peptides, including glutathione.[1] Thiazolidines are oxidized to symmetrical disulfides with iodine;[2] they do not react with thiocyanogen in a neutral solution.[3]

Formation[4]

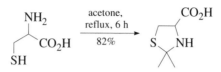

1. [Reaction scheme: cysteine (NH_2, CO_2H, SH) → acetone, reflux, 6 h, 82% → 2,2-dimethylthiazolidine-4-carboxylic acid (S, NH, CO_2H)]

Cleavage

1. HCl, H_2O, CH_3OH, 25°, 3 days, high yield.[4]
2. $HgCl_2$, H_2O, 25°, 2 days or 60–70°, 15 min; H_2S, 20 min, 30–40% yield.[4]
3. *N*-BOC thiazolidines can be cleaved with ScmCl (methoxycarbonylsulfenyl chloride) (AcOH, DMF, H_2O) to afford the Scm derivative in >90% yield.[5]
4. [Reaction scheme: N-BOC, Ph-substituted thiazolidine bearing H and alkenyl $(CH_2)_4CO_2H$ chain → Na, NH_3, 95% → BocHN, HS, H, $(CH_2)_4CO_2H$]

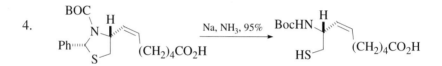

1. F. E. King, J. W. Clark-Lewis, G. R. Smith, and R. Wade, *J. Chem. Soc.*, 2264 (1959).
2. S. Ratner and H. T. Clarke, *J. Am. Chem. Soc.*, **59**, 200 (1937).
3. R. G. Hiskey and W. P. Tucker, *J. Am. Chem. Soc.*, **84**, 4789 (1962).
4. J. C. Sheehan and D.-D. H. Yang, *J. Am. Chem. Soc.*, **80**, 1158 (1958).
5. D. S. Kemp and R. I. Carey, *J. Org. Chem.*, **54**, 3640 (1989).

S-Acetamidomethyl Thioacetal (Acm–SR): $RSCH_2NHCOCH_3$ (Chart 7)

Formation

1. $AcNHCH_2OH$, concd. HCl, pH 0.5, 25°, 1–2 days, 52% yield.[1]
2. $AcNHCH_2OH$, TFA.[2]

Cleavage

1. $Hg(OAc)_2$, pH 4, 25°, 1 h; H_2S; air, 98% yield of cystine.[1] An *S*-acetamidomethyl group is hydrolyzed by the strongly acidic (6 *N* HCl, 110°, 6 h) or strongly basic conditions used to cleave amide bonds. The group is stable to anhydrous trifluoroacetic acid and to hydrogen fluoride (0°, 1 h; 18°, 1 h, 10% cleaved). It is stable to zinc in acetic acid and to hydrazine in acetic acid or methanol.[1] If the Acm group is oxidized, there is no satisfactory method to liberate the cysteine. Cleavage of the sulfoxide with HF/anisole or CH_3SO_3H/anisole affords Cys(C_6H_4OMe).[3]

2. $2\text{-}NO_2C_6H_4SCl$, AcOH; $HO(CH_2)_2SH$ or $NaBH_4$, quant.[4]
3. PhSH. This reagent affords the phenyl disulfide.[3]
4. $ClSCO_2Me$, MeOH, 80% yield.[5]

NHBOC, CO_2H, SCH_2NHAc —[$MeOC(O)SCl$, $CHCl_3$]→ NHBOC, CO_2H, S–SCO_2Me

These conditions convert the Acm group to a methyl *S*-sulfenylthiocarbonate group (Scm group), which can be cleaved with dithiothreitol.[6]

5. ClCOSCl, $CHCl_3$; PhNHMe.[6]

NHBOC, CO_2H, SCH_2NHAc —[1. $ClC(O)SCl$, $CHCl_3$; 2. $PhNHCH_3$; 98%]→ NHBOC, CO_2H, S–S–C(=O)–N(Ph)(CH_3)

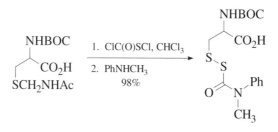

The ***S*-(*N*-methyl-*N*-phenylcarbamoyl)sulfenyl** group (Snm group) produced under these conditions is stable to HF or CF_3SO_3H. Since there are few acid-stable–SH protective groups, the Snm group should prove useful where strong acids are encountered in synthesis.

6. $MeSiCl_3$, Ph_2SO, TFA, 4°, 30 min, 93% yield. These conditions also cleave the Tacm, Bam (benzamidomethyl), *t*-Bu, MeOBn, and MeBn groups in high yield.[7]
7. AgTFA, TFA/anisole (95:5), 3 h, rt; H_2S.[8]
8. $Tl(TFA)_3$, TFA, anisole, 1 h, 66% yield.[9]
9. $AgBF_4$, anisole, TFA, 4°, 1 h, 93% yield. The benzamidomethyl (Bam), 4-methoxybenzyl, and 2,4,6-trimethylbenzyl (Tmb) groups are only partially cleaved under these conditions (87%, 87%, and 73% respectively).[10]
10. I_2, Met, Tyr, His, and Trp are susceptible to overoxidation with iodine if the reaction conditions are not carefully controlled.[11]
11. TFA, triisopropylsilane, 70% yield.[12]

1. D. F. Veber, J. D. Milkowski, S. L. Varga, R. G. Denkewalter, and R. Hirschmann, *J. Am. Chem. Soc.*, **94**, 5456 (1972); J. D. Milkowski, D. F. Veber, and R. Hirschmann, *Org. Synth., Collect., Vol. VI*, 5 (1988).
2. P. Marbach and J. Rudinger, *Helv. Chim. Acta*, **57**, 403 (1974).
3. H. Yajima, K. Akaji, S. Funakoshi, N. Fujii, and H. Irie, *Chem. Pharm. Bull.*, **28**, 1942 (1980).

4. L. Moroder, F. Marchiori, G. Borin, and E. Schoffone, *Biopolymers*, **12**, 493 (1973); A. Fontana, *J. Chem. Soc., Chem. Commun.*, 976 (1975).
5. R. G. Hiskey, N. Muthukumaraswamy, and R. R. Vunnam, *J. Org. Chem.*, **40**, 950 (1975).
6. A. L. Schroll and G. Barany, *J. Org. Chem.*, **54**, 244 (1989).
7. K. Akaji, T. Tatsumi, M. Yoshida, T. Kimura, Y. Fujiwara, and Y. Kiso, *J. Chem. Soc., Chem. Commun.*, 167 (1991).
8. Z. Chen and B. Hemmasi, *Biol. Chem. Hoppe-Seyler*, **374**, 1057 (1993).
9. N. Fujii, A. Otaka, S. Funakoshi, K. Bessho, and H. Yajima, *J. Chem. Soc., Chem. Commun.*, 163 (1987); C. Garcia-Echeverria, M. A. Molins, F. Alberico, M. Pons, and E. Giralt, *Int. J. Pept. Protein Res.*, **35**, 434 (1990).
10. M. Yoshida, K. Akaji, T. Tatsumi, S. IInuma, Y. Fujiwara, T. Kimura, and Y. Kiso, *Chem. Pharm. Bull.*, **38**, 273 (1990).
11. B. Kamber, *Helv. Chim. Acta*, **54**, 927 (1971); B. Kamber, A. Hartmann, K. Eisler, B. Riniker, H. Rink, P. Sieber, and W. Rittle, *Helv. Chim. Acta*, **63**, 899 (1980).
12. P. R. Singh, M. Rajopadhye, S. L. Clark, and N. E. Williams, *Tetrahedron Lett.*, **37**, 4117 (1996).

S-Trimethylacetamidomethyl Thioacetal (Tacm–SR): $(CH_3)_3CCONHCH_2SR$

Formation[1,2]

1. $(CH_3)_3CCONHCH_2OH$, TFA, rt, 1 h, >85% yield.

Cleavage

1. I_2, AcOH, EtOH, 25°, 1 h, 100% yield.[1,2] These conditions can result in methionine oxidation.[3]
2. $Hg(OAc)_2$, TFA, 0°, 30 min. The Tacm group is stable to HF (0° 1 h); to 1 *M* CF_3COOH, $PhSCH_3$ (0°, 1 h); to 0.5 *M* NaOH/MeOH (0°, 1 h); to NH_2NH_2; to MeOH; and to Zn/AcOH.[1,2] It is not stable to 25% HBr/AcOH, 2 h, rt.[1] This group was reported to be more useful than the Acm group, because it was less susceptible to by-product formation and oxidation.[2] The Pim (phthalimidomethyl) group is stable under these conditions.[4]
3. $AgBF_4$, anisole, 0°, 1 h, quant. These conditions also cleave the Acm group.[3]

1. Y. Kiso, M. Yoshida, Y. Fujiwara, T. Kimura, M. Shimokura, and K. Akaji, *Chem. Pharm. Bull.*, **38**, 673 (1990).
2. Y. Kiso, M. Yoshida, T. Kimura, Y. Fujiwara, and M. Shimokura, *Tetrahedron Lett.*, **30**, 1979 (1989).
3. M. Yoshida, K. Akaji, T. Tatsumi, S. Iinuma, Y. Fujiwara, T. Kimura, and Y. Kiso, *Chem. Pharm. Bull.*, **38**, 273 (1990).
4. Y.-D. Gong and N. Iwasawa, *Chem. Lett.*, 2139 (1994).

S-Benzamidomethyl Thioacetal (Bam–SR): $RSCH_2NHCOC_6H_5$

S-Benzamidomethyl-*N*-methylcysteine has been prepared as a crystalline derivative ($HOCH_2NHCOC_6H_5$, anhydr. CF_3CO_2H, 25°, 45 min, 88% yield as the trifluoroacetate salt) and cleaved (100% yield) by treatment with mercury(II) acetate (pH 4, 25°, 1 h) followed by hydrogen sulfide. Attempted preparation of *S*-acetamidomethyl-*N*-methylcysteine resulted in noncrystalline material, shown by TLC to be a mixture.[1] The Bam–SR group is also cleaved with $AgBF_4$/TFA, 4°, >1 h[2] and $MeSiCl_3/Ph_2SO$, 4°, 30 min, 100% cleavage.[3] The latter conditions also cleave the Acm, Tacm, *t*-Bu, 4-methoxybenzyl, and 4-methylbenzyl groups.

1. P. K. Chakravarty and R. K. Olsen, *J. Org. Chem.*, **43**, 1270 (1978).
2. M. Yoshida, T. Tatsumi, Y. Fujiwara, S. Iinuma, T. Kimura, K. Akaji, and Y. Kiso, *Chem. Pharm. Bull.*, **38**, 1551 (1990).
3. K. Akaji, T. Tatsumi, M. Yoshida, T. Kimura, Y. Fujiwara, and Y. Kiso, *J. Chem. Soc., Chem. Commun.*, 167 (1991).

S-Allyloxycarbonylaminomethyl Thioacetal (Allocam–SR): $CH_2{=}CHCH_2OC(O)NHCH_2SR$

Formation/Cleavage[1]

RSH + $HOCH_2NHC(O)OCH_2CH{=}CH_2$ —(TFA, CH_2Cl_2, rt; 70–95%, yield)→ $RSCH_2NHC(O)OCH_2CH{=}CH_2$

$RSCH_2NHC(O)OCH_2CH{=}CH_2$ —(1. Bu_3SnH, $PdCl_2(Ph_3P)_2$, AcOH, CH_2Cl_2; 2. I_2, 65–100%)→ RSSR

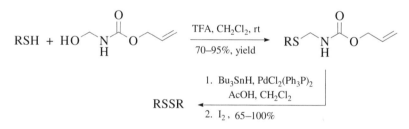

1. A. M. Kimbonguila, A. Merzouk, F. Guibe, and A. Loffet, *Tetrahedron Lett.*, **35**, 9035 (1995).

S-Phenylacetamidomethyl Thioacetal (Phacm–SR): $C_6H_5CH_2C(O)NHCH_2SR$

Formation

The Phacm group is introduced by the same methodology as the Acm group[1] [$PhCH_2C(O)NHCH_2OH$, TFMSA].[2]

Cleavage

1. Penicillin G. acylase, pH 7.8 buffer, 35°, 30 min to 2 h. These conditions result in isolation of the disulfide, but if *β*-mercaptoethanol is included in the reaction mixture, the thiol can be isolated.[2]

2. I_2, 80% aq. AcOH. The disulfide is isolated.[2]

The Phacm group is stable to the following conditions: DIEA–CH_2Cl_2, TFA–CH_2Cl_2, piperidine–DMF, 0.1 *M* TBAF–DMF, and DBU–DMF for 24 h at rt; to HF-anisole or *p*-cresol (9:1) at 0° for 1 h; and to TFA-scavengers (phenol, $HSCH_2CH_2SH$, *p*-cresol, anisole) for 2 h at 25°. It is partially stable (>80%) to TFMSA–TFA–*p*-cresol for 2 h at 25°. These stability characteristics make the group compatible with BOC- or Fmoc-based peptide synthesis.[2]

1. F. Albericio, A. Grandas, A. Porta, E. Pedroso, and E. Giralt, *Synthesis*, 271 (1987).
2. M. Royo, J. Alsina, E. Giralt, U. Slomcyznska, and F. Albericio, *J. Chem. Soc., Perkin Trans. 1*, 1095 (1995).

S-Phthalimidomethyl Thioacetal (Pim–SR): O NCH$_2$SR O

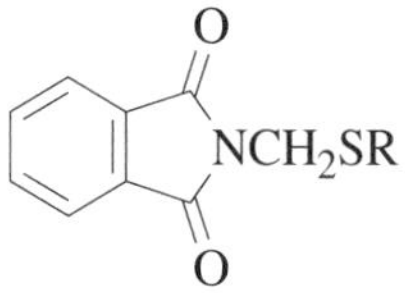

Formation[1]

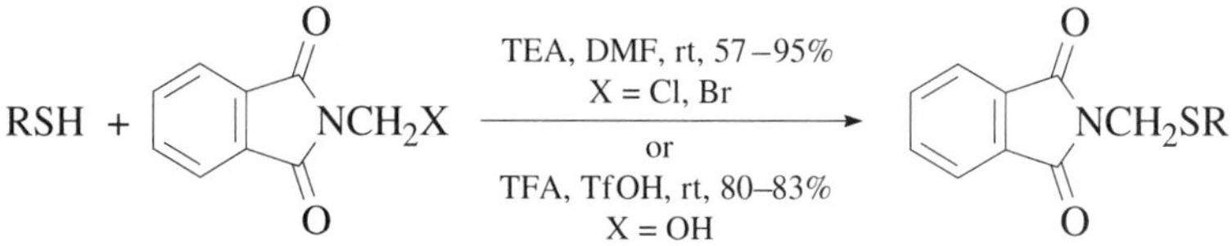

RSH + (phthalimide) NCH$_2$X → NCH$_2$SR

TEA, DMF, rt, 57–95%, X = Cl, Br; or TFA, TfOH, rt, 80–83%, X = OH

Cleavage[1]

1. NH_2NH_2, H_2O, MeOH, 0°–rt 1–2 h; $Hg(OAc)_2$, 2–3 h or $Cu(OAc)_2$, 3–24 h; $HSCH_2CH_2OH$, 71–92% yield. These conditions return the free thiol. The use of $Hg(OAc)_2$ cleaves the Acm (acetamidomethyl) group in the presence of the Pim group.
2. NH_2NH_2, H_2O, MeOH, 0°–rt, 1–2 h; I_2, rt, 1–2 h, 79–89% yield. The disulfide is formed.

1. Y.-D. Gong and N. Iwasawa, *Chem. Lett.*, 2139 (1994).

S-Acetyl-, S-Carboxy-, and S-Cyanomethyl Thioethers: $ArSCH_2X$

X = $-COCH_3$, $-CO_2H$, –CN (Chart 7)

In an attempt to protect thiophenols during electrophilic substitution reactions on the aromatic ring, the three substituted thioethers were prepared. After acetylation of the aromatic ring (with moderate yields), the protective group was converted to the disulfide in moderate yields, 50–60%, by oxidation with hydrogen peroxide/boiling mineral acid, nitric acid, or acidic potassium permanganate. [1]

1. D. Walker, *J. Org. Chem.*, **31**, 835 (1966).

Substituted *S*-Ethyl Derivatives

A thiol, usually under basic catalysis, can undergo Michael addition to an activated double bond, resulting in protection of the sulfhydryl group as a substituted *S*-ethyl derivative. Displacement of an ethyl tosylate by thiolate also affords an *S*-ethyl derivative.

S-(2-Nitro-1-phenyl)ethyl Thioether: $RSCH(C_6H_5)CH_2NO_2$ (Chart 7)

Formation

1. $PhCH{=}CHNO_2$, *N*-methylmorpholine, pH 7–8, 10 min, 70% yield.[1]

Cleavage

The protective group is removed by mildly alkaline conditions that do not cleave methyl or benzyl esters. The group is stable to CF_3COOH, HCl–AcOH, and HBr–AcOH. A polymer-bound version of this group has also been developed.[2]

S-2-(2,4-Dinitrophenyl)ethyl Thioether (Dnpe-SR):

Formation

2-(2,4-Dinitrophenyl)ethyl tosylate, DIPEA, DMF, 63% yield.[3]

Cleavage

Piperidine, DMF (1:1), 30 min, 25°, 57–90% yield.[3]

1. G. Jung, H. Fouad, and G. Heusel, *Angew. Chem., Int. Ed. Engl.*, **14**, 817 (1975).
2. G. Heusel and G. Jung, *Liebigs Ann. Chem.*, 1173 (1979).
3. M. Royo, C. Garcia-Echeverria, E. Giralt, R. Eritja, and F. Albericio, *Tetrahedron Lett.*, **33**, 2391 (1992).

S-2-(4′-Pyridyl)ethyl Thioether: $C_4H_4NCH_2CH_2SR$

Formation[1]*/Cleavage*[2]

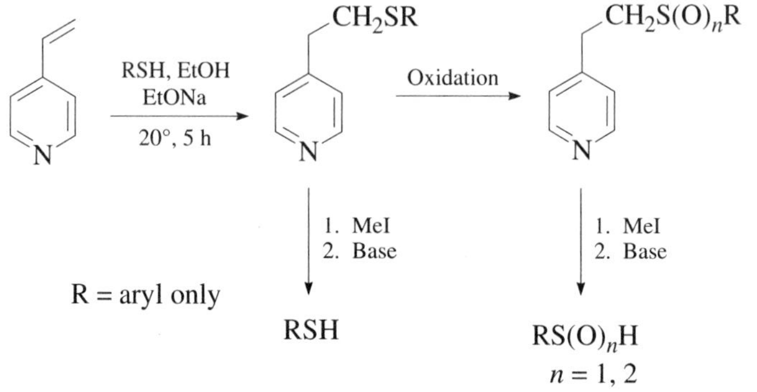

RSH, EtOH, EtONa, 20°, 5 h; Oxidation; 1. MeI 2. Base

R = aryl only

RSH

$RS(O)_nH$, $n = 1, 2$

The intermediate sulfides can be oxidized to the corresponding sulfoxides and sulfones and then liberated to give sulfenic and sulfinic acids.

1. A. R. Katritzky, I. Takahashi, and C. M. Marson, *J. Org. Chem.*, **51**, 4914 (1986).
2. A. R. Katritzky, G. R. Khan, and O. A. Schwarz, *Tetrahedron Lett.*, **25**, 1223 (1984).

S-2-Cyanoethyl Thioether: $NCCH_2CH_2SR$

Formation

1. $BrCH_2CH_2CN$, K_2CO_3, DMF.[1]

Cleavage

1. The 2-cyanoethyl group was cleaved from an aromatic sulfide with $K_2CO_3/NaBH_4$ (DMF, 135°, 70% yield).[2]

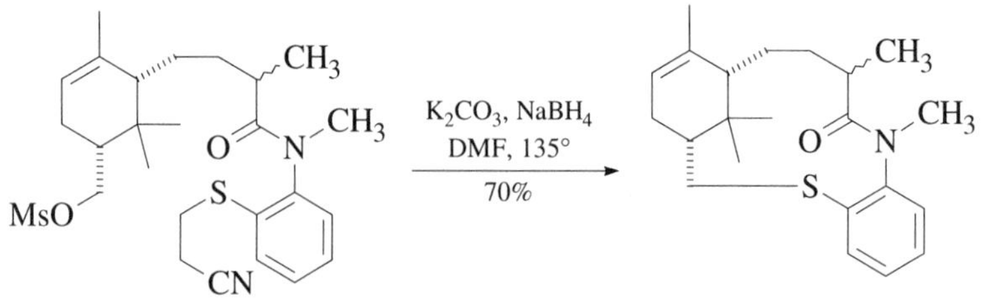

K_2CO_3, $NaBH_4$, DMF, 135°, 70%

2. Concd. NH_4OH, rt, quant.[1]
3. *t*-BuOK, DMF, 50–94%.[3]

1. M. S. Christopherson and A. D. Broom, *Nucleic Acids Res.*, **19**, 5719 (1991).

2. Y. Ohtsuka and T. Oishi, *Tetrahedron Lett.*, **27**, 203 (1986).
3. A. Kakehi, S. Ito, N. Yamada, and K. Yamaguchi, *Bull. Chem. Soc. Jpn.*, **63**, 829 (1990).

S-2-(Trimethylsilyl)ethyl Thioether: $TMSCH_2CH_2SR$

Cleavage

1. $Bu_4N^+F^-$, 3Å, THF, rt, >53% yield.[1]
2. $MeSS^+Me_2\,BF_4^-$ forms a disulfide in 92% yield that is cleaved to the thiol with $Ph_3P/MeOH/H_2O$ in 90% yield.[2]

1. M. Koreeda and W. Yang, *J. Am. Chem. Soc.*, **116**, 10793 (1994).
2. M. B. Anderson, M. G. Ranasinghe, J. T. Palmer, and P. L. Fuchs, *J. Org. Chem.*, **53**, 3125 (1988).

S-2,2-Bis(carboethoxy)ethyl Thioether: $RSCH_2CH(COOC_2H_5)_2$ (Chart 7)

Formation

1. $CH_2{=}C(CO_2Et)_2$, EtOH, 1 h, 74% yield.[1]

Cleavage

1. 1 *N* KOH, EtOH, 20°, 5–10 min, 80% yield. *S*-2,2-Bis(carboethoxy)ethyl thioether, stable to acidic reagents such as trifluoroacetic acid and hydrogen bromide/acetic acid, has been used in a synthesis of glutathione.[1]

1. T. Wieland and A. Sieber, *Justus Liebigs Ann. Chem.*, **722**, 222 (1969); *idem, ibid.*, **727**, 121 (1969).

S-(1-*m*-Nitrophenyl-2-benzoyl)ethyl Thioether: $ArSCH(C_6H_4\text{–}m\text{-}NO_2)CH_2COC_6H_5$

An *S*-(1-*m*-nitrophenyl-2-benzoyl)ethyl thioether was used to protect thiophenols during electrophilic substitution reactions of the benzene ring.[1]

Formation

1. $PhCOCH{=}CHC_6H_4\text{–}m\text{-}NO_2$, piperidine, benzene, 96% yield.[1]

Cleavage

1. $Pb(OAc)_2$, EtOH, pH 8–10; dil HCl, 77% yield.[1]

1. A. H. Herz and D. S. Tarbell, *J. Am. Chem. Soc.*, **75**, 4657 (1953).

S-2-Phenylsulfonylethyl Thioether and *S*-1-(4-Methylphenylsulfonyl)-2-methylprop-2-yl Thioether: $PhSO_2CH_2CH_2SR$ and 4-$CH_3C_6H_4SO_2CH_2C(CH_3)_2SR$

Formation/Cleavage[1,2]

$$\mathrm{RSH} \underset{t\text{-BuOK, THF, DME or } t\text{-BuOH, 80–100\%}}{\overset{\mathrm{PhSO_2CH{=}CH_2,\ (Et_3N,\ THF)\ \ or\ \ MeONa,\ MeOH}}{\rightleftarrows}} \mathrm{RSCH_2CH_2SO_2Ph}\ \ 84\text{–}100\%$$

1. Y. Kuroki and R. Lett, *Tetrahedron Lett.*, **25**, 197 (1984).
2. L. Horner and H. Lindel, *Phosphorus Sulfur*, **15**, 1 (1983).

Silyl Thioethers

Silyl-derived protective groups are also used to mask the thiol function. A complete compilation is not given here, since silyl derivatives are described in the section on alcohol protection. The formation and cleavage of silyl thioethers proceed analogously to those of simple alcohols. The Si–S bond is weaker than the Si–O bond, and therefore, sulfur derivatives are more susceptible to hydrolysis. For the most part, silyl ethers are rarely used to protect the thiol function, because of their instability. Silyl ethers have been used for *in situ* protection of the –SH group during amide formation.[1] The use of the sterically demanding and thus more stable **triisopropylsilyl** thioether may prove worthwhile.[2]

1. E. W. Abel, *J. Chem. Soc.*, 4933 (1961); L. Birkofer, W. Konkol, and A. Ritter, *Chem. Ber.*, **94**, 1263 (1961).
2. J. C. Arnould, M. Didelot, C. Cadilhac, and M. J. Pasquet, *Tetrahedron Lett.*, **37**, 4523 (1996).

THIOESTERS

S-Acetyl Derivative: $RSCOCH_3$

S-Benzoyl Derivative: $RSCOC_6H_5$ (Chart 7)

Formation

1. Ac_2O, $KHCO_3$, 55% yield.[1]
2. BzCl, NaOH, $KHCO_3$, 0–5°, 30 min., 50% yield.[2]

$$\begin{array}{c} CH_2SH \\ | \\ CHOH \\ | \\ CH_2OH \end{array} \xrightarrow[\text{MeOH, 25°, 1.5 h, 54\%}]{\text{PhCSSMe, cat. NaOMe}} \begin{array}{c} CH_2SCSPh \\ | \\ CHOH \\ | \\ CH_2OH \end{array}$$

The base-catalyzed reaction of thiothreitol with methyl dithiobenzoate selectively protects a thiol group as an *S*-thiobenzoyl derivative in the presence of a hydroxyl group.[2]

Cleavage

1. 0.2 *N* NaOH, N_2, 20°, 2–15 min, 100% yield.[1]
2. Aqueous NH_3, N_2, 20°, 95–100% yield.[1]
3. HBr, AcOH, 25°, 30 min, 5% to a substantial amount.[1]
4. CF_3CO_2H, phenol, reflux, 30 min, 2–5% yield. In this case, an *S*–Cbz group is removed.[1]
5. $Fe(NO_3)_3$–Clayfen.[3]
6. NaSMe, MeOH, 23°, 81–95% yield.[4] This procedure is chemoselective for removal of a thioacetate in the presence of an acetate.

Two disadvantages are associated with the use of *S*-acetyl or *S*-benzoyl derivatives in peptide syntheses: (a) base-catalyzed hydrolysis of *S*-acetyl- and *S*-benzoylcysteine occurs with *β*-elimination to give olefinic side products, CH_2=C–(NHPG)CO–;[5] (b) the yields of peptides formed by coupling an unprotected amino group in an *S*-acylcysteine are low because of prior S–N acyl migration.[6]

An *S*-acetyl group is stable to oxidation of a double bond by ozone (–20°, 5.5 h, 73% yield).[7]

S-Trifluoroacetyl Derivative: $RSCOCF_3$

Formation

1. $CF_3COSC_6F_5$, Pyr, DMF, 75% yield.[8]

1. L. Zervas, I. Photaki, and N. Ghelis, *J. Am. Chem. Soc.*, **85**, 1337 (1963).
2. E. J. Hedgley and N. H. Leon, *J. Chem. Soc. C*, 467 (1970).
3. H. M. Meshram, *Tetrahedron Lett.*, **34**, 2521 (1993).
4. O. B. Wallace and D. M. Springer, *Tetrahedron Lett.*, **39**, 2693 (1998).
5. R. G. Hiskey, R. A. Upham, G. M. Beverly, and W. C. Jones, Jr., *J. Org. Chem.*, **35**, 513 (1970).
6. R. G. Hiskey, T. Mizoguchi, and T. Inui, *J. Org. Chem.*, **31**, 1192 (1966).
7. I. Ernest, J. Gosteli, C. W. Greengrass, W. Holick, D. E. Jackman, H. R. Pfaendler, and R. B. Woodward, *J. Am. Chem. Soc.*, **100**, 8214 (1978).

8. L. M. Gayo and M. J. Suto, *Tetrahedron Lett.*, **37**, 4915 (1996).

S-N-[[(p-Biphenylyl)isopropoxy]carbonyl]-N-methyl-γ-aminothiobutyrate: $BpocN(CH_3)CH_2CH_2CH_2COSR$ **and S-N-(t-Butoxycarbonyl)-N-methyl-γ-aminothiobutyrate:** $BOCN(CH_3)CH_2CH_2CH_2COSR$

Formation/Cleavage[1]

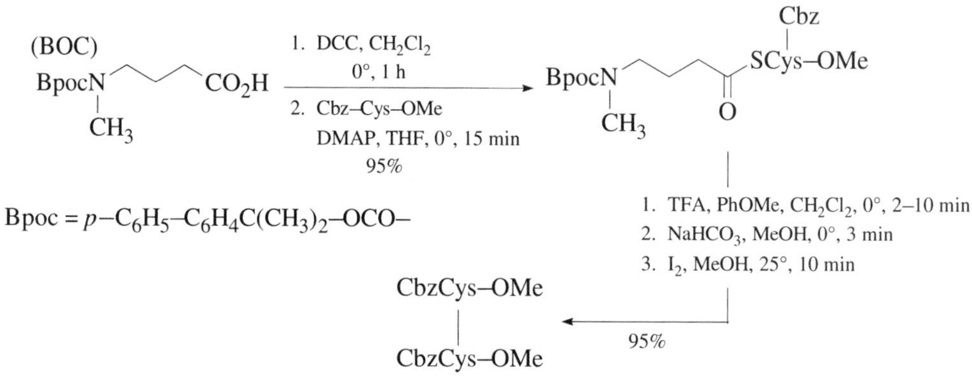

(BOC)BpocN(CH_3)$CH_2CH_2CH_2CO_2H$ → [1. DCC, CH_2Cl_2, 0°, 1 h; 2. Cbz–Cys–OMe, DMAP, THF, 0°, 15 min; 95%] → BpocN(CH_3)$CH_2CH_2CH_2$C(=O)SCys(Cbz)–OMe

→ [1. TFA, PhOMe, CH_2Cl_2, 0°, 2–10 min; 2. $NaHCO_3$, MeOH, 0°, 3 min; 3. I_2, MeOH, 25°, 10 min; 95%] → CbzCys–OMe | CbzCys–OMe

Bpoc = $p–C_6H_5–C_6H_4C(CH_3)_2–OCO–$

Deprotection is effected only by step 1 (TFA, PhOMe, CH_2Cl_2, 0°, 2–10 min).

1. N. G. Galakatos and D. S. Kemp, *J. Org. Chem.*, **50**, 1302 (1985).

Thiocarbonate Derivatives

When cysteine reacts with an alkyl or aryl chloroformate, both the –SH and –NH groups are protected, as a thiocarbonate and as a carbamate, respectively. Selective or simultaneous removal of the protective groups is possible. (See cleavage conditions 3–6 for an *S*-benzyloxycarbonyl derivative, page 485.)

S-2,2,2-Trichloroethoxycarbonyl Derivative: $RSCOOCH_2CCl_3$

Cleavage

1. Electrolysis, –1.5 V, $LiClO_4$, CH_3OH, 90% yield. The conditions can be adjusted to form either the sulfide or disulfide.[1]

1. M. F. Semmelhack and G. E. Heinsohn, *J. Am. Chem. Soc.*, **94**, 5139 (1972).

S-t-Butoxycarbonyl Derivative (BOC–SR): $RSCOOC(CH_3)_3$

t-Butyl chloroformate reacts with cysteine to protect both the amine and thiol groups; as with *N,S*-bis(benzyloxycarbonyl)cysteine, selective or simultaneous removal of the *N*- or *S*-protective groups can be effected.[1] Treatment with HCl/EtOAc efficiently cleaves the S–BOC group.[2]

1. M. Muraki and T. Mizoguchi, *Chem Pharm. Bull.*, **19**, 1708 (1971).
2. F. S. Gibson, S. C. Bergmeier, and H. Rapoport, *J. Org. Chem.*, **59**, 3216 (1994).

S-Benzyloxycarbonyl Derivative (RS-Cbz, RS-Z): $RSCOOCH_2C_6H_5$

Formation[1]

NH_2 / CO_2H / SCbz ← aq. $NaHCO_3$, CbzCl, 0°, 1 h; 10°, 1 h, 67% — NH_2 / CO_2H / SH — aq. NaOH, CbzCl, 30 min, 71% → NHCbz / CO_2H / SCbz

Cleavage

1. Concd. NH_4OH, 25°, 1 h, 90% yield.[1]
2. Na, NH_3, 62% yield.[1]
3. 0.1 *N* $NaOCH_3$, CH_3OH, N_2, 30 min–3 h, 100% yield.[2] An *S*-benzoyl group is removed (95–100% yield) in 5–10 min.
4. CF_3COOH, reflux, 30 min, ca. quant.[2] An *N*-Cbz group is also removed under these conditions.
5. 2 *N* HBr, AcOH, 25°, 30 min.[2,3] The *S*-Cbz group is removed slowly under these conditions, but the *N*-Cbz group is completely cleaved, thus providing some selectivity in the protection scheme for cysteine.
6. Electrolysis, −2.6 V, $R_4N^+X^-$, DMF.[4] Both an *N*-Cbz group and an *S*-Cbz group are removed under these conditions.

1. A. Berger, J. Noguchi, and E. Katchalski, *J. Am. Chem. Soc.*, **78**, 4483 (1956).
2. L. Zervas, I. Photaki, and N. Ghelis, *J. Am. Chem. Soc.*, **85**, 1337 (1963).
3. M. Sokolovsky, M. Wilchek, and A. Patchornik, *J. Am. Chem. Soc.*, **86**, 1202 (1964).
4. V. G. Mairanovsky, *Angew. Chem., Int. Ed. Engl.*, **15**, 281 (1976).

S-p-Methoxybenzyloxycarbonyl Derivative: $RSCOOCH_2C_6H_4$-*p*-OCH_3

S-p-Methoxybenzyloxycarbonylcysteine has been prepared in low yield (30%). It has been used in peptide syntheses, but is very labile to acids and bases.[1]

1. I. Photaki, *J. Chem. Soc. C*, 2687 (1970).

Thiocarbamate Derivatives

Thiocarbamates, formed by reaction of a thiol with an isocyanate, are stable in acidic and neutral solutions and are readily cleaved by basic hydrolysis. The

β-elimination that can occur when an *S*-acyl group is removed with base from a cysteine derivative does not occur under the conditions needed to cleave a thiocarbamate.[1]

S-(N-Ethylcarbamate): $RSCONHC_2H_5$ (Chart 7)

Formation[1]

1. EtN=C=O, pH 1→ pH 6, 20°, 70 h, 67% yield.

Cleavage

1. 1 *N* NaOH, 20°, 20 min, 100% yield.[1]
2. NH_3 or NH_2NH_2, methanol, 20°, 2 h, 100% yield.[1]
3. Na/NH_3, –30°, 3 min, 100% yield.[1]
4. $Hg(OAc)_2$, H_2O, CH_3OH, 30 min; H_2S, 4 h, 79% yield.[2]
5. $AgNO_3$, H_2O, CH_3OH; concd. HCl, 3 h, 62% yield.[2]

This protective group is stable to acidic hydrolysis (4.5 *N* HBr/HOAc; 1 *N* HCl; CF_3CO_2H, reflux). There is no evidence of S → N acyl migration in *S-(N-*ethylcarbamates) (RS = cysteinyl).[1] Oxidation of *S-(N-*ethylcarbamoyl)cysteine with performic acid yields cysteic acid.[2]

1. St. Guttmann, *Helv. Chim. Acta*, **49**, 83 (1966).
2. H. T. Storey, J. Beacham, S. F. Cernosek, F. M. Finn, C. Yanaihara, and K. Hofmann, *J. Am. Chem. Soc.*, **94**, 6170 (1972).

S-(N-Methoxymethylcarbamate): $RSCONHCH_2OCH_3$

Formation[1]

1. $CH_3OCH_2N{=}C{=}O$, pH 4–5, 2 min, 100% yield.

At pH 4–5, the reaction is selective for the protection of thiol groups in the presence of α- or ε-amino groups.

Cleavage[1]

1. At pH 9.6, a cysteine derivative is cleaved in 100% yield and a glutathione derivative in 80% yield.

1. H. Tschesche and H. Jering, *Angew. Chem., Int. Ed. Engl.*, **12**, 756 (1973).

MISCELLANEOUS DERIVATIVES

Unsymmetrical Disulfides

A thiol can be protected by oxidation (with O_2; H_2O_2; I_2; ···) to the corresponding symmetrical disulfide, which subsequently can be cleaved by reduction: [Sn/HCl; Na/xylene, Et_2O, or NH_3; $LiAlH_4$; $NaBH_4$; or thiols such as $HO(CH_2)_2SH$]. Unsymmetrical disulfides have also been prepared and are discussed.

***S*-Ethyl Disulfide:** $RSSC_2H_5$ (Chart 7)

Formation

1. EtS(O)SEt, –70°, 1 h, 80–90% yield.[1]

Cleavage

1. PhSH, >50° or $HSCH_2CO_2H$, 45°, 15 h, quant.[2] The *S*-ethyl disulfide is stable to acid-catalyzed hydrolysis (CF_3CO_2H) of carbamates and to ammonolysis (25% NH_3/CH_3OH).[2]

1. D. A. Armitage, M. J. Clark, and C. C. Tso, *J. Chem. Soc., Perkin Trans. 1*, 680 (1972).
2. N. Inukai, K. Nakano, and M. Murakami, *Bull. Chem. Soc. Jpn.*, **40**, 2913 (1967).

***S*-*t*-Butyl Disulfide:** $RSSC(CH_3)_3$

Formation

1. $CH_3OC(O)SCl$, 0–5°, 1.5 h; *t*-BuSH, MeOH, 5 days, 97% crude, 46% pure.[1] The reaction proceeds through an *S*-sulfenyl thiocarbonate.
2. *t*-BuO_2CNHN(S-*t*-Bu)CO_2–*t*-Bu, H_2O.[2]

Cleavage

1. $NaBH_4$.[3]
2. Bu_3P, trifluoroethanol/water (95/5).[4]

1. L. Field and R. Ravichandran, *J. Org. Chem.*, **44**, 2624 (1979).
2. E. Wünsch, L. Moroder, and S. Romani, *Hoppe-Seyler's Z. Physiol. Chem.*, **363**, 1461 (1982).
3. E. Wünsch and R. Spangenberg, in *Peptides, 1969*, E. Schoffone, Ed., North Holland, Amsterdam, p. 1971.

4. R. Ramage and A. S. J. Stewart, *J. Chem. Soc., Perkin Trans. 1*, 1947 (1993).

Substituted *S*-Phenyl Disulfides: $RSSC_6H_4$–Y

Three substituted *S*-phenyl unsymmetrical disulfides have been prepared, **i**,[1] **ii**,[2] and **iii**[3] — compounds **i** and **ii** by reaction of a thiol with a sulfenyl halide, compound **iii** from a thiol and an aryl thiosulfonate ($ArSO_2SAr$). The disulfides are cleaved by reduction ($NaBH_4$) or by treatment with excess thiol ($HSCH_2CH_2OH$).

RSS–C_6H_3–2-NO_2-4-R′ RSS–C_6H_4–2-N=N–C_6H_5 RSS–C_6H_4–2-COOH

i (R′ = H, NO_2) **ii** **iii**

1. A. Fontana, E. Scoffone, and C. A. Benassi, *Biochemistry*, **7**, 980 (1968); A. Fontana, *J. Chem. Soc., Chem. Commun.*, 976 (1975).
2. A. Fontana, F. M. Veronese, and E. Scoffone, *Biochemistry*, **7**, 3901 (1968).
3. L. Field and P. M. Giles, Jr., *J. Org. Chem.*, **36**, 309 (1971).

Sulfenyl Derivatives

S-Sulfonate Derivative: $RSSO_3^-$

Formation

1. Na_2SO_3, cat. cysteine, O_2, pH 7–8.5, 1 h, quant.[1]

Cleavage

1. $HSCH_2CH_2OH$, pH 7.5, 25°, 2 h, 100% yield.[1]
2. $NaBH_4$.[1] *S*-Sulfonates are stable at pH 1–9; they are unstable in hot acidic solutions and in 0.1 *N* sodium hydroxide.

1. W. W.-C. Chan, *Biochemistry*, **7**, 4247 (1968).

S-Sulfenylthiocarbonate: RSSCOOR′

A number of *S*-sulfenylthiocarbonates have been prepared to protect thiols. A benzyl derivative, R′=CH_2Ph, is stable to trifluoroacetic acid (25°, 1 h), but not to HBr/AcOH, and provides satisfactory protection during peptide syntheses;[1] a *t*-butyl derivative, R′ = *t*-Bu, is too labile in base to provide protection.[1] A methyl derivative, R′=CH_3, has been used to protect a cysteine fragment that is subsequently converted to a cystine.[2]

NHBOC / CO_2H / STr $\xrightarrow[\text{2 days, 60\%}]{\text{BnOC(O)SCl, 0°, 2 h, –25°}}$ NHBOC / CO_2H / $SSCO_2Bn$ $\xrightarrow{\text{TFA, rt, 1 h}}$ NH_2 / CO_2H / $SSCO_2Bn$

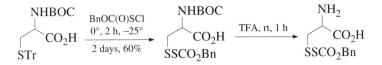

1. K. Nokihara and H. Berndt, *J. Org. Chem.*, **43**, 4893 (1978).
2. R. G. Hiskey, N. Muthukumaraswamy, and R. R. Vunnam, *J. Org. Chem.*, **40**, 950 (1975).

S-3-Nitro-2-pyridinesulfenyl Sulfide (Npys–SR): 3-NO_2–C_5H_3NSSR

These sulfides are prepared from other sulfur-protected cysteine derivatives by reaction with the sulfenyl chloride.[1] The Npys group can also be introduced directly by treatment of the thiol with NpysCl.[2]

Conversion of Conventional *S*-Protective Groups into the NpysSR Derivative[1]

Starting Material	Npys–X, Eq.	Conditions	% Yield
Boc–Cys(Bn)–OH	Cl, 1.2	rt, 24 h, CH_2Cl_2	No reaction
Boc–Cys(MeOBn)–OH[3]	Cl, 1.2	0°, 30 min, CH_2Cl_2	92
Boc–Cys(Me_2Bn)–OH	Cl, 1.2	0°, 30 min, CH_2Cl_2	90
Z–Cys(MeOBn)–Phe–Phe–Gln–Asn–O–*t*-Bu	Cl, 1.2	rt, 30 min, CH_2Cl_2, CF_3COOH (1 : 1)	85
Fmoc–Cys(*t*-Bu)–OH	Cl, 1.2	0°, 30 min, CH_2Cl_2	80
Boc–Cys(Tr)–OH	Cl, 1.2	–30°, 3 h, CH_2Cl_2	91
Boc–Cys(Acm)–OH	Cl, 1.2	0°, 30 min, AcOH	63
Z–Cys(Bn)–OH	Br, 2.0	rt, 10 h, CH_2Cl_2	21
Z–Cys(Bn)–OH	Cl, 2.0	rt, 5 h, CF_3CH_2OH	61
Z–Cys(Bn)–OH	Br, 2.4	rt, 3 h, CF_3CH_2OH, AcOH (10 : 1)	73
Z–Cys(Bn)–Pro–Leu–GlyNH_2	Br, 2.4	rt, 3 h, CF_3CH_2OH, AcOH (10 : 1)	70

The Npys group can be cleaved reductively with Bu_3P, H_2O, or mercaptoethanol. It has also been cleaved with 2-mercaptopyridine, 2-mercaptomethylimidazole, or 2-mercaptoacetic acid in methanol/acetic acid. Selective cleavage of the *O*-Npys bond over the *S*-Npys bond can be achieved with the aromatic thiols.[4] The Npys group is stable to CF_3COOH (24 h), 4 *M* HCl/dioxane (24 h), and HF (1 h).[2] The related reagent, 2-pyridinesulfenyl chloride, has also been proposed as a useful reagent for the deprotection of the *S*-trityl, *S*-diphenylmethyl, *S*-acetamidomethyl, *S*-*t*-butyl, and ***S*-*t*-butylsulfenyl** groups, but it is very susceptible to hydrolysis.[5]

1. R. Matsueda, S. Higashida, R. J. Ridge, and G.R. Matsueda, *Chem. Lett.*, 921 (1982).
2. R. Matsueda, T. Kimura, E. T. Kaiser, and G. R. Matsueda, *Chem. Lett.*, 737 (1981).

3. O. Ploux, M. Caruso, G. Chassaing, and A. Marquet, *J. Org. Chem.*, **53**, 3154 (1988).
4. O. Rosen, S. Rubinraut, and M. Fridkin, *Int. J. Pept. Protein Res.*, **35**, 545 (1990).
5. J. V. Castell and A. Tun-Kyi, *Helv. Chim. Acta*, **62**, 2507 (1979).

S-[Tricarbonyl[1,2,3,4,5-η]-2,4-cyclohexadien-1-yl]-iron(1+) Thioether: $[(\eta\text{-}^5C_6H_7)Fe(CO)_3]SR$

Formation

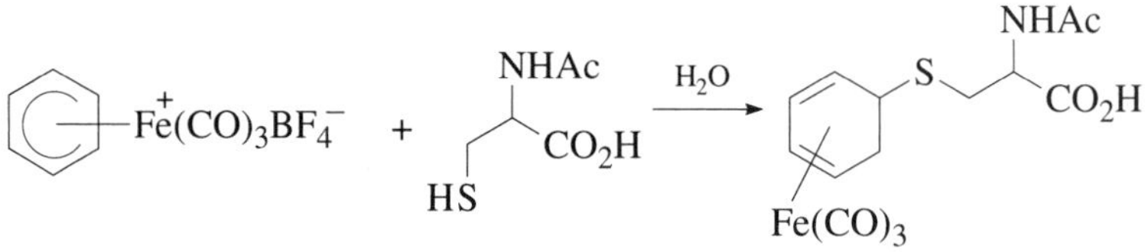

$\text{C}_6\text{H}_7\overset{+}{\text{Fe}}(CO)_3BF_4^-$ + HS–CH₂CH(NHAc)CO₂H $\xrightarrow{H_2O}$ [(η⁵-cyclohexadienyl)Fe(CO)₃]–S–CH₂CH(NHAc)CO₂H

Cleavage

Treatment with HBF_4 in $CHCl_3$ liberates the thiol and returns the derivatizing agent, $[(\eta\text{-}^5C_6H_7)Fe(CO)_3]^+ BF_4^-$ [tricarbonyl[1,2,3,4,5-η]-2,4-cyclohexadien-1-yl-iron(1+) tetrafluoroborate] as a precipitate.[1]

1. S. Fu, J. A. Carver, and L. A. P. Kane-Maguire, *J. Organomet. Chem.*, **454**, C11 (1993).

Oxathiolones: 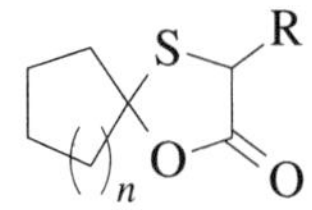 $n = 1, 2$

Oxathiolones are formed by heating a ketone with the mercaptocarboxylic acid in the presence of TsOH. They are cleaved by either acid (TFA, H_2O, THF) or base (NaOH, acetone) hydrolysis.[1]

1. L. M. Gustavson, D. S. Jones, J. S. Nelson, and A. Srinivasan, *Synth. Commun.*, **21**, 249 (1991).

Protection for Dithiols

Dithio Acetals and Ketals

S,S′-Methylene (i), *S, S′*-Isopropylidene (ii), and *S, S′*-Benzylidene (iii), Derivatives

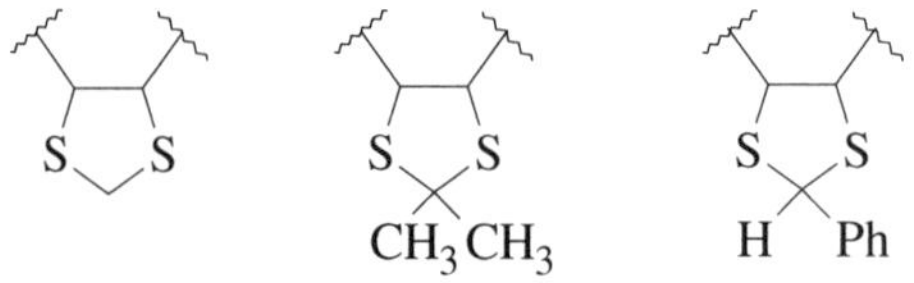

S S (i); S S, CH_3 CH_3 (ii); S S, H Ph (iii)

i **ii** **iii**

Dithiols, like diols, have been protected as *S,S′*-methylene,[1] *S,S′*-isopropylidene,[2] and *S,S′*-benzylidene[3] derivatives, formed by reaction of the dithiol with formaldehyde, acetone, or benzaldehyde, respectively. The methylene and benzylidene derivatives are cleaved by reduction with sodium/ammonia. The isopropylidene[2] and benzylidene[3] derivatives are cleaved by mercury(II) chloride; with sodium/ammonia, the isopropylidene derivative is converted to a monothio ether, $HSCHRCHRSCHMe_2$.[1]

1. E. D. Brown, S. M. Igbal, and L. N. Owen, *J. Chem. Soc.*, *C* 415 (1966).
2. E. P. Adams, F. P. Doyle, W. H. Hunter, and J. H. C. Nayler, *J. Chem. Soc.*, 2674 (1960).
3. L. W. C. Miles and L. N. Owen, *J. Chem. Soc.*, 2938 (1950).

S,S′-p-Methoxybenzylidene Derivative: $(RS)_2CHC_6H_4\text{–}4\text{-}OCH_3$

Formation [1]

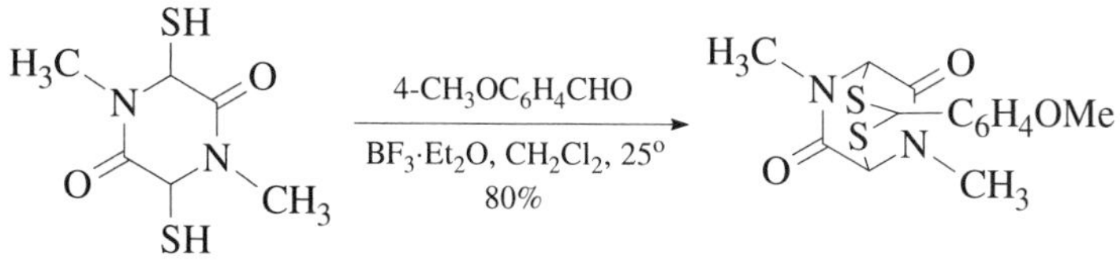

4-$CH_3OC_6H_4CHO$, $BF_3{\cdot}Et_2O$, CH_2Cl_2, 25°, 80%

Cleavage[1]

1. MCPBA, CH_2Cl_2, 0°
2. $BF_3{\cdot}Et_2O$ or BCl_3 or H_2SO_4 or $HClO_4$

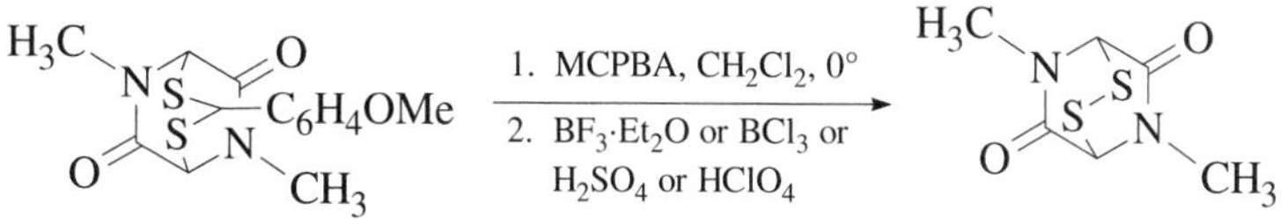

The preceding epidithioketopiperazine is present in natural products, including the gliotoxins and sporidesmins.[1]

1. Y. Kishi, T. Fukuyama, and S. Nakatusuka, *J. Am. Chem. Soc.*, **95**, 6490 (1973).

Protection for Sulfides

Since sulfides tend to react with electrophiles, a method for protecting sulfides could be quite useful. Sulfoxides can be used to protect sulfides and are easily formed by a variety of oxidants. Sulfides can be regenerated with thiols,[1] $SiCl_4$ (0°, 15 min, TFA, anisole),[2] $LiBH_4/Me_3SiCl$,[3] and $DMF{\cdot}SO_3/HSCH_2CH_2SH$ (DMF, Pyr, rt, 85% yield).[4]

Sulfides can also be protected as sulfonium salts.

S-Methylsulfonium Salt: $R_2S^+CH_3\ X^-$

Formation

1. $CH_3OSO_2CF_3$, CH_2Cl_2, 99% yield.[5]
2. MeOTs, EtOAc, rt, 4 days, 85% yield.[6]

Cleavage

1. DMF, Et_3N, $HSCH_2CH_2OH$, rt, 78% yield.[6]
2. $LiAlH_4$, THF.[5]

   A methylsulfonium salt is stable to NH_3/MeOH and to TFA, but not to hydrogenolysis (H_2/Pd–C).[6]

S-Benzyl- and *S*-4-Methoxybenzylsulfonium Salt: $R_2S^+CH_2Ph\ X^-$

Formation

1. $C_6H_5CH_2OTf$, CH_3CN.[7]
2. 4-$MeOC_6H_4CH_2Cl$, $AgBF_4$, CH_3CN, 97–99% yield.[8]

Cleavage

The benzylsulfonium salt is cleaved by hydrogenolysis (H_2/Pd–C, MeOH);[7] the 4-methoxybenzylsulfonium salt is cleaved by methylamine (100%).[8]

S-1-(4-Phthalimidobutyl)sulfonium Salt

Formation/Cleavage[8]

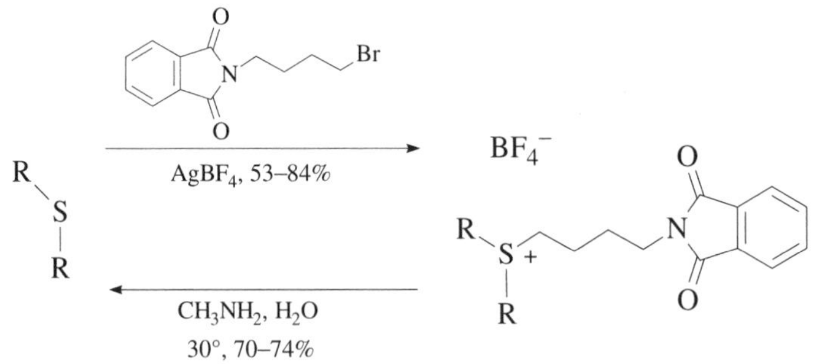

1. N. Fujii, A. Otaka, S. Funakoshi, H. Yajima, O. Nishimura, and M. Fujino, *Chem. Pharm. Bull.*, **34**, 869 (1986).
2. Y. Kiso, M. Yoshida, T. Fujisaki, T. Mimoto, T. Kimura, and M. Shimokura, *Pept. Chem., 1986*, **24th**, 205 (1987); *Chem. Abstr.*, **108:** 112924j (1988).

3. A. Giannis and K. Sandhoff, *Angew. Chem., Int. Ed. Engl.*, **28**, 218 (1989).
4. S. Futaki, T. Taike, T. Yagami, T. Akita, and K. Kitagawa, *Tetrahedron Lett.*, **30**, 4411 (1989).
5. V. Cere, A. Guenzi, S. Pollicino, E. Sandri, and A. Fava, *J. Org. Chem.*, **45**, 261 (1980).
6. M. Bodanszky and M. A. Bednareck, *Int. J. Pept. Protein Res.*, **20**, 408 (1982).
7. R. C. Roemmele and H. Rapoport, *J. Org. Chem.*, **54**, 1866 (1989).
8. J. T. Doi and G. W. Luehr, *Tetrahedron Lett.*, **26**, 6143 (1985).

S–P Derivatives

S-(Dimethylphosphino)thioyl Group (Mpt–SR): $(CH_3)_2P(S)SR$

S-(Diphenylphosphino)thioyl Group (Ppt-SR): $Ph_2P(S)SR$

Formation

1. MptCl, $(i\text{-Pr})_2EtN$, $CHCl_3$, 79% yield. The Mpt group on the nitrogen in cysteine can be selectively removed with HCl/Ph_3P, leaving the *S*–Mpt group intact.[1]

Cleavage

1. $AgNO_3$, H_2O, Pyr, 0°, 1 h; H_2S, 100% yield.[1]
2. KF, 18-crown-6 or $Bu_4N^+F^-$, CH_3CN, MeOH, 88% yield.[2]

   The related *S*-(diphenylphosphino)thioyl group (Ppt group) has also been cleaved using these conditions.[3] The Mpt derivative of cysteine is not stable to DBU; it forms dehydroalanine. The Mpt group is stable to TFA and to 1 *M* HCl, but not to HBr/AcOH or 6 *M* HCl.[1]
3. $Bu_4N^+F^-$, THF, AcOH, >76% yield.[4]

1. M. Ueki and K. Shinozaki, *Bull. Chem. Soc. Jpn.*, **56**, 1187 (1983).
2. M. Ueki and K. Shinozaki, *Bull. Chem. Soc. Jpn.*, **57**, 2156 (1984).
3. L. Horner, R. Gehring, and H. Lindel, *Phosphorus Sulfur*, **11**, 349 (1981).
4. M. Ueki, H. Takeshita, A. Sacki, H. Komatsu, and T. Katoh, *Pept. Chem. 1994, 32nd*, 173 (1995), *Chem. Abstr.*, **123:** 257332j (1995).

7

PROTECTION FOR THE AMINO GROUP

A great many protective groups have been developed for the amino group, including carbamates (>NCO_2R), used for the protection of amino acids in peptide and protein syntheses,[1] and amides (>NCOR), used more widely in syntheses of alkaloids and for the protection[2] of the nitrogen bases adenine, cytosine, and guanine in nucleotide syntheses.

Carbamates are formed from an amine with a wide variety of reagents, the chloroformate being the most common; amides are formed from the acid chloride. *n*-Alkyl carbamates are cleaved by acid-catalyzed hydrolysis; *N*-alkylamides are cleaved by acidic or basic hydrolysis at reflux and by ammonolysis, conditions that cleave peptide bonds.

In this chapter, detailed information is provided for the more useful protective groups (some of which are included in Reactivity Charts 8–10); structures and references are given for protective groups that seem to have more limited use.[3]

CARBAMATES

Carbamates can be used as protective groups for amino acids to minimize racemization in peptide synthesis. Racemization occurs during the base-catalyzed coupling reaction of an *N*-protected, carboxyl-activated amino acid and takes place in the intermediate oxazolone that forms readily from an *N*-acyl-protected amino acid (R′ = alkyl, aryl):

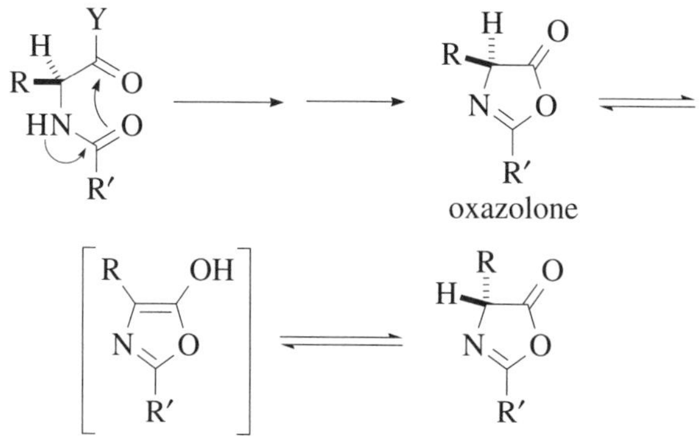

oxazolone

To minimize racemization, the use of nonpolar solvents, a minimum of base, low-reaction temperatures, and carbamate protective groups (R′ = *O*-alkyl or *O*-aryl) are effective.

Many carbamates have been used as protective groups. They are arranged in this chapter in order of increasing complexity of structure. The most useful compounds (not necessarily the simplest structures) are *t*-butyl (BOC), readily cleaved by acidic hydrolysis; benzyl (Cbz or Z), cleaved by catalytic hydrogenolysis; 2,4-dichlorobenzyl, stable to the acid-catalyzed hydrolysis of benzyl and *t*-butyl carbamates; 2-(biphenylyl)isopropyl, cleaved more easily than *t*-butyl carbamate by dilute acetic acid; 9-fluorenylmethyl, cleaved by *β*-elimination with base; isonicotinyl, cleaved by reduction with zinc in acetic acid; 1-adamantyl, readily cleaved by trifluoroacetic acid; and allyl, readily cleaved by Pd-catalyzed isomerization.

1. See reference 22 (**Peptides**) in Chapter 1.
2. See reference 23 (**Oligonucleotides**) in Chapter 1. See also C. B. Reese, *Tetrahedron*,

34, 3143 (1978); V. Amarnath and A. D. Broom, *Chem. Rev.*, **77**, 183 (1977).

3. See also E. Wünsch, "Blockierung und Schutz der α-Amino-Funktion," in *Methoden der Organischen Chemie* (*Houben-Weyl*), Georg Thieme Verlag, Stuttgart, 1974, Vol. 15/1, pp. 46–305; J. W. Barton, "Protection of N–H Bonds and NR_3," in *Protective Groups in Organic Chemistry*, J. F. W. McOmie, Ed., Plenum Press, New York and London, 1973, pp. 43–93; L. A. Carpino, *Acc. Chem. Res.*, **6**, 191–198 (1973); Y. Wolman, "Protection of the Amino Group," in *The Chemistry of the Amino Group*, S. Patai, Ed., Wiley-Interscience, New York, 1968, Vol. 4, pp. 669–699; E. Gross and J. Meienhofer, Eds., *The Peptides: Analysis, Synthesis, Biology, Vol. 3: Protection of Functional Groups in Peptide Synthesis*, Academic Press, New York, 1981; P. J. Kocienski, *Protecting Groups*, Thieme Medical Publishers, New York, 1994, Chapter 6.

Methyl and Ethyl Carbamate: $CH_3OC(O)NR_2$ (Chart 8)

Formation

1. CH_3OCOCl, K_2CO_3, reflux 12 h.[1]
2. CO, O_2, MeOH, HCl, $PdCl_2$, $CuCl_2$.[2]
3. CO, EtOH, O_2, KI, Pd–C[3] or $Pd(OAc)_2$.[4] Electrochemical oxidation has also been used (55–99% yield).[5]
4. CO, O_2, Co(tpp), NaI, EtOH, 68 atm, 3 h, 180°.[6]
5. R_2NH → R_2NCO_2Et (reagent: EtO_2CN-thiazolidine-2-thione, CH_2Cl_2, rt, 93–97%) Ref. 7
6. *N*-[(Methoxy)carbonyloxy]succinimide, Pyr, rt, >89% yield.[8]
7. CO_2, $HC(OEt)_3$, 40 h, 120° 45 atm, 83% yield.[9]
8. CO_2, TEA, RCl, 20–76% yield.[10]
9. From a thiocarbamate: NaOMe, MeOH, reflux, 43 h, 90% yield.[11]

Cleavage

1. *n*-PrSLi, 0°, 8.5 h, 75–80% yield.[12]
2. Me_3SiI, 50°, 70% yield.[13,14]

TMSI, CH_2Cl_2

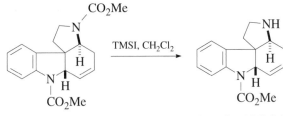

CO_2Me, N, H, N, H, CO_2Me → NH, H, N, H, CO_2Me

Contains 10% fully deprotected material Ref. 15

3. KOH, H_2O, ethylene glycol, 100°, 12 h, 88% yield.[16]
4. HBr, AcOH, 25°, 18 h.[17,18]
5. $NaAlH_2(OCH_2CH_2OCH_3)_2$, benzene, rt, 80% yield.[19]
6. $Ba(OH)_2$, H_2O, MeOH, 110°, 12 h.[20]
7. K_2CO_3, MeOH, 67% yield. These conditions were used to cleave a methyl carbamate from an aziridine.[21]
8. $NH_2NH_2 \cdot H_2O$, KOH, 98% yield.[22]
9. Dimethyl sulfide, methanesulfonic acid, 5°, 58–100% yield.[23]
10. NaHTe, 45–83% yield.[24]
11. TMSOK, MeOH, reflux, 48 h, 67% yield.[25]
12. MeLi, THF, 0°.[26]
13. AcCl, NaI, CH_3CN, 16 h, 60°, 52% yield.[27]
14. NaOH, MeOH, rt, 80% yield.[28] Cleavage occurs under such mild conditions because the N–O nitrogen in this case is a much better leaving group than the typical aliphatic amine.

Sugar–O–N(EtO_2C)–[HO, BnO, O, OCH_3 sugar ring] $\xrightarrow[\text{rt, 80\%}]{\text{NaOH, MeOH}}$ Sugar–O–N(H)–[HO, BnO, O, OCH_3 sugar ring]

1. E. J. Corey, M. G. Bock, A. P. Kozikowski, A. V. Rama Rao, D. Floyd, and B. Lipshutz, *Tetrahedron Lett.*, 1051 (1978).
2. H. Alper and F. W. Hartstock, *J. Chem. Soc., Chem. Commun.*, 1141 (1985).
3. S. Fukuoka, M. Chono, and M. Kohno, *J. Org. Chem.*, **49**, 1458 (1984).
4. T. Pri-Bar and J. Schwartz, *J. Org. Chem.*, **60**, 8124 (1995).
5. F. W. Hartstock, D. G. Herrington, and L. B. McMahon, *Tetrahedron Lett.*, **35**, 8761 (1994).
6. T. W. Leung and B. D. Dombek, *J. Chem. Soc., Chem. Commun.*, 205 (1992).
7. L. C. Chen and S. C. Yang, *J. Chin. Chem. Soc. (Taipei)*, **33**, 347 (1986).
8. S. B. Rollins and R. M. Williams, *Tetrahedron Lett.*, **38**, 4033 (1997).
9. S. Ishii, H. Nakayama, Y. Hoshida, and T. Yamashita, *Bull. Chem. Soc. Jpn.*, **62**, 455 (1989).
10. W. McGhee, D. Riley, K. Christ, Y. Pan, and B. Parnas, *J. Org. Chem.*, **60**, 2820 (1995).
11. S. K. Tandel, S. Ragappa, and S. V. Pansare, *Tetrahedron*, **49**, 7479 (1993).
12. E. J. Corey, L. O. Weigel, D. Floyd, and M. G. Bock, *J. Am. Chem. Soc.*, **100**, 2916 (1978).
13. R. S. Lott, V. S. Chauhan, and C. H. Stammer, *J. Chem. Soc., Chem. Commun.*, 495 (1979).
14. S. Raucher, B. L. Bray, and R. F. Lawrence, *J. Am. Chem. Soc.*, **109**, 442 (1987).
15. V. H. Rawal and C. Michoud, *J. Org. Chem.*, **58**, 5583 (1993).

16. E. Wenkert, T. Hudlicky, and H. D. H. Showalter, *J. Am. Chem. Soc.*, **100**, 4893 (1978).
17. M. C. Wani, H. F. Campbell, G. A. Brine, J. A. Kepler, M. E. Wall, and S. G. Levine, *J. Am. Chem. Soc.*, **94**, 3631 (1972).
18. P. Magnus, J. Rodrigues-Lôpez, K. Mulholland, and I. Matthews, *J. Am. Chem. Soc.*, **114**, 382 (1992).
19. G. R. Lenz, *J. Org. Chem.*, **53**, 4447 (1988).
20. P. M. Wovkulich and M. R. Uskokovic, *Tetrahedron*, **41**, 3455 (1985).
21. K. F. McClure and S. J. Danishefsky, *J. Am. Chem. Soc.*, **115**, 6094 (1993).
22. T. Shono, Y. Matsumura, K. Uchida, K. Tsubata, and A. Makino, *J. Org. Chem.*, **49**, 300 (1984).
23. H. Irie, H. Nakanishi, N. Fujii, Y. Mizuno, T. Fushimi, S. Funakoshi, and H. Yajima, *Chem. Lett.*, 705 (1980).
24. X.-J. Zhou and Z.-Z. Huang, *Synth. Commun.*, **19**, 1347 (1989).
25. S. J. Hays, P. M. Novak, D. F. Ortwine, C. F. Bigge, N. L. Colbry, G. Johnson, L. J. Lescosky, T. C. Malone, A. Michael, M. D. Reily, L. L. Coughenour, L. J. Brahce, J. L. Shillis, and A. Probert, Jr., *J. Med. Chem.*, **36**, 654 (1993).
26. M. Tius and M. A. Keer, *J. Am. Chem. Soc.*, **114**, 5959 (1992).
27. M. Ihara, A. Hirabayashi, N. Taniguchi, and K. Fukumoto, *Heterocycles*, **33**, 851 (1992).
28. D. Yang, S.-H. Kim, and D. Kahne, *J. Am. Chem. Soc.*, **113**, 4715 (1991).

9-Fluorenylmethyl Carbamate (Fmoc–NR$_2$) (Chart 8):

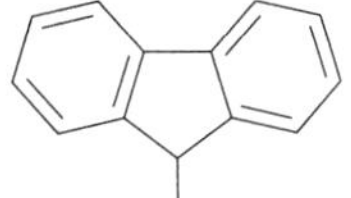

$CH_2OC(O)NR_2$

One major advantage of the Fmoc protective group is that it has excellent acid stability; thus, BOC and benzyl-based groups can be removed in its presence. Other advantages are that it is readily cleaved, nonhydrolytically, by simple amines, and the protected amine is liberated as its free base.[1] The Fmoc group is generally considered to be stable to hydrogenation conditions, but it has been shown that, under some circumstances, it can be cleaved with H_2/Pd–C, AcOH, MeOH ($t_{1/2}$ = 3–33 h).[2]

Formation

1. Fmoc–Cl, $NaHCO_3$, aq. dioxane, 88–98% yield.[3] Diisopropylethylamine is reported to suppress dipeptide formation during Fmoc introduction with Fmoc–Cl.[4]
2. Fmoc–N_3, $NaHCO_3$, aq. dioxane, 88–98% yield.[3,5] This reagent reacts more slowly with amino acids than does the acid chloride. It is not the safest method for Fmoc introduction, because of the azide.
3. Fmoc–OBt (Bt = benzotriazol-1-yl).[6,7]
4. Fmoc–OSu (Su = succinimidyl), H_2O, CH_3CN.[6–9] The advantage of Fmoc–OSu

is that little or no oligopeptides are formed when amino acid derivatives are prepared.[10]

5. Fmoc–OC_6F_5, $NaHCO_3$, H_2O, acetone, rt, 64–99% yield.[11]

Cleavage

1. The Fmoc group is cleaved under mild conditions with an amine base to afford the free amine and dibenzofulvene. The accompanying table gives the approximate half-lives for the deprotection of Fmoc–ValOH by a variety of amine bases in DMF.[10] The half-lives shown in the table will vary, depending on the structure of the Fmoc–amine derivative.

Amine	$t_{1/2}$
20% Piperidine	6 sec
5% Piperidine	20 sec
50% Morpholine	1 min
50% Dicyclohexylamine	35 min
10% *p*-Dimethylaminopyridine	85 min
50% Diisopropylethylamine[12]	10.1 h

2. $Bu_4N^+F^-$, DMF, rt, 2 min.[13]
3. $Bu_4N^+F^-$, *n*-$C_8H_{17}SH$, 92–100% yield.[14] The thiol is used to scavenge the liberated fulvene
4. Piperazine attached to a polymer has also been used to cleave the Fmoc group.[15]
5. Tris(2-aminoethyl)amine, CH_2Cl_2. This amine acts as the deblocking agent and the scavenger for the dibenzofulvene and does not cause the formation of precipitates or emulsions, which sometimes occur.[1b]
6. Direct conversion of an Fmoc group to a Cbz group: KF, TEA, DMF, *N*-benzyloxycarbonyloxy-5-norbornene-2,3-dicarboximide, 7–12 h, 83–99% yield.[16]

9-(2-Sulfo)fluorenylmethyl Carbamate

Because of the electron-withdrawing sulfonic acid substituent, cleavage occurs under milder conditions than are needed for the Fmoc group (0.1 *N* NH_4OH; 1% Na_2CO_3, 45 min).[17]

9-(2,7-Dibromo)fluorenylmethyl Carbamate:

Br Br

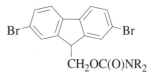

$CH_2OC(O)NR_2$

Because of the two electron-withdrawing bromine groups, pyridine can be used to cleave this derivative from its parent amine.[18]

17-Tetrabenzo[*a,c,g,i*]fluorenylmethyl Carbamate (Tbfmoc–NR_2):

CH_2OCONR_2

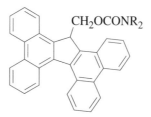

This Fmoc analog is prepared from the chloroformate, *O*-succinimide, or *p*-nitrophenyl carbonate and is cleaved with 10% piperidine in 1:1 6 *M* guanidine/IPA.[19] It was designed to interact strongly on a column of porous graphitized carbon so as to aid in the purification of peptides after cleavage from the resin.

2-Chloro-3-indenylmethyl Carbamate (Climoc–NR_2) and Benz[*f*]inden-3-ylmethyl Carbamate (Bimoc–NR_2):

Cl O O NR_2 O O NR_2

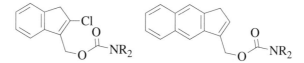

These base-sensitive protective groups were introduced from the chloroformate or azidoformate. They are more sensitive to base than is the Fmoc group. Cleavage times with 0.2 mL of piperidine to 0.1 mmole of urethane in 5 mL of $CHCl_3$ at rt occur as follows: Climoc, <10 min; Bimoc, <14 h; Fmoc, 18 h.[20]

1. (a) L. A. Carpino, *Acc. Chem. Res.*, **20**, 401 (1987); (b) L. A. Carpino, D. Sadat-Aalaee, and M. Beyermann, *J. Org. Chem.*, **55**, 1673 (1990).
2. E. Atherton, C. Bury, R. C. Sheppard, and B. J. Williams, *Tetrahedron Lett.*, 3041 (1979).
3. L. A. Carpino and G. Y. Han, *J. Org. Chem.*, **37**, 3404 (1972).
4. F. M. F. Chen and N. L. Benoiton, *Can. J. Chem.*, **65**, 1224 (1987).
5. M. Tessier, F. Albericio, E. Pedroso, A. Grandas, R. Eritja, E. Giralt, C. Granier, and J. Van Rietschoten, *Int. J. Pept. Protein Res.*, **22**, 125 (1983).
6. A. Paquet, *Can. J. Chem.*, **60**, 976 (1982).
7. G. F. Sigler, W. D. Fuller, N. C. Chaturvedi, M. Goodman, and M. Verlander, *Biopolymers*, **22**, 2157 (1983).
8. R. C. de L. Milton, E. Becker, S. C. F. Milton, J. E. H. Baxter, and J. F. Elsworth, *Int. J. Pept. Protein Res.*, **30**, 431 (1987).
9. L. Lapatsanis, G. Milias, K. Froussios, and M. Kolovos, *Synthesis*, 671 (1983).
10. For a review of the use of Fmoc protection in peptide synthesis, see E. Atherton and R. C. Sheppard, "The Fluorenylmethoxycarbonyl Amino Protecting Group," in *The*

Peptides, S. Udenfriend and J. Meienhofer, Eds., Academic Press, New York, 1987, Vol. 9, p. 1.
11. I. Schoen and L. Kisfaludy, *Synthesis*, 303 (1986).
12. T. Hoeg-Jensen, M. H. Jakobsen, and A. Holm, *Tetrahedron Lett.*, **32**, 6387 (1991).
13. M. Ueki and M. Amemiya, *Tetrahedron Lett.*, **28**, 6617 (1987).
14. M. Ueki, N. Nishigaki, H. Aoki, T. Tsurusaki, and T. Katoh, *Chem. Lett.*, 721 (1993).
15. L. A. Carpino, E. M. E. Mansour, and J. Knapczyk, *J. Org. Chem.*, **48**, 666 (1983).
16. W. R. Li, J. Jiang, and M. M. Joullié, *Synlett*, 362 (1993).
17. R. B. Merrifield and A. E. Bach, *J. Org. Chem.*, **43**, 4808 (1978).
18. L. A. Carpino, *J. Org. Chem.*, **45**, 4250 (1980).
19. A. R. Brown, S. L. Irving, R. Ramage, and G. Raphy, *Tetrahedron*, **57**, 11815 (1995); R. Ramage and G. Raphy, *Tetrahedron Lett.*, **33**, 385 (1992).
20. L. A. Carpino, B. J. Cohen, Y. Z. Lin, K. E. Stephens, Jr., and S A. Triolo, *J. Org. Chem.*, **55**, 251 (1990).

2,7-Di-*t*-butyl[9-(10,10-dioxo-10,10,10,10-tetrahydrothioxanthyl)]methyl Carbamate (DBD–Tmoc–NR_2):

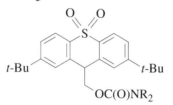

O O S *t*-Bu *t*-Bu $OC(O)NR_2$

Formation

1. DBD–TmocCl, $NaHCO_3$, H_2O, dioxane.[1]

Cleavage[1]

The DBD–Tmoc group is stable to TFA and HBr/AcOH.

1. 50–75° in DMSO, 4.5–16 h, 100% yield.
2. Pd–C, HCO_2NH_4, MeOH.
3. Pyridine. The Fmoc group is stable to pyridine.

1. L. A. Carpino, H.-S. Gao, G.-S. Ti, and D. Segev, *J. Org. Chem.*, **54**, 5887 (1989).

1,1-Dioxobenzo[*b*]thiophene-2-ylmethyl Carbamate (Bsmoc–NR_2):

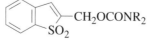

CH_2OCONR_2 SO_2

During the cleavage of the Fmoc group with base, dibenzofulvene is liberated and must be scavenged to prevent its reaction with the liberated peptides during peptide synthesis. The Bsmoc group was designed so that the cleavage agent [(tris(2-aminoethyl)amine] also serves as the scavenging agent.

The Bsmoc derivative is formed from the chloroformate or the *N*-hydroxysuccinimide ester. It is cleaved rapidly by a Michael addition with tris(2-aminoethyl)amine at a rate that leaves Fmoc derivatives intact. More hindered bases, such as *N*-methylcyclohexylamine or diisopropylamine, do not react with the Bsmoc group, but do cleave the Fmoc group, illustrating the importance of steric effects in additions to Michael acceptors.

The Bsmoc group is stable to TFA, HCl/EtOAc at rt for 24 h, to tertiary amines, and to hydrogenolysis, but it is not stable to HBr/AcOH. It is readily cleaved by RSH and base (DIPEA).[1]

1. L. A. Carpino, M. Philbin, M. Ismail, G. A. Truran, E. M. E. Mansour, S. Iguchi, D. Ionescu, A. El-Faham, C. Riemer, R. Warrass, and M. S. Weiss, *J. Am. Chem. Soc.*, **119**, 9915 (1997).

Substituted Ethyl Carbamates

2,2,2-Trichloroethyl Carbamate (Troc–NR_2): $Cl_3CCH_2OC(O)NR_2$) (Chart 8)

Formation

1. Cl_3CCH_2OCOCl, Pyr or aq. NaOH, 25°, 12 h.[1,2]
2. Silylate with $Me_3SiN{=}C(OSiMe_3)CH_3$, then treat with Cl_3CCH_2OCOCl.[3]
3. Cl_3CCH_2OCO–O-succinimidyl, 1 *N* NaOH or 1 *N* Na_2CO_3, dioxane, 77–96% yield.[4,5] This method does not result in oligopeptide formation when it is used to prepare amino acid derivatives.
4. Treatment of a tertiary benzylamine also affords the Troc derivative with cleavage of the benzyl group (Cl_3CCH_2OCOCl, CH_3CN, 93% yield).[6]
5. $CH_3OC(O)OC(O)OCH_2CCl_3$, CH_2Cl_2, rt, 3.5 h, 90–97% yield.[7]

Cleavage

1. Zn, THF, H_2O, pH 4.2, 30 min, 86% yield or pH 5.5–7.2, 18 h, 96% yield.[8] Under these conditions, the Troc group can be cleaved in the presence of the BOC, benzyl, and trifluoroacetamido groups, and these groups can in turn be cleaved individually in the presence of a Troc group.[9]
2. Electrolysis at a Hg cathode, 1.7 V (SCE), DMF, >72% yield.[10]
3. Electrolysis, −1.7 V, 0.1 *M* $LiClO_4$, 85% yield.[11]
4. Zn–Pb couple, 4:1 THF/1 *M* NH_4OAc.[12]

5. Cd, AcOH.[13] These conditions were reported to be superior to the use of Zn/AcOH. The authors also report that the Troc group is not stable to hydrogenation with Pd–C (TsOH, DMF, H_2), but is stable to hydrogenation with Ru–C or Pt–C.
6. Cd–Pb, AcOH, 89–94% yield.[14] This reagent also cleaves trichloroethyl esters and carbonates.

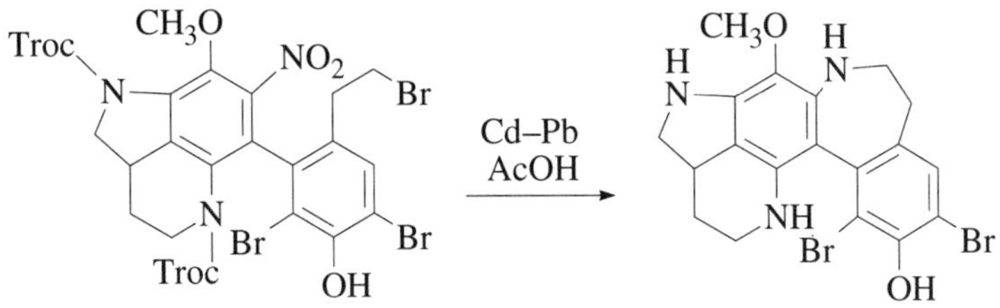

Cd–Pb
AcOH

7. Cobalt(I) phthalocyanine.[15]

8. 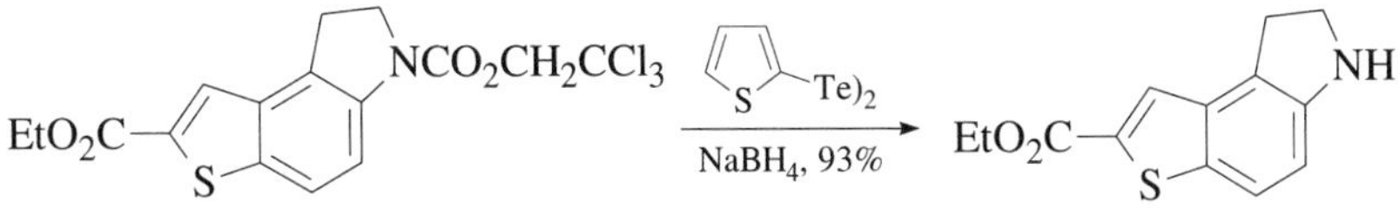 EtO_2C ... $NCO_2CH_2CCl_3$ → (2-thienyl-Te$)_2$, $NaBH_4$, 93% → EtO_2C ... NH

Ref. 16

1. T. B. Windholz and D. B. R. Johnston, *Tetrahedron Lett.*, 2555 (1967).
2. J. F. Carson, *Synthesis*, 268 (1981).
3. S. Raucher and D. S. Jones, *Synth. Commun.*, **15**, 1025 (1985).
4. A. Paquet, *Can. J. Chem.*, **60**, 976 (1982).
5. L. Lapatsanis, G. Milias, K. Froussios, and M. Kolovos, *Synthesis*, 671 (1983).
6. V. H. Rawal, R. J. Jones, and M. P. Cava, *J. Org. Chem.*, **52**, 19 (1987).
7. Y. Kita, J.-i. Haruta, H. Yasuda, K. Fukunaga, Y. Shirouchi, and Y. Tamura, *J. Org. Chem.*, **47**, 2697 (1982).
8. G. Just and K. Grozinger, *Synthesis*, 457 (1976).
9. R. J. Bergeron and J. S. McManis, *J. Org. Chem.*, **53**, 3108 (1988).
10. L. Van Hijfte and R. D. Little, *J. Org. Chem.*, **50**, 3940 (1985).
11. M. F. Semmelhack and G. E. Heinsohn, *J. Am. Chem. Soc.*, **94**, 5139 (1972).
12. L. E. Overman and R. L. Freerks, *J. Org. Chem.*, **46**, 2833 (1981).
13. G. Hancock, I. J. Galpin, and B. A. Morgan, *Tetrahedron Lett.*, **23**, 249 (1982).
14. Q. Dong, C. E. Anderson, and M. A. Ciufolini, *Tetrahedron Lett.*, **36**, 5681 (1995).
15. H. Eckert and I. Ugi, *Liebigs Ann. Chem.*, 278 (1979).
16. M. V. Lakshmikantham, Y. A. Jackson, R. J. Jones, G. J. O'Malley, K. Ravichandran, and M. P. Cava, *Tetrahedron Lett.*, **27**, 4687 (1986).

2-Trimethylsilylethyl Carbamate (Teoc–NR_2):
$(CH_3)_3SiCH_2CH_2OC(O)NR_2$ (Chart 8)

Formation

1. Teoc–O-succinimidyl, $NaHCO_3$ or TEA, dioxane, H_2O, rt, overnight, 43–96% yield.[1,2] The use of Teoc–OSu for the protection of amino acids proceeds without oligopeptide formation. Teoc–O-Benzotriazolyl was also examined, but was inferior to the succinimide derivative.
2. Teoc–OC_6H_4–4-NO_2, NaOH, *t*-BuOH, 66–89% yield.[3,4] The phenyl carbamate can be converted to a Teoc derivative.[5]
3. Teoc–Cl or Teoc–N_3.[6]
4. The Teoc derivative can be prepared by cleavage of an N–Bn bond with Teoc–Cl in THF. This is a general method for the removal of benzyl groups from nitrogen.[7] Methyl and ethyl groups are also cleaved, but more slowly (24 h versus 4 h) and in lower yield.

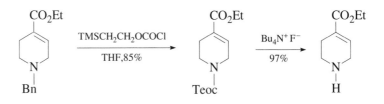

Cleavage

1. $Bu_4N^+F^-$, $KF{\cdot}2H_2O$, CH_3CN, 50°, 8 h, 93% yield or 28°, 70 h, 93% yield.[8]
2. CF_3COOH, 0°, 90% yield.[6]
3. $ZnCl_2$, CH_3NO_2 or $ZnCl_2$, CF_3CH_2OH.[4] These conditions cause partial BOC cleavage. The BOC group can be removed in the presence of a Teoc group with TsOH.[4]
4. Tris(dimethylamino)sulfonium difluorotrimethylsilicate (TAS-F), DMF, >76% yield.[9]
5. $Bu_4N^+Cl^-$, $KF{\cdot}2H_2O$, CH_3CN, 45°.[10]

1. R. E. Shute and D. H. Rich, *Synthesis*, 346 (1987).
2. A. Paquet, *Can. J. Chem.*, **60**, 976 (1982).
3. E. Wuensch, L. Moroder, and O. Keller, *Hoppe-Seyler's Z. Physiol. Chem.*, **362**, 1289 (1981).
4. A. Rosowsky and J. E. Wright, *J. Org. Chem.*, **48**, 1539 (1983).
5. A. I. Meyers, K. A. Babiak, A. L. Campbell, D. L. Comins, M. P. Fleming, R. Henning, M. Heuschmann, J. P. Hudspeth, J. M. Kane, P. J. Reider, D. M. Roland, K. Shimizu, K. Tomioka, and R. D. Walkup, *J. Am. Chem. Soc.*, **105**, 5015 (1983).

6. L. A. Carpino, J.-H. Tsao, H. Ringsdorf, E. Fell, and G. Hettrich, *J. Chem. Soc., Chem. Commun.*, 358 (1978).
7. A. L. Campbell, D. R. Pilipauskas, I. K. Khanna, and R. A. Rhodes, *Tetrahedron Lett.*, **28**, 2331 (1987).
8. L. A. Carpino and A. C. Sau, *J. Chem. Soc., Chem. Commun.*, 514 (1979).
9. W. R. Roush, D. S. Coffey, and D. J. Madar, *J. Am. Chem. Soc.*, **119**, 11331 (1997).
10. J. H. Van Maarseveen and H. W. Scheeren, *Tetrahedron*, **49**, 2325 (1993).

2-Phenylethyl Carbamate (hZ–NR_2): $R_2NCO_2CH_2CH_2Ph$

The 2-phenylethyl carbamate ("homo Z" = homobenzyloxycarbonyl derivative) is prepared from the chloroformate and can be cleaved with H_2/Pd–C if the catalyst is freshly prepared [$Pd(OAc)_2$, HCO_2NH_4). This derivative is stable to CF_3COOH, HBr/AcOH, HCl/Et_2O, and normal hydrogenation with Pd/C (1 atm). Hydrogenolysis of the hZ group is slower than that of the Fmoc group, which in turn is slower than hydrogenolysis of the Z group (Cbz).[1]

1. L. A. Carpino and A. Tunga, *J. Org. Chem.*, **51**, 1930 (1986).

1-(1-Adamantyl)-1-methylethyl Carbamate (Adpoc–NR_2):
(1-Adamantyl) $C(Me)_2OC(O)NR_2$

The 1-Adpoc derivative is cleaved by CF_3COOH (0°, 4–5 min) 10^3 times faster than the *t*-BOC derivative.[1]

1. H. Kalbacher and W. Voelter, *Angew. Chem., Int. Ed. Engl.*, **17**, 944 (1978); *idem., J. Chem. Soc., Chem. Commun.*, 1265 (1980).

2-Chloroethyl Carbamate: $ClCH_2CH_2OC(O)NR_2$

Cleavage

1. SmI_2, THF, 70°, 7 h, 70% yield.[1]

1,1-Dimethyl-2-haloethyl Carbamate:
$XCH_2C(CH_3)_2OC(O)NR_2$, X = Br, Cl (Chart 8)

Formation[2]

1. $XCH_2C(CH_3)_2OCOCl$, THF, Et_3N, H_2O, $CHCl_3$, 0°, 1.5 h (X = Br, 41–79% yield; X = Cl, 60–86% yield). These halo-substituted *t*-butyl chloroformates are more stable than an unsubstituted *t*-butyl chloroformate.

Cleavage[2]

1. CH_3OH, reflux, 1 h.
2. $BF_3 \cdot Et_2O$, CF_3COOH, 25°.
3. 4 *N* HBr, AcOH, 25°, 1 h.
4. Na, NH_3.

1,1-Dimethyl-2,2-dibromoethyl Carbamate (DB-*t*-BOC–NR_2): $Br_2CHC(CH_3)_2OC(O)NR_2$

The DB-*t*-BOC group is introduced with the chloroformate and can be cleaved solvolytically in hot ethanol or by HBr/AcOH. It is stable to CF_3COOH, 24 h; HCl, $MeNO_2$, 24 h; HCl, AcOH, 24 h; HBr, $MeNO_2$, 5 h.[3]

1,1-Dimethyl-2,2,2-trichloroethyl Carbamate (TCBOC–NR_2): $Cl_3CC(CH_3)_2OC(O)NR_2$

The TCBOC group is stable to the alkaline hydrolysis of methyl esters and to the acidic hydrolysis of *t*-butyl esters. It is rapidly cleaved by the supernucleophile lithium cobalt(I)-phthalocyanine, by zinc in acetic acid,[4] by Cd/Pb in NH_4OAc,[5] and by cobalt phthalocyanine (0.1 eq. $NaBH_4$, EtOH, 77–90% yield).[6]

1. T. P. Ananthanarayan, T. Gallagher, and P. Magnus, *J. Chem. Soc., Chem. Commun.*, 709 (1982).
2. T. Ohnishi, H. Sugano, and M. Miyoshi, *Bull. Chem. Soc. Jpn.*, **45**, 2603 (1972).
3. L. A. Carpino, N.W. Rice, E. M. E. Mansour, and S. A. Triolo, *J. Org. Chem.*, **49**, 836 (1984).
4. H. Eckert, M. Listl, and I. Ugi, *Angew. Chem., Int. Ed. Engl.*, **17**, 361 (1978).
5. R. M. Rzasa, H. A. Shea, and D. Romo, *J. Am. Chem. Soc.*, **120**, 591 (1998).
6. H. Eckert and Y. Kiesel, *Synthesis*, 947 (1980).

1-Methyl-1-(4-biphenylyl)ethyl Carbamate (Bpoc–NR_2): *p*-$PhC_6H_4C(Me)_2OC(O)NR_2$ (Chart 8)

Formation

1. Bpoc-N_3, 35–80% yield.[1]

Cleavage

1. This derivative is readily cleaved by acidic hydrolysis (dil. CF_3COOH, CH_2Cl_2, 10 min, quant.). It is cleaved 3000 times faster than the *t*-BOC

derivative because of stablization of the cation by the biphenyl group.[1] BnSH was found to be the most effective scavenger for $PhC_6H_4C^+Me_2$ when deblocking is performed in 0.5% TFA/CH_2Cl_2.[2]

2. Tetrazole, trifluoroethanol, 24 h, 95% yield.[3] These conditions also cleave the *N*-trityl group. If deprotection is performed in the presence of an acylating agent, acylation proceeds directly.
3. *N*-Hydroxybenzotriazole, trifluoroethanol, rt.[4] Trityl and Nps (2-nitrophenylsulfenyl) groups are also cleaved under these conditions.

1. R. S. Feinberg and R. B. Merrifield, *Tetrahedron*, **28**, 5865 (1972).
2. D. S. Kemp, N. Fotouhi, J. G. Boyd, R. I. Carey, C. Ashton, and J. Hoare, *Int. J. Pept. Protein Res.*, **31**, 359 (1988).
3. M. Bodanszky, A. Bodanszky, M. Casaretto, and H. Zahn, *Int. J. Pept. Protein Res.*, **26**, 550 (1985).
4. M. Bodanszky, M. A. Bednarek, and A. Bodanszky, *Int. J. Pept. Protein Res.*, **20**, 387 (1982).

1-(3,5-Di-*t*-butylphenyl)-1-methylethyl Carbamate (*t*-Bumeoc–NR_2):

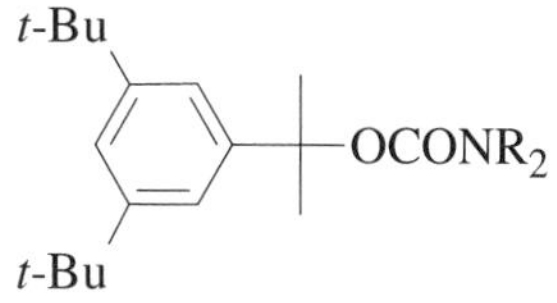

t-Bu

$OCONR_2$

t-Bu

Formation[1]

The *t*-Bumeoc adduct is prepared from the acid fluoride or the mixed carbonate in dioxane, H_2O, NaOH.

Cleavage[1]

Cleavage occurs with acid. The following tables give relative rate data that are useful for comparing other, more commonly employed, derivatives of phenylalanine (Phe).

Half-life of *t*-Bumeoc–Phe–OH with Different Acids

Acid	Half-life (min)	Complete cleavage (min)
3% TFA/CH_2Cl_2	0.07	0.6
80% AcOH/H_2O	2.1	18.8
AcOH/HCO_2H/H_2O (7:1:2)	22.0	167.0

Comparison of Cleavage Rates for Various Carbamate Protective Groups

Group	k_{rel}[a]	k_{rel}[b]
Boc	1	1
Ppoc[c]	700	750
Adpoc[d]	2400	600
Bpoc[e]	2800	2000
t-Bumeoc	4000	8000

a. 80% $AcOH/H_2O$. b. $AcOH/HCO_2H/H_2O$ (7:1:2). c. Ppoc = 1-Methyl-1-(triphenylphosphonio)ethyl. d. Adpoc = 1-Methyl-1-(1-adamantyl)ethyl. e. Bpoc = 1-Methyl-1-(4-biphenylyl) ethyl.

1. W. Voelter and J. Mueller, *Liebigs Ann. Chem.*, 248 (1983).

2-(2′- and 4′-Pyridyl)ethyl Carbamate (Pyoc–NR_2)

Formation/Cleavage[1,2]

H_2N–CH(R)–CO_2Na → (4-pyridyl)–$CH_2CH_2OCO_2C_6H_4$–4-NO_2, pH 11, 76–82% → PyocNH–CH(R)–CO_2H

Reverse: 1. MeI; 2. Morpholine; 90–95%

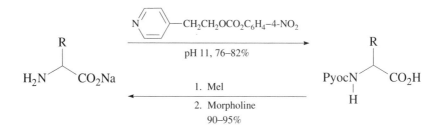

The Pyoc derivative is not affected by H_2/Pd–C or TFA.

1. H. Kunz and S. Birnbach, *Tetrahedron Lett.*, **25**, 3567 (1984).
2. H. Kunz and R. Barthels, *Angew. Chem., Int. Ed. Engl.*, **22**, 783 (1983).

2,2-Bis(4′-nitrophenyl)ethyl Carbamate (Bnpeoc–NR_2): (4-O_2N–$C_6H_4)_2CHCH_2OCONR_2$

The Bnpeoc group was developed as a base-labile protective group for solid-phase peptide synthesis. The carbamate is formed from the *O*-succinimide (DMF, 10% Na_2CO_3 or 5% $NaHCO_3$) and is cleaved using DBN, DBU, DBU/AcOH, or piperidine.[1]

1. R. Ramage, A. J. Blake, M. R. Florence, T. Gray, G. Raphy, and P. L. Roach, *Tetrahedron*, **47**, 8001 (1991).

N-(2-Pivaloylamino)-1,1-dimethylethyl Carbamate:

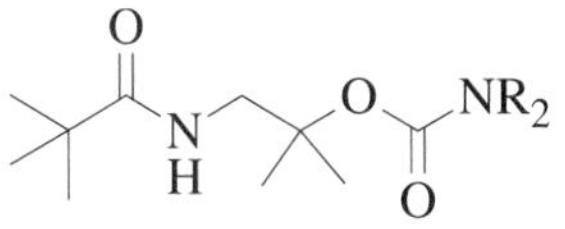

This group and a related series of amides were developed as a BOC-like protective group that, upon cleavage, does not release the *t*-butyl cation. The group is designed so that when it is cleaved, the cation is scavenged internally to give 4,5-dihydrooxazoles. Cleavage with TFA or HBr/AcOH occurs more slowly than with the normal BOC group, but the addition of $CaCl_2$ to TFA increases the rate.[1]

This protective group is formed from the alcohol and an isocyanate as follows:

$$(CH_3)_3CC(O)NHCH_2C(CH_3)_2OH \xrightarrow{RN=C=O} (CH_3)_3CC(O)NHCH_2C(CH_3)_2OC(O)NR_2$$

1. M. Gormanns and H. Ritter, *Tetrahedron*, **49**, 6965 (1993).

2-[(2-Nitrophenyl)dithio]-1-phenylethyl Carbamate (NpSSPeoc–NR_2):

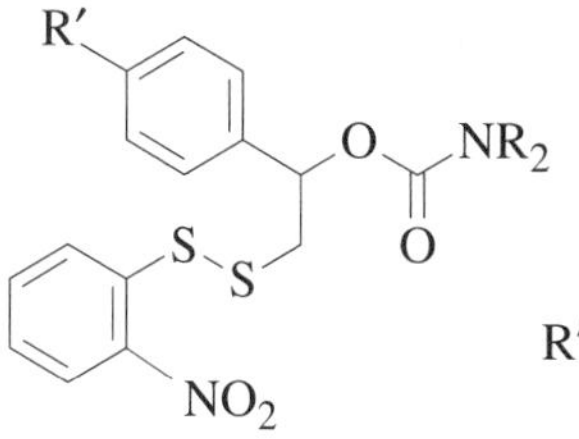

R′ = H, NO_2

This protective group was designed as part of the development of an affinity chromatography method for the purification of hydrophobic peptides. S–S cleavage in peptides containing the group, followed by attachment of the resulting S–S moiety to an affinity support (e.g., iodoacetamide resin), was found to be the simplest and highest yielding method. When R′ = H, the protective group is efficiently cleaved with TFA, and when R = NO_2, TfOH in TFA must be used.[1]

1. I. Sucholeiki and P. T. Lansbury, Jr., *J. Org. Chem.*, **58**, 1318 (1993).

2-(*N*,*N*-Dicyclohexylcarboxamido)ethyl Carbamate:

$(C_6H_{11})_2NC(O)CH_2CH_2OCONR_2$

Formation

1. $(C_6H_{11})_2NC(O)CH_2CH_2OCOCl$, diisopropylethylamine, CH_2Cl_2, 0°, 15 min.[1]

Cleavage

1. *t*-BuOK, *t*-BuOH, 18-crown-6, THF, 0°, 30 min, 100% yield. This protective group is stable to $LiAlH_4$; 3 *N* NaOH, MeOH, rt; H_2, RaNi, 1500 psi, 100°, EtOH; and TFA.[1]

1. T. Fukuyama, L. Li, A. A. Laird, and R. K. Frank, *J. Am. Chem. Soc.*, **109**, 1587 (1987).

t-Butyl (BOC) Carbamate: $(CH_3)_3COC(O)NR_2$ (Chart 8)

The BOC group is used extensively in peptide synthesis for amine protection.[1] It is not hydrolyzed under basic conditions and is inert to many other nucleophilic reagents.

Formation

1. $(BOC)_2O$, NaOH, H_2O, 25°, 10–30 min, 75–95% yield.[2] This is one of the more common methods for introduction of the BOC group. It has the advantage that the by-products are innocuous and easily removed.
2. $(BOC)_2O$, TEA, MeOH or DMF, 40–50°, 87–99% yield. These nonaqueous conditions were used in the protection of O^{17}-labeled amino acids so that the label would not be lost because of exchange with water.[3]
3. $(BOC)_2O$, EtOH or MeOH, $NaHCO_3$, ultrasound, 84–100% yield.[4]
4. $(BOC)_2O$, $Me_4NOH{\cdot}5H_2O$, CH_3CN, 88–100% yield. These conditions were found to be very good for sterically hindered amino acids.[5]
5. Sterically hindered amines often tend to form ureas with $(BOC)_2O$, because of isocyanate formation.[6,7] The problem can be avoided by reacting the amine with NaHMDS and then with $(BOC)_2O$.[8] The isocyanates can also be converted to the BOC group by heating with *t*-BuOH. When other alcohols are used, the corresponding carbamate is produced.[9]
6. BOC−ON=C(CN)Ph, Et_3N, 25°, several hours, 72–100% yield.[10] This reagent selectively protects primary amines in the presence of secondary amines.[11]
7. BOC−ONH_2.[12] This reagent reacts with amines 1.5–2.5 times faster than $(BOC)_2O$. Hydroxylamine can be used catalytically in the presence of $(BOC)_2O$ to generate this reagent *in situ*.
8. BOC−OCH(Cl)CCl_3 (1,2,2,2-tetrachloroethyl *tert*-butyl carbonate, BOC−OTCE), THF, K_2CO_3 or dioxane, H_2O, Et_3N, 60–91% yield.[13] This reagent is a cheap solid that can be distilled and that has the effectiveness of $(BOC)_2O$.
9. *t*-BuO−C(O)−S−(2-pyridyl) $\xrightarrow{RNH_2}$ RNHBOC Ref. 14

10. BOC–N_3, DMSO, 25°.[15]
11. BOC–$OC_6H_4S^+Me_2$ $MeSO_4^-$, H_2O.[16] This is a water-soluble reagent for the introduction of the BOC group.
12. 1-(*t*-Butoxycarbonyl)benzotriazole, NaOH, dioxane, 20°, 88–96% yield.[17]
13. BOC derivatives can be prepared directly from azides by hydrogenation in the presence of $(BOC)_2O$.[18]

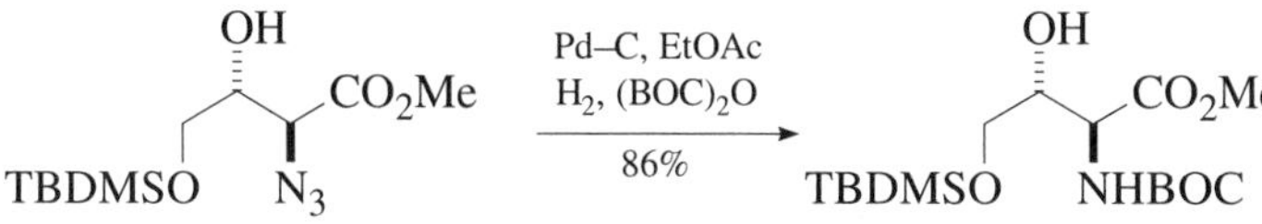

Pd–C, EtOAc, H_2, $(BOC)_2O$, 86%

14. Derivatization of cyclic urethanes with $(BOC)_2O$ makes the urethane carbonyl susceptible to hydrolysis under mild conditions and leaves the amine protected as a BOC derivative.[19]

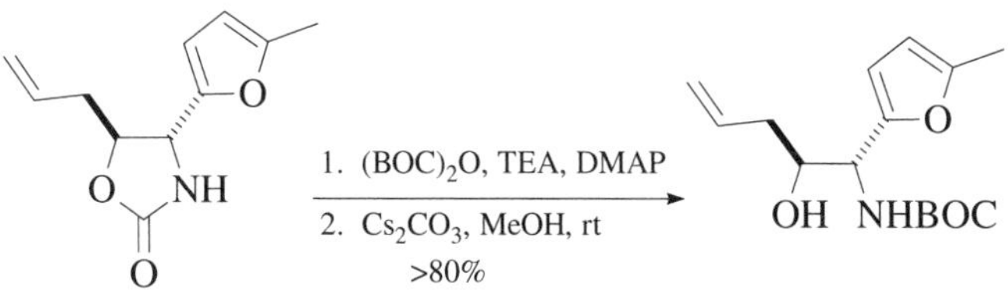

1. $(BOC)_2O$, TEA, DMAP
2. Cs_2CO_3, MeOH, rt
>80%

15. From an acetamide or benzamide: $(BOC)_2O$, THF, DMAP; hydrazine, 70–94% yield.[20]
16. 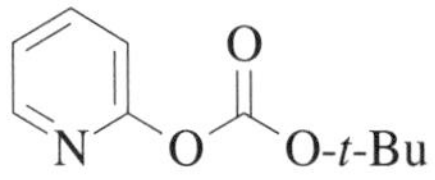 dioxane, H_2O, Et_3N, 70–94% yield.[21]
17. 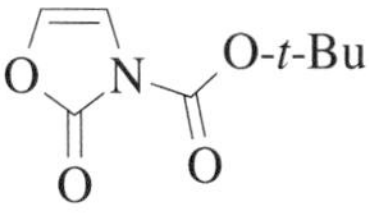 50% acetone, H_2O, DMAP, TEA, 85–95% yield.[22]

This method was also used to prepare the benzyl, methyl, ethyl, and *p*-methoxybenzyl derivatives. A polymeric version of the reagent was described as well.

18. 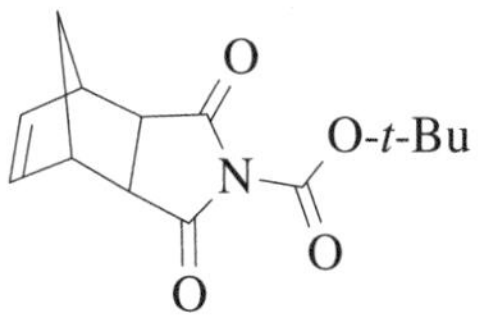 This reagent is useful for the selective protection of primary amines.[23]
19. Monoprotection of small diamines [$H_2N(CH_2)_xNH_2$, $x = 2–6$] is achieved by reacting an excess of the amine with $(BOC)_2O$ in dioxane (75–90% yield).[24]

20. *t*-BuOCOF.[25,26]
21. Directly from a carbobenzoxy-protected amine: 1,4-cyclohexadiene, Pd/C, $(BOC)_2O$, EtOH, rt, 86–96% yield.[27]
22. Directly from a Fmoc-protected amine: TEA, $(BOC)_2O$, KF, DMF.[28]
23. Directly from an azide: $(BOC)_2O$, Et_3SiH, 20% $Pd(OH)_2$–C, EtOH, 75–99% yield.[29]

Cleavage

1. 3 *M* HCl, EtOAc, 25°, 30 min, 96% yield.[30] With MeOH as the solvent, a diphenylmethyl ester is not affected.[31] The combination of HCl/EtOAc leaves TBDMS and TBDPS ethers[32] and *t*-butyl esters and nonphenolic ethers[33] intact during BOC cleavage, but *S*–BOC derivatives are cleaved.

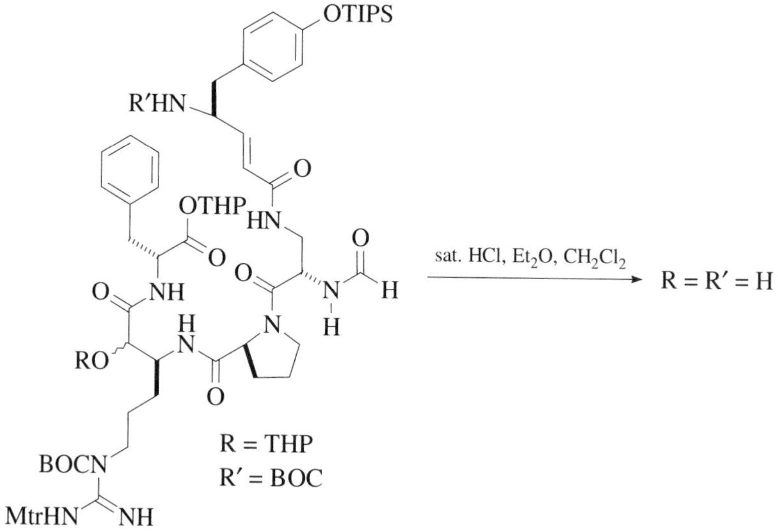

OTIPS R′HN O OTHP HN O O NH O N H O RO H N N H BOC N MtrHN NH O R = THP R′ = BOC

sat. HCl, Et_2O, CH_2Cl_2 → R = R′ = H

Ref. 34

2. AcCl, MeOH, 95–100% yield. This is a convenient method for generating anhydrous HCl in methanol. These conditions are also used to prepare methyl esters from carboxylic acids and for the formation of amine hydrochlorides.[35]
3. CF_3COOH, PhSH, 20°, 1 h, 100% yield.[36] Thiophenol is used to scavenge the liberated *t*-butyl cations, thus preventing alkylation of methionine or tryptophan. Other scavengers, such as anisole, thioanisole, thiocresol, cresol, and dimethyl sulfide, have also been used.[37] TBDPS[38] and TBDMS[39] groups are stable to TFA during BOC cleavage.
4. TsOH, THF, CH_2Cl_2, 5 min. This method was developed for solid-phase peptide synthesis as a safe large-scale alternative to the use of TFA, which is expensive, corrosive, and a waste problem on a large scale.[40]

5. 10% H_2SO_4, dioxane.[41] These conditions are similar to the use of 50% TFA/CH_2Cl_2 and are considered safer for large-scale use than the use of the volatile, corrosive, and expensive TFA. The authors provide a comparison of many of the acidic methods for BOC cleavage.
6. Me_3SiI, $CHCl_3$ or CH_3CN, 25°, 6 min, 100% yield.[42,43] Me_3SiI also cleaves carbamates, esters, ethers, and ketals under neutral, nonhydrolytic conditions. Some selectivity can be achieved by control of the reaction conditions.
7. $AlCl_3$, $PhOCH_3$, CH_2Cl_2, CH_3NO_2, 0–25°, 2–5 h, 73–88% yield.[44]
8. The BOC group can be removed thermally, either neat (185°, 20–30 min, 97% yield)[45] or in diphenyl ether.[46]
9. Bromocatecholborane.[47]

ZHN–CH(CO_2Et)–$(CH_2)_3$–NHBOC → (catecholBBr) → ZHN–CH(CO_2Et)–$(CH_2)_3$–NH_2

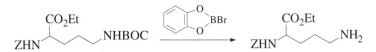

10. Me_3SiCl, PhOH, CH_2Cl_2, 20 min, 100% yield.[48] Under these conditions, benzyl groups are not cleaved and thus provide marked improvement over the conventional 50% TFA/CH_2Cl_2 used in peptide synthesis.
11. 1 *M* $SiCl_4$, 3 *M* phenol, CH_2Cl_2, 10 min. The Fmoc, Cbz, and Bn ester and ether groups were not noticeably cleaved even after 18 h of exposure.[49]
12. Trimethylsilyl triflate (TMSOTf), $PhSCH_3$, CF_3COOH.[50] These conditions also cleave the following protective groups used in peptide synthesis: (MeO)Z–, Bn–, Ts–, $Cl_2C_6H_3CH_2$–, BOM (benzyloxymethyl)–, Mts–, MBS–, *t*-Bu–SR, and Ad–SR. The rate of cleavage is reported to be faster than with TfOH/TFA. The conditions do not cleave a BnSR, Acm, or Arg(NO_2) group.
13. The BOC group can be cleaved with TBDMSOTf and the intermediate silyl carbamate converted to other nitrogen-protective groups by treatment with fluoride followed by a suitable alkylating agent.[51]

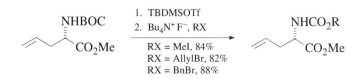

NHBOC / CO_2Me → 1. TBDMSOTf; 2. $Bu_4N^+F^-$, RX → $NHCO_2R$ / CO_2Me

RX = MeI, 84%
RX = AllylBr, 82%
RX = BnBr, 88%

14. 0.05 *M* $MeSO_3H$, dioxane, CH_2Cl_2 (1:9).[52,53] This reagent also cleaves the Moz (4-methoxybenzyloxycarbonyl) group.
15. In an amine bearing two BOC groups, 2 eq. of TFA in CH_2Cl_2 will cleave only one BOC, leaving a monoprotected primary amine.[54]
16. $Mg(ClO_4)_2$, >67% yield.[55] These conditions cleave one of the two BOC groups on a primary amine.

17. $BF_3 \cdot Et_2O$, 4Å ms, CH_2Cl_2, rt, 20 h, 77–98% yield.[56]

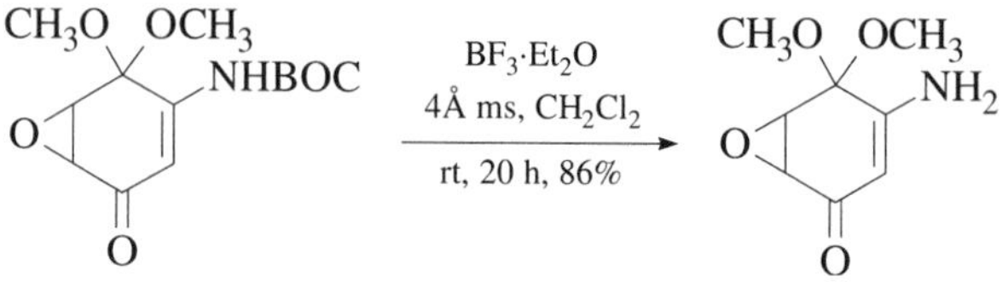

$BF_3 \cdot Et_2O$, 4Å ms, CH_2Cl_2, rt, 20 h, 86%

18. $SnCl_4$, AcOH, THF, CH_2Cl_2, toluene or CH_3CN, 82–98% yield. This method was developed because acid-based methods were incompatible with the presence of a thioamide peptide bond.[57] Guanidines were cleanly deprotected.[58]
19. From RN(Ts)BOC: DMF, 100–120°, 24 h, >69–75% yield.[59]
20. Silica gel, heat under vacuum, 80–92% yield.[60] These conditions selectively remove only the indole BOC group from a fully *t*-Bu-based, protected tryptophan.
21. Migration of a BOC is normally not observed, but in the following case a BOC group moved to a hydroxyl. The stabilizing effect of the tosyl group makes this possible.[61]

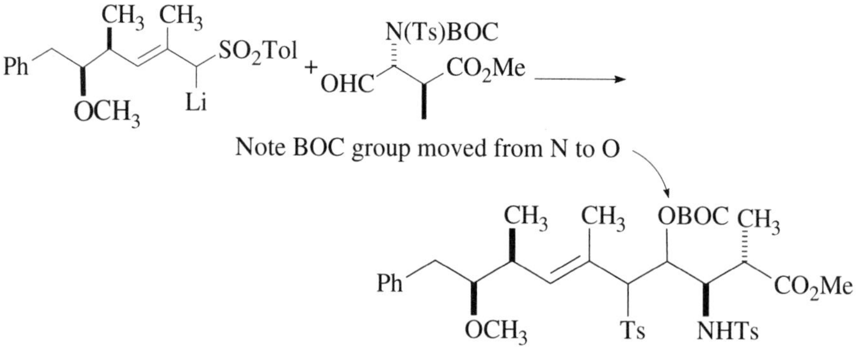

Note BOC group moved from N to O

22. 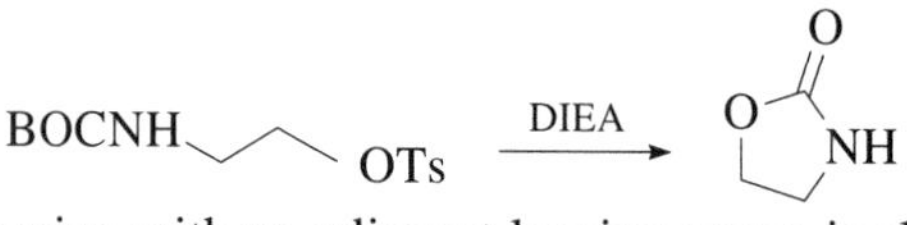 BOCNH~~OTs —DIEA→ oxazolidone. A BOC-protected primary amine with an adjacent leaving group is slowly converted to an oxazolidone.[62]
23. The conversion of the BOC group to other carbamates is achieved by heating the alcohol, $Ti(O\text{–}i\text{-}Pr)_4$ in toluene. Teoc-, Cbz-, and Alloc-protected primary amines have been prepared in this fashion. The reaction is selective for a primary BOC derivative.[63]
24. CAN, CH_3CN, 90–99% yield.[64]
25. $ZnBr_2$, CH_2Cl_2, 89–94% yield. These conditions selectively cleave the BOC group from secondary amines in the presence of the primary derivatives.[65]

1. M. Bodanszky, *Principles of Peptide Chemistry*, Springer-Verlag; New York, 1984, p. 99.
2. D. S. Tarbell, Y. Yamamoto, and B. M. Pope, *Proc. Natl. Acad. Sci., USA*, **69**, 730 (1972).
3. E. Ponnusamy, U. Fotadar, A. Spisni, and D. Fiat, *Synthesis*, 48 (1986).
4. J. Einhorn, C. Einhorn, and J. L. Luche, *Synlett*, 37 (1991).
5. E. M. Khalil, N. L. Subasinghe, and R. L. Johnson, *Tetrahedron Lett.*, **37**, 3441 (1996).
6. H.-J. Knölker, T. Braxmeier, and G. Schlechtingen, *Angew. Chem., Int. Ed. Engl.* **34**, 2497 (1995).
7. H.-J. Knölker, T. Braxmeier, and G. Schlechtingen, *Synlett*, 502 (1996).
8. T. A. Kelly and D. W. McNeil, *Tetrahedron Lett.*, **35**, 9003 (1994).
9. H.-J. Knölker and T. Braxmeier, *Tetrahedron Lett.*, **37**, 5861 (1996).
10. M. Itoh, D. Hagiwara, and T. Kamiya, *Bull. Chem. Soc. Jpn.*, **50**, 718 (1977).
11. G. M. Cohen, P. M. Cullis, J. A. Hartley, A. Mather, M. C. R. Symons, and R. T. Wheelhouse, *J. Chem. Soc., Chem. Commun.*, 298 (1992).
12. R. B. Harris and I. B. Wilson, *Tetrahedron Lett.*, **24**, 231 (1983).
13. G. Barcelo, J. P. Senet, and G. Sennyey, *J. Org. Chem.*, **50**, 3951 (1985).
14. S. Kim, J. I. Lee, and K. Y. Yi, *Bull. Chem. Soc. Jpn.*, **58**, 3570 (1985).
15. J. B. Hansen, M. C. Nielsen, U. Ehrbar, and O. Buchardt, *Synthesis*, 404 (1982).
16. I. Azuse, H. Okai, K. Kouge, Y. Yamamoto, and T. Koizumi, *Chem. Express*, **3**, 45 (1988).
17. A. R. Katritzky, C. N. Fali, J. Li, D. J. Ager, and I. Prakash, *Synth. Commun.*, **27**, 1623 (1997).
18. S. Saito, H. Nakajima, M. Inaba, and T. Moriwake, *Tetrahedron Lett.*, **30**, 837 (1989).
19. T. Ishizuka and T. Kunieda, *Tetrahedron Lett.*, **28**, 4185 (1987).
20. M. J. Burk and J. G. Allen, *J. Org. Chem.*, **62**, 7054 (1997).
21. F. Effenberger and W. Brodt, *Chem. Ber.*, **118**, 468 (1985).
22. T. Kunieda, T. Higuchi, Y. Abe, and M. Hirobe, *Chem. Pharm. Bull.*, **32**, 2174 (1984).
23. I. Grapsas, Y. J. Cho, and S. Mobashery, *J. Org. Chem.*, **59**, 1918 (1994).
24. A. P. Krapcho and C. S.Kuell, *Synth. Commun.*, **20**, 2559 (1990).
25. L. A. Carpino, K. N. Parameswaran, R. K. Kirkley, J. W. Spiewak, and E. Schmitz, *J. Org. Chem.*, **35**, 3291 (1970).
26. For an improved preparation of this reagent, see V. A. Dang, R. A. Olofson, P. R. Wolf, M. D. Piteau, and J.-P. G. Senet, *J. Org. Chem.*, **55**, 1847 (1990).
27. J. S. Bajwa, *Tetrahedron Lett.*, **33**, 2955 (1992).
28. W.-R. Li, J. Jiang, and M. J. Joullie, *Tetrahedron Lett.*, **34**, 1413 (1993).
29. H. Kotsuki, T. Ohishi, and T. Araki, *Tetrahedron Lett.*, **38**, 2129 (1997).
30. G. L. Stahl, R. Walter, and C. W. Smith, *J. Org. Chem.*, **43**, 2285 (1978).
31. Z. Tozuka, H. Takasugi, and T. Takaya, *J. Antibiotics*, **36**, 276 (1983).
32. F. Cavelier and C. Enjalbal, *Tetrahedron Lett.*, **37**, 5131 (1996).

33. F. S. Gibson, S. C. Bergmeier, and H. Rapoport, *J. Org. Chem.*, **59**, 3216 (1994).
34. P. Wipf and H. Kim, *J. Org. Chem.*, **58**, 5592 (1993).
35. A. Nudelman, Y. Bechor, E. Falb, B. Fischer, B. A. Wexler, and A. Nudelman, *Synth. Commun.*, **28**, 471 (1998).
36. B. F. Lundt, N. L. Johansen, A. Vølund, and J. Markussen, *Int. J. Pept. Protein Res.*, **12**, 258 (1978).
37. See, for example, M. Bodanszky and A. Bodanszky, *Int. J. Pept. Protein Res.*, **23**, 565 (1984); Y. Masui, N. Chino, and S. Sakakibara, *Bull. Chem. Soc. Jpn.*, **53**, 464 (1980).
38. P. A. Jacobi, S. Murphree, F. Rupprecht, and W. Zheng, *J. Org. Chem.*, **61**, 2413 (1996).
39. J. Deng, Y. Hamada, and T. Shioiri, *J. Am. Chem. Soc.*, **117**, 7824 (1995).
40. H. R. Brinkman, J. J. Landi, Jr., J. B. Paterson, Jr., and P. J. Stone, *Synth. Commun.*, **21**, 459 (1991).
41. R. A. Houghten, A. Beckman, and J. M. Ostresh, *Int. J. Pept. Protein Res.*, **27**, 653 (1986).
42. R. S. Lott, V. S. Chauhan, and C. H. Stammer, *J. Chem. Soc., Chem. Commun.*, 495 (1979).
43. For a review on the use of Me_3SiI, see G. A. Olah and S. C. Narang, *Tetrahedron*, **38**, 2225 (1982).
44. T. Tsuji, T. Kataoka, M. Yoshioka, Y. Sendo, Y. Nishitani, S. Hirai, T. Maeda, and W. Nagata, *Tetrahedron Lett.*, 2793 (1979).
45. V. H. Rawal, R. J. Jones, and M. P. Cava, *J. Org. Chem.*, **52**, 19 (1987).
46. H. H. Wasserman, G. D. Berger, and K. R. Cho, *Tetrahedron Lett.*, **23**, 465 (1982).
47. R. K. Boeckman, Jr., and J. C. Potenza, *Tetrahedron Lett.*, **26**, 1411 (1985).
48. E. Kaiser, Sr., T. Kubiak, J. P. Tam, and R. B. Merrifield, *Tetrahedron Lett.*, **29**, 303 (1988); E. Kaiser, Sr., F. Picart, T. Kubiak , J. P. Tam, and R. B. Merrifield, *J. Org. Chem.*, **58**, 5167 (1993).
49. K. M. Sivanandaiah, V. V. S. Babu, and B. P. Gangadhar, *Tetrahedron Lett.*, **37**, 5989 (1996).
50. N. Fujii, A. Otaka, O. Ikemura, K. Akaji, S. Funakoshi, Y. Hayashi, Y. Kuroda, and H. Yajima, *J. Chem. Soc., Chem. Commun.*, 274 (1987).
51. M. Sakaitani and Y. Ohfune, *Tetrahedron Lett.*, **26**, 5543 (1985); *idem.*, *J. Org. Chem.*, **55**, 870 (1990).
52. Y. Kiso, A. Nishitani, M. Shimokura, Y. Fujiwara, and T. Kimura, *Pept. Chem., 1987*, 291 (1988), *Chem. Abstr.*, **109** : 190837t (1988).
53. Y. Kiso, Y. Fujiwara, T. Kimura, A. Nishitani, and K. Akaji, *Int. J. Pept. Protein, Res.*, **40**, 308 (1992).
54. R. A. T. M. van Benthem, H. Hiemstra, and W. N. Speckamp, *J. Org. Chem.*, **57**, 6083 (1992).
55. F. Burkhart, M. Hoffmann, and H. Kessler, *Angew. Chem., Int. Ed. Engl.*, **36**, 1191 (1997).
56. E. F. Evans, N. J. Lewis, I. Kapfer, G. Macdonald, and R. J. K. Taylor, *Synth. Commun.*, **27**, 1819 (1997).

57. R. Frank and M. Schutkowski, *J. Chem. Soc., Chem. Commun.*, 2509 (1996).
58. H. Miel and S. Rault, *Tetrahedron Lett.*, **38**, 7865 (1997).
59. K. E. Krakowiak and J. S. Bradshaw, *Synth. Commun.*, **26**, 3999 (1996).
60. T. Apelqvist and D. Wensbo, *Tetrahedron Lett.*, **37**, 1471 (1996).
61. R. J. Valentekovich and S. L. Schreiber, *J. Am. Chem. Soc.*, **117**, 9069 (1995).
62. T. P. Curran, M. P. Pollastri, S. M. Abelleira, R. J. Messier, T. A. McCollum, and C. G. Rowe, *Tetrahedron Lett.*, **35**, 5409 (1994).
63. G. Shapiro and M. Marzi, *J. Org. Chem.*, **62**, 7096 (1997).
64. J. R. Hwu, M. L. Jain, S.-C. Tsay, and G. H. Hakimelahi, *Tetrahedron Lett.*, **37**, 2035 (1996).
65. S. C. Nigam, A. Mann, M. Taddei, and C.-G. Wermuth, *Synth. Commun.*, **19**, 3139 (1989).

1-Adamantyl Carbamate (1-Adoc–NR_2): 1-Adamantyl–OC(O)NR_2 (Chart 8)

The 1-Adoc group is very similar to the *t*-BOC group in its sensitivity to acid, but often provides more crystalline derivatives of amino acids.

Formation

1. 1-Adoc–Cl, NaOH, 27–98% yield.[1]
2. 1-Adoc–O–2-pyridyl, 70–95% yield.[2]
3. 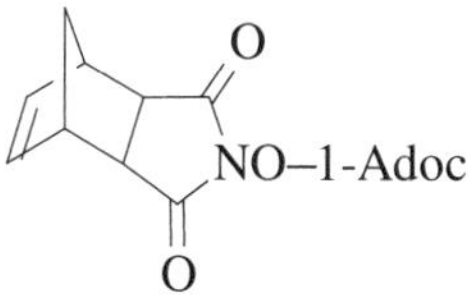 O O NO–1-Adoc Dioxane, H_2O, 2–3 h, 82–84% yield.[3]
4. 1-Adoc–F, 84–94% yield.[4,5] The solvolytic decomposition of this reagent has been examined.[6]

Cleavage

1. CF_3CO_2H, 25°, 15 min, 100% yield.[1]

2-Adamantyl Carbamate (2-Adoc–NR_2): 2-Adamantyl–OC(O)NR_2

Formation

1. 2-Adamantyl chloroformate.[7]

Cleavage

1. Trifluoromethanesulfonic acid or anhydrous HF at 0°. The 2-Adoc group is stable to HCl/dioxane, TFA, 25% HBr/AcOH, and TMSBr/thioanisole/TFA for up to 24 h.[7]

1. W. L. Haas, E. V. Krumkalns, and K. Gerzon, *J. Am. Chem. Soc.*, **88**, 1988 (1966).
2. F. Effenberger and W. Brodt, *Chem. Ber.*, **118**, 468 (1985).
3. P. Henklein, H.-U. Heyne, W.-R. Halatsch, and H. Niedrich, *Synthesis*, 166 (1987).
4. L. Moroder, L. Wackerle, and E. Wuensch, *Hoppe-Seyler's Z. Physiol. Chem.*, **357**, 1647 (1976).
5. R. Presentini and G. Antoni, *Int. J. Pept. Protein Res.*, **27**, 123 (1986).
6. D. N. Kevill and J. B. Kyong, *J. Org. Chem.*, **57**, 258 (1992).
7. Y. Nishiyama, N. Shintomi, Y. Kondo, and Y. Okada, *J. Chem. Soc., Perkin Trans.1*, 3201 (1994).

Vinyl Carbamate (Voc–NR_2): CH_2=CHOC(O)NR_2 (Chart 8)

The olefin of the Voc group is very susceptible to electrophilic reagents and thus is readily cleaved by reaction with bromine or mercuric acetate.

Formation

1. CH_2=CHOCOCl, MgO, H_2O, dioxane, pH 9–10, 90% yield.[1]
2. CH_2=CHOCOSPh, Et_3N, dioxane or DMF, H_2O, 25°, 16 h, 50–80% yield.[2]

Cleavage

1. Anhydrous HCl, dioxane, 25°, 97% yield.[1]
2. HBr, AcOH, 94% yield.[1]
3. Br_2, CH_2Cl_2, then MeOH, 95% yield.[1]
4. $Hg(OAc)_2$, AcOH, H_2O, 25°, 97% yield.[1]

1. R. A. Olofson, Y. S. Yamamoto, and D. J. Wancowicz, *Tetrahedron Lett.*, 1563 (1977).
2. A. J. Duggan and F. E. Roberts, *Tetrahedron Lett.*, 595 (1979). For an improved preparation of this reagent, see R. A. Olofson and J. Cuomo, *J. Org. Chem.*, **45**, 2538 (1980).

Allyl Carbamate (Aloc–NR_2 or Alloc–NR_2): CH_2=$CHCH_2OC(O)NR_2$ (Chart 8)

Formation

1. CH_2=$CHCH_2OCOCl$, Pyr.[1]
2. $(CH_2{=}CHCH_2OCO)_2O$, dioxane, H_2O, reflux or CH_2Cl_2, 1 h, rt, 67–96% yield.[2]
3. CH_2=$CHCH_2OC(O)$–O–benzotriazolyl.[3]
4. $(CH_2{=}CHCH_2OCO)_2O$, phosphate buffer pH 8, Subtilisin.[4]

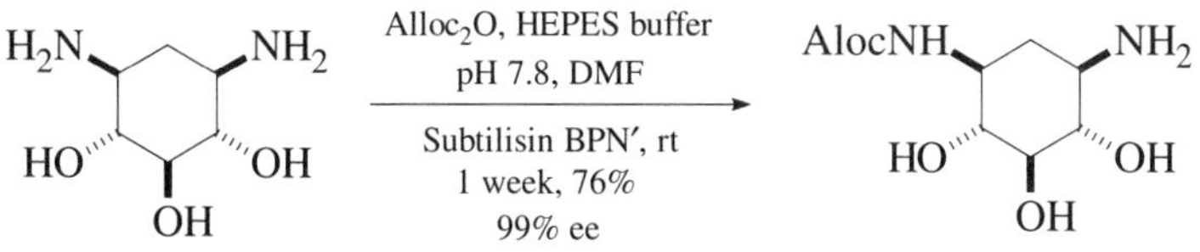

Alloc$_2$O, HEPES buffer pH 7.8, DMF; Subtilisin BPN′, rt 1 week, 76% 99% ee

5. Allyl bromide, CO_2, 18-crown-6, 9–55% yield.[5]
6. (a) $NO_2C_6H_4OCOCl$, (b) allyl alcohol, CH_3CN, 3 h, rt, 88% yield.[6]

Cleavage

1. $Ni(CO)_4$ (Caution: very toxic!), DMF, H_2O (95:5), 55°, 4 h, 83–95% yield[1].
2. $Pd(Ph_3P)_4$, Bu_3SnH, AcOH, 70–100% yield.[7]
3. $Pd(Ph_3P)_4$, Me_2NTMS, 89–100% yield as the TMS carbamate that is easily hydrolyzed. This method was developed to suppress allylamine formation.[8]
4. $Pd(Ph_3P)_4$, dimedone, THF, 88–95% yield.[9] The catalyst is not poisoned by the presence of thioethers such as methionine. Diethyl malonate has also been used as a nucleophile to trap the π-allylpalladium intermediate and regenerate Pd(0).[10]
5. $Pd(Ph_3P)_2Cl_2$, Bu_3SnH, p-$NO_2C_6H_4OH$, CH_2Cl_2, 70–100% yield.[7,11] This reaction works best in the presence of acids. AcOH and pyridinium acetate are also effective.
6. $Pd_2(dba)_3 \cdot CHCl_3$, [tris(dibenzylideneacetone)dipalladium-(chloroform)], HCO_2H, 74–100% yield.[12]
7. The Aloc group can be converted to a silyl carbamate that is readily hydrolyzed.[13,14]

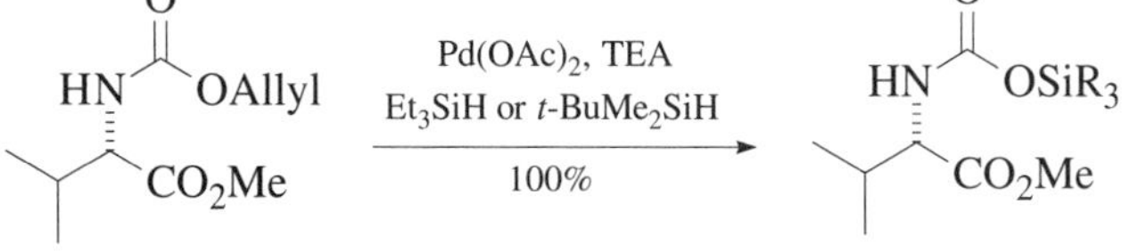

$Pd(OAc)_2$, TEA; Et_3SiH or t-$BuMe_2SiH$; 100%

8. $Pd(Ph_3P)_4$, 2-ethylhexanoic acid.[15]

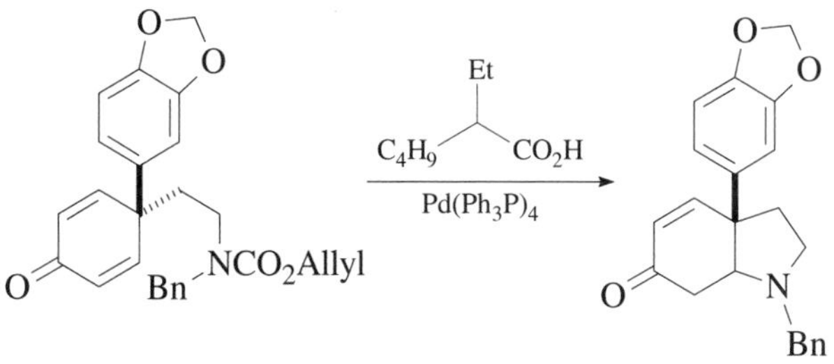

$Pd(Ph_3P)_4$

Ref. 16

9. $Pd(OAc)_2$, TPPTS, CH_3CN, H_2O, Et_2NH, 30 min, 89–99% yield. Deprotection can be achieved in the presence of a prenyl or cinnamyl ester, but as the reaction times increase, these esters are also cleaved.[17,18] Prenyl carbamates and allyl carbonates are cleaved similarly.
10. $Pd(Ph_3P)_4$ and Bu_3SnH convert the Alloc group to other amine derivatives when electrophiles such as $(BOC)_2O$, AcCl, TsCl, or succinic anhydride are added. Hydrolysis of the stannyl carbamate with acetic acid gives the free amine.[19]
11. $Pd(Ph_3P)_4$, *N,N′*-dimethylbarbituric acid, 92% yield.[20]
12. $Pd(Ph_3P)_4$, HCO_2H, TEA[21] or AcOH, NMO.[22]
13. $Pd(Ph_3P)_4$, $NaBH_4$, THF, 91% yield. If $(BOC)_2O$ or CbzOSu is included in the reaction, transprotection to the BOC or Cbz derivative is achieved.[23]
14. $Pd(dba)_2$, dppb, Et_2NH or 2-mercaptobenzoic acid, THF or EtOH, 15–120 min, 99–100% yield.[24] Allyl carbonates and allyl ethers are cleaved similarly.
15. Pd(0), water soluble phosphine, H_2O, CH_3CN, 10 min, 100% yield.[25]

1-Isopropylallyl Carbamate: (Ipaoc–NR$_2$): $CH_2=CHCH(CH(CH_3)_2)OC(O)NR_2$

This group was developed to minimize the problem of nitrogen allylation during the deprotection step, because deprotection proceeds with *β*-hydride elimination. The derivative is stable to TFA and 6 *N* HCl.[26]

Formation / Cleavage

RNH_2 → [$CH_2=CHCH(CH(CH_3)_2)OC(O)$S-pyrimidine, 90–95%] → $CH_2=CHCH(CH(CH_3)_2)OC(O)NHR$

← [$Pd_2(dba)_3-CHCl_3$, Ph_3P, dioxane]

Cinnamyl Carbamate (Coc–NR$_2$): $PhCH=CHCH_2OC(O)NR_2$ (Chart 8)

Formation

1. $PhCH=CHCH_2OCO$-O-benzotriazolyl, Et_3N, dioxane or DMF, rt, 16 h, 71–100% yield.[27]

Cleavage

1. $Pd(Ph_3P)_4$, THF, Pyr, HCO_2H, heat, 4 min.[27]

2. $Hg(OAc)_2$, CH_3OH, HNO_3, 23°, 2–4 h, then KSCN, H_2O, 23°, 12–16 h.[28]

4-Nitrocinnamyl Carbamate (Noc–NR$_2$):
4-$NO_2C_6H_4CH{=}CHCH_2OC(O)NR_2$

The Noc group, developed for amino acid protection, is introduced with the acid chloride (Et_3N, H_2O, dioxane, 2 h, 20°, 61–95% yield). It is cleaved with $Pd(Ph_3P)_4$ (THF, *N,N*-dimethylbarbituric acid, 8 h, 20°, 80% yield). It is not isomerized by Wilkinson's catalyst, thus allowing selective removal of the allyl ester group.[29]

3-(3′-Pyridyl)prop-2-enyl Carbamate (Paloc–NR$_2$):
3-C_5H_4N–$CH{=}CHCH_2OC(O)NR_2$

The Paloc group was developed as an amino acid protective group that is introduced with the *p*-nitrophenyl carbonate (H_2O, dioxane, 68–89% yield). It is exceptionally stable to TFA and to rhodium-catalyzed allyl isomerization, but it is conveniently cleaved with $Pd(Ph_3P)_4$ (methylaniline, THF, 20°, 10 h, 74–89% yield).[30]

1. E. J. Corey and J. W. Suggs, *J. Org. Chem.*, **38**, 3223 (1973).
2. G. Sennyey, G. Barcelo, and J.-P. Senet, *Tetrahedron Lett.*, **28**, 5809 (1987).
3. Y. Hayakawa, H. Kato, M. Uchiyama, H. Kajino, and R. Noyori, *J. Org. Chem.*, **51**, 2400 (1986).
4. B. Orsat, P. B. Alper, W. Moree, C.-P. Mak, and C.-H. Wong, *J. Am. Chem. Soc.*, **118**, 712 (1996).
5. M. Aresta and E. Quaranta, *Tetrahedron*, **48**, 1515 (1992).
6. N. Choy, K. Y. Moon, C. Park, Y. C. Son, W. H. Jung, H. Choi, C. S. Lee, C. R. Kim, S. C. Kim, and H. Yoon, *Org. Prep. Proced. Int.*, **28**, 173 (1996).
7. O. Dangles, F. Guibé, G. Balavoine, S. Lavielle, and A. Marquet, *J. Org. Chem.*, **52**, 4984 (1987).
8. A. Merzouk and F. Guibe, *Tetrahedron Lett.*, **33**, 477 (1992).
9. H. Kunz and C. Unverzagt, *Angew. Chem., Int. Ed. Engl.*, **23**, 436 (1984).
10. P. Boullanger and G. Descotes, *Tetrahedron Lett.*, **27**, 2599 (1986).
11. P. Four and F. Guibé, *Tetrahedron Lett.*, **23**, 1825 (1982).
12. I. Minami, Y. Ohashi, I. Shimizu, and J. Tsuji, *Tetrahedron Lett.*, **26**, 2449 (1985).
13. M. Sakaitani, N. Kurokawa, and Y. Ohfune, *Tetrahedron Lett.*, **27**, 3753 (1986).
14. M. Sakaitani and Y. Ohfune, *J. Org. Chem.*, **55**, 870 (1990).
15. P. D. Jeffrey and S. W. McCombie, *J. Org. Chem.*, **47**, 587 (1982).
16. S. F. Martin and C. L. Campbell, *J. Org. Chem.*, **53**, 3184 (1988).
17. S. Lemaire-Audoire, M. Savignac, E. Blart, G. Pourcelot, J. P. Genêt, and J.-M. Bernard, *Tetrahedron Lett.*, **35**, 8783 (1994); S. Lemaire-Audoire, M. Savignac, E. Blart, J.-M. Bernard, and J. P. Genêt, *Tetrahedron Lett.*, **38**, 2955 (1997).

18. J. P. Genêt, E. Blart, M. Savignac, S. Lemeune, and J.-M. Paris, *Tetrahedron Lett.*, **34**, 4189 (1993).
19. E. C. Roos, P. Bernabe, H. Hiemstra, W. N. Speckamp, B. Kaptein, and W. H. J. Boesten, *J. Org. Chem.*, **60**, 1733 (1995).
20. C. Unverzagt and H. Kunz, *Biorg. Med. Chem.*, **2**, 1189 (1994).
21. Y. Kanda, H. Arai, T. Ashizawa, M. Morimoto, and M. Kasai, *J. Med. Chem.*, **35**, 2781 (1992).
22. J. Lee, J. H. Griffin, and T. I. Nicas, *J. Org. Chem.*, **61**, 3983 (1996).
23. R. Beugelmans, L. Neuville, M. Bois-Choussy, J. Chastanet, and J. Zhu, *Tetrahedron Lett.*, **36**, 3129 (1995).
24. J. P. Genêt, E. Blart, M. Savignac, S. Lemeune, S. Lemaire-Audoire, and J. M. Bernard, *Synlett*, 680 (1993).
25. J. P. Genêt, E. Blart, M. Savignac, S. Lemeune, S. Lemaire-Audoire, J. M. Paris, and J. M. Bernard, *Tetrahedron*, **50**, 497 (1994).
26. I. Minami, M. Yukara, and J. Tsuji, *Tetrahedron Lett.*, **28**, 2737 (1987).
27. H. Kinoshita, K. Inomata, T. Kameda, and H. Kotake, *Chem. Lett.*, 515 (1985).
28. E. J. Corey and M. A. Tius, *Tetrahedron Lett.*, 2081 (1977).
29. H. Kunz and J. März, *Angew. Chem., Int. Ed. Engl.*, **27**, 1375 (1988).
30. K. von dem Bruch and H. Kunz, *Angew. Chem., Int. Ed. Engl.*, **29**, 1457 (1990).

8-Quinolyl Carbamate (Chart 8): N O_2CNR_2

Formation/Cleavage[1]

An 8-quinolyl carbamate is cleaved under neutral conditions by Cu(II)- or Ni(II)-catalyzed hydrolysis.

1. E. J. Corey and R. L. Dawson, *J. Am. Chem. Soc.*, **84**, 4899 (1962).

***N*-Hydroxypiperidinyl Carbamate** (Chart 8): N–$OCONR_2$

A piperidinyl carbamate, stable to aqueous alkali and to cold acid (30% HBr, 25°, several hours) is best cleaved by reduction.

Formation[1]

1. 1-Piperidinyl-OCOX (X = 2,4,5-trichlorophenyl,...), Et_3N, 55–85% yield.

Cleavage[1]

1. H_2, Pd–C, AcOH, 20°, 30 min, 95% yield.
2. Electrolysis, 200 mA, 1 *N* H_2SO_4, 20°, 90 min, 90–93% yield.

3. $Na_2S_2O_4$, AcOH, 20°, 5 min, 93% yield.
4. Zn, AcOH, 20°, 10 min, 94% yield.

1. D. Stevenson and G. T. Young, *J. Chem. Soc. C*, 2389 (1969).

Alkyldithio Carbamate: $R_2NCOSSR'$

Alkyldithio carbamates are prepared from the acid chloride (Et_3N, EtOAc, 0°) and amino acid, either free or as the *O*-silyl derivatives (70–88% yield).[1] The *N*-(*i*-propyldithio) carbamate has been used in the protection of proline during peptide synthesis.[2] Alkyldithio carbamates can be cleaved with thiols, NaOH, Ph_3P/TsOH. They are stable to acid. Cleavage rates are a function of the size of the alkyl group, as illustrated in the following table:

Relative Rates of Cleavage of Alkyldithio Carbamates

Alkyl Group (R′)	$HSCH_2CH_2OH$	NaOH
CH_3	100	100
Et	33	32
i-Pr	1.4	1.3
t-Bu	0.0002	—
Ph	460	500

The rates were determined using the proline derivative as a substrate.[3] The *t*-Bu carbamate was unusually stable in NaOH.

1. E. Wünsch, L. Moroder, R. Nyfeler, and E. Jäger, *Hoppe-Seyler's Z. Physiol. Chem.*, **363**, 197 (1982).
2. F. Albericio and G. Barany, *Int. J. Pept. Protein Res.*, **30**, 177 (1987).
3. G. Barany, *Int. J. Pept. Protein Res.*, **19**, 321 (1982).

Benzyl Carbamate (Cbz– or Z–NR$_2$): $PhCH_2OC(O)NR_2$ (Chart 8)

Formation

1. $PhCH_2OCOCl$, Na_2CO_3, H_2O, 0°, 30 min, 72% yield.[1] Alpha–omega diamines can be protected somewhat selectively with this reagent at a pH between 3.5 and 4.5, but the selectivity decreases as the chain length increases [$H_2N(CH_2)_nNH_2$, $n = 2$, 71% mono; $n = 7$, 29% mono].[2] Hindered amino acids are protected in DMSO (DMAP, TEA, heat, 47–82% yield). These conditions also convert a carboxylic acid to the benzyl ester.[3]
2. $PhCH_2OCOCl$, MgO, EtOAc, 3 h, 70° to reflux, 60% yield.[4]
3. $(PhCH_2OCO)_2O$, dioxane, H_2O, NaOH or Et_3N.[5,6] This reagent was reported to give better yields in preparing amino acid derivatives than when $PhCH_2OCOCl$ was used. The reagent decomposes at 50°.

4. $PhCH_2OCO_2$–C(OMe)=CH_2, 90–98% yield.[7]
5. $PhCH_2OCO_2$–succinimidyl, >70% yield.[8] This reagent avoids the formation of amino acid dimers and is a stable, easily handled solid.
6. $PhCH_2OCO$–benzotriazolyl, NaOH, dioxane, rt.[9]
7. $PhCH_2OCOCN$, CH_2Cl_2, CH_3CN or 1,2-dimethoxyethane.[10]
8. $PhCH_2OCO$–imidazolyl, 4-dimethylaminopyridine, 16 h, rt, 76% yield.[11] Two primary amines were protected in the presence of a secondary amine.

9. [Scheme: bicyclic imide NOCOCl reagent → ROH, Et_3N, dioxane, 64–91%; R'CH(NH_2)CO_2H → RO–C(=O)–NH–CH(R')–CO_2H]

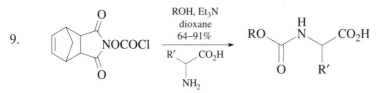

ROH = *t*-BuOH, BnOH, FmOH, AdamantylOH, $PhC_6H_4C(Me)_2OH$, $CH_3SO_2CH_2CH_2OH$

This method is suitable for the preparation of BOC-, Fmoc-, Adoc-, and Bpoc-protected amino acids. The acid chloride is a stable, storable solid.[12]

10. CO_2, BnCl, DMF, Cs_2CO_3, 58–96% yield.[13]
11. [Structure: 5-methyl-1,3,4-thiadiazol-2-yl–S–C(=O)–OBn] TEA, 88% yield.[14]
12. 4-$NO_2C_6H_4OCO_2Bn$, Pyr, DMF, 26°, 24 h, 74% yield. Primary amines are selectively protected over secondary amines, but anilines are insufficiently nucleophilic to react with this reagent.[15]
13. 1,3-Bis(benzyloxycarbonyl)-3,4,5,6-tetrahydropyrimidine-2-thione, refluxing dioxane, 7 h.[16]
14. [4-(Benzyloxycarbonyloxy)phenyl]dimethylsulfonium methyl sulfate, NaOH, H_2O, 51–95% yield.[17] This is a water-soluble reagent for benzyloxy carbamate formation. Analogous reagents for the introduction of BOC and Fmoc were also prepared and give the respective derivatives in similar high yields.

Cleavage

1. H_2/Pd–C.[1] If hydrogenation is carried out in the presence of $(BOC)_2O$, the released amine is directly converted to the BOC derivative.[18] The formation of *N*-methylated lysines during the hydrogenolysis of a Z group has been observed with MeOH/DMF as the solvent.[19] Formaldehyde derived oxidatively from methanol is the source of the methyl carbon.[20]
2. H_2/Pd–C, NH_3, –33°, 3–8 h, quantitative.[21] When ammonia is used as the solvent, cysteine or methionine units in a peptide do not poison the

catalyst. Additionally, amines inhibit the reduction of BnO ethers; thus, selectivity can be achieved for the Z group.[22]

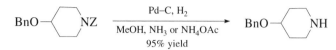

BnO–(piperidine)NZ —[Pd–C, H_2 / MeOH, NH_3 or NH_4OAc / 95% yield]→ BnO–(piperidine)NH

3. Pd–C or Pd black, hydrogen donor, solvent, 25° or reflux in EtOH, 15 min to 2 h, 80–100% yield. Several hydrogen donors, including cyclohexene,[23] 1,4-cyclohexadiene,[24] formic acid,[25,26] *cis*-decalin,[27] and HCO_2NH_4,[28] have been used for catalytic transfer hydrogenation, which is, in general, a more rapid reaction than catalytic hydrogenation. The use of this technique in the presence of $(BOC)_2O$ converts a Z-protected amine to a BOC-protected amine.[29]
4. $CaNi_5$, H_2, MeOH, H_2O.[30] The catalyst is a hydrogen storage alloy and is partially consumed by the reaction of Ca with water or methanol.
5. Et_3SiH, cat. Et_3N, cat. $PdCl_2$, reflux, 3 h, 80% yield.[31] If the reaction is performed in the presence of *t*-$BuMe_2SiH$, the *t*-butyldimethylsilyl carbamate can be isolated because of its greater stability.[32] *S*-Benzyl groups are stable to these conditions, but benzyl esters and benzyl ethers are cleaved.[31] A similar procedure has been published, but in this case the benzyl ether was stable to the cleavage conditions.[18]

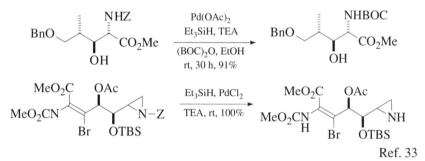

BnO–CH(…)(NHZ)(OH)–CO_2Me —[$Pd(OAc)_2$, Et_3SiH, TEA / $(BOC)_2O$, EtOH, rt, 30 h, 91%]→ BnO–…(NHBOC)(OH)–CO_2Me

MeO_2C, OAc, MeO_2CN, Br, OTBS, N–Z —[Et_3SiH, $PdCl_2$ / TEA, rt, 100%]→ MeO_2C, OAc, MeO_2CNH, Br, OTBS, NH

Ref. 33

6. *t*-$BuMe_2SiH$, $Pd(OAc)_2$, TEA, CH_2Cl_2, rt, 95–100% yield. In this case, the relatively stable TBDMS carbamate is isolated.[34]
7. Me_3SiI, CH_3CN, 25°, 6 min, 100% yield.[35,36]
8. TMSBr, PhSMe, TFA, 0°, 1 h, 70% yield.[37]
9. Pd-Poly(ethylenimine), HCO_2H.[38] This catalyst system was reported to be better than Pd/C or Pd black for Z removal.
10. $AlCl_3$, $PhOCH_3$, 0–25°, 5 h, 73% yield.[39] These conditions are compatible with β-lactams.
11. BBr_3, CH_2Cl_2, −10°, 1 h → 25°, 2 h, 80–100% yield.[40] Benzyl carbamates of larger peptides can be cleaved by boron tribromide in trifluoroacetic acid, since the peptides are more soluble in acid than in methylene chloride.[41]

12. BCl_3, CH_2Cl_2, rt.[42]
13. 253.7 nm, *hν*, 55°, 4 h, CH_3OH, H_2O, 70% yield.[43,44]
14. Electrolysis: −2.9 V, DMF, $R_4N^+X^-$, 70–80% yield[45] or Pd/graphite cathode, MeOH, AcOH, 2.5% $NaClO_4$ (0.5 mole/L), 99% yield.[46] Benzyl ethers and tosylates are stable to these conditions, but benzyl esters are cleaved.
15. Benzyl carbamates of pyrrole-type nitrogens can be cleaved with nucleophilic reagents such as hydrazine; hydrogenation and HF treatment are also effective.[47]
16. Benzyl carbamates are readily cleaved under strongly acidic conditions: HBr, AcOH;[48] 50% CF_3COOH (25°, 14 days, partially cleaved);[49] 70% HF, Pyr;[50] CF_3SO_3H;[51] FSO_3H,[52] or CH_3SO_3H.[52] In cleaving benzyl carbamates from peptides, 0.5 *M* 4-(methylmercapto)phenol in CF_3CO_2H has been recommended to suppress Bn^+ additions to aromatic amino acids.[53] To achieve deprotection via an S_N-2 mechanism, which also reduces the problem of Bn^+ addition, HF–Me_2S–*p*-cresol (25:65:10, v/v) has been recommended for peptide deprotection.[54]
17. Na/NH_3.[55]
18. 0.15 *M* $Ba(OH)_2$, heat, 40 h, 3:2 glyme/H_2O, 75% yield.[56]

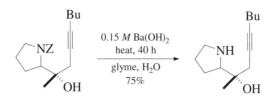

In this case, the following reagents failed to afford clean deprotection because of destruction of the acetylene: Me_3SiI, BBr_3, Me_2BBr, BF_3/EtSH, $AlCl_3$/EtSH, MeLi/LiBr, KOH/EtOH.

19. $BF_3{\cdot}Et_2O$, CH_3SCH_3, CH_2Cl_2, 92% yield.[57]
20. $BF_3{\cdot}Et_2O$, EtSH, CH_2Cl_2, rt. 76–96% yield.[58] It is possible to achieve some selectivity for a secondary derivative over a primary one when the reaction is conducted under more dilute conditions.

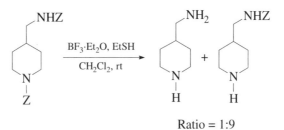

21. 40% KOH, MeOH, H_2O, 85–94% yield.[59]

22. $K_3[Co(CN)_5]$, H_2, MeOH, 20°, 3 h.[60] Benzyl ethers are not cleaved under these conditions.
23. $LiBH_4$ or $NaBH_4$, Me_3SiCl, THF, 24 h, 88–95%.[61] This combination of reagents also reduces all functional groups that can normally be reduced with diborane.
24. Catecholborane halide cleaves benzyl carbamates in the presence of ethyl and benzyl esters and TBDMS ethers.[62]
25. 6 *N* HCl, reflux, 1 h, 92% yield.[63]
26. AcCl and NaI transform a Z-protected amine into an acetamide (84% yield).[64]
27. $CCl_3C(=NH)OBn$ (TFA, heat, 46–56% yield) will exchange the Teoc and BOC groups for the Z group.[65]

1. M. Bergmann and L. Zervas, *Ber.*, **65**, 1192 (1932).
2. G. J. Atwell and W. A. Denny, *Synthesis*, 1032 (1984).
3. D. B. Berkowitz and M. L. Pedersen, *J. Org. Chem.*, **59**, 5476 (1994).
4. M. Dymicky, *Org. Prep. Proced. Int.*, **21**, 83 (1989).
5. W. Graf, O. Keller, W. Keller, G. Wersin, and E. Wuensch, *Peptides 1986: Proc. 19th Eur. Pept. Symp.*, D. Theodoropoulos, Ed., de Gruyter, New York, 1987, p. 73.
6. G. Sennyey, G. Barcelo, and J.-P. Senet, *Tetrahedron Lett.*, **27**, 5375 (1986).
7. Y. Kita, J. Haruta, H. Yasuda, K. Fukunaga, Y. Shirouchi, and Y. Tamura, *J. Org. Chem.*, **47**, 2697 (1982).
8. A. Paquet, *Can. J. Chem.*, **60**, 976 (1982).
9. E. Wuensch, W. Graf, O. Keller, W. Keller, and G. Wersin, *Synthesis*, 958 (1986).
10. S. Murahashi, T. Naota, and N. Nakajima, *Chem. Lett.*, 879 (1987).
11. S. K. Sharma, M. J. Miller, and S. M. Payne, *J. Med. Chem.*, **32**, 357 (1989).
12. P. Henklein, H.-U. Heyne, W.-R. Halatsch, and H. Niedrich, *Synthesis*, 166 (1987).
13. K. J. Butcher, *Synlett*, 825 (1994).
14. M. Allainmat, P. L'Haridon, L. Toupet, and D. Plusquellec, *Synthesis*, 27 (1990).
15. D. R. Kelly and M. Gingell, *Chem. Ind. (London)*, 888 (1991).
16. N. Matsumura, A. Noguchi, A. Kitayoshi, and H. Inoue, *J. Chem. Soc., Perkin Trans. 1*, 2953 (1995).
17. I. Azuse, M. Tamura, K. Kinomura, H. Okai, K. Kouge, F. Hamatsu, and T. Koizumi, *Bull. Chem. Soc. Jpn.*, **62**, 3103 (1989).
18. M. Sakaitani, K. Hori, and Y. Ohfune, *Tetrahedron Lett.*, **29**, 2983 (1988).
19. J.-P. Mazaleyrat, J. Xie, and M. Wakselman, *Tetrahedron Lett.*, **33**, 4301 (1992).
20. N. L. Benoiton, *Int. J. Pept. Protein Res.*, **41**, 611 (1993).
21. J. Meienhofer and K. Kuromizu, *Tetrahedron Lett.*, 3259 (1974).
22. H. Sajiki, *Tetrahedron Lett.*, **36**, 3465 (1995).
23. A. E. Jackson and R. A. W. Johnstone, *Synthesis*, 685 (1976); G. M. Anantharamaiah and K. M. Sivanandaiah, *J. Chem. Soc., Perkin Trans. 1*, 490 (1977).

24. A. M. Felix, E. P. Heimer, T. J. Lambros, C. Tzougraki, and J. Meienhofer, *J. Org. Chem.*, **43**, 4194 (1978).
25. K. M. Sivanandaiah and S. Gurusiddappa, *J. Chem. Res., Synop.*, 108 (1979); B. ElAmin, G. M. Anantharamaiah, G. P. Royer, and G. E. Means, *J. Org. Chem.*, **44**, 3442 (1979).
26. M. J. O. Anteunis, C. Becu, F. Becu, and M. F. Reyniers, *Bull. Soc. Chim. Belg.*, **96**, 775 (1987).
27. Y. Okada and N. Ohta, *Chem. Pharm. Bull.*, **30**, 581 (1982).
28. M. Makowski, B. Rzeszotarska, L. Smelka, and Z. Kubica, *Liebigs Ann. Chem.*, 1457 (1985).
29. J. S. Bajwa, *Tetrahedron Lett.*, **33**, 2955 (1992).
30. Y. Kawasaki, H. Konishi, M. Morita, M. Kawanari, S. i. Dosako, and I. Nakajima, *Chem. Pharm. Bull.*, **42**, 1238 (1994).
31. L. Birkofer, E. Bierwirth, and A. Ritter, *Chem. Ber.*, **94**, 821 (1961).
32. M. Sakaitani, N. Kurokawa, and Y. Ohfune, *Tetrahedron Lett.*, **27**, 3753 (1986).
33. R. S. Coleman and A. J. Carpenter, *J. Org. Chem.*, **57**, 5813 (1992).
34. M. Sakaitani and Y. Ohfune, *J. Org. Chem.*, **55**, 870 (1990).
35. R. S. Lott, V. S. Chauhan, and C. H. Stammer, *J. Chem. Soc., Chem. Commun.*, 495 (1979).
36. M. Ihara, N. Taniguchi, K. Nogochi, K. Fujumoto, and T. Kametani, *J. Chem. Soc., Perkin Trans. 1*, 1277 (1988).
37. U. Schmidt, V. Leitenberger, H. Griesser, J. Schmidt, and R. Meyer, *Synthesis*, 1248 (1992).
38. D. R. Coleman and G. P. Royer, *J. Org. Chem.*, **45**, 2268 (1980).
39. T. Tsuji, T. Kataoka, M. Yoshioka, Y. Sendo, Y. Nishitani, S. Hirai, T. Maeda, and W. Nagata, *Tetrahedron Lett.*, 2793 (1979).
40. A. M. Felix, *J. Org. Chem.*, **39**, 1427 (1974).
41. J. Pless and W. Bauer, *Angew Chem., Int. Ed. Engl.*, **12**, 147 (1973).
42. T. Fukuyama, R. K. Frank, and C. F. Jewell, Jr., *J. Am. Chem. Soc.*, **102**, 2122 (1980).
43. S. Hanessian and R. Masse, *Carbohydr. Res.*, **54**, 142 (1977).
44. For a review of photochemically labile protective groups, see V. N. R. Pillai, *Synthesis*, 1 (1980).
45. V. G. Mairanovsky, *Angew Chem., Int. Ed. Engl.*, **15**, 281 (1976).
46. M. A. Casadei and D. Pletcher, *Synthesis*, 1118 (1987).
47. M. Chorev and Y. S. Klausner, *J. Chem. Soc., Chem. Commun.*, 596 (1976).
48. D. Ben-Ishai and A. Berger, *J. Org. Chem.*, **17**, 1564 (1952).
49. A. R. Mitchell and R. B. Merrifield, *J. Org. Chem.*, **41**, 2015 (1976).
50. S. Matsuura, C.-H. Niu, and J. S. Cohen, *J. Chem. Soc., Chem. Commun.*, 451 (1976).
51. H. Yajima, N. Fujii, H. Ogawa, and H. Kawatani, *J. Chem. Soc., Chem. Commun.*, 107 (1974).
52. H. Yajima, H. Ogawa, and H. Sakurai, *J. Chem. Soc., Chem. Commun.*, 909 (1977).
53. M. Bodanszky and A. Bodanszky, *Int. J. Pept. Protein Res.*, **23**, 287 (1984).

54. J. P. Tam, W. F. Heath, and R. B. Merrifield, *J. Am. Chem. Soc.*, **105**, 6442 (1983).
55. I. Schon, T. Szirtes, T. Uberhardt, A. Rill, A. Csehi, and B. Hegedus, *Int. J. Pept. Protein Res.*, **22**, 92 (1983).
56. L. E. Overman and M. J. Sharp, *Tetrahedron Lett.*, **29**, 901 (1988).
57. I. H. Sanchez, F. J. López, J. J. Soria, M. I. Larraza, and H. J. Flores, *J. Am. Chem. Soc.*, **105**, 7640 (1983).
58. D. S. Bose and D. E. Thurston, *Tetrahedron Lett.*, **31**, 6903 (1990).
59. S. R. Angle and D. O. Arnaiz, *Tetrahedron Lett.*, **30**, 515 (1989).
60. G. Losse and H. U. Stiehl, *Z. Chem.*, **21**, 188 (1981).
61. A. Giannis and K. Sandhoff, *Angew. Chem., Int. Ed. Engl.*, **28**, 218 (1989).
62. R. K. Boeckman, Jr., and J. C. Potenza, *Tetrahedron Lett.*, **26**, 1411 (1985).
63. G. Chelucci, M. Falorni, and G. Giacomelli, *Synthesis*, 1121 (1990).
64. M. Ihara, A. Hirabayashi, N. Taniguchi, and K. Fukumoto, *Heterocycles*, **33**, 851 (1992).
65. A. G. M. Barrett and D. Pilipauskas, *J. Org. Chem.*, **55**, 5170 (1990).

p-Methoxybenzyl Carbamate (Moz–NR$_2$): *p*-MeOC$_6$H$_4$CH$_2$OC(O)NR$_2$

Formation

1. Moz-ON=C(CN)Ph, H_2O, Et_3N, rt, 6 h, 90% yield.[1]
2. MozN$_3$.[2,3]

Cleavage

The Moz group is more readily cleaved by acid than is either the benzyloxycarbonyl or BOC group.[4,5] The section on benzyl carbamates should be consulted, since many of the methods for formation and cleavage described there should be applicable to the Moz group as well.

1. TsOH, CH_3CN, acetone, rt.[6,7]

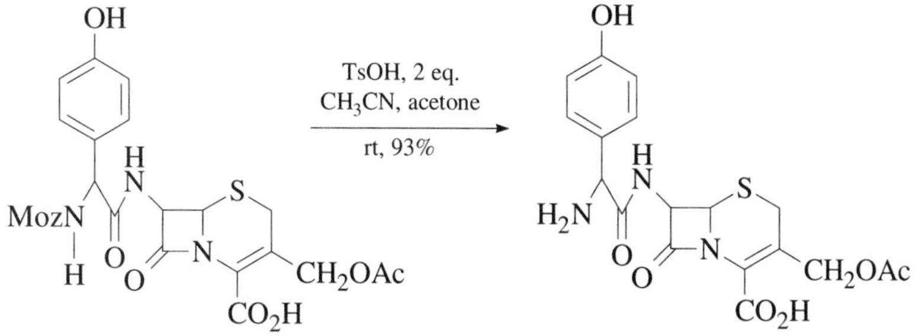

Ref. 5

2. 10% CF_3COOH, CH_2Cl_2, 100% yield.[1,4]
3. CH_3SO_3H, *m*-cresol, CH_2Cl_2. The addition of *m*-cresol greatly accelerates the rate of cleavage.[8]

1. S.-T. Chen, S.-H. Wu, and K.-T. Wang, *Synthesis*, 36 (1989).
2. J. M. Kerr, S. C. Banville, and R. N. Zuckermann, *J. Am. Chem. Soc.*, **115**, 2529 (1993).
3. S. S. Wang, S. T. Chen, K. T. Wang, and R. B. Merrifield, *Int. J. Pept. Protein Res.*, **30**, 662 (1987).
4. F. Weygand and K. Hunger, *Chem. Ber.*, **95**, 1 (1962).
5. S. S. Wang, S. T. Chen, K. T. Wang, and R. B. Merrifield, *Int. J. Pept. Protein Res.*, **30**, 662 (1987).
6. H. Yajima, H. Ogawa, N. Fujii, and S. Funakoshi, *Chem. Pharm. Bull.*, **25**, 740 (1977).
7. H. Yamada, H. Tobiki, N. Tanno, H. Suzuki, K. Jimpo, S. Ueda, and T. Nakagome, *Bull. Chem. Soc. Jpn.*, **57**, 3333 (1984).
8. H. Tamamura, J. Nakamura, K. Noguchi, S. Funakoshi, and N. Fujii, *Chem. Pharm. Bull.*, **41**, 954 (1993).

p-Nitrobenzyl Carbamate (PNZ–NR$_2$): *p*-$NO_2C_6H_4CH_2OC(O)NR_2$ (Chart 8)

Formation[1]

1. *p*-$NO_2C_6H_4CH_2OCOCl$, base, 0°, 1.5 h, 78% yield.

Cleavage

1. H_2/Pd–C, 10 h, 87% yield.[1] A nitrobenzyl carbamate is more readily cleaved by hydrogenolysis than is a benzyl carbamate; it is also more stable to acid-catalyzed hydrolysis than a benzyl carbamate, and therefore, selective cleavage is possible.
2. 4 *N* HBr, AcOH, 60°, 2 h, 68% yield.[1]
3. $Na_2S_2O_4$, NaOH.[2] This method was used for deprotection of a glucosamine.[3] Cleavage occurs by reduction to the amine, which then undergoes a 1,6-elimination.
4. Electrolysis, −1.2 V, DMF, $R_4N^+X^-$.[4]

1. J. E. Shields and F. H. Carpenter, *J. Am. Chem. Soc.*, **83**, 3066 (1961); S. Hashiguchi, H. Natsugari, and M. Ochiai, *J. Chem. Soc., Perkin Trans. 1*, 2345 (1988).
2. P. J. Romanovskis, P. Henklein, J. A. Benders, I. V. Siskov, and G. I. Chipens, in *Fifth All-Union Symposium on Protein and Peptide Chemistry and Physics, Abstracts*, Baku (1980), p. 229 (in Russian); P. J. Romanovskis, I. V. Siskov, I. K. Liepkaula, E. A. Porunkevich, M. P. Ratkevich, A. A. Skujins, and G. I. Chipens, "Linear and Cyclic Analogs of ACTH Fragments: Synthesis and Biological Activity," in *Peptides: Synthesis, Structure, Function: Proceedings of the Seventh American Peptide Symposium*, University of Wisconsin, Madison (1981), D. H. Rich and E. Gross, Eds., Pierce Chem. Co., Rockford, IL, 1981, pp. 229–232.

3. X. Qian and O. Hindsgaul, *J. Chem. Soc., Chem. Commun.*, 1059 (1997).
4. V. G. Mairanovsky, *Angew. Chem., Int. Ed. Engl.*, **15**, 281 (1976); H. L. S. Maia, M. J. Medeiros, M. I. Montenegro, and D. Pletcher, *Port. Electrochim. Acta*, **5**, 187 (1987); *ibid., Chem. Abstr.*, **109**: 118114n (1989).

Benzyl carbamates substituted with one or more halogens are much more stable to acidic hydrolysis than are the unsubstituted benzyl carbamates.[12] For example, the 2,4-dichlorobenzyl carbamate is 80 times more stable to acid than is the simple benzyl derivative.[3] Halobenzyl carbamates can also be cleaved by hydrogenolysis with Pd–C.[3] The following halobenzyl carbamates have been found to be useful when increased acid stability is required:

p-Bromobenzyl Carbamate[4]

p-Chlorobenzyl Carbamate[1,2]

2,4-Dichlorobenzyl Carbamate[3] (Chart 8)

1. K. Noda, S. Terada, and N. Izumiya, *Bull. Chem. Soc. Jpn.*, **43**, 1883 (1970).
2. B. W. Erickson and R. B. Merrifield, *J. Am. Chem. Soc.*, **95**, 3757 (1973).
3. Y. S. Klausner and M. Chorev, *J. Chem. Soc., Perkin Trans. 1*, 627 (1977).
4. D. M. Channing, P. B. Turner, and G. T. Young, *Nature*, **167**, 487 (1951).

4-Methylsulfinylbenzyl Carbamate (Msz–NR_2): $CH_3S(O)C_6H_4CH_2OCONR_2$

The Msz group is stable to TFA/anisole, NaOH, and hydrazine.

Formation

1. Msz–O-succinimidyl, CH_3CN, H_2O, Et_3N, 45% yield.[1]

Cleavage

1. $SiCl_4$, TFA, anisole.[1] $SiCl_4$ serves to reduce the sulfoxide prior to acid-catalyzed cleavage. Other sulfoxide-reducing agents could probably be used.

1. Y. Kiso, T. Kimura, M. Yoshida, M. Shimokura, K. Akaji, and T. Mimoto, *J. Chem. Soc., Chem. Commun.*, 1511 (1989).

9-Anthrylmethyl Carbamate (Chart 8): $CH_2OC(O)NR_2$

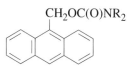

Formation

1. 9-Anthryl–$CH_2OCO_2C_6H_4$–*p*-NO_2, DMF, 25°, 86% yield.[1]

Cleavage[1]

1. CH_3SNa, DMF, –20°, 1–7 h, 77–91% yield or 25°, 4 min, 86% yield.
2. CF_3COOH, CH_2Cl_2, 0°, 5 min, 88–92% yield. The anthrylmethyl carbamate is stable to 0.01 *N* lithium hydroxide (25°, 6 h), to 0.1 *N* sulfuric acid (25°, 1 h), and to 1 *M* trifluoroacetic acid (25°, 1 h, dioxane).

1. N. Kornblum and A. Scott, *J. Org. Chem.*, **42**, 399 (1977).

Diphenylmethyl Carbamate: $Ph_2CHOC(O)NR_2$ (Chart 8)

The diphenylmethyl carbamate, prepared from the azidoformate, is readily cleaved by mild acid hydrolysis (1.7 *N* HCl, THF, 65°, 10 min, 100% yield).[1]

1. R. G. Hiskey and J. B. Adams, *J. Am. Chem. Soc.*, **87**, 3969 (1965).

Assisted Cleavage

Several protective groups have been prepared that rely on a *β*-elimination to effect cleavage. Often, the protective group must first be activated to increase the acidity of the *β*-hydrogen. In general, the derivatives are prepared by standard procedures, from either the chloroformate or mixed carbonate.

2-Methylthioethyl Carbamate: $MeSCH_2CH_2OC(O)NR_2$

A 2-methylthioethyl carbamate is cleaved by 0.01 *N* NaOH after alkylation to $Me_2S^+CH_2CH_2–$ or by 0.1 *N* NaOH after oxidation to the sulfone.[1]

2-Methylsulfonylethyl Carbamate: $MeSO_2CH_2CH_2OC(O)NR_2$

This is the oxidized form of the preceding methylthio derivative. It is stable to catalytic hydrogenolysis and does not poison the catalyst. It is also stable to liq. HF (30 min), but is cleaved in 5 sec with 1 *N* NaOH.[2,3]

2-(*p*-Toluenesulfonyl)ethyl Carbamate: $4\text{-}CH_3–C_6H_4SO_2CH_2CH_2OC(O)NR_2$

This derivative is similar to the methylsulfonylethyl derivative. It is cleaved by 1 *M* NaOH, <1 h.[4] The related **4-chlorobenzenesulfonylethyl carbamate** has also been used as a protective group that can be cleaved with DBU or tetramethylguanidine.[5]

[2-(1,3-Dithianyl)]methyl Carbamate (Dmoc–NR_2): S, S (1,3-dithiane ring)–$CH_2OC(O)NR_2$

Cleavage occurs by prior activation with peracetic acid to the bissulfone, followed by mild base treatment.[6]

4-Methylthiophenyl Carbamate (Mtpc–NR_2): 4-$MeSC_6H_4OC(O)NR_2$[7]

2,4-Dimethylthiophenyl Carbamate (Bmpc–NR_2):
2,4-$(MeS)_2C_6H_3OC(O)NR_2$[8]

After activation with peracetic acid and base treatment, derivatives of primary amines form the isocyanate, which can be trapped with water to effect hydrolysis or with an alcohol to form other carbamates.

2-Phosphonioethyl Carbamate (Peoc–NR_2): $R_3P^+CH_2CH_2OC(O)NR'_2\ X^-$

This derivative is stable to trifluoroacetic acid; it is cleaved by mild bases (pH 8.4; 0.1 *N* NaOH, 1 min, 100% yield).[9]

1-Methyl-1-(triphenylphosphonio)ethyl (2-Triphenylphosphonioisopropyl) Carbamate (Ppoc–NR_2): $Ph_3P^+CH_2CH(CH_3)OC(O)NR_2\ X^-$

This derivative is similar to the Peoc group, except that it is four times more stable to base and is not as susceptible to side reactions as is the Peoc group.[10]

1,1-Dimethyl-2-cyanoethyl Carbamate:
$(CN)CH_2C(CH_3)_2OC(O)NR_2$ (Chart 8)

This derivative is stable to trifluoroacetic acid and is cleaved by aqueous K_2CO_3 or Et_3N, 25°, 6 h, 90% yield.[11]

2-Dansylethyl Carbamate (Dnseoc–NR_2):

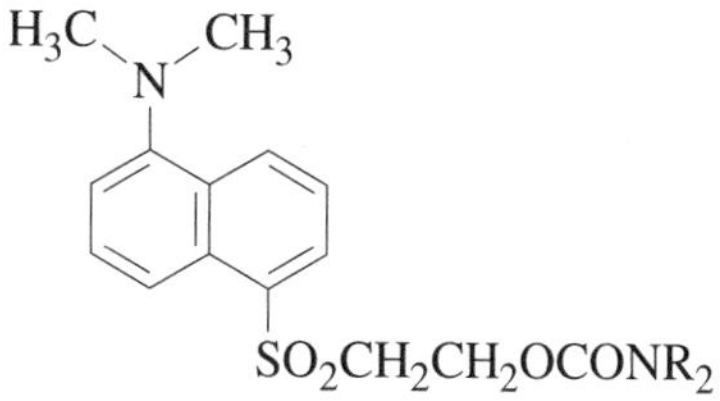

H_3C CH_3
N
$SO_2CH_2CH_2OCONR_2$

The Dnseoc group was developed as a base-labile protective group for the 5′-hydroxyl in oligonucleotide synthesis. It is cleaved with DBU in aprotic solvents. The condensation of oligonucleotide synthesis can be monitored by UV detection at 350 nm or by fluorescence at 530 nm of the liberated vinylsulfone.[12]

2-(4-Nitrophenyl)ethyl Carbamate (Npeoc–NR_2):
4-$NO_2C_6H_4CH_2CH_2OCONR_2$

The Npeoc group was introduced for protection of the exocyclic amino functions of nucleic acid bases, but has also been used for simple amines.

Formation

1. 4-$NO_2C_6H_4CH_2CH_2OCOCl$.[13]

2. [imidazolium chloride carbonate structure: Cl^-, $-N^+$ imidazole $N-C(=O)-O-CH_2CH_2-C_6H_4NO_2$] DMAP, DMF, 75–97% yield.[14]

Cleavage

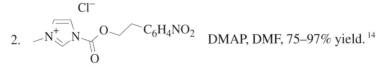

1. DBU, CH_3CN or Pyr.[15]
2. Photolysis, for ***N-o*-nitrodiphenylmethoxycarbonyl** compounds.[16]

1. H. Kunz, *Chem. Ber.*, **109**, 3693 (1976).
2. G. I. Tesser and I. C. Balver-Geers, *Int. J. Pept. Protein Res.*, **7**, 295 (1975).
3. D. Filippov, G.A. van der Marel, E. Kuyl-Yeheskiely, and J. H. van Boom, *Synlett*, 922 (1994).
4. A. T. Kader and C. J. M. Sterling, *J. Chem. Soc.*, 258 (1964).
5. V. V. Samukov, A. N. Sabirov, and M. L. Troshkov, *Zh. Obshch. Khim.*, **58**, 1432 (1988); *Chem. Abstr.*, **110**: 76008u (1989).
6. H. Kunz and R. Barthels, *Chem. Ber.*, **115**, 833 (1982); R. Barthels and H. Kunz, *Angew. Chem., Int. Ed. Engl.*, **21**, 292 (1982).
7. H. Kunz and K. Lorenz, *Angew. Chem., Int. Ed. Engl.*, **19**, 932 (1980).
8. H. Kunz and H.-J. Lasowski, *Angew. Chem., Int. Ed. Engl.*, **25**, 170 (1986).
9. H. Kunz, *Angew. Chem., Int. Ed. Engl.*, **17**, 67 (1978).
10. H. Kunz and G. Schaumloeffel, *Liebigs Ann. Chem.*, 1784 (1985).
11. E. Wünsch and R. Spangenberg, *Chem. Ber.*, **104**, 2427 (1971).
12. F. Bergmann and W. Pfleiderer, *Helv. Chim. Acta*, **77**, 203 (1994).
13. H. Sigmund and W. Pfleiderer, *Helv. Chim. Acta*, **77**, 1267 (1994).
14. G. Walcher and W. Pfleiderer, *Helv. Chim. Acta*, **79**, 1067 (1996).
15. F. Himmelsbach, B. S. Schulz, T. Trichtinger, R. Charubala, and W. Pfleiderer, *Tetrahedron*, **40**, 59 (1984).
16. J. A. Baltrop, P. J. Plant, and P. Schofield, *J. Chem. Soc., Chem. Commun.*, 822 (1966).

A series of carbamates has been prepared that are cleaved by the liberation of a phenol, which, when treated with base, cleaves the carbamate by quinone methide formation through a 1,6-elimination.[1]

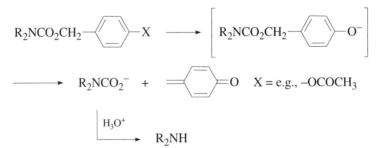

$$R_2NCO_2CH_2-C_6H_4-X \longrightarrow [R_2NCO_2CH_2-C_6H_4-O^-] \longrightarrow R_2NCO_2^- + CH_2{=}C_6H_4{=}O \quad X = \text{e.g., } -OCOCH_3$$

$$R_2NCO_2^- \xrightarrow{H_3O^+} R_2NH$$

4-Phenylacetoxybenzyl Carbamate (PhAcOZ–NR_2):
4-($C_6H_5CH_2CO_2$)$C_6H_4CH_2OCONR_2$

Preparation of PhAcOZ amino acids proceeds from the chloroformate, and cleavage is accomplished enzymatically with penicillin G acylase (pH 7 phosphate buffer, 25°, $NaHSO_3$, 40–88% yield).[2,3] In a related approach, the 4-acetoxy derivative is used, but in this case deprotection is achieved using the lipase, acetyl esterase, from oranges (pH 7, NaCl buffer, 45°, 57–70% yield).[4]

4-Azidobenzyl Carbamate (ACBZ–NR_2): 4-$N_3C_6H_4CH_2OCONR_2$

The carbamate, prepared from the 4-nitrophenyl carbonate, is cleaved by reduction with dithiothreitol (DTT) and TEA to give the aniline, which triggers fragmentation, releasing the amine.[5]

4-Azidomethoxybenzyl Carbamate: $N_3CH_2OC_6H_4CH_2OC(O)NR_2$

Amino acids are protected as the 4-nitrophenyl carbonate (H_2O, dioxane, 54–85% yield) and cleaved by reduction of the azide with $SnCl_2$. The group is stable to the conditions normally used to cleave a BOC group, but it is not expected to be stable to a large number of strongly reducing conditions.[6]

***m*-Chloro-*p*-acyloxybenzyl Carbamate:** [7,8]

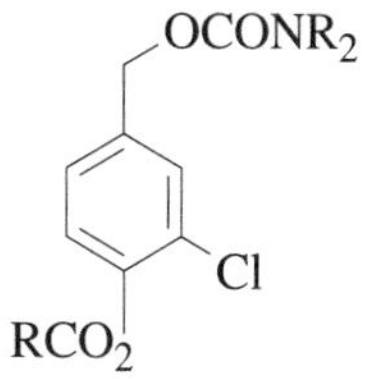

OCONR$_2$ / Cl / RCO$_2$

Cleavage

1. $NaHCO_3/Na_2CO_3$ or H_2O_2/NH_3, $NaHSO_3$, 1 h.
2. 0.1 *N* NaOH, 10 min, 100% yield.
3. H_2/Pd–C.
4. HBr, AcOH.

***p*-(Dihydroxyboryl)benzyl Carbamate (Dobz–NR_2):**

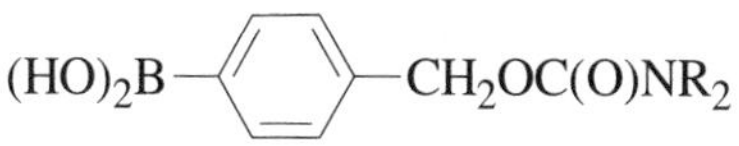

$(HO)_2B$–C_6H_4–$CH_2OC(O)NR_2$

Formation[9]

1. 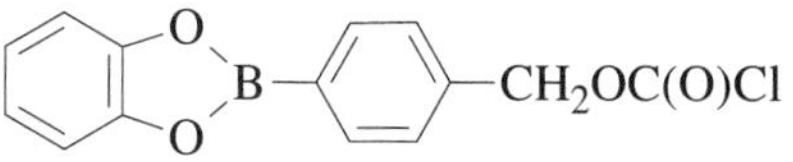 (catechol boronate)B–C_6H_4–$CH_2OC(O)Cl$ aqueous base; aqueous acid

Cleavage[10]

1. H_2O_2, pH 9.5, 25°, 5 min, 90% yield.
2. H_2, Pd–C.
3. HBr, AcOH.

5-Benzisoxazolylmethyl Carbamate (Bic–NR_2) (Chart 8):

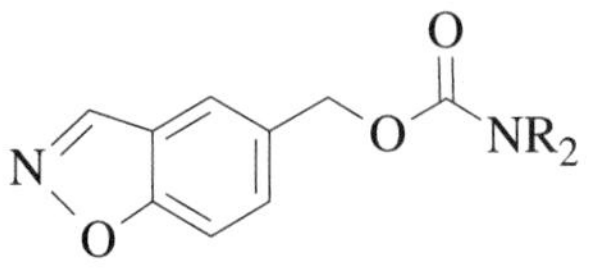

Formation

1. $ClCO_2CH_2$–5-benzisoxazole, pH 8.5–9.0, CH_3CN, 0°, 1 h, 63% yield.[10]

Cleavage[10]

1. Et_3N, CH_3CN or DMF, 25°, 30 min; Na_2SO_3, EtOH, H_2O, 40°, 3 h, pH 7, 92% yield or CF_3COOH, 90 min, 95% yield.
2. H_2, Pd–C.
3. HBr, AcOH. This derivative is stable to trifluoroacetic acid.

2-(Trifluoromethyl)-6-chromonylmethyl Carbamate (Tcroc–NR_2):[11,12]

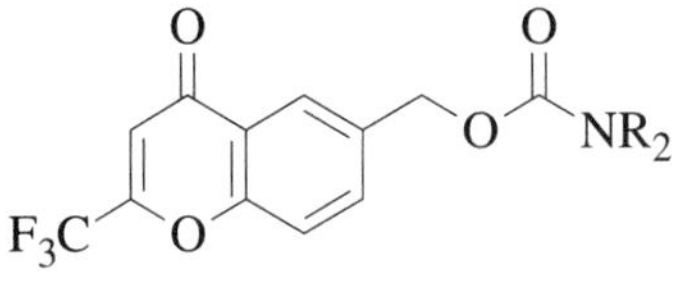

Cleavage

1. $PrNH_2$ or hydrazine. The Tcroc group resists cleavage by CF_3COOH.

1. M. Wakselman, *Nouv. J. Chim.*, **7**, 439 (1983).
2. T. Pohl and H. Waldmann, *Angew. Chem., Int. Ed. Engl.*, **35**, 1720 (1996).
3. D. Sebastian, A. Heuser, S. Schulze, and H. Waldemann, *Synthesis*, 1098 (1997).
4. H. Waldmann and E. Nägele, *Angew. Chem., Int. Ed. Engl.*, **34**, 2259 (1995).

5. A. Mitchinson, B. T. Golding, R. J. Griffin, and M. C. O'Sullivan, *J. Chem. Soc., Chem. Commun.*, 2613 (1994).
6. B. Loubinoux and P. Gerardin, *Tetrahedron Lett.*, **32**, 351 (1991).
7. M. Wakselman and E. G.-Jampel, *J. Chem. Soc., Chem. Commun.*, 593 (1973).
8. G. Le Corre, E. G.-Jampel, and M. Wakselman, *Tetrahedron*, **34**, 3105 (1978).
9. D. S. Kemp and D. C. Roberts, *Tetrahedron Lett.*, 4629 (1975).
10. D. S. Kemp and C. F. Hoyng, *Tetrahedron Lett.*, 4625 (1975).
11. D. S. Kemp and G. Hanson, *J. Org. Chem.*, **46**, 4971 (1981).
12. D. S. Kemp, D. R. Bolin, and M. E. Parham, *Tetrahedron Lett.*, **22**, 4575 (1981).

Photolytic Cleavage

The following carbamates can be cleaved by photolysis.[1] They can be prepared either from the chloroformate or from the mixed carbonate.

1. *m*-Nitrophenyl Carbamate[2]

2. 3,5-Dimethoxybenzyl Carbamate[3]

3. 1-Methyl-1-(3,5-dimethoxyphenyl)ethyl Carbamate (Ddz–NR_2). The carbamate, prepared in 80% yield from the azidoformate or pentachlorophenyl carbonate, is cleaved by photolysis and, as expected, by acidic hydrolysis (TFA, 20°, 8 min, 100% yield).[4,5]

4. α-Methylnitropiperonyl Carbamate (Menpoc–NR_2).[6] The half-life for the photochemical cleavage is on the order of 20–30 sec for a variety of amino acid derivatives. This rate is substantially faster than that of the 2-nitrobenzyl carbamate.

5. *o*-Nitrobenzyl Carbamate[7]

6. 3,4-Dimethoxy-6-nitrobenzyl Carbamate[7,8] (Chart 8)

7. Phenyl(*o*-nitrophenyl)methyl Carbamate (Npeoc–NR_2)[9]

8. 2-(2-Nitrophenyl)ethyl Carbamate.[10] The photolytic removal of this group occurs twice as fast as does the 2-nitrobenzyl carbamate. Additionally, substitution at the alpha carbon increases the rate of cleavage even more.

9. 6-Nitroveratryl Carbamate (Nvoc–NR_2)[11]

10. 4-Methoxyphenacyl Carbamate (Phenoc–NR_2):[12]

Formation/Cleavage[12]

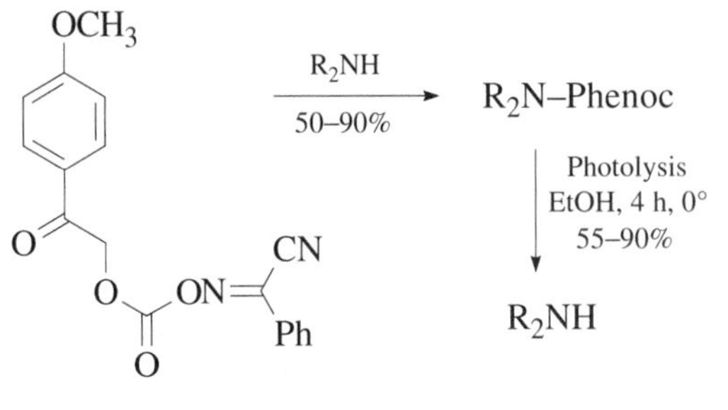

OCH$_3$ / O / O / ON= / CN / Ph / O —(R_2NH, 50–90%)→ R_2N–Phenoc —(Photolysis, EtOH, 4 h, 0°, 55–90%)→ R_2NH

This group is stable to 50% TFA/CH_2Cl_2, NaOH, and 20% piperidine/DMF.

11. 3′,5′-Dimethoxybenzoin Carbamate (DMBOCONR$_2$)

Formation/Cleavage[13]

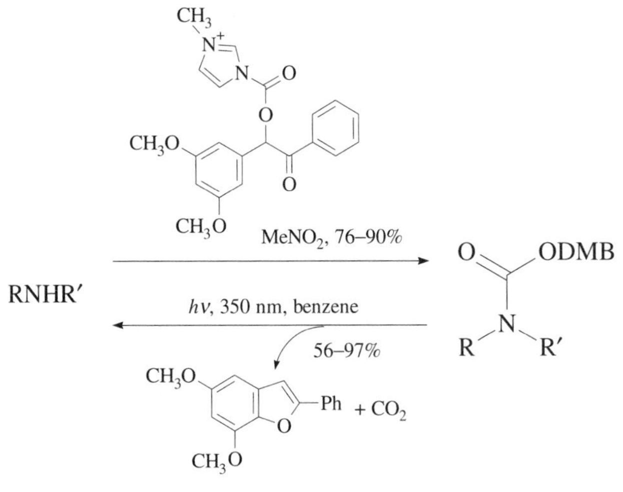

RNHR′ —(CH_3-imidazolium carbonate of 3′,5′-dimethoxybenzoin (CH_3O, CH_3O), $MeNO_2$, 76–90%)→ O=C(ODMB)–N(R)(R′)

O=C(ODMB)–NRR′ —(hv, 350 nm, benzene, 56–97%)→ RNHR′ + 5,7-dimethoxy-2-phenylbenzofuran (CH_3O, CH_3O, Ph) + CO_2

The DMB carbamate can also be introduced through the 4-nitrophenyl carbonate.[14] It has been prepared from an isocyanate and 3′,5′-dimethoxybenzoin.[15] The synthesis of a number of other substituted benzoins as possible protective groups has been described.[16]

1. For a review of photochemically labile protective groups, see V. N. R. Pillai, *Synthesis*, 1 (1980).
2. Th. Wieland, Ch. Lamperstorfer, and Ch. Birr, *Makromol. Chem.*, **92**, 277 (1966).
3. J. W. Chamberlin, *J. Org. Chem.*, **31**, 1658 (1966).
4. C. Birr, W. Lochinger, G. Stahnke, and P. Lang, *Justus Liebigs Ann. Chem.*, **763**, 162 (1972); C. Birr, *Justus Liebigs Ann. Chem.*, 1652 (1973).

5. J. F. Cameron and J. M. J. Fréchet, *J. Org. Chem.*, **55**, 5919 (1990).
6. C. P. Holmes and B. Kiangsoontra, *Pept.: Chem., Struct. Biol., Proc. Am. Pept. Symp., 13th*, 110 (1994).
7. B. Amit, U. Zehavi, and A. Patchornik, *J. Org. Chem.*, **39**, 192 (1974).
8. K. Burgess, S. E. Jacutin, D. Lim, and A. Shitangkoon, *J. Org. Chem.*, **62**, 5165 (1997).
9. J. A. Baltrop, P. J. Plant, and P. Schofield, *J. Chem. Soc., Chem. Commun.*, 822 (1966).
10. A. Hasan, K.-P. Stengele, H. Giegrich, P. Cornwell, K. R. Isham, R. A. Sachleben, W. Pfleiderer, and R. S. Foote, *Tetrahedron*, **53**, 4247 (1997).
11. S. B. Rollins and R. M. Williams, *Tetrahedron Lett.*, **38**, 4033 (1997).
12. G. Church, J.-M. Ferland, and J. Gauthier, *Tetrahedron Lett.*, **30**, 1901 (1989).
13. M. C. Pirrung and C.-Y. Huang, *Tetrahedron Lett.*, **36**, 5883 (1995).
14. G. Papageorgiou and J. E. T. Corrie, *Tetrahedron*, **53**, 3917 (1997).
15. J. F. Cameron, C. G. Willson, and J. M. J. Fréchet, *J. Chem. Soc., Chem. Commun.*, 923 (1995).
16. J. F. Cameron, C. G. Willson, and J. M. J. Fréchet, *J. Chem. Soc., Perkin Trans. 1*, 2429 (1997).

Urea-Type Derivatives

Phenothiazinyl-(10)-carbonyl Derivative:

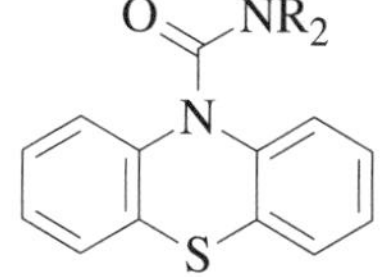

This derivative is prepared in 51–82% yield and is cleaved with $Ba(OH)_2$ or NaOH in 52–96% yield after oxidation of the sulfur with hydrogen peroxide. The derivative is stable to CF_3COOH and NaOH.[1]

1. J. Gante, W. Hechler, and R. Weitzel in *Peptites 1986: Proc. 19th Eur. Pept. Symp.*, 1986, D. Theodoropoulos, Ed., de Gruyter, Berlin, 1987, pp. 87–90.

***N′-p*-Toluenesulfonylaminocarbonyl Derivative:** $R_2NCONHSO_2C_6H_4$-*p*-CH_3

This sulfonyl urea, prepared from an amino acid and *p*-tosyl isocyanate in 20–80% yield, is cleaved by alcohols (95% aq. EtOH, *n*-PrOH, or *n*-BuOH, 100°, 1 h, 95% yield). It is stable to dilute base, to acids (HBr/AcOH or cold CF_3CO_2H), and to hydrazine.[1]

1. B. Weinstein, T. N.-S. Ho, R. T. Fukura, and E. C. Angell, *Synth. Commun.*, **6**, 17 (1976).

N′-Phenylaminothiocarbonyl Derivative: $R_2NCSNHC_6H_5$ (Chart 8)

This thiourea, prepared from an amino acid and phenyl isothiocyanate,[1] is cleaved by anhydrous trifluoroacetic acid (an N–$COCF_3$ group is stable)[2] and by oxidation (*m*-$ClC_6H_4CO_3H$, 0°, 1.5 h, 73% yield; H_2O_2/AcOH, 80°, 80 min, 44% yield).[3]

1. P. Edman, *Acta Chem. Scand.*, **4**, 277 (1950).
2. F. Borrás and R. E. Offord, *Nature*, **227**, 716 (1970).
3. J. Kollonitsch, A. Hajós, and V. Gábor, *Chem. Ber.*, **89**, 2288 (1956).

Miscellaneous Carbamates

The following carbamates have seen little use since the preparation of the first edition of this book; they are listed here for completeness. For the most part, they are variations of the BOC and benzyl carbamates, with the exception of the azo derivatives, which are highly colored. The differences between them are largely in the strength of the acid required for their cleavage.

***t*-Amyl Carbamate**[1]

***S*-Benzyl Thiocarbamate**[2] (Chart 8)

Butynyl Carbamate[3]

***p*-Cyanobenzyl Carbamate**[4]

Cyclobutyl Carbamate[5] (Chart 8). The half-life for cleavage in neat CF_3CO_2H is > 300 min.

Cyclohexyl Carbamate[6]

Cyclopentyl Carbamate[6]

Cyclopropylmethyl Carbamate[5]

***p*-Decyloxybenzyl Carbamate**[7]

Diisopropylmethyl Carbamate[6]

2,2-Dimethoxycarbonylvinyl Carbamate[8]

***o*-(*N*,*N*-Dimethylcarboxamido)benzyl Carbamate**[9]

1,1-Dimethyl-3-(*N*,*N*-dimethylcarboxamido)propyl Carbamate[9]

1,1-Dimethylpropynyl Carbamate[10] (Chart 8)

Di(2-pyridyl)methyl Carbamate[9]

2-Furanylmethyl Carbamate[11]

2-Iodoethyl Carbamate[12]

Isobornyl Carbamate[13]

Isobutyl Carbamate[14]

Isonicotinyl Carbamate[15] (Chart 8)

***p*-(*p*′-Methoxyphenylazo)benzyl Carbamate**[16]

1-Methylcyclobutyl Carbamate[5] (Chart 8)

1-Methylcyclohexyl Carbamate[5] (Chart 8). The half-life for cleavage in neat CF_3CO_2H is 2 min and 180 min in formic acid.

1-Methyl-1-cyclopropylmethyl Carbamate[6]

1-Methyl-1-(*p*-phenylazophenyl)ethyl Carbamate[17]

1-Methyl-1-phenylethyl Carbamate[18] (Chart 8)

1-Methyl-1-(4′-pyridyl)ethyl Carbamate[9]

Phenyl Carbamate[19]

***p*-(Phenylazo)benzyl Carbamate**[16]

2,4,6-Tri-*t*-butylphenyl Carbamate[20]

4-(Trimethylammonium)benzyl Carbamate[21]

2,4,6-Trimethylbenzyl Carbamate[22]

1. S. Sakakibara, I. Honda, K. Takada, M. Miyoshi, T. Ohnishi, and K. Okumura, *Bull. Chem. Soc. Jpn.*, **42**, 809 (1969); S. Matsuura, C.-H. Niu, and J. S. Cohen, *J. Chem. Soc., Chem. Commun.*, 451 (1976).
2. H. B. Milne, S. L. Razniak, R. P. Bayer, and D. W. Fish, *J. Am. Chem. Soc.*, **82**, 4582 (1960).
3. S. Miyazawa, K. Okano, T. Kawahara, Y. Machida, and I. Yamatsu, *Chem. Pharm. Bull.*, **40**, 762 (1992).
4. K. Noda, S. Terada, and N. Izumiya, *Bull. Chem. Soc. Jpn.*, **43**, 1883 (1970).
5. S. F. Brady, R. Hirschmann, and D. F. Veber, *J. Org. Chem.*, **42**, 143 (1973).
6. F. C. McKay and N. F. Albertson, *J. Am. Chem. Soc.*, **79**, 4686 (1957).
7. H. Brechbühler, H. Büchi, E. Hatz, J. Schreiber, and A. Eschenmoser, *Helv. Chim. Acta*, **48**, 1746 (1965).
8. A. Gomez-Sanchez, P. B. Moya, and J. Bellanato, *Carbohydr. Res.*, **135**, 101 (1984).
9. S. Coyle, O. Keller, and G. T. Young, *J. Chem. Soc., Perkin Trans. 1*, 1459 (1979).
10. G. L. Southard, B. R. Zaborowsky, and J. M. Pettee, *J. Am. Chem. Soc.*, **93**, 3302 (1971).
11. G. Losse and K. Neubert, *Tetrahedron Lett.*, 1267 (1970).
12. J. Grimshaw, *J. Chem. Soc.*, 7136 (1965).

13. M. Fujino, S. Shinagawa, O. Nishimura, and T. Fukuda, *Chem. Pharm. Bull.*, **20**, 1017 (1972).
14. R. L. Letsinger and P. S. Miller, *J. Am. Chem. Soc.*, **91**, 3356 (1969).
15. D. F. Veber, W. J. Paleveda, Y. C. Lee, and R. Hirschmann, *J. Org. Chem.*, **42**, 3286 (1977).
16. R. Schwyzer, P. Sieber, and K. Zatsko, *Helv. Chim. Acta*, **41**, 491 (1958).
17. A. T.-Kyi and R. Schwyzer, *Helv. Chim. Acta*, **59**, 1642 (1976).
18. B. E. B. Sandberg and U. Ragnarsson, *Int. J. Pept. Protein Res.*, **6**, 111 (1974); H. Franzen and U. Ragnarsson, *Acta Chem. Scand. B*, **33**, 690 (1979).
19. J. D. Hobson and J. G. McCluskey, *J. Chem. Soc. C*, 2015 (1967); R. W. Adamiak and J. Stawinski, *Tetrahedron Lett.*, 1935 (1977).
20. D. Seebach and T. Hassel, *Angew. Chem., Int. Ed. Engl.*, **17**, 274 (1978).
21. Y. Zhang, X. Wang, L. Li, and P. Zhang, *Sci. Sin. Ser. B*, **29**, 1009 (1986); *Chem. Abstr.*, **108**: 6354p (1988).
22. Y. Isowa, M. Ohmori, M. Sato, and K. Mori, *Bull. Chem. Soc. Jpn.*, **50**, 2766 (1977).

AMIDES

Simple amides are generally prepared from the acid chloride or the anhydride. They are exceptionally stable to acidic or basic hydrolyis and are classically hydrolyzed by brute force by heating in strongly acidic or basic solutions. Among simple amides, hydrolytic stability increases from formyl to acetyl to benzoyl. Lability of the haloacetyl derivatives to mild acid hydrolysis increases with substitution: acetyl < chloroacetyl < dichloroacetyl < trichloroacetyl < trifluoroacetyl.[1] Note that amide hydrolysis under acidic or basic[2] conditions is greatly facilitated in the presence of a neighboring hydroxyl group that can participate in the hydrolysis.[3] Although a number of imaginative amide-derived protective groups have been developed, most are not commonly used because they contain other reactive functionality or are not commercially available or because other more easily introduced and cleaved groups, such as the BOC, Alloc, and Cbz groups, serve more adequately for amine protection. Amide derivatives of the nucleotides are not discussed in this section, since their behavior is atypical of amides. They are generally more easily hydrolyzed than the typical amide. Several review articles discuss amides as –NH protective groups.[4–7]

1. R. S. Goody and R. T. Walker, *Tetrahedron Lett.*, 289 (1967).
2. B. F. Cain, *J. Org. Chem.*, **41**, 2029 (1976).
3. See, for example, C. K. Lai, R. S. Buckanin, S. J. Chen, D. F. Zimmerman, F. T. Sher, and G. A. Berchtold, *J. Org. Chem.*, **47**, 2364 (1982); E. R. Koft, P. Dorff, and R. Kullnig, *J. Org. Chem.*, **54**, 2936 (1989).
4. E. Wünsch, "Blockierung und Schutz der α-Amino-Function," in *Methoden der Organischen Chemie (Houben-Weyl)*, Georg Thieme Verlag, Stuttgart, 1974, Vol. 15/1, pp. 164–203, 250–264.

5. J. W. Barton, "Protection of N–H Bonds and NR_3," in *Protective Groups in Organic Chemistry*, J. F. W. McOmie, Ed., Plenum Press, New York and London, 1973, pp. 46–56.
6. Y. Wolman, "Protection of the Amino Group," in *The Chemistry of the Amino Group*, S. Patai, Ed., Wiley-Interscience, New York, 1968, Vol. 4, pp. 669–682.
7. E. Gross and J. Meienhofer, Eds., *The Peptides: Analysis: Synthesis, Biology*, Vol. 3: *Protection of Functional Groups in Peptide Synthesis*, Academic Press, New York, 1981.

Formamide: R_2NCHO (Chart 9)

Formation

1. 98% HCO_2H, Ac_2O, 25°, 1 h, 78–90% yield.[1,2] The use of formic acetic anhydride for esterification and amide formation has been reviewed.[3]
2. HCO_2H, DCC, Pyr, 0°, 4 h, 87–90% yield.[4] These conditions produce *N*-formyl derivatives of *t*-butyl amino acid esters with a minimum of racemization.
3. HCO_2H, $EtN{=}C{=}N(CH_2)_3NMe_2 \cdot HCl$, 0°, 15 min; then *N*-methylmorpholine, 5°, 20 h, 65–96% yield. This method can be used with amine hydrochlorides.[5]
4. C_6F_5OCHO, $CHCl_3$, rt, 5–30 min, 85–99% yield.[6]
5. 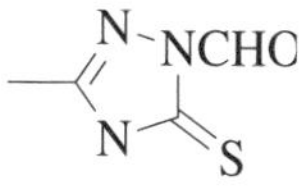 [structure: N-NCHO, N, S] This reagent also formylates alcohols in the presence of added base.[7]
6. *t*-$BuMe_2SiCl$, DMAP, Et_3N, DMF, 35–60°, 65–85% yield.[8]
7. DMF, silica gel, heat, 5 h, 100% yield,[9] or DMF, ZrO, heat, 5 h, 92% yield.[10]
8. HCO_2Et, heat.[11]
9. Triethyl orthoformate, 50–100% yield.[12]
10. HCO_2CH_2CN, CH_2Cl_2, rt, 12 h, 62–97% yield.[13]
11. Vinyl formates readily react with amines, alcohols, and phenols to give the formamide or ester.[14]

Cleavage

1. HCl, H_2O, dioxane, 25°, 48 h, or reflux, 1 h, 80–95% yield.[1]
2. Hydrazine, EtOH, 60°, 4 h, 60–80% yield.[15]
3. H_2/Pd–C, THF, HCl, 25°, 5–7 h, quant.[16]
4. 15% H_2O_2, H_2O, 60°, 2 h, 80% yield.[17]
5. AcCl, $PhCH_2OH$, 20°, 24 h, or 60°, 3 h, good yields.[18]

6. $h\nu$, 254 nm, CH_3CN, 100% yield.[19]
7. NaOH, H_2O, reflux, 18 h, 85% yield.[20]

1. J. C. Sheehan and D.-D. H. Yang, *J. Am. Chem. Soc.*, **80**, 1154 (1958).
2. E. G. E. Jahngen, Jr., and E. F. Rossomando, *Synth. Commun.*, **12**, 601 (1982).
3. P. Strazzolini, A. G. Giumanini, and S. Cauci, *Tetrahedron*, **46**, 1018 (1990).
4. M. Waki and J. Meienhofer, *J. Org. Chem.*, **42**, 2019 (1977).
5. F. M. F. Chen and N. L. Benoiton, *Synthesis*, 709 (1979).
6. L. Kisfaludy and L. Ötvös, Jr., *Synthesis*, 510 (1987).
7. H. Yazawa and S. Goto, *Tetrahedron Lett.*, **26**, 3703 (1985).
8. S. W. Djuric, *J. Org. Chem.*, **49**, 1311 (1984).
9. European Patent to Japan Tobacco, Inc., EP. 271093, June 15, 1988.
10. K. Takahashi, M. Shibagaki, and H. Matsushita, *Agric. Biol. Chem.*, **52**, 853 (1988).
11. H. Schmidhammer and A. Brossi, *Can. J. Chem.*, **60**, 3055 (1982).
12. T. Chancellor and C. Morton, *Synthesis*, 1023 (1994).
13. W. Duczek, J. Deutsch, S. Vieth, and H.-J. Niclas, *Synthesis*, 37 (1996).
14. M. Neveux, C. Bruneau, and P. H. Dixneuf, *J. Chem. Soc., Perkin Trans. 1*, 1197 (1991).
15. R. Geiger and W. Siedel, *Chem. Ber.*, **101**, 3386 (1968).
16. G. Losse and D. Nadolski, *J. Prakt. Chem.*, **24**, 118 (1964).
17. G. Losse and W. Zönnchen, *Justus Liebigs Ann. Chem.*, **636**, 140 (1960).
18. J. O. Thomas, *Tetrahedron Lett.*, 335 (1967).
19. B. K. Barnett and T. D. Roberts, *J. Chem. Soc., Chem. Commun.*, 758 (1972).
20. U. Hengartner, A. D. Batcho, J. F. Blount, W. Leimgruber, M. E. Larscheid, and J. W. Scott, *J. Org. Chem.*, **44**, 3748 (1979).

Acetamide: R_2NAc (Chart 9)

Formation

The simplest method for acetamide preparation involves reaction of the amine with acetic anhydride or acetyl chloride with or without added base. The following are some other methods:

1. C_6F_5OAc, DMF, 25°, 1–12 h, 78–91% yield.[1] These conditions allow selective acylation of amines in the presence of alcohols. If triethylamine is used in place of DMF, alcohols are also acylated (75–85% yield).
2. Ac_2O, 18-crown-6, Et_3N, 98% yield.[2] The crown ether forms a complex with a primary amine, thus allowing selective acylation of a secondary amine.
3. $AcOC_6H_4$-p-NO_2, pH 11.[3]

$$H_2N\text{-}CH(CO_2H)(CH_2)_3NH_2 \xrightarrow[\text{pH 11}]{O_2N\text{-}C_6H_4\text{-}OAc} H_2N\text{-}CH(CO_2H)(CH_2)_3NHAc$$

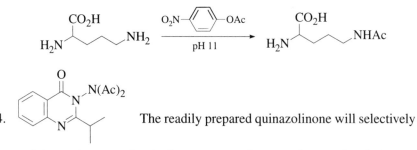

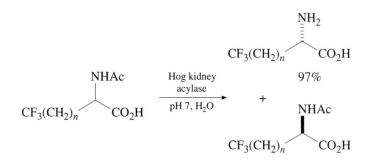

4. [quinazolinone structure: 3-N(Ac)$_2$-2-isopropyl-4(3H)-quinazolinone] The readily prepared quinazolinone will selectively acylate a primary amine in the presence of a secondary amine, but more uniquely, it will selectively acylate a pyrrolidine over a piperidine with 3:1 selectivity and dimethylamine over diethylamine with 9:1 selectivity.[4]
5. Vinyl acetate or diethyl carbonate, $Cp_2Sm(THF)_2$, 80–99% yield.[5] Aniline fails to react under these conditions.
6. *N,N*-Diacetyl-2-trifluoromethylaniline, organic solvents, 3–24 h, rt or reflux, 54–99% yield. Acylation selectivity is a very sensitive function of steric effects; this reagent selectively acylates pyrrolidine over piperidine (15:1). It is more selective than the diacetylaminoquinazolinones.[6]
7. Ac_2NOMe selectively acylates a primary amine of a spermidine.[7]

Cleavage

1. 1.2 *N* HCl, reflux, 9 h, 61–77% yield.[8]
2. 85% Hydrazine, 70°, 15 h, 68% yield.[9]
3. $Et_3O^+BF_4^-$, CH_2Cl_2, 25°, 1–2 h, 90% yield, then aq. $NaHCO_3$, satisfactory yields.[10]
4. Hog kidney acylase, pH 7, H_2O, 36°, 35 h.[11,12] In this case, deprotection also proceeds with resolution, since only one enantiomer is cleaved.

$$CF_3(CH_2)_nCH(NHAc)CO_2H \xrightarrow[\text{pH 7, } H_2O]{\text{Hog kidney acylase}} CF_3(CH_2)_nCH(NH_2)CO_2H \ (97\%) + CF_3(CH_2)_nCH(NHAc)CO_2H$$

5. Enzymatic hydrolysis with Aspergillis Acylase, pH 8.5, 75% yield.[13]
6. Simple amides that are difficult to cleave can first be converted to a BOC derivative by an exchange process that relies on the reduced electrophilicity of the carbamate as well as its increased steric bulk.[14–16]

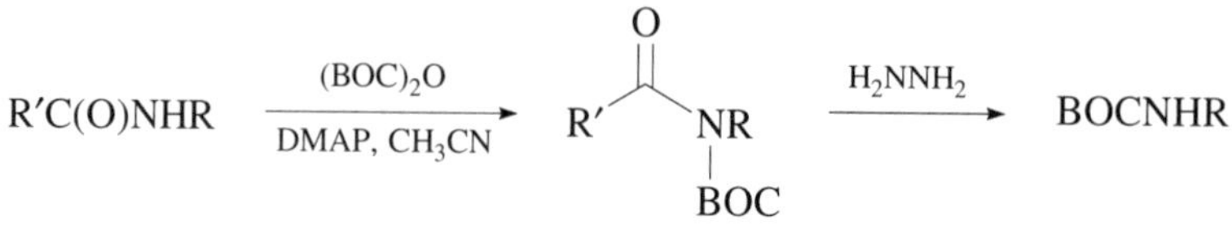

$$R'C(O)NHR \xrightarrow[\text{DMAP, } CH_3CN]{(BOC)_2O} R'C(O)N(BOC)R \xrightarrow{H_2NNH_2} BOCNHR$$

7. Na, BuOH, 120°, 62% yield.[17]
8. Ca, NH_3, DME, EtOH, 4 h, 96% yield.[18]
9. For most common amides cleavage is quite difficult, but in the case of an aziridine (which exhibited reduced participation significantly in amide resonance because of the nonplanar amide moiety[19]) hydrolysis is much simpler, as shown in the following illustration:[20]

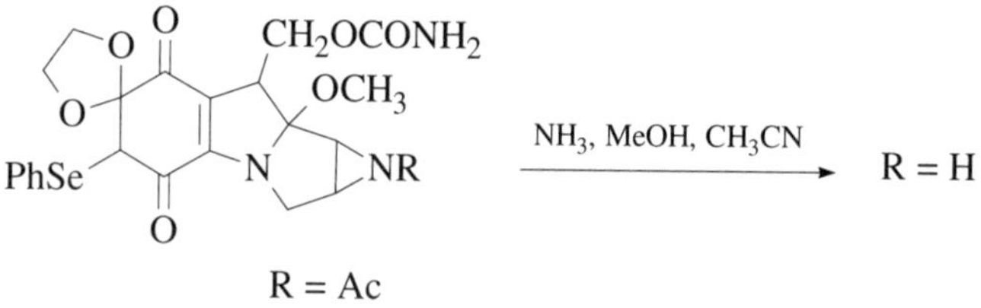

$$\text{R = Ac} \xrightarrow{NH_3,\ MeOH,\ CH_3CN} \text{R = H}$$

   As the lone pair and the carbonyl group become more orthogonal, reducing the level of resonance, the rate of amide hydrolysis increases.[21,22]
10. In a diacetamide, one acetamide is easily cleaved by hydrolysis with NaOMe and MeOH,[23] which is consistent with the use of *N,N*-diacetylaminoquinazoline,[4] 2-trifluoromethyl-*N,N*-diacetylaniline,[6] and *N*-methoxydiacetamide as amidating agents.[7]

1. I. Kisfaludy, T. Mohacsi, M. Low, and F. Drexler, *J. Org. Chem.*, **44**, 654 (1979).
2. A. G. M. Barrett and J. C. A. Lana, *J. Chem. Soc., Chem. Commun.*, 471 (1978).
3. F. Kanai, T. Kaneko, H. Morishima, K. Isshiki, T. Takita, T. Takeuchi, and H. Umezawa, *J. Antibiotics (Tokyo)*, **38**, 39 (1985).
4. R. S. Atkinson, E. Barker, and M. J. Sutcliffe, *J. Chem. Soc., Chem. Commun.*, 1051 (1996).
5. Y. Ishii, M. Takeno, Y. Kawasaki, A. Muromachi, Y. Nishiyama, and S. Sakaguchi, *J. Org. Chem.*, **61**, 3088 (1996).
6. Y. Murakami, K. Kondo, K. Miki, Y. Akiyama, T. Watanabe, and Y. Yokoyma, *Tetrahedron Lett.*, **38**, 3751 (1997).
7. Y. Kikugawa, K. Mitsui, and T. Sakamoto, *Tetrahedron Lett.*, **31**, 243 (1990).
8. G. A. Dilbeck, L. Field, A. A. Gallo, and R. J. Gargiulo, *J. Org. Chem.*, **43**, 4593 (1978).
9. D. D. Keith, J. A. Tortora, and R. Yang, *J. Org. Chem.*, **43**, 3711 (1978).
10. S. Hanessian, *Tetrahedron Lett.*, 1549 (1967).

11. T. Tsushima, K. Kawada, S. Ishihara, N. Uchida, O. Shiratori, J. Higaki, and M. Hirata, *Tetrahedron*, **44**, 5375 (1988).
12. R. J. Cox, W. A. Sherwin, L. K. P. Lam, and J. C. Vederas, *J. Am. Chem. Soc.*, **118**, 7449 (1996).
13. F. VanMiddlesworth, C. Dufresne, F. E. Wincott, R. T. Mosley, and K. E. Wilson, *Tetrahedron Lett.*, **33**, 297 (1992).
14. L. Grehn, K. Gunnarsson, and U. Ragnarsson, *J. Chem. Soc., Chem. Commun.*, 1317 (1985).
15. L. Grehn, K. Gunnarsson, and U. Ragnarsson, *Acta Chem. Scand. Ser. B.*, **B40**, 745 (1986).
16. D. J. Kempf, *Tetrahedron Lett.*, **30**, 2029 (1989).
17. M. Obayashi and M. Schlosser, *Chem. Lett.*, 1715 (1985).
18. A. J. Pearson and D. C. Rees, *J. Am. Chem. Soc.*, **104**, 1118 (1982).
19. G. V. Shustov, G. K. Kadorkina, S. V. Varlamov, A. V. Kachanov, R. G. Kostyanovsky, and A. Rauk, *J. Am. Chem. Soc.*, **114**, 1616 (1992).
20. H. Arai and M. Kasai, *J. Org. Chem.*, **58**, 4151 (1993).
21. A. J. Bennet, Q.-P. Wang, H. Stebocka-Tilk, V. Somayaji, and R. C. Brown, *J. Am. Chem. Soc.*, **112**, 6383 (1990).
22. A. J. Kirby, I. V. Konarov, P. D. Wothers, and N. Feeder, *Angew. Chem., Int. Ed. Engl.*, **37**, 785 (1998).
23. J. C. Castro-Palomino and R. R. Schmidt, *Tetrahedron Lett.*, **36**, 6871 (1995).

Chloroacetamide: R_2NCOCH_2Cl (Chart 9)

Monochloroacetamides are cleaved (by "assisted removal") by reagents that contain two nucleophilic groups (e.g., *o*-phenylenediamine,[1] thiourea,[2,3] 1-piperidinethiocarboxamide,[4] 3-nitropyridine-2-thione,[5] 2-aminothiophenol[6]):

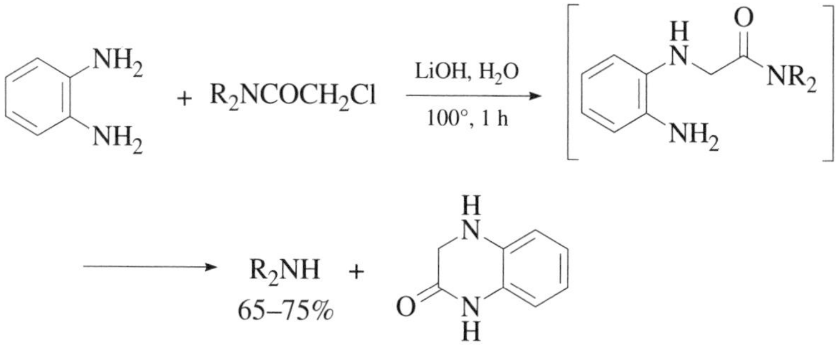

The chloroacetamide can also be cleaved by first converting it to the pyridiniumacetamide (Pyr, 90°, 1 h, 70–90% yield) followed by mild basic hydrolysis (0.1 *N* NaOH, 25°)[7] or by acidic hydrolysis (4 *N* HCl, 60°, 8 h).[8]

1. R. W. Holley and A. D. Holley, *J. Am. Chem. Soc.*, **74**, 3069 (1952).
2. M. Masaki, T. Kitahara, H. Kurita, and M. Ohta, *J. Am. Chem. Soc.*, **90**, 4508 (1968).

3. J. E. Baldwin, M. Otsuka, and P. M. Wallace, *Tetrahedron*, **42**, 3097 (1986); T. Allmendinger, G. Rihs, and H. Wetter, *Helv. Chim. Acta*, **71**, 395 (1988).
4. W. Steglich and H.-G. Batz, *Angew Chem., Int. Ed. Engl.*, **10**, 75 (1971).
5. K. Undheim and P. E. Fjeldstad, *J. Chem. Soc., Perkin Trans. I*, 829 (1973).
6. J. D. Glass, M. Pelzig, and C. S. Pande, *Int. J. Pept. Protein Res.*, **13**, 28 (1979).
7. C. H. Gaozza, B. C. Cieri, and S. Lamdan, *J. Heterocycl. Chem.*, **8**, 1079 (1971).
8. B. E. Ledford and E. M. Carreira, *J. Am. Chem. Soc.*, **117**, 11811 (1995).

Trichloroacetamide: $R_2NCOCCl_3$ (Chart 9)

Formation

1. $Cl_3CCOCCl_3$, hexane, 65°, 90 min, 65–97% yield.[1]

Cleavage

1. $NaBH_4$, EtOH, 1 h, 65% yield.[2]

1. B. Sukornick, *Org. Synth., Collect. Vol. V*, 1074 (1973).
2. F. Weygand and E. Frauendorfer, *Chem. Ber.*, **103**, 2437 (1970).

Trifluoroacetamide (TFA): R_2NCOCF_3 (Chart 9)

Formation

1. CF_3CO_2Et, Et_3N, CH_3OH, 25°, 15–45 h, 75–95% yield.[1] A polymeric version of this approach has also been developed.[2] This reagent selectively protects a primary amine in the presence of a secondary amine.[3]
2. $(CF_3CO)_2O$, 18-crown-6, Et_3N, 95% yield.[4] Complex formation of a primary amine with 18-crown-6 allows selective acylation of a secondary amine.
3. CF_3COO–succinimidyl, CH_2Cl_2, 0°, 85% yield.[5] These conditions selectively introduced the TFA group onto a primary amine in the presence of a secondary amine.
4. (Trifluoroacetyl)benzotriazole, THF, rt, 85–100% yield.[6] This reagent can be used to prepare trifluoroacetate esters.
5. TFA, Ph_3P, NBS, CH_2Cl_2, Pyr, 81–99% yield. This methodology can be used for the preparation of other amides from simple carboxylic acids.[7]
6. $(CF_3CO)_2O$, Pyr, CH_2Cl_2.[8]
7. $CF_3CO_2C_6F_5$, Pyr, DMF, 52–92% yield.[9]
8. CF_3CO_2H, MeOH, rt, 97% yield.[10]
9. 2-Trifluoroacetoxypyridine, ether, 20°, 30 min, 93% yield.[11]

Cleavage

The trifluoroacetamide is one of the more easily cleaved amides.

1. K_2CO_3 or Na_2CO_3, MeOH, H_2O, rt, 55–95% yield.[5,12] Note that the trifluoroacetamide has been cleaved in the presence of a methyl ester, which illustrates the ease of hydrolysis of the trifluoroacetamide group.[13]

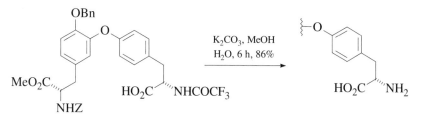

OBn O MeO_2C NHZ HO_2C $NHCOCF_3$ → K_2CO_3, MeOH, H_2O, 6 h, 86% → O HO_2C NH_2

2. NH_3, MeOH.[14]
3. Lewatit 500, MeOH, 96% yield.[10]
4. By phase-transfer hydrolysis: KOH, $Et_3BnN^+Br^-$, H_2O, CH_2Cl_2 or ether, 75–95% yield.[15]
5. 0.2 *N* $Ba(OH)_2$, CH_3OH, 25°, 2 h, 79% yield.[16]
6. $NaBH_4$, EtOH, 20°, or 60°, 1 h, 60–100% yield.[17,18]
7. $PhCH_2N^+Et_3OH^-$, CH_2Cl_2, –40°, 48 h.[8]
8. HCl, MeOH, 65°, 24 h.[19]

1. T. J. Curphey, *J. Org. Chem.*, **44**, 2805 (1979).
2. P. I. Svirskaya, C. C. Leznoff, and M. Steinman, *J. Org. Chem.*, **52**, 1362 (1987).
3. D. Xu, K. Prasad, O. Repic, and T. J. Blacklock, *Tetrahedron Lett.*, **36**, 7357 (1995); M. C. O'Sullivan and D. M. Dalrymple, *Tetrahedron Lett.*, **36**, 3451 (1995).
4. A. G. M. Barrett and J. C. A. Lana, *J. Chem. Soc., Chem. Commun.*, 471 (1978).
5. R. J. Bergeron and J. J. McManis, *J. Org. Chem.*, **53**, 3108 (1988).
6. A. R. Katritzky, B. Yang, and D. Semenzin, *J. Org. Chem.*, **62**, 726 (1997).
7. P. Froyen, *Tetrahedron Lett.*, **38**, 5359 (1997).
8. S. G. Pyne, *Tetrahedron Lett.*, **28**, 4737 (1987).
9. L. M. Gayo and M. J. Suto, *Tetrahedron Lett.*, **37**, 4915 (1996).
10. L. F. Tietze, C. Schneider, and A. Grote, *Chem.—Eur. J.*, **2**, 139 (1996).
11. T. Keumi, M. Shimada, T. Morita, and H. Kitajima, *Bull. Chem. Soc. Jpn.*, **63**, 2252 (1990).
12. H. Newman, *J. Org. Chem.*, **30**, 1287 (1965); J. Quick and C. Meltz, *J. Org. Chem.*, **44**, 573 (1979); M. A. Schwartz, B. F. Rose, and B. Vishnuvajjala, *J. Am. Chem. Soc.*, **95**, 612 (1973).
13. D. L. Boger and D. Yohannes, *J. Org. Chem.*, **54**, 2498 (1989).

14. M. Imazawa and F. Eckstein, *J. Org. Chem.*, **44**, 2039 (1979).
15. D. Albanese, F. Corcella, D. Landini, A. Maia, and M. Penso, *J. Chem. Soc., Perkin Trans. 1*, 247 (1997).
16. F. Weygand and W. Swodenk, *Chem. Ber.*, **90**, 639 (1957).
17. F. Weygand and E. Frauendorfer, *Chem. Ber.*, **103**, 2437 (1970).
18. Z. H. Kudzin, P. Lyzwa, J. Luczak, and G. Andrijewski, *Synthesis*, 44 (1997).
19. S. B. King and B. Ganem, *J. Am. Chem. Soc.*, **116**, 562 (1994).

Phenylacetamide: $R_2NCOCH_2C_6H_5$

This amide, readily formed from an amine and the anhydride[1] or enzymatically using penicillin amidase,[2] is readily cleaved by penicillin acylase (pH 8.1, *N*-methylpyrrolidone, 65–95% yield). This deprotection procedure works on peptides,[3–5] phosphorylated peptides,[6] and oligonucleotides,[7] as well as on non-peptide substrates.[8,9] The deprotection of racemic phenylacetamides with penicillin acylase can result in enantiomer enrichment of the cleaved amine and the remaining amide.[10] An immobilized form of penicillin G acylase has been developed.[11]

1. A. R. Jacobson, A. N. Makris, and L. M. Sayre, *J. Org. Chem.*, **52**, 2592 (1987).
2. C. Ebert, L. Gardossi, and P. Linda, *Tetrahedron Lett.*, **37**, 9377 (1996).
3. R. Didziapetris, B. Drabnig, V. Schellenberger, H.-D. Jakubke, and V. Svedas, *FEBS Lett.*, **287**, 31 (1991).
4. For a review on the use of enzymes in protective group manipulation in peptide chemistry, see J. D. Glass in *The Peptides*, S. Undenfriend and J. Meienhofer, Eds., Academic Press, 1987, Vol. 9, p. 167.
5. D. Sebastian, A. Heuser, S. Schulze, and H. Waldmann, *Synthesis*, 1098 (1997).
6. H. Waldmann, A. Heuser, and S. Schulze, *Tetrahedron Lett.*, **37**, 8725 (1996).
7. H. Waldmann and A. Reidel, *Angew. Chem., Int. Ed. Engl.*, **36**, 647 (1997).
8. H. Waldmann, *Tetrahedron Lett.*, **29**, 1131 (1988) and references cited therein.
9. V. M. Vrudhula, H. P. Svensson, and P. D. Senter, *J. Med. Chem.*, **38**, 1380 (1995).
10. A. L. Margolin, *Tetrahedron Lett.*, **34**, 1239 (1993).
11. T. F. Favino, G. Fronza, C. Fuganti, D. Fuganti, P. Grasselli, and A. Mele, *J. Org. Chem.*, **61**, 8975 (1996).

3-Phenylpropanamide: $R_2NCOCH_2CH_2C_6H_5$ (Chart 9)

A 3-phenylpropanamide, prepared from a nucleoside, is hydrolyzed under mild conditions by α-chymotrypsin (37°, pH 7, 2–12 h).[1]

1. H. S. Sachdev and N. A. Starkovsky, *Tetrahedron Lett.*, 733 (1969).

Pent-4-enamide: CH_2=$CHCH_2CH_2C(O)NR_2$

Formation

1. $(CH_2$=$CHCH_2CH_2CO)_2O$, Pyr, CH_2Cl_2, MeOH, H_2O, 90–99% yield.[1]
2. CH_2=$CHCH_2CH_2CO_2CH_2CN$, 3-methyl-3-pentanol, Subtilisin Carlsberg. These conditions were used to resolve a chiral amine (43% yield, 97% ee).[2]

Cleavage

1. I_2, THF, H_2O, 83–94% yield.[1,2]

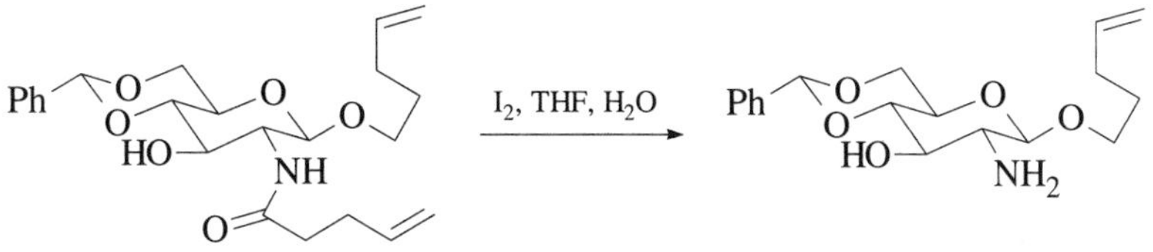

I_2, THF, H_2O

1. R. Madsen, C. Roberts, and B. Fraser-Reid, *J. Org. Chem.*, **60**, 7920 (1995).
2. S. Takayama, W. J. Moree, and C.-H. Wong, *Tetrahedron Lett.*, **37**, 6287 (1996).

Picolinamide: R_2NCO–2-pyridyl (Chart 9)

The picolinamide is prepared in 95% yield from picolinic acid/DCC and an amino acid and is hydrolyzed in 75% yield by aqueous $Cu(OAc)_2$[1] or by electrochemical reduction (sulfuric acid, MeOH, 20°, 20–94% yield).[2]

3-Pyridylcarboxamide: R_2NCO–3-pyridyl

The 3-pyridylcarboxamide, prepared from the anhydride (Pyr, 99% yield), is cleaved (55–86% yield) by basic hydrolysis (0.5 *M* NaOH, rt) after quaternization of the pyridine nitrogen with methyl iodide.[3]

1. A. K. Koul, B. Prashad, J. M. Bachhawat, N. S. Ramegowda, and N. K. Mathur, *Synth. Commun.*, **2**, 383 (1972).
2. N. Auzeil, G. Dutruc-Rosset, and M. Largeron, *Tetrahedron Lett.*, **38**, 2283 (1997).
3. S. Ushida, *Chem. Lett.*, 59 (1989).

***N*-Benzoylphenylalanyl Derivative:** $R_2NCOCH(NHCOC_6H_5)CH_2C_6H_5$

This derivative, prepared from an amino acid and the acyl azide, is selectively cleaved in 80% yield by chymotrypsin.[1]

1. R. W. Holley, *J. Am. Chem. Soc.*, **77**, 2552 (1955).

Benzamide: $R_2NCOC_6H_5$ (Chart 9)

Formation

1. PhCOCl, Pyr, 0°, high yield.[1]
2. PhCOCN, CH_2Cl_2, −10°, 92% yield.[2] This reagent readily acylates amines in the presence of alcohols.
3. $PhCOCF(CF_3)_2$, $Me_2NCH_2CH_2NMe_2$ (TMEDA), 25°, 30 min, high yield.[3]
4. 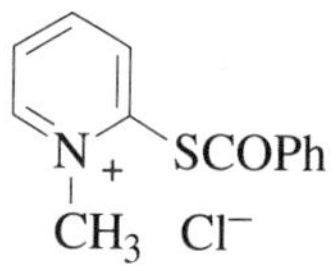 aq. $NaHCO_3$ or aq. NaOH, good yields.[4]
5. $(PhCO)_2NOCH_3$, DMF, H_2O, or dioxane, 3–26 h, 66–89%.[5] The reagent is selective for primary amines.

Cleavage

1. 6 *N* HCl, reflux, 48 h or HBr, AcOH, 25°, 72 h, 80% yield.[6]
2. $(HF)_n$/Pyr, 25°, 60 min, 100% yield.[7] Polyhydrogen fluoride/pyridine cleaves most of the protective groups used in peptide synthesis.
3. Electrolysis, −2.3 V, $Me_4N^+X^-$, CH_3OH, 70 min, 60–90% yield.[8]
4. $(Me_2CHCH_2)_2AlH$, $PhCH_3$, −78°, 80% yield.[9] Since the *N*-benzoyl group in this substrate could not be removed by hydrolysis, a less selective reductive cleavage with diisobutylaluminum hydride was used.
5. Hydrazine, EtOH, 85% yield.[10] Note that the cleavage of an anilide and a benzoylpyrrole is much more facile than that of a typical aliphatic benzamide.

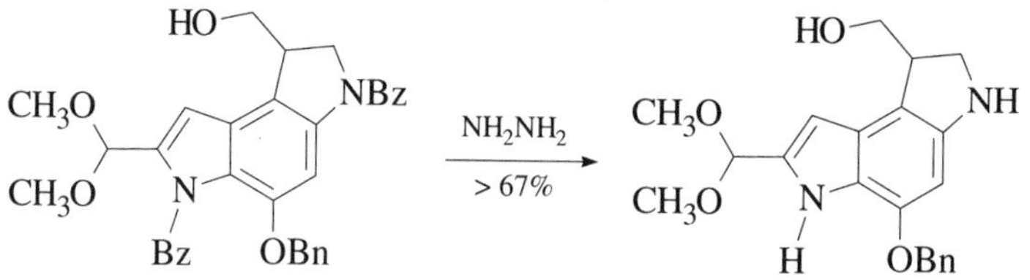

1. E. White, *Org. Synth., Collect Vol. V*, 336 (1973).
2. S.-I. Murahashi, T. Naota, and N. Nakajima, *Tetrahedron Lett.*, **26**, 925 (1985).
3. N. Ishikawa and S. Shin-ya, *Chem. Lett.*, 673 (1976).
4. M. Yamada, Y. Watabe, T. Sakakibara, and R. Sudoh, *J. Chem. Soc., Chem. Commun.*, 179 (1979).
5. Y. Kikugawa, K. Mitsui, T. Sakamoto, M. Kawase, and H. Tamiya, *Tetrahedron Lett.*, **31**, 243 (1990).

6. D. Ben-Ishai, J. Altman, and N. Peled, *Tetrahedron*, **33**, 2715 (1977); P. Hughes and J. Clardy, *J. Org. Chem.*, **53**, 4793 (1988).
7. S. Matsuura, C.-H. Niu, and J. S. Cohen, *J. Chem. Soc., Chem. Commun.*, 451 (1976).
8. L. Horner and H. Neumann, *Chem. Ber.*, **98**, 3462 (1965).
9. J. Gutzwiller and M. Uskokovic, *J. Am. Chem. Soc.*, **92**, 204 (1970).
10. D. L. Boger and K. Machiya, *J. Am. Chem. Soc.*, **114**, 10056 (1992); D. L. Boger, J. A. McKie, T. Nishi, and T. Ogiku, *J. Am. Chem. Soc.*, **119**, 311 (1997).

p-Phenylbenzamide: $R_2NCOC_6H_4$–*p*-C_6H_5

The phenylbenzamide is prepared from the acid chloride in the presence of Et_3N (86% yield) and can be cleaved with 3% Na(Hg) (MeOH, 25°, 4 h, 81% yield).[1] Most amides react only slowly with Na(Hg). Phenylbenzamides are generally crystalline compounds, an aid in their purification.[2]

1. R. B. Woodward and 48 co-workers, *J. Am. Chem. Soc.*, **103**, 3210 (1981).
2. R. M. Scribner, *Tetrahedron Lett.*, 3853 (1976).

Assisted Cleavage

A series of amides has been prepared as protective groups that are cleaved by intramolecular cyclization after activation, by reduction of a nitro group, or by activation by other chemical means. These groups have not found much use. A significant consideration in examining the use of any of these amides is that the nature of the amine will have a substantial effect on the rate of deprotection. Amines, such as those of the nucleobases and many anilines whose basicity is much reduced compared with that of typical primary and secondary aliphatic amines, tend to be cleaved at a much faster rate. Structural effects, such as the "trimethyl lock"[1] effect, exert a considerable influence on the effectiveness of the deprotection. The concept of assisted cleavage is generalized in the following scheme:

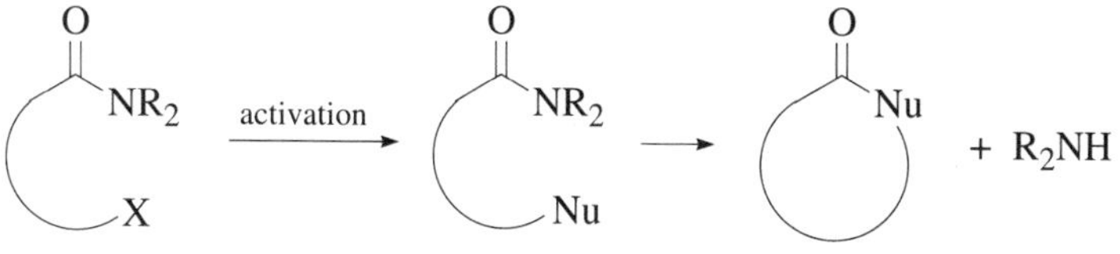

O NR2 X —activation→ O NR2 Nu → O Nu + R_2NH

Amide Cleavage Induced by Nitro Group Reduction

In this series of compounds, any reagent that is capable of reducing a nitro group should be capable of initiating deprotection.

1. ***o*-Nitrophenylacetamide**[2] (Chart 9)

2. ***o*-Nitrophenoxyacetamide**[3] (Chart 9)

3. **3-(*o*-Nitrophenyl)propanamide**[4]

4. **2-Methyl-2-(*o*-nitrophenoxy)propanamide**[2,5] (Chart 9)

5. **3-Methyl-3-nitrobutanamide**[6]

6. ***o*-Nitrocinnamide**[7] (Chart 9)

7. ***o*-Nitrobenzamide**[8]

8. **3-(4-*t*-Butyl-2,6-dinitrophenyl)-2,2-dimethylpropanamide**[9]

Amide Cleavage Induced by Release of an Alcohol

In this series of amides, hydrolysis or aminolysis of a simple ester, cleavage of a silyl group, a *cis/trans* isomerization, or reduction of a quinone to a hydroquinone exposes an alcohol that then induces deprotection by intramolecular addition to the amide carbonyl.

1. ***o*-(Benzoyloxymethyl)benzamide (BMB).**[10] Cleavage is initiated by ester hydrolysis.

2. **2-(Acetoxymethyl)benzamide (AMB).**[11,12] Cleavage is initiated by ester hydrolysis.

3. **2-[(*t*-Butyldiphenylsiloxy)methyl]benzamide (SiOMB).**[13,14] Cleavage is induced by silyl ether cleavage.

4. **3-(3′,6′-Dioxo-2′,4′,5′-trimethylcyclohexa-1′,4′-diene)-3,3-dimethylpropionamide (Q).** The application of this well-known acid [3-(3′,6′-dioxo-2′,4′,5′-trimethylcyclohexa-1′,4′-diene)-3,3-dimethylpropionic acid] to protection of the amino function for peptide synthesis has been examined. Reduction of the quinone with sodium dithionite causes rapid "trimethyl lock"[1]-facilitated ring closure with release of the amine.[15,16]

5. ***o*-Hydroxy-*trans*-cinnamide.** Cleavage is initiated by photolysis at 365 nm, which favors trans- to cis-isomerization. Intramolecular lactonization releases the amine.[17]

Amides Cleaved by Other Chemical Reactions

1. **2-Methyl-2-(*o*-phenylazophenoxy)propanamide**[18] (Chart 9). Cleaved by reduction.

2. 4-Chlorobutanamide[19] (Chart 9). Cleaved by cyclization induced with silver ion.

3. Acetoacetamide[20] (Chart 9). Cleaved with hydrazine.

4. 3-(*p*-Hydroxyphenyl)propanamide[21] (Chart 9). Cleaved by oxidation with NBS.

5. (*N*′-Dithiobenzyloxycarbonylamino)acetamide[22] Cleaved by TFA induced cyclization.

6. *N*-Acetylmethionine Derivative[23] (Chart 9). Cleaved by alkylation of the thioether with iodoacetamide followed by cyclization.

1. B. Wang, M. G. Nicolaou, S. Liu, and R. T. Borchardt, *Biorg. Chem.*, **24**, 39 (1996).
2. F. Cuiban, *Rev. Roum. Chim.*, **18**, 449 (1973).
3. R. W. Holley and A. D. Holley, *J. Am. Chem. Soc.*, **74**, 3069 (1952).
4. I. D. Entwistle, *Tetrahedron Lett.*, 555 (1979).
5. C. A. Panetta, *J. Org. Chem.*, **34**, 2773 (1969).
6. T.-L. Ho, *Synth. Commun.*, **10**, 469 (1980).
7. G. Just and G. Rosebery, *Synth. Commun.*, **3**, 447 (1973).
8. A. K. Koul, J. M. Bachhawat, B. Rashad, N. S. Ramegowda, A. K. Mathur, and N. K. Mathur, *Tetrahedron*, **29**, 625 (1973).
9. F. Johnson, I. Habus, R. G. Gentles, S. Shibutani, H.-C. Lee, C. R. Iden, and R. Rieger, *J. Am. Chem. Soc.*, **114**, 4923 (1992).
10. B. F. Cain, *J. Org. Chem.*, **41**, 2029 (1976).
11. W. H. A. Kuijpers, J. Huskens, and C. A. A. van Boeckel, *Tetrahedron Lett.*, **31**, 6729 (1990).
12. W. H. A. Kuijpers, E. Kuyl-Yeheskiely, J. H. van Boom, and C. A. A. van Boeckel, *Nucleic Acids Res.*, **21**, 3493 (1993).
13. C. M. Dreef-Tromp, E. M. A. Van Dam, H. Van den Elst, J. E. Van den Boogaart, G. A. Van der Marel, and J. H. Van Boom, *Recl. Trav. Chim. Pays-Bas*, **110**, 378 (1991).
14. C. M. Dreef-Tromp, E. M. A. Van Dam, H. Van den Elst, G. A. Van der Marel, and J. H. Van Boom, *Nucleic Acids Res.*, **18**, 6491 (1990).
15. L. A. Carpino and F. Nowshad, *Tetrahedron Lett.*, **34**, 7009 (1993).
16. B. Wang, S. Liu, and R.T. Borchardt, *J. Org. Chem.*, **60**, 539 (1995).
17. B. Wang and A. Zheng, *Chem. Pharm. Bull.*, **45**, 715 (1997).
18. C. A. Panetta and A.-U. Rahman, *J. Org. Chem.*, **36**, 2250 (1971).
19. H. Peter, M. Brugger, J. Schreiber, and A. Eschenmoser, *Helv. Chim. Acta*, **46**, 577 (1963).
20. C. Di Bello, F. Filira, V. Giormani, and F. D′Angeli, *J. Chem. Soc. C*, 350 (1969).
21. G. L. Schmir and L. A. Cohen, *J. Am. Chem. Soc.*, **83**, 723 (1961); L. Farber and L. A. Cohen, *Biochemistry*, **5**, 1027 (1966).

22. F. E. Roberts, *Tetrahedron Lett.*, 325 (1979).
23. W. B. Lawson, E. Gross, C. M. Foltz, and B. Witkop, *J. Am. Chem. Soc.*, **84**, 1715 (1962).

A number of protective groups have been developed that simultaneously protect both sites of a primary nitrogen. These may prove to be useful in cases where acidic hydrogens on nitrogen cannot be tolerated.

4,5-Diphenyl-3-oxazolin-2-one (Chart 8):

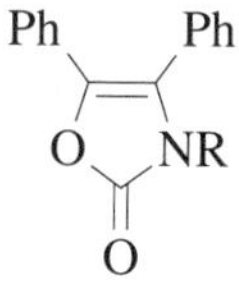

Ph Ph O NR O

Formation[1]

1. 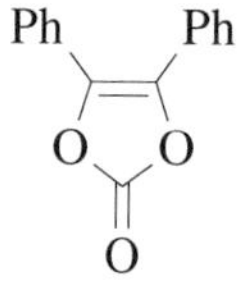 Ph Ph O O O — DMF, 0.5 h; CF_3CO_2H, 67–85% yield.

Cleavage

1. H_2/Pd–C, aq. HCl, 25°, 12 h, quantitative.[1,2]
2. Na/NH_3, 75–85% yield.[1]
3. *m*-$ClC_6H_4CO_3H$, then water, 70% yield.[1]
4. O_2, photolysis, –30°, then Zn, AcOH, quant.[3]

Cyclic Imide Derivatives

***N*-Phthalimide** (Chart 9):

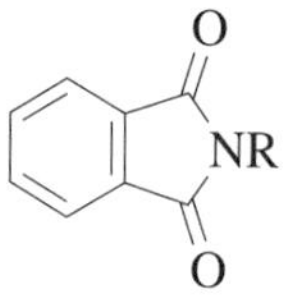

O NR O

Formation

1. Phthalic anhydride, $CHCl_3$, 70°, 4 h, 85–93% yield.[4]
2. Phthalic anhydride, $TaCl_5$-SiO_2, 5 min, 88–92% yield.[5]
3. Phthalic anhydride, HMDS, rt, 1 h, then reflux with $ZnBr_2$ 1 h, 94% yield.[6]
4. *o*-$(CH_3OOC)C_6H_4COCl$, Et_3N, THF, 0°, 2 h, 90–95% yield.[7]
5. Phthalimide–NCO_2Et, aq. Na_2CO_3, 25°, 10–15 min, 85–95% yield.[8] This

reagent can be used to protect selectively primary amines in the presence of secondary amines.[9]

6. 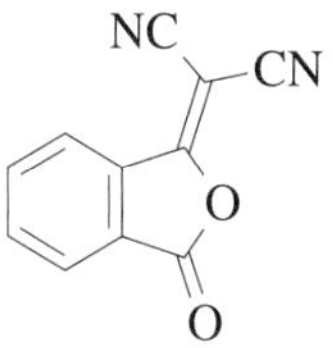 (2-(dicyanomethylene)phthalide: NC, CN, O, O) 65–91% yield.[10]

7. 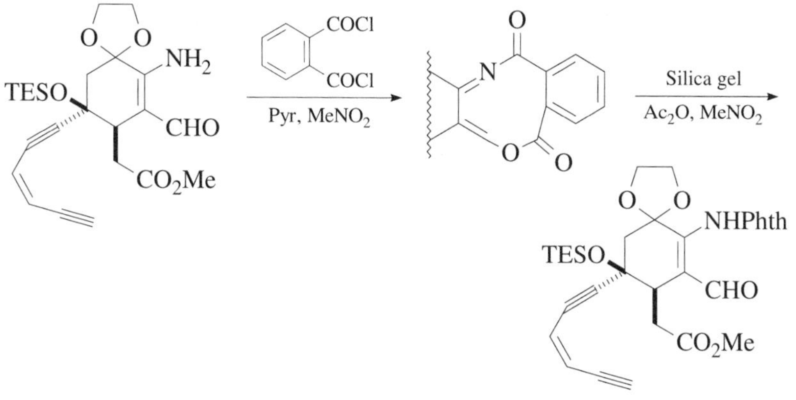 TESO, O, O, NH_2, CHO, CO_2Me → (phthaloyl chloride: COCl, COCl; Pyr, $MeNO_2$) → N, O, O, O → (Silica gel; Ac_2O, $MeNO_2$) → TESO, O, O, NHPhth, CHO, CO_2Me

Ref. 11

Cleavage

1. Hydrazine, EtOH, 25°, 12 h; H_3O^+, 76% yield.[4,12]
2. $PhNHNH_2$, *n*-Bu_3N, reflux, 2 h, 83% yield.[13]
3. $Na_2S{\cdot}H_2O$, H_2O, THF, 68–90% yield; DCC(—H_2O), 67–97% yield; hydrazine; dil. HCl, 55–95% yield.[14] This method is used to cleave *N*-phthalimido penicillins; hydrazine attacks an intermediate phthalisoimide instead of the azetidinone ring.
4. $NaBH_4$, 2-propanol, H_2O (6:1); AcOH, pH 5, 80°, 5–8 h.[15,16] This method was reported to be superior in cases where hydrazine proved to be inefficient.
5. $MeNH_2$, EtOH, rt, 5 min, then heat, 2.5 h, 89% yield.[17] Butylamine has also been used.[18]
6. (a) Base, H_2O, CH_3CN; (b) 0.2 pH 8 buffer, phthalyl amidase.[19]
7. $Me_2NCH_2CH_2CH_2NH_2$, MeOH, TEA, 5°, 24 h, 60% yield.[20]
8. $HONH_2$, MeONa, MeOH, >72% yield.[21]
9. Hydrazine acetate, MeOH, reflux, >82% yield.[22]
10. $MeNHNH_2$. This reagent was used to prevent diimide formation, which resulted in acetylene reduction.[23]

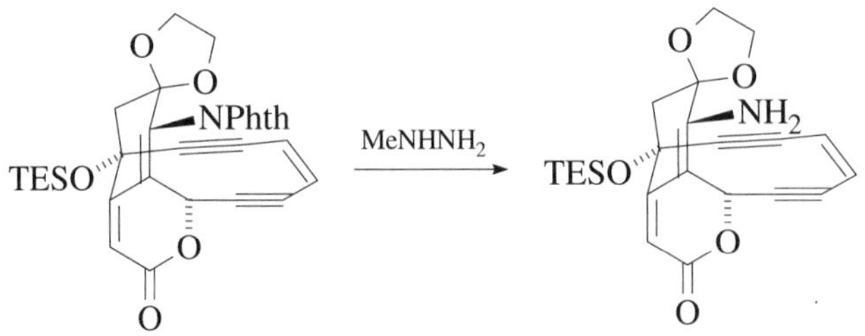

11. The phthalimido group is susceptible to basic reagents and thus must occasionally be protected. This is accomplished by treatment with pyrrolidine to open the ring (>90%), which can be closed by treatment with HF, $B(OH)_3$, THF, H_2O, 73–99% yield.[24]
12. MsOH, HCO_2H.[25]
13. Ethylenediamine, butanol, 90°, 67–96% yield.[26] These conditions were used when heating with butylamine failed to give clean conversions.
14. Diaion WA-20, EtOH, H_2O, 80–90°, 1 h, 87–92% yield.[27]

N-Tetrachlorophthalimide (TCP)

The use of this group was developed to improve the quality and mildness of the cleavage reaction in the synthesis of complex amino sugars.[28] The **dichlorophthalimide** group has also been examined in this context, but little is known of its advantages or disadvantages.[29]

Formation

1. Tetrachlorophthalic anhydride, microwaves, 90% yield.[30]
2. Tetrachlorophthalic anhydride, TEA; Ac_2O, Pyr.[31]

Cleavage

1. Ethylenediamine, CH_3CN, THF, EtOH, 60°.[30,32] The phthalimide group and *O*-acetate are not cleaved with this reagent.[33]

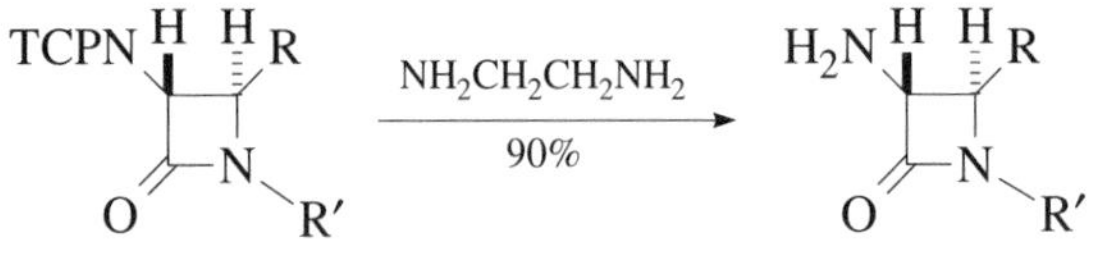

2. Polymer–$NH(CH_2)_xNH_2$, (x = 2, 4, 6), BuOH, 85°, 92–96% yield. The polymer-supported amine helps in the final purification of oligosaccharides that have used the TCP group for $–NH_2$ protection.[34]
3. (a) $NaBH_4$ (b) AcOH, >60–80% yield.[35,36]

4-Nitro-*N*-phthalimide

The 4-nitro-*N*-phthalimide, prepared by heating the amine with the anhydride to 130° for 30 min, is cleaved with $MeNHCH_2CH_2NH_2$ (71–92% yield). These cleavage conditions were compatible with cephalosporins, where the phthalimide was removed in 92% yield at −50° in 30 min.[37]

N-Dithiasuccinimide (Dts–NR) (Chart 9):

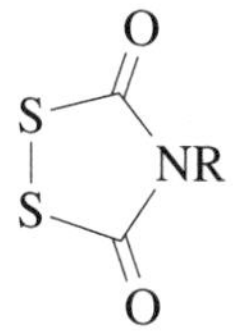

Formation

1. $EtOCS_2CH_2CO_2H$ or $EtOCS_2CSOEt$; ClSCOCl, 0–45°, 70–90% yield.[38–40]
2. PEG(2000)–$OCS_2CH_2CONH_2$; TMSNH(CO)NHTMS; ClCOSCl.[38]

Cleavage

The Dts group is cleaved by treatment with a thiol and base, e.g., $HOCH_2CH_2SH$, Et_3N, 25°, 5 min, $HSCH_2C(O)NHMe$, Pyr, 5 min.[41] Dithiothreitol (DIPEA, CH_2Cl_2, 87–98% yield) seems to be the most trouble-free method for Dts deprotection.[40b] In the presence of an azide, the Dts group can be removed with $NaBH_4$[42] or with $HSCH_2CH_2CH_2SH$ (DIPEA, CH_2Cl_2, 94% yield).[43] The use of Zn (AcOH, Ac_2O, THF, 80–87% yield) cleaves the Dts group in the presence of the extremely sensitive pentafluorophenyl ester.[40a]

The Dts group, stable to acidic cleavage of *t*-butyl carbamates (12 *N* HCl, AcOH, reflux; HBr, AcOH), to mild base ($NaHCO_3$), and to photolytic cleavage of *o*-nitrobenzyl carbamates, can be used in orthogonal schemes for protection of peptides.[41] Merrifield defines an orthogonal system as a set of completely independent classes of protective groups wherein each class of protective groups can be removed in any order and in the presence of all other classes.[41]

N-2,3-Diphenylmaleimide:

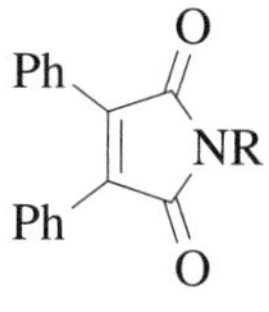

The diphenylmaleimide is prepared from the anhydride, 33–87% yield, and cleaved by hydrazinolysis, 65–75% yield. It is stable to acid (HBr, AcOH, 48 h) and to mercuric cyanide. It is colored and easily located during chromatography, and has been prepared to protect steroidal amines and amino sugars.[41]

N-2,5-Dimethylpyrrole:

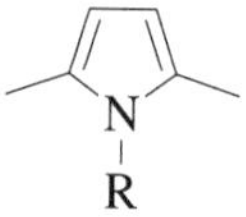

Formation

1. $CH_3C(O)CH_2CH_2C(O)CH_3$, AcOH, 88% yield.[44,45]

Cleavage

1. $H_2NOH{\cdot}HCl$, EtOH, H_2O, 73% yield.[44,46]
2. Ozone, –78°, MeOH; $NaBH_4$; HCl, MeOH, H_2O.[47,48]

N-2,5-Bis(triisopropylsiloxy)pyrrole (BIPSOP)

These derivatives are formed from the succinimide by silylation (TIPSOTf, TEA, CH_2Cl_2, 0°–rt, 68–87% yield). Deprotection is achieved by hydrolysis of the silyl groups followed by succinimide cleavage with hydrazine (EtOH, H_2O, reflux, 72% yield).[49] The succinimides were prepared by heating the amine with succinic anhydride followed by ring closure with AcCl or Ac_2O/NaOAc. They may also be prepared by reacting succinic anhydride with the amine and HMDS followed by ring closure with $ZnBr_2$ (reflux, 1 h).[6]

N-1,1,4,4-Tetramethyldisilylazacyclopentane adduct (STABASE)

Formation/Cleavage[50–53]

1. 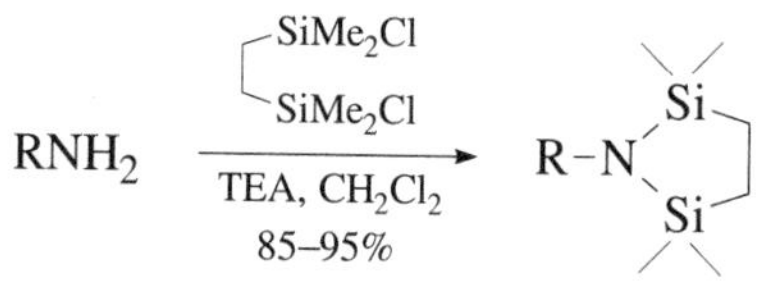 RNH_2 → (ClMe$_2$SiCH$_2$CH$_2$SiMe$_2$Cl; TEA, CH_2Cl_2, 85–95%) → R–N(SiMe$_2$CH$_2$CH$_2$SiMe$_2$)

2. $Me_2NSi(Me)_2CH_2CH_2Si(Me)_2NMe_2$, ZnI_2, 140°, 8 h, 72% yield.[54] The amine adducts are stable to the following reagents: *n*-BuLi (THF, –25°), *s*-BuLi (Et_2O, –25°); lithium diisopropylamide; saturated aqueous ammonium chloride; H_2O; MeOH; 2 *N* $NaHCO_3$; pyridinium dichromate, CH_2Cl_2; $KF{\cdot}2H_2O$, THF, H_2O; saturated aqueous sodium dihydrogen phosphate. The derivative is not stable to strong acid or base; to pyridinium chlorochromate, CH_2Cl_2; or to $NaBH_4$, EtOH.

1,1,3,3-Tetramethyl-1,3-disilaisoindoline (Benzostabase, BSB):

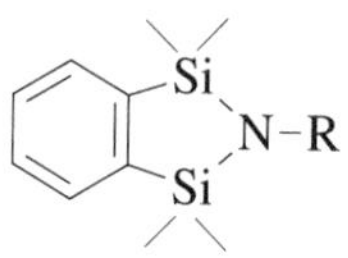

Formation

1. 1,2-Bisdimethylsilylbenzene, $Rh(Ph_3P)_3Cl$, toluene, 120°, 71–92% yield.[55]
2. 1,2-Bisdimethylsilylbenzene, CsF, HMPA, 71–92% yield.[55]
3. 1,2-Bisdimethylsilylbenzene, $PdCl_2$, toluene, rt, 69–87% yield.[56]
4. 1,2-Bis(diethylsilyl)benzene, $PdCl_2$ or CsF, DMPU, 50–86% yield. The **tetraethyl analogue (TEDI)** was found to be more stable to acid than was the tetramethyl derivative. Exposure of BnNBSB and BnNTEDI to a phosphate buffer at pH 2.5 resulted in a cleavage half-life of <0.4 min for the BSB derivative and a half-life of ~30 min for the TEDI analogue. The TEDI group can also be introduced with the dibromide and TEA.[57]
5. A difluorinated analogue was found to be somewhat more stable to acid than was the BSB derivative but overall it showed no major advantage to the original Benzostabase.[58]

Cleavage

1. Cleavage is achieved by simple acid hydrolysis. The Benzostabase group is reasonably stable to base (KOH, MeOH).[58]

N-5-Substituted 1,3-Dimethyl-1,3,5-triazacyclohexan-2-one and *N*-5-Substituted 1,3-Dibenzyl-1,3,5-triazacyclohexan-2-one

Formation[59]

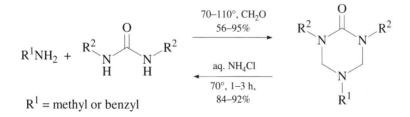

R^1NH_2 + R^2NH–CO–NHR^2

70–110°, CH_2O
56–95%

aq. NH_4Cl
70°, 1–3 h,
84–92%

R^1 = methyl or benzyl

Cleavage

1. Aqueous NH_4Cl, 70°, 1–3 h, 84–92% yield.[59]
2. $HN(CH_2CH_2OH)_3$.[60]
3. 1 *N* HCl, 23°, >84% yield.[61]

The triazone is stable to $LiAlH_4$; PtO_2/H_2, EtOH, 48 h; Pd-black/H_2,THF, 1 h; *n*-BuLi, THF,–40°, 30 min; PhMgBr, THF,–78°, 30 min; Wittig reagents; DIBAL, THF, rt, 3 h; $LiBH_4$, THF, 40°; acylation, silylation, and anhydrous acids ($TiCl_4$,

CH_2Cl_2, −78°, 30 min; TsOH, toluene, 12 h; neat CF_3CO_2H, 15 min). Extended exposure (48 h) of a triazone to neat CF_3CO_2H results in cleavage.[62]

1-Substituted 3,5-Dinitro-4-pyridone

Formation/Cleavage[63]

$$\mathrm{RNH_2} \underset{\substack{\mathrm{MeNH_2,\ PrNH_2,\ or\ hexylNH_2} \\ \mathrm{H_2O,\ Pyr,\ 0.5\text{–}2\ h} \\ 83\text{–}97\%}}{\overset{\substack{\text{4-}\mathrm{NO_2\text{–}C_6H_4\text{–}N}(\text{3,5-dinitro-4-pyridone}) \\ \mathrm{H_2O,\ Pyr,\ rt,\ 2\text{–}24\ h} \\ 72\text{–}100\%}}{\rightleftarrows}} \text{1-R-3,5-dinitro-4-pyridone}$$

1,3,5-Dioxazine: (N—R)

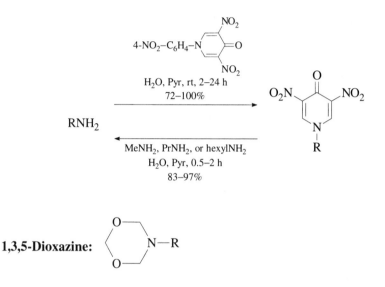

The reaction of a cepham primary amine with 20 eq. of 37% formalin produces the dioxazine in 75% yield. The dioxazine is sufficiently stable to allow the formation of Wittig reagents and to carry out an olefination with formaldehyde. Treatment of the dioxazine with 6 *N* HCl in CH_2Cl_2 releases the amine in excellent yield.[64]

1. J. C. Sheehan and F. S. Guziec, *J. Org. Chem.*, **38**, 3034 (1973).
2. S. V. Pansare and J. C. Vederas, *J. Org. Chem.*, **52**, 4804 (1987); U. Sreenivasan, R. K. Mishra, and R. L. Johnson, *J. Med. Chem.*, **36**, 256 (1993).
3. F. S. Guziec, Jr., and E. T. Tewes, *J. Heterocycl. Chem.*, **17**, 1807 (1980).
4. T. Sasaki, K. Minamoto, and H. Itoh, *J. Org. Chem.*, **43**, 2320 (1978).
5. S. Chandrasekhar, M. Takhi, and G. Uma, *Tetrahedron Lett.*, **38**, 8089 (1997).
6. P. Y. Reddy, S. Kondo, T. Toru, and Y. Ueno, *J. Org. Chem.*, **62**, 2652 (1997).
7. D. A. Hoogwater, D. N. Reinhoudt, T. S. Lie, J. J. Gunneweg, and H. C. Beyerman, *Recl. Trav. Chim. Pays-Bas*, **92**, 819 (1973).
8. G. H. L. Nefkins, G. I. Tesser, and R. J. F. Nivard, *Recl. Trav. Chim. Pays-Bas*, **79**, 688 (1960); C. R. McArthur, P. M. Worster, and A. U. Okon, *Synth. Commum.*, **13**, 311 (1983).

9. G. Sosnovsky and J. Lukszo, *Z. Naturforsch. B*, **41B**, 122 (1986).
10. J. A. Moore and J.-H. Kim, *Tetrahedron Lett.*, **32**, 3449 (1991).
11. K. C. Nicolaou, *Angew. Chem., Int. Ed. Engl.*, **32**, 1377 (1993).
12. For a mechanistic study of this reaction, see M. N. Khan, *J. Org. Chem.*, **60**, 4536 (1995).
13. I. Schumann and R. A. Boissonnas, *Helv. Chim. Acta*, **35**, 2235 (1952).
14. S. Kukolja and S. R. Lammert, *J. Am. Chem. Soc.*, **97**, 5582 (1975).
15. F. Dasgupta and P. J. Garegg, *J. Carbohydr. Chem.*, **7**, 701 (1988).
16. J. O. Osby, M. G. Martin, and B. Ganem, *Tetrahedron Lett.*, **25**, 2093 (1984).
17. M. S. Motawia, J. Wengel, A. E. S. Abdel-Megid, and E. B. Pedersen, *Synthesis*, 384 (1989).
18. P. L. Durette, E. P. Meitzner, and T. Y. Shen, *Tetrahedron Lett.*, 4013 (1979).
19. C. A. Costello, A. J. Kreuzman, and M. J. Zmijewski, *Tetrahedron Lett.*, **37**, 7469 (1996).
20. T. Kamiya, M. Hashimoto, O. Nakaguchi, and T. Oku, *Tetrahedron*, **35**, 323 (1979).
21. D. R. Mootoo and B. Fraser-Reid, *Tetrahedron Lett.*, **30**, 2363 (1989).
22. H. H. Lee, D. A. Schwartz, J. F Harris, J. P. Carver, and J. J. Krepinsky, *Can. J. Chem.*, **64**, 1912 (1986).
23. A. L. Smith, C.-K. Hwang, E. Pitsinos, G. R. Scarlato, and K. C. Nicolaou, *J. Am. Chem. Soc.*, **114**, 3134 (1992).
24. B. Astleford and L. O. Weigel, *Tetrahedron Lett.*, **32**, 3301 (1991).
25. S. Kotha, D. Anglos, and A. Kuki, *Tetrahedron Lett.*, **33**, 1569 (1992).
26. O. Kanie, S. C. Crawley, M. M. Palcic, and O. Hindsgaul, *Carbohydr. Res.*, **243**, 139 (1993).
27. M.Kuriyama, Y. Inoue, and K. Kitagawa, *Synthesis*, 735 (1990).
28. For a review of the use of TCP in amino sugar synthesis, see J. Debenham, R. Rodebaugh and B. Fraser-Reid, *Liebigs Ann./Recl.*, 791 (1997).
29. H. Shimizu, Y. Ito, Y. Matsuzaki, H. Iijima, and T. Ogawa, *Biosci., Biotech., Biochem.*, **60**, 73 (1996).
30. A. K. Bose, M. Jayaraman, A. Okawa, S. S. Bari, E. W. Robb, and M. S. Manhas, *Tetrahedron Lett.*, **37**, 6989 (1996).
31. J. S. Debenham, S. D. Debenham, and B. Fraser-Reid, *Bioorg. Med. Chem.*, **4**, 1909 (1996).
32. J. S. Debenham, R. Rodebaugh, and B. Frasier-Reid, *J. Org. Chem.*, **61**, 6478 (1996).
33. J. S. Debenham, R. Madsen, C. Roberts, and B. Fraser-Reid, *J. Am. Chem. Soc.*, **117**, 3302 (1995).
34. P. Stangier and O. Hindsgaul, *Synlett*, 179 (1996).
35. B. A. Roe, C. G. Boojamra, J. L. Griggs, and C. R. Bertozzi, *J. Org. Chem.*, **61**, 6442 (1996).
36. J. C. Castro-Palomino and R. R. Schmidt, *Tetrahedron Lett.*, **36**, 5343 (1995).
37. H. Tsubouchi, K. Tsuji, and H. Ishikawa, *Synlett*, 63 (1994).
38. S. Zalipsky, F. Albericio, U. Slomczynska, and G. Barany, *Int. J. Pept. Protein Res.*, **30**, 748 (1987).

39. U. Zehavi, *J. Org. Chem.*, **42**, 2819 (1977).
40. For an application in glucosamine chemistry, see (a) K. J. Jensen, P. R. Hansen, D. Venugopal, and G. Barany, *J. Am. Chem. Soc.*, **118**, 3148 (1996); (b) E. Meinjohanns, M. Meldal, H. Paulsen, and K. Bock, *J. Chem. Soc., Perkin Trans. 1*, 405 (1995).
41. G. Barany and R. B. Merrifield, *J. Am. Chem. Soc.*, **99**, 7363 (1977); *idem*, **102**, 3084 (1980).
42. I. Christiansen-Brams, M. Meldal, and K. Bock, *J. Chem. Soc., Perkin Trans. 1*, 1461 (1993).
43. E. Meinjohanns, M. Meldal, T. Jensen, O. Werdelin, L. Galli-Stampino, S. Mouritsen, and K. Bock, *J. Chem. Soc., Perkin Trans. 1*, 871 (1997).
44. S. P. Bruekelman, S. E. Leach, G. D. Meakins, and M. D. Tirel, *J. Chem. Soc., Perkin Trans. 1*, 2801 (1984).
45. J. E. Macor, B. L. Chenard, and R. J. Post, *J. Org. Chem.*, **59**, 7496 (1994).
46. S. P. Breukelman, G. D. Meakins, and M. D. Tirel, *J. Chem. Soc., Chem. Commun.*, 800 (1982).
47. C. Kashima, T. Maruyama, Y. Fujioka, and K. Harada, *J. Chem. Soc., Perkin Trans. 1*, 1041 (1989).
48. A. P. Davis and T. J. Egan, *Tetrahedron Lett.*, **33**, 8125 (1992).
49. S. F. Martin and C. Limberakis, *Tetrahedron Lett.*, **38**, 2617 (1997).
50. S. Djuric, J. Venit, and P. Magnus, *Tetrahedron Lett.*, **22**, 1787 (1981).
51. T. Högberg, P. Ström, and U. H. Lindberg, *Acta Chem. Scand., Ser. B.*, **B39**, 414 (1985).
52. T. L. Guggenheim, *Tetrahedron Lett.*, **25**, 1253 (1984).
53. M. J. Sofia, P. K. Chakravarty, and J. A. Katzenellenbogen, *J. Org. Chem.*, **48**, 3318 (1983).
54. K. Deshayes, R. D. Broene, I. Chao, C. B. Knobler, and F. Diederich, *J. Org. Chem.*, **56**, 6787 (1991).
55. R. P. Bonar-Law, A. P. Davis, and B. J. Dorgan, *Tetrahedron Lett.*, **31**, 6721 (1990). *idem*, *Tetrahedron*, **49**, 9855 (1993).
56. R. P. Boner-Law, A. P. Davis, B. J. Dorgan, M. T. Reetz, and A. Wehrsig, *Tetrahedron Lett.*, **31**, 6725 (1990).
57. A. P. Davis and P. J. Gallagher, *Tetrahedron Lett.*, **36**, 3269 (1995).
58. F. Cavelier-Frontin, R. Jacquier, J. Paladino, and J. Verducci, *Tetrahedron*, **47**, 9807 (1991).
59. S. Knapp, J. J. Hale, M. Bastos, and F. S. Gibson, *Tetrahedron Lett.*, **31**, 2109 (1990).
60. S. Knapp and J. J. Hale, *J. Org. Chem.*, **58**, 2650 (1993).
61. S. R. Angle, J. M. Fevig, S. D. Knight, R. W. Marguis, Jr., and L. E. Overman, *J. Am. Chem. Soc.*, **115**, 3966 (1993).
62. S. Knapp, J. J. Hale, M. Bastos, A. Molina, and K. Y. Chen, *J. Org. Chem.*, **57**, 6239 (1992).
63. E. Matsumura, H. Kobayashi, T. Nishikawa, M. Ariga, Y. Tohda, and T. Kawashima, *Bull. Chem. Soc. Jpn.*, **57**, 1961 (1984); E. Matsumura, M. Ariga, Y. Tohda, and T. Kawashima, *Tetrahedron Lett.*, **22**, 757 (1981).
64. Y. Katsura and M. Aratani, *Tetrahedron Lett.*, **35**, 9601 (1994).

SPECIAL –NH PROTECTIVE GROUPS

N-Alkyl and *N*-Aryl Amines

N-Methylamine: CH_3NR_2

Formation

1. Methylamines are commonly formed by reacting the amine with a methylating agent such as MeI or dimethyl sulfate.
2. Preparation from an amine and $TMSCHN_2$ (HBF_4, CH_2Cl_2, H_2O) has also been explored.
3. For primary aromatic amines: dimethyl carbonate, Y-Zeolite, 130–150°, 72–93% yields.[1]
4. HCHO, HCO_2H, 5° then reflux, 12 h, 91% yield.[2,3]

Cleavage

1. The cleavage of a methylamine can be accomplished photochemically in the presence of an electron acceptor such as 9,10-dicyanoanthracene.[4]

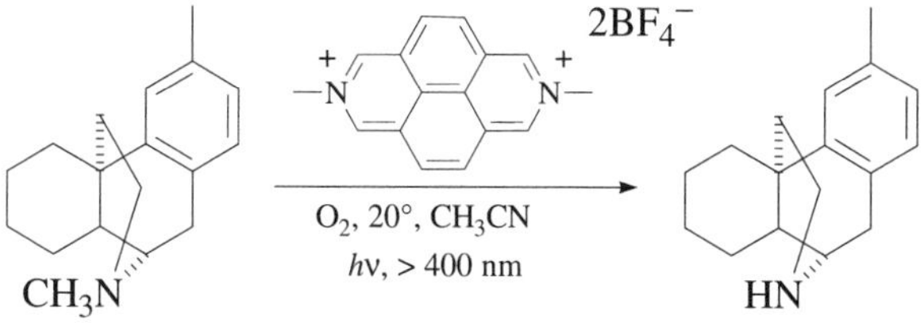

CH_3N … $\xrightarrow[O_2,\ 20°,\ CH_3CN;\ h\nu,\ >400\ nm]{2BF_4^-}$ … HN

2. Photolysis with visible light, DAP^{+2}; TMSCN. The photochemical reaction generates an iminium ion that is trapped with cyanide.[5]
3. CH_2=CHOCOCl, K_2CO_3, CH_2Cl_2.[6] The *N*-methyl group of a tertiary amine is converted to a vinyl carbamate that is easily hydrolyzed.
4. I_2, CaO, THF, MeOH. A dimethylaniline is converted to a monomethylaniline.[7]
5. CS_2, MeI, THF, 6 h, 30°, 97% yield. *N*-Methylpiperidine is converted to a dithiocarbamate.[8]
6. 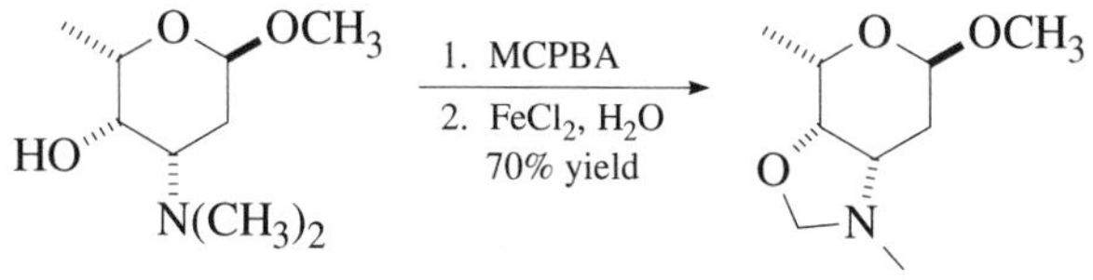 HO … OCH_3 … $N(CH_3)_2$ $\xrightarrow[2.\ FeCl_2,\ H_2O\ \ 70\%\ yield]{1.\ MCPBA}$ O … OCH_3 … N

Ref. 9

7. *t*-BuOOH, $RuCl_2(Ph_3P)_2$, benzene, rt, 3 h, 83% yield. The methyl group is

converted to t-BuOOCH$_2$NR$_2$, which can then be hydrolyzed, releasing the secondary amine.[10] The oxidation of amines has been reviewed.[11]

8. PhSeH, 160°, 5 days, 68% yield.[12]
9. RuCl$_3$, H$_2$O$_2$, MeOH, 55–80% yield.[13] These conditions convert the methyl to a MOM group that can be removed by hydrolysis.

1. M. Selva, A. Bomben, and P. Tundo, *J. Chem. Soc., Perkin Trans. 1*, 1041 (1997).
2. G. Chelucci, M. Falorni, and G. Giacomelli, *Synthesis*, 1121 (1990).
3. For a review of the Leukart reaction, see M. L. Moore, *Org. React.*, **5**, 301 (1949).
4. J. Santamaria, R. Ouchabane, and J. Rigaudy, *Tetrahedron Lett.*, **30**, 2927 (1989).
5. J. Santamaria, M. T. Kaddachi, and J. Rigaudy, *Tetrahedron Lett.*, **31**, 4735 (1990).
6. J. R. Ferguson, K. W. Lumbard, F. Scheinmann, A. V. Stachulski, P. Stjernlöf, and S. Sundell, *Tetrahedron Lett.*, **36**, 8867 (1995); R. A. Olofson, R. C. Schnur, L. Bunes, and J. P. Pepe, *ibid.*, 1567 (1977).
7. K. Acosta, J. W. Cessac, P. N. Rao, and H. K. Kim, *J. Chem. Soc., Chem. Commun.*, 1985 (1994).
8. M. D. Pujol and G. Guillaumet, *Synth. Commun.*, **22**, 1231 (1992)
9. J. P. Gesson, J. C. Jacquesy, and M. Mondon, *Synlett*, 669 (1990).
10. S.-I. Murahashi, T. Naota, and K. Yonemura, *J. Am. Chem. Soc.*, **110**, 8256 (1988).
11. S.-I. Murahashi, *Angew. Chem., Int. Ed. Engl.*, **34**, 2443 (1995).
12. R. P. Polniaszek and L. W. Dillard, *J. Org. Chem.*, **57**, 4103 (1992).
13. S.-I. Murahashi, T. Naota, N. Miyaguchi, and T. Nakato, *Tetrahedron Lett.*, **33**, 6991 (1992).

N-t-Butylamine: $(CH_3)_3CNR_2$

The t-butyl group can be cleaved from a cyclopropylamine upon prolonged heating in acid (H_3O^+, reflux, 3–5 days).[1]

1. N. De Kimpe, P. Sulmon, and P. Brunet, *J. Org. Chem.*, **55**, 5777 (1990).

N-Allylamine: CH_2=CHCH$_2$NR$_2$ (Chart 10)

Formation

1. Allyl chloride, Cu(0), Cu(ClO$_4$)$_2$·6H$_2$O, Et$_2$O, 97% yield.[1]
2. AllylOAc, Pd(Ph$_3$P)$_4$, diisopropylamine.[2]
3. Ni(cod)$_2$, Bu$_4$N$^+$PF$_6^-$, dppb, THF, 50°.[3]
4. Allyl bromide, K$_2$CO$_3$, THF, heat, 75% yield.[4] This is a fairly general method that has been used widely for the preparation of allylamines.

Cleavage

1. Isomerization to the enamine (t-BuOK, DMSO), followed by hydrolysis.[5]
2. Rhodium-catalyzed isomerization.[6] Ru(cod)(cot) has been used to convert an allylamine into an enamine.[7]

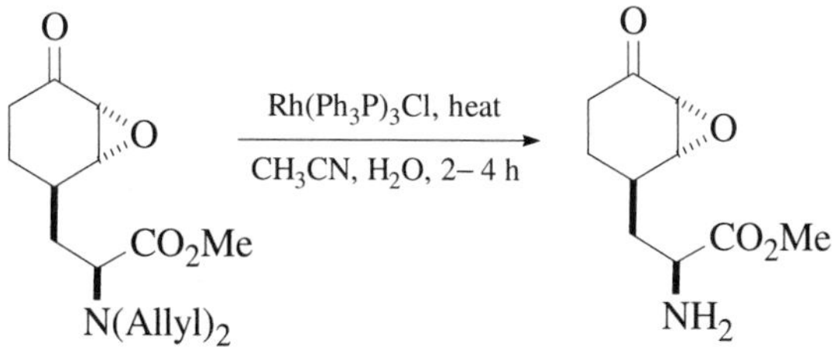

Rh(Ph$_3$P)$_3$Cl, heat

CH$_3$CN, H$_2$O, 2–4 h

CO$_2$Me N(Allyl)$_2$ → CO$_2$Me NH$_2$

In the presence of a nearby hydroxyl, the aminal is formed.[8]

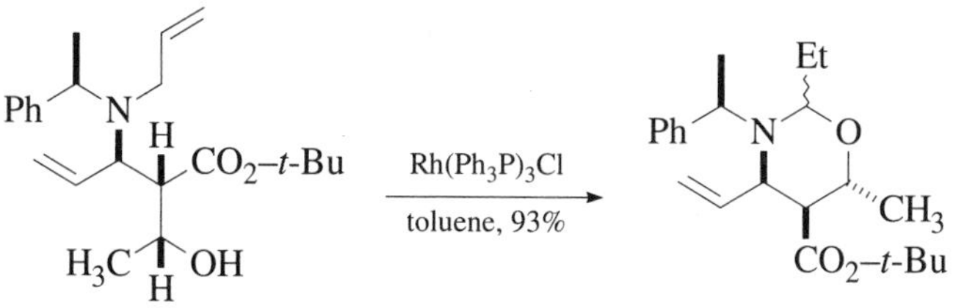

Rh(Ph$_3$P)$_3$Cl

toluene, 93%

Ref. 8

The use of Pd(Ph$_3$P)$_4$ and *N,N*-dimethylbarbituric acid removed the allyl group in 98% yield.

3. Pd(Ph$_3$P)$_4$, and *N,N*-dimethylbarbituric acid, 30°, 1.5–3 h, 91–100% yield.[2]
4. Pd/C, MsOH, H_2O, 82% yield.[9] In certain heterocyclic systems this method failed, but it was successful when MsOH was replaced with $BF_3 \cdot Et_2O$.[10]
5. Pd(Ph$_3$P)$_4$, RSO$_2$Na, CH$_2$Cl$_2$ or THF-MeOH, 70–99% yield. These conditions were shown to be superior to the use of sodium 2-ethylhexanoate. Methallyl, crotyl, allyl and cinnamyl ethers, the alloc group, and allyl esters are all efficiently cleaved by this method.[11]
6. Cp$_2$Zr, then water, 66% yield.[12] *O*-Allyl ethers are cleaved at a faster rate; THP, acetonide, Bn ethers, and benzoates are stable.
7. Pd(dba)DPPB, 2-thiobenzoic acid, THF, 70–100% yield.[13] Tertiary allylamines are cleaved efficiently at 20°, but secondary allylamines require heating to 60° to achieve cleavage. Thus, it is possible to monodeallylate a diallylamine by running the cleavage at 20°.[14]
8. CH$_3$CHCl(OCOCl), then methanolysis with MeOH, 74% yield.[15]

1. J. B. Baruah and A. G. Samuelson, *Tetrahedron*, **47**, 9449 (1991).
2. F. Garro-Helion, A. Merzouk, and F. Guibé, *J. Org. Chem.*, **58**, 6109 (1993).
3. H. Bricout, J.-F. Carpentier, and A. Mortreux, *J. Chem. Soc., Chem. Commun.*, 1863 (1995).
4. G. A. Molander and P. J. Nichols, *J. Org. Chem.*, **61**, 6040 (1996).

5. R. Gigg and R. Conant, *J. Carbohydr. Chem.*, **1**, 331 (1983).
6. B. C. Laguzza and B. Ganem, *Tetrahedron Lett.*, **22**, 1483 (1981).
7. T. Mitsudo, S.-W. Zhang, N. Satake, T. Kondo, and Y. Watanabe, *Tetrahedron Lett.*, **33**, 5533 (1992).
8. S. G. Davies and D. R. Fenwick, *J. Chem. Soc., Chem. Commun.*, 565 (1997).
9. Q. Liu, A. P. Marchington, N. Boden, and C. M. Rayner, *J. Chem. Soc., Perkin Trans. 1*, 511 (1997).
10. S. Jaime-Figueroa, Y. Liu, J. M. Muchowski, and D. G. Putman, *Tetrahedron Lett.*, **39**, 1313 (1998).
11. M. Honda, H. Morita, and I. Nagakura, *J. Org. Chem.*, **62**, 8932 (1997).
12. H. Ito, T. Taguchi, and Y. Hanzawa, *J. Org. Chem.*, **58**, 774 (1993).
13. S. Lemaire-Audoire, M. Savignac, J. P. Genêt, and J.-M. Bernard, *Tetrahedron Lett.*, **36**, 1267 (1995); W. F. Bailey and X.-L. Jiang, *J. Org. Chem.*, **61**, 2596 (1996).
14. S. Lemaire-Audoire, M. Savignac, C. Dupuis, and J. P. Genêt, *Bull. Soc. Chim. Fr.*, **132**, 1157 (1995).
15. P. Magnus and L. S. Thurston, *J. Org. Chem.*, **56**, 1166 (1991).

N-[2-(Trimethylsilyl)ethoxy]methylamine (SEM–NR$_2$): $(CH_3)_3SiCH_2CH_2OCH_2–NR_2$

The SEM derivative of a secondary aromatic amine, prepared from SEMCl (NaH, DMF, 0°, 100% yield), can be cleaved with HCl (EtOH, >88% yield).[1]

1. Z. Zeng and S. C. Zimmerman, *Tetrahedron Lett.*, **29**, 5123 (1988).

N-3-Acetoxypropylamine: $R_2NCH_2CH_2CH_2OCOCH_3$ (Chart 10)

Formation

NH → N–(CH$_2$)$_3$–OAc

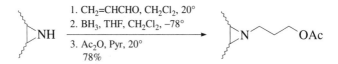

1. CH_2=CHCHO, CH_2Cl_2, 20°
2. BH_3, THF, CH_2Cl_2, –78°
3. Ac_2O, Pyr, 20°

78%

Cleavage

N–(CH$_2$)$_3$–OAc → NH

1. NaOMe, MeOH, 20°
2. DMSO, DCC, CF_3CO_2H, Pyr, 20°
3. $HClO_4$, $PhNMe_2$, 20°

35%

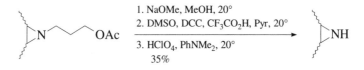

A 3-acetoxypropyl group was used to protect an aziridine –NH group during the synthesis of mitomycins A and C; acetyl, benzoyl, ethoxycarbonyl, and methoxymethyl groups were unsatisfactory.[1]

2. T. Fukuyama, F. Nakatsubo, A. J. Cocuzza, and Y. Kishi, *Tetrahedron Lett.*, 4295 (1977).

N-Cyanomethylamine: $NCCH_2NR_2$

The cyanomethylamine, formed from the amine and bromoacetonitrile (DMF, TEA, 86–96% yield), is cleaved by reduction of the nitrile followed by hydrolysis (PtO_2, H_2, EtOH, 96–98% yield)[1] or with $AgNO_3$/EtOH (92% yield).[2]

1. A. Benarab, S. Boyé, L. Savelon, and G. Guillaumet, *Tetrahedron Lett.*, **34**, 7567 (1993).
2. L. E. Overman and J. Shim, *J. Org. Chem.*, **56**, 5005 (1991).

N-(1-Isopropyl-4-nitro-2-oxo-3-pyrrolin-3-yl)amine

Formation/Cleavage[1]

RNH_2 → (1-*i*-Pr-4-nitro-3-ethoxy-3-pyrrolin-2-one, pH 8–9, 82%) → O_2N/NHR pyrrolinone (N-*i*-Pr)

Cleavage: NH_3, MeOH, 2 h, rt, 56%

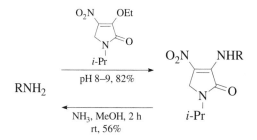

1. P. L. Southwick, G. K. Chin, M. A. Koshute, J. R. Miller, K. E. Niemela, C. A. Siegel, R. T. Nolte, and W. E. Brown, *J. Org. Chem.*, **49**, 1130 (1984).

N-2,4-Dimethoxybenzylamine (Dmb–NR_2): 2,4-$(CH_3O)_2C_6H_3CH_2NR_2$

The dimethoxybenzyl group was used for backbone protection of the pseudopeptides of the form Xaaψ(CH_2N)Gly (Xaa = amino acid). It is introduced by reductive alkylation with the aldehyde and $NaCNBH_3$. Acidolysis with TFMSA in TFA/thioanisole is used to remove it from the amine, but the efficiency is dependent upon the peptide sequence.[1]

1. Y. Sasaki and J. Abe, *Chem. Pharm. Bull.*, **45**, 13 (1997).

2-Azanorbornenes:

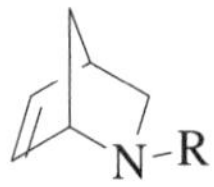

N-R

A primary amine, protected by reaction of the amine with cyclopentadiene and formaldehyde (H_2O, rt, 3 h),[1] is cleaved by trapping cyclopentadiene with *N*-methylmaleimide (H_2O, 2.5 h, 23–50°, 61–97% yield),[2] $CuSO_4$ (EtOH or MeOH, 70°, 74–99%), or Bio-Rad AG 50W-X2 acid ion-exchange resin, 82–98% yield.[3]

1. S. D. Larsen and P. A. Grieco, *J. Am. Chem. Soc.*, **107**, 1768 (1985).
2. P. A. Grieco, D. T. Parker, W. F. Forbare, and R. Ruckle, *J. Am. Chem. Soc.*, **109**, 5859 (1987); P. A. Grieco and B. Bahsas, *J. Org. Chem.*, **52**, 5746 (1987).
3. P. A. Grieco and J. D. Clark, *J. Org. Chem.*, **55**, 2271 (1990).

***N*-2,4-Dinitrophenylamine:** $2,4\text{-}(NO_2)_2C_6H_3NR_2$

The DNP derivative, prepared from 2,4-dinitrofluorobenzene[1–3] is released from the nitrogen with an anionic ion exchange resin.[4,5] When used for histidine protection, the DNP group has been observed to migrate to nearby lysine residues during Fmoc cleavage.[6]

1. P. F. Lloyd and M. Stacey, *Tetrahedron*, **9**, 116 (1960).
2. K. Izawa, T. Ineyama, K. Fujii, and T. Suami, *Carbohydr. Res.*, **205**, 415 (1990).
3. Y. Nakamura, A. Ito, and C.-g. Shin, *Bull. Chem. Soc. Jpn.*, **67**, 2151 (1994).
4. H. Tsunoda, J. Inokuchi, K. Yamagishi, and S. Ogawa, *Liebigs Ann. Chem.*, 279 (1995).
5. T. E. Nicolas and R. W. Franck, *J. Org. Chem.*, **60**, 6904 (1995).
6. J.-C. Gesquiere, J. Najib, T. Letailleur, P. Maes, and A. Tartar, *Tetrahedron Lett.*, **34**, 1921 (1993).

Quaternary Ammonium Salts: $R_3N^+CH_3\ I^-$ (Chart 10)

Formation

1. CH_3I, CH_3OH, $KHCO_3$, 20°, 24 h, 85–95% yield. These salts are generally used to protect tertiary amines during oxidation reactions. Under the conditions cited, quaternary salts are formed from primary, secondary, or tertiary amines, including amino acids, in the presence of hydroxyl or phenol groups.[1]

Cleavage

1. PhSNa, 2-butanone, reflux, 24–36 h, 85% yield.[2]

1. F. C. M. Chen and N. L. Benoiton, *Can. J. Chem.*, **54**, 3310 (1976).
2. M. Shamma, N. C. Deno, and J. F. Remar, *Tetrahedron Lett.*, 1375 (1966).

N-Benzylamine (R_2N–Bn): R_2NCH_2Ph (Chart 10)

Formation

1. BnCl, aq. K_2CO_3, reflux, 30 min; H_2, Pd–C, 77% yield.[1]

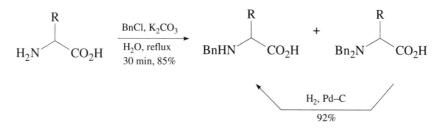

H_2N–CH(R)–CO_2H → (BnCl, K_2CO_3; H_2O, reflux, 30 min, 85%) → BnHN–CH(R)–CO_2H + Bn_2N–CH(R)–CO_2H

Bn_2N–CH(R)–CO_2H → (H_2, Pd–C, 92%) → BnHN–CH(R)–CO_2H

2. BnBr, EtOH, Na_2CO_3, H_2O, CH_2Cl_2, reflux.[2]
3. BnBr, Et_3N, CH_3CN.[3] Examples 2 and 3 produce dibenzyl derivatives from primary amines.
4. $PhCHN_2$, HBF_4, –40°, CH_2Cl_2, 57–68% yield.[4] $SnCl_2 \cdot H_2O$ has been used to catalyze this transformation.[5]
5. PhCHO, 6 *M* HCl in MeOH, MeOH, $NaCNBH_3$.[6]
6. PhCHO, PhSeSePh, $NaBH_4$, EtOH, 1.5 h, 25°, 90% yield.[7]
7. PhCHO, $CHCl_3$, 3Å molecular sieves; $NaBH_4$ alcohol solvent, 66% yield. These conditions were used to protect selectively the terminal ends of a polyamine.[8]

Cleavage

1. Pd–C, 4.4% HCOOH, CH_3OH, 25°, 10 h, 80–90% yield.[3,9] The cleavage of benzylamines with H_2/Pd–C is often very slow.[10] Note in example 1 above that one of the benzyl groups can be selectively removed from a dibenzyl derivative.
2. Pd–C, ROH, HCO_2NH_4,[11] hydrazine or sodium hypophosphite, 42–91% yield.[12] 2-Benzylaminopyridine and benzyladenine were stable to these reaction conditions. Lower yields occurred because of the water solubility of the product, thus hampering isolation. Cyclohexene can be used as a hydrogen source in the transfer hydrogenation.[13]

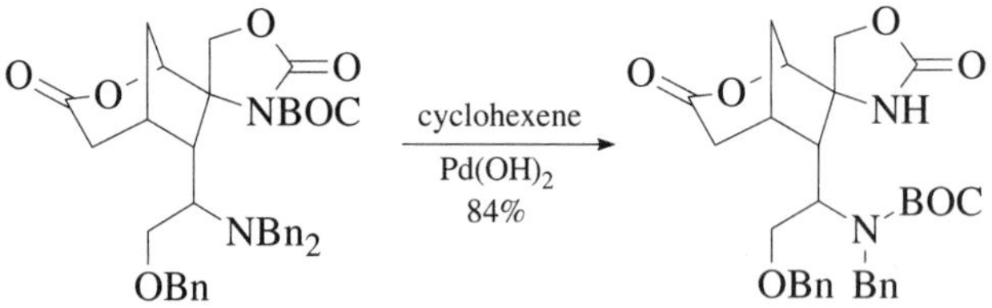

cyclohexene, $Pd(OH)_2$, 84%

Note that the OBn group is retained and that the BOC group has migrated

3. 20% $Pd(OH)_2$, EtOH, H_2, 55 psi, 19 h. A benzyl ether was not cleaved.[14]
4. Na, NH_3, excellent yields.[15]
5. $h\nu$, 405 nm ($CuSO_4$: NH_3 solution filter), CH_3CN, H_2O, 9,10-dicyanoanthracene, 6–10 h, 78–90% yield.[16]

   Benzyl groups, as well as other alkyl groups, can be converted to various carbamates as a variation of the von Braun reaction.[17,18] The carbamates can then be cleaved by conditions that are outlined in the section on carbamates.
6. CCl_3CH_2OCOCl, CH_3CN, 93%.[19]
7. (a) $ClCO_2Et$, CH_2Cl_2, reflux; (b) $PhNEt_2$–BI_3, 25°, 85–89% yield.[20]
8. $Me_3SiCH_2CH_2OCOCl$, THF, −50°, then 25°, overnight, 78–91% yield.[21]
9. α-Chloroethyl chloroformate, NaOH.[22,23] The 4-methoxybenzyl group is selectively cleaved with this reagent, and the benzyl group is cleaved in preference to the 4-nitrobenzyl group.[24]

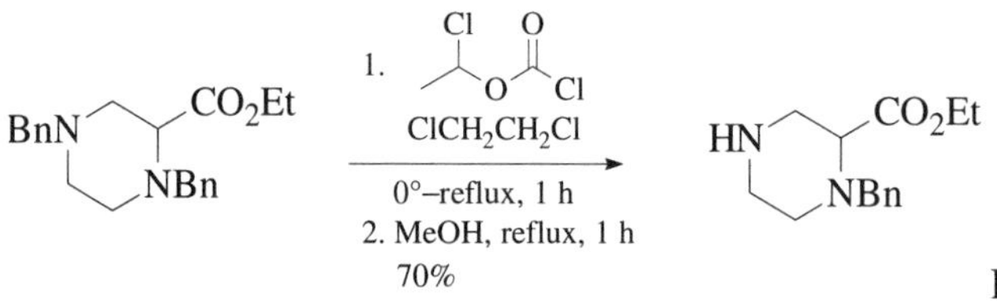

1. $ClCH_2CH_2Cl$, 0°–reflux, 1 h; 2. MeOH, reflux, 1 h; 70%

Ref. 25

10. Vinyl chloroformate is reported to be the best reagent for dealkylation of tertiary alkyl amines.[26]
11. Allyl chloroformate, CH_2Cl_2, >80% yield.[27] In this case, the benzylamine was converted to an alloc carbamate.

    Oxidative methods include the following:
12. RuO_4, NH_3, H_2O, 70% yield.[28]
13. *m*-Chloroperoxybenzoic acid followed by $FeCl_2$, −10°, 6–80% yield.[29]
14. Co(II)L, *t*-BuOOH, DMSO, 40°; H_2O, 90–97% yield.[30]
15. *t*-BuOLi, $CuBr_2$, 20 min, THF, rt, 99%.[31]
16. TPAP, NMO, rt, CH_3CN, 89% yield.[32]

N-4-Methoxybenzylamine (MPM–NR$_2$): $CH_3OC_6H_4CH_2NR_2$

Formation

$MeOC_6H_4CH_2Br$, KI, K_2CO_3, DMF, 92% yield.[33]

Cleavage

1. Pd/C, HCl, MeOH, H_2.[34]
2. $Pd(OH)_2$, H_2. A hydroxamic acid is stable to these conditions.[35]
3. α-Chloroethyl chloroformate, THF, 89–98% yield.[24]
4. DDQ is often used to remove the MPM group from alcohols and can be used to cleave it from an amine, but in the following case overoxidation also occurs:[36]

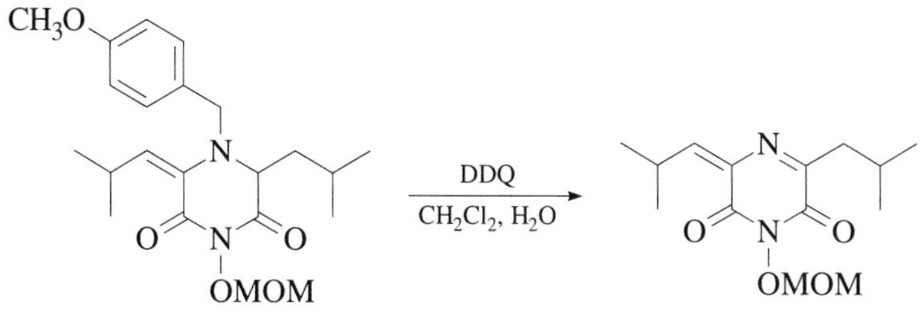

DDQ, CH_2Cl_2, H_2O

N-2,4-Dimethoxybenzylamine (DMPM–NR$_2$): $2,4\text{-}(CH_3O)_2C_6H_3CH_2NR_2$

Cleavage of the DMPM group is achieved by conversion with trifluoroacetic anhydride to the amide, which is then removed with $NaBH_4$/EtOH (93–97% yield).[37]

N-2-Hydroxybenzylamine (HBn–NR$_2$): $2\text{-}(HO)C_6H_4CH_2NR_2$

Amino acids were protected by reductive alkylation with salicylaldehyde ($NaBH_4$, KOH, aq. EtOH). The amine is released by treatment with CF_3SO_3H (TFA, EDT, PhSMe, 2 h, >75% yield).[38]

1. L. Velluz, G. Amiard, and R. Heymes, *Bull. Soc. Chim. Fr.*, 1012 (1954).
2. N. Yamazaki and C. Kibayashi, *J. Am. Chem. Soc.*, **111**, 1397 (1989).
3. B. D. Gray and P. W. Jeffs, *J. Chem. Soc., Chem. Commun.*, 1329 (1987).
4. L. J. Liotta and B. Ganem, *Tetrahedron Lett.*, **30**, 4759 (1989).
5. L. J. Liotta and B. Ganem, *Isr. J. Chem.*, **31**, 215 (1991).
6. C. M. Cain, R. P. C. Cousins, G. Coumbarides, and N. S. Simpkins, *Tetrahedron*, **46**, 523 (1990).

7. A. Guy and J. F. Barbetti, *Synth. Commun.*, **22**, 853 (1992).
8. J. A. Sclafani, M. T. Maranto, T. M. Sisk, and S. A. Van Arman, *J. Org. Chem.*, **61**, 3221 (1996).
9. B. ElAmin, G. M. Anantharamaiah, G. P. Royer, and G. E. Means, *J. Org. Chem.*, **44**, 3442 (1979).
10. W. H. Hartung and R. Simonoff, *Org. React.*, **VII**, 263 (1953).
11. S. Ram and L. D. Spicer, *Tetrahedron Lett.*, **28**, 515 (1987); *idem*, *Synth. Commun.*, **17**, 415 (1987).
12. B. M. Adger, C. O'Farrell, N. J. Lewis, and M. B. Mitchell, *Synthesis*, 53 (1987).
13. A. S. Kende, K. Liu, and K. M. J. Brands, *J. Am. Chem. Soc.*, **117**, 10597 (1995).
14. R. C. Bernotas and R. V. Cube, *Synth. Commun.*, **20**, 1209 (1990).
15. V. du Vigneaud and O. K. Behrens, *J. Biol. Chem.*, **117**, 27 (1937).
16. G. Pandey and K. S. Rani, *Tetrahedron Lett.*, **29**, 4157 (1988).
17. H. A. Hageman, *Org. React.*, **7**, 198 (1953).
18. For a review, see J. H. Cooley and E. J. Evain, *Synthesis*, 1 (1989).
19. V. H. Rawal, R. J. Jones, and M. P. Cava, *J. Org. Chem.*, **52**, 19 (1987).
20. J. V. B. Kanth, C. K. Reddy, and M. Periasamy, *Synth. Commun.*, **24**, 313 (1994).
21. A. L. Campbell, D. R. Pilipauskas, I. K. Khanna, and R. A. Rhodes, *Tetrahedron Lett.*, **28**, 2331 (1987).
22. R. A. Olofson, J. T. Martz, J.-P. Senet, M. Piteau, and T. Malfroot, *J. Org. Chem.*, **49**, 2081 (1984).
23. P. DeShong and D. A. Kell, *Tetrahedron Lett.*, **27**, 3979 (1986).
24. B. V. Yang, D. O'Rourke, and J. Li, *Synlett*, 195 (1993).
25. S. Gubert, C. Braojos, A. Sacristan, and J. A. Ortiz, *Synthesis*, 318 (1991).
26. R. A. Olofson, R. C. Schnur, L. Bunes, and J. P. Pepe, *Tetrahedron Lett.*, 1567 (1977).
27. E. Magnier, Y. Langlois, and C. Mérienne, *Tetrahedron Lett.*, **36**, 9475 (1995).
28. X. Gao and R. A. Jones, *J. Am. Chem. Soc.*, **109**, 1275 (1987).
29. T. Monkovic, H. Wong, and C. Bachand, *Synthesis*, 770 (1985).
30. K. Maruyama, T. Kusukawa, Y. Higuchi, and A. Nishinaga, *Chem. Lett.*, 1093 (1991).
31. J. Yamaguchi and T. Takeda, *Chem. Lett.*, 1933 (1992).
32 A. Goti and M. Romani, *Tetrahedron Lett.*, **35**, 6567 (1994).
33. M. Yamato, Y. Takeuchi, and Y. Ikeda, *Heterocycles*, **26**, 191 (1987).
34. B. M. Trost, M. J. Krische, R. Radinov, and G. Zanoni, *J. Am. Chem. Soc.*, **118**, 6297 (1996).
35. M. Rowley, P. D. Leeson, B. J. Williams, K. W. Moore, and R. Baker, *Tetrahedron*, **48**, 3557 (1992).
36. S. B. Singh, *Tetrahedron Lett.*, **36**, 2009 (1995).
37. P. Nussbaumer, K. Baumann, T. Dechat, and M. Harasek, *Tetrahedron*, **47**, 4591 (1991).
38. T. Johnson and M. Quibell, *Tetrahedron Lett.*, **35**, 463 (1994).

N-(Diphenylmethyl)amine (DPM–NR$_2$): Ph_2CHNR_2

Formation

1. By reduction of a benzophenone imine with $NaCNBH_3$, pH 6, 25°.[1,2]
2. (Diphenylmethyl)amine is used as a convenient protected source of ammonia.[3]

Cleavage

1. Et_3SiH, TFA, 86% yield.[4]
2. Pd/C, cyclohexene, 1 *M* HCl, EtOH, 83% yield.[5] Ammonium formate can also be used as a source of hydrogen.[2]
3. $Pd(OH)_2$, H_2, MeOH, 20 bar, 40°, 8 h, 90% yield.[6]

N-Bis(4-methoxyphenyl)methylamine: $(4\text{-}MeOC_6H_4)_2CHNR_2$ (Chart 10)

This derivative has been used to protect the amines of amino acids [$(4\text{-}MeOC_6H_4)_2CHCl$, Et_3N, 0→20°, 20 h, 67% yield]. It is easily cleaved with 80% AcOH (80°, 5 min, 73% yield).[7]

N-5-Dibenzosuberylamine (DBS–NR$_2$):

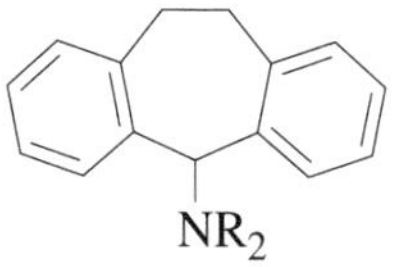

NR$_2$

The dibenzosuberylamine is prepared in quantitative yield from an amine or amino acid and suberyl chloride; this chloride has also been used to protect hydroxyl, thiol, and carboxyl groups. Although the dibenzosuberylamine is stable to 5 *N* HCl/dioxane (22°, 16 h) and to refluxing HBr (1 h), it is completely cleaved by some acids (HCOOH, CH_2Cl_2, 22°, 2 h; CF_3COOH, CH_2Cl_2, 22°, 0.5 h; BBr_3, CH_2Cl_2, 22°, 0.5 h; 4 *N* HBr, AcOH, 22°, 1 h; 60% AcOH, reflux, 1 h) and by reduction (H_2, Pd–C, CH_3OH, 22°, 1 h, 100% cleaved).[8] Hydrogenolysis in the presence of formaldehyde converts the DBS group to a methylamine.[9]

N-Triphenylmethylamine (Tr–NR$_2$): Ph_3CNR_2 (Chart 10)

The bulky triphenylmethyl group has been used to protect a variety of amines such as amino acids, penicillins, and cephalosporins. Esters of *N*-trityl α-amino acids are shielded from hydrolysis and require forcing conditions for cleavage. The α-proton is also shielded from deprotonation, which means that esters elsewhere in the molecule can be selectively deprotonated.

Formation

1. TrCl, Et_3N, 25°, 4 h.[10]
2. TrBr, $CHCl_3$, DMF, rt, 0.5–1 h; Et_3N, rt, 50 min.[11] These conditions also lead to tritylation of carboxyl groups in the amino acids, but the protective

groups can be selectively hydrolyzed. This method was considered to be an improvement over the standard methods of *N*-tritylation of amino acids.

3. (i) Silylation of $-CO_2H$ with Me_3SiCl, Et_3N; (ii) TrCl, Et_3N; (iii) MeOH, 65–92% yield.[12]
4. To effect *N*-tritylation of serine, Me_2SiCl_2 should be used in the silylation step.

Cleavage

1. HCl, acetone, 25°, 3 h, 80% yield.[10]
2. H_2, Pd black, EtOH, 45°, 92% yield.[13] If the hydrogenolysis is performed in the presence of $(BOC)_2O$ or Fmoc–OSu, the released amine is converted to the BOC and Fmoc derivatives *in situ*.[14]
3. Na, NH_3.[15]
4. Hydroxybenzotriazole (HOBT), trifluoroethanol, rt.[16]
5. 1-Hydroxy-7-azabenzotriazole, TMSCl, in trifluoroethanol or TMSCl in trifluoroethanol, quant.[17]
6. 0.2% TFA, 1% H_2O, CH_2Cl_2.[17]

N-[(4-Methoxyphenyl)diphenylmethyl]amine (MMTr–NR_2): $(4\text{-}CH_3O\text{–}C_6H_4)(C_6H_5)_2C\text{–}NR_2$ (Chart 10)

The MMTr derivative is easily prepared from amino acids[18] and is readily cleaved by acid hydrolysis (5% CCl_3CO_2H, 4°, 5 min, 100% yield)[19] or ($CHCl_2CO_2H$, anisole, CH_2Cl_2, rt, 1 h).[18]

1. K. M. Czerwinski, L. Deng, and J. M. Cook, *Tetrahedron Lett.*, **33**, 4721 (1992).
2. E. D. Cox, L. K. Hamaker, J. Li, P. Yu, K. M. Czerwinski, L. Deng, D. W. Bennett, J. M. Cook, W. H. Watson, and M. Krawiec, *J. Org. Chem.*, **62**, 44 (1997).
3. M. E. Jung and Y. M. Choi, *J. Org. Chem.*, **56**, 6729 (1991).
4. W. L. Neumann, M. M. Rogic, and T. J. Dunn, *Tetrahedron Lett.*, **32**, 5865 (1991).
5. L. E. Overman, L. T. Mendelson, and E. J. Jacobsen, *J. Am. Chem. Soc.*, **105**, 6629 (1983).
6. E. Bacqué, J.-M. Paris, and S. Le Bitoux, *Synth. Commun.*, **25**, 803 (1995).
7. R. W. Hanson and H. D. Law, *J. Chem. Soc.*, 7285 (1965).
8. J. Pless, *Helv. Chim. Acta*, **59**, 499 (1976).
9. C. Y. Hong, L. E. Overman, and A. Romero, *Tetrahedron Lett.*, **38**, 8439 (1997).
10. H. E. Applegate, C. M. Cimarusti, J. E. Dolfini, P. T. Funke, W. H. Koster, M. S. Puar, W. A. Slusarchyk, and M. G. Young, *J. Org. Chem.*, **44**, 811 (1979).
11. M. Mutter and R. Hersperger, *Synthesis*, 198 (1989).
12. K. Barlos, D. Papaioannou, and D. Theodoropoulos, *J. Org. Chem.*, **47**, 1324 (1982).
13. L. Zervas and D. M. Theodoropoulos, *J. Am. Chem. Soc.*, **78**, 1359 (1956).

14. C. Dugave and A. Menez, *J. Org. Chem.*, **61**, 6067 (1996).
15. H. Nesvadba and H. Roth, *Monatsh. Chem.*, **98**, 1432 (1967).
16. M. Bodanszky, M. A. Bednarek, and A. Bodanszky, *Int. J. Pept. Protein Res.*, **20**, 387 (1982).
17. J. Alsina, E. Giralt, and F. Albericio, *Tetrahedron Lett.*, **37**, 4195 (1996).
18. G. M. Dubowchik and S. Radia, *Tetrahedron Lett.*, **38**, 5257 (1997).
19. Y. Lapidot, N. de Groot, M. Weiss, R. Peled, and Y. Wolman, *Biochim. Biophys. Acta*, **138**, 241 (1967).

N-9-Phenylfluorenylamine (Pf–NR$_2$): 9-(C_6H_5)–($C_{13}H_8$)–NR_2

Formation

1. 9-Pf–Br, $Pb(NO_3)_2$, CH_3CN, rt, 28 h, >80% yield.[1,2]
2. 9-Pf–Br, K_3PO_4, CH_3NO_2. This method avoids the use of lead nitrate.[3]

Cleavage

This group was reported to be 6000 times more stable to acid than the trityl group because of destabilization of the cation by the fluorenyl group.[4]

1. CF_3COOH, CH_3CN, H_2O, 0°, 1 h $\rightarrow$ rt, 1 h.[5]
2. H_2, Pd/C, EtOAc, AcOH.[6,7]
3. Li, NH_3, THF, 76% yield.[8]

1. P. L. Feldman and H. Rapoport, *J. Org. Chem.*, **51**, 3882 (1986).
2. B. D. Christie and H. Rapoport, *J. Org. Chem.*, **50**, 1239 (1985).
3. S. C. Bergmeier, A. A. Cobas, and H. Rapoport, *J. Org. Chem.*, **58**, 2369 (1993).
4. R. Bolton, N. B. Chapman, and J. Shorter, *J. Chem. Soc.*, 1895 (1964).
5. J. P. Whitten, D. Muench, R. U. Culse, P. L. Nyce, B. M. Baron, and I. A. McDonald, *Bioorg. Med. Chem. Lett.*, **1**, 441 (1991).
6. H.-G. Lombart and W. D. Lubell, *J. Org. Chem.*, **61**, 9437 (1996).
7. J. A. Campbell, W. K. Lee, and H. Rapoport, *J. Org. Chem.*, **60**, 4602 (1995).
8. W. D. Lubbel, T. F. Jamison, and H. Rapoport, *J. Org. Chem.*, **55**, 3511 (1990).

N-Ferrocenylmethylamine (Fcm–NR$_2$): $C_{10}H_{10}FeCH_2NR_2$

The Fcm derivative is prepared from amino acids on treatment with formylferrocene and Pd-phthalocyanine by reductive alkylation (60–89% yield). It is cleaved with 2-thionaphthol/CF_3COOH. Its primary advantage is its color, making it easily detected.[1]

1. H. Eckert and C. Seidel, *Angew. Chem., Int. Ed. Engl.*, **25**, 159 (1986).

_N_-2-Picolylamine _N′_-Oxide: R_2NCH_2–2-pyridyl _N′_-Oxide (Chart 10)

_N_-2-Picolylamine _N′_-oxide, used in oligonucleotide syntheses, is cleaved by acetic anhydride at 22°, followed by methanolic ammonia (85–95% yield).[1]

1. Y. Mizuno, T. Endo, T. Miyaoka, and K. Ikeda, _J. Org. Chem._, **39**, 1250 (1974).

Imine Derivatives

A number of imine derivatives have been prepared as amine protective groups, but most of these have not seen extensive use. The most widely used are the benzylidene and diphenylmethylene derivatives. The less used derivatives are listed, for completeness, with their references at the end of this section. For the most part, they are prepared from the aldehyde and the amine by water removal; cleavage is effected by acid hydrolysis.

_N_-1,1-Dimethylthiomethyleneamine: $(MeS)_2C{=}NR$

This group was used to protect the nitrogen of glycine in a synthesis of amino acids.[1]

Formation

1. CS_2, TEA, $CHCl_3$, 20–40°, 1 h; MeI, reflux, 1 h, 77% yield.[2]
2. CS_2, NaOH, benzene; MeI, benzene, TEBA, 20°, 39–86% yield.[3]
3. CS_2, TEA, $BrCH_2CH_2Br$, 70–75% yield.[4]

Cleavage

1. H_2O_2, HCO_2H, TsOH, 0° → 20°, 90% yield.[2]
2. HCl, H_2O, THF, rt, 100% yield.[2,5]

MeS, R, H, O, N, N, CH_2OMOM, MeS, $MOMOCH_2$ —(1 N HCl)→ R, H, H_2N, CO_2H

3. Direct conversion to other protective groups is possible.[6]

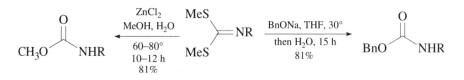

CH_3O–C(=O)–NHR ←($ZnCl_2$, MeOH, H_2O, 60–80°, 10–12 h, 81%)— $(MeS)_2C{=}NR$ —(BnONa, THF, 30°, then H_2O, 15 h, 81%)→ BnO–C(=O)–NHR

***N*-Benzylideneamine:** RN=CHPh (Chart 10)

Most applications of this derivative have been for the preparation and modification of amino acids, although some applications in the area of carbohydrates have been reported. The derivative is stable to *n*-butyllithium, lithium diisopropylamide, and *t*-BuOK.[7]

Formation

1. PhCHO, Et_3N, 80–90% yield.[8]
2. PhCHO, Na_2SO_4, benzene, rt, 99% yield.[9] A primary amine is protected in the presence of a secondary amine.[10]
3. PhCHO, trimethyl orthoformate, 89–100% yield.[11]

Cleavage

1. 1 *N* HCl, 25°, 1 h.[1,12]
2. H_2, Pd–C, CH_3OH.[13]
3. Hydrazine, EtOH, reflux, 6 h, 70% yield.[14]
4. Girard-T Reagent, >75% yield.[15]

***N-p*-Methoxybenzylideneamine:** 4-$MeOC_6H_4CH$=NR

The *N-p*-methoxybenzylideneamine has been used to protect glucosamines.[16]

Formation

1. 4-$MeOC_6H_4CHO$, benzene, pyridine, heat, >72% yield.[17]

Cleavage

1. MeOH, 10% aq. AcOH, $TsNHNH_2$, >81% yield.[13,18]
2. 5 *N* HCl.[19]

***N*-Diphenylmethyleneamine:** RN=CPh_2

The derivative of glycine, prepared from benzophenone (cat. $BF_3 \cdot Et_2O$, xylene, reflux, 82% yield), has found considerable use in the preparation of amino acids. It can also be prepared by an exchange reaction with benzophenonimine (Ph_2C=NH, CH_2Cl_2, rt).[20] It is stable to DIBAH, Grignard reagents, strong base,[21] and osmium oxidations.[22] When used for the protection of serine, it increases the nucleophilicity of the hydroxyl group and improves β-*O*-glycosylation.[23] Benzophenonimine has been used as a protective group for ammonia in the amination of aromatic rings.[24] The fluorene analogue, prepared from fluorenone ($TiCl_4$, toluene, 0°), has also been used to protect a primary amine.[22]

Cleavage

1. Concd. HCl, reflux, 6 h, or aq. citric acid, 12 h.[25]
2. H_2, Pd–C, MeOH, rt, 14 h, 90% yield.[26]
3. NH_2OH, 3 min, pH 4–6.[27,28]

***N*-[(2-Pyridyl)mesityl]methyleneamine:** $(C_5H_4N)(Me_3C_6H_2)C{=}NR$[29]

The imine, prepared from an amine and $(C_5H_4N)(Me_3C_6H_2)CO$ ($TiCl_4$, toluene, reflux, 12 h; NaOH, 80% yield), can be cleaved with concd. HCl (reflux). The protective group was used to direct α-alkylation of amines.

***N*-(*N′N′*-Dimethylaminomethylene)amine *N*-(*N′,N′*-Dimethylformamidine):** $RN{=}CHN(CH_3)_2$

The formamidine is prepared by heating the primary amine in DMF-dimethylacetal (81–100% yield). Deprotection is effected by heating in EtOH with $ZnCl_2$.[30] $LiAlH_4$ (Et_2O, reflux), hydrazine (AcOH, MeOH), KOH (MeOH, reflux),[31] dilute ammonia (high yield),[32] and concd. HCl (reflux, 65–90% yield)[33] are also known to cleave the formamidine group.

***N*-(*N′,N′*-Dibenzylaminomethylene)amine *N*-(*N′,N′*-Dibenzylformamidine):** $(C_6H_5CH_2)_2NCH{=}NR$

Heating a primary amine with dibenzylformamide–dimethyl acetal in CH_3CN gives the formamidine in 49–99% yield. It is cleaved by hydrogenolysis ($Pd(OH)_2$, MeOH, H_2O, H_2, 52–99% yield).[34]

***N*-(*N′-t*-Butylaminomethylene)amine *N*-(*N′-t*-Butylformamidine):** $(CH_3)_3CN{=}CH\text{-}NR_2$

The *t*-butylformamidine was used to protect and direct the course of metalation of secondary amines. It is formed from *N,N*-dimethyl-*N′-t*-butylformamidine by an acid-catalyzed exchange reaction or from the *N-t*-butylimidate tetrafluoroborate salt and is cleaved with hydrazine.[35]

***N,N′*-Isopropylidenediamine** (Chart 10):[36]

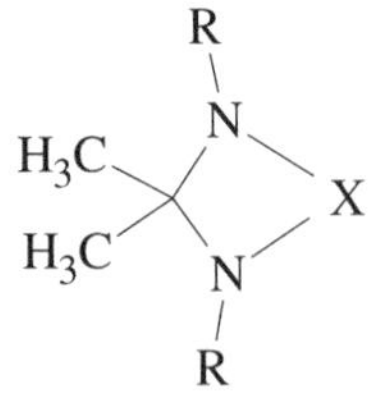

***N-p*-Nitrobenzylideneamine:** $4\text{-}NO_2C_6H_4CH{=}NR$[37] (Chart 10)

***N*-Salicylideneamine:** $2\text{-}HO\text{–}C_6H_4CH{=}NR$[38] (Chart 10)

***N*-5-Chlorosalicylideneamine:** 2-HO–5-ClC_6H_3CH=NR[39]

***N*-(5-Chloro-2-hydroxyphenyl)phenylmethyleneamine:**
RN=C(Ph)C_6H_3–2-OH-5-Cl[40,41]

***N*-(2-Chlorobenzylidene)cyclohexylamine:** C_6H_{11}N=CHR[42]

This imine is stable to the $Fe(acac)_3$-catalyzed Grignard coupling of aryl halides.

***N-t*-Butylideneamine:** $(CH_3)_3$CCH=NR[43]

1. S. Ikegami, T. Hayama, T. Katsuki, and M. Yamaguchi, *Tetrahedron Lett.*, **27**, 3403 (1986); S. Ikegama, H. Uchiyama, T. Hayama, T. Katsuki, and M. Yamaguchi, *Tetrahedron*, **44**, 5333 (1988).
2. D. Hoppe and L. Beckmann, *Liebigs Ann. Chem.*, 2066 (1979).
3. C. Alvarez-Ibarra, M. L. Quiroga, E. Martinez-Santos, and E. Toledano, *Org. Prep. Proced. Int.*, **23**, 611 (1991).
4. S. Hanessian and Y. L. Bennani, *Tetrahedron Lett.*, **31**, 6465 (1990).
5. W. Oppolzer, R. Moretti, and S. Thomi, *Tetrahedron Lett.*, **30**, 6009 (1989).
6. M. Anbazhagan, T. I. Reddy, and S. Rajappa, *J. Chem. Soc., Perkin Trans. 1*, 1623 (1997).
7. N. De Kimpe and P. Sulmon, *Synlett*, 161 (1990).
8. P. Bey and J. P. Vevert, *Tetrahedron Lett.*, 1455 (1977).
9. B.W. Metcalf and P. Casara, *Tetrahedron Lett.*, 3337 (1975).
10. J. D. Prugh, L. A. Birchenough, and M. S. Egbertson, *Synth. Commun.*, **22**, 2357 (1992).
11. G. C. Look, M. M. Murphy, D. A. Campbell, and M. A. Gallop, *Tetrahedron Lett.*, **36**, 2937 (1995).
12. D. Ferroud, J. P. Genet, and R. Kiolle, *Tetrahedron Lett.*, **27**, 23 (1986).
13. R. A. Lucas, D. F. Dickel, R. L. Dziemian, M. J. Ceglowski, B. L. Hensle, and H. B. MacPhillamy, *J. Am. Chem. Soc.*, **82**, 5688 (1960).
14. G. W. J. Fleet and I. Fleming, *J. Chem. Soc. C*, 1758 (1969).
15. T. Watanabe, S. Sugawara, and T. Miyadera, *Chem. Pharm Bull.*, **30**, 2579 (1982).
16. A. Marra and P. Sinay, *Carbohydr. Res.*, **200**, 319 (1990).
17. D. R. Mootoo and B. Fraser-Reid, *Tetrahedron Lett.*, **30**, 2363 (1989).
18. F. Baumberger, A. Vasella, and R. Schauer, *Helv. Chim. Acta*, **71**, 429 (1988).
19. M. Bergmann and L. Zervas, *Ber.*, **64**, 975 (1931).
20. T. Hvidt, W. A. Szarek, and D. B. Maclean, *Can. J. Chem.*, **66**, 779 (1988); M. A. Peterson and R. Polt, *J. Org. Chem.*, **58**, 4309 (1993).
21. R. Polt and M. A. Peterson, *Tetrahedron Lett.*, **31**, 4985 (1990).
22. E. J. Corey, A. Guzman-Perez, and M. C. Noe, *J. Am. Chem. Soc.*, **117**, 10805 (1995).

23. L. Szabo, Y. Li, and R. Polt, *Tetrahedron Lett.*, **32**, 585 (1991).
24. J. P. Wolfe, J. Ahman, J. P. Sadighi, R. A. Singer, and S. L. Buchwald, *Tetrahedron Lett.*, **38**, 6367 (1997).
25. M. J. O'Donnell, J. M. Boniece, and S. E. Earp, *Tetrahedron Lett.*, 2641 (1978).
26. L. Wessjohann, G. McGaffin, and A. de Meijere, *Synthesis*, 359 (1989).
27. K.-J. Fasth, G. Antoni, and B. Langström, *J. Chem. Soc., Perkin Trans. 1*, 3081 (1988).
28. M. Lögers, L. E. Overman, and G. S. Welmaker, *J. Am. Chem. Soc.*, **117**, 9139 (1995).
29. J. M. Hornback and B. Murugaverl, *Tetrahedron Lett.*, **30**, 5853 (1989).
30. D. Toste, J. McNulty, and I. W. J. Still, *Synth. Commun.*, **24**, 1617 (1994).
31. A. I. Meyers, P. D. Edwards, W. F. Rieker, and T. R. Bailey, *J. Am. Chem. Soc.*, **106**, 3270 (1984); A. I. Meyers, *Aldrichimica Acta*, **18**, 59 (1985).
32. J. Zemlicka, S. Chládek, A. Holy, and J. Smrt, *Collect. Czech. Chem. Commun.*, **31**, 3198 (1966).
33. J. J. Fitt and H. W. Gschwend, *J. Org. Chem.*, **42**, 2639 (1977).
34. S. Vincent, S. Mons, L. Lebeau, and C. Mioskowki, *Tetrahedron Lett.*, **38**, 7527 (1997).
35. A. I. Meyers, P. D. Edwards, W. F. Rieker, and T. R. Bailey, *J. Am. Chem. Soc.*, **106**, 3270 (1984).
36. P. M. Hardy and D. J. Samworth, *J. Chem. Soc., Perkin Trans. 1*, 1954 (1977).
37. J. L. Douglas, D. E. Horning, and T. T. Conway, *Can. J. Chem.*, **56**, 2879 (1978).
38. J. N. Williams and R. M. Jacobs, *Biochem. Biophys. Res. Commun.*, **22**, 695 (1966).
39. J. C. Sheehan and V. J. Grenada, *J. Am. Chem. Soc.*, **84**, 2417 (1962).
40. B. Halpern and A. P. Hope, *Aust. J. Chem.*, **27**, 2047 (1974).
41. A. Abdipranoto, A. P. Hope, and B. Halpern, *Aust. J. Chem.*, **30**, 2711 (1977).
42. L. N. Pridgen, L. Snyder, and J. Prol, Jr., *J. Org. Chem.*, **54**, 1523 (1989).
43. S. Kanemasa, O. Uchida, and E. Wada, *J. Org. Chem.*, **55**, 4411 (1990).

Enamine Derivative

***N*-(5,5-Dimethyl-3-oxo-1-cyclohexenyl)amine** (Chart 10):

NHR

O

This vinylogous amide has been prepared in 70% yield to protect amino acid esters. It is cleaved by treatment with either aqueous bromine[1] or nitrous acid (90% yield).[2]

1. B. Halpern and L. B. James, *Aust. J. Chem.*, **17**, 1282 (1964).
2. B. Halpern and A. D. Cross, *Chem. Ind. (London)*, 1183 (1965).

N-2,7-Dichloro-9-fluorenylmethyleneamine:

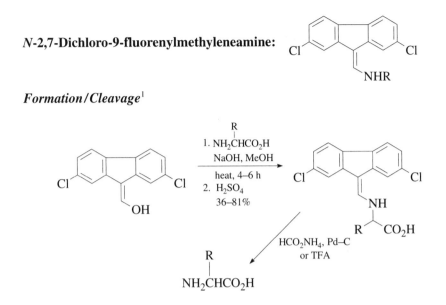

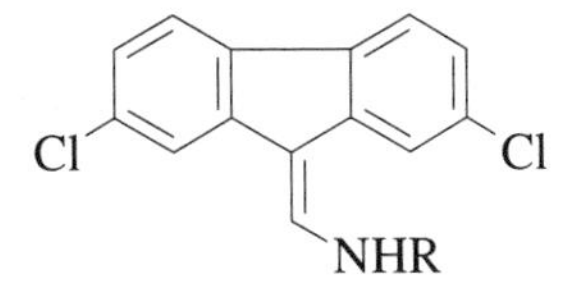

Cl Cl NHR

Formation/Cleavage[1]

R
1. NH_2CHCO_2H
NaOH, MeOH
heat, 4–6 h
2. H_2SO_4
36–81%

Cl Cl OH → Cl Cl NH R CO_2H

HCO_2NH_4, Pd–C
or TFA

R
NH_2CHCO_2H

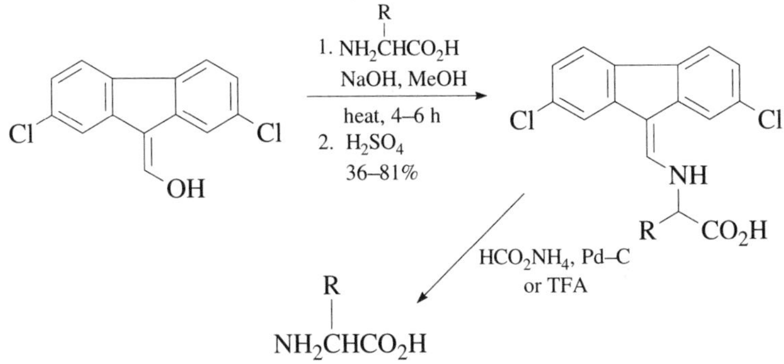

1. L. A. Carpino, H. G. Chao, and J.-H. Tien, *J. Org. Chem.*, **54**, 4302 (1989).

N-2-(4,4-Dimethyl-2,6-dioxocyclohexylidene)ethylamine (Dde–NR$_2$):

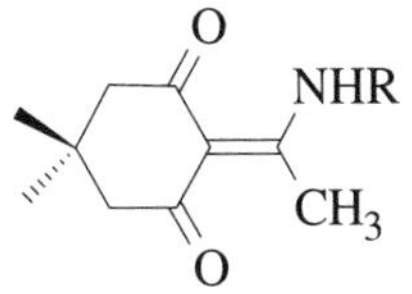

O NHR CH$_3$ O

The Dde group was developed for amine protection in solid-phase peptide synthesis. It is formed from 2-acetyldimedone in DMF and cleaved using 2% hydrazine in DMF,[1,2] or ethanolamine.[3] Hydrazinolysis of the Dde group in the presence of the Aloc group was found to be troublesome because of hydrogenation of the allyl group, unless allyl alcohol was included in the deprotection mixture.[4] A number of structurally similar analogues employing the concept of stabilization through conjugation and intramolecular hydrogen bonding have been prepared for the same purpose.[5–9]

1. B. W. Bycroft, W. C. Chan, S. R. Chhabra, and N. D. Hone, *J. Chem. Soc., Chem. Commun.*, 778 (1993).
2. I. A. Nash, B. W. Bycroft, and W. C. Chan, *Tetrahedron Lett.*, **37**, 2625 (1996).
3. J.-C. Truffert, O. Lorthioir, U. Asseline, N. T. Thuong, and A. Brack, *Tetrahedron Lett.*, **35**, 2353 (1994).
4. B. Rohwedder, Y. Mutti, P. Dumy, and M. Mutter, *Tetrahedron Lett.*, **39**, 1175 (1998).

5. M. de G. García Martin, C. Gasch, and A. Gómez-Sánchez, *Carbohydr. Res.*, **199**, 139 (1990).
6. J. Svete, M. Aljaz-Rozic, and B. Stanovnik, *J. Heterocycl. Chem.*, **34**, 177 (1997).
7. M. Abarbri, A. Guignard, and M. Lamant, *Helv. Chim. Acta*, **78**, 109 (1995).
8. M. A. Pradera, D. Olano, and J. Fuentes, *Tetrahedron Lett.*, **36**, 8653 (1995).
9. S. R. Chhabra, B. Hothi, D. J. Evans, P. D. White, B. W. Bycroft, and W. C. Chan, *Tetrahedron Lett.*, **39**, 1603 (1998).

N-4,4,4-Trifluoro-3-oxo-1-butenylamine (Tfav–NR_2):

H R O N F_3C

This group was developed for the protection of amino acids. It is formed from 4-ethoxy-1,1,1-trifluoro-3-buten-2-one in aqueous sodium hydroxide (70–94% yield). Primary amino acids form the *Z*-enamines, whereas secondary amines such as proline form the *E*-enamines. Deprotection is achieved with 1–6 *N* aqueous HCl in dioxane at rt.[1,2]

1. M. G. Gorbunova, I. I. Gerus, S. V. Galushko, and V. P. Kukhar, *Synthesis*, 207 (1991).
2. I. I. Gerus, M. G. Gorbunova, and V. P. Kukhar, *J. Fluorine Chem.*, **69**, 195 (1994).

N-(1-Isopropyl-4-nitro-2-oxo-3-pyrrolin-3-yl)amine

Formation/Cleavage[1]

O_2N OEt N O *i*-Pr

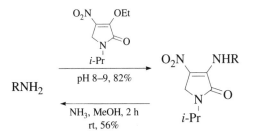

RNH_2 → pH 8–9, 82% → O_2N NHR N O *i*-Pr

← NH_3, MeOH, 2 h rt, 56%

1. P. L. Southwick, G. K. Chin, M. A. Koshute, J. R. Miller, K. E. Niemela, C. A. Siegel, R. T. Nolte, and W. E. Brown, *J. Org. Chem.*, **49**, 1130 (1984).

N-Hetero Atom Derivatives

Six categories of *N*-hetero atom derivatives are considered: N-M (M = boron, copper); N-N (e.g., *N*-nitro, *N*-nitroso); *N*-oxides (used to protect tertiary amines); N-P (e.g., phosphinamides, phosphonamides); N-SiR_3 (R = CH_3), and N-S (e.g., sulfonamides, sulfenamides).

N-Metal Derivatives

N-Borane Derivatives: $R_3N^+BH_3^-$

Aminoboranes can be prepared from diborane to protect a tertiary amine during oxidation.[1,2]

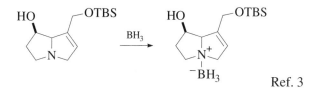

HO OTBS $\xrightarrow{BH_3}$ HO OTBS N^+ $^-BH_3$ Ref. 3

They are cleaved by refluxing in ethanol,[4] methanolic sodium carbonate,[5] TFA,[6] or ammonium chloride.[7] The aminoborane was found to be stable to LDA and KHMDS.[7]

N-Diphenylborinic Acid Derivative

Formation/Cleavage[8,9]

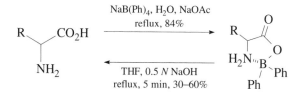

R CO₂H NH₂ — $NaB(Ph)_4$, H_2O, NaOAc, reflux, 84% → ; ← THF, 0.5 *N* NaOH, reflux, 5 min, 30–60%

This derivative is stable to acetic acid and CF_3CO_2H.[9]

N-Diethylborinic Acid Derivative

Formation

The diethylborinic acid derivative has been prepared from triethylborane (THF, reflux).[10]

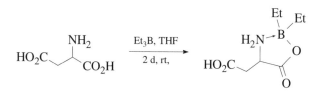

HO_2C NH_2 CO_2H — Et_3B, THF, 2 d, rt, →

After esterification of the remaining carboxyl group, the boron was removed with HCl(g) (Et_2O, rt, 15 min, >80% yield).[10,11]

N-Difluoroborinic acid

Formation

HO–CH₂–CH(NH₂)–CO₂H $\xrightarrow[40\text{–}50°,\ 2\text{–}4\ h]{BF_3 \cdot Et_2O,\ THF}$ (H₂N→BF₂–O–C(=O) chelate)

These water-sensitive derivatives can be used to form cleanly the *t*-butyl ethers of serine and threonine. They are cleaved with aqueous acid or base.[12]

N,N′-3,5-Bis(trifluoromethyl)phenylboronic acid

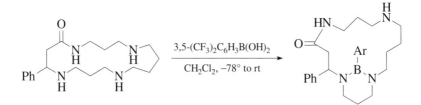

$\xrightarrow[CH_2Cl_2,\ -78°\ \text{to rt}]{3,5\text{-}(CF_3)_2C_6H_3B(OH)_2}$

The free amine can be monoacylated. Without this protection, only the bisacylated derivative is obtained.[13]

1. J. L. Brayer, J. P. Alazard, and C. Thal, *Tetrahedron*, **46**, 5187 (1990).
2. C. J. Swain, C. Kneen, R. Herbert, and R. Baker, *J. Chem. Soc., Perkin Trans. 1*, 3183 (1990).
3. J. D. White, J. C. Amedio, Jr., S. Gut, and L. Jayasinghe, *J. Org. Chem.*, **54**, 4268 (1989).
4. A. Picot and X. Lusinchi, *Bull. Soc. Chim. Fr.*, 1227 (1977).
5. M. A. Schwartz, B. F. Rose, and B. Vishnuvajjala, *J. Am. Chem. Soc.*, **95**, 612 (1973).
6. S. Choi, I. Bruce, A. J. Fairbanks, G. W. J. Fleet, A. H. Jones, R. J. Nash, and L. E. Fellows, *Tetrahedron Lett.*, **32**, 5517 (1991).
7. V. Ferey, P. Vedrenne, L. Toupet, T. Le Gall, and C. Mioskowski, *J. Org. Chem.*, **61**, 7244 (1996).
8. I. Staatz, U. H. Granzer, A. Blume, and H. J. Roth, *Liebigs Ann. Chem.*, 127 (1989).
9. G. H. L. Nefkens and B. Zwanenburg, *Tetrahedron*, **39**, 2995 (1983).
10. F. Albericio, E. Nicolás, J. Rizo, M. Ruiz-Gayo, E. Pedroso, and E. Giralt, *Synthesis*, 119 (1990).
11. J. Robles, E. Pedroso, and A. Grandas, *Synthesis*, 1261 (1993).
12. J. Wang, Y. Okada, W. Li, T. Yokoi, and J. Zhu, *J. Chem. Soc., Perkin Trans. 1*, 621 (1997).

13. K. Ishihara, Y. Kuroki, N. Hanaki, S. Ohara, and H. Yamamoto, *J. Am. Chem. Soc.*, **118**, 1569 (1996).

N-[Phenyl(pentacarbonylchromium- or -tungsten)carbenyl]amine:

$$(CO)_5M{=}C(NR_2)R'$$

$R' = Ph$, or CH_3; $M = Cr$ or W

These transition metal carbenes, prepared in 66–97% yield from amino acid esters, are cleaved by acid hydrolysis (CF_3CO_2H, 20°, 80% yield; 80% AcOH; M = W; BBr_3, –25°).[1]

1. K. Weiss and E. O. Fischer, *Chem. Ber.*, **109**, 1868 (1976).

N-Copper or N-Zinc Chelate: $RNH_2\cdots M\cdots OH$

M = Cu(II), Zn(II)

Formation/Cleavage

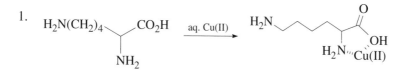

1. $H_2N(CH_2)_4CH(NH_2)CO_2H$ —aq. Cu(II)→ $H_2N(CH_2)_4CH(NH_2)C(O)OH$ chelated through H_2N and OH to Cu(II)

   A copper chelate selectively protects the α-NH_2 group in lysine. The chelate is cleaved by 2 *N* HCl or by EDTA, $(HO_2CCH_2)_2NCH_2CH_2N(CH_2CO_2H)_2$.[1] This mode of protection is sufficient to allow alkylation of a copper-protected tyrosine at the phenol (75% yield).[2]
2. In an aminoglycoside, a vicinal amino hydroxy group can be protected as a Cu(II) chelate. After acylation of other amine groups, the chelate is cleaved by aqueous ammonia.[3] The copper chelate can also be cleaved with $Bu_2NC(S)NHBz$ (EtOH, reflux, 2 h).[4]
3. After examination of the complexing ability of Ca(II), Cr(III), Mn(II), Fe(III), Co(II), Ni(II), Cu(II), Zn(II), Ru(III), Ag(I), and Sn(IV), the authors decided that Zn(II) provides the best protection for vicinal amino hydroxy groups during trifluoroacetylation of other amino groups in the course of some syntheses of kanamycin derivatives.[5]

1. R. Ledger and F. H. C. Stewart, *Aust. J. Chem.*, **18**, 933 (1965).

2. K. Nakanishi, R. Goodnow, K. Konno, M. Niwa, R. Bukownik, T. A. Kallimopoulos, P. Usherwood, A. T. Eldefrawi, and M. E. Eldefrawi, *Pure Appl. Chem.*, **62**, 1223 (1990).
3. S. Hanessian and G. Patil, *Tetrahedron Lett.*, 1035 (1978).
4. K. H. König, L. Kaul, M. Kuge, and M. Schuster, *Liebigs Ann. Chem.*, 1115 (1987).
5. T. Tsuchiya, Y. Takagi, and S. Umezawa, *Tetrahedron Lett.*, 4951 (1979).

18-Crown-6 Derivative

The primary amine of an amino acid as its tosylate salt can be protected by coordination with a crown ether. The protection scheme was sufficient to allow the HOBt/DDC coupling of amino acids. The crown is removed by treatment with diisopropylethylamine or KCl solution.[1,2]

1. P. Botti, H. L. Ball, E. Rizzi, P. Lucietto, M. Pinori, and P. Mascagni, *Tetrahedron*, **51**, 5447 (1995).
2. C. B. Hyde and P. Mascagni, *Tetrahedron Lett.*, **31**, 399 (1990).

N–N Derivatives

***N*-Nitroamine:** R_2NNO_2 (Chart 10)

Formation

H_2N–C(=NH)–NH–$(CH_2)_3$–CH(NHPG)–CO_2H $\xrightarrow[0°, 1\ h,\ 80\%]{HNO_3,\ H_2SO_4}$ H_2N–C(=NNO_2)–NH–$(CH_2)_3$–CH(NHPG)–CO_2H

An *N*-nitro derivative is used primarily to protect the guanidino group in arginine; it is cleaved by reduction: H_2/Pd–C, AcOH/CH_3OH, ~80% yield;[1] 10% Pd–C/cyclohexadiene, 25°, 2 h, good yields;[2] Pd–C/4% HCO_2H–CH_3OH, 5 h, 100% yield;[3] $TiCl_3$/pH 6, 25°, 45 min, 70–98% yield;[4] $SnCl_2$/60% HCO_2H, 63% yield;[5] electrolysis, 1 *N* H_2SO_4, 1–6 h, 85–95% yield.[6]

1. K. Hofmann, W. D. Peckham, and A. Rheiner, *J. Am. Chem. Soc.*, **78**, 238 (1956).
2. A. M. Felix, E. P. Heimer, T. J. Lambros, C. Tzougraki, and J. Meienhofer, *J. Org. Chem.*, **43**, 4194 (1978).
3. B. ElAmin, G. M. Anantharamaiah, G. P. Royer, and G. E. Means, *J. Org. Chem.*, **44**, 3442 (1979).
4. R. M. Freidinger, R. Hirschmann, and D. F. Veber, *J. Org. Chem.*, **43**, 4800 (1978).
5. T. Hayakawa, Y. Fujiwara, and J. Noguchi, *Bull. Chem. Soc. Jpn.*, **40**, 1205 (1967).

6. P. M. Scopes, K. B. Walshaw, M. Welford, and G. T. Young, *J. Chem. Soc.*, 782 (1965).

N-Nitrosoamine: R_2NNO

N-Nitroso derivatives, prepared from secondary amines and nitrous acid, are cleaved by reduction (H_2/Raney Ni, EtOH, 28°, 3.5 h[1]; CuCl/concd. HCl[2]). Since many *N*-nitroso compounds are carcinogens, and because some racemization and cyclodehydration of *N*-nitroso derivatives of *N*-alkyl amino acids occur during peptide syntheses,[3,4] *N*-nitroso derivatives are of limited value as protective groups.

1. M. Harfenist and E. Magnein, *J. Am. Chem. Soc.*, **79**, 2215 (1957).
2. C. F. Koelsch, *J. Am. Chem. Soc.*, **68**, 146 (1946).
3. P. Quitt, R. O. Studer, and K. Vogler, *Helv. Chim. Acta*, **47**, 166 (1964).
4. F. H. C. Stewart, *Aust. J. Chem.*, **22**, 2451 (1969).

Amine *N*-Oxide: $R_3N \rightarrow O$ (Chart 10)

Amine oxides, prepared to protect tertiary amines during methylation[1,2] and to prevent their protonation in diazotized aminopyridines,[3] can be cleaved by reduction (e.g., SO_2/H_2O, 1 h, 22°, 63% yield[1]; H_2/Pd–C, AcOH, Ac_2O, 7 h, 91% yield;[2] Zn/HCl, 30% yield,[3] reduction with RaNi[4]). Photolytic reduction of an aromatic amine oxide has been reported [i.e., 4-nitropyridine *N*-oxide, 300 nm, $(MeO)_3PO/CH_2Cl_2$, 15 min, 85–95% yield].[5]

1. F. N. H. Chang, J. F. Oneto, P. P. T. Sah, B. M. Tolbert, and H. Rapoport, *J. Org. Chem.*, **15**, 634 (1950).
2. J. A. Berson and T. Cohen, *J. Org. Chem.*, **20**, 1461 (1955).
3. F. Koniuszy, P. F. Wiley, and K. Folkers, *J. Am. Chem. Soc.*, **71**, 875 (1949).
4. K. Toshima, Y. Nozaki, S. Mukaiyama, T. Tamai, M. Nakata, K. Tatsuta, and M. Kinoshita, *J. Am. Chem. Soc.*, **117**, 3717 (1995).
5. C. Kaneko, A. Yamamoto, and M. Gomi, *Heterocycles*, **12**, 227 (1979).

Triazene Derivative:

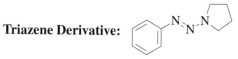

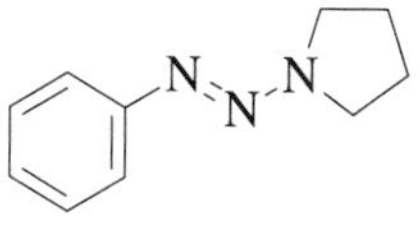

Protection of primary aryl amines as the triazene is accomplished by diazotization of the amine followed by reaction with pyrrolidine in aq. KOH. This group is stable to metalation of the aromatic ring by metal halogen exchange. The amine is recovered by reductive cleavage with Ni–Al alloy (aq. KOH, rt, 37–68% yield).[1]

1. M. L. Gross, D. H. Blank, and W. M. Welch, *J. Org. Chem.*, **58**, 2104 (1993).

N–P Derivatives

Diphenylphosphinamide (Dpp–NR$_2$): $Ph_2P(O)NR_2$ (Chart 10)

Phosphinamides are stable to catalytic hydrogenation, used to cleave benzyl-derived protective groups, and to hydrazine.[1] The rate of hydrolysis of phosphinamides is a function of the steric and electronic factors around the phosphorus.[2] This derivative has largely been used for the protection of amino acids and occurs few, if any, times in the general synthetic literature.

Formation

1. Ph_2POCl, *N*-methylmorpholine, 0°, 60–90% yield.[3]

Cleavage

1. The Dpp group is cleaved by the following acidic conditions: AcOH, HCOOH, H_2O, 24 h, 100% yield; 80% CF_3COOH, ca. quant; 0.4 *M* HCl, 90% CF_3CH_2OH, ca. quant.; *p*-TsOH, H_2O–CH_3OH, ca. quant.; 80% AcOH, 3 days, not completely cleaved.[3] The Dpp group is slightly less stable to acid than the BOC group.[2,3]
2. MeOH, $BF_3·Et_2O$, CH_2Cl_2, 0°–rt, 81–93% yield.[4]
3. Bu_2CuLi, PhLi, or Ph_2CuLi cleaved the Dpp group from an aziridine (63–83% yield), but Me_2CuLi resulted in ring opening.[4]

Dimethyl- and Diphenylthiophosphinamide (Mpt–NR$_2$ and Ppt–NR$_2$): $(CH_3)_2P(S)NR_2$ (Chart 10) and $Ph_2P(S)NR_2$

The Mpt and Ppt derivatives can be prepared from an amino acid and the thiophosphinyl chloride [$Me_2P(S)Cl$ or $Ph_2P(S)Cl$, respectively, 41–78% yield; lysine gives 16% yield].[5] The Mpt group is cleaved with HCl or $Ph_3P·HCl$[6] and is cleaved 60 times faster than the BOC group. The Ppt group is the more stable of the two groups.

Dialkyl Phosphoramidates: $(RO)_2P(O)NR_2$

Formation

1. $(EtO)_2P(O)H$, CCl_4, aq. NaOH, $PhCH_2N^+Et_3\ Cl^-$, 0°, 1 h → 22°, 1 h, 75–90% yield.[7,8]
2. $(BuO)_2P(O)H$, Et_3N, CCl_4.[9]

3. (i-PrO)$_2$P(O)Cl, 73–93% yield.[10]

Cleavage

Phosphoramidates are cleaved with HCl-saturated THF (70–94% yield). Their stability is dependent upon the alkyl group, the methyl derivative being the least stable. They also have good stability to organic acids and Lewis acids.[10]

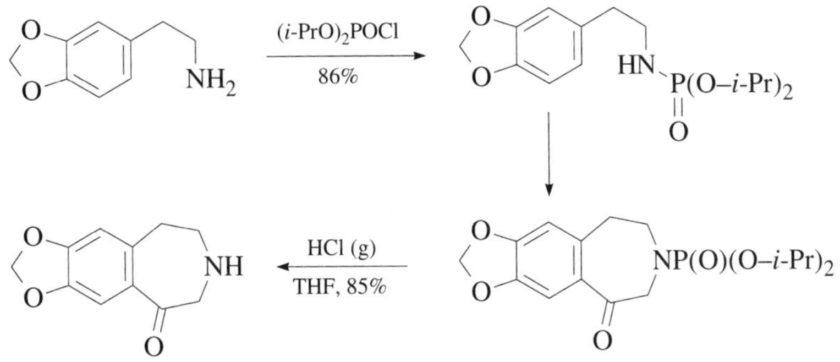

(i-PrO)$_2$POCl, 86%

HCl (g), THF, 85%

NP(O)(O–i-Pr)$_2$

Dibenzyl and Diphenyl Phosphoramidate: (BnO)$_2$P(O)NR$_2$ and (PhO)$_2$P(O)NR$_2$

Dibenzyl phosphoramidates have been prepared from amino acids and the phosphoryl chloride, (BnO)$_2$P(O)Cl.[11] A diphenyl phosphoramidate has been prepared from a glucosamine; it was converted by transesterification into a dibenzyl derivative to facilitate cleavage.[12]

1. G. W. Kenner, G. A. Moore, and R. Ramage, *Tetrahedron Lett.*, 3623 (1976).
2. R. Ramage, B. Atrash, D. Hopton, and M. J. Parrott, *J. Chem. Soc., Perkin Trans. 1*, 1217 (1985).
3. R. Ramage, D. Hopton, M. J. Parrott, G. W. Kenner, and G. A. Moore, *J. Chem. Soc., Perkin Trans. 1*, 1357 (1984).
4. H. M. I. Osborn and J. B. Sweeney, *Synlett*, 145 (1994).
5. S. Ikeda, F. Tonegawa, E. Shikano, K. Shinozaki, and M. Ueki, *Bull. Chem. Soc. Jpn.*, **52**, 1431 (1979).
6. M. Ueki, T. Inazu, and S. Ikeda, *Bull. Chem. Soc. Jpn.*, **52**, 2424 (1979).
7. A. Zwierzak, *Synthesis*, 507 (1975).
8. A. Zwierzak and K. Osowska, *Synthesis*, 223 (1984).
9. Y.-F. Zhao, S.-K. Xi, A.-T. Song, and G.-J. Ji, *J. Org. Chem.*, **49**, 4549 (1984).
10. Y. F. Zhao, G. J. Ji, S. K. Xi, H. G. Tang, A. T. Song, and S. Z. Wei, *Phosphorus Sulfur*, **18**, 155 (1983).
11. A. Cosmatos, I. Photaki, and L. Zervas, *Chem. Ber.*, **94**, 2644 (1961).
12. M. L. Wolfrom, P. J. Conigliaro, and E. J. Soltes, *J. Org. Chem.*, **32**, 653 (1967).

Iminotriphenyphosphorane: $Ph_3P{=}NR$

This derivative is most conveniently prepared by reacting an azide with triphenylphosphine. It was used because of its stability towards Ph_2PLi. Its aqueous hydrolysis is well documented.[1,2]

1. S.-T. Liu and C.-Y. Liu, *J. Org. Chem.*, **57**, 6079 (1992).
2. M. Campbell and M. J. McLeish, *J. Chem. Res., Synop.*, 148 (1993).

N–Si Derivatives

For the most part, silyl derivatives such as **trimethylsilylamines** have not been used extensively for amine protection because of their high reactivity to moisture, although they do provide satisfactory protection when prepared and used under anhydrous conditions.[1,2] The more stable and sterically demanding ***t*-butyldiphenylsilyl** group has been used to protect primary amines in the presence of secondary amines, thus allowing selective acylation or alkylation of the secondary amine.[3] Silylamines are reported not to be stable to oxidative conditions.[3] Silylamines are readily cleaved in the presence of silyl ethers.[4] For a more thorough discussion of silylating reagents, the section on alcohol protection should be consulted, since many of the reagents described there will also silylate amines.

1. J. R. Pratt, W. D. Massey, F. H. Pinkerton, and S. F. Thames, *J. Org. Chem.*, **40**, 1090 (1975).
2. A. B. Smith, III, M. Visnick, J. N. Haseltine, and P. A. Sprengeler, *Tetrahedron*, **42**, 2957 (1986).
3. L. E. Overman, M. E. Okazaki, and P. Mishra, *Tetrahedron Lett.*, **27**, 4391 (1986).
4. T. P. Mawhinney and M. A. Madson, *J. Org. Chem.*, **47**, 3336 (1982).

N–S Derivatives

N-Sulfenyl Derivatives

Sulfenamides, R_2NSR', prepared from an amine and a sulfenyl halide,[1,2] are readily cleaved by acid hydrolysis and have been used in syntheses of peptides, penicillins, and nucleosides. They are also cleaved by nucleophiles[3] and by Raney nickel desulfurization.[4] The synthesis and application of sulfenamides have been reviewed.[5]

Benzenesulfenamide: $R_2NSC_6H_5$, **A** (Chart 10)

2-Nitrobenzenesulfenamide (Nps–NR_2): $R_2NSC_6H_4$–*o*-NO_2, **B** (Chart 10)

The 2-nitrobenzenesulfenamide has been used for the protection of amino acids[6,7] and nucleosides.[8]

Formation

1. *o*-$NO_2C_6H_4SCl$, NaOH, dioxane, 79% yield.[9]
2. *o*-$NO_2C_6H_4SSCN$, $AgNO_2$.[10]

Cleavage

1. Sodium iodide, CH_3OH, CH_2Cl_2, AcOH, 0°, 20 min, 53% yield.[11]
2. Acidic hydrolysis: HCl/Et_2O or EtOH, 0°, 1 h, 95% yield.[12]
3. By nucleophiles: 13 reagents, 5 min–12 h, 90% cleaved.[3]
4. PhSH or $HSCH_2CO_2H$, 22°, 1 h.[13]
5. $CH_3C_6H_4SH$, TsOH, CH_2Cl_2, 84% yield.[14,15]

MeC_6H_4SH, TsOH / CH_2Cl_2, 84% → $\overset{+}{N}H_3\overset{-}{O}Ts$

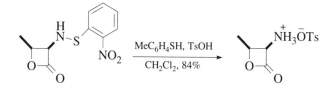

6. 2-Mercaptopyridine/CH_2Cl_2, 1 min, 100% yield.[16]
7. NH_4SCN, 2-methyl-1-indolylacetic acid.[7]
8. HOBt, aniline, DMF. These conditions give the amine as the HOBt salt, which may be acylated without the addition of a tertiary amine.[14]
9. Catalytic desulfurization: Raney Ni/DMF, column, few hours, satisfactory yields.[4]
10. 2-Acylthiomercaptobenzotriazoles, PPTS, 52–80% yield. In this case, the amide is formed rather than the free amine.[17]

2,4-Dinitrobenzenesulfenamide: $R_2NSC_6H_3$–2,4-$(NO_2)_2$, **C**

The 2,4-dinitrobenzenesulfenamide is cleaved with *p*-thiocresol/TsOH.[18]

Pentachlorobenzenesulfenamide: $R_2NSC_6Cl_5$, **D**

Benzenesulfenamide and a number of substituted benzenesulfenamides (compounds **B**, **C**, and **D**) have been prepared to protect the 7-amino group in cephalosporins. They are cleaved by sodium iodide (CH_3OH, CH_2Cl_2, AcOH, 0°, 20 min, 53% yield from sulfenamide **B**).[11]

The *o*-nitrobenzenesulfenamide has been used for the protection of amino acids.[6,7] *o*-Nitrobenzenesulfenamides, **B**, are also cleaved by acidic hydrolysis (HCl/Et_2O or EtOH, 0°, 1 h, 95% yield);[12] by nucleophiles (13 reagents, 5 min–12 h, 90% cleaved);[3] by PhSH or $HSCH_2CO_2H$, 22°, 1 h;[13] by 2-mercaptopyridine/CH_2Cl_2, 1 min, 100% yield);[16] by NH_4SCN, 2-methyl-1-indolylacetic acid;[7] and by catalytic desulfurization (Raney Ni/DMF, column, a few hours, satisfactory yield).[4]

The 2,4-dinitrobenzenesulfenamide, **C**, is cleaved with *p*-thiocresol/TsOH.[18]

2-Nitro-4-methoxybenzenesulfenamide: $R_2NSC_6H_3$–2-NO_2-4-OCH_3

This sulfenamide, prepared from an amino acid, the sulfenyl chloride, and sodium bicarbonate, is cleaved by acid hydrolysis (HOAc/dioxane, 22°, 30 min, 95% yield).[19]

Triphenylmethylsulfenamide: $R_2NSC(C_6H_5)_3$

The tritylsulfenamide can be prepared from an amine and the sulfenyl chloride (Na_2CO_3, THF, H_2O or Pyr, CH_2Cl_2, 64–96% yield);[20] it is cleaved by hydrogen chloride in ether or ethanol (0°, 1 h, 90% yield),[12] $CuCl_2$ (THF, EtOH, 58–67% yield), Me_3SiI (77–96% yield),[20] I_2 (0.1 *M*, THF, collidine, H_2O, 97% yield),[21] Bu_3SnH, 115°, toluene, 5 min, 82% yield.[22] The tritylsulfenamide is stable to 1 *N* HCl, base, $NaCNBH_3$, $LiAlH_4$, *m*-chloroperoxybenzoic acid, pyridinium chlorochromate, Jones Reagent, Collins oxidation, and Moffat oxidation. The stability of this group is largely due to steric hindrance.

***N*-1-(2,2,2-Trifluoro-1,1-diphenyl)ethylsulfenamide (TDE):**
$CF_3C(Ph)_2S$–NR_2

The sulfenamide is prepared from the sulfenyl chloride (Na_2CO_3, THF, H_2O, rt, 95–100% yield or CH_2Cl_2, TEA, 87–96% yield). It is cleaved with Na/NH_3, (67–94% yield) or with HCl/Et_2O (80–98% yield). In the latter method, the sulfenyl chloride can be recovered. The TDE group is stable to strong aqueous HCl, NaOH, $NaBH_4$, $LiAlH_4/Et_2O$ at 0°, Bu_3SnH (toluene, 90°), $Pd(OH)_2/H_2$, and Ac_2O/Pyr.[23]

3-Nitro-2-pyridinesulfenamide (Npys–NR_2)

This group, which is more stable than the 2-nitrobenzenesulfenamide, has been developed to protect amino acids. It is readily introduced with the sulfenyl chloride[24] (52–74% yield).

Cleavage

1. Triphenylphosphine, pentachlorophenol, or 2-thiopyridine *N*-oxide. The group is stable to CF_3COOH, but can be cleaved with 0.1 *M* HCl.[25]
2. 2-Mercaptopyridine and 2-mercapto-1-methylimidazole.[26]
3. 2-Mercaptopyridine *N*-oxide, CH_2Cl_2. A thousand fold excess of this reagent is required to achieve good yields for cleavage in solid-phase peptide synthesis.[27]

1. For other methods of preparation, see F. A. Davis and U. K. Nadir, *Org. Prep. Proced. Int.*, **11**, 33 (1979).

2. For a review of sulfenamides, see L. Craine and M. Raban, *Chem. Rev.*, **89**, 689 (1989).
3. W. Kessler and B. Iselin, *Helv. Chim. Acta*, **49**, 1330 (1966).
4. J. Meienhofer, *Nature*, **205**, 73 (1965).
5. I. V. Koval, *Russ. Chem. Rev.*, **65**, 421 (1996).
6. S. Romani, G. Bovermann, L. Moroder, and E. Wünsch, *Synthesis*, 512 (1985).
7. I. F. Luescher and C. H. Schneider, *Helv. Chim. Acta*, **66**, 602 (1983).
8. M. Sekine, *J. Org. Chem.*, **54**, 2321 (1989).
9. Y. Pu, F. M. Martin, and J. C. Vederas, *J. Org. Chem.*, **56**, 1280 (1991).
10. J. Savrda and D. H. Veyrat, *J. Chem. Soc. C*, 2180 (1970).
11. T. Kobayashi, K. Iino, and T. Hiraoka, *J. Am. Chem. Soc.*, **99**, 5505 (1977).
12. L. Zervas, D. Borovas, and E. Gazis, *J. Am. Chem. Soc.*, **85**, 3660 (1963).
13. A. Fontana, F. Marchiori, L. Moroder, and E. Schoffone, *Tetrahedron Lett.*, 2985 (1966).
14. M. A. Bednarek and M. Bodanszky, *Int. J. Pept. Protein. Res.*, **45**, 64 (1995).
15. Y. Pu, C. Lowe, M. Sailer, and J. C. Vederas, *J. Org. Chem.*, **59**, 3643 (1994).
16. M. Stern, A. Warshawsky, and M. Fridkin, *Int. J. Pept. Protein Res.*, **13**, 315 (1979).
17. M. N. Rao, A. G. Holkar, and N. R. Ayyangar, *J. Chem. Soc., Chem. Commun.*, 1007 (1991).
18. E. M. Gordon, M. A. Ondetti, J. Pluscec, C. M. Cimarusti, D. P. Bonner, and R. B. Sykes, *J. Am. Chem. Soc.*, **104**, 6053 (1982).
19. Y. Wolman, *Isr. J. Chem.*, **5**, 231 (1967).
20. B. P. Branchaud, *J. Org. Chem.*, **48**, 3538 (1983).
21. H. Takaku, K. Imai, and M. Nagai, *Chem. Lett.*, 857 (1988).
22. M. Sekine and K. Seio, *J. Chem. Soc., Perkin Trans.1*, 3087 (1993).
23. T. Netscher and T. Wellar, *Tetrahedron*, **47**, 8145 (1991).
24. For a one-pot preparation of the reagent, see M. Ueki, M. Honda, Y. Kazama, and T. Katoh, *Synthesis*, 21 (1994).
25. R. Matsueda and R. Walter, *Int. J. Pept. Protein Res.*, **16**, 392 (1980).
26. O. Rosen, S. Rubinraut, and M. Fridkin, *Int. J. Pept. Protein Res.*, **35**, 545 (1990).
27. S. Rajagopalan, T. J. Heck, T. Iwamoto, and J. M. Tomich, *Int. J. Pept. Protein Res.*, **45**, 173 (1995).

N-Sulfonyl Derivatives: R_2NSO_2R'

Sulfonamides are prepared from an amine and a sulfonyl chloride in the presence of pyridine or aqueous base.[1] The sulfonamide is one of the most stable nitrogen protective groups. Most arylsulfonamides are stable to alkaline hydrolysis and to catalytic reduction; they are cleaved by Na/NH_3,[2] Na/butanol,[3] sodium naphthalenide,[4] or sodium anthracenide,[5] and by refluxing in acid (48% HBr/cat. phenol).[6] Sulfonamides of less basic amines such as pyrroles and indoles are much easier to cleave than those of the more basic alkyl amines. In fact, sulfonamides of the less basic amines (pyrroles, indoles, and imidazoles) can be cleaved by basic hydrolysis, which is almost impossible for the alkyl amines. Because of the inherent differences between the aromatic –NH group and simple aliphatic amines, the protection of these compounds (pyrroles, indoles, and imidazoles) will be described in a separate section. One appealing property of sulfonamides is that the derivatives are more crystalline than amides or carbamates.

1. E. Fischer and W. Lipschitz, *Ber.*, **48**, 360 (1915).
2. V. du Vigneaud and O. K. Behrens, *J. Biol. Chem.*, **117**, 27 (1937).
3. G. Wittig, W. Joos, and P. Rathfelder, *Justus Liebigs Ann. Chem.*, **610**, 180 (1957).
4. S. Ji, L. B. Gortler, A. Waring, A. Battisti, S. Bank, W. D. Closson, and P. Wriede, *J. Am. Chem. Soc.*, **89**, 5311 (1967).
5. K. S. Quaal, S. Ji, Y. M. Kim, W. D. Closson, and J. A. Zubieta, *J. Org. Chem.*, **43**, 1311 (1978).
6. H. R. Synder and R. E. Heckert, *J. Am. Chem. Soc.*, **74**, 2006 (1952).

p-Toluenesulfonamide (TsNR$_2$): *p*-$CH_3C_6H_4SO_2NR_2$ (Chart 10)

Benzenesulfonamide: $PhSO_2NR_2$

In general, the benzenesulfonyl group is somewhat more reactive than the tosyl group, in both its formation and ease of cleavage.

Formation

1. Tosylates are generally formed from an amine and tosyl chloride in an inert solvent such as CH_2Cl_2 with an acid scavenger such as pyridine or triethylamine. They may also be prepared using the Schotten–Baumann reaction.
2. 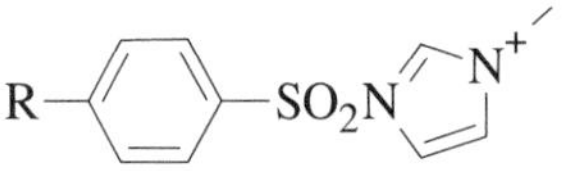 R–C_6H_4–SO_2N N^+ TfO^- This reagent is effective for the formation of sulfonamides of hindered amines.[1]
3. 1-Phenylsulfonylbenzotriazole, THF, 1-methylimidazole, reflux, 64–99% yield.[2] The reagent also benzenesulfonates phenols (51–99% yield).

Cleavage

1. HBr, AcOH, 70°, 8 h, 45–50% yield.[3] During the synthesis of L-2-amino-3-oxalylaminopropionic acid, a neurotoxin, cleavage with Na/NH_3 or $[C_{10}H_8^{\bullet}]^- Na^+$ gave a complex mixture of products.
2. HBr, P, reflux, 24 h, 74–88% yield. An *N*-benzyl group survived these brutal conditions.[4]

3. HF–Pyr, anisole, rt, >62% yield.[5]
4. $NaAlH_2(OCH_2CH_2OCH_3)_2$, benzene or toluene, reflux, 20 h, 65–75% yield.[6] Note that $LiAlH_4$ does not cleave sulfonamides of primary amines; those from secondary amines must be heated to 120°.
5. Electrolysis, $Me_4N^+Cl^-$, 5°, 65–98% yield.[7–9] Acylation of a tosylated amine with BOC or benzoyl reduces the potential required for electrolytic cleavage so that these aryltosyl groups can be selectively removed in the presence of a simple tosylamide.[10]
6. Electrolysis, ascorbic acid, anthracene, $Et_4N^+BF_4^-$, DMF.[11]
7. Sodium naphthalenide.[12,13] This reagent has been used to remove the tosyl group from an amide.[14]

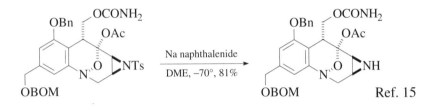

Ref. 15

It is possible to retain a Bn group when using this reagent.[16]

8. Sodium anthracenide, DME, 85% yield.[17]
9. Li, catalytic naphthalene, –78°, THF, 65–99% yield.[18]
10. Li, NH_3, 75% yield[19] or Na, NH_3.[20,21]
11. Na, IPA.[22]
12. SmI_2, DMPU, 50–97% yield.[23,24] The reaction works well for alkyl-substituted aziridines; benzenesulfonamides react faster than tosyl amides.
13. 48% HBr, phenol, 30 min, heat, 85% yield.[9,25] 4-Hydroxybenzoic acid has been used in place of phenol to aid in the isolation process. The addition of water to the reaction mixture caused most of the hydroxybenzoic acid derivatives to precipitate, thus greatly simplifying the isolation.[26]
14. $HClO_4$, AcOH, 100°, 1 h, 30–75% yield.[27]
15. $h\nu$, Et_2O, 6–20 h, 85–90% yield.[28,29]
16. $h\nu$, EtOH, H_2O, $NaBH_4$, 1,2-dimethoxybenzene.[30] This is a photosensitized electron-transfer reaction. Other reductants, such as hydrazine and $BH_3 \cdot NH_3$ are also effective.

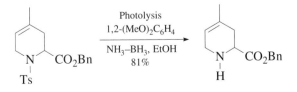

17. $h\nu$, β-naphthoxide anion, $NaBH_4$, quantitative.[31]
18. Na(Hg), Na_2HPO_4.[32,33]

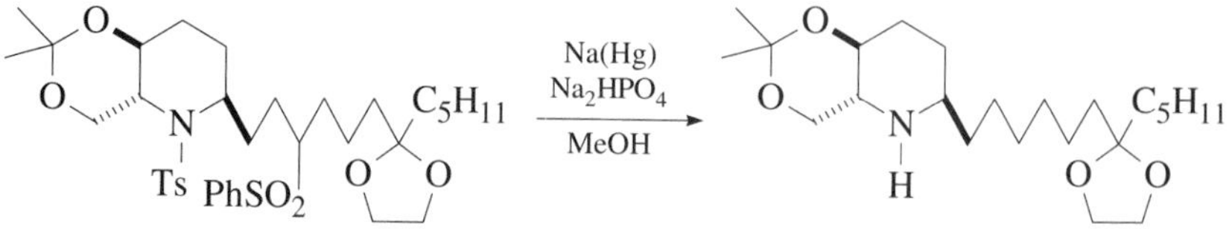

Na(Hg), Na_2HPO_4, MeOH

19. In this example, the enone was not reduced.[34]

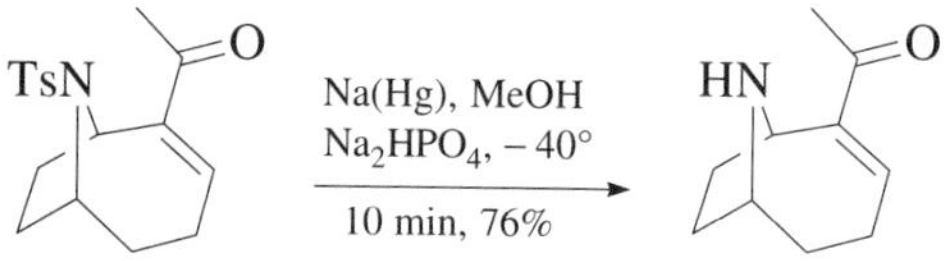

Na(Hg), MeOH, Na_2HPO_4, − 40°, 10 min, 76%

20. SMEAH, *o*-xylene, reflux, 91% yield.[35]
21. During attempted acetonide formation of an amino alcohol derivative, smooth tosyl cleavage was observed. The reaction is general for those cases having a carboxyl group, as in the following example, but fails for simple amino alcohol derivatives that lack this functionality.[36]

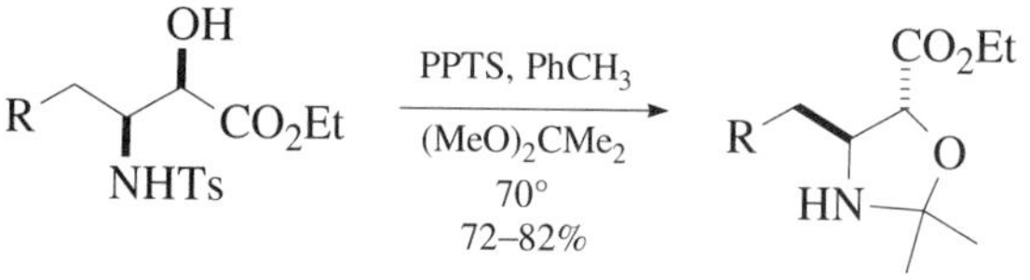

PPTS, $PhCH_3$, $(MeO)_2CMe_2$, 70°, 72–82%

1. J. F. O'Connell and H. Rapoport, *J. Org. Chem.*, **57**, 4775 (1992).
2. A. R. Katritzky, G. Zhang, and J. Wu, *Synth. Commun.*, **24**, 205 (1994).
3. B. E. Haskell and S. B. Bowlus, *J. Org. Chem.*, **41**, 159 (1976).
4. U. Jordis, F. Sauter, S. M. Siddiqi, B. Küenburg, and K. Bhattacharya, *Synthesis*, 925 (1990).
5. W. Oppolzer, H. Bienaymé, and A. Genevois-Borella, *J. Am. Chem. Soc.*, **113**, 9660 (1991).
6. E. H. Gold and E. Babad, *J. Org. Chem.*, **37**, 2208 (1972).
7. L. Horner and H. Neumann, *Chem. Ber.*, **98**, 3462 (1965).
8. T. Moriwake, S. Saito, H. Tamai, S. Fujita, and M. Inaba, *Heterocycles*, **23**, 2525 (1985).
9. R. C. Roemmele and H. Rapoport, *J. Org. Chem.*, **53**, 2367 (1988).
10. L. Grehn, L. S. Maia, L. S. Monteiro, M. I. Montenegro, and U. Ragnarsson, *J. Chem. Res., Synop.*, 144 (1991).
11. K. Oda, T. Ohnuma, and Y. Ban, *J. Org. Chem.*, **49**, 953 (1984).

12. J. M. McIntosh and L. C. Matassa, *J. Org. Chem.*, **53**, 4452 (1988).
13. C. H. Heathcock, T. A. Blumenkopf, and K. M. Smith, *J. Org. Chem.*, **54**, 1548 (1989).
14. H. Nagashima, N. Ozaki, M. Washiyama, and K. Itoh, *Tetrahedron Lett.*, **26**, 657 (1985); J. R. Henry, L. R. Marcin, M. C. McIntosh, P. M. Scola, G. D. Harris, Jr., and S. M. Weinreb, *Tetrahedron Lett.*, **30**, 5709 (1989).
15. T. Katoh, E. Itoh, T. Yoshino, and S. Terashima, *Tetrahedron Lett.*, **37**, 3471 (1996).
16. W.-S. Zhou, W.-G. Xie, Z.-H. Lu, and X. F. Pan, *Tetrahedron Lett.*, **36**, 1291 (1995).
17. P. Magnus, M. Giles, R. Bonnert, C. S. Kim, L. McQuire, A. Merritt, and N. Vicker, *J. Am. Chem. Soc.*, **114**, 4403 (1992).
18. E. Alonso, D. J. Ramón, and M. Yus, *Tetrahedron*, **53**, 14355 (1997).
19. C. H. Heathcock, K. M. Smith, and T. A. Blumenkopf, *J. Am. Chem. Soc.*, **108**, 5022 (1986).
20. A. G. Schultz, P. J. McCloskey, and J. J. Court, *J. Am. Chem. Soc.*, **109**, 6493 (1987).
21. N. Yamazaki and C. Kibayashi, *J. Am. Chem. Soc.*, **111**, 1396 (1989).
22. J. S. Bradshaw, K. E. Krakowiak, and R. M. Izatt, *Tetrahedron*, **48**, 4475 (1992).
23. E. Vedejs and S. Lin, *J. Org. Chem.*, **59**, 1602 (1994).
24. For glucosamines; See D. C. Hill, L. A. Flugge, and P. A. Petillo, *J. Org. Chem.*, **62**, 4864 (1997).
25. R. S. Compagnone and H. Rapoport, *J. Org. Chem.*, **51**, 1713 (1986).
26. C. J. Opalka, T. E. D'Ambra, J. J. Faccone, G. Bodson, and E. Cossement, *Synthesis*, 766 (1995).
27. D. P. Kudav, S. P. Samant, and B. D. Hosangadi, *Synth. Commun.*, **17**, 1185 (1987).
28. A. Abad, D. Mellier, J. P. Pète, and C. Portella, *Tetrahedron Lett.*, 4555 (1971).
29. W. Yuan, K. Fearson, and M. H. Gelb, *J. Org. Chem.*, **54**, 906 (1989).
30. T. Hamada, A. Nishida, and O. Yonemitsu, *J. Am. Chem. Soc.*, **108**, 140 (1986).
31. J. F. Art, J. P. Kestemont, and J. P. Soumillion, *Tetrahedron Lett.*, **32**, 1425 (1991).
32. T. N. Birkinshaw and A. B. Holmes, *Tetrahedron Lett.*, **28**, 813 (1987).
33. F. Chavez and A. D. Sherry, *J. Org. Chem.*, **54**, 2990 (1989).
34. P. Somfai and J. Åhman, *Tetrahedron Lett.*, **33**, 3791 (1992).
35. M. Ishizaki, O. Hoshino, and Y. Iitaka, *J. Org. Chem.*, **57**, 7285 (1992).
36. S. Chandrasekhar and S. Mohapatra, *Tetrahedron Lett.*, **39**, 695 (1998).

2,3,6-Trimethyl-4-methoxybenzenesulfonamide (Mtr–NR_2)[1]

2,4,6-Trimethoxybenzenesulfonamide (Mtb–NR_2)[1] (Chart 10)

2,6-Dimethyl-4-methoxybenzenesulfonamide (Mds–NR_2)[2]

Pentamethylbenzenesulfonamide (Pme–NR_2)[2]

2,3,5,6-Tetramethyl-4-methoxybenzenesulfonamide (Mte–NR_2)[2]

4-Methoxybenzenesulfonamide (Mbs–NR_2)[2]

2,4,6-Trimethylbenzenesulfonamide (Mts–NR_2)[3]

2,6-Dimethoxy-4-methylbenzenesulfonamide (iMds–NR$_2$)[3]

3-Methoxy-4-*t*-butylbenzenesulfonamide[4]

These sulfonamides have been used to protect the guanidino group of arginine.[5] Their acid stablity, as determined by TFA cleavage of the N^G-Arg derivative (25°, 60 min), is as follows: Mtr (52%) > Mds (22%) ≈ Mtb (20%) > Pme (2%) > Mte (1.6%) > Mts ≈ Mbs > iMbs. The Mtr group has been used to protect the ε-nitrogen of lysine. The following table gives the % cleavage of Lys(Mtr) in various acids (MSA = methanesulfonic acid):[6]

	0.15 *M* MSA TFA, PhSMe (9:1) 20°	0.3 *M* MSA TFA, PhSMe (9:1) 20°	TFA, PhSMe (9:1) 50°	HF, PhSMe 0°	MSA, PhSMe 20°	TFA 20°
1 h	80.7	95.1	15.1	3.6	2.3	0
2 h	91.9	99.3	33.6	—	—	0

The rate of cleavage is four to five times faster if dimethyl sulfide is included in the TFA/PhSMe mixture.[7]

The use of 1 *M* HBF$_4$ in TFA/thioanisole was found to give significant rate accelerations during cleavage of the Mtr group.[8] Sulfuric acid at 90° has also been used to cleave the Mtr group.[9]

2,2,5,7,8-Pentamethylchroman-6-sulfonamide (Pmc–NR$_2$):

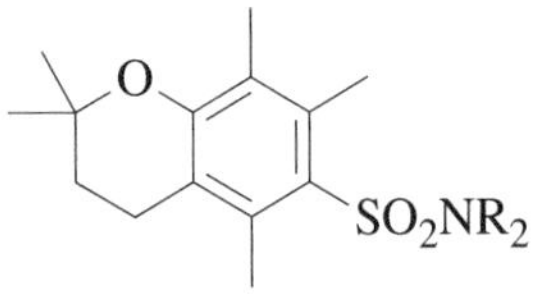

SO$_2$NR$_2$

This group was developed for the protection of N^G-Arg. It is effectively an analogue of the Mtr group, but has the useful property that it is cleaved in TFA/PhSMe in only 20 min. The enhanced rate of cleavage is attributed to the forced overlap of the oxygen electrons with the incipient cation during cleavage. The Pmc group can also be cleaved with 50% TFA/CH$_2$Cl$_2$, which does not cleave the benzyloxy carbamate.[10,11] One problem associated with the Pmc group is that it tends to migrate to other amino acids, such as tryptophan, during acidolysis. This problem, which cannot be completely suppressed with the usual scavenging agents,[12] is also sequence dependent.[13] Another problem observed with both the Mtr and Pmc groups when serine and threonine are present is that of *O*-sulfonation, which was best supressed by the addition of 5% water to the cleavage mixture,[14] but the addition of water was not always effective.[15]

Attempts to develop a more acid labile protecting group than the Pmc group[16] has led to the preparation of the related **Pbf** group, which was shown to to be 1.2–1.4 times as sensitive to TFA as the Pmc group.[17]

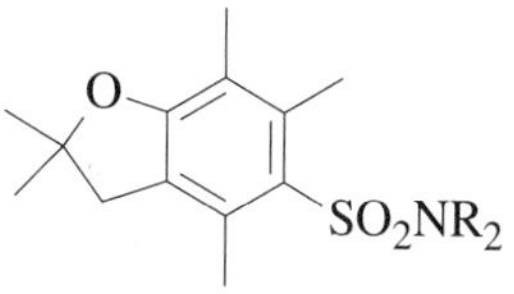

1. E. Atherton, R. C. Sheppard, and J. D. Wade, *J. Chem. Soc., Chem. Commun.*, 1060 (1983).
2. M. Wakimasu, C. Kitada, and M. Fujino, *Chem. Pharm. Bull.*, **29**, 2592 (1981).
3. H. Yajima, K. Akaji, K. Mitani, N. Fujii, S. Funakoshi, H. Adachi, M. Oishi, and Y. Akazawa, *Int. J. Pept. Protein Res.*, **14**, 169 (1979).
4. S. S. Ali, K. M. Khan, H. Echner, W. Voelter, M Hasan, and Atta-ur-Rahman, *J. Prakt. Chem./Chem.-Ztg.*, **337**, 12 (1995).
5. M. Fujino, M. Wakimasu, and C. Kitada, *Chem. Pharm. Bull.*, **29**, 2825 (1981); M. Fujino, O. Nishimura, M. Wakimasu, and C. Kitada, *J. Chem. Soc., Chem. Commun.*, 668 (1980).
6. M. Wakimasu, C. Kitada, and M. Fujino, *Chem. Pharm. Bull.*, **30**, 2766 (1982).
7. K. Saito, T. Higashijima, T. Miyazawa, M. Wakimasu, and M. Fujino, *Chem. Pharm. Bull.*, **32**, 2187 (1984).
8. K. Akaji, M. Yoshida, T. Tatsumi, T. Kimura, Y. Fujiwara, and Y. Kiso, *J. Chem. Soc., Chem. Commun.*, 288 (1990).
9. T. J. McMurry, M. Brechbiel, C. Wu, and O. A. Gansow, *Bioconjugate Chem.*, **4**, 236 (1993).
10. R. Ramage and J. Green, *Tetrahedron Lett.*, **28**, 2287 (1987).
11. J. Green, O. M. Ogunjobi, R. Ramage, A. S. J. Stewart, S. McCurdy, and R. Noble, *Tetrahedron Lett.*, **29**, 4341 (1988).
12. C. G. Fields and G. B. Fields, *Tetrahedron Lett.*, **34**, 6661 (1993).
13. A. Stierandova, N. F. Sepetov, G. V. Nikiforovich, and M. Lebl, *Int. J. Pept. Protein Res.*, **43**, 31 (1994).
14. E. Jaeger, H. Remmer, G. Jung, J. Metzger, W. Oberthür, K. P. Rücknagel, W. Schäfer, J. Sonnenbichler, and I. Zetl, *Biol. Chem. Hoppe-Seyler*, **374**, 349 (1993).
15. A. G. Beck-Sickinger, G. Schnorrenberg, J. Metzger, and G. Jung, *Int. J. Pept. Protein Res.*, **38**, 25 (1991).
16. I. M. Eggleston, J. H. Jones, and P. Ward, *J. Chem. Res., Synop.*, 286 (1991).
17. H. N. Shroff, L. A. Carpino, H. Wenschuh, E. M. E. Mansour, S. A. Triolo, G. W. Griffin, and F. Albericio, *Pept.: Chem., Struct., Biol., Proc. Am. Pept. Symp., 13th*, 121 (1994); L. A. Carpino, H. N. Shroff, S. A. Triolo, E. M. E. Mansour, H. Wenschuh, and F. Albericio, *Tetrahedron Lett.*, **34**, 7829 (1993).

2- or 4-Nitrobenzenesulfonamide (Nosyl–NR_2 or Ns–NR_2):

Formation

1. NsCl, TEA, CH_2Cl_2, 97% yield.[1]

2. The Schotten–Baumann protocol can also be used.

Cleavage

The nosylamide is stable to strong acid and strong base.

1. K_2CO_3 or Cs_2CO_3, DMF or CH_3CN, PhSH, 88–96% yield.[1] This process is not always selective for *p*-nosylate cleavage. Some amines, especially cyclic ones, tend to form 4-phenyl thioethers by nitro displacement as by-products of the cleavage process. This seems to be true only for the *p*-nosylate.[2]
2. LiOH, DMF, $HSCH_2CO_2H$, 93–98% yield. This method has the advantage that the thioether by-products can be washed out by acid/base extraction.[1]
3. Electrolysis, DMF.[3] In the case of primary nosylates, –NH deprotonation competes with cleavage.
4. DBU, DMF, $HSCH_2CH_2OH$, >48% yield. These conditions were used to remove the nosyl group from *N*-methylated peptides.[4]

Nosylaziridines can be opened with a variety of nucleophiles, in preference to nucleophilic cleavage of the nosylate.[5]

2,4-Dinitrobenzenesulfonamide (DNs–NR_2)

Formation[6]

1. 2,4-Dinitrobenzenesulfonyl chloride, Pyr or lutidine, CH_2Cl_2.

Cleavage[6]

1. Propylamine 20 eq., CH_2Cl_2, 20°, 10 min, 88–93% yield.
2. $HSCH_2CO_2H$, TEA, CH_2Cl_2, 23°, 5 min, 91–98% yield. Since the rate of cleavage of the DNs group is much greater than that of the Ns group, it can be cleaved preferentially. DNs derivatives of primary amines under strongly basic conditions can rearrange to give an aniline with loss of SO_2. A similar process occurs for Ns derivatized primary amines, but much harsher conditions are required.[7]
3. Cleavage with thioacids (RCOSH) results in the formation of amides, $R'_2NC(O)R$.[8] The concept was extended to the formation of ureas, thioureas, and thioamides.[9]

1. T. Fukuyama, C.-K. Jow, and M. Cheung, *Tetrahedron Lett.*, **36**, 6373 (1995).
2. P. G. M. Wuts and J. M. Northuis, *Tetrahedron Lett.*, **39**, 3889 (1998).
3. N. R. Stradiotto, M. V. B. Zanoi, O. R. Nascimento, and E. F. Koury, *J. Chim. Phys. Phys.–Chim. Biol.*, **91**, 75 (1994).
4. S. C. Miller and T. S. Scanlan, *J. Am. Chem. Soc.*, **119**, 2301 (1997).

5. P. E. Maligres, M. M. See, D. Askin, and P. J. Reider, *Tetrahedron Lett.*, **38**, 5253 (1997).
6. T. Fukuyama, M. Cheung, C.-K. Jow, Y. Hidai, and T. Kan, *Tetrahedron Lett.*, **38**, 5831 (1997).
7. P. Müller and N.-T. M. Phuong, *Helv. Chim. Acta*, **62**, 494 (1979).
8. T. Messeri, D. D. Sternbach, and N. C. O. Tomkinson, *Tetrahedron Lett.*, **39**, 1669 (1998).
9. T. Messeri, D. D. Sternbach, and N. C. O. Tomkinson, *Tetrahedron Lett.*, **39**, 1673 (1998).

Benzothiazole-2-sulfonamide (Betsyl–NR_2 or Bts–NR_2):

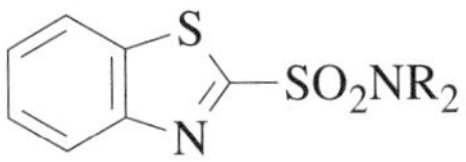

Formation

1. The Bts derivative is formed from the sulfonyl chloride, either using aprotic conditions for simple amines or by the Schotten–Baumann protocol for amino acids (87–97% yield). The primary drawback of this reagent is that its stability depends on its quality. It can on occasion rapidly and exothermically lose SO_2 to give 2-chlorobenzothiazole.[1,2]

Cleavage

1. Zn, AcOH, EtOH.[1]
2. Al–Hg, ether, H_2O.[1]
3. Slow addition of excess H_3PO_2 to 1 *M* DMF solution of substrate at 50°.[1]
4. PhSH, DIPEA, DMF.[2]

1. E. Vedejs, S. Lin, A. Klapars, and J. Wang, *J. Am. Chem. Soc.*, **118**, 9796 (1996).
2. P. G. M. Wuts, R. L. Gu, and J. M. Northuis, *Tetrahedron Lett.*, **38**, 9155 (1998).

Pyridine-2-sulfonamide

Formation

1. Pyridine-2-sulfonyl chloride, aq. K_2CO_3, ether, 64–98% yield.[1]

Cleavage

1. SmI_2, THF or DMPU, rt, 76–94% yield.[1] Deprotection of the pyridinesulfonamide in the presence of a cinnamoyl group was possible when done without a proton source. BOC, *N*-benzyl, *N*-allyl, and trifluoroacetamido groups were all stable to these conditions.[2]
2. Electrolysis.[1]

1. C. Goulaouic-Dubois, A. Guggisberg, and M. Hesse, *J. Org. Chem.*, **60**, 5969 (1995).
2. C. Goulaouic-Dubois, A. Guggisberg, and M. Hesse, *Tetrahedron*, **51**, 12573 (1995).

Methanesulfonamide (Ms–NR$_2$): $CH_3SO_2NR_2$

The mesylate group, introduced with methanesulfonyl chloride, can be cleaved with lithium aluminum hydride and dissolving metal reduction (Na, *t*-BuOH, HMPT, NH_3, 64% yield).[1]

2-(Trimethylsilyl)ethanesulfonamide (SES–NR$_2$): $Me_3SiCH_2CH_2SO_2NR_2$

Formation

1. SES–Cl, Et_3N, DMF, 0°, 88–95% yield.[2]

Cleavage

The SES group is stable to TFA, hot 6 *M* HCl, THF; $LiBH_4$, CH_3CN, $BF_3{\cdot}Et_2O$, 40% HF/EtOH.

1. DMF, CsF, 95°, 9–40 h, 80–93% yield.[2]
2. $Bu_4N^+F^-$, CH_3CN, reflux, >85% yield.[2,3]
3. CsF, DMF, 95°.[4]
4. CsF, DMF, $(BOC)_2O$, 50°, 6 h, 0.01 *M*, 96% yield. The amine is converted to a BOC derivative, which prevents diketopiperazine formation.[5]

9-Anthracenesulfonamide

This group was used to protect the guanidine nitrogen of arginine. It is cleaved by hydrogenation (H_2, Pd–C, 24 h), SmI_2 (THF, *t*-BuOH), Al(Hg) (H_2O, pH 7),[6] photolysis, or TFA/anisole.[7] It was reported that the anthracenesulfonamide can be cleaved by reduction with $HSCH_2CH_2CH_2SH$/DIPEA,[8] but treatment with PhSH/DIPEA/DMF gives cleavage by an addition–elimination mechanism in which 9-phenylthioanthracene is isolated as the only by-product.[9]

4-(4′,8′-Dimethoxynaphthylmethyl)benzenesulfonamide (DNMBS–NR$_2$):

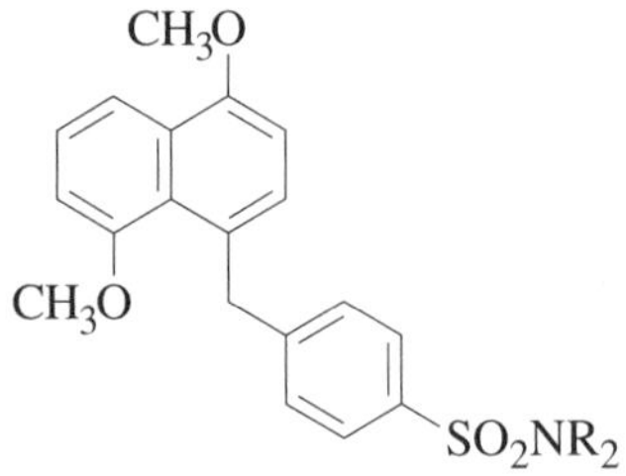

CH_3O

CH_3O

SO_2NR_2

The DNMBS derivative, readily prepared from an amine and the sulfonyl chloride, is efficiently ($\phi = 0.65$) cleaved photochemically ($h\nu$ >300 nm, EtOH, $NH_3 \cdot BH_3$, 77–91% yield).[10] A water-soluble version of this group has been prepared and its photolytic cleavage examined.[11]

1. P. Merlin, J. C. Braekman, and D. Daloze, *Tetrahedron Lett.*, **29**, 1691 (1988).
2. S. M. Weinreb, D. M. Demko, T. A. Lessen, and J. P. Demers, *Tetrahedron Lett.*, **27**, 2099 (1986).
3. R. S. Garigipati and S. M. Weinreb, *J. Org. Chem.*, **53**, 4143 (1988).
4. N. Matzanke, R. J. Gregg, and S. M. Weinreb, *J. Org. Chem.*, **62**, 1920 (1997).
5. D. L. Boger, J.-H. Chen, and K. W. Saionz, *J. Am. Chem. Soc.*, **118**, 1629 (1996).
6. A. J. Robinson and P. B. Wyatt, *Tetrahedron*, **49**, 11329 (1993).
7. H. B. Argens and D. S. Kemp, *Synthesis*, 32 (1988).
8. J. Y. Roberge, X. Beebe, and S. J. Danishefsky, *Science*, **269**, 202 (1995).
9. P. G. M. Wuts and R. L. Gu, unpublished results.
10. T. Hamada, A. Nishida, and O. Yonemitsu, *Tetrahedron Lett.*, **30**, 4241 (1989).
11. J. E. T. Corrie and G. Papageorgiou, *J. Chem. Soc., Perkin Trans. 1*, 1583 (1996).

Benzylsulfonamide: $C_6H_5CH_2SO_2NR_2$ (Chart 10)

Benzylsulfonamides, prepared in 40–70% yield, are cleaved by reduction (Na, NH_3, 75% yield; H_2-Raney Ni, 65–85% yield, but not by H_2-PtO_2) and by acid hydrolysis (HBr or HI, slow).[1] They are also cleaved by photolysis (2–4 h, 40–90% yield).[2] The similar ***p*-methylbenzylsulfonamide (PMS–NR$_2$)** has been prepared to protect the ε-amino group in lysine; it is quantitatively cleaved by anhydrous hydrogen fluoride/anisole (–20°, 60 min).[3] Another example of this seldom-used group is illustrated in the following example.[4]

Formation

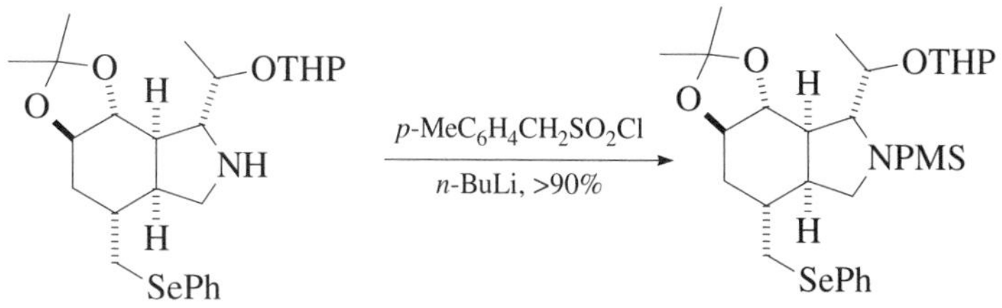

p-$MeC_6H_4CH_2SO_2Cl$, n-BuLi, >90%

Cleavage

HF, anisole, rt, 2 h

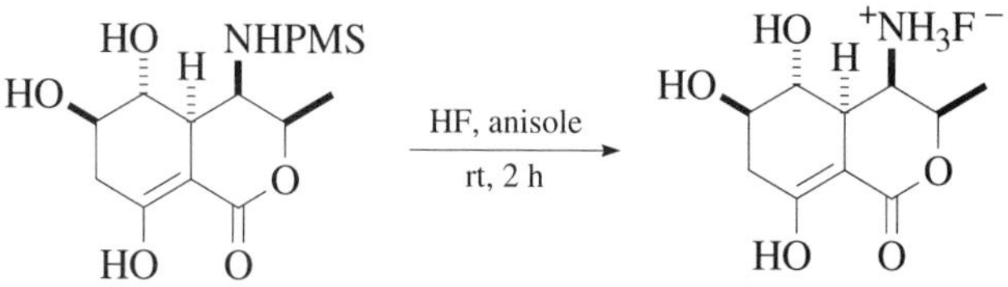

1. H. B. Milne and C.-H. Peng, *J. Am. Chem. Soc.*, **79**, 639, 645 (1957).
2. J. A. Pinock and A. Jurgens, *Tetrahedron Lett.*, 1029 (1979).
3. T. Fukuda, C. Kitada, and M. Fujino, *J. Chem. Soc., Chem. Commun.*, 220 (1978).
4. M. Yoshioka, H. Nakai, and M. Ohno, *J. Am. Chem. Soc.*, **106**, 1133 (1984).

Trifluoromethylsulfonamide: $R_2NSO_2CF_3$ (Chart 10)

A trifluoromethylsulfonamide can be prepared from a primary amine to allow monoalkylation of that amine.[1] The triflamide is not stable to strong base, which causes elimination to an imine,[2] but when used to protect an indole, it is cleaved with K_2CO_3 in refluxing methanol.[6]

Formation

1. $(CF_3SO_2)_2O$, CH_2Cl_2, –78°, ~quant.[1]

Cleavage

1. $NaAlH_2(OCH_2CH_2OCH_3)_2$, benzene, reflux, few min, 95% yield.[1]
2. 4-Br–$C_6H_4COCH_2Br$, K_2CO_3, acetone, 12 h; H_3O^+, 80% yield.[3]
3. $LiAlH_4$, Et_2O, reflux, 90–95% yield.[1,4]
4. Na (NH_3, *t*-BuOH, THF).[5]
5. $BH_3 \cdot THF$, >3h.[6]

1. J. B. Hendrickson and R. Bergeron, *Tetrahedron Lett.*, 3839 (1973).
2. S. Bozec-Ogor, V. Salou-Guiziou, J. J. Yaouanc, and H. Handel, *Tetrahedron Lett.*, **36**, 6063 (1995).
3. J. B. Hendrickson, R. Bergeron, A. Giga, and D. Sternbach, *J. Am. Chem. Soc.*, **95**, 3412 (1973).
4. K. E. Bell, D. W. Knight, and M. B. Gravestock, *Tetrahedron Lett.*, **36**, 8681 (1995).
5. M. L. Edwards, D. M. Stemerick, and J. R. McCarthy, *Tetrahedron Lett.*, **31**, 3417 (1990); D. F. Taber and Y. Wang, *J. Am. Chem. Soc.*, **119**, 22 (1997).
6. M. Lögers, L. E. Overman, and G. S. Welmaker, *J. Am. Chem. Soc.*, **117**, 9139 (1995).

Phenacylsulfonamide: $R_2NSO_2CH_2COC_6H_5$ (Chart 10)

Like the trifluoromethylsulfonamides, phenacylsulfonamides are used to prevent dialkylation of primary amines. Phenacylsulfonamides are prepared in 91–94% yield from the sulfonyl chloride and are cleaved in 66–77% yield by Zn/AcOH/trace HCl.[1]

1. J. B. Hendrickson and R. Bergeron, *Tetrahedron Lett.*, 345 (1970).

t-Butylsulfonamide (Bus–NR_2)

Since *t*-butylsulfonyl chloride is unstable, the sulfonamide must be prepared by an indirect method. It is prepared by reacting the amine with the sulfinyl chloride to give the sulfinamide, which is oxidized with MCPBA or $RuCl_3/NaIO_4$ (CH_2Cl_2, CH_3CN, H_2O, 0°, 75–97% yield) to give the sulfonamide. It is cleaved with TFA or, more efficiently, with TfOH (CH_2Cl_2, anisole, 88–100% yield). Secondary Bus derivatives can be cleaved in the presence of a primary derivative.[1]

1. P. Sun and S. M. Weinreb, *J. Org. Chem.*, **62**, 8604 (1997).

PROTECTION FOR IMIDAZOLES, PYRROLES, INDOLES, AND OTHER AROMATIC HETEROCYCLES

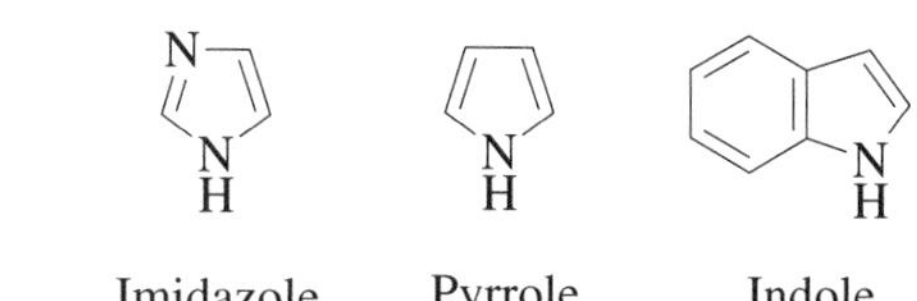

Imidazole Pyrrole Indole

The protective group chemistry of these amines has been separated from that of the simple amines because, chemically, the two classes behave quite differently with respect to protective group cleavage. The increased acidity of these aromatic amines makes it easier to cleave the various amide, carbamate, and sulfonamide groups that are used to protect this class. A similar situation arises in the deprotection of nucleoside bases (e.g., the isobutanamide is cleaved with methanolic ammonia),[1] again, because of the increased acidity of the NH group.

N-Sulfonyl Derivatives

N,N-Dimethylsulfonamide: $R_2N–SO_2NMe_2$

Formation

1. Imidazole, Me_2NSO_2Cl, Et_3N, PhH, 16 h, 95% yield.[2]

Cleavage

1. 2 *M* HCl, reflux, 4 h.[2–4]
2. 2% KOH, H_2O, reflux, 12 h, 64–92% yield.[3] This group is more stable to *n*-BuLi than is the benzyl group in the protection of imidazoles.

Mesitylenesulfonamide (Mts–NR$_2$): $R_2N–SO_2–C_6H_2–2,4,6-(CH_3)_3$

Formation/Cleavage[5]

$$\text{MeOZTrp–OBn} \xrightarrow[\text{Cetyl(Me)}_3\text{N}^+\text{Cl}^-]{\text{MtsCl, NaOH, CH}_2\text{Cl}_2} \text{Z(OMe)Trp(N}^{\text{in}}\text{-Mts)–OBn}$$

BuLi and MtsCl (84% yield) can also be used to protect an indole.[6]
The Mts group is stable to CF_3COOH, 1 *N* NaOH, hydrazine, 4 *N* HCl, 25% HBr–AcOH, and H_2–Pd, but is cleaved with 1 *M* CF_3SO_3H/CF_3COOH/thioanisole or CH_3SO_3H/CF_3COOH/thioanisole. Thioanisole is required to obtain clean conversions. The Mts group is not efficiently cleaved by HF.

***p*-Methoxyphenylsulfonamide (Mps–NR$_2$):** $R_2N–SO_2–C_6H_4–4-OCH_3$

Formation

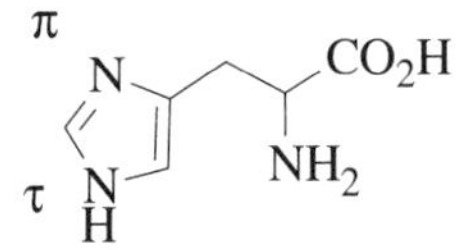

π N, τ N–H, CO_2H, NH_2

Histidine (His)

1. *p*-MeO-$C_6H_4SO_2Cl$, imidazole = His.[7,8]

Cleavage

1. CF_3COOH, Me_2S, 40–60 min, 100%, imidazole = His(Mps).[9]
2. Hydrazine, 1 *N* NaOH, HOBT and HF. [9] The Mps group on histidine is stable to CF_3COOH/anisole and to 25% HBr/AcOH.
3. Mg, MeOH, 60% yield.[10]

Benzenesulfonamide (Bs–NR$_2$): $R_2N–SO_2C_6H_5$

***p*-Toluenesulfonamide (Ts–NR$_2$):** $R_2N–SO_2C_6H_4–4-CH_3$

Formation

1. Imidazole, *p*-toluenesulfonyl chloride, Et_3N.[11,12]
2. Pyrrole, benzenesulfonyl chloride, NaH, DMF, 60% yield.[13]

Cleavage

1. Ac_2O, Pyr; H_2O or trifluoroacetic anhydride, Pyr, 0.5–16 h, 95–100% yield, imidazole = His(Tos).[7,12]

2. 1-Hydroxybenzotriazole (HOBT), THF, 1 h, imidazole = His(Tos).[8]
3. Pyr/HCl, DMF, imidazole = His(Tos).[14]
4. CF_3CO_2H, Me_2S, 40–60 min, 100% yield, imidazole = His(Tos).[15] The related benzenesulfonyl group has been used to protect pyrroles and indoles and is cleaved with NaOH/H_2O/dioxane, rt, 2 h.[16,17]
5. KOH, MeOH, 98% yield (indole deprotection).[18,19] Sodium hydroxide can also be used (pyrrole deprotection).[13]
6. Mg, MeOH, sonication 20–40 min, 100% yield.[20] Sulfonamide-protected amides are also efficiently cleaved by this method.[21]
7. Mg, MeOH, NH_4Cl, benzene, rt.[22]
8. PhSH, AIBN, benzene, reflux, 2 h, 90% yield.[23]
9. The very similar benzenesulfonamide is cleaved with TBAF (THF, reflux, 38–100% yield).[24]

Carbamates

2,2,2-Trichloroethyl Carbamate (Troc–NR_2): $R_2NCO_2CH_2CCl_3$

Formation/Cleavage[25]

$$\text{BOC–TrpOBn} \xrightarrow{\text{TrocCl, NaOH, } Bu_4NH^+SO_4^-} \text{BOC–Trp}(N^{in}\text{–Troc})\text{OBn}$$

The Troc group on tryptophan is stable to CF_3COOH, CF_3SO_3H, and H_2–Pd, but can be cleaved with 0.01 *M* NaOH/MeOH, hydrazine/MeOH/H_2O, or Cd/AcOH/DMF. Cleavage with Zn/AcOH is only partially complete.

2-(Trimethylsilyl)ethyl Carbamate (Teoc–NR_2): $R_2NCO_2CH_2CH_2Si(CH_3)_3$

The Teoc group is introduced onto pyrroles or indoles with 4-nitrophenyl 2-(trimethylsilyl)ethyl carbonate and NaH in 61–64% yield. It can be removed with $Bu_4N^+F^-$ in CH_3CN.[26]

t-Butyl Carbamate (BOC–NR_2): $R_2N–CO_2$–*t*-C_4H_9

Formation

The BOC group has been introduced onto the imidazole nitrogen of histidine with BOCF, pH 7–8;[27] $BOCN_3$, MgO,[28] or $(BOC)_2O$.[26,29] It can be introduced onto pyrroles and indoles with phenyl *t*-butyl carbonate and NaH, 67–91% yield,[30] or with NaH, $BOCN_3$.[31]

Cleavage

The N^{im}-BOC group can be removed under the usual conditions for removing the BOC group: CF_3COOH and HF. It can also be removed with hydrazine and NH_3/MeOH. NaOMe/MeOH/THF has been used to remove the BOC group from pyrroles in 66–99% yield.[31] Thermolysis at 180° cleaves the BOC group from indoles and pyrroles in 92–99% yield.[32]

2,4-Dimethylpent-3-yl Carbamate (Doc–NR$_2$): $[(CH_3)_2CH]_2CHOC(O)NR_2$

The Doc group, introduced with the chloroformate and either DMAP or *t*-BuOK, is quite acid stable, but can be cleaved with TFMSA–thioanisole–EDT–TFA (10 min, rt) or with *p*-cresol-HF (1 h, 0°).[33] The Doc group was found to be suitable for tryptophan protection in *t*-Bu-based peptide synthesis, since no *t*-butylation of tryptophan was observed during acid deprotection.

Cyclohexyl Carbamate (Hoc–NR$_2$): $C_6H_{11}OCONR_2$

The Hoc group was developed for tryptophan protection to minimize alkylation during BOC-mediated peptide synthesis. It is introduced with the chloroformate (NaOH, CH_2Cl_2, $Bu_4N^+HSO_4^-$) and can be cleaved with HF without the need to include thiols in the cleavage mixture.[34]

1,1-Dimethyl-2,2,2-trichloroethyl Carbamate (TcBOC–NR$_2$): $R_2NCO_2C(CH_3)_2CCl_3$

Formation/Cleavage[35]

Indole (R, R′, N–H) → $Cl_3CC(Me)_2OCOCl$, TEA, CH_2Cl_2, 80% → N–C(O)OC(Me)$_2$CCl$_3$ indole (R, R′)

Reverse: Zn, AcOH, MeOH, 20 min, 76%

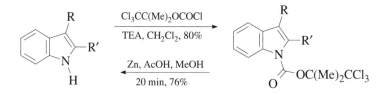

1-Adamantyl Carbamate (1-Adoc–NR$_2$): R_2NCO_2–1-adamantyl

Formation

1. 1-AdocCl, histidine, NaOH, Na_2CO_3, H_2O, 86% yield; forms N$^{\alpha}$,N^{im}-(Adoc)$_2$–HisOH.[36]

Cleavage

The 1-Adoc group can be cleaved by the same methods used to cleave the BOC group.[36] The 1-Adoc group is somewhat more stable than the BOC group to acid.

2-Adamantyl Carbamate (2-Adoc–NR$_2$): R_2NCO_2–2-adamantyl

Formation

2-Adoc–Cl, aq. NaOH, dioxane, 76% yield for His isolated as the cyclohexylamine salt.[37]

Cleavage

The 2-Adoc group is stable to TFA, but cleaved completely within 10 min with 25% HBr/AcOH, HF, and TFMSA/thioanisole/TFA. Under basic conditions, the group is slowly cleaved in 10% aq. TEA or 20% piperidine/DMF, but rapidly cleaved in 2 mol dm^{-3} aq. NaOH.[37]

N-Alkyl and *N*-Aryl Derivatives

***N*-Vinylamine:** CH_2=CH–NR$_2$

The vinyl group has been used to protect the nitrogen of benzimidazole during metallation with lithium diisopropylamide. It is introduced with vinyl acetate [Hg(OAc)$_2$, H_2SO_4, reflux, 24 h] or dibromoethane (TEA, reflux; 10% aq. NaOH reflux)[38] and cleaved by ozonolysis (MeOH, –78°)[39] or $KMnO_4$ (acetone, reflux, 99% yield).[38]

***N*-2-Chloroethylamine:** $R_2NCH_2CH_2Cl$

Formation/Cleavage[40]

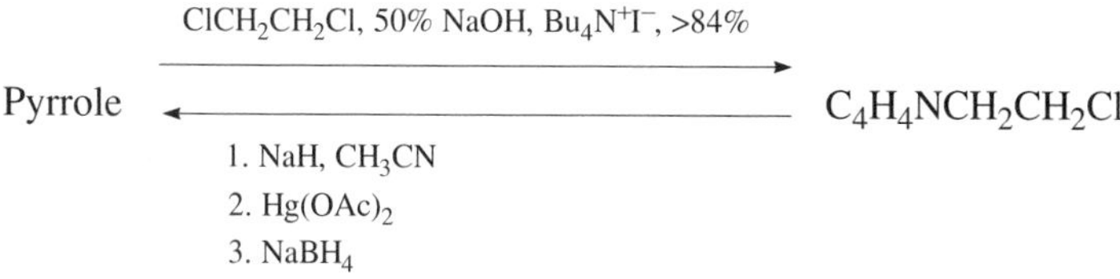

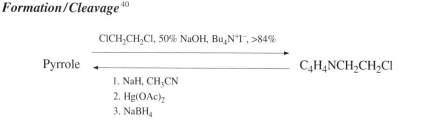

***N*-(1-Ethoxy)ethylamine, EE–NR$_2$:** $R_2NCH(OCH_2CH_3)CH_3$

Formation/Cleavage[41]

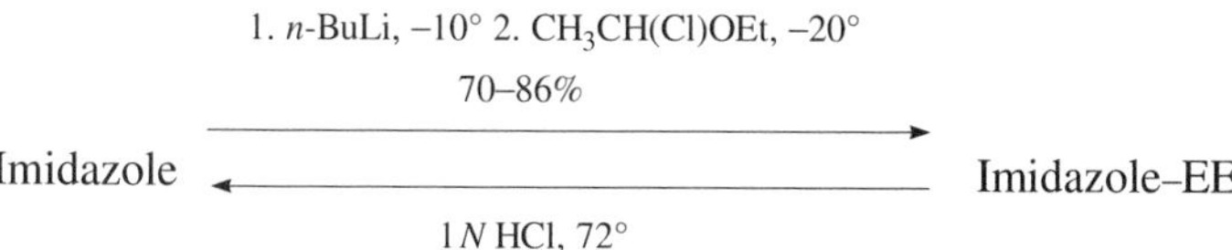

N-2-(2′-Pyridyl)ethyl- and N-2-(4′-Pyridyl)ethylamine:
$R_2NCH_2CH_2$–2-(C_5H_4N) and $R_2NCH_2CH_2$–4-(C_5H_4N)

Formation/Cleavage

$$=NH \underset{\text{AlCl}_3\text{, ClCH}_2\text{CH}_2\text{Cl; NaOH, 18–93\%; or MeI, acetone, 25°; NaOH}^{43}}{\overset{\text{2- or 4-vinylpyridine, AcOH, 22–94\%}^{42}\text{ or Na, 4-vinylpyridine}^{43}}{\rightleftarrows}} =NCH_2CH_2Pyr$$

A series of substituted benzimidazoles and pyrroles was protected and deprotected using this methodology.

N-2-(4-Nitrophenyl)ethylamine (PNPE–NR_2): 4-$NO_2C_6H_4CH_2CH_2NR_2$

The PNPE group is cleaved from a pyrrole with DBU (CH_3CN, rt, 81% yield).[44, 45]

N-Trialkylsilylamines: $R_2N–SiR'_3$

Pyrroles and indoles can be protected with the **t-butyldimethylsilyl** group by treatment with TBDMSCl and n-BuLi or NaH.[46] **Triisopropylsilyl chloride** (NaH, DMF, 0°–rt, 73% yield) has been used to protect the pyrrole nitrogen in order to direct electrophilic attack to the 3-position.[47] It has also been used to protect an indole.[48,49] This derivative can be prepared from the silyl chloride and K.[50] The silyl protective group is cleaved with $Bu_4N^+F^-$, THF, rt or with CF_3COOH.

N-Allylamine: CH_2=$CHCH_2NR_2$

Guanine is catalytically protected at the 9-position with allyl acetate [($Pd(Ph_3P)_4$, Cs_2CO_3, DMSO, 68% yield)].[51] The τ-nitrogen of BOC-protected histidine is protected by bisalkylation with allyl bromide followed by removal of the N-π allyl group with $Pd(Ph_3P)_4$ (Et_2NH, $NaHCO_3$ or $PhSiH_3$, 80–85% yield). Removal of the allyl group is achieved by palladium-catalyzed transfer of the allyl group to N,N′-dimethylbarbituric acid.[52]

N-Benzylamine (Bn–NR_2): $PhCH_2$–NR_2

Formation

1. BnCl, NH_3, Na.[53]

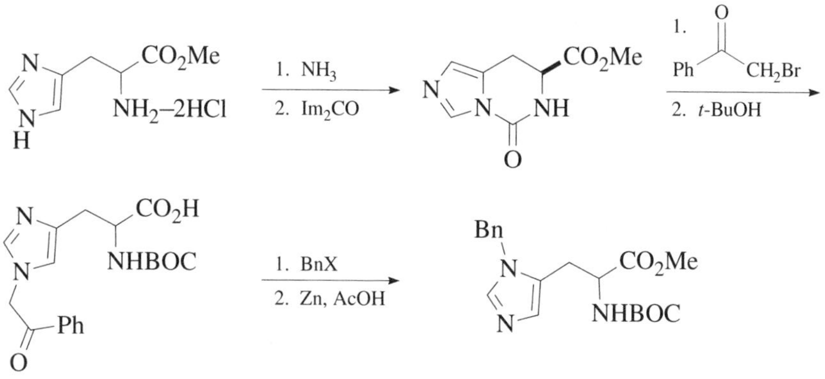

The following benzyl halides were used: $PhCH_2Br$, 82% yield; $PhCH(CH_3)Br$, 33% yield; $(Ph)_2CHBr$, 50% yield; 3,4-$(MeO)_2C_6H_3CH_2Cl$, 52% yield.[54]

2. From an electron-deficient sodium imidazolide: $PhCH_2OP^+(NMe_2)_3\ PF_6^-$, DMF, 24 h, heat, 40% yield.[55]
3. From indole: dibenzyl oxylate, *t*-BuOK, DMF, reflux, 86% yield.[56]
4. MeLi, BnBr, THF, –40° to rt, 39–74% yield.[57]
5. BnBr, NaH, DMF or DMSO, rt–50°, 57–75% yield.[58]

Cleavage

1. Cyclohexadiene, Pd-black, 25°, 100% yield, imidazole = His(Bn).[59] With H_2/Pd–C, the normal conditions for benzyl group removal, it is difficult to remove the benzyl group on histidine without also causing the reduction of other aromatic groups that may be present.[60]
2. $AlCl_3$, benzene or anisole, reflux, 25–91% yield, cleaved from a pyrido-[2,3-*b*]indole[61] and indole.[58]

N-*p*-Methoxybenzylamine (MPM–NR_2): $R_2N–CH_2C_6H_4–4\text{-}OCH_3$

The MPM group was used in the preparation of a variety of triazoles,[62] imidazoles,[63] and pyrazoles.[64] It is readily cleaved with CF_3COOH at 65° (52–100% yield). It is also cleaved from a pyrido[2,3-*b*]indole with DDQ, 88% yield.[61]

N-3,4-Dimethoxybenzylamine: 3,4-$(MeO)_2C_6H_3CH_2NR_2$

A 3,4-dimethoxybenzyl derivative, cleaved by acid (concd. H_2SO_4/anhyd. CF_3COOH, anisole), was used to protect a pyrrole –NH group during the synthesis of a tetrapyrrole pigment precursor. Neither an *N*-benzyl nor an *N*-*p*-methoxybenzyl derivative could be cleaved satisfactorily. Hydrogenolysis of the benzyl derivatives led to cyclohexyl compounds; acidic cleavage resulted in migration of the benzyl groups to the free α-position.[65]

N-3-Methoxybenzylamine and N-3,5-Dimethoxybenzylamine: 3-(MeO)$C_6H_4CH_2NR_2$ and 3,5-$(MeO)_2C_6H_3CH_2NR_2$

These benzylamines have been used for the protection of adenine and can be cleaved by photolysis at 254 nm.[66]

N-2-Nitrobenzylamine (ONB–NR_2): $R_2N–CH_2C_6H_4$–2-NO_2 (Chart 10)

Formation

1. BOC–His(N^{im}Ag)OMe, 2-NO_2–$C_6H_4CH_2Br$, PhH, 4 h, reflux.[67]

Cleavage

1. *hv*, dioxane, 1 h, 100% yield.[67,68] The ONB group is stable to CF_3COOH, to HCl–AcOH, and to NaOH–MeOH, but is slowly cleaved by hydrogenation.

The related **4-nitrobenzyl** group, used to protect a benzimidazole, can be cleaved with H_2O_2 (EtOH, NaOH, 50°, 72% yield).[69]

N-2,4-Dinitrophenylamine (DNP–NR_2): 2,4-$(NO_2)_2$–$C_6H_3NR_2$ (Chart 10)

The dinitrophenyl group has been used to protect the imidazole –NH group in histidines (45% yield)[70] by reaction with 2,4-dinitrofluorobenzene and potassium carbonate. Imidazole –NH groups, but not α-amino acid groups, are quantitatively regenerated by reaction with 2-mercaptoethanol (22°, pH 8, 1 h).[71] The 2,4-dinitrophenyl group on the N^{im} of histidine reduces racemization in peptide synthesis because of its electron-withdrawing character.[72] In Fmoc-based, peptide synthesis, the DNP group is not stable, because it migrates to the ε-NH_2 group of lysine.[73]

N-Phenacylamine: $R_2NCH_2COC_6H_5$ (Chart 10)

The phenacyl group is stable to HBr–AcOH, CF_3COOH, and CF_3SO_3H.[74] It is used to protect the π-nitrogen in histidine in order to reduce racemization during peptide-bond formation.[75]

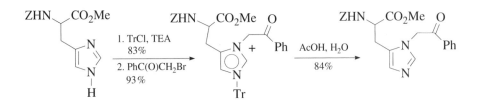

ZHN–CH(CO_2Me)–CH$_2$–(imidazolyl, N–H) → 1. TrCl, TEA, 83%; 2. PhC(O)CH_2Br, 93% → ZHN–CH(CO_2Me)–CH$_2$–(imidazolium, N^+–CH_2C(O)Ph, N–Tr) → AcOH, H_2O, 84% → ZHN–CH(CO_2Me)–CH$_2$–(imidazolyl, N–CH_2C(O)Ph)

N-Triphenylmethylamine, (Tr–NR$_2$) and N-Diphenylmethylamine (Dpm –NR$_2$): R_2NCPh_3 and R_2NCHPh_2

Formation

1. BOC–His, TrCl, Pyr.[76]
2. From a tetrazole: TrCl, CH_2Cl_2, TBAB, NaOH, H_2O.[77]

Cleavage

The trityl group can be cleaved with HBr–AcOH, 2 h; CF_3COOH, 30 min; formic acid, 2 min, and by hydrogenation.[78] The trityl group in BOC–His(Tr)OH is stable to 1 *M* HCl/AcOH, rt, 20 h. The **diphenylmethyl** group was introduced in the same manner as the trityl group. It is more stable to acid than the trityl group, but not significantly.[76,78] The trityl group has also been used to protect simple imidazoles.[79]

The following table gives the comparative stabilities of the N^{α}-Tr, N^{Im}-Tr and *N*-BOC groups of Tr–His(Tr)–Lys(BOC)–OMe to various acidic conditions:[80]

Cleavage Conditions	% Cleavage		
	N^{α}-Tr	N^{Im}-Tr	*N*-BOC
5% HCO_2H, $ClCH_2CH_2Cl$, 8 min, 20°	100	1	0
$ClCH_2CH_2Cl$, MeOH, TEA, 5 min, 20°	100	<1	0
2.5 eq. HCl in 90% AcOH, 1 min, 20°	100	<1	<1
1 *N* HCl in 90% AcOH, 20 min, 20°	100	<1	100
90% AcOH, 1.5 h, 60°	100	100	<1
5% Pyr·HCl, in MeOH, 2 h, 60°	100	100	<1
95% TFA, 1 h, 20°	100	100	100

N-(Diphenyl-4-pyridylmethyl)amine (Dppm–NR$_2$):
$R_2N–C(Ph)_2–4-(C_5H_4N)$ (Chart 10)

Formation

1. Ph_2–4-(C_5H_4N)CCl, Et_3N, $CHCl_3$, (Z)- or (BOC)-HisOMe.[81,82]

Cleavage

The diphenyl-4-pyridylmethyl group is cleaved by Zn/AcOH, 1.5 h, 91% yield; H_2/Pd–C, 91% yield; or by electrolytic reduction, 2.5 h, 0°, 87% yield. The Dppm group is stable to trifluoroacetic acid.[81,83]

N-(N′,N′-Dimethyl)hydrazine: $R_2N–NMe_2$

The dimethylamine group can be cleaved from a pyrrole in low yield with chromous acetate.[84]

Amino Acetal Derivatives

***N*-Hydroxymethylamine:** $HOCH_2–NR_2$

Formation/Cleavage[85]

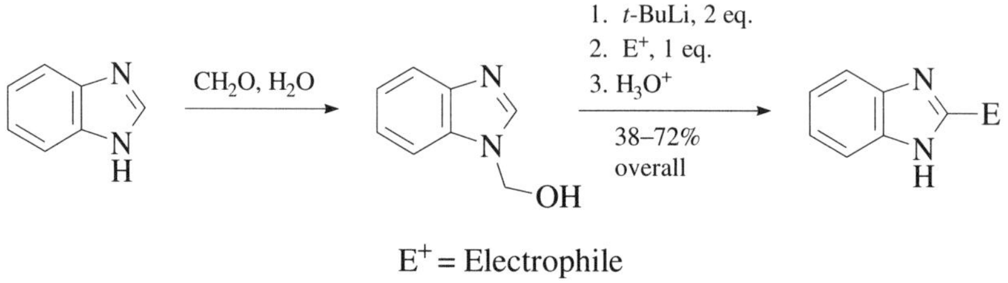

1. *t*-BuLi, 2 eq.
2. E^+, 1 eq.
3. H_3O^+

CH_2O, H_2O

38–72% overall

E^+ = Electrophile

***N*-Methoxymethylamine (MOM–NR$_2$):** $R_2NCH_2OCH_3$ (Chart 10)

The MOM group is introduced onto an indole through the sodium salt (NaOH, DMSO, 0°, 0.5 h; MOMCl, 22°, 0.5 h, 90% yield). It is removed with $BF_3·Et_2O$ (Ac_2O, LiBr, 20°, 48 h, 86% yield).[86]

Removal of the the related **ethoxymethyl** group from an imidazole with 6 *N* HCl at reflux is slow and low yielding.[87] Small structural effects at a site seemingly remote from the MOM group can have a significant influence on the deprotection process. The MOM group in compound **a** is easily removed with acid, but the cleavage with HCl in compound **b** proved quite difficult.[88]

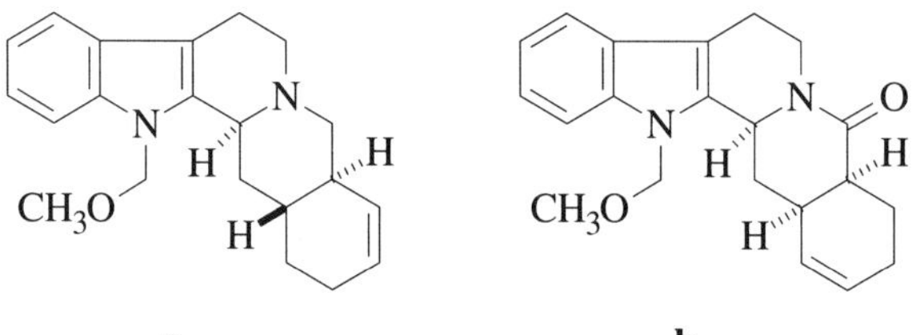

a **b**

***N*-Diethoxymethylamine (DEM–NR$_2$):** $(EtO)_2CH–NR_2$

Formation/Cleavage[89,90]

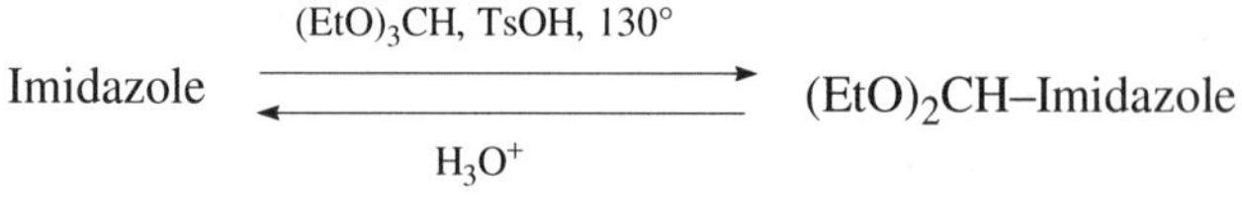

Imidazole —[$(EtO)_3CH$, TsOH, 130°]→ / ←[H_3O^+]— $(EtO)_2CH$–Imidazole

DEM protection of an indole is also effective (46–82% yield), and cleavage occurs efficiently with 2 *N* HCl (EtOH, rt, 0.5 h, 86–93% yield).[91]

_N_-(2-Chloroethoxy)methylamine: $R_2NCH_2OCH_2CH_2Cl$

This derivative has been prepared from an indole, the chloromethyl ether, and potassium hydride in 50% yield; it is cleaved in 84% yield by potassium cyanide/18-crown-6 in refluxing acetonitrile.[92]

_N_-[2-(Trimethylsilyl)ethoxy]methylamine (SEM–NR$_2$):
$R_2NCH_2OCH_2CH_2Si(CH_3)_3$

Formation

1. Imidazole, indole or pyrrole, NaH, SEMCl, 50–85% yield.[93–95]

Cleavage

1. 1 *M* $Bu_4N^+F^-$, THF, reflux, 45 min, 46–90% yield or dil. HCl.[93,94]
2. $BF_3 \cdot Et_2O$; base.[96,97]
3. $Bu_4N^+F^-$, ethylenediamine (ethylenediamine was used as a formaldehyde scavenger), 45–98% yield.[96] Neat TBAF under vacuum has been used (90% yield).[98]
4. 3 *M* HCl, EtOH, reflux, 1 h, 95% yield.[99]
5. PPTS, MeOH, 24 h.[100]

_N-t_-Butoxymethylamine (Bum–NR$_2$): R_2NCH_2O–*t*-C_4H_9

The Bum derivative has been used to protect the π-nitrogen of histidine to prevent racemization during peptide bond formation.[101] The related **1-adamantyloxymethylamine** has been used similarly for histidine protection.[102]

_N-t_-Butyldimethylsiloxymethylamine

The N-9 position of adenine was protected by formylation with basic formalin followed by silylation with TBDMSCl in Pyr, 86% yield. This group is removed with TFA/H_2O, 20°, 2 h.[103]

_N_-Pivaloyloxymethylamine (POM–NR$_2$): $R_2NCH_2OCOC(CH_3)_3$ (Chart 10)

The POM group is introduced onto pyrroles and indoles by treatment with NaH, $(CH_3)_3CCO_2CH_2Cl$[104] in THF at rt in 65–78% yield.[105] It is removed by hydrolysis with MeOH, NaOH[105], or NH_3, MeOH (25°, 4 h, 30–80% yield).[106]

_N_-Benzyloxymethylamine (BOM–NR$_2$): $R_2NCH_2OCH_2C_6H_5$ (Chart 10)

The BOM group is introduced onto an indole with the chloromethyl ether and sodium hydride in 80–90% yield. It is cleaved in 92% yield by catalytic reduction followed by basic hydrolysis,[107,108] or by CF_3COOH, HBr or 6 *M* HCl at

110°.[109] It has been used to protect the π-nitrogen of histidine, preventing racemization during peptide bond formation. It has also been used to protect the τ-nitrogen of histidine ($BnOCH_2Cl$, Et_2O; Et_3N, MeOH).[110] During protective group cleavage of BOM-protected histidine, the formaldehyde that is liberated can react with *N*-terminal cysteine residues to form thiazolidines.[111,112]

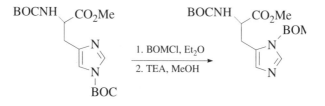

BOCNH CO₂Me … 1. BOMCl, Et_2O 2. TEA, MeOH → BOCNH CO₂Me BOM

N-Dimethylaminomethylamine: $(CH_3)_2NCH_2NR_2$

An indole protected by a Mannich reaction with formaldehyde and dimethylamine is stable to lithiation. The protective group is removed with $NaBH_4$ (EtOH, THF, reflux).[113] The related piperidine analogue has been used similarly for the protection of a triazole.[114]

N-2-Tetrahydropyranylamine (THP–NR$_2$):
R_2N–2-tetrahydropyranyl (Chart 10)

The THP derivative of the imidazole nitrogen in purines has been prepared by treatment with dihydropyran (TsOH, 55°, 1.5 h, 50–85% yield). It is cleaved by acid hydrolysis.[115] The THP group is useful for the protection of 1,2,4-triazoles.[116]

Amides

Carbon Dioxide: CO_2

The *in situ* generation of the carbon dioxide adduct of an indole provides sufficient protection and activation of an indole for metalation at C-2 with *t*-butyllithium. The lithium reagent can be quenched with an electrophile, and quenching of the reaction with water releases the carbon dioxide.[117, 118]

Formamide: R_2N–CHO

Formation[119]***/Cleavage***[120]

$$\text{Tryptophan} \xrightarrow{HCO_2H,\ HCl} \text{Tryptophan } (N^{in}\text{–CHO})$$

The formyl group is cleaved with HF/anisole/$(CH_2SH)_2$.[120] It is also cleaved at pH 9–10.[119]

N,N-Diethylureide: $(CH_3CH_2)_2NC(O)NR_2$

The ureide, which is stable to BuLi, was used for the protection of indole. It is cleaved with 25% NaOH in EtOH, reflux.[121]

Dichloroacetamide: $Cl_2CHCONR_2$

The dichloroacetamide of indole, formed by refluxing a mixture of dichloroacetyl chloride in dichloroethane, is cleaved upon treatment with TEA (CH_2Cl_2, rt).[122]

Pivalamide: $(CH_3)_3CCONR_2$

A pivalamide of an indole, introduced with PvCl (NaH, DMF, 0°, 1 h, 96% yield), is efficiently cleaved with MeSNa (MeOH, 20°, 2 h, 96% yield).[123]

Diphenylthiophosphinamide: $Ph_2P(S)–NR_2$

This group was used to protect the tryptophan nitrogen.

Formation

$Ph_2P(S)Cl$, $NaHSO_4$, NaOH, CH_2Cl_2, 0°, 88% yield.[124]

Cleavage

1. 0.25 *M* Methanesulfonic acid, thioanisole in CF_3COOH, 0°, 90 min.[124]
2. 0.25 *M* Trifluoromethanesulfonic acid, 0.25 *M* thioanisole in CF_3COOH, 0°, 50 min.[124]
3. 0.1 *M* $Bu_4N^+F^-$, DMSO or DMF, 25°, 10 min.[124,125]
4. 0.5 *M* KF, 18-crown-6, CH_3CN, 25°, 3 h.[124]

1. H. Büchi and H. G. Khorana, *J. Mol. Biol.*, **72**, 251 (1972).
2. D. J. Chadwick and R. I. Ngochindo, *J. Chem. Soc., Perkin Trans. 1*, 481 (1984).
3. A. J. Carpenter and D. J. Chadwick, *Tetrahedron*, **42**, 2351 (1986).
4. S. Harusawa, Y. Murai, H. Moriyama, T. Imazu, H. Ohishi, R. Yoneda, and T. Kurihara, *J. Org. Chem.*, **61**, 4405 (1996).
5. N. Fujii, S. Futaki, K. Yasumura, and H. Yajima, *Chem. Pharm. Bull.*, **32**, 2660 (1984).
6. L. W. Boteju, K. Wegner, X. Qian, and V. J. Hruby, *Tetrahedron*, **50**, 2391 (1994).
7. J. M. van der Eijk, R. J. M. Nolte, and J. W. Zwikker, *J. Org. Chem.*, **45**, 547 (1980).
8. T. Fujii and S. Sakakibara, *Bull. Chem. Soc. Jpn.*, **47**, 3146 (1974).
9. K. Kitagawa, K. Kitade, Y. Kiso, T. Akita, S. Funakoshi, N. Fujii, and H. Yajima, *J. Chem. Soc., Chem. Commun.*, 955 (1979).

10. B. Danieli, G. Lesma, M. Martinelli, D. Passarella, and A. Silvani, *J. Org. Chem.*, **62**, 6519 (1997).
11. S. Sakakibara and T. Fujii, *Bull. Chem. Soc. Jpn.*, **42**, 1466 (1969).
12. E. Wuensch in *Methoden der Organischen Chemie* (*Houben-Weyl*), E. Mueller, Ed., Georg Thieme Verlag, Stuttgart, 1974, Vol. 15/1, p. 223.
13. C. F. Masaguer, E. Ravina, and J. Fueyo, *Heterocycles*, **34**, 1303 (1992).
14. H. C. Beyerman, J. Hirt, P. Kranenburg, J. L. M. Syrier, and A. Van Zon, *Recl. Trav. Chim. Pays-Bas*, **93**, 256 (1974).
15. K. Kitagawa, K. Kitade, Y. Kiso, T. Akita, S. Funakoshi, N. Fujii, and H. Yajima, *Chem. Pharm. Bull.*, **28**, 926 (1980).
16. J. Rokach, P. Hamel, M. Kakushima, and G. M. Smith, *Tetrahedron Lett.*, **22**, 4901 (1981).
17. W. A. Remers, R. H. Roth, G. J. Gibs, and M. J. Weiss, *J. Org. Chem.*, **36**, 1232 (1971).
18. A. P. Kozikowski and Y.-Y. Chen, *J. Org. Chem.*, **46**, 5248 (1981).
19. M. G. Saulnierand and G. W. Gribble, *J. Org. Chem.*, **47**, 2810 (1982).
20. Y. Yokoyama, T. Matsumoto, and Y. Murakami, *J. Org. Chem.*, **60**, 1486 (1995).
21. B. Nyasse, L. Grehn, and U. Ragnarsson, *J. Chem. Soc., Chem. Commun.*, 1017 (1997).
22. H. Ishibashi, T. Tabata, K. Hanaoka, H. Iriyama, S. Akamatsu, and M. Ikeda, *Tetrahedron Lett.*, **34**, 489 (1993).
23. S. H. Kim, I. Figueroa, and P. L. Fuchs, *Tetrahedron Lett.*, **38**, 2601 (1997).
24. A. Yasuhara and T. Sakamoto, *Tetrahedron Lett.*, **39**, 595 (1998).
25. Y. Kiso, M. Inai, K. Kitagawa, and T. Akita, *Chem. Lett.*, 739 (1983).
26. L. Grehn and U. Ragnarsson, *Angew. Chem., Int. Ed. Engl.*, **23**, 296 (1984).
27. E. Schnabel, H. Herzog, P. Hoffmann, E. Klauke, and I. Ugi, *Justus Liebigs Ann. Chem.*, **716**, 175 (1968); E. Schnabel, J. Stoltefuss, H. A. Offe, and E. Klauke, *Justus Liebigs Ann. Chem.*, **743**, 57 (1971).
28. M. Fridkin and H. J. Goren, *Can. J. Chem.*, **49**, 1578 (1971).
29. V. F. Pozdnev, *Zh. Obshch. Khim.*, **48**, 476 (1978); *Chem. Abstr.*, **89**: 24739m (1978).
30. D. Dhanak and C. B. Reese, *J. Chem. Soc., Perkin Trans. 1*, 2181 (1986).
31. I. Hasan, E. R. Marinelli, L.-C. C. Lin, F. W. Fowler, and A. B. Levy, *J. Org. Chem.*, **46**, 157 (1981).
32. V. H. Rawal and M. P. Cava, *Tetrahedron Lett.*, **26**, 6141 (1985).
33. A. Karström and A. Undén, *J. Chem. Soc., Chem. Commun.*, 1471 (1996).
34. Y. Nishiuchi, H. Nishio, T. Inui, T. Kimura, and S. Sakakibara, *Tetrahedron Lett.*, **37**, 7529 (1996).
35. S. Raucher, J. E. Macdonald, and R. F. Lawrence, *J. Am. Chem. Soc.*, **103**, 2419 (1981).
36. W. L. Haas, E. V. Kromkalns, and K. Gerzon, *J. Am. Chem. Soc.*, **88**, 1988 (1966).
37. Y. Nishiyama, N. Shintomi, Y. Kondo, T. Izumi, and Y. Okada, *J. Chem. Soc., Perkin Trans. 1*, 2309 (1995).

38. D. J. Hartley and B. Iddon, *Tetrahedron Lett.*, **38**, 4647 (1997).
39. Y. L. Chen, K. G. Hedberg, and K. J. Guarino, *Tetrahedron Lett.*, **30**, 1067 (1989).
40. C. Gonzalez, R. Greenhouse, R. Tallabs, and J. M. Muchowski, *Can. J. Chem.*, **61**, 1697 (1983).
41. T. S. Manoharan and R. S. Brown, *J. Org. Chem.*, **53**, 1107 (1988).
42. M. Ichikawa, C. Yamamoto, and T. Hisano, *Chem. Pharm. Bull.*, **29**, 3042 (1981).
43. A. R. Katritzky, G. R. Khan, and C. M. Marson, *J. Heterocycl. Chem.*, **24**, 641 (1987).
44. B. Santiago, C. R. Dalton, E. W. Huber, and J. M. Kane, *J. Org. Chem.*, **60**, 4947 (1995).
45. E. D. Edstrom and Y. Wei, *J. Org. Chem.*, **60**, 5069 (1995).
46. B. H. Lipshutz, B. Huff, and W. Hagen, *Tetrahedron Lett.*, **29**, 3411 (1988).
47. J. M. Muchowski and D. R. Solas, *Tetrahedron Lett.*, **24**, 3455 (1983).
48. P. J. Beswick, C. S. Greenwood, T. J. Mowlem, G. Nechvatal, and D. A. Widdowson, *Tetrahedron*, **44**, 7325 (1988).
49. M. Iwao, *Heterocycles*, **36**, 29 (1993).
50. K. P. Stefan, W. Schuhmann, H. Parlar, and F. Korte, *Chem. Ber.*, **122**, 169 (1989).
51. L. L. Gundersen, T. Benneche, F. Rise, A. Gogoll, and K. Undheim, *Acta Chem. Scand.*, **46**, 761 (1992).
52. A. M. Kimbonguila, S. Boucida, F. Guibé, and A. Loffet, *Tetrahedron*, **53**, 12525 (1997).
53. V. du Vigneaud and O. K. Behrens, *J. Biol. Chem.*, **117**, 27 (1937).
54. C. J. Chivikas and J. C. Hodges, *J. Org. Chem.*, **52**, 3591 (1987).
55. M. Searcey, J. B. Lee, and P. L. Pye, *Chem. Ind.* (*London*), 569 (1989).
56. J. Bergman, P. Ola Norrby, and P. Sand, *Tetrahedron*, **46**, 6113 (1990).
57. H. Suzuki, A. Tsukuda, M. Kondo, M. Aizawa, Y. Senoo, M. Nakajima, T. Watanabe, Y. Yokoyama, and Y. Murakami, *Tetrahedron Lett.*, **36**, 1671 (1995).
58. T. Watanabe, A. Kobayashi, M. Nishiura, H. Takahashi, T. Usui, I. Kamiyama, N. Mochizuki, K. Noritake, Y. Yokoyama, and Y. Murkami, *Chem. Pharm. Bull.*, **39**, 1152 (1991).
59. A. M. Felix, E. P. Heimer, T. J. Lambros, C. Tzougraki, and J. Meienhofer, *J. Org. Chem.*, **43**, 4194 (1978).
60. E. C. Jorgensen, G. C. Windridge, and T. C. Lee, *J. Med. Chem.*, **13**, 352 (1970).
61. I. T. Forbes, C. N. Johnson, and M. Thompson, *J. Chem. Soc., Perkin Trans. 1*, 275 (1992).
62. D. R. Buckle and C. J. M. Rockell, *J. Chem. Soc., Perkin Trans. 1*, 627 (1982).
63. T. Kamijo, R. Yamamoto, H. Hirada, and K. Iizuka, *Chem. Pharm. Bull.*, **31**, 1213 (1983).
64. C. Subramanyam, *Synth. Commun.*, **25**, 761 (1995).
65. M. I. Jones, C. Froussios, and D. A. Evans, *J. Chem. Soc., Chem. Commun.*, 472 (1976).
66. A. Er-Rhaimini, N. Mohsinaly, and R. Mornet, *Tetrahedron Lett.*, **31**, 5757 (1990).
67. S. M. Kalbag and R. W. Roeske, *J. Am. Chem. Soc.*, **97**, 440 (1975).

68. T. Voelker, T. Ewell, J. Joo, and E. D. Edstrom, *Tetrahedron Lett.*, **39**, 359 (1998).
69. R. Balasuriya, S. J. Chandler, M. J. Cook, and D. J. Hardstone, *Tetrahedron Lett.*, **24**, 1385 (1983).
70. E. Siepmann and H. Zahn, *Biochim. Biophys. Acta*, **82**, 412 (1964).
71. S. Shaltiel, *Biochem. Biophys. Res. Commun.*, **29**, 178 (1967).
72. M. C. Lin, B. Gutte, D. G. Caldi, S. Moore, and R. B. Merrifield, *J. Biol. Chem.*, **247**, 4768 (1972).
73. J.-C. Gesquière, J. Najib, T. Letailleur, P. Maes, and A. Tartar, *Tetrahedron Lett.*, **34**, 1921 (1993).
74. A. R. Fletcher, J. H. Jones, W. I. Ramage, and A. V. Stachulski, in *Peptides 1978*, I. Z. Siemion and G. Kupryszewski, Eds., Wroclaw University Press, Wroclaw, Poland, 1979, pp. 168–171.
75. A. R. Fletcher, J. H. Jones, W. I. Ramage, and A. V. Stachulski, *J. Chem Soc., Perkin Trans. 1*, 2261 (1979).
76. G. Losse and U. Krychowski, *J. Prakt. Chem.*, **312**, 1097 (1970).
77. B. E. Huff, M. E. LeTourneau, M. A. Staszak, and J. A. Ward, *Tetrahedron Lett.*, **37**, 3655 (1996).
78. G. Losse and U. Krychowski, *Tetrahedron Lett.*, 4121 (1971).
79. N. J. Curtis and R. S. Brown, *J. Org. Chem.*, **45**, 4038 (1980); K. L. Kirk, *J. Org. Chem.*, **43**, 4381 (1978); J. L. Kelley, C. A. Miller, and E. W. McLean, *J. Med. Chem.*, **20**, 721 (1977).
80. P. Sieber and B. Riniker, *Tetrahedron Lett.*, **28**, 6031 (1987).
81. S. Coyle and G. T. Young, *J. Chem. Soc., Chem. Commun.*, 980 (1976).
82. S. Coyle, O. Keller, and G. T. Young, *J. Chem. Soc., Chem. Commun.*, 939 (1975).
83. S. Coyle, A. Hallett, M. S. Munns, and G. T. Young, *J. Chem. Soc., Perkin Trans. 1*, 522 (1981).
84. G. R. Martinez, P. A. Grieco, E. Williams, K.-i. Kanai, and C. V. Srinivasan, *J. Am. Chem. Soc.*, **104**, 1436 (1982).
85. A. R. Katritzky and K. Akutagawa, *J. Org. Chem.*, **54**, 2949 (1989).
86. R. J. Sundberg and H. F. Russell, *J. Org. Chem.*, **38**, 3324 (1973).
87. T. P. Demuth, Jr., D. C. Lever, L. M. Gorgos, C. M. Hogan, and J. Chu, *J. Org. Chem.*, **57**, 2963 (1992).
88. A. I. Meyers, T. K. Highsmith, and P. T. Bounora, *J. Org. Chem.*, **56**, 2960 (1991).
89. N. J. Curtis and R. S. Brown, *J. Org. Chem.*, **45**, 4038 (1980).
90. S. Ohta, M. Matsukawa, N. Ohashi, and K. Nagayama, *Synthesis*, 78 (1990).
91. P. Gmeiner, J. Kraxner, and B. Bollinger, *Synthesis*, 1196 (1996).
92. A. J. Hutchison and Y. Kishi, *J. Am. Chem. Soc.*, **101**, 6786 (1979).
93. J. P. Whitten, D. P. Matthews, and J. R. McCarthy, *J. Org. Chem.*, **51**, 1891 (1986).
94. B. H. Lipshutz, W. Vaccaro, and B. Huff, *Tetrahedron Lett.*, **27**, 4095 (1986).
95. M. P. Edwards, A. M. Doherty, S. V. Ley, and H. M. Organ, *Tetrahedron*, **42**, 3723 (1986).
96. J. M. Muchowski and D. R. Solas, *J. Org. Chem.*, **49**, 203 (1984).
97. C. R. Dalton, J. M. Kane, and D. Rampe, *Tetrahedron Lett.*, **33**, 5713 (1992).

98. O. A. Moreno and Y. Kishi, *J. Am. Chem. Soc.*, **118**, 8180 (1996).
99. D. P. Matthews, J. P. Whitten, and J. R. McCarthy, *J. Heterocycl. Chem.*, **24**, 689 (1987).
100. J. G. Phillips, L. Fadnis, and D. R. Williams, *Tetrahedron Lett.*, **38**, 7835 (1997).
101. R. Colombo, F. Colombo, and J. H. Jones, *J. Chem. Soc., Chem. Commun.*, 292 (1984).
102. Y. Okada, J. Wang, T. Yamamoto, Y. Mu, and T. Yokoi, *J. Chem. Soc., Perkin Trans.1*, 2139 (1996); Y. Okada, J. Wang, T. Yamamoto, and Y. Mu, *Chem. Pharm. Bull.*, **44**, 871 (1996).
103. G. C. Magnin, J. Dauvergne, A. Burger, and J.-F. Biellmann, *Tetrahedron Lett.*, **37**, 7833 (1996).
104. For a preparation of the chloride and iodide, see P. P. Iyer, D. Yu, N.-h. Ho, and S. Agrawal, *Synth. Commun.*, **25**, 2739 (1995).
105. D. Dhanak and C. B. Reese, *J. Chem. Soc., Perkin Trans. 1*, 2181 (1986).
106. M. Rasmussen and N. J. Leonard, *J. Am. Chem. Soc.*, **89**, 5439 (1967).
107. H. J. Anderson and J. K. Groves, *Tetrahedron Lett.*, 3165 (1971).
108. J. E. Macor, J. T. Forman, R. J. Post, and K. Ryan, *Tetrahedron Lett.*, **38**, 1673 (1997).
109. T. Brown, J. H. Jones, and J. D. Richards, *J. Chem. Soc., Perkin Trans. 1*, 1553 (1982).
110. T. Brown and J. H. Jones, *J. Chem. Soc., Chem. Commun.*, 648 (1981).
111. J.-C. Gesquiere, E. Diesis, and A. Tartar, *J. Chem. Soc., Chem. Commun.*, 1402 (1990).
112. M. A. Mitchell, T. A. Runge, W. R. Mathews, A. K. Ichhpurani, N. K. Harn, P. J. Dobrowolski, and F. M. Eckenrode, *Int. J. Pept. Protein Res.*, **36**, 350 (1990).
113. A. R. Katritzky, P. Lue, and Y.-X. Chen, *J. Org. Chem.*, **55**, 3688 (1990).
114. A. R. Katritzky, P. Lue, and K. Yannakopoulou, *Tetrahedron*, **46**, 641 (1990).
115. R. K. Robins, E. F. Godefroi, E. C Taylor, L. R. Lewis, and A. Jackson, *J. Am. Chem. Soc.*, **83**, 2574 (1961).
116. J. S. Bradshaw, C. W. McDaniel, K. E. Krakowiak, and R. M. Izatt, *J. Heterocycl. Chem.*, **27**, 1477 (1990).
117. R. L. Hudkins, J. L. Diebold, and F. D. Marsh, *J. Org. Chem.*, **60**, 6218 (1995).
118. A. R. Katritsky and K. Akutagawa, *Tetrahedron Lett.*, **26**, 5935 (1985).
119. A. Previero, M. A. Coletti-Previero, and J. C. Cavadore, *Biochim. Biophys. Acta*, **147**, 453 (1967).
120. G. R. Matsueda, *Int. J. Pept. Protein Res.*, **20**, 26 (1982).
121. J. Castells, Y. Troin, A. Diez, M. Rubiralta, D. S. Grierson, and H. P. Husson, *Tetrahedron*, **37**, 7911 (1991).
122. W. G. Rajeswaran and L. A. Cohen, *Tetrahedron Lett.*, **38**, 7813 (1997).
123. K. Teranishi, S.-i. Nakatsuka, and T. Goto, *Synthesis*, 1018 (1994).
124. Y. Kiso, T. Kimura, M. Shimokura, and T. Narukami, *J. Chem. Soc., Chem. Commun.*, 287 (1988).
125. Y. Kiso, T. Kimura, Y. Fujiwara, M. Shimokura, and A. Nishitani, *Chem. Pharm. Bull.*, **36**, 5024 (1988).

PROTECTION FOR THE AMIDE –NH

Protection of the amide –NH is an area of protective group chemistry that has received little attention, and as a consequence, few good methods exist for amide –NH protection. Most of the cases found in the literature do not represent protective groups in the true sense, in that the protective group is often incorporated as a handle to introduce nitrogen into a molecule, rather than installed to protect a nitrogen that at some later time is deblocked. For this reason, many of the following examples deal primarily with removal, instead of formation and cleavage.

Amides

N-Allylamide: $CH_2{=}CHCH_2{-}NRCO{-}$

Formation

The allyl group was used to protect the nitrogen in a β-lactam synthesis, but was removed in a four-step sequence.[1]

1. $CH_2{=}CHCH_2Cl$, CsF, DMF.[2] The use of allyl iodide gives *O*-alkylation.
2. $CH_2{=}CHCH_2Br$, P4 base, THF, −100° to −78°.[3]
3. NaH, LiBr, DME, DMF, allyl bromide, 88% yield.[4]
4. Allyl chloride, $Pd(Ph_3P)_4$, TEA, 89% yield.[5]

allyl chloride, $Pd(Ph_3P)_4$, TEA, 89%

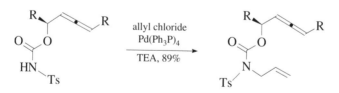

Cleavage

1. $Rh(Ph_3P)_3Cl$, toluene, reflux, 81% to the enamide; O_3, MeOH; DMS; $NaHCO_3$, 87% yield.[6,7]
2. $Pd(Ph_3P)_4$, HCO_2H, TEA, dioxane, reflux, 80% yield. Cleavage is from an imide.[8]

N-*t*-Butylamide (*t*-Bu-NRCO–)

The *t*-butyl group is introduced as a *t*-butylamine and is cleaved with strong acid (70–97% yield).[9]

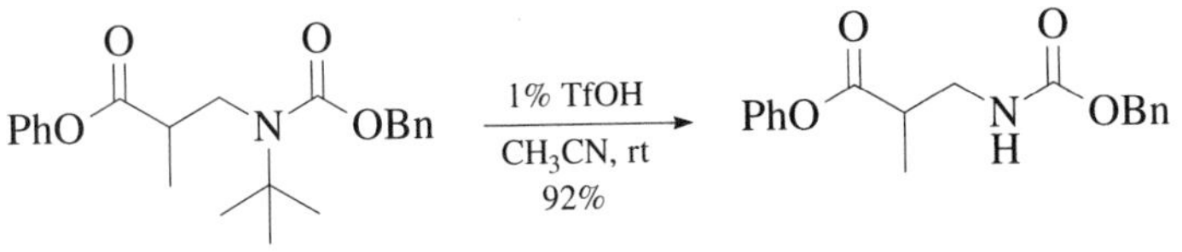

$$\text{PhO-CO-CH(CH}_3\text{)-CH}_2\text{-N}(t\text{-Bu})\text{-CO-OBn} \xrightarrow[\text{CH}_3\text{CN, rt}\ 92\%]{1\%\ \text{TfOH}} \text{PhO-CO-CH(CH}_3\text{)-CH}_2\text{-NH-CO-OBn}$$

N-Dicyclopropylmethylamide (Dcpm–NRCO–): $(C_3H_5)_2CH$–NRCO–

Half-lives for Cleavage of CH_3CONHR in Neat TFA at rt

R	$t_{1/2}$ (min)
Dicyclopropylmethyl	19
Dimethylcyclopropylmethyl	1–2
Me_2PhC–	15
$MePh_2C$–	<1

Cleavage is achieved by acidolysis in neat TFA. *N*-Cyclopropylmethyl, *N*-*t*-butyl, *N*-*t*-adamantyl, and *N*-(1-methylcyclohexyl)acetamide were not affected by these conditions.[10]

N-Methoxymethylamide (MOM–NRCO–): CH_3OCH_2–NRCO–

Formation

1. MOMCl, *t*-BuOK, DMSO.[11]

Cleavage

1. BBr_3, 31% yield.[11]

N-Methylthiomethylamide (MTM–NRCO–): CH_3SCH_2–NRCO–

Cleavage

1. $SOCl_2$; $NaHCO_3$, H_2O; heat to 120° under vacuum, 80% yield.[12]

N-Benzyloxymethylamide (BOM–NRCO–): $C_6H_5CH_2OCH_2$–NRCO–

Cleavage

1. The BOM group can be cleaved with $H_2/Pd(OH)_2$–C, MeOH, which also removes the BOM group from alcohols.[13]
2. BBr_3, 25°, toluene or $AlCl_3$, toluene, reflux.[14]

N-2,2,2-Trichloroethoxymethylamide: $Cl_3CCH_2OCH_2$–NRCO–

Formation

1. $Cl_3CCH_2OCH_2Cl$, KH, THF, 0° to rt, 20 min, 93% yield.[15]

Cleavage

1. 5% Na(Hg), Na_2HPO_4, MeOH, 67% yield.[15]

N-t-Butyldimethylsiloxymethylamide: t-$C_4H_9(CH_3)_2SiOCH_2$–NRCO–

Formation

1. $TBDMSOCH_2Cl$, TEA, CH_2Cl_2, −78°, rt, 24 h, >89% yield.[16]

Cleavage

1. $Bu_4N^+F^-$, THF, rt, 30 min, 70% yield.[16]

N-Pivaloyloxymethylamide: $(CH_3)_3CCO_2CH_2$–NRCO–

Formation

1. NaH, DMF, $PvOCH_2Cl$, rt, 12 h, 80% yield.[17]

Cleavage

1. NaOH, THF, rt, 4 days, 48% yield.[17]

N-Cyanomethylamide: $NCCH_2$–NRCO–

Formation

1. $BrCH_2CN$, EtONa, DMF, 82–85% yield.[18] Phenols and amines have also been protected by this method.

Cleavage

1. H_2, PtO_2, EtOH, 85–95% yield.[18]

N-Pyrrolidinomethylamide

Formation

1. HCHO, pyrrolidine, 93% yield.[19,20]

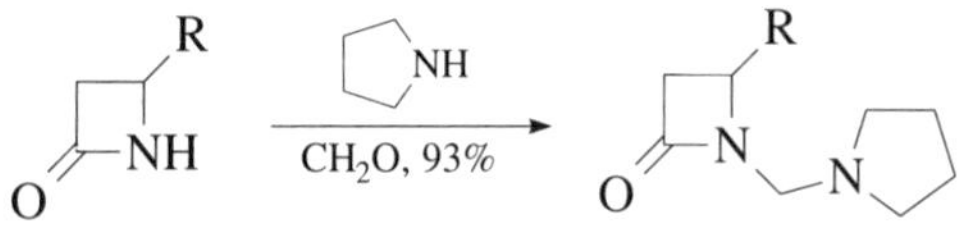

R NH CH$_2$O, 93% R O NH O N N

Cleavage

1. MeOH, 1% HCl, or 1:9 THF, 1% HCl, >52–85% yield.[20] This group was used to protect a β-lactam amide nitrogen during deprotonation of the α-position.

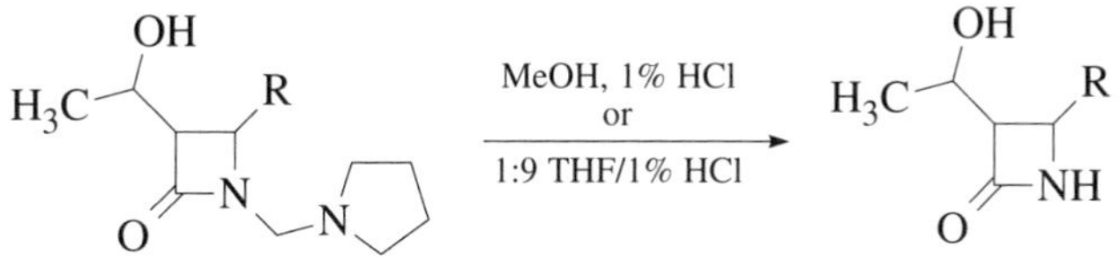

OH H_3C R O N N MeOH, 1% HCl or 1:9 THF/1% HCl OH H_3C R O NH

N-Methoxyamide: MeO–NRCO–

The methoxy group on a β-lactam nitrogen was cleaved by reduction with Li ($EtNH_2$, *t*-BuOH, THF, −40°, 71% yield). A benzyloxy group was stable to these cleavage conditions.[21]

N-Benzyloxyamide (BnO–NRCO–): $C_6H_5CH_2O$–NRCO–

The benzyloxy group on a β-lactam nitrogen was cleaved by hydrogenolysis (H_2, Pd–C) or by $TiCl_3$ [MeOH, H_2O, $(NH_4)_2CO_3$, Na_2CO_3].[22]

N-Methylthioamide: MeS–NRCO–

Formation

1. LDA, HMPA, $CH_3SSO_2CH_3$, −78° → 0°, 94% yield.[23]

Cleavage

1. 2-Pyridinethiol, Et_3N, CH_2Cl_2, 95% yield. The methylthioamide group is stable to 2.5 *N* NaOH, THF, H_2O and to 10% H_2SO_4, MeOH, H_2O.[23] The section on sulfenamides should be consulted for a related approach to nitrogen protection. Some of the derivatives presented there may also be applicable to amides.

N-Triphenylmethylthioamide: Ph_3CS–NRCO–

Cleavage

1. Bu_3P, EtOH, THF, 115°, 48 h, 75% yield.[24]

2. Me_3SiI, CH_2Cl_2, 25°, 7 h, 81% yield.[24]
3. Li, NH_3.[24]
4. W2 Raney Ni.[24] Li/NH_3 and Raney Ni also cleave benzylic C–N bonds.

N-t-Butyldimethylsilylamide (TBDMS–NRCO–): t-$C_4H_9(CH_3)_2$Si–NRCO–

Formation

1. TBDMSCl, Et_3N, CH_2Cl_2, 98% yield.[25–27] This methodology is also used to protect the BOCNH derivatives.[28]
2. TBDMSOTf, collidine.[29]

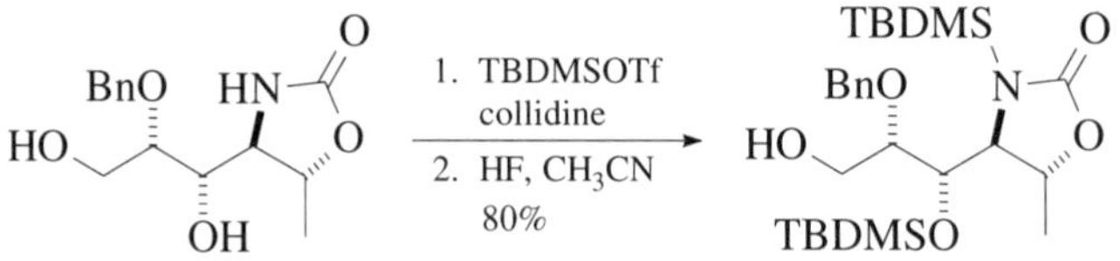

1. TBDMSOTf collidine
2. HF, CH_3CN
80%

Silylation of both the primary and secondary hydroxyl groups is followed by selective deprotection to regenerate the primary hydroxyl group.

3. 10% Pd/C, t-$BuMe_2SiH$, hexane, CH_2Cl_2, rt, 2 h, 80% yield.[30] These conditions also silylate alcohols, amines, and carboxylic acids.

Cleavage

1. 1 *N* HCl, MeOH, rt, 91% yield.[32] The TBDMS derivative of a β-lactam nitrogen is reported to be stable to lithium diisopropylamide, citric acid, Jones oxidation, and BH_3–diisopropylamine, but not to $Pb(OAc)_4$ oxidation.
2. MeSNa, THF, H_2O, >38% yield.[33]
3. Aq. HF, CH_3CN, DBU or t-BuOK.[31]

N-Triisopropylsilylamide (TIPS–NRCO–): (i-Pr)$_3$Si–NRCO–

Formation

1. TIPSOTf, DBU, CH_3CN.[34]

Cleavage

1. HF–Pyr, TBAF or NaOAc in DMSO/H_2O at 65°.[35]

N-4-Methoxyphenylamide: 4-CH_3O–C_6H_4–NRCO–

This group has been used extensively in β-lactam syntheses, to introduce the nitrogen as *p*-anisidine. The direct introduction of the 4-methoxyphenyl group has not been demonstrated, but the introduction of a simple phenyl group has, and these methods could perhaps be extended to the 4-methoxyphenyl group.[36]

Cleavage

1. Electrolysis, CH_3CN, H_2O, $LiClO_4$, 1.5 V, rt, 60–95% yield.[37] The released quinone is removed by forming the bisulfite adduct, which can be washed out with water.
2. Ceric ammonium nitrate, CH_3CN, H_2O, 0°, 95% yield.[38,39] In the presence of chloride ion, cleavage fails.[40]
3. Ozonolysis, then reduction with $Na_2S_2O_4$ at 50°, 57% yield.[41] The **3,4-dimethoxyphenyl derivative** was cleaved in 71% yield using these conditions. Ceric ammonium nitrate was reported not to work in this example.

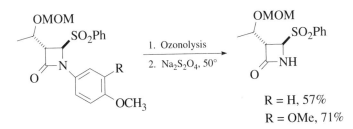

OMOM, SO_2Ph, N, R, O, OCH_3 — 1. Ozonolysis; 2. $Na_2S_2O_4$, 50° → OMOM, SO_2Ph, NH, O

R = H, 57%
R = OMe, 71%

4. $(NH_4)_2S_2O_8$, $AgNO_3$, CH_3CN, H_2O, 60°, 57–62% yield.[42]

N-4-(Methoxymethoxy)phenylamide ($MOMOC_6H_4$–NRCO–): $4\text{-}MeOCH_2OC_6H_4$–NRCO–

This group was developed for a case in which direct oxidation of the methoxyphenyl group with CAN was not very efficient. Prior removal of the MOM group [HCl, $HC(OMe)_3$, MeOH], followed by oxidation with CAN, was reported to be more effective.[43]

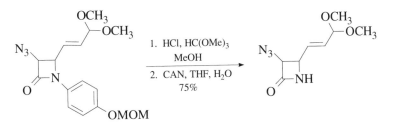

N_3, OCH_3, OCH_3, N, O, OMOM — 1. HCl, $HC(OMe)_3$, MeOH; 2. CAN, THF, H_2O, 75% → N_3, OCH_3, OCH_3, NH, O

N-2-Methoxy-1-naphthylamide: $2\text{-}CH_3O\text{–}C_{10}H_6$–NRCO–

This group was removed from a cyclic urethane with CAN.[44]

N-Benzylamide (Bn–NRCO–): $C_6H_5CH_2$–NRCO–

Formation

1. BnCl, KH, THF, rt, 100% yield.[45]
2. $Et_3BuN^+Br^-$, toluene, H_2O, BnCl, K_2CO_3, reflux.[46]
3. PhCHO, Pd/C, Na_2SO_4, H_2, 40 bar, 100°, 93% yield.[47]
4. BnBr neat, 120°.[48] This reaction also works with Ph_2CHBr to give the **diphenylmethylamide** derivative.

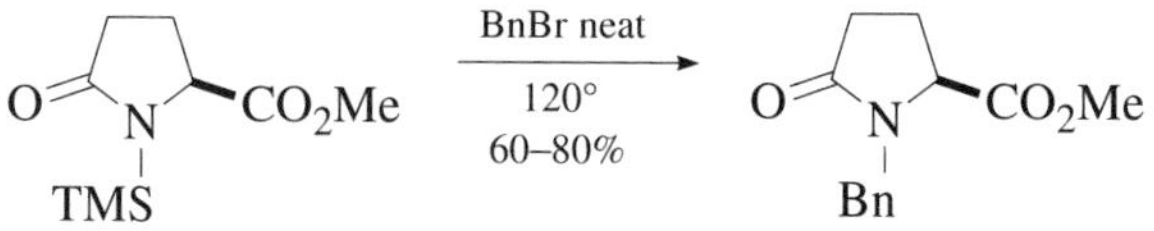

BnBr neat, 120°, 60–80%

5. BnCl, CsF, DMF, 83% yield.[2]
6. Treatment of an amide with $BnOC(=NH)CCl_3$ (TMSOTf, CH_2Cl_2, 85–88% yield) protects the amide by *O*-alkylation.[49]

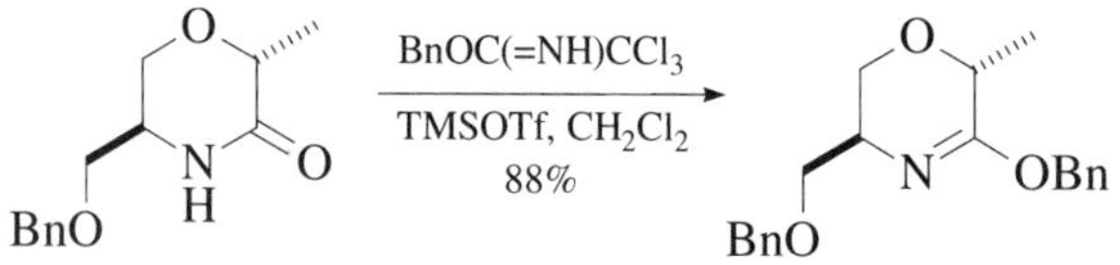

$BnOC(=NH)CCl_3$, TMSOTf, CH_2Cl_2, 88%

Cleavage

1. H_2, Pd-C, AcOH, 2 days.[50] Debenzylation of a benzylacetamide by hydrogenolyis is much slower than hydrogenolysis of a benzyl oxygen bond. Hydroxyl groups protected with benzyl groups or benzylidene groups are readily cleaved without affecting amide benzyl groups. It is often impossible to remove the benzyl group on an amide by hydrogenolysis.
2. Na or Li and ammonia, excellent yields.[51] This is a very good method for removing a benzyl group from an amide and will usually work when hydrogenolysis does not. A dissolving metal reduction can be effected without cleavage of a sulfur–carbon bond. Note also the unusual selectivity in the following cleavage. This was attributed to steric compression.[52]

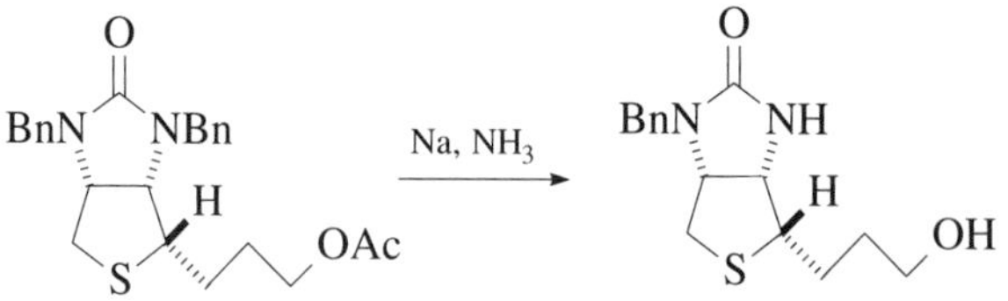

Na, NH_3

Primary benzyl amides are not cleaved under these conditions.[53]

3. Li, catalytic naphthalene, –78°, THF, 97–99% yield. In addition, tosylamides and mesylamides are cleaved with similar efficiency.[54]
4. *t*-BuLi, THF, –78°; O_2 or MoOPH, [oxodiperoxymolybdenum-(hexamethylphosphorictriamide)(pyridine)], 30–68% yield.[55] This method uses the amide carbonyl to direct benzylic metalation.
5. *t*-BuOK, DMSO, O_2, 20°, 20 min.[56]
6. Sunlight, $FeCl_3$, H_2O, acetone, 21% yield.[57]
7. 95% HCO_2H, 50–60°, 74–91% yield.[58] This method was used to remove the α-methylbenzyl group from an amide.
8. Aqueous HBr, 85% yield.[59]
9. Orthophosphoric acid, phenol, 53% yield.[60] Methods 7 and 8 were used to remove the benzyl group from a biotin precursor.

N-4-Methoxybenzylamide (PMB–NRCO–): 4-$CH_3OC_6H_4CH_2$–NRCO–

Formation

1. NaH, 4-MeO–$C_6H_4CH_2Br$, DMF, rt, 12 h, 62% yield.[61]
2. 4-MeO–$C_6H_4CH_2Cl$, DBU, CH_3CN, 45°, 6 h, 92% yield.[62]
3. 4-MeO–$C_6H_4CH_2Cl$, Ag_2O.[63]

Cleavage

1. Ceric ammonium nitrate (CAN), CH_3CN, H_2O, rt, 12 h, 96% yield.[64,65] Benzylamides are not cleaved under these conditions. Some of the methods used to cleave the benzyl group should also be effective for cleavage of the PMB group. Ceric ammonium nitrate is also used to cleave the PMB group from a sulfonamide nitrogen.[66]
2. *t*-BuLi, THF, –78°, O_2, 60% yield.[67,68]

R OTBDMS O OCH_3 PMBN S O —— *t*-BuLi / THF, –78° / O_2, 60% ——> R OTBDMS O OCH_3 HN S O

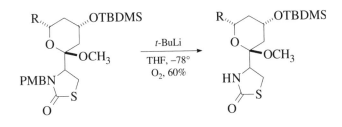

3. H_2, $PdCl_2$, EtOAc, AcOH, rt, 90% yield.[69]
4. $AlCl_3$, anisole, rt, 81–96% yield. An acetonide survived these conditions.[70]
5. TFA, reflux.[71]

N-2,4-Dimethoxybenzylamide and N-3,4-Dimethoxybenzylamide:
2,4- and 3,4-$(CH_3O)_2$–$C_6H_3CH_2$–NRCO–

Cleavage

1. TFA, 85% yield.[72,73]

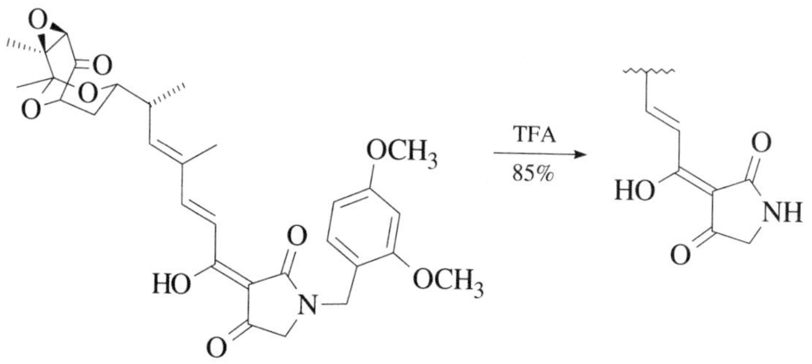

2. TFA, anisole, 75% yield.[74] Thioanisole has been used in this cleavage reaction to scavenge the benzyl cation.[75] Its absence results in considerable alkylation of the indolocarbazole nucleus.[76]

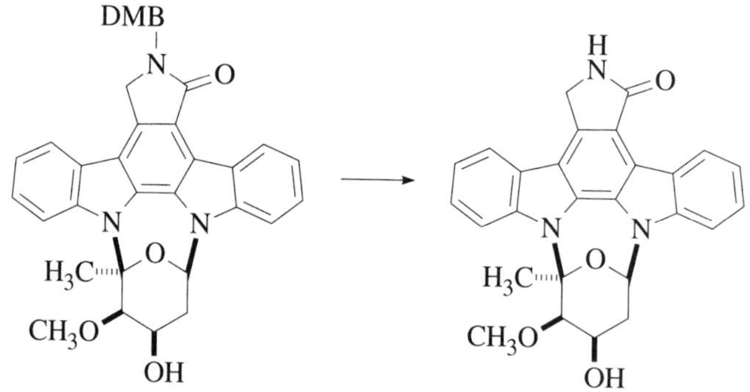

3. DDQ, $CHCl_3$, H_2O.[77] The 3,4-dimethoxybenzyl group could be cleaved from a sulfonamide with DDQ (8–50% yield).[78]
4. Ceric ammonium nitrate, CH_3CN, H_2O, 78% yield.[79]
5. The 3,4-dimethoxybenzyl group has been cleaved from an amide with Na/NH_3, 82% yield.[80]

N-2-Acetoxy-4-methoxybenzylamide (AcHmb–NRCO–):
2-Ac–4-MeO$C_6H_3CH_2$–NRCO–

This group is used for peptide backbone protection. The acetoxy group makes it stable to TFA that is used to cleave the BOC group during peptide synthesis.

When the Ac group is removed (20% piperidine/DMF or 5% hydrazine/DMF), it becomes the Hmb group that is used to improve solubility and prevent aspartamide formation[81–83] and is readily cleaved with TFA.[84] The related **2-Fmoc-4-methoxybenzyl** group has also been prepared and used in peptide synthesis.[85]

***N-o*-Nitrobenzylamide (–OCRN–ONB):** 2-$NO_2C_6H_4CH_2$–NRCO–

Cleavage[86]

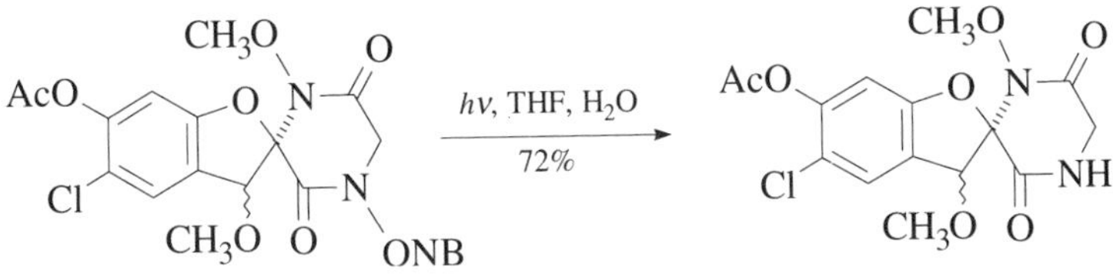

AcO, Cl, CH_3O, O, N, CH_3O, O, N, ONB —(hv, THF, H_2O; 72%)→ AcO, Cl, CH_3O, O, N, CH_3O, O, NH

***N*-Bis(4-methoxyphenyl)methylamide (DAM–NRCO–):** (4-$MeOC_6H_4)_2$CH–NRCO–

The DAM (dianisylmethyl) group, used to protect the –NH group of a β-lactam, can be cleaved with ceric ammonium nitrate (H_2O, CH_3CN, 0°, 91% yield)[87] or HCl (IPA, 60°, 4 h).[88]

Cleavage

1. $(NH_4)_2Ce(NO_3)_6$, CH_3CN, H_2O, 63% yield.[89]
2. TFA, $BF_3{\cdot}Et_2O$, anisole, Et_3SiH.[90]

***N*-Bis(4-methoxyphenyl)phenylmethylamide (DMTr–NRCO–):** (4-$MeOC_6H_4)_2$PhC–NRCO–

Formation

1. The DMTr group was selectively introduced into a biotin derivative.[91]

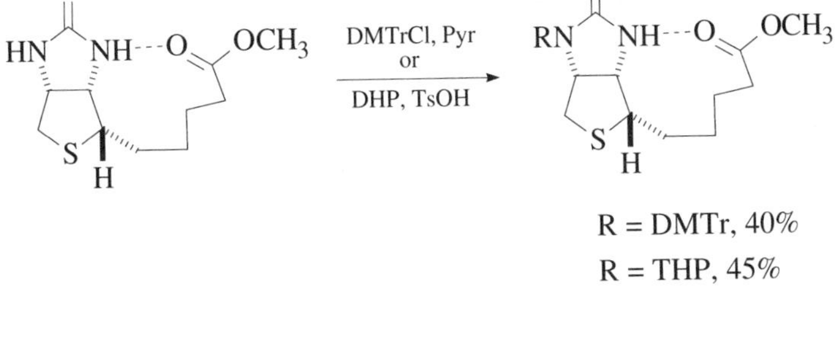

O, HN, NH, O, OCH_3, S, H —(DMTrCl, Pyr or DHP, TsOH)→ O, RN, NH, O, OCH_3, S, H

R = DMTr, 40%

R = THP, 45%

N-Bis(4-methylsulfinylphenyl)methylamide: $(4\text{-MeS(O)C}_6\text{H}_4)_2\text{CH–NRCO–}$

This group was developed for the protection of primary amides of amino acids. It is introduced by amide bond formation with the benzhydrylamine. It is cleaved with 1 *M* $SiCl_4$/anisole/TFA/0° or 1 *M* TMSOTf/thioanisole/TFA, 0°. Cleavage occurs by initial sulfoxide reduction followed by acidolysis.[92]

N-Triphenylmethylamide (Tr-NRCO–): $(C_6H_5)_3C\text{–NRCO–}$

The trityl group was introduced on a primary amide, $RCONH_2$, in the presence of a secondary amide with TrOH, Ac_2O, H_2SO_4, AcOH, 60°, 75% yield. It is stable to BOC removal with 1 *N* HCl in 50% isopropyl alcohol, 30 min, 50°, but can be cleaved with TFA.[93] The following table gives the cleavage rates with TFA for a number of protected primary amides.

Compound	$t_{1/2}$ (min)	
Fmoc–Asn(Tr)–OH	8	
Fmoc–Gln(Tr)–OH	2	
Fmoc–Gln(Tmob)–OH	9	Tmob = 2,4,6-trimethoxybenzyl
Fmoc–Gln(Mbh)–OH	27	Mbh = 4,4′-dimethoxybenzyhydryl
Ac–Pro–Asn(Tr)–Gly–Phe–OH	9	

N-9-Phenylfluorenylamide (Pf–NRCO–)

Cleavage

1. TFA, CH_2Cl_2, 84% yield.[94]

N-Bis(trimethylsilyl)methylamide: $(TMS)_2CH\text{–NRCO–}$

Cleavage

1. $(NH_4)_2Ce(NO_3)_6$, CH_3CN, H_2O, rt, 3 h, 84–95% yield. These conditions gave a *β*-lactam formimide that was then hydrolyzed with $NaHCO_3$, Na_2CO_3, H_2O, rt, 2 h, 78–95% yield.[95,96]
2. (i) TBAF, CH_3CHO, (ii) ozonolysis, DMS, (iii) $NaHCO_3$.[96]

N-*t*-Butoxycarbonylamide (BOC–NRCO–): $t\text{-}C_4H_9OCO\text{–NRCO–}$

Formation

1. $(BOC)_2O$, Et_3N, DMAP, 25°, 15 h, 78–96% yield.[97,98] The rate of reaction of $(BOC)_2O$ with an amide –NH is a function of the acidity of the –NH when steric factors are the same. The more acidic the –NH the faster is the reaction. For example, 4-thiazolidinone, pKa = 18.3, reacts in 2 min, whereas pyrrolidinone, pKa = 24.2, requires 2 h to reach completion.[99] If the amide

is sufficiently acidic, the same methodology can be used to prepare the methyl and benzyl carbamates.

2. BuLi, $(BOC)_2O$.[100]
3. $(BOC)OCO_2(BOC)$, DMAP.[101]
4. The very similar **1-Adoc** derivative of amides can be prepared from $(Adoc)_2O$/DMAP in CH_3CN. It is a little more reactive than $(BOC)_2O$.[101]

Cleavage

1. It should be noted that when a BOC-protected amide is subjected to nucleophilic reagents such as MeONa, hydrazine, and LiOH, the amide bond is cleaved in preference to the BOC group (85–96% yield) because of the difference in steric factors.[102] The BOC group can be removed by the methods used to remove it from simple amines.
2. $Mg(ClO_4)_2$, CH_3CN, 99% yield.[103] These conditions do not cleave a *t*-butyl ester or *t*-butyl carbamate.
3. TMSOTf, CH_2Cl_2.[104]

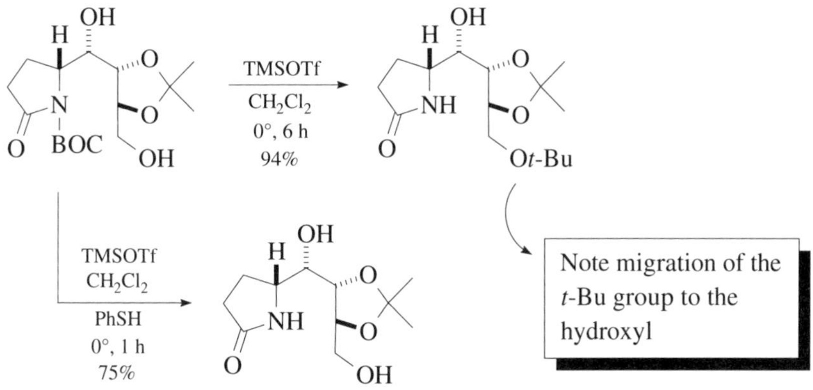

TMSOTf, CH_2Cl_2, 0°, 6 h, 94%

TMSOTf, CH_2Cl_2, PhSH, 0°, 1 h, 75%

Note migration of the *t*-Bu group to the hydroxyl

4. $Mg(OMe)_2$, MeOH, 82–90% yield.[105] This method is also effective for the Cbz and MeOCO derivatives, giving 78 and 86% yields, respectively.
5. Sm, I_2, MeOH, reflux 24 h, 95% yield.[106] This reagent also cleaves the Cbz group and other carbamates and esters.

N-Benzyloxycarbonylamide (Cbz–NRCO–)

Formation

1. *n*-BuLi, THF, –78°; CbzCl, –78° to 0°, 87–92% yield.[107]
2. $(BnO_2C)_2O$, DMAP, CH_3CN, 90% yield.[99]

Cleavage

1. Aqueous LiOH, dioxane, 86–92% yield.[107]

N-Methoxy- and *N*-Ethoxycarbonylamide (MeOCO–NRCO–)

Formation

1. $(MeO_2C)_2O$, DMAP, CH_3CN, 5 min, 71% yield. It appears that only amides having a fairly acidic NH are acylated under these conditions. δ-Valerolactam fails to react.[99]
2. $4\text{-}NO_2C_6H_4OCO_2Me$, DMAP, 92% yield.[108]
3. 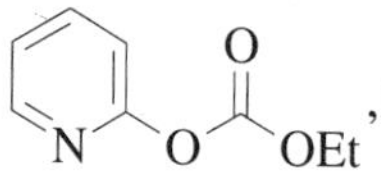 2-Pyridyl-O-C(=O)OEt, K_2CO_3, CH_3CN, reflux, 94% yield.[109]

N-p-Toluenesulfonylamide (Ts–NRCO–)

Cleavage

1. Sodium naphthalenide, DME, 0° → 20°, 6 h, 59–94%.[110] A benzyl ether was stable to these reductive conditions.[111]

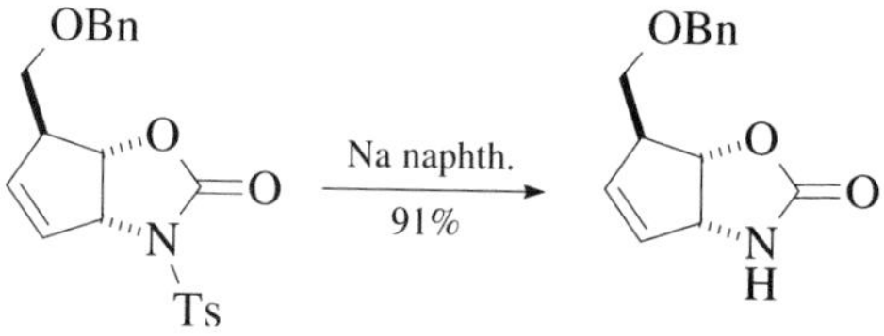

Na naphth., 91%

Ref. 111

2. Sodium anthracenide.[112]

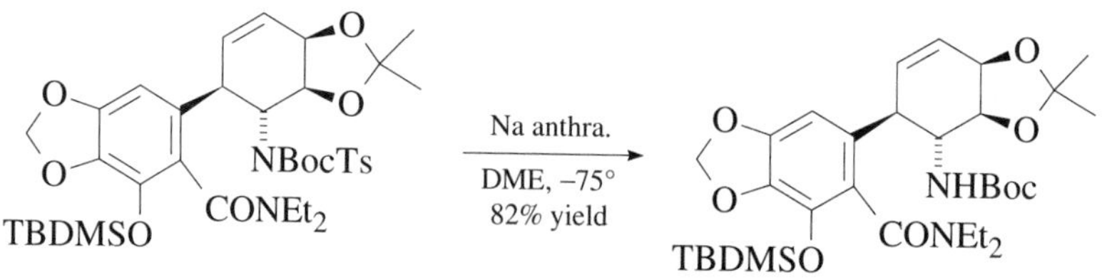

Na anthra., DME, −75°, 82% yield

3. Bu_3SnH, AIBN, toluene, 35–94% yield.[113]
4. Electrolysis, TFA, DMF, Hg cathode, 70–98% yield.[114] A number of other sulfonamides are cleaved similarly.[115]
5. Photolysis, CH_3CN, 300 nm, 86% yield.[116]
6. Mg, MeOH, sonication, 20–40 min, 93–100% yield. The benzenesulfonyl, cyanophenylsulfonyl, 4-methoxybenzenesulfonyl, and the 4-bromosulfonyl

groups were all efficiently removed. The reaction is not compatible with the nosyl and troc groups. The troc group is converted to a dichloroethoxycarbonyl group.[117]

7. Li, catalytic naphthalene, −78°, THF, 97–99% yield. In addition, benzylamides and methanesulfonamides are efficiently cleaved.[54]

***N,O*-Isopropylidene Ketals:**

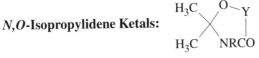

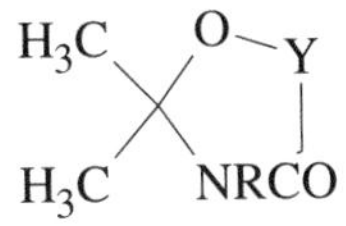

Formation

1. 2-Methoxypropene, $BF_3 \cdot Et_2O$, CH_2Cl_2, rt, 0.5 h, 84% yield.[118]
2. 2,2-Dimethoxypropane, toluene, TsOH, rt, 18 h, > 65% yield.[121]
3. $(CH_3)_2C(OCH_3)_2$, acetone, TsOH, rt, 97% yield.[119]
4. For the related **cyclohexylidene acetal**: cyclohexanone, TsOH, benzene, reflux 40 h with Soxhlet containing 4Å molecular sieves, 82% yield.[120]

Cleavage

1. Aqueous AcOH, 3 h, >65% yield.[121]
2. Pyridinium chlorochromate. In this case, the alcohol that is cleaved is simultaneously oxidized to give a ketone.[118]

***N,O*-Benzylidene Acetals:**

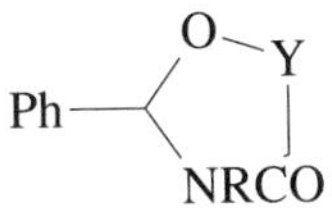

Formation

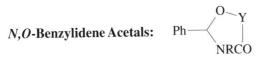

1. $PhCH(OMe)_2$, $BF_3 \cdot Et_2O$, 72% yield.[122]

Cleavage

1. Acid hydrolysis.[123]
2. Hydrogenolysis, Pd–C, hydrazine, MeOH, 95% yield.[124]

***N,O*-Formylidene Acetals:**

Cleavage[125]

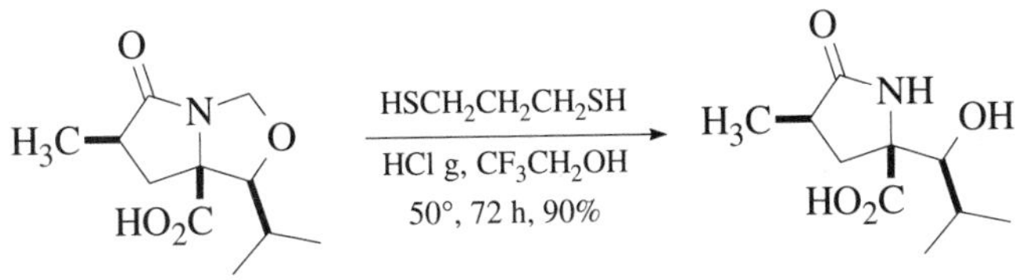

$HSCH_2CH_2CH_2SH$, HCl g, CF_3CH_2OH, 50°, 72 h, 90%

N-Butenylamide: $CH_3CH_2CH{=}CH{-}NRCO{-}$

Formation

1. Butanal, P_2O_5, toluene, reflux.[126]
2. Butanal, TsOH, toluene, 70% yield.[127]

Cleavage

1. $Et_3O^+BF_4^-$; H_2O; pH 8, 67% yield.[127]
2. $KMnO_4$, acetone, H_2O, 0°, 10 min, 78–90% yield. These conditions are used for the related **ethenyl** group.[128]
3. THF, 1% aq. HCl (9:2), reflux, 36 h; THF, H_2O (1:1), Na_2CO_3, reflux, 1 h, 62% yield.[128]
4. 4-$NO_2C_6H_4CO_3H$, THF, H_2O, HCO_2H, (10:10:1), 25°, 80% yield.[129]

N-[(E)-2-(Methoxycarbonyl)vinyl]amide: $MeO_2CC{=}CH{-}NRCO{-}$

Formation

1. Methyl propiolate, DMAP, rt, <10 min.[130]

Cleavage

1. Pyrrolidine, CH_3CN, rt, <2 h, >98% yield.[130]
2. CSA·2 H_2O, MeOH, reflux, 1.5 h, >92% yield.[130]

N-Diethoxymethylamide (DEM–NRCO–): $(EtO)_2CH{-}NRCO{-}$

Formation

1. $CH(OEt)_3$, 160°, 25–78% yield.[131]

Cleavage

1. TFA, CH_2Cl_2, rt, 1 h; 2 *N* NaOH, rt, 0.5 h, 37–90% yield.[131]

N-(1-Methoxy-2,2-dimethylpropyl)amide

Formation

1.

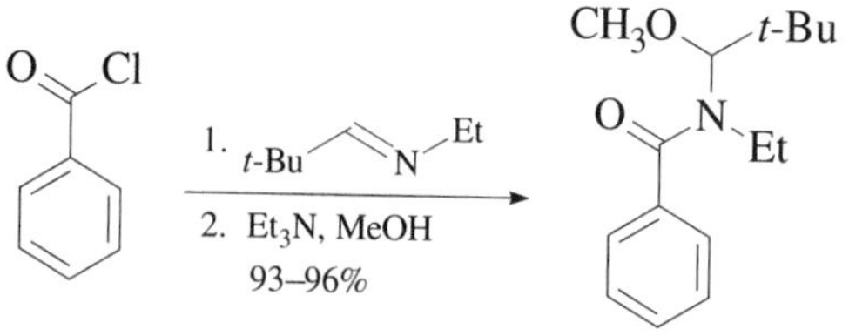

O Cl
1. t-Bu N Et
2. Et_3N, MeOH
93–96%
CH_3O t-Bu
O N Et

This protective group was used to improve the directed ortho metallation.[132]

Cleavage

1. HCl, dioxane, >71–82% yield.[132]

N-2-(4-Methylphenylsulfonyl)ethylamide: 4-$CH_3C_6H_4SO_2CH_2CH_2$–NRCO–

Formation

1. (4-Methylphenylsulfonyl)ethylamine was used to introduce the nitrogen in a β-lactam synthesis.[133]

Cleavage

1. By β-elimination with *t*-BuOK, THF, 1,5 h, –35° → 0°C, 72% yield.[133]

PROTECTION FOR THE SULFONAMIDE –NH

N-Diphenylmethylsulfonamide (DPM–$NRSO_2R'$)

Cleavage

1. Hydrogenation, H_2, 1 atm, $Pd(OH)_2/C$, CH_3OH, THF, Et_3N, 18 h, 87–99% yield.[134] In this case, the use of benzyl, 2,4-dimethoxybenzyl, 3,4-dimethoxybenzyl and 4-nitrobenzyl protective groups was unsatisfactory because of ring saturation during the hydrogenolysis. Oxidative cleavage of 2,4- and 3,4-dimethoxybenzyl groups led to complex mixtures.

N-2,4,6-Trimethoxybenzylsulfonamide (Tmob–$NRSO_2R'$)

Formation

1. The Tmob group is introduced by reaction of the sulfonyl chloride with 2,4,6-trimethoxybenzylamine.[135]

Cleavage

1. TFA, CH_2Cl_2, CH_3SCH_3, 92% yield.[135]

N-4-Hydroxy-2-methyl-3(2*H*)-isothiazolone 1,1-Dioxide[136]

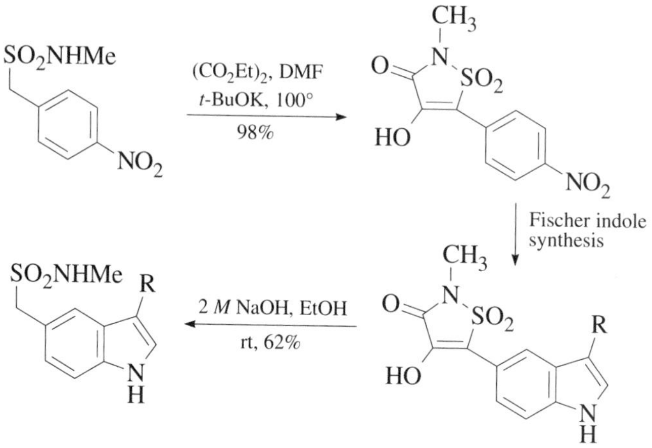

When the benzylic position was protected, an indole could be prepared without side products.

1. T. Fukuyama, A. A. Laird, and C. A. Schmidt, *Tetrahedron Lett.*, **25**, 4709 (1984).
2. T. Sato, K. Yoshimatsu, and J. Otera, *Synlett*, 845 (1995).
3. T. Pietzonkz and D. Seebach, *Angew. Chem., Int. Ed. Engl.*, **31**, 1481 (1992).
4. H. Liu, S.-B. Ko, H. Josien, and D. P. Curran, *Tetrahedron Lett.*, **36**, 8917 (1995).
5. M. Kimura, K. Fugami, S. Tanaka, and Y. Tamaru, *J. Org. Chem.*, **57**, 6377 (1993).
6. T. A. Lessen, D. M. Demko, and S. M. Weinreb, *Tetrahedron Lett.*, **31**, 2105 (1990).
7. B. Moreau, S. Lavielle, and A. Marquet, *Tetrahedron Lett.*, 2591 (1977).
8. T. Koch and M. Hesse, *Synthesis*, 931 (1992); *idem, ibid.*, 251 (1995).
9. M. J. Earle, R. A. Fairhurst, H. Heaney, and G. Papageorgiou, *Synlett*, 621 (1990).
10. L. A. Carpino, H.-G. Chao, S. Ghassemi, E. M. E. Mansour, C. Riemer, R. Warrass, D. Sadat-Aalaee, G. A. Truran, H. Imazumi, A. El-Faham, D. Ionescu, M. Ismail, T. L. Kowaleski , C. H. Han, H. Wenschuh, M. Beyermann, M. Bienert, H. Shroff, F. Albericio, S. A. Triolo, N. A. Sole, and S. A. Kates, *J. Org. Chem.*, **60**, 7718 (1995).
11. G. W. Kirby, D. J. Robins, and W. M. Stark, *J. Chem. Soc., Chem. Commun.*, 812 (1983).
12. E. D. Edstrom, X. Feng, and S. Tumkevicius, *Tetrahedron Lett.*, **37**, 759 (1996).
13. S. Hanessian, *Trends in Synthetic Carbohydrate Chemistry*, ACS Symp. Ser. **386** (1989), p. 64.
14. E. Zara-Kaczian and P. Matyus, *Heterocycles*, **36**, 519 (1993).
15. A. B. Smith, III, J. Barbosa, W. Wong, and J. L. Wood, *J. Am. Chem. Soc.*, **118**, 8316 (1996).

16. T. Benneche, L. L. Gundersen, and K. Undheim, *Acta Chem. Scand., Ser. B*, **B42**, 384 (1988).
17. E. C. Taylor and W. B. Young, *J. Org. Chem.*, **60**, 7947 (1995).
18. A. Benarab, S. Boyé, L. Savelon, and G. Guillaumet, *Tetrahedron Lett.*, **34**, 7567 (1993).
19. G. Cignarella, G. F. Cristiani, and E. Testa, *Justus Liebigs Ann. Chem.*, **661**, 181 (1963).
20. A. B. Hamlet and T. Durst, *Can. J. Chem.*, **61**, 411 (1983).
21. F. Shirai and T. Nakai, *Tetrahedron Lett.*, **29**, 6461 (1988).
22. P. G. Mattingly and M. J. Miller, *J. Org. Chem.*, **46**, 1557 (1981).
23. N. V. Shah and L. D. Cama, *Heterocycles*, **25**, 221 (1987).
24. D. A. Burnett, D. J. Hart, and J. Liu, *J. Org. Chem.*, **51**, 1929 (1986).
25. P. J. Reider and E. J. J. Grabowski, *Tetrahedron Lett.*, **23**, 2293 (1982).
26. H. Hiemstra, W. J. Klaver, and W. N. Speckamp, *Tetrahedron Lett.*, **27**, 1411 (1986).
27. D. J. Hart, C.-S. Lel, W. H. Pirkle, M. H. Hyon, and A. Tsipouras, *J. Am. Chem. Soc.*, **108**, 6054 (1986).
28. J. Roby and N. Voyer, *Tetrahedron Lett.*, **38**, 191 (1997).
29. D. E. Ward and B. F. Kaller, *Tetrahedron Lett.*, **34**, 407 (1993).
30. K. Yamamoto and M. Takemae, *Bull. Chem. Soc. Jpn.*, **62**, 2111 (1989).
31. S. Knapp, A. T. Levorse, and J. A. Potenza, *J. Org. Chem.*, **53**, 4773 (1988).
32. R. W. Ratcliffe, T. N. Salzmann, and B. G. Christensen, *Tetrahedron Lett.*, **21**, 31 (1980).
33. H.-O. Kim, C. Lum, and M. S. Lee, *Tetrahedron Lett.*, **38**, 4935 (1997).
34. R. C. F. Jones and A. D. Bates, *Tetrahedron Lett.*, **27**, 5285 (1986).
35. S. F. Vice, W. R. Bishop, S. W. McCombie, H. Dao, E. Frank, and A. K. Ganguly, *Bioorg. Med. Chem. Lett.*, **4**, 1333 (1994).
36. P. López-Alvarado, C. Avendaño, and J. C. Menéndez, *Tetrahedron Lett.*, **33**, 6875 (1992); M. S. Akhtar, W. J. Brouillette, and D. V. Waterhous, *J. Org. Chem.*, **55**, 5222 (1990).
37. E. G. Corley, S. Karady, N. L. Abramson, D. Ellison, and L. M. Weinstock, *Tetrahedron Lett.*, **29**, 1497 (1988).
38. D. R. Kronenthal, C. Y. Han, and M. K. Taylor, *J. Org. Chem.*, **47**, 2765 (1982).
39. D.-C. Ha and D. J. Hart, *Tetrahedron Lett.*, **28**, 4489 (1987).
40. J. Fetter, L. T. Giang, T. Czuppon, K. Lempert, M. Kajtar-Peredy, and G. Czira, *Tetrahedron*, **50**, 4188 (1994).
41. H. Yanagisawa, A. Ando, M. Shiozaki, and T. Hiraoka, *Tetrahedron Lett.*, **24**, 1037 (1983).
42. K. Bhattarai, G. Cainelli, and M. Panunzio, *Synlett*, 229 (1990).
43. T. Fukuyama, R. K. Frank, and C. F. Jewell, Jr., *J. Am. Chem. Soc.*, **102**, 2122 (1980).
44. B. M. Trost and A. A. Sudhakar, *J. Am. Chem. Soc.*, **110**, 7933 (1988).
45. Y. Xia and A. P. Kozikowski, *J. Am. Chem. Soc.*, **111**, 4116 (1989).

46. U. R. Kalkote, A. R. Choudhary, and N. R. Ayyangar, *Org. Prep. Proced. Int.*, **24**, 83 (1992).
47. F. Fache, L. Jacquot, and M. Lemaire, *Tetrahedron Lett.*, **35**, 3313 (1994).
48. F. Effenberger, W. Müller, R. Keller, W. Wild, and T. Ziegler, *J. Org. Chem.*, **55**, 3064 (1990).
49. J. Danklmaier and H. Hoenig, *Synth. Commun.*, **20**, 203 (1990).
50. R. Gigg and R. Conant, *Carbohydr. Res.*, **100**, C5 (1982).
51. T. Ohgi and S. M. Hecht, *J. Org. Chem.*, **46**, 1232 (1981); M. Y. Kim, J. E. Starrett, Jr., and S. M. Weinreb, *J. Org. Chem.*, **46**, 5383 (1981); S. Sugasawa and T. Fujii, *Chem. Pharm. Bull.*, **6**, 587 (1958); F. X. Webster, J. G. Millar, and R. M. Silverstein, *Tetrahedron Lett.*, **27**, 4941 (1986).
52. G. F. Field, *J. Org. Chem.*, **43**, 1084 (1978).
53. P. A. Jacobi, H. L. Brielmann, and S. I. Hauck, *Tetrahedron Lett.*, **36**, 1193 (1995).
54. E. Alonso, D. J. Ramón, and M. Yus, *Tetrahedron*, **53**, 14355 (1997).
55. R. M. Williams and E. Kwast, *Tetrahedron Lett.*, **30**, 451 (1989).
56. R. Gigg and R. Conant, *J. Chem. Soc., Chem. Commun.*, 465 (1983).
57. M. Barbier, *Heterocycles*, **23**, 345 (1985).
58. J. E. Semple, P. C. Wang, Z. Lysenko, and M. M. Joullié, *J. Am. Chem. Soc.*, **102**, 7505 (1980).
59. E. G. Baggiolini, H. L. Lee, G. Pizzolato, and M. R. Uskokovic, *J. Am. Chem. Soc.*, **104**, 6460 (1982).
60. G. F. Field, W. J. Zally, L. H. Sternbach, and J. F. Blout, *J. Org. Chem.*, **41**, 3853 (1976).
61. M. Yamaura, T. Suzuki, H. Hashimoto, J. Yoshimura, and C. Shin, *Chem. Lett.*, 1547 (1984).
62. T. Akiyama, H. Nishimoto, and S. Ozaki, *Bull. Chem. Soc. Jpn.*, **63**, 3356 (1990).
63. Y. Takahashi, H. Yamashita, S. Kobayashi, and M. Ohno, *Chem. Pharm. Bull.*, **34**, 2732 (1986).
64. M. Yamaura, T. Suzuki, H. Hashimoto, J. Yoshimura, T. Okamoto, and C. Shin, *Bull. Chem. Soc. Jpn.*, **58**, 1413 (1985).
65. J. Yoshimura, M. Yamaura, T. Suzuki, and H. Hashimoto, *Chem. Lett.*, 1001 (1983).
66. J. Morris and D. G. Wishka, *J. Org. Chem.*, **56**, 3549 (1991).
67. A. B. Smith, III, I. Noda, S. W. Remiszewski, N. J. Liverton, and R. Zibuck, *J. Org. Chem.*, **55**, 3977 (1990); R. M. Williams, T. Glinka, E. Kwast, H. Coffman, and J. K. Stille, *J. Am. Chem. Soc.*, **112**, 808 (1990).
68. J. H. Rigby and M. E. Mateo, *J. Am. Chem. Soc.*, **119**, 12655 (1997).
69. J. H. Rigby and V. Gupta, *Synlett*, 547 (1995).
70. T. Akiyama, Y. Takesue, M. Kumegawa, H. Nishimoto, and S. Ozaki, *Bull. Chem. Soc. Jpn*, **64**, 2266 (1991).
71. G. M. Brooke, S. Mohammed, and M. C. Whiting, *J. Chem. Soc., Chem. Commun.*, 1511 (1997).
72. R. H. Schlessinger, G. R. Bebernitz, P. Lin, and A. Y. Poss, *J. Am. Chem. Soc.*, **107**, 1777 (1985).
73. P. DeShong, S. Ramesh, V. Elango, and J. J. Perez, *J. Am. Chem. Soc.*, **107**, 5219

(1985); S. S. Shimshock, R. E. Waltermire, and P. DeShong, *J. Am. Chem. Soc.*, **113**, 8791 (1991).

74. J. L. Wood, B. M. Stoltz, and S. N. Goodman, *J. Am. Chem. Soc.*, **118**, 10656 (1996).
75. J. L. Wood, B. M. Stoltz, and H.-J. Dietrich, *J. Am. Chem. Soc.*, **117**, 10413 (1995).
76. J. L. Wood, B. M. Stoltz, H-J. Dietrich, D. A. Pflum, and D. T. Petsch, *J. Am. Chem. Soc.*, **119**, 9641 (1997).
77. S. Mori, H. Iwakura, and S. Takechi, *Tetrahedron Lett.*, **29**, 5391 (1988).
78. E. Grunder-Klotz and J. D. Ehrhardt, *Tetrahedron Lett.*, **32**, 751 (1991).
79. L. E. Overman and T. Osawa, *J. Am. Chem. Soc.*, **107**, 1698 (1985).
80. T. G. Back, K. Brunner, P. W. Codding, and A. W. Roszak, *Heterocycles*, **28**, 219 (1989).
81. L. C. Packman, *Tetrahedron Lett.*, **36**, 7523 (1995); C. Hyde, T. Johnson, D. Owen, M. Quibell, and R. C. Sheppard, *Int. J. Pept. Protein Res.*, **43**, 431 (1994).
82. T. Johnson, L. C. Packman, C. B. Hyde, D. Owen, and M. Quibell, *J. Chem. Soc., Perkin Trans 1*, 719 (1996).
83. E. Nicolás, M. Pujades, J. Bacardit, E. Giralt, and F. Albericio, *Tetrahedron Lett.*, **38**, 2317 (1997).
84. M. Quibell, W. G. Turnell, and T. Johnson, *Tetrahedron Lett.*, **35**, 2237 (1994).
85. T. Johnson, M. Quibell, D. Owen, and R. C. Sheppard, *J. Chem. Soc., Chem. Commun.*, 369 (1993).
86. G. F. Miknis and R. M. Williams, *J. Am. Chem. Soc.*, **115**, 537 (1993).
87. T. Kawabata, Y. Kimura, Y. Ito, and S. Terashima, *Tetrahedron Lett.*, **27**, 6241 (1986).
88. Y. Kobayashi, Y. Takemoto, Y. Ito, and S. Terashima, *Tetrahedron Lett.*, **31**, 3031 (1990).
89. C. Palomo, J. M. Aizpurua, J. M. García, M. Iturburu, and J. M. Odriozola, *J. Org. Chem.*, **59**, 5184 (1994).
90. Y. Kobayashi, Y. Ito, and S. Terashima, *Bull. Chem. Soc. Jpn.*, **62**, 3041 (1989).
91. A. M. Alves, D. Holland, and M. D. Edge, *Tetrahedron Lett.*, **30**, 3089 (1989).
92. M. Patek and M. Lebl, *Collect. Czech. Chem. Commun.*, **57**, 508 (1992); *idem.*, *Tetrahedron Lett.*, **31**, 5209 (1990).
93. P. Sieber and B. Riniker, *Tetrahedron Lett.*, **32**, 739 (1991).
94. E. Fernández-Megía and F. J. Sardina, *Tetrahedron Lett.*, **38**, 673 (1997).
95. C. Palomo, J. M. Aizpurua, M. Legido, and R. Galarza, *J. Chem. Soc., Chem. Commun.*, 233 (1997); C. Palomo, J. M. Aizpurua, M. Legido, A. Mielgo, and R. Galarza, *Chem.—Eur. J.*, **3**, 1432 (1997).
96. C. Palomo, J. M. Aizpurua, J. M. García, R. Galarza, M. Legido, R. Urchegui, P. Román, A. Luque, J. Server-Carrió, and A. Linden, *J. Org. Chem.*, **62**, 2070 (1997).
97. D. L. Flynn, R. E. Zelle, and P. A. Grieco, *J. Org. Chem.*, **48**, 2424 (1983).
98. Y. Ohfune and M. Tomita, *J. Am. Chem. Soc.*, **104**, 3511 (1982).
99. M. M. Hansen, A. R. Harkness, D. S. Coffey, F. G. Bordwell, and Y. Zhao, *Tetrahedron Lett.*, **36**, 8949 (1995).

100. A. Giovannini, D. Savoia, and A. Umani-Ronchi, *J. Org. Chem.*, **54**, 228 (1989).
101. A. Könnecke, L. Grehn, and U. Ragnarsson., *Tetrahedron Lett.*, **31**, 2697 (1990).
102. M. J. Burk and J. G. Allen, *J. Org. Chem.*, **62**, 7054 (1997).
103. J. A. Stafford, M. F. Brackeen, D. S. Karanewsky, and N. L. Valvano, *Tetrahedron Lett.*, **34**, 7873 (1993).
104. G. Casiraghi, F. Ulgheri, P. Spanu, G. Rassu, L. Pinna, G. G. Fava, M. B. Ferrari, and G. Pelosi, *J. Chem. Soc., Perkin Trans. 1*, 2991 (1993).
105. Z.-Y. Wei and E. E. Knaus, *Tetrahedron Lett.*, **35**, 847 (1994).
106. R. Yanada, N. Negoro, K. Bessho, and K. Yanada, *Synlett*, 1261 (1995).
107. R. W. Hungate, J. L. Chen, K. E. Starbuck, S. A. Macaluso, and R. S. Rubino, *Tetrahedron Lett.*, **37**, 4113 (1996).
108. M. J. Crossley and R. C. Reid, *J. Chem. Soc., Chem. Commun.*, 2237 (1994).
109. M. A. Neanwell, S. Y. Sit, J. Gao, H. S. Wong, Q. Gao, D. R. St. Laurent, and N. Balasubramanian, *J. Org. Chem.*, **60**, 1565 (1995).
110. J. Martens and M. Scheunemann, *Tetrahedron Lett.*, **32**, 1417 (1991).
111. B. M. Trost and D. L. Van Vranken, *J. Am. Chem. Soc.*, **115**, 444 (1993).
112. T. Hudlicky, X. Tian, K. Königsberger, R. Maurya, J. Rouden, and B. Fan, *J. Am. Chem. Soc.*, **118**, 10,752 (1996).
113. A. F. Parsons and R. M. Pettifer, *Tetrahedron Lett.*, **37**, 1667 (1996).
114. M. A. Casadei, A. Gessner, A. Inesi, W. Jugelt, and F. M. Moracci, *J. Chem. Soc., Perkin Trans. 1*, 2001 (1992).
115. B. Nyasse, L. Grehn, U. Ragnarsson, H. L. S. Maia, L. S. Monteiro, I. Leito, I. Koppel, and J. Koppel, *J. Chem. Soc., Perkin Trans. 1*, 2025 (1995).
116. C. Li and P. L. Fuchs, *Tetrahedron Lett.*, **34**, 1855 (1993).
117. B. Nyasse, L. Grehn, and U. Ragnarsson, *J. Chem. Soc., Chem. Commun.*, 1017 (1997).
118. R. Camerini, M. Panumzio, G. Bonanomi, D. Donati, and A. Perboni, *Tetrahedron Lett.*, **37**, 2467 (1996).
119. T. Yoshino, Y. Nagata, E. Itoh, M. Hashimoto, T. Katoh, and S. Terashima, *Tetrahedron Lett.*, **37**, 3475 (1996); K. Mori and H. Matsuda, *Liebigs Ann. Chem.*, 131 (1992).
120. L. Williams, Z. Zhang, X. Ding, and M. M. Joullié, *Tetrahedron Lett.*, **36**, 7031 (1995).
121. D. Favara, A. Omodei-Salè, P. Consonni, and A. Depaoli, *Tetrahedron Lett.*, **23**, 3105 (1982).
122. H. Cheng, P. Keitz, and J. B. Jones, *J. Org. Chem.*, **59**, 7671 (1994).
123. E. Didier, E. Fouque, I. Taillepied, and A. Commerçon, *Tetrahedron Lett.*, **35**, 2349 (1994).
124. Y. Hamada, A. Kawai, Y. Kohno, O. Hara, and T. Shioiro, *J. Am. Chem. Soc.*, **111**, 1524 (1989).
125. E. J. Corey and G. A. Reichard, *Tetrahedron Lett.*, **34**, 6973 (1993).
126. T. W. Kwon, P. F. Keusenkothen, and M. B. Smith, *J. Org. Chem.*, **57**, 6169 (1992).
127. M. B. Smith, C. J. Wang, P. F. Keusenkothen, B. T. Dembofsky, J. G. Fay, C. A. Zezza, T. W. Kwon, J. Sheu, Y. C. Son, and R. F. Menezes, *Chem. Lett.*, 247 (1992).

128. G. I. Georg, P. He, J. Kant, and J. Mudd, *Tetrahedron Lett.*, **31**, 451 (1990).
129. J. V. Heck and B. G. Christensen, *Tetrahedron Lett.*, **22**, 5027 (1981).
130. M. Faja, X. Ariza, C. Galvez, and J. Vilarrasa, *Tetrahedron Lett.*, **36**, 3261 (1995).
131. P. Gmeiner and B. Bollinger, *Synthesis*, 168 (1995).
132. D. P. Phillion and D. M. Walker, *J. Org. Chem.*, **60**, 8417 (1995).
133. D. DiPietro, R. M. Borzilleri, and S. M. Weinreb, *J. Org. Chem.*, **59**, 5856 (1994).
134. M. A. Poss and J. A. Reid, *Tetrahedron Lett.*, **33**, 7291 (1992).
135. G. Videnov, B. Aleksiev, M. Stoev, T. Paipanova, and G. Jung, *Liebigs Ann. Chem.* 941 (1993).
136. P. Remuzon, C. Dussy, J. P. Jacquet, M. Soumeillant, and D. Bouzard, *Tetrahedron Lett.*, **36**, 6227 (1995).

8

PROTECTION FOR THE ALKYNE –CH

Protection of an acetylenic hydrogen is often necessary because of its acidity. The bulk of a silane can protect an acetylene against catalytic hydrogenation because of rate differences between an olefin (primary or secondary) vs. the more hindered protected alkyne.[1] Trialkylsilylacetylenes are often used as a convenient method for introducing an acetylenic unit because they tend to be easily handled liquids or solids, as opposed to gaseous acetylene.

Formation

1. Trialkylsilanes are usually formed by the addition of a lithium or Grignard reagent to the silyl chloride,[2] and thus, discussions related to the formation of the silyl acetylene bond will be kept to a minimum. Silyl acetylenes are prepared from the alkynylcopper(I) reagents in the presence of PPh_3, Zn or TMEDA in CH_3CN at 100°, 36–98% yield.[3] It is interesting to note that the

reaction can be reversed to give the alkynylcopper(I) reagent in the presence of CuCl and 1,3-dimethyl-2-imidazolidinone.[4]

R—≡—Cu ⇄ R—≡—SiR_3 (forward: R_3SiCl, CH_3CN, PPh_3, Zn, TMEDA, 100°, 36–98%; reverse: CuCl, DMI)

Trimethylsilylalkyne (TMS-alkyne)

Cleavage

1. KF, MeOH, 50°, 89% yield.[5,6]
2. $AgNO_3$, 2,6-lutidine, 90% yield.[7]
3. $AgNO_2$, MeOH, H_2O, 24°, cool to 0°, add KCN, then HCl, 96% yield.[8,9] The reduced electron density of the propargylic alkyne directs the electrophilic silver to the other alkyne and activates it for cleavage.

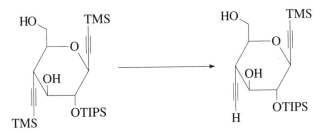

HO, TMS, O, OH, TMS, OTIPS → HO, TMS, O, OH, H, OTIPS

4. $Bu_4N^+ F^-$, THF, rt, quant.[10]
5. K_2CO_3, MeOH[10] or KOH, MeOH, 76%, 99% yield.[11,12] Under the basic conditions shown in the example, the more electron-deficient silylalkyne will be cleaved faster.[13]

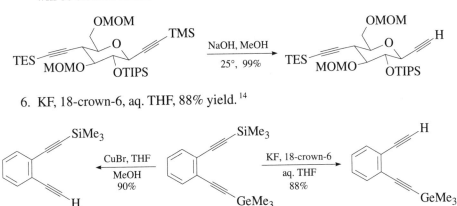

OMOM, TMS, O, TES, MOMO, OTIPS —NaOH, MeOH, 25°, 99%→ OMOM, H, O, TES, MOMO, OTIPS

6. KF, 18-crown-6, aq. THF, 88% yield.[14]

(H, SiMe$_3$ product) ←CuBr, THF, MeOH, 90%— SiMe$_3$ / GeMe$_3$ substrate —KF, 18-crown-6, aq. THF, 88%→ H / GeMe$_3$ product

In a similar example, a trimethylsilyl group was cleaved with NaOH, MeOH, H_2O in the presence of a triethylgermyl group.[15] The latter group can also be cleaved with methanolic $HClO_4$; the rate increases with increasing electron density.[16]

7. $Bu_4N^+ F^-$, 0.4 eq., THF, MeOH, –20° to –10°, 98% yield.[17]

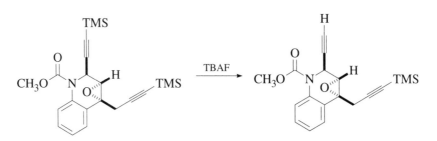

TMS, CH_3O, O, N, O, H, TMS —TBAF→ H, CH_3O, O, N, O, H, TMS

8. $Na(MeO)_3BH$, THF, H_2O, –20°, 2.5 h, 60% yield + 20% starting material (SM).[6]

HO, TMS, TMS —$Na(MeO)_3BH$, THF, H_2O, –20°, 2.5 h, 60% yield, 20% SM→ HO, H, TMS

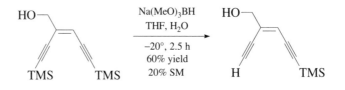

9. MeLi/LiBr.[18]
10. Amberlyst basic resin, MeOH, 80–98% yield.[19]
11. LiOH, THF, H_2O, 1 h, 98% yield. A TIPS-alkyne is stable to these conditions.[20]

Triethylsilylalkyne (TES-alkyne)

The relative rates of cleavage in aqueous, methanolic alkali at 29.4° for the following silanes are $PhC{\equiv}CSiMe_3$/$PhC{\equiv}CSiEtMe_2$/$PhC{\equiv}CSiEt_2Me$/ $PhC{\equiv}C\ SiEt_3$/ $PhC{\equiv}CSiPh_3$, 277:49:7.4:1:11.8.[21] A TES group can be cleaved selectively in the presence of a TBDMS group (*t*-BuOK, MeOH, 40°, 65%).[8]

t-Butyldimethylsilylalkyne and Thexyldimethylsilylalkyne (TBDMS- and TDS-alkyne)

Formation

For the TBDMS group, KHMDS, THF, TBDMSOTf, –78°, 98% yield.[8] The TDS group behaves similarly, except that it is slightly more hindered.

Cleavage

$Bu_4N^+ F^-$, THF, –23°, 75% yield.[22,23]

Dimethyl[1,1-dimethyl-3-(tetrahydro-2*H*-pyran-2-yloxy)propylsilylalkyne)] (DOPS-alkyne)

Cleavage

THF, 0.1 eq. BuLi, –78°, 2.5 h; –20°, 2 h.[10]

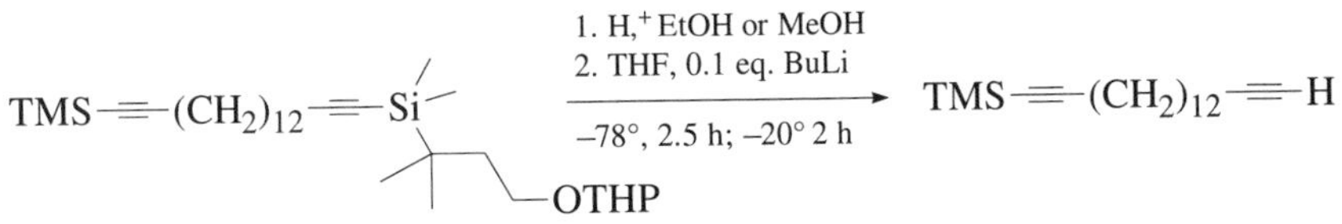

TMS–≡–$(CH_2)_{12}$–≡–Si(Me)$_2$–C(Me)$_2$–CH_2CH_2–OTHP → (1. H,$^+$ EtOH or MeOH; 2. THF, 0.1 eq. BuLi, –78°, 2.5 h; –20° 2 h) → TMS–≡–$(CH_2)_{12}$–≡–H

Protection of the OH with an alcohol-protective group gives this approach considerable versatility

Biphenyldimethylsilylalkyne (BDMS-alkyne)

Formation

BuLi, BDMSCl, THF, 75–98% yield. The advantage of this group is that many of the derivatives tend to be crystalline and thus provide a safe alternative for purification. Some smaller silylalkynes have been reported to explode upon distillation.[24]

Cleavage

K_2CO_3, MeOH, 72–98% yield. Cleavage occurs selectively in the presence of biphenyldiisopropylalkyne.[24]

Triisopropylsilylalkyne (TIPS-alkyne)

Cleavage

TBAF, THF, H_2O, 20°, 99% yield.[25]

Biphenyldiisopropylsilylalkyne (BDIPS-alkyne)

Formation

BuLi, BDIPSCl, THF, 81% yield.[24]

Cleavage

The cleavage of this group is reported to be similar to that of the triisopropylsilyl analogue.[24]

2-(2-Hydroxypropyl)alkyne [($Me_2C(OH)$-alkyne]

Formation

In this case, the low-cost 2-methyl-2-hydroxy-3-butyne is used as a convenient source of acetylene.

Cleavage

NaOH, benzene, reflux, > 96% yield.[26–28]

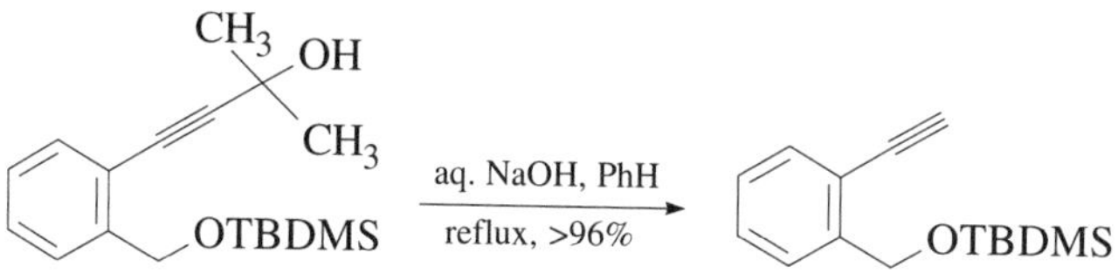

aq. NaOH, PhH
reflux, >96%

Ref. 26

1. C. J. Palmer and J. E. Casida, *Tetrahedron Lett.*, **31**, 2857 (1990).
2. For a review of the synthesis of silyl and germanyl alkynes, see W. E. Davidsohn and M. C. Henry, *Chem. Rev.*, **67**, 73 (1967).
3. H. Sugita, Y. Hatanaka, and T. Hiyama, *Chem. Lett.*, 379 (1996).
4. H. Ito, K. Arimoto, H.-o. Senusui, and A. Hosomi, *Tetrahedron Lett.*, **38**, 3977 (1997).
5. T. Saito, M. Morimoto, C. Akiyama, T. Matsumoto, and K. Suzuki, *J. Am. Chem. Soc.*, **117**, 10757 (1995).
6. A. G. Myers, P. M. Harrington, and E. Y. Kuo, *J. Am. Chem. Soc.*, **113**, 694 (1991).
7. E. M. Carreira and J. Du Bois, *J. Am. Chem. Soc.*, **117**, 8106 (1995).
8. J. Alzeer and A. Vasella, *Helv. Chim. Acta*, **78**, 177 (1995).
9. E. J. Corey and H. A. Kirst, *Tetrahedron Lett.*, 5041 (1968).
10. C. Cai and A. Vasella, *Helv. Chim. Acta*, **78**, 732 (1995).
11. L. T. Scott, M. J. Cooney, and D. Johnels, *J. Am. Chem. Soc.*, **112**, 4054 (1990).
12. Y.-F. Lu and A. G. Fallis, *Tetrahedron Lett.*, **34**, 3367 (1993).
13. C. Eaborn, R. Eastmond, and D. R. M. Walton, *J. Chem. Soc. B*, 127 (1971); J. Alzeer and A. Vasella, *Helv. Chim. Acta*, **78**, 1219 (1996).
14. A. Ernst, L. Gobbi, and A. Vasella, *Tetrahedron Lett.*, **37**, 7959 (1996).
15. R. Eastmond and D. R. M. Walton, *Tetrahedron*, **28**, 4591 (1972).
16. C. Eaborn, R. Eastmond, and D. R. M. Walton, *J. Chem. Soc. B*, 752 (1970).
17. T. Nishikawa, A. Ino, and M. Isobe, *Tetrahedron*, **50**, 1449 (1994).
18. L. Birkoffer, A. Ritter, and H. Dickopp, *Chem. Ber.*, **96**, 1473 (1963).
19. J. Bach, R. Berenguer, J. Garcia, T. Loscertales, and J. Vilarrasa, *J. Org. Chem.*, **61**, 9021 (1996).
20. Y. Tobe, N. Utsumi, K. Kawabata, and K. Naemura, *Tetrahedron Lett.*, **37**, 9325 (1996).

21. C. Eaborn and D. R. M. Walton, *J. Organomet. Chem.*, **4**, 217 (1965).
22. D. Elbaum, T. B. Nguyen, W. L. Jorgensen, and S. L. Schreiber, *Tetrahedron*, **50**, 1503 (1994).
23. A. G. Myers, N. J. Tom, M. E. Fraley, S. B. Cohen, and D. J. Madar, *J. Am. Chem. Soc.*, **119**, 6072 (1997).
24. J. Anthony and F. Diederich, *Tetrahedron Lett.*, **32**, 3787 (1991).
25. F. Diederich, Y. Rubin, O. L. Chapman, and N. S. Goroff, *Helv. Chim. Acta*, **77**, 1441 (1994).
26. C. S. Swindell, W. Fan, and P. G. Klimko, *Tetrahedron Lett.*, **35**, 4959 (1994).
27. S. J. Harris and D. R. M. Walton, *Tetrahedron*, **34**, 1037 (1978).
28. J. G. Rodriquez, R. Martin-Villamil, F. H. Cano, and I. Fonseca, *J. Chem. Soc., Perkin Trans 1*, 709 (1997).

9

PROTECTION FOR THE PHOSPHATE GROUP

"Phosphate esters and anhydrides dominate the living world."[1] Major areas of synthetic interest include oligonucleotides[2] (polymeric phosphate diesters), phosphorylated peptides, phospholipids, glycosyl phosphates, and inositol phosphates.[2b,3]

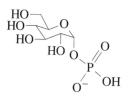

a glycosyl phosphate

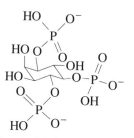

D-*myo*-inositol 1,4,5-triphosphate

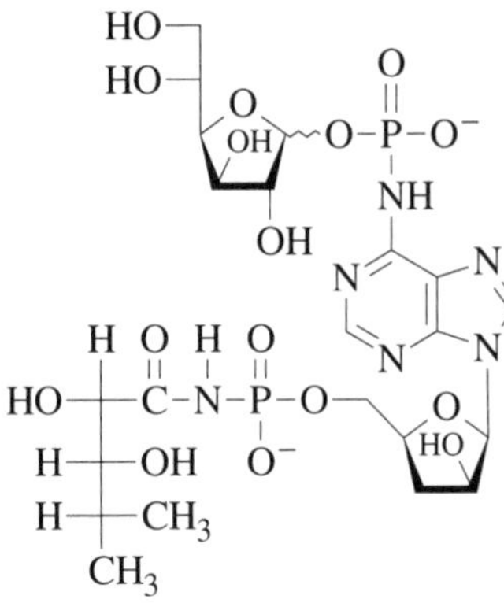

Agrocin 84[4]

The steps involved in automated oligonucleotide synthesis illustrate the current use of protective groups in phosphate chemistry (Scheme 1). Oligonucleotide synthesis involves the protection and deprotection of the 5′-OH, the amino groups on adenine, guanine, and cytosine, and –OH groups on phosphorus.

A difference in the problems associated with the protection and deprotection of phosphoric acid species, compared with the other functionalities in this book (alcohols, phenols, aldehydes and ketones, carboxylic acids, amines, and thiols), lies in the fact that phosphoric acid is tribasic (pK_1 = 2.12, pK_2 = 7.21, pK_3 = 12.66). These large differences in pKa's are reflected in large differences in rates of alkaline hydrolysis of the corresponding esters [e.g., $t_{1/2}$ at 1 *M* NaOH in water, 35°: $(CH_3O)_3PO$, 30 minutes; $(CH_3O)_2PO_2^-$, 11 years].[5] Large differences are often found in the rates of successive removal of blocking groups from phosphate derivatives, especially under nonacidic conditions. Phosphate esters are also hydrolyzed by acid,[5] but here the relative rates are closer together.

A consequence of the tribasic nature of phosphoric acid (three –OH groups attached to phosphorus) is the increased number of options available in the overall process of conversion of alcohol to protected phosphate. The conversion might be carried out by the sequence

$$R\text{–}O\text{–}H \rightarrow R\text{–}O\text{–}PO_3H_2 \rightarrow R\text{–}O\text{–}P(O)(OH)\text{–}O\text{–}PG$$

or by the formation of the R–O–P attachment *after* the formation of P–O–PG, i.e., introduction of the phosphate moiety in a form that is already protected. Another major difference in protection (and deprotection) in the phosphorus area lies in the availability of two major valence states, P(III) and P(V), of this second-row element. Both of these aspects [the order of formation of the bonds to P and the use of P(III) as well as P(V)] are important in current phosphate protection practice.

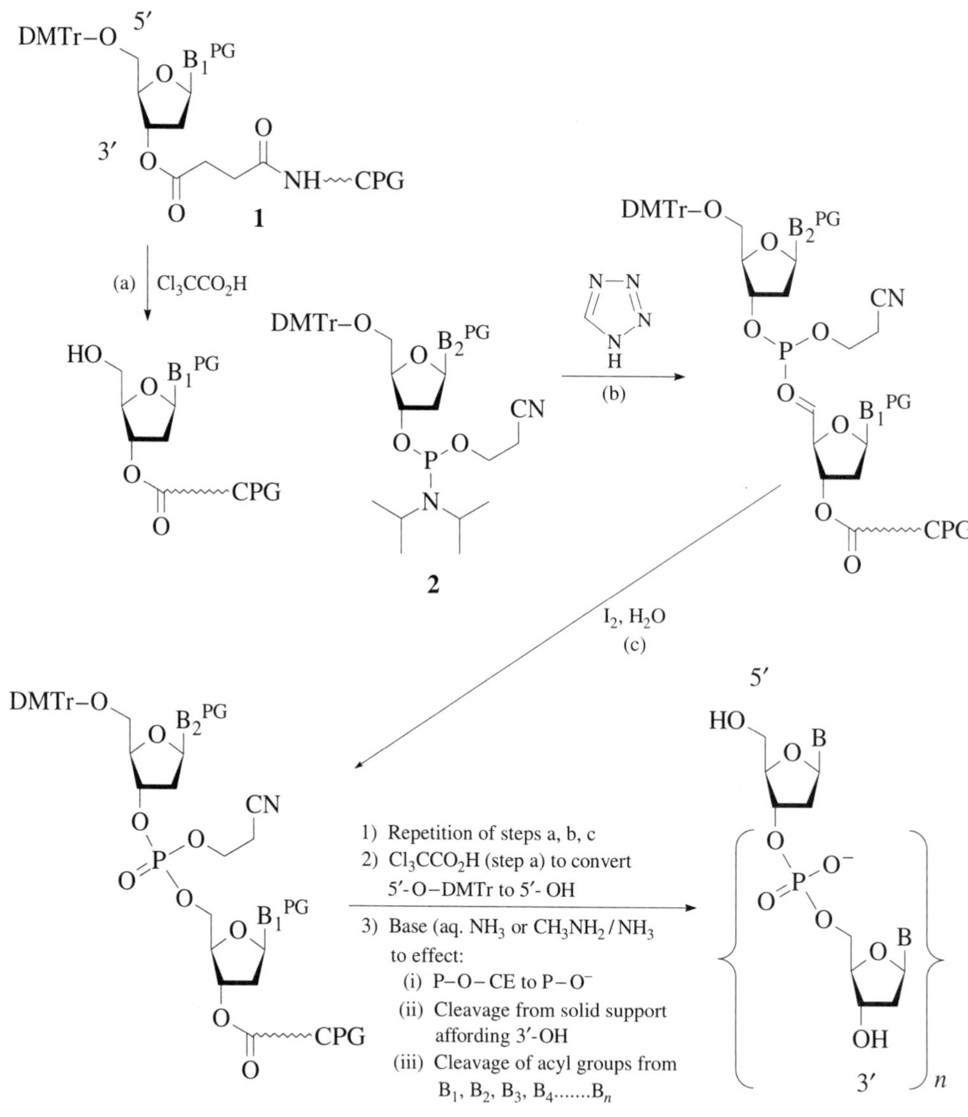

Scheme 1. Automated Synthesis of Oligonucleotides. Synthetic Cycle for the Phosphoroamidite Method.

Phosphate protection may begin at the stage of phosphoryl chloride (phosphorus oxychloride). A protective group may be introduced by reaction of this acid chloride with an alcohol[6] to afford an ester with the desired combination of stability to certain conditions and lability to others:

$$\mathrm{POCl_3 + ROH \rightarrow \underset{\text{a phosphorodichloridate}}{RO\text{-}P(O)Cl_2} \xrightarrow{\text{slower}} \underset{\text{a phosphorochloridate}}{(RO)_2P(O)Cl} \xrightarrow{\text{slower}} (RO)_3PO}$$

A disadvantage of phosphoryl chloride reagents is that they are not very reactive. In the mid-1970s, Letsinger and co-workers introduced a new paradigm that makes use of the more reactive phosphorus(III) reagents.[7] In this approach a monoprotected phosphorodichloridite ($ROPCl_2$)[8,9] is coupled with an alcohol, followed by a second condensation with another alcohol, to produce a triester. Oxidation with aqueous iodine affords a phosphate:[2,10]

$$\mathrm{ROPCl_2 + R'OH \rightarrow ROP(OR')Cl \xrightarrow{R''OH} ROP(OR')(OR'')}$$

$$\mathrm{\xrightarrow{I_2,\ H_2O} ROP(O)(OR')(OR'')}$$

The disadvantage of this method is that the dichloridites and monochloridites are sensitive to water and thus could not be used readily in automated oligonucleotide synthesis. This problem was overcome by Beaucage and Caruthers, who developed the phosphoramidite approach. In this method, derivatives of the form $R'OP(NR_2)_2$ react with one equivalent of an alcohol (catalyzed by species such as 1*H*-tetrazole) to form diesters, $R'OP(OR'')NR_2$, which usually are stable, easily handled solids. These phosphoroamidites are easily converted to phosphite triesters by reaction with a second alcohol (catalyzed by 1*H*-tetrazole). Here, again, oxidation of the phosphite triester with aqueous iodine affords the phosphate triester. Over the years, numerous protective groups and amines have been examined for use in this approach. Much of the work has been reviewed.[2,10]

SOME GENERAL METHODS FOR PHOSPHATE ESTER FORMATION

1. Phosphoric acids may be esterified using an alcohol and an activating agent:
   (a) carbodiimides, e.g., DCC.[11,12]
   (b) arylsulfonyl chloride and a base (TPS, Pyr).[13]
   (c) Various sulfonamido derivatives ($ArSO_2$–Z, Z = 1-imidazolyl, 1-triazolyl, 1-tetrazolyl).[2j,14,15]
   (d) CCl_3CN.[16–18]
   (e) $SOCl_2$, DMF, −20°, 70–90% yield:[19] $RP(O)(OH)_2 \rightarrow RP(O)(OH)OR$.
   (f) $[(Me_2N)_3PBr]^+PF_6^-$, DIEPA, CH_2Cl_2.[20]
2. Nucleophilic (S_N2) reactions for the formation of benzyl, allyl, and certain alkyl phosphates [e.g., Me_4N^+ $(RO)_2P(O)O^-$ and an alkyl halide in refluxing DME].[21,22]

3. Reaction of a phosphoric acid with a diazoalkane [CH_2N_2,[23,17] $ArCHN_2$, (N-oxido-α-pyridyl)CHN_2, Ar_2CN_2].[24]
4. Primary alcohols may be phosphorylated by use of the Mitsunobu reaction (Ph_3P, DEAD, HBF_4, Pyr). Of several salts examined, the potassium salt of the phosphate was the best.
5. One of the most widely used methods for the formation of phosphate esters involves the conversion of a P–N bond of a phosphorus(III) compound to a P–O bond by ROH, catalyzed by 1*H*-tetrazole, followed by oxidation to the phosphorus(V) derivative:[2]

(a) $R'OH + (R''O)P(NR_2)_2 \xrightarrow{1H\text{-tetrazole}} (R'O)P(NR_2)OR''$

phosphorodiamidite → phosphoroamidite

(b) $(R'O)P(NR_2)OR'' \xrightarrow{I_2,\ H_2O} (R'O)P(O)(NR_2)OR''$

phosphoroamidite → phosphoroamidate

6. Preparation of $(MeO)_2P–O–R$: ROH, $(MeO)_3P$, CBr_4, Pyr, 70–98% yield.[25] The alkyl dimethyl phosphite may then be oxidized to the corresponding phosphate by aq. iodine, *t*-butyl hydroperoxide, or peracid.

REMOVAL OF PROTECTIVE GROUPS FROM PHOSPHORUS

All the approaches for deblocking protective groups described earlier in this book have found application in the removal of protective groups from phosphorus derivatives. Because phosphate protection and deprotection are commonly associated with compounds that contain acid-sensitive sites (e.g., glycosidic linkages and DMTr–O– groups of nucleotides), the most widely used protective groups on phosphorus are those that are deblocked by base.

In the following list, "P^V–O–" stands for phosphorus(V) derivatives — usually $(R^1O)P(O)(OR^2)$–O–, in which R^1 and R^2 are not specified:

$$R^1O-\overset{\overset{\displaystyle O}{\|}}{\underset{\underset{\displaystyle R^2O}{|}}{P}}-O-(\text{Protective group}) = \text{“}P^VO-(\text{Protective group})\text{”}$$

1a or **1b**

1a $R^1 = R^2$ = alkyl or aryl

1b R^1 = H, R^2 = alkyl or aryl

1. Groups removed by base (in one step or in the second of two steps):
   (a) One-step removal via β-elimination of various β-substituted ethyl derivatives:

(i) $P^v–O–CH_2CH_2CN + TEA \rightarrow P^v–O^- + CH_2{=}CHCN$ Ref. 26

(ii) $P^v–O–CH_2CH_2–SiMe_3 + Bu_4N^+F^-$, THF $\rightarrow P^v–O^-$ Ref. 4

(b) Two-step removal:

(i) oxidation–elimination

$$P^V–O–CH_2CH_2–S–R \xrightarrow{\text{Oxid'n}} P^V–OCH_2CH_2–SO_2R$$

$$\xrightarrow{\text{base}} P^V–O^- \qquad \text{Ref. 14}$$

(ii) reduction-elimination

$P^V–O–CH_2$–(2-anthryl-9,10-quinone) $\xrightarrow{\text{Red'n}}$ corresponding hydro-

quinone $\xrightarrow{\text{base}} P^v–O^-$ Ref. 27

(c) Aryl phosphates and strong base. As stated earlier, dialkyl phosphates are quite stable to base. The P^v–O–aryl moiety is more labile to base than the P^v–O–alkyl moiety (hydroxide attack at P and ejection of $Ar–O^-$):

$$P^v–O–Aryl + OH^- \rightarrow P^v–O^- + ArO^- \qquad \text{Ref. 28}$$

2. Hydrogenolysis: $P^v–O–CH_2Ph$, H_2, Pd.[29]
3. Reduction: $P^v–O–CH_2CCl_3$, Zn/Cu, DMF.[30]
4. S_N2 displacement:
   (a) $P^v–O–CH_2Ph$ + NaI, $CH_3CN \rightarrow P^v–OH$ (or $P^v–O^-$) Ref. 31
   (b) $P^v–O–CH_3 + PhS^-$, DMF $\rightarrow P^v–O^-$ + PhSMe Ref. 32
5. Acid: $P^v–O–t\text{-}Bu + H^+ \rightarrow P^v–OH$ Ref. 33
6. Photolysis: $P^v–O–R \xrightarrow{h\nu} P^v–OH$ (or $P^v–O^-$) Ref. 34
   R = 3,5-dinitrophenyl, 2-nitrobenzyl, 3,5-dimethoxybenzyl, pyrenylmethyl, desyl, 4-methoxybenzoylmethyl
7. Oxidation: $P^v–O–C_6H_4–p$-NHTr, I_2, acetone, NH_4OAc.[35]
8. Metal ion catalysis: P^v–O–8-quinolinyl, $CuCl_2$, DMSO, $H_2O \rightarrow P^v–O^-$ Ref. 36
9. TMSCl, TMSBr or TMSI: $P^v–O–CH_3$, TMSI, CH_3CN.[37]
10. Cleavage of P^v–NHR to P^v–OH: P^v–NH–Ph, isoamyl nitrite, HOAc.[38]
11. Cleavage of P^v–S–R:
    (a) P^v–S–Et, I_2, Pyr $\rightarrow P^v–O^-$ Ref. 39
    (b) P^v–S–Ph, Zn $\rightarrow P^v–O^-$ Ref. 40
12. Transesterification: conversion of P^v–O–R to P^v–O–R′.
    (a) Transesterification–hydrogenolysis:

$$P^v–O–Ph + Bn–O^- \rightarrow P^v–O–Bn \xrightarrow{H_2,\ Pd} P^v–OH \text{ (or } P^v–O^-) \qquad \text{Ref. 41}$$

(b) Transesterification–elimination:

$$P^{v}\text{–O–R} + R'\text{–CH=N–O}^{-} \rightarrow P^{v}\text{–O–N=CHR} \xrightarrow{\text{base}} P^{v}\text{–O}^{-} + R'CN \quad \text{Ref. 42}$$

13. Electrolysis (has seen little use):

$$P^{V}\text{–O–}CH_2CCl_3 \xrightarrow{\text{electrolytic reduction}} P^{V}\text{– O}^{-}$$

Ref. 43

The following sections primarily describe many of the methods used for the cleavage of some of the more common phosphate protective groups. Since most of these groups are introduced by either the phosphate or phosphite method, little information is included here about their formation. The cited references generally describe the means that were used to introduce the protective group. In some cases, methods of formation are described, but this is done only when alternative methods to the phosphate or phosphite procedure were used.

1. F. W. Westheimer, "Why Nature Chose Phosphates," *Science*, **235**, 1173–1178 (1987).
2. Reviews: (a) S. L. Beaucage and R. P. Iyer, *Tetrahedron*, **48**, 2223–2311 (1992); (b) S. L. Beaucage and R. P. Iyer, *Tetrahedron*, **49**, 10441–10488 (1993); (c) S. L. Beaucage and R. P. Iyer, *Tetrahedron*, **49**, 6123–6194 (1993); (d) R. Cosstick in *Rodd's Chemistry of Carbon Compounds*, Supplement to the 2d ed., Suppl. Vol. IV, Part L, M. F. Ansell, Ed., Elsevier, 1988, pp. 61–128; (e) H. Kossel and H. Seliger, *Fortschr. Chem. Org. Naturst.*, **32**, 298–508 (1975); (f) G. C. Crockett, *Aldrichimica Acta*, **16**, 47 (1983); (g) J. W. Engels and E. Uhlmann, *Angew. Chem., Int. Ed. Engl.*, **28**, 716 (1989); (h) V. Amarnath and A. D. Broom, *Chem. Rev.*, **77**, 183–217 (1977); (i) F. Eckstein, "Protection of Phosphoric and Related Acids," in *Protective Groups in Organic Chemistry*, J. F. W. McOmie, Ed., Plenum Press, New York and London, 1973 pp. 217–234; (j) E. Sonveaux, "The Organic Chemistry Underlying DNA Synthesis," *Bioorg. Chem.*, **14**, 274–325 (1986); (k) S. L. Beaucage and M. H. Caruthers, "The Chemical Synthesis of DNA/RNA," in *Bioorganic Chemistry: Nucleic Acids*, S. M. Hecht, Ed., Oxford University Press, New York, 1996, Chapter 2, pp. 36–74.
3. B. V. L. Potter and D. Lampe, *Angew. Chem., Int. Ed. Engl.*, **34**, 1933 (1995).
4. T. Moriguchi, T. Wada, and M. Sekine, *J. Org. Chem.*, **61**, 9223 (1996).
5. J. R. Cox, Jr., and J. O. B. Ramsay, "Mechanisms of Nucleophilic Substitutions in Phosphate Esters," *Chem. Rev.*, **64**, 317 (1964).
6. A. M. Modro and T. A. Modro, *Org. Prep. Proced. Int.*, **24**, 57 (1992).
7. R. L. Letsinger, J. L. Finnan, G. A. Heavner, and W. B. Lunsford, *J. Am. Chem. Soc.*, **97**, 3278 (1975).
8. C. A. A. Claesen, R. P. A. M. Segers, and G. I. Tesser, *Recl. Trav. Chim. Pays-Bas*, **104**, 119 (1985).
9. K. K. Ogilvie, N. Y. Theriault, J. M. Seifert, R. T. Pon, and M. J. Nemer, *Can. J. Chem.*, **58**, 2686 (1980).
10. C. A. A. Claesen, R. P. A. M. Segers, and G. I. Tesser, *Recl. Trav. Chim. Pays-Bas*, **104**, 209 (1985).

11. A. Burger and J. J. Anderson, *J. Am. Chem. Soc.*, **79**, 3575 (1957).
12. W. F. Gilmore and H. A. McBride, *J. Pharm. Sci.*, **63**, 965 (1974).
13. E. Ohtsuka, H. Tsuji, T. Miyake, and M. Ikehara, *Chem. Pharm. Bull.*, **25**, 2844 (1977).
14. C. B. Reese, *Tetrahedron*, **34**, 3143 (1978); R. W. Adamiak, M. Z. Barciszewska, E. Biala, K. Grzeskowiak, R. Kierzek, A. Kraszewski, W. T. Markiewicz, and M. Wiewiorowski, *Nucleic Acids Res.*, **3**, 3397 (1976); H. Takaku, M. Kato, and S. Ishikawa, *J. Org. Chem.*, **46**, 4062 (1981).
15. B. L. Gaffney and R. A. Jones, *Tetrahedron Lett.*, **23**, 2257 (1982); M. Sekine, J.-i. Matsuzaki, and T. Hata, *ibid.*, **23**, 5287 (1982).
16. C. Wasielewski, M. Hoffmann, E. Witkowska, and J. Rachon, *Rocz. Chem.*, **50**, 1613 (1976).
17. J. Szewczyk, J. Rachon, and C. Wasielewski, *Pol. J. Chem.*, **56**, 477 (1982).
18. J. Szewczyk and C. Wasielewski, *Pol. J. Chem.*, **55**, 1985 (1981).
19. M. Hoffmann, *Synthesis*, 557 (1986).
20. N. Galéotti, J. Coste, P. Bedos, and P. Jouin, *Tetrahedron Lett.*, **37**, 3997 (1996).
21. M. Kluba, A. Zwierzak, and R. Gramze, *Rocz. Chem.*, **48**, 227 (1974).
22. A. Zwierzak and M. Kluba, *Tetrahedron*, **27**, 3163 (1971).
23. M. Hoffmann, *Pol. J. Chem.*, **53**, 1153 (1979).
24. G. Lowe and B. S. Sproat, *J. Chem. Soc., Perkin Trans. 1*, 1874 (1981).
25. V. B. Oza and R. C. Corcoran, *J. Org. Chem.*, **60**, 3680 (1995).
26. H. M. Hsiung, *Tetrahedron Lett.*, **23**, 5119 (1982).
27. N. Balgobin, M. Kwiatkowski, and J. Chattopadhyaya, *Chem. Scr.*, **20**, 198 (1982).
28. G. De Nanteuil, A. Benoist, G. Remond, J.-J. Descombes, V. Barou, and T. J. Verbeuren, *Tetrahedron Lett.*, **36**, 1435 (1995).
29. M. M. Sim, H. Kondo, and C.-H. Wong, *J. Am. Chem. Soc.*, **115**, 2260 (1993).
30. J. H. Van Boom, P. M. J. Burgers, R. Crea, G. van der Marel, and G. Wille, *Nucleic Acids Res.*, **4**, 747 (1977).
31. K. H. Scheit, *Tetrahedron Lett.*, 3243 (1967).
32. B. H. Dahl, K. Bjergaarde, L. Henriksen, and O. Dahl, *Acta Chem. Scand.*, **44**, 639 (1990).
33. J. W. Perich, P. F. Alewood, and R. B. Johns, *Aust. J. Chem.*, **44**, 233 (1991).
34. For a review of phosphate ester photochemistry, see R. S. Givens and L. W. Kueper, III, *Chem. Rev.*, **93**, 55 (1993).
35. E. Ohtsuka, S. Morioka, and M. Ikehara, *J. Am. Chem. Soc.*, **95**, 8437 (1973).
36. H. Takaku, Y. Shimada, and T. Hata, *Chem. Lett.*, 873 (1975).
37. J. Vepsäläinen, H. Nupponen, and E. Pohjala, *Tetrahedron Lett.*, **34**, 4551 (1993).
38. E. Ohtsuka, T. Ono, and M. Ikehara, *Chem. Pharm. Bull.*, **33**, 3274 (1981).
39. E. Heimer, M. Ahmad, S. Roy, A. Ramel, and A. L. Nussbaum, "Nucleoside *S*-Alkyl Phosphorothioates: VI. Synthesis of Deoxyribonucleotide Oligomers," *J. Am. Chem. Soc.*, **94**, 1707 (1972).
40. M. Sekine, K. Hamaoki, and T. Hata, *Bull. Chem. Soc. Jpn.*, **54**, 3815 (1981).
41. D. C. Billington, R. Baker, J. J. Kulagowski, and I. M. Mawer, *J. Chem. Soc., Chem. Commun.*, 314 (1987).

42. S. S. Jones and C. B. Reese, *J. Am. Chem. Soc.*, **101**, 7399 (1979).
43. J. Engels, *Angew. Chem., Int. Ed. Engl.*, **18**, 148 (1979).

ALKYL PHOSPHATES

Methyl: CH_3–

Formation

1. A phosphonic acid can be esterified with CH_2N_2 in 88–100% yield.[1,2]

Cleavage

1. 2-Mercaptobenzothiazole, *N*-methylpyrrolidone, DIPEA. The reagent has the advantage that it is odorless and does not lead to internucleotide cleavage, but the cleavage rate is 10 times slower than when thiophenol is used.[3]
2. Thiophenol, TEA, DMF or dioxane.[4] In the case of dimethyl phosphonates, this method can be used to remove selectively only one methyl group.[5] Lithium thiophenoxide is also effective.[6]
3. $H_2NC(=O)(N{\equiv}C)C{=}C(SNa)_2$, DMF. This odorless and easily prepared reagent is relatively nonbasic ($pK_B = 8.4$) and cleaves the methyl group about four times faster than thiophenol. It is also used to remove the 2,4-dichlorobenzyl group from phosphates and dithiophosphates.[4]
4. Ammonia. Cleavage is not as clean as with thiophenol.[7]
5. 10% Me_3SiBr, CH_3CN, 1–2 h, 25°, >97% yield.[8,9] This reagent is also useful for the cleavage of ethyl phosphates[10] and phosphonates.[11]
6. 1 *M* Me_3SiBr, thioanisole, TFA.[8,12]
7. 45% HBr, AcOH.[13,14,15] This method and the use of TMSI were not suitable for the deprotection of phosphorylated serines.[16] Diethyl phosphates are cleaved very slowly.[17]
8. TMSI, CH_3CN.[14,15,18] *In situ*–generated TMSI is also effective.[19]
9. Aqueous pyridine.[20]
10. NaI, acetone.[21,22]
11. The use of TMSOTf and thioanisole results in rapid ($t_{1/2} = 7$ min) cleavage of one methyl in a dimethyl phosphate, whereas the second methyl is cleaved only slowly ($t_{1/2} = 12$ h).[23] The method has been further refined for peptide synthesis.[24]
12. Fmoc chemistry is compatible with methyl phosphates when methanolic

K_2CO_3 is used to remove the Fmoc group instead of the usual amines.[25]

13. Dimethyl sulfide, CH_3SO_3H. Methyl phosphates are selectively cleaved in the presence of other alkyl phosphates.[26]
14. *t*-Butylamine, 46°, 15 h.[27]

Ethyl: C_2H_5-

Cleavage

1. Ethyl phosphates are usually cleaved by acid hydrolysis.[28]
2. TMSBr, CH_3CN.[29]
3. NH_4OH, MeOH.[29] These conditions result in cleavage of only one ethyl group of a diethyl phosphonate. Selective monodeprotection of a number of alkyl-protected phosphates is fairly general for cases where cleavage occurs by the release of phosphate or phosphonate anions.
4. LiBr has been used to cleave the ethyl group.[30]

4-(*N*-Trifluoroacetylamino)butyl: $CF_3C(O)NH(CH_2)_4-$

Ammonia treatment removes the TFA group, which then releases the phosphate and pyrrolidine through intramolecular cyclization. The analogous pentyl derivative was also prepared.[31]

Isopropyl: $(CH_3)_2CH-$

A diisopropyl phosphonate is cleaved with TMSBr, TEA, CH_2Cl_2, rt.[32] Dioxane can also be used as solvent.[33]

Cyclohexyl (cHex): $C_6H_{11}-$

Cleavage

1. The cyclohexyl phosphate, used in the protection of phosphorylated serine derivatives, is introduced by the phosphoramidite method and cleaved with TFMSA/MTB/*m*-cresol/1,2-ethanedithiol/TFA, 4 h, 0° to rt.[34]
2. Monocyclohexyl phosphates and phosphonates can be cleaved by a two-step process in which the ester is treated with an epoxide such as propylene oxide to form another ester, which, upon treatment with base, releases the cyclohexyl alcohol.[35]

***t*-Butyl:** $(CH_3)_3C-$

t-Butyl phosphates are acid sensitive.[36] They are not stable to Zn/AcOH.[37]

Cleavage

1. 1 *M* HCl, dioxane, 4 h.[16,38]
2. TFA, thiophenol[12] or thioanisole.[39]
3. TMSCl, TEA, CH_3CN, 75°, 2 h.[40]

1-Adamantyl:

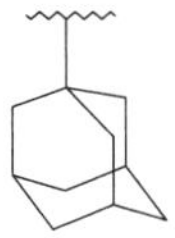

An adamantyl phosphonate, prepared from adamantyl bromide and Ag_2O, is easily cleaved with TFA in CH_2Cl_2.[41]

Allyl: CH_2=$CHCH_2$–

Cleavage

1. $Rh(Ph_3P)_3Cl$, acetone, H_2O, reflux, 2 h, 86% yield.[42]
2. $Pd(Ph_3P)_4$, Ph_3P, RCO_2K, EtOAc, 25°, 83% yield.[42,43] Diethylammonium formate,[44] NH_3,[45] and $BuNH_2$[46,47] have also been used as allyl scavengers in this process. In a diallyl phosphate, deprotection results in the cleavage of only a single allyl group.[48]
3. $Pd_2(dba)_3$–$CHCl_3$, Ph_3P, butylamine, formic acid, THF, 50°, 0.5–1 h.[49]
4. Concd. ammonia, 70°.[50]
5. $PdCl_2(Ph_3P)_2$, Bu_3SnH; $ClB(OR)_2$, then aqueous hydrolysis.[51]
6. NaI.[52]

2-Trimethylsilylprop-2-enyl (TMSP): CH_2=C(TMS)CH_2–

This derivative is stable to AcOH and methanolic ammonia, but not to 0.5 *N* aq. NaOH.

Cleavage

1. H_2, Pd–C, EtOH.[53]
2. $Et_4N^+F^-$, CH_3CN, 48 h, reflux. TMSF and allene are formed in the cleavage reaction. These conditions are not compatible with phenyl phosphates, which are cleaved preferentially with fluoride.[53] Cleavage of a bis TMSP phosphate results in the cleavage of only one of the TMSP groups.

3-Pivaloyloxy-1,3-dihydroxypropyl Derivative:

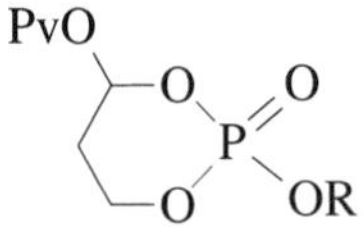

PvO O O P O OR

This group was designed as an enzymatically cleavable protective group. Cleavage is achieved using an esterase present in mouse plasma or hog liver carboxylate esterase.[54]

1. M. Hoffmann, *Pol. J. Chem.*, **53**, 1153 (1979).
2. J. Szewdzyk, J. Rachon, and C. Wasielewski, *Pol. J. Chem.*, **56**, 477 (1982).
3. A. Andrus and S. L. Beaucage, *Tetrahedron Lett.*, **29**, 5479 (1988).
4. B. H. Dahl, K. Bjergaarde, L. Henriksen, and O. Dahl, *Acta Chem. Scand.*, **44**, 639 (1990).
5. B. Müller, T. J. Martin, C. Schaub, and R. R. Schmidt, *Tetrahedron Lett.*, **29**, 509 (1998).
6. G. W. Daud and E. E. van Tamelen, *J. Am. Chem. Soc.*, **99**, 3526 (1977).
7. T. Tanaka and R. L. Letsinger, *Nucleic Acids Res.*, **10**, 3249 (1982).
8. R. M. Valerio, J. W. Perich, E. A. Kitas, P. F. Alewood, and R. B. Johns, *Aust. J. Chem.*, **42**, 1519 (1989).
9. C. E. McKenna, M. T. Higa, N. H. Cheung, and M.-C. McKenna, *Tetrahedron Lett.*, 155 (1977).
10. A. Holy, *Collect. Czech. Chem. Comm.*, **54**, 446 (1989).
11. L. Qiao and J. C. Vederas, *J. Org. Chem.*, **58**, 3480 (1993).
12. E. A. Kitas, R. Knorr, A. Trzeciak, and W. Bannwarth, *Helv. Chim. Acta*, **74**, 1314 (1991).
13. P. Kafarski, B. Lejczak, P. Mastalerz, J. Szweczyk, and C. Wasielewski, *Can. J. Chem.*, **60**, 3081 (1982).
14. J. Zygmunt, P. Kafarski, and P. Mastalerz, *Synthesis*, 609 (1978).
15. P. Kafarski and M. Soroka, *Synthesis*, 219 (1982).
16. J. W. Perich, P. F. Alewood, and R. B. Johns, *Aust. J. Chem.*, **44**, 233 (1991).
17. R. M. Valerio, P. F. Alewood, R. B. Johns, and B. E. Kemp, *Int. J. Pept. Protein Res.*, **33**, 428 (1989).
18. For a general review on the use of TMSI, see G. A. Olah and S. C. Narang, *Tetrahedron*, **38**, 2225 (1982).
19. J. Vepsäläinen, H. Nupponen, and E. Pohjala, *Tetrahedron Lett.*, **34**, 4551 (1993).
20. H. Vecerkova and J. Smrt, *Collect. Czech. Chem. Comm.*, **48**, 1323 (1983).
21. D. V. Patel, E. M. Gordon, R. J. Schmidt, H. N. Weller, M. G. Young, R. Zahler, M. Barbacid, J. M. Carboni, J. L. Gullo-Brown, L. Hunihan, C. Ricca, S. Robinson, B. R. Seizinger, A. V. Tuomari, and V. Manne, *J. Med. Chem.*, **38**, 435 (1995).
22. J. M. Delfino, C. J. Stankovic, S. L. Schreiber, and F. M. Richards, *Tetrahedron Lett.*, **28**, 2323 (1987).
23. E. A. Kitas, J. W. Perich, G. W. Tregear, and R. B. Johns, *J. Org. Chem.*, **55**, 4181 (1990).
24. A. Otaka, K. Miyoshi, M. Kaneko, H. Tamamura, N. Fujii, M. Momizu, T. R. Burke, Jr., and P. P. Roller, *J. Org. Chem.*, **60**, 3967 (1995).
25. W. H. A. Kuijpers, J. Huskens, L. H. Koole, and C. A. A. Van Boekel, *Nucleic Acids Res.*, **18**, 5197 (1990).
26. L. Jacob, M. Julia, B. Pfeiffer, and C. Rolando, *Synthesis*, 451 (1983).
27. D. J. H. Smith, K. K. Ogilvie, and M. F. Gillen, *Tetrahedron Lett.*, **21**, 861 (1980).
28. J. L. Kelley, E. W. McLean, R. C. Crouch, D. R. Averett, and J. V. Tuttle, *J. Med. Chem.*, **38**, 1005 (1995).
29. J. Matulic-Adamic, P. Haeberli, and N. Usman, *J. Org. Chem.*, **60**, 2563 (1995).

30. H. Krawczyk, *Synth. Commun.*, **27**, 3151 (1997).
31. A. Wilk, K. Srinivasachar, and S. L. Beaucage, *J. Org. Chem.*, **62**, 6712 (1997).
32. J.-L. Montchamp, L. T. Piehler, and J. W. Frost, *J. Am. Chem. Soc.*, **114**, 4453 (1992).
33. C. J. Salomon and E. Breuer, *Tetrahedron Lett.*, **36**, 6759 (1995).
34. T. Wakamiya, K. Saruta, J.-i. Yasuoka, and S. Kusumoto, *Bull. Chem. Soc. Jpn.*, **68**, 2699 (1995).
35. M. Sprecher, R. Oppenheimer, and E. Nov, *Synth. Commun.*, **23**, 115 (1993).
36. J. W. Perich and E. C. Reynolds, *Synlett*, 577 (1991).
37. G. Shapiro and D. Buechler, *Tetrahedron Lett.*, **35**, 5421 (1994).
38. J. W. Perich and R. B. Johns, *Synthesis*, 142 (1988).
39. J. M. Lacombe, F. Andriamanampisoa, and A. A. Pavia, *Int. J. Pept. Protein Res.*, **36**, 275 (1990).
40. M. Sekine, S. Iimura, and T. Nakanishi, *Tetrahedron Lett.*, **32**, 395 (1991).
41. A. Yiotakis, S. Vassiliou, J. Jiracek, and V. Dive, *J. Org. Chem.*, **61**, 6601 (1996).
42. M. Kamber and G. Just, *Can. J. Chem.*, **63**, 823 (1985).
43. D. B. Berkowitz and D. G. Sloss, *J. Org. Chem.*, **60**, 7047 (1995).
44. Y. Hayakawa, H. Kato, T. Nobori, R. Noyori, and J. Imai, *Nucl. Acids Symp. Ser.*, **17**, 97 (1986).
45. W. Bannwarth and E. Küng, *Tetrahedron Lett.*, **30**, 4219 (1989).
46. Y. Hayakawa, M. Uchiyama, H. Kato, and R. Noyori, *Tetrahedron Lett.*, **26**, 6505 (1985).
47. T. Pohl and H. Waldmann, *J. Am. Chem. Soc.*, **119**, 6702 (1997).
48. A. Sawabe, S. A. Filla, and S. Masamune, *Tetrahedron Lett.*, **33**, 7685 (1992).
49. Y. Hayakawa, S. Wakabayashi, H. Kato, and R. Noyori, *J. Am. Chem. Soc.*, **112**, 1691 (1990).
50. F. Bergmann, E. Kueng, P. Iaiza, and W. Bannwarth, *Tetrahedron*, **51**, 6971 (1995).
51. H. X. Zhang, F. Guibé, and G. Balavoine, *Tetrahedron Lett.*, **29**, 623 (1988).
52. Y. Hayakawa, M. Hirose, and R. Nyori, *Nucleosides Nucleotides*, **8**, 867 (1989).
53. T.-H. Chan and M. Di Stefano, *J. Chem. Soc., Chem. Commun.*, 761 (1978).
54. D. Farquhar, S. Khan, M. C. Wilkerson, and B. S. Andersson, *Tetrahedron Lett.*, **36**, 655 (1995).

2-Substituted Ethyl Phosphates

2-Cyanoethyl: $NCCH_2CH_2$–

Formation

1. $NCCH_2CH_2OH$, triisopropylbenzenesulfonyl chloride, Pyr, rt, 15 h.[1]
2. $NCCH_2CH_2OH$, DCC, Pyr.[2]
3. $NCCH_2CH_2OH$, 8-quinolinesulfonyl chloride, 1-methylimidazole, Pyr, rt.[3]
4. For monoprotection of a phosphonic acid: $NCCH_2CH_2OH$, Cl_3CCN, 74–93% yield.[4]

Cleavage

1. Aqueous ammonia, dioxane.[5]
2. Alkaline hydrolysis.[2]
3. TMSCl, DBU, CH_2Cl_2, 25°. The presence of TMSCl allows for complete deprotection of a biscyanoethyl phosphate. Without TMSCl, only one cyanoethyl group was cleaved.[6]
4. $Bu_4N^+F^-$, THF, 30 min.[7]
5. In a study of the use of various amines for the deprotection of the cyanoethyl group, it was found that primary amines are the most effective in achieving rapid cleavage. The following times for complete cleavage of the cyanoethyl group in phosphate **1** were obtained: TEA, 180 min; DIPA, 60 min; Et_2NH, 30 min; *s*-$BuNH_2$, 20 min; *t*-$BuNH_2$, 10 min; *n*-$PrNH_2$, 2 min.[8] Further study showed that *t*-$BuNH_2$ was most suitable because it did not react with protected nucleobases. Methylamine/ammonia was also a fast (5 min), effective reagent for deprotection.[9]

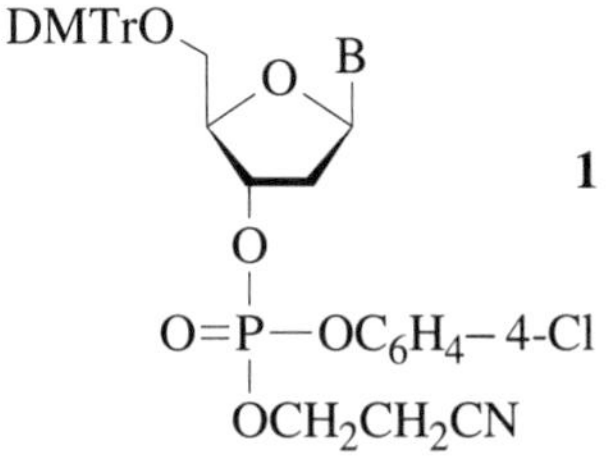

DMTrO
O B
1
O
O=P—OC_6H_4–4-Cl
OCH_2CH_2CN

2-Cyano-1,1-dimethylethyl (CDM): $CNCH_2C(CH_3)_2$–

Cleavage

1. Ammonia.[10]
2. DBU, *N,O*-bis(trimethylsilyl)acetamide.[11] Thiophosphorylated derivatives are cleaved more rapidly than the phosphorylated counterpart.
3. 0.2 *N* NaOH, dioxane, CH_3OH.[10]
4. Guanidine, tetramethylguanidine, or $Bu_4N^+OH^-$.[12]

4-Cyano-2-butenyl

This is a vinylogous analogue of the cyanoethyl group that is removed by δ-elimination with ammonium hydroxide.[13]

***N*-(4-Methoxyphenyl)hydracrylamide, *N*-Phenylhydracrylamide, and *N*-Benzylhydracrylamide Derivatives:** $ArNHC(O)CH_2CH_2$–

These derivatives, used for 5′-phosphate protection, are prepared by using the DCC coupling protocol and are cleaved with 2 *N* NaOH at rt.[14] The protected phosphates can be purified using benzoylated DEAE–Cellulose.

2-(Methyldiphenylsilyl)ethyl (DPSE): $(C_6H_5)_2CH_3SiCH_2CH_2$–

2-(Trimethylsilyl)ethyl (TSE): $(CH_3)_3SiCH_2CH_2$–

These groups, along with a number of other trialkylsilylethyl derivatives, were examined for protection of phosphorothioates. Only the phenyl-substituted silyl derivative was useful, because simple trialkylsilyl derivatives were prone to acid-catalyzed thiono–thiolo rearrangement.[15] Other trialkylsilylethyl derivatives also suffer from inherent instability upon storage,[16] but the trimethylsilylethyl group has been used successfully in the synthesis of the very sensitive agrocin 84[17] and for internucleotide phosphate protection with the phosphoramidite approach.[18]

Formation

1. The ester is introduced by means of the phosphoramidite method.[15,19]

Cleavage

1. Ammonium hydroxide, rt, 1 h.[15,19,20]
2. $Bu_4N^+F^-$ THF, Pyr, H_2O.[15,21,22]
3. Methylamine, H_2O.[15]
4. SiF_4, CH_3CN, H_2O, 20 min.[23]
5. NH_4F, methanol, 60°. One of two DPSE groups is cleaved.[24]
6. HF, CH_3CN, H_2O. In this case, both DPSE groups are removed.[24] This method effectively removes the trimethylsilylethyl group.[25]
7. TFA, CH_2Cl_2 or TFA, phenol, 30 min.[16]

2-(Triphenylsilyl)ethyl: $(C_6H_5)_3SiCH_2CH_2$–

This group, used for 5′-phosphate protection, has hydrophobicity similar to that of the dimethoxytrityl group and thus was expected to assist in reverse-phase HPLC purification of product from failure sequences in oligonucleotide synthesis. The group is cleaved with $Bu_4N^+F^-$ in DMSO at 70°.[26]

2-(*S*-Acetylthio)ethyl (SATE): $CH_3C(O)SCH_2CH_2$–

Formation

1. The SATE ester is formed from a phosphite using PvCl activation followed by oxidation to the phosphate with I_2/H_2O.[27]

Cleavage

1. Enzymatic hydrolysis exposes the sulfide, which undergoes episulfide formation releasing the phosphate.[27] This method was developed for

intracellular delivery of a monophosphate. This concept was also extended to the use of an *S*-glucoside that could be activated by a glucosidase to release the thiol.[28]

2. Treatment of $(EtO)_2P(S)SCH_2CH_2SC(O)R$ (R = Bz was preferred) with ammonia gives $(EtO)_2P(S)S^-$.[29]

2-(4-Nitrophenyl)ethyl (NPE): $4\text{-}NO_2C_6H_4CH_2CH_2-$

The use of this group in nucleotide and nucleoside synthesis has been reviewed.[30,31]

Cleavage

1. 0.5 *M* DBU in Pyr or CH_3CN. In this study,[32] the cleavage of a series of **2-(pyrazin-2-yl)ethyl phosphates** was compared with that of the NPE group. The former group was found to be cleaved with DBU in CH_3CN.[32–34] The related **2-(2-chloro-4-nitrophenyl)ethyl** ester is cleaved with the weaker base TEA in CH_3CN.[35] The addition of thymine during DBU deprotection improves the yield, because thymine scavenges the released 4-nitrostyrene.[36] The **2-(2-nitrophenyl)ethyl** group is cleaved about six times more slowly with DBU as the base.[37] Upon DBU treatment, a **bis-2-(4-nitrophenyl)ethyl** phosphate releases only a single Npe group.[38]

2-(2′-Pyridyl)ethyl (Pyet)

Cleavage

1. NaOMe, MeOH, Pyr or *t*-BuOK, Pyr, *t*-BuOH.[39] This group is reasonably stable to aqueous NaOH, ammonia, and 80% acetic acid.
2. MeI, CH_3CN.[40]
3. PhOCOCl, CH_3CN, 20°, 6 h; ammonia, Pyr.[41]

2-(4′-Pyridyl)ethyl

The 4′-pyridylethyl group was found to be more effective for internucleotide phosphate protection than the 2′-pyridylethyl group, because its cleavage proceeded with greater efficiency. It is cleaved in a two-step process: acylation with PhOCOCl increases the acidity of the benzylic protons, facilitating E-2 elimination by ammonia.[42]

2-(3-Arylpyrimidin-2-yl)ethyl

Cleavage of this ester with DBU is faster than cleavage of the Npes group; it can also be cleaved with the weaker base TEA/Pyr.[43]

2-(Phenylthio)ethyl: $C_6H_5SCH_2CH_2$–

Formation

1. From $ROP(O)(OH)_2$: $PhSCH_2CH_2OH$, DCC.[44]

2. Cyclic phosphate RO–P(=O)(OCH₂CH₂O) $\xrightarrow[\text{MeCN}]{\text{PhSH, Base}}$ RO–P(=O)(OCH_2CH_2SPh)(O⁻)   Ref. 45

3. $PhSCH_2CH_2OH$, triisopropylbenzenesulfonyl chloride, DMF, HMPA, rt, 8 h, 65–70% yield.[46]

Cleavage

1. $NaIO_4$, 1 h, rt; 2 *N* NaOH, 30 min, rt.[44,45]
2. *N*-Chlorosuccinimide; 1 *N* NaOH.[47] With this method, the sulfide is oxidized completely to the sulfone, which is cleaved with hydroxide more readily than the sulfoxide formed by periodate oxidation. It has been reported that oxidation of the sulfide leads to oxidation of adenine and guanine.[48] However, see the discussion of the TPTE group below.

2-(4′-Nitrophenyl)thioethyl (PTE)

This group is stable to TEA and morpholine in pyridine at 20°. It is cleaved by oxidation with MCPBA followed by elimination with TEA in Pyr, 10 min, 20°.[49] The rate of cleavage is proportional to the strength of the electron-withdrawing group on the phenyl ring.[50]

2-(4′-Triphenylmethylphenylthio)ethyl (TPTE): 2-[4-$(C_6H_5)_3CC_6H_4S$]CH_2CH_2–

The TPTE group, an analogue of the 2-(phenylthio)ethyl group, was developed to impart lipophilicity to protected oligonucleotides so that they could be isolated by solvent extraction. It is formed from the phosphoric acid and the alco-hol using either DCC or TPS as coupling agents. Cleavage is effected by base treatment after oxidation with $NaIO_4$ or NCS.[51]

2-[2′-(Monomethoxytrityloxy)ethylthio]ethyl

This easily prepared lipophilic 5′-phosphate protective group is cleaved by NCS oxidation (dioxane, triethylammonium hydrogen carbonate, 2 h, rt) followed by ammonia-induced β-elimination.[3]

Dithiodiethanol Derivative (DTE): $HOCH_2CH_2SSCH_2CH_2$–

Cleavage

1. Reduction of the disulfide by a reductase exposes the thiol, which then closes to give an episulfide, releasing the phosphate.[27]

2-(*t*-Butylsulfonyl)ethyl (B*ᵗ*SE): $(CH_3)_3CSO_2CH_2CH_2$–

The B^{t}SE group was used for internucleotide protection and is removed with ammonia, also used to remove *N*-acyl protective groups. Compared with the methylsulfonylethyl group,[52] the B^{t}SE group has better solubility properties for solution phase synthesis.[53]

2-(Phenylsulfonyl)ethyl (PSE): $C_6H_5SO_2CH_2CH_2$–

The use of this group avoids the problems associated with the oxidation of the phenylthioethyl group.

Cleavage

1. TEA, Pyr, 20°, <3 h.[48,54]

2-(Benzylsulfonyl)ethyl

This group is cleaved with 2 eq. of TEA in Pyr at a rate somewhat slower than that of the phenylsulfonylethyl group.[55]

1. E. Ohtsuka, H. Tsuji, T. Miyake, and M. Ikehara, *Chem. Pharm. Bull.*, **25**, 2844 (1977).
2. G. M. Tener, *J. Am. Chem. Soc.*, **83**, 159 (1961).
3. K. Kamaike, T. Ogawa, and Y. Ishido, *Nucleosides Nucleotides*, **12**, 1015 (1993).
4. J. Szewdzyk, J. Rachon, and C. Wasielewski, *Pol. J. Chem.*, **56**, 477 (1982).
5. J. Robles, E. Pedroso, and A. Grandas, *J. Org. Chem.*, **59**, 2482 (1994).
6. D. A. Evans, J. R. Gage, and J. L. Leighton, *J. Org. Chem.*, **57**, 1964 (1992).
7. K. K. Ogilvie, S. L. Beaucage, and D. W. Entwistle, *Tetrahedron Lett.*, 1255 (1976).
8. H. M. Hsiung, *Tetrahedron Lett.*, **23**, 5119 (1982).
9. M. P. Reddy, N. B. Hanna, and F. Farooqui, *Tetrahedron Lett.*, **35**, 4311 (1994).
10. J. E. Marugg, C. E. Dreef, G. A. Van der Marel, and J. H. Van Boom, *Recl.: J. R. Neth. Chem. Soc.*, **103**, 97 (1984).
11. M. Sekine, H. Tsuruoka, S. Iimura, and T. Wada, *Nat. Prod. Lett.*, **5**, 41 (1994).
12. Yu. V. Tumanov, V. V. Gorn, V. K. Potapov, and Z. A. Shabarova, *Dokl. Akad. Nauk SSSR*, **270**, 1130 (1983); *Chem. Abstr.* **99:** 212865e (1983).
13. V. T. Ravikumar, Z. S. Cheruvallath, and D. L. Cole, *Tetrahedron Lett.*, **37**, 6643 (1996).

14. S. A. Narang, O. S. Bhanot, J. Goodchild, and J. Michniewicz, *J. Chem. Soc., Chem. Commun.*, 516 (1970).
15. A. H. Krotz, P. Wheeler, and V. T. Ravikumar, *Angew. Chem., Int. Ed. Engl.*, **34**, 2406 (1995).
16. H.-G. Chao, M. S. Bernatowicz, P. D. Reiss, and G. R. Matsueda, *J. Org. Chem.*, **59**, 6687 (1994).
17. T. Moriguchi, T. Wada, and M. Sekine, *J. Org. Chem.*, **61**, 9223 (1996).
18. T. Wada, M. Tobe, T. Nagayama, K. Furusawa, and M. Sekine, *Nucleic Acids Symp. Ser.*, **29**, 9 (1993).
19. V. T. Ravikumar, H. Sasmor, and D. L. Cole, *Bioorg. Med. Chem. Lett.*, **3**, 2637 (1993).
20. V. T. Ravikumar, T. K. Wyrzykiewicz, and D. L. Cole, *Tetrahedron*, **50**, 9255 (1994).
21. S. Honda and T. Hata, *Tetrahedron Lett.*, **22**, 2093 (1981).
22. T. Wada and M. Sekine, *Tetrahedron Lett.*, **35**, 757 (1994).
23. V. T. Ravikumar and D. L. Cole, *Gene*, **149**, 157 (1994); V. T. Ravikumar, *Synth. Commun.*, **25**, 2164 (1995).
24. K. C. Ross, D. L. Rathbone, W. Thomson, and S. Freeman, *J. Chem. Soc., Perkin Trans. 1*, 421 (1995).
25. A. Sawabe, S. A. Filla, and S. Masamune, *Tetrahedron Lett.*, **33**, 7685 (1992).
26. J. E. Celebuski, C. Chan, and R. A. Jones, *J. Org. Chem.*, **57**, 5535 (1992).
27. C. Périgaud, G. Gosselin, I. Lefebvre, J. L. Girardet, S. Benzaria, I. Barber, and J. L. Imbach, *Bioorg. Med. Chem. Lett.*, **3**, 2521 (1993).
28. N. Schlienger, C. Perigaud, G. Gosselin, and J.-L. Imbach, *J. Org. Chem.*, **62**, 7216 (1997).
29. W. T. Wiesler and M. H. Caruthers, *J. Org. Chem.*, **61**, 4272 (1996).
30. W. Pfleiderer, F. Himmelsbach, R. Charubala, H. Schirmeister, A. Beiter, B. Schultz, and T. Trichtinger, *Nucleosides Nucleotides*, **4**, 81 (1985).
31. F. Himmelsbach, B. S. Schulz, T. Trichtinger, R. Charubala, and W. Pfleiderer, *Tetrahedron*, **40**, 59 (1984).
32. W. Pfleiderer, H. Schirmeister, T. Reiner, M. Pfister, and R. Charubala, "Biophosphates and Their Analogs–Synthesis, Structure, Metabolism and Activity," *Bioact. Mol.* **3**, 133 (1987).
33. For a brief review, see W. Pfleiderer, M. Schwarz, and H. Schirmeister, *Chem. Scr.*, **26**, 147–154 (1986).
34. E. Uhlmann and W. Pfleiderer, *Helv. Chim. Acta*, **64**, 1688 (1981).
35. E. Uhlmann and W. Pfleiderer, *Nucl. Acids Res., Spec. Publ.*, **4**, 25 (1978).
36. A. M. Avino and R. Eritja, *Nucleosides Nucleotides*, **13**, 2059 (1994).
37. E. Uhlmann and W. Pfleiderer, *Tetrahedron Lett.*, **21**, 1181 (1980).
38. E. Uhlmann, R. Charubala, and W. Pfleiderer, *Nucl. Acids Symp. Ser.*, **9**, 131 (1981).
39. W. Freist, R. Helbig, and F. Cramer, *Chem. Ber.*, **103**, 1032 (1970).
40. H. Takaku, S. Hamamoto, and T. Watanabe, *Chem. Lett.*, 699 (1986).
41. S. Hamamoto, N. Shishido, and H. Takaku, *Nucl. Acids Symp. Ser.*, **17**, 93 (1986).
42. S. Hamamoto, Y. Shishido, M. Furuta, H. Takaku, M. Kawashima, and M. Takaki, *Nucleosides Nucleotides*, **8**, 317 (1989).

43. T. Reiner and W. Pfleiderer, *Nucleosides Nucleotides*, **6**, 533 (1987).
44. R. H. Wightman, S. A. Narang, and K. Itakura, *Can. J. Chem.*, **50**, 456 (1972).
45. N. T. Thuong, M. Chassignol, U. Asseline, and P. Chabrier, *Bull. Soc. Chim. Fr.*, II-51 (1981).
46. S. A. Narang, K. Itakura, C. P. Bahl, and N. Katagiri, *J. Am. Chem. Soc.*, **96**, 7074 (1974).
47. K. L. Agarwal, M. Fridkin, E. Jay, and H. G. Khorana, *J. Am. Chem. Soc.*, **95**, 2020 (1973).
48. N. Balgobin, S. Josephson, and J. B. Chattopadhyaya, *Tetrahedron Lett.*, **22**, 1915 (1981).
49. N. Balgobin and J. Chattopadhyaya, *Chem. Scr.*, **20**, 144 (1982).
50. N. Balgobin, C. Welch, and J. B. Chattopadhyaya, *Chem. Scr.*, **20**, 196 (1982).
51. K. L. Agarwal, Y. A. Berlin, H.-J. Fritz, M. J. Gait, D. G. Kleid, R. G. Lees, K. E. Norris, B. Ramamoorthy, and H. G. Khorana, *J. Am. Chem. Soc.*, **98**, 1065 (1976).
52. C. Claesen, G. I. Tesser, C. E. Dreef, J. E. Marugg, G. A. van der Marel, and J. H. van Boom, *Tetrahedron Lett.*, **25**, 1307 (1984).
53. C. A. A. Claesen, C. J. M. Daemean, and G. I. Tesser, *Recl. Trav. Chim. Pays-Bas*, **105**, 116 (1986).
54. S. Josephson and J. B. Chattopadhyaya, *Chem. Scr.*, **18**, 184 (1981).
55. E. Felder, R. Schwyzer, R. Charubala, W. Pfleiderer, and B. Schulz, *Tetrahedron Lett.*, **25**, 3967 (1984).

Haloethyl Phosphates

2,2,2-Trichloroethyl: Cl_3CCH_2O-

Myoinositol bis(trichloroethyl)phosphates were not as stable to pyridine at 20°, as were the related benzyl analogs.[1]

Formation

1. Trichloroethanol, DCC, Pyr, rt, 15 h.[2]
2. A phosphonic acid was monoesterified with trichloroethanol, CCl_3CN in Pyr at 100°.[3]

Cleavage

1. Electrolysis at a Hg cathode, −1.2 V (Ag wire), CH_3CN, DMF, $Bu_4N^+BF_4^-$, 2,6-lutidine.[4] LiCl or $LiClO_4$ have been used as electrolytes in the electrochemical removal of haloethyl phosphates.[5]
2. Zn, acetylacetone, DMF, Pyr.[6,7] Chelex resin can be used to remove the zinc from these deprotections.[8]
3. Na, ammonia.[9] These conditions also remove cyanoethyl- and benzyl-protective groups. Phosphorothioates are similarly deprotected.

4. Zn(Cu), DMF.[10,11]
5. NaOH, aqueous dioxane.[12]
6. The trichloroethyl group is stable to Pd-catalyzed hydrogenolysis in AcOH/TFA, but when hydrogenolysis was attempted using EtOAc/MeOH as solvent, partial removal of the trichloroethyl group occurred along with Fmoc cleavage. Clean cleavage was observed in aqueous ethanol as solvent.[13,14]
7. Hydrogenolysis: Pd, Pyr.[15]
8. $Bu_4N^+F^-$, THF.[16]
9. Zn, anthranilic acid. Anthranilic acid was used to prevent complexation of the zinc with the oligonucleotides.[17]

2,2,2-Trichloro-1,1-dimethylethyl(TCB): $Cl_3CC(CH_3)_2O-$

Formation

1. The ester is introduced as the bis-TCB monochlorophosphate.[18]

Cleavage

1. Cobalt(I)-phthalocyanine, CH_3CN, 48 h. In a phosphate with two TCB groups, the first is cleaved considerably faster than the second.[18,19]
2. Bu_3P, DMF, TEA, 80°, quant.[20,21] Trichloroethyl phosphates are also cleaved.
3. Zn, acac, TEA, CH_3CN.[22]

2,2,2-Tribromoethyl: Br_3CCH_2-

Formation

1. $(RO)(Cl_3CCH_2O)P(O)Cl$, Br_3CCH_2OH.

Cleavage

1. Electrolysis at a Hg cathode, −0.5 to −0.6 V, $LiClO_4$, CH_3CN, Pyr. The trichloroethyl ester, which requires a greater reduction potential for cleavage, is retained under these conditions.[4]
2. Zn(Cu), DMF, 20°.[23]
3. Zn(Cu), Bu_3N, H_3PO_4, Pyr, rt.[24]

2,3-Dibromopropyl: $BrCH_2CHBrCH_2-$

Treatment of this protective group with KI/DMF for 24 h results in complete cleavage. The group is stable to Pyr/TEA/H_2O, but not to 7 *M* NH_4OH/MeOH.[25]

2,2,2-Trifluoroethyl: CF_3CH_2–

The trifluoroethyl group was used as an activating group in the phosphotriester approach to oligonucleotide synthesis, as well as a protective group that could be removed with 4-nitrobenzaldoxime (tetramethylguanidine, dioxane, H_2O).[26]

1,1,1,3,3,3-Hexafluoro-2-propyl: $(CF_3)_2CH$–

Cleavage of this group is achieved with tetramethylguanidinium *syn*-2-pyridinecarboxaldoxime.[27,28] Tris(hexafluoro-2-propyl) phosphites are sufficiently reactive to undergo transesterification with alcohols in a stepwise fashion.[29]

1. T. Desai, A. Fernandez-Mayoralas, J. Gigg, R. Gigg, and S. Payne, *Carbohydr. Res.*, **234**, 157 (1992).
2. E. Ohtsuka, H. Tsuji, T. Miyake, and M. Ikehara, *Chem. Pharm. Bull.*, **25**, 2844 (1977).
3. J. Szewczyk and C. Wasielewski, *Pol. J. Chem.*, **55**, 1985 (1981).
4. J. Engels, *Angew. Chem., Int. Ed. Engl.*, **18**, 148 (1979).
5. J. Engels, *Liebigs Ann. Chem.*, 557 (1980).
6. M. Sekine, K. Hamaoki, and T. Hata, *Bull. Chem. Soc. Jpn.*, **54**, 3815 (1981).
7. R. W. Adamiak, E. Biala, K. Grzeskowiak, R. Kierzek, A. Kraszewski, W. T. Markiewicz, J. Stawinski, and M. Wiewiorowski, *Nucleic Acids Res.*, **4**, 2321 (1977).
8. Y. Ichikawa and Y. C. Lee, *Carbohydr. Res.*, **198**, 235 (1990).
9. N. J. Noble, A. M. Cooke, and B. V. L. Potter, *Carbohydr. Res.*, **234**, 177 (1992).
10. F. Eckstein, *Chem. Ber.*, **100**, 2236 (1967).
11. M. Heuer, K. Hohgardt, F. Heinemann, H. Kühne, W. Dietrich, D. Grzelak, D. Müller, P. Welzel, A. Markus, Y. van Heijenoort, and J. van Heijenoort, *Tetrahedron*, **50**, 2029 (1994).
12. T. Neilson and E. S. Werstiuk, *Can. J. Chem.*, **49**, 3004 (1971).
13. A. Paquet, *Int. J. Pept. Protein Res.*, **39**, 82 (1992).
14. N. Mora, J. M. Lacombe, and A. A. Pavia, *Int. J. Pept. Protein Res.*, **45**, 53 (1995).
15. K. Grzeskowiak, R. W. Adamiak, and M. Wiewiorowski, *Nucleic Acids Res.*, **8**, 1097 (1980).
16. K. K. Ogilvie, S. L. Beaucage, and D. W. Entwistle, *Tetrahedron Lett.*, 1255 (1976).
17. A. Wolter and H. Köster, *Tetrahedron Lett.*, **24**, 873 (1983).
18. H. A. Kellner, R. G. K. Schneiderwind, H. Eckert, and I. K Ugi, *Angew. Chem., Int. Ed. Engl.*, **20**, 577 (1981).
19. P. Lemmen, K. M. Buchweitz, and R. Stumpf, *Chem. Phys. Lipids*, **53**, 65 (1990).
20. R. L. Letsinger, E. P. Groody, and T. Tanaka, *J. Am. Chem. Soc.*, **104**, 6805 (1982).
21. R. L. Letsinger, E. P. Groody, N. Lander, and T. Tanaka, *Tetrahedron*, **40**, 137 (1984).
22. A. B. Kazi and J. Hajdu, *Tetrahedron Lett.*, **33**, 2291 (1992).
23. J. H. Van Boom, P. M. J. Burgers, R. Crea, G. van der Marel, and G. Wille, *Nucleic Acids Res.*, **4**, 747 (1977).
24. L. Desaubry, I. Shoshani, and R. A. Johnson, *Tetrahedron Lett.*, **36**, 995 (1995).

25. A. Kraszewski and J. Strawinski, *Nucleic Acids Symp. Ser.*, **9**, 135 (1981).
26. H. Takaku, H. Tsuchiya, K. Imai, and D. E. Gibbs, *Chem. Lett.*, 1267 (1984).
27. S. Yamakage, M. Fujii, H. Takaku, and M. Uemura, *Tetrahedron*, **45**, 5459 (1989).
28. H. Takaku, T. Watanabe, and S. Hamamoto, *Tetrahedron Lett.*, **29**, 81 (1988).
29. T. Watanabe, H. Sato, and H. Takaku, *J. Am. Chem. Soc.*, **111**, 3437 (1989).

BENZYL PHOSPHATES

Benzyl (Bn): $C_6H_5CH_2$–

Formation

1. From a tributylstannyl phosphate: BnBr, $Et_4N^+Br^-$, CH_3CN, reflux. Phenacyl, 4-nitrobenzyl, and simple alkyl derivatives were similarly prepared. Yields are substrate and alkylating-agent dependent.[1]
2. Diphenyl phosphates are converted by transesterification to dibenzyl phosphates upon treatment with BnONa in THF at 25° in 83% yield.[2]

Cleavage

1. Pd–C, H_2, formic acid.[3]
2. Pd–C, EtOH, $NaHCO_3$, H_2.[4] Hydrogenolysis in the presence of NH_4OAc cleaves only one benzyl group of a dibenzyl phosphate.[5]
3. Na, ammonia.[6,7] Cyanoethyl and trichloroethyl phosphates are also deprotected.
4. 1 *M* TFMSA in TFA, thioanisole.[8] Dibenzyl phosphates are only partially labile to TFA alone.
5. TFA, thiophenol.[9]
6. A dibenzyl phosphate is monodeprotected with TFA–CH_2Cl_2.[10]
7. LiSPh, THF, HMPA, 30 min, >95% yield.[11]
8. NaI, CH_3CN.[12]
9. TMSBr, Pyr, CH_2Cl_2, rt, 1.5 h.[13] Phenolic phosphates were stable to this reagent.[14]
10. With dibenzyl phosphates or phosphonates, treatment with refluxing *N*-methylmorpholine results in monodebenzylation (60–100% yield).[15]
11. Quinuclidine, toluene, reflux.[16] In dibenzyl phosphates, only one benzyl group is removed.

***o*-Nitrobenzyl:** 2-NO_2-$C_6H_4CH_2$–

Formation

1. *o*-Nitrobenzyl alcohol, DCC, rt, 2 days. Pyridine slowly reacts to displace the nitrobenzyl ester, forming a 2-nitrobenzylpyridinium salt.[17]

Cleavage

1. Photolysis.[18–20]
2. Cleavage of an *S*-2-nitrobenzyl phosphorothioate is achieved with thiophenoxide in 5 min.[21]

4-Nitrobenzyl: 4-$NO_2C_6H_4CH_2$–

The 4-nitrobenzyl group, used in the synthesis of phosphorylated serine, is introduced by the phosphoramidite method and can be cleaved with TFMSA/MTB/*m*-cresol/1,2-ethanedithiol/TFA, 4 h, 0° to rt.[22] *N*-Methylmorpholine at 80° also cleaves a 4-nitrobenzyl phosphate triester.[23]

2,4-Dinitrobenzyl: 2,4-$(NO_2)_2$–$C_6H_3CH_2$–

Formation

This group has been used for the protection of a phosphorodithioate and is cleaved with 4-methylthiophenol and TEA.[24]

4-Chlorobenzyl: 4-$ClC_6H_4CH_2$–

Cleavage

1. Hydrogenolysis: Pd–C, *t*-BuOH, NaOAc, H_2O.[25–27]
2. From a phosphorothioate: TFMSA, *m*-cresol, thiophenol, TFA. These conditions minimized the migration of the benzyl group to the thione.[28]

4-Chloro-2-nitrobenzyl: 4-Cl–2-$NO_2C_6H_3CH_2$–

The 4-chloro-2-nitrobenzyl group was useful in the synthesis of dithymidine phosphorothioates. It could be cleaved with a minimun of side reactions with PhSH, TEA, Pyr.[29]

4-Acyloxybenzyl: 4-$RCO_2C_6H_4CH_2$–

4-Acyloxybenzyl esters were designed to be released under physiological conditions. Porcine liver carboxyesterase efficiently releases the phosphate by acetate hydrolysis and quinonemethide formation. In a diester, the first ester is cleaved faster than the second.[30]

1-Oxido-4-methoxy-2-picolyl:

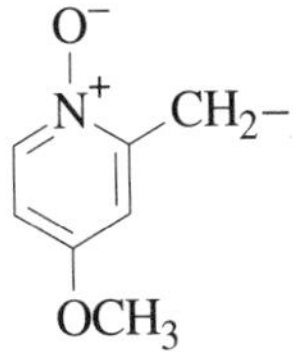

O^-
N^+ CH_2–
OCH_3

The oxidopicolyl group increases the rate and efficiency of internucleotide phosphodiester synthesis.[31] It is cleaved with piperidine.[32]

Fluorenyl-9-methyl (Fm): CH_2-

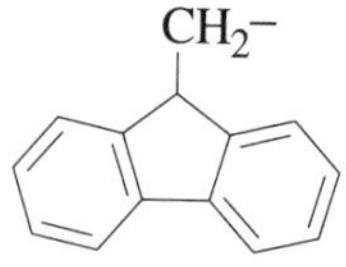

Formation

1. 5′-Nucleoside phosphates are protected using triisopropylbenzenesulfonyl chloride in Pyr.[33]

Cleavage

1. TEA, Pyr, 20°, 2 h.[34] These conditions were developed for use with 2-chlorophenyl protection at the internucleotide junctions.
2. TEA, CH_3CN, 14 h, rt.[35]
3. 0.1 *M* NaOH, 0°, 10 min.[33]
4. Concd. NH_4OH, 50°, 2 h.[33]

The fluorenyl-9-methyl group has been shown to be of particular value in studies of deoxynucleoside dithiophosphates.[36]

Pyrenylmethyl: CH_2-

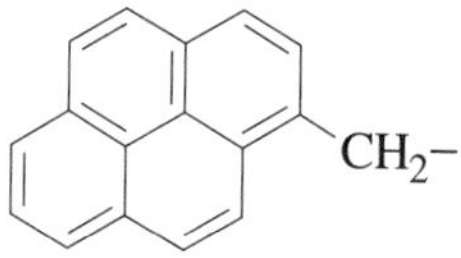

This derivative, synthesized by a silver oxide–promoted condensation of pyrenylmethyl chloride and a dialkyl phosphate (92% yield), is quantitatively cleaved by photolysis at >300 nm in 60 min.[37]

2-(9,10-Anthraquinonyl)methyl or 2-Methyleneanthraquinone (MAQ):

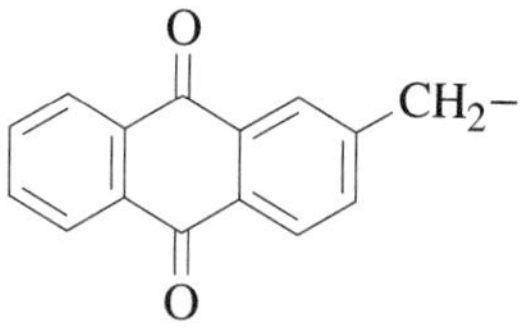

O

CH_2-

O

This group is stable to TEA/Pyr and to 80% acetic acid. It is cleaved by reduction with sodium dithionite at pH 7.3.[38]

5-Benzisoxazolylmethylene (Bim): N O CH_2O-

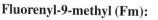

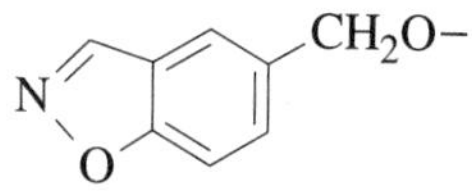

This group was effective in the synthesis of oligonucleotides using the phosphotriester approach.

Cleavage

1. TEA, Pyr, < 2 h.[39]

Cleavage Rates of Various Arylmethyl Phosphates

The following table compares the cleavage rates for a variety of benzyl phosphates using thiols or pyridine for the reaction[40,41]

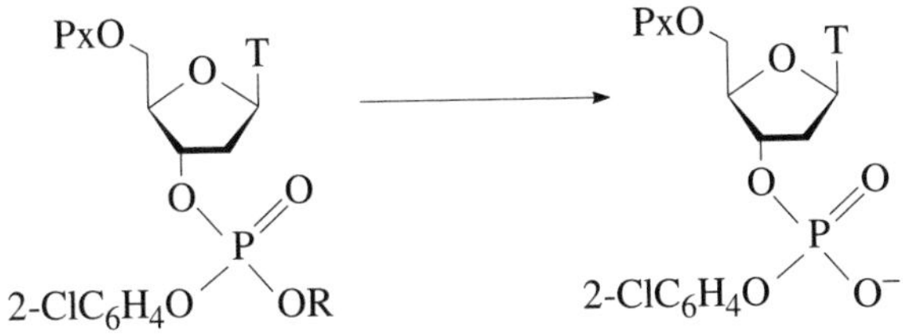

PxO, O, T; O, O, P; 2-ClC$_6$H$_4$O, OR → PxO, O, T; O, O, P; 2-ClC$_6$H$_4$O, O$^-$

Px = 9-phenylxanthen-9-yl (pixyl)

Substrate R =	*p*-Thiocresol/TEA/ACN $t_{1/2}$ (min)	(min)	Pyridine $t_{1/2}$ (h)	Ratio of half-lives (Pyr/RSH)
CH_3–	45	—	12	16
Bn–	30	—	12	24
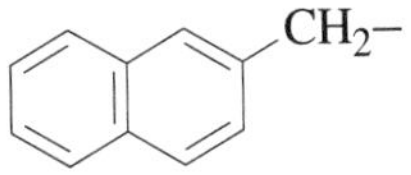 CH_2–	5	60	5	60
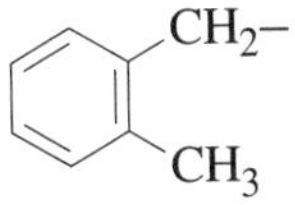 CH_2–, CH_3	7	90	3	26
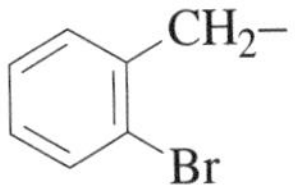 CH_2–, Br	4	45	10	150
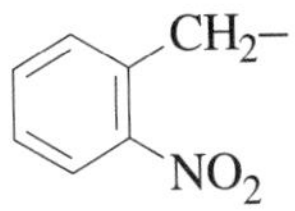 CH_2–, NO_2	5	60	68	820
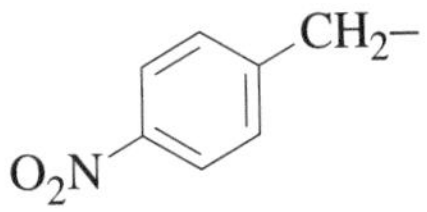 CH_2–, O_2N	2	20	40	1200
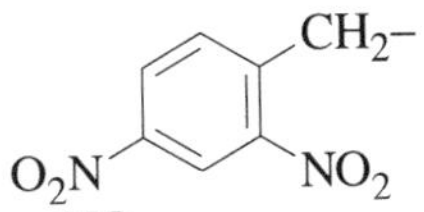 CH_2–, O_2N, NO_2	~10 sec	~1	120	~43,000
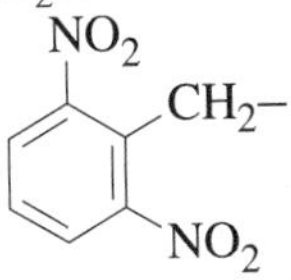 NO_2, CH_2–, NO_2	~10 sec	~1	45	~16,000

Diphenylmethyl (Dpm): $(C_6H_5)_2CH-$

The reaction of phosphoric acid with diphenyldiazomethane in dioxane gives the triphosphate.[42,43]

Cleavage

1. $(DpmO)_3PO$, upon reaction with NaI, Pyr at 100°, gives $(DpmO)_2P(O)ONa$ quantitatively. $Bu_3N^+HI^-$ can also be used to remove a single Dpm group.[42]
2. H_2, Pd-C, aqueous methanol.[42]
3. Trifluoroacetic acid.[43]

***o*-Xylene Derivative:** [cyclic o-xylylene phosphate structure: benzo-fused ring –O–P(=O)(OR)–O–]

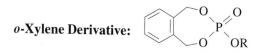

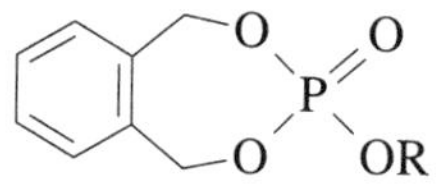

Cleavage

1. Hydrogenolysis: H_2, Pd–C, rt, 17 h.[44–46]

Benzoin Derivative: Ph–C(=O)–CH(Ph)–

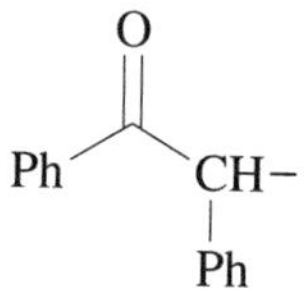

Formation

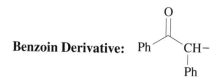

1. From $(EtO)_2P(O)Cl$: benzoin, Ag_2O.[37]
2. Bu_3NH–cAMP, desyl bromide.[47]

Cleavage

1. Photolysis, >300 nm.[37,48]

3′,5′-Dimethoxybenzoin Derivative (3′,5′-DMB)

The phosphate ester, prepared either through phosphoramidite or phosphoryl chloride protocols, is cleavable by photolysis (350 nm, benzene, 83–87% yield).[49–51]

4-Hydroxyphenacyl: $4\text{-}HOC_6H_4C(O)CH_2-$

The 4-hydroxyphenacyl group is removed by photolysis (300 nm, CH_3CN, tris buffer).[52,53]

4-Methoxyphenacyl: $4\text{-}CH_3OC_6H_4C(O)CH_2-$

Introduced with α-diazo-4-methoxyacetophenone, the phenacyl group is cleaved by photolysis with Pyrex-filtered mercury light in 74–86% yield.[54]

1. H. Ayukawa, S. Ohuchi, M. Ishikawa, and T. Hata, *Chem. Lett.*, 81 (1995).
2. D. C. Billington, R. Baker, J. J. Kulagowski, and I. M. Mawer, *J. Chem. Soc., Chem. Commun.*, 314 (1987).
3. J. W. Perich, P. F. Alewood, and R. B. Johns, *Aust. J. Chem.*, **44**, 233 (1991).
4. M. M. Sim, H. Kondo, and C.-H. Wong, *J. Am. Chem. Soc.*, **115**, 2260 (1993).
5. J. Scheigetz, M. Gilbert, and R. Zamboni, *Org. Prep. Proced. Int.*, **29**, 561 (1997).
6. N. J. Noble, A. M. Cooke, and B. V. L. Potter, *Carbohydr. Res.*, **234**, 177 (1992).
7. A. M. Riley, P. Guedat, G. Schlewer, B. Spiess, and B. V. L. Potter, *J. Org. Chem.*, **63**, 295 (1998).
8. T. Wakamiya, K. Saruta, S. Kusumoto, K. Nakajima, K. Yoshizawa-Kumagaye, S. Imajoh-Ohmi, and S. Kanegasaki, *Chem. Lett.*, 1401 (1993).
9. E. A. Kitas, R. Knorr, A. Trzeciak, and W. Bannwarth, *Helv. Chim. Acta*, **74**, 1314 (1991).
10. Z. Tian, C. Gu, R. W. Roeske, M. Zhou, and R. L. Van Etten, *Int. J. Pept. Protein Res.*, **42**, 155 (1993).
11. G. W. Daud and E. E. van Tamelen, *J. Am. Chem. Soc.*, **99**, 3526 (1977).
12. K. H. Scheit, *Tetrahedron Lett.*, 3243 (1967).
13. P. M. Chouinard and P. A. Bartlett, *J. Org. Chem.*, **51**, 75 (1986).
14. S. Lazar and G. Guillaumet, *Synth. Commun.*, **22**, 923 (1992); H.-G. Chao, M. S. Bernatowicz, C. E. Klimas, and G. R. Matsueda, *Tetrahedron Lett.*, **34**, 3377 (1993).
15. M. Saady, L. Lebeau, and C. Mioskowski, *J. Org. Chem.*, **60**, 2946 (1995).
16. M. Saady, L. Lebeau, and C. Mioskowski, *Tetrahedron Lett.*, **36**, 4785 (1995).
17. E. Ohtsuka, H. Tsuji, T. Miyake, and M. Ikehara, *Chem. Pharm. Bull.*, **25**, 2844 (1977).
18. M. Rubenstein, B. Amit, and A. Patchornik, *Tetrahedron Lett.*, 1445 (1975).
19. J. W. Walker, G. P. Reid, J. A. McCray, and D. R. Trentham, *J. Am. Chem. Soc.*, **110**, 7170 (1988).
20. E. Ohtsuka, T. Tanaka, S. Tanaka, and M. Ikehara, *J. Am. Chem. Soc.*, **100**, 4580 (1978).
21. Z. J. Lesnikowski and M. M. Jaworska, *Tetrahedron Lett.*, **30**, 3821 (1989).
22. T. Wakamiya, K. Saruta, J.-i. Yasuoka, and S. Kusumoto, *Bull. Chem. Soc. Jpn.*, **68**, 2699 (1995).
23. J. Smrt, *Collect. Czech. Chem. Commun.* **37**, 1870 (1972).
24. G. M. Porritt and C. B. Reese, *Tetrahedron Lett.*, **31**, 1319 (1990).
25. A. H. van Oijen, C. Erkelens, J. H. Van Boom, and R. M. J. Liskamp, *J. Am. Chem. Soc.*, **111**, 9103 (1989).
26. H. B. A. de Bont, J. H. Van Boom, and R. M. J. Liskamp, *Recl. Trav. Chim. Pays-Bas*, **109**, 27 (1990).
27. A. H. van Oijen, H. B. A. de Bont, J. H. van Boom, and R. M. J. Liskamp, *Tetrahedron Lett.*, **32**, 7723 (1991).
28. D. B. A. de Bont, W. J. Moree, J. H. van Boom, and R. M. J. Liskamp, *J. Org. Chem.*, **58**, 1309 (1993).
29. A. Püschl, J. Kehler, and O. Dahl, *Nucleosides Nucleotides*, **16**, 145 (1997).
30. A. G. Mitchell, W. Thomson, D. Nicholls, W. J. Irwin, and S. Freeman, *J. Chem. Soc., Perkin Trans. 1*, 2345 (1992).

31. T. Szabo, A. Kers, and J. Stawinski, *Nucleic Acids Res.*, **23**, 893 (1995).
32. N. N. Polushin, I. P. Smirnov, A. N. Verentchikov, and J. M. Coull, *Tetrahedron Lett.*, **37**, 3227 (1996).
33. N. Katagiri, C. P. Bahl, K. Itakura, J. Michniewicz, and S. A. Narang, *J. Chem. Soc., Chem. Commun.*, 803 (1973).
34. C. Gioeli and J. Chattopadhyaya, *Chem. Scr.*, **19**, 235 (1982).
35. Y. Watanabe and M. Nakatomi, *Tetrahedron Lett.*, **39**, 1583 (1998).
36. P. H. Seeberger, E. Yau, and M. H. Caruthers, *J. Am. Chem. Soc.*, **117**, 1472 (1995).
37. T. Furuta, H. Torigai, T. Osawa, and M. Iwamura, *Chem. Lett.*, 1179 (1993).
38. N. Balgobin, M. Kwiatkowski, and J. Chattopadhyaya, *Chem. Scr.*, **20**, 198 (1982).
39. N. Balgobin and J. Chattopadhyaya, *Chem. Scr.*, **20**, 142 (1982).
40. C. Christodoulou and C. B. Reese, *Nucl. Acids Symp. Ser.*, **11**, 33 (1982).
41. C. Christodoulou and C. B. Reese, *Tetrahedron Lett.*, **24**, 951 (1983).
42. G. Lowe and B. S. Sproat, *J. Chem. Soc., Perkin Trans. 1*, 1874 (1981).
43. M. Hoffmann, *Pol. J. Chem.*, **59**, 395 (1985).
44. Y. Watanabe, T. Shinohara, T. Fujimoto, and S. Ozaki, *Chem. Pharm. Bull.*, **38**, 562 (1990).
45. S. Ozaki, Y. Kondo, N. Shiotani, T. Ogasawara, and Y. Watanabe, *J. Chem. Soc., Perkin Trans. 1*, 729 (1992).
46. Y. Watanabe, Y. Komoda, K. Ebisuya, and S. Ozaki, *Tetrahedron Lett.*, **31**, 255 (1990).
47. R. S. Givens, P. S. Athey, L. W. Kueper, III, B. Matuszewski, and J.-y. Xue, *J. Am. Chem. Soc.*, **114**, 8708 (1992).
48. R. S. Givens and B. Matuszewski, *J. Am. Chem. Soc.*, **106**, 6860 (1984).
49. M. C. Pirrung and S. W. Shuey, *J. Org. Chem.*, **59**, 3890 (1994).
50. J. E. Baldwin, A. W. McConnaughie, M. G. Moloney, A. J. Pratt, and S. B. Shim, *Tetrahedron*, **46**, 6879 (1990).
51. For a review of phosphate ester photochemistry, see R. S. Givens and L. W. Kueper, III, *Chem. Rev.*, **93**, 55 (1993).
52. R. S. Givens and C.-H. Park, *Tetrahedron Lett.*, **37**, 6259 (1996).
53. C.-H. Park and R. S. Givens, *J. Am. Chem. Soc.*, **119**, 2453 (1997).
54. W. W. Epstein and M. Garrossian, *J. Chem. Soc., Chem. Commun.*, 532 (1987).

PHENYL PHOSPHATES

Phenyl: C_6H_5–

Cleavage

1. PtO_2 (stoichiometric), TFA, AcOH, H_2, 91% yield.[1,2] This method cannot be used in substrates that contain a tyrosine, because tyrosine is easily reduced in the acidic medium. Neutral conditions fail to cleave phenyl phosphates.[3]
2. Aqueous HCl, reflux.[4]

3. $Bu_4N^+F^-$, THF, Pyr, H_2O, rt, 30 min.[5] These conditions result in the formation of a mixture of fluorophosphate and phosphate. In the case of oligonucleotides, some internucleotide bond cleavage is observed with this reagent.
4. NaOH, THF[6] or LiOH, dioxane.[7]
5. See the discussion of the cleavage of 2-chlorophenyl (below) for oximate rate comparisons.

2-Methylphenyl and 2,6-Dimethylphenyl

These groups were more effective than the phenyl group for protection of phosphoserine during peptide synthesis. They are cleaved by hydrogenolysis with stoichiometric PtO_2 in AcOH.[8]

2-Chlorophenyl: 2-Cl-C_6H_4–

Cleavage

1. Tetramethylguanidinium 4-nitrobenzaldoxime, dioxane, H_2O, 20°, 22 h.[9] This reagent cleaves the 2-chlorophenyl ester 2.5 times faster than the 4-chlorophenyl ester and 25 times faster than the phenyl ester. The use of *syn*-2-nitrobenzaldoxime increases the rate an additional 2.5 to 4 times.[10] Oximate cleavage proceeds by nucleophilic addition–elimination to give an oxime ester that, with base, undergoes another elimination to give a nitrile and phosphate anion.[11]
2. NaOH, Pyr, H_2O, 0°.[12]
3. *syn*-Pyridine-2-aldoxime, tetramethylguanidine, dioxane, Pyr, H_2O.[13]

4-Chlorophenyl: 4-Cl-C_6H_4–

Halogen-substituted phenols were originally introduced for phosphate protection to minimize internucleotide bond cleavage during deprotection.[14]

1. NH_4OH, 55°, 3 h.[15]
2. Treatment of an internucleotide 4-chlorophenyl ester with CsF and an alcohol (MeOH, EtOH, neopentylOH) results in transesterification.[16]

2,4-Dichlorophenyl: 2,4-$Cl_2C_6H_3$–

1. 4-Nitrobenzaldoxime, tetramethylguanidine, THF.[17]
2. Aqueous ammonia, dioxane, 12 h, 60°.[18]

2,5-Dichlorophenyl: 2,5-$Cl_2C_6H_3$–

Cleavage

1. 4-Nitrobenzaldoxime, TEA, dioxane, H_2O.[19] Cleavage occurs in the presence of 4-nitrophenylethyl phosphate.

2. Pyridine-2-carbaldoxime, TEA, H_2O, dioxane. The 2-(1-methyl-2-imidazolyl)phenyl group is not removed under these conditions.[20]

2,6-Dichlorophenyl: 2,6-$Cl_2C_6H_3$–

Cleavage

1. 4-Nitrobenzaldoxime, TEA, dioxane, H_2O.[21]

2-Bromophenyl: 2-BrC_6H_4–

Cleavage of the bromophenyl group is achieved with $Cu(OAc)_2$ in Pyr, H_2O. The 2-chlorophenyl group is stable to these conditions.[22]

4-Nitrophenyl (PNP): 4-$NO_2C_6H_4$–

Cleavage

1. *p*-Thiocresol, TEA, CH_3CN.[9] The 4-nitrophenyl group is removed in the presence of a 2-chlorophenyl group.
2. 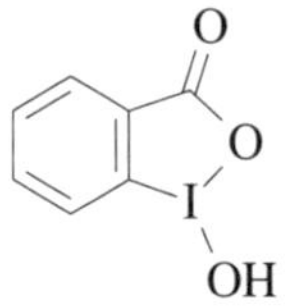 (Organoindinane), aqueous micellar cetyltrimethylammonium chloride, pH 8.[23]
3. Tetrabutylammonium acetate, 20 h, 20°. For comparison, the 2,4-dichlorophenyl group was removed in 100 h.[24]
4. *syn*-4-Nitrobenzaldoxime, tetramethylguanidine, dioxane, CH_3CN, 16 h.[24]
5. 0.125 *N* NaOH, dioxane.[24]
6. 4-Nitrophenyl phosphonates are transesterified in the presence of DBU and an alcohol.[25]

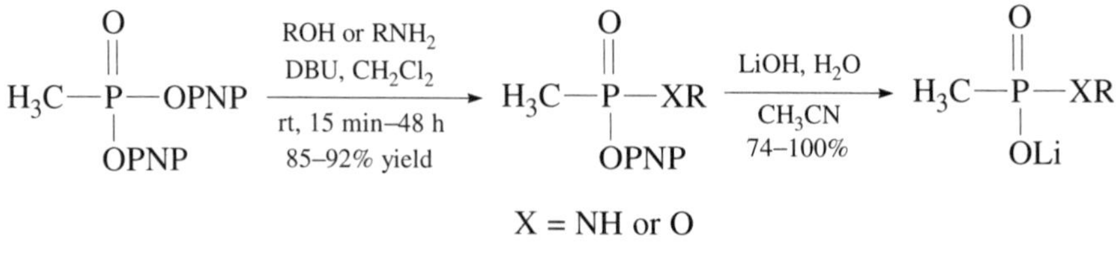

$$H_3C{-}P(=O)(OPNP){-}OPNP \xrightarrow[\text{rt, 15 min–48 h; 85–92\% yield}]{\text{ROH or RNH}_2\text{, DBU, CH}_2\text{Cl}_2} H_3C{-}P(=O)(OPNP){-}XR \xrightarrow[\text{CH}_3\text{CN; 74–100\%}]{\text{LiOH, H}_2\text{O}} H_3C{-}P(=O)(OLi){-}XR$$

X = NH or O

7. Zr^{+4}, H_2O, pH 3.5, 37°.[26]

3,5-Dinitrophenyl: 3,5-$(NO_2)_2C_6H_3$–

Photolysis through a Pyrex filter in Pyr, EtOH, H_2O cleaves this phosphate ester.[27] The rate increases with increasing pH.

4-Chloro-2-nitrophenyl: 4-Cl-2-$NO_2C_6H_3$–

Cleavage is achieved with refluxing NaOH (15 min), but some deamination occurs with deoxyriboadenosine-5′-phosphate.[28] The ester is formed using the DCC protocol for phosphate ester formation.

2-Chloro-4-triphenylmethylphenyl

The lipophilicity of this phosphate protective group helps in the chromatographic purification of oligonucleotides. It is removed by the oximate method.[29]

2-Methoxy-5-nitrophenyl

This ester is cleaved by photolysis at >300 nm in basic aqueous acetonitrile.[30]

1,2-Phenylene:

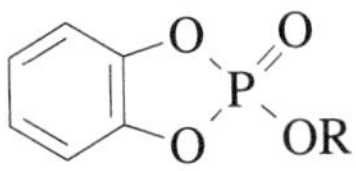

The phenylene group is removed oxidatively with $Pb(OAc)_4$ in dioxane.[31]

4-Triphenylmethylaminophenyl: 4-[$(C_6H_5)_3$CNH]C_6H_4–

Formation

1. $TrNHC_6H_4OH$, DCC, Pyr.

Cleavage

1. Iodine, acetone or DMF, ammonium acetate, rt, 2 h. The tritylaminophenyl group is stable to isoamyl nitrite/acetic acid.[32]

4-Benzylaminophenyl: 4-[$C_6H_5CH_2NH$]C_6H_4–

Cleavage

1. Electrolysis: 0.6–1.0 V, 3 h, DMF, H_2O, $NaClO_4$.[33] The related 4-tritylaminophenyl and 4-methoxyphenyl groups were not cleanly cleaved.

1-Methyl-2-(2-hydroxyphenyl)imidazole Derivative:

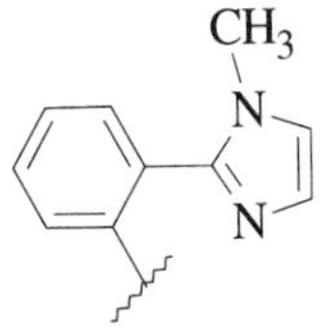

The rate of oligonucleotide synthesis by the triester method using mesitylenesulfonyl chloride was increased five- to tenfold when this group was used as a protective group during internucleotide bond formation. It was removed with concd. NH_4OH at 60° for 12 h[18] or by the oximate method.[20]

8-Quinolyl

This group is stable to acid and alkali. It has been used as a copper-activated leaving group for triphosphate protection.[34]

Formation

1. Ph_3P, 2,2′-dipyridyl disulfide, Pyr, rt, 6 h.[35]
2. $(PhO)_3P$, 2,2′-dipyridyl diselenide, Pyr, rt, 12 h.[36]

Cleavage

1. $CuCl_2$, DMSO, H_2O, 40–45°, 5 h.[35]

5-Chloro-8-quinolyl

Formation

1. 5-Chloro-8-hydroxyquinoline, $POCl_3$, Pyr, 92% yield.[37]
2. 2,2′-Dipyridyl diselenide, $(PhO)_3P$, Pyr, rt, 12 h, 80–85% yield.[38]

Cleavage

1. Aqueous ammonia, 2 days, 27°.[39]
2. $Zn(OAc)_2$, Pyr, H_2O, 28 h, 98% yield.[12]
3. 2-Pyridinecarboxaldoxime, tetramethylguanidine, dioxane, H_2O, 90% yield.[12]
4. $ZnCl_2$, aq. Pyr, rt, 12 h.[38,40]
5. Pyr, *t*-$BuNH_2$, H_2O. Cleavage occurs in the presence of the 2,6-dichlorophenyl phosphate.[41]
6. The 5-chloro-8-quinolyl group can also be activated with $CuCl_2$ under anhydrous conditions and used in triphosphate formation.[42,43]

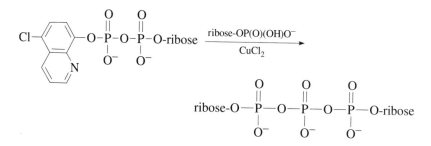

5-Cl-8-quinolyl–O–P(O)(O⁻)–O–P(O)(O⁻)–O-ribose $\xrightarrow[\text{CuCl}_2]{\text{ribose-OP(O)(OH)O}^-}$ ribose-O–P(O)(O⁻)–O–P(O)(O⁻)–O–P(O)(O⁻)–O-ribose

Thiophenyl: C_6H_5S–

The phosphorodithioate is stable to heating at 100°, 80% acetic acid (1 h), dry or aqueous pyridine (days) and refluxing methanol, ethanol or isopropyl alcohol for 1 h.

Formation

1. $(ArS)_2P(O)O^-$ $C_6H_{11}NH_3^+$ is prepared from the phosphinic acid with TMSCl, TEA, PhSSPh in THF at rt, 20 h, in 83% yield.[44]

Cleavage

1. Treatment of $ROP(O)(SPh)_2$ (**1**) with 0.2 *N* NaOH (dioxane, rt, 15 min)[44] or pyridinium phosphinate (Pyr, TEA)[45] quantitatively gives $ROP(O)(SPh)O^-$ (**2**).

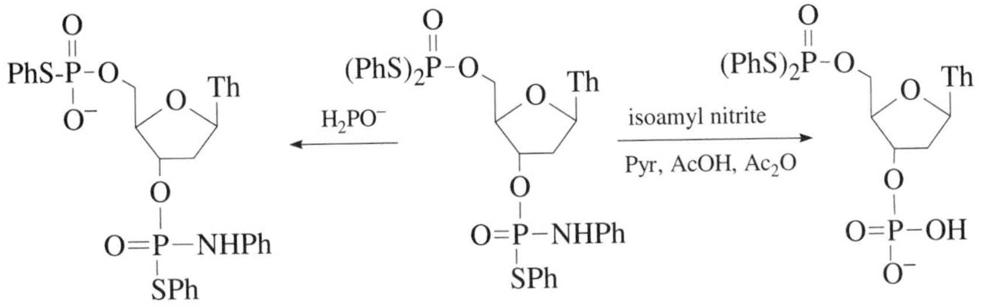

H_2PO^- ← **1** → isoamyl nitrite, Pyr, AcOH, Ac_2O

2 **1** **3**

Ref. 45

2. AgOAc (Pyr, H_2O) cleaves both thioates of **1** to give a phosphate.[44]
3. Treatment of **2** with I_2 or AgOAc also gives the phosphate.[44]
4. Treatment of **1** with Zn (acetylacetone, Pyr, DMF) gives the phosphate.[44]
5. Treatment of **1** with phosphinic acid and triazole gives **2**.[44]
6. Treatment of $(RO)_2P(O)SPh$ with Bu_3SnOMe converts it to $(RO)_2P(O)$-OMe.[46, 47]
7. $(Bu_3Sn)_2O$; TMSCl; H_2O.[48,49]
8. Treatment of $ROP(O)(SPh)_2$ with H_3PO_3/Pyr gives ROP(O)(SPh)OH.[50]
9. Phosphorothioates, when activated with $AgNO_3$ under anhydrous conditions in the presence of monophosphates, are converted into diphosphates.[51]
10. Tributylstannyl 2-pyridine-*syn*-carboxaldoxime, Pyr.[48]

Salicylic Acid Derivative

Salicylic acid was used for phosphite protection in the synthesis of glycosyl phosphites and phosphates. This derivative is very reactive and readily forms a phosphite upon treatment with an alcohol or a phosphonic acid upon aqueous hydrolysis.[52]

1. W. H. A. Kuijpers, J. Huskens, L. H. Koole, and C. A. A. Van Boekel, *Nucleic Acids Res.*, **18**, 5197 (1990).

2. J. W. Perich, P. F. Alewood, and R. B. Johns, *Aust. J. Chem.*, **44**, 233 (1991).
3. Y. Ichikawa and Y. C. Lee, *Carbohydr. Res.*, **198**, 235 (1990).
4. C. C. Tam, K. L. Mattocks, and M. Tishler, *Synthesis*, 188 (1982).
5. K. K. Ogilvie and S. L. Beaucage, *Nucleic Acids Res.*, **7**, 805 (1979).
6. G. De Nanteuil, A. Benoist, G. Remond, J.-J. Descombes, V. Barou, and T. J. Verbeuren, *Tetrahedron Lett.*, **36**, 1435 (1995).
7. R. Plourde and M. d'Alarcao, *Tetrahedron Lett.*, **31**, 2693 (1990).
8. M. Tsukamoto, R. Kato, K. Ishiguro, T. Uchida, and K. Sato, *Tetrahedron Lett.*, **32**, 7083 (1991).
9. S. S. Jones and C. B. Reese, *J. Am. Chem. Soc.*, **101**, 7399 (1979).
10. C. B. Reese and L. Zard, *Nucleic Acids Res.*, **9**, 4611 (1981).
11. C. B. Reese and L. Yau, *Tetrahedron Lett.*, 4443 (1978).
12. K. Kamaike, S. Ueda, H. Tsuchiya, and H. Takaku, *Chem. Pharm. Bull.*, **31**, 2928 (1983).
13. T. Tanaka, T. Sakata, K. Fujimoto, and M. Ikehara, *Nucleic Acids Res.*, **15**, 6209 (1987).
14. J. H. van Boom, P. M. J. Burgers, P. H. van Deursen, R. Arentzen, and C. B. Reese, *Tetrahedron Lett.*, 3785 (1974).
15. E. Ohtsuka, T. Tanaka, T. Wakabayashi, Y. Taniyama, and M. Ikehara, *J. Chem. Soc., Chem. Commun.*, 824 (1978).
16. U. Asseline, C. Barbier, and N. T. Thuong, *Phosphorus Sulfur*, **26**, 63 (1986).
17. B. Mlotkowska, *Liebigs Ann. Chem.*, 1361 (1991).
18. B. C. Froehler and M. D. Matteucci, *J. Am. Chem. Soc.*, **107**, 278 (1985).
19. E. Uhlmann and W. Pfleiderer, *Helv. Chim. Acta*, **64**, 1688 (1981).
20. B. S. Sproat, P. Rider, and B. Beijer, *Nucleic Acids Res.*, **14**, 1811 (1986).
21. H. Takaku, S. Hamamoto, and T. Watanabe, *Chem. Lett.*, 699 (1986).
22. Y. Stabinsky, R. T. Sakata, and M. H. Caruthers, *Tetrahedron Lett.*, **23**, 275 (1982).
23. R. A. Moss, B. Wilk, K. Krogh-Jespersen, J. T. Blair, and J. D. Westbrook, *J. Am. Chem. Soc.*, **111**, 250 (1989).
24. J. A. J. Den Hartog and J. H. Van Boom, *Recl.: J. R. Neth. Chem. Soc.*, **100**, 285 (1981).
25. D. S. Tawfik, Z. Eshhar, A. Bentolila, and B. S. Green, *Synthesis*, 968 (1993).
26. R. A. Moss, J. Zhang, and K. G. Ragunathan, *Tetrahedron Lett.*, **39**, 1529 (1998).
27. A. J. Kirby and A. G. Varvoglis, *J. Chem. Soc., Chem. Commun.*, 406 (1967).
28. S. A. Narang, O. S. Bhanot, J. Goodchild, and R. Wightman, *J. Chem. Soc., Chem. Commun.*, 91 (1970).
29. J. J. Vasseur, B. Rayner, and J. L. Imbach, *Tetrahedron Lett.*, **24**, 2753 (1983).
30. N. R. Graciani, D. S. Swanson, and J. W. Kelly, *Tetrahedron*, **51**, 1077 (1995).
31. L.-d. Liu and H.-w. Liu, *Tetrahedron Lett.*, **30**, 35 (1989).
32. E. Ohtsuka, S. Morioka, and M. Ikehara, *J. Am. Chem. Soc.*, **95**, 8437 (1973).
33. E. Ohtsuka, T. Miyake, M. Ikehara, A. Matsumuto, and H. Ohmori, *Chem. Pharm. Bull.*, **27**, 2242 (1979).
34. K. Fukuoka, F. Suda, M. Ishikawa, and T. Hata, *Nucl. Acids Symp. Ser.*, **29**, 35 (1993).

35. H. Takaku, Y. Shimada, and T. Hata, *Chem. Lett.*, 873 (1975).
36. H. Takaku, R. Yamaguchi and T. Hata, *J. Chem. Soc., Perkin Trans. 1*, 519 (1978).
37. H. Takaku, K. Kamaike, and M. Suetake, *Chem. Lett.*, 111 (1983).
38. H. Takaku, R. Yamaguchi, and T. Hata, *Chem. Lett.*, 5 (1979).
39. S. C. Srivastava and A. L. Nussbaum, *J. Carbohydr., Nucleosides, Nucleotides*, **8**, 495 (1981).
40. H. Takaku, R. Yamaguchi, T. Nomoto, and T. Hata, *Tetrahedron Lett.*, 3857 (1979).
41. H. Takaku, K. Imai, and M. Nagai, *Chem. Lett.*, 857 (1988).
42. K. Fukuoka, F. Suda, R. Suzuki, M. Ishikawa, H. Takaku, and T. Hata, *Nucleosides Nucleotides*, **13**, 1557 (1994).
43. K. Fukuoka, F. Suda, R. Suzuki, H. Takaku, M. Ishikawa, and T. Hata, *Tetrahedron Lett.*, **35**, 1063 (1994).
44. M. Sekine, K. Hamaoki, and T. Hata, *Bull. Chem. Soc. Jpn.*, **54**, 3815 (1981).
45. T. Hata, T. Kamimura, K. Urakami, K. Kohno, M. Sekine, I. Kumagai, K. Shinozaki, and K. Miura, *Chem. Lett.*, 117 (1987).
46. S. Ohuchi, H. Ayukawa, and T. Hata, *Chem. Lett.*, 1501 (1992).
47. Y. Watanabe and T. Mukaiyama, *Chem. Lett.*, 389 (1979).
48. M. Sekine, H. Tanimura, and T. Hata, *Tetrahedron Lett.*, **26**, 4621 (1985).
49. H. Tanimura, M. Sekine, and T. Hata, *Tetrahedron*, **42**, 4179 (1986); H. Tanaka, H. Hayakawa, K. Obi, and T. Miyasaka, *idem*, 4187 (1986).
50. M. Sekine, K. Hamaoki, and T. Hata, *J. Org. Chem.*, **44**, 2325 (1979).
51. K. Fukuoka, F. Suda, R. Suzuki, M. Ishikawa, H. Takaku, and T. Hata, *Nucleosides Nucleotides*, **13**, 1557 (1994).
52. J. P. G. Hermans, E. De Vroom, C. J. J. Elie, G. A. Van der Marel, and J. H. Van Boom, *Recl. Trav. Chim. Pays-Bas*, **105**, 510 (1986).

AMIDATES

```
              O–
              |
R– or Ar–NH-P—O–
              ||
              O
```

Anilidate: $C_6H_5NH–$

A polymeric version of this group has been developed for terminal phosphate protection in ribooligonucleotide synthesis.[1]

Formation

1. Ph_3P, 2,2′-dipyridyl disulfide, aniline, 60% yield.[2]

Cleavage

1. Isoamyl nitrite, Pyr, acetic acid.[3,4]

4-Triphenylmethylanilidate: 4-$(C_6H_5)_3CC_6H_4NH$–

This highly lipophilic group is cleaved with isoamyl nitrite in Pyr/AcOH.[5] The use of a lipophilic 5′-phosphate protective group aids in reverse-phase HPLC purification of oligonucleotides.

[*N*-(2-Trityloxy)ethyl]anilidate: $(C_6H_5)_3COCH_2CH_2$-C_6H_4-N–

This lipophilic group, developed for 5′-phosphate protection in oligonucleotide synthesis, is removed with 80% AcOH in 1 h.[6,7] The related trityloxyethylamino group has been used in a similar capacity for phosphate protection and is also cleaved with 80% AcOH.[8]

***p*-(*N,N*-Dimethylamino)anilidate:** *p*-$(CH_3)_2NC_6H_4NH$–

This group was developed to aid in the purification of polynucleotides by adsorbing the phosphoroanilidates on an acidic ion-exchange resin.[9] Derivatives containing this group as a terminal phosphate protective group could be adsorbed on an acid ion-exchange resin for purification. The group is removed with 80% acetic acid at 80° for 3 h.[10]

Formation

1. DCC, *N,N*-dimethyl-*p*-phenylenediamine.[10]

Cleavage

1. 80% acetic acid, 80°, 3 h.[10]
2. Isoamyl nitrite, Pyr, AcOH.[11]

3-(*N,N*-Diethylaminomethyl)anilidate: 3-$[(C_2H_5)_2NCH_2]C_6H_4NH$–

Cleavage is effected with isoamyl nitrite in Pyr/AcOH.[12,13]

***p*-Anisidate:** *p*-$CH_3OC_6H_4NH$–

Cleavage

1. Pyr, AcOH, isoamyl nitrite.[14,15]
2. $Bu_4N^+NO_2^-$, Ac_2O, Pyr, rt, 10 min.[16]

2,2′-Diaminobiphenyl Derivative

Formation

1. 2,2′-Diaminobiphenyl, Ph_3P, $(PyS)_2$.[17]

Cleavage

1. Isoamyl nitrite, Pyr, AcOH, AgOAc, benzoic anhydride.[17]

n-Propylamine and *i*-Propylamine Derivatives

These derivatives provide effective protection for phosphotyrosine in Fmoc-based peptide synthesis. They are cleaved with 95% TFA.[18]

N,N′-Dimethyl-(*R,R*)-1,2-diaminocyclohexyl

This group was used as a protective group and chiral directing group for the asymmetric synthesis of α-aminophosphonic acids. It is cleaved by acid hydrolysis.[19]

Morpholino Derivative

Morpholine has been used for 5′-phosphate protection in oligonucleotide synthesis and can be cleaved with 0.01 *N* HCl without significant depurination of bases having free exocyclic amino functions.[20,21]

1. E. Ohtsuka, S. Morioka, and M. Ikehara, *J. Am. Chem. Soc.*, **94**, 3229 (1972).
2. E. Ohtsuka, H. Tsuji, T. Miyake, and M. Ikehara, *Chem. Pharm. Bull.*, **25**, 2844 (1977).
3. M. Sekine and T. Hata, *Tetrahedron Lett.*, **24**, 5741 (1983).
4. E. Ohtsuka, T. Ono, and M. Ikehara, *Chem. Pharm. Bull.*, **29**, 3274 (1981).
5. K. L. Agarwal, A. Yamazaki, and H. G. Khorana, *J. Am. Chem. Soc.*, **93**, 2754 (1971).
6. T. Tanaka, Y. Yamada, S. Tamatsukuri, T. Sakata, and M. Ikehara, *Nucl. Acids Symp. Ser.*, **17**, 85 (1986).
7. T. Tanaka, Y. Yamada, and M. Ikehara, *Tetrahedron Lett.*, **27**, 3267 (1986).
8. T. Tanaka, Y. Yamada, and M. Ikehara, *Tetrahedron Lett.*, **27**, 5641 (1986).
9. K. Tajima and T. Hata, *Bull. Chem. Soc. Jpn.*, **45**, 2608 (1972).
10. T. Hata, K. Tajima, and T. Mukaiyama, *J. Am. Chem. Soc.*, **93**, 4928 (1971).
11. K. Tajima and T. Hata, "Simple Protecting Group Protection–Purification Handle for Polynucleotide Synthesis, II." *Bull. Chem. Soc. Jpn.*, **45**, 2608 (1972).
12. T. Hata, I. Nakagawa, and N. Takebayashi, *Tetrahedron Lett.*, 2931 (1972).
13. T. Hata, I. Nakagawa, and Y. Nakada, *Tetrahedron Lett.*, 467 (1975).
14. S. Iwai, M. Asaka, H. Inoue, and E. Ohtsuka, *Chem. Pharm. Bull.*, **33**, 4618 (1985).
15. E. Ohtsuka, M. Shin, Z. Tozuka, A. Ohta, K. Kitano, Y. Taniyama, and M. Ikehara, *Nucl. Acids Symp. Ser.*, **11**, 193 (1982).
16. S. Nishino, Y. Nagato, Y. Hasegawa, K. Kamaike, and Y. Ishido, *Nucl. Acids Symp. Ser.*, **20**, 73 (1988).
17. M. Nishizawa, T. Kurihara, and T. Hata, *Chem. Lett.*, 175 (1984).
18. M. Ueki, J. Tachibana, Y. Ishii, J. Okumura, and M. Goto, *Tetrahedron Lett.*, **37**, 4953 (1996).
19. S. Hanessian and Y. L. Bennani, *Synthesis*, 1272 (1994).

20. C. van der Marel, G. Veeneman, and J. H. van Boom, *Tetrahedron Lett.*, **22**, 1463 (1981).
21. A. Kondo, Y. Uchimura, F. Kimizuka, and A. Obayashi, *Nucl. Acids Symp. Ser.*, **16**, 161 (1985).

MISCELLANEOUS DERIVATIVES

Ethoxycarbonyl: EtO_2C–

The ethoxycarbonyl group was developed for the protection of phosphonates. The derivative is prepared by reaction of tris(trimethylsilyl) phosphite with ethyl chloroformate and can be cleaved by hydrolysis of the ester followed by silylation with bistrimethylsilylacetamide.[1]

(Dimethylthiocarbamoyl)thio: $(CH_3)_2NC(S)S$–

This group, used for internucleotide protection, is introduced with 8-quinolinesulfonyl chloride, $[(CH_3)_2NC(S)S]_2$, and Ph_3P and is cleaved with BF_3 (dioxane, H_2O, rt).[2]

1. M. Sekine, H. Mori, and T. Hata, *Bull. Chem. Soc. Jpn.*, **55**, 239 (1982).
2. H. Takaku, M. Kato, and S. Ishikawa, *J. Org. Chem.*, **46**, 4062 (1981).

10

REACTIVITIES, REAGENTS, AND REACTIVITY CHARTS

REACTIVITIES

In the selection of a protective group, it is of paramount importance to know the reactivity of the resulting protected functionality toward various reagents and reaction conditions. The number of reagents available to the organic chemist is large; approximately 8000 reagents are reviewed in the excellent series of books by the Fiesers.[1] In an effort to assess the effect of a wide variety of standard types of reagents and reaction conditions on the different possible protected functionalities, 108 prototype reagents have been selected and grouped into 16 categories:[2]

A. Aqueous
B. Nonaqueous Bases
C. Nonaqueous Nucleophiles
D. Organometallic
E. Catalytic Reduction
F. Acidic Reduction
G. Basic or Neutral Reduction
H. Hydride Reduction
I. Lewis Acids
J. Soft Acids
K. Radical Addition

L. Oxidizing Agents
M. Thermal Reactions
N. Carbenoids
O. Miscellaneous
P. Electrophiles

These 108 reagents are used in the Reactivity Charts that have been prepared for each class of protective groups. The reagents and some of their properties are described on the following pages.

REAGENTS

A. AQUEOUS

1. pH < 1, 100°	Refluxing HBr
2. pH < 1	1 *N* HCl
3. pH 1	0.1 *N* HCl
4. pH 2–4	0.01 *N* HCl; 1–0.01 *N* HOAc
5. pH 4–6	0.1 *N* H_3BO_3; phosphate buffer; HOAc–NaOAc
6. pH 6–8.5	H_2O
7. pH 8.5–10	0.1 *N* HCO_3^-; 0.1 *N* OAc^-; satd. $CaCO_3$
8. pH 10–12	0.1 *N* CO_3^{2-}; 1–0.01 *N* NH_4OH; 0.01 *N* NaOH; satd $Ca(OH)_2$
9. pH > 12	1–0.1 *N* NaOH
10. pH > 12, 150°	

B. NONAQUEOUS BASES

11. NaH	
12. $(C_6H_5)_3CNa$	$pK_a = 32$
13. $[C_{10}H_8]^- \cdot Na^+$	$pK_a \cong 37$
14. $CH_3SOCH_2^-Na^+$	$pK_a = 35$
15. KO–*t*-C_4H_9	$pK_a = 19$
16. $LiN(i\text{-}C_3H_7)_2$	(LDA) $pK_a = 36$
17. Pyridine; Et_3N	$pK_a = 5; 10$
18. $NaNH_2$; NaNHR	$pK_a = 36$

C. NONAQUEOUS NUCLEOPHILES

19. $NaOCH_3/CH_3OH$, 25°	$pK_a = 16$
20. Enolate anion	$pK_a = 20$
21. NH_3; RNH_2; RNHOH	$pK_a = 10$

22. RS^-; N_3^-; SCN^-	
23. OAc^-; X^-	$pK_a = 4.5$
24. NaCH, pH 12	
25. HCN, cat. CN^-, pH 6	$pK_a = 9$. For cyanohydrin formation

D. ORGANOMETALLIC

26. RLi	
27. RMgX	
28. Organozinc	Reformatsky reaction. Similar: R_2Cu; R_2Cd
29. Organocopper	R_2CuLi
30. Wittig; ylide	Includes sulfur ylides

E. CATALYTIC REDUCTION

31. H_2/Raney Ni	
32. H_2/Pt, pH 2–4	
33. H_2/Pd–C	
34. H_2/Lindlar	
35. H_2/Rh–C or H_2/Rh–Al_2O_3	Avoids hydrogenolysis of benzyl ethers

F. ACIDIC REDUCTION

36. Zn/HCl
37. Zn/HOAc; $SnCl_2$/HCl
38. Cr(II), pH 5

G. BASIC OR NEUTRAL REDUCTION

39. Na/*l* NH_3
40. Al(Hg)
41. $SnCl_2$/Py
42. H_2S or HSO_3^-

H. HYDRIDE REDUCTION

43. $LiAlH_4$	
44. Li–*s*-Bu_3BH, −50°	Li-Selectride
45. $[(CH_3)_2CHCH(CH_3)]_2BH$	Disiamylborane
46. B_2H_6, 0°	
47. $NaBH_4$	
48. $Zn(BH_4)_2$	Neutral reduction
49. $NaBH_3CN$, pH 4–6	
50. $(i\text{-}C_4H_9)_2AlH$, −60°	Dibal
51. $Li(O\text{–}t\text{-}C_4H_9)_3AlH$, 0°	

I. LEWIS ACIDS (ANHYDROUS CONDITIONS)

52. $AlCl_3$, 80°	
53. $AlCl_3$, 25°	
54. $SnCl_4$, 25°; $BF_3{\cdot}Et_2O$	
55. $LiClO_4$; $MgBr_2$	For epoxide rearrangement
56. TsOH, 80°	Catalytic amount
57. TsOH, 0°	Catalytic amount

J. SOFT ACIDS

58. Hg(II)	
59. Ag(I)	
60. Cu(II)/Py	For example, for Glaser coupling

K. RADICAL ADDITION

61. HBr/initiator	"Acidic" HX addition; acidity ≅ TsOH, 0°
62. HX/initiator	Neutral HX addition; X = P, S, Se, Si
63. NBS/CCl_4, $h\nu$ or heat	Allylic bromination
64. $CHBr_3$; $BrCCl_3$; CCl_4/In·	Carbon-halogen addition

L. OXIDIZING AGENTS

65. OsO_4	
66. $KMnO_4$, 0°, pH 7	
67. O_3, −50°	
68. RCO_3H, 0°	Epoxidation of olefins; prototype for H_2O_2/H^+
69. RCO_3H, 50°	Baeyer-Villiger oxidation of hindered ketones
70. CrO_3/Py	Collins oxidation
71. CrO_3, pH 1	Jones oxidation
72. H_2O_2/OH^-, pH 10–12	
73. Quinone	Dehydrogenation
74. $^1O_2$	Singlet oxygen
75. CH_3SOCH_3, 100°	(DMSO); HCO_3^- may be added to maintain neutrality
76. NaOCl, pH 10	
77. Aq. NBS	Nonradical conditions
78. I_2	
79. C_6H_5SCl; C_6H_5SeX	
80. Cl_2; Br_2	
81. MnO_2/CH_2Cl_2	
82. $NaIO_4$, pH 5–8	
83. SeO_2, pH 2–4	
84. SeO_2/Py	In EtOH/cat. Py

85. $K_3Fe(CN)_6$, pH 7–10	Phenol coupling
86. Pb(IV), 25°	Glycol and α-hydroxy acid cleavage
87. Pb(IV), 80°	Oxidative decarboxylation
88. $Tl(NO_3)_3$, pH 2	Oxidative rearrangement of olefins

M. THERMAL REACTIONS

89. 150°	Some Cope rearrangements and Cope eliminations
90. 250°	Claisen or Cope rearrangement
91. 350°	Ester cracking; Conia "ene" reaction

N. CARBENOIDS

92. $:CCl_2$	
93. $N_2CHCO_2C_2H_5/Cu$, 80°	
94. CH_2I_2/Zn–Cu	Simmons–Smith addition

O. MISCELLANEOUS

95. *n*-Bu_3SnH/initiator	
96. $Ni(CO)_4$	
97. CH_2N_2	
98. $SOCl_2$	
99. Ac_2O, 25°	Acetylation
100. Ac_2O, 80°	Dehydration
101. DCC	Dicylohexylcarbodiimide, $C_6H_{11}N{=}C{=}NC_6H_{11}$
102. CH_3I	
103. $(CH_3)_3O^+BF_4^-$	Or CH_3OSO_2F=Magic Methyl: **SEVERE POISON**
104. 1. LiN–*i*-Pr_2; 2. MeI	For C-alkylation
105. 1. K_2CO_3; 2. MeI	For *O*-alkylation

P. ELECTROPHILES

106. RCHO	
107. RCOCl	
108. C^+ion/olefin	For cation-olefin cyclization

REACTIVITY CHARTS

One requirement of a protective group is stability to a given reaction. The charts that follow were prepared as a guide to relative reactivities and thereby as an aid in the choice of a protective group. The reactivities in the charts were estimated by the individual and collective efforts of a group of synthetic chemists. *It is important to realize that not all the reactivities in the charts have been determined experimentally and considerable conjecture has been exercised.* For those cases in which a literature reference was available concerning the use of a protective group and one of the 108 prototype reagents, the reactivity is printed in italic type. However, an exhaustive search for such references has not been made; therefore, the absence of italic type does not imply an experimentally unknown reactivity.

There are four levels of reactivity in the charts:

"H" (high) indicates that under the conditions of the prototype reagent, the protective group is readily removed to regenerate the original functional group.

"M" (marginal) indicates that the stability of the protected functionality is marginal and depends on the exact parameters of the reaction. The protective group may be stable, may be cleaved slowly, or may be unstable to the conditions. Relative rates are always important, as illustrated in the following example[5] (in which a monothioacetal is cleaved in the presence of a dithiane), and may have to be determined experimentally.

CH_3SCH_2O Me OSiMe$_2$–*t*-Bu S S HO

$\xrightarrow[\text{CH}_3\text{CN–H}_2\text{O (4:1)}, 25°, 1–2 \text{ h}]{\text{HgCl}_2/\text{CaCO}_3{}^5}$

HO Me OSiMe$_2$–*t*-Bu S S HO

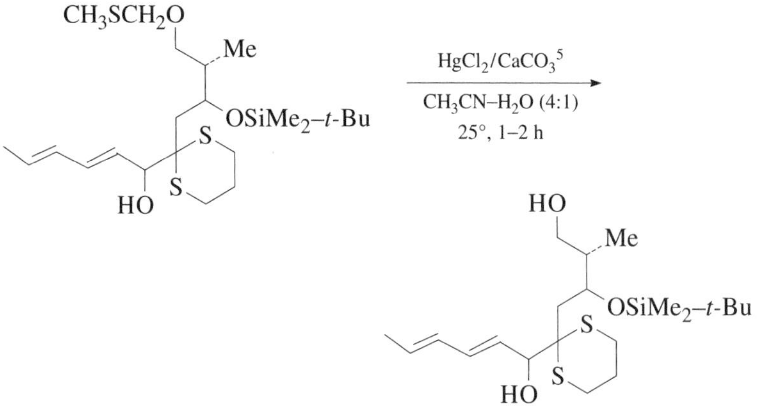

"L" (low) indicates that the protected functionality is stable under the reaction conditions.

"R" (reacts) indicates that the protected compound reacts readily, but that the original functional group is not restored. The protective group may be changed to a new protective group (eq. 1) or to a reactive intermediate (eq. 2), or the protective group may be unstable to the reaction conditions and react further (eq. 3).

$$(1)\ \text{ROCOC}_6\text{H}_4\text{–}p\text{-NO}_2 \xrightarrow{\text{H}_2\,/\,\text{Pd-C}} \text{ROCOC}_6\text{H}_4\text{–}p\text{-NH}_2$$

(2) $RCONR'_2 \xrightarrow{Me_3O^+ BF_4^-} [RC(OMe){=}N^+R'_2\ BF_4^-]$

(3) $RCH(OR')_2 \xrightarrow{pH < 1,\ 100°} [RCHO] \longrightarrow$ condensation products

The reactivities in the charts refer *only* to the protected functionality, not to atoms adjacent to the functional group; for example, RCOOEt $\xrightarrow{LDA}$: "L" (low) reactivity of PG(Et). However, if the protected functionality is $R_2CHCOOEt$, this substrate obviously *will* react with LDA. Reactivity of the entire substrate must be evaluated by the chemist.

Five reagents [#25: HCN, pH 6; #88: $Tl(NO_3)_3$; #103: $Me_3O^+ BF_4^-$; #104: LiN–*i*-Pr_2/MeI; and #105: K_2CO_3/MeI] were added after some of the charts had been completed; reactivities to these reagents are not included for all charts.

The number used to designate a protective group (PG) in a Reactivity Chart is the same as that used in the body of the text in the *first* edition.

Protective group numbers in the Reactivity Charts are not continuous, since not all of the protective groups described in the text are included in the charts. The protective groups that are included in the Reactivity Charts are, in general, those that have been used most widely; consequently, considerable experimental information is available for them.

The Reactivity Charts were prepared in collaboration with the following chemists, to whom we are most grateful: John O. Albright, Dale L. Boger, Dr. Daniel J. Brunelle, Dr. David A. Clark, Dr. Jagabandhu Das, Herbert Estreicher, Anthony L. Feliu, Dr. Frank W. Hobbs, Jr., Paul B. Hopkins, Dr. Spencer Knapp, Dr. Pierre Lavallée, John Munroe, Jay W. Ponder, Marcus A. Tius, Dr. David R. Williams, and Robert E. Wolf, Jr.

[1] L. F. Fieser and M. Fieser, *Reagents for Organic Synthesis*, Wiley-Interscience, New York, 1967, Vol. 1; M. Fieser and L. F. Fieser, Vols. 2–7, 1969–1979; M. Fieser, Vols. 8–17, 1980–1994.

[2] The categories and prototype reagents used in this study are an expansion of an earlier set of 11 categories and 60 prototype reagents,[3] originally compiled for use in LHASA[4] (Logic and Heuristics Applied to Synthetic Analysis), a long-term research program at Harvard University for Computer-Assisted Synthetic Analysis.

[3] E. J. Corey, H. W. Orf, and D. A. Pensak, *J. Am. Chem. Soc.*, **98**, 210 (1976).

[4] Selected references include E. J. Corey, *Quart. Rev., Chem. Soc.*, **25**, 455 (1971); H. W. Orf, Ph. D. Thesis, Harvard University, 1976.

[5] E. J. Corey and M.G. Bock, *Tetrahedron Lett.*, 2643 (1975).

Reactivity Chart 1. Protection for Hydroxyl Group: Ethers

1. Methyl Ether
2. Methoxymethyl Ether (MOM)
3. Methylthiomethyl Ether (MTM)
6. 2-Methoxyethoxymethyl Ether (MEM)
8. Bis(2-chloroethoxy)methyl Ether
9. Tetrahydropyranyl Ether (THP)
11. Tetrahydrothiopyranyl Ether
12. 4-Methoxytetrahydropyranyl Ether
13. 4-Methoxytetrahydrothiopyranyl Ether
15. Tetrahydrofuranyl Ether
16. Tetrahydrothiofuranyl Ether
17. 1-Ethoxyethyl Ether
18. 1-Methyl-1-methoxyethyl Ether
21. 2-(Phenylselenyl)ethyl Ether
22. *t*-Butyl Ether
23. Allyl Ether
26. Benzyl Ether
28. *o*-Nitrobenzyl Ether
35. Triphenylmethyl Ether
36. α-Naphthyldiphenylmethyl Ether
37. *p*-Methoxyphenyldiphenylmethyl Ether
41. 9-(9-Phenyl-10-oxo)anthryl Ether (Tritylone)
43. Trimethylsilyl Ether (TMS)
45. Isopropyldimethylsilyl Ether
46. *t*-Butyldimethylsilyl Ether (TBDMS)
48. *t*-Butyldiphenylsilyl Ether
51. Tribenzylsilyl Ether
53. Triisopropylsilyl Ether

(See chart, pp. 709–711.)

Reactivity Chart 1. Protection for the Hydroxyl Group: Ethers

PG	pH<1, 100°	pH<1	pH 1	pH 2-4	pH 4-6	pH 6-8.5	pH 8.5-10	pH 10-12	pH>12	pH>12, 150°	NaH	Ph_3CNa	$(C_{10}H_8)^{\cdot-}Na^+$	$MeSOCH_2^-Na^+$	KO-_t_-Bu	LiN-_i_-Pr_2	Py; R_3N	$NaNH_2$	NaOMe	Enolate	NH_3; RNH_2	RS^-; N_3^-; SCN^-	OAc^-; X^-	NaCN, pH 12	HCN, pH 6	RLi	RMgX	Organozinc	Organocopper	Wittig; ylide	H_2/Raney (Ni)	H_2/Pt pH 2-4	H_2/Pd	H_2/Lindlar	H_2/Rh	Zn/HCl	Zn/HOAc	Cr(II), pH 5
	1	2	3	4	5	6	7	8	9	10	11	12	13	14	15	16	17	18	19	20	21	22	23	24	25	26	27	28	29	30	31	32	33	34	35	36	37	38
	A. AQUEOUS										B. BASIC								C. NUCLEOPHILIC							D. ORGANOMET.					E. CAT. REDN.					F.		
1	H	M	L	L	L	L	L	L	L	L	L	L	L	L	L	L	L	L	L	L	L	L	L	L	L	L	L	L	L	L	L	L	L	L	L	L	L	L
2	H	H	H	M	L	L	L	L	L	M	L	L	L	L	L	L	L	L	L	L	L	L	L	L	L	L	L	L	L	L	L	M	L	L	L	H	M	L
3	H	H	M	L	L	L	L	L	L	M	L	M	R	M	L	M	L	L	L	L	L	L	L	L	L	L	L	L	L	L	R	R	R	L	R	R	R	M
6	H	H	L	L	L	L	L	L	L	L	L	L	L	L	L	L	L	L	L	L	L	L	L	L	L	L	L	L	L	L	L	L	L	L	L	M	L	L
8	H	H	H	H	M	L	L	L	M	H	L	R	R	R	R	R	L	R	R	R	M	R	L	R		R	R	M	R	L	R	R	R	L	L	H	H	M
9	H	H	H	H	M	L	L	L	L	L	L	L	L	L	L	L	L	L	L	L	L	L	L	L	M	L	L	L	L	L	L	H	L	L	L	H	H	M
11	H	H	H	H	L	L	L	L	L	M	L	M	R	R	L	M	L	L	L	L	L	L	L	L		L	L	L	L	L	R	R	R	L	R	H	H	M
12	H	H	H	H	M	L	L	L	L	L	L	L	L	L	L	L	L	L	L	L	L	L	L	L		L	L	L	L	L	L	H	L	L	L	H	H	M
13	H	H	H	H	M	L	L	L	L	M	L	L	M	L	L	L	L	L	L	L	L	L	L	L		M	L	L	L	L	R	R	R	L	R	H	H	M
15	H	H	H	H	H	L	L	L	L	L	L	L	L	L	L	L	L	L	L	L	L	L	L	L		L	L	L	L	L	L	H	L	L	L	H	H	H
16	H	H	H	M	L	L	L	L	L	M	L	M	R	R	L	M	L	L	L	L	L	L	L	L		L	L	L	L	L	R	R	R	L	R	H	H	M
17	H	H	H	H	L	L	L	L	L	L	L	L	L	L	L	L	L	L	L	L	L	L	L	L		L	L	L	L	L	L	H	L	L	L	H	H	L
18	H	H	H	H	M	L	L	L	L	L	L	L	L	L	L	L	L	L	L	L	L	L	L	L		L	L	L	L	L	L	H	L	L	L	H	H	M
21	H	H	M	L	L	L	L	L	L	M	L	R	R	R	M	R	L	R	M	L	L	R	L	L		R	L	L	L	L	R	R	R	L	R	H	H	M
22	H	H	L	L	L	L	L	L	L	L	L	L	L	L	L	L	L	L	L	L	L	L	L	L	L	L	L	L	L	L	L	L	L	L	L	L	L	L
23	H	H	L	L	L	L	L	L	M	R	L	L	L	L	R	L	L	L	R	L	L	L	L	L	L	L	L	L	L	L	R	R	R	L	R	L	L	L
26	H	H	L	L	L	L	L	L	L	L	L	L	R	L	L	L	L	L	L	L	L	L	L	L	L	L	L	L	L	L	H	H	H	L	L	L	L	L
28	H	H	L	L	L	L	L	L	L	H	R	M	R	R	L	R	L	R	L	L	L	L	L	L	L	M	R	R	R	L	H	H	H	L	R	R	R	R
35	H	H	H	H	L	L	L	L	L	L	L	L	H	L	L	L	L	L	L	L	L	L	L	L		L	L	L	L	L	H	H	H	L	L	H	H	L
36	H	H	H	H	L	L	L	L	L	M	L	L	H	L	L	L	L	L	L	L	L	L	L	L		L	L	L	L	L	H	H	H	L	L	H	H	L
37	H	H	H	H	M	L	L	L	L	L	L	L	H	L	L	L	L	L	L	L	L	L	L	L		L	L	L	L	L	H	H	H	L	L	H	H	M
41	H	H	L	L	L	L	L	L	L	L	L	L	R	L	L	L	L	L	L	M	R	L	L	L		R	R	R	L	R	R	R	R	L	R	R	R	R
43	H	H	H	H	H	H	H	H	H	H	H	H	H	H	L	L	L	H	H	L	H	H	L	H	H	H	H	L	L	L	L	H	H	L	L	H	H	H
45	H	H	H	H	M	L	L	L	H	H	L	L	L	L	L	L	L	L	L	L	L	L	L	L	L	H	H	L	L	H	M	H	M	L	L	H	H	M
46	H	H	H	H	L	L	L	L	H	H	L	L	L	L	L	L	L	L	L	L	L	L	L	L		L	L	L	L	L	L	H	L	L	L	H	L	L
48	H	M	M	L	L	L	L	L	H	H	L	L	L	L	L	L	L	L	L	L	L	L	L	L		L	L	L	L	L	L	L	L	L	L	M	L	L
51	H	H	H	H	L	L	L	L	H	H	L	L	H	L	L	L	L	L	L	L	L	L	L	L		L	L	L	L	L	L	H	L	L	L	H	H	L
53	H	H	H	H	L	L	L	L	H	H	L	L	L	L	L	L	L	L	L	L	L	L	L	L		L	L	L	L	L	L	H	L	L	L	H	H	L

Reactivity Chart 1. Protection for the Hydroxyl Group: Ethers (*Continued*)

PG	G. Na/NH_3 39	Al(Hg) 40	$SnCl_2$/Py 41	HSO_3^-; H_2S 42	H. HYDRIDE REDN. $LiAlH_4$ 43	Li-s-Bu_3BH 44	$(C_5H_{11})_2BH$ 45	B_2H_6, 0° 46	$NaBH_4$ 47	$Zn(BH_4)_2$ 48	$NaBH_3CN$ pH 4-6 49	i-Bu_2AlH 50	Li(OtBu)$_3$AlH 51	I. $AlCl_3$, 80° 52	$AlCl_3$, 25° 53	$SnCl_4$; BF_3 54	$LiClO_4$; $MgBr_2$ 55	TsOH, 80° 56	TsOH, 0° 57	J. Hg(II) 58	Ag(I) 59	Cu(II)/Py 60	K. HBr/In· 61	HX/In· 62	NBS/CCl_4 63	Br_3CCl/In· 64	L. OXIDANTS OsO_4 65	$KMnO_4$, pH 7,0° 66	O_3, -50° 67	RCO_3H, 0° 68	RCO_3H, 50° 69	CrO_3/Py 70	CrO_3, pH 1 71	H_2O_2 pH 10-12 72	Quinone 73	$^1O_2$ 74	DMSO, 100° 75	NaOCl pH 10 76	aq NBS 77
1	L	L	L	L	L	L	L	L	L	L	L	L	L	H	H	H	L	L	L	L	L	L	L	L	M	L	L	L	L	L	L	L	L	L	L	L	L	L	L
2	L	L	L	L	L	L	L	L	L	L	L	L	L	H	H	H	L	H	L	L	L	L	H	L	R	L	L	L	R	L	M	L	L	L	L	L	L	L	R
3	R	L	M	L	L	L	L	L	L	L	L	L	L	H	H	M	M	M	L	H	H	L	M	L	M	M	L	R	R	R	R	L	H	L	L	R	M	R	R
6	L	L	L	L	L	L	L	L	L	L	L	L	L	H	H	H	H	H	L	L	L	L	H	L	R	L	L	L	R	L	L	L	M	L	L	L	L	L	M
8	R	M	M	L	M	L	L	L	L	L	L	M	L	H	H	H	L	H	M	L	R	L	H	L	R	R	L	L	R	L	M	L	H	L	L	L	L	L	M
9	L	L	L	L	L	L	L	L	L	L	L	L	L	H	H	H	M	H	L	L	L	L	H	L	M	L	L	L	H	L	H	L	H	L	L	L	L	L	L
11	R	L	L	L	L	L	L	L	L	L	L	L	L	H	M	M	L	H	L	H	H	L	M	L	M	L	L	R	R	R	R	L	H	L	L	R	M	R	R
12	L	L	L	L	L	L	L	L	L	L	M	L	L	H	H	H	L	H	L	L	L	L	H	L	M	L	L	L	R	L	H	L	H	L	L	L	L	L	R
13	R	L	L	L	L	L	L	L	L	L	M	L	L	H	H	H	L	H	L	L	L	L	H	L	M	L	L	R	R	R	R	L	H	L	L	R	M	R	R
15	L	L	L	L	L	L	L	L	L	L	H	L	L	H	H	H	M	H	L	L	L	L	H	L	M	L	L	L	R	L	H	L	H	L	L	L	L	L	R
16	R	L	L	L	L	L	L	L	L	L	L	L	L	H	M	M	L	H	L	H	H	L	M	L	M	L	L	R	R	R	R	L	H	L	L	R	M	R	R
17	L	L	L	L	L	L	L	L	L	L	L	L	L	H	H	H	L	H	L	L	L	L	M	L	R	L	L	L	R	L	H	L	H	L	L	L	L	L	M
18	L	L	L	L	L	L	L	L	L	L	M	L	L	H	H	H	M	H	M	L	L	L	H	L	M	L	L	L	L	H	R	L	R	L	L	L	L	L	M
21	R	M	L	L	L	L	L	L	L	L	L	L	L	H	M	M	L	M	L	L	L	L	R	L	R	L	R	R	R	R	R	R	R	R	L	R	M	R	R
22	L	L	L	L	L	L	L	L	L	L	L	L	L	H	H	H	L	H	L	L	L	L	M	L	L	L	L	L	L	L	L	L	L	L	L	L	L	L	L
23	H	L	L	L	L	L	R	R	L	L	L	R	L	H	H	H	L	L	L	R	L	L	R	R	R	R	R	R	R	R	R	L	M	L	L	R	L	L	R
26	H	L	L	L	L	L	L	L	L	L	L	L	L	H	H	H	L	L	L	L	L	L	R	L	R	L	L	L	L	L	L	L	L	L	L	L	L	L	L
28	R	R	L	L	R	L	L	L	L	L	L	L	L	H	L	L	L	L	L	L	L	L	R	L	R	L	L	L	L	L	L	L	L	L	L	L	L	L	L
35	H	M	H	L	L	L	L	L	L	L	M	L	L	H	H	H	L	H	H	L	L	L	H	L	L	L	L	L	L	L	H	L	H	L	L	L	L	L	L
36	H	M	L	L	L	L	L	L	L	L	M	L	L	H	H	H	L	H	H	L	L	L	H	L	L	L	L	L	L	L	H	L	H	L	L	L	L	L	L
37	H	M	L	L	L	L	L	L	L	L	M	L	L	H	H	H	L	H	H	L	L	L	H	L	L	L	L	L	L	M	H	L	H	L	L	L	L	L	L
41	R	R	L	R	R	R	R	L	R	R	L	R	M	H	H	H	M	H	L	L	L	L	H	L	L	L	L	L	L	L	R	L	L	L	L	L	L	L	L
43	H	L	M	H	H	H	L	H	H	H	H	H	H	H	M	M	L	H	L	L	L	L	H	M	R	M	L	H	L	L	H	L	H	H	L	L	H	H	H
45	L	L	L	L	L	L	L	L	L	L	M	L	L	H	L	L	L	H	M	L	L	L	H	L	M	L	L	L	L	L	H	L	H	L	L	L	L	L	L
46	L	L	L	L	L	L	L	L	L	L	L	L	L	H	M	M	L	H	L	L	L	L	H	L	M	L	L	L	L	L	L	L	L	L	L	L	L	L	L
48	L	L	L	L	L	L	L	L	L	L	L	L	L	H	L	L	L	M	L	L	L	L	L	L	M	L	L	L	L	L	L	L	M	L	L	L	L	L	L
51	R	R	L	L	L	L	L	L	L	L	L	L	L	H	L	L	L	H	L	L	L	L	R	L	R	L	L	L	L	L	L	L	H	L	L	L	L	L	L
53	L	L	L	L	L	L	L	L	L	L	L	L	L	H	M	M	L	H	L	L	L	L	H	L	M	L	L	L	L	L	L	L	H	L	L	L	L	L	L

Reactivity Chart 1. Protection for the Hydroxyl Group: Ethers (*Continued*)

	I_2	PhSeX; PhSCl	Br_2; Cl_2	MnO_2/CH_2Cl_2	$NaIO_4$ pH 5-8	SeO_2 pH 2-4	SeO_2/Py	$K_3Fe(CN)_6$, pH 8	Pb(IV), 25°	Pb(IV), 80°	$Tl(NO_3)_3$	150°	250°	350°	$:CCl_2$	N_2CHCO_2R/Cu	$CH_2I_2/Zn(Cu)$	$R_3SnH/In\cdot$	$Ni(CO)_4$	CH_2N_2	$SOCl_2$	Ac_2O, 25°	Ac_2O, 80°	DCC	MeI	$Me_3O^+BF_4^-$	1.LDA 2.MeI	1.K_2CO_3 2.MeI	RCHO	RCOCl	C^+/olefin
PG	78	79	80	81	82	83	84	85	86	87	88	89	90	91	92	93	94	95	96	97	98	99	100	101	102	103	104	105	106	107	108
	L. OXIDANTS											M.			N.			O. MISCELLANEOUS											P.		
1	L	L	M	L	L	L	L	L	L	L	L	L	L	L	L	L	L	L	L	L	L	L	L	L	L	L	L	L	L	L	L
2	L	L	R	L	L	H	L	L	L	H	H	L	L	R	L	L	L	L	L	L	L	L	L	L	L	M	L	L	L	L	H
3	L	L	R	L	R	H	M	L	R	R	R	L	M	R	M	R	L	M	L	L	L	L	L	L	R	R	M	R	L	L	M
6	L	L	M	L	L	L	L	L	L	M	M	L	L	R	L	L	L	L	L	L	L	L	L	L	L	M	L	L	L	L	H
8	L	L	M	L	L	H	L	L	M	H		L	L	R	L	M	R	R	L	L	L	L	M	L	L	M					
9	L	L	M	L	L	H	L	L	M	H	H	L	M	H	L	L	L	L	L	L	L	L	L	L	L	M	L	L	L	L	R
11	L	L	R	L	R	H	M	M	H	H		M	M	H	M	M	L	M	L	L	L	L	L	L	R	R	R	R			
12	L	L	R	L	L	H	L	L	H	H		L	M	H	L	L	L	L	L	L	L	L	M	L	L	M					
13	L	L	R	L	R	H	M	M	R	H		L	M	H	M	M	L	M	L	L	L	L	M	L	R	R	R	R			
15	L	L	R	L	M	H	L	L	H	H		L	M	R	L	L	L	L	L	L	L	L	H	L	L	M					
16	L	L	R	L	R	H	M	M	H	H		R	R	R	M	M	L	M	L	L	L	L	L	L	R	R	R	R			
17	L	L	M	L	L	H	L	L	M	H		L	L	R	L	L	L	L	L	L	L	L	L	L	L	M					
18	L	L	M	L	M	H	L	L	M	H		L	M	R	L	L	L	L	L	L	L	L	M	L	L	M					
21	R	L	R	R	R	R	R	L	R	R		R	R	R	M	M	M	M	L	L	L	L	L	L	R	R					
22	L	L	L	L	L	L	L	L	L	L	M	L	R	R	L	L	L	L	L	L	L	L	L	L	L	L	L	L	L	L	H
23	M	R	R	L	L	R	R	L	L	R	R	L	R	R	R	R	R	M	M	L	L	L	L	L	L	M	L	L	L	L	H
26	L	L	M	L	L	L	L	L	L	L	L	L	M	R	L	L	L	L	L	L	L	L	L	L	L	M	L	L	L	L	L
28	L	L	M	L	L	L	L	L	L	L	L	L	M	R	L	L	L	L	L	L	L	L	L	L	L	M	M	L	L	L	L
35	L	L	L	L	L	H	L	L	L	H		L	M	R	L	L	L	L	L	L	L	L	M	L	L	L				L	
36	L	L	L	L	L	H	L	L	L	H		L	M	R	L	L	L	L	L	L	L	L	L	L	L	L					
37	L	L	L	L	M	H	L	L	M	H		L	M	R	L	L	L	L	L	L	L	L	M	L	L	L					
41	L	L	L	L	L	L	L	L	L	L		L	R	R	L	L	L	M	L	L	L	L	M	L	L	L	L				
43	L	L	L	L	H	H	L	H	H	H	H	L	L	R	L	L	L	L	L	L	L	M	H	L	L	H	M	L	L	L	L
45	L	L	L	L	M	H	L	L	H	H		L	L	R	L	L	L	L	L	L	L	L	H	L	L	M					
46	L	L	L	L	L	H	L	L	M	L	M	L	L	R	L	L	L	L	L	L	L	L	M	L	L	M	L	L	L	L	L
48	L	L	L	L	L	L	L	L	L	M		L	L	R	L	L	L	L	L	L	L	L	L	L	L	L				L	
51	L	L	M	L	L	H	L	L	M	H		L	L	R	L	L	L	L	L	L	L	L	M	L	L	L					
53	L	L	L	L	L	H	L	L	M	H		L	L	R	L	L	L	L	L	L	L	L	M	L	L	M					

Reactivity Chart 2. Protection for Hydroxyl Group: Esters

1. Formate Ester
3. Acetate Ester
6. Trichloroacetate Ester
10. Phenoxyacetate Ester
19. Isobutyrate Ester
22. Pivaloate Ester
23. Adamantoate Ester
27. Benzoate Ester
31. 2,4,6-Trimethylbenzoate (Mesitoate) Ester
34. Methyl Carbonate
36. 2,2,2-Trichloroethyl Carbonate
39. Allyl Carbonate
41. *p*-Nitrophenyl Carbonate
42. Benzyl Carbonate
46. *p*-Nitrobenzyl Carbonate
47. *S*-Benzyl Thiocarbonate
48. *N*-Phenylcarbamate
51. Nitrate Ester
53. 2,4-Dinitrophenylsulfenate Ester

(See chart, pp. 713–715.)

Reactivity Chart 2. Protection for the Hydroxyl Group: Esters

PG	pH<1, 100°	pH<1	pH 1	pH 2-4	pH 4-6	pH 6-8.5	pH 8.5-10	pH 10-12	pH>12	pH>12, 150°	NaH	Ph_3CNa	$(C_{10}H_8)^{\dot{-}}Na^+$	$MeSOCH_2^-Na^+$	KO-t-Bu	LiN-i-Pr_2	Py; R_3N	$NaNH_2$	NaOMe	Enolate	NH_3; RNH_2	RS^-; N_3^-; SCN^-	OAc^-; X^-	NaCN, pH 12	HCN, pH 6	RLi	RMgX	Organozinc	Organocopper	Wittig; ylide	H_2/Raney (Ni)	H_2/Pt pH 2-4	H_2/Pd	H_2/Lindlar	H_2/Rh	Zn/HCl	Zn/HOAc	Cr(II), pH 5
	1	2	3	4	5	6	7	8	9	10	11	12	13	14	15	16	17	18	19	20	21	22	23	24	25	26	27	28	29	30	31	32	33	34	35	36	37	38
	A. AQUEOUS										B. BASIC								C. NUCLEOPHILIC							D. ORGANOMET.					E. CAT. REDN.					F.		
1	H	H	H	M	L	L	H	H	H	H	H	L	H	H	H	H	L	H	H	H	H	H	L	M		H	H	M	M	H	M	M	M	L	L	M	M	L
3	H	M	L	L	L	L	M	H	H	H	R	H	H	H	H	H	L	R	H	R	M	L	L	H	L	H	H	L	L	L	L	L	L	L	L	L	L	L
6	H	M	L	L	L	L	H	H	H	H	L	H	H	H	H	L	L	L	H	R	H	H	L	R		H	H	H	H	L	R	R	R	L	R	R	R	H
10	H	M	L	L	L	L	M	H	H	H	M	M	H	H	H	H	L	R	H	R	M	H	L	L		H	H	L	L	L	L	L	L	L	L	L	L	L
19	H	M	L	L	L	L	L	M	H	H	H	H	H	H	L	M	L	H	M	M	M	L	L	L		H	H	L	L	L	L	L	L	L	L	L	L	L
22	H	M	L	L	L	L	L	L	M	H	L	L	H	M	H	L	L	M	M	M	H	L	L	L	L	H	M	L	L	L	L	L	L	L	L	L	L	L
23	H	H	L	L	L	L	L	L	M	H	L	L	H	M	L	L	L	L	L	L	L	L	L	L		L	L	L	L	L	L	L	L	L	L	L	L	L
27	H	M	L	L	L	L	L	M	H	H	L	L	H	H	L	L	L	H	M	M	M	L	L	L	L	H	H	L	L	L	L	L	L	L	L	L	L	L
31	H	M	L	L	L	L	L	L	L	H	L	L	H	L	H	L	L	L	L	L	L	L	L	L		L	L	L	L	L	L	L	L	L	L	L	L	L
34	H	M	L	L	L	L	L	H	H	H	L	L	H	H	M	L	L	H	M	M	M	L	L	L	M	R	R	L	H	M	R	L	L	L	L	R	H	H
36	H	M	L	L	L	L	M	H	H	H	L	H	H	H	H	H	L	H	R	R	M	H	L	R		R	R	L	H	M	R	L	L	L	R	H	H	H
39	H	M	L	L	L	L	L	H	H	H	L	H	H	H	M	H	L	R	M	M	M	L	L	L		R	R	L	H	L	R	R	R	L	R	L	L	L
41	H	M	L	L	L	L	H	H	H	H	L	L	H	H	R	L	L	H	R	R	M	L	L	M		R	R	M	M	M	R	R	R	L	R	R	R	R
42	H	M	L	L	L	L	L	H	H	H	L	H	H	H	M	H	L	H	M	M	M	L	L	L	M	R	R	L	L	L	H	H	H	L	M	L	L	L
46	H	M	L	L	L	L	L	H	H	H	L	H	H	H	R	H	L	H	R	R	M	L	L	L		R	R	L	L	L	H	H	H	L	R	R	R	R
47	H	L	L	L	L	L	L	H	H	H	L	H	H	H	R	H	L	H	R	R	M	L	L	L		R	R	M	M	H	R	R	R	L	H	L	L	L
48	H	L	L	L	L	L	L	L	M	H	R	H	H	H	R	R	L	H	M	L	M	L	L	L		R	R	L	L	L	L	L	L	L	L	L	L	L
51	H	H	M	L	L	L	L	L	H	H	L	L	H	H	H	M	L	H	H	H	H	M	L	L		H	H	M	M	M	H	H	H	L	H	H	H	H
53	H	M	M	L	L	L	L	H	H	H	L	L	H	H	H	H	L	H	H	H	H	H	H	H		M	M	H	H	H	H	H	H	L	H	H	H	H

Reactivity Chart 2. Protection for the Hydroxyl Group: Esters (*Continued*)

	Na/NH_3	Al(Hg)	$SnCl_2/Py$	HSO_3^-; H_2S	$LiAlH_4$	Li-s-Bu_3BH	$(C_5H_{11})_2BH$	B_2H_6, 0°	$NaBH_4$	$Zn(BH_4)_2$	$NaBH_3CN$ pH 4-6	i-Bu_2AlH	Li(OtBu)$_3$AlH	$AlCl_3$, 80°	$AlCl_3$, 25°	$SnCl_4$; BF_3	$LiClO_4$; $MgBr_2$	TsOH, 80°	TsOH, 0°	Hg(II)	Ag(I)	Cu(II)/Py	HBr/In·	HX/In·	NBS/CCl_4	Br_3CCl/In·	OsO_4	$KMnO_4$, pH 7, 0°	O_3, −50°	RCO_3H, 0°	RCO_3H, 50°	CrO_3/Py	CrO_3, pH 1	H_2O_2 pH 10-12	Quinone	$^1O_2$	DMSO, 100°	NaOCl pH 10	aq NBS
PG	39	40	41	42	43	44	45	46	47	48	49	50	51	52	53	54	55	56	57	58	59	60	61	62	63	64	65	66	67	68	69	70	71	72	73	74	75	76	77
	G.				H. HYDRIDE REDN.									I.						J.			K.				L. OXIDANTS												
1	H	L	L	L	H	H	M	M	M	M	M	H	M	H	H	L	L	H	L	L	L	L	M	L	L	L	L	L	L	L	L	L	L	H	L	L	M	H	L
3	H	L	L	L	H	M	L	L	L	L	L	H	L	L	L	L	L	M	L	L	L	L	L	L	L	L	L	L	L	L	L	L	L	L	L	L	L	L	L
6	H	H	L	L	H	H	L	L	L	M	L	H	L	R	R	L	L	M	L	L	R	L	R	L	L	L	L	L	L	L	L	L	L	H	L	L	H	H	L
10	H	H	L	L	H	M	L	L	L	L	L	H	L	R	R	L	L	M	L	L	L	L	M	L	L	L	L	L	M	L	L	L	L	L	L	L	L	M	L
19	H	H	L	L	H	L	L	L	L	L	L	H	L	L	L	L	L	M	L	L	L	L	L	L	L	L	L	L	L	L	L	L	L	L	L	L	L	L	L
22	H	L	L	L	H	L	L	L	L	L	L	L	L	L	L	L	L	L	L	L	L	L	L	L	L	L	L	L	L	L	L	L	L	L	L	L	L	L	L
23	H	L	L	L	H	L	L	L	L	L	L	L	L	L	L	L	L	L	L	L	L	L	L	L	L	L	L	L	L	L	L	L	L	L	L	L	L	L	L
27	H	L	L	L	H	L	L	L	L	L	L	H	L	L	L	L	L	M	L	L	L	L	L	L	L	L	L	L	L	L	L	L	L	L	L	L	L	L	L
31	H	L	L	L	H	L	L	L	L	L	L	L	L	L	L	L	L	L	L	L	L	L	L	L	R	L	L	L	L	L	L	L	L	L	L	L	L	L	L
34	H	L	L	L	H	L	L	L	L	L	L	H	L	L	L	L	L	M	L	L	L	L	L	L	L	L	L	L	L	L	L	L	L	H	L	L	L	L	L
36	H	R	L	L	H	M	L	L	L	L	L	H	L	R	R	L	L	M	L	L	R	L	M	L	L	L	L	L	L	L	L	L	L	H	L	L	M	M	L
39	H	L	L	L	H	L	H	R	L	L	L	H	L	R	R	H	L	M	L	M	L	L	R	R	R	R	H	H	H	H	H	L	L	R	L	R	L	H	R
41	H	H	L	L	H	M	L	L	L	M	L	H	R	R	M	L	L	M	L	L	L	L	L	L	L	L	L	L	L	L	L	L	L	H	L	L	M	H	L
42	H	L	L	L	H	L	L	L	L	L	L	H	L	R	R	L	L	M	L	L	L	L	L	L	R	L	L	L	L	L	L	L	L	L	L	L	L	L	L
46	H	R	L	L	H	L	L	L	L	L	L	H	L	R	M	L	L	M	L	L	L	L	L	L	R	L	L	L	L	L	L	L	L	L	L	L	L	L	L
47	H	L	L	L	H	H	L	L	M	M	L	H	M	R	M	L	L	M	L	R	M	L	L	L	L	L	L	R	R	M	R	L	R	H	L	L	L	H	R
48	H	L	L	L	H	L	M	H	L	L	L	H	L	R	L	L	L	M	L	L	L	L	L	L	L	L	L	L	L	L	L	L	L	L	L	L	L	L	L
51	H	H	H	M	H	H	H	H	R	H	H	H	H	R	R	R	L	M	L	L	L	L	H	H	H	H	L	L	L	L	L	L	M	L	L	L	L	L	L
53	H	H	H	H	H	H	H	H	H	H	H	H	H	H	H	H	L	M	L	M	M	L	R	R	R	R	L	R	R	R	R	L	R	H	L	M	H	H	H

Reactivity Chart 2. Protection for the Hydroxyl Group: Esters (*Continued*)

PG	I_2	PhSeX; PhSCl	Br_2; Cl_2	MnO_2/CH_2Cl_2	$NaIO_4$ pH 5-8	SeO_2 pH 2-4	SeO_2/Py	$K_3Fe(CN)_6$, pH 8	Pb(IV), 25°	Pb(IV), 80°	$Tl(NO_3)_3$	150°	250°	350°	$:CCl_2$	N_2CHCO_2R/Cu	$CH_2I_2/Zn(Cu)$	$R_3SnH/In\cdot$	$Ni(CO)_4$	CH_2N_2	$SOCl_2$	Ac_2O, 25°	Ac_2O, 80°	DCC	MeI	$Me_3O^+BF_4^-$	1. LDA 2. MeI	1. K_2CO_3 2. MeI	RCHO	RCOCl	C^+/olefin
	78	79	80	81	82	83	84	85	86	87	88	89	90	91	92	93	94	95	96	97	98	99	100	101	102	103	104	105	106	107	108
	L. OXIDANTS											M.			N.			O. MISCELLANEOUS											P.		
1	L	L	L	L	L	M	*M*	M	L	L		L	M	H	L	L	L	L	L	L	M	*L*	L	L	L	*L*	R	M	L	L	L
3	L	L	*L*	*L*	*L*	*L*	*L*	L	L	*L*	*L*	*L*	M	*H*	L	*L*	L	*L*	L	*L*	*L*	*L*	*L*	*L*	*L*						
6	L	L	L	L	L	L	L	H	L	L		M	H	H	L	L	R	*R*	*M*	*L*	L	L	L	L	L						
10	L	L	R	L	L	L	M	L	L	L		L	L	H	L	L	L	*L*	L	L	L	L	L	L	L						
19	L	L	L	L	L	L	L	L	L	L		L	M	H	L	L	L	L	L	L	L	L	L	L	L						
22	L	L	L	L	L	L	L	L	L	L	*L*	L	M	H	L	L	L	L	L	L	L	L	L	L	L	L	L	L	L	L	L
23	L	L	L	L	L	L	L	L	L	L		L	L	H	R	L	L	*L*	L	L	L	L	L	L	L						
27	L	L	L	L	L	L	L	L	L	L	*L*	*L*	M	H	L	L	L	*L*	L	*L*	L	*L*	L	L	L	L	L	L	L	L	L
31	L	L	L	L	L	*L*	*L*	L	L	*L*		L	M	H	L	L	L	L	L	*L*	L	*L*	L	L	L						
34	L	L	L	L	L	L	L	L	*L*	L	*L*	M	*H*	H	L	L	L	L	L	L	*L*	L	L	L	L	R	L	M	L	L	L
36	L	L	L	L	L	L	L	L	L	L	*L*	M	H	H	L	L	H	*R*	*M*	*L*	L	L	L	L	L	R	R	H	L	L	L
39	*M*	R	R	L	L	H	*H*	L	L	R		M	H	H	R	R	R	*H*	*H*	L	L	L	L	L	L						
41	L	L	L	L	L	L	*M*	L	L	L		M	H	H	L	L	L	*R*	L	L	L	L	L	L	L						
42	L	L	L	L	L	*M*	*M*	L	L	L	*L*	M	H	H	L	L	L	L	*L*	L	L	L	L	L	L	R	M	M	L	L	L
46	L	L	L	L	L	H	*H*	L	L	L		M	H	H	L	L	L	*H*	L	L	*L*	L	L	L	L						
47	M	L	H	L	L	L	L	L	L	R		M	H	H	L	L	L	*R*	L	*L*	L	*L*	*L*	L	L						
48	L	L	L	L	L	L	L	L	L	L		M	M	H	L	L	L	L	L	L	L	*R*	L	L	*L*		*R*	*L*			
51	L	L	L	L	L	L	L	L	L	L		H	H	H	L	L	L	H	L	L	H	*L*	L	*L*	L						
53	M	L	R	L	*M*	*M*	*M*	L	R	R		H	H	H	L	L	M	H	*M*	L	M	*L*	L	*L*	H						

Reactivity Chart 3. Protection for 1,2- and 1,3-Diols

1. Methylenedioxy Derivative
2. Ethylidene Acetal
6. Acetonide Derivative
11. Benzylidene Acetal
13. *p*-Methoxybenzylidene Acetal
18. Methoxymethylene Acetal
20. Dimethoxymethylenedioxy Derivative
28. Cyclic Carbonates
29. Cyclic Boronates

(See chart, pp. 717–719.)

Reactivity Chart 3. Protection for 1,2- and 1,3-Diols

	A. AQUEOUS										B. BASIC								C. NUCLEOPHILIC							D. ORGANOMET.					E. CAT. REDN.					F.		
	pH<1, 100°	pH<1	pH 1	pH 2–4	pH 4–6	pH 6–8.5	pH 8.5–10	pH 10–12	pH>12	pH>12, 150°	NaH	Ph_3CNa	$(C_{10}H_8)^{\dot{-}}Na^+$	$MeSOCH_2^-Na^+$	KO-t-Bu	LiN-i-Pr_2	Py; R_3N	$NaNH_2$	NaOMe	Enolate	NH_3; RNH_2	RS^-; N_3^-; SCN^-	OAc^-; X^-	NaCN, pH 12	HCN, pH 6	RLi	RMgX	Organozinc	Organocopper	Wittig; ylide	H_2/Raney (Ni)	H_2/Pt pH 2–4	H_2/Pd	H_2/Lindlar	H_2/Rh	Zn/HCl	Zn/HOAc	Cr(II), pH 5
PG	1	2	3	4	5	6	7	8	9	10	11	12	13	14	15	16	17	18	19	20	21	22	23	24	25	26	27	28	29	30	31	32	33	34	35	36	37	38
1	H	H	L	*L*	L	L	L	L	L	L	L	L	L	L	L	L	L	L	L	L	L	L	L	L		L	L	L	L	L	L	L	L	L	L	L	L	L
2	H	H	H	M	L	L	L	L	L	L	L	L	L	L	L	L	L	L	L	L	L	L	L	*L*		L	L	L	L	L	L	M	L	L	L	H	M	L
6	H	H	*H*	M	L	L	L	*L*	*L*	*L*	*L*	L	*L*	L	*L*	L	*L*	L	*L*	*L*	*L*	*L*	*L*	*L*	L	L	*L*	L	*L*	*L*	*L*	M	*L*	L	L	H	M	*L*
11	H	H	H	*H*	L	L	L	L	*L*	M	*L*	L	R	L	L	L	*L*	L	*L*	L	*L*	L	*L*	L		L	L	L	*L*	L	H	H	*H*	L	L	H	H	L
12	H	H	H	H	M	L	L	L	L	M	L	L	R	L	L	L	L	L	L	L	L	L	L	L		L	L	L	L	L	H	H	H	L	L	H	H	M
18	H	H	H	*H*	*H*	L	L	L	L	M	L	L	L	M	L	L	L	L	L	L	L	L	L	L		M	H	H	L	L	L	H	L	L	L	H	H	H
20	H	H	H	H	H	L	L	L	L	*M*	L	L	L	*M*	*L*	L	L	L	L	*L*	L	*L*	L	L		*M*	*H*	H	L	*L*	L	H	L	L	L	H	H	H
28	H	*L*	*L*	L	L	L	L	*H*	*H*	H	L	L	*H*	*H*	*L*	L	*L*	*H*	*M*	*M*	M	M	M	*L*	L	H	H	L	L	L	L	L	*L*	L	*L*	L	L	L
29	H	H	H	H	H	*H*	H	H	H	H	L	L	*H*	*H*	*H*	L	*L*	*H*	*H*	*H*	*L*	*L*	*L*	*H*	M	H	H	H	H	H	L	H	L	L	*L*	H	H	M

Reactivity Chart 3. Protection for 1,2- and 1,3-Diols (*Continued*)

PG	Na/NH_3	Al(Hg)	$SnCl_2/Py$	HSO_3^-; H_2S	$LiAlH_4$	Li-s-Bu_3BH	$(C_5H_{11})_2BH$	B_2H_6, 0°	$NaBH_4$	$Zn(BH_4)_2$	$NaBH_3CN$ pH 4-6	i-Bu_2AlH	Li(OtBu)$_3$AlH
	39	40	41	42	43	44	45	46	47	48	49	50	51
	G.				H. HYDRIDE REDN.								
1	L	L	L	L	L	L	L	L	L	L	*L*	L	L
2	L	L	L	L	L	L	L	L	*L*	L	*L*	L	L
6	*L*	*L*	L	L	*L*	L	*L*	L	*L*	L	*L*	*L*	L
11	*H*	L	L	L	L	L	L	L	*L*	L	*L*	L	L
12	H	L	L	L	L	L	L	L	L	L	*L*	L	L
18	L	L	L	L	R	L	L	L	L	L	*M*	L	*L*
20	L	L	L	L	R	L	L	L	L	L	*M*	L	*L*
28	H	L	L	*L*	H	*L*	*L*	L	L	*L*	L	H	L
29	H	M	L	*L*	H	*H*	*H*	H	H	*H*	H	H	H

PG	$AlCl_3$, 80°	$AlCl_3$, 25°	$SnCl_4$; BF_3	$LiClO_4$; $MgBr_2$	TsOH, 80°	TsOH, 0°	Hg(II)	Ag(I)	Cu(II)/Py	HBr/In·	HX/In·	NBS/CCl_4	$Br_3CCl/In·$
	52	53	54	55	56	57	58	59	60	61	62	63	64
	I.						J.			K.			
1	H	H	*H*	L	M	L	L	L	L	R	L	L	L
2	H	H	H	L	M	L	L	L	L	M	L	L	L
6	H	H	H	L	M	L	*L*	*L*	L	M	L	*L*	L
11	H	H	*H*	L	H	*L*	*L*	L	L	H	L	*R*	L
12	H	H	H	L	M	L	L	L	L	H	L	R	L
18	H	H	H	M	*R*	H	L	L	L	H	L	H	L
20	H	H	H	M	R	H	L	L	L	H	L	L	L
28	H	H	L	L	M	L	L	L	L	L	L	L	L
29	H	H	*L*	L	*L*	L	H	H	L	L	L	L	L

PG	OsO_4	$KMnO_4$, pH 7,0°	O_3, -50°	RCO_3H, 0°	RCO_3H, 50°	CrO_3/Py	CrO_3, pH 1	H_2O_2 pH 10-12	Quinone	$^1O_2$	DMSO, 100°	NaOCl pH 10	aq NBS
	65	66	67	68	69	70	71	72	73	74	75	76	77
	L. OXIDANTS												
1	L	L	R	L	L	L	L	L	L	L	L	L	*L*
2	L	L	R	L	*M*	L	H	L	L	L	L	L	*L*
6	*L*	L	*L*	*L*	*L*	*L*	H	*L*	*L*	L	*L*	L	*L*
11	L	L	*R*	L	H	*L*	H	L	L	L	*L*	L	*L*
12	L	L	R	M	H	L	H	L	L	L	L	L	*M*
18	L	M	L	H	H	L	H	L	L	L	L	L	*M*
20	L	M	L	H	H	L	H	L	L	L	L	L	*M*
28	L	*L*	L	*L*	L	*L*	L	H	L	L	L	M	L
29	L	H	L	*H*	*H*	L	H	H	L	L	*L*	H	H

Reactivity Chart 3. Protection for 1,2- and 1,3-Diols (*Continued*)

PG	78 I_2	79 PhSeX; PhSCl	80 Br_2; Cl_2	81 MnO_2/CH_2Cl_2	82 $NaIO_4$ pH 5–8	83 SeO_2 pH 2–4	84 SeO_2/Py	85 $K_3Fe(CN)_6$, pH 8	86 Pb(IV), 25°	87 Pb(IV), 80°	88 $Tl(NO_3)_3$	89 150°	90 250°	91 350°	92 $:CCl_2$	93 N_2CHCO_2R/Cu	94 $CH_2I_2/Zn(Cu)$
	L. OXIDANTS											M.			N.		
1	L	L	*L*	L	L	L	L	L	L	L	L	L	L	M	L	L	L
2	L	L	L	L	*L*	M	L	L	L	M		L	L	M	L	L	L
6	*L*	L	*L*	*L*	*L*	M	L	L	*L*	*M*	H	*L*	*L*	L	L	L	L
11	L	L	R	L	*L*	H	L	L	L	M		L	M	H	L	L	L
12	L	L	R	L	L	H	L	L	L	M		L	M	H	*L*	*M*	L
18	L	L	L	L	M	H	L	L	*L*	*H*		L	M	H	L	L	L
20	L	L	L	L	M	R	L	L	*L*	*H*		M	H	H	L	L	L
28	L	L	*L*	L	L	L	L	L	*L*	L	L	L	M	H	L	L	L
29	L	L	*L*	*L*	*H*	H	L	H	H	H	H	L	H	H	L	L	L

PG	95 $R_3SnH/In\cdot$	96 $Ni(CO)_4$	97 CH_2N_2	98 $SOCl_2$	99 Ac_2O, 25°	100 Ac_2O, 80°	101 DCC	102 MeI	103 $Me_3O^+BF_4^-$	104 1.LDA 2.MeI	105 1.K_2CO_3 2.MeI	106 RCHO	107 RCOCl	108 C^+/olefin
	O. MISCELLANEOUS											P.		
1	L	L	*L*	L	L	*M*	L	L	M	L	L	L	L	H
2	L	L	*L*	L	*L*	*M*	L	L	M					
6	*L*	L	*L*	*L*	*L*	*M*	*L*	*L*	M	L	L	*L*	*L*	*H*
11	*L*	L	*L*	L	L	*M*	L	*L*	M				*L*	
12	L	L	L	L	L	*M*	L	L	R					
18	L	L	L	L	M	*R*	L	L	R					
20	L	L	L	L	M	*R*	L	L	R					
28	L	L	L	*L*	L	*L*	L	L	L	L	H	L	L	L
29	M	*L*	*L*	M	M	H	L	L	R	L	H	L	M	H

Reactivity Chart 4. Protection for Phenols and Catechols

Phenols

1. Methyl Ether
2. Methoxymethyl Ether
3. 2-Methoxyethoxymethyl Ether
4. Methylthiomethyl Ether
6. Phenacyl Ether
7. Allyl Ether
8. Cyclohexyl Ether
9. *t*-Butyl Ether
10. Benzyl Ether
11. *o*-Nitrobenzyl Ether
12. 9-Anthrylmethyl Ether
13. 4-Picolyl Ether
15. *t*-Butyldimethylsilyl Ether
16. Aryl Acetate
17. Aryl Pivaloate
18. Aryl Benzoate
19. Aryl 9-Fluorenecarboxylate
20. Aryl Methyl Carbonate
21. Aryl 2,2,2-Trichloroethyl Carbonate
22. Aryl Vinyl Carbonate
23. Aryl Benzyl Carbonate
25. Aryl Methanesulfonate

Catechols

27. Methylenedioxy Derivative
28. Acetonide Derivative
30. Diphenylmethylenedioxy Derivative
31. Cyclic Borates
32. Cyclic Carbonates

(See chart, pp. 721–723.)

Reactivity Chart 4. Protection for Phenols and Catechols

PG	A. AQUEOUS										B. BASIC								C. NUCLEOPHILIC							D. ORGANOMET.					E. CAT. REDN.					F.		
	1 pH<1, 100°	2 pH<1	3 pH 1	4 pH 2-4	5 pH 4-6	6 pH 6-8.5	7 pH 8.5-10	8 pH 10-12	9 pH>12	10 pH>12, 150°	11 NaH	12 Ph_3CNa	13 $(C_{10}H_8)^{\cdot-}Na^+$	14 $MeSOCH_2^-Na^+$	15 KO-t-Bu	16 LiN-i-Pr_2	17 Py; R_3N	18 $NaNH_2$	19 NaOMe	20 Enolate	21 NH_3; RNH_2	22 RS^-; N_3^-; SCN^-	23 OAc^-; X^-	24 NaCN, pH 12	25 HCN, pH 6	26 RLi	27 RMgX	28 Organozinc	29 Organocopper	30 Wittig; ylide	31 H_2/Raney (Ni)	32 H_2/Pt pH 2-4	33 H_2/Pd	34 H_2/Lindlar	35 H_2/Rh	36 Zn/HCl	37 Zn/HOAc	38 Cr(II), pH 5
1	H	M	*L*	L	L	L	*L*	*L*	*L*	*L*	*L*	L	L	*L*	*L*	*L*	L	*L*	*L*	*L*	*L*	*H*	*L*	*H*	L	*L*	*L*	L	L	*L*	*L*	*L*	*L*	L	L	*L*	*L*	L
2	H	H	*H*	*M*	L	L	L	L	L	L	*L*	L	L	L	L	L	L	L	*L*	*L*	L	M	L	*L*	*L*	*L*	*L*	L	L	*L*	L	M	*L*	L	L	H	M	L
3	H	*H*	L	L	L	L	L	L	L	L	*L*	L	L	L	L	L	L	L	L	L	L	M	L	L	L	L	L	L	L	L	L	M	L	L	L	L	L	L
4	H	H	M	*M*	L	L	L	L	L	L	L	L	R	L	L	L	L	L	*L*	L	L	M	L	L	L	M	*L*	L	L	L	R	R	R	L	R	H	H	M
6	H	L	*L*	L	L	L	L	L	*L*	H	R	R	R	R	M	R	L	R	R	R	L	M	L	M	L	R	R	M	M	R	M	R	R	L	L	H	*H*	H
7	H	L	L	L	L	L	L	L	M	H	L	L	R	L	R	L	L	L	R	L	L	M	L	M	L	R	M	L	M	L	R	R	R	L	R	M	L	L
8	H	*H*	L	L	L	L	L	L	*L*	L	L	L	L	L	L	L	L	L	L	L	*L*	*L*	L	L	L	L	L	L	L	L	L	L	L	L	L	M	L	L
9	H	*H*	*H*	L	L	L	L	L	L	L	L	L	L	L	L	L	*L*	L	L	L	L	L	L	L	L	L	L	L	L	L	L	L	L	L	L	H	M	L
10	H	*H*	L	L	L	L	L	*L*	*L*	L	*L*	L	R	L	L	L	L	L	L	L	L	M	L	L	L	M	L	*L*	L	L	H	H	*H*	L	H	*L*	L	L
11	H	*L*	L	L	L	L	L	L	*L*	R	L	R	R	R	L	R	L	R	L	L	*L*	M	L	L	L	R	R	M	R	L	H	H	H	L	H	R	R	R
12	H	M	L	L	L	L	L	L	L	L	L	L	R	L	L	L	L	L	L	L	*L*	*H*	L	M	L	R	M	L	L	L	H	H	H	L	H	L	L	L
13	H	L	L	L	L	L	L	L	L	L	L	L	R	L	L	L	L	L	L	L	L	M	L	L	L	R	M	L	L	L	H	H	H	L	H	M	L	L
15	H	H	H	H	M	L	L	L	M	H	*L*	L	L	L	L	L	L	L	L	L	L	L	L	L	L	L	L	L	L	L	L	H	L	L	L	H	M	M
16	H	H	H	M	L	L	*H*	H	*H*	H	R	R	R	H	R	R	L	H	H	H	*H*	M	L	H	L	R	R	L	L	L	L	M	L	L	L	L	L	L
17	H	M	L	L	L	L	L	L	*M*	H	L	L	R	M	L	L	L	M	M	M	L	M	L	M	L	L	L	L	L	L	L	L	L	L	L	L	L	L
18	H	H	*H*	L	L	L	L	M	*H*	H	L	L	R	H	L	L	L	H	M	M	M	M	L	H	L	R	R	L	L	L	L	L	*L*	L	L	L	L	L
19	H	M	L	L	L	L	L	M	H	H	R	R	R	R	R	R	L	H	M	M	M	M	L	H	L	R	R	L	L	L	L	L	L	L	L	L	L	L
20	H	H	*H*	*H*	L	L	L	M	*H*	H	L	L	R	H	M	L	L	H	M	R	M	H	L	H	M	H	H	L	L	L	L	H	L	L	L	H	H	M
21	H	*M*	*L*	L	L	L	M	H	H	H	R	R	R	R	R	R	L	H	R	R	M	M	L	H	L	H	H	M	R	M	R	*L*	*L*	L	R	H	*H*	H
22	*H*	*M*	*L*	L	L	L	L	*H*	H	H	L	L	R	H	L	L	L	H	M	R	M	M	L	H	L	H	H	L	L	L	R	R	R	L	R	H	L	L
23	H	H	M	L	L	L	L	M	H	H	L	L	R	H	M	L	L	H	M	M	M	M	L	H	L	H	H	L	L	L	H	H	*H*	L	L	M	L	L
25	L	L	*L*	L	L	L	L	*L*	*H*	H	R	R	R	H	L	R	L	R	M	M	M	L	L	M	*L*	*H*	*M*	L	R	*M*	R	L	*L*	L	L	L	*L*	L
27	H	M	L	L	L	L	*L*	*L*	L	L	*L*	*L*	L	L	*L*	L	*L*	*L*	*L*	L	*L*	L	L	L	L	L	M	L	L	L	*L*	L	*L*	L	L	L	L	L
28	H	H	*H*	*M*	L	L	L	L	L	L	L	L	L	L	L	L	*L*	L	L	L	*L*	L	L	L	*L*	*L*	L	L	L	L	L	M	L	L	L	H	M	L
30	H	H	H	*H*	L	L	L	L	L	L	L	L	R	L	L	L	L	L	L	L	L	L	L	L	L	L	*L*	L	L	L	H	H	*H*	L	L	H	M	M
31	H	H	*H*	L	L	L	L	L	H	H	R	R	H	H	R	R	L	R	H	H	H	H	L	H	M	H	H	H	H	H	L	H	L	L	L	H	H	M
32	H	H	H	L	L	L	M	*H*	H	H	L	L	R	H	L	L	L	H	M	M	M	L	L	H	L	R	R	L	L	L	L	L	L	L	L	M	L	L

Reactivity Chart 4. Protection for Phenols and Catechols (*Continued*)

PG	G. 39 Na/NH_3	40 Al(Hg)	41 $SnCl_2/Py$	42 HSO_3^-; H_2S	H. HYDRIDE REDN. 43 $LiAlH_4$	44 Li-s-Bu_3BH	45 $(C_5H_{11})_2BH$	46 B_2H_6, 0°	47 $NaBH_4$	48 $Zn(BH_4)_2$	49 $NaBH_3CN$ pH 4-6	50 i-Bu_2AlH	51 Li(OtBu)$_3$AlH	I. 52 $AlCl_3$, 80°	53 $AlCl_3$, 25°	54 $SnCl_4$; BF_3	55 $LiClO_4$; $MgBr_2$	56 TsOH, 80°	57 TsOH, 0°	J. 58 Hg(II)	59 Ag(I)	60 Cu(II)/Py	K. 61 HBr/In·	62 HX/In·	63 NBS/CCl_4	64 Br_3CCl/In·	L. OXIDANTS 65 OsO_4	66 $KMnO_4$, pH 7, 0°	67 O_3, −50°	68 RCO_3H, 0°	69 RCO_3H, 50°	70 CrO_3/Py	71 CrO_3, pH 1	72 H_2O_2 pH 10-12	73 Quinone	74 $^1O_2$	75 DMSO, 100°	76 NaOCl pH 10	77 aq NBS
1	R	L	L	L	L	L	L	L	L	L	L	L	L	H	M	L	L	L	L	L	L	L	L	L	L	L	L	L	L	L	L	L	L	L	L	L	L	L	L
2	R	L	L	L	L	L	L	L	L	L	L	L	L	H	H	H	M	H	L	L	L	L	H	L	R	L	L	L	L	L	M	L	R	L	L	L	L	L	L
3	R	L	L	L	L	L	L	L	L	L	L	L	L	H	M	L	L	H	L	L	L	L	M	L	R	L	L	L	L	L	L	L	M	L	L	L	L	L	L
4	R	L	L	L	L	L	L	L	L	L	L	M	L	H	H	M	L	M	L	H	M	L	H	L	R	L	L	R	R	R	R	L	R	L	L	R	L	R	R
6	R	R	L	L	R	R	R	R	R	R	R	R	R	H	H	L	L	M	L	L	L	L	L	L	L	L	L	L	L	L	R	L	L	L	L	L	L	L	L
7	R	L	L	L	L	L	R	R	L	L	L	L	L	H	H	L	L	L	L	R	L	L	R	R	R	R	R	R	R	R	R	L	L	L	L	R	L	L	R
8	R	L	L	L	L	L	L	L	L	L	L	L	L	M	L	L	L	L	L	L	L	L	M	L	L	L	L	L	L	L	L	L	L	L	L	L	L	L	L
9	R	L	L	L	L	L	L	L	L	L	L	L	L	H	H	H	L	H	L	L	L	L	M	L	L	L	L	L	L	L	L	L	R	L	L	L	L	L	L
10	R	L	L	L	L	L	L	L	L	L	L	L	L	H	M	L	L	L	L	L	L	L	R	R	R	R	L	L	L	L	M	L	L	L	L	L	L	L	L
11	R	R	L	L	R	R	L	L	L	L	L	R	L	H	M	L	L	L	L	L	L	L	M	M	M	M	L	L	L	L	M	L	L	L	L	L	L	L	L
12	R	L	L	L	L	L	L	L	L	L	L	L	L	H	M	L	L	M	L	L	L	L	R	R	R	R	L	L	L	L	M	L	L	L	L	L	L	L	L
13	R	L	L	L	L	L	L	L	L	L	L	L	L	H	H	L	L	M	L	L	L	L	R	R	R	R	L	R	R	R	R	L	M	L	L	L	L	L	L
15	R	L	L	L	L	L	L	L	L	L	M	M	L	H	M	L	L	H	L	L	L	L	H	L	L	L	L	L	L	L	M	L	R	L	L	L	L	L	L
16	R	L	L	L	H	H	L	L	M	L	L	H	M	R	R	M	L	H	L	L	L	L	L	L	L	L	L	L	L	L	L	L	L	H	L	L	L	H	L
17	R	L	L	L	H	L	L	L	L	L	L	M	L	R	L	L	L	M	L	L	L	L	L	L	L	L	L	L	L	L	L	L	L	L	L	L	L	L	L
18	R	L	L	L	H	M	L	L	L	L	L	H	M	R	R	M	L	H	L	L	L	L	L	L	L	L	L	L	L	L	L	L	L	H	L	L	L	M	L
19	R	L	L	L	H	H	L	L	L	L	L	H	M	H	H	L	L	H	L	L	L	L	R	L	R	L	L	L	L	L	L	L	L	M	L	L	L	M	L
20	R	L	L	L	H	L	L	L	L	L	M	H	M	H	H	L	L	H	M	L	L	L	H	L	L	L	L	L	L	L	L	L	H	M	L	L	L	M	L
21	R	M	L	L	H	M	L	L	M	L	L	H	M	H	H	L	L	H	L	L	R	L	M	L	L	L	L	L	L	L	L	L	L	R	L	L	L	R	L
22	R	L	L	L	H	L	R	R	L	L	L	H	M	H	H	L	L	H	L	H	L	L	R	R	R	R	R	R	R	R	R	L	R	R	L	R	L	R	R
23	R	L	L	L	H	L	L	L	L	L	L	H	M	H	H	L	L	H	L	L	L	L	M	L	R	L	L	L	L	L	M	L	M	M	L	L	L	L	L
25	R	L	L	L	L	L	L	L	L	L	L	L	L	H	L	L	L	L	L	L	L	L	L	L	L	L	L	L	L	L	L	L	L	L	L	L	L	L	L
27	R	L	L	L	L	L	L	L	L	L	L	L	L	H	H	L	L	M	L	L	L	L	M	L	R	L	L	L	L	L	L	L	M	L	L	L	L	L	L
28	R	L	L	L	L	L	L	L	L	L	M	L	L	H	H	M	L	M	L	L	L	L	H	L	L	L	L	L	L	L	L	L	R	L	L	L	L	L	L
30	R	M	L	L	L	L	L	L	L	L	L	L	L	H	H	M	L	M	L	L	L	L	H	L	L	L	L	L	L	L	L	L	R	L	L	L	L	L	L
31	R	M	L	L	H	H	H	H	H	H	H	H	H	H	H	H	L	H	L	L	L	L	H	L	L	L	L	L	L	L	R	L	R	L	L	L	L	L	L
32	R	L	L	L	H	L	L	L	L	L	L	H	M	H	H	L	L	H	L	L	L	L	H	L	L	L	L	L	L	L	L	L	R	R	L	L	L	M	L

Reactivity Chart 4. Protection for Phenols and Catechols (*Continued*)

PG	I_2	PhSeX; PhSCl	Br_2; Cl_2	MnO_2/CH_2Cl_2	$NaIO_4$ pH 5-8	SeO_2 pH 2-4	SeO_2/Py	$K_3Fe(CN)_6$, pH 8	Pb(IV), 25°	Pb(IV), 80°	$Tl(NO_3)_3$	150°	250°	350°	$:CCl_2$	N_2CHCO_2R/Cu	CH_2I_2/Zn(Cu)	R_3SnH/In·	$Ni(CO)_4$	CH_2N_2	$SOCl_2$	Ac_2O, 25°	Ac_2O, 80°	DCC	MeI	$Me_3O^+BF_4^-$	1.LDA 2.MeI	1.K_2CO_3 2.MeI	RCHO	RCOCl	C^+/olefin
	78	79	80	81	82	83	84	85	86	87	88	89	90	91	92	93	94	95	96	97	98	99	100	101	102	103	104	105	106	107	108
	L. OXIDANTS											M.			N.			O. MISCELLANEOUS											P.		
1	L	L	L	L	L	L	L	L	L	L	L	L	L	L	L	L	L	L	L	L	L	L	L	L	L	L	L	L	L	L	L
2	L	L	L	L	L	M	L	L	M	R	R	L	M	H	L	L	L	L	L	L	L	L	L	L	L	L	L	L	L	L	H
3	L	L	L	L	L	L	L	L	L	L	L	L	M	H	L	L	L	L	L	L	L	L	L	L	L	L	L	L	L	L	L
4	L	L	R	L	R	M	L	L	R	R	R	L	M	H	M	R	L	M	L	L	L	L	L	L	R	R	L	L	L	L	M
6	L	M	M	L	L	R	L	L	M	R	R	L	L	M	R	R	L	L	L	L	L	L	L	L	L	M	R	M	L	L	L
7	L	R	R	L	L	H	M	L	L	R	R	M	R	R	R	R	R	M	M	L	L	L	L	L	L	M	L	L	L	L	R
8	L	L	L	L	L	L	L	L	L	L	L	L	L	L	L	L	L	L	L	L	L	L	L	L	L	L	L	L	L	L	L
9	L	L	L	L	L	L	L	L	L	M	L	L	M	H	L	L	L	L	L	L	L	L	L	L	L	L	L	L	L	L	H
10	L	L	L	L	L	L	L	L	L	L	L	L	L	L	L	L	L	L	L	L	L	L	L	L	L	L	L	L	L	L	L
11	L	L	L	L	L	L	L	L	L	L	L	L	L	L	L	L	L	L	L	L	L	L	L	L	L	L	L	L	L	L	L
12	L	L	M	L	L	L	L	L	L	L	L	L	L	L	L	L	L	L	L	L	L	L	L	L	L	L	L	L	L	L	L
13	L	L	L	L	L	L	L	L	L	L	L	L	L	L	L	L	L	L	L	L	L	L	L	L	R	R	L	L	L	L	L
15	L	L	L	L	L	M	L	L	M	R	R	L	L	L	L	L	L	L	L	L	L	L	M	L	L	L	L	L	L	L	L
16	L	L	L	L	L	M	L	R	L	L	M	L	L	L	L	L	L	L	L	L	L	L	L	L	L	L	R	H	L	L	L
17	L	L	L	L	L	L	L	L	L	L	L	L	L	H	L	L	L	L	L	L	L	L	L	L	L	L	L	L	L	L	L
18	L	L	L	L	L	L	L	L	L	L	L	L	L	M	L	L	L	L	L	L	L	L	L	L	L	L	L	M	L	L	L
19	L	L	L	L	L	L	L	L	L	L	L	L	L	M	L	L	L	L	L	L	L	L	L	L	L	L	R	M	L	L	L
20	L	L	L	L	L	H	L	L	L	M	R	L	M	H	L	L	L	L	L	L	L	L	L	L	L	R	L	M	L	L	M
21	L	L	L	L	L	L	L	M	L	L	L	L	M	H	L	L	R	R	M	L	L	L	L	L	L	M	R	H	L	L	L
22	M	R	R	L	L	L	L	L	M	R	R	L	M	H	R	R	R	R	R	L	L	L	L	L	L	R	L	H	L	L	R
23	L	L	L	L	L	L	L	L	L	L	R	L	M	H	L	L	L	L	L	L	L	L	L	L	L	R	L	M	L	L	L
25	L	L	L	L	L	L	L	L	L	L	L	L	L	M	L	L	L	L	L	L	L	L	L	L	L	L	R	L	L	L	L
27	L	L	L	L	L	L	L	L	L	L	L	L	L	L	L	L	L	L	L	L	M	L	L	L	L	L	L	L	L	L	L
28	L	L	L	L	L	M	L	L	L	M	M	L	L	H	L	L	L	L	L	L	L	L	L	L	L	L	L	L	L	L	M
30	L	L	L	L	L	R	L	L	L	R	R	L	M	H	L	L	L	L	L	L	L	L	L	L	L	L	L	L	L	L	M
31	L	L	L	L	L	L	L	L	L	R	R	L	M	H	L	L	L	L	L	L	R	R	R	L	R	R	R	R	R	R	H
32	L	L	L	L	L	L	L	M	L	L	M	L	L	M	L	L	L	L	L	L	L	L	L	L	L	R	L	L	L	L	L

Reactivity Chart 5. Protection for the Carbonyl Group

1. Dimethyl Acetals and Ketals
3. Bis(2,2,2-trichloroethyl) Acetals and Ketals
5. 1,3-Dioxanes
6. 5-Methylene-1,3-dioxanes
7. 5,5-Dibromo-1,3-dioxanes
8. 1,3-Dioxolanes
9. 4-Bromomethyl-1,3-dioxolanes
10. 4-*o*-Nitrophenyl-1,3-dioxolanes
11. *S,S′*-Dimethyl Acetals and Ketals
19. 1,3-Dithianes
20. 1,3-Dithiolanes
24. 1,3-Oxathiolanes
26. *O*-Trimethylsilyl Cyanohydrins
29. *N,N*-Dimethylhydrazones
30. 2,4-Dinitrophenylhydrazones
33. *O*-Phenylthiomethyl Oximes
34. Substituted Methylene Derivatives
43. Bismethylenedioxy Derivatives

(See chart, pp. 725–727.)

Reactivity Chart 5. Protection for the Carbonyl Group

PG	1 pH<1, 100°	2 pH<1	3 pH 1	4 pH 2-4	5 pH 4-6	6 pH 6-8.5	7 pH 8.5-10	8 pH 10-12	9 pH>12	10 pH>12, 150°	11 NaH	12 Ph_3CNa	13 $(C_{10}H_8)^{\dot{-}}Na^+$	14 $MeSOCH_2^-Na^+$	15 KO-t-Bu	16 LiN-i-Pr_2	17 Py; R_3N	18 $NaNH_2$	19 NaOMe	20 Enolate	21 NH_3; RNH_2	22 RS^-; N_3^-; SCN^-	23 OAc^-; X^-	24 NaCN, pH 12	25 HCN, pH 6	26 RLi	27 RMgX	28 Organozinc	29 Organocopper	30 Wittig; ylide	31 H_2/Raney (Ni)	32 H_2/Pt pH 2-4	33 H_2/Pd	34 H_2/Lindlar	35 H_2/Rh	36 Zn/HCl	37 Zn/HOAc	38 Cr(II), pH 5
	A. AQUEOUS										B. BASIC								C. NUCLEOPHILIC							D. ORGANOMET.					E. CAT. REDN.					F.		
1	H	*H*	*H*	L	*L*	L	L	L	*L*	L	*L*	L	L	L	*L*	*L*	L	L	*L*	*L*	*L*	*L*	L	*L*		*L*	*L*	L	*L*	*L*	L	L	*L*	*L*	*L*	H	H	L
3	H	H	H	L	L	L	L	L	M	R	R	R	H	R	R	R	L	R	R	R	*M*	R	R	R		R	R	R	R	R	R	R	R	L	*M*	H	R	M
5	H	H	*H*	L	L	L	L	L	L	L	L	L	L	L	L	*L*	L	L	L	L	L	L	L	L		L	L	L	L	L	L	M	L	L	L	H	H	M
6	H	H	H	L	L	L	L	L	L	R	L	*L*	H	R	R	L	L	L	L	L	L	L	L	L		*L*	L	L	L	L	R	R	R	*L*	*R*	H	H	M
7	H	H	H	L	L	L	L	M	R	R	R	R	H	R	R	*R*	L	R	R	R	M	R	R	R		R	R	R	R	R	R	R	R	L	*M*	H	R	M
8	H	H	*H*	M	L	L	*L*	*L*	*L*	*L*	*L*	L	L	*L*	*L*	*L*	*L*	*L*	*L*	*L*	*L*	*L*	*L*	*L*		*L*	*L*	L	*L*	*L*	*L*	*R*	L	*L*	*L*	H	*L*	L
9	H	H	H	M	L	L	L	L	R	R	R	R	H	R	R	R	L	R	R	R	*L*	R	R	R		*L*	R	R	M	L	R	R	R	L	L	H	H	M
10	H	H	H	M	L	L	L	L	L	H	L	*M*	H	H	L	R	L	*M*	L	M	L	L	L	L		R	R	*M*	R	L	R	R	R	*M*	*R*	H	R	R
11	R	R	L	L	L	L	L	L	L	L	*L*	L	L	L	L	L	L	L	*L*	L	*L*	L	L	L		*L*	L	L	L	L	R	R	R	*H*	*R*	L	L	L
19	R	M	*L*	*L*	L	L	L	L	*L*	L	*L*	L	L	L	*L*	*L*	*L*	L	*L*	*L*	*L*	L	L	L		*L*	*L*	L	L	*L*	*R*	R	R	*H*	*R*	L	L	L
20	R	*R*	L	L	L	L	L	L	*L*	*H*	L	L	L	L	*L*	L	*L*	L	*L*	L	*L*	L	*L*	L		*L*	L	L	L	*L*	*R*	R	*L*	*H*	*R*	L	L	L
24	H	H	*H*	*H*	L	L	*L*	L	L	R	L	L	L	L	L	L	L	L	L	L	L	L	*L*	*L*		L	*L*	L	L	L	*R*	R	R	*H*	*R*	H	M	L
26	H	H	H	*H*	H	*H*	H	*H*	H	H	L	L	R	R	M	*L*	L	R	H	*L*	L	H	*L*	L		M	M	H	L	R	R	R	R	L	R	H	R	R
29	H	L	L	L	L	L	L	L	L	H	L	L	*M*	L	L	*R*	L	L	L	*L*	*R*	L	L	L		*L*	L	L	L	L	R	R	R	L	R	H	H	M
30	H	*L*	L	L	L	L	L	L	L	*H*	R	R	R	R	R	R	L	R	M	M	L	L	L	L		R	R	R	L	R	R	R	R	L	R	R	*R*	*R*
33	H	H	M	L	L	L	L	L	L	R	L	R	*H*	L	L	M	L	L	*L*	L	L	L	L	L		R	L	L	L	L	R	R	R	L	R	R	R	M
34	R	M	*L*	L	L	L	L	L	H	H	*H*	*H*	R	R	R	R	L	R	H	R	M	L	L	L		R	R	R	R	R	R	R	*L*	L	R	M	M	M
43	*H*	H	M	M	L	L	L	L	L	*L*	L	L	L	L	L	L	L	L	*L*	L	*L*	L	L	L		L	*L*	L	L	L	*L*	L	L	L	L	H	H	L

Reactivity Chart 5. Protection for the Carbonyl Group (*Continued*)

PG	Na/NH_3	Al(Hg)	$SnCl_2$/Py	HSO_3^-; H_2S	$LiAlH_4$	Li-s-Bu_3BH	$(C_5H_{11})_2BH$	B_2H_6, 0°	$NaBH_4$	$Zn(BH_4)_2$	$NaBH_3CN$ pH 4-6	i-Bu_2AlH	Li(OtBu)$_3$AlH	$AlCl_3$, 80°	$AlCl_3$, 25°	$SnCl_4$; BF_3	$LiClO_4$; $MgBr_2$	TsOH, 80°	TsOH, 0°	Hg(II)	Ag(I)	Cu(II)/Py	HBr/In·	HX/In·	NBS/CCl_4	Br_3CCl/In·	OsO_4	$KMnO_4$, pH 7, 0°	O_3, -50°	RCO_3H, 0°	RCO_3H, 50°	CrO_3/Py	CrO_3, pH 1	H_2O_2 pH 10-12	Quinone	$^1O_2$	DMSO, 100°	NaOCl pH 10	aq NBS
	39	40	41	42	43	44	45	46	47	48	49	50	51	52	53	54	55	56	57	58	59	60	61	62	63	64	65	66	67	68	69	70	71	72	73	74	75	76	77
	G.				H. HYDRIDE REDN.									I.						J.			K.				L. OXIDANTS												
1	L	L	L	L	L	L	L	M	L	L	L	L	L	H	R	H	L	H	M	L	L	L	H	L	L	L	L	L	L	L	L	L	H	L	L	L	L	L	L
3	R	M	L	L	R	L	L	M	L	L	L	L	L	H	H	H	L	H	M	L	R	L	R	R	R	R	L	L	L	L	L	L	H	L	L	L	H	L	L
5	L	L	L	L	L	L	L	M	L	L	L	L	L	H	H	H	L	L	L	L	L	L	H	L	L	L	L	L	L	L	L	L	M	L	L	L	L	L	L
6	R	L	L	L	L	L	R	R	L	L	L	L	L	H	H	H	L	H	M	R	L	L	H	R	H	R	R	R	R	R	R	L	H	L	L	M	L	L	R
7	R	H	M	L	R	L	L	L	L	L	L	M	M	H	H	H	L	L	L	L	R	L	H	H	H	H	L	L	L	L	L	L	H	L	L	L	H	L	L
8	L	L	L	L	L	L	L	M	L	L	L	L	L	H	H	H	L	L	L	L	L	L	H	L	L	L	L	L	L	L	L	L	H	L	L	L	L	L	L
9	L	H	L	L	M	L	L	M	L	L	L	L	L	H	H	H	L	L	L	L	R	L	H	H	L	L	L	L	L	L	L	L	L	L	L	L	R	L	L
10	R	R	L	L	R	R	L	M	L	L	L	L	L	H	H	H	L	L	L	L	L	L	H	H	L	L	L	L	L	L	L	L	H	L	L	L	L	L	L
11	R	L	L	L	L	L	L	L	L	L	L	L	L	H	L	L	L	L	L	H	H	H	H	L	L	L	L	R	R	R	R	L	M	L	L	R	M	R	R
19	R	L	L	L	L	L	L	L	L	L	L	L	L	H	L	L	L	L	L	H	H	H	H	L	L	L	L	R	R	R	R	L	M	L	L	R	M	R	R
20	R	L	L	L	L	L	L	L	L	L	L	L	L	H	L	L	L	L	L	H	H	H	H	L	L	L	L	R	R	R	R	L	M	L	L	R	M	R	R
24	R	L	L	L	L	L	L	L	L	L	L	L	L	H	H	M	L	H	L	H	H	H	H	L	L	L	L	R	R	R	R	L	H	L	L	R	M	R	R
26	R	R	M	L	R	R	R	R	M	M	L	R	R	H	H	H	M	H	H	L	L	L	R	L	H	L	L	R	H	R	R	L	R	R	L	L	R	R	R
29	L	L	R	L	R	L	L	R	L	L	R	R	L	H	L	L	L	L	L	L	L	R	R	M	L	M	L	H	H	H	H	H	H	H	L	H	L	H	H
30	R	R	M	L	R	R	L	L	L	M	R	R	M	L	L	L	L	L	L	L	L	H	R	L	L	L	L	L	H	H	H	L	L	L	L	M	L	H	R
33	R	R	M	L	L	L	L	R	L	L	R	R	M	H	M	L	L	L	L	R	R	M	R	L	R	L	L	R	H	R	R	L	R	L	L	R	M	M	R
34	R	R	M	L	R	R	R	R	M	L	R	R	M	M	L	L	L	L	L	L	L	L	R	R	R	R	R	R	H	R	R	L	M	L	L	M	M	M	M
43	L	L	L	L	L	L	L	L	L	L	L	L	L	R	R	M	M	L	L	L	L	L	M	L	L	L	L	L	L	L	L	L	L	L	L	L	L	L	L

Reactivity Chart 5. Protection for the Carbonyl Group (*Continued*)

PG	I_2	PhSeX; PhSCl	Br_2; Cl_2	MnO_2/CH_2Cl_2	$NaIO_4$ pH 5–8	SeO_2 pH 2–4	SeO_2/Py	$K_3Fe(CN)_6$, pH 8	Pb(IV), 25°	Pb(IV), 80°	$Tl(NO_3)_3$	150°	250°	350°	$:CCl_2$	N_2CHCO_2R/Cu	$CH_2I_2/Zn(Cu)$	$R_3SnH/In\cdot$	$Ni(CO)_4$	CH_2N_2	$SOCl_2$	Ac_2O, 25°	Ac_2O, 80°	DCC	MeI	$Me_3O^+BF_4^-$	1. LDA 2. MeI	1. K_2CO_3 2. MeI	RCHO	RCOCl	C^+/olefin
	78	79	80	81	82	83	84	85	86	87	88	89	90	91	92	93	94	95	96	97	98	99	100	101	102	103	104	105	106	107	108
	L. OXIDANTS											M.			N.			O. MISCELLANEOUS											P.		
1	L	*L*	L	*L*	*L*	M	L	L	L	L	*L*	*L*	H	*H*	L	*L*	L	L	L	L	L	*L*	L	L	*L*	M			*L*	*L*	M
3	L	L	L	L	L	M	L	L	L	L		*L*	M	R	*L*	*M*	H	R	M	L	L	L	L	L	L	M			L	L	M
5	L	L	R	L	L	M	L	L	*L*	L		L	*L*	R	L	*L*	L	L	L	L	L	L	L	L	L	*L*			L	*L*	*L*
6	R	R	R	L	*R*	M	R	L	R	R		L	L	R	R	R	R	L	R	L	L	L	L	L	L	M			L	L	*R*
7	L	L	R	L	L	M	L	L	L	*L*		L	L	R	*L*	*M*	H	R	M	L	L	L	L	L	L	M			L	L	M
8	*L*	*L*	*R*	L	*L*	H	*L*	*L*	*L*	L		*L*	*L*	R	*L*	*L*	*L*	L	L	*L*	*L*	*L*	*L*	L	*L*	*L*			L	L	*R*
9	L	L	R	L	M	H	L	L	L	*L*		L	L	R	*L*	*L*	H	R	L	L	L	L	L	L	L	M			L	L	M
10	L	L	R	L	M	H	L	L	L	M		L	L	R	*L*	*L*	L	R	L	L	L	L	L	L	L	M			L	L	M
11	*H*	L	R	M	*R*	M	L	L	R	R		L	*L*	R	M	M	L	M	L	L	L	L	L	L	R	*R*			L	L	M
19	L	*L*	R	M	R	M	L	L	*L*	M		L	L	R	M	M	L	M	L	L	L	*L*	L	L	*H*	*L*			L	*L*	M
20	*H*	*L*	R	M	R	M	L	*L*	R	R		*L*	*L*	R	M	M	L	M	L	*L*	L	L	L	L	*H*	*R*			*L*	*L*	*L*
24	L	L	R	L	R	M	L	L	R	R		*L*	*L*	R	M	M	L	M	L	L	L	L	M	L	*R*	R			L	L	M
26	L	L	R	R	R	R	L	R	R	R		L	M	R	L	R	L	L	L	L	L	*R*	R	L	L	M			L	L	*R*
29	L	L	*H*	*M*	*H*	*L*	*M*	L	*R*	R		L	R	R	R	R	R	L	L	L	L	L	*L*	L	*R*	R			*L*	L	M
30	L	L	*L*	M	H	*L*	*L*	L	R	R		L	M	R	R	R	R	L	L	L	L	*L*	L	L	M	R			L	L	L
33	L	L	R	L	R	M	L	L	R	R		L	R	R	R	R	R	M	L	L	L	L	L	L	R	R			L	L	M
34	L	L	R	R	L	L	L	L	L	L		L	M	R	R	R	R	R	L	L	L	L	*L*	L	L	M			L	L	*R*
43	L	L	*L*	L	*L*	L	L	L	*L*	M		L	L	L	L	L	L	*L*	L	*L*	*L*	L	M	L	L	M			L	L	L

Reactivity Chart 6. Protection for the Carboxyl Group

Esters

1. Methyl Ester
2. Methoxymethyl Ester
3. Methylthiomethyl Ester
4. Tetrahydropyranyl Ester
7. Benzyloxymethyl Ester
8. Phenacyl Ester
13. *N*-Phthalimidomethyl Ester
15. 2,2,2-Trichloroethyl Ester
16. 2-Haloethyl Ester
21. 2-(*p*-Toluenesulfonyl)ethyl Ester
23. *t*-Butyl Ester
27. Cinnamyl Ester
30. Benzyl Ester
31. Triphenylmethyl Ester
33. Bis(*o*-nitrophenyl)methyl Ester
34. 9-Anthrylmethyl Ester
35. 2-(9,10-Dioxo)anthrylmethyl Ester
42. Piperonyl Ester
45. Trimethylsilyl Ester
47. *t*-Butyldimethylsilyl Ester
50. *S*-*t*-Butyl Ester
59. 2-Alkyl-1,3-oxazolines

Amides and Hydrazides

64. *N*,*N*-Dimethylamide
68. *N*-7-Nitroindoylamide
71. Hydrazides
72. *N*-Phenylhydrazide
73. *N*,*N*′-Diisopropylhydrazide

(See chart, pp. 729–731.)

Reactivity Chart 6. Protection for the Carboxyl Group

PG	pH<1, 100°	pH<1	pH 1	pH 2-4	pH 4-6	pH 6-8.5	pH 8.5-10	pH 10-12	pH>12	pH>12, 150°	NaH	Ph_3CNa	$(C_{10}H_8)^{\cdot-}Na^+$	$MeSOCH_2^-Na^+$	KO-t-Bu	LiN-i-Pr_2	Py; R_3N	$NaNH_2$	NaOMe	Enolate	NH_3; RNH_2	RS^-; N_3^-; SCN^-	OAc^-; X^-	NaCN, pH 12	HCN, pH 6	RLi	RMgX	Organozinc	Organocopper	Wittig; ylide	H_2/Raney (Ni)	H_2/Pt pH 2-4	H_2/Pd	H_2/Lindlar	H_2/Rh	Zn/HCl	Zn/HOAc	Cr(II), pH 5
	1	2	3	4	5	6	7	8	9	10	11	12	13	14	15	16	17	18	19	20	21	22	23	24	25	26	27	28	29	30	31	32	33	34	35	36	37	38
	A. AQUEOUS										B. BASIC								C. NUCLEOPHILIC							D. ORGANOMET.					E. CAT. REDN.					F.		
1	H	H	L	L	L	L	L	M	H	H	L	L	R	R	L	L	L	L	L	R	M	H	L	L		R	R	L	L	L	L	L	L	L	L	L	L	L
2	H	H	H	M	L	L	L	L	H	H	L	L	R	R	L	L	L	L	R	R	M	L	L	L		R	R	L	L	L	L	L	L	L	L	H	M	L
3	H	M	L	L	L	L	L	L	M	H	L	L	R	R	L	L	L	L	R	R	M	M	L	L		R	R	L	L	L	R	R	R	L	R	H	M	M
4	H	H	H	H	L	L	L	M	H	H	L	L	R	R	L	L	L	L	R	R	M	L	L	L		R	R	L	L	L	L	H	L	L	L	H	H	L
7	H	H	H	M	L	L	L	L	H	H	L	L	R	R	L	L	L	L	R	R	M	L	L	L		R	R	L	L	L	H	H	H	L	M	H	M	L
8	H	L	L	L	L	L	L	M	H	H	R	R	R	R	M	R	L	R	R	R	M	H	L	H		R	R	R	L	R	H	H	H	L	M	H	H	H
13	H	H	L	L	L	L	H	H	H	H	L	L	R	R	L	L	L	L	R	R	H	L	L	L		R	R	L	L	L	L	M	L	L	M	H	R	L
15	R	L	L	L	L	L	M	H	H	H	R	R	R	R	R	L	L	R	L	R	M	M	L	M		R	R	M	R	L	R	R	L	L	L	H	H	M
16	H	H	M	L	L	L	L	H	H	H	R	R	R	R	R	R	R	R	R	R	M	R	L	H		R	R	R	R	R	R	R	R	L	L	M	M	M
21	H	H	L	L	L	L	M	H	H	H	R	R	R	R	R	R	R	R	R	R	M	R	M	H		R	R	L	L	H	R	L	L	L	L	M	M	M
23	H	H	H	H	L	L	L	L	H	H	L	L	R	R	L	L	L	L	L	L	L	L	L	H		R	R	L	L	L	L	L	L	L	L	H	L	L
27	H	H	H	L	L	L	L	H	H	H	L	L	R	R	L	L	L	L	R	R	M	L	L	L		R	R	L	H	L	R	R	R	L	R	M	L	L
30	H	M	L	L	L	L	L	H	H	H	L	L	R	R	L	L	L	L	R	R	M	M	L	L		R	R	L	L	L	H	H	H	L	L	L	L	L
31	H	H	H	H	M	L	L	M	H	H	L	L	R	R	L	L	L	L	R	R	L	L	L	L		R	R	L	L	L	H	H	H	L	M	H	H	M
33	H	M	L	L	L	L	L	M	H	H	L	L	R	R	L	L	L	L	R	R	M	L	L	L		R	R	L	L	L	R	R	R	L	R	R	R	R
34	H	H	M	L	L	L	L	L	H	H	L	L	R	R	L	L	L	L	R	R	M	H	L	L		R	R	L	L	L	H	H	H	L	M	M	L	L
35	H	H	L	L	L	L	L	M	H	H	L	L	R	R	L	L	L	L	R	R	M	M	L	L		R	R	M	L	M	H	H	H	L	M	R	R	L
42	H	H	H	L	L	L	L	M	H	H	L	L	R	R	L	L	L	L	R	R	M	M	L	L		R	R	L	L	L	H	H	H	L	M	H	M	L
45	H	H	H	H	H	H	H	H	H	H	L	L	R	R	L	L	L	L	R	R	H	H	H	H		R	R	H	H	H	H	H	H	H	H	H	H	H
47	H	H	H	H	H	M	H	H	H	H	L	L	R	R	L	L	L	L	R	R	M	L	L	L		R	R	L	L	L	L	H	L	L	L	H	H	H
50	H	H	M	L	L	L	L	L	M	H	L	L	R	R	L	L	L	L	R	R	R	R	M	H		R	R	R	R	L	R	R	R	L	R	H	L	L
59	H	H	M	L	L	L	L	L	M	H	L	L	M	L	L	L	L	L	L	L	L	R	L	L		L	L	L	L	L	R	R	R	L	R	R	R	R
64	H	L	L	L	L	L	L	L	L	H	L	L	M	M	L	L	L	L	L	L	L	L	L	L		R	R	L	L	L	M	L	L	L	L	L	L	L
68	H	M	L	L	L	L	L	M	H	H	L	L	R	H	H	L	L	L	L	M	L	L	L	L		R	R	L	L	L	R	R	R	R	R	R	R	R
71	H	H	M	L	L	L	L	L	M	H	R	R	R	R	R	R	L	R	R	L	L	L	L	L		R	R	L	L	L	H	H	M	L	L	M	L	L
72	H	H	M	L	L	L	L	L	M	H	R	R	R	R	L	R	L	R	R	L	L	L	L	L		R	R	L	L	L	H	H	M	L	L	M	L	L
73	H	H	L	L	L	L	L	L	L	H	L	L	H	R	L	R	L	R	L	L	L	L	L	L		R	R	L	L	L	H	H	M	L	L	M	L	L

Reactivity Chart 6. Protection for the Carboxyl Group (*Continued*)

PG	Na/NH_3	Al(Hg)	$SnCl_2$/Py	HSO_3^-; H_2S	$LiAlH_4$	Li-*s*-Bu_3BH	$(C_5H_{11})_2BH$	B_2H_6, 0°	$NaBH_4$	$Zn(BH_4)_2$	$NaBH_3CN$ pH 4-6	*i*-Bu_2AlH	Li(O*t*Bu)$_3$AlH	$AlCl_3$, 80°	$AlCl_3$, 25°	$SnCl_4$; BF_3	$LiClO_4$; $MgBr_2$	TsOH, 80°	TsOH, 0°
	39	40	41	42	43	44	45	46	47	48	49	50	51	52	53	54	55	56	57
		G.							H. HYDRIDE REDN.							I.			
1	*R*	*L*	L	*L*	*R*	M	*L*	*L*	*L*	*L*	*L*	R	M	R	M	L	L	M	*L*
2	R	L	L	L	R	M	L	L	L	L	L	R	M	H	H	L	L	H	L
3	R	L	L	L	R	M	L	L	*L*	L	L	R	M	H	H	M	L	R	L
4	R	L	L	L	R	M	L	L	L	L	L	R	M	H	H	*H*	L	H	*M*
7	R	L	L	L	R	M	L	L	L	L	L	R	M	R	R	M	L	H	L
8	R	R	L	L	R	R	R	R	M	M	L	R	M	R	R	M	L	R	*L*
13	R	L	L	L	R	M	R	R	L	L	L	R	L	R	L	L	L	L	*L*
15	R	M	L	*L*	R	M	L	*L*	L	L	L	R	M	R	R	*L*	L	L	L
16	R	M	M	L	R	M	L	L	L	L	R	R	M	R	R	L	L	L	L
21	R	M	M	L	R	M	L	L	L	L	R	R	M	R	L	L	L	L	L
23	R	L	L	L	R	L	L	*L*	*L*	L	*L*	*L*	L	R	*H*	*H*	L	M	L
27	R	L	L	L	R	M	R	R	L	L	L	R	M	R	R	R	L	H	L
30	*H*	L	L	*L*	R	M	L	*L*	*L*	L	L	R	M	R	*R*	M	L	*L*	*L*
31	R	L	L	L	R	L	L	*L*	L	L	M	R	L	R	R	L	L	H	M
33	H	R	L	L	R	M	L	L	L	L	L	R	M	R	R	L	L	L	L
34	H	L	L	L	R	M	L	L	L	L	L	R	M	R	R	M	L	L	L
35	H	R	M	L	R	R	M	M	M	M	L	R	M	R	R	M	L	L	L
42	H	L	L	L	R	M	L	L	L	L	L	R	M	R	R	M	L	R	L
45	R	H	L	H	R	H	H	R	H	H	H	R	H	H	H	H	H	H	M
47	R	L	L	L	R	M	L	*L*	L	L	H	R	M	H	H	H	L	H	L
50	R	L	L	L	R	M	L	*L*	L	L	L	R	M	R	*M*	L	L	*H*	L
59	R	M	L	L	*L*	L	*M*	*L*	*L*	*L*	*M*	*M*	L	*R*	*L*	*L*	*L*	*R*	L
64	R	L	L	L	*R*	L	*R*	R	L	L	L	*R*	*L*	*L*	L	*L*	L	*L*	L
68	R	R	L	L	R	L	R	R	L	L	L	R	L	*M*	L	L	L	*L*	L
71	*L*	L	L	L	R	L	R	R	L	L	L	R	L	R	L	L	L	L	L
72	R	L	L	L	*R*	L	R	R	L	L	L	R	L	*M*	L	L	L	L	L
73	R	L	L	L	R	L	R	R	L	L	L	R	L	*M*	L	L	L	L	L

PG	Hg(II)	Ag(I)	Cu(II)/Py	HBr/In·	HX/In·	NBS/CCl_4	Br_3CCl/In·	OsO_4	$KMnO_4$, pH 7, 0°	O_3, −50°	RCO_3H, 0°	RCO_3H, 50°	CrO_3/Py	CrO_3, pH 1	H_2O_2 pH 10-12	Quinone	$^1O_2$	DMSO, 100°	NaOCl pH 10	aq NBS
	58	59	60	61	62	63	64	65	66	67	68	69	70	71	72	73	74	75	76	77
		J.			K.								L. OXIDANTS							
1	*L*	*L*	L	*L*	*L*	*L*	*L*	*L*	L	*L*	*L*	L	*L*	L	M	*L*	*L*	*L*	M	*L*
2	L	L	L	L	L	L	L	L	L	M	L	H	L	H	L	L	L	L	L	L
3	*H*	*H*	L	L	L	*H*	L	L	R	R	R	R	L	R	L	L	R	M	R	R
4	L	L	L	M	L	*M*	L	L	L	M	L	H	L	H	M	L	L	L	H	*L*
7	L	L	L	L	L	R	L	L	L	M	L	H	L	H	L	L	L	L	L	L
8	L	L	L	L	L	L	L	L	L	L	L	R	L	L	R	L	L	L	M	R
13	L	L	L	L	L	L	L	L	L	L	L	L	L	L	H	L	L	H	H	L
15	*L*	R	L	R	R	R	R	L	L	L	*L*	L	L	L	M	L	L	M	H	*L*
16	L	M	L	L	L	L	L	L	L	L	L	L	L	M	H	L	L	M	H	L
21	L	L	L	L	L	R	L	L	L	L	L	L	L	L	H	L	L	M	H	L
23	L	*L*	L	L	L	*L*	L	L	L	L	*L*	H	L	H	L	L	L	L	L	L
27	*R*	L	L	R	R	R	R	R	R	R	R	R	L	H	R	L	R	L	H	R
30	L	L	L	L	L	R	L	*L*	*L*	*L*	*L*	L	*L*	*L*	M	L	L	L	M	L
31	*L*	L	L	M	L	L	L	L	L	L	M	H	L	H	M	L	L	H	M	L
33	L	L	L	L	L	R	L	L	L	L	L	L	L	L	M	L	L	L	M	L
34	L	L	L	L	L	R	L	L	L	L	L	L	L	M	L	L	R	L	L	L
35	L	L	L	L	L	R	L	L	L	L	L	L	L	L	M	L	L	L	M	L
42	L	L	L	L	L	R	L	L	L	L	L	L	L	H	M	L	L	L	M	L
45	H	H	L	H	L	*L*	L	L	H	L	H	H	L	H	H	L	L	H	H	H
47	L	L	L	L	L	L	L	L	M	L	H	H	L	H	H	L	L	H	H	M
50	*R*	*R*	*R*	L	L	L	L	L	R	*R*	L	M	L	M	R	L	R	M	H	H
59	L	L	L	*L*	*L*	*L*	*L*	R	R	*L*	R	R	*L*	R	L	L	L	L	L	L
64	*L*	L	L	L	L	L	L	L	L	L	*L*	L	L	L	L	L	L	L	L	L
68	R	L	L	R	R	R	R	R	M	R	L	M	L	L	M	L	*R*	L	L	M
71	*R*	L	H	R	L	R	L	R	R	R	*R*	R	R	M	R	R	R	R	*R*	*H*
72	H	*L*	*H*	L	L	R	L	R	R	R	R	R	R	H	R	R	R	R	H	H
73	H	*L*	H	L	L	R	L	R	R	R	R	R	R	*H*	R	R	R	R	H	*H*

Reactivity Chart 6. Protection for the Carboxyl Group (*Continued*)

PG	78	79	80	81	82	83	84	85	86	87	88	89	90	91	92	93	94	95	96	97	98	99	100	101	102	103	104	105	106	107	108
	L. OXIDANTS											M.			N.			O. MISCELLANEOUS											P.		
	I_2	PhSeX; PhSCl	Br_2; Cl_2	MnO_2/CH_2Cl_2	$NaIO_4$ pH 5-8	SeO_2 pH 2-4	SeO_2/Py	$K_3Fe(CN)_6$, pH 8	Pb(IV), 25°	Pb(IV), 80°	$Tl(NO_3)_3$	150°	250°	350°	$:CCl_2$	N_2CHCO_2R/Cu	$CH_2I_2/Zn(Cu)$	$R_3SnH/In\cdot$	$Ni(CO)_4$	CH_2N_2	$SOCl_2$	Ac_2O, 25°	Ac_2O, 80°	DCC	MeI	$Me_3O^+BF_4^-$	1.LDA 2.MeI	1.K_2CO_3 2.MeI	RCHO	RCOCl	C^+/olefin
1	*L*	L	*L*	*L*	*L*	*L*	*L*	L	L	*L*		*L*	*L*	*L*	*L*	*L*	L	*L*	*L*	*L*	*L*	*L*	*L*	*L*	*L*	*L*			*L*	*L*	*L*
2	L	L	R	L	L	L	L	L	L	H		L	L	R	L	L	L	L	L	L	L	L	L	L	L	L			L	L	M
3	L	L	R	L	R	M	L	L	R	R		L	L	R	M	M	L	M	L	L	L	L	L	L	*R*	R			L	L	L
4	L	L	M	L	L	H	L	L	L	H		*M*	*H*	R	L	L	L	L	L	L	*M*	M	M	L	L	L			L	*L*	M
7	L	L	R	L	L	M	L	L	L	H		L	L	R	L	L	L	L	L	L	L	L	M	L	L	L			L	L	M
8	L	R	R	L	L	L	L	L	L	L		L	L	R	L	L	L	L	L	L	L	L	L	L	L	*L*			L	L	L
13	L	L	L	L	L	L	L	H	L	L		L	M	R	L	L	L	L	L	L	L	L	L	L	L	R			L	L	L
15	L	L	L	L	*L*	L	L	M	L	L		M	H	R	L	M	R	R	M	L	L	*L*	L	*L*	L	L			L	L	L
16	L	L	L	L	L	L	L	L	L	L		L	M	R	L	L	R	R	L	L	L	L	L	L	L	L			L	L	L
21	L	L	L	L	L	L	L	L	L	M		M	H	R	L	L	L	L	L	L	L	L	L	L	L	L			L	L	L
23	*L*	*L*	L	*L*	L	H	L	L	*L*	H		*M*	*H*	H	L	*L*	L	L	L	*L*	*L*	*L*	M	L	L	L			*L*	*L*	H
27	R	R	R	L	L	M	L	L	L	M		R	R	R	R	R	R	L	R	L	L	L	L	L	L	L			L	L	R
30	L	L	*L*	*L*	*L*	M	L	L	L	L		L	L	R	L	L	L	*L*	L	L	*L*	L	L	*L*	L	L			L	*L*	L
31	L	L	L	L	L	H	L	L	L	H		M	H	R	L	L	L	L	L	L	L	L	M	L	L	L			L	L	H
33	L	L	L	L	L	L	L	L	L	L		L	L	R	L	L	L	L	L	L	L	L	L	L	L	L			L	*L*	L
34	L	L	L	L	L	M	L	L	L	L		L	L	R	L	L	L	L	L	L	L	L	L	L	L	L			L	L	L
35	L	L	M	L	L	M	L	L	L	L		L	L	R	L	L	L	L	L	L	L	L	L	L	L	L			L	L	L
42	L	L	M	L	L	M	L	L	L	R		L	L	R	L	*M*	L	L	L	L	L	L	L	L	L	L			L	L	R
45	H	L	L	L	H	H	L	H	H	H		*L*	*L*	R	L	M	L	L	L	L	*L*	H	H	L	L	H			L	M	L
47	H	L	L	L	M	H	L	H	H	H		L	L	R	L	L	L	L	L	L	L	L	H	L	L	L			L	M	L
50	L	L	R	*L*	L	L	L	L	L	R		M	H	R	M	M	L	L	L	L	L	*L*	L	L	L	R			L	L	L
59	L	L	*R*	L	L	M	M	L	L	R		L	L	R	L	L	L	L	L	L	*L*	*L*	*L*	*L*	*R*	*R*			*L*	*R*	R
64	L	L	L	L	L	L	L	L	L	*L*		*L*	L	L	L	L	L	L	L	L	L	L	L	L	L	R			L	L	*L*
68	L	L	M	L	L	L	L	L	L	M		L	L	M	R	R	R	L	L	M	L	L	L	L	L	R			L	L	*M*
71	*R*	*R*	*R*	R	R	R	M	*R*	*H*	R		M	*R*	*R*	L	R	L	L	L	L	R	R	R	L	R	R			*R*	*R*	*R*
72	H	R	R	H	H	R	M	H	H	R		M	R	R	L	M	L	L	L	L	R	*R*	R	L	R	R			R	R	R
73	H	R	R	H	*H*	R	M	H	*H*	R		M	R	R	L	M	L	L	L	L	R	R	R	L	R	R			R	*R*	R

Reactivity Chart 7. Protection for the Thiol Group

1. *S*-Benzyl Thioether
3. *S*-*p*-Methoxybenzyl Thioether
5. *S*-*p*-Nitrobenzyl Thioether
6. *S*-4-Picolyl Thioether
7. *S*-2-Picolyl *N*-Oxide Thioether
8. *S*-9-Anthrylmethyl Thioether
9. *S*-Diphenylmethyl Thioether
10. *S*-Di(*p*-methoxyphenyl)methyl Thioether
12. *S*-Triphenylmethyl Thioether
15. *S*-2,4-Dinitrophenyl Thioether
16. *S*-*t*-Butyl Thioether
19. *S*-Isobutoxymethyl Monothioacetal
20. *S*-2-Tetrahydropyranyl Monothioacetal
23. *S*-Acetamidomethyl Aminothioacetal
25. *S*-Cyanomethyl Thioether
26. *S*-2-Nitro-1-phenylethyl Thioether
27. *S*-2,2-Bis(carboethoxy)ethyl Thioether
30. *S*-Benzoyl Derivative
36. *S*-(*N*-Ethylcarbamate)
38. *S*-Ethyl Disulfide

(See chart, pp. 733–735.)

Reactivity Chart 7. Protection for the Thiol Group

	pH<1, 100°	pH<1	pH 1	pH 2-4	pH 4-6	pH 6-8.5	pH 8.5-10	pH 10-12	pH>12	pH>12, 150°	NaH	Ph_3CNa	$(C_{10}H_8)^{\dot{-}}Na^+$	$MeSOCH_2^-Na^+$	KO-t-Bu	LiN-i-Pr_2	Py; R_3N	$NaNH_2$	NaOMe	Enolate	NH_3; RNH_2	RS^-; N_3^-; SCN^-	OAc^-; X^-	NaCN, pH 12	HCN, pH 6	RLi	RMgX	Organozinc	Organocopper	Wittig; ylide	H_2/Raney (Ni)	H_2/Pt pH 2-4	H_2/Pd	H_2/Lindlar	H_2/Rh	Zn/HCl	Zn/HOAc	Cr(II), pH 5
PG	1	2	3	4	5	6	7	8	9	10	11	12	13	14	15	16	17	18	19	20	21	22	23	24	25	26	27	28	29	30	31	32	33	34	35	36	37	38
	A. AQUEOUS										B. BASIC								C. NUCLEOPHILIC							D. ORGANOMET.					E. CAT. REDN.					F.		
1	H	*L*	L	*L*	L	L	L	L	L	M	L	R	R	R	*L*	R	L	L	L	L	L	*L*	L	L		*R*	R	L	L	L	R	R	R	*M*	R	L	L	L
3	H	*H*	*M*	*L*	L	L	L	L	*L*	M	L	R	R	R	L	R	L	L	L	L	L	L	L	L		R	R	L	L	L	R	R	R	*M*	*R*	M	L	L
5	H	L	L	L	L	L	L	L	M	H	M	R	R	R	L	R	L	R	M	L	L	M	L	M		R	R	L	R	L	R	R	*R*	*M*	R	R	R	R
6	H	*L*	L	L	L	L	L	L	*L*	M	L	R	R	R	L	R	L	L	L	L	L	*L*	L	L		R	R	L	L	L	R	R	R	*M*	R	L	L	L
7	H	*H*	*M*	L	L	L	L	L	L	M	L	R	R	R	L	R	L	L	L	L	L	L	L	L		R	R	L	L	L	R	R	R	*R*	R	R	R	R
8	H	*M*	L	L	L	L	L	L	L	M	L	R	R	R	L	R	L	L	L	L	L	*H*	L	L		R	R	L	L	L	R	R	R	*M*	R	L	L	L
9	*H*	*L*	L	L	L	L	L	L	L	M	L	R	R	R	L	R	L	*M*	*L*	L	L	M	L	L		R	R	L	L	L	R	R	R	*M*	R	L	L	L
10	H	H	L	L	L	L	L	L	L	M	L	R	R	R	L	R	L	*M*	L	L	L	M	L	L		R	R	L	L	L	R	R	R	*M*	R	L	L	L
12	H	*H*	*M*	*L*	L	L	L	L	L	M	L	*L*	R	L	L	*L*	L	L	*L*	L	L	*R*	L	L		L	L	L	L	L	R	R	R	*R*	R	M	L	L
15	H	L	L	L	L	M	H	H	H	H	*R*	*R*	R	R	*R*	R	L	*R*	*M*	M	M	H	L	*H*		R	R	L	R	L	R	R	R	*R*	R	R	R	R
16	H	*M*	*L*	L	L	L	L	*L*	*L*	M	L	L	L	L	L	*L*	*L*	L	*L*	*L*	*L*	L	L	L		*L*	L	L	L	L	R	R	M	*L*	M	L	L	L
19	H	H	L	L	L	L	L	L	M	H	L	L	R	H	L	L	L	L	L	L	L	H	L	L		L	L	L	L	L	R	R	R	L	R	H	L	L
20	H	H	L	L	L	L	L	L	M	H	L	L	R	H	L	L	L	L	L	L	L	R	L	L		L	L	L	L	L	R	R	R	L	R	H	L	L
23	*H*	*M*	*L*	*L*	*L*	*L*	*L*	*L*	*L*	*H*	R	R	R	R	L	R	L	R	R	L	*L*	L	L	L		R	R	L	L	L	R	R	R	*M*	R	L	*L*	L
25	H	R	L	L	L	L	L	M	M	H	R	R	R	R	M	R	L	R	R	L	*L*	*L*	L	L		R	R	L	L	L	R	R	R	*L*	R	R	R	L
26	H	*L*	L	L	L	L	H	H	*H*	R	H	H	R	H	H	H	H	H	H	R	H	M	*L*	*M*		R	R	L	R	L	R	R	R	*R*	R	R	R	R
27	H	*L*	L	L	L	L	L	H	*H*	R	H	H	R	H	H	H	L	H	R	R	R	M	L	*M*		R	R	*R*	R	L	R	R	R	*L*	R	L	L	L
30	H	L	L	L	L	L	M	H	H	H	L	L	H	H	H	H	H	H	H	H	H	H	L	H		H	H	H	L	H	R	R	R	L	R	H	L	L
36	H	L	L	L	L	M	H	H	H	H	R	R	R	H	H	R	H	H	H	H	H	H	L	L		H	H	L	L	L	R	R	R	L	R	H	L	L
38	H	L	L	L	L	L	L	L	H	H	R	R	R	H	R	R	L	R	R	R	L	M	L	H		H	H	H	M	H	R	R	R	H	R	H	H	H

Reactivity Chart 7. Protection for the Thiol Group (*Continued*)

PG	G. 39 Na/NH_3	40 Al(Hg)	41 $SnCl_2$/Py	42 HSO_3^-; H_2S	H. HYDRIDE REDN. 43 $LiAlH_4$	44 Li-s-Bu_3BH	45 $(C_5H_{11})_2BH$	46 B_2H_6, 0°	47 $NaBH_4$	48 $Zn(BH_4)_2$	49 $NaBH_3CN$ pH 4-6	50 i-Bu_2AlH	51 Li(OtBu)$_3$AlH	I. 52 $AlCl_3$, 80°	53 $AlCl_3$, 25°	54 $SnCl_4$; BF_3	55 $LiClO_4$; $MgBr_2$	56 TsOH, 80°	57 TsOH, 0°
1	*H*	L	L	L	L	L	L	*L*	L	L	L	L	L	L	L	L	L	M	L
3	*H*	L	L	L	L	L	L	L	L	L	L	L	L	M	L	L	L	M	L
5	H	R	L	L	R	*M*	L	L	L	L	L	L	L	L	L	L	L	M	L
6	H	L	L	L	L	L	L	L	L	L	L	L	L	M	L	L	L	M	L
7	H	R	R	*R*	*R*	*R*	L	R	M	L	L	R	*R*	H	M	L	L	L	L
8	H	L	L	L	L	L	L	L	L	L	L	L	L	M	L	L	L	M	L
9	*H*	L	L	L	L	L	L	L	L	L	L	L	L	H	M	L	L	M	L
10	*H*	L	L	L	L	L	L	L	L	L	L	L	L	H	M	M	L	M	L
12	*H*	L	L	L	L	L	L	*H*	L	L	L	L	L	H	H	M	L	M	L
15	R	R	M	L	R	*M*	*M*	*M*	L	L	L	*R*	L	L	L	L	L	M	L
16	L	L	L	L	*L*	L	L	L	L	L	L	L	L	H	M	L	L	M	L
19	H	L	L	L	L	L	L	L	L	L	L	L	L	H	H	M	L	M	L
20	M	L	L	L	L	L	L	L	L	L	L	L	L	H	H	M	L	M	L
23	L	L	L	L	R	L	R	R	L	L	L	R	L	L	L	L	L	M	L
25	R	L	L	L	R	*M*	L	R	L	L	L	R	L	M	L	L	L	M	L
26	R	R	L	L	R	*M*	L	L	L	L	L	L	*R*	L	L	L	L	M	L
27	R	L	L	L	R	R	L	*L*	L	L	L	R	L	M	L	L	L	M	L
30	H	L	L	L	H	H	L	M	H	M	L	H	L	R	M	L	L	M	L
36	H	L	L	L	H	M	M	H	L	L	L	H	L	R	H	L	L	M	L
38	H	H	H	H	H	H	H	H	H	H	M	H	H	R	H	L	L	M	L

PG	J. 58 Hg(II)	59 Ag(I)	60 Cu(II)/Py	K. 61 HBr/In·	62 HX/In·	63 NBS/CCl_4	64 Br_3CCl/In·	L. OXIDANTS 65 OsO_4	66 $KMnO_4$, pH 7, 0°	67 O_3, -50°	68 RCO_3H, 0°	69 RCO_3H, 50°	70 CrO_3/Py	71 CrO_3, pH 1	72 H_2O_2 pH 10-12	73 Quinone	74 $^1O_2$	75 DMSO, 100°	76 NaOCl pH 10	77 aq NBS
1	*L*	M	L	R	L	R	R	L	R	R	R	R	L	R	L	L	*R*	L	R	*R*
3	*R*	R	L	R	L	R	R	L	R	R	R	R	L	R	L	L	*R*	L	R	R
5	L	L	L	R	L	R	R	L	R	M	R	R	L	R	L	L	*R*	L	R	R
6	L	L	L	R	L	R	R	L	R	R	R	R	L	R	L	L	*M*	M	R	R
7	L	L	L	R	L	R	R	L	R	R	R	R	L	R	L	L	*M*	L	R	R
8	M	M	L	R	L	R	R	L	R	R	R	R	L	R	L	L	*R*	L	R	R
9	*M*	*M*	L	R	L	R	R	L	R	R	R	R	L	R	L	L	*R*	L	R	R
10	R	R	L	R	L	R	R	L	R	R	R	R	L	R	L	L	*R*	L	R	R
12	*R*	*R*	L	L	L	L	L	L	R	R	R	R	L	R	L	L	*M*	L	R	R
15	L	L	L	L	L	L	L	L	R	L	M	R	L	R	R	L	*M*	R	R	R
16	*R*	R	L	L	L	L	L	L	R	R	R	R	L	R	L	L	*M*	L	R	R
19	R	R	L	L	L	R	L	L	R	R	R	R	M	R	L	L	M	L	R	R
20	R	R	L	L	L	R	L	L	R	R	R	R	M	R	L	L	M	L	R	R
23	*R*	R	L	L	L	L	L	L	R	R	R	R	L	R	L	L	*R*	L	R	R
25	L	L	L	R	L	R	L	L	R	R	R	R	L	R	R	L	*R*	L	R	R
26	L	L	L	R	L	R	R	L	R	R	R	R	L	R	R	L	*R*	M	R	R
27	L	L	L	L	L	R	L	L	R	R	R	R	L	R	R	L	*M*	L	R	R
30	R	R	L	L	L	M	L	L	R	M	R	R	L	R	H	L	R	L	R	R
36	R	R	L	L	L	M	L	L	R	R	R	R	L	R	H	L	R	M	R	R
38	R	R	M	R	R	R	R	M	R	R	R	R	R	R	L	R	R	L	R	R

Reactivity Chart 7. Protection for the Thiol Group (*Continued*)

PG	I_2	PhSeX; PhSCl	Br_2; Cl_2	MnO_2/CH_2Cl_2	$NaIO_4$ pH 5-8	SeO_2 pH 2-4	SeO_2/Py	$K_3Fe(CN)_6$, pH 8	Pb(IV), 25°	Pb(IV), 80°	$Tl(NO_3)_3$	150°	250°	350°	$:CCl_2$	N_2CHCO_2R/Cu	CH_2I_2/Zn(Cu)
	78	79	80	81	82	83	84	85	86	87	88	89	90	91	92	93	94
	L. OXIDANTS											M.			N.		
1	*L*	L	R	M	R	R	M	L	R	R		L	L	M	M	R	L
3	M	L	R	M	R	R	M	L	R	R		L	L	M	M	R	L
5	R	L	R	L	R	R	L	L	R	R		L	L	M	M	R	L
6	R	L	R	L	R	R	M	L	R	R		L	L	M	M	R	L
7	R	L	R	L	R	R	M	L	R	R		L	L	M	R	R	L
8	R	L	R	M	R	R	R	L	R	R		L	L	M	M	R	L
9	*M*	L	R	M	R	R	R	L	R	R		L	L	M	M	R	L
10	M	L	R	M	R	R	R	L	R	R		L	L	M	M	R	L
12	*R*	L	M	L	R	L	L	L	R	R		L	L	M	M	R	L
15	R	L	M	L	L	M	L	R	R	R		L	L	M	M	R	L
16	*L*	L	M	L	R	L	L	L	R	R		L	L	R	M	R	L
19	R	L	R	L	R	M	L	L	R	R		L	M	R	M	R	L
20	R	L	R	L	R	M	L	L	R	R		L	M	R	M	R	L
23	*R*	L	R	L	R	M	L	L	R	R		L	L	M	M	R	L
25	R	L	R	L	R	M	L	M	R	R		L	L	M	M	R	L
26	R	L	R	L	R	*M*	*L*	*H*	R	R		M	H	H	M	R	L
27	R	L	R	L	R	M	L	H	R	R		M	H	H	M	R	L
30	L	L	R	L	L	L	L	H	L	M		M	R	R	M	M	L
36	L	L	R	L	R	L	L	H	L	M		R	R	R	M	M	L
38	R	L	R	R	R	R	R	R	R	R		H	H	R	M	R	R

PG	R_3SnH/In·	$Ni(CO)_4$	CH_2N_2	$SOCl_2$	Ac_2O, 25°	Ac_2O, 80°	DCC	MeI	$Me_3O^+BF_4^-$	1.LDA 2.MeI	1.K_2CO_3 2.MeI	RCHO	RCOCl	C^+/olefin
	95	96	97	98	99	100	101	102	103	104	105	106	107	108
	O. MISCELLANEOUS											P.		
1	*R*	L	L	L	L	L	*L*	R	R	R	R	*L*	*L*	M
3	R	L	L	L	L	L	*L*	R	R	R	R	L	L	M
5	R	L	L	L	L	L	L	M	M	R	M	L	L	L
6	R	L	*L*	L	L	L	*L*	R	R	R	R	L	L	L
7	R	L	L	R	*R*	*R*	L	R	R	R	R	L	H	L
8	R	L	L	L	L	L	L	R	R	R	R	L	L	M
9	R	L	L	L	L	L	L	R	R	R	R	L	L	M
10	R	L	L	L	L	L	L	R	R	R	R	L	L	M
12	L	L	L	L	L	L	L	L	L	L	L	L	L	M
15	R	L	L	L	L	L	L	L	L	R	L	L	L	L
16	L	L	L	L	*L*	L	L	M	M	M	M	*L*	L	M
19	M	L	L	L	L	L	L	R	R	M	R	L	L	M
20	M	L	L	L	L	L	L	R	R	M	R	L	L	M
23	L	L	L	L	L	M	L	R	R	R	R	L	L	L
25	*M*	L	L	L	L	L	L	R	R	R	R	L	L	L
26	R	L	L	L	L	L	L	R	R	R	R	L	L	L
27	L	L	L	L	L	L	L	R	R	R	R	L	L	L
30	L	L	L	L	L	L	L	L	R	L	L	L	L	L
36	R	L	L	L	L	M	L	L	R	L	L	L	L	L
38	R	R	L	L	L	L	L	M	R	R	M	L	L	R

Reactivity Chart 8. Protection for the Amino Group: Carbamates

1. Methyl Carbamate
5. 9-Fluorenylmethyl Carbamate
8. 2,2,2-Trichloroethyl Carbamate
11. 2-Trimethylsilylethyl Carbamate
16. 1,1-Dimethylpropynyl Carbamate
20. 1-Methyl-1-phenylethyl Carbamate
22. 1-Methyl-1-(4-biphenylyl)ethyl Carbamate
24. 1,1-Dimethyl-2-haloethyl Carbamate
26. 1,1-Dimethyl-2-cyanoethyl Carbamate
28. *t*-Butyl Carbamate
30. Cyclobutyl Carbamate
31. 1-Methylcyclobutyl Carbamate
35. 1-Adamantyl Carbamate
37. Vinyl Carbamate
38. Allyl Carbamate
39. Cinnamyl Carbamate
44. 8-Quinolyl Carbamate
45. *N*-Hydroxypiperidinyl Carbamate
47. 4,5-Diphenyl-3-oxazolin-2-one
48. Benzyl Carbamate
53. *p*-Nitrobenzyl Carbamate
55. 3,4-Dimethoxy-6-nitrobenzyl Carbamate
58. 2,4-Dichlorobenzyl Carbamate
65. 5-Benzisoxazolylmethyl Carbamate
66. 9-Anthrylmethyl Carbamate
67. Diphenylmethyl Carbamate
71. Isonicotinyl Carbamate
72. *S*-Benzyl Carbamate
75. *N*-(*N′*-Phenylaminothiocarbonyl) Derivative

(See chart, pp. 737–739.)

Reactivity Chart 8. Protection for the Amino Group: Carbamates

PG	pH<1, 100°	pH<1	pH 1	pH 2-4	pH 4-6	pH 6-8.5	pH 8.5-10	pH 10-12	pH>12	pH>12, 150°	NaH	Ph_3CNa	$(C_{10}H_8)^{\cdot-}Na^+$	$MeSOCH_2^-Na^+$	KO-t-Bu	LiN-i-Pr_2	Py; R_3N	$NaNH_2$
	1	2	3	4	5	6	7	8	9	10	11	12	13	14	15	16	17	18
	A. AQUEOUS										B. BASIC							
1	H	*H*	*L*	L	L	L	*L*	L	M	*H*	L	L	L	L	L	L	*L*	L
5	H	M	L	L	L	L	L	L	M	H	M	H	L	L	L	L	*M*	H
8	H	H	*L*	L	L	L	L	L	*M*	M	R	R	R	R	R	R	L	R
11	H	*H*	H	M	L	L	L	L	*L*	*M*	L	L	L	L	L	L	L	L
16	H	H	H	L	L	L	L	L	L	M	L	R	R	R	L	R	L	R
20	H	H	M	L	L	L	L	L	L	M	L	L	L	L	L	L	L	L
22	H	H	H	M	L	L	L	L	L	M	L	L	L	L	L	L	L	L
24	H	*H*	M	L	L	L	L	L	L	H	L	L	H	L	L	L	L	M
26	H	*L*	L	L	L	L	M	*H*	H	H	M	R	M	H	H	H	H	H
28	H	*H*	M	M	L	L	L	L	*L*	M	L	L	L	L	L	L	L	L
30	H	*H*	M	L	L	L	L	L	L	H	L	L	L	L	L	L	L	L
31	H	*H*	H	*M*	L	L	L	L	L	H	L	L	L	L	L	L	L	L
35	H	H	H	M	L	L	L	L	L	M	L	L	L	L	L	L	L	L
37	H	*H*	H	H	M	L	L	*L*	*L*	M	L	L	L	L	L	L	L	L
38	H	H	M	M	L	L	L	L	L	M	L	L	L	L	L	L	L	L
39	H	H	M	M	L	L	L	L	L	M	L	L	L	L	L	L	L	L
44	H	H	M	L	L	L	L	L	H	H	L	L	L	L	L	L	L	L
45	H	*L*	L	L	L	L	L	L	*L*	M	L	L	M	L	L	L	L	L
47	H	*L*	L	L	L	L	L	L	L	M	L	L	H	L	L	L	L	L
48	H	*H*	*L*	*L*	L	L	*L*	*L*	*L*	*H*	*L*	R	L	L	*L*	*L*	L	H
53	H	*H*	L	L	L	L	L	L	L	M	L	R	R	L	L	L	H	H
55	H	M	*L*	L	L	L	L	L	L	M	L	L	L	L	L	L	L	L
58	H	M	L	L	L	L	L	L	L	M	L	L	L	L	L	L	H	M
65	H	*H*	M	L	L	L	*L*	L	M	H	L	R	H	H	H	H	*R*	H
66	H	H	*L*	L	L	L	L	*L*	L	M	L	R	L	L	L	L	L	L
67	H	*H*	*H*	M	L	L	L	L	L	M	L	R	L	L	L	L	L	L
71	H	*L*	*L*	*L*	L	L	L	L	L	M	L	R	L	L	L	L	L	L
72	H	M	L	L	L	L	L	L	*H*	H	L	R	H	L	L	L	L	L
75	H	*H*	M	L	L	L	L	L	H	H	L	L	L	L	L	L	L	L

PG	NaOMe	Enolate	NH_3; RNH_2	RS^-; N_3^-; SCN^-	OAc^-; X^-	NaCN, pH 12	HCN, pH 6	RLi	RMgX	Organozinc	Organocopper	Wittig; ylide	H_2/Raney (Ni)	H_2/Pt pH 2-4	H_2/Pd	H_2/Lindlar	H_2/Rh	Zn/HCl	Zn/HOAc	Cr(II), pH 5
	19	20	21	22	23	24	25	26	27	28	29	30	31	32	33	34	35	36	37	38
	C. NUCLEOPHILIC							D. ORGANOMET.					E. CAT. REDN.					F.		
1	L	L	*L*	H	*L*	L	*L*	H	*L*	L	*L*	*L*	L	L	L	L	L	L	L	L
5	M	L	*H*	L	L	M	L	H	H	M	L	L	L	L	*M*	L	R	L	L	L
8	*L*	R	M	M	L	L	L	H	H	R	R	R	L	L	*L*	*L*	L	*H*	*H*	H
11	L	L	*L*	L	L	L	L	H	H	L	L	L	L	M	*L*	L	L	M	L	L
16	L	L	L	L	L	L	L	H	H	R	M	R	H	H	*H*	*H*	R	H	H	L
20	L	L	L	L	L	L	L	H	H	L	L	L	M	H	L	L	R	H	H	L
22	L	L	L	L	L	L	L	H	H	L	L	L	M	H	L	L	R	H	H	H
24	R	R	M	L	L	L	L	H	H	M	R	M	M	R	R	L	L	H	H	H
26	H	H	H	H	M	H	L	H	H	H	M	M	R	R	R	L	R	R	R	L
28	L	L	*L*	*L*	L	L	L	H	H	L	L	L	L	H	*L*	*L*	L	H	H	M
30	L	L	L	L	L	L	L	H	H	L	L	L	L	M	L	L	L	H	M	L
31	L	L	L	L	L	L	L	H	H	L	L	L	L	M	L	L	L	H	M	L
35	L	L	L	L	L	L	L	H	H	L	L	L	L	M	L	L	L	H	H	L
37	L	L	L	L	L	L	L	H	H	L	L	L	H	H	H	L	R	H	H	M
38	L	L	H	L	L	L	L	H	H	L	L	L	H	H	H	L	R	H	H	L
39	L	L	H	L	L	L	L	H	H	L	L	L	H	H	H	L	R	H	H	L
44	L	L	L	L	L	L	L	H	H	L	L	L	L	L	L	L	L	H	H	L
45	L	L	L	L	L	L	L	H	H	L	L	L	L	*H*	H	L	H	H	*H*	L
47	M	M	L	L	L	L	L	H	H	L	L	L	R	H	*H*	L	R	L	L	L
48	L	L	*H*	*L*	L	L	L	H	H	L	L	L	H	*H*	*H*	L	R	M	*L*	L
53	M	M	H	L	L	L	L	H	H	L	R	L	H	H	*H*	L	R	R	R	R
55	M	M	M	L	L	L	L	H	H	L	R	L	H	H	H	L	R	R	R	R
58	M	M	H	H	L	L	L	H	H	M	L	L	H	H	*H*	L	R	L	L	L
65	H	H	H	H	L	M	L	H	H	L	L	L	H	H	*H*	L	R	H	M	L
66	L	L	*L*	*H*	L	L	L	H	H	L	L	L	H	H	H	L	R	M	M	L
67	L	L	H	L	L	L	L	H	H	M	L	L	H	H	H	L	R	H	M	L
71	L	L	M	L	L	L	L	H	H	L	L	L	H	H	*H*	L	R	*H*	*H*	L
72	H	M	H	L	L	M	L	H	H	L	L	L	*H*	R	H	L	R	L	L	L
75	L	L	L	L	L	L	L	H	H	M	L	L	R	R	R	L	R	H	H	R

Reactivity Chart 8. Protection for the Amino Group: Carbamates (*Continued*)

	Na/NH_3	Al(Hg)	$SnCl_2$/Py	HSO_3^-; H_2S	$LiAlH_4$	Li-*s*-Bu_3BH	$(C_5H_{11})_2BH$	B_2H_6, 0°	$NaBH_4$	$Zn(BH_4)_2$	$NaBH_3CN$ pH 4-6	*i*-Bu_2AlH	Li(O*t*Bu)$_3$AlH	$AlCl_3$, 80°	$AlCl_3$, 25°	$SnCl_4$; BF_3	$LiClO_4$; $MgBr_2$	TsOH, 80°	TsOH, 0°	Hg(II)	Ag(I)	Cu(II)/Py	HBr/In·	HX/In·	NBS/CCl_4	Br_3CCl/In·	OsO_4	$KMnO_4$, pH 7, 0°	O_3, −50°	RCO_3H, 0°	RCO_3H, 50°	CrO_3/Py	CrO_3, pH 1	H_2O_2 pH 10-12	Quinone	$^1O_2$	DMSO, 100°	NaOCl pH 10	aq NBS
PG	39	40	41	42	43	44	45	46	47	48	49	50	51	52	53	54	55	56	57	58	59	60	61	62	63	64	65	66	67	68	69	70	71	72	73	74	75	76	77
	G.				H. HYDRIDE REDN.									I.						J.			K.				L. OXIDANTS												
1	L	*L*	*L*	L	*R*	L	L	L	L	L	L	*R*	L	R	L	*L*	L	L	L	L	L	L	L	L	*L*	L	*L*	L	L	L	*L*	L	H	*L*	L	*L*	*L*	L	*L*
5	H	L	L	L	M	L	L	L	L	L	L	M	L	R	L	L	L	M	L	L	L	H	R	R	R	L	L	L	L	L	L	H	L	L	L	L	L	L	L
8	R	M	H	L	R	L	L	*L*	L	L	L	M	L	R	R	R	L	L	L	L	L	L	M	M	M	M	L	L	*L*	*L*	L	*L*	*L*	L	L	L	R	L	L
11	L	L	L	L	R	L	L	L	L	L	L	M	L	R	L	L	L	H	H	L	L	L	L	L	L	L	L	L	L	L	L	L	H	L	L	L	L	L	L
16	H	L	L	L	R	L	R	R	L	L	L	M	L	R	H	L	L	R	H	R	L	R	R	R	L	R	R	R	R	M	R	L	H	L	L	L	L	L	R
20	R	L	L	L	R	L	L	L	L	L	L	M	L	R	H	M	L	R	H	L	L	L	L	L	L	L	L	L	L	L	L	L	H	L	L	L	L	L	L
22	R	L	L	L	R	L	L	L	L	L	M	M	L	R	H	M	L	R	H	L	L	L	H	L	L	L	L	L	L	L	M	L	H	L	L	L	L	L	L
24	*H*	L	H	L	R	L	L	L	L	L	L	M	L	R	*H*	M	L	M	L	L	M	L	M	L	L	L	L	L	L	L	L	L	M	L	L	L	R	L	L
26	H	L	L	L	R	M	L	M	L	L	L	M	L	R	H	M	L	M	L	L	L	H	L	L	L	L	L	L	L	L	L	H	L	R	L	L	M	M	L
28	*L*	*L*	L	*L*	M	L	L	L	*L*	L	M	*L*	L	R	*H*	M	L	R	*H*	L	*L*	L	H	L	L	L	L	*L*	*L*	*L*	M	L	H	L	L	L	L	L	L
30	L	L	L	L	M	L	L	L	L	L	L	M	L	R	M	L	L	M	L	L	L	L	M	L	L	L	L	L	L	L	L	L	M	L	L	L	L	L	L
31	L	L	L	L	M	L	L	L	L	L	L	M	L	R	M	M	L	R	M	L	L	L	M	L	L	L	L	L	L	L	L	L	H	L	L	L	L	L	L
35	L	L	L	L	M	L	L	L	L	L	L	M	L	R	L	L	L	L	L	L	L	L	M	L	L	L	L	L	L	L	L	L	H	L	L	L	L	L	L
37	L	L	L	L	R	L	R	R	L	L	M	M	L	R	L	L	L	R	L	*R*	L	L	R	R	R	R	R	R	R	R	R	L	H	L	L	R	L	L	R
38	L	L	L	L	R	L	R	R	L	L	L	M	L	R	H	H	L	R	L	R	L	L	R	R	R	R	R	R	R	R	R	L	M	L	L	M	L	L	R
39	H	L	L	L	R	L	R	R	L	L	L	M	L	R	H	H	L	R	L	*R*	L	L	R	R	R	R	R	R	R	R	R	L	M	L	L	R	L	L	R
44	H	L	L	L	M	L	L	L	L	L	L	M	L	R	L	L	L	L	L	L	M	*H*	R	L	L	L	L	L	L	R	R	L	L	L	L	L	L	L	L
45	H	L	M	L	M	L	L	L	L	L	L	M	L	R	L	L	L	L	L	L	L	L	L	L	L	L	L	L	L	R	R	L	L	L	L	L	L	L	L
47	*H*	L	L	L	M	L	M	R	L	L	L	M	L	R	L	L	L	M	L	L	L	L	R	R	R	R	M	M	M	*H*	H	L	M	L	L	R	L	L	R
48	*H*	L	L	L	R	L	L	*L*	*L*	L	L	*L*	L	R	L	*L*	L	R	*L*	L	*L*	H	H	R	R	L	*L*	L	L	L	L	*L*	*L*	*L*	L	L	L	L	*L*
53	*H*	R	L	L	R	L	L	L	L	L	L	M	L	R	L	L	L	L	L	L	L	H	M	L	R	L	L	L	L	L	L	H	L	L	L	L	L	R	L
55	H	R	L	L	R	L	L	L	L	L	L	M	L	R	R	L	L	R	L	L	L	L	M	L	R	L	L	L	L	L	L	L	L	L	L	L	L	M	L
58	H	L	L	L	R	L	L	L	L	L	L	M	L	R	H	L	L	M	L	L	L	H	M	L	R	L	L	L	L	L	L	H	L	L	L	L	L	L	L
65	H	L	M	L	M	L	L	L	L	L	L	M	L	R	H	L	L	R	L	L	L	H	H	R	R	L	R	R	R	R	R	R	M	L	L	M	L	L	R
66	H	L	L	L	M	L	L	L	L	L	L	M	L	R	H	L	L	R	L	L	L	L	R	R	R	R	L	L	L	L	L	L	L	L	L	R	L	L	L
67	H	L	L	L	M	L	L	L	L	L	L	M	L	R	H	L	L	R	L	L	L	L	H	R	R	L	L	L	L	L	M	L	H	L	L	L	L	L	L
71	H	L	L	L	M	L	L	L	L	L	L	M	L	R	M	L	L	R	L	L	L	L	M	M	R	L	L	L	L	R	R	L	L	L	L	L	L	L	L
72	*H*	L	M	L	M	L	L	L	L	L	L	M	L	R	H	L	L	R	L	H	H	L	R	R	R	L	L	R	R	*H*	H	L	L	M	L	L	L	R	R
75	H	R	L	L	M	L	L	L	L	L	L	M	L	R	M	M	L	L	L	*R*	M	M	R	R	R	L	L	L	R	*H*	H	L	M	*H*	L	R	R	R	R

Reactivity Chart 8. Protection for the Amino Group: Carbamates (*Continued*)

PG	I_2	PhSeX; PhSCl	Br_2; Cl_2	MnO_2/CH_2Cl_2	$NaIO_4$ pH 5-8	SeO_2 pH 2-4	SeO_2/Py	$K_3Fe(CN)_6$, pH 8	Pb(IV), 25°	Pb(IV), 80°	$Tl(NO_3)_3$	150°	250°	350°	$:CCl_2$	N_2CHCO_2R/Cu	CH_2I_2/Zn(Cu)	R_3SnH/In·	$Ni(CO)_4$	CH_2N_2	$SOCl_2$	Ac_2O, 25°	Ac_2O, 80°	DCC	MeI	$Me_3O^+BF_4^-$	1.LDA 2.MeI	1.K_2CO_3 2.MeI	RCHO	RCOCl	C^+/olefin
	78	79	80	81	82	83	84	85	86	87	88	89	90	91	92	93	94	95	96	97	98	99	100	101	102	103	104	105	106	107	108
	L. OXIDANTS											M.			N.			O. MISCELLANEOUS											P.		
1	L	L	L	*L*	L	L	L	L	L	L	L	L	*L*	M	L	*L*	L	L	L	*L*	L	*L*	L	L	L	R	L	L	*L*	L	L
5	L	L	L	L	L	M	L	L	L	L	L	L	M	H	L	L	L	L	L	L	L	L	L	L	L	R	L	L	L	L	L
8	L	L	L	L	L	L	L	L	L	L	L	L	M	H	L	L	R	R	M	L	*L*	*L*	L	L	L	R	*L*	L	L	L	R
11	L	L	L	L	L	M	L	L	L	L	L	L	M	H	L	L	L	L	L	L	L	L	L	L	L	R	L	L	L	L	L
16	L	R	R	L	L	R	R	L	R	R	R	L	M	H	R	R	R	R	R	L	L	L	L	L	L	R	R	L	L	L	R
20	L	L	L	L	L	H	L	L	L	L	L	L	M	H	L	L	L	L	L	L	L	L	L	L	L	R	L	L	L	L	M
22	L	L	L	L	L	H	L	L	L	L	L	L	M	H	L	L	L	L	L	L	L	L	L	L	L	R	L	L	L	L	M
24	L	L	L	L	L	L	L	L	L	L	L	H	H	H	L	L	M	R	L	L	L	L	L	L	L	R	L	L	L	L	H
26	L	L	L	L	L	L	H	M	L	L	L	M	H	H	L	L	L	L	L	L	L	L	L	L	L	R	R	H	L	L	M
28	L	L	L	L	L	H	*L*	L	*L*	L	L	*H*	H	H	L	L	L	L	L	*L*	L	L	L	*L*	L	R	*L*	L	*L*	*L*	H
30	L	L	L	L	L	M	L	L	L	L	L	L	M	H	L	L	L	L	L	L	L	L	L	L	L	R	L	L	L	L	M
31	L	L	L	L	L	M	L	L	L	L	L	L	M	H	L	L	L	L	L	L	L	L	L	L	L	R	L	L	L	L	M
35	L	L	L	L	L	M	L	L	L	L	L	L	L	M	L	L	L	L	L	L	L	L	L	L	L	R	L	L	L	L	H
37	L	R	*R*	L	L	H	L	L	M	R	R	L	M	H	R	R	R	R	H	L	L	L	L	L	L	R	L	L	L	L	R
38	L	R	R	L	L	M	R	L	M	R	R	L	M	R	R	R	R	R	*H*	L	L	L	L	L	L	R	L	L	L	L	R
39	L	R	R	L	L	M	M	L	M	R	R	L	M	R	R	R	R	R	H	L	L	L	L	L	L	R	L	L	L	L	R
44	L	L	L	L	L	L	L	L	L	L	L	L	L	M	L	L	R	L	H	L	L	L	L	L	R	R	L	R	L	L	M
45	L	L	L	L	L	L	L	L	L	L	L	L	M	H	L	L	L	L	L	L	L	L	L	L	R	R	L	R	L	L	L
47	L	L	R	L	L	L	L	L	M	R	M	L	L	M	R	R	R	R	R	L	L	L	L	L	L	R	L	L	L	L	L
48	L	*L*	L	L	*L*	L	L	L	L	L	L	L	L	M	L	L	L	L	L	L	L	L	L	*L*	L	R	L	L	L	*L*	M
53	L	L	L	L	L	L	R	L	L	L	L	L	L	M	L	L	L	R	L	L	L	L	L	L	L	R	L	L	L	L	M
55	L	L	L	L	L	L	M	L	L	L	L	L	L	M	M	M	M	R	L	L	L	L	L	L	L	R	L	L	L	L	M
58	L	L	L	L	L	L	R	L	L	L	L	L	L	M	L	L	L	R	L	L	L	L	L	L	L	R	L	L	L	L	L
65	L	L	R	L	L	L	R	L	M	R	M	L	L	M	M	M	M	M	L	L	L	L	L	L	R	R	L	L	L	L	H
66	L	L	L	L	L	L	L	L	L	L	L	L	L	M	L	L	L	L	L	L	L	L	L	L	L	R	L	L	L	L	M
67	L	L	L	L	L	L	L	L	L	L	L	L	L	M	L	L	L	L	L	L	L	L	L	L	L	R	L	L	L	L	H
71	L	L	L	L	L	L	L	L	L	L	L	L	L	M	L	L	M	L	R	L	L	L	L	L	R	R	L	R	L	L	L
72	L	L	R	L	R	L	L	L	L	M	L	L	L	M	L	L	L	R	L	L	L	L	L	L	L	R	L	L	L	L	L
75	R	R	R	L	R	M	L	L	M	R	M	L	M	H	R	R	R	R	L	R	L	L	M	L	R	R	R	R	L	L	M

Reactivity Chart 9. Protection for the Amino Group: Amides

1. *N*-Formyl
2. *N*-Acetyl
3. *N*-Chloroacetyl
5. *N*-Trichloroacetyl
6. *N*-Trifluoroacetyl
7. *N*-*o*-Nitrophenylacetyl
8. *N*-*o*-Nitrophenoxyacetyl
9. *N*-Acetoacetyl
12. *N*-3-Phenylpropionyl
13. *N*-3-(*p*-Hydroxyphenyl)propionyl
15. *N*-2-Methyl-2-(*o*-nitrophenoxy)propionyl
16. *N*-2-Methyl-2-(*o*-phenylazophenoxy)propionyl
17. *N*-4-Chlorobutyryl
19. *N*-*o*-Nitrocinnamoyl
20. *N*-Picolinoyl
21. *N*-(*N′*-Acetylmethionyl)
23. *N*-Benzoyl
29. *N*-Phthaloyl
31. *N*-Dithiasuccinoyl

(See chart, pp. 741–743.)

Reactivity Chart 9. Protection for the Amino Group: Amides

PG	pH<1, 100°	pH<1	pH 1	pH 2-4	pH 4-6	pH 6-8.5	pH 8.5-10	pH 10-12	pH>12	pH>12, 150°	NaH	Ph_3CNa	$(C_{10}H_8)^{\dot{-}}Na^+$	$MeSOCH_2^-Na^+$	KO-t-Bu	LiN-i-Pr_2	Py; R_3N	$NaNH_2$	NaOMe	Enolate	NH_3; RNH_2	RS^-; N_3^-; SCN^-	OAc^-; X^-	NaCN, pH 12	HCN, pH 6	RLi	RMgX	Organozinc	Organocopper	Wittig; ylide	H_2/Raney (Ni)	H_2/Pt pH 2-4	H_2/Pd	H_2/Lindlar	H_2/Rh	Zn/HCl	Zn/HOAc	Cr(II), pH 5
	1	2	3	4	5	6	7	8	9	10	11	12	13	14	15	16	17	18	19	20	21	22	23	24	25	26	27	28	29	30	31	32	33	34	35	36	37	38
	A. AQUEOUS										B. BASIC								C. NUCLEOPHILIC							D. ORGANOMET.					E. CAT. REDN.					F.		
1	H	*H*	*H*	*L*	L	L	L	M	H	*H*	*H*	L	R	L	L	L	L	L	L	L	*H*	L	L	M	L	H	H	L	L	L	L	*H*	L	L	L	H	L	L
2	*H*	M	L	*L*	L	L	L	*L*	M	H	*L*	R	R	R	L	*R*	*L*	R	*L*	L	*H*	*L*	*L*	L	L	H	M	L	L	L	*L*	L	*L*	L	L	M	L	L
3	*H*	M	L	L	L	L	L	L	*H*	H	M	R	R	R	L	R	*L*	R	R	L	H	R	L	R	L	H	M	M	R	L	M	R	R	L	L	H	H	*R*
5	H	L	L	L	L	L	M	H	H	H	L	L	R	L	L	R	*L*	L	R	R	H	M	L	R	L	H	M	M	R	M	R	R	R	L	R	H	H	M
6	H	*L*	*L*	*L*	L	L	M	*H*	*H*	H	L	L	R	L	L	L	*L*	L	R	R	*H*	M	L	R	L	H	M	L	M	M	R	R	*L*	L	L	H	H	M
7	H	L	L	L	L	L	L	L	M	H	M	R	R	R	L	R	L	R	L	L	H	L	L	L	L	H	H	M	L	L	R	*H*	R	L	R	H	H	R
8	H	L	L	L	L	L	L	L	M	H	M	R	R	R	L	L	L	R	M	M	M	M	L	M	L	H	R	L	L	L	R	R	*R*	L	R	H	H	R
9	H	*L*	L	L	L	L	L	L	H	H	R	R	R	R	R	R	L	R	R	R	*H*	L	L	L	R	H	R	R	R	R	M	R	L	L	M	L	L	L
12	H	L	L	L	L	L	L	L	M	H	M	R	R	R	L	R	L	R	L	L	M	L	L	L	L	H	M	L	L	L	L	L	L	L	R	L	L	L
13	H	L	L	L	L	L	L	L	M	H	R	R	R	R	R	R	L	R	R	L	M	L	L	L	L	H	R	L	L	L	L	L	L	L	R	L	L	L
15	H	L	L	L	L	L	L	L	M	H	L	L	R	L	L	L	L	L	L	L	M	L	L	L	L	H	R	L	L	L	R	*H*	R	L	R	H	H	R
16	H	L	L	L	L	L	L	L	H	H	L	L	R	M	L	H	L	L	L	L	M	L	L	L	L	H	R	R	L	L	R	H	R	R	R	H	H	H
17	H	L	L	L	L	L	L	L	*L*	H	M	R	R	R	R	R	L	R	R	L	M	R	L	R	L	H	M	L	R	L	M	R	R	L	L	L	L	L
19	H	L	L	L	L	L	L	L	M	H	L	L	R	R	L	L	L	L	M	R	M	M	L	M	L	H	R	R	L	L	R	*H*	R	L	R	H	H	R
20	H	L	L	L	L	L	L	L	M	H	L	L	*R*	L	L	L	L	L	L	L	H	L	L	L	L	H	M	L	L	L	L	L	L	L	R	L	L	L
21	H	L	L	L	L	L	L	L	M	H	R	R	R	R	R	R	L	R	L	L	H	L	L	L	L	H	R	L	L	L	R	R	R	R	R	L	L	L
23	H	*H*	L	L	L	L	L	*L*	*H*	H	L	L	*R*	L	*L*	*L*	L	L	L	L	*H*	L	L	M	L	H	M	L	L	L	L	L	*L*	L	R	L	L	L
29	*H*	L	*L*	*L*	L	L	L	L	*R*	H	L	*L*	*R*	L	*L*	L	L	L	*L*	*L*	*H*	*L*	L	L	L	H	M	*L*	L	L	L	L	*L*	L	R	L	L	L
31	M	*L*	L	L	L	L	*L*	M	H	H	L	L	R	R	L	L	L	L	H	H	H	*H*	L	H	L	H	R	R	M	H	R	R	R	R	R	R	R	R

Reactivity Chart 9. Protection for the Amino Group: Amides (*Continued*)

PG	Na/NH_3	Al(Hg)	$SnCl_2/Py$	HSO_3^-; H_2S	$LiAlH_4$	Li-*s*-Bu_3BH	$(C_5H_{11})_2BH$	B_2H_6, 0°	$NaBH_4$	$Zn(BH_4)_2$	$NaBH_3CN$ pH 4-6	*i*-Bu_2AlH	$Li(O$*t*$Bu)_3AlH$	$AlCl_3$, 80°	$AlCl_3$, 25°	$SnCl_4$; BF_3	$LiClO_4$; $MgBr_2$	TsOH, 80°	TsOH, 0°
	39	40	41	42	43	44	45	46	47	48	49	50	51	52	53	54	55	56	57
	G.				H. HYDRIDE REDN.									I.					
1	R	L	L	L	*R*	L	H	R	L	L	L	H	L	L	L	L	L	L	L
2	R	L	L	L	*R*	L	H	R	*L*	L	*L*	H	L	L	*L*	*L*	L	L	L
3	R	M	L	L	R	M	H	R	L	L	L	H	L	M	M	L	L	L	L
5	R	M	L	L	H	M	H	R	*H*	M	M	H	M	M	M	L	L	L	L
6	R	L	L	L	H	M	H	R	*H*	M	M	H	M	L	*L*	L	L	*L*	L
7	R	R	L	L	R	M	H	R	L	L	L	H	L	L	L	L	L	L	L
8	R	R	L	L	R	M	H	R	L	L	L	H	L	L	L	L	L	L	L
9	R	L	L	L	R	R	H	R	R	R	R	H	R	R	R	M	L	L	L
12	R	L	L	L	R	L	H	R	L	L	L	H	L	L	L	L	L	L	L
13	R	L	L	L	R	L	H	R	L	L	L	H	L	L	L	L	L	L	L
15	R	R	L	L	R	M	H	R	L	L	L	H	L	L	L	L	L	L	L
16	R	*H*	R	L	R	L	H	R	R	R	L	H	M	R	L	L	L	L	L
17	R	L	L	L	R	M	H	R	L	L	L	H	L	M	M	L	L	L	L
19	R	R	L	L	R	M	H	R	L	L	L	H	L	M	M	L	L	L	L
20	R	L	L	L	R	L	H	R	L	L	L	H	L	L	L	L	L	L	L
21	R	L	L	L	R	L	H	R	L	L	L	H	L	L	L	L	L	L	L
23	R	*L*	L	L	*R*	L	H	R	*L*	L	L	*H*	L	L	L	L	L	*L*	L
29	*R*	L	L	L	R	L	H	R	*L*	L	L	H	L	L	L	L	L	L	L
31	R	R	R	L	R	R	H	R	R	L	L	H	R	R	R	L	L	R	L

PG	Hg(II)	Ag(I)	Cu(II)/Py	HBr/In·	HX/In·	NBS/CCl_4	Br_3CCl/In·	OsO_4	$KMnO_4$, pH 7, 0°	O_3, −50°	RCO_3H, 0°	RCO_3H, 50°	CrO_3/Py	CrO_3, pH 1	H_2O_2 pH 10-12	Quinone	$^1O_2$	DMSO, 100°	NaOCl pH 10	aq NBS
	58	59	60	61	62	63	64	65	66	67	68	69	70	71	72	73	74	75	76	77
	J.			K.				L. OXIDANTS												
1	*L*	L	L	L	L	L	L	L	L	*L*	L	M	L	H	M	L	L	L	L	L
2	*L*	L	L	L	L	L	L	L	L	L	L	*L*	L	*L*	*L*	L	*L*	L	L	L
3	L	H	L	L	L	L	L	L	L	L	L	L	L	L	L	L	L	R	L	L
5	L	H	L	R	R	R	R	L	L	L	L	L	L	L	L	L	L	L	R	L
6	L	M	L	L	L	L	L	L	L	L	*L*	R	L	L	L	L	L	L	R	L
7	L	L	L	L	L	M	L	L	L	L	L	L	L	L	L	L	L	L	L	L
8	L	L	L	L	L	L	L	L	L	L	L	L	L	L	L	L	L	L	L	L
9	L	L	*L*	L	L	L	L	L	*H*	L	L	M	L	*H*	L	L	L	L	R	R
12	L	L	L	L	L	R	L	L	L	L	L	L	L	L	L	L	L	L	L	L
13	L	L	L	R	R	R	R	L	R	L	L	L	M	R	M	M	L	L	R	*H*
15	L	L	L	L	L	L	L	L	L	L	L	L	L	L	L	L	L	L	L	L
16	L	R	L	R	R	R	R	L	L	R	M	R	L	R	R	L	R	L	L	L
17	L	*H*	L	L	L	L	L	L	L	L	L	L	L	L	L	L	L	R	L	L
19	L	L	L	R	R	R	R	R	R	R	L	M	L	L	R	L	L	L	L	R
20	H	L	H	L	L	R	L	L	L	L	M	R	M	M	L	L	L	L	L	L
21	L	L	L	L	L	L	L	L	R	R	R	R	L	R	L	L	R	L	R	R
23	*L*	L	L	L	L	L	L	*L*	L	L	L	L	*L*	L	L	L	L	*L*	L	L
29	L	L	*H*	L	L	L	L	L	L	*L*	L	*L*	L	L	L	L	L	L	L	L
31	R	M	L	R	R	R	R	M	R	R	R	R	L	R	R	R	R	R	R	R

Reactivity Chart 9. Protection for the Amino Group: Amides (*Continued*)

	I_2	PhSeX; PhSCl	Br_2; Cl_2	MnO_2/CH_2Cl_2	$NaIO_4$ pH 5–8	SeO_2 pH 2–4	SeO_2/Py	$K_3Fe(CN)_6$, pH 8	Pb(IV), 25°	Pb(IV), 80°	$Tl(NO_3)_3$	150°	250°	350°	$:CCl_2$	N_2CHCO_2R/Cu	$CH_2I_2/Zn(Cu)$	$R_3SnH/In\cdot$	$Ni(CO)_4$	CH_2N_2	$SOCl_2$	Ac_2O, 25°	Ac_2O, 80°	DCC	MeI	$Me_3O^+BF_4^-$	1.LDA 2.MeI	1.K_2CO_3 2.MeI	RCHO	RCOCl	C^+/olefin
PG	78	79	80	81	82	83	84	85	86	87	88	89	90	91	92	93	94	95	96	97	98	99	100	101	102	103	104	105	106	107	108
	L. OXIDANTS											M.			N.			O. MISCELLANEOUS											P.		
1	L	L	M	L	L	L	L	L	L	L	L	*L*	L	L	L	L	L	L	L	L	L	L	L	*L*	L	R	*R*	L	L	*H*	L
2	L	L	L	L	L	L	L	L	L	*L*	L	*L*	*L*	L	L	L	L	L	L	*L*	L	*L*	*L*	*L*	L	*R*	*R*	L	L	*L*	L
3	L	L	L	L	L	L	L	L	L	L	L	L	L	L	L	L	M	R	L	L	L	L	*L*	L	L	R	R	L	L	L	L
5	L	L	L	L	L	L	L	M	L	L	L	*L*	M	R	L	L	H	R	M	L	L	L	*L*	L	L	R	L	M	L	L	L
6	L	L	L	L	M	L	L	M	L	L	L	L	M	R	L	L	L	R	L	L	L	L	*L*	*L*	L	R	L	L	L	*L*	L
7	L	L	L	L	L	L	L	L	L	L	*L*	L	L	L	L	L	L	L	L	L	L	L	*L*	L	L	R	R	L	L	L	L
8	L	L	L	L	L	L	L	L	L	L	L	L	L	L	L	L	L	L	L	L	L	L	*L*	L	L	R	R	L	L	L	L
9	L	L	R	L	*M*	*H*	*M*	L	M	R	M	L	L	M	L	M	L	*R*	L	M	L	M	M	L	L	R	R	R	L	M	M
12	L	L	L	L	L	L	L	L	L	L	L	L	L	L	L	L	L	L	L	L	L	L	*L*	L	L	R	R	L	L	L	L
13	M	L	*H*	M	M	M	M	R	M	R	L	L	L	L	L	L	L	L	L	R	L	R	R	R	L	R	R	R	L	R	L
15	L	L	L	L	L	L	L	L	L	L	L	L	L	L	L	L	L	L	L	L	L	L	*L*	L	L	R	L	L	L	L	L
16	L	L	L	L	L	L	L	R	L	M	R	R	R	R	M	R	R	R	L	L	L	L	*L*	L	L	R	L	M	L	L	R
17	L	L	L	L	L	L	L	L	L	L	L	R	R	R	L	L	R	R	L	L	L	L	*L*	L	L	R	R	L	L	L	L
19	L	L	M	L	L	L	L	L	L	R	M	L	L	L	R	R	R	L	L	L	L	L	*L*	L	L	R	L	L	L	L	*M*
20	L	L	L	L	L	L	L	L	L	L	L	L	L	L	L	M	M	L	L	L	L	L	*L*	L	M	R	L	M	L	L	L
21	L	L	R	L	R	M	L	L	R	R	L	M	H	H	M	M	L	L	L	L	L	L	*L*	L	R	R	R	R	L	L	L
23	L	L	L	L	L	L	L	*L*	L	L	L	L	*L*	L	L	L	L	L	L	*L*	L	*L*	*L*	L	L	R	L	L	L	*L*	L
29	L	L	*L*	L	L	L	L	L	L	L	L	L	L	L	L	L	L	L	L	L	*L*	L	*L*	*L*	L	R	L	L	L	L	L
31	M	M	R	R	M	R	*M*	R	R	R	L	L	M	H	M	M	H	R	R	L	L	L	*L*	L	M	R	M	L	L	L	L

Reactivity Chart 10. Protection for the Amino Group: Special –NH Protective Groups

1. *N*-Allyl
2. *N*-Phenacyl
3. *N*-3-Acetoxypropyl
5. Quaternary Ammonium Salts
6. *N*-Methoxymethyl
8. *N*-Benzyloxymethyl
9. *N*-Pivaloyloxymethyl
12. *N*-Tetrahydropyranyl
13. *N*-2,4-Dinitrophenyl
14. *N*-Benzyl
16. *N*-*o*-Nitrobenzyl
17. *N*-Di(*p*-methoxyphenyl)methyl
18. *N*-Triphenylmethyl
19. *N*-(*p*-Methoxyphenyl)diphenylmethyl
20. *N*-Diphenyl-4-pyridylmethyl
21. *N*-2-Picolyl *N′*-Oxide
24. *N,N′*-Isopropylidene
25. *N*-Benzylidene
27. *N*-*p*-Nitrobenzylidene
28. *N*-Salicylidene
33. *N*-(5,5-Dimethyl-3-oxo-1-cyclohexenyl)
37. *N*-Nitro
39. *N*-Oxide
40. *N*-Diphenylphosphinyl
41. *N*-Dimethylthiophosphinyl
47. *N*-Benzenesulfenyl
48. *N*-*o*-Nitrobenzenesulfenyl
55. *N*-2,4,6-Trimethylbenzenesulfonyl
56. *N*-Toluenesulfonyl
57. *N*-Benzylsulfonyl
59. *N*-Trifluoromethylsulfonyl
60. *N*-Phenacylsulfonyl

(See chart, pp. 745–747.)

Reactivity Chart 10. Protection for the Amino Group: Special —NH Protective Groups

PG	pH<1, 100°	pH<1	pH 1	pH 2-4	pH 4-6	pH 6-8.5	pH 8.5-10	pH 10-12	pH>12	pH>12, 150°	NaH	Ph_3CNa	$(C_{10}H_8)^{\cdot-}Na^+$	$MeSOCH_2^-Na^+$	KO-t-Bu	LiN-i-Pr_2	Py; R_3N	$NaNH_2$	NaOMe	Enolate	NH_3; RNH_2	RS^-; N_3^-; SCN^-	OAc^-; X^-	NaCN, pH 12	HCN, pH 6	RLi	RMgX	Organozinc	Organocopper	Wittig; ylide	H_2/Raney (Ni)	H_2/Pt pH 2-4	H_2/Pd	H_2/Lindlar	H_2/Rh	Zn/HCl	Zn/HOAc	Cr(II), pH 5
	1	2	3	4	5	6	7	8	9	10	11	12	13	14	15	16	17	18	19	20	21	22	23	24	25	26	27	28	29	30	31	32	33	34	35	36	37	38
	A. AQUEOUS										B. BASIC								C. NUCLEOPHILIC							D. ORGANOMET.					E. CAT. REDN.					F.		
1	H	L	L	L	L	L	L	L	L	R	L	R	R	R	*R*	R	L	R	L	L	L	L	L	L	L	L	L	L	L	L	R	R	H	L	R	L	L	L
2	H	*L*	L	L	L	L	L	L	*L*	R	R	R	R	R	R	R	L	R	M	M	*M*	L	L	R	R	R	R	M	L	R	L	*R*	R	L	R	H	*H*	L
3	H	R	L	L	L	L	M	R	R	R	L	R	R	R	L	R	L	R	*R*	R	R	L	L	L	L	R	R	L	L	L	L	L	*L*	L	L	L	L	L
5	H	L	L	L	L	L	L	L	L	R	L	R	L	M	M	R	L	R	L	L	L	*H*	L	M	L	R	R	*L*	L	R	L	L	L	L	L	H	L	L
6	H	H	L	L	L	L	L	L	L	L	L	L	L	L	L	L	L	L	L	L	L	L	L	L	L	L	L	L	L	L	L	L	L	L	L	L	L	L
8	H	H	*L*	L	L	L	L	L	*L*	L	L	L	L	L	L	L	L	L	L	L	L	L	L	L	L	L	L	*L*	L	L	H	H	*R*	L	L	L	L	L
9	H	H	M	L	L	L	M	H	H	H	L	L	R	R	L	L	L	H	H	M	*H*	L	L	L	L	R	R	L	L	L	L	L	L	L	L	M	L	L
12	H	H	*H*	H	H	L	L	L	L	*L*	L	L	L	L	L	L	L	L	L	L	L	L	L	L	L	L	L	L	L	L	L	H	L	L	L	H	H	H
13	H	L	*L*	L	L	L	L	L	H	H	L	M	R	L	L	L	L	H	M	M	*L*	*H*	L	R	L	R	R	M	M	L	R	R	R	L	R	R	R	R
14	H	L	*L*	L	L	L	L	L	L	L	L	L	L	L	L	L	L	L	L	L	L	L	L	L	L	L	L	L	L	L	H	M	M	L	*H*	L	L	L
16	H	L	L	L	L	L	L	L	L	R	L	M	R	L	L	L	L	L	L	L	L	L	L	R	L	H	H	M	M	L	H	H	H	L	R	R	R	R
17	H	H	H	*H*	L	L	L	L	*L*	L	L	L	L	L	L	L	L	L	L	L	L	L	L	L	L	L	L	L	L	L	H	H	H	L	L	H	H	L
18	H	*H*	*H*	*H*	M	L	L	L	L	L	L	L	L	L	L	L	L	L	L	L	L	L	L	L	L	L	L	L	L	L	H	H	H	L	L	H	H	L
19	H	H	*H*	H	M	L	L	L	L	L	L	L	L	L	L	L	L	L	L	L	L	L	L	L	L	L	L	L	L	L	H	H	H	L	L	H	H	L
20	M	*L*	L	L	L	L	L	L	L	L	L	L	L	L	L	L	L	L	L	L	L	L	L	L	L	R	R	L	L	L	H	H	*H*	L	L	H	*H*	L
21	H	L	L	L	L	L	L	L	L	R	L	L	R	L	L	*H*	L	*H*	L	L	L	L	L	L	L	R	R	L	L	L	R	R	R	R	R	R	R	R
24	H	H	H	M	L	L	L	L	L	H	L	L	R	L	L	L	L	L	L	L	H	L	L	L	R	R	R	R	M	R	R	R	R	L	R	R	R	R
25	H	H	*H*	M	L	L	L	L	M	H	L	L	R	L	L	L	L	L	L	L	*H*	L	L	L	R	R	R	R	M	R	*R*	R	*H*	L	R	R	R	R
27	H	H	H	M	L	L	L	L	L	H	L	M	R	L	L	*L*	L	L	L	L	H	L	L	R	R	R	R	R	M	M	R	R	R	L	R	R	R	R
28	H	H	*H*	M	L	L	L	L	M	H	R	R	R	R	R	R	L	R	R	R	H	*L*	L	R	R	R	R	R	R	R	*R*	R	R	L	R	R	R	R
33	H	H	*L*	L	L	L	L	L	L	H	L	R	R	R	R	R	L	R	H	R	H	R	L	R	L	R	R	R	M	M	R	R	R	L	R	R	R	R
37	H	*L*	*L*	L	L	L	L	L	*L*	H	L	L	R	L	L	L	H	H	H	H	R	*L*	*L*	*L*	*L*	H	H	H	M	L	R	H	*H*	L	H	H	H	H
39	H	M	L	L	L	L	L	L	L	H	L	L	R	L	L	M	L	L	L	L	L	L	L	L	L	R	R	L	L	L	H	*H*	*H*	H	H	*H*	H	H
40	H	*H*	*H*	*M*	L	L	L	L	L	H	L	L	L	L	L	L	L	L	L	L	*L*	L	L	L	L	L	L	L	L	L	L	M	L	L	L	H	M	L
41	H	*H*	*H*	*L*	L	L	L	L	L	H	L	L	R	L	L	L	L	L	L	L	L	L	L	L	L	M	M	M	L	L	R	R	R	L	R	H	L	L
47	H	M	L	L	L	L	L	L	M	H	L	L	R	L	L	L	L	L	L	R	H	H	*H*	H	H	L	L	L	L	L	R	R	R	L	R	L	L	L
48	H	H	*H*	*H*	L	L	L	L	M	H	L	M	R	L	L	L	L	L	L	R	H	*H*	H	H	*H*	R	R	M	M	L	*R*	R	R	L	R	H	H	R
55	*H*	L	L	L	L	L	L	L	*L*	*L*	L	L	H	L	L	L	L	L	L	L	*L*	L	L	L	L	H	H	H	L	L	L	L	L	L	L	H	L	L
56	*H*	L	L	L	L	L	L	L	L	*L*	L	L	*H*	L	L	L	L	L	L	L	*L*	*L*	L	L	L	H	H	H	L	L	*L*	*L*	L	L	L	*H*	*L*	L
57	H	M	L	L	L	L	L	L	L	H	R	R	R	R	R	R	L	R	R	R	L	R	L	L	L	H	H	H	L	L	*H*	H	H	L	L	H	M	M
59	H	L	L	L	L	L	L	L	L	H	L	L	H	L	L	L	L	L	L	L	L	L	L	L	L	H	H	H	H	L	L	L	L	L	L	L	L	M
60	H	L	L	L	L	L	L	*L*	R	H	R	R	R	R	R	R	L	R	R	R	R	R	L	R	R	H	H	H	R	R	H	H	H	L	R	H	*H*	M

Reactivity Chart 10. Protection for the Amino Group: Special —NH Protective Groups (*Continued*)

PG	Na/NH_3	Al(Hg)	$SnCl_2/Py$	HSO_3^-; H_2S	$LiAlH_4$	Li-*s*-Bu_3BH	$(C_5H_{11})_2BH$	B_2H_6, 0°	$NaBH_4$	$Zn(BH_4)_2$	$NaBH_3CN$ pH 4-6	*i*-Bu_2AlH	Li(O*t*Bu)$_3$AlH	$AlCl_3$, 80°	$AlCl_3$, 25°	$SnCl_4$; BF_3	$LiClO_4$; $MgBr_2$	TsOH, 80°	TsOH, 0°
	39	40	41	42	43	44	45	46	47	48	49	50	51	52	53	54	55	56	57
	G.				H. HYDRIDE REDN.									I.					
1	R	L	L	L	L	L	R	R	L	L	L	L	L	L	L	L	L	L	L
2	*R*	R	L	R	R	R	R	R	R	R	M	R	R	L	L	L	L	L	L
3	R	L	L	L	R	L	L	L	L	L	L	R	M	R	L	L	L	R	L
5	L	L	L	L	L	L	L	L	L	L	L	L	L	R	L	L	L	L	L
6	L	L	L	L	L	L	L	L	L	L	L	L	L	H	H	H	L	M	L
8	H	L	L	L	L	L	L	L	L	L	L	L	L	H	H	H	L	M	L
9	R	L	L	L	R	L	L	L	L	L	L	R	M	H	H	M	L	R	L
12	L	L	L	L	L	L	L	L	L	L	L	L	L	H	H	H	L	M	L
13	R	R	R	L	R	L	L	L	L	L	L	R	M	L	L	L	L	L	L
14	H	M	L	L	*L*	L	L	L	L	L	L	L	L	L	L	L	L	L	L
16	H	R	L	L	R	L	L	L	L	L	M	R	M	L	L	L	L	L	L
17	H	L	L	L	L	L	L	L	L	L	L	L	L	R	R	L	M	L	L
18	*H*	L	L	L	L	L	L	L	L	L	L	L	L	L	L	L	L	L	L
19	H	L	L	L	L	L	L	L	L	L	L	L	L	R	R	L	M	L	L
20	H	L	L	L	L	L	L	L	L	L	L	L	L	L	L	L	L	L	L
21	R	R	R	L	R	M	R	R	M	L	L	M	M	R	L	L	L	L	L
24	R	M	L	M	R	L	R	R	L	L	R	R	R	L	L	L	L	M	L
25	R	M	L	*R*	R	L	R	R	M	M	R	R	R	L	L	L	L	M	L
27	R	R	L	R	R	L	R	R	M	M	R	R	R	L	L	L	L	M	L
28	R	M	L	M	R	L	R	R	M	M	R	R	R	R	R	R	L	R	L
33	R	L	L	R	R	R	R	R	L	L	M	R	R	L	L	L	L	H	L
37	R	R	R	L	R	L	R	R	L	L	R	*L*	R	L	L	L	L	L	L
39	H	H	H	*H*	H	M	R	R	M	L	M	H	H	R	L	L	L	L	L
40	H	L	L	L	M	L	L	L	L	L	L	L	L	L	L	L	L	M	L
41	H	L	L	L	M	L	L	L	L	L	L	L	L	L	L	L	L	M	L
47	H	H	M	M	H	M	R	R	R	R	M	H	H	*H*	L	L	L	M	L
48	H	H	M	*H*	H	H	R	R	R	R	M	H	H	L	L	L	L	M	L
55	H	H	L	L	L	L	L	L	L	L	L	L	L	L	L	L	L	L	L
56	*H*	H	L	L	*L*	L	L	L	L	L	L	L	L	L	L	L	L	L	L
57	*H*	H	L	L	L	L	L	L	L	L	L	L	L	L	L	L	L	L	L
59	M	L	L	L	*R*	L	L	L	L	L	L	L	L	L	L	L	L	L	L
60	H	H	L	L	R	R	R	R	M	M	M	R	R	L	L	L	L	L	L

PG	Hg(II)	Ag(I)	Cu(II)/Py	HBr/In·	HX/In·	NBS/CCl_4	Br_3CCl/In·	OsO_4	$KMnO_4$, pH 7, 0°	O_3, −50°	RCO_3H, 0°	RCO_3H, 50°	CrO_3/Py	CrO_3, pH 1	H_2O_2 pH 10-12	Quinone	$^1O_2$	DMSO, 100°	NaOCl pH 10	aq NBS
	58	59	60	61	62	63	64	65	66	67	68	69	70	71	72	73	74	75	76	77
	J.			K.				L. OXIDANTS												
1	R	L	L	R	R	R	R	R	R	R	R	R	R	L	R	R	R	M	M	R
2	L	L	L	L	L	L	L	R	R	R	R	R	R	L	R	R	R	M	M	R
3	L	L	L	L	L	L	L	R	R	R	R	R	R	L	R	R	R	M	R	R
5	L	M	L	L	L	L	L	L	L	L	L	L	L	L	L	L	L	H	L	L
6	L	L	L	L	L	L	L	R	R	R	R	R	R	L	R	R	R	M	M	R
8	L	L	L	L	L	R	L	R	R	R	R	R	R	L	R	R	R	M	M	R
9	L	L	L	L	L	L	L	R	R	R	R	R	R	M	R	R	R	M	R	R
12	L	L	L	L	L	L	L	R	R	R	R	R	R	H	R	R	R	M	M	R
13	L	L	L	L	L	L	L	*L*	*M*	*M*	*M*	*H*	*L*	L	*L*	*L*	L	*L*	*M*	*L*
14	L	L	L	L	L	R	L	R	R	R	R	R	R	L	R	R	R	M	M	R
16	L	L	L	L	L	R	L	R	R	R	R	R	R	L	R	R	R	M	M	R
17	L	L	L	L	L	R	L	R	R	R	R	R	R	H	R	R	R	M	M	R
18	L	L	L	L	L	L	L	R	R	R	R	R	R	H	R	R	L	M	M	R
19	L	L	L	L	L	L	L	R	R	R	R	R	R	H	R	R	L	M	M	R
20	L	L	L	L	L	L	L	R	R	R	R	R	R	L	R	R	L	M	M	R
21	L	L	R	L	L	R	L	R	R	R	R	R	R	L	R	R	R	R	M	R
24	M	M	H	R	R	R	R	R	R	R	R	R	R	H	M	R	*L*	R	R	R
25	M	M	H	R	R	R	R	R	R	R	R	R	R	H	M	R	*L*	R	R	R
27	M	M	H	R	R	R	R	R	R	R	R	R	R	H	M	R	*L*	R	*R*	R
28	M	M	H	R	R	R	R	R	R	R	R	R	R	H	M	R	*M*	R	R	R
33	L	M	H	R	R	R	R	R	R	R	R	R	*L*	L	*L*	R	R	*L*	*L*	R
37	L	L	H	R	R	R	R	R	R	R	R	R	R	R	R	R	R	R	R	R
39	L	L	R	L	L	L	L	L	L	L	L	L	L	L	L	L	L	H	L	L
40	L	L	L	L	L	L	L	L	L	L	L	L	L	H	L	L	L	L	L	L
41	H	H	H	L	L	L	L	L	L	L	L	L	L	H	L	L	L	L	L	L
47	H	H	H	L	L	*R*	L	R	R	R	R	R	R	L	L	L	R	*L*	R	R
48	H	H	H	L	L	L	L	R	R	R	R	R	R	L	L	L	M	*L*	R	R
55	L	L	L	L	L	R	L	L	L	L	L	L	L	L	L	L	L	L	M	L
56	L	L	L	L	L	R	L	L	L	L	L	L	L	L	L	L	L	L	M	L
57	L	L	L	L	L	M	L	L	L	L	L	L	L	L	L	L	L	L	M	*L*
59	L	L	L	L	L	L	L	L	L	L	L	L	L	L	L	L	L	L	M	L
60	L	L	L	L	L	L	L	L	L	L	L	L	L	L	L	L	R	L	M	R

Reactivity Chart 10. Protection for the Amino Group: Special —NH Protective Groups (*Continued*)

PG	I_2	PhSeX; PhSCl	Br_2; Cl_2	MnO_2/CH_2Cl_2	$NaIO_4$ pH 5–8	SeO_2 pH 2–4	SeO_2/Py	$K_3Fe(CN)_6$, pH 8	Pb(IV), 25°	Pb(IV), 80°	$Tl(NO_3)_3$	150°	250°	350°	:CCl_2	N_2CHCO_2R/Cu	CH_2I_2/Zn(Cu)
	78	79	80	81	82	83	84	85	86	87	88	89	90	91	92	93	94
	L. OXIDANTS											M.			N.		
1	L	R	R	R	R	R	R	R	R	R	R	L	L	M	R	R	R
2	L	M	R	R	R	R	R	R	R	R	R	L	L	M	L	L	L
3	L	L	R	R	R	M	L	R	R	R	R	L	L	R	L	L	L
5	L	L	L	L	L	L	L	L	L	L	L	R	R	R	L	L	L
6	L	L	R	R	R	M	L	R	R	R	R	L	L	M	L	L	L
8	L	L	R	R	R	R	M	R	R	R	R	L	L	M	L	L	L
9	L	L	R	R	R	M	L	R	R	R	R	L	L	H	L	L	L
12	L	L	R	R	R	M	L	R	R	R	R	L	L	H	L	L	L
13	L	L	*M*	*L*	*L*	*L*	L	*L*	*L*	*L*	*L*	L	M	H	L	L	L
14	L	L	R	R	R	R	M	R	R	R	R	L	L	M	L	L	L
16	L	L	R	R	R	M	L	R	R	R	R	L	L	R	L	L	L
17	L	L	R	R	R	R	M	R	R	R	R	L	L	M	L	L	L
18	L	L	R	R	R	M	L	R	R	R	R	L	L	M	L	L	L
19	L	L	R	R	R	M	L	R	R	R	R	L	L	M	L	L	L
20	L	L	R	R	R	M	L	R	R	R	R	L	L	M	L	L	L
21	L	L	R	R	R	R	M	R	R	R	R	R	R	R	L	L	L
24	R	R	R	R	R	R	R	R	R	R	R	L	M	H	R	R	R
25	R	R	R	R	R	R	M	R	R	R	R	L	M	H	R	R	R
27	R	R	R	R	R	M	M	R	R	R	R	L	M	H	R	R	R
28	R	R	R	R	R	R	M	R	R	R	R	L	M	H	R	R	R
33	L	*R*	R	L	R	R	R	M	R	R	R	L	L	M	R	R	R
37	R	R	R	R	R	R	R	R	R	R	R	L	R	R	R	R	R
39	L	L	L	L	L	L	L	L	L	L	L	H	R	R	L	L	L
40	L	L	L	L	L	M	L	L	L	L	L	L	L	M	L	L	L
41	L	L	L	L	R	M	L	L	L	L	L	L	L	M	L	L	L
47	R	L	R	*R*	R	R	L	R	R	R	R	L	M	H	*H*	L	L
48	R	L	R	R	R	M	L	M	R	R	R	L	M	R	L	L	L
55	L	L	L	L	L	M	L	L	L	L	L	L	M	H	L	L	L
56	L	L	L	L	L	M	L	L	L	L	L	L	M	H	L	L	L
57	L	R	*M*	L	L	R	M	L	L	L	L	L	M	H	L	L	L
59	L	L	L	L	L	L	L	L	L	L	L	L	M	R	L	L	L
60	L	R	R	L	L	R	R	L	L	L	L	L	M	H	L	L	L

PG	R_3SnH/In·	$Ni(CO)_4$	CH_2N_2	$SOCl_2$	Ac_2O, 25°	Ac_2O, 80°	DCC	MeI	$Me_3O^+BF_4^-$	1.LDA 2.MeI	1.K_2CO_3 2.MeI	RCHO	RCOCl	C^+/olefin
	95	96	97	98	99	100	101	102	103	104	105	106	107	108
	O. MISCELLANEOUS											P.		
1	L	R	L	L	L	L	L	R	R	L	R	L	*L*	R
2	R	L	R	L	L	L	*L*	R	R	R	R	L	L	R
3	L	L	L	L	L	L	L	R	R	M	R	L	L	R
5	L	L	L	L	L	L	L	L	L	L	L	L	*R*	L
6	L	L	L	L	L	L	L	R	R	L	R	L	L	R
8	L	L	L	L	L	*L*	L	R	R	L	R	L	L	R
9	L	L	L	L	L	L	L	R	R	L	R	L	L	R
12	L	L	L	L	L	L	L	R	R	L	R	L	L	R
13	L	L	L	L	L	L	L	M	R	L	M	L	L	R
14	L	L	*L*	L	L	L	L	R	R	*L*	R	L	L	R
16	L	L	L	L	L	L	L	M	R	L	M	L	L	R
17	L	L	L	L	L	L	L	R	R	L	R	L	L	R
18	L	L	L	L	L	L	L	R	R	L	R	L	L	R
19	L	L	L	L	L	L	L	R	R	L	R	L	L	R
20	L	M	L	L	R	R	L	R	R	L	R	L	L	R
21	M	M	M	H	*H*	H	L	R	R	L	R	L	R	R
24	L	L	L	R	R	R	*L*	M	R	L	M	L	L	R
25	R	L	L	L	*L*	H	L	M	R	*L*	M	L	L	R
27	R	L	L	L	M	R	L	M	R	L	M	L	L	R
28	R	L	M	R	R	R	R	M	R	R	R	L	R	R
33	L	L	L	L	M	H	*L*	L	R	R	L	L	L	R
37	R	R	R	L	M	H	*L*	M	R	R	M	R	R	R
39	L	L	L	H	H	H	L	L	R	L	L	L	R	R
40	L	L	L	L	L	L	L	R	R	L	R	L	L	M
41	L	L	L	L	L	L	L	R	R	L	R	L	L	M
47	*H*	L	L	*L*	M	R	L	R	R	R	R	L	L	R
48	L	L	L	*L*	R	R	L	R	R	R	R	L	L	R
55	L	L	L	L	L	L	L	L	R	R	R	L	L	M
56	L	L	L	*L*	L	L	*L*	L	R	R	R	L	L	M
57	L	L	L	L	L	L	L	L	R	R	M	L	L	M
59	L	L	L	L	L	L	L	L	R	R	R	L	L	M
60	R	L	R	R	R	R	L	M	R	R	R	L	L	M

INDEX